Molecular Mechanisms of Cardiac Hypertrophy and Failure

Molecular Mechanisms of Cardiac Hypertrophy and Failure

Edited by

Richard A Walsh MD
John H Hord Professor and Chairman,
Department of Medicine
Case Western Reserve University and
University Hospitals of Cleveland
Cleveland, Ohio, USA

Associate Editors

Michael Schneider, Stephen Vatner, Eduardo Marban,
Jon Seidman, Christine Seidman

Taylor & Francis
Taylor & Francis Group

LONDON AND NEW YORK

© 2005 Taylor & Francis, an imprint of the Taylor & Francis Group

First published in the United Kingdom in 2005 by Taylor & Francis, an imprint of the Taylor & Francis Group, 2 Park Square, Milton Park, Abingdon, Oxon OX14 4RN

Tel.:   +44 (0)20 7017 6000
Fax.:   +44 (0)20 7017 6699
E-mail: info.medicine@tandf.co.uk
Website: http://www.tandf.co.uk/medicine

A CIP record for this book is available from the British Library.

Library of Congress Cataloging-in-Publication Data

Data available on application

ISBN 1 84214 248 8
ISBN 978-1-84214-248-6

Distributed in North and South America by

Taylor & Francis
2000 NW Corporate Blvd
Boca Raton, FL 33431, USA

Within Continental USA
Tel: 800 272 7737; Fax: 800 374 3401
Outside Continental USA
Tel: 561 994 0555; Fax: 561 361 6018
E-mail: orders@crcpress.com

Distributed in the rest of the world by
Thomson Publishing Services
Cheriton House
North Way
Andover, Hampshire SP10 5BE, UK
Tel.: +44 (0)1264 332424
E-mail: salesorder.tandf@thomsonpublishingservices.co.uk

Composition by Phoenix Photosetting, Chatham, Kent, UK
Printed and bound in Great Britain by CPI, Bath

Dedication
To Donna, Ciara and Maeve

Contents

Section I: Mechanisms for cardiac hypertrophy

Contents

Section IV: Genetic basis for cardiomyopathy

Contents

Contributors

Hiroshi Akazawa MD
Assistant Professor
Division of Cardiovascular Pathophysiology
Chiba University Graduate School of Medicine
Chiba
Japan

Piero Anversa MD
Professor of Medicine, Microbiology and
    Immunology, and Pathology
Vice-Chairman, Department of Medicine
Director of the Cardiovascular Research Institute
New York Medical College
Valhalla, New York
USA

Carl V Barnes MD
Attending Physician, Internal Medicine
Denver VA Medical Center
Ambulatory Care, Internal Medicine
Denver, Colorado
USA

Rhonda Bassel-Duby PhD
Associate Professor of Molecular Biology
Department of Molecular Biology
University of Texas Southwestern Medical Center
Dallas, Texas
USA

Craig T Basson MD, PhD
Professor of Medicine
Greenberg Division of Cardiology
Weill Medical College of Cornell University
New York Presbyterian Hospital
Cornell Medical Center
New York, New York,
USA

Jonathan M Bella MD
Associate Professor of Medicine
Division of Cardiology
Bronx-Lebanon Hospital Center / Albert Einstein
    College of Medicine
Bronx, New York
USA

Donald M Bers PhD
Professor and Chair of Physiology
Department of Physiology
Loyola University Chicago
Stritch School of Medicine
Maywood, Illinois
USA

Penelope A Boyden PhD
Professor of Pharmacology
Department of Pharmacology
Columbia University
New York, New York
USA

Michael R Bristow MD, PhD
Co-Director, CU CVI; S. Golbert Blount Professor
    of Medicine (Cardiology)
Division of Cardiology
University of Colorado Health Sciences Center
Denver, Colorado
USA

Jana Burchfield PhD
Post-doctoral Research Fellow
Winters Center for Heart Failure Research
Section of Cardiology
Baylor College of Medicine
Texas Heart Institute and St Luke's Episcopal
    Hospital
Houston, Texas
USA

Kevin P Campbell PhD
Department Chair, Physiology and Biophysics
HHMI Investigator
Carver College of Medicine
University of Iowa
Howard Hughes Medical Institute
Iowa City, Iowa
USA

Angela Clerk PhD
Reader in Biochemistry and Cell Biology
National Heart and Lung Institute (NHLI)
    Division
Faculty of Medicine
Imperial College London
London
UK

Wilson S Colucci MD
Thomas J. Ryan Professsor of Medicine
Chief, Cardiovascular Medicine
Boston University Medical center
Boston, Massachusetts
USA

Michael T Crow PhD
Associate Professor of Medicine
Johns Hopkins University
School of Medicine
Baltimore, Maryland
USA

Christophe Depre MD, PhD
Associate Professor
Department of Cell Biology and Molecular
    Medicine
Cardiovascular Research Institute
University of Medicine and Dentistry of New
    Jersey
New Jersey Medical School
Newark, New Jersey
USA

Anne M Deschamps BS
Division of Cardiothoracic Surgery Research
Medical University of South Carolina
Charleston, South Carolina
USA

Richard B Devereux MD
Professor of Medicine
Greenberg Division of Cardiology
Weill Medical College of Cornell University
New York Presbyterian Hospital
Cornell Medical Center
New York, New York,
USA

Mark H Drazner MD, MSc
Medical Director, Parkland Memorial Hospital
    CHF Clinic
Associate Professor of Medicine
Donald W. Reynolds Cardiovascular Clinical
    Research Center
Division of Cardiology
University of Texas Southwestern Medical Center
Dallas, Texas
USA

Wen Dun PhD
Associate Research Scientist
Department of Pharmacology
Columbia University
New York, New York
USA

Loren J Field PhD
Professor of Medicine and Pediatrics
Krannert Institute of Cardiology and Herman B
    Wells Center for Pediatric Research
Indianapolis University School of Medicine
Indiana
USA

Jens Fielitz MD
Department of Molecular Biology
University of Texas Southwestern Medical Center
Dallas, Texas
USA

Harry A Fozzard MD
Otho S.A. Sprague Distinguished Service
    Professor of Medical Sciences, Emeritus
Cardiac Electrophysiology Laboratories
Department of Medicine
The University of Chicago
Chicago, Illinois
USA

William H Frishman MD
Rosenthal Professor and Chairman, Department
    of Medicine
Professor of Pharmacology / Director of Medicine
New York Medical College and Westchester
    Medical Center
New York, New York
USA

Barry Greenberg MD
Professor of Medicine
University of California
San Diego, California
USA

Gerd Hasenfuss MD
Professor of Medicine
Chair of Cardiology and Pneumology
University of Göttingen
Heart Center
Department of Cardiology and Pneumology
Göttingen
Germany

Brian D Hoit MD
Professor of Medicine, Physiology and Biophysics
Case Western Reserve University and University
    Hospitals of Cleveland
Cleveland, Ohio
USA

Charles L Hoppel MD
Professor of Pharmacology, Medicine and
    Oncology
Case Western Reserve University School of
    Medicine
Louis Stokes VA Medical Center
Cleveland, Ohio
USA

Seigo Izumo MD
Vice President & Global Head of Cardiovascular
    Research
Novartis Institutes for BioMedical Research
Japan

Arundathi Jayatilleke MD
Department of Medicine
Duke University Medical Center
Durham, North Carolina
USA

Stefan Kääb MD, PhD
Ludwig Maximilian's University Munich
Department of Medicine 1, Cardiology
Klinikum Grosshadern, Munich
Germany

Richard N Kitsis MD
Professor of Medicine and Cell Biology
The Gerald and Myra Dorros Chair in
    Cardiovascular Disease
Chief, Division of Cardiology
Director, Cardiovascular Research Center
Albert Einstein College of Medicine
Bronx, New York
USA

Brian Kobilka MD
Professor of Molecular and Cellular Physiology
Stanford University School of Medicine
Stanford, California
USA

Issei Komuro
Professor, Department of Cardiovascular Science
    and Medicine
Chiba University Graduate School of Medicine
Chiba
Japan

Evangelia G Kranais PhD
Professor and Director, Cardiovascular Biology
Department of Pharmacology and Cell Biophysics
University of Cincinnati College of Medicine
Cincinnati, Ohio
USA

Stephen B Liggett MD
Professor of Medicine and Physiology
Director, Cardiopulmonary Genomics Program
University of Maryland School of Medicine
Baltimore, Maryland
USA

Jonathan C Makielski MD
Professor of Medicine and Physiology
Division of Cardiovascular Medicine
Department of Medicine
University of Wisconsin
Madison, Wisconsin
USA

Kartik Mani MB BS
Fellow in Cardiology
Albert Einstein College of Medicine
Bronx, New York
USA

Douglas L Mann MD
Professor of Medicine, Molecular Physiology and
    Biophysics
Winters Center for Heart Failure Research
Section of Cardiology
Baylor College of Medicine
Texas Heart Institute and St Luke's Episcopal
    Hospital
Houston, Texas
USA

Kenneth B Margulies MD
Associate Professor of Medicine, Research
    Director
Hearth Failure and Transplant Program
University of Pennsylvania
Philadelphia, Pennsylvania
USA

Barry J Marron MD
Director, Hypertrophic Cardiomyopathy Center
Minneapolis Heart Institute Foundation
Minneapolis, Minnesota
USA

Sunil N Matiwala MD
Heart Failure Fellow
Division of Cardiology
Boston University Medical Center
Boston, Massachusetts
USA

Julie R McMullen BSc(Hons), PhD
Senior Research Officer and Laboratory Head
Division of Cardiology
Baker Heart Research Institute
Melbourne, Victoria
Australia

Elizabeth M McNally MD, PhD
Associate Professor
Department of Medicine, Section of Cardiology
Department of Human Genetics
The University of Chicago
Chicago, Illinois
USA

Luisa Mestroni MD, FACC, FESC
Associate Professor of Medicine / Cardiology
Director, Molecular Genetics Program,
    Cardiovascular Institute and Adult Medical
    Genetics Program, Department of Internal
    Medicine
University of Colorado Health Sciences Center
Aurora, Colorado
USA

Jeanne Mialet-Perez PhD
Post-doctoral Fellow
Cardiopulmonary Research Center
University of Cincinnati College of Medicine
Cincinnati, Ohio
USA

Daniel E Michele PhD
Assistant Professor of Physiology
Department of Molecular and Integrative
    Physiology
Department of Internal Medicine
University of Michigan
Ann Arbor, Michigan
USA

Michael Näbauer MD, PhD
Ludwig Maximilian's University Munich
Department of Medicine 1, Cardiology
Klinikum Grosshadern, Munich
Germany

Timothy M Olsen MD
Associate Professor of Medicine and Pediatrics
Divisions of Cardiovascular Diseases and Pediatric
    Cardiology
Mayo Clinic College of Medicine
Rochester, Minnesota
USA

Eric N Olsen PhD
Professor and Chairman of Molecular Biology
University of Texas Southwestern Medical Center
Dallas, Texas
USA

Cam Patterson MD, FACC
Craige Distunguished Professor of Cardiovascular
    Medicine
Director, Division of Cardiology and Carolina
    Cardiovascular Biology Center
Chapel Hill, North Carolina
USA

David R Pimentel MD
Assistant Professor of Medicine
Boston University Medical Center
Boston, Massachusetts
USA

J David Port PhD
Professor of Medicine and Pharmacology
Division of Cardiology
University of Colorado Health Sciences Center
Denver, Colorado
USA

R S Ramagadran MD
Post-doctoral Research Fellow
Winters Center for Heart Failure Research
Section of Cardiology
Baylor College of Medicine
Texas Heart Institute and St Luke's Episcopal
    Hospital
Houston, Texas
USA

Charles Redwood PhD
British Heart Foundation Basic Science Lecturer
University of Oxford
Oxford
UK

Howard A Rockman MD
Professor of Medicine, Cell Biology and Molecular
    Genetics
Division of Cardiology
Duke University Medical Center
Durham, North Carolina
USA

Junichi Sadoshima MD, PhD
Professor, Department of Cell Biology and
    Molecular Medicine
Cardiovascular Research Institute
UMDNJ, New Jersey Medical School
Newark, New Jersey
USA

Jeffrey E Saffitz MD, PhD
Mallinckrodt Professor of Pathology
Harvard Medical School
Chairman, Department of Pathology
Beth Israel Deaconess Medical Center
Boston, Massachusetts
USA

Allen M Samarel MD
William B. Knapp Professor of Medicine and
    Physiology
Cardiovascular Institute
Loyola University Chicago
Stritch School of Medicine
Maywood, Illinois
USA

Douglas B Sawyer MD, PhD
Associate Professor of Medicine
Boston University Medical Center
Boston, Massachusetts
USA

Michael D Schneider MD
Professor, Baylor College of Medicine
Co-Director, Center for Cardiovascular
    Development
Houston, Texas
USA

J G Seidman PhD
Henrietta B. and Frederick H. Bugher Professor of
    Cardiovascular Genetics
Department of Genetics
Harvard Medical School
Boston, Massachusetts
USA

Christine E Seidman MD
Professor of Medicine and Genetics
Harvard Medical School
Brigham and Women's Hospital
Boston, Massachusetts
USA

Christopher Semsarian MB BS, PhD
Head, Molecular Cardiology Group
Cardiologist, Centenary Institute and Royal
    Prince Alfred Hospital
Newtown, NSW
Australia

Kersten M Small PhD
Associate Professor
Cardiopulmonary Research Center
University of Cincinnati College of Medicine
Cincinnati, Ohio
USA

R John Solaro PhD
University Professor and Head of Physiology and
    Biophysics
Department of Physiology and Biophysics
University of Illinois at Chicago
College of Medicine
Chicago, Illinois
USA

Edmund H Sonnenblick MD
Department of Medicine
New York Medical College
Valhalla, New York
USA

Madison S Spach MD
Department of Pediatrics
Duke University Medical Center
Durham, North Carolina
USA

Francis G Spinale MD, PhD
Professor of Surgery and Physiology
Division of Cardiothoracic Surgery
Medical University of South Carolina
Charleston, South Carolina
USA

Peter M Spooner PhD
Associate Professor of Cardiology and Executive
    Director, D.W. Reynolds Cardiovascular
    Center
Division of Cardiology
Department of Medicine
Johns Hopkins University
Baltimore, Maryland
USA

Brian A Stanley MS
Department of Cardiology
Queen's University
Kingston, Ontario
Canada
Department of Medicine
Johns Hopkins University
Baltimore, Maryland
USA

Peter H Sugden PhD
Professor of Cellular Biochemistry
National Heart and Lung Institute (NHLI)
    Division
Faculty of Medicine
Imperial College London
London
UK

Matthew R G Taylor MD, PhD
Assistant Professor
Adult Medical Genetics Program
University of Colorado Health Sciences Center
Aurora, Colorado
USA

Jennifer E Van Eyk PhD
Associate Professor
Departments of Medicine, Biological Chemistry
    and Biomedical Engineering
Johns Hopkins University
Baltimore, Maryland
USA

Stephen F Vatner MD
University Professor and Department Chair
Director of CVRI
Department of Cell Biology and Molecular
    Medicine
Cardiovascular Research Institute
University of Medicine and Dentistry of New
    Jersey
New Jersey Medical School
Newark, New Jersey
USA

Lynne E Wagoner MD
Associate Professor of Medicine
Division of Cardiology
University of Cincinnati College of Medicine
Cincinnati, Ohio
USA

Richard A Walsh MD
John H. Hord Professor and Chairman
Department of Medicine
Case Western University and University Hospitals
    of Cleveland
Cleveland, Ohio
USA

Hugh Watkins MD, PhD
Field Marshal Alexander Professor of
    Cardiovascular Medicine
University of Oxford
Oxford
UK

Arthur B Zinn MD, PhD
Associate Professor of Genetics and Pediatrics
Center for Human Genetics
Case Western Reserve University
University Hospitals of Cleveland
Cleveland, Ohio
USA

Preface

The major impetus for the design and execution of *Molecular Mechanisms for Cardiac Hypertrophy and Failure* has been the remarkable insights into the pathogenesis of abnormal cardiac growth and function that have been derived from the application of molecular and cellular approaches to this problem over the past decade. The syndrome of congestive heart failure is a major and growing public health problem in both developed and developing countries. In developed countries, the aging of our populations has allowed more protracted impacts of genetic and environmental factors that have resulted in the continued increase in incidence and prevalence of this condition. Another major contributory factor has been the substantial improvements in the management of acute coronary syndromes. The advent of pharmacologic and interventional therapies has significantly improved mortality but at the expense of a large and growing population of patients who have variably damaged left ventricular function. Fifty years ago congestive heart failure was viewed as a cardiocirculatory disturbance and therapy focused on improving cardiac and renal function using inotropic and diuretic agents. Subsequently biochemical insights elucidated the important role of abnormal activation of various neurohormonal systems in the development and progression of congestive heart failure. Most recently molecular approaches coupled with traditional biochemical and physiologic phenotypic insights are beginning to provide important insights for novel approaches to therapy. This book is designed to summarize and analyze critically the current advances in this area for a broad audience of cardiologists, clinical and basic research trainees and established scientists who find it increasingly challenging to have a general perspective on the state-of-the-art knowledge in this area. We are fortunate to have recruited contributors who have made pivotal contributions in this area and who are acknowledged leaders in their respective fields.

The book has been divided into four sections that encompass the major pathophysiologic areas which contribute to the development and evolution of cardiac hypertrophy and failure.

Section I edited by Michael Schneider focuses on mechanisms for cardiac hypertrophy. The relative roles of normal cardiovascular growth and development, cell cycle control, apoptosis and aberrant signal transduction are explored critically.

Section II edited by Steven Vatner explores the mechanisms for abnormal cardiac function in heart failure. The authors provide state-of-the-art analyses of abnormalities in excitation contraction coupling, calcium homeostasis, myocardial ischemia and its variants in the production of contractile depression and abnormalities of cardiac relaxation.

Section III edited by Eduardo Marban is focused on the molecular mechanisms responsible for arrhythmogenesis and provides ideas for potential novel targets for drug cell and gene therapy.

Finally, Section IV edited by Jon and Christine Seidman highlights the current state of the molecular genetic basis for cardiomyopathy.

Richard A Walsh, M.D.

Acknowledgements

The design and implementation of this book would not have been possible without the able assistance of a number of individuals in addition to the Associate Editors and chapter contributors. I am grateful for the guidance provided by our Managing Editor, Jonathan Gregory. I am particularly thankful for the expert administrative and secretarial assistance provided by Ann Smisek and Jill Pritchett. Finally, as always, I appreciate the understanding and support of my wife Donna.

Richard A Walsh, M.D.

Abbreviations

AC	adenylyl cyclase	ARB	angiotensin receptor blocker
2-DLC	two-dimensional liquid chromatography	ARC	apoptosis repressor with a caspase recruitment domain
3'UTR	3' untranslated region	ARE	A+U-rich nucleotide element
3-DLD	three-dimensional liquid chromatography	ARIC	Atherosclerosis Risk in Communities Study
4E-BP1	4E binding protein 1	α-sk-actin	alpha-skeletal actin
AC	alternating current	βARK	β-adrenergic receptor kinase
ACE	angiotensin-converting enzyme	ARVC	arrhythmogenic right ventricular cardiomyopathy
ACEI	ACE inhibitor		
AD	autosomal dominant	ARVD	arrhythmogenic right ventricular dysplasia
AD EDMD	autosomal dominant Emery–Dreifuss muscular dystrophy	ASC	apoptosis-associated speck-like protein containing a capsase recruitment domain
AF	atrial fibrillation		
AFLP	acute fatty liver of pregnancy	AVB	atrioventricular block
AI	aortic insufficiency	AVN	atrioventricular node
AK	adenylate kinase	AVR	aortic valve replacement
AKAP	A-kinase-anchoring protein	Bap31	B-cell receptor-associated protein 31
Akt/AKT	see PKB		
ALP	actinin-associated LIM protein	Bcl-x$_L$	Bcl-x protein long isoform
aLQTS	acquired long QT syndrome	BEST	Beta-blocker Extends Survival Trial
AMPK	AMP-activated protein kinase		
ANF	atrial natriuretic factor	BH3	Bcl-2 homology domain 3
Ang I	angiotensin I	bHLH	basic helix loop helix
Ang II	angiotensin II	BIR	*baculovirus* inhibitor of apoptosis repeats
ANOVA	analysis of variance		
ANP	atrial natriuretic peptide	BK	bradykinin
ANT	adenine nucleotide translocase	BMD	Becker's muscular dystrophy
AP	action potential	BMP	bone morphogenetic protein
Apaf	apoptotic protease activating factor	BNip3	Bcl-2/adenovirus E1B nineteen kD-interacting protein 3
APD	action potential duration	BNP	brain natriuretic peptide
AR	adrenergic receptor	BP	blood pressure
AR	autosomal recessive	BRF1	butyrate response factor 1

ca	constitutively active	CRT	cardiac resynchronization therapy
CABG	coronary artery bypass graft	CsA	cyclosporin A
CACT	carnitine-acylcarnitine translocase	CSQ	calsequestrin
CAF	cyclophosphamide	CT	carnitine transporter
CA-IEF	carrier ampholyte isoelectric focusing	CT-1	cardiotrophin-1
CaM	calmodulin	cTnC	cardiac troponin C
CAM	constitutive active mutation	cTnI	cardiac troponin I
CaMK	calcium/calmodulin-dependent protein kinase	CV	cardiovascular
		CVB	coxsackievirus B
CARD	caspase recruitment domain	Cx	connexin
CARDIA	Coronary Artery Risk Development in Young Adults (study)	DAD	delayed afterdepolarization
		DAG	diacylglycerol
		DC	direct current
		DCM	dilated cardiomyopathy
CARP	cardiac ankryin repeat protein	DD	death domain
CBB	Coomassie brilliant blue	DDC	diethyldithiocarbamic acid
CCD	charge-coupled device	DED	death effector domain
CCT	chaperonin-containing T-complex (also known as TCP-1)	Del	developmentally regulated endothelial cell locus
CCU	coronary care unit	DG	diacylglycerol
CEC	cation exchange chromatography	DGC	dystrophin–glycoprotein complex
CHF	congestive heart failure	DGC	dystrophin–glycoprotein complex
CHIP	carboxyl-terminus of Hsp70-interacting protein	DHPR	dihydropyridine receptor (L-type Ca^{2+} channel)
CHP	Chinese hamster ovary	DHT	5-dehydrotestosterone
CHS	Cardiovascular Health Study	DIABLO	direct IAP-binding protein with low pI
CI	confidence interval		
CICR	Ca^{2+}-induced Ca^{2+} release	DISC	death-inducing signaling complex
CIP	CDK-interacting protein	DMD	Duchenne muscular dystrophy
CK	creatine kinase	dn	dominant negative
CM	cardiomyopathic Syrian (hamster)	DPMK	myotonic dystrophy protein kinase
		DSP	desmoplakin
CN	calcineurin	DSS	Dahl salt-sensitive (rats)
CNBD	cyclic nucleotide-binding domain	EAD	early afterdepolarization
CoQ_{10}	2,3-dimethoxy-5-methyl-6-decaprenyl benzoquinone (coenzyme Q_{10})	EBNA-LP	Epstein–Barr virus nuclear antigen-leader protein
		EBV	Epstein–Barr virus
COUP-TF	chicken ovalbumin upstream promoter transcription factor	E–C	excitation–contraction (coupling)
		ECG	electrocardiogram
COX	cytochrome c oxidase	ECM	extracellular matrix
COX-2	cyclooxygenase-2	ECSOD	extracellular superoxide dismutase
CPT I	carnitine palmitoyltransferase I	EF	ejection fraction
CPT II	carnitine palmitoyltransferase II	EGF	epidermal growth factor
CPVT	catecholaminergic polymorphic ventricular tachycardia	EGFR	epidermal growth factor receptor
		EIF	eukaryotic translation initiation factor
CRE	cAMP-responsive element		
CREB	cAMP response element-binding protein	eIF4E	eukaryotic initiation factor 4E
		EMCV	encephalomyocarditis virus

EMT	epithelial–mesenchymal transformation	G6PD	glucose 6-phosphate dehydrogenase
EndoG	endonuclease G	GAP	GTPase-activating protein
ENMC	European Neuromuscular Center	GAPDH	Glyceraldehyde-3-phosphate dehydrogenase
eNOS	endothelial nitric oxide synthase		
EPC	endothelial progenitor cell	GATA4	GATA-binding protein 4
EPHESUS	Eplerenone Post-AMI Heart Failure Efficacy and Survival (study)	GEF	guanine nucleotide exchange factor
EPR	electron paramagnetic resonance spectroscopy	GH	growth hormone
		Gp130	glycoprotein 130
ER	endoplasmic reticulum	GPCR	G protein-coupled receptor
ERK	extracellular signal-regulated kinase	GPx	glutathione peroxidase
		GRK	G protein receptor kinase
ES	embryonic stem (cell)	GSH	reduced glutathione
ESI	electrospray ionization	GSK	glycogen synthase kinase
ESP	end-systolic pressure	HAT	histone acetyltransferase
EST	expressed sequence tag	HCM	hypertrophic cardiomyopathy
ET-1	endothelin-1	HCN	hyperpolarization-activated cyclic nucleotide-modulated (channel)
ETF	electron transfer flavoprotein		
ETF DH	electron transfer flavoprotein dehydrogenase	HDAC	histone deacetylase
		HELLP	hemolysis, elevated liver enzymes, and low platelets
FADD	Fas-associated death domain protein		
		HF	heart failure
FAF	familial 'lone' atrial fibrillation	HIF	hypoxia-inducible factor
FAK	focal adhesion kinase	HPLC	high-pressure liquid chromatography
FCMD	Fukuyama congenital muscular dystrophy		
		HRT	hairy-related transcription factor
FDA	Food and Drug Administration	HSC	hematopoietic stem cell
FDAR	frequency-dependent acceleration of relaxation	HSF1	heat shock factor-1
		HSP	heat-shock protein
FFR	force–frequency response	Htr	high temperature requirement protein
FGF (Fgf)	fibroblast growth factor		
FHC	familial hypertrophic cardiomyopathy	HyperGEN	Hypertension Genetic Epidemiology Network
FHL1	four-and-a-half LIM domain protein-1	IAP	inhibitor of apoptosis proteins
		ICAM	intercellular adhesion molecule
FKBP	FK-506-binding protein	ICAT	isotope-coded affinity tags
FKRP	fukutin-related protein	ICD	implantable cardio defibrillator
FLIP$_L$	Fas-associated death domain protein-like-interleukin-1β-converting enzyme	ICER	inducible cyclic AMP early repressor
		IDC	idiopathic dilated cardiomyopathy
FPVT	familial polymorphic ventricular tachycardia	IEF	isoelectric focusing
		IGF	insulin-like growth factor
FSH	follicle-stimulating hormone	IgG	immunoglobulin G
FSHD	facioscapulohumeral muscular dystrophy	IL	interleukin
		IMM	inner mitochondrial membrane
FT-MS	Fourier transform mass spectrometer	iNOS	inducible isoform of NO synthase
		InsP$_3$/IP$_3$	inositol 1,4,5-trisphosphate

Ip	inhibitory peptide	MERRF	myoclonic epilepsy and ragged red fibers
IPG	immobilized pH gradient	MHC	myosin heavy chain
IRK1	inward rectifier K^+ channel	MI	myocardial infarction
ISA	intrinsic sympathomimetic activity	MIBG	^{123}I-metaiodobenzylguanidine
Iso	isoprenaline	MKK	MAPK kinase
IVF	idiopathic ventricular fibrillation	MKKK	MKK kinase
JAK	janus kinase	MLC	myosin light chain
JNK	c-jun N-terminal kinase	MLC-2v	ventricle-specific myosin light chain
LA	left atrium		
LAMP-2	lysosome-associated membrane protein-2	MLCK	myosin light chain kinase
		MLCP	myosin light chain phosphatase
LAP 2β	lamin-associated protein 2β	MLP	muscle LIM protein
LC	liquid chromatography	MMP	matrix metalloproteinase
LCAD	long chain acyl coenzyme A dehydrogenase	MPTP	mitochondrial permeability transition pore
LCAHD	long chain acyl-CoA dehydrogenase deficiency	MRFIT	Multiple Risk Factor Intervention Trial
LDL	low density lipoprotein	MRI	magnetic resonance imaging
LGMD	limb girdle muscular dystrophy	MRI	magnetic resonance imaging
LIF	leukemia inhibitory factor	mRNA	messenger ribonucleic acid
LIFR	LIF receptor	MRTF	myocardin-related transcription factor
LQTS	long QT syndrome		
LV	left ventricle (ventricular)	MS	mass spectrometer/spectrometry
LVAD	left ventricular assist device	MS/MS	tandem mass spectrometry
LVEF	left ventricular ejection fraction	MSC	mesenchymal stem cell
LVH	left ventricular hypertrophy	MT-MMP	membrane-type MMP
LVMI	left ventricular mass index	mTOR	mammalian target of rapamycin
MAD	mandibuloacral dysplasia	MudPIT	multidimensional protein identification technology
MADS	MCM1, agamous, deficiens, and SRF		
		MuRF1	muscle RING finger protein 1
MALDI	matrix-assisted laser desorption ionization	MyBP-C	myosin-binding protein C
		NARP	neuropathy, ataxia, and retinitis pigmentosa
MAO	monoamine oxidase		
MAP	mitogen-activated protein	NCX	sodium–calcium exchanger
MAPC	multipotential adult progenitor cell	NE	norepinephrine
		NEP	neutral endopeptidase
MAPK	mitogen-activated protein kinase	Nesprin	nuclear envelope spectrin repeat protein
MCAD	medium chain acyl coenzyme A dehydrogenase		
		NFAT	nuclear factor of activated T cells
MCIP	modulatory calcineurin-interacting protein/myocyte-enriched calcineurin-interacting protein	NF-ATc	nuclear factor of activated T-cells
		NF-κB	nuclear factor-κB
		NHLBI	National Heart Lung and Blood Institute
Mcl-1	myeloid cell leukemia sequence 1		
MEB	muscle-eye-brain disease	Nix	Nip3-like protein X
MEF	myocyte enhancer factor	NMR	nuclear magnetic resonance
MELAS	mitochondrial encephalomyopathy, lactic acidosis, and stroke-like episodes	nNOS	neuronal isoform of NO synthase
		NO	nitric oxide

NOMAS	The Northern Manhattan Study
NOS	nitric oxide synthase
NRVM	neonatal rat ventricular myocytes
NSF	N-ethylmaleimide-sensitive factor
NYHA	New York Heart Association
OMM	outer mitochondrial membrane
OPMD	oculopharyngeal muscular dystrophy
Or-SUDS	Oregon Sudden Unexpected Death (study)
OXPHOS	oxidative phosphorylation
PA	phosphatidic acid
PAGE	polyacrylamide gel electrophoresis
PAK	p21-activated kinase
PARP	poly(ADP-ribose) polymerase
PBS	phosphate-buffered saline
PCA	primary cardiac arrest
PCR	polymerase chain reaction
PDC	pyruvate dehydrogenase complex
PDE	phosphodiesterase
PDE	phosphodiesterase
PDGFR	platelet-derived growth factor receptor
PDZ	PSD-95/Dlg/ZO-1
PE	phenylephrine
PESP	post-extrasystolic potentiation
PGDF	Platelet-derived growth factor
$PGF_{2\alpha}$	prostaglandin $F_{2\alpha}$
PH	pleckstrin homology (domain)
Pi	inorganic phosphate
pI	isoelectric point
PI3K	phosphoinositide 3-kinase
PICP	carboxyterminal propeptide of type I procollagen
PINP	aminoterminal propeptide of type I procollagen
PIP_2	phosphatidyl 4,5-biphosphate
PKA	protein kinase A
PKB	protein kinase B (also known as Akt)
PKC	protein kinase C
PLB	phospholamban
PLB	phospholipase B
PLC	phospholipase C
PLC	propionyl-L-carnitine
PLD	phospholipase D
PLN	phospholamban
PMA	phorbol 12-myristate 13-acetate

PMF	peptide mass fingerprinting
POH	pressure overload hypertrophy
PP-1	protein phosphatase-1
PP2A	Protein phosphatase-2A
PPH	phosphatidate phosphohydrolase
PTCA	percutaneous transluminal coronary angioplasty
$PtdIns(4,5)P_2$	phosphatidylinositol 4,5-bisphosphate
PWT	posterior wall thickness
PYD	pyrin domain
RAAS	renin–angiotensin–aldosterone
rAAV	recombinant adeno-associated virus
RACK1	receptor for activated C kinase1
RAS	renin–angiotensin system
RB	retinoblastoma
RBBB	right bundle branch block
RCC	rapid cooling contracture
RIP	receptor-interacting serine-threonine protein kinase
RMD	rippling muscle disease
ROK	rho-dependent kinase
ROS	reactive oxygen species
RP-HPLC	reversed phase HPLC
RPTK	receptor protein tyrosine kinase
rRNA	ribosomal RNA
RTK	receptor tyrosine kinase
RV	right ventricle
RVOT	right ventricular outflow tract tachycardia
RyR	ryanodine receptor
SA	sinoatrial
SAC	stretch-activated ion channel
SAECG	signal-averaged ECG
SAN	sinoatrial node
SAVE	Survival and Ventricular Enlargement (trial)
Sca-1	stem cell antigen-1
SCAD	short chain acyl coenzyme A dehydrogenase
SCD	sudden cardiac death
SDS-PAGE	SDS polyacrylamide gel electrophoresis
SEC	size exclusion chromatography
SELDI	surface-enhanced laser desorption-ionization (mass spectrometry)
SEPN1	Selenoprotein N1

SERCA	sarcoplasmic reticulum Ca^{2+} ATPase		TOF-MS	'time-of-flight' mass spectrometer
SHR	spontaneously hypertensive		TOT	tropomodulin-overexpressing transgenic
SIDS	sudden infant death syndrome		TPPE	thiamine pyrophosphate effect
SL	sarcomere length		TR	thyroid hormone receptor
Smac	second mitochondria-derived activator of caspase		TRADD	TNF receptor superfamily 1A-associated via death domain
SNP	single nucleotide polymorphism		TRAF	TNF receptor-associated factor
SOD	superoxide dismutase		tRNA	transfer RNA
SOLVD	Studies of Left Ventricular Dysfunction		TSC-22	transforming growth factor β-stimulating clone
SP	'side population'		TSH	thyroid-stimulating hormone
SR	sarcoplasmic reticulum		TTP	tristetraprolin
SRF	serum response factor		UBF	upstream binding factor
ssTnI	slow skeletal TnI		UCP2	uncoupling protein 2
STAT	signal transducer and activator of transcription		UV	ultraviolet
			VACR	voltage-activated Ca^{2+} release
SWOP	second window of preconditioning		VALIANT	Valsartan in Acute Myocardial Infarction (study)
TAC	transverse aortic constriction			
TACE	tumor necrosis factor-alpha converting enzyme		VCAM	vascular cell adhesion molecule
			VDAC	voltage-dependent anion channel
TACE	TNF-α-converting enzyme		VEGF	vascular endothelial growth factor
TBRIII	transforming growth factor-beta type III receptor		VF	ventricular fibrillation
			VLCAD	very long chain acyl coenzyme A dehydrogenase
TCA	tricarboxylic acid			
TCP-1	T-complex polypeptide 1 (also known as CCT)		VOH	volume overload hypertrophy
			VSMC	vascular smooth muscle cells
TEF	transcription enhancer factor		VSRM	voltage-sensitive release mechanism
Tfam	mitochondrial transcription factor A			
			VT	ventricular tachycardia
TFP	trifunctional protein		WT	wild-type
TG	transgenic		WT-1	Wilms' tumor-1
TGF	transforming growth factor		WWII	Second World War
TGF-β	transforming growth factor-β		WWS	Walker–Warburg syndrome
TIMI	Thrombolysis In Myocardial Infarction (trial)		X-GAL	5-bromo-4-chloro-3-indolyl-β-D-galactoside
TIMP	tissue inhibitor of metalloproteinases		XLDCM	X-linked dilated cardiomyopathy
			XL-EDMD	X-linked Emery–Dreifuss muscular dystrophy
Tm	tropomyosin			
TM	transmembrane (receptor)		XO	xanthine oxidase
TNF	tumor necrosis factor		YY1	yin yang-1
TNF-α	tumor necrosis factor-α		ZASP	Z line-associated proteins
TnI	troponin I		ZO	zonula occludens

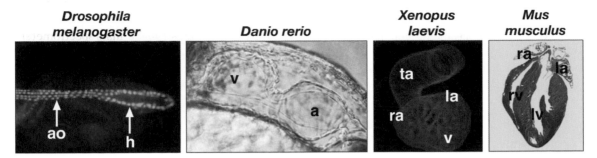

Figure 1.1. *Heart structures. Shown are hearts of* Drosophila melanogaster *(fruit fly) at stage 16;* Danio rerio *(zebrafish) at 48 hours post-fertilization;* Xenopus laevis *(tadpole) at stage 46; and* Mus musculus *(mouse) at 6 weeks of age. The* Drosophila melanogaster *heart shows the cardiac enhancer of the* Hand *gene driving green fluorescent protein (a gift from Zhe Han at University of Texas Southwestern); the zebrafish heart was modified[79] with permission from Dider Stainier at University of California, San Francisco; and the* Xenopus laevis *heart, stained with cardiac troponin T was obtained from Xenbase (www.xenbase.org/atlas/atlas.html). a, atrium; ao, aorta; h, heart; la, left atrium, lv, left ventricle; ra, right atrium; rv, right ventricle; ta, truncus arteriosis; v, ventricle.*

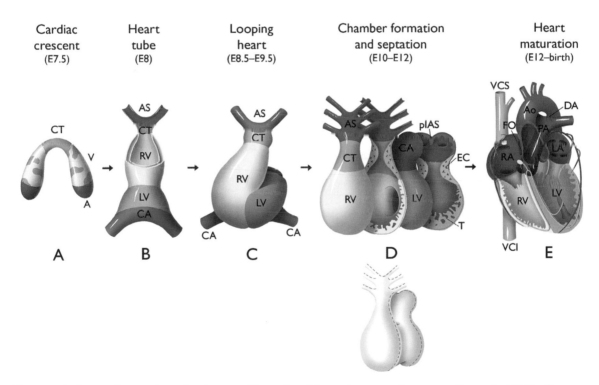

Figure 1.2. *Stages of murine heart development. Illustration of five major stages of mouse cardiogenesis. (A) Cardiac crescent formation at E7.5; (B), heart tube at E8.0; (C), looping of the heart at E8.5–9.5; (D), chamber formation and septation at E10–12, the grey image below shows the whole heart with broken red lines marking the site of sectioning to display the internal regions of the heart; (E), chamber maturation, septation and valve-formation at E12 to birth; coronary arteries are shown in red and conduction system (sinoatrial node, atrioventricular node, His-bundle and branches) in yellow. A, atrium; Ao, aorta; AS, atria sinus venosa; CA, common atrium; CT, conotruncus; DA, ductus arteriosus; EC, endocardial cushion; FO, foramen ovale; LA, left atrium; LV, left ventricle; PA, pulmonary artery; pIAS, primary intra-atrial septum; RA, right atrium; RV, right ventricle; T, trabeculation; V, ventricles; VCS, vena cava superior; and VCI, vena cava inferior. (Illustrated by Ryan Carre, University of Texas. Southwestern Medical Center).*

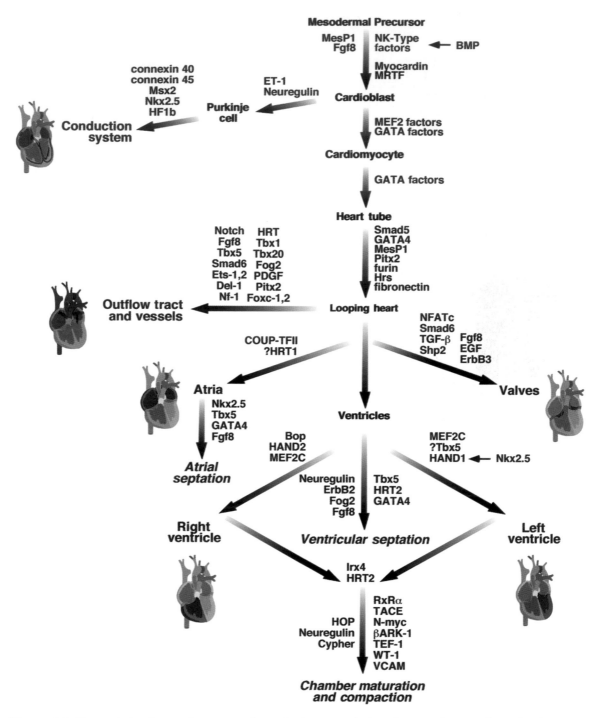

Figure 1.3. *Transcription factors determining heart development. A flow diagram depicting heart development and highlighting the transcription factors involved at each stage. Modified[109] with permission from Deepak Srivastava and Eric N Olson (University of Texas Southwestern Medical Center).*

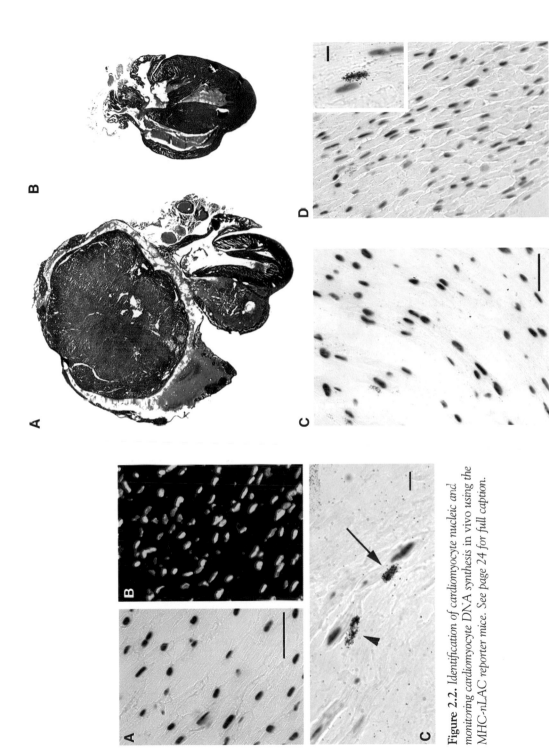

Figure 2.3. *Examples of genetic modifications that enhance cardiomyocyte cell cycle activity in vivo. See page 29 for full caption.*

Figure 2.2. *Identification of cardiomyocyte nucleic and monitoring cardiomyocyte DNA synthesis in vivo using the MHC-nLAC reporter mice. See page 24 for full caption.*

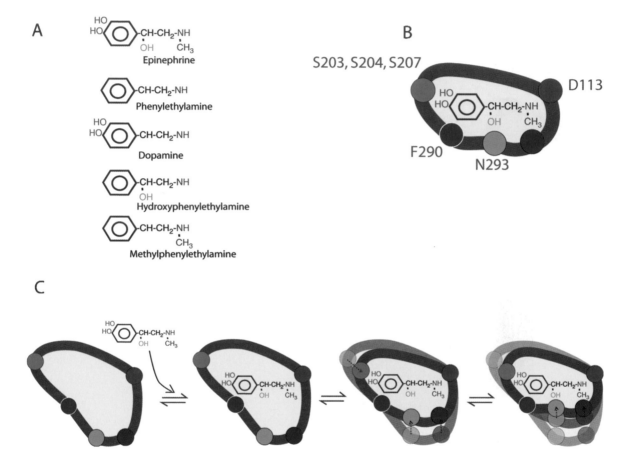

Figure 5.8. *Models of agonist binding and activation. (A) Structures of epinephrine, phenylethylamine, dopamine, hydroxyphenylethylamine and methylphenylethylamine. (B) Lock-and-key-type agonist-binding model. Receptor sites that interact with specific substituents of the ligand are shown as colored circles. (C) Sequential agonist binding model. The models are based on a study by Liapakis et al,[84] and previously published in an accompanying commentary: Kobilka B. Agonist binding: a multistep process. Mol Pharmacol 2004; 65: 1060–2.*

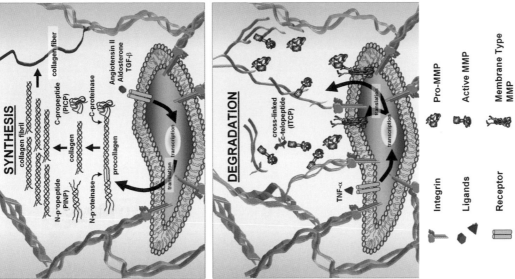

Figure 7.2. *The balance between synthesis and degradative rates determines the steady state levels of fibrillar collagen within the myocardial interstitium. See page 106 for full caption.*

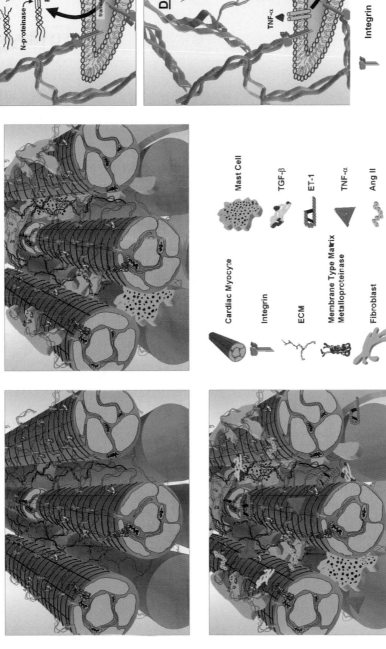

Figure 7.1. *While historically considered a static structure, it is now becoming recognized that the myocardial extracellular matrix (ECM) is a complex microenvironment containing a large portfolio of matrix proteins, signaling molecules, proteases, and cell types. See pages 102–103 for full caption.*

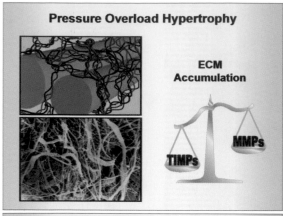

Figure 7.3. *Significant differences in ECM composition and structure occur in pressure (POH) or volume overload (VOH) hypertrophy. See page 110 for full caption.*

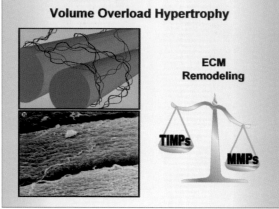

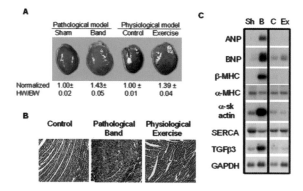

Figure 8.2. *Pathological and physiological cardiac hypertrophy. Please see page 119 for full caption.*

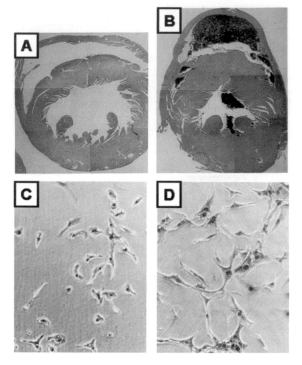

Figure 16.1. *Hypertrophic effects of IL-6. Coronal section of (A) littermate control and (B) double transgenic mice overexpressing IL-6 and IL-6R, showing the development of cardiac hypertrophy in the double transgenic mice. Morphologic appearance of cultured neonatal cardiac myocytes in the absence (C) and presence (D) of IL-6 and IL-6R, showing increased cell size in the cells that were treated with IL-6 and IL-6R. Reproduced with permission from Hirota H, Yoshida K, Kishimoto T et al. Continuous activation of gp130, a signal-transducing receptor component for interleukin 6-related cytokines, causes myocardial hypertrophy in mice. Proc Natl Acad Sci USA 1995; 92: 4862–6.*

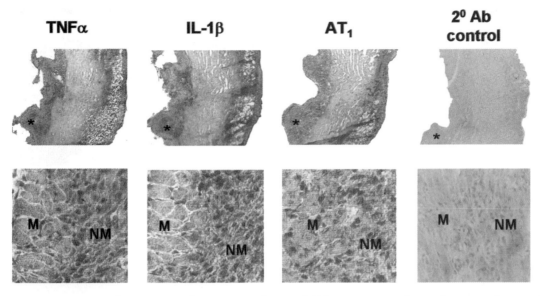

Figure 15.1B. *Association between pro-inflammatory cytokines and AT₁ receptor upregulation. Association between the appearance of pro-inflammatory cytokines and AT₁ upregulation in the post-MI heart. Sections were obtained from an adult male rat heart 7 days following coronary artery ligation. A large infarct in the anterior wall of the left ventricle is noted. Immunohistochemical staining demonstrates the appearance of both TNF-α and IL-1β in the peri-infarction zone. There is also evidence of increased AT₁ receptor density on non-myocytes (but not on myocytes) that have infiltrated into this region. M, myocytes; NM, non-myocytes.*

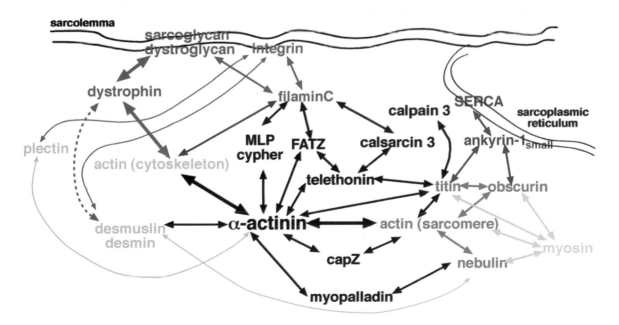

Figure 19.3. *The Z line. See page 316 for full caption.*

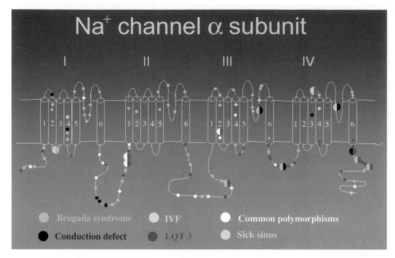

Figure 23.1a. *The voltage dependent Na⁺ channel (SCN5A) structure and gating kinetics. (A) Topological diagram of the transmembrane structure of SCN5A, with the extracellular side at the top. The locations of arrhythmia-causing mutations and common (>0.5%) polymorphisms are shown. (B) Kinetic state diagram for SCN5A. See text for further explanation. IVF, idiopathic ventricular fibrillation.*

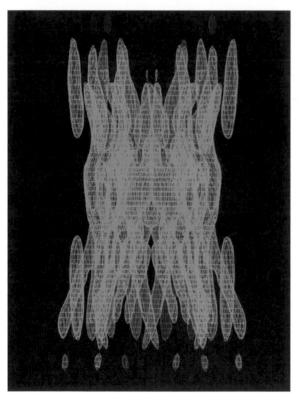

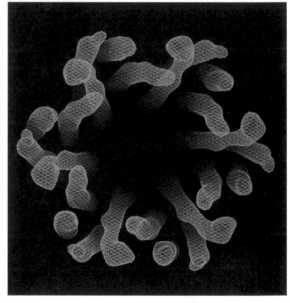

Figure 25.3. *Left panel: side view of the three-dimensional density map of a recombinant cardiac gap junction channel. The outer boundary of the MAP shows that the diameter of the channel is ~80 Å within the two membrane regions and then narrows in the extracellular gap to a diameter of ~50 Å. The rod-like features in the MAP corresponding to membrane α-helices. Right panel: a top view looking toward the extracellular gap of the three-dimensional map of recombinant gap junction channel. For clarity, only the cytoplasmic and most of the membrane-spanning regions of one connexon are shown. The rod-shaped features reveal the packing of 24 transmembrane α-helices within each connexin. Reprinted from Unger VM, Kumar NM, Gilula NB et al. Three-dimensional structure of a recombinant gap junction membrane channel. Science 1999; 283: 1176–80.*

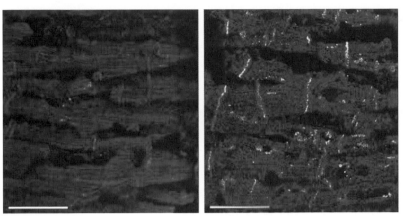

Figure 25.8. *Confocal images of control human myocardium (left) and failing human myocardium (right) stained with an anti-Cx45 antibody showing upregulation of Cx45 in the failing heart. Bar = 20 μm. Reprinted with permission from Yamada KA, Rogers JW, Sundset R et al. Upregulation of connexin45 in heart failure. J Cardiovasc Electrophysiol 2003; 14: 1205–12.*

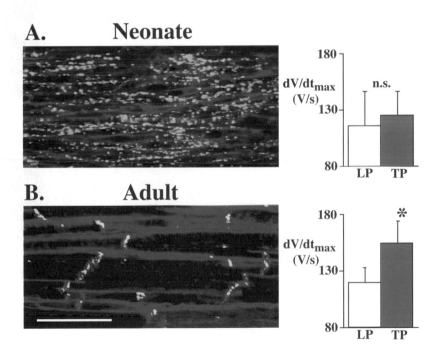

Figure 25.9. *Connexin43 distribution in neonatal and adult canine ventricular muscle (left) and the associated anisotropic changes in dV/dt$_{max}$ (right). Gap junctions (green or yellow) are labeled with antibodies to Cx43. The sarcolemma (red) is labeled with wheat germ agglutinin.[111] (A) Neonatal distribution of gap junctions typical of that in canine hearts from birth to 2 months of age. In the neonatal ventricular muscle, there is no significant difference in mean LP dV/dt$_{max}$ and mean TP dV/dt$_{max}$, which also occurs in synthetic neonatal cellular monolayers.[112] (B) Mature canine ventricular muscle. Bar = 50 μm. In adult ventricular muscle, mean TP dV/dt$_{max}$ is significantly greater than mean LP dV/dt$_{max}$. Data presented as mean ±1 SD; *P < 0.001 (n = 24). Modified with permission from Spach MS, Heidlage JF, Dolber PC et al. Electrophysiological effects of remodeling cardiac gap junctions and cell size: experimental and model studies of normal cardiac growth. Circ Res 2000; 86: 302–11.*

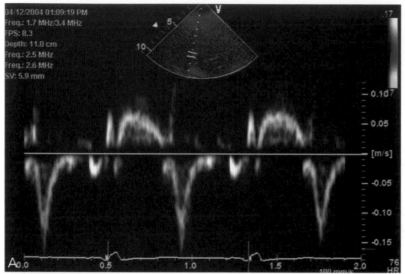

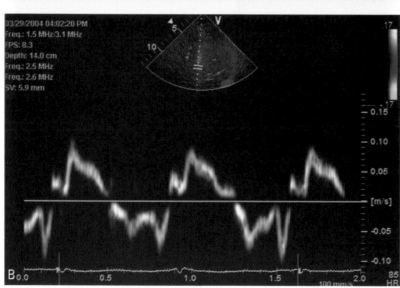

Figure 28.3. *Tissue Doppler echocardiographic images of the interventricular septum in (A) a normal individual, and (B) a hypertrophic cardiomyopathy patient.*

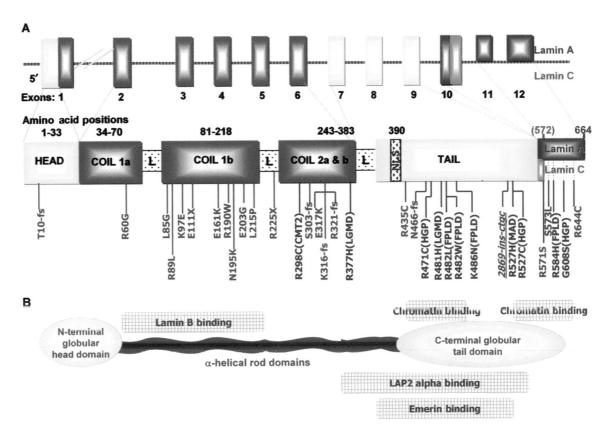

Figure 33.1. *The LMNA gene. (A) The exons encoding the domains of lamin A/C are shown; alternative splicing in exon 10 results in the two principal products. The location of mutations reported in dilated cardiomyopathy (red) and other laminopathies (blue) are indicated by their corresponding amino acid residue position and the abbreviation for the amino acid. CMT2, Charcot–Marie–Tooth type 2; LGMD, limb girdle muscular dystrophy; HGP, Hutchinson–Gilford progeria; MAD, mandibuloacral dysplasia; FPLD, familial partial lipodystrophy; fs, frameshift; NLS, nuclear localization signal region, L-linker segment. All mutation numbers refer to amino acid positions except for the italicized entry in the tail region representing a 4 bp insertion at mRNA position 2869/amino acid residue 525. (B) The general structure of the lamin A or C protein product with the various reported binding sites (to other proteins and chromatin) is depicted.*

A

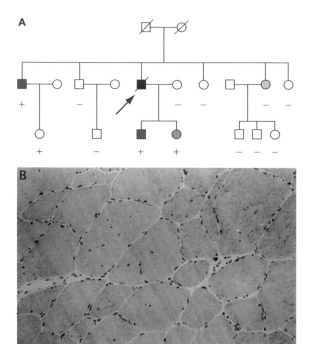

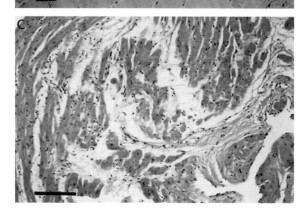

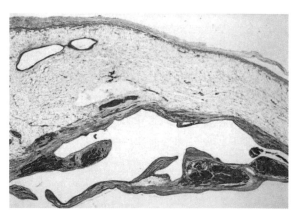

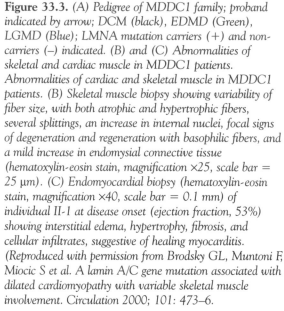

Figure 33.3. (A) Pedigree of MDDC1 family; proband indicated by arrow; DCM (black), EDMD (Green), LGMD (Blue); LMNA mutation carriers (+) and non-carriers (–) indicated. (B) and (C) Abnormalities of skeletal and cardiac muscle in MDDC1 patients. Abnormalities of cardiac and skeletal muscle in MDDC1 patients. (B) Skeletal muscle biopsy showing variability of fiber size, with both atrophic and hypertrophic fibers, several splittings, an increase in internal nuclei, focal signs of degeneration and regeneration with basophilic fibers, and a mild increase in endomysial connective tissue (hematoxylin-eosin stain, magnification ×25, scale bar = 25 μm). (C) Endomyocardial biopsy (hematoxylin-eosin stain, magnification ×40, scale bar = 0.1 mm) of individual II-1 at disease onset (ejection fraction, 53%) showing interstitial edema, hypertrophy, fibrosis, and cellular infiltrates, suggestive of healing myocarditis. (Reproduced with permission from Brodsky GL, Muntoni F, Miocic S et al. A lamin A/C gene mutation associated with dilated cardiomyopathy with variable skeletal muscle involvement. Circulation 2000; 101: 473–6.

Figure 33.6. Histological findings in ARVD/C showing fatty tissue replacement of myocardial muscle in right free ventricular wall (×2.5). Reproduced with permission from Severini GM, Krajinovic M, Pinamonti B et al. A new locus for arrhythmogenic right ventricular dysplasia on the long arm of chromosome 14. Genomics 1996; 31: 193–200.

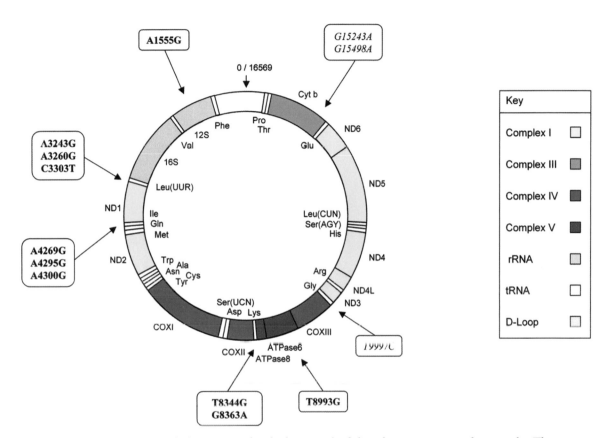

Figure 34.5. *Morbidity map of the human mitochondrial genome for defects that may cause cardiomyopathy. The mitochondrial genome contains two circular strands of DNA, an outer strand and an inner strand. The two strands encode for a total of 37 genes, 2 ribosomal RNAs (rRNAs), 22 transfer RNAs (tRNAs) and 13 messenger RNAs (mRNAs). For simplicity, the two strands have been shown as a single circular DNA. The 2 rRNA genes and the 22 tRNA genes are indicated inside the circle. The rRNA genes are designated by the sedimentation rate of the product they encode (12S and 16S), whereas the tRNA genes are designated by the standard three-letter abbreviation for the amino acids they transfer. The 13 mRNA genes are designated by the standard abbreviations for the OXPHOS subunits they encode: complex I (ND, NADH dehydrogenase), ND1, ND2, ND3, ND4, ND4L, ND5, ND6 and ND7; complex III cytochrome b, (cyt b); complex IV (COX, cytochrome c oxidase), COXI, COXII and COXIII; and complex V (ATPase, ATP synthase), ATPase6 and ATPase8. The origin of replication is located in the displacement loop (D-loop); the nucleotides of the mtDNA sequence also begin and end in the displacement loop. Mutations that have been confirmed to cause cardiomyopathy are indicated in bold type, while mutations that have been implicated provisionally (i.e. only demonstrated by one laboratory) are indicated in italics.*[24]

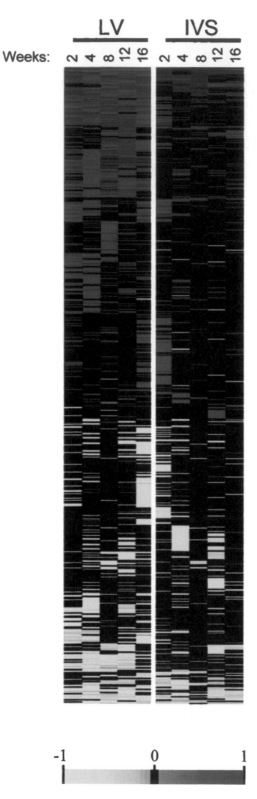

Figure 36.3. *Heat map display of clustered gene expression patterns of 731 clones that display differential expression in rat myocardial infarction (MI). Each row represents a different cDNA, and columns pertain to data collected at five time points (weeks) after surgery from left ventricular infarct zone (LV) and non-infarct zone (IVS). Normalized data values, displayed in shades of red and blue, represent elevated and repressed expression, respectively, in MI tissue relative to control tissue (scale is shown at bottom). Insignificant differential expression values between 1.4 and −1.4 (before normalization) are set to 0 and shown in black. Genes with similar expression patterns are clustered together in 58 different clusters, and the clusters were arranged by nearest similarity to other clusters. Reproduced with permission from Stanton LW, Garrard LJ, Damm D et al. Altered patterns of gene expression in response to myocardial infarction. Circ Res 2000; 86: 939–45.*

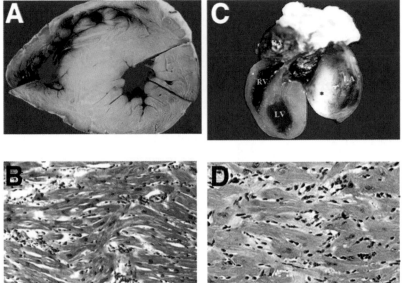

Figure 37.2. *Comparison of mouse and human HCM. Postmortem human heart sample showing gross hypertrophy and reduction in left ventricular chamber size (A) associated with myocyte hypertrophy, myofiber disarray and interstitial fibrosis (B). Postmortem heart sample form mouse expressing the Arg403Gln mutation in the myosin heavy chain gene, showing myocardial hypertrophy and left atrial enlargement (C) associated with typical histopathological features of HCM (D). LA, left atrium; LV, left ventricle; RV, right ventricle.*

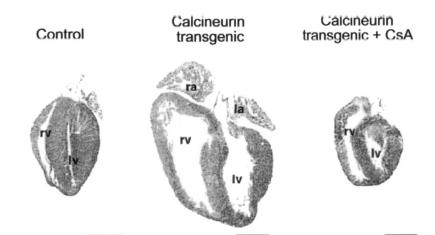

Figure 39.4. *Inhibition of calcineurin signaling prevents the development of DCM in activated calcineurin-expressing transgenic mice. This transgenic model implicates calcineurin signaling as a novel pathway for the development of cardiac hypertrophy and DCM. Hematoxylin and eosin-stained histological section of hearts from non-transgenic mice (left) and calcineurin transgenic mice treated with vehicle (center) or CsA (right). Mice expressing activated calcineurin show DCM at 25 days (center), while CsA treatment attentuates cardiac chamber dilation. la, left atrium; lv, left ventricle; ra, right atrium; rv, right ventricle. From Molkentin JD, Lu JR, Antos CL et al. A calcineurin-dependent transcriptional pathway for cardiac hypertrophy. Cell 1998; 93: 215–28, with permission.*

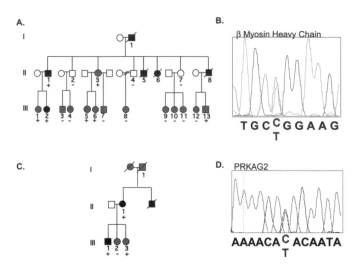

Figure 40.1. *Genetic definition of the molecular basis for cardiomyopathy in two families. Pedigrees indicated sex (circles denote women, boxes denote men) and clinical status (solid fill, affected; open, unaffected, hatched uncertain status). Genotypes are provided: + mutation present; − mutation absent. (A) Sudden death in multiple individuals (I-1, II-5, -6, -8) prompted clinical evaluation of surviving family members. Clinical diagnosis was uncertain in II-3 who had prior thoracic surgery and poor imaging studies, and in younger family members (generation IV). Genetic analyses of affected individual II-1 defined a disease-causing mutation that allowed definitive genotype assignment of all family members. Note that genotype clarified diagnosis of II-3 and identified individuals at risk death from HCM. (B) An arginine to glutamine substitution at residue 719 in the β myosin heavy chain gene causes HCM. DNA traces show both the normal and mutant myosin sequences. (C) A small family with cardiac hypertrophy (II-1 and III-1) and electrographic abnormalities (II-1, first degree atrioventricular block and bradycardia; III-1 ventricular pre-excitation pattern). Pedigree structure is consistent with dominant inheritance. (D) DNA analyses of the PRKAG2 gene revealed a tyrosine to histidine missense mutation in residue 477, indicating glycogen storage as the cause of cardiac hypertrophy. Individual II-1 subsequently developed conduction system disease necessitating pacemaker implantation.*

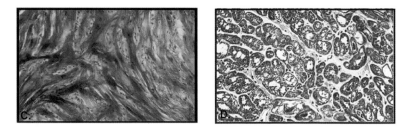

Figure 40.2c, d. *Genetic causes of hypertrophy: HCM and PRKAG2 cardiomyopathy. Comparison of cross-sectional echocardiogram images in HCM (A) and PRKAG2 cardiomyopathy (B). Note the comparable and significant LVH in each. Images were obtained late in diastole. (C) Mason trichrome stain of LV tissue reveals classifc HCM histopathology with myocyte (red) enlargement and disarray, and also significant interstitial fibrosis (blue). (D) Histopathology in PRKAG2 cardiomyopathy also shows myocyte hypertrophy but without disarray and with prominent vacuoles that appear empty, due to the loss of glycogen during tissue fixation. Note the minimal interstitial fibrosis compared to HCM specimen (C).*

Mechanisms for Cardiac Hypertrophy

CHAPTER 1

Mechanisms of normal cardiovascular growth and development

Jens Fielitz, Eric N Olson, Rhonda Bassel-Duby

Heart development involves a wondrous and precisely orchestrated series of molecular and morphogenic events, and even subtle perturbation of this process can have catastrophic consequences in the form of congenital heart disease. Many details of the morphological formation of the heart are known; however, many of the molecules and mechanisms that regulate heart development and growth are still being elucidated. Identification of the genes that orchestrate heart formation is a nec-

essary prerequisite for development of therapeutic strategies for cardiac repair and regeneration in the settings of congenital and acquired cardiovascular diseases. Over the past few decades, hearts of fruit flies, zebrafish, tadpoles, chicken and mice have served as model systems to elucidate the signaling factors involved in heart formation (Fig. 1.1). Strikingly, although the Drosophila 'heart' is a simple linear tubular structure and the vertebrate heart is a multichamber complex organ, many of

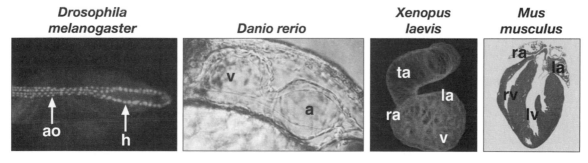

Figure 1.1. *Heart structures. Shown are hearts of* Drosophila melanogaster *(fruit fly) at stage 16;* Danio rerio *(zebrafish) at 48 hours post-fertilization;* Xenopus laevis *(tadpole) at stage 46; and* Mus musculus *(mouse) at 6 weeks of age. The* Drosophila melanogaster *heart shows the cardiac enhancer of the* Hand *gene driving green fluorescent protein (a gift from Zhe Han at University of Texas Southwestern); the zebrafish heart was modified[79] with permission from Dider Stainier at University of California, San Francisco; and the Xenopus laevis heart, stained with cardiac troponin T was obtained from Xenbase (www.xenbase.org/atlas/atlas.html). a, atrium; ao, aorta; h, heart; la, left atrium, lv, left ventricle; ra, right atrium; rv, right ventricle; ta, truncus arteriosis; v, ventricle. See color plate section.*

the transcription factors that regulate heart development are evolutionarily conserved. Based on morphological, biochemical and genetic information garnered from various organisms, a clear picture of the mechanisms of heart development is starting to emerge. This chapter highlights some of the transcription factors and mechanisms involved in heart formation, starting at commitment of a mesodermal precursor cell to a cardiac cell lineage onto heart chamber maturation. Elucidation of the mechanisms governing heart development will provide necessary information to understand the intricacies of heart formation in humans, and offers critical contributions to the design of therapeutics to combat congenital and adult heart disease.

Progression of mesoderm to cardiomyocytes to heart tube

The embryonic plate, initially possessing two layers, is ovoid, and is formed at the union between the yolk sac and the amniotic cavity. The primitive streak is located in the midline of the long axis of the oval disc, with the node at its cranial end. Cells migrate from the upper layer through the primitive streak, by a process called gastrulation, to form the three germ layers of the embryo proper: the ectoderm, the endoderm, and the mesoderm. Commitment of mesodermal cells to the cardiogenic lineage is established during and shortly after gastrulation, upon which the cardiac mesoderm, head mesoderm and foregut endoderm ingress through the primitive streak and migrate to the anterior of the embryo, condensing into the cardiac crescent (Fig. 1.2A).

In *Drosophila melanogaster* (fruit fly), as in vertebrate embryos, the initial commitment of mesodermal progenitor cells to a cardiac fate is dependent on signaling between adjacent tissues. Entry of cells into the cardiac lineage in response to the appropriate signals is coupled to the expression of a set of transcription factors that initiate the program for cardiac gene expression (Fig. 1.3). Members of the bone morphogenetic protein (BMP) family play positive roles in establishment of the cardiac lineage, while members of the Wnt family, *wingless* in *Drosophila*, are involved in both

induction and suppression effects of cardiogenesis, depending on the Wnt isoform and Wnt mediator.[1] Overexpression of Wnt3A and Wnt8 proteins in frog embryos or exposure of mesoderm from the heart field of chick embryos to Wnt3A and Wnt1 blocks cardiac differentiation.[2,3] In mice, endodermal ablation of *beta-catenin*, a downstream effector of the canonical Wnt signaling pathway, results in the formation of multiple hearts.[4]

A major breakthrough in our understanding of heart development was the discovery of the NK-class homeobox gene *tinman*, which encodes a transcription factor that is required for the formation of the primitive heart in *Drosophila* and marks the commitment of the precursor cells to the cardiac lineage.[5] A mammalian ortholog of *tinman*, called *Nkx2.5/Csx*, is expressed from the onset of embryonic heart formation, initially in the cardiac crescent, until adulthood.[6,7] Although highly conserved and restricted to the cardiac lineage like *tinman*, mammalian *Nkx2.5* is dispensable for establishment of the cardiac lineage. Instead, mice lacking *Nkx2.5* die during mid-embryogenesis from abnormalities in growth and development of the left ventricular chamber of the heart, which suggests that *Nkx2.5* serves a different function than *tinman*, or that other genes, such as *Nkx2.3*, *Nkx2.6*, *Nkx2.7* and *Nkx2.8*, have redundant functions in specification of cardiac cell fate.[8]

Expressed concomitantly with Nkx2.5 is myocardin, a potent activator of serum response factor.[9] Studies in *Xenopus* showed that expression of a dominant negative form of myocardin ablates heart formation, and ectopic expression of myocardin induces cardiac gene expression.[10] These findings implicate a role for myocardin in early differentiation of cardiac cells and show that myocardin directs cells to a cardiac lineage. In mice lacking myocardin, a severe vascular defect was seen at embryonic day 10.5 (E10.5) resulting in embryonic lethality.[11] However, no morphological abnormalities were observed in the developing heart in mice lacking myocardin suggesting that the myocardin-related transcription factors (MRTFs), MRTF-A and MRTF-B, expressed in cardiac cells play a redundant role in specification of cardiac cells.

The GATA family of zinc finger-containing genes has been shown to be involved in the pro-

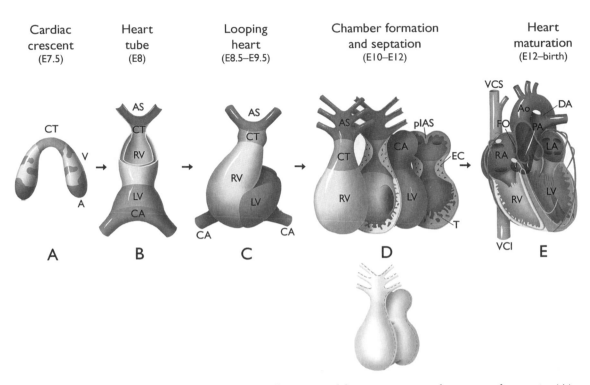

Figure 1.2. *Stages of murine heart development. Illustration of five major stages of mouse cardiogenesis. (A) Cardiac crescent formation at E7.5; (B), heart tube at E8.0; (C), looping of the heart at E8.5–9.5; (D), chamber formation and septation at E10–12, the grey image below shows the whole heart with broken lines marking the site of sectioning to display the internal regions of the heart; (E), chamber maturation, septation and valve-formation at E12 to birth. A, atrium; Ao, aorta; AS, atria sinus venosa; CA, common atrium; CT, conotruncus; DA, ductus arteriosus; EC, endocardial cushion; FO, foramen ovale; LA, left atrium; LV, left ventricle; PA, pulmonary artery; pIAS, primary intra-atrial septum; RA, right atrium; RV, right ventricle; T, trabeculation; V, ventricles; VCS, vena cava superior; and VCI, vena cava inferior. (Illustrated by Ryan Carre, University of Texas Southwestern Medical Center). See color plate section.*

gression of differentiation of the mesodermal progenitor cells to a cardiac fate (Fig. 3).[12] In *Drosophila*, the GATA4 homolog *pannier* is required for normal proliferation of cardiogenic precursors,[13] and in zebrafish a mutation of the GATA5 homolog, *faust*, causes impaired cardiac differentiation and cardia bifida.[14] In mice, three GATA factors, GATA4, GATA5 and GATA6, are expressed in the developing heart. Using tetraploid embryo complementation it has been shown that clonal E9.5 GATA4 knockout mice embryos display heart defects characterized by disrupted looping morphogenesis and septation, and a hypoplastic ventricular myocardium.[15] However, GATA6 seems to be dispensable for early development of the heart, septum transversum mesenchyme, and vasculature.[16] Unlike the phenotype seen in *faust* (GATA5) zebrafish mutants, mice lacking GATA5 showed no obvious cardiac defects, raising the possibility that during evolution different functions may have arisen for these two GATA factors.[17,18] However, GATA5 knockout mice may express a truncated form of GATA5, containing both zinc fingers and the C-terminal activation domain, which would still be transcriptionally active, thus a role for GATA5 during murine cardiogenesis

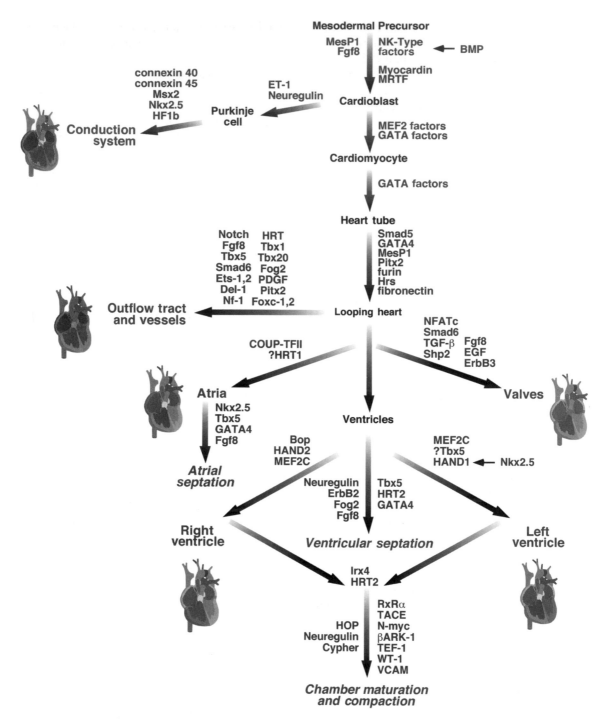

Figure 1.3. *Transcription factors determining heart development. A flow diagram depicting heart development and highlighting the transcription factors involved at each stage. Modified[109] with permission from Deepak Srivastava and Eric N Olson (University of Texas Southwestern Medical Center). See color plate section.*

cannot yet be excluded.[19] Notably, in humans, it has been shown that haplo-insufficiency of GATA4 can result in cardiac septal defects.[20] In addition, GATA factors have been shown to be involved in the migration of the cardiac progenitor cells to form the heart tube and in controlling cell survival in cardiomyocytes.[14,18,21,22]

MesP1, a basic helix-loop-helix (bHLH) transcription factor has been shown to be required for migration of cardiac precursor cells. Mice lacking *MesP1* lack migratory activity of the *MesP1*-expressing heart precursor cells; however, the cells eventually acquire migratory activity and give rise to an abnormal heart tube. Mice mutated in both the *MesP1* and *MesP2* genes have a complete block in migration of mesoderm from the primitive streak, resulting in a complete lack of cardiac cells.[23]

The myocyte enhancer factor-2 (MEF2) factors, MEF2A, MEF2B, MEF2C and MEF2D, are all expressed in cardiac cells and serve as transcription factors regulating other transcription factors necessary for heart formation.[24] MEF2 also autoregulates its expression and activates a myriad of cardiac-specific genes. Ablation of the *MEF2 Drosophila* homolog, *dmef*, results in a loss of any muscle formation, establishing *dmef* as a nodal transcription factor of muscle development. Cardiogenesis is initiated in mice lacking *Mef2c*, however, the regions of the heart corresponding to both the left ventricle and the right ventricle are severely hypoplastic.[25]

Fibroblast growth factor (Fgf) 8 is expressed in the cardiac mesoderm and later in the heart ventricle. In zebrafish, mutation of the *Fgf8* gene severely reduces the expression of *nkx2.5* and *gata4* and results in an abnormally small heart with a reduction of ventricular size. Fgf8 also plays a role in development of all the pharyngeal arches and in neural crest cell survival. Neural crest cells contribute to the formation of the aorto-pulmonary septum at the arterial pole of the heart. Mice lacking *Fgf8* have defects in cardiovascular patterning, including hypoplastic arch arteries and deficient outflow tract septation, primarily resulting in transposition with double outlet right ventricle and, less frequently, persistent truncus arteriosus.[26] Other cardiac defects include atrial and ventricular septal defects, as well as atrioventricular valve defects.

Cells of the cardiac crescent in the anterior lateral mesoderm, called the primary heart field, form the tubular heart. The rapid elongation of the tubular heart occurs through the addition of cells to its growing poles. At the venous pole, cells in the caudal region of the cardiac crescent give rise to myocardium of the atrioventricular canal, the atria and the inflow tract. At the arterial pole, the heart tube is extended by the anterior, or secondary, heart field; extracardiac cells situated in the pharyngeal mesoderm that contribute to the myocardium giving rise to the outflow tract and outlet region of the right ventricle.[27–29] The anterior heart field appears to be added to the heart at the time of cardiac looping.[27–29] The identification of at least two distinct populations of cells contributing to the heart provides a likely explanation for the spatial restriction of gene expression and function that has often been observed in the heart.[30]

The mechanisms controlling growth and differentiation processes within, and the crosstalk between, the primary and secondary heart field are just beginning to be understood. The LIM-homeodomain transcription factor Isl1 was one of the first transcription factors within the secondary heart field shown to be involved in transcriptional control of outflow tract and right ventricular development.[31] Furthermore, the forkhead transcription factor Foxh1 is also expressed in the secondary heart field, and *Foxh1* mutant embryos, like embryos lacking *Isl1* display defects in the outflow tract and right ventricle.[32] Isl1 and Foxh1 directly activate transcription of *Mef2c* in the anterior heart field by activating two independent cardiac enhancers in collaboration with GATA factors and Nkx2.5, respectively.[33] Furthermore, *Bop* is a direct target of MEF2C during anterior heart field development, implying that *Bop* is indirectly regulated by Isl1/GATA factors and Foxh1/Nkx2.5.[34] Thus, Isl1/GATA factors and Foxh1/Nkx2.5 appear to act at the top of a cascade of cardiac transcription factors in the anterior heart field.

The heart tube and the looping of the heart

The four-chambered heart develops from a single heart tube (Fig. 1.2). The linear heart tube is composed of two concentric layers: an outer muscular layer (the myocardium) and an inner endothelial layer (the endocardium), and is subdivided into two major chambers, with unidirectional blood flow directed from the posterior atrium to the anterior ventricle (Fig. 1.2B). However, in the linear heart tube the atrium is not connected to the right ventricle, and the left ventricle is not connected to the outflow tract. The right and left halves of the heart are brought into parallel position by looping of the heart tube (Fig. 1.2B and 1.2C), transforming the linear heart tube into a complex structure with two atria and two ventricles, assuring unidirectional blood flow with two separate, parallel circulations. The looping of the bulboventricular segment of the heart will cause the atrium and sinus venosus to become dorsal to the heart loop. Prior to looping, the tubular heart forms constrictions defined from cranial to caudal: bulbus cordis, ventricle and atrium (Fig. 1.2B). The bulbus cordis is continuous with the truncus arteriosus which connects cranially to the aortic sac. The truncus arteriosus gives rise to the aorta and pulmonary artery; the bulbus cordis forms the ventricular outflow tract; the primitive ventricle forms the inlet and future ventricles; the primitive atrium forms the right and left atria, and the sinus venosus forms the posterior right and left atria walls, coronary sinus, inferior and superior vena cava, umbilical and pulmonary veins (Fig. 1.2E). The truncus arteriosus is outside the cranial end of the pericardial sac, and the primitive atrium and sinus venosus lay outside the caudal end of the pericardial sac. As cardiac looping proceeds, the paired atria form a common chamber and move into the pericardial sac.

Misfolding of the heart tube due to the failure of migration and/or proliferation of cells involved in ventrolateral morphogenesis leads to cardia bifida, characterized by two lateral myocardial clusters instead of one single midline heart. Several gene knockout mice exhibit cardia bifida, including *fibronectin*,[35] GATA4,[18] *Furin*,[36] *Hrs*,[37] and *Smad5*[38] knockout mice, implicating involvement of these genes in formation of the looping heart (Fig. 1.3).

Tbx5, a member of the T-box transcription factor family, is required for correct development of the atria and the sinus venosus–atrial junctional region. Mice lacking *Tbx5* display a severely hypoplastic sinus venosus, atrial, and primitive ventricle segments together with a failure of looping by E8.5.[39] TBX5 mutations in humans have been implicated in Holt–Oram syndrome, a rare inherited disease characterized by congenital heart and upper limb defects. Mice lacking one copy of *Tbx5* that survive to adulthood display large ostium secundum atrial septal defects, whereas those that die as fetuses have ventricular septal defects. As in Holt–Oram syndrome patients, electrocardiographic analysis of these mice shows sinus pauses and both first and second-degree atrioventricular blocks, indicating abnormalities in the conduction system of the heart.

Disruptions of the *Tbx1* gene in mice produce aortic arch anomalies together with outflow tract and truncal septation defects resembling those seen in DiGeorge syndrome.[40] Clinical studies have shown that *Tbx1* mutations are responsible for different phenotypes observed in DiGeorge syndrome, confirming that *Tbx1* is a major genetic determinant of the syndrome.[41]

Tbx20 is expressed exclusively in endocardial cushions in the atrioventricular canal and outflow tract regions in mice at E9.5. By E10.5, *Tbx20* transcripts are detected in the myocardium of the atria and the ventricular wall, not in the trabeculae. In the adult, *Tbx20* expression remains stronger in the atria than the ventricles and occurs in the vena cava and the sinoatrial node. Mice lacking *Tbx20* die in mid-gestation with severely malformed, underdeveloped hearts in which heart chamber formation has failed.[42–44] Studies of the mutant *Tbx20* mice showed that Tbx20 regulates cell proliferation in a region-specific manner by repressing *Tbx2*, which in turn represses *Nmyc1*, revealing how Tbx20 coordinates the transcriptional programs that control heart tissue specification and growth.

Zebrafish studies of mutants with morphogenetic defects have revealed several genes that are essential for cardiac fusion and heart tube assembly, including *casanova*, *bonnie and clyde*, *gata5*, *one-eyed pinhead*, *hand2*, *miles apart*, and *heart and soul*. In addition, analysis of the *jekyll* gene has indicated its important role during the morphogenesis of the atrioventricular valve. Further studies are necessary to address the role of these factors in cardiac patterning, cardiac morphogenesis, and the relationship between these processes.[45]

Formation of the ventricles

Soon after cardiac muscle cells are specified, they become segregated into different populations based on their positions in the linear heart tube. Precursors of the right ventricular chamber are positioned anteriorly to those that give rise to the left ventricular chamber (Fig. 1.2C). As the heart loops and twists, these two populations of cells come to lie side by side and, through growth that is analogous to ballooning of the different regions of the heart tube, the two ventricular chambers are formed (Fig. 1.2C and 1.2D).[46] Newborns are often seen in which one ventricular chamber develops normally but the other is severely underdeveloped or even missing, so-called single outlet left- or right ventricle. The clinical observation that the growth of one ventricle or the other can be specifically perturbed suggests that the ventricular chambers develop independently. The notion that the heart develops in this type of modular manner is supported by the phenotypes of mouse and zebrafish mutants that exhibit abnormalities in specific segments of the heart, leaving the remainder of the heart intact.[47]

An important insight into the genetic circuitry of ventricular development was provided by the discovery of two chamber-restricted bHLH transcription factors, HAND1 (also known as eHAND/Thing-1/Hxt) and HAND2 (also known as dHAND/Hed), which are expressed in the left and right ventricular chambers, respectively.[48] During mouse heart development, the Hand genes are expressed in the cardiac crescent (Fig. 1.2A), but their expression patterns subsequently demarcate distinct regions of the developing heart tube. Hand2 is expressed throughout the linear heart tube before becoming restricted predominantly to the right ventricular chamber.[49] In contrast, as early as the linear heart tube stage, Hand1 is expressed in the specific segments that give rise to the left ventricle and conotruncus, and its expression remains restricted to these regions as the heart

develops.[50] The highly specific patterns of expression of these genes indicate that they respond to positional information and signals that are unique to different regions of the developing heart. Knockout mice lacking Hand2 fail to form a right ventricle, and mice lacking both genes do not form ventricles. These mutant mice, the first to demonstrate that a single gene mutation could result in the ablation of an entire segment of the heart, supported the notion that the heart is assembled in a modular fashion in which each compartment is governed by unique genetic programs.

Mice lacking the Hand1 gene die at E8.5 from placental and extra-embryonic abnormalities which has precluded analysis of its potential role in later stages of heart development. Using a conditional Hand1-null allele and cardiac-specific gene excision together with Hand2 knockout mice revealed specific functions of Hand1 in growth and maturation of the heart that are distinct from those of Hand2. In this regard, mouse embryos homozygous for the cardiac Hand1 gene deletion survived until the perinatal period, when they died from a spectrum of cardiac abnormalities, such as membranous ventricular septal defects, overriding aorta, hyperplastic atrioventricular valves, double outlet right ventricle and outflow tract defects.[51] However, this cardiac phenotype in Hand1 knockout mice is much less severe than that of Hand2 mutant embryos in which the entire right ventricular region of the heart is absent.[52] These differences in ventricular phenotypes are likely to reflect important distinctions in the expression patterns of Hand1 and Hand2 described above. Therefore, in the absence of Hand1, residual Hand2 expression in the LV and outflow tract may partially compensate for the loss of Hand1. By contrast, in the absence of Hand2 there is a more complete lack of HAND factors in the presumptive right ventricle (RV).[52]

The zebrafish genome has only a single hand gene that is most similar to hand2. Interestingly, the zebrafish heart has only a single ventricle, and the zebrafish mutant hands off, containing a mutation in the HAND2 gene, fails to form a ventricular chamber.[53]

Another family of bHLH transcription factors with important functions in cardiovascular development is the HRT (hairy-related transcription factors) family, also known as Hey, Hesr, CHF and Herp factors.[54] HRT genes show highly specific expression patterns that demarcate regions of the developing heart, vasculature, pharyngeal arches, and somites. HRT1 and HRT2 are expressed in atrial and ventricular precursors, respectively, and in the cardiac outflow tract, aortic arch arteries and pharyngeal arches.[54] In many regions of the embryo, the pattern of HRT gene expression mirrors that of Notch signaling-related molecules. The transmembrane receptor Notch participates in an evolutionarily conserved cell-interaction system that plays fundamental roles in embryonic patterning and development.[55] Upon activation by ligands such as Delta or Jagged on neighboring cells, the intracellular domain of the Notch receptor (Notch IC) is cleaved and translocated to the nucleus together with Suppressor of Hairless (Su(H)),[56] or related molecules. Su(H) provides DNA binding specificity, whereas Notch IC functions as an activation domain. In response to Notch signaling, Su(H) activates transcription of the HES genes, whose products repress transcription, thereby preventing undifferentiated precursors from achieving differentiated phenotypes.

HRT genes contain binding sites for Su(H) in their regulatory regions that mediate transcriptional activation by Notch signaling.[56] The existence of a negative feedback loop in which HRT proteins interfere with Notch-dependent activation of HRT2 expression has been demonstrated. This type of negative autoregulation may function to terminate Notch signaling, thereby resulting in a transient or periodic signal, as is characteristic of Notch signaling in many cell types. While it is clear that HRTs can function as transcriptional repressors, at least in part by recruiting histone deacetylases and the mSin3 and N-CoR corepressors, their specific target genes have not been defined. HRT proteins have also been shown to dimerize with HAND factors, which can modulate the expression of target genes for both sets of factors.[57]

Homozygous *Hrt2* null mice die during the perinatal period and show a spectrum of cardiac malformations including ventricular septal defects, tetralogy of Fallot, and tricuspid atresia.[58–60] Zebrafish embryos harboring a mutation in *gridlock*, an ortholog of hrt2, also show localized abnormal-

ities of the aorta that resemble human coarctation of the aorta.[61] Consistent with this finding, Notch and Delta ligands are specifically expressed in the vasculature, and Notch mutant embryos show embryonic lethality from vascular defects. Moreover, human mutations in *Jagged1*, which encodes a Notch ligand, cause Alagille syndrome, which is a dominantly inherited multisystem disorder that particularly affects the cardiovascular system, especially the outflow tract and aortic arch artery derivatives.[62]

In zebrafish, reduction of *hrt2/gridlock* expression results in downregulation of arterial markers and upregulation of venous markers.[63] Similarly, interference with Notch signaling reduces artery-specific markers. These findings suggest that the Notch–HRT2 pathway governs the choice between arterial and venous cell fates by actively repressing the venous gene program.

Expression of the Iroquois homeobox gene 4 (*Irx4*), a member of the Iroquois family of homeodomain-containing transcription factor genes, is restricted to the ventricular myocardium during heart development. In particular, Irx4 expression is more concentrated in the outer curvature of the myocardium which gives rise to ballooning chambers. In the absence of *Irx4*, the genes encoding atrium-specific slow myosin heavy chain and atrial natriuretic factor are upregulated in the ventricles, implicating Irx4 as a repressor of atria-specific gene expression in the ventricle. Studies showed that in fact Irx4 does not bind directly to the promoter region of the gene for atrium-specific slow myosin heavy chain but that it may repress by interacting with retinoic acid receptor/vitamin D protein complexes.[64] Mice lacking *Irx4* have a partial disturbance of ventricle-specific gene expression with impaired cardiac function.[65]

Pitx2 is a homeodomain transcription factor that plays a role in directing cardiac asymmetric morphogenesis. *Pitx2c* is expressed on the left side of the cardiac crescent both in mice and chickens. As heart development progresses, Pitx2c is restricted to the left side of the heart tube. With further development Pitx2c remains on the ventricular side of the looping heart. *Pitx2c* expression is confined to the left region of the inflow tract and atrioventricular canal, and *Pitx2c* expression in the outflow tract shifts from a ventral to a left position, spanning from the ventricular junction to the aortic arches.[66] Expression of *Pitx2c* declines in later fetal stages and is not detected past E16.5. Mice lacking *Pitx2* display severe cardiovascular defects, such as atrial isomerism, double inlet left ventricle, transposition of the great arteries, persistent truncus arteriosus and abnormal aortic arch remodeling.[67]

Hop, a homeodomain-only protein, is expressed at about E8.0 during mouse development, and by E9.5 is strongly expressed in the myocardium and regions in the pharyngeal arches.[68,69] *Hop* is a downstream target gene of Nkx2.5 during myocardial development based on the absence of Hop expression in *Nkx2.5* null mice and Nkx2.5 binding sites in a cardiac enhancer region of the *Hop* gene. Ablation of *Hop* in mice results in half of the mice dying in mid-gestation due to ventricular hypocellularity, thinning and perforation of the myocardium. However, some of the *Hop*-deficient mice are viable but have an increase in the number of cardiac myocytes, suggesting delayed withdrawal from the cell cycle that normally occurs with birth, and indicating a role for Hop in normal withdrawal from the cell cycle in the perinatal period.

Myocardial compaction

Ventricular trabeculation starts in the apical region of the left and right ventricle soon after looping in both mouse and human (Fig. 1.2D). The most likely reason for myocardial trabeculation is to increase myocardial oxygenation in the absence of coronary circulation. Shortly after looping, the trabeculae provide most of the ventricular mass in the embryonic human and mouse heart. Concomitant with ventricular septation, the trabeculae start to compact at their base adjacent to the outer compact myocardium, adding substantially to its thickness. This process is fairly rapid (in the mouse, between E13 and E14, and in the human between 10 and 12 weeks), coinciding with establishment of the coronary circulation. Non-compaction of the myocardium results in the persistence of multiple prominent ventricular trabeculations and deep intertrabecular recesses, and has serious functional consequences for the heart.[70]

A complete lack of myocardial compaction has been found in mice with a mutated gene for retinoid X-receptor (RxR)-α, underscoring the importance of RxR for myocardial compaction.[71] In humans, ventricular non-compaction mainly affects the left ventricle and is often associated with depressed systolic and diastolic cardiac function, systemic embolism, development of ventricular tachyarrhythmias and sudden cardiac death, both in adult and pediatric populations.[72–74] Furthermore, it has been shown that the epicardium and epicardially derived cells play a crucial role in the development of the compact layer of the ventricular myocardial walls. In this regard, models with perturbed epicardial development showed inhibition of ventricular compaction. In addition, a 'thin-myocardium syndrome' has been observed in vascular cell adhesion molecule-1 (VCAM-1) knockout mice, and has been described for other transcription factors like RxR-α, N-myc, transcription enhancer factor-1 (TEF-1), arginyltransferase 1 (Ate-1), and Wilms' tumor-1 (WT-1).[75–80] Furthermore, the β-adrenergic receptor kinase 1 (β ARK1), a member of the G protein-coupled receptor kinase family, plays a role in cardiac development and appears to be important in myocardial compaction. β *Ark1* mutant mouse embryos die before E15.5 showing a pronounced ventricular hypoplasia identical to 'thin myocardium syndrome', and a decreased cardiac ejection fraction.[81]

The tumor necrosis factor-α converting enzyme (TACE) plays an essential regulatory role during ventricular compaction and cardiac development. Mice deficient in *Tace* exhibit markedly enlarged fetal hearts with increased myocardial trabeculation and reduced cell compaction, mimicking the pathological changes of non-compaction of ventricular myocardium. In addition, larger cardiomyocyte cell size and increased cell proliferation are observed in the ventricles of *Tace* knockout mouse hearts.[82]

In human studies, mutations in *Cypher/ZASP*, a component of the Z-line in both skeletal and cardiac muscle, have been associated with isolated non-compaction of the left ventricular myocardium.[83]

Formation of the atria

The fate of the atrioventricular myocardium is very important, because it is involved in changing the connection of atria to ventricles from being serial to parallel and unidirectional, and it restricts the electrical continuity of atria and ventricles to the atrioventricular node, which is specialized for electrical conduction. If non-specialized (working) myocardium remains at the atrioventricular junction, propagation of the electrical current through the heart is abnormal and may result in arrhythmias, as seen in pre-excitation syndromes such as Wolff-Parkinson–White syndrome.

Atrial septation starts when the common atrium becomes indented externally by the bulbus cordis and truncus arteriosus. This indentation corresponds internally with the septum primum and its orifice the ostium primum, growing from the posterosuperior wall and extending towards the endocardial cushion of the atrioventricular canal. As the superior and inferior endocardial cushions fuse, the concave lower edge of the septum primum fuses with it, obliterating the ostium primum. However, just before this happens fenestrations appear in the posterosuperior part of the septum forming the ostium secundum, thus maintaining communication between the right and left atrium. Meanwhile, the septum secundum develops on the anterosuperior wall of the right atrium covering the ostium secundum.[84]

Little is known about the molecular mechanism of atrial development. The orphan nuclear receptor chicken ovalbumin upstream promoter transcription factor-2 (COUP-TFII) is expressed in atrial precursors and is required for atrial, but not ventricular, growth.[85] Retinoid signaling has also been implicated in atrial specification and in the regulation of the atrioventricular border along the anterior–posterior axis of the heart tube.[86] How this border is established remains unclear. A pathological state resulting in abnormal localization of the atrioventricular borders is seen in Epstein's anomaly in humans, in which the tricuspid valve is displaced inferiorly into the right ventricle, resulting in 'atrialization' of the right ventricular myocardium. It will be interesting to determine whether the gene responsible for Epstein's anomaly is involved in atrioventricular specification.

Valvuloseptal development

The single primitive atrium is linked to the single primitive ventricle (Fig. 1.2B), and thus in order to form the four-chambered heart, which ensures a unidirectional blood flow, septa and valves, respectively, must develop between the two ventricles, two atria, and between the atria and ventricles (Fig. 1.2E). In this process an acellular matrix, referred to as the cardiac jelly, is involved. As the heart loops, the cardiac jelly accumulates in the junction between the atria and ventricles, the atrioventricular junction and the developing outflow tract, forming the endocardial cushion tissues (Fig. 1.2D).

The major atrioventricular cushion-derived tissue forms a mesenchymal 'bridge' from the posterior wall of the atrioventricular canal to the anterior wall of the heart and later forms the membranous atrioventricular septum, the septal leaflet of the tricuspid valve, and the aortic leaflet of the mitral valve.

Fusion of cushion-derived tissues and epicardially derived mesenchyme leads to the interruption of the continuity between atrial and myocardial ventricular myocardium, ensuring functional separation and electrical insulation between atrial and ventricular working myocardium. The only place where this separation does not take place is where the atrioventricular node and proximal His bundle develop (Fig. 1.2E). These components of the atrioventricular conduction system relay the cardiac impulse from atria to ventricles. The next step in valvuloseptal development is myocardial delaminiation, which leads to the formation of freely moveable leaflets of the atrioventricular valves. During this process the lower part of the atrioventricular cushions, with its associated myocardial layer, separates from adjacent ventricular myocardium by an as-yet undetermined mechanism. This finally results in the formation of fibrous leaflets that are attached to either the ventricular walls or the interventricular septum through the developing papillary muscles.

Connexin45 has been shown to play a role in development of endocardial cushions. *Connexin45*-deficient mice die of heart failure at around E10. The cardiac walls of the *connexin45*-deficient mice show an endocardial cushion defect, while the cardiac jelly is present.[87] Heart contractions of connexin45 mutant mice are initiated but conduction block appears within 24 hours after the first few contractions. These results indicate a requirement for gap junction channels during early cardiogenesis.[87,88]

After the formation of the cardiac jelly within the two distinct regions, the atrioventricular canal and the ventricular outflow tract, the cardiac jelly thickens to form the endocardial cushions. Within the ensuing day, endocardial cells that line the atrioventricular and outflow tract cushions respond to extrinsic molecular signals and transform into mesenchyme as they invade the underlying cardiac jelly, a process called epithelial–mesenchymal transformation (EMT).[89–91]

Transforming growth factor-β type III receptor (TBRIII) is expressed in cardiac endothelial cells that undergo EMT. Antagonizing this receptor in explanted chick atrioventricular cushions leads to inhibition of both mesenchyme formation and migration. Expression of the TBRIII in normally unresponsive endothelial cells results in mesenchyme formation in response to transforming growth factor-β2 (TGF-β2), and supports the role for the TBRIII in mediating atrioventricular cushion transformation.[92] Furthermore, a genome-wide screen using short tandem repeat polymorphic markers to search for a genetic locus involved in atrioventricular cushion defects in humans, linked these defects to a region of chromosome 1 near the gene encoding the TBRIII.[93]

Studies of genetic interaction between the epidermal growth factor receptor (EGFR) and the protein-tyrosine-phosphatase Shp2 showed that EGFR and Shp2 are components of a common growth-factor signaling pathway, and that Shp2 is required for EGFR signaling *in vivo*. In particular, EGFR and Shp2 are required for semilunar valvulogenesis, since mutant mice develop mild aortic stenosis and regurgitation. Deletion of the gene encoding platelet-derived growth factor receptor α (PDGFRα) in mice harboring the *Patch* mutation causes defects in cardiac outflow tract formation, suggesting a functional role of PDGFRα in normal cardiac outflow tract development.

The Ca^{2+}/calcineurin/nuclear factor of activated T-cells (NFATc) signaling pathway is essential for normal cardiac valve and septum morphogenesis. NFATc is first expressed in the heart tube of mouse embryos at E7.5, and is restricted to the specialized endocardium that gives rise to the valves and septum. Between E9 and E11.5, NFATc expression was observed in the endocardium of the ventricles, atrium, the outflow tract and the apex of the ventricular septum. NFATc is activated in endocardial cells adjacent to the interface with the cardiac jelly and myocardium, which are thought to give the inductive stimulus to the valve primordia. Mice with a disruption in the *Nfatc* gene fail to develop normal cardiac valves and septa, and die *in utero* from congestive heart failure between E13.5 and E17.5.[94,95] *Nfatc* mutant mouse embryos show hypertrophied ventricular walls and small chambers.[95] In addition, the proximal region of the outflow tract is narrowed or occluded in *Nfatc* mutant mouse embryos. By E13.5, *Nfatc* mutant embryos display abnormalities in the structure of the aortic and pulmonary valves, including defective atrioventricular valve and septum formation.

Neurofibromin (Nf), a negative regulator of Ras-signaling, normally acts to modulate EMT and proliferation in the developing heart.[96] Beginning at E8.0, Nf is expressed ubiquitously in the developing mouse embryo. At E11.5, Nf is expressed in the endocardial cushion, in the atrioventricular region and in the outflow tract. Expression in the endocardial cushion continues through E13.5. In the absence of Nf, embryonic mouse hearts develop overabundant endocardial cushions due to hyperproliferation and lack of apoptosis. The *Nf* mutant cushions fail to condense and thin, in order to form mature valve leaflets by E13.5, leading to a compromised inflow into the ventricles. *Nf* mutant mice also show hypoplastic left and right ventricles and a double-outlet right ventricle. In explant cultures, *Nf* deficiency is reproduced by activation of Ras-signaling pathways, and the *Nf* null mutant phenotype is prevented by inhibiting Ras *in vitro*,[96] highlighting the importance of Ras-signaling in normal cardiac development.

The $\alpha v\beta3$ integrin receptor ligand, Del1 (developmentally regulated endothelial cell locus) is associated with normal development of vascular-like structures. Del1 is a matrix protein that promotes adhesion of endothelial cells through interaction with the $\alpha v \beta 3$ integrin receptor. Embryonic endothelial-like yolk sac cells expressing recombinant Del1 protein, or grown on an extracellular matrix containing Del1 protein, are inhibited from forming vascular-like structures. Expression of Del1 protein in the chick chorioallantoic membrane leads to loss of vascular integrity and promotes vessel remodeling.[97]

Neuregulins, a family of structurally diverse glycoproteins, and their receptors are expressed during embryonic development and are involved in cell fate determination. Neuregulins share a common EGF-like domain that is a critical part of the molecule, as it alone can stimulate tyrosine phosphorylation of ErbB2. Neuregulin signals through a family of protein tyrosine kinases of the class I EGFR family, namely ErbB2, ErbB3 and ErbB4. While ErbB3 on its own lacks any biological activity, transactivation of ErbB3 with ErbB2 generates a potent signal. *Neuregulin-* and *ErbB2-* deficient mouse embryos die at E10.5 due to a lack of cardiac ventricular myocyte differentiation, leading to a thinner ventricular wall, absence of trabeculation and ventricular dilation. ErbB3-deficient embryos survive until E13.5, exhibiting cardiac cushion abnormalities, with lack of mesenchyme and thinning of the cushion, leading to blood reflux through defective and hypoplastic valves.[98]

Friend-of-GATA-2 (*fog2*)-deficient mouse embryos die at mid-gestation with a cardiac defect characterized by a thin ventricular myocardium, markedly dilated and thinned atria, and a common atrioventricular canal. In addition, *fog2*-deficient embryos exhibit defects seen in the human congenital malformation tetralogy of Fallot with atrial septal defect, subaortic ventricular septal defect, overriding aorta and subpulmonic stenosis. Remarkably, coronary vasculature is absent in *fog2*-deficient mouse hearts.[80] Despite formation of an intact epicardial layer and expression of epicardium-specific genes, the markers of cardiac vessel development intercellular adhesion molecule-2 (ICAM-2) and FLK-1 are not expressed, indicative of failure to activate their expression and/or to initiate the EMT of epicardial cells. The

phenotype of *fog2*-deficient hearts with a thin myocardium and coronary vascular defects is rescued using heart-specific overexpression of fog2, demonstrating that fog2 is required and sufficient for myocardial and coronary vessel development.[99]

The winged-helix transcription factors Foxc1 and 2 are required synergistically for cardiovascular development. Embryos lacking either *Foxc1* or *2*, and most compound heterozygotes, die pre- or perinatally with similar abnormal phenotypes, including defects in the cardiovascular system, suggesting that the genes have similar, dose-dependent functions, and compensate for each other in early development. However, there are phenotypic differences between compound homo- and heterozygous *Foxc1* and *2* mutants, suggesting that the functions of the two genes are not completely identical and/or that there are quantitative differences in the level and onset of their expression. Compound homozygous *Foxc1/Foxc2* mutant embryos die earlier (at E9.0–9.5), and with much more severe defects than single homozygotes alone. Significantly, they have profound abnormalities in the first and second branchial arches, the early remodeling of blood vessels, a smaller heart and an enlarged pericardial sac. Analysis of gene expression in *Foxc1/Foxc2* mutants suggests that Foxc1 and 2 proteins interact with the Notch signaling pathway.[100]

The homeobox transcription factor Msx1 is expressed by a number of non-myocardial cell populations, including cells undergoing EMT. *Msx2* expression is restricted to a distinct subpopulation of myocardial cells undergoing a specialization into cardiac conduction system-specific cells and can therefore be used to identify precursor cells that give rise to the conduction system.[101]

The transcription factor Smad6 belongs to a group of intracellular mediators that are initiated by TGF-β ligands and acts as a negative regulator. Smad6 is strongly expressed in endocardial cells at E9.5 and is largely restricted to the heart and blood vessels. Deletion of *Smad6* demonstrates its involvement in endocardial cushion transformation. In this regard, *Smad6* mutants show hyperplastic thickening of cardiac valves and defects in outflow tract septation, resulting in an extremely narrow ascending aorta accompanied by a large pulmonary trunk, or vice versa. The role of Smad6 in homeostasis of the adult cardiovascular system is indicated by the development of ossification of the cardiac outflow tracts and elevated blood pressure in viable *Smad6* null mutants.[102]

The *Sox4* gene, a member of the Sox (for Sry-box) family of transcription factors, plays an important role in development of the endocardially derived tissue responsible for development of the atrioventricular junction. Homozygous *Sox4*-deficient mouse embryos develop normally until E12–13, the onset of valvular morphogenesis in the arterial pole of the heart, and die due to cardiac failure at E14, showing rapidly developing generalized edema. Dying *Sox4* mutant embryos typically display valvular malfunction. In this regard, the development of the endocardial ridges into the semilunar valves and the outlet portion of the muscular ventricular septum, leading to a common arterial trunk, are affected. *Sox4* deficiency leads to malformations ranging from infundibular septum defect to complete transposition of the great arteries.[103,104]

Formation of the conduction system

Each heartbeat and synchronized round of cardiac contraction and relaxation is controlled by a specialized cluster of cells in the right atrium, called the sino-atrial node, which functions as the pacemaker (Fig. 1.2E, shown in yellow in the color plate section). Electrical impulses are propagated through the heart by the cardiac conduction system, consisting of Purkinje cells, and by direct cell–cell coupling of cardiac myocytes. The final result of the formation of the conduction system is an efficient and coordinated contraction of both atria and ventricles. The embryonic origins of cardiac conduction system cell lineages have been unclear until recently, when it was shown that Purkinje cells are derived from a subpopulation of ventricular cardiomyocytes in response to signaling by endothelin-1 and neuregulins.[105] Fate mapping of cardiac conduction lineages in the mouse has yielded insights into the transcriptional pathways responsible for the formation of this specialized cardiac structure, and the molecular basis of cardiac conduction defects.

Transcription factors like HF1b/SP4 and Nkx2.5 control conduction system lineages. It has been shown that HF1b plays a critical role in formation of the conduction system lineage, and the loss of *HF1b* leads to a confused electrophysiological identity in Purkinje and ventricular cell lineages, resulting in marked tachy- and brady-arrhythmias, and sudden cardiac death. Mice which have a neural crest-restricted knockout of *HF1b* show marked arrhythmogenesis and conduction system defects, implicating neural crest cues in conduction system development and disease. Mice harboring a ventricular-restricted knockout of *Nkx2.5* display completely normal conduction at birth, but develop a hypoplastic atrioventricular node.[106] During maturation, progressive complete heart block ensues, associated with a selective dropout of distal atrioventricular nodal cell lineages at the boundaries of the penetrating His bundle. In humans, *NKX2.5* has been identified through genetic linkage analysis of pedigrees with non-syndromic congenital heart disease.[107] Mutations in *NKX2.5* have been identified in individuals with cardiac septal defects and conduction abnormalities, indicating that Nkx2.5 is important for regulation of septation during cardiac morphogenesis and for maturation and maintenance of atrioventricular node function throughout life.

Formation of the coronary arteries

The increase in myocardial function during development requires a separate vasculature of the myocardial walls, ensuring an adequate supply of blood to, and transport of metabolites from, the myocardium. The formation of the coronary system is preceded by the outgrowth of the proepicardial organ over the heart to form the epicardium, followed by migration of endothelial cells from the septum transversum into the subepicardial space to form a capillary network. Subsequently, epicardial cells transdifferentiate and migrate to form the subepicardial and subendocardial mesenchyme. These mesenchymal cells then differentiate into smooth muscle cells, and adventitial fibroblasts of the coronary vessel walls differentiate into myocardial interstitial fibroblasts, subendocardial cells, and cells in the cushion mesenchyme.

The role of Ets-1 and Ets-2 in the normal development of coronary arteries and myocardial development has been shown in chicken embryos. Ablation of *Ets1* and *Ets2* causes the absence of one or both coronary orifices in 40% of the mutant embryos, and underdevelopment of the coronary arteries.[108] Furthermore, in the ventricular periphery, the developing coronary arteries were not regularly organized, but showed an irregular distribution over the free-wall myocardium. In addition, the ventricular wall was attenuated and trabeculae were broader and fewer in number than in the hearts of control animals, especially in the right ventricle. Furthermore, the ventricular septum did not close in 25% of the antisense *Ets*-treated embryos, resulting in ventricular septal defects.[108]

Concluding remarks

Remarkable progress has been made over the past few years in determining the molecular mechanisms involved during heart development. Exploration of vertebrate and invertebrate model systems reveals a high degree of conservation among many of the transcription factors regulating heart formation, allowing elucidation of these factors in a multifaceted approach. The sensitivity of human and animal model hearts to perturbations during embryogenesis points to a precise orchestration between myocardial, endothelial, epicardial and neural crest precursor cells and their specific pathways required to form the functioning mature four-chambered heart. However, the specific interactions of the different signaling pathways and the communication between different cell types in formation and maturation of the cardiovascular system still remains to be determined. Many studies use mouse knockout models to determine the role of transcription factors during cardiogenesis, but little is known about the activators, inhibitors and downstream target genes of these transcription factors. Identification of the genes responsible for various cardiovascular diseases enables development of genetic testing as a diagnostic tool to

screen and assess the risk of various cardiac disorders, including congenital heart disease.[109] From a clinical perspective, we can anticipate that the growing knowledge of the molecular pathways during cardiogenesis will be increasingly useful for the understanding, diagnosis and treatment of heart disease.

References

1. Wu X, Golden K, Bodmer R. Heart development in *Drosophila* requires the segment polarity gene wingless. Dev Biol 1995; 169: 619–28.
2. Marvin MJ, Di Rocco G, Gardiner A, Bush SM, Lassar AB. Inhibition of Wnt activity induces heart formation from posterior mesoderm. Genes Dev 2001; 15: 316–27.
3. Schneider VA, Mercola M. Wnt antagonism initiates cardiogenesis in *Xenopus laevis*. Genes Dev 2001; 15: 304–15.
4. Lickert H, Kutsch S, Kanzler B et al. Formation of multiple hearts in mice following deletion of beta-catenin in the embryonic endoderm. Dev Cell 2002; 3: 171–81.
5. Bodmer R. The gene tinman is required for specification of the heart and visceral muscles in Drosophila. Development 1993; 118: 719–29.
6. Lints TJ, Parsons LM, Hartley L et al. Nkx-2.5: a novel murine homeobox gene expressed in early heart progenitor cells and their myogenic descendants. Development 1993; 119: 419–31.
7. Komuro I, Izumo S. Csx: a murine homeobox-containing gene specifically expressed in the developing heart. Proc Natl Acad Sci U S A 1993; 90: 8145–9.
8. Lyons I, Parsons LM, Hartley L et al. Myogenic and morphogenetic defects in the heart tubes of murine embryos lacking the homeobox gene Nkx2–5. Genes Dev 1995; 9: 1654–66.
9. Wang D, Chang PS, Wang Z et al. Activation of cardiac gene expression by myocardin, a transcriptional cofactor for serum response factor. Cell 2001; 105: 851–62.
10. Small EM, Warkman AS, Wang DZ et al. Myocardin is sufficient and necessary for cardiac gene expression in *Xenopus*. Development 2005; 132: 987–97.
11. Li S, Wang DZ, Wang Z et al. The serum response factor coactivator myocardin is required for vascular smooth muscle development. Proc Natl Acad Sci U S A 2003; 100: 9366–70.
12. Evans T. Regulation of cardiac gene expression by

GATA-4/5/6. Trends Cardiovasc Med 1997; 7: 75–83.

13. Gajewski K, Fossett N, Molkentin JD et al. The zinc finger proteins Pannier and GATA4 function as cardiogenic factors in *Drosophila*. Development 1999; 126: 5679–88.

14. Reiter JF, Alexander J, Rodaway A et al. Gata5 is required for the development of the heart and endoderm in zebrafish. Genes Dev 1999; 13: 2983–95.

15. Watt AJ, Battle MA, Li J, Duncan SA. GATA4 is essential for formation of the proepicardium and regulates cardiogenesis. Proc Natl Acad Sci U S A 2004; 101: 12573–8.

16. Zhao R, Watt AJ, Li J et al. GATA6 is essential for embryonic development of the liver but dispensable for early heart formation. Mol Cell Biol 2005; 25: 2622–31.

17. Kikuchi Y, Verkade H, Reiter JF et al. Notch signaling can regulate endoderm formation in zebrafish. Dev Dyn 2004; 229: 756–62.

18. Molkentin JD, Lin Q, Duncan SA et al. Requirement of the transcription factor GATA4 for heart tube formation and ventral morphogenesis. Genes Dev 1997; 11: 1061–72.

19. Nemer G, Qureshi ST, Malo D, Nemer M. Functional analysis and chromosomal mapping of Gata5, a gene encoding a zinc finger DNA-binding protein. Mamm Genome 1999; 10: 993–9.

20. Garg V, Kathiriya IS, Barnes R et al. GATA4 mutations cause human congenital heart defects and reveal an interaction with TBX5. Nature 2003; 424: 443–7.

21. Kuo CT, Morrisey EE, Anandappa R et al. GATA4 transcription factor is required for ventral morphogenesis and heart tube formation. Genes Dev 1997; 11: 1048–60.

22. Suzuki YJ, Evans T. Regulation of cardiac myocyte apoptosis by the GATA-4 transcription factor. Life Sci 2004; 74: 1829–38.

23. Kitajima S, Takagi A, Inoue T et al. MesP1 and MesP2 are essential for the development of cardiac mesoderm. Development 2000; 127: 3215–26.

24. McKinsey TA, Zhang CL, Olson EN. MEF2: a calcium-dependent regulator of cell division, differentiation and death. Trends Biochem Sci 2002; 27: 40–7.

25. Lin Q, Schwarz J, Bucana C et al. Control of mouse cardiac morphogenesis and myogenesis by transcription factor MEF2C. Science 1997; 276: 1404–7.

26. Abu-Issa R, Smyth G, Smoak I et al. Fgf8 is required for pharyngeal arch and cardiovascular development in the mouse. Development 2002; 129: 4613–25.

27. Kelly RG, Brown NA, Buckingham ME. The arterial pole of the mouse heart forms from Fgf10-expressing cells in pharyngeal mesoderm. Dev Cell 2001; 1: 435–40.

28. Mjaatvedt CH, Nakaoka T, Moreno-Rodriguez R et al. The outflow tract of the heart is recruited from a novel heart-forming field. Dev Biol 2001; 238: 97–109.

29. Waldo KL, Kumiski DH, Wallis KT et al. Conotruncal myocardium arises from a secondary heart field. Development 2001; 128: 3179–88.

30. RJ Schwartz and EN Olson. Building the heart piece by piece: modularity of cis-elements regulating Nkx2–5 transcription. Development 1999; 126: 4187–92.

31. Cai CL, Liang X, Shi Y et al. Isl1 identifies a cardiac progenitor population that proliferates prior to differentiation and contributes a majority of cells to the heart. Dev Cell 2003; 5: 877–89.

32. von Both I, Silvestri C, Erdemir T et al. Foxh1 is essential for development of the anterior heart field. Dev Cell 2004; 7: 331–45.

33. Dodou E, Verzi MP, Anderson JP, Xu SM, Black BL. Mef2c is a direct transcriptional target of ISL1 and GATA factors in the anterior heart field during mouse embryonic development. Development 2004; 131: 3931–42.

34. Phan D, Rasmussen TL, Nakagawa O et al. BOP, a regulator of right ventricular heart development, is a direct transcriptional target of MEF2C in the developing heart. Development 2005; 132: 2669–78.

35. George EL, Georges-Labouesse EN, Patel-King RS et al. Defects in mesoderm, neural tube and vascular development in mouse embryos lacking fibronectin. Development 1993; 119: 1079–91.

36. Roebroek AJ, Umans L, Pauli IG et al. Failure of ventral closure and axial rotation in embryos lacking the proprotein convertase Furin. Development 1998; 125: 4863–76.

37. Komada M, Soriano P. Hrs, a FYVE finger protein localized to early endosomes, is implicated in vesicular traffic and required for ventral folding morphogenesis. Genes Dev 1999; 13: 1475–85.

38. Chang H, Huylebroeck D, Verschueren K et al. Smad5 knockout mice die at mid-gestation due to multiple embryonic and extraembryonic defects. Development 1999; 126: 1631–42.

39. Bruneau BG, Nemer G, Schmitt JP et al. A murine model of Holt–Oram syndrome defines roles of the T-box transcription factor Tbx5 in cardiogenesis and disease. Cell 2001; 106: 709–21.

40. Epstein JA. Developing models of DiGeorge syndrome. Trends Genet 2001; 17: S13–17.

41. Yagi H, Furutani Y, Hamada H et al. Role of TBX1 in human del22q11.2 syndrome. Lancet 2003; 362: 1366–73.

42. Stennard FA, Costa MW, Lai D et al. Murine T-box transcription factor Tbx20 acts as a repressor during heart development, and is essential for adult heart integrity, function and adaptation. Development 2005; 132: 2451–62.

43. Takeuchi JK, Mileikovskaia M, Koshiba-Takeuchi K et al. Tbx20 dose-dependently regulates transcription factor networks required for mouse heart and motoneuron development. Development 2005; 132: 2463–74.

44. Cai CL, Zhou W, Yang L et al. T-box genes coordinate regional rates of proliferation and regional specification during cardiogenesis. Development 2005; 132: 2475–87.

45. Yelon D. Cardiac patterning and morphogenesis in zebrafish. Dev Dyn 2001; 222: 552–63.

46. van den Hoff MJ, Kruithof BP, Moorman AF et al. Formation of myocardium after the initial development of the linear heart tube. Dev Biol 2001; 240: 61–76.

47. Fishman MC, Olson EN. Parsing the heart: genetic modules for organ assembly. Cell 1997; 91: 153–6.

48. Firulli AB. A HANDful of questions: the molecular biology of the heart and neural crest derivatives (HAND)-subclass of basic helix-loop-helix transcription factors. Gene 2003; 312: 27–40.

49. Srivastava D, Thomas T, Lin Q et al. Regulation of cardiac mesodermal and neural crest development by the bHLH transcription factor, dHAND. Nat Genet 1997; 16: 154–60.

50. Cserjesi P, Brown D, Lyons GE et al. Expression of the novel basic helix-loop-helix gene eHAND in neural crest derivatives and extraembryonic membranes during mouse development. Dev Biol 1995; 170: 664–78.

51. McFadden DG, Barbosa AC, Richardson JA, Schneider MD, Srivastava D, Olson EN. The Hand1 and Hand2 transcription factors regulate expansion of the embryonic cardiac ventricles in a gene dosage-dependent manner. Development 2005; 132: 189–201.

52. Srivastava D, Cserjesi P, Olson EN. A subclass of bHLH proteins required for cardiac morphogenesis. Science 1995; 270: 1995–9.

53. Yelon D, Ticho B, Halpern ME et al. The bHLH transcription factor hand2 plays parallel roles in zebrafish heart and pectoral fin development. Development 2000; 127: 2573–82.

54. Fischer A, Gessler M. Hey genes in cardiovascular development. Trends Cardiovasc Med 2003; 13: 221–6.

55. Artavanis-Tsakonas S, Rand MD, Lake RJ. Notch signaling: cell fate control and signal integration in development. Science 1999; 284: 770–6.

56. Iso T, Kedes L, Hamamori Y. HES and HERP families: multiple effectors of the Notch signaling pathway. J Cell Physiol 2003; 194: 237–55.

57. Firulli BA, Hadzic DB, McDaid JR et al. The basic helix-loop-helix transcription factors dHAND and eHAND exhibit dimerization characteristics that suggest complex regulation of function. J Biol Chem 2000; 275: 33567–73.

58. Sakata Y, Kamei CN, Nakagami H et al. Ventricular septal defect and cardiomyopathy in mice lacking the transcription factor CHF1/Hey2. Proc Natl Acad Sci U S A 2002; 99: 16197–202.

59. Gessler M, Knobeloch KP, Helisch A et al. Mouse gridlock: no aortic coarctation or deficiency, but fatal cardiac defects in Hey2 –/– mice. Curr Biol 2002; 12: 1601–4.

60. Donovan J, Kordylewska A, Jan YN et al. Tetralogy of fallot and other congenital heart defects in Hey2 mutant mice. Curr Biol 2002; 12: 1605–10.

61. Zhong TP, Rosenberg M, Mohideen MA et al. Gridlock, an HLH gene required for assembly of the aorta in zebrafish. Science 2000; 287: 1820–4.

62. Oda T, Elkahloun AG, Pike BL et al. Mutations in the human Jagged1 gene are responsible for Alagille syndrome. Nat Genet 1997; 16: 235–42.

63. Zhong TP, Childs S, Leu JP et al. Gridlock signalling pathway fashions the first embryonic artery. Nature 2001; 414: 216–20.

64. Wang GF, Nikovits W, Jr., Bao ZZ et al. Irx4 forms an inhibitory complex with the vitamin D and retinoic X receptors to regulate cardiac chamber-specific slow MyHC3 expression. J Biol Chem 2001; 276: 28835–41.

65. Bruneau BG, Bao ZZ, Fatkin D et al. Cardiomyopathy in Irx4-deficient mice is preceded by abnormal ventricular gene expression. Mol Cell Biol 2001; 21: 1730–6.

66. Campione M, Ros MA, Icardo JM et al. Pitx2 expression defines a left cardiac lineage of cells: evidence for atrial and ventricular molecular isomerism in the iv/iv mice. Dev Biol 2001; 231: 252–64.

67. Lu MF, Pressman C, Dyer R, Johnson RL, Martin JF. Function of Rieger syndrome gene in left-right asymmetry and craniofacial development. 1999; 401: 276–8.

68. Chen F, Kook H, Milewski R et al. *Hop* is an unusual

homeobox gene that modulates cardiac development. Cell 2002; 110: 713–23.

69. Shin CH, Liu ZP, Passier R et al. Modulation of cardiac growth and development by HOP, an unusual homeodomain protein. Cell 2002; 110: 725–35.

70. Richardson P, McKenna W, Bristow M et al. Report of the 1995 World Health Organization/International Society and Federation of Cardiology Task Force on the Definition and Classification of Cardiomyopathies. Circulation 1996; 93: 841–2.

71. Gruber PJ, Kubalak SW, Pexieder T et al. RXR alpha deficiency confers genetic susceptibility for aortic sac, conotruncal, atrioventricular cushion, and ventricular muscle defects in mice. J Clin Invest 1996; 98: 1332–43.

72. Oechslin EN, Attenhofer Jost CH, Rojas JR et al. Long-term follow-up of 34 adults with isolated left ventricular noncompaction: a distinct cardiomyopathy with poor prognosis. J Am Coll Cardiol 2000; 36: 493–500.

73. Neudorf UE, Hussein A, Trowitzsch E et al. Clinical features of isolated noncompaction of the myocardium in children. Cardiol Young 2001; 11: 439–42.

74. Ichida F, Hamamichi Y, Miyawaki T et al. Clinical features of isolated noncompaction of the ventricular myocardium: long-term clinical course, hemodynamic properties, and genetic background. J Am Coll Cardiol 1999; 34: 233–40.

75. Kwee L, Baldwin HS, Shen HM et al. Defective development of the embryonic and extraembryonic circulatory systems in vascular cell adhesion molecule (VCAM-1) deficient mice. Development 1995; 121: 489–503.

76. Kastner P, Grondona JM, Mark M et al. Genetic analysis of RXR alpha developmental function: convergence of RXR and RAR signaling pathways in heart and eye morphogenesis. Cell 1994; 78: 987–1003.

77. Charron J, Malynn BA, Fisher P et al. Embryonic lethality in mice homozygous for a targeted disruption of the N-myc gene. Genes Dev 1992; 6: 2248–57.

78. Chen Z, Friedrich GA, Soriano P. Transcriptional enhancer factor 1 disruption by a retroviral gene trap leads to heart defects and embryonic lethality in mice. Genes Dev 1994; 8: 2293–301.

79. Kwon YT, Kashina AS, Davydov IV et al. An essential role of N-terminal arginylation in cardiovascular development. Science 2002; 297: 96–9.

80. Kreidberg JA, Sariola H, Loring JM et al. WT-1 is required for early kidney development. Cell 1993; 74: 679–91.

81. Jaber M, Koch WJ, Rockman H et al. Essential role of beta-adrenergic receptor kinase 1 in cardiac development and function. Proc Natl Acad Sci U S A 1996; 93: 12974–9.

82. Shi W, Chen H, Sun J et al. TACE is required for fetal murine cardiac development and modeling. Dev Biol 2003; 261: 371–80.

83. Vatta M, Mohapatra B, Jimenez S et al. Mutations in Cypher/ZASP in patients with dilated cardiomyopathy and left ventricular non-compaction. J Am Coll Cardiol 2003; 42: 2014–27.

84. Wenink AC, Gittenberger-de Groot AC. The role of atrioventricular endocardial cushions in the septation of the heart. Int J Cardiol 1985; 8: 25–44.

85. Pereira FA, Qiu Y, Zhou G et al. The orphan nuclear receptor COUP-TFII is required for angiogenesis and heart development. Genes Dev 1999; 13: 1037–49.

86. Dyson E, Sucov HM, Kubalak SW et al. Atrial-like phenotype is associated with embryonic ventricular failure in retinoid X receptor alpha –/– mice. Proc Natl Acad Sci U S A 1995; 92: 7386–90.

87. Gitler AD, Lu MM, Jiang YQ et al. Molecular markers of cardiac endocardial cushion development. Dev Dyn 2003; 228: 643–50.

88. Kumai M, Nishii K, Nakamura K et al. Loss of connexin45 causes a cushion defect in early cardiogenesis. Development 2000; 127: 3501–12.

89. Mjaatvedt CH, Markwald RR. Induction of an epithelial-mesenchymal transition by an in vivo adheron-like complex. Dev Biol 1989; 136: 118–28.

90. Eisenberg LM, Markwald RR. Molecular regulation of atrioventricular valvuloseptal morphogenesis. Circ Res 1995; 77: 1–6.

91. Schroeder JA, Jackson LF, Lee DC et al. Form and function of developing heart valves: coordination by extracellular matrix and growth factor signaling. J Mol Med 2003; 81: 392–403.

92. Brown CB, Boyer AS, Runyan RB et al. Requirement of type III TGF-beta receptor for endocardial cell transformation in the heart. Science 1999; 283: 2080–2.

93. Sheffield VC, Pierpont ME, Nishimura D et al. Identification of a complex congenital heart defect susceptibility locus by using DNA pooling and shared segment analysis. Hum Mol Genet 1997; 1: 117–21.

94. Ranger AM, Grusby MJ, Hodge MR et al. The transcription factor NF-ATc is essential for cardiac valve formation. Nature 1998; 392: 186–90.

95. de la Pompa JL, Timmerman LA, Takimoto H et al. Role of the NF-ATc transcription factor in morphogenesis of cardiac valves and septum. Nature 1998; 392: 182–6.

96. Lakkis MM, Epstein JA. Neurofibromin modulation of ras activity is required for normal endocardial-mesenchymal transformation in the developing heart. Development 1998; 125: 4359–67.

97. Hidai C, Zupancic T, Penta K et al. Cloning and characterization of developmental endothelial locus-1: an embryonic endothelial cell protein that binds the alphavbeta3 integrin receptor. Genes Dev 1998; 12: 21–33.

98. Erickson SL, O'Shea KS, Ghaboosi N et al. ErbB3 is required for normal cerebellar and cardiac development: a comparison with ErbB2-and heregulin-deficient mice. Development 1997; 124: 4999–5011.

99. Svensson EC, Huggins GS, Lin H et al. A syndrome of tricuspid atresia in mice with a targeted mutation of the gene encoding Fog-2. Nat Genet 2000; 25: 353–6.

100. Kume T, Jiang H, Topczewska JM et al. The murine winged helix transcription factors, Foxc1 and Foxc2, are both required for cardiovascular development and somitogenesis. Genes Dev 2001; 15: 2470–82.

101. Chan-Thomas PS, Thompson RP, Robert B et al. Expression of homeobox genes Msx-1 (Hox-7) and Msx-2 (Hox-8) during cardiac development in the chick. Dev Dyn 1993; 197: 203–16.

102. Galvin KM, Donovan MJ, Lynch CA et al. A role for smad6 in development and homeostasis of the cardiovascular system. Nat Genet 2000; 24: 171–4.

103. Schilham MW, Oosterwegel MA, Moerer P et al. Defects in cardiac outflow tract formation and pro-B-lymphocyte expansion in mice lacking Sox-4. Nature 1996; 380: 711–14.

104. Ya J, Schilham MW, de Boer PA et al. Sox4-deficiency syndrome in mice is an animal model for common trunk. Circ Res 1998; 83: 986–94.

105. Gourdie RG, Harris BS, Bond J et al. Development of the cardiac pacemaking and conduction system. Birth Defects Res C Embryo Today 2003; 69: 46–57.

106. Tevosian SG, Deconinck AE, Tanaka M et al. FOG-2, a cofactor for GATA transcription factors, is essential for heart morphogenesis and development of coronary vessels from epicardium. Cell 2000; 101: 729–39.

107. Schott JJ, Benson DW, Basson CT et al. Congenital heart disease caused by mutations in the transcription factor NKX2–5. Science 1998; 281: 108–11.

108. Lie-Venema H, Gittenberger-de Groot AC, van Empel LJ et al. Ets-1 and Ets-2 transcription factors are essential for normal coronary and myocardial development in chicken embryos. Circ Res 2003; 92: 749–56.

109. Srivastava D, Olsen EN. A genetic blueprint for cardiac development. Nature 2000; 407: 221–26.

CHAPTER 2

Cardiomyocyte cell cycle control

Loren J Field

Introduction

Growth of the heart is traditionally divided into two phases. The initial phase of growth occurs during embryonic development and is characterized by the proliferation of differentiated, albeit immature, cardiomyocytes. The second phase of growth occurs after birth, and is characterized by hypertrophic growth of cardiomyocytes that have for the most part exited the cell cycle. Although cardiomyocyte DNA synthesis is observed in the adult heart, the magnitude and origin of these activities is debated. This chapter provides a brief review of mammalian cell cycle regulation, followed by a description of cardiomyocyte proliferation during development. Studies examining cardiomyocyte cell cycle activity in normal and injured adult hearts are then discussed. This is followed by a brief summary of efforts to enhance cardiomyocyte cell cycle activity using biochemical (i.e. addition of exogenous growth factors, cytokines, etc) or genetic (i.e. altering gene expression via gene transfer or gene ablation approaches, etc) interventions.

The mammalian cell cycle

The mammalian cell cycle is divided into four distinct phases (Fig. 2.1).[1, 2] During the G1 (or gap 1) phase, cells are poised to initiate cell cycle activity, waiting until favorable environmental conditions are encountered. During this period, members of the retinoblastoma (RB) protein family exist in a hypo-phosphorylated state. Hypo-phosphorylated RB binds to and renders inactive members of the E2F family (transcription factors that activate the expression of genes required for cell cycle progression). When conditions favorable for cell cycle progression are encountered (as for example the presence of appropriate mitogenic growth factors), G1 cells commit to a new round of division and transit through the so called 'restriction point' of the cell cycle. This process entails the transcriptional activation of D-type cyclins (cyclin D1, D2 or D3). Newly translated D-type cyclins form a binary complex with cyclin-dependent kinase 4 or 6 (CDK4 or 6), thereby rendering the kinase active. The cyclin D/CDK complex phosphorylates RB, which in turn results in the release of E2F proteins and the transcriptional activation of genes required for DNA synthesis and cell cycle progression. The initiation of DNA replication constitutes entry into the S (or synthesis) phase of the cell cycle.

Similar cyclin/CDK complexes regulate other cell cycle checkpoints. For example, cyclin E/CDK2 activity ensures continued phosphorylation of RB family members during late G1 phase, while cyclin A/CDK2 activity regulates checkpoints at early S phase. The INK family of CDK inhibitors (p15, p16, p18 and p19) can negatively

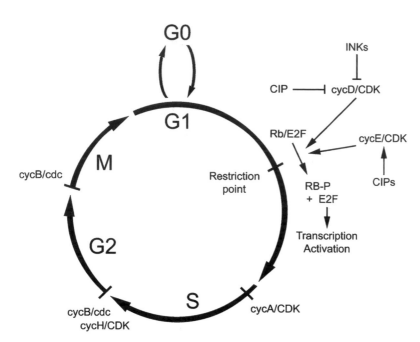

Figure 2.1. *Schematic diagram of the mammalian cell cycle. G0, gap 0; G1, gap 1; S, synthesis phase; G2, gap 2; M, mitosis; cyc, cyclin; CDK, cyclin-dependent kinase; CIP, CDK-interacting protein; INK, cyclin-dependent kinase inhibitor; Rb, retinoblastoma protein; cdc, cell division cycle.*

regulate restriction point transit and S phase entry by binding to CDK4 and 6, thereby preventing phosphorylation of RB family members. The CIP/KIP family of CDK inhibitors (p21, p27 and p57) negatively regulates other cyclin/CDK complexes, in particular cyclin E/CDK. Interestingly, at other stages of the cell cycle p21 and p27 appear to provide a molecular scaffold that facilitates cyclin D/CDK complex formation, and in this regard are positive regulators of cell cycle progression.[3]

After completion of chromosome replication, cells enter the G2 (or gap 2) phase of the cell cycle. G2 is characterized by an increase in cell growth prior to the initiation of the M (or mitosis) phase. M phase in turn is characterized by the orderly segregation of the newly replicated chromosomes to daughter cells, followed by cytokinesis. M phase cells are recognized by the characteristic appearance of chromosomes as they segregate to the daughter cells (particularly during metaphase, where chromosomes are aligned along the equatorial plate of the cell). Transit through S/G2 and M phase checkpoints requires cyclin B/cdc2 activity. There are a number of additional checkpoints during G2 and M phases to ensure proper chromosomal replication and segregation. Following mitosis,

the newly formed daughter cells enter the G1 phase and are once again able to initiate a new round of replication, provided that favorable conditions exist. Alternatively, the cells can either permanently or transiently withdraw from the cell cycle; such cells are frequently said to be in the G0, or gap zero, phase.

Monitoring cardiomyocyte cell cycle activity in the heart

The first factor to consider when monitoring cardiomyocyte proliferation is the experimental endpoint used to score cell cycle activity. In instances where the rate of proliferation is high, the simplest and arguably the most straightforward endpoint for cell cycle activity entails direct counting of cardiomyocytes. This can readily be accomplished using enzymatically (i.e. collagenase),[4] or chemically (i.e. potassium hydroxide)[5] dispersed cell preparations. It is somewhat more difficult to accurately document changes in cardiomyocyte number in histological sections, as developmentally or injury-induced changes in heart shape and cellular content need to be taken into account. In addition, issues pertaining to representative sampling of tissue sections must also be considered.

Cardiomyocyte cell cycle activity is more frequently quantitated by monitoring some aspect of DNA replication in histological sections (i.e. induction of S-phase-associated proteins, incorporation of labeled deoxyribonucleotides, quantitation of nuclear DNA content, presence of mitotic figures, etc). Unfortunately, adult cardiomyocytes can undergo polyploidization and karyokinesis in the absence of cytokinesis;[6] consequently these hallmarks of DNA replication can also be present in cells that ultimately do not divide. Moreover many markers associated with cell cycle progression in cultured or rapidly proliferating cells can also be induced in terminally differentiated cardiomyocytes in the absence of chromosomal replication. Thus, it is very difficult to document that cardiomyocyte cell division actually occurs via histological analysis of the adult heart, particularly if the process is occurring at a low frequency. The advantages and disadvantages of various assays used to monitor cardiomyocyte cell cycle activity have been critically reviewed.[7]

The second factor to consider when monitoring cardiomyocyte proliferation in the heart is the criteria used to distinguish cardiomyocytes from non-myocytes. Once again, this is easily accomplished in dispersed cell preparations. While this would also seem to be a somewhat trivial task in histological sections, analyses are complicated by the fact that cardiomyocytes comprise less than 20% of the total number of cells present in the adult heart (although they constitute more than 90% of the heart's mass). Cardiac fibroblasts are the most prevalent cell type within the adult heart. In addition, cells comprising the hematopoietic, vascular and autonomic nervous systems also are quite abundant.

This point is illustrated by histochemical analysis of transgenic mice that express a nuclear localized beta-galactosidase reporter gene under the transcriptional control of the alpha-cardiac myosin heavy chain (MHC) promoter. These animals (designated MHC-nLAC)[8] exhibit nuclear beta-galactosidase in cardiomyocytes, but not in other cell types. Beta-galactosidase activity is detected by incubating sections prepared from transgenic hearts with a chromogenic beta-galactosidase substrate (5-bromo-4-chloro-3-indolyl-β-D-galactoside, or X-GAL). Incubation with X-GAL results in the formation of an intense dark blue signal over cardiomyocyte nuclei (Fig. 2.2A). The section shown was also stained with Hoechst 33342, a fluorescent dye that avidly binds to DNA. Examination of the same microscopic field under fluorescent illumination reveals the presence of many nuclei that failed to stain blue in the presence of X-GAL (i.e. non-cardiomyocytes, Fig. 2.2B). These data demonstrate that the vast preponderance of nuclei (and therefore cells) present in the adult myocardium are non-cardiomyocytes.

Accurate identification of cardiomyocytes (and more specifically, cardiomyocyte nuclei) is of critical importance if DNA replication is to be used as a surrogate marker for cardiomyocyte cell cycle activity. In situations where injured tissues are examined, issues pertaining to cell type identification are particularly critical, as high rates of non-cardiomyocyte proliferation are present. Immune cytology analyses for cardiomyocyte-restricted proteins (as for example structural proteins comprising the myofiber) are frequently used to identify cardiomyocytes in histological sections. However, in light of the prevalence of non-cardiomyocyte nuclei in a typical histological section of the heart (Fig. 2.2A and B), a certain degree of subjectivity is encountered when using immune cytology analysis of myofiber proteins as the criteria to establish the identity of a given nucleus. This subjectivity is further complicated if immune cytology read-outs are utilized both for cell-type identification (i.e. myofiber protein expression) and for monitoring cell cycle progression (i.e. BrdU incorporation or expression of S phase proteins), as the simultaneous use of multiple immune cytology assays increases the possibility of encountering non-specific staining.

Cardiomyocyte cell cycle activity during normal development

In the mouse, differentiated, contracting cardiomyocytes are first detected at embryonic day 7.5. Prior to their differentiation, cardiomyocyte precursors in the lateral plate mesoderm exhibit

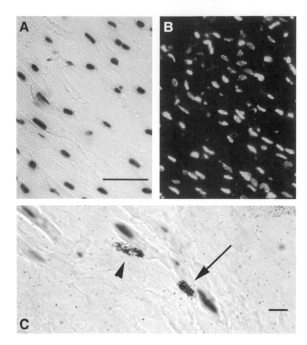

Figure 2.2. *Identification of cardiomyocyte nucleic and monitoring cardiomyocyte DNA synthesis* in vivo *using the MHC-nLAC reporter mice. (A) Image of the interventricular septum of an MHC-nLAC transgenic mouse heart following X-GAL staining (10 μm cryosection). Blue staining indicates the presence of cardiomyocyte nuclei. The section was also stained with Hoechst 33342. Scale bar (for A and B) = 50 μm. (B) Image of the same microscopic field depicted in (A) under fluorescent illumination. Note the high number of non-cardiomyocyte nuclei present in the section (identified by light blue staining). The X-GAL reaction product quenches Hoechst fluorescence, consequently the cardiomyocyte nuclei are not visible under fluorescent illumination. (C) Image of the infarct border zone in an MHC-nLAC transgenic mouse heart. Two weeks after myocardial infarction, the mouse received a single injection of tritiated thymidine and was killed 4 hours later. The heart was harvested, cryosectioned at 10 μm, stained with X-GAL and processed for autoradiography. DNA synthesis can be seen in a cardiomyocyte nucleus (arrow) and in a fibroblast nucleus (arrowhead). Scale bar = 10 μm. See color plate section.*

exceedingly high rates of cell cycle activity, with as many as 70% of the cells labeled following a single injection of tritiated thymidine. A transient reduction in the tritiated thymidine labeling index is observed upon cardiomyogenic differentiation, however by embryonic day 11 the cardiomyocyte labeling index increases to approximately 45%.[8] As heart development progresses, a transmural gradient in cell cycle activity develops, with cardiomyocytes in the compact zone (located near the outer surface of the heart) exhibiting an approximately 2-fold greater rate of cell division compared to those in the inner trabeculae.

This gradient in cell cycle activity probably underlies the formation of transmural, cone-shaped growth units previously identified by retroviral- and transgene-based lineage mapping experiments in the chick and mouse, respectively.[9–12] The bases of the growth units are localized along the compact zone of the ventricular wall, while the apices project towards the inner trabeculae. Interestingly, analysis of cell cycle activity in individual clones of embryonic cardiomyocytes *in vitro* indicates that the daughter cells from a given progenitor have an intrinsic propensity to undergo a similar number of cell divisions. Similar proliferation rates in daughter cells ensure that the cone-shaped growth units adopt a symmetrical morphology.

The overall rate of cardiomyocyte proliferation gradually declines at later stages of embryogenesis. In the mouse, cardiomyocyte tritiated thymidine incorporation levels drop to zero at birth, and then recover to approximately 10% by neonatal day 4. Pulse-chase labeling experiments indicate that the preponderance of cardiomyocyte DNA synthesis occurring during neonatal life in the mouse contributes to the formation of multinucleated cells, rather than to an increase in cell number.[13] The onset of multinucleation is associated with a marked increase in hypertrophic cardiomyocyte growth, as well as a marked increase in cell cycle activity of non-cardiomyocytes (particularly in fibroblasts and vascular components). By neonatal day 20, cardiomyocyte DNA synthesis is essentially absent in the mouse heart. Similar patterns of developmental growth, multinucleation and cell cycle withdrawal have been reported in the rat.[14,15]

Although it is generally agreed that the adult mammalian heart exhibits some degree of cardiomyocyte cell cycle activity, the level at which this occurs has been debated.[7,16] Issues discussed above concerning the methods employed to monitor cell cycle activity and cardiomyocyte identification undoubtedly have contributed at least in part to the discrepant results reported in the literature. Other factors, including potential species and strain differences, must also be taken into consideration. In an effort to non-subjectively monitor cardiomyocyte DNA synthesis in the mouse, the MHC-nLAC animals described above have been used in conjunction with a DNA synthesis assay.[17] The transgenic mice are given an injection of tritiated thymidine and humanely killed four hours later. The hearts are then isolated and sectioned. The resulting sections are incubated in the presence of X-GAL (to stain the cardiomyocyte nuclei blue) and then processed for autoradiography (to determine which nuclei were synthesizing DNA at the time of isotope injection). Cardiomyocytes synthesizing DNA are identified by the appearance of silver grains over blue nuclei, while non-cardiomyocytes synthesizing DNA are identified by the presence of silver grains in the absence of blue nuclear staining. Using this approach, only 0.0005% of the cardiomyocytes in a normal adult heart (i.e. one nucleus of 180 000 screened) are actively synthesizing DNA.[17] These data support the notion that the rate of cardiomyocyte cell cycle activity is exceedingly low in uninjured adult mouse hearts.

As indicated above, since cardiomyocytes are frequently multinucleated and/or polyploid, the presence of DNA replication does not necessarily indicate that the cells will undergo cytokinesis. In this regard there is an interesting discrepancy between the low frequency of cardiomyocyte DNA synthesis seen with the X-GAL/tritiated thymidine assay described above compared with the comparably high frequency of cardiomyocyte mitotic indices observed by some groups in normal adult hearts.[18] Assuming that cardiomyocyte misidentification does not account for these discrepancies, the data would suggest that some karyokinetic and/or cytokinetic events attributed to cell cycle induction might alternatively result from nuclear division in polyploid cells and/or cellular division in multinucleated cells, both occurring in the absence of de novo cardiomyocyte DNA synthesis.

Finally, it is of interest to note that several studies suggest the presence of adult stem cells with cardiomyogenic capability. For example, beta-galactosidase-expressing cardiomyocytes are seen following myocardial infarction in hearts of non-transgenic mice which have previously undergone bone marrow transplantation with donor cells prepared from transgenic animals with a ubiquitously expressed reporter gene.[19] In agreement with this, cardiomyocytes containing Y chromosomes can be seen in male transplant patients who receive hearts from female donors, although the frequency at which this occurs appears to be highly variable (ranging over many orders of magnitude in the studies reported to date).[20–24] Interpretation of such 'transdifferentiation' events is particularly difficult as it is now clear that marrow-derived cells have the capacity to fuse with adult cardiomyocytes.[25] Other studies suggest that specific hematopoietic progenitor cells may possess cardiomyogenic potential,[26] however several recent studies using either lineage-restricted or ubiquitously expressed reporter transgenes have failed to reproduce these results.[27–29] Finally, stem cells with apparent cardiomyogenic and/or cardiofusogenic activity have recently been identified in the adult heart.[30–32] In light of these observations, it is possible that some of the cardiomyocyte cell cycle activity detected in the adult heart might reflect de novo cardiomyogenic differentiation, or stem cell/cardiomyocyte fusion, and subsequent proliferation. The reader is referred to several recent review articles that discuss recent advances and relevant issues concerning myogenic stem cells.[33,34]

Cardiomyocyte cell cycle activity during pathophysiology

Many forms of cardiovascular disease are associated with cardiomyocyte death. In mammalian hearts dead cardiomyocytes are typically replaced by collagen deposition, and in instances where large numbers of cardiomyocytes are lost (as in myocardial infarction), overt scar formation

occurs while the cells are still actively dividing, rather than after terminal differentiation. This is an important consideration if the ultimate goal is to identify genetic pathways with which to induce regenerative growth in the adult heart, as traditional transgenic and knockout approaches rely on the intrinsic regulatory properties of the promoter or gene under study, respectively. Fortunately, recent variations of both techniques now permit a high degree of control over the temporal induction or ablation of any gene of interest within the myocardium of intact animals.[47,48] These latter approaches thus permit assessment of altered gene expression after cardiomyocyte terminal differentiation, which is a more appropriate substrate in which to test the regenerative capacity of a given genetic intervention.

An early example of enhanced cardiomyocyte cell cycle activity *in vivo* is demonstrated by mice carrying a transgene comprised of the atrial natriuretic factor promoter and sequences encoding the SV40 large T antigen oncoprotein (ANF-TAG mice).[49] T antigen is able to induce proliferation by binding to and altering the activity of genes that normally regulate cell cycle control (as for example members of the RB family or the p53 tumor suppressor protein). T antigen expression is limited to the atria in adult ANF-TAG mice, and results in the formation of right atrial tumors (compare the right atria in an ANF-TAG mouse to that of a non-transgenic control mouse, Fig. 2.3A and B, respectively). Similar results are obtained when T antigen expression is targeted to the ventricles.[50] Tumors from the ANF-TAG mice have been used to generate cardiomyocyte cell lines, which in turn have been used to identify cardiomyocyte T antigen-binding proteins.[51,52] One of these proteins, designated p193, appears to play an important role in cardiomyocyte cell cycle regulation as expression of a dominant negative version of p193 allows ventricular cardiomyocytes to re-enter the cell cycle following myocardial infarction.[35]

Targeted expression of the D-Type cyclins provides a more subtle (and perhaps more clinically relevant) example of enhanced cardiomyocyte cell cycle activity. Expression of cyclin D1, D2 or D2 under the regulation of the MHC promoter results in mild developmental hyperplasia, with adult transgenic hearts approximately 20–30% larger that those of their non-transgenic littermates.[53] The ventricular myocardium of adult cyclin D transgenic mice has an increased cardiomyocyte content (compare the X-GA- stained section of the interventricular septum of an MHC-nLAC mouse to that of an MHC-cycD/MHC-nLAC double transgenic mouse, Fig. 2.3C and D, respectively). Moreover, low levels of cardiomyocyte cell cycle activity persist in the adult transgenic mice (Fig. 2.3D, insert). Interestingly, recent studies have revealed that mice expressing cyclin D2 retain cardiomyocyte cell cycle activity following a variety of myocardial injuries, whereas mice expressing cyclin D1 or D3 do not. Sustained cardiomyocyte cell cycle activity in the cyclin D2 mice is associated with a progressive regression of infarct size following permanent coronary artery occlusion, suggesting that this pathway might be particularly suitable for inducing regenerative cardiac growth after myocardial infarction.

Summary

The studies summarized above indicate that cardiomyocytes in the adult mammalian heart retain at least a limited capacity for cell cycle activation, and that this activity is increased in response to some forms of injury. Other studies indicate that mobilization of cardiomyogenic stem cells (resident in the heart or derived from a non-cardiac source) might contribute to de novo cardiomyogenesis with concomitant cell cycle activity in the resulting cells. Despite these activities, the intrinsic regenerative capacity of the heart is insufficient to effect appreciable myocardial regeneration following injury. In this regard it is quite encouraging that there are ample experimental data indicating that cardiomyocyte cell cycle activity can be enhanced with both biochemical and genetic approaches. With further refinement, application of these approaches to native or stem cell-derived cardiomyocytes might result in sufficient cell cycle activity to antagonize ongoing cardiomyocyte loss, or alternatively reverse myocyte loss following myocardial infarction.

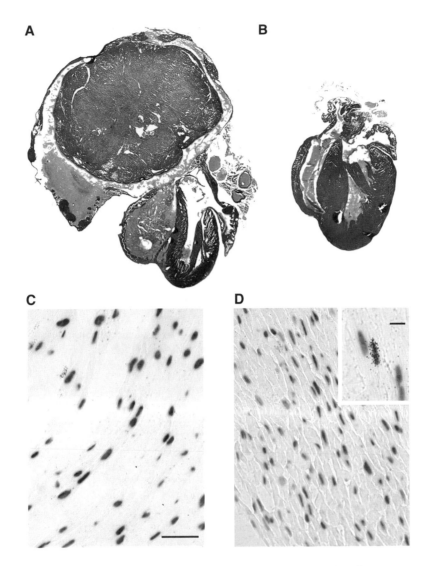

Figure 2.3. *Examples of genetic modifications that enhance cardiomyocyte cell cycle activity* in vivo. *(A) Survey micrograph of a section prepared from an ANF-TAG transgenic mouse heart, stained with hematoxylin and eosin. A prominent cardiomyocyte tumor is present in the right atrium of the heart. (B) Survey micrograph of a section prepared from the heart of a non-transgenic littermate of the animal depicted in (A), stained with hematoxylin and eosin. (C) Image of the interventricular septum of an MHC-nLAC transgenic mouse heart following X-GAL staining. Scale bar (for C and D) = 50 μm. (D) Image of the interventricular septum of an MHC-nLAC / MHC-cycD double transgenic mouse heart following X-GAL staining (the heart shown is from a littermate of the animal depicted in (C)). Note the increased number of cardiomyocyte nuclei in the heart of the double transgenic mouse. The insert shows cardiomyocyte DNA synthesis in an adult MHC-cycD transgenic mouse heart (as evidenced by the presence of silver grains over a blue nucleus). The mouse received a single injection of tritiated thymidine and was killed 4 hours later. The heart was harvested, cryosectioned at 10 μm, stained with X-GAL and processed for autoradiography. Insert scale bar = 10 μm. See color plate section.*

References

1. Schang LM. The cell cycle, cyclin-dependent kinases, and viral infections: new horizons and unexpected connections. Prog Cell Cycle Res 2003; 5: 103–24.

2. Quinn LM, Richardson H. Bcl-2 in cell cycle regulation. Cell Cycle. 2004; 3: 7–9.

3. Sherr CJ, Roberts JM. CDK inhibitors: positive and negative regulators of G1-phase progression. Genes Dev 1999; 13: 1501–12.

4. Soonpaa MH, Field LJ. Assessment of cardiomyocyte DNA synthesis during hypertrophy in adult mice. Am J Physiol Heart Circ Physiol 1994; 266: H1439–45.

5. Gerdes AM, Onodera T, Tamura T et al. New method to evaluate myocyte remodeling from formalin-fixed biopsy and autopsy material. J Card Fail 1998; 4: 343–8.

6. Rumiantsev PP. Growth and hyperplasia of cardiac muscle cells. London, New York: Harwood Academic Publishers; 1991.

7. Soonpaa MH, Field LJ. Survey of studies examining mammalian cardiomyocyte DNA synthesis. Circ Res 1998; 83: 15–26.

8. Soonpaa MH, Koh GY, Klug MG, Field LJ. Formation of nascent intercalated disks between grafted fetal cardiomyocytes and host myocardium. Science 1994; 264: 98–101.

9. Mikawa T, Borisov A, Brown AM, Fischman DA. Clonal analysis of cardiac morphogenesis in the chicken embryo using a replication-defective retrovirus: I. Formation of the ventricular myocardium. Dev Dyn 1992; 193: 11–23.

10. Mikawa T, Cohen-Gould L, Fischman DA. Clonal analysis of cardiac morphogenesis in the chicken embryo using a replication-defective retrovirus. III: Polyclonal origin of adjacent ventricular myocytes. Dev Dyn 1992; 195: 133–41.

11. Mikawa T, Fischman DA, Dougherty JP, Brown AM. In vivo analysis of a new lacZ retrovirus vector suitable for cell lineage marking in avian and other species. Exp Cell Res 1991; 195: 516–23.

12. Meilhac SM, Kelly RG, Rocancourt D et al. A retrospective clonal analysis of the myocardium reveals two phases of clonal growth in the developing mouse heart. Development 2003; 130: 3877–89.

13. Soonpaa MH, Kim KK, Pajak L, Franklin M, Field LJ. Cardiomyocyte DNA synthesis and binucleation during murine development. Am J Physiol Heart Circ Physiol 1996; 271: H2183–9.

14. Clubb FJ Jr, Bishop SP. Formation of binucleated myocardial cells in the neonatal rat. An index for growth hypertrophy. Lab Invest 1984; 50: 571–7.

15. Li F, Wang X, Capasso JM, Gerdes AM. Rapid transition of cardiac myocytes from hyperplasia to hypertrophy during postnatal development. J Mol Cell Cardiol 1996; 28: 1737–46.

16. Anversa P, Kajstura J. Ventricular myocytes are not terminally differentiated in the adult mammalian heart. Circ Res 1998; 83: 1–14.

17. Soonpaa MH, Field LJ. Assessment of cardiomyocyte DNA synthesis in normal and injured adult mouse hearts. Am J Physiol Heart Circ Physiol 1997; 272: H220–6.

18. Limana F, Urbanek K, Chimenti S et al. bcl-2 overexpression promotes myocyte proliferation. Proc Natl Acad Sci U S A 2002; 99: 6257–62.

19. Jackson KA, Majka SM, Wang H et al. Regeneration of ischemic cardiac muscle and vascular endothelium by adult stem cells. J Clin Invest 2001; 107: 1395–402.

20. Quaini F, Urbanek K, Beltrami AP et al. Chimerism of the transplanted heart. N Engl J Med 2002; 346: 5–15.

21. Laflamme MA, Myerson D, Saffitz JE, Murry CE. Evidence for cardiomyocyte repopulation by extracardiac progenitors in transplanted human hearts. Circ Res 2002; 90: 634–40.

22. Deb A, Wang S, Skelding KA et al. Bone marrow-derived cardiomyocytes are present in adult human heart: A study of gender-mismatched bone marrow transplantation patients. Circulation 2003; 107: 1247–9.

23. Muller P, Pfeiffer P, Koglin J et al. Cardiomyocytes of noncardiac origin in myocardial biopsies of human transplanted hearts. Circulation 2002; 106: 31–5.

24. Glaser R, Lu MM, Narula N, Epstein JA. Smooth muscle cells, but not myocytes, of host origin in transplanted human hearts. Circulation 2002; 106: 17–19.

25. Alvarez-Dolado M, Pardal R, Garcia-Verdugo JM et al. Fusion of bone-marrow-derived cells with Purkinje neurons, cardiomyocytes and hepatocytes. Nature 2003; 425: 968–73.

26. Orlic D, Kajstura J, Chimenti S et al. Bone marrow cells regenerate infarcted myocardium. Nature 2001; 410: 701–5.

27. Murry CE, Soonpaa MH, Reinecke H et al. Haematopoietic stem cells do not transdifferentiate into cardiac myocytes in myocardial infarcts. Nature 2004; 428: 664–8.

28. Balsam LB, Wagers AJ, Christensen JL et al. Haematopoietic stem cells adopt mature haematopoietic fates in ischaemic myocardium. Nature 2004; 428: 668–73.

29. Nygren JM, Jovinge S, Breitbach M et al. Bone marrow-derived hematopoietic cells generate cardio-

myocytes at a low frequency through cell fusion, but not transdifferentiation. Nat Med 2004; 10: 494–501.

30. Beltrami AP, Barlucchi L, Torella D et al. Adult cardiac stem cells are multipotent and support myocardial regeneration. Cell 2003; 114: 763–76.

31. Hierlihy AM, Seale P, Lobe CG, Rudnicki MA, Megeney LA. The post-natal heart contains a myocardial stem cell population. FEBS Lett 2002; 530: 239–43.

32. Oh H, Bradfute SB, Gallardo TD et al. Cardiac progenitor cells from adult myocardium: homing, differentiation, and fusion after infarction. Proc Natl Acad Sci U S A 2003; 100: 12313–18.

33. Dowell JD, Rubart M, Pasumarthi KB, Soonpaa MH, Field LJ. Myocyte and myogenic stem cell transplantation in the heart. Cardiovasc Res 2003; 58: 336–50.

34. Hassink RJ, Dowell JD, Brutel de la Riviere A, Doevendans PA, Field LJ. Stem cell therapy for ischemic heart disease. Trends Mol Med 2003; 9: 436–41.

35. Nakajima H, Nakajima HO, Tsai SC, Field LJ. Expression of mutant p193 and p53 permits cardiomyocyte cell cycle reentry after myocardial infarction in transgenic mice. Circ Res 2004; 94: 1606–14.

36. Bader D, Oberpriller J. Autoradiographic and electron microscopic studies of minced cardiac muscle regeneration in the adult newt, notophthalmus viridescens. J Exp Zool 1979; 208: 177–93.

37. Bader D, Oberpriller JO. Repair and reorganization of minced cardiac muscle in the adult newt (Notophthalmus viridescens). J Morphol 1978; 155: 349–57.

38. Poss KD, Wilson LG, Keating MT. Heart regeneration in zebrafish. Science 2002; 298: 2188–90.

39. Pasumarthi KB, Field LJ. Cardiomyocyte cell cycle regulation. Circ Res 2002; 90: 1044–54.

40. Dowell JD, Field LJ, Pasumarthi KB. Cell cycle regulation to repair the infarcted myocardium. Heart Fail Rev 2003; 8: 293–303.

41. Svensson EC, Marshall DJ, Woodard K et al. Efficient and stable transduction of cardiomyocytes after intramyocardial injection or intracoronary perfusion with recombinant adeno-associated virus vectors. Circulation 1999; 99: 201–5.

42. Metcalfe BL, Huentelman MJ, Parilak LD et al. Prevention of cardiac hypertrophy by angiotensin II type 2 receptor gene transfer. Hypertension 2004; 43: 1233–8.

43. Pasumarthi KB, Tsai SC, Field LJ. Coexpression of mutant p53 and p193 renders embryonic stem cell-derived cardiomyocytes responsive to the growth-promoting activities of adenoviral E1A. Circ Res 2001; 88: 1004–11.

44. Pasumarthi KB, Field LJ. Cardiomyocyte enrichment in differentiating ES cell cultures: strategies and applications. Methods Mol Biol 2002; 185: 157–68.

45. Pasumarthi KBS, Field LJ. Genetic models of heart disease. In: McManus B, Braunwald E (eds). Atlas of Cardiovascular Pathology for the Physician. Philadelphia: Current Medicine Inc.; 2001: 1–17.

46. Field LJ. Modulation of the cardiomyocyte cell cycle in genetically altered animals. Ann N Y Acad Sci 2004; 1015: 160–70.

47. Redfern CH, Coward P, Degtyarev MY et al. Conditional expression and signaling of a specifically designed Gi-coupled receptor in transgenic mice. Nat Biotechnol 1999; 17: 165–9.

48. Sohal DS, Nghiem M, Crackower MA et al. Temporally regulated and tissue-specific gene manipulations in the adult and embryonic heart using a tamoxifen-inducible Cre protein. Circ Res 2001; 89: 20–5.

49. Field LJ. Atrial natriuretic factor-SV40 T antigen transgenes produce tumors and cardiac arrhythmias in mice. Science 1988; 239: 1029–33.

50. Katz EB, Steinhelper ME, Delcarpio JB et al. Cardiomyocyte proliferation in mice expressing alpha-cardiac myosin heavy chain-SV40 T-antigen transgenes. Am J Physiol Heart Circ Physiol 1992; 262: H1867–76.

51. Daud AI, Lanson NA, Jr., Claycomb WC, Field LJ. Identification of SV40 large T-antigen-associated proteins in cardiomyocytes from transgenic mice. Am J Physiol Heart Circ Physiol 1993; 264: H1693–1700.

52. Tsai SC, Pasumarthi KB, Pajak L et al. Simian virus 40 large T antigen binds a novel Bcl-2 homology domain 3-containing proapoptosis protein in the cytoplasm. J Biol Chem 2000; 275: 3239–46.

53. Soonpaa MH, Koh GY, Pajak L et al. Cyclin D1 overexpression promotes cardiomyocyte DNA synthesis and multinucleation in transgenic mice. J Clin Invest 1997; 99: 2644–54.

CHAPTER 3

Molecular mechanisms of cardiac myocyte death

Kartik Mani, Michael T Crow, Richard N Kitsis

Introduction

Cardiac myocyte death occurs by apoptosis, necrosis, and perhaps autophagy. These death forms exhibit distinct morphologies, but appear to be regulated by partially overlapping molecular circuits. Apoptosis, the best understood form of cell death, is mediated by two central pathways, one involving cell surface receptors and the other cytoplasmic organelles including mitochondria and the endoplasmic reticulum. The mechanisms of necrosis are less well defined, although recent work in lower organisms suggests that this process may not be as 'accidental' as previously believed. Each of these types of cell death occurs in the heart under stressful conditions and during disease. In experimental models, cell death has been causally linked to the pathogenesis of myocardial infarction and heart failure. These data suggest that the regulated nature of certain forms of cardiac myocyte death may be exploited to provide novel therapies for these most common cardiac syndromes. This chapter reviews recent advances in the understanding of cell death mechanisms and how they relate to heart disease.

Many ways to die

Apoptosis, autophagy, and necrosis are defined by morphological characteristics. In addition, several hybrid (e.g. 'aponecrosis'[1]) and variant (e.g. caspase-independent cell death that retains apoptotic morphology) forms of cell death appear to exist. As mechanisms become more precisely delineated, it should be possible to redefine cell death processes according to molecular criteria.

Apoptosis

Apoptosis, or type 1 cell death, is an ancient suicide process that is hardwired into all metazoan cells.[2] Apoptosis can result in the deletion of single cells from an organism, but often large numbers of contiguous cells are involved. Apoptotic cells exhibit loss of cytoplasm ('shrinkage necrosis',[3] an early term for apoptosis), membrane blebbing, chromatin condensation and margination against the inner surface of the nuclear membrane, and fragmentation of cytoplasm, organelles, and nuclei into membrane-enclosed apoptotic bodies.[4] These are phagocytosed by macrophages or neighboring cells, avoiding the dumping of intracellular contents into the extracellular space and the concomitant inflammatory response.[5] During apoptosis, the morphology of cytoplasmic organelles remains grossly intact until late in the process, although abnormalities are clearly present in mitochondria.[6] Biochemical events in apoptosis include the translocation of cytochrome c and other apoptogens from mitochondria to cytosol, activation of caspases, externalization of plasma membrane phosphatidyl serine, and fragmentation of DNA.

Apoptosis is essential for the structural and functional prenatal development of multiple tissues. In postnatal tissues, this death process plays equally important roles in maintaining tissue homeostasis (e.g. appropriate cell number and composition), tissue remodeling, surveillance for malignant transformation, and immunity. Given these critical physiological roles, it is not surprising that disease can result from either deficient or excessive apoptosis.[7] For example, defects in apoptosis are involved in the pathogenesis of some cancers, while induction of apoptosis contributes to stroke, myocardial infarction, and heart failure. A wide variety of stimuli induce apoptosis in cardiac myocytes. These include deprivation of growth/survival factors,[8,9] glucose,[10,11] and oxygen[12] (particularly when followed by reoxygenation[13]), acidosis,[14] reactive oxygen species,[15] stretch,[16] angiotensin II,[17] β1-adrenergic agonists,[18–20] tumor necrosis factor-α (TNF-α),[21] Fas ligand (FasL),[22,23] and anthracyclines.[24] Cardiac myocytes undergo apoptosis *in vivo* during myocardial infarction,[25–27] particularly when followed by reperfusion,[28,29] heart failure and cardiomyopathy,[30–35] myocarditis,[36] and allograft rejection.[37]

The evidence linking cardiac myocyte apoptosis with disease is most convincing for ischemia–reperfusion and heart failure. In both humans and animal models of ischemia–reperfusion, up to 30% of cardiac myocytes die by apoptosis.[25,29,38,39] This death takes place in the first 24 hours and is largely localized to the infarct itself, although the border zone is often involved.[25,27,29,38,39] Little apoptosis occurs in the remote myocardium at these early time points. The magnitude and temporal spatial distribution of cardiac myocyte apoptosis in heart failure, however, is markedly different from that of ischemia–reperfusion injury. First, the rate of apoptosis in failing hearts is extremely low (0.023% in rodents and 0.08–0.25% in humans compared with 0.001–0.002% in controls).[32–34,40] Second, while most cardiac myocyte apoptosis in ischemia–reperfusion takes place in the first 24 hours, low levels of cell death continue for months in heart failure. Third, cardiac myocyte apoptosis occurs diffusely throughout the myocardium in the case of non-ischemic dilated cardiomyopathy and is localized primarily to non-infarcted segments in cardiomyopathy resulting from prior myocardial infarction.

Ultimately, the consequence of cell death in these syndromes is to reduce the number of functioning cardiac myocytes. Since quantification of steady-state numbers of cells is time consuming, however, most studies do not directly measure this parameter. Rather, assays of apoptotic markers (e.g. DNA fragmentation *in situ*) are used as surrogates to generate rates of cell death. Extrapolating these rates to provide estimates of steady cardiac myocyte numbers is difficult, however, because information is lacking concerning the length of time that the cell death marker remains positive in this cell type. Moreover, even if cardiac myocyte numbers could be derived from apoptotic rates, this information by itself would not be sufficient to determine whether cardiac myocyte apoptosis plays a causal role in the pathogenesis. To make this assessment, it is necessary to determine the effect of inhibiting cardiac myocyte apoptosis during a pathological process on the expected changes in cardiac structure and function. For these reasons, investigators have used genetic and pharmacologic interventions to manipulate apoptosis. In the case of ischemia–reperfusion injury, multiple independent mutations in apoptotic signaling molecules (discussed below) reduce infarct size ~50–70% in mice, and attenuate cardiac dysfunction.[22,41–46] Although attempts to translate this work using a variety of caspase inhibitors have produced mixed results, 21–52% decreases in infarct size have been observed in some studies.[47–50] Thus, cardiac myocyte apoptosis is critical for ischemia–reperfusion injury in these experimental rodent models.

The major question concerning the role of cardiac myocyte apoptosis in heart failure is whether rates of cell death are consistent with the known time course of the disease. The earliest studies reported unrealistically high rates, possibly due to technical issues. Given the protracted time course of heart failure, these high rates of cardiac myocyte loss would be inconsistent with the continued existence of the heart – although some have postulated that replacement of myocytes from progenitor cells might mitigate these large losses. On the other hand, questions have also been raised as to whether the very low, but abnormal, rates of cardiac myocyte apoptosis measured in more recent heart failure studies are large enough to be important in pathogenesis. This issue was addressed directly in one study that showed that rates of cardiac myocyte apoptosis as low as 0.023% were sufficient to cause a lethal, dilated cardiomyopathy over 2–6 months in transgenic mice with heart-restricted expression of an inducible caspase-8 allele.[40] In addition, this phenotype could be prevented by administration of caspase inhibitors. Since the rates of cardiac myocyte apoptosis in patients with dilated cardiomyopathy are actually 5–10-fold higher than in this transgenic model,[32–34] these data suggest that apoptosis may be a causal mechanism in human heart failure as well. Independent support for a mechanistic role for apoptosis in heart failure is provided by the demonstration that caspase inhibition decreases apoptosis and, most significantly, abolishes the 30% mortality resulting from peripartum cardiomyopathy in transgenic mice with heart-specific G$_{\alpha q}$ over-expression.[51] These and other studies indicate that cardiac myocyte apoptosis is a causal component in the pathogenesis of heart failure.[52–54]

Activation of the apoptotic program (and caspases) triggers the deconstruction of the cell by cleaving multiple proteins.[55] Included among these in cardiac myocytes are components of the sarcomere (e.g. troponin T, α-actin, and α-actinin).[56–58] Thus, the apoptotic program may mediate cardiac myocyte dysfunction even if it fails to execute the cell.[44,53] Since rescue of cardiac function correlates with inhibition of cell death in many models,[40,44,51] it is likely that anti-apoptotic interventions (e.g. caspase inhibitors) work mainly through inhibiting cell death. Inhibition of contractile dysfunction, however, may provide an additional mechanism for their beneficial effects.

Although a substantial body of work indicates that cardiac myocyte apoptosis is critical in the pathogenesis of ischemia–reperfusion injury and heart failure, it should be remembered that these investigations were performed primarily in rodent models. Additional studies in larger animals and humans are imperative to assess the applicability of these results to human disease. If cardiac myocyte apoptosis is as important in the human conditions, this death program may constitute a target for novel therapies directed against myocardial infarction and heart failure.

Autophagy

Autophagy is a fundamental cellular process in which double-membrane vacuoles capture cytoplasm and organelle fragments and deliver them to lysosomes for degradation and subsequent recycling to the cytoplasm.[59] In addition, in chaperone-mediated autophagy, a specialized variant of this process, proteins are transported to the lysosome for degradation by an ubiquitin-like system without involvement of a vacuole. Autophagy is thought to carry out several important cellular functions. First, it is likely to be the major mechanism for the degradation of long-lived cellular proteins.[60] Second, it is a process by which proteins and organelles can be recycled to provide the cell with chemical sources of energy.[59] This mechanism is upregulated during conditions of nutrient deprivation in yeast and serves an important survival function. The available data suggest that this is also the case in mammalian cells. Third, autophagy appears to be important for tumor suppression, although the mechanism is incompletely understood.[61–63]

Autophagy is also hypothesized to mediate a distinct form of cell death (type 2 cell death), which has garnered considerable attention in recent years. Cells undergoing autophagic cell death may exhibit enlargement of the mitochondria, endoplasmic reticulum, and Golgi apparatus.[64] Autophagic cell death is postulated to occur in many cellular contexts. For example, autophagy is a prominent feature in some paradigms of neuronal cell death as well as neurodegenerative disorders such as Parkinson's disease and Huntington's disease.[1] As will be discussed below, however, it remains quite controversial whether the autophagy observed in these syndromes is a cause of, an effect of, or unrelated to the accompanying cell death.

Autophagy occurs in the myocardium in diverse situations. It is a key component of lysosomal storage disorders such as Danon's cardiomyopathy, which is caused by a deficiency of lysosome-associated membrane protein-2 (LAMP-2).[65,66] In addition, numerous studies have demonstrated regulation of autophagy by humoral or mechanical loads. Isoproterenol and aortic constriction decrease autophagy,[67,68] while propanolol and verapamil increase it.[69] These data have been interpreted as demonstrating an inverse relationship between cardiac work and autophagy. Analogous to the situation in yeast, starvation induces myocardial autophagy in animals.[70] Similarly, myocardial autophagy is stimulated by hypoxia,[71] hypoxia-reoxygenation,[72] and treatment with non-metabolizable sugars.[73] While cardiac myocyte death is elicited by some of these stimuli, cell death was not noted to occur in conjunction with autophagy in the studies cited above. In contrast, both autophagy and cell death have been observed in cardiac myocytes of failing human hearts.[74–76] Moreover, autophagy and necrosis were felt to be more important than apoptosis as the mode of cardiac myocyte demise in this setting.[75,76] Additional studies will be important to further test this conclusion.

It is reasonable to hypothesize that massive organelle destruction involves cell death. Accordingly, the concept of autophagic cell death is further bolstered by the association of autophagy with large-scale tissue destruction during development, metamorphosis in insects and amphibians, and in some disease states.[1] In addition, molecular data provide further indirect support. For example, molecules such as caspases, the death ligand TRAIL, and type I phosphatidylinositol 3′-kinases, which play key roles in apoptotic signaling, also regulate autophagy.[77–79] Despite these arguments, definitive proof that autophagy causes cell death in any system is currently lacking. Moreover, the existing data are consistent with a non-death function. For example, the correlation between autophagy and cell death may simply reflect a 'clean-up' role for autophagy in cell death carried out by another process (e.g. apoptosis[1]). In addition, increasing evidence that autophagy is important for survival in mammalian cells raises the possibility that it is activated in stressful situations as a defense to injury rather than a pathological process.

To test directly the relationship between autophagy and cell death, it is necessary to inhibit autophagy. Genetic approaches in yeast and mammalian cells have begun to dissect pathways that control this process. Transfection of Beclin 1 into breast cancer cells stimulates autophagy and suppresses tumorigenesis.[61] Importantly, however, Beclin 1 does not influence cell death in this setting. Interestingly, Beclin 1 is a Bcl-2 (B cell leukemia/lymphoma-2 protein)-binding protein, but the relationship of this interaction to Beclin 1's functions is not known.[80] Inactivation of one *beclin 1* allele in the mouse reduces Beclin 1 abundance, decreases autophagy, and predisposes to a variety of tumors in adult tissues, consistent with the observed mono-allelic deletion of Beclin 1 in several human cancers.[62,63] Although Beclin 1[−/−] embryonic stem (ES) cells exhibit markedly decreased autophagy under conditions of nutrient deprivation, they undergo cell death at rates similar to wild-type ES cells in response to serum deprivation or UV light. Thus, Beclin 1 is critical for autophagy and tumor suppression but not for cell death. These observations provide the strongest evidence to date dissociating autophagy from cell death. The molecular dissection of autophagy is in its early stages, however. It will be important to study the effects of mutations in additional autophagy genes to further assess the relationship between autophagy and cell death.

Necrosis

Necrosis (oncosis; type 3 cell death) is morphologically characterized by cell swelling and loss of plasma membrane integrity leading to a marked inflammatory response that often results in collateral damage to neighboring cells.[64] In contrast, plasma membrane integrity is maintained in classical apoptosis, although it should be noted that secondary necrosis can follow apoptosis in situations where apoptotic corpses are not cleared efficiently (e.g. in cell culture). Necrosis is often thought of as 'pathological' cell death, and in many instances, this is correct. However, necrosis can also occur in physiological contexts. For example, although most developmental cell deaths occur by apoptosis, necrosis has also been noted.[81]

Necrosis is an important cause of cardiac myocyte death in disease. During myocardial infarction in the rat, necrosis appears later than apoptosis and to be of smaller magnitude.[26] In addition, 0.5–1.2% of cardiac myocytes in explanted hearts from patients with ischemic and non-ischemic dilated cardiomyopathies showed necrosis while 0.08–0.18% were apoptotic.[34] These estimates indicate that cardiac myocyte death during human myocardial infarction and heart failure occurs by both necrosis and apoptosis. While many instances of necrosis in these syndromes are probably primary events, the kinetics of the two death forms in myocardial infarction raise the possibility that some instances of necrosis may be secondary events following on the heels of apoptosis.

The mechanism of necrosis is obscure. While its *sine qua non* is ATP depletion and a failure of plasma membranes leading to a breakdown in cellular homeostasis, even the chronology of these events is not known with certainty. Moreover, as noted above, necrosis has traditionally been regarded as an 'accidental' and unregulated form of cell death. Recent work in *C. elegans* challenges this view (Fig. 3.1). Gain of function mutations in amiloride-sensitive Na^+ channels (DEG-1 and MEC-4) induce necrosis-like cell death in neurons, which is rescued by loss of function mutations in the ryanodine (UNC-68) and IP_3 (ITR-1) receptors (sarcoplasmic/endoplasmic reticulum Ca^{2+} release channels) or calreticulin (CRT-1) and calnexin (CNX-1) (endoplasmic reticulum (ER) proteins that regulate Ca^{2+}).[82,83] Na^+ channel (MEC-4)-induced necrosis was restored in the face of calreticulin deficiency by thapsigargin, which augments cytoplasmic Ca^{2+}.[83] These data suggest that necrosis may be an active process that is mediated by specific cellular programs. Moreover, they imply that an increase in the cytoplasmic calcium

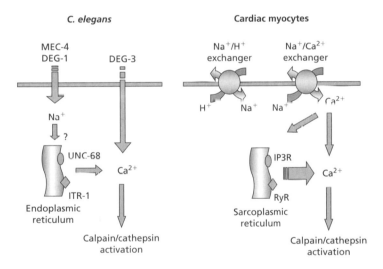

Figure 3.1. *Potential necrosis pathways. Left, necrosis pathways in C. elegans. In the nematode, gain of function mutations in the plasma membrane Na^+ channels MEC-4 and DEG-1 induce necrotic death. This is hypothesized to be mediated by Ca^{2+} release through UNC-68 and ITR-1, endoplasmic reticulum Ca^{2+} release channels. Elevations in intracellular Ca^{2+} are thought to activate calpains, which cleave and activate cathepsins. Necrosis is also induced by a gain of function mutation in the plasma membrane Ca^{2+} channel DEG-3. This pathway is also genetically linked with calpains and cathepsins, but not with the ER Ca Ca^{2+} release channels. Right, hypothetical necrosis pathway in mammalian cardiac myocytes. Ischemia produces intracellular acidosis. Na^+ enters the cell through the sarcolemma Na^+/H^+ exchanger. Ca^{2+} is exchanged for Na^+ through the sarcolemma Na^+/Ca^{2+} exchanger. Intracellular Ca^{2+} induces Ca^{2+}-mediated Ca^{2+} release through the ryanodine receptor (RyR) and inositol 1,4,5-trisphosphate receptor (IP3R), sarcoplasmic reticulum Ca^{2+} release channels. Increased intracellular Ca^{2+} activates calpains and cathepsins as above. Figure adapted with permission from Yuan J, Lipinski M, Degterev A. Diversity in the mechanisms of neuronal cell death. Neuron 2003; 40: 401–13.*

concentration may play an important role in necrosis, a hypothesis that has also been proposed for mammalian cells.

Increases in Ca^{2+} are known to activate calpains, which are non-lysosomal, non-caspase, cysteine proteases. Loss of function mutations in certain *C. elegans* calpains (CLP-1 and TRA-3), as well as cathepsin aspartyl proteases (ASP-3 and ASP-4), rescue necrosis induced by Na^+ channel activation, suggesting that these proteases are situated downstream of Na^+ channels in a necrotic pathway.[84] Interestingly, a gain of function mutation of a plasma membrane Ca^{2+} channel (DEG-3) also causes necrosis, which is rescued by deficiencies of these proteases but not by mutations in the above endoplasmic reticulum (ER) proteins.[84] This suggests that the plasma membrane Ca^{2+} channel is part of a pathway that signals in parallel to the same downstream proteases as those activated by the activated Na^+ channels. These studies in the nematode suggest a role for Ca^{2+} as a mediator of necrosis that signals downstream to calpains and cathepsins.[1]

Although the relevance of these *C. elegans* studies to the mammalian heart is not yet clear, it is worth remembering that the essential blueprint for apoptosis (discussed below) is conserved from worms to humans. In addition, there are some immediate parallels that can be drawn between necrosis pathways in *C. elegans* and the ischemic heart. During ischemia, intracellular pH decreases, stimulating Na^+–H^+ exchange at the plasma membrane.[85] The subsequent accumulation of cytosolic Na^+ results in cytosolic Ca^{2+} overload through plasma membrane Na^+–Ca^{2+} exchange. Inhibition of the amiloride-sensitive Na^+–H^+ exchanger with cariporide,[86] or of the Na^+–Ca^{2+} exchanger with KB-R7943,[87] decreases infarct size following ischemia–reperfusion. Consistent with these data, complete knockout of *NHE1*,[88] or heterozygous deletion of *NCX1*,[89] the primary cardiac Na^+–H^+ and Na^+–Ca^{2+} exchangers respectively, reduce infarct size following ischemia–reperfusion. Inhibition of the Na^+–H^+ exchanger with EMD-87580 also inhibits myocardial remodeling and improves cardiac function following myocardial infarction.[90] Since markers were not consistently evaluated, it is difficult to know in each case

whether decreases in apoptosis or necrosis were primarily responsible for reductions in infarct size. Other studies have implicated calpains in these processes. For example, calpains are activated during myocardial ischemia, and calpain inhibition reduces infarct size.[91] The involvement of the Na^+–H^+ and Na^+–Ca^{2+} exchangers and calpains suggest the existence of a deliberate necrotic program in mammalian cells that appears to be functionally conserved back to the worm.

Core apoptotic pathways

Of the pathways that mediate cell death, the molecular regulation of apoptosis is best understood. For this reason and because of its relevance to myocardial biology, the remainder of this chapter will discuss the core apoptotic pathways in detail and their role in cardiac pathophysiology. Apoptosis is mediated by two pathways (Fig. 3.2): the extrinsic (or death receptor) pathway and the intrinsic (or mitochondrial) pathway.[92] The goals of each pathway are the same: to activate caspases and destroy mitochondrial function. Caspases are a subfamily of cysteine proteases that hydrolyze peptide bonds' carboxyl to aspartic acid residues.[93] These proteases play critical roles in apoptosis by transmitting upstream death signals and by cleaving cellular proteins to disassemble the cell. Humans possess approximately a dozen caspases, each encoded by an individual gene. Human caspases-2, -8, -9, -10, and -12, which are upstream (apical, signaling) caspases, relay death signals to downstream caspases. Downstream (effector, executioner) caspases-3, -6, and -7, on the other hand, cleave cellular proteins to kill the cell. Because caspases are lethal molecules, they are synthesized and exist in a dormant state as largely inactive procaspases. The primary structure of procaspases consists of an N-terminal prodomain of variable length followed by ~20 kDa (p20) and ~10 kDa (p10) domains, with the catalytic activity residing in the latter two domains. Upstream procaspases are activated by dimerization,[94–96] which usually occurs when they are recruited into multiprotein complexes (discussed below). On the other hand, downstream procaspases, which already exist as inactive dimers,

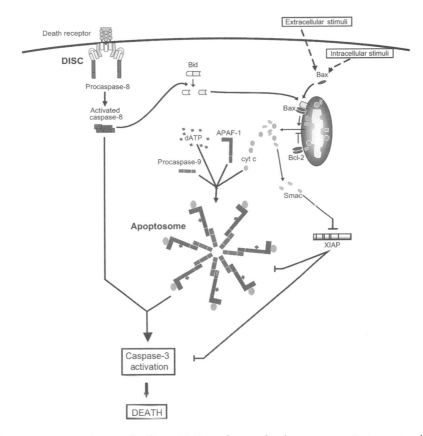

Figure 3.2. *Core apoptosis pathways. In the extrinsic pathway, death receptor activation stimulates death-inducing signaling complex (DISC) assembly (see text) leading to procaspase-8 activation. Activated caspase-8 then cleaves downstream procaspase-3, which proteolyses cellular substrates killing the cell. In the intrinsic pathway, intracellular and extracellular death signals are transmitted to the mitochondria through BH3-only proteins (e.g. Bid) and Bax, which translocates to the outer mitochondrial membrane. Bax and Bak (not shown) stimulate the release of cytochrome c (cyt c), Smac/DIABLO, and other apoptogens (not shown). Apoptogen release is opposed by Bcl-2 and Bcl-xL (not shown). Cytochrome c, dATP, Apaf-1, and procaspase-9 assemble into the apoptosome (see text) leading to procaspase-9 activation, which subsequently activates procaspase-3. XIAP inhibits activation of procaspase-9 as well as active caspases-9, -3, and -7 (not shown). Smac/DIABLO and Omi/HtrA2 (not shown) bind XIAP relieving this inhibition. Bid, a direct substrate of caspase-8, connects the extrinsic and intrinsic pathways. Following cleavage, its C-terminus translocates to and inserts into the outer mitochondrial membrane, triggering activation of Bax and Bak and cytochrome c release. Figure reproduced with permission from Crow MT, Mani K, Nam Y-J, Kitsis RN. The mitochondrial death pathway and cardiac myocyte apoptosis. Circ Res 2004; 95: 957–70.*

are activated when cleaved by upstream caspases. The cleavage sites are located in linkers between the prodomain, p20 and p10 subunits.[97] In this way, cleavage separates these domains, allowing the non-covalent reassociation of two p20 and two p10 subunits into the active caspases. Following acti-

vation, downstream caspases carry out the proteolytic destruction of the cell.[55]

Extrinsic pathway

The extrinsic pathway is activated by a death ligand, which may be a soluble protein (e.g. tumor necrosis factor (TNF)-α) or a membrane protein on another cell (e.g. Fas (CD95/Apo-1) ligand (FasL)).[98] The death ligand binds as a trimer to its cognate cell surface death receptor, also a preformed trimer.[99] Ligand binding is presumed to induce a conformational change in the death receptor that triggers the assembly of the death-inducing signaling complex (DISC),[100–102] one of several multiprotein complexes that play critical roles in apoptotic signaling. As illustrated by the DISC formed following Fas activation, the liganded death receptor recruits the adaptor protein FADD (Fas-associated death domain protein) which, in turn, recruits procaspases-8 and -10. As in a number of other apoptotic protein complexes, interactions between proteins in the DISC are mediated by death-fold motifs. Death-fold motifs include death domains (DD), death effector domains (DED), caspase recruitment domains (CARD), and pyrin domains (PYD).[103,104] While the amino acid sequences of these domains differ significantly, their three dimensional structures are similar, each consisting of six antiparallel α-helices.[105–107] Death-fold motifs engage only in monovalent interactions. In addition, they are thought to bind in a homotypic manner (DD with DD; DED with DED; CARD with CARD etc), although this view has been recently challenged.[108] In the case of the Fas DISC, liganded Fas binds FADD through DDs in each protein.[109,110] Upstream procaspases have long prodomains that contain DEDs or CARDs, which are used for their recruitment into complexes. Accordingly, FADD recruits procaspase-8 through DED–DED interactions.[101,102] Thus, death receptor activation brings procaspase-8 molecules into close proximity with each other.

Recruitment of procaspase-8 into the DISC leads to its activation. Even in the unprocessed state, procaspase-8 (and some other upstream procaspases) possesses a small, but significant, amount of caspase activity. For this reason, procaspase-8 activation in the DISC was initially thought to involve autocleavage due to forced proximity.[111] As discussed above, however, recent biochemical studies have demonstrated that activation of procaspase-8 and other upstream procaspases is mediated instead by dimer formation induced by forced proximity.[94–96] Proteolytic processing following this

event may stabilize dimer formation but is not the activating event.[96] Following its activation, caspase-8 transmits the apoptotic signal downstream by proteolytically activating procaspase-3.[97] In addition, activation of the extrinsic pathway can also trigger the intrinsic pathway through caspase-8-mediated cleavage of Bid (BH3 (B cell leukemia/lymphoma-2 (Bcl-2) homology domain 3) interacting domain death agonist),[112–114] a pro-apoptotic Bcl-2 protein (discussed below).

Fine control is critical for any biological process with profound implications. Accordingly, the pathways that drive apoptosis are balanced by ones that oppose it. Several endogenous inhibitors, which are present at high levels in cardiac and skeletal muscle, antagonize death receptor signaling at the level of the DISC. FLIP$_L$ (Fas-associated death domain protein-like-interleukin-1β-converting enzyme inhibitory protein) is a DED containing protein similar to procaspase-8, but enzymatically dead because of several missense mutations. FLIP$_L$ inhibits procaspase-8 recruitment by binding to the DEDs of FADD and procaspase-8, presumably precluding their association.[115] FLIP$_S$, an isoform generated by alternative splicing is comprised solely of two DEDs and is thought to act through the same mechanism. ARC (Apoptosis Repressor with a CARD) binds to Fas and FADD through CARD–DD interactions, and to procaspase-8 through CARD-DED interactions to preclude formation of the DISC.[108,116] Of course, despite the expression of these endogenous inhibitors, cells including cardiac myocytes do die in response to activation of the extrinsic pathway (e.g. FasL). This apparent contradiction may be explained by the rapid decreases in the levels of these inhibitory proteins during apoptosis.[117–119] In the case of FLIP, this has been shown to be mediated by increased degradation via the ubiquitin-proteasome pathway,[120] a mechanism that modulates the abundance of many apoptosis regulators.

TNF, an important mediator in cardiovascular biology, is an unusual death ligand as it can stimulate both cell death and survival. Initially, the TNF was thought to assemble a DISC similar to that of Fas with one minor modification: TRADD (TNF receptor superfamily 1A-associated via death domain) was believed to function as a linker between TNFR1 (TNF receptor 1) and FADD. A significant revision to this model has recently been proposed, however.[121] Rather than recruiting a DISC directly to TNFR1, two sequential multi-protein complexes form. Complex I consists of TNFR1, TRADD, TRAF2 (TNF receptor-associated factor 2), and RIP (receptor-interacting serine-threonine protein kinase). Complex II, which is thought to form following an internalization event, consists of TRADD, FADD, RIP, TRAF2, and procaspases-8 and/or -10. This model explains the inability to detect FADD in complex with TNFR1. In addition, it provides a framework for understanding how TNFR1, in contrast to Fas, can signal life as well as death: $FLIP_L$ plays a dual role in this scheme by inhibiting procaspases-8 and -10 in Complex II and activating NF-ΚB (nuclear factor-ΚB), which transactivates the expression of survival proteins, in collaboration with TRAF2.

The importance of death receptor signaling in cardiac myocyte apoptosis is well established. Hypoxia-induced death in these cells can be antagonized by a FADD dominant negative mutant,[122] FLIP,[123] and CrmA (cytokine response modifier A; a cowpox virus serpin that primarily inhibits caspases-8 and -1).[124] Although hypoxia is classically thought to be an activator of the intrinsic pathway, these observations and the fact that death ligands are secreted during hypoxia[23] suggest that hypoxia-induced death may also be mediated through an autocrine/paracrine loop that activates the extrinsic pathway. The importance of the extrinsic pathway to cardiac myocyte apoptosis is illustrated by *lpr* mice, which lack Fas and exhibit marked reductions in infarct size following ischemia–reperfusion.[22,23] In addition, overexpression of ARC decreases infarct size in isolated, perfused hearts subjected to ischemia–reperfusion,[125] and pathological post-infarct remodeling in intact animals.[126]

In contrast to Fas, genetic evidence suggests that the net effect of TNF signaling during ischemia–reperfusion is cell survival.[127] This is based on data showing that, while single deletions of TNFR1 and TNFR2 do not alter infarct size following permanent coronary artery occlusion, the double TNFR1/TNFR2 knockout exhibits larger infarcts. The basis for these observations is probably related to the duality noted above in TNF life–death signaling, but the mechanisms are incompletely understood.

In addition to apoptosis, death receptor signaling also mediates non-apoptotic processes. Both Fas and TNF stimulation in the presence of caspase inhibition have been shown to cause necrosis in non-cardiac cells.[128,129] This effect is mediated by FADD and RIP.[130] While RIP is not needed for apoptosis, its serine-threonine kinase domain appears to be essential for necrosis. The substrates for this kinase remain to be defined. Neither the extent to which death receptor signaling regulates cardiac myocyte necrosis nor the relationship between this pathway and the Na^+–Ca^{2+}–calpain necrotic pathways discussed above are currently understood.

Fas signaling has also been demonstrated to stimulate cellular proliferation in T lymphocytes through FADD.[131,132] Perhaps related to this growth effect, long-term overexpression of FasL in cardiac myocytes elicits hypertrophy,[133] while Fas-deficient *lpr* mice fail to undergo hypertrophy in response to pressure overload.[134] In contrast, short-term cardiac overexpression of high levels of FasL causes cardiac myocyte apoptosis.[22] An understanding of mechanistic relationships between Fas-mediated growth and death will require further work. Also unexplained is the embryonic lethal 'thin myocardial' phenotype resulting from germ line deletion of both pro- and anti-apoptotic mediators of the extrinsic pathway including FADD,[135] procaspase-8,[136] and FLIP.[137] Where examined, changes in myocyte apoptosis and proliferation do not appear to account for these observations. Whether this phenotype is myocyte-autonomous, and its underlying basis remain to be determined.

Intrinsic pathway: pre-mitochondrial events

The intrinsic pathway is the most ancient and fundamental death mechanism and is responsible for transducing the vast majority of extra- and intracellular death signals including: loss of survival/growth factors, nutrient deprivation, hypoxia, oxidative stress, radiation, drugs, DNA damage, and ER stress. In the intrinsic pathway, death signals transduced by a variety of peripheral pathways are relayed first to BH3 (Bcl-2 homology domain 3)-only proteins and, from there, to multidomain pro-apoptotic Bcl-2 proteins (see below). The latter trigger mitochondrial dysfunction and the release of mitochondrial apoptogens into the cytosol. These apoptogens subsequently stimulate caspase activation or cause direct cellular damage.

The Bcl-2 family includes both pro- and anti-apoptotic proteins.[92] The pro-apoptotics can be categorized as: BH3-only proteins such as Bid and Bad (Bcl-2-antagonist of cell death); and multidomain proteins such as Bax (Bcl-2-associated X protein) and Bak (Bcl-2-antagonist/killer). Antiapoptotic Bcl-2 proteins include Bcl-2 itself and Bcl-x$_L$ (Bcl-x protein long isoform). The multidomain pro-apoptotics and anti-apoptotics are characterized by the presence of BH1–3 domains. Most anti-apoptotic Bcl-2 proteins also contain a BH4 domain that is required for cytoprotection. The primary structure of the BH3-only proteins is dissimilar to the other groups and each other except in the short (10–16 amino acid) BH3 domain which, in most cases, is required for killing. Despite these differences, the various Bcl-2 proteins each possess 8–9 α-helices such that the three-dimensional structures for the three groups are similar.[138]

BH3-only proteins

BH3-only proteins transduce upstream death signals to multidomain pro-apoptotics Bax and Bak. BH3-only proteins differ both in terms of the specific stimuli that activate them and the mechanisms by which they are activated. For example, Bid,[139] which is activated through proteolytic cleavage, senses signals emanating from the extrinsic pathway.[112–114] In contrast, Bad, which is activated by dephosphorylation, responds to growth factor deprivation.[140–143] As there are multiple BH3-only proteins, the discussion to follow will focus on ones that are central to apoptotic signaling or play known roles in cardiac myocytes.

In most cells (type II cells), killing via the extrinsic pathway is amplified by secondary activation of the intrinsic pathway. Bid provides a critical link through which this is accomplished.[112–114] Bid activation involves a cleavage event that removes its N-terminus to reveal its BH3 domain. This cleavage event is carried out by caspase-8 following activation of the extrinsic pathway. Bid can

also be cleaved by calpain,[144] granzyme B,[145,146] and caspase-3, the latter functioning as a feed-forward mechanism.[147] Following cleavage, the truncated carboxy fragment of Bid (tBid) moves to the mitochondria. At the mitochondria, tBid inserts into the outer membrane via its C-terminus such that its BH3 domain faces the cytosol.[114,118] The BH3 domain binds to and activates Bak,[148] which also resides in the outer mitochondrial membrane, displacing Bak's inhibitor VDAC2 (voltage-dependent anion channel 2, discussed below).[149] In addition, the BH3 domain binds Bax,[139] and enhances its insertion into the outer mitochondrial membrane during apoptosis.[150] Bid is important in the pathogenesis of myocardial ischemia–reperfusion injury as mice that lack this protein exhibit a 53% reduction in infarct size (P Lee, SJ Korsmeyer, RN Kitsis, unpublished data).

In healthy cells, Bad is held in an inactive state by serine phosphorylation events that sequester it at the 14-3-3 protein.[141–143] Relevant kinases include Akt (v-akt (AKR thymoma viral oncogene) homolog serine-threonine kinase)/PKB (protein kinase B), PKA (protein kinase A), p90RSK (p90 ribosomal S6 kinase), and p70S6K (p70 ribosomal S6 kinase). Bad is dephosphorylated in the absence of survival factors such as insulin-like growth factor (IGF)-1, which releases it from the 14-3-3 protein allowing it to translocate to the mitochondria. Similar to tBid, Bad activates Bak by displacing VDAC2.[149] The role of Bad in cardiac myocyte apoptosis remains to be determined. It is notable, however, that the susceptibility of these cells to apoptotic stimuli is increased by expression of dominant negative 14-3-3.[151] Since 14-3-3 binds several pro-apoptotic proteins, however, direct experiments will be required to delineate the precise role of Bad.

BNip3 (Bcl-2/adenovirus E1B nineteen kD-interacting protein 3),[152,153] and Nix (Nip3-like protein X)/BNip3L (BNip3-like protein) are members of the BH3-only family that have recently been recognized to play important roles in cardiac myocyte apoptosis.[52,154,155] The abundance of these proteins is transcriptionally regulated by HIF-1α (hypoxia-inducible factor-1α).[152,153] The expression of BNip3 is induced by hypoxia and acidosis in cardiac myocytes and heart failure,[154,155] and

BNip3 is required for cardiac myocyte apoptosis due to hypoxia-acidosis.[154] Similarly, the expression of Nix/BNip3L is increased by hemodynamic overload, and transgenic mice that overexpress the hypertrophy-inducing protein $G_{\alpha q}$.[52,156] $G_{\alpha q}$ transgenic mice develop a lethal peripartum cardiomyopathy,[156] and Nix/BNip3L is critical for its pathogenesis.[52] Similar to Bid and Bad, BNip3 and Nix/BNip3L are targeted to the outer mitochondrial membrane. Unlike Bid and Bad, however, the 'BH3-like' domains of BNip3 and Nix/BNip3L are not required for cytotoxicity.[157] Rather, their transmembrane domains are essential,[157] suggesting that death is mediated through other mechanisms.

Multidomain pro-apoptotic Bcl-2 proteins

Bax and Bak are the major multidomain pro-apoptotic Bcl-2 proteins.[158–161] Either suffices to transmit upstream death signals to the mitochondria; conversely, the absence of both proteins blocks cell death mediated by the intrinsic pathway.[162] In healthy cells, Bax is cytosolic. In response to activators of the intrinsic pathway, it translocates to the mitochondria, oligomerizes, and inserts into the outer mitochondrial membrane via its most carboxy α-helix (α9).[163–166] Studies demonstrating that an N-terminal neo-epitope is unmasked during apoptosis suggest that conformational changes are important in Bax activation.[165] Although the details of this conformational activation are not known, the solution structure showing that α9 is buried in a hydrophobic cleft when Bax is inactive suggests that conformational activation may involve release of this helix.[166] The mechanisms that regulate Bax activation and translocation are also poorly understood, but several proteins have been implicated. Ku70,[167] Humanin,[168] and ARC[108,169] bind Bax and inhibit its conformational activation, while ASC (apoptosis-associated speck-like protein containing a CARD),[170] another Bax-binding protein, does the opposite. The tumor suppressor p53, which mediates apoptosis in response to multiple stresses, also stimulates Bax activation, but the mechanism may be indirect as p53 and Bax do not appear to interact.[171] In addition to conformationally activating Bax, p53 also induces its expression through a transcriptional mechanism.[172] Knockdown of caspase-2 blocks Bax

translocation, although the mechanism is unclear.[173] Bax can also be activated by cleavage of its inhibitory N-terminus by calpain,[174,175] a mechanism that may be a point of overlap between necrotic (calpain) and apoptotic (Bax) signaling. Once targeted to the mitochondria, Bax triggers the release of cytochrome c through mechanisms that will be considered in detail below. A role for Bax in ischemia–reperfusion injury has been established by experiments showing that its absence reduces infarct size by 50% in isolated hearts.[46] These data suggest that, although Bax and Bak perform redundant functions in many cell systems,[162] Bax appears to be critical for cardiac myocyte apoptosis induced by certain stresses. Bak is also conformationally regulated.[148,176] In contrast to Bax, however, Bak is constitutively localized at the outer mitochondrial membrane where it is held in place by its C-terminus. Thus, unlike Bax, Bak cannot be silenced by its C-terminus. Rather, Bak is inhibited by Mcl-1 (myeloid cell leukemia sequence 1 isoform 2), an anti-apoptotic Bcl-2 protein,[177] and VDAC 2, an outer mitochondrial membrane protein that may be part of the mitochondrial permeability transition pore (MPTP, discussed below).[149] Pro-apoptotic tBid and Bad activate Bak by displacing VDAC2, while p53 displaces Mcl-1. Once activated, Bak oligomerizes in the membrane and stimulates cytochrome c release (see below).

Anti-apoptotic Bcl-2 proteins

Anti-apoptotic Bcl-2 proteins, such as Bcl-2,[178–180] and Bcl-x_L,[181] antagonize the intrinsic pathway. Although they have long been recognized to inhibit cytochrome c release,[182,183] their precise mechanisms of action are not well understood. Both proteins are constitutively localized at the outer mitochondrial membrane, ER, and nuclear envelope,[184] while a portion of Bcl-x_L is cytosolic.[163] Bcl-2 and Bcl-x_L were first thought to regulate multidomain pro-apoptotics, such as Bax and Bak, by direct binding (rheostat model).[158] Other work, however, has questioned whether these interactions are artifactual.[185] Furthermore, knockout experiments suggest that Bcl-2 and Bax are functionally independent.[186] An alternative model that takes these data into consideration postulates that Bcl-2 and Bcl-x_L serve as 'sinks' for BH3-only proteins, such as tBid, thus preventing their access to Bax and Bak.[187] In this model, other BH3-only proteins, such as Bad, may promote apoptosis by displacing tBid from Bcl-2.[188] The importance of anti-apoptotic Bcl-2 proteins in cardiac myocytes is illustrated by 48–64% reductions in infarct size following ischemia–reperfusion in transgenic mice that overexpress Bcl-2 in the heart.[41,42]

p53

The transcription factor p53 mediates apoptosis induced by multiple cellular stresses including oxidative stress, hypoxia, and DNA damage.[189] Both transcriptional and non-transcriptional mechanisms mediate p53-induced apoptosis. The non-transcriptional effects of p53 on Bax and Bak have been discussed previously.[171,177] Transcriptional targets include genes encoding Noxa, Puma, and Bid (BH3-only proteins), Bax, ASC (which induces Bax activation as described above), Apaf-1 (a component of the apoptosome), caspase-6, and Fas.[170,189] Hypoxia increases p53 levels in cultured cardiac myocytes, and p53 is sufficient to induce apoptosis in normoxic cardiac myocytes.[190] In contrast, cardiac myocyte apoptosis induced by ischemia–reperfusion appears to be p53 independent.[14,38] It should be noted, however, that the effects of p53 on infarct size *per se* have not yet been assessed. Stretch-induced apoptosis in cultured cardiac myocytes is p53 dependent, and p53's transcriptional induction of angiotensinogen and the type 1 angiotensin II receptor is believed to be involved.[191,192]

ER pathway

Recent work has shown that the ER is a critical component of the intrinsic pathway in mediating apoptosis induced by certain stimuli including ceramide, arachidonic acid, and oxidative stress.[193] A major mechanism by which the ER modulates apoptosis is by influencing intracellular Ca^{2+}. Significant pools of Bax, Bak, and Bcl-2 exist at the ER membrane,[184,194] and modulate ER Ca^{2+} stores.[193,195,196] Bax and Bak increase intraluminal Ca^{2+} stores through unknown mechanisms, resulting in more marked release of Ca^{2+} in response to an apoptotic signal. Bcl-2, on the other hand, decreases ER Ca^{2+} stores, which blunts death stimulus-induced Ca^{2+} release. Intracellular Ca^{2+} can trigger cell death

through several mechanisms. The first is mitochondrial toxicity, including that mediated by MPTP opening,[197] which can result in cytochrome c release. Cytochrome c release can potentiate this pathway in a feedback loop by binding the inositol 1,4,5-trisphosphate (IP_3) receptor, an ER Ca^{2+} release channel, further augmenting Ca^{2+} release.[198] Second, ER Ca^{2+} release can induce cell death by activating calpain. Among other substrates, calpain can cleave and activate procaspase-12.[199] Following cleavage, caspase-12 may translocate from the ER membrane to the cytoplasm and activate procaspase-9, through a mechanism that does not require mitochondria or the apoptosome (defined below).[200,201] Caspase-9 can then activate procaspase-3 and induce apoptosis. Of note, experiments in knockout mice demonstrate that caspase-12 is critical for apoptosis resulting from ER stress.[202] The role of caspase-12 in humans is less clear, however, as some populations express polymorphisms with non-sense mutations.[203] As previously noted, Ca^{2+}-activated calpain can also cleave Bid, triggering apoptosis.[144] In addition, the extrinsic death pathway may be linked with ER signaling through Bap31 (B-cell receptor-associated protein 31), an integral ER membrane protein that is cleaved by caspase-8 resulting in ER Ca^{2+} release.[204] It is notable that the actions of Ca^{2+} on the mitochondria, caspases, and calpain provide multiple potential points of convergence between apoptotic and necrotic signaling. As intracellular Ca^{2+} concentrations increase in cardiac myocytes by an order of magnitude with each contraction (10^{-7} to 10^{-6} M), it remains to be determined why these cells do not succumb under normal conditions.[205]

Intrinsic pathway: intramitochondrial events

Mitochondria amplify upstream death signals by undergoing structural and functional remodeling and releasing apoptogens. The release of these proteins triggers apoptotic events in the cytosol and nucleus (see below). Apoptogens include cytochrome c.[183,206] Smac (second mitochondria-derived activator of caspase/DIABLO (direct IAP binding protein with low pI),[207,208] Omi/HtrA$_2$ (high temperature requirement protein A$_2$),[209,210] AIF,[211] and EndoG (endonuclease G).[212]

Apoptogen release

Cytochrome c was the first mitochondrial apoptogen to be identified.[83,206,213] Its release was noted to be stimulated by pro-apoptotic and inhibited by anti-apoptotic Bcl-2 proteins.[183,213,214] Despite this, the precise mechanism(s) by which it moves to the cytosol remains an enigma and one of the most important unresolved issues in the cell death field. Cytochrome c release was initially hypothesized to be regulated by the MPTP (Fig. 3.3A). Under stress conditions (e.g. oxidative stress, Ca^{2+} overload), this non-specific pore allows molecules <1.5 kDa to cross the usually impermeable inner mitochondrial membrane.[197] Thus, opening of the MPTP would allow the influx of water into the solute-rich mitochondrial matrix, causing

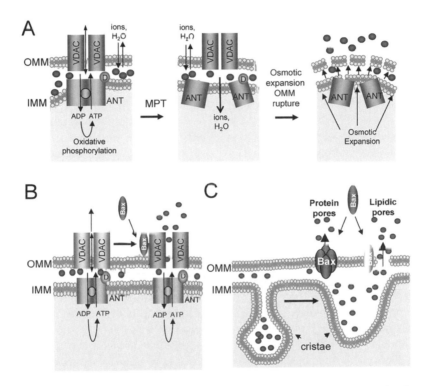

Figure 3.3. *Models of cytochrome c release. (A) MPT model. In healthy cells (left panel), the MPTP functions as a nucleotide exchanger. In response to stress signals, ANT undergoes a conformational change that converts it from a ligand-specific translocase to a non-specific pore that collapses the inner mitochondrial membrane (IMM) potential and allows the entry of water down its osmotic gradient (middle panel). The subsequent swelling of the IMM ruptures the outer mitochondrial membrane (OMM) releasing cytochrome c (right panel). D, cyclophilin D; (B) Bax-VDAC model. In healthy cells (left), VDAC functions as a component of the nucleotide exchange complex and restricts the passage of cytochrome c. During apoptosis (right), Bax binds VDAC changing its properties to permit cytochrome c release. (C) Bax pore models. In healthy cells (left), ~15% of cytochrome c is in the mitochondrial intermembrane space and ~85% in cristae that are sequestered from the intermembrane space because of narrow junctions. During apoptosis, remodeling of cristae widens the junctions allowing the majority of cytochrome c to access the intermembrane space. Bax is postulated to form protein pores or to stimulate the formation of lipid pores in the OMM, through which cytochrome c translocates into the cytosol. Figure reproduced with permission from Crow MT, Mani K, Nam Y-J, Kitsis RN. The mitochondrial death pathway and cardiac myocyte apoptosis. Circ Res 2004; 95: 957–70.* ● = cytochrome c

expansion of the convoluted inner mitochondrial membrane and rupture of the outer mitochondrial membrane. It was hypothesized that cytochrome c release resulted from disruption of the outer membrane. Consistent with a role for the MPTP in cytochrome c release, pro-apoptotic and anti-apoptotic Bcl-2 proteins have been reported to bind ANT (adenine nucleotide translocase) and VDAC, which are believed to be components of the MPTP.[149,215–218] The functional effects of these interactions remain unclear, however (see below).

Disruption of the outer mitochondrial membrane may be one mechanism of cytochrome c release during cell death associated with mito-chondrial calcium overload (e.g. ischemia–reperfusion).[219] Several important observations contradict this model, however. First, in many instances of apoptosis, cytochrome c release occurs before or in the absence of mitochondrial swelling.[220] Second, caspase inhibitors antagonize MPTP opening, but do not effect cytochrome c release.[221] These data suggest that MPTP opening is not the primary mechanism for cytochrome c release, although it may serve to amplify this process through a feedback loop in which acti-vated caspases induce the mitochondrial perme-ability transition.[222]

Other models to explain cytochrome c release emphasize the potential roles of Bax and Bak in altering the permeability of the outer mitochon-drial membrane. Although the mechanism by which Bax/Bak would accomplish this is not known, three models have been entertained: (1) Bax/Bak induce other proteins (e.g. VDAC) to form pores; (2) Bax/Bak themselves form pores; or (3) Bax/Bak induce lipid pores.

As discussed above, Bax interacts with VDAC,[218] and this interaction converts VDAC to a non-selective pore large enough to pass cytochrome c in reconstituted liposomes (Fig. 3.3B).[217] The relevance of this mechanism has been challenged, however, by experiments showing that cytochrome c release is not influenced by VDAC deletion in yeast.[223] The possibility that VDAC serves as an outer mitochondrial membrane pore also needs to be squared with VDAC2's inhibition of apoptosis through its interaction with Bak.[149]

In the second model, pores in the outer mito-chondrial membrane have been proposed to con-sist of Bcl-2 proteins themselves (Fig. 3.3C). The three-dimensional structure of Bcl-x_L's central α-helices resembles portions of diphtheria toxin and the colicins, suggesting that Bcl-2 proteins can form channels.[224] This proved to be the case as Bax, Bcl-2, and Bcl-x_L form ion channels in artifi-cial membranes.[225–227] Moreover, consistent with its anti-apoptotic properties, Bcl-2 antagonizes Bax channels.[225] Furthermore, Bax can form pores large enough to permit the passage of cytochrome c.[228]

Although it is clear that Bcl-2 proteins by themselves can form channels under some condi-tions, other data suggest that outer mitochondrial membrane pores may not consist primarily of these proteins. For example, Bax-induced chan-nels lack the discrete conductance states expected of a protein-lined channel.[229] In addi-tion, Bax-induced channels appear to be influ-enced by the lipid composition of the membrane.[230] For these reasons, a third model, Bax-induced lipid pores, has been postulated (Fig. 3.3C). Bax monomers and tBID together or oligomerized Bax alone induce pores in outer mitochondrial membranes and liposomes that are large enough to accommodate 2 MDa mole-cules.[230] Lipids from outer mitochondrial mem-branes, especially cardiolipin, are more potent than lipids from microsomal membranes at pro-moting Bax-induced pores. These experiments are consistent with Bax-induced lipid pores playing a role in outer mitochondrial membrane permeabi-lization during apoptosis.

Cytochrome c is the smallest of the mitochon-drial apoptogens. Little work has been directed to understanding the mechanisms by which the other apoptogens gain access to the cytosol during apop-tosis. Some data suggest, however, that these mechanisms may differ. For example, caspase inhi-bition does not affect cytochrome c release in most systems,[221] while it blocks the release of AIF and EndoG.[231,232] Contrary to initial reports,[233] the release of Smac/DIABLO and Omi/HtrA$_2$ also appear to be caspase-independent.[232] Additional studies will be required to understand the under-lying release mechanisms that are differentially sensitive to caspase inhibition.

Mechanisms of mitochondrial remodeling during apoptosis

Since the mitochondrial intermembrane space contains only 15% of the cytochrome c,[234] permeabilization of the outer membrane alone cannot explain the near total release of cytochrome c in many apoptotic models.[235] The remainder of the cytochrome c, which is located on the outer aspect of the inner membrane in the cristae, cannot access the intermembrane space under normal conditions, because of the narrow width of the cristae junctions. During apoptosis, the width of these junctions increase from 19 to 57 nm (Fig. 3.3C).[6] Moreover, this remodeling process augments connections among cristae. The net result is that most mitochondrial cytochrome c can move to the intermembrane space, rendering it available for release following permeabilization of the outer mitochondrial membrane. Thus, mitochondrial remodeling is as important in the intramitochondrial events of apoptosis as the apoptogen release mechanism itself.

The mechanisms underlying outer membrane permeabilization and mitochondrial remodeling differ.[6] When tBid is employed to stimulate cytochrome c release, its BH3 domain as well as Bax and/or Bak are required for outer membrane permeabilization.[148,162] In contrast, remodeling occurs independently of the BH3 domain and Bax and Bak. Interestingly, the remodeling process can be inhibited by cyclosporin A, an inhibitor of MPTP opening. MPTP opening does take place during remodeling, but only transiently, so that mitochondrial swelling does not occur. How MPTP opening contributes to the remodeling process is not known. Although mitochondria in hydrogen peroxide-stimulated cardiac myocytes undergo remodeling,[235] the kinetics and extent of cytochrome c release may not be as marked as that in some other systems.[6,236]

Intrinsic pathway: post-mitochondrial events
Post-mitochondrial caspase activation: the apoptosome

The apoptosome is a multiprotein complex in which procaspase-9 is activated. It consists of procaspase-9, the adaptor protein Apaf-1 (apoptotic protease activating factor-1), cytochrome c, and dATP. The release of cytochrome c to the cytosol is the trigger that stimulates apoptosome formation. Once cytosolic, cytochrome c binds WD-40 repeats in Apaf-1.[237,238] In addition, dATP, which already resides in the cytoplasm, interacts with Apaf-1's nucleotide binding domain. Together, these events stimulate Apaf-1 homo-oligomerization (through its nucleotide binding domain) and recruitment of procaspase-9 (through CARDs in Apaf-1 and procaspase-9).[107,238–247] The resulting complex, which exceeds 1 MDa, resembles a wheel with heptad symmetry (Fig. 3.2).[243,247] The recruitment of procaspase-9 leads to its dimerization and activation. Dimerization, rather than proteolytic cleavage, is the critical activating mechanism,[94–96] analogous to procaspase-8 activation in the DISC. Procaspase-3 is also recruited into the apoptosome where it is activated by caspase-9-mediated cleavage.[248] The pathway is amplified by a feedback loop in which caspase-3 cleaves and activates procaspase-9.[249] Deletion of cytochrome c,[250] Apaf-1,[251] or procaspase-9[252,253] markedly decreases apoptosis induced by activators of the intrinsic pathway. The importance of the mitochondrial pathway in myocardial ischemia–reperfusion injury is underscored by 53–68% reductions in infarct size in transgenic mice overexpressing either of two dominant negative procaspase-9 alleles in the heart (C-F Peng, G Tremp, A Silberstein, RN Kitsis, unpublished data).

Post-mitochondrial inhibition of apoptosis by inhibitors of apoptosis proteins

The post-mitochondrial portion of the intrinsic pathway is opposed by IAPs (inhibitor of apoptosis proteins) that guard against the potentially disastrous consequences of inadvertent caspase activation.[254] Members of the IAP family, which includes XIAP (X-linked inhibitor of apoptosis protein) and cIAP1 and cIAP2 (cellular inhibitor of apoptosis proteins 1 and 2), contain *baculovirus* inhibitor of apoptosis repeats (BIR). IAPs binds to and inhibit already activated upstream caspase-9 and downstream caspases-3 and -7 by blocking substrate access.[255–258] XIAP also inhibits the activation of procaspase-9 in the apoptosome by interfering with its dimerization.[259] Thus, IAPs inhibit the intrinsic pathway even after cytochrome c release.

To allow apoptosis to proceed efficiently, it is necessary to relieve the inhibition exerted by the IAPs. When released into the cytoplasm, the mitochondrial apoptogens Smac/DIABLO and Omi/HtrA$_2$ bind to IAPs and neutralize them by displacing caspases.[207,208,210] In addition, Omi/HtrA$_2$, which possesses a serine protease activity, inactivates XIAP irreversibly through proteolysis.[260] Moreover, reduction in XIAP is critical for cell death to occur in some cell types.[261]

In addition to the functions discussed above, IAPs also possess ubiquitin E3 ligase activity.[262] In keeping with their anti-apoptotic function, IAPs mediate the degradation of pro-apoptotic proteins including caspases-3 and -7 and Smac.[263–266] In response to some apoptotic stresses, however, IAPs degrade themselves,[262] perhaps to allow apoptosis to proceed without delay. Currently, the mechanisms that regulate IAP E3 ligase activity are not clear.

Caspase-independent apoptosis: AIF

AIF is a mitochondrial apoptogen that translocates to the nucleus and promotes DNA degradation.[211] AIF is not an endonuclease but is believed to co-operate with one in the initial phase of DNA degradation (50 kb fragments) during apoptosis. In addition to its nuclear effects, AIF feeds back on mitochondria to trigger cytochrome c release. Although AIF release may be caspase dependent,[231,232] its actions in the cytosol and nucleus do not require caspases or components of the apoptosome.[211,267] Two genetic models of AIF deficiency have yielded paradoxical results. Consistent with its pro-apoptotic functions, AIF-deficient embryonic stem cells exhibit defects in apoptosis.[267] In contrast, the 80% reductions in AIF in the naturally occurring *harlequin* mouse mutant confer increased sensitivity to oxidative stress.[268] These data have been interpreted as showing that AIF plays roles both in the promotion of apoptosis and the handling of oxidative stress.[269] AIF translocation to the nucleus is dependent on PARP (poly(ADP-ribose) polymerase) activation, and AIF is required for PARP-induced cell death.[270] PARP knockout mice subjected to ischemia–reperfusion exhibit decreased infarct size, raising the possibility that AIF is implicated in this process.[271]

Concluding remarks

This chapter reviews progress over the past decade in delineating the regulation of cell death and its role in myocardial disease. The majority of work has focused on the pathways that activate caspases during apoptosis. Less attention has been given to understanding how cleavage of various cellular substrates results in the deconstruction of the cell or the subsequent engulfment of apoptotic corpses.[55,272] Numerous experiments have demonstrated that cardiac myocyte apoptosis plays a causal role in ischemia–reperfusion injury and heart failure in rodents. Studies that determine whether these paradigms hold true in larger animals and humans will be critical in evaluating whether apoptosis will provide a novel therapeutic target for heart disease. Finally, another emerging frontier is the molecular regulation of necrosis, a process which may be more 'programmed' than originally thought.

Acknowledgements

This work was funded by grants from the NIH (R01s HL73935 to MTC and HL60665 and HL61550 to RNK). RNK is The Dr Gerald and Myra Dorros Chair in Cardiovascular Disease of the Albert Einstein College of Medicine and the recipient of the Monique Weill-Caulier Career Scientist Award. We acknowledge Chang-Fu Peng, Alan Thoms-Chesley, and Eric Bind for discussions.

References

1. Yuan J, Lipinski M, Degterev A. Diversity in the mechanisms of neuronal cell death. Neuron 2003; 40: 401–13.

2. Metzstein MM, Stanfield GM, Horvitz HR. Genetics of programmed cell death in C. elegans: past, present and future. Trends Genet 1998; 14: 410–16.

3. Kerr JF. Shrinkage necrosis: a distinct mode of cellular death. J Pathol 1971; 105: 13–20.

4. Kerr JF, Wyllie AH, Currie AR. Apoptosis: a basic biological phenomenon with wide-ranging implications in tissue kinetics. Br J Cancer 1972; 26: 239–57.

5. Savill J, Fadok V. Corpse clearance defines the meaning of cell death. Nature 2000; 407: 784–8.

6. Scorrano L, Ashiya M, Buttle K et al. A distinct pathway remodels mitochondrial cristae and mobilizes cytochrome c during apoptosis. Dev Cell 2002; 2: 55–67.

7. Thompson CB. Apoptosis in the pathogenesis and treatment of disease. Science 1995; 267: 1456–62.

8. Fujio Y, Kunisada K, Hirota H, Yamauchi-Takihara K, Kishimoto T. Signals through gp130 upregulate bcl-x gene expression via STAT1-binding cis-element in cardiac myocytes. J Clin Invest 1997; 99: 2898–905.

9. Sheng Z, Knowlton K, Chen J, Hoshijima M, Brown JH, Chien KR. Cardiotrophin 1 (CT-1) inhibition of cardiac myocyte apoptosis via a mitogen-activated protein kinase dependent pathway. Divergence from downstream CT-1 signals for myocardial cell hypertrophy. J Biol Chem 1997; 272: 5783–91.

10. Bialik S, Cryns VL, Drincic A et al. The mitochondrial apoptotic pathway is activated by serum and glucose deprivation in cardiac myocytes. Circ Res 1999; 85: 403–14.

11. Malhotra R, Brosius FC, 3rd. Glucose uptake and glycolysis reduce hypoxia induced apoptosis in cultured neonatal rat cardiac myocytes. J Biol Chem 1999; 274: 12567–75.

12. Tanaka M, Ito H, Adachi S et al. Hypoxia induces apoptosis with enhanced expression of Fas antigen messenger RNA in cultured neonatal rat cardiomyocytes. Circ Res 1994; 75: 426–33.

13. Kang PM, Haunstetter A, Aoki H, Usheva A, Izumo S. Morphological and molecular characterization of adult cardiomyocyte apoptosis during hypoxia and reoxygenation. Circ Res 2000; 87: 118–25.

14. Webster KA, Discher DJ, Kaiser S, Hernandez O, Sato B, Bishopric NH. Hypoxia-activated apoptosis of cardiac myocytes requires reoxygenation or a pH shift and is independent of p53. J Clin Invest 1999; 104: 239–52.

15. von Harsdorf R, Li PF, Dietz R. Signaling pathways in reactive oxygen species induced cardiomyocyte apoptosis. Circulation 1999; 99: 2934–41.

16. Cheng W, Li B, Kajstura J, Li P et al. Stretch-induced programmed myocyte cell death. J Clin Invest 1995; 96: 2247–59.

17. Kajstura J, Cigola E, Malhotra A et al. Angiotensin II induces apoptosis of adult ventricular myocytes in vitro. J Mol Cell Cardiol 1997; 29: 859–70.

18. Shizukuda Y, Buttrick PM, Geenen DL et al. Beta-adrenergic stimulation causes cardiocyte apoptosis: influence of tachycardia and hypertrophy. Am J Physiol Heart Circ Physiol 1998; 275: H961–8.

19. Communal C, Singh K, Pimentel DR, Colucci WS. Norepinephrine stimulates apoptosis in adult rat ventricular myocytes by activation of the beta-adrenergic pathway. Circulation 1998; 98: 1329–34.

20. Xiao RP. Beta-adrenergic signaling in the heart: dual coupling of the beta2-adrenergic receptor to G(s) and G(i) proteins. Sci STKE 2001; 2001: RE15.

21. Krown KA, Page MT, Nguyen C et al. Tumor necrosis factor alpha-induced apoptosis in cardiac myocytes. Involvement of the sphingolipid signaling cascade in cardiac cell death. J Clin Invest 1996; 98: 2854–65.

22. Lee P, Sata M, Lefer DJ et al. Fas pathway is a critical mediator of cardiac myocyte death and MI during ischemia–reperfusion in vivo. Am J Physiol Heart Circ Physiol 2003; 284: H456–63.

23. Jeremias I, Kupatt C, Martin-Villalba A et al. Involvement of CD95/Apo1/Fas in cell death after myocardial ischemia. Circulation 2000; 102: 915–20.

24. Wang L, Ma W, Markovich R, Chen JW, Wang PH. Regulation of cardiomyocyte apoptotic signaling by insulin-like growth factor I. Circ Res 1998; 83: 516–22.

25. Olivetti G, Quaini F, Sala R et al. Acute myocardial infarction in humans is associated with activation of programmed myocyte cell death in the surviving portion of the heart. J Mol Cell Cardiol 1996; 28: 2005–16.

26. Kajstura J, Cheng W, Reiss K et al. Apoptotic and necrotic myocyte cell deaths are independent contributing variables of infarct size in rats. Lab Invest 1996; 74: 86–107.

27. Saraste A, Pulkki K, Kallajoki M et al. Apoptosis in human acute myocardial infarction. Circulation 1997; 95: 320–3.

28. Gottlieb RA, Burleson KO, Kloner RA, Babior BM, Engler RL. Reperfusion injury induces apoptosis in rabbit cardiomyocytes. J Clin Invest 1994; 94: 1621–8.

29. Fliss H, Gattinger D. Apoptosis in ischemic and reperfused rat myocardium. Circ Res 1996; 79: 949–56.

30. Narula J, Haider N, Virmani R et al. Apoptosis in myocytes in end-stage heart failure. N Engl J Med 1996; 335: 1182–9.

31. Condorelli G, Morisco C, Stassi G et al. Increased cardiomyocyte apoptosis and changes in proapoptotic and antiapoptotic genes bax and bcl-2 during left ventricular adaptations to chronic pressure overload in the rat. Circulation 1999; 99: 3071–8.

32. Olivetti G, Abbi R, Quaini F et al. Apoptosis in the failing human heart. N Engl J Med 1997; 336: 1131–41.

33. Saraste A, Pulkki K, Kallajoki M et al. Cardiomyocyte apoptosis and progression of heart failure to transplantation. Eur J Clin Invest 1999; 29: 380–6.

34. Guerra S, Leri A, Wang X et al. Myocyte death in the failing human heart is gender dependent. Circ Res 1999; 85: 856–66.

35. Mallat Z, Tedgui A, Fontaliran F et al. Evidence of apoptosis in arrhythmogenic right ventricular dysplasia. N Engl J Med 1996; 335: 1190–6.

36. Saraste A, Arola A, Vuorinen T et al. Cardiomyocyte apoptosis in experimental coxsackievirus B3 myocarditis. Cardiovasc Pathol 2003; 12: 255–62.

37. Narula J, Acio ER, Narula N et al. Annexin-V imaging for noninvasive detection of cardiac allograft rejection. Nat Med 2001; 7: 1347–52.

38. Bialik S, Geenen DL, Sasson IE et al. Myocyte apoptosis during acute myocardial infarction in the mouse localizes to hypoxic regions but occurs independently of p53. J Clin Invest 1997; 100: 1363–72.

39. Palojoki E, Saraste A, Eriksson A et al. Cardiomyocyte apoptosis and ventricular remodeling after myocardial infarction in rats. Am J Physiol Heart Circ Physiol 2001; 280: H2726–31.

40. Wencker D, Chandra M, Nguyen K et al. A mechanistic role for cardiac myocyte apoptosis in heart failure. J Clin Invest 2003; 111: 1497–504.

41. Brocheriou V, Hagege AA, Oubenaissa A et al. Cardiac functional improvement by a human Bcl-2 transgene in a mouse model of ischemia/reperfusion injury. J Gene Med 2000; 2: 326–33.

42. Chen Z, Chua CC, Ho YS et al. Overexpression of Bcl-2 attenuates apoptosis and protects against myocardial I/R injury in transgenic mice. Am J Physiol Heart Circ Physiol 2001; 280: H2313–20.

43. Miao W, Luo Z, Kitsis RN, Walsh K. Intracoronary, adenovirus-mediated Akt gene transfer in heart limits infarct size following ischemia–reperfusion injury in vivo. J Mol Cell Cardiol 2000; 32: 2397–402.

44. Matsui T, Tao J, del Monte F et al. Akt activation preserves cardiac function and prevents injury after transient cardiac ischemia in vivo. Circulation 2001; 104: 330–5.

45. Peng CF, Lee P, DeGuzman A et al. Multiple independent mutations in apoptotic signaling pathways markedly decrease infarct size due to myocardial ischemia–reperfusion. Circulation (Suppl) 2001; 104: II–187.

46. Hochhauser E, Kivity S, Offen D et al. Bax ablation protects against myocardial ischemia–reperfusion injury in transgenic mice. Am J Physiol Heart Circ Physiol 2003; 284: H2351–9.

47. Yaoita H, Ogawa K, Maehara K, Maruyama Y. Attenuation of ischemia/reperfusion injury in rats by a caspase inhibitor. Circulation 1998; 97: 276–81.

48. Holly TA, Drincic A, Byun Y et al. Caspase inhibition reduces myocyte cell death induced by myocardial ischemia and reperfusion in vivo. J Mol Cell Cardiol 1999; 31: 1709–15.

49. Huang JQ, Radinovic S, Rezaiefar P, Black SC. In vivo myocardial infarct size reduction by a caspase inhibitor administered after the onset of ischemia. Eur J Pharmacol 2000; 402: 139–42.

50. Yang W, Guastella J, Huang JC et al. MX1013, a dipeptide caspase inhibitor with potent in vivo antiapoptotic activity. Br J Pharmacol 2003; 140: 402–12.

51. Hayakawa Y, Chandra M, Miao W et al. Inhibition of cardiac myocyte apoptosis improves cardiac function and abolishes mortality in the peripartum cardiomyopathy of Galpha(q) transgenic mice. Circulation 2003; 108: 3036–41.

52. Yussman MG, Toyokawa T, Odley A et al. Mitochondrial death protein Nix is induced in cardiac hypertrophy and triggers apoptotic cardiomyopathy. Nat Med 2002; 8: 725–30.

53. Yarbrough WM, Mukherjee R, Escobar GP et al. Pharmacologic inhibition of intracellular caspases after myocardial infarction attenuates left ventricular remodeling: a potentially novel pathway. J Thorac Cardiovasc Surg 2003; 126: 1892–9.

54. Hirota H, Chen J, Betz UA et al. Loss of a gp130 cardiac muscle cell survival pathway is a critical event in the onset of heart failure during biomechanical stress. Cell 1999; 97: 189–98.

55. Fischer U, Janicke RU, Schulze-Osthoff K. Many cuts to ruin: a comprehensive update of caspase substrates. Cell Death Differ 2003; 10: 76–100.

56. Laugwitz KL, Moretti A, Weig HJ et al. Blocking caspase-activated apoptosis improves contractility in failing myocardium. Hum Gene Ther 2001; 12: 2051–63.

57. Moretti A, Weig HJ, Ott T et al. Essential myosin light chain as a target for caspase-3 in failing myocardium. Proc Natl Acad Sci USA 2002; 99: 11860–5.

58. Communal C, Sumandea M, de Tombe P et al. Functional consequences of caspase activation in cardiac myocytes. Proc Natl Acad Sci USA 2002; 99: 6252–6.

59. Klionsky DJ, Emr SD. Autophagy as a regulated pathway of cellular degradation. Science 2000; 290: 1717–21.

60. Ohsumi Y. Molecular dissection of autophagy: two ubiquitin-like systems. Nat Rev Mol Cell Biol 2001; 2: 211–16.

61. Liang XH, Jackson S, Seaman M et al. Induction of autophagy and inhibition of tumorigenesis by beclin 1. Nature 1999; 402: 672–6.

62. Qu X, Yu J, Bhagat G et al. Promotion of tumorigenesis by heterozygous disruption of the beclin 1 autophagy gene. J Clin Invest 2003; 112: 1809–20.

63. Yue Z, Jin S, Yang C, Levine AJ, Heintz N. Beclin 1, an autophagy gene essential for early embryonic development, is a haplo insufficient tumor suppressor. Proc Natl Acad Sci USA 2003; 100: 15077–82.

64. Schweichel JU, Merker HJ. The morphology of various types of cell death in prenatal tissues. Teratology 1973; 7: 253–66.

65. Tanaka Y, Guhde G, Suter A et al. Accumulation of autophagic vacuoles and cardiomyopathy in LAMP-2-deficient mice. Nature 2000; 406: 902–6.

66. Nishino I, Fu J, Tanji K et al. Primary LAMP-2 deficiency causes X-linked vacuolar cardiomyopathy and myopathy (Danon disease). Nature 2000; 406: 906–10.

67. Dammrich J, Pfeifer U. Acute effects of isoproterenol on cellular autophagy. Inhibition in myocardium but stimulation in liver parenchyma. Virchows Arch B Cell Pathol Incl Mol Pathol 1981; 38: 209–18.

68. Dammrich J, Pfeifer U. Cardiac hypertrophy in rats after supravalvular aortic constriction. II. Inhibition of cellular autophagy in hypertrophying cardiomyocytes. Virchows Arch B Cell Pathol Incl Mol Pathol 1983; 43: 287–307.

69. Bahro M, Pfeifer U. Short-term stimulation by propranolol and verapamil of cardiac cellular autophagy. J Mol Cell Cardiol 1987; 19: 1169–78.

70. de Waal EJ, Vreeling-Sindelarova H, Schellens JP, James J. Starvation-induced microautophagic vacuoles in rat myocardial cells. Cell Biol Int Rep 1986; 10: 527–33.

71. David H, Behrisch D, Vivar Flores OD. Postnatal development of myocardial cells after oxygen deficiency in utero. Pathol Res Pract 1985; 179: 370–6.

72. Decker RS, Wildenthal K. Lysosomal alterations in hypoxic and reoxygenated hearts. I. Ultrastructural and cytochemical changes. Am J Pathol 1980; 98: 425–44.

73. Wildenthal K, Dees JH, Buja LM. Cardiac lysosomal derangements in mouse heart after long-term exposure to nonmetabolizable sugars. Circ Res 1977; 40: 26–35.

74. Shimomura H, Terasaki F, Hayashi T et al. Autophagic degeneration as a possible mechanism of myocardial cell death in dilated cardiomyopathy. Jpn Circ J 2001; 65: 965–8.

75. Hein S, Arnon E, Kostin S et al. Progression from compensated hypertrophy to failure in the pressure-overloaded human heart: structural deterioration and compensatory mechanisms. Circulation 2003; 107: 984–91.

76. Knaapen MW, Davies MJ, De Bie M et al. Apoptotic versus autophagic cell death in heart failure. Cardiovasc Res 2001; 51: 304–12.

77. Martin DN, Baehrecke EH. Caspases function in autophagic programmed cell death in Drosophila. Development 2004; 131: 275–84.

78. Mills KR, Reginato M, Debnath J, Queenan B, Brugge JS. Tumor necrosis factor related apoptosis-inducing ligand (TRAIL) is required for induction of autophagy during lumen formation in vitro. Proc Natl Acad Sci USA 2004; 101: 3438–43.

79. Petiot A, Ogier-Denis E, Blommaart EF, Meijer AJ, Codogno P. Distinct classes of phosphatidylinositol 3′-kinases are involved in signaling pathways that control macroautophagy in HT-29 cells. J Biol Chem 2000; 275: 992–8.

80. Liang XH, Kleeman LK, Jiang HH et al. Protection against fatal Sindbis virus encephalitis by beclin, a novel Bcl-2-interacting protein. J Virol 1998; 72: 8586–96.

81. Chu-Wang IW, Oppenheim RW. Cell death of motoneurons in the chick embryo spinal cord. II. A quantitative and qualitative analysis of degeneration in the ventral root, including evidence for axon outgrowth and limb innervation prior to cell death. J Comp Neurol 1978; 177: 59–85.

82. Hall DH, Gu G, Garcia-Anoveros J et al. Neuropathology of degenerative cell death in Caenorhabditis elegans. J Neurosci 1997; 17: 1033–45.

83. Xu K, Tavernarakis N, Driscoll M. Necrotic cell death in C. elegans requires the function of calreticulin and regulators of Ca²⁺ release from the endoplasmic reticulum. Neuron 2001; 31: 957–71.

84. Syntichaki P, Xu K, Driscoll M, Tavernarakis N. Specific aspartyl and calpain proteases are required for neurodegeneration in C. elegans. Nature 2002; 419: 939–44.

85. Karmazyn M, Gan XT, Humphreys RA, Yoshida H, Kusumoto K. The myocardial Na(+)-H(+) exchange: structure, regulation, and its role in heart disease. Circ Res 1999; 85: 777–86.

86. Klein HH, Pich S, Bohle RM, Wollenweber J, Nebendahl K. Myocardial protection by Na⁺-H⁺ exchange inhibition in ischemic, reperfused porcine hearts. Circulation 1995; 92: 912–7.

87. Inserte J, Garcia-Dorado D, Ruiz-Meana M et al. Effect of inhibition of Na^+/Ca^{2+} exchanger at the time of myocardial reperfusion on hypercontracture and cell death. Cardiovasc Res 2002; 55: 739–48.

88. Wang Y, Meyer JW, Ashraf M, Shull GE. Mice with a null mutation in the NHE1 Na^+-H^+ exchanger are resistant to cardiac ischemia–reperfusion injury. Circ Res 2003; 93: 776–82.

89. Ohtsuka M, Takano H, Suzuki M et al. Role of Na^+-Ca^{2+} exchanger in myocardial ischemia/reperfusion injury: evaluation using a heterozygous Na^+-Ca^{2+} exchanger knockout mouse model. Biochem Biophys Res Commun 2004; 314: 849–53.

90. Chen L, Chen CX, Gan XT et al. Inhibition and reversal of myocardial infarction-induced hypertrophy and heart failure by NHE-1 inhibition. Am J Physiol Heart Circ Physiol 2004; 286: H381–7.

91. Chen M, Won DJ, Krajewski S, Gottlieb RA. Calpain and mitochondria in ischemia/reperfusion injury. J Biol Chem 2002; 277: 29181–6.

92. Danial NN, Korsmeyer SJ. Cell death: critical control points. Cell 2004; 116: 205–19.

93. Thornberry NA, Lazebnik Y. Caspases: enemies within. Science 1998; 281: 1312–6.

94. Stennicke HR, Deveraux QL, Humke EW et al. Caspase-9 can be activated without proteolytic processing. J Biol Chem 1999; 274: 8359–62.

95. Renatus M, Stennicke HR, Scott FL et al. Dimer formation drives the activation of the cell death protease caspase 9. Proc Natl Acad Sci USA 2001; 98: 14250–5.

96. Boatright KM, Renatus M, Scott FL et al. A unified model for apical caspase activation. Mol Cell 2003; 11: 529–41.

97. Stennicke HR, Jurgensmeier JM, Shin H et al. Procaspase-3 is a major physiologic target of caspase-8. J Biol Chem 1998; 273: 27084–90.

98. Ashkenazi A, Dixit VM. Death receptors: signaling and modulation. Science 1998; 281: 1305–8.

99. Siegel RM, Frederiksen JK, Zacharias DA et al. Fas preassociation required for apoptosis signaling and dominant inhibition by pathogenic mutations. Science 2000; 288: 2354–7.

100. Kischkel FC, Hellbardt S, Behrmann I et al. Cytotoxicity-dependent APO-1 (Fas/CD95)-associated proteins form a death-inducing signaling complex (DISC) with the receptor. EMBO J 1995; 14: 5579–88.

101. Boldin MP, Goncharov TM, Goltsev YV, Wallach D. Involvement of MACH, a novel MORT1/FADD-interacting protease, in Fas/APO-1- and TNF receptor-induced cell death. Cell 1996; 85: 803–15.

102. Muzio M, Chinnaiyan AM, Kischkel FC et al. FLICE, a novel FADD-homologous ICE/CED-3-like protease, is recruited to the CD95 (Fas/APO-1) death-inducing signaling complex. Cell 1996; 85: 817–27.

103. Hofmann K, Bucher P, Tschopp J. The CARD domain: a new apoptotic signalling motif. Trends Biochem Sci 1997; 22: 155–6.

104. Weber CH, Vincenz C. The death domain superfamily: a tale of two interfaces? Trends Biochem Sci 2001; 26: 475–81.

105. Huang B, Eberstadt M, Olejniczak ET et al. NMR structure and mutagenesis of the Fas (APO-1/CD95) death domain. Nature 1996; 384: 638–41.

106. Eberstadt M, Huang B, Chen Z et al. NMR structure and mutagenesis of the FADD (Mort1) death-effector domain. Nature 1998; 392: 941–5.

107. Vaughn DE, Rodriguez J, Lazebnik Y, Joshua-Tor L. Crystal structure of Apaf-1 caspase recruitment domain: an alpha-helical Greek key fold for apoptotic signaling. J Mol Biol 1999; 293: 439–47.

108. Nam Y-J, Mani K, Ashton AW et al. Inhibition of both the extrinsic and intrinsic death pathways through nonhomotypic death-fold interactions. Mol Cell 2004; 15: 901–12.

109. Chinnaiyan AM, O'Rourke K, Tewari M, Dixit VM. FADD, a novel death domain containing protein, interacts with the death domain of Fas and initiates apoptosis. Cell 1995; 81: 505–12.

110. Boldin MP, Varfolomeev EE, Pancer Z et al. A novel protein that interacts with the death domain of Fas/APO1 contains a sequence motif related to the death domain. J Biol Chem 1995; 270: 7795–8.

111. Muzio M, Stockwell BR, Stennicke HR, Salvesen GS, Dixit VM. An induced proximity model for caspase-8 activation. J Biol Chem 1998; 273: 2926–30.

112. Luo X, Budihardjo I, Zou H, Slaughter C, Wang X. Bid, a Bcl2 interacting protein, mediates cytochrome c release from mitochondria in response to activation of cell surface death receptors. Cell 1998; 94: 481–90.

113. Li H, Zhu H, Xu CJ, Yuan J. Cleavage of BID by caspase 8 mediates the mitochondrial damage in the Fas pathway of apoptosis. Cell 1998; 94: 491–501.

114. Gross A, Yin XM, Wang K et al. Caspase cleaved BID targets mitochondria and is required for cytochrome c release, while BCL-XL prevents this release but not tumor necrosis factor-R1/Fas death. J Biol Chem 1999; 274: 1156–63.

115. Irmler M, Thome M, Hahne M et al. Inhibition of death receptor signals by cellular FLIP. Nature 1997; 388: 190–5.

116. Koseki T, Inohara N, Chen S, Nunez G. ARC, an inhibitor of apoptosis expressed in skeletal muscle and heart that interacts selectively with caspases. Proc Natl Acad Sci USA 1998; 95: 5156–60.

117. Rasper DM, Vaillancourt JP, Hadano S et al. Cell death attenuation by 'Usurpin', a mammalian DED-caspase homologue that precludes caspase-8 recruitment and activation by the CD-95 (Fas, APO-1) receptor complex. Cell Death Differ 1998; 5: 271–88.

118. Ekhterae D, Lin Z, Lundberg MS et al. ARC inhibits cytochrome c release from mitochondria and protects against hypoxia-induced apoptosis in heart-derived H9c2 cells. Circ Res 1999; 85: e70–7.

119. Neuss M, Monticone R, Lundberg MS et al. The apoptotic regulatory protein ARC (apoptosis repressor with caspase recruitment domain) prevents oxidant stress-mediated cell death by preserving mitochondrial function. J Biol Chem 2001; 276: 33915–22.

120. Fukazawa T, Fujiwara T, Uno F et al. Accelerated degradation of cellular FLIP protein through the ubiquitin–proteasome pathway in p53-mediated apoptosis of human cancer cells. Oncogene 2001; 20: 5225–31.

121. Micheau O, Tschopp J. Induction of TNF receptor I-mediated apoptosis via two sequential signaling complexes. Cell 2003; 114: 181–90.

122. Chao W, Shen Y, Li L, Rosenzweig A. Importance of FADD signaling in serum deprivation- and hypoxia-induced cardiomyocyte apoptosis. J Biol Chem 2002; 12: 12.

123. Davidson SM, Stephanou A, Latchman DS. FLIP protects cardiomyocytes from apoptosis induced by simulated ischemia/reoxygenation, as demonstrated by short hairpin-induced (shRNA) silencing of FLIP mRNA. J Mol Cell Cardiol 2003; 35: 1359–64.

124. Gurevich RM, Regula KM, Kirshenbaum LA. Serpin protein CrmA suppresses hypoxia-mediated apoptosis of ventricular myocytes. Circulation 2001; 103: 1984–91.

125. Gustafsson AB, Sayen MR, Williams SD, Crow MT, Gottlieb RA. TAT protein transduction into isolated perfused hearts: TAT-apoptosis repressor with caspase recruitment domain is cardioprotective. Circulation 2002; 106: 735–9.

126. Chatterjee S, Bish LT, Jayasankar V et al. Blocking the development of postischemic cardiomyopathy with viral gene transfer of the apoptosis repressor with caspase recruitment domain. J Thorac Cardiovasc Surg 2003; 125: 1461–9.

127. Kurrelmeyer KM, Michael LH, Baumgarten G et al. Endogenous tumor necrosis factor protects the adult cardiac myocyte against ischemic-induced apoptosis in a murine model of acute myocardial infarction. Proc Natl Acad Sci USA 2000; 97: 5456–61.

128. Vercammen D, Brouckaert G, Denecker G et al. Dual signaling of the Fas receptor: initiation of both apoptotic and necrotic cell death pathways. J Exp Med 1998; 188: 919–30.

129. Matsumura H, Shimizu Y, Ohsawa Y et al. Necrotic death pathway in Fas receptor signaling. J Cell Biol 2000; 151: 1247–56.

130. Holler N, Zaru R, Micheau O et al. Fas triggers an alternative, caspase-8-independent cell death pathway using the kinase RIP as effector molecule. Nat Immunol 2000; 1: 489–95.

131. Zhang J, Cado D, Chen A, Kabra NH, Winoto A. Fas-mediated apoptosis and activation-induced T-cell proliferation are defective in mice lacking FADD/Mort1. Nature 1998; 392: 296–300.

132. Newton K, Harris AW, Bath ML, Smith KG, Strasser A. A dominant interfering mutant of FADD/MORT1 enhances deletion of autoreactive thymocytes and inhibits proliferation of mature T lymphocytes. EMBO J 1998; 17: 706–18.

133. Nelson DP, Setser E, Hall DG et al. Proinflammatory consequences of transgenic fas ligand expression in the heart. J Clin Invest 2000; 105: 1199–208.

134. Badorff C, Ruetten H, Mueller S et al. Fas receptor signaling inhibits glycogen synthase kinase 3 beta and induces cardiac hypertrophy following pressure overload. J Clin Invest 2002; 109: 373–81.

135. Yeh WC, Pompa JL, McCurrach ME et al. FADD: essential for embryo development and signaling from some, but not all, inducers of apoptosis. Science 1998; 279: 1954–8.

136. Varfolomeev EE, Schuchmann M, Luria V et al. Targeted disruption of the mouse caspase 8 gene ablates cell death induction by the TNF receptors, Fas/Apo1, and DR3 and is lethal prenatally. Immunity 1998; 9: 267–76.

137. Yeh WC, Itie A, Elia AJ et al. Requirement for Casper (c-FLIP) in regulation of death receptor-induced apoptosis and embryonic development. Immunity 2000; 12: 633–42.

138. Petros AM, Olejniczak ET, Fesik SW. Structural biology of the Bcl-2 family of proteins. Biochim Biophys Acta 2004; 1644: 83–94.

139. Wang K, Yin XM, Chao DT, Milliman CL, Korsmeyer SJ. BID: a novel BH3 domain-only death agonist. Genes Dev 1996; 10: 2859–69.

140. Yang E, Zha J, Jockel J et al. Bad, a heterodimeric partner for Bcl-XL and Bcl-2, displaces Bax and promotes cell death. Cell 1995; 80: 285–91.

141. Datta SR, Dudek H, Tao X et al. Akt phosphorylation of BAD couples survival signals to the cell-intrinsic death machinery. Cell 1997; 91: 231–41.

142. Datta SR, Katsov A, Hu L et al. 14-3-3 proteins and survival kinases cooperate to inactivate BAD by BH3 domain phosphorylation. Mol Cell 2000; 6: 41–51.

143. Datta SR, Ranger AM, Lin MZ et al. Survival factor-mediated BAD phosphorylation raises the mitochondrial threshold for apoptosis. Dev Cell 2002; 3: 631–43.

144. Chen M, He H, Zhan S et al. Bid is cleaved by calpain to an active fragment in vitro and during myocardial ischemia/reperfusion. J Biol Chem 2001; 276: 30724–8.

145. Barry M, Heibein JA, Pinkoski MJ et al. Granzyme B short-circuits the need for caspase 8 activity during granule-mediated cytotoxic T-lymphocyte killing by directly cleaving Bid. Mol Cell Biol 2000; 20: 3781–94.

146. Sutton VR, Davis JE, Cancilla M et al. Initiation of apoptosis by granzyme B requires direct cleavage of bid, but not direct granzyme B-mediated caspase activation. J Exp Med 2000; 192: 1403–14.

147. Slee EA, Keogh SA, Martin SJ. Cleavage of BID during cytotoxic drug and UV radiation-induced apoptosis occurs downstream of the point of Bcl-2 action and is catalysed by caspase-3: a potential feedback loop for amplification of apoptosis-associated mitochondrial cytochrome c release. Cell Death Differ 2000; 7: 556–65.

148. Wei MC, Lindsten T, Mootha VK et al. tBID, a membrane-targeted death ligand, oligomerizes BAK to release cytochrome c. Genes Dev 2000; 14: 2060–71.

149. Cheng EH, Sheiko TV, Fisher JK, Craigen WJ, Korsmeyer SJ. VDAC2 inhibits BAK activation and mitochondrial apoptosis. Science 2003; 301: 513–17.

150. Eskes R, Desagher S, Antonsson B, Martinou JC. Bid induces the oligomerization and insertion of Bax into the outer mitochondrial membrane. Mol Cell Biol 2000; 20: 929–35.

151. Xing H, Zhang S, Weinheimer C, Kovacs A, Muslin AJ. 14-3-3 proteins block apoptosis and differentially regulate MAPK cascades. EMBO J 2000; 19: 349–58.

152. Sowter HM, Ratcliffe PJ, Watson P, Greenberg AH, Harris AL. HIF-1-dependent regulation of hypoxic induction of the cell death factors BNIP3 and NIX in human tumors. Cancer Res 2001; 61: 6669–73.

153. Bruick RK. Expression of the gene encoding the proapoptotic Nip3 protein is induced by hypoxia. Proc Natl Acad Sci USA 2000; 97: 9082–7.

154. Kubasiak LA, Hernandez OM, Bishopric NH, Webster KA. Hypoxia and acidosis activate cardiac myocyte death through the Bcl-2 family protein BNIP3. Proc Natl Acad Sci USA 2002; 11: 11.

155. Regula KM, Ens K, Kirshenbaum LA. Inducible expression of BNIP3 provokes mitochondrial defects and hypoxia-mediated cell death of ventricular myocytes. Circ Res 2002; 91: 226–31.

156. Adams JW, Sakata Y, Davis MG et al. Enhanced Galphaq signaling: a common pathway mediates cardiac hypertrophy and apoptotic heart failure. Proc Natl Acad Sci USA 1998; 95: 10140–5.

157. Ray R, Chen G, Vande Velde C et al. BNIP3 heterodimerizes with Bcl-2/Bcl-X(L) and induces cell death independent of a Bcl-2 homology 3 (BH3) domain at both mitochondrial and nonmitochondrial sites. J Biol Chem 2000; 275: 1439–48.

158. Oltvai ZN, Milliman CL, Korsmeyer SJ. Bcl-2 heterodimerizes in vivo with a conserved homolog, Bax, that accelerates programmed cell death. Cell 1993; 74: 609–19.

159. Farrow SN, White JH, Martinou I et al. Cloning of a bcl-2 homologue by interaction with adenovirus E1B 19K. Nature 1995; 374: 731–3.

160. Chittenden T, Harrington EA, O'Connor R et al. Induction of apoptosis by the Bcl-2 homologue Bak. Nature 1995; 374: 733–6.

161. Kiefer MC, Brauer MJ, Powers VC et al. Modulation of apoptosis by the widely distributed Bcl-2 homologue Bak. Nature 1995; 374: 736–9.

162. Wei MC, Zong WX, Cheng EH et al. Proapoptotic BAX and BAK: a requisite gateway to mitochondrial dysfunction and death. Science 2001; 292: 727–30.

163. Hsu YT, Wolter KG, Youle RJ. Cytosol-to-membrane redistribution of Bax and Bcl-X(L) during apoptosis. Proc Natl Acad Sci USA 1997; 94: 3668–72.

164. Wolter KG, Hsu YT, Smith CL, Nechushtan A, Xi XG, Youle RJ. Movement of Bax from the cytosol to mitochondria during apoptosis. J Cell Biol 1997; 139: 1281–92.

165. Nechushtan A, Smith CL, Hsu YT, Youle RJ. Conformation of the Bax C-terminus regulates subcellular location and cell death. EMBO J 1999; 18: 2330–41.

166. Suzuki M, Youle RJ, Tjandra N. Structure of Bax: coregulation of dimer formation and intracellular localization. Cell 2000; 103: 645–54.

167. Sawada M, Sun W, Hayes P, Leskov K, Boothman DA, Matsuyama S. Ku70 suppresses the apoptotic translocation of Bax to mitochondria. Nat Cell Biol 2003; 5: 320–9.

168. Guo B, Zhai D, Cabezas E et al. Humanin peptide suppresses apoptosis by interfering with Bax activation. Nature 2003; 423: 456–61.

169. Gustafsson AB, Tsai JG, Logue SE, Crow MT, Gottlieb RA. Apoptosis repressor with caspase recruitment domain protects against cell death by interfering with Bax activation. J Biol Chem 2004; 279: 21233–8.

170. Ohtsuka T, Ryu H, Minamishima YA et al. ASC is a Bax adaptor and regulates the p53-Bax mitochondrial apoptosis pathway. Nat Cell Biol 2004; 6: 121–8.

171. Chipuk JE, Kuwana T, Bouchier-Hayes L et al. Direct activation of Bax by p53 mediates mitochondrial membrane permeabilization and apoptosis. Science 2004; 303: 1010–4.

172. Miyashita T, Reed JC. Tumor suppressor p53 is a direct transcriptional activator of the human bax gene. Cell 1995; 80: 293–9.

173. Lassus P, Opitz-Araya X, Lazebnik Y. Requirement for caspase-2 in stressinduced apoptosis before mitochondrial permeabilization. Science 2002; 297: 1352–4.

174. Wood DE, Newcomb EW. Cleavage of Bax enhances its cell death function. Exp Cell Res 2000; 256: 375–82.

175. Gao G, Dou QP. N-terminal cleavage of bax by calpain generates a potent proapoptotic 18-kDa fragment that promotes bcl-2-independent cytochrome C release and apoptotic cell death. J Cell Biochem 2000; 80: 53–72.

176. Griffiths GJ, Dubrez L, Morgan CP et al. Cell damage-induced conformational changes of the pro-apoptotic protein Bak in vivo precede the onset of apoptosis. J Cell Biol 1999; 144: 903–14.

177. Leu JI, Dumont P, Hafey M, Murphy ME, George DL. Mitochondrial p53 activates Bak and causes disruption of a Bak-Mcl1 complex. Nat Cell Biol 2004; 6: 443–50.

178. Bakhshi A, Jensen JP, Goldman P et al. Cloning the chromosomal breakpoint of t(14; 18) human lymphomas: clustering around JH on chromosome 14 and near a transcriptional unit on 18. Cell 1985; 41: 899–906.

179. Cleary ML, Sklar J. Nucleotide sequence of a t(14; 18) chromosomal breakpoint in follicular lymphoma and demonstration of a breakpoint-cluster region near a transcriptionally active locus on chromosome 18. Proc Natl Acad Sci USA 1985; 82: 7439–43.

180. Tsujimoto Y, Cossman J, Jaffe E, Croce CM. Involvement of the bcl-2 gene in human follicular lymphoma. Science 1985; 228: 1440–3.

181. Boise LH, Gonzalez-Garcia M, Postema CE et al. bcl-x, a bcl-2-related gene that functions as a dominant regulator of apoptotic cell death. Cell 1993; 74: 597–608.

182. Kharbanda S, Pandey P, Schofield L et al. Role for Bcl-xL as an inhibitor of cytosolic cytochrome C accumulation in DNA damage-induced apoptosis. Proc Natl Acad Sci USA 1997; 94: 6939–42.

183. Kluck RM, Bossy-Wetzel E, Green DR, Newmeyer DD. The release of cytochrome c from mitochondria: a primary site for Bcl-2 regulation of apoptosis. Science 1997; 275: 1132–6.

184. Krajewski S, Tanaka S, Takayama S et al. Investigation of the subcellular distribution of the bcl-2 oncoprotein: residence in the nuclear envelope, endoplasmic reticulum, and outer mitochondrial membranes. Cancer Res 1993; 53: 4701–14.

185. Hsu YT, Youle RJ. Nonionic detergents induce dimerization among members of the Bcl-2 family. J Biol Chem 1997; 272: 13829–34.

186. Knudson CM, Korsmeyer SJ. Bcl-2 and Bax function independently to regulate cell death. Nat Genet 1997; 16: 358–63.

187. Cheng EH, Wei MC, Weiler S et al. BCL-2, BCL-X(L) sequester BH3 domain-only molecules preventing BAX- and BAK-mediated mitochondrial apoptosis. Mol Cell 2001; 8: 705–11.

188. Letai A, Bassik MC, Walensky LD et al. Distinct BH3 domains either sensitize or activate mitochondrial apoptosis, serving as prototype cancer therapeutics. Cancer Cell 2002; 2: 183–92.

189. Fridman JS, Lowe SW. Control of apoptosis by p53. Oncogene 2003; 22: 9030–40.

190. Long X, Boluyt MO, Hipolito ML et al. p53 and the hypoxia-induced apoptosis of cultured neonatal rat cardiac myocytes. J Clin Invest 1997; 99: 2635–43.

191. Leri A, Claudio PP, Li Q et al. Stretch-mediated release of angiotensin II induces myocyte apoptosis by activating p53 that enhances the local renin–angiotensin system and decreases the Bcl-2-to-Bax protein ratio in the cell. J Clin Invest 1998; 101: 1326–42.

192. Leri A, Fiordaliso F, Setoguchi M et al. Inhibition of p53 function prevents renin–angiotensin system activation and stretch-mediated myocyte apoptosis. Am J Pathol 2000; 157: 843–57.

193. Scorrano L, Oakes SA, Opferman JT et al. BAX and BAK regulation of endoplasmic reticulum Ca^{2+}: a control point for apoptosis. Science 2003; 300: 135–9.

194. Zong WX, Li C, Hatzivassiliou G, Lindsten T, Yu QC, Yuan J, Thompson CB. Bax and Bak can localize to the endoplasmic reticulum to initiate apoptosis. J Cell Biol 2003; 162: 59–69.

195. Pinton P, Ferrari D, Magalhaes P et al. Reduced loading of intracellular Ca^{2+} stores and downregulation of capacitative Ca^{2+} influx in Bcl-2-overexpressing cells. J Cell Biol 2000; 148: 857–62.

196. Foyouzi-Youssefi R, Arnaudeau S, Borner C et al. Bcl-2 decreases the free Ca^{2+} concentration within the endoplasmic reticulum. Proc Natl Acad Sci USA 2000; 97: 5723–8.

197. Halestrap AP, McStay GP, Clarke SJ. The permeability transition pore complex: another view. Biochimie 2002; 84: 153–66.

198. Boehning D, Patterson RL, Sedaghat L et al. Cytochrome c binds to inositol (1,4,5) trisphosphate receptors, amplifying calcium dependent apoptosis. Nat Cell Biol 2003; 5: 1051–61.

199. Nakagawa T, Yuan J. Cross-talk between two cysteine protease families. Activation of caspase-12 by calpain in apoptosis. J Cell Biol 2000; 150: 887–94.

200. Morishima N, Nakanishi K, Takenouchi H, Shibata T, Yasuhiko Y. An endoplasmic reticulum stress-specific caspase cascade in apoptosis. Cytochrome c-independent activation of caspase-9 by caspase-12. J Biol Chem 2002; 277: 34287–94.

201. Rao RV, Castro-Obregon S, Frankowski H et al. Coupling endoplasmic reticulum stress to the cell death program. An Apaf-1-independent intrinsic pathway. J Biol Chem 2002; 277: 21836–42.

202. Nakagawa T, Zhu H, Morishima N et al. Caspase-12 mediates endoplasmic-reticulum-specific apoptosis and cytotoxicity by amyloid-beta. Nature 2000; 403: 98–103.

203. Saleh M, Vaillancourt JP, Graham RK et al. Differential modulation of endotoxin responsiveness by human caspase-12 polymorphisms. Nature 2004; 429: 75–9.

204. Breckenridge DG, Stojanovic M, Marcellus RC, Shore GC. Caspase cleavage product of BAP31 induces mitochondrial fission through endoplasmic reticulum calcium signals, enhancing cytochrome c release to the cytosol. J Cell Biol 2003; 160: 1115–27.

205. Marks AR. Calcium and the heart: a question of life and death. J Clin Invest 2003; 111: 597–600.

206. Liu X, Kim CN, Yang J, Jemmerson R, Wang X. Induction of apoptotic program in cell-free extracts: requirement for dATP and cytochrome c. Cell 1996; 86: 147–57.

207. Du C, Fang M, Li Y, Li L, Wang X. Smac, a mitochondrial protein that promotes cytochrome c-dependent caspase activation by eliminating IAP inhibition. Cell 2000; 102: 33–42.

208. Verhagen AM, Ekert PG, Pakusch M et al. Identification of DIABLO, a mammalian protein that promotes apoptosis by binding to and antagonizing IAP proteins. Cell 2000; 102: 43–53.

209. Faccio L, Fusco C, Chen A et al. Characterization of a novel human serine protease that has extensive homology to bacterial heat shock endoprotease HtrA and is regulated by kidney ischemia. J Biol Chem 2000; 275: 2581–8.

210. Suzuki Y, Imai Y, Nakayama H et al. A serine protease, HtrA2, is released from the mitochondria and interacts with XIAP, inducing cell death. Mol Cell 2001; 8: 613–21.

211. Susin SA, Lorenzo HK, Zamzami N et al. Molecular characterization of mitochondrial apoptosis inducing factor. Nature 1999; 397: 441–6.

212. Li LY, Luo X, Wang X. Endonuclease G is an apoptotic DNase when released from mitochondria. Nature 2001; 412: 95–9.

213. Yang J, Liu X, Bhalla K et al. Prevention of apoptosis by Bcl-2: release of cytochrome c from mitochondria blocked. Science 1997; 275: 1129–32.

214. Rosse T, Olivier R, Monney L et al. Bcl-2 prolongs cell survival after Bax-induced release of cytochrome c. Nature 1998; 391: 496–9.

215. Marzo I, Brenner C, Zamzami N et al. Bax and adenine nucleotide translocator cooperate in the mitochondrial control of apoptosis. Science 1998; 281: 2027–31.

216. Belzacq AS, Vieira HL, Verrier F et al. Bcl-2 and Bax modulate adenine nucleotide translocase activity. Cancer Res 2003; 63: 541–6.

217. Shimizu S, Narita M, Tsujimoto Y. Bcl-2 family proteins regulate the release of apoptogenic cytochrome c by the mitochondrial channel VDAC. Nature 1999; 399: 483–7.

218. Narita M, Shimizu S, Ito T et al. Bax interacts with the permeability transition pore to induce permeability transition and cytochrome c release in isolated mitochondria. Proc Natl Acad Sci USA 1998; 95: 14681–6.

219. Kim J-S, He L, Lemasters JJ. Mitochondrial permeability transition: a common pathway to necrosis and apoptosis. Biochem Biophys Res Commun 2003; 304: 463–70.

220. Scorrano L, Ashiya M, Buttle K et al. A distinct pathway remodels mitochondrial cristae and mobilizes cytochrome c during apoptosis. Developmental Cell 2002; 2: 55–67.

221. Bossy-Wetzel E, Newmeyer DD, Green DR. Mitochondrial cytochrome c release in apoptosis

occurs upstream of DEVD-specific caspase activation and independently of mitochondrial transmembrane depolarization. EMBO J 1998; 17: 37–49.

222. Ricci JE, Munoz-Pinedo C, Fitzgerald P et al. Disruption of mitochondrial function during apoptosis is mediated by caspase cleavage of the p75 subunit of complex I of the electron transport chain. Cell 2004; 117: 773–86.

223. Priault M, Chaudhuri B, Clow A, Camougrand N, Manon S. Investigation of bax-induced release of cytochrome c from yeast mitochondria permeability of mitochondrial membranes, role of VDAC and ATP requirement. Eur J Biochem 1999; 260: 684–91.

224. Muchmore SW, Sattler M, Liang H et al. X-ray and NMR structure of human Bcl-xL, an inhibitor of programmed cell death. Nature 1996; 381: 335–41.

225. Antonsson B, Conti F, Ciavatta A et al. Inhibition of Bax channel-forming activity by Bcl-2. Science 1997; 277: 370–2.

226. Schlesinger PH, Gross A, Yin XM et al. Comparison of the ion channel characteristics of proapoptotic BAX and antiapoptotic BCL-2. Proc Natl Acad Sci USA 1997; 94: 11357–62.

227. Minn AJ, Velez P, Schendel SL et al. Bcl-x(L) forms an ion channel in synthetic lipid membranes. Nature 1997, 385: 353 7.

228. Saito M, Korsmeyer SJ, Schlesinger PH. BAX-dependent transport of cytochrome c reconstituted in pure liposomes. Nat Cell Biol 2000; 2: 553–5.

229. Basanez G, Nechushtan A, Drozhinin O et al. Bax, but not Bcl-xL, decreases the lifetime of planar phospholipid bilayer membranes at subnanomolar concentrations. Proc Natl Acad Sci USA 1999; 96: 5492–7.

230. Kuwana T, Mackey MR, Perkins G et al. Bid, Bax, and lipids cooperate to form supramolecular openings in the outer mitochondrial membrane. Cell 2002; 111: 331–42.

231. Arnoult D, Parone P, Martinou JC et al. Mitochondrial release of apoptosis-inducing factor occurs downstream of cytochrome c release in response to several proapoptotic stimuli. J Cell Biol 2002; 159: 923–9.

232. Arnoult D, Gaume B, Karbowski M et al. Mitochondrial release of AIF and EndoG requires caspase activation downstream of Bax/Bak-mediated permeabilization. EMBO J. 2003; 22: 4385–99.

233. Adrain C, Creagh EM, Martin SJ. Apoptosis-associated release of Smac/DIABLO from mitochondria requires active caspases and is blocked by Bcl-2. EMBO J 2001; 20: 6627–36.

234. Bernardi P, Azzone GF. Cytochrome c as an electron shuttle between the outer and inner mitochondrial membranes. J Biol Chem 1981; 256: 7187–92.

235. Akao M, O'Rourke B, Teshima Y, Seharaseyon J, Marban E. Mechanistically distinct steps in the mitochondrial death pathway triggered by oxidative stress in cardiac myocytes. Circ Res 2003; 92: 186–94.

236. Goldstein JC, Waterhouse NJ, Juin P, Evan GI, Green DR. The coordinate release of cytochrome c during apoptosis is rapid, complete and kinetically invariant. Nature Cell Biology 2000; 2: 156–62.

237. Zou H, Henzel WJ, Liu X, Lutschg A, Wang X. Apaf-1, a human protein homologous to C. elegans CED-4, participates in cytochrome c-dependent activation of caspase-3. Cell 1997; 90: 405–13.

238. Li P, Nijhawan D, Budihardjo I et al. Cytochrome c and dATP-dependent formation of Apaf-1/caspase-9 complex initiates an apoptotic protease cascade. Cell 1997; 91: 479–89.

239. Srinivasula SM, Ahmad M, Fernandes-Alnemri T, Alnemri ES. Autoactivation of procaspase-9 by Apaf-1-mediated oligomerization. Mol Cell 1998; 1: 949–57.

240. Hu Y, Ding L, Spencer DM, Nunez G. WD-40 repeat region regulates Apaf-1 selfassociation and procaspase-9 activation. J Biol Chem 1998; 273: 33489–94.

241. Saleh A, Srinivasula SM, Acharya S, Fishel R, Alnemri ES. Cytochrome c and dATP-mediated oligomerization of Apaf-1 is a prerequisite for procaspase-9 activation. J Biol Chem 1999; 274: 17941–5.

242. Zou H, Li Y, Liu X, Wang X. An APAF-1 cytochrome c multimeric complex is a functional apoptosome that activates procaspase-9. J Biol Chem 1999; 274: 11549–56.

243. Qin H, Srinivasula SM, Wu G et al. Structural basis of procaspase-9 recruitment by the apoptotic protease-activating factor 1. Nature 1999; 399: 549–57.

244. Zhou P, Chou J, Olea RS, Yuan J, Wagner G. Solution structure of Apaf-1 CARD and its interaction with caspase-9 CARD: a structural basis for specific adaptor/caspase interaction. Proc Natl Acad Sci USA 1999; 96: 11265–70.

245. Day CL, Dupont C, Lackmann M, Vaux DL, Hinds MG. Solution structure and mutagenesis of the caspase recruitment domain (CARD) from Apaf-1. Cell Death Differ 1999; 6: 1125–32.

246. Jiang X, Wang X. Cytochrome c promotes caspase-9 activation by inducing nucleotide binding to Apaf-1. J Biol Chem 2000; 275: 31199–203.

247. Acehan D, Jiang X, Morgan DG et al. Three-dimensional structure of the apoptosome: implications for assembly, procaspase-9 binding, and activation. Mol Cell 2002; 9: 423–32.

248. Bratton SB, Walker G, Srinivasula SM et al. Recruitment, activation and retention of caspases-9 and -3 by Apaf-1 apoptosome and associated XIAP complexes. EMBO J 2001; 20: 998–1009.

249. Slee EA, Harte MT, Kluck RM et al. Ordering the cytochrome c initiated caspase cascade: hierarchical activation of caspases-2, -3, -6, -7, -8, and -10 in a caspase-9-dependent manner. J Cell Biol 1999; 144: 281–92.

250. Li K, Li Y, Shelton JM et al. Cytochrome c deficiency causes embryonic lethality and attenuates stress-induced apoptosis. Cell 2000; 101: 389–99.

251. Yoshida H, Kong YY, Yoshida R et al. Apaf1 is required for mitochondrial pathways of apoptosis and brain development. Cell 1998; 94: 739–50.

252. Hakem R, Hakem A, Duncan GS et al. Differential requirement for caspase 9 in apoptotic pathways in vivo. Cell 1998; 94: 339–52.

253. Kuida K, Haydar TF, Kuan CY et al. Reduced apoptosis and cytochrome c-mediated caspase activation in mice lacking caspase 9. Cell 1998; 94: 325–37.

254. Liston P, Fong WG, Korneluk RG. The inhibitors of apoptosis: there is more to life than Bcl2. Oncogene 2003; 22: 8568–80.

255. Sun C, Cai M, Meadows RP et al. NMR structure and mutagenesis of the third Bir domain of the inhibitor of apoptosis protein XIAP. J Biol Chem 2000; 275: 33777–81.

256. Chai J, Shiozaki E, Srinivasula SM et al. Structural basis of caspase-7 inhibition by XIAP. Cell 2001; 104: 769–80.

257. Huang Y, Park YC, Rich RL et al. Structural basis of caspase inhibition by XIAP: differential roles of the linker versus the BIR domain. Cell 2001; 104: 781–90.

258. Riedl SJ, Renatus M, Schwarzenbacher R et al. Structural basis for the inhibition of caspase-3 by XIAP. Cell 2001; 104: 791–800.

259. Shiozaki EN, Chai J, Rigotti DJ et al. Mechanism of XIAP-mediated inhibition of caspase-9. Mol Cell 2003; 11: 519–27.

260. Yang QH, Church-Hajduk R, Ren J et al. Omi/HtrA2 catalytic cleavage of inhibitor of apoptosis (IAP) irreversibly inactivates IAPs and facilitates caspase activity in apoptosis. Genes Dev 2003; 17: 1487–96.

261. Potts PR, Singh S, Knezek M, Thompson CB, Deshmukh M. Critical function of endogenous XIAP in regulating caspase activation during sympathetic neuronal apoptosis. J Cell Biol 2003; 163: 789–99.

262. Yang Y, Fang S, Jensen JP, Weissman AM, Ashwell JD. Ubiquitin protein ligase activity of IAPs and their degradation in proteasomes in response to apoptotic stimuli. Science 2000; 288: 874–7.

263. Huang H, Joazeiro CA, Bonfoco E et al. The inhibitor of apoptosis, cIAP2, functions as a ubiquitin-protein ligase and promotes in vitro monoubiquitination of caspases 3 and 7. J Biol Chem 2000; 275: 26661–4.

264. Suzuki Y, Nakabayashi Y, Takahashi R. Ubiquitin-protein ligase activity of X-linked inhibitor of apoptosis protein promotes proteasomal degradation of caspase-3 and enhances its anti-apoptotic effect in Fas-induced cell death. Proc Natl Acad Sci USA 2001; 98: 8662–7.

265. MacFarlane M, Merrison W, Bratton SB, Cohen GM. Proteasome-mediated degradation of Smac during Apoptosis: XIAP promotes Smac Ubiquitination in Vitro. J Biol Chem 2002; 277: 36611–16.

266. Hu S, Yang X. Cellular inhibitor of apoptosis 1 and 2 are ubiquitin ligases for the apoptosis inducer Smac/DIABLO. J Biol Chem 2003; 278: 10055–60.

267. Joza N, Susin SA, Daugas E et al. Essential role of the mitochondrial apoptosis-inducing factor in programmed cell death. Nature 2001; 410: 549–54.

268. Klein JA, Longo-Guess CM, Rossmann MP et al. The harlequin mouse mutation downregulates apoptosis inducing factor. Nature 2002; 419: 367–74.

269. Lipton SA, Bossy-Wetzel E. Dueling activities of AIF in cell death versus survival: DNA binding and redox activity. Cell 2002; 111: 147–50.

270. Yu SW, Wang H, Poitras MF et al. Mediation of poly(ADP-ribose) polymerase-1-dependent cell death by apoptosis-inducing factor. Science 2002; 297: 259–63.

271. Pieper AA, Walles T, Wei G et al. Myocardial post-ischemic injury is reduced by polyADPribose polymerase-1 gene disruption. Mol Med 2000; 6: 271–82.

272. Ravichandran KS. 'Recruitment signals' from apoptotic cells: invitation to a quiet meal. Cell 2003; 113: 817–20.

CHAPTER 4

Transcription factors and hypertrophy

Hiroshi Akazawa, Issei Komuro

Introduction

Recent studies have identified several transcription pathways that are crucially involved in generation and progression of cardiac hypertrophy. Especially, nuclear factor of activated T cells and cardiac transcription factors such as myocyte enhancer factor 2 transcription factors and GATA family transcription factors are key regulators for transcriptional regulation during cardiac hypertrophy. Divergent intracellular signals mediate extracellular hypertrophic stimulation and activate these transcription factors. Pharmacological and genetic inhibition have highlighted the significant role of these transcription factors in cardiac hypertrophy, and raised the possibility of the transcriptional pathways as novel therapeutic targets for pathological hypertrophy. In this section, regulatory roles of these transcription factors during development of cardiac hypertrophy are discussed.

Transcriptional adaptation during cardiac hypertrophy

In a variety of pathological conditions (e.g. hypertension, valvular heart disease, myocardial infarction, and cardiomyopathy) that impose hemodynamic burden on the heart, cardiomyocytes undergo hypertrophic growth. Although cardiac hypertrophy *per se* is initially compensatory and beneficial, prolongation of this process leads to deleterious outcomes such as congestive heart failure, arrhythmia, and sudden death.[1] Accordingly, the elucidation of molecular mechanisms underlying key processes generating cardiac hypertrophy is an important subject of intense research from a clinical point of view.

When placed in an unusual environment, specialized cells in a multicellular organism alter their properties by changing the patterns of gene expression. Although gene expression is regulated at several levels in the pathway from DNA to protein, transcriptional control is, in general, the most critical point for expression of each molecule. In response to alterations in hemodynamic workload, cardiomyocytes promptly adjust their contractile performance and simultaneously enter into hypertrophic growth. Cellular responses characteristic of cardiac hypertrophy include an increase in cell size due to accelerated synthesis of sarcomeric and structural proteins, and reprogramming of the fetal cardiac genes.[2,3] It is well known that some of the contractile proteins, ion channels and metabolic enzymes have both fetal and adult isoforms, which have similar, but not identical, functions. Transcriptional control during cardiac hypertrophy involves switching of gene expression from normally expressed adult isoforms to fetal isoforms. A shift from adult isoform to fetal isoform of sarcomeric myosin heavy chain (MHC) gives rise to a decrease in the initial speed of sarcomeric shortening, but improves the efficacy of contraction for an equivalent stroke work.[2] Therefore, this kind of transcriptional switch is generally regarded as an adaptation of myocardium to increased workload. Likewise, expression of the gene for atrial natriuretic peptide (ANP) is differentially regulated before and after birth. ANP is expressed in both the atria and ventricles during embryogenesis, but ventricular expression is silenced postnatally. In hypertrophied hearts, however, ventricular expression of ANP is reactivated to a great extent. In so much as ANP has potent natriuretic, diuretic and vasodilatory effects, upregulation of the ANP gene during cardiac hypertrophy is interpreted to be an adaptive response as well. In addition, recent high-throughput expression profilings have identified a plethora of genes other than fetal isoforms or natriuretic peptides that are upregulated in hypertrophied hearts, including genes encoding extracellular matrix proteins, growth factors, signaling molecules and metabolic enzymes.[4] The questions at issue are how extracellular hypertrophic stimulation is perceived and converted into intracellular signals, and how these signals change the transcriptional program. A more intractable problem is how altered gene expression eventually leads to hypertrophic cell growth.

With regard to transcriptional adaptation induced by hypertrophic stimulation, several transcription factors have been identified to play critical roles in extracellular signal-regulated gene programs. These transcription factors include nuclear factor of activated T cells (NFAT) and cardiac transcription factors such as myocyte enhancer factor 2 (MEF2) transcription factors and GATA family transcription factors. NFAT is especially involved in calcium (Ca^{2+})-triggered transcriptional activation. Cardiac transcription factors essentially regulate expressions of myriad cardiac genes both during cardiogenesis and during the adaptive process in response to hemodynamic stresses.

The NFAT family of transcription factors

It has been well established that NFAT is a transcriptional activator of the gene encoding interleukin-2 (IL2), which moves from the cytoplasm into the nucleus in response to Ca^{2+} signaling. Increased intracellular Ca^{2+} binds to and activates Ca^{2+}-binding proteins, including calmodulin (CaM), which regulates several downstream effectors such as calcineurin and Ca^{2+}/CaM-dependent protein kinases (CaMKs) (Fig. 4.1). Activated calcineurin dephosphorylates serine residues of NFAT leading to nuclear import of NFAT by unmasking its nuclear localization signals and, thereby, allows NFAT to transactivate calcineurin-responsive genes (Table 4.1).[5] The NFAT family of transcription factors consists of five NFAT proteins: NFATc1 (or NFATc, NFAT2), NFATc2 (or NFATp, NFAT1), NFATc3 (or NFATx, NFAT4), NFATc4 (or NFAT3), and NFAT5, and shares a highly homologous DNA binding domain related to the Rel domain in the NF-κB family of transcription factors.

A growing body of evidence has suggested that Ca^{2+} signaling plays a crucial role in the generation of cardiac hypertrophy. Mechanical stretch or stimulation by agonists such as angiotensin II (Ang II), phenylephrine (PE), and endothelin-1 (ET-1) evokes a signaling cascade leading to an elevation of Ca^{2+} concentration in cardiomyocytes.[3,6] Initial insights into a role of the calcineurin–NFAT pathway in mediating cardiac hypertrophy in addition to an immune response originated from the observation that NFATc4 forms a complex with a zinc finger transcription factor GATA4 and synergistically transactivates the gene encoding brain natriuretic peptide (BNP).[7] Indeed, the calcineurin–NFAT pathway is activated in response to hypertrophic stimulation. In cultured cardiomyocytes, stimulation by Ang II or PE upregulates an NFAT-dependent reporter, and hypertrophic responses as well as NFAT activation are attenuated by pharmacological inhibition of calcineurin using cyclosporin A (CsA) or FK506.[7] In the hearts of mice, calcineurin activity is elevated by pressure overload,[8] and β-adrenergic stimulation.[9] Recently, transgenic mice expressing an NFAT-dependent luciferase reporter were generated to monitor the activity of calcineurin–NFAT signaling in murine hearts.[10] In pressure-overloaded hypertrophy of the heart, upregulation of NFAT-luciferase activity precedes the onset of hypertrophic cell growth and

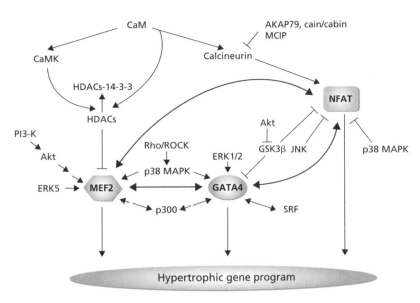

Figure 4.1. *Roles of transcription factors in regulation of cardiac gene program during cardiac hypertrophy. In response to Ca^{2+} signaling evoked by hypertrophic stimulation, NFAT is dephosphorylated by calcineurin and translocates into the nucleus, where it activates gene expression partly through forming a complex with GATA4. The calcineurin–NFAT pathway is counter-regulated by NFAT phosphorylation by GSK3β, p38 MAPK, and JNK. Transcriptional activity of GATA4 is stimulated through phosphorylation by ERK1/2 and p38 MAPK, although phosphorylation by GSK3β suppresses GATA4 activity through nuclear export of GATA4. In addition, the transcriptional activity of GATA4 is regulated through physical interaction with NFAT, MEF2, SRF, or a coactivator, p300. MEF2 transcriptional activity is enhanced through phosphorylation by p38 MAPK and ERK5, and physical interaction with GATA4, NFAT, and coactivator p300. In addition, MEF2 might be involved in the PI3-K/Akt-mediated hypertrophic signal. Most important, MEF2 functions as an important effector of Ca^{2+} signaling. Activated calcineurin or CaMK enhances MEF2 activity in vivo. Activation of MEF2 is dependent on dissociation from class II HDACs. Signal-mediated phosphorylation of HDACs recruits chaperones 14-3-3, and consequently MEF2 is dissociated from HDAC–MEF2 complex, although the endogenous HDAC kinases have not been determined.*

is sustained thereafter. In contrast, NFAT-luciferase activity shows no significant change in hearts undergoing physiological hypertrophy induced by exercise or insulin-like growth factor stimulation, suggesting that the calcineurin–NFAT pathway plays a distinctive role in pathological hypertrophy, but not in physiological hypertrophy.

Activation of the calcineurin–NFAT pathway is sufficient to induce cardiac hypertrophy, because transgenic mice that express constitutively active forms of calcineurin or NFATc4 in the heart exhibit severe cardiac hypertrophy and, occasionally, congestive heart failure (Table 4.2).[7] To validate whether calcineurin is required for hypertrophic growth in the hearts, the effects of pharmacological

inhibition of calcineurin have been investigated in numerous animal models of cardiac hypertrophy. Although conflicting results have been reported in studies using CsA and FK506, possibly due to differences in experimental protocols, most studies have demonstrated that inhibition of calcineurin suppresses development of cardiac hypertrophy.[11] The importance of calcineurin in cardiac hypertrophy has been strengthened by genetic approaches. Calcineurin Aβ (CnAβ)-deficient mice showed smaller hearts, and hypertrophic responses after aortic constriction, Ang II-infusion and isoproterenol (Iso)-infusion were attenuated.[12] In line with this phenotype, transgenic mice overexpressing a dominant-negative mutant of *CnA* displayed

Table 4.1. *Post-translational modification by protein phosphorylation or dephosphorylation during cardiac hypertrophy*

Transcription factor	Kinases or phosphatases	Effect	Reference numbers
NFAT[a]	Calcineurin	Nuclear import	8
	GSK3β[b]	Nuclear export	21, 22
	p38 MAPK[c]	Nuclear export	24
	JNK[d]	Nuclear export	25
MEF2[e]	p38 MAPK	Activation?	27, 28
	ERK5[f]	Activation?	29, 30
GATA	ERK1/2	Activation	40, 41
	p38 MAPK	Activation?	44, 45
	GSK3β	Nuclear export	23

[a] NFAT, nuclear factor of activated T cells.
[b] GSK3β, glycogen synthase kinase 3β.
[c] p38 MAPK, p38 mitogen-activated protein kinase.
[d] JNK, c-jun N-terminal kinase.
[e] MEF2, myocyte enhancer factor 2.
[f] ERK, extracellular signal-regulated kinase.

Table 4.2. *Genetically engineered mice revealing postnatal roles of NFAT, MEF2, and GATA transcription factors in the hearts*

Transcription factor		Phenotype	Reference numbers
NFAT[a]			
NFATc4	CA[b]-TG	Cardiac hypertrophy	8
NFATc3	KO[c]	Attenuated cardiac hypertrophy in response to pressure overload, Ang II stimulation, and calcineurin activation	19
NFATc4	KO	No effect on cardiac hypertrophy	19
MEF2[d]			
MEF2-C	DN[e]-TG[f]	Attenuated postnatal growth of the myocardium	27
GATA			
GATA4	TG	Cardiac hypertrophy	42

[a] NFAT, nuclear factor of activated T cells.
[b] CA, constitutively active form.
[c] KO, knockout mice.
[d] MEF2, myocyte enhancer factor 2.
[e] DN, dominant-negative form.
[f] TG, transgenic mice.

compromised hypertrophic responses induced by pressure overload.[13] In the hearts of transgenic mice, an increase in calcineurin enzymatic activity after aortic banding was significantly suppressed, although the basal activity of calcineurin was comparable to that in wild-types. Likewise, over-expression of endogenous calcineurin inhibitors, AKAP79 or Cain/Cabin, attenuated hypertrophic responses in cultured cardiomyocytes after Ang II- and PE-stimulation,[14] and in murine hearts after pressure overload and Iso-stimulation.[15] MCIP (modulatory calcineurin-interacting protein) 1 and MCIP 2 are also calcineurin-inhibitory proteins, and transgenic overexpression of MCIP1 in murine hearts blunted hypertrophy induced by pressure overload and Iso stimulation as well as hypertrophy observed in transgenic mice expressing constitutively active calcineurin.[16,17]

Among the transcription factors whose expressions are regulated by calcineurin such as NFAT, MEF2, NF-κB, and Elk-1, NFAT plays a central role in calcineurin-mediated cardiac hypertrophy. Whereas only a few direct downstream targets for NFAT have been identified such as BNP and adenylosuccinate synthetase I, activation of NFATc4 is sufficient for generation of cardiac hypertrophy in murine hearts (Table 4.2).[7] In addition, adenoviral delivery of a dominant-negative mutant of NFAT inhibited calcineurin-, ET-1, and cardiotrophin-1-induced hypertrophic responses in cultured cardiomyocytes.[18] Genetic evidence has also been provided that NFAT functions as a necessary downstream effector of calcineurin in development of cardiac hypertrophy. Targeted disruption of NFATc3 attenuates cardiac hypertrophy induced by pressure overload or Ang II-stimulation, and by overexpression of constitutively active calcineurin, while disruption of NFATc4 has no significant effect on development of hypertrophy (Table 4.2).[19]

Recent studies have suggested that nuclear accumulation of NFAT is opposed by phosphorylation through NFAT kinases such as glycogen synthase kinase 3β (GSK3β), p38 mitogen-activated protein kinase (p38 MAPK), and c-jun N-terminal kinase (JNK) (Table 4.1). GSK3β is a serine/threonine kinase that is catalytically active under basal conditions, and diverse cellular responses are controlled through inactivation of GSK3β by multiple signaling pathways such as the phosphoinositide 3-kinase (PI3-K)-Akt pathway and the Wnt pathway.[20] In cultured cardiomyocytes, adenoviral transfer of constitutively active GSK3β accelerates nuclear export of NFAT, and attenuates hypertrophic responses in response to ET-1 and PE.[21] Moreover, transgenic overexpression of constitu-

tively active GSK3β in murine hearts prevented generation of cardiac hypertrophy in response to activation of calcineurin, β-adrenergic stimulation, and pressure overload, and calcineurin-induced nuclear import of NFAT is attenuated by activated GSK3β.[22] A recent study has demonstrated that the anti-hypertrophic effects of GSK3β are also mediated in part by other substrates such as GATA4.[23] In addition, the p38 MAPK and JNK pathways, consisting of the ternary branches of the MAPK cascades, affect the NFAT activities by regulating nucleocytoplasmic shuttling of NFAT. Transient transfection of a dominant-negative form of p38 MAPK upregulates an NFAT-dependent luciferase reporter, and induces significant nuclear translocation of NFAT in cultured cardiomyocytes.[24] Consistently, transgenic mice overexpressing a dominant-negative form of p38 MAPK showed accelerated cardiac hypertrophy in response to pressure overload and Ang II-, Iso-, and PE-stimulation. More importantly, transgenic mice expressing an NFAT-dependent luciferase reporter reveal upregulation of NFAT-luciferase activity in the hearts of dominant-negative p38 MAPK transgenic mice.[74] Similarly, wild-type JNK1 and JNK2 antagonize calcineurin-induced nuclear import of NFAT, and dominant-negative mutants of JNK1 and JNK2 enhance NFAT activities both in cultured cardiomyocytes and in murine hearts.[25] Transgenic overexpression of dominant-negative mutants of JNK1 and JNK2 in murine hearts induces spontaneous cardiac hypertrophy with aging, and enhances hypertrophy in response to pressure overload.[25] These results suggest that the calcineurin–NFAT signaling mediating cardiac hypertrophy is counteracted by phosphorylation of NFAT through GSK3β, p38, MAPK, and JNKs.

MEF2 transcription factors

MEF2 transcription factors consist of four members (MEF2-A, MEF2-B, MEF2-C, and MEF2-D) in vertebrates, and share a MADS (MCM1, agamous, deficiens, and SRF) domain and a MEF2-specific domain in the N-terminus in common, which together allow dimerization and binding to their cognate DNA sequence.[26] The MEF2-binding

DNA sequences have been identified within the promoter regions of numerous cardiac genes,[26] and it is now established that MEF2 family members are involved in transcriptional regulation in the heart as well as skeletal muscle, immune system and neurons. In particular, MEF2-C is involved in transcriptional regulation in postnatal hearts, because transgenic mice overexpressing a dominant-negative MEF2-C show attenuated postnatal growth of the myocardium (Table 4.2).[27]

Consistent with the implications of MEF2 in transcriptional regulation during myocardial cell hypertrophy, the MEF2-binding site within the myosin light chain (MLC2) promoter is required during PE-mediated and ET-1-mediated hypertrophy, and the MEF2 DNA-binding activity is increased in the hearts of rats subjected to pressure overload or volume overload.[26] Recent studies have dissected complex signaling pathways that link hypertrophic stimulation to MEF2 activation (Fig. 4.1). First, MEF2 is phosphorylated by p38 MAPK (Table 4.1).[27,28] Although activation of p38 MAPK induces hypertrophic growth in cultured cardiomyocytes, and p38 MAPK phosphorylates MEF2 in hypertrophied heart,[28] the pathophysiological significance of the p38 MAPK–MEF2 pathway during cardiac hypertrophy is challenged by the observation that targeted inhibition of p38 MAPK promotes cardiac hypertrophy in murine hearts.[24] Second, MEF2 is activated through phosphorylation by extracellular signal-regulated kinase (ERK) 5 (Table 4.1).[29] In cultured cardiomyocytes, ERK5 is activated by PE, leukemia inhibitory factor, and oxidative and osmotic stress.[30] A dominant-negative form of MEK5, the MAPK kinase for ERK5, inhibits leukemia inhibitory factor-induced hypertrophy in cultured cardiomyocytes, and transgenic overexpression of a constitutively active MEK5 in the heart leads to eccentric hypertrophy.[30] These results suggest a significant role for the ERK5–MEF2 pathway in generation of cardiac hypertrophy, whereas it remains to be determined whether MEF2 is an essential downstream effector for ERK5-induced cardiac hypertrophy. During skeletal muscle differentiation, MEF2 activity is enhanced through the PI3-K–Akt pathway.[31] Interestingly, transgenic mice overexpressing a constitutively active form of either PI3-K or Akt exhibit physiological cardiac hypertrophy.[32] Although the transcriptional activity of MEF2 has not been examined in these transgenic mice, it may be possible that MEF2 is involved in PI3-K/Akt-mediated myocardial cell hypertrophy.

In addition, the MEF2 factors are defined as important effectors that function downstream in the binary pathways of Ca^{2+} signaling (Table 4.1). MEF2 activity is stimulated by CaMK, as indicated by LacZ expression in the hearts of double transgenic mice expressing activated CaMKIV and a MEF2-dependent LacZ reporter.[33] Activation of MEF2 by CaMK is mediated mainly through the phosphorylation of transcriptional repressors, the histone deacetylases (HDACs).[34] In particular, class II HDACs (HDAC-4, HDAC-5, HDAC-7, and HDAC-9) associate with MEF2 to repress MEF2-induced gene expression.[31] In general, transcriptional activity is largely influenced by the state of histone acetylation, which is balanced through opposing activities of HDACs and histone acetyltransferases (HATs). HDACs repress gene expression through intrinsic deacetylase activity and recruitment of a transcriptional corepressor COOH-terminal-binding protein.[35] Recent studies have demonstrated that phosphorylation of HDACs by CaMKs results in the recruitment of intracellular chaperones 14-3-3 to dissociate the HDAC–MEF2 formation.[36] Consequently, HDACs are exported and sequestered in the cytoplasm by the nucleocytoplasmic shuttling mechanism, and MEF2 is released from HDACs in the nucleus and transcriptionally activated through binding to coactivators harboring intrinsic HAT activity, such as p300 and CBP.[36] HDAC-4 has a CaM-binding domain that overlaps the MEF2-binding domain, and dissociation of MEF2 from HDACs is also regulated by CaM,[37] indicating that the HDAC–MEF2 complex is controlled by a series of mediators in the Ca^{2+} signaling pathway.

The importance of class II HDACs during cardiac hypertrophy is underscored by a genetic approach. HDAC9-deficient mice show spontaneous cardiac hypertrophy and are predisposed to severe hypertrophic growth after pressure overload.[38] In cultured cardiomyocytes, overexpression of class II HDACs with mutations of two conserved CaMK phosphorylation sites blocks hypertrophic features, including agonist-induced gene expression of ANP and β-MHC and histone acetylation of the promoter regions of these genes.[38] These data indicate the repressive role of class II HDACs in generation of cardiac hypertrophy. Although HDAC kinase activity is enhanced in cardiac extracts from hypertrophied hearts of mice, and CaMKs are capable of phosphorylating HDACs, it remains unclear whether CaMKs are the functional HDAC kinases that are responsive to hypertrophic stimulation, because HDAC kinase activity in *in vitro* kinase assays is only partially blocked by CaMK inhibitors.[38] Instead, HDAC kinase activity is enhanced by calcineurin signaling, in as much as HDAC9-deficient mice harboring the activated calcineurin transgene show more severe cardiac hypertrophy together with increased transcriptional activity of MEF2.[38] Calcineurin-mediated dephosphorylation of MEF2 is observed in skeletal muscle and neurons.[31] The precise role of calcineurin in the activation of MEF2 during cardiac hypertrophy remains to be determined.

Collectively, MEF2 activity is enhanced in response to hypertrophic stimulation, and MEF2 functions as an essential effector of divergent intracellular signaling pathways mediating hypertrophic features.

GATA transcription factors

GATA transcription factors are characterized by the conserved double zinc fingers that are required for binding to the consensus DNA sequence. Among six GATA transcription factors in vertebrates, GATA4, GATA5, and GATA6 are involved in transcriptional regulation of cardiac genes.[39] Functional analysis of the *cis*-regulatory elements has revealed that GATA4 directly regulates basal expression of numerous cardiac genes.[39] In addition to maintaining the basal transcription levels of these cardiac genes, GATA4 is critically involved in inducible gene expression evoked by hypertrophic stimulations. For example, GATA-binding elements are required for the upregulation of the genes for β-MHC or the Ang II type 1a receptor in response to aortic constriction, and GATA-binding elements are responsible for

inducible gene expression of *BNP* in the hearts of bilaterally nephrectomized rats.[39]

Intriguingly, the expression levels of *GATA4* are not affected by hypertrophic stimulation induced by pressure overload, α-adrenergic agonists, or ET-1. (reviewed in reference 39). Recent studies have suggested that GATA4 is activated through posttranscriptional modification in response to hypertrophic stimulation, on the basis of an increase in DNA-binding activity of GATA4 in response to pressure overload or neurohumoral stimulations (Figure 4.1).[39] Indeed, GATA4 activation induced by PE stimulation is coupled with serine phosphorylation of GATA4.[40,41] In particular, ERK2 directly phosphorylates GATA4 *in vitro*, and PE-induced phosphorylation and activation of GATA4 are inhibited either by incubation with an ERK kinase (MEK1) inhibitor, or by adenoviral transfection of a dominant-negative MEK1 (Table 4.1). Furthermore, a dominant-negative GATA4 inhibited MEK1-induced hypertrophic responses in cultured cardiomyocytes,[42] suggesting that GATA4 may function as a transcriptional effector acting downstream to the ERK signaling pathway, a key biochemical signal mediating hypertrophic responses.[43] GATA4 is also activated through direct serine phosphorylation by the p38 MAPK pathway (Table 4.1).[44,45] Pharmacological inhibition of p38 MAPK attenuated ET-1-induced protein synthesis, as well as transcriptional activity and phosphorylation of GATA4.[45] Further experiments using genetic approaches will be required for elucidation of the pathological role of the p38 MAPK–GATA4 pathway, like the p38 MAPK–MEF2 pathway, in hypertrophic cell growth in the hearts. A recent report has suggested that Rho and ROCK, a target of Rho, are linked to PE-induced GATA4 activation through the ERK pathway.[46] Moreover, the potentiation of GATA4 transcriptional activity through p38 MAPK is induced by RhoA,[44] a member of the Rho family of GTPases, which regulates diverse cellular events such as transcriptional regulation, cell growth control, and membrane trafficking, as well as cytoskeletal organization. In cardiomyocytes, Rho is critically involved in mediating hypertrophic responses.[47] Collectively, these observations highlight the role of GATA4 as an essential transcriptional effector on which divergent protein phosphorylation pathways converge during the generation of cardiac hypertrophy.

The transcriptional activity of GATA4 is regulated through its nucleocytoplasmic shuttling mechanism. GSK3β directly phosphorylates GATA4 and thereby decreases basal and β-adrenergic-stimulated GATA4 expression in the nucleus by activating the nuclear export system (Table 4.1).[23] Therefore, GSK3β may inhibit cardiac hypertrophy by interfering with transcriptional activity of GATA4 as well as NFAT.

In addition to phosphorylation, the transcriptional activity of GATA4 is regulated through interaction with cofactors such as p300.[48] p300 interacts with GATA transcription factors to enhance the promoter activation of the genes encoding ANP,[48,49] and β-MHC,[48] which is dependent on the HAT activity of p300 (Fig. 4.1).

Consistent with the essential role of GATA4 in activating the gene program in response to hypertrophic stimulation, the overexpression of GATA4 generated cardiac hypertrophy both in cultured cardiomyocytes,[42,44] and in the hearts of mice.[42] These results suggest that GATA4 is a sufficient transcriptional regulator for the generation of cardiac hypertrophy. Moreover, the overexpression of a dominant-negative GATA4 by adenoviral gene transfer inhibited an agonist-induced increase in protein synthesis and hypertrophic gene expression in cultured cardiomyocytes.[42] Targeting disruption of GATA4 in mice resulted in embryonic lethality that was due to failure in the formation of a ventrally fused heart tube,[50] and conditional gene targeting of GATA4 will be informative to elucidate the role of GATA4 in postnatal hearts.

Taken together, GATA4 transcriptional activity is positively regulated by multiple signaling pathways in response to hypertrophic stimulation. GATA4 plays an essential role in transcriptional regulation during the generation of cardiac hypertrophy.

Protein–protein interactions among transcription factors during cardiac hypertrophy

Protein–protein interactions among transcription factors have been postulated to be important to enable the transcriptional regulation to be fine-tuned in diverse cellular responses including hypertrophic cell growth (Fig. 4.1).[51] The interaction between NFAT and GATA4 is particularly noteworthy,[7] because both play critical roles in activating the hypertrophic gene program. In T lymphocytes or skeletal muscle, activated calcineurin promotes complex formation between NFAT and MEF2 to synergistically transactivate downstream target genes.[31] Furthermore, MEF2 and GATA4 synergistically activate the transcription of several cardiac genes, such as those encoding ANP, BNP, α-MHC, and cardiac α-actin.[52] In addition, the transcriptional activity of GATA4 is regulated through protein–protein interaction with SRF.[51] SRF is a transcriptional regulator of a wide variety of cardiac-specific genes, and cardiac overexpression of SRF induces hypertrophic features in mice.

Conceivably, combinations of transcription factors execute regulatory decisions in response to hypertrophic stimulation. Although transcriptional synergy is implicated in controlling the expression of several cardiac genes, an important issue remains unsolved – how much the cooperative transcriptional regulation weighs with the generation and progression of cardiac hypertrophy. It is also undetermined how the mutual interactions are regulated in response to hypertrophic stimulation. Functional analysis of the individual transcriptional regulators and elucidation of their interactive roles will be required.

Conclusion

Functional roles of transcription factors during cardiac hypertrophy have been considerably deciphered. Mechanistic insights have been provided into the signaling pathways that enhance the activities of these transcription factors. Furthermore, phenotypical analysis of genetically engineered mice has revealed sufficient and necessary roles of transcription factors in the development of cardiac hypertrophy. However, a difficult problem remains unsolved – how the transcription factors affect protein synthesis and myocardial cell size. An increased capacity of protein synthesis underlying hypertrophic growth is facilitated not by increased translational efficiency, but by ribosome accumulation resulting from increased transcription of ribosomal DNA by the nucleolar factor upstream binding factor (UBF).[53,54] Interestingly, UBF expression is significantly increased in murine hearts after pressure overload,[55] and adenoviral introduction of UBF antisense RNA into cultured cardiomyocytes abolished an increase in general protein synthesis and hypertrophic cell growth in response to α-adrenergic and contraction stimulation, but had little effect on fetal gene expression.[54] It has not been clarified how the UBF activity is transcriptionally regulated during cardiac hypertrophy.

Finally, transcription factors may be potential therapeutic targets for prevention and treatment of cardiac hypertrophy. Although compensatory cardiac hypertrophy is beneficial in some pathological conditions, evidence-based studies have suggested that the regression of cardiac hypertrophy in patients leads to better prognosis.[1] It is an ideal adaptation to excessive workload to enhance myocardial contractility without a pathological increase in left ventricular mass. Interestingly, cardiac overexpression of MCIP1 attenuates hypertrophic growth without deteriorating cardiac performance, even at 3 months after pressure overload.[17] These findings confirm that the calcineurin–NFAT pathway is coupled to generation of pathological hypertrophy, but not physiological hypertrophy,[10] and provide an experimental basis that novel therapeutic interventions directing hypertrophic signaling are feasible. Further investigations are required to understand the precise molecular mechanisms for how the transcription factors regulate progression of cardiac hypertrophy, and to target these molecules for therapeutic purposes.

Acknowledgements

This work was supported in part by grants from the Japanese Ministry of Education, Science, Sports, and Culture; Japan Health Sciences Foundation; Takeda Medical Research Foundation; Takeda Science Foundation; Uehara Memorial Foundation; Kato Memorial Trust for Nambyo Research; Japan Medical Association (to IK); Japanese Heart Foundation/Pfizer Japan Grant on Cardiovascular Disease Research (to HA). The New Energy and Industrial Technology Development Organization is acknowledged for support of this work.

References

1. Lorell BH, Carabello BA. Left ventricular hypertrophy: pathogenesis, detection, and prognosis. Circulation 2000; 102: 470–9.

2. Komuro I, Yazaki Y. Control of cardiac gene expression by mechanical stress. Annu Rev Physiol 1993; 55: 55–75.

3. Sadoshima J, Izumo S. The cellular and molecular response of cardiac myocytes to mechanical stress. Annu Rev Physiol 1997; 59: 551–71.

4. Friddle CJ, Koga T, Rubin EM et al. Expression profiling reveals distinct sets of genes altered during induction and regression of cardiac hypertrophy. Proc Natl Acad Sci U S A 2000; 97: 6745–50.

5. Hogan PG, Chen L, Nardone J et al. Transcriptional regulation by calcium, calcineurin, and NFAT. Genes Dev 2003; 17: 2205–32.

6. Kudoh S, Akazawa H, Takano H et al. Stretch-modulation of second messengers: effects on cardiomyocyte ion transport. Prog Biophys Mol Biol 2003; 82: 57–66.

7. Molkentin JD, Lu JR, Antos CL et al. A calcineurin-dependent transcriptional pathway for cardiac hypertrophy. Cell 1998; 93: 215–28.

8. Shimoyama M, Hayashi D, Takimoto E et al. Calcineurin plays a critical role in pressure overload-induced cardiac hypertrophy. Circulation 1999; 100: 2449–54.

9. Zou Y, Yao A, Zhu W et al. Isoproterenol activates extracellular signal-regulated protein kinases in cardiomyocytes through calcineurin. Circulation 2001; 104: 102–18.

10. Wilkins BJ, Dai YS, Bueno OF et al. Calcineurin/NFAT coupling participates in pathological, but not physiological, cardiac hypertrophy. Circ Res 2004; 94: 110 18.

11. Molkentin JD. Calcineurin and beyond: cardiac hypertrophic signaling. Circ Res 2000; 87: 731–8.

12. Bueno OF, Wilkins BJ, Tymitz KM et al. Impaired cardiac hypertrophic response in Calcineurin Abeta -deficient mice. Proc Natl Acad Sci U S A 2002; 99: 4586–91.

13. Zou Y, Hiroi Y, Uozumi H et al. Calcineurin plays a critical role in the development of pressure overload-induced cardiac hypertrophy. Circulation 2001; 104: 97–101.

14. Taigen T, De Windt LJ, Lim HW et al. Targeted inhibition of calcineurin prevents agonist-induced

cardiomyocyte hypertrophy. Proc Natl Acad Sci U S A 2000; 97: 1196–201.

15. De Windt LJ, Lim HW, Bueno OF et al. Targeted inhibition of calcineurin attenuates cardiac hypertrophy in vivo. Proc Natl Acad Sci U S A 2001; 98: 3322–7.

16. Rothermel BA, McKinsey TA, Vega RB et al. Myocyte-enriched calcineurin-interacting protein, MCIP1, inhibits cardiac hypertrophy in vivo. Proc Natl Acad Sci U S A 2001; 98: 3328–33.

17. Hill JA, Rothermel B, Yoo KD et al. Targeted inhibition of calcineurin in pressure-overload cardiac hypertrophy. Preservation of systolic function. J Biol Chem 2002; 277: 10251–5.

18. van Rooij E, Doevendans PA, de Theije CC et al. Requirement of nuclear factor of activated T-cells in calcineurin-mediated cardiomyocyte hypertrophy. J Biol Chem 2002; 277: 48617–26.

19. Wilkins BJ, De Windt LJ, Bueno OF et al. Targeted disruption of NFATc3, but not NFATc4, reveals an intrinsic defect in calcineurin-mediated cardiac hypertrophic growth. Mol Cell Biol 2002; 22: 7603–13.

20. Hardt SE, Sadoshima J. Glycogen synthase kinase-3beta: a novel regulator of cardiac hypertrophy and development. Circ Res 2002; 90: 1055–63.

21. Haq S, Choukroun G, Kang ZB et al. Glycogen synthase kinase-3beta is a negative regulator of cardiomyocyte hypertrophy. J Cell Biol 2000; 151: 117–30.

22. Antos CL, McKinsey TA, Frey N et al. Activated glycogen synthase-3 beta suppresses cardiac hypertrophy in vivo. Proc Natl Acad Sci U S A 2002; 99: 907–12.

23. Morisco C, Seta K, Hardt SE et al. Glycogen synthase kinase 3beta regulates GATA4 in cardiac myocytes. J Biol Chem 2001; 276: 28586–97.

24. Braz JC, Bueno OF, Liang Q et al. Targeted inhibition of p38 MAPK promotes hypertrophic cardiomyopathy through upregulation of calcineurin-NFAT signaling. J Clin Invest 2003; 111: 1475–86.

25. Liang Q, Bueno OF, Wilkins BJ et al. c-Jun N-terminal kinases (JNK) antagonize cardiac growth through cross-talk with calcineurin-NFAT signaling. EMBO J 2003; 22: 5079–89.

26. Black BL, Olson EN. Transcriptional control of muscle development by myocyte enhancer factor-2 (MEF2) proteins. Annu Rev Cell Dev Biol 1998; 14: 167–96.

27. Kolodziejczyk SM, Wang L, Balazsi K et al. MEF2 is upregulated during cardiac hypertrophy and is required for normal post-natal growth of the myocardium. Curr Biol 1999; 9: 1203–6.

28. Han J, Molkentin JD. Regulation of MEF2 by p38 MAPK and its implication in cardiomyocyte biology. Trends Cardiovasc Med 2000; 10: 19–22.

29. Kato Y, Kravchenko VV, Tapping RI et al. BMK1/ERK5 regulates serum-induced early gene expression through transcription factor MEF2C. EMBO J 1997; 16: 7054–66.

30. Nicol RL, Frey N, Pearson G et al. Activated MEK5 induces serial assembly of sarcomeres and eccentric cardiac hypertrophy. EMBO J 2001; 20: 2757–67.

31. McKinsey TA, Zhang CL, Olson EN. MEF2: a calcium-dependent regulator of cell division, differentiation and death. Trends Biochem Sci 2002; 27: 40–7.

32. Sugden PH. Ras, Akt, and mechanotransduction in the cardiac myocyte. Circ Res 2003; 93: 1179–92.

33. Passier R, Zeng H, Frey N et al. CaM kinase signaling induces cardiac hypertrophy and activates the MEF2 transcription factor in vivo. J Clin Invest 2000; 105: 1395–406.

34. McKinsey TA, Zhang CL, Lu J et al. Signal-dependent nuclear export of a histone deacetylase regulates muscle differentiation. Nature 2000; 408: 106–11.

35. Zhang CL, McKinsey TA, Olson EN. Association of class II histone deacetylases with heterochromatin protein 1: potential role for histone methylation in control of muscle differentiation. Mol Cell Biol 2002; 22: 7302–12.

36. McKinsey TA, Zhang CL, Olson EN. Signaling chromatin to make muscle. Curr Opin Cell Biol 2002; 14: 763–72.

37. Youn HD, Grozinger CM, Liu JO. Calcium regulates transcriptional repression of myocyte enhancer factor 2 by histone deacetylase 4. J Biol Chem 2000; 275: 22563–7.

38. Zhang CL, McKinsey TA, Chang S et al. Class II histone deacetylases act as signal-responsive repressors of cardiac hypertrophy. Cell 2002; 110: 479–88.

39. Liang Q, Molkentin JD. Divergent signaling pathways converge on GATA4 to regulate cardiac hypertrophic gene expression. J Mol Cell Cardiol 2002; 34: 611–16.

40. Morimoto T, Hasegawa K, Kaburagi S et al. Phosphorylation of GATA-4 is involved in alpha 1-adrenergic agonist-responsive transcription of the endothelin-1 gene in cardiac myocytes. J Biol Chem 2000; 275: 13721–6.

41. Liang Q, Wiese RJ, Bueno OF et al. The transcription factor GATA4 is activated by extracellular signal-regulated kinase 1- and 2-mediated phosphorylation of serine 105 in cardiomyocytes. Mol Cell Biol 2001; 21: 7460–9.

42. Liang Q, De Windt LJ, Witt SA et al. The transcription factors GATA4 and GATA6 regulate cardiomyocyte hypertrophy in vitro and in vivo. J Biol Chem 2001; 276: 30245–53.

43. Bueno OF, Molkentin JD. Involvement of extracellular signal-regulated kinases 1/2 in cardiac hypertrophy and cell death. Circ Res 2002; 91: 776–81.

44. Charron F, Tsimiklis G, Arcand M et al. Tissue-specific GATA factors are transcriptional effectors of the small GTPase RhoA. Genes Dev 2001; 15: 2702–19.

45. Kerkela R, Pikkarainen S, Majalahti-Palviainen T et al. Distinct roles of mitogen-activated protein kinase pathways in GATA-4 transcription factor-mediated regulation of B-type natriuretic peptide gene. J Biol Chem 2002; 277: 13752–60.

46. Yanazume T, Hasegawa K, Wada H et al. Rho/ROCK pathway contributes to the activation of extracellular signal-regulated kinase/GATA-4 during myocardial cell hypertrophy. J Biol Chem 2002; 277: 8618–25.

47. Clerk A, Sugden PH. Small guanine nucleotide-binding proteins and myocardial hypertrophy. Circ Res 2000; 86: 1019–23.

48. Dai YS, Markham BE. p300 Functions as a coactivator of transcription factor GATA-4. J Biol Chem 2001; 276: 37178–85.

49. Kakita T, Hasegawa K, Morimoto T et al. p300 protein as a coactivator of GATA-5 in the transcription of cardiac-restricted atrial natriuretic factor gene. J Biol Chem 1999; 274: 34096–102.

50. Molkentin JD. The zinc finger-containing transcription factors GATA-4, -5, and -6. Ubiquitously expressed regulators of tissue-specific gene expression. J Biol Chem 2000; 275: 38949–52.

51. Akazawa H, Komuro I. Roles of cardiac transcription factors in cardiac hypertrophy. Circ Res 2003; 92: 1079–88.

52. Morin S, Charron F, Robitaille L et al. GATA-dependent recruitment of MEF2 proteins to target promoters. EMBO J 2000; 19: 2046–55.

53. Hannan RD, Rothblum LI. Regulation of ribosomal DNA transcription during neonatal cardiomyocyte hypertrophy. Cardiovasc Res 1995; 30: 501–10.

54. Brandenburger Y, Jenkins A, Autelitano DJ et al. Increased expression of UBF is a critical determinant for rRNA synthesis and hypertrophic growth of cardiac myocytes. FASEB J 2001; 15: 2051–3.

55. Brandenburger Y, Arthur JF, Woodcock EA et al. Cardiac hypertrophy in vivo is associated with increased expression of the ribosomal gene transcription factor UBF. FEBS Lett 2003; 548: 79–84.

CHAPTER 5

G protein-coupled receptor activation

Brian Kobilka

Introduction

G protein-coupled receptors (GPCRs) represent the single largest class of membrane proteins in the human genome and they are the largest group of targets for the pharmaceutical industry. GPCRs share a common structural signature of seven hydrophobic segments predicted to be membrane-spanning domains with an extracellular amino terminus and an intracellular carboxyl terminus (Fig. 5.1). While the vast majority of GPCRs have been shown to activate one or more cytoplasmic heterotrimeric GTP-binding proteins (G proteins), there is now considerable evidence that some GPCRs can activate signaling pathways that don't involve heterotrimeric G proteins.[1,2] For this reason, the terms seven transmembrane (TM) receptor or heptahelical receptor have also been used in place of GPCR.

Eukaryotic GPCRs have been classified by sequence similarity into five classes (A–F or 1–5);[3,4] however, not all of these classes are represented in the human genome. Detailed analysis of the human genome reveals at least 800 unique GPCRs; of these, approximately 460 are predicted to be olfactory receptors.[5] These human receptors could be clustered into five families based on sequence similarity within the seven TM segments:[5] the rhodopsin family (701 members), the adhesion family (24 members), the frizzled/taste family (24 members), the glutamate family (15 members), and the secretin family (15 members). We do not yet know the physiological function of a large fraction of the 800 GPCRs, and they are therefore referred to as orphan GPCRs. However, the deorphanizing of non-olfactory GPCRs is an ongoing process in academic labs and within the pharmaceutical industry,[6] and the actual number of orphan GPCRs continues to decline.

The structural and functional similarity of GPCRs as a family of proteins stands in contrast to the structural diversity of the natural agonists.[7] These range from subatomic particles (a photon), to ions (H^+ and Ca^{2+}), to small organic molecules, to peptides and proteins. The locations of the binding domains for many GPCRs have been determined (Fig. 5.1).[7] Many small organic agonists bind to sites formed by the TM segments. Peptide hormones and proteins often bind to the amino terminus and extracellular sequences joining the TM domains, as well as within the TM segments. However, size of the ligand alone cannot be used to predict the location of the binding site. Glycoprotein hormones, glutamate, and Ca^{2+} all activate their respective receptors by binding to relatively large amino terminal domains.[7,8] It is interesting to note that for many GCPRs that bind their native agonists on extracellular loops or the amino terminus, it has been possible to identify small molecular weight allosteric modulators that bind within the TM domains.[9–13]

In contrast to the diversity in the size of native

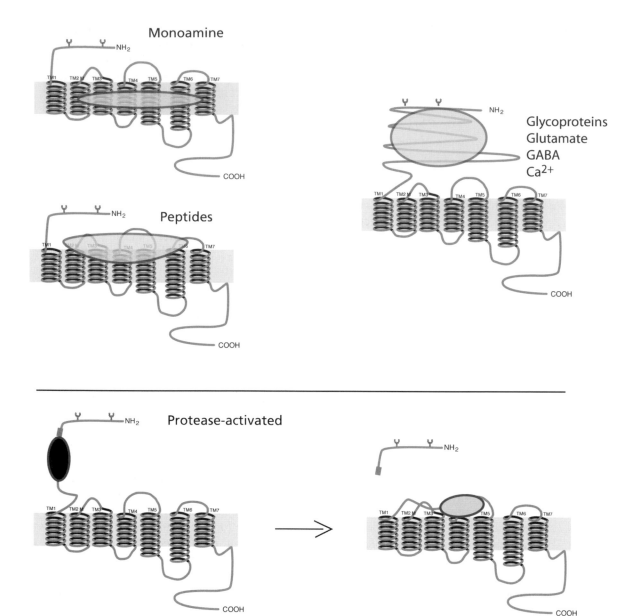

Figure 5.1. *Location of agonist binding sites for different GPCRs.*

agonists and the location of the ligand binding sites, the vast majority of known GPCRs have been shown to activate one or more of the 16 known G protein α-subunits.[14] G proteins are structurally homologous, and the mechanism by which different G proteins are activated by GPCRs is likely to be similar. Therefore, it is likely that the diverse modes of agonist binding to extracellular structures and transmembrane domains result in similar structural changes in cytoplasmic domains that interface with heterotrimeric G proteins.

This chapter will review what is known about the mechanism of transmembrane signaling by GPCRs, specifically, the process by which agonist binding leads to conformational changes necessary for G protein activation. There is a paucity of experimental data that directly address this process; however, several recent studies are beginning to provide mechanistic insight. These studies suggest that a lock and key model of agonist binding does not apply to GPCRs. That is, there is no readily available agonist binding site in the non-liganded receptor. For agonists to bind, intramolecular interactions that keep the receptor in an inactive state must be broken. Evidence suggests that agonists bind in stages involving one or more conformational intermediates. If correct, the mechanism will have implications for drug design and in silico ligand docking experiments.

Evidence for structural and mechanistic homology among GPCRs

Before addressing details about the mechanism of GPCR activation I will briefly review evidence that GPCRs are structurally homologous and probably undergo similar structural changes when activating G proteins.

Rhodopsin

Much is known about the structure and mechanism of activation of rhodopsin. Rhodopsin is a highly specialized GPCR in which *cis*-retinal behaves as a covalently bound inverse agonist that is converted to a full agonist upon photoisomerization to all *trans*-retinal. Therefore the mechanism of agonist activation is highly specialized in contrast to the vast majority of GPCRs that are activated by diffusible agonists. Nevertheless, there is evidence that the structural changes that occur in rhodopsin upon activation by light are similar to structural changes observed in other GPCRs. Therefore, I will briefly discuss rhodopsin structure and what is known about light-induced conformational changes. This area has been the subject of several excellent reviews.[15–18]

Rhodopsin is the only GPCR for which a high-resolution crystal structure is available. Structures from both two-dimensional (2D),[19,20] and 3D[21–24] crystals have been obtained. The most recent 3D structure of rhodopsin has a resolution of 2.6 Å.[23]

While the recent 2D structure determined by cryo-electron microsopy of 2D crystals is of lower resolution (5.6 Å), it gives information about the orientation of TM segments relative to the lipid bilayer that cannot be obtained from 3D crystals.

The most detailed information about structural changes associated with activation of a GPCR come from studies of rhodopsin. This is in part due to the natural abundance and biochemical stability relative to other GPCRs. Electron paramagnetic resonance spectroscopy (EPR) studies provide evidence that photoactivation of rhodopsin involves a rotation and tilting of transmembrane segment 6 (TM6) relative to TM3.[25] Further support for motion of TM6 during rhodopsin activation was provided by chemical reactivity measurements and fluorescence spectroscopy,[26] as well as by ultra-violet absorbance spectroscopy,[27] and by zinc crosslinking of histidines.[28] Light-induced conformational changes have also been observed in the cytoplasmic domain spanning TM1 and TM2, and the cytoplasmic end of TM7.[29–31]

GPCRs activated by diffusible agonists

While rhodopsin has long been used as a model system for GPCR activation, it is unique among GPCRs because of the presence of a covalent linkage between the receptor and its ligand, retinal. Thus, the dynamic processes of agonist association and dissociation common to GPCRs for hormones and neurotransmitters are not part of the activation mechanism of rhodopsin. High-resolution structures for other GPCRs have not yet been obtained; however, there is indirect evidence that some rhodopsin family members are structurally very similar to rhodopsin. Ballesteros and coworkers found that structural insights obtained from mutagenesis data and substituted cysteine accessibility studies on monoamine receptors were consistent with the high-resolution structure of rhodopsin, suggesting that rhodopsin serves as a good template for molecular model building.[32] However, Archer and colleagues found that rhodopsin might not be a good template for models of more distantly related family rhodopsin family members such as the cholecystokinin CCK1 receptor.[33]

In spite of the remarkable diversity of ligands and ligand-binding domains in the family of GPCRs, there is considerable evidence for a common mechanism of activation. When comparing sequences, GPCRs are most similar at the cytoplasmic ends of the transmembrane segments adjacent to the second and third cytoplasmic domains, the regions known to interact with cytoplasmic G proteins.[34] Members of the large family of GPCRs transduce signals by activating one or more members of the relatively small family of highly homologous heterotrimeric G proteins. For example, the thyroid-stimulating hormone (TSH) receptor is activated by a large glycoprotein hormone that binds to the amino terminus while the β_2 adrenoceptor is activated by adrenaline (approximately the size of a single amino acid) that binds to the TM segments; yet both of these receptors activate the same G protein (G_s), indicating that the structural changes in the cytoplasmic domains of these two receptors must be very similar. Moreover, many GPCRs exhibit promiscuous coupling to more than one G protein. For example, rhodopsin preferentially couples to transducin while the β_2AR preferentially couples to G_s; however, both are capable of activating G_i.[35]

Additional evidence that GPCRs undergo similar conformational changes within TM segments and cytoplasmic domains comes from biophysical and biochemical studies. Fluorescence spectroscopic studies of β_2 adrenoceptors labeled with fluorescent probes demonstrate movement in both TM3 and TM6 upon activation.[36] More recent studies of β_2AR labeled with fluorescent probes at the cytoplasmic end of TM6 provide evidence that agonists induce a rotation or tilting movement of the cytoplasmic end of TM6 similar to that observed in rhodopsin.[37,38] Additional support for movement of TM3 and TM6 in the β_2AR comes from zinc crosslinking studies,[39] and chemical reactivity measurements in constitutively active β_2AR mutants.[40,41] Cysteine crosslinking studies on the M3 muscarinic receptor provide evidence for the movement of the cytoplasmic ends of TM5 and TM6 toward each other upon agonist activation.[42]

GPCR oligomers

There is a growing body of evidence that GPCRs exist as dimers (or oligomers) and that dimers may be important for G protein activation for at least some GPCRs.[43,44] Recent cryoelectron microscopy images suggest that rhodopsin may exist as homodimers in rod outer segment membranes.[45] Dimerization is clearly an important mechanism of receptor activation for the glutamate family of GPCRs,[8] where ligand-induced changes in the dimer interface of the amino terminal ligand binding domain has been demonstrated by crystallography.[46,47] However, the role of dimerization in the activation of rhodopsin family members is less clear. The effect of agonist binding on the formation or disruption of dimers is not consistent among the rhodopsin family members that have been examined.[43] Moreover, ligands interact with individual receptor monomers, and there is currently no evidence that ligands span the interface between receptor dimers. If changes in dimerization occur, it is probably a secondary consequence of ligand-induced changes in the arrangement of the TM segments. Therefore, while dimers may be important for receptor activation, we must first understand the agonist-induced structural changes that occur in the context of individual GPCR monomers.

Conformational states

Proteins are often thought of as rigid structures. The classic model of receptors is the lock-and-key analogy where the agonist fits precisely into a complementary pocket in the receptor protein. However, it is known the proteins are dynamic molecules that exhibit rapid, small-scale structural fluctuations. One of the best ways to discuss protein conformations is in terms of an energy diagram (Fig. 5.2). The basal conformational state is the lowest energy state of the protein in a particular environment. The width of the energy well reflects the conformational flexibility. The probability that a protein will undergo transitions to other conformational states is a function of the energy difference between the two states and the height of the energy barrier between the two states. In the case of a receptor, the ligand-binding energy may change both the depth of an energy well and the height of an energy barrier. The simple diagram

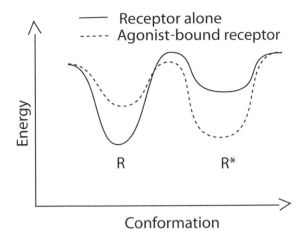

Figure 5.2. *Energy landscape diagram describing a possible mechanism of GPCR activation by an agonist.*

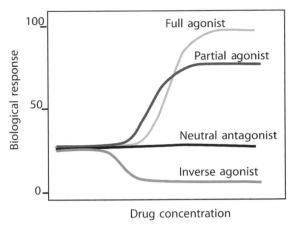

Figure 5.3. *Ligand efficacy. The effect of different classes of drugs on a GPCR that has some detectable basal activity. The response is plotted as a percentage of the maximum response.*

shown in Fig. 5.2 shows only two conformational states, however, as discussed below, there is evidence that GPCRs may exist in multiple conformational states.

Basal activity and ligand efficacy

With the exception of rhodopsin, most GPCRs do not behave as simple bimodal switches. Rhodopsin has virtually no detectable basal activity in the absence of light, but can be fully activated by a single photon. For many GPCRs there is a considerable amount of basal, agonist-independent activity. The activity of receptors can be either increased or decreased by different classes of ligands (Fig. 5.3). The term 'efficacy' is used to describe the effect of a ligand on the functional properties of the receptor (for a more complete discussion of efficacy refer to reference 48). Agonists are defined as ligands that fully activate the receptor. Partial agonists induce submaximal activation of the G protein even at saturating concentrations. Inverse agonists inhibit basal activity. Antagonists have no effect on basal activity, but competitively block access of other ligands. Therefore, based on functional behavior, GPCRs behave more like rheostats than simple bimodal switches. Ligands can 'dial in' virtually any level of activity from fully active to fully inactive.

Multiple agonist-specific states

A number of kinetic models have been developed to explain ligand efficacy based on indirect measures of receptor conformation such as ligand-binding affinity, and the activation of G proteins and effector enzymes.[49 52] A simple two-state model can describe much of the functional behavior of GPCRs.[50] The two-state model proposes that a receptor exists primarily in two states, the inactive state (R) and the active state (R*). In the absence of ligands, the level of basal receptor activity is determined by the equilibrium between R and R*. The efficacy of ligands reflects their ability to alter the equilibrium between these two states. Full agonists bind to and stabilize R*, while inverse agonists bind to and stabilize R. Partial agonists have some affinity for both R and R*, and are therefore less effective in shifting the equilibrium towards R*.

While the two-state model can explain the spectrum of responses to ligands of different efficacy in simple experimental systems consisting of one receptor and one G protein, there is a growing body of experimental evidence for multiple conformational states.[53] Thus each ligand may induce or stabilize a unique conformational state that can be distinguished by the activity of that state towards different signaling molecules (G proteins, kinases, arrestins). In the case of the β_2AR it has been

75

possible to monitor directly ligand-specific states using fluorescence spectroscopy.[54,55] β_2ARs were labeled at C265 on the cytoplasmic end of TM6 adjacent to the G protein coupling domain. Conformational changes in the G protein-coupling domain change the molecular environment around the fluorophore, leading to changes in fluorescence intensity and fluorescence lifetime. Fluorescence lifetime analysis can detect discrete conformational states in a population of molecules, while fluorescence intensity measurements reflect the weighted average of one or more discrete states. A single broad distribution of fluorescent lifetimes is observed in the absence of ligands, suggesting that this domain oscillates around a single detectable conformation (Fig. 5.4A). Antagonist binding reduces the width of the distribution, but does not change the mean lifetime, suggesting that the conformation does not change, but the domain's flexibility is reduced (Fig. 5.4A). However, when the β_2AR is bound to a full agonist, two lifetime distributions are observed (Fig. 5.4B and C), suggesting two distinct conformations. Moreover, the conformations induced by a full agonist can be distinguished from those induced by partial agonists (Fig. 5.4C).[55]

Defining the 'active state'

As we learn more about the complexity of GPCR signaling it is becoming more difficult to define exactly what is meant by activation and the 'active state'. Recent work from several labs have shown that GPCRs can activate signaling pathways by G protein-independent mechanisms such as through arrestin (and possibly other signaling molecules),[2,56] and that the 'active state' for receptor activation of arrestin or other G protein-independent pathways may differ from that for receptor activation of a G protein.[2] Thus a drug classified as an inverse agonist when monitoring receptor activation of a G protein-dependent signaling pathway, may behave as a partial agonist for a G protein-independent signaling pathway.[2,56] Moreover, the fully active state my differ for different G proteins.[53] Thus, for a GPCR capable of activating more than one G protein, a drug may act as a full agonist towards one G protein and as a partial agonist towards another. Therefore, to simplify the discus-

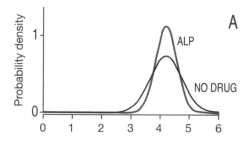

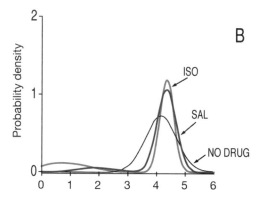

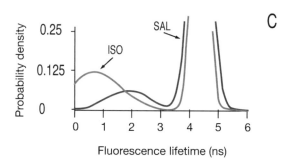

Figure 5.4. *Fluorescence lifetime distributions of β_2AR labeled at Cys265 with fluorescein maleimide.[55] (A) Single lifetime distributions are observed for unliganded receptor and receptor bound to the neutral antagonist alprenolol (ALP). (B) and (C) Two lifetime distributions are observed for β_2AR bound to the full agonist isoproterenol (ISO) and the partial agonist salbutamol (SAL). The short lifetime distribution for ISO is different from that for SAL, consistent with a different active conformation. Points plotted are probability density.*

sion, I will define the activity of a particular conformational state of a GPCR by the effect that conformational state has on the activity of the receptor's cognate (or preferred) G protein. Thus a full agonist maximally activates the cognate G protein, while an inverse agonist maximally inhibits any basal activation of the G protein by the receptor.

Agonist activation

Insights from constitutively active mutants

To understand the process of receptor activation we must first understand the basal or non-liganded state of the receptor. Some GPCRs such as rhodopsin and the follicle-stimulating hormone (FSH)[57] receptor have little or no detectable basal activity while other GPCRs such as the canabinoid

receptors exhibit a high degree of basal activity.[58,59] Basal activity could reflect inherent flexibility of the receptor, and the tendency to exist in more than one conformational state in the absence of ligands. It could also reflect a highly constrained state that has a relatively high affinity for a G protein. The concept of basal activity and receptor activation can be considered in terms of an energy landscape (Fig. 5.5). In the case of a receptor with low basal activity, in the absence of agonist, the receptor may be relatively constrained to one inactive conformational state having a deep energy well (Fig. 5.5A). High basal activity might be explained by smaller energy differences between the inactive and active states with a lower energy barrier (Fig. 5.5B). This might also be thought of as a receptor with greater conformational flexibility (fewer conformational constraints). Alternatively, it is possible that a receptor may exist predominantly in one constrained state that has intermediate activity

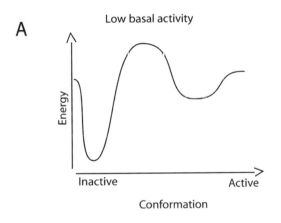

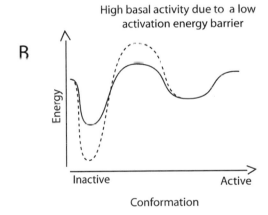

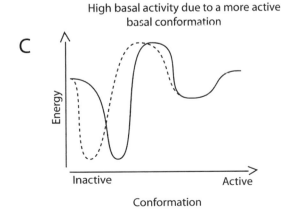

Figure 5.5. *Energy landscapes used to explain possible mechanisms of elevated basal activity. (A) Receptor with low basal activity. (B) and (C) Receptors with high basal activity, relative to (A) (shown as a broken line). (B) The basal activity is elevated because of a reduced energy barrier separating the basal and active conformational states. (C) The basal activity is higher because the basal conformational state has a higher activity towards the G protein.*

77

towards its G protein (Fig. 5.5C). While both of these mechanisms may apply to different receptors, there is experimental evidence linking conformational flexibility and structural instability to elevated basal activity.[60]

TM domains are held in the basal state by intervening loops and non-covalent interactions between side chains. The non-covalent interactions appear to play a greater role in determining the specific basal arrangement of the TM segments relative to each other than do some of the intervening loop structures. Evidence for this comes from proteolysis and split receptor studies. For example co-transfecting a plasmid encoding the amino terminus through TM5 with a plasmid encoding TM6 through the carboxyl terminus generates a functional β_2AR.[61] These fragments assemble and are held together by non-covalent interactions. Similar observations have been made for the muscarinic receptor. Schoneberg and colleagues were able to generate functional M3 muscarinic receptors with discontinuity within the loop connecting TM3 and TM4, the loop connecting TM4 and TM5, and the loop connecting TM5 and TM6.[62] Similarly, both the α_2 adrenergic receptor,[63] and the β_2AR (G Swaminath and BK Kobilka, unpublished data) are capable of binding ligands after proteolytic cleavage of loop structures.

The degree of basal activity can be dramatically enhanced by single point mutations in a variety of structural domains.[64] These constitutive active mutations (CAMs) provide insight into the structural basis of basal activity and of receptor activation. The fact that CAMs can be generated in virtually any structural domain suggests that in the inactive state the receptor structure is constrained by intramolecular interactions that link TM segments or link TM segments with inter-TM segment loops.[64] Mutations that disrupt these interactions would increase the 'flexibility' of the protein (movement of TM domains relative to each other), and the probability that the receptor can assume an active conformation.

One of the best-characterized examples of CAMs are those that disrupt the highly conserved (D/E)R(Y/W) sequence. In rhodopsin there is a network of interactions between E134 and R135 at the cytoplasmic end of TM3, and E257 and T251 at the cytoplasmic end of TM6.[22] Disruption of this network by mutating E134 to glutamine leads to constitutive activity in opsin.[65] Experimental evidence indicates that E134 is protonated during activation of rhodopsin, demonstrating that disruption of this network is part of the normal activation process.[66] Amino acids constituting this bridge are conserved in other GPCRs, and mutations of the acidic amino acid have been reported to increase basal activity in a number of other GPCRs including the β_2AR,[41] the H_2 histamine receptor,[67] the α_{1b} adrenoceptor,[68,69] and the angiotensin (AT_1) receptor.[70] The β_2AR becomes constitutively active when the pH is reduced from 7.5 to 6.5, presumably due to the disruption of this and/or other intramolecular interaction as a result of protonation of an acidic amino acid.[71]

One might predict that mutations that lead to enhanced basal activity by disrupting intramolecular interactions could also lead to decreased structural stability. Mutation of L272 at the cytoplasmic end of TM6 in the β_2AR to alanine results in elevated basal activity,[72] as well as biochemical instability.[60] Purified L272Aβ_2AR denatures 2–3 times faster than wild-type receptor.[60] The increased basal activity observed in the native β_2AR at reduced pH is also associated with an increased rate of denaturation.[71] Denaturation can be attenuated by both agonists and antagonists.[60] Instability has also been reported in constitutive active mutants of the β_1 adrenoceptor,[73] and the H_2 histamine receptor.[67] It is worth mentioning that ligands, both agonists and antagonists, can stabilize the receptor against denaturation and act as biochemical chaperones,[73,74] suggesting that they form stabilizing bridges between TM segments.

Agonists act as reversible CAMs to alter the arrangement of TM segments

So what do CAMs tell us about GPCR activation? Active states can be achieved by destabilizing the normal arrangement of TM domains by mutations at several different sites. As discussed above, TM domains are held in the basal state primarily by a network of non-covalent interactions between side chains. Thus, any compound that disrupts the network of intramolecular interactions that characterize the basal state could have agonist activity. In the case of the monamine receptors, side chains of TM domains also form the ligand-binding site. Thus, by binding within the TM segments, agonists may act like reversible CAMs and disrupt the normal packing of TM domains. Figure 5.6 shows two possible ways that ligands may influence the arrangement of TM domains. Ligands may serve as bridges that create new interactions between TM domains (Fig. 5.6A). These ligands may move specific TM domains closer to each other, push them further apart or rotate one relative to the other. At the other end of the spectrum, ligands may act by simply disrupting existing intramolecular interactions (Fig. 5.6B). An example of an agonist binding to and displacing stabilizing interactions comes from the AT_1 receptor. Evidence suggests that Asn111 interacts with Tyr4 of angiotensin.[75] Moreover, mutation of Asn111 to Ala leads to constitutive activity.[76] Experimental evidence suggests that Asn111 interacts with Asn295 in TM7 to stabilize the inactive state of the receptor.[77] It is likely that a combination of the mechanisms shown in Fig. 5.7 may be operable for any given ligand, particularly for larger ligands such as peptide agonists.

Agonist binding and activation is multistep process

In both of the models shown in Fig. 5.6, the ability of the ligand to bind depends on the dynamic nature of the non-covalent interactions between TM segments. That is, agonist activation cannot be explained by a simple lock-and-key model. There is no preformed binding site for the agonist. In the basal state, sites of contact for the ligand are either not optimally aligned to bind the agonist (Fig. 5.6A) or are involved in intramolecular interactions (Fig. 5.6B). Thus, interactions must break and reform on a time scale compatible with rapid binding and activation of receptors. In the case of the model shown in Fig. 5.6A, ligands may first bind to one interacting site, and be poised to bind to a subsequent site upon disruption of intramolecular interactions between TM segments. This would involve the formation of one or more intermediate conformational states.

Insights from biophysical studies

Evidence in support of a multistep process for agonist binding comes from catecholamine binding to the β_2AR.[54,55] Fluorescence lifetime studies reveal the existence of at least one intermediate conformational state in the presence of the full agonist isoproterenol or partial agonists dobutamine and

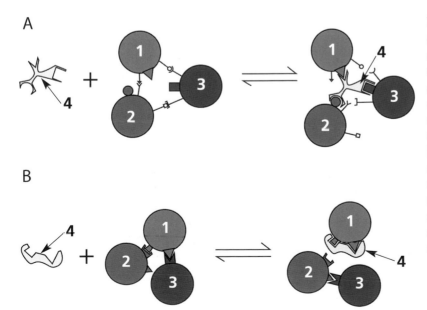

A

B

Figure 5.6. *Mechanisms by which agonist binding may change the relative arrangement of TM segments. (A) The agonist binding requires disruption of intramolecular interactions and the formation of new interactions with the ligand. (B) The agonist binds directly to amino acids involved in forming stabilizing intramolecular interactions. 1, 2 and 3 represent three of the seven transmembrane domains of a GPCR. 4 is the agonist.*

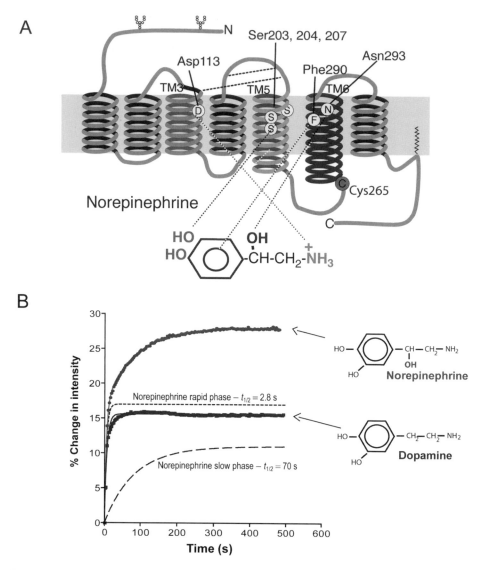

Figure 5.7. *Agonist-induced conformational changes in the* β_2*AR. (A) Sites of interaction between norepinephrine and the* β_2*AR identified by site-directed mutagenesis. The catecholamine nitrogen interacts with Asp113 in TM3.[80] Hydroxyls on the catechol ring interact with serines 203,[82] 204 and 207[80] in TM6. The chiral* β*-hydroxyl interacts with Asn293 in TM6,[81] and the aromatic ring interacts with Phe290 in TM6.[80] Also shown is the relative position of Cys265, the labeling site for tetramethylrhodamine. (B) Agonist-induced conformational changes in purified* β_2*AR labeled with tetramethylrhodamine at Cys265. Conformational responses to norepinephrine and dopamine were examined by monitoring changes in fluorescence intensity as a function of time. The response to norepinephrine was best fitted with a two-site exponential association function, while there was no significant difference between a one-site and a two-site fit for the response to dopamine. The rapid and slow components of the biphasic response to norepinephrine are shown as broken lines.*

salbutamol.[55] The existence of an intermediate conformational state can also be demonstrated kinetically in the β_2AR. The β_2AR is a good model system for studying agonist binding because much is known about the sites of interactions between catecholamine ligands and the receptor. β_2AR was labeled at Cys265 at the cytoplasmic end of TM6 with tetramethylrhodamine (TMR-β_2AR). An environmentally sensitive fluorophore covalently bound to Cys265 is well positioned to detect agonist-induced conformational changes relevant to G protein activation. Based on homology with rhodopsin,[22] Cys265 is located in the third intracellular loop (IC3) at the cytoplasmic end of the transmembrane 6 (TM6). Mutagenesis studies have shown this region of IC3 to be important for G protein coupling.[78,79] Moreover, TM6, along with TM3 and TM5, contain amino acids that form the agonist-binding site. The sites of interaction between catecholamines and the β_2AR have been extensively characterized,[80–82] and are summarized in Fig. 5.7A. The amine nitrogen interacts with Asp 113 in TM3,[80] the catechol hydroxyls interact with serines in TM5 [80–82] Interactions with the aromatic ring and the chiral β-hydroxyl have both been mapped to TM6.[81]

Agonist-induced conformational changes can be observed by monitoring fluorescence intensity of TMR-β_2AR over time. Binding of the catecholamine norepinephrine results in curve that is best fit by a two-site exponential association function (Fig. 5.7B).[54] This suggests that upon catecholamine binding, β_2ARs undergo transitions to two kinetically distinguishable conformational states. Using a panel of chemically related catechol derivatives, we identified the specific chemical groups on the agonist responsible for the rapid and slow conformational changes in the receptor. Only rapid conformational changes were observed in the presence of dopamine (Fig. 5.7B). In contrast, both rapid and slow conformational changes were observed upon binding to norepinephrine, epinephrine and isoproterenol. These results suggest that formation of interactions between the catechol ring and the receptor, and the amine nitrogen and the receptor occur rapidly, while interactions between the β hydroxyl and Asn293 occur more slowly, possibly due to the need to overcome a strong stabilizing intramolecular interaction.

The conformational changes observed in these biophysical assays were correlated with biological responses in cellular assays. Dopamine, which induces only a rapid conformational change, is efficient at activating G_s, but not receptor internalization. In contrast, norepinephrine and epinephrine, which induce both rapid and slow conformational changes, are efficient at activating G_s and receptor internalization. Evidence for intermediate conformational states has also been observed in a peptide receptor. Time-resolved peptide-binding studies on the neurokinin receptor revealed that an agonist peptide binds with biphasic kinetics. The rapid binding component was associated with a cellular calcium response while the slow component was required for cAMP signaling.[83] These results support a mechanistic model for GPCR activation where contacts between the receptor and structural determinants of the agonist stabilize a succession of conformational states with distinct cellular functions.

Insights from binding studies

These biophysical studies on the β_2AR are supported by an elegantly simple series of experiments examining the binding properties and efficacy of a panel of ligands that represent components of the catecholamine epinephrine (Fig. 5.8A).[84] By comparing the affinity of these different compounds it is possible to determine the contribution of each chemical substituent (catechol hydroxyls, β-hydroxyl, N-methyl) of the agonist to binding affinity and efficacy. The results suggest that there is no preformed binding site for the agonist epinephrine in the unliganded β_2AR. Fig. 5.8B illustrates the concept of a preformed binding site. In the simple lock-and-key type model, all of the amino acid side chains that contribute to the binding site for epinephrine are in the optimal position to engage the chemical substituents of the ligand. The contribution of each substituent to the binding affinity would depend on the type of interaction between the substituent and the receptor (aromatic-interactions, van der Waals' forces, hydrogen bonding, ion pairs). However, one would expect that each interaction would make a significant contribution

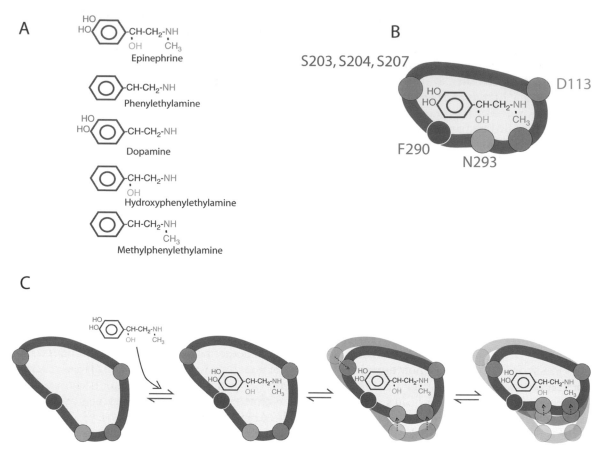

Figure 5.8. *Models of agonist binding and activation. (A) Structures of epinephrine, phenylethylamine, dopamine, hydroxyphenylethylamine and methylphenylethylamine. (B) Lock-and-key-type agonist-binding model. Receptor sites that interact with specific substituents of the ligand are shown as colored circles. (C) Sequential agonist binding model. The models are based on a study by Liapakis et al,[84] and previously published in an accompanying commentary: Kobilka B. Agonist binding: a multistep process. Mol Pharmacol 2004; 65: 1060–2. See color plate section.*

to ligand affinity that would be relatively independent of the presence of the other substituents. Based on these assumptions, there should be an increase in binding affinity by adding the catechol hydroxyls to phenylethylamine to make dopamine, or by adding the *N*-methyl to phenylethylamine to make methylphenylethylamine, or by adding the β-hydroxyl to phenylethylamine to make hydroxyphenylethylamine. Yet, the authors found similar binding affinities for the four compounds. That is, adding any one of the substituents (catechol hydroxyls, β-hydroxyl or the *N*-methyl) to

phenylethylamine did not increase the affinity. However, the authors observed that the catechol hydroxyls, the β-hydroxyl or *N*-methyl group increased the affinity by ~10-fold when added to a ligand having any one of the other three substituents (considering the two catechol hydroxyls as one substituent). Moreover, each of these groups contributed to a 60- to 120-fold increase in binding affinity in ligands having both of the other two substituents.

These binding studies do not support the simple model shown in Fig 5.8C, but suggest that the binding pocket changes during the process of binding. One possible way to explain these results is a model whereby the agonist binds through a series of conformational intermediates as shown in Fig. 5.8B. In the unliganded state there is a minimal, low-affinity binding site that forms interactions between the receptor and a few structural features on the agonist (e.g. the aromatic ring and the amine). Binding to this site increases the probability of a conformational transition that is stabilized by an interaction between the receptor and the catechol hydroxyls. The binding energy gained by interactions between the receptor and the catechol hydroxyls pays for the energetic costs of the conformational change. This conformational transition increases the probability of yet another conformational change stabilized by interactions between receptor and the β-hydroxyl and/or the N-methyl. Thus, the receptor becomes activated through a series of conformational intermediates, and the energetic costs of receptor activation are paid in instalments.

Insights from native and engineered metal ion-binding sites

Engineered metal ion-binding sites have proven useful for providing intramolecular distance constraints between adjacent TM domains. Successful generation of a metal ion-binding site is usually detected when a metal ion competes with the native ligand in a binding assay.[85] The few examples where engineered or native metal ion-binding sites lead to receptor activation have provided insight about the mechanism of receptor activation.[86,87] Elling and colleagues engineered a metal ion-binding site into the β_2AR by mutating Asp113 to histidine and Asn312 to cysteine. The mutant receptor was no longer activated by catecholamines; however, it could be partially activated by Zn^{2+}.[86] The results demonstrate that stabilizing interactions between specific amino acids on TM3 and TM7 by a bridging metal ion stabilized an active state of the receptor.

Concluding remarks

This review has addressed activation of GPCRs using data from only a small subset of rhodopsin family members in which the agonist-binding site is formed by the TM segments. At the other end of the spectrum are receptors for glycoprotein hormones, and the glutamate family of receptors in which the ligand binding site is found within a large amino terminal domain. The mechanism linking ligand binding to receptor activation for these receptors is likely to be more complex.[8,88] Nevertheless, glycoprotein hormone receptors can be activated by mutations within TM segments, and glutamate receptor activity can be modulated by small organic compounds that bind within the TM segments,[8] suggesting that agonist binding ultimately leads to the disruption of interactions that stabilize the arrangement of the TM segments. In the case of the glycoprotein hormones, evidence suggests that part of the amino terminus interacts with the sequence between TM4 and TM5 to stabilize the inactive state, and that this interaction is disrupted by agonist binding.[88,89]

Glutamate family receptors are homodimers held together by the amino terminal Venus flytrap domain. Agonist binding leads to large structural changes in the Venus flytrap motifs and would be predicted to alter the relative arrangement of the 7TM segments from each monomer.[8,90] This may in turn alter orientation of the TM segments within each monomer.

While we have learned a great deal about GPCR structure and the cellular signaling pathways activated by GPCRs over the past 20 years, much remains to be learned about the mechanism of activation of this fascinating family of membrane proteins. A better understanding of the complex process of agonist binding and activation may facilitate the design of more effective and selective pharmaceuticals.

References

1. Luttrell LM, Lefkowitz RJ. The role of beta-arrestins in the termination and transduction of G-protein-coupled receptor signals. J Cell Sci 2002; 115: 455–65.
2. Azzi M, Charest PG, Angers S et al. Beta-arrestin-mediated activation of MAPK by inverse agonists reveals distinct active conformations for G protein-coupled receptors. Proc Natl Acad Sci USA 2003; 100: 11406–11.

3. Attwood TK, Findlay JB. Fingerprinting G-protein-coupled receptors. Protein Eng 1994; 7: 195–203.

4. Kolakowski LF, Jr. GCRDb: a G-protein-coupled receptor database. Recept Channels 1994; 2: 1–7.

5. Fredriksson R, Lagerstrom MC, Lundin LG et al. The G-protein-coupled receptors in the human genome form five main families. Phylogenetic analysis, paralogon groups, and fingerprints. Mol Pharmacol 2003; 63: 1256–72.

6. Howard AD, McAllister G, Feighner SD et al. Orphan G-protein-coupled receptors and natural ligand discovery. Trends Pharmacol Sci 2001; 22: 132–40.

7. Ji TH, Grossmann M, Ji I. G protein-coupled receptors. I. Diversity of receptor-ligand interactions. J Biol Chem 1998; 273: 17299–302.

8. Pin JP, Galvez T, Prezeau L. Evolution, structure, and activation mechanism of family 3/C G-protein-coupled receptors. Pharmacol Ther 2003; 98: 325–54.

9. Knoflach F, Mutel V, Jolidon S et al. Positive allosteric modulators of metabotropic glutamate 1 receptor: characterization, mechanism of action, and binding site. Proc Natl Acad Sci USA 2001; 98: 13402–7.

10. Carroll FY, Stolle A, Beart PM et al. BAY36-7620: a potent non-competitive mGlu1 receptor antagonist with inverse agonist activity. Mol Pharmacol 2001; 59: 965–73.

11. Pagano A, Ruegg D, Litschig S et al. The non-competitive antagonists 2-methyl-6-(phenylethynyl) pyridine and 7-hydroxyiminocyclopropan[b]chromen-1a-carboxylic acid ethyl ester interact with overlapping binding pockets in the transmembrane region of group I metabotropic glutamate receptors. J Biol Chem 2000; 275: 33750–8.

12. Litschig S, Gasparini F, Ruegg D et al. CPCCOEt, a noncompetitive metabotropic glutamate receptor 1 antagonist, inhibits receptor signaling without affecting glutamate binding. Mol Pharmacol 1999; 55: 453–61.

13. Ray K, Northup J. Evidence for distinct cation and calcimimetic compound (NPS 568) recognition domains in the transmembrane regions of the human Ca^{2+} receptor. J Biol Chem 2002; 277: 18908–13.

14. Sprang SR. G protein mechanisms: insights from structural analysis. Annu Rev Biochem 1997; 66: 639–78.

15. Sakmar TP. Structure of rhodopsin and the superfamily of seven-helical receptors: the same and not the same. Curr Opin Cell Biol 2002; 14: 189–95.

16. Sakmar TP, Franke RR, Khorana HG. The role of the retinylidene Schiff base counterion in rhodopsin in determining wavelength absorbance and Schiff base pKa. Proc Natl Acad Sci USA 1991; 88: 3079–83.

17. Ridge KD, Abdulaev NG, Sousa M et al. Photo-transduction: crystal clear. Trends Biochem Sci 2003; 28: 479–87.

18. Hubbell WL, Altenbach C, Hubbell CM et al. Rhodopsin structure, dynamics, and activation: a perspective from crystallography, site-directed spin labeling, sulfhydryl reactivity, and disulfide cross-linking. Adv Protein Chem 2003; 63: 243–90.

19. Schertler GF, Villa C, Henderson R. Projection structure of rhodopsin. Nature 1993; 362: 770–2.

20. Krebs A, Edwards PC, Villa C et al. The three-dimensional structure of bovine rhodopsin determined by electron cryomicroscopy. J Biol Chem 2003; 278: 50217–25.

21. Okada T, Le Trong I, Fox BA et al. X-Ray diffraction analysis of three-dimensional crystals of bovine rhodopsin obtained from mixed micelles. J Struct Biol 2000; 130: 73–80.

22. Palczewski K, Kumasaka T, Hori T et al. Crystal structure of rhodopsin: A G protein-coupled receptor. Science 2000; 289: 739–45.

23. Okada T, Fujiyoshi Y, Silow M et al. Functional role of internal water molecules in rhodopsin revealed by X-ray crystallography. Proc Natl Acad Sci USA 2002; 99: 5982–7.

24. Teller DC, Okada T, Behnke CA et al. Advances in determination of a high-resolution three-dimensional structure of rhodopsin, a model of G-protein-coupled receptors (GPCRs). Biochemistry 2001; 40: 7761–72.

25. Farrens DL, Altenbach C, Yang K et al. Requirement of rigid-body motion of transmembrane helices for light activation of rhodopsin. Science 1996; 274: 768–70.

26. Dunham TD, Farrens DL. Conformational changes in rhodopsin. Movement of helix f detected by site-specific chemical labeling and fluorescence spectroscopy. J Biol Chem 1999; 274: 1683–90.

27. Lin SW, Sakmar TP. Specific tryptophan UV-absorbance changes are probes of the transition of rhodopsin to its active state. Biochemistry 1996; 35: 11149–59.

28. Sheikh SP, Zvyaga TA, Lichtarge O et al. Rhodopsin activation blocked by metal-ion-binding sites linking transmembrane helices C and F. Nature 1996; 383: 347–50.

29. Altenbach C, Klein-Seetharaman J, Cai K et al. Structure and function in rhodopsin: mapping light-dependent changes in distance between residue 316 in helix 8 and residues in the sequence 60–75, covering the cytoplasmic end of helices TM1 and TM2 and their connection loop CL1. Biochemistry 2001; 40: 15493–500.

30. Altenbach C, Klein-Seetharaman J, Hwa J et al. Structural features and light-dependent changes in the sequence 59–75 connecting helices I and II in rhodopsin: a site-directed spin-labeling study. Biochemistry 1999; 38: 7945–9.

31. Altenbach C, Cai K, Khorana HG et al. Structural features and light-dependent changes in the sequence 306–322 extending from helix VII to the palmitoylation sites in rhodopsin: a site-directed spin-labeling study. Biochemistry 1999; 38: 7931–7.

32. Ballesteros JA, Shi L, Javitch JA. Structural mimicry in G protein-coupled receptors: implications of the high-resolution structure of rhodopsin for structure-function analysis of rhodopsin-like receptors. Mol Pharmacol 2001; 60: 1–19.

33. Archer E, Maigret B, Escrieut C et al. Rhodopsin crystal: new template yielding realistic models of G-protein-coupled receptors? Trends Pharmacol Sci 2003; 24: 36–40.

34. Mirzadegan T, Benko G, Filipek S et al. Sequence analyses of G-protein-coupled receptors: similarities to rhodopsin. Biochemistry 2003; 42: 2759–67.

35. Cerione RA, Staniszewski C, Benovic JL et al. Specificity of the functional interactions of the β-adrenergic receptor and rhodopsin with guanine nucleotide regulatory proteins reconstituted in phospholipid vesicles. J Biol Chem 1985; 260: 1493–500.

36. Gether U, Lin S, Ghanouni P et al. Agonists induce conformational changes in transmembrane domains III and VI of the beta2 adrenoceptor. EMBO J 1997; 16: 6737–47.

37. Jensen AD, Guarnieri F, Rasmussen SG et al. Agonist-induced conformational changes at the cytoplasmic side of TM 6 in the beta 2 adrenergic receptor mapped by site-selective fluorescent labeling. J Biol Chem 2001; 276: 9279–90.

38. Ghanouni P, Steenhuis JJ, Farrens DL et al. Agonist-induced conformational changes in the G-protein-coupling domain of the beta 2 adrenergic receptor. Proc Natl Acad Sci USA 2001; 98. 5997 6002.

39. Sheikh SP, Vilardarga JP, Baranski TJ et al. Similar structures and shared switch mechanisms of the beta2-adrenoceptor and the parathyroid hormone receptor. Zn(II) bridges between helices III and VI block activation. J Biol Chem 1999; 274: 17033–41.

40. Javitch JA, Fu D, Liapakis G et al. Constitutive activation of the beta2 adrenergic receptor alters the orientation of its sixth membrane-spanning segment. J Biol Chem 1997; 272: 18546–9.

41. Rasmussen SG, Jensen AD, Liapakis G et al. Mutation of a highly conserved aspartic acid in the beta2 adrenergic receptor: constitutive activation, structural insta-bility, and conformational rearrangement of transmembrane segment 6. Mol Pharmacol 1999; 56: 175–84.

42. Ward SD, Hamdan FF, Bloodworth LM et al. Conformational changes that occur during M3 muscarinic acetylcholine receptor activation probed by the use of an in situ disulfide cross- linking strategy. J Biol Chem 2002; 277: 2247–57.

43. Angers S, Salahpour A, Bouvier M. Dimerization: an emerging concept for G protein-coupled receptor ontogeny and function. Annu Rev Pharmacol Toxicol 2002; 42: 409–35.

44. Devi LA. Heterodimerization of G-protein-coupled receptors: pharmacology, signaling and trafficking. Trends Pharmacol Sci 2001; 22: 532–7.

45. Liang Y, Fotiadis D, Filipek S et al. Organization of the G protein-coupled receptors rhodopsin and opsin in native membranes. J Biol Chem 2003; 278: 21655–62.

46. Kunishima N, Shimada Y, Tsuji Y et al. Structural basis of glutamate recognition by a dimeric metabotropic glutamate receptor. Nature 2000; 407: 971–7.

47. Tsuchiya D, Kunishima N, Kamiya N et al. Structural views of the ligand-binding cores of a metabotropic glutamate receptor complexed with an antagonist and both glutamate and Gd^{3+}. Proc Natl Acad Sci USA 2002; 99: 2660–5.

48. Kenakin T. Efficacy at G-protein-coupled receptors. Nat Rev Drug Discov 2002; 1: 103–10.

49. Lefkowitz RJ, Cotecchia S, Samama P et al. Constitutive activity of receptors coupled to guanine nucleotide regulatory proteins. Trends Pharmacol Sci 1993; 14(8):303–7.

50. Leff P. The two-state model of receptor activation [see comments]. Trends Pharmacol Sci 1995; 16: 89–97.

51. Weiss JM, Morgan PH, Lutz MW et al. The cubic ternary complex receptor-occupancy model. III. Resurrecting efficacy. J Theor Biol 1996; 181: 381–97.

52. Kenakin T. Inverse, protean, and ligand-selective agonism: matters of receptor conformation. FASEB J 2001; 15: 598–611.

53. Kenakin T. Ligand-selective receptor conformations revisited: the promise and the problem. Trends Pharmacol Sci 2003; 24: 346–54.

54. Swaminath G, Xiang Y, Lee TW et al. Sequential binding of agonists to the beta2 adrenoceptor: kinetic evidence for intermediate conformational states. J Biol Chem 2004; 279: 686–691.

55. Ghanouni P, Gryczynski Z, Steenhuis JJ et al. Functionally different agonists induce distinct conformations in the G protein coupling domain of the beta 2 adrenergic receptor. J Biol Chem 2001; 276: 24433–6.

56. Baker JG, Hall IP, Hill SJ. Agonist actions of 'beta-

blockers' provide evidence for two agonist activation sites or conformations of the human beta1-adrenoceptor. Mol Pharmacol 2003; 63: 1312–21.

57. Kudo M, Osuga Y, Kobilka BK et al. Transmembrane regions V and VI of the human luteinizing hormone receptor are required for constitutive activation by a mutation in the third intracellular loop. J Biol Chem 1996; 271: 22470–8.

58. Sharma S, Sharma SC. An update on eicosanoids and inhibitors of cyclooxygenase enzyme systems. Indian J Exp Biol 1997; 35: 1025–31.

59. Nie J, Lewis DL. Structural domains of the CB1 cannabinoid receptor that contribute to constitutive activity and G-protein sequestration. J Neurosci 2001; 21: 8758–64.

60. Gether U, Ballesteros JA, Seifert R et al. Structural instability of a constitutively active G protein-coupled receptor. Agonist-independent activation due to conformational flexibility. J Biol Chem 1997; 272: 2587–90.

61. Kobilka BK, Kobilka TS, Daniel K et al. Chimeric alpha 2-,beta 2-adrenergic receptors: delineation of domains involved in effector coupling and ligand binding specificity. Science 1988; 240: 1310–6.

62. Schoneberg T, Liu J, Wess J. Plasma membrane localization and functional rescue of truncated forms of a G protein-coupled receptor. J Biol Chem 1995; 270: 18000–6.

63. Wilson AL, Guyer CA, Cragoe EJ et al. The hydrophobic tryptic core of the porcine alpha 2-adrenergic receptor retains allosteric modulation of binding by Na$^+$, H$^+$, and 5-amino-substituted amiloride analogs. J Biol Chem 1990; 265: 17318–22.

64. Parnot C, Miserey-Lenkei S, Bardin S et al. Lessons from constitutively active mutants of G protein-coupled receptors. Trends Endocrinol Metab 2002; 13: 336–43.

65. Kim JM, Altenbach C, Thurmond RL et al. Structure and function in rhodopsin: rhodopsin mutants with a neutral amino acid at E134 have a partially activated conformation in the dark state. Proc Natl Acad Sci USA 1997; 94: 14273–8.

66. Arnis S, Fahmy K, Hofmann KP et al. A conserved carboxylic acid group mediates light-dependent proton uptake and signaling by rhodopsin. J Biol Chem 1994; 269: 23879–81.

67. Alewijnse AE, Timmerman H, Jacobs EH et al. The effect of mutations in the DRY motif on the constitutive activity and structural instability of the histamine H(2) receptor. Mol Pharmacol 2000; 57: 890–8.

68. Scheer A, Fanelli F, Costa T et al. Constitutively active mutants of the alpha 1B-adrenergic receptor:

role of highly conserved polar amino acids in receptor activation. EMBO J 1996; 15: 3566–78.

69. Scheer A, Fanelli F, Costa T et al. The activation process of the alpha1B-adrenergic receptor: potential role of protonation and hydrophobicity of a highly conserved aspartate. Proc Natl Acad Sci USA 1997; 94: 808–13.

70. Gaborik Z, Jagadeesh G, Zhang M et al. The role of a conserved region of the second intracellular loop in AT1 angiotensin receptor activation and signaling. Endocrinology 2003; 144: 2220–8.

71. Ghanouni P, Schambye H, Seifert R et al. The effect of pH on beta(2) adrenoceptor function. Evidence for protonation-dependent activation. J Biol Chem 2000; 275: 3121–7.

72. Samama P, Cotecchia S, Costa T et al. A mutation-induced activated state of the beta 2-adrenergic receptor. Extending the ternary complex model. J Biol Chem 1993; 268: 4625–36.

73. McLean AJ, Zeng FY, Behan D et al. Generation and analysis of constitutively active and physically destabilized mutants of the human beta(1)-adrenoceptor. Mol Pharmacol 2002; 62: 747–55.

74. Petaja-Repo UE, Hogue M, Bhalla S et al. Ligands act as pharmacological chaperones and increase the efficiency of delta opioid receptor maturation. EMBO J 2002; 21: 1628–37.

75. Noda K, Feng YH, Liu XP et al. The active state of the AT1 angiotensin receptor is generated by angiotensin II induction. Biochemistry 1996; 35: 16435–42.

76. Groblewski T, Maigret B, Larguier R et al. Mutation of Asn111 in the third transmembrane domain of the AT1A angiotensin II receptor induces its constitutive activation. J Biol Chem 1997; 272: 1822–6.

77. Balmforth AJ, Lee AJ, Warburton P et al. The conformational change responsible for AT1 receptor activation is dependent upon two juxtaposed asparagine residues on transmembrane helices III and VII. J Biol Chem 1997; 272: 4245–51.

78. O'Dowd BF, Hnatowich M, Regan JW et al. Site-directed mutagenesis of the cytoplasmic domains of the human β_2-adrenergic receptor. Localization of regions involved in G protein-receptor coupling. J Biol Chem 1988; 263: 15985–92.

79. Liggett SB, Caron MG, Lefkowitz RJ et al. Coupling of a mutated form of the human beta 2-adrenergic receptor to Gi and Gs. Requirement for multiple cytoplasmic domains in the coupling process. J Biol Chem 1991; 266: 4816–21.

80. Strader C, Sigal I, Dixon R. Structural basis of β-adrenergic receptor function. FASEB J 1989; 3: 1825–32.

81. Wieland K, Zuurmond HM, Krasel C et al. Involvement of Asn-293 in stereospecific agonist recognition and in activation of the beta 2-adrenergic receptor. Proc Natl Acad Sci USA 1996; 93: 9276–81.

82. Liapakis G, Ballesteros JA, Papachristou S et al. The forgotten serine. A critical role for Ser-2035.42 in ligands binding to and activation of the beta 2-adrenergic receptor. J Biol Chem 2000; 275: 37779–88.

83. Palanche T, Ilien B, Zoffmann S et al. The neurokinin A receptor activates calcium and cAMP responses through distinct conformational states. J Biol Chem 2001; 276: 34853–61.

84. Liapakis G, Chan WC, Papadokostaki M et al. Synergistic contributions of the functional groups of epinephrine to its affinity and efficacy at the beta2 adrenergic receptor. Mol Pharmacol 2004; 65: 1181–90.

85. Elling CE, Thirstrup K, Nielsen SM et al. Engineering of metal-ion sites as distance constraints in structural and functional analysis of 7TM receptors. Fold Des 1997; 2:S76–80.

86. Elling CE, Thirstrup K, Holst B et al. Conversion of agonist site to metal-ion chelator site in the beta(2)-adrenergic receptor. Proc Natl Acad Sci USA 1999; 96: 12322–7.

87. Holst B, Elling CE, Schwartz TW. Metal ion-mediated agonism and agonist enhancement in melanocortin MC1 and MC4 receptors. J Biol Chem 2002; 277: 47662–70.

88. Yi CS, Song YS, Ryu KS et al. Common and differential mechanisms of gonadotropin receptors. Cell Mol Life Sci 2002; 59: 932–40.

89. Nishi S, Nakabayashi K, Kobilka B et al. The ectodomain of the luteinizing hormone receptor interacts with exoloop 2 to constrain the transmembrane region: studies using chimeric human and fly receptors. J Biol Chem 2002; 277: 3958–64.

90. Tateyama M, Abe H, Nakata H et al. Ligand-induced rearrangement of the dimeric metabotropic glutamate receptor 1alpha. Nat Struct Mol Biol 2004; 11: 637–42.

CHAPTER 6

Mechanotransduction in cardiomyocyte hypertrophy

Allen M Samarel

Introduction

Mechanotransduction refers to the cellular mechanisms by which load-bearing cells sense physical forces, transduce the forces into biochemical signals, and generate appropriate responses leading to alterations in cellular structure and function. The physical forces encountered by living cells include membrane stretch, gain and loss of adhesion, and compression due to an increase in pressure. The signal transduction pathways that are activated in response to physical forces include many unique components, as well as elements shared by other, more traditional signaling pathways. Mechanotransduction in the heart affects the beat-to-beat regulation of cardiac performance, but also profoundly affects the proliferation, differentiation and survival of the cellular components that comprise the human myocardium. Understanding the molecular basis for mechanotransduction is therefore important to our overall understanding of growth regulation during cardiac hypertrophy and failure.

Muscle cells are unique in that they respond to externally applied mechanical forces, and also generate internal loads that are transmitted to adjacent cells and their surrounding extracellular matrix (ECM). Cardiomyocytes (the individual muscle cells that comprise the atrial and ventricular force-generating compartments of the mammalian heart) are particularly sensitive to externally applied and intrinsically generated mechanical load. The process of mechanotransduction is critical to many aspects of cardiomyocyte function, wherein alterations in mechanical load produce profound changes in cardiomyocyte contractility, shape, gene expression, and growth.

A relatively simple way to quantify mechanical loading of cardiomyocytes *in vivo* is to equate mechanical load to the wall stress of a sphere. Based on LaPlace's Law, the wall stress (σ) is the mechanical force per unit area, and is proportional to the product of the pressure inside the sphere (P) and its radius of curvature r, divided by the wall thickness (h):

$$6.1 \quad \sigma = (P \times r)/2h$$

Wall stress and, by inference, the mechanical load on individual cardiomyocytes, varies continuously throughout the cardiac cycle, constituting both passive filling of the chambers during atrial and ventricular diastole, and the isovolumic and ejection phases during atrial and ventricular systole. Wall stress also varies in the different chambers of the heart. In general, systolic and diastolic pressures are lower in the atria than in the ventricles, whereas systolic pressures are considerably higher on the left as compared to the right side of the heart. These regional differences in pressure (and by inference, wall stress) account in part for developmental differences in chamber dimensions, wall thickness, and protein turnover rates that occur during the transition from the fetal to the adult pattern of circulation. Alterations in wall stress are also ultimately responsible for changes in chamber geometry that accompany pathological remodeling of the adult heart in response to hemodynamic overload. Here, increased wall stress, due to pressure or volume overload, or segmental loss of functioning myocardium, is transduced into biochemical signals that increase the rate of protein synthesis, alter cell shape, and increase the transcription rate of genes normally expressed predominantly during fetal life. Although the normalization of wall stress by hypertrophic growth may prove to be dispensable in maintaining cardiac performance under some circumstances,[1] a comprehensive understanding of cardiomyocyte mechanotransduction remains critically important to our understanding of the molecular mechanisms of adaptive cardiac hypertrophy. Indeed, the regulation of cardiomyocyte growth and atrophy are primarily dependent on the interpretation of, and response to, mechanical stimuli.

Observational studies conducted during the 1970s addressing the hypertrophic growth response of human myocardium to pathological changes in systolic and diastolic wall stress fostered the development of experimental model systems in which to explore cardiomyocyte mechanotransduction *in vivo* and *in vitro*. These model systems now span the breadth of experimental cardiology, from complex large animal models of human cardiovascular disease, to studies conducted on isolated cardiomyocyte membranes subjected to mechanical deformation *in vitro*. Many studies have relied upon cultured cardiomyocytes in order to identify potential mechanosensors in cardiomyocytes, and to define their downstream effectors and molecular targets. Research has also been greatly aided by molecular techniques to over-express and inhibit specific components of the putative mechanochemical signaling pathways by gene targeting and related approaches conducted in isolated cells and transgenic animals. These experimental approaches have provided significant insights into cardiomyocyte mechanotransduction that have already proved useful in preventing or ameliorating adverse ventricular remodeling during the progression of cardiac hypertrophy and its transition to heart failure.

Experimental paradigms to explore cardiomyocyte mechanotransduction *in vitro*

The heart increases its mass as one of several ways to compensate for an increase in hemodynamic load. The increase in mass is predominantly the result of an increase in the volume of individual cardiomyocytes, although other cellular and extracellular components of ventricular myocardium may contribute. The plasticity of hypertrophying cardiomyocytes to alterations in hemodynamic load allows for long-lasting adaptation to mechanical stress as a consequence of increased pressure, volume, or segmental loss of viable myocardium. However, attempts to experimentally alter cardiomyocyte mechanotransduction *in vivo* are problematic, as cardiomyocytes not only sense mechanical load, but are also needed to generate mechanical force in order to ensure the survival of the entire organism. This paradox has led investigators to attempt to simulate the effects of *in vivo* mechanical loading of ventricular myocardium using freshly isolated cardiomyocytes, and cardiomyocytes maintained in primary culture. These derivative model systems have intrinsic problems of their own, including the presence of contaminating non-muscle cells, the absence of an authentic three-dimensional ECM for normal cell attachment, and the elimination of normal cell-to-cell communication afforded by the complex three-dimensional architecture of the myocardial tissue. Other drawbacks often include limited viability, the absence of normal excitation–contraction coupling mechanisms and neuroendocrine factors that may be critical for mechanochemical signaling, and the inability to precisely model mechanical load as it changes throughout the cardiac cycle. Nevertheless, *in vitro* models have the advantage of allowing for the study of mechanical load on cardiomyocyte signal transduction and growth in the absence of other, confounding variables.

The vast majority of *in vitro* studies of load-induced cardiomyocyte hypertrophy have relied upon primary cultures of neonatal or adult ventricular myocytes grown on an elastic substratum subjected to static or cyclic stretch. In a pioneering study, Mann and colleagues first demonstrated that isolated adult feline cardiomyocytes increased rates of protein and RNA synthesis in response to application of a 10% linear stretch.[2] These results were obtained in randomly oriented cells attached to a silastic membrane coated with laminin, but correlated well with earlier studies performed with cultured skeletal muscle myotubes,[3] isolated papillary muscles,[4] and the isolated, perfused rat heart.[5-8] Since then, investigators have devised a variety of clever experimental approaches to ascertain the relative contributions of passive stretch versus active tension development,[9-12] cell orientation,[13-15] and three-dimensionality of cultured cardiomyocytes to the process of mechanotransduction.[16,17] These in vitro systems, in combination with additional studies performed in intact animals, have helped to identify the putative mechanosensors in cardiomyocytes, and to characterize their downstream signal transduction pathways resulting in load-induced cardiomyocyte hypertrophy.

What are the mechanosensors in cardiomyocytes? (Fig. 6.1)

Cardiomyocytes rely on several intracellular components to sense mechanical load, and convert mechanical stimuli into biochemical events that affect cellular structure and function. These sensors include protein components within the myofilaments and Z-discs, stretch-activated ion channels, and other membrane-associated proteins that link the ECM to the cytoskeleton. Information regarding each component's role in modulating cardiomyocyte growth is limited. However, it is likely that one or more of the stretch sensors are primarily responsible for the induction of cardiomyocyte hypertrophy in response to increased wall stress.

Myofilament stretch receptors

Cardiac muscle displays the fundamental property of length-dependent activation, in which myofilament force generation at any given concentration of activator calcium ion concentration increases with increasing sarcomere length. This property, in part responsible for the Frank–Starling law of the heart, requires that one or more myofilament proteins assume the role of the length sensor. Recent studies have suggested that the giant cytoskeletal protein titin is the responsible element. This 700 kDa protein spans the entire distance from the Z-disc to the M-line, and is well-accepted as the major cytoskeletal protein responsible for generating intrinsic passive stiffness of the sarcomere during linear stretch. More importantly, selective removal of titin from myofilaments reduced length-dependent activation, suggesting that titin, or other

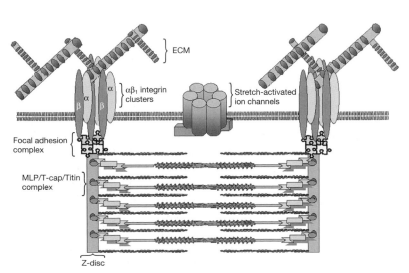

Figure 6.1. *Identity and schematic localization of putative mechanosensors in the cardiomyocyte.*

titin-interacting proteins are responsible.[18] Although titin may control lattice spacing, and actin–myosin interactions that result in length-dependent changes in force generation,[19] there is no specific information currently available regarding a direct role for titin in mechanotransduction leading to cardiac hypertrophy. Nevertheless, titin's interaction with multiple cytoskeletal proteins within the Z-disc (see below), and the recent discovery of alterations in titin isoforms during cardiac development,[20] suggest that it may function as an important stress-transducer during both short- and long-term adaptation to mechanical overload.

Stretch sensors within the Z-disc

Knoll and colleagues recently proposed that defects in the muscle LIM protein (MLP)/T-cap/titin complex lead to defects in stretch receptor function resulting in dilated cardiomyopathy.[21] MLP is a member of a large family of proteins that contains one or more double zinc finger structures (LIM domains) mediating specific contacts between proteins that participate in the formation of multiprotein complexes.[22] Mice deficient in MLP developed a progressive cardiomyopathy characterized by left ventricular chamber dilatation, decreased ejection fraction, and premature death from congestive heart failure.[23] Isolated papillary muscles from juvenile MLP–/– mice demonstrated increased passive stretch properties, suggesting an intrinsic abnormality in titin function. Importantly, neonatal cardiomyocytes isolated from MLP–/– mice failed to increase BNP and ANF mRNA levels in response to 10% biaxial passive stretch, but responded normally to $G_{\alpha q}$-coupled receptor stimulation with phenylephrine, indicating a specific defect in mechanotransduction. MLP interacts with multiple proteins within the Z-disc, including α-actinin and the titin-binding protein telethonin (T-cap). Absence of MLP caused the displacement of T-cap in a small minority of cardiomyocytes, suggesting that the loss of MLP led to a destabilization of the anchoring of the Z disc to the proximal end of the T-cap/titin complex, thereby affecting titin's role in mechanical stretch receptor function. Exactly how the loss of MLP interferes with downstream signals for cardiomyocyte survival and hypertrophy remains

unknown. However, other LIM proteins are known to directly interact with focal adhesion proteins, including integrins, to regulate cell shape, spreading, and elongation.[22,24,25] MLP itself has been proposed to translocate to the nucleus and to serve as a transcriptional co-activator in muscle cells.[21]

Stretch-activated ion channels

The presence of stretch-activated ion channels (SACs) in striated muscle was initially described by Guharay and Sachs in 1984.[26] Since then, SACs have been described in a variety of different cell types, including neonatal and adult cardiomyocytes isolated from several species. SACs are ion channels that are directly gated by mechanical deformation of the plasma membrane. A variety of approaches have been used to demonstrate SACs in cardiomyocyte membranes, including direct stretch of the cell, hypotonic swelling, and application of positive pressure to the voltage-clamping pipette. Mechanical deformation by these various techniques has revealed the presence of multiple types of SACs with differing ion selectivity, and differing sensitivity to specific blocking agents. Cardiomyocyte SACs appear to be K^+ selective, cation selective, or chloride selective, and, in the case of the cation SACs, sensitive to blockade by Gd^{3+} or streptomycin. Indeed, Gd^{3+} is often used to test whether a particular response to mechanical loading is due to the activation of SACs, although its specificity has recently been questioned.[27] For example, Gd^{3+}, but not diltiazem or NiCl, inhibited stretch-induced atrial natriuretic factor (ANF) secretion by isolated rat atria,[28] suggesting an important role of SACs, rather than voltage-gated ion channels, in the regulation of stretch-induced ANF release. SACs may also be important in augmentation of contractility produced by increased coronary perfusion pressure, and in stretch-induced arrythmogenesis. However, their role in mechanotransduction leading to cardiomyocyte hypertrophy remains unclear. For instance, Sadoshima et al. showed that Gd^{3+} failed to inhibit stretch-induced c-fos expression, or stretch-induced increases in protein synthesis in neonatal cardiomyocytes.[29] Similarly, Yamazaki et al. showed that neither Gd^{3+} nor streptomycin prevented stretch-induced mitogen-activated protein kinase (MAPK) activation in neonatal rat ventricular myocytes (NRVM).[30] In contrast, an inhibitor of the Na^+–H^+ exchanger partially attenuated stretch-induced cellular growth, indicating a potential role for this exchanger, rather than SACs, in hypertrophic mechanotransduction. SACs may contribute to stretch-induced increases in intracellular Ca^{2+} ($[Ca^{2+}]_I$), although L-type Ca^{2+} channels, and Ca^{2+}-induced Ca^{2+} release (CICR) are also involved.[31] Indeed, $[Ca^{2+}]_I$ transients are critical for many aspects of the hypertrophic phenotype induced by growth factors and mechanical stretch in cultured cardiomyocytes.[32–35] SAC-dependent increases in $[Na^+]_I$ could stimulate local increases in $[Ca^{2+}]_I$ via Na^+–Ca^{2+} exchange, and thereby produce sufficient activator Ca^{2+} to stimulate CICR and, downstream, Ca^{2+}-dependent signaling pathways leading to hypertrophic growth.

Stretch receptors within the ECM–integrin–cytoskeletal complex

Perhaps the most convincing evidence for the presence of a specific mechanosensor in cardiomyocytes comes from studies of the ECM–integrin–cytoskeletal complex. Cardiomyocytes attach to ECM proteins via cell surface receptors known as integrins. Integrins are heterodimeric integral membrane proteins consisting of single α and β chains. At least 18 α and 8 β subunits have been identified, with more than 24 paired integrin receptors expressed.[36] Additional complexity arises from the existence of multiple isoforms of individual α and β chains generated by alternative splicing of α and β integrin hnRNA transcripts. Cardiomyocytes express α_1, α_3, α_5, α_6, α_7, α_9, and α_{10}, whereas the predominant β subunit expressed is the ubiquitous β_{1A} subunit, and the striated muscle-specific β_{1D} isoform.[37] The relative expression of the various α chains varies throughout development, accounting in large part for the differences in adhesive properties of immature versus adult cardiomyocytes. For instance, NRVM express integrins that predominantly bind to fibronectin and type I collagen, whereas adult rat ventricular cells express predominantly laminin receptors. Integrin receptor subtypes also change in response to hemodynamic overload, suggesting an important regulatory role of integrins in mechanotransduction.[38,39]

Cardiomyocyte integrins are not randomly distributed on the cell surface, but rather are found within the sarcolemmal membrane directly adjacent to costameres. Costameres are band-like structures that link the Z-discs to the sarcolemmal membrane, and are therefore considered important sites for bidirectional communication of mechanical forces.[40] In addition to integrins, costameres contain many of the same cytoskeletal proteins that are found within focal adhesions of adherent non-muscle cells. Indeed, cardiomyocytes placed into culture on a two-dimensional surface develop typical focal adhesions over time, containing vinculin, α-actinin, paxillin, and other focal adhesion proteins. The physical interaction between integrin cytoplasmic domains and adaptor proteins within the cytoskeleton generates a submembrane adhesion plaque that appears critical for transmitting mechanical force between the ECM and the actin cytoskeleton. Furthermore, the cytoplasmic domains of β-integrin subunits play a direct role in these connections, as at least four focal adhesion proteins (talin, α actinin, filamin and tensin) can link integrins directly to actin filaments.[41] With the inclusion of additional protein–protein interactions, it is apparent that focal adhesions, and by inference costameres, are sites of close interaction between the ECM and the intracellular environment.

In addition to their structural role, costameres (and their focal adhesion counterparts in cultured cardiomyocytes) are also sites for the localization of signaling molecules that are important in cardiomyocyte survival and growth. These include protein tyrosine kinases such as FAK, PYK2, Src and Csk, and serine-threonine protein kinases such as ILK, PKCε, and PAK. Other adapter proteins, such as Crk, DOCK180, and Cas, can link focal adhesion proteins to other downstream signaling cascades that may be important in both mechanotransduction, and growth factor signaling. In general, there appears to be substantial crosstalk between integrin and growth factor receptors in many cell types, suggesting that integrin- and growth factor-mediated cellular responses are locally coordinated within focal adhesions.[42]

Whereas adhesion of cardiomyocytes has long been considered an important factor in regulating their growth and survival, the responsible molecular mechanisms are only now being identified. Many cell types, including cardiomyocytes, normally require integrin-mediated adhesion in order to stay alive, and the downstream signals resulting from integrin engagement and clustering within focal adhesion complexes cooperate with those from growth factor receptors to prevent apoptosis.[43] Apoptosis mediated by the loss of cell attachment to the ECM is referred to as 'anoikis' from the Greek word for 'homelessness'. This form of apoptosis contributes to normal tissue homeostasis by ensuring that cells remain in their proper tissue environment. Although the significance of anoikis to human cardiac development and disease is presently unknown, Ding et al. have recently suggested that it may be a contributing factor to disease progression in some forms of cardiac hypertrophy and heart failure.[44]

Attachment is clearly one way in which costameres and focal adhesions contribute to mechanotransduction, but there is also substantial evidence to indicate that mechanical forces (generated by passive stretch and active tension development of cultured cardiomyocytes) are 'sensed' by focal adhesion complexes, and are transduced into biochemical signals leading to sarcomeric assembly and altered gene expression characteristic of *in vivo* cardiomyocyte hypertrophy. Stretch-induced deformation of cardiomyocyte integrins triggers the recruitment and activation of several signaling kinases, including FAK, Src, ROCK, and MAPKs, to the cytoplasmic face of the focal adhesion complex, where they participate in downstream signaling to the nucleus and other organelles.[45] Results obtained in mechanically stressed, cultured cells complement elegant studies performed in pressure-overloaded, intact myocardium,[46–50] and also support the close interaction between the ECM–integrin–cytoskeletal complex and growth factor receptor signaling during cardiomyocyte hypertrophy.[51–57]

Exactly how the ECM–integrin–cytoskeletal complex senses mechanical stimuli remains somewhat of a mystery. Seminal observations by Ingber and colleagues,[58] using a magnetic twisting device to transfer force directly from integrins to the local cytoskeleton, suggest that mechanical deformation of one or more adhesion plaque proteins is the proximal step in an intracellular signaling cascade that leads to global cytoskeletal rearrangements and mechanotransduction at multiple, distant sites within the cell. FAK is one candidate protein that may be responsible for integrin-mediated mechanotransduction within focal adhesions. FAK is a non-receptor protein tyrosine kinase that associates with the cytoplasmic tail of β-integrins via an autoinhibitory, FERM domain located in its N-terminus. The C-terminal region of FAK (the so-called focal adhesion targeting sequence, or FAT) also binds directly to paxillin and talin, two cytoskeletal proteins that localize to sites of integrin clustering. Once localized, FAK phosphorylates itself at a single tyrosine residue (Y_{397}) during integrin engagement and clustering. This autophosphorylation site serves as a high-affinity binding site (pYAEI motif) for the SH2 domain of Src-family protein tyrosine kinases. Once bound to FAK, Src can then phosphorylate FAK at residues Y_{576} and Y_{577} within the catalytic domain (which augments FAK kinase activity toward exogenous substrates), and at Y_{861} and Y_{925} near its C-terminus. The Y_{925} phosphorylation site promotes the binding to FAK of Grb2, and other adapter proteins and kinases containing SH2 domains. The FAK–Src complex also directly tyrosine-phosphorylates paxillin, and other cytoskeletal proteins involved in focal adhesion and stress fiber formation. FAK, and a highly homologous protein tyrosine kinase known as PYK2, are both expressed in neonatal and adult cardiomyocytes, where they are activated in response to mechanical loading,[45,59,60] and agonists (e.g. phenylephrine, angiotensin II, endothelin) that stimulate Gq-coupled receptors.[51–57,61] Thus, focal adhesion kinases (and other protein kinases bound to FAK during integrin clustering) can activate downstream signaling pathways that regulate integrin-mediated mechanotransduction at local as well as distant sites within the cell.

A stretch sensor within caveolae?

In another recent report, Kawamura et al. proposed that mechanical force might be transduced into cellular signaling pathways in cardiomyocytes via specific proteins within caveolae.[62] Caveolae are cholesterol- and sphingolipid-enriched sarcolemmal invaginations that contain a variety of important signal transduction molecules, including receptors, heterotrimeric G proteins, and the small GTPases Rac1 and RhoA. Biaxial stretch of NRVM caused the time-dependent activation of Rac1 and RhoA. Stretch also caused the redistribution of both GTPases from lighter cell fractions enriched in caveolin-3 into heavier cell fractions enriched in focal adhesion proteins. In contrast, stretch had no effect on the distribution of caveolin-3, suggesting that mechanical loading caused the release of the GTPases from caveolar microdomains.[62] As stretch may flatten caveolar invaginations and expose the GTPases to guanine nucleotide exchange factors, the authors propose that mechanical deformation might directly result in activation of downstream signaling that requires RhoA and Rac1 activation. In support of this argument, methyl β-cyclodextrin treatment, which disrupted caveolar microdomains, prevented stretch-induced RhoA and Rac1 activation, and also inhibited stretch-induced increases in cell size, actin myofilament organization, and ANF expression. Interestingly, other stretch-responsive events, including activation of ERKs, were unaffected, indicating that other stretch sensors must remain operative in caveolae-depleted cardiomyocytes.[62]

Biochemical signaling pathways activated in response to mechanical load (Fig. 6.2)

Mechanical loading activates a number of downstream signaling pathways in cardiomyocytes, and nearly all these pathways have been implicated in hypertrophic growth in response to other, exogenous stimuli. However, the nature of the mechanical stimulus (i.e. adhesion, compression, static/cyclic stretch, stimulated contraction) may have some influence on which pathways are activated, and what phenotype is ultimately observed. Simple adhesion of neonatal or adult cardiomyocytes to appropriate ECM proteins is critical for cell survival, and for hypertrophic growth in response to

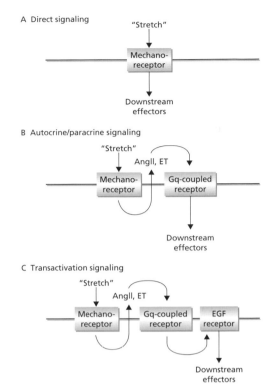

Figure 6.2. *Potential mechanisms for converting mechanosensory signals into biochemical signals resulting in cardiomyocyte hypertrophy. (A) Mechanical signals are directly converted by the mechanosensors into biochemical signals that induce cardiomyocyte hypertrophy. (B) Mechanical signals are 'sensed' by mechanosensors in the cell membrane or other locations in the cell, that then stimulate the release of growth factors which are responsible for the induction of cardiomyocyte hypertrophy. (C) Mechanical signals are 'sensed' by mechanosensors, that directly or indirectly transactivate adjacent growth factor receptors that induce cardiomyocyte hypertrophy.*

soluble growth factors. Adhesion-dependent cell signaling has not been as extensively studied in cardiomyocytes as in other cell types. However, there is good evidence to indicate that integrin-mediated cardiomyocyte attachment directly stimulates growth, and is also necessary for the hypertrophic response to neurohormonal agonists such as phenylephrine.[51] FAK appears to play a central role, as it is strongly activated by adhesion of cardiomyocytes to purified ECM components (M Cvejic, MC Heidkamp, AM Samarel, unpublished data), as well as neurohormonal agonists that induce hypertrophy of cultured neonatal cardiomyocytes.[52–55] Similarly, integrin and FAK levels (probably reflecting the overall number of focal adhesions) increase substantially in response to intrinsic mechanical loading due to contractile activity of adherent, high-density NRVM.[60,63] Static stretch of NRVM grown on silastic membranes also rapidly activates FAK, and causes the assembly of a multicomponent signaling complex at the site of integrin clustering.[45]

Another immediate effect of stretch on cardiomyocyte mechanotransduction is the release into the extracellular space of peptide growth factors. Sadoshima et al. first demonstrated that uniaxial, static stretch of NRVM caused the release into the culture medium of preformed angiotensin II, and the released peptide growth factor was necessary and sufficient for the induction of stretch-induced cardiomyocyte hypertrophy.[64] Since then, other growth factors, including endothelin-1, VEGF, TGF-β, and cardiotrophin-1 have also been shown to be released into the medium of cultured cardiomyocytes in response to mechanical loading. Stretch-induced release of angiotensin II and endothelin-1 also occurs in intact, adult feline papillary muscle, leading to the rapid activation of Na^+–H^+ exchange via a protein kinase C-dependent signaling pathway.[65] Indeed, there are numerous downstream effectors activated during autocrine–paracrine release of peptide growth factors, which probably amplify the initial growth stimulus triggered by mechanical stretch.[40] Autocrine–paracrine activation of downstream signaling from one or more of these growth factor receptors may be required for the full development of hypertrophy during mechanical overload.

A third potential consequence of mechanical overload is the transactivation of growth factor-receptor tyrosine kinases.[66] Transactivation may occur by autocrine–paracrine release of angiotensin II or endothelin-1, activation of their cognate receptors, which is then followed by intracellular tyrosine phosphorylation of the epidermal growth factor (EGF) receptor.[67] Alternatively, receptor tyrosine kinases

may undergo activation more directly in response to integrin engagement and clustering, wherein integrin-dependent FAK, PYK2, and/or Src activation play an intermediary role. A third possibility involves the autocrine/paracrine release by cardiomyocytes or cardiac fibroblasts of matrix metalloproteinases in response to mechanical overload. The metalloproteinases, in turn, release heparin-binding growth factors such as HB-EGF from the ECM, and thereby stimulate downstream signaling.[68]

Consequences of mechanotransduction in cardiomyotyes (Fig. 6.3)

As inferred by the aforementioned complex, interacting signaling web that connects mechanosensory and growth factor-dependent signal transduction pathways, it is not surprising that investigators have identified many different downstream effectors that are activated in response to mechanical loading. These include the well-described MAPK signaling cascades (i.e. ERKs, JNKs, and p38[MAPK]), cell survival pathways that involve phosphoinositol-3-kinase and Akt, the JAK–STAT signaling pathways, and Ca^{2+}-dependent pathways involved in the activation of calcineurin and CAMKs. Their regulation by mechanical load is similar to their activation by neurohormonal agonists, and has been extensively reviewed in the recent literature.[40,69–71] Thus, mechanosensory pathways that lead to cardiomyocyte hypertrophy share critical com-

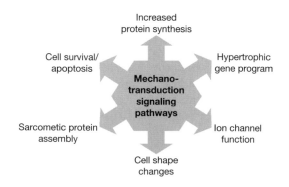

Figure 6.3. *Consequences of mechanotransduction during cardiomyocyte hypertrophy.*

ponents with signaling pathways activated in response to norepinephrine, endothelin-1, angiotensin II, and other peptide growth factors and cytokines. Major responses related to the growth of cardiomyocytes include an increase in the overall rate of cardiomyocyte protein synthesis (including the contractile proteins myosin and actin), enhanced sarcomeric protein assembly, and a generalized decrease in the susceptibility of sarcomeric proteins to intracellular proteolysis. In response to certain pathological forms of hemodynamic overload, mechanosensory pathways also increase the transcription rates of cardiomyocyte-specific genes that are normally expressed predominantly during fetal life (i.e. activation of the 'hypertrophic' or 'fetal' gene program). Re-activation of the fetal gene program improves energy utilization in the face of reduced myocardial flow reserve, but also alters the contractile properties of the adult myocardium, resulting in reduced force generation, abnormal calcium handling, and slowed myocardial relaxation. Mechanosensory pathways activated in response to hemodynamic overload also affect normal cell-to-cell communication and cell shape, and produce local alterations in the actin cytoskeleton that affect ion channel gating and responsiveness to β-adrenergic receptor signaling. Finally, there is some evidence to suggest that mechanical overload of cardiomyocytes can, directly or indirectly, lead to the induction of apoptosis.

Summary and future directions

The cardiomyocyte was once naively viewed as a 'black box', in which mechanical factors mysteriously regulated the size of the box via alterations in the rate of synthesis and degradation of its relatively homogenous protein components. It should now be apparent from this brief review that a great deal of new information has been acquired over the past 30 years regarding mechanotransduction in cardiomyocyte hypertrophy. We now know that mechanical factors influence virtually every step in the process of myocardial gene expression, from the transcription of genes encoding specific components of the myofibril, to their ultimate disassembly and breakdown. Many of the signaling pathways that convert mechanical stimuli into biochemical pathways that modulate cardiomyocyte growth have been identified. Ongoing studies using combined molecular, genetic and pharmacological approaches should further characterize the responsible mechanosensors and signaling effectors. However, specific questions regarding the complex nature of the mechanical signals, and the manner in which these signals are interpreted remain unanswered. For instance, what are the cellular mechanisms responsible for load-dependent, physiological versus pathological hypertrophy? Are there intrinsic differences in the mechanical signals *per se* that lead to cardiomyocyte hypertrophy with normal or enhanced function, versus pathological hypertrophy with its aberrant contraction and relaxation? Furthermore, are there specific mechanosenors that are responsible for concentric versus eccentric cardiomyocyte hypertrophy? Can mechanosensory pathways be interrupted or inhibited in order to reduce left ventricular hypertrophy and its transition to heart failure? The answers to these and other questions await the development of better model systems with which to study cardiomyocyte mechanotransduction in isolation. Once available, the results from these derivative systems can be applied to complex *in vivo* models of human cardiomyocyte hypertrophy and heart failure, with the ultimate goal of reducing the incidence and prevalence of cardiac hypertrophy and heart failure.

Acknowledgements
The author is supported by NIH RO1 grants HL34328, HL63711, and HL64956, and a grant to the Cardiovascular Institute from the Ralph and Marian Falk Trust for Medical Research.

References

1. Esposito G, Rapacciuolo A, Naga Prasad SV et al. Genetic alterations that inhibit in vivo pressure-overload hypertrophy prevent cardiac dysfunction despite increased wall stress. Circulation 2002; 105: 85–92.

2. Mann DL, Kent RL, Cooper G, 4th. Load regulation of the properties of adult feline cardiocytes: growth induction by cellular deformation. Circ Res 1989; 64: 1079–90.

3. Vandenburgh H, Kaufman S. In vitro model for

stretch-induced hypertrophy of skeletal muscle. Science 1979; 203: 265–8.

4. Peterson MB, Lesch M. Protein synthesis and amino acid transport in the isolated rabbit right ventricular papillary muscle. Effect of isometric tension development. Circ Res 1972; 31: 317–27.

5. Schreiber SS, Morkin E. Protein synthesis in cardiac hypertrophy. Circ Res 1972; 31: 629–30.

6. Takala T. Protein synthesis in the isolated perfused rat heart. Effects of mechanical work load, diastolic ventricular pressure and coronary pressure on amino acid incorporation and its transmural distribution into left ventricular protein. Basic Res Cardiol 1981; 76: 44–61.

7. Smith DM, Sugden PH. Stimulation of left-atrial protein-synthesis rates by increased left-atrial filling pressures in the perfused working rat heart in vitro. Biochem J 1983; 216: 537–42.

8. Kira Y, Kochel PJ, Gordon EE et al. Aortic perfusion pressure as a determinant of cardiac protein synthesis. Am J Physiol Cell Physiol 1984; 246: C247–58.

9. Ivester CT, Kent RL, Tagawa H et al. Electrically stimulated contraction accelerates protein synthesis rates in adult feline cardiocytes. Am J Physiol Heart Circ Physiol 1993; 265: H666–74.

10. Samarel AM, Engelmann GL. Contractile activity modulates myosin heavy chain-β expression in neonatal rat heart cells. Am J Physiol Heart Circ Physiol 1991; 261: H1067–77.

11. Sharp WW, Terracio L, Borg TK et al. Contractile activity modulates actin synthesis and turnover in cultured neonatal rat heart cells. Circ Res 1993; 73: 172–83.

12. Yamamoto K, Dang QN, Maeda Y et al. Regulation of cardiomyocyte mechanotransduction by the cardiac cycle. Circulation 2001; 103: 1459–64.

13. Simpson DG, Terracio L, Terracio M et al. Modulation of cardiac myocyte phenotype in vitro by the composition and orientation of the extracellular matrix. J Cell Physiol 1994; 161: 89–105.

14. Simpson DG, Majeski M, Borg TK et al. Regulation of cardiac myocyte protein turnover and myofibrillar structure in vitro by specific directions of stretch. Circ Res 1999; 85: e59–69.

15. Gopalan SM, Flaim C, Bhatia SN et al. Anisotropic stretch-induced hypertrophy in neonatal ventricular myocytes micropatterned on deformable elastomers. Biotechnol Bioeng 2003; 81: 578–87.

16. Zile MR, Cowles MK, Buckley JM et al. Gel stretch method: a new method to measure constitutive properties of cardiac muscle cells. Am J Physiol Heart Circ Physiol 1998; 274: H2188–202.

17. Mansour H, de Tombe P, Samarel AM et al. Restoration of resting sarcomere length after uniaxial static strain is regulated by protein kinase C epsilon and focal adhesion kinase. Circ Res 2004; 94: 642–9.

18. Cazorla O, Wu Y, Irving TC et al. Titin-based modulation of calcium sensitivity of active tension in mouse skinned cardiac myocytes. Circ Res 2001; 88: 1028–35.

19. Konhilas JP, Irving TC, de Tombe PP. Frank–Starling law of the heart and the cellular mechanisms of length-dependent activation. Pflugers Arch 2002; 445: 305–10.

20. Lahmers S, Wu Y, Call DR et al. Developmental control of titin isoform expression and passive stiffness in fetal and neonatal myocardium. Circ Res 2004; 94: 505–13.

21. Knoll R, Hoshijima M, Hoffman HM et al. The cardiac mechanical stretch sensor machinery involves a Z disc complex that is defective in a subset of human dilated cardiomyopathy. Cell 2002; 111: 943–55.

22. Wixler V, Geerts D, Laplantine E et al. The LIM-only protein DRAL/FHL2 binds to the cytoplasmic domain of several alpha and beta integrin chains and is recruited to adhesion complexes. J Biol Chem 2000; 275: 33669–78.

23. Hongo M, Ryoke T, Schoenfeld J et al. Effects of growth hormone on cardiac dysfunction and gene expression in genetic murine dilated cardiomyopathy. Basic Res Cardiol 2000; 95: 431–41.

24. McGrath MJ, Mitchell CA, Coghill ID et al. Skeletal muscle LIM protein 1 (SLIM1/FHL1) induces $\alpha_5\beta_1$-integrin-dependent myocyte elongation. Am J Physiol Cell Physiol 2003; 285: C1513–26.

25. Robinson PA, Brown S, McGrath MJ et al. Skeletal muscle LIM protein 1 regulates integrin-mediated myoblast adhesion, spreading, and migration. Am J Physiol Cell Physiol 2003; 284: C681–95.

26. Guharay F, Sachs F. Stretch-activated single ion channel currents in tissue-cultured embryonic chick skeletal muscle. J Physiol 1984; 352: 685–701.

27. Lacampagne A, Gannier F, Argibay J et al. The stretch-activated ion channel blocker gadolinium also blocks L-type calcium channels in isolated ventricular myocytes of the guinea-pig. Biochim Biophys Acta 1994; 1191: 205–8.

28. Laine M, Arjamaa O, Vuolteenaho O et al. Block of stretch-activated atrial natriuretic peptide secretion by gadolinium in isolated rat atrium. J Physiol 1994; 480: 553–61.

29. Sadoshima J, Takahashi T, Jahn L et al. Roles of mechano-sensitive ion channels, cytoskeleton, and contractile activity in stretch-induced immediate-early

gene expression and hypertrophy of cardiac myocytes. Proc Natl Acad Sci USA 1992; 89: 9905–9.

30. Yamazaki T, Komuro I, Kudoh S et al. Role of ion channels and exchangers in mechanical stretch-induced cardiomyocyte hypertrophy. Circ Res 1998; 82: 430–7.

31. Ruwhof C, van Wamel JT, Noordzij LA et al. Mechanical stress stimulates phospholipase C activity and intracellular calcium ion levels in neonatal rat cardiomyocytes. Cell Calcium 2001; 29: 73–83.

32. Sadoshima J, Qiu Z, Morgan JP et al. Angiotensin II and other hypertrophic stimuli mediated by G protein-coupled receptors activate tyrosine kinase, mitogen-activated protein kinase, and 90-kD S6 kinase in cardiac myocytes. The critical role of Ca^{2+}-dependent signaling. Circ Research 1995; 76: 1–15.

33. Eble DM, Qi M, Waldschmidt S et al. Contractile activity is required for sarcomeric assembly in phenylephrine-induced cardiac myocyte hypertrophy. Am J Physiol Cell Physiol 1998; 274: C1226–37.

34. Molkentin JD, Lu JR, Antos CL et al. A calcineurin-dependent transcriptional pathway for cardiac hypertrophy. Cell 1998; 93: 215–28.

35. Cadre BM, Qi M, Eble DM et al. Cyclic stretch down-regulates calcium transporter gene expression in neonatal rat ventricular myocytes. J Mol Cell Cardiol 1998; 30: 2247–59.

36. Ross RS, Borg TK. Integrins and the myocardium. Circ Res 2001; 88: 1112–19.

37. Shai SY, Harpf AE, Babbitt CJ et al. Cardiac myocyte-specific excision of the β_1 integrin gene results in myocardial fibrosis and cardiac failure. Circ Res 2002; 90: 458–64.

38. Terracio L, Rubin K, Gullberg D et al. Expression of collagen binding integrins during cardiac development and hypertrophy. Circ Res 1991; 68: 734–44.

39. Babbitt CJ, Shai SY, Harpf AE et al. Modulation of integrins and integrin signaling molecules in the pressure-loaded murine ventricle. Histochem Cell Biol 2002; 118: 431–9.

40. Sussman MA, McCulloch A, Borg TK. Dance band on the Titanic: biomechanical signaling in cardiac hypertrophy. Circ Res 2002; 91: 888–98.

41. Bershadsky AD, Balaban NQ, Geiger B. Adhesion-dependent cell mechanosensitivity. Annu Rev Cell Dev Biol 2003; 19: 677–95.

42. Eliceiri BP. Integrin and growth factor receptor crosstalk. Circ Res 2001; 89: 1104–10.

43. Wang P, Valentijn AJ, Gilmore AP et al. Early events in the anoikis program occur in the absence of caspase activation. J Biol Chem 2003; 278: 19917–25.

44. Ding B, Price RL, Goldsmith EC et al. Left ventricular hypertrophy in ascending aortic stenosis mice: anoikis and the progression to early failure. Circulation 2000; 101: 2854–62.

45. Torsoni AS, Constancio SS, Nadruz W, Jr. et al. Focal adhesion kinase is activated and mediates the early hypertrophic response to stretch in cardiac myocytes. Circ Res 2003; 93: 140–7.

46. Kuppuswamy D, Kerr C, Narishige T et al. Association of tyrosine-phosphorylated c-Src with the cytoskeleton of hypertrophying myocardium. J Biol Chem 1997; 272: 4500–8.

47. Laser M, Willey CD, Jiang W et al. Integrin activation and focal complex formation in cardiac hypertrophy. J Biol Chem 2000; 275: 35624–30.

48. Domingos PP, Fonseca PM, Nadruz W, Jr. et al. Load-induced focal adhesion kinase activation in the myocardium: role of stretch and contractile activity. Am J Physiol Heart Circ Physiol 2002; 282: H556–64.

49. Bayer AL, Heidkamp MC, Patel N et al. PYK2 expression and phosphorylation increases in pressure over-load-induced left ventricular hypertrophy. Am J Physiol Heart Circ Physiol 2002; 283: H695–706.

50. Torsoni AS, Fonseca PM, Crosara-Alberto DP et al. Early activation of p160^ROCK by pressure overload in rat heart. Am J Physiol Cell Physiol 2003; 284: C1411–19.

51. Ross RS, Pham C, Shai SY et al. β_1 integrins participate in the hypertrophic response of rat ventricular myocytes. Circ Res 1998; 82: 1160–72.

52. Eble DM, Strait JB, Govindarajan G et al. Endothelin-induced cardiac myocyte hypertrophy: role for focal adhesion kinase. Am J Physiol Heart Circ Physiol 2000; 278: H1695–707.

53. Taylor JM, Rovin JD, Parsons JT. A role for focal adhesion kinase in phenylephrine-induced hypertrophy of rat ventricular cardiomyocytes. J Biol Chem 2000; 275: 19250–7.

54. Pham CG, Harpf AE, Keller RS et al. Striated muscle-specific β_{1D} integrin and FAK are involved in cardiac myocyte hypertrophic response pathway. Am J Physiol Heart Circ Physiol 2000; 279: H2916–26.

55. Kovacic-Milivojevic B, Roediger F, Almeida EA et al. Focal adhesion kinase and p130^Cas mediate both sarcomeric organization and activation of genes associated with cardiac myocyte hypertrophy. Mol Biol Cell 2001; 12: 2290–307.

56. Heidkamp MC, Bayer AL, Kalina JA et al. GFP-FRNK disrupts focal adhesions and induces anoikis in neonatal rat ventricular myocytes. Circ Res 2002; 90: 1282–9.

57. Heidkamp MC, Bayer AL, Scully BT et al. Activation of focal adhesion kinase by protein kinase Cϵ in

neonatal rat ventricular myocytes. Am J Physiol Heart Circ Physiol 2003; 285: H1684–96.

58. Wang N, Butler JP, Ingber DE. Mechanotransduction across the cell surface and through the cytoskeleton. Science 1993; 260: 1124–7.

59. Seko Y, Takahashi N, Tobe K et al. Pulsatile stretch activates mitogen-activated protein kinase (MAPK) family members and focal adhesion kinase (p125^{FAK}) in cultured rat cardiac myocytes. Biochem Biophys Res Commun 1999; 259: 8–14.

60. Eble DM, Qi M, Strait J et al. Contraction-dependent hypertrophy of neonatal rat venticular myocytes: potential role for focal adhesion kinase. In: Takeda N, Dhalla NS (eds). The Hypertrophied Heart. Boston: Kluwer Academic Publishers; 2000: 91–107.

61. Bayer AL, Ferguson AG, Lucchesi PA et al. Pyk2 expression and phosphorylation in neonatal and adult cardiomyocytes. J Mol Cell Cardiol 2001; 33: 1017–30.

62. Kawamura S, Miyamoto S, Brown JH. Initiation and transduction of stretch-induced RhoA and Rac1 activation through caveolae: cytoskeletal regulation of ERK translocation. J Biol Chem 2003; 278: 31111–17.

63. Sharp WW, Simpson DG, Borg TK et al. Mechanical forces regulate focal adhesion and costamere assembly in cardiac myocytes. Am J Physiol Heart Circ Physiol 1997; 273: H546–56.

64. Sadoshima J, Xu Y, Slayter HS et al. Autocrine release of angiotensin II mediates stretch-induced hypertrophy of cardiac myocytes in vitro. Cell 1993; 75: 977–84.

65. Cingolani HE, Alvarez BV, Ennis IL et al. Stretch-induced alkalinization of feline papillary muscle: an autocrine-paracrine system. Circ Res 1998; 83: 775–80.

66. Saito Y, Berk BC. Transactivation: a novel signaling pathway from angiotensin II to tyrosine kinase receptors. J Mol Cell Cardiol 2001; 33: 3–7.

67. Kodama H, Fukuda K, Takahashi T et al. Role of EGF receptor and Pyk2 in endothelin-1-induced ERK activation in rat cardiomyocytes. J Mol Cell Cardiol 2002; 34: 139–50.

68. Shah BH, Catt KJ. A central role of EGF receptor transactivation in angiotensin II-induced cardiac hypertrophy. Trends Pharmacol Sci 2003; 24: 239–44.

69. Sadoshima J, Izumo S. The cellular and molecular response of cardiac myocytes to mechanical stress. Annu Rev Physiol 1997; 59: 551–71.

70. Molkentin JD, Dorn IG, 2nd. Cytoplasmic signaling pathways that regulate cardiac hypertrophy. Annu Rev Physiol 2001; 63: 391–426.

71. Sugden PH. Ras, Akt, and mechanotransduction in the cardiac myocyte. Circ Res 2003; 93: 1179–92.

CHAPTER 7

Extracellular matrix

Anne M Deschamps, Francis G Spinale

Introduction

Myocardial remodeling occurs in response to a prolonged cardiovascular stress and is characterized by a cascade of compensatory structural events within the myocardium. This remodeling process has been observed following myocardial infarction (MI), with hypertrophy, or cardiomyopathic disease. Progressive left ventricular (LV) dilation in patients with congestive heart failure (CHF) is associated with a greater incidence of morbidity and mortality. Furthermore, pharmacological interventions that provide a beneficial effect on survival in CHF patients attenuate the rate and extent of LV dilation. These observational data suggest that interventions that directly alter the LV myocardial remodeling process hold therapeutic promise in the setting of CHF. A number of cellular and extracellular factors probably contribute to the complex process of myocardial remodeling. For example, myocardial remodeling following MI includes changes in coronary vascular structure and function, myocyte loss, hypertrophy of remaining myocytes, and increased size and number of non-myocyte cells. Collectively, these alterations result in non-uniform changes in LV myocardial wall geometry. While myocardial remodeling is accompanied by modifications in the cellular constituents of the LV myocardium, most notably those of the remaining viable myocytes, significant alterations in the structure and composition of the extracellular matrix (ECM) occur. The goals of this chapter are to revisit and revise the concepts regarding structure and function of the myocardial ECM and then to place these concepts in reference to hypertrophy and the progression to failure.

The myocardial interstitium: structure and function

A new perspective

It has become increasingly evident that the myocardial ECM is not a static structure, but rather a dynamic entity, which may play a fundamental role in myocardial adaptation to a pathological stress, and thereby facilitate the remodeling process. Therefore, identification and understanding of the biological systems responsible for ECM synthesis and degradation within the myocardium hold particular relevance in the progression of CHF.

The view of the myocardial interstitium has changed over the past several years. The myocardial ECM was previously thought to serve solely as a means to align cells and provide structure to the tissue – the scaffolding effect. While this is an important function of the ECM, emerging evidence suggests that the ECM plays a complex and divergent role in influencing cell behavior. For example, the ECM plays a functional role in cell migration, proliferation, adhesion and cell-to-cell signaling. In light of this, the myocardial ECM is not regarded as merely a static structure, but is rather viewed as a complex system of dynamic interactions between matrix molecules, signaling proteins, resident cells, and transmembrane proteins. The composition of these ECM constituents and interactions are shown schematically in Fig. 7.1.

Macromolecules
Basement membrane
Matrix molecules localized in the myocardial ECM can be organized into two separate entities, one being the basement membrane and the other the collagen network. The basement membrane forms a dense sheet-like, specialized type of ECM that separates the stroma from the cell membrane. Collagen type IV is the most abundant protein found within the basement membrane and it forms a covalently stabilized polygonal framework. A second polymer network of laminin self-assembles and is bridged to the collagen type IV network by nidogen, a dumbbell-shaped sulfated glycoprotein. Minor components of the myocardial basement membrane include fibronectin and proteoglycans. Myocytes from cardiomyopathic tissue displayed a 50% reduction in adhesion to basement membrane proteins when compared to control tissue, suggesting that changes within the basement membrane can contribute to LV remodeling during cardiomyopathic disease.

Collagen network
Collagen is a major structural protein of the ECM. In the myocardium, fibroblasts typically secrete collagen polypeptide chains known as α chains. Three α chains are then wound tightly around each other to form a tensile, triple helical super structure. Collagen type I and type III constitute the majority of the fibrillar collagen content of the myocardium and form a highly organized network. This network is composed of three distinct units: (1) endomysial collagen, which surrounds and connects individual myocytes to each other and to capillaries; (2) perimysial collagen, which surrounds and connects bundles of cardiomyocytes; and (3) epimysial collagen, which forms an outer layer.

Alterations in the content and quality of the collagen network lead to changes in LV structure and function. Increases in collagen, as seen with pressure overload hypertrophy, increase wall stiffness and yield a less compliant LV with a hypertrophic phenotype. Degradation of existing collagen, as seen with MI, decreases wall stiffness and yields a more compliant LV with a dilated phenotype. In both cases, the net result is an abnormally functioning LV.

Resident cells
Cardiac myocytes
The interaction of the ECM with the myocyte forms a critical pathway with respect to the alignment of myocytes with the LV free wall, coordinates sarcomere shortening with muscle fiber shortening, and provides the interface for mechanical signaling necessary to regulate myocyte growth. The ECM influences changes in myocyte size, structure, and function. In adaptive hypertrophy, structural alterations in the ECM must occur in order to facilitate myocyte growth. Although myocytes account for three-quarters of the myocardial volume, this cell type only accounts for a third of the cell population.

Myocardial fibroblasts
Approximately two-thirds of the myocardium is composed of non-myocytes, a majority of which are

Figure 7.1. *While historically considered a static structure, it is now becoming recognized that the myocardial extracellular matrix (ECM) is a complex microenvironment containing a large portfolio of matrix proteins, signaling molecules, proteases, and cell types. (A) Schematic of myocytes arranged in a linear fashion to demonstrate the interdigitations of the fibrillar collagen weave and the interaction/location of the transmembrane proteins, the integrins. The integrins are critical for myocyte adhesion to the ECM and multiple isoforms of these integrins exist (for review see Ross and Borg, 2001[5]). Engagement of the integrins provides the mechanical fulcrum for the transduction of sarcomere shortening to myocyte shortening, and in turn facilitates muscle fiber shortening. In addition, the integrins form an important intracellular signaling role by activation of intracellular*

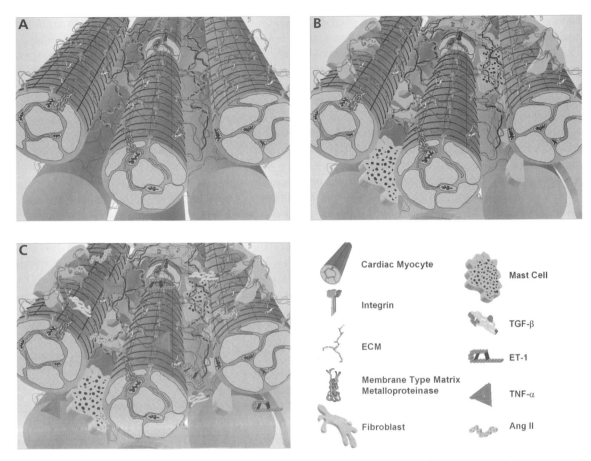

Legend:
- Cardiac Myocyte
- Integrin
- ECM
- Membrane Type Matrix Metalloproteinase
- Fibroblast
- Mast Cell
- TGF-β
- ET-1
- TNF-α
- Ang II

pathways such as MAP kinases. An important enzyme system responsible for ECM degradation/remodeling is the matrix metalloproteinases (MMPs), and a transmembrane subfamily of MMPs are the membrane-type MMPs that can be co-localized with integrins. The activation of the MT-MMPs can cause local ECM degradation and in turn alter the degree of integrin engagement and ultimately intracellular signaling pathways. It is likely that this pathway of outside-in/inside-out signaling is operative in the hypertrophic response. (B) A number of cell types exist within the ECM and include the fibroblast, smooth muscle cells (compartmentalized to the vascular wall) and mast cells. The fibroblasts are not static within the ECM and different fibroblast phenotypes exist which are dependent upon the stimulus and region of the myocardium. It is very likely that the fibroblast influences ECM structure and function significantly, but also interacts through multiple signaling mechanisms with the myocyte. The mast cell is becoming recognized as an important cellular constituent of the ECM and can release a number of proteases as well as activate endogenous protease pathways such as MMPs. It is likely that changes in the phenotype and synthetic profile of the mast cell changes during the hypertrophic response and in turn, alters ECM structure and composition. (C) Superimposed on the ECM structure and interstitial cells are a number of signaling molecules which are locally released. These signaling molecules are part of the highly dynamic microenvironment that exists within the ECM. While not inclusive, representative signaling molecules for the growth factor pathway (TGF), bioactive peptides (ET-1), inflammatory (TNF) and neurohormonal pathways (Ang II) have been placed into the interstitial space to emphasize the highly diverse and complex domain that makes up the myocardial ECM. Changes in the levels of these bioactive molecules probably occur during the hypertrophic response, and in turn change the synthetic/degradative pathways for ECM proteins. Illustrations created by Anne Deschamps. See color plate section.

fibroblasts. The fibroblast is a member of the connective-tissue family of cells whose main function is to secrete ECM proteins. The myocardial fibroblast responds to an injury by increasing proliferation and synthesis of ECM components. In the myocardium, the majority of excreted ECM components include laminin, collagens I, III, and IV, and fibronectin. In addition, a conversion of interstitial fibroblasts to myofibroblasts, which possess contractile components and differences in secretory products typically found in smooth muscle cells, can occur in response to mechanical tension and neurohormonal stimulation.

Mast cells

Mast cells are nucleated hematopoietic cells found in most connective tissues including the myocardium. Mast cells are mainly located around blood vessels and between myocytes in all sections of the heart. Whereas other granulocytes mature in the bone marrow and then circulate in the blood, mast cells have a unique function in which they mature in vascularized tissue. Known for their granulatory properties due to high histamine content, mast cells have also been shown to release several cytokines and growth factors that influence ECM remodeling, including tumor necrosis factor-α (TNF-α) and transforming growth factor-β (TGF-β).[1] Cardiac mast cells, by synthesizing and secreting pro-hypertrophic cytokines and pro-fibrotic growth factors, participate in the induction and maintenance of cardiac hypertrophy and fibrosis.[2] Mast cell density and release of mast cell constituents can influence the biology of the myocardial ECM.[2] Moreover, experimental studies have implicated that mast cell proliferation and degradation contribute to the remodeling process in hypertrophy.[3]

Transmembrane proteins
Integrins

The integrins are a family of transmembrane heterodimeric proteins that traverse the cell membrane and link the extracellular matrix to its cytoplasmic actin cytoskeleton. Integrins are comprised of an α and β subunit in which both contain a single transmembrane domain. To date, approximately 25 different integrins can form due to a variety of combinations between α and β subunits.[4] However, the number of α and β subunits expressed in the myocardium is relatively small.[5] Integrins were initially thought to function as an ECM–cytoskeleton attachment molecule, but have since been shown to play numerous roles in signal transduction, differentiation, proliferation, and migration.[5] An important feature of the integrin family is their ability to elicit a signaling cascade as a result of changes in mechanical stress induced by pressure or volume overload or in post-MI remodeling. Integrin-mediated signaling involves classical mitogen-activated protein kinase pathways and plays a critical role in cardiomyocyte hypertrophy.[6] Furthermore, overexpression of certain integrins has been shown to have an impact on the hypertrophic response in myocytes.[7] Aside from the extracellular environment eliciting a response, or what has been coined outside-in signaling, there is a phenomenon of inside-out signaling that alters the binding of the integrin to ECM components.

Membrane-bound matrix metalloproteinases

Matrix metalloproteinases (MMPs) are a family of pro-enzymes that are activated to degrade all components of the ECM. Notably, however, there is class within this family that resides within the cell membrane and has a major influence within the extracellular space. Structurally, membrane-type MMPs (MT-MMPs) consist of a signal sequence, a pro-domain, a catalytic domain, and a hemopexin-like domain. Four of the six members of the MT family have an additional transmembrane domain and a small cytoplasmic tail, while the two remaining members are attached to the plasma membrane by a glycosylphosphatidyl-inositol anchor. These MT-MMPs are expressed in a number of myocardial cell types including myocytes, fibroblasts, and infiltrating inflammatory cells. The functions of these unique MMPs are discussed in detail in a later section.

Signaling proteins
Transforming growth factor-β

The transforming growth factor (TGF superfamily) comprises more than 30 members that have a wide range of functions during development, cell cycle control and growth, extracellular matrix regulation, and immune response.[8] TGF-β_1 is the most abundant and well studied of the three TGF-β isoforms, and will be the focus of this section. After myocardial injury, inflammatory cells can secrete TGF-β and induce a signaling cascade through binding their serine-threonine kinase receptors, TGF-β_{RI} and TGF-β_{RII}. Binding of TGF-β to the type I receptor causes the type II receptor to translocate and form a heteromeric complex. The type I receptor is phosphorylated by the type II receptor which in turn phosphorylates R-Smads. These proteins subsequently dimerize and translocate to the nucleus, in turn inducing transcriptional activation of specific target genes.

One downstream response is the induction of ECM synthesis. In transgenic mice that overexpress the TGF-β_1 isoform, a significant cardiac hypertrophy along with interstitial fibrosis resulted, which was not present in non-transgenic animals.[9] TGF-β also stimulates the production of collagen in myocardial fibroblasts.[10] Further, tranilast, a nonspecific TGF-β inhibitor, showed attenuated LV hypertrophy and fibrosis in a rat model of hypertension.[11]

Tumor necrosis factor-α

Tumor necrosis factor-alpha (TNF-α) is a pleiotropic cytokine that has both adaptive and maladaptive effects on the heart. It is produced by a number of cell types including neutrophils, monocytes and macrophages, T cells, mast cells, fibroblasts, vascular smooth muscle cells, and cardiac myocytes. TNF-α is initially synthesized as a 26 kDa type II transmembrane protein. It is subsequently cleaved and shed by TNF-α converting enzyme (TACE), also known as a metalloproteinase and disintegrin-17 (ADAM-17). In the extracellular space, three 17 kDa monomers combine to produce a biologically active homotrimer, which binds a TNF receptor and elicits a signaling cascade. These signaling cascades include the mitogen-activated protein kinases (ERK, JNK and p38) and the transcription factors AP-1 and NFκb. Increased levels of TNF-α have been reported in patients with advanced cardiac failure.[12] In mice, the cardiac-specific overexpression of TNF-α led to progressive LV dilation and remodeling within 4–12 weeks, partly due to the activation of the MMP family.[13]

Angiotensin II

Studies using microdialysis to sample the myocardial extracellular fluid have characterized peptides such as angiotensin II (Ang II) and endothelin (ET-1). These studies have demonstrated the dynamic and changing environment of the myocardial extracellular space. Ang II is an important mediator in cardiac remodeling associated with LV hypertrophy, following MI, and in the setting of CHF. The octapeptide, Ang II, is produced by the proteolytic removal of two residues from angiotensin I (Ang I) by angiotensin-converting enzyme (ACE). *In vivo* evidence suggests that there is a high capacity of Ang I to Ang II conversion in the interstitial fluid space of the myocardium.[14] Ang II can then bind to one of four known receptors, of which the AT1 and AT2 subtypes are most prevalent in the myocardium. Signaling cascades elicited by the binding of Ang II to its receptor can result in numerous effects including cardiac hypertrophy, fibrosis, and systemic vasoconstriction.

Aside from its potent vasoconstricting properties, Ang II has been shown to be an efficient inducer of fibrosis. Through chronic infusion of Ang II in the rat, increased fibronectin expression was associated with interstitial fibrosis.[15] In addition, a study conducted by Kawano and coworkers discovered multiple profibrotic effects of Ang II on human cardiac fibroblasts.[16] Fibroblasts stimulated with 100 nmol/l of Ang II showed a 2-fold increase in laminin and fibronection mRNA levels at 24 hours.[16] Additionally, expression of plasminogen activator inhibitor-1, an inhibitor of ECM degradation, was enhanced.[16]

Endothelin-1

The mature 21 amino acid peptide ET-1 is synthesized from a 38 amino acid precursor by ET-converting enzyme. ET-1 has equal affinity for two receptor subtypes (ET$_A$ and ET$_B$) and produces diverse physiological effects. Although ET-1 was originally discovered as a product of endothelial cells, and named aptly, both cardiac myocytes and smooth muscle cells also release this bioactive peptide. Therefore, the production and release of ET-1 can occur locally within the vascular and myocardial compartments, to regulate a number of physiological processes. Increased synthesis and release of ET-1 exacerbates LV pump dysfunction in a number of cardiovascular diseases.[17] Moreover, ET-1 exposure can induce changes in fibrillar-collagen synthesis and degradation pathways.[18,19]

Determinants of matrix synthesis

Fibrillar collagen synthesis

While a number of matrix components are synthesized by cardiac cellular components, the mechanism of fibrillar collagen production is unique. The fibrillar collagens are the most abundant proteins in the extracellular matrix and provide the scaffold for tissue structure and function. Collagen fibrils are assembled in the extracellular space within an infolding of the plasma membrane. First, procollagen is translated into a triple helical protein with N- and C-propeptides. The N-propeptide is trimeric while the C-propeptide takes on a globular shape. These propeptides are subsequently cleaved by N- and C-proteinases to produce the procollagen

fragments: aminoterminal propeptide of type I procollagen (PINP) and carboxyterminal propeptide of type I procollagen (PICP). Quantification of these small peptides has been used to determine the synthesis rate of collagen. After removal of the propeptides, the collagen spontaneously self-assembles into fibrils that are stabilized by covalent cross-linking. Aggregation of collagen fibrils forms a collagen fiber. A schematic of fibrillar collagen synthesis is shown in Fig. 7.2 (top). Net synthesis of fibrillar collagen is dependent upon extrinsic stimuli such as TGF-β and Ang II for transcription and translation of procollagen chains. Alterations in the quantity and quality of fibrillar collagen have been observed in overload hypertrophy.

Determinants of matrix degradation

Matrix metalloproteinases (MMPs)

The matrix metalloproteinases (MMPs) are a family of more than 25 species of zinc-dependent proteases that are essential for normal tissue remodeling in processes such as bone growth, wound healing, and reproduction. MMPs are responsible for turnover of the ECM, which in turn

Figure 7.2. *The balance between synthesis and degradative rates determines the steady state levels of fibrillar collagen within the myocardial interstitium. (A) A number of extracellular stimuli will influence fibroblast collagen synthesis rates. These extracellular signals include angiotensin II, aldosterone, and growth factors such as TGF-β. Mechanical signals transduced through the integrins also contribute to stimuli which affect collagen synthesis rates. Intracellular signals generated by biochemical and/or physical stimuli result in transcription and translation of nascent collagen proteins which occur in the formation of the mature collagen fibril. These steps include cleavage of extension peptides by specific proteinases and then self-assembly of these collagen molecules into a collagen fibril. These collagen structures are highly vulnerable to degradation by MMPs and other proteases. The final maturation of collagen fibrils is through the formation of covalent cross-links that make up the mature collagen fiber. (B) The degradation of the collagen matrix within the myocardium involves a number of biochemical events requiring a number of protease systems. This schematic demonstrates the MMP system, but serine proteases and other proteases also can degrade particular components of the myocardial ECM. Extracellular signals, either biochemical or physical can induce the transcription and translation of MMPs. The two major types of MMPS are the soluble type, and the membrane type (MT-MMP). The soluble types are released into the interstitium, and upon activation will degrade the collagen fibrils. This is not a random process, but rather degradation is dependent upon the type of MMPs synthesized and the exposure of specific peptide sequences of the collagen molecule. The MT-MMPs are proteolytically active once transported to the cell membrane, and degrade a wide portfolio of ECM proteins. The small peptide fragments released during degradation are detectable in the bloodstream and can be utilized for quantifying the relative degree of collagen degradation. Illustrations created by Anne Deschamps. See color plate section.*

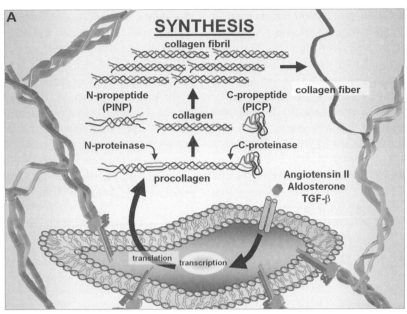

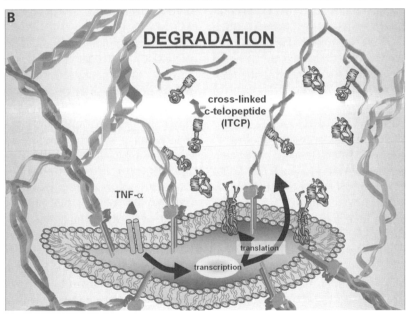

facilitates tissue remodeling.[20,21] MMPs are synthesized as inactive zymogens and are secreted into the extracellular space as pro-enzymes.[20,21] The pro-MMP binds specific ECM proteins and remains enzymatically quiescent until the propeptide domain is cleaved. Cleavage results in exposure of the zinc active site in the catalytic domain and subsequent activation. The disruption of the cysteine–zinc interaction in the MMP active site is essential in the activation of MMPs. This 'cysteine switch' hypothesis provides a mechanism for the activation of secreted MMPs.[22] Instead of a sporadic distribution of pro-MMPs throughout the ECM, there is a specific allotment of these proteolytic enzymes within the extracellular space. Moreover, a large reservoir of recruitable MMPs exists, which upon activation can result in a rapid surge of ECM proteolytic activity. As shown in Fig. 7.2 (bottom), upon activation of these MMPs, degradation of collagen fibrils and other ECM components takes place. Small peptide fragments, such as the cross-linked c-telopeptide (ITCP), during the degradation process are detectable in the blood stream and can be utilized for quantifying the relative degree of collagen degradation.

MMP classification

MMPs are classified into subgroups according to *in vitro* substrate specificity and/or structure. Interstitial collagenase (MMP-1), neutrophil collagenase (MMP-8), and collagenase-3 (MMP-13) possess high substrate specificity for fibrillar collagens, whereas the gelatinases (MMP-2 and MMP-9) demonstrate high substrate affinity for denatured fibrillar collagen and basement membrane proteins. The substrate portfolio for stromelysin (MMP-3) includes important myocardial ECM proteins such as aggrecan, fibronectin, and fibrillar collagens. Stromelysins also participate in the MMP activational cascade by directly processing pro-MMP species.[23,24] For example, Murphy and colleagues reported a 12-fold increase in the conversion of pro-MMP-1 to active MMP-1 in the presence of MMP-3.[23] In addition, other MMPs such as MMP-1, MMP-2, and the membrane-type MMPs (MT-MMPs) can also activate pro-MMPs.[25] MT-MMPs contain a transmembrane domain, and are probably activated intracellularly

through a pro-protein convertase pathway.[25] Thus, unlike the secretable MMPs, MT-MMPs are inserted into the cell membrane already activated. MT1-MMP degrades fibrillar collagens and a wide range of ECM components, as well as proteolytically processing pro-MMP-2 and pro-MMP-13. There is emerging evidence to suggest that MMPs can also degrade non-matrix substrates such as bioactivate peptides, in turn affecting cell proliferation, migration, and apoptosis.[26] A table of MMPs identified within the myocardium and ECM substrates can be found in Table 7.1.

MMP activation/inhibition

Due to the interaction of MMPs with other MMP family members, as well as members of other protease families, the activation of a few key enzymes can initiate a cascade of proteolytic activity in a feed-forward manner. For example, pro-MMPs are activated not only by MMP-1, MMP-2, and the MT-MMPs, but also by the serine proteases chymotrypsin, trypsin, and plasmin. Thus, the regulation of these upstream processes and proteolytic cascades may directly affect MMP activity. Because MMPs degrade various components of the ECM, it is important that MMPs be tightly controlled in order to prevent excessive tissue degradation. Therefore, a group of endogenous proteins, the tissue inhibitors of matrix metalloproteinases (TIMPs), exists to regulate activity of the MMPs. Currently, there are four known TIMP species.[27,28] TIMPs bind to the active site of MMPs and block access to extracellular matrix substrates. The MMP–TIMP complex is formed in a stoichiometric 1:1 molar ratio and forms an important endogenous system for regulating MMP activity *in vivo*.[27,28] The importance of inhibitory control of MMPs within the myocardium has been demonstrated through the genetic deletion of TIMP-1 expression.[29] TIMP-1 null mice display basal LV remodeling in the absence of a pathological stimulus.[29] By four months of age, these mice had increased LV end-diastolic volume, mass, and wall stress, but reduced fibrillar collagen content occured.[29] TIMPs have been identified in every cell examined. TIMP-4 showed the highest level of expression in human myocardial tissue and may be the predominant cardiac isoform.[27]

Table 7.1. *Classes of MMPs that have been identified within the myocardium. The MMP numbering and nomenclature is shown along with ECM substrates*

Name	Number	Substrate/function
Collagenase		
Interstitial collagenase	MMP-1	Collagens I, II, III, VII and basement membrane components
Collagenase 3	MMP-13	Collagens I, II, III
Neutrophil collagenase	MMP-8	Collagens I, II, III and basement membrane components
Gelatinase		
Gelatinase A	MMP-2	Gelatins, collagens I, IV, V, VII and basement membrane components
Gelatinase B	MMP-9	Gelatins, collagens IV, V XIV and basement membrane components
Stromelysin		
Stromelysin 1	MMP-3	Fibronectin, laminin, collagens III, IV, IX and MMP activation
Membrane-type MMP		
MT1-MMP	MMP-14	Collagens I, II, III, fibronectin, laminin-1, activates pro-MMP-2 and pro-MMP-13

ECM changes in overload hypertrophy

Two distinct patterns of left ventricular hypertrophy (LVH) occur in response to persistent load: pressure overload hypertrophy (POH) and volume overload hypertrophy (VOH). Changes in LV wall stress drive the adaptation to hypertrophy, with different manifestations to the different types of overload. POH results in a concentric hypertrophy, whereby wall thickness is increased while LV diameter remains the same or is decreased. A myocyte undergoes concentric hypertrophy by adding sarcomeres in parallel to achieve an increase in width. In contrast, VOH results in an eccentric hypertrophy whereby wall thickness remains the same or is decreased, while LV diameter increases. Eccentric hypertrophy involves the addition of sarcomeres in series to achieve an increase in myocyte length.

In POH, accumulation of myocardial fibrillar collagen influences passive compliance properties of the LV and results in diastolic dysfunction.[30,31] Increases in ACE and TGF-β are just some of the factors that contribute to increases in collagen I

and III expression during POH.[32] In VOH, collagen phenotypes are altered and the degree of these changes is related to the degree of LV wall stress and degree of regurgitant volume.[33–35] In addition, cardiac fibroblasts have been shown to produce abnormal proportions of non-collagen ECM, specifically fibronectin, in volume overload.[36] In both of these hypertrophic remodeling processes, significant myocardial matrix remodeling occurs and facilitates changes in myocyte shape. Fig. 7.3 depicts the significant differences in ECM composition and structure that occur in POH and VOH.

POH and MMPs
Human studies
Past studies have documented reduced plasma levels of MMPs in patients with systemic hypertension and LVH.[33,37] Decreased plasma levels of MMP-1 were reported in hypertensive patients with hypertrophy, and reduced plasma levels of MMP-9 were observed in untreated hypertensive patients.[33,37] The ratio of TIMP-1 to MMP-1 was also markedly increased in patients with severe LV hypertrophy (left ventricular mass index ≥ 159).[33,37,38] Taken

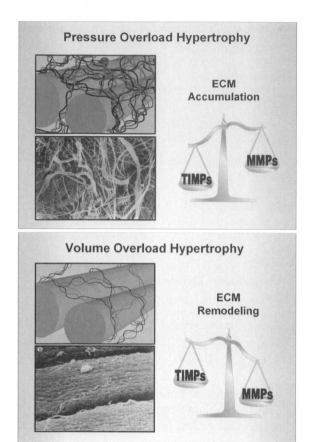

Figure 7.3. *Significant differences in ECM composition and structure occur in pressure (POH) or volume overload (VOH) hypertrophy. In POH, ECM accumulation occurs and appears to be related to the degree of the pressure overload stimulus. In marked contrast, VOH is associated with ECM degradation and reformation which apparently facilitates changes in myocyte length and eventually LV dilation. The schematic demonstrates the changes in ECM structure with POH or VOH, and representative scanning electron micrographs from myocardial samples are shown. With respect to POH, the stoichiometric relationship between MMPs and TIMPs shifts to one of decreased net proteolytic activity, and thereby matrix accumulation. In contrast, the stoichiometric balance shifts in favor of increased proteolytic activity in VOH, and thereby accelerated ECM degradation and remodeling. The precise signaling pathways which translate these two overload states to mutually distinct synthetic/degradative processes is not fully understood, but probably entails the cells and signaling molecules illustrated in Fig. 7.1. The scanning electron micrographs were extracted with permission from Abrahams C, Janicki JS, Weber KT. Myocardial hypertrophy in Macaca fascicularis. Structural remodeling of the collagen matrix. Lab Invest 1987; 56: 676–83; and Dell'italia LJ, Balcells E, Meng QC et al. Volume-overload cardiac hypertrophy is unaffected by ACE inhibitor treatment in dogs. Am J Physiol Heart Circ Physiol 1997; 273: H961–70, and were reproduced with permission from the publisher. See color plate section.*

together, these studies suggest an initial decrease in MMPs and increase in TIMPs during POH, to favor matrix accumulation. In contrast to patterns of collagen accumulation in POH, LV remodeling with volume overload is characterized by excessive ECM degradation and LV dilation.

Animal models

As hypertrophy progresses to decompensation and eventual LV failure, time-dependent changes in the activation of myocardial MMPs are likely to occur. In fact, several animal studies have demonstrated such changes in myocardial MMP levels throughout the development of POH.[39,40] In the spontaneously hypertensive rat, the development of compensated hypertrophy is associated with an increase in TIMP-4 levels.[41] However, over time MMP-2 and MMP-9 levels increase and TIMP levels fall, favoring myocardial remodeling and changes in the ECM related to decompensation and failure.[39,40] Additionally, Iwanaga et al. reported disparate MMP and TIMP levels in the compensation stage of LVH compared to the transition to CHF.[42] While net MMP-2 and TIMP-2 and -4 activity remained unchanged at the compensation stage, MMP-2 activity increased significantly, and surpassed that of the TIMPs as LVH decompensation progressed to LV failure.[42]

Both physical and chemical stimuli influence LV remodeling in POH in a time-dependent manner.[42,43] Myocardial MMP-9 zymographic activity was increased 3-fold in a dog model of acute pressure overload.[43] The increase in MMP-9 zymographic activity was not paralleled by increases in MMP-9 or TIMP-1 protein levels, suggestive of diminished inhibitory control.[43] With prolonged pressure overload, the zymographic activity to protein ratio along with the zymographic activity to TIMP-1 ratio returned to basal values.[43] Furthermore, it has been demonstrated in a sheep model of POH that relieving the pressure of the overload stimulus directly affected MMP and TIMP levels.[44]

VOH and MMPs
Human studies
Unlike collagen accumulation in POH, chronic VOH states display disruption of normal fibrillar collagens, which may be due to enhanced proteolytic activity of myocardial MMPs.[33–36,43]

However, the specific profiles and time-dependent changes of MMPs and TIMPs in patients with VOH are not well understood. Some insight can be gleaned from these and future studies.[33–36,43]

Animal models
In aortic regurgitation, cardiac fibroblasts produce abnormal proportions of non-collagenous ECM, while little change in collagen synthesis occurs.[36] In addition, myocardial MMP levels and zymographic activity for MMPs-2 and -9 are increased.[43,45,46] Rat models of volume overload have shown not only an increase in MMP-2 and MMP-9 zymographic activity, which was associated with changes in LV function and geometry, but also an attenuation in LVH and adverse LV remodeling following MMP inhibition with a concomitant maintenance of normal LV function (Fig. 7.4).[45,47]

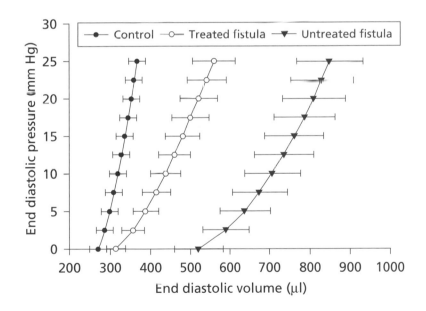

Figure 7.4. *Aorto-caval fistulas were created in rats in order to produce an LV volume overload stimulus. LV pressure–volume curves were generated ex vivo in order to assess the relative degree of LV chamber remodeling at 8 weeks, following creation of the volume overload. A significant shift to the right occurred in the fistula rats consistent with LV remodeling and dilation. Treatment with an MMP inhibition from weeks 2–8 resulted in a shift to the left in the LV pressure–volume relation, indicating a reduction in the degree of LV remodeling – despite an equivalent volume overload stimulus. Error bars are standard errors. Reproduced with permission from Chancey AL, Brower GL, Peterson JT et al. Effects of matrix metalloproteinase inhibition on ventricular remodeling due to volume overload. Circulation 2002; 105: 1983–8.*

Taken together, these LVH studies suggest that: (1) changes in myocardial MMP and TIMP levels probably contribute to the LV remodeling process given a pressure or volume overload stimulus; (2) different patterns of MMP and TIMP expression occur depending on the type and duration of overload states; and (3) increased levels of MMPs and TIMPs may be decreased with pharmacological inhibition or alleviation of the overload stimulus.

Myocardial interstitium in failure

The pathways by which LV hypertrophy leads to occult CHF remain incompletely understood. However, clear defects in both LV systolic and diastolic function occur that are accompanied by changes in myocardial geometry and structure. The previous sections outlined how the myocardial interstitium may facilitate this remodeling process. The most widely studied process with respect to maladaptive remodeling and changes in the myocardial interstitium is that of post-MI remodeling. However, the disease state, which uniformly gives rise to CHF and is accompanied by gross changes in LV geometry and structure, is dilated cardiomyopathy (DCM). A number of studies have been performed in patients with end-stage DCM with respect to specific ECM degradative pathways. Thus, reviewing what has been observed in this particular cardiac disease state may provide insight and a useful paradigm for myocardial ECM remodeling.

A number of studies have examined relative MMP and TIMP expression in end-stage human DCM. Increased *in vitro* myocardial MMP zymographic activity in DCM samples has been observed.[48] Also, it has been demostrated that myocardial MMP-9 was increased due to either ischemic or non-ischemic origins, and MMP-3 was increased in non-ischemic DCM myocardial extracts.[49,50] A selective upregulation of certain MMP species has been observed in DCM. For example, the collagenase MMP-13 is expressed at low levels in normal myocardium but is significantly increased in end-stage DCM (Fig 7.5).[49,50] Interstitial collagenase (MMP-1), in contrast, is

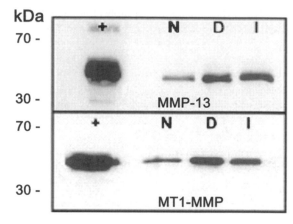

Figure 7.5. *Myocardial levels of MMP-13 and MT1-MMP were determined using immunoblotting techniques in normal and DCM myocardium. MMP-13 was expressed at low levels in normal human myocardium (N) but increased significantly in non-ischemic DCM (D) and ischemic cardiomyopathy (I). MT1-MMP was present in normal myocardium (N) and increased by over 3-fold in cardiomyopathic samples (D, I). MMP-13 and MT1-MMP were localized to positive controls (+). Reproduced with permission from Spinale FG, Coker ML, Heung LJ et al. A matrix metalloproteinase induction/activation system exists in the human left ventricular myocardium and is upregulated in heart failure. Circulation 2000; 102: 1944–9.*

decreased.[49,50] In addition, because TIMP-1 binds with less affinity to MMP-13 than to MMP-1, there is a loss of endogenous control as MMP-13 levels increase.[51] Furthermore, increases in MT1-MMP of over 3-fold have been shown in cardiomyopathic samples (Fig. 7.5).[50] Because of its biologically diverse actions, MT1-MMP is not only able to degrade surrounding matrix but also is capable of initiating cascades of proteolytic activity.[25,52] Therefore, the emergence of certain MMP species within the DCM myocardium may contribute to the increased susceptibility of the myocardial fibrillar collagen network to degradation, and subsequent maladaptive myocardial remodeling.

Myocardial TIMPs are variably expressed in end-stage DCM.[49,53] Thomas and colleagues found an increased abundance of TIMP-1 and TIMP-2 in DCM myocardium samples.[49] In a study by Li and co-workers, TIMP-1 and TIMP-3 levels were reduced in cardiomyopathic samples, whereas TIMP-2 levels were unchanged when compared to control myocardium.[53] Previous studies of cardiomyopathic disease have suggested that changes occur in the stoichiometric ratio of MMPs to TIMPs.[50,53] For example, the MMP-1–TIMP-1 complex in both ischemic and non-ischemic DCM, was reduced.[50] Both MMPs and TIMPs are encoded by unique genes and differ in promoter regions. Thus, extracellular stimuli may induce differential expression of TIMPs in DCM. Nevertheless, the quantitative reduction of MMP/TIMP complexes within the myocardium in end-stage DCM has significant relevance to the LV remodeling process. As observed post-MI, a change in the MMP/TIMP stoichiometric ratio in DCM could favor persistent MMP activation, resulting in uncontrolled ECM turnover and progressive LV remodeling.

Conclusions and future directions

Remodeling in hypertension/hypertrophy is characterized by excessive matrix accumulation and abnormalities in compliance due to changes in wall stress. The factors that drive ECM remodeling are numerous. However, a specific pattern of MMPs and TIMPs appear to occur in pressure and volume overload states. Therefore, cellular and molecular triggers, which in turn give rise to ECM changes, probably hold therapeutic promise.

Modifying MMPs in myocardial hypertrophy

Current pharmacological approaches are focused upon global neurohormonal pathways such as the renin–angiotensin–aldosterone and adrenergic systems. Targeting of therapies to the myocardial ECM, however, would provide a selective and specific strategy to inhibit myocardial hypertrophy. A recent study by Badenhorst and colleagues used collagen cross-link breakers to modify myocardial

compliance properties.[54] Furthermore, the selective activation of particular MMPs may alleviate the accumulation of the ECM in hypertrophy. For example, serine protease activation of MMPs has been shown to reduce fibrillar collagen and decrease elastic stiffness.[55] MMP intervention in LVH will probably be entirely different from intervention in remodeling post-MI and in cardiomyopathy. The previous two disease states suggest that inhibition of selective MMPs will attenuate LV remodeling, whereas in LVH, MMP activation may be necessary to restore the normal balance between ECM synthesis and degradation. Although directly modifying MMP levels may have beneficial effects on myocardial compliance, more research is needed to determine the optimum conditions to favorably alter the myocardium without introducing unintended side-effects.

Future directions

Developing pharmacological strategies that interfere with upstream signaling cascades involved in MMP transcription will add to our understanding of the complex myocardial remodeling process and the specific role of MMPs. For example, cytokine interruption, as with TNF-α inhibitors, may be a useful pharmacological tool to identify the signaling pathways obligatory for MMP species induction.[56] Ideally, through targeting specific bioactive molecules or blocking nuclear binding sites, MMPs implicated in adverse remodeling could be inhibited while necessary basal expression of beneficial MMP levels could continue. Gene delivery systems provide the potential for local modification of MMPs and TIMPs. Emerging evidence also suggests that non-matrix substrates of MMPs influence cell function. For example, the ability of particular MMPs to cleave multiple non-matrix substrates, such as integrins, cytokines, and growth factors, renders regulation by MMPs specific and significant to numerous pathways.[26] Thus, more clinically applicable strategies specific to each disease state and time course can be realized through acknowledging and targeting upstream signals of MMPs, as well as their various non-matrix substrates. In consideration of the divergent physical and biochemical pathways involved in LV remodeling, defining the molecular triggers of

MMP and TIMP expression and targeting the upstream mechanisms responsible may prove to be an important therapeutic paradigm for heart failure treatment.

Acknowledgements

This study was supported by NIH grants HL59165, P01 HL48788–08, and a Career Development Award from the Ralph H Johnson Veterans' Association Medical Center.

References

1. Wedemeyer J, Tsai M, Galli SJ. Roles of mast cells and basophils in innate and acquired immunity. Curr Opin Immunol 2000; 12: 624–31.

2. Shiota N, Rysa J, Kovanen PT et al. A role for cardiac mast cells in the pathogenesis of hypertensive heart disease. J Hypertens 2003; 21: 1935–44.

3. Stewart JA, Jr., Wei CC, Brower GL et al. Cardiac mast cell- and chymase-mediated matrix metalloproteinase activity and left ventricular remodeling in mitral regurgitation in the dog. J Mol Cell Cardiol 2003; 35: 311–19.

4. Humphries MJ. Integrin structure. Biochem Soc Trans 2000; 28: 311–39.

5. Ross RS, Borg TK. Integrins and the myocardium. Circ Res 2001; 88: 1112–19.

6. MacKenna DA, Dolfi F, Vuori K et al. Extracellular signal-regulated kinase and c-Jun NH2-terminal kinase activation by mechanical stretch is integrin-dependent and matrix-specific in rat cardiac fibroblasts. J Clin Invest 1998; 101: 301–10.

7. Ross RS, Pham C, Shai SY et al. Beta1 integrins participate in the hypertrophic response of rat ventricular myocytes. Circ Res 1998; 82: 1160–72.

8. Topper JN. TGF-beta in the cardiovascular system: molecular mechanisms of a context-specific growth factor. Trends Cardiovasc Med 2000; 10: 132–7.

9. Rosenkranz S, Flesch M, Amann K et al. Alterations of beta-adrenergic signaling and cardiac hypertrophy in transgenic mice overexpressing TGF-beta(1). Am J Physiol Heart Circ Physiol 2002; 283: H1253–62.

10. Sigusch HH, Campbell SE, Weber KT. Angiotensin II-induced myocardial fibrosis in rats: role of nitric oxide, prostaglandins and bradykinin. Cardiovasc Res 1996; 31: 546–54.

11. Pinto YM, Pinto-Sietsma SJ, Philipp T et al. Reduction in left ventricular messenger RNA for transforming growth factor beta(1) attenuates left ventricular fibrosis and improves survival without lowering blood pressure in the hypertensive TGR(mRen2)27 Rat. Hypertension 2000; 36: 747–54.

12. Torre-Amione G, Kapadia S, Lee J et al. Tumor necrosis factor-alpha and tumor necrosis factor receptors in the failing human heart. Circulation 1996; 93: 704–11.

13. Sivasubramanian N, Coker ML, Kurrelmeyer KM et al. Left ventricular remodeling in transgenic mice with cardiac restricted overexpression of tumor necrosis factor. Circulation 2001; 104: 826–31.

14. Wei CC, Meng QC, Palmer R et al. Evidence for angiotensin-converting enzyme- and chymase-mediated angiotensin II formation in the interstitial fluid space of the dog heart in vivo. Circulation 1999; 99: 2583–9.

15. Crawford DC, Chobanian AV, Brecher P. Angiotensin II induces fibronectin expression associated with cardiac fibrosis in the rat. Circ Res 1994; 74: 727–39.

16. Kawano H, Do YS, Kawano Y et al. Angiotensin II has multiple profibrotic effects in human cardiac fibroblasts. Circulation 2000; 101: 1130–7.

17. Schiffrin EL, Intengan HD, Thibault G et al. Clinical significance of endothelin in cardiovascular disease. Curr Opin Cardiol 1997; 12: 354–67.

18. Seccia TM, Belloni AS, Kreutz R et al. Cardiac fibrosis occurs early and involves endothelin and AT-1 receptors in hypertension due to endogenous angiotensin II. J Am Coll Cardiol 2003; 41: 666–73.

19. Saam T, Ehmke H, Haas C et al. Effect of endothelin blockade on early cardiovascular remodeling in the one-clip-two-kidney hypertension of the rat. Kidney Blood Press Res 2003; 26: 325–32.

20. Visse R, Nagase H. Matrix metalloproteinases and tissue inhibitors of metalloproteinases: structure, function, and biochemistry. Circ Res 2003; 92: 827–39.

21. Creemers EE, Cleutjens JP, Smits JF et al. Matrix metalloproteinase inhibition after myocardial infarction: a new approach to prevent heart failure? Circ Res 2001; 89: 201–10.

22. Van Wart HE, Birkedal-Hansen H. The cysteine switch: a principle of regulation of metalloproteinase activity with potential applicability to the entire matrix metalloproteinase gene family. Proc Natl Acad Sci USA 1990; 87: 5578–82.

23. Murphy G, Cockett MI, Stephens PE et al. Stromelysin is an activator of procollagenase. A study with natural and recombinant enzymes. Biochem J 1987; 248: 265–8.

24. Nagase H. Activation mechanisms of matrix metalloproteinases. Biol Chem 1997; 378: 151–60.

25. Hernandez-Barrantes S, Bernardo M, Toth M et al. Regulation of membrane type-matrix metalloproteinases. Semin Cancer Biol 2002; 12: 131–8.

26. McCawley LJ, Matrisian LM. Matrix metalloproteinases: they're not just for matrix anymore! Curr Opin Cell Biol 2001; 13: 534–40.

27. Greene J, Wang M, Liu YE et al. Molecular cloning and characterization of human tissue inhibitor of metalloproteinase 4. J Biol Chem 1996; 271: 30375–80.

28. Brew K, Dinakarpandian D, Nagase H. Tissue inhibitors of metalloproteinases: evolution, structure and function. Biochim Biophys Acta 2000; 1477: 267–83.

29. Roten L, Nemoto S, Simsic J et al. Effects of gene deletion of the tissue inhibitor of the matrix metalloproteinase-type 1 (TIMP-1) on left ventricular geometry and function in mice. J Mol Cell Cardiol 2000; 32: 109–20.

30. Zile MR, Brutsaert DL. New concepts in diastolic dysfunction and diastolic heart failure: Part II: causal mechanisms and treatment. Circulation 2002; 105: 1503–8.

31. Diez J, Querejeta R, Lopez B et al. Losartan-dependent regression of myocardial fibrosis is associated with reduction of left ventricular chamber stiffness in hypertensive patients. Circulation 2002; 105: 2512–7.

32. Fielitz J, Hein S, Mitrovic V et al. Activation of the cardiac renin-angiotensin system and increased myocardial collagen expression in human aortic valve disease. J Am Coll Cardiol 2001; 37: 1443–9.

33. Laviades C, Varo N, Fernandez J et al. Abnormalities of the extracellular degradation of collagen type I in essential hypertension. Circulation 1998; 98: 535–40.

34. Grossman W, Jones D, McLaurin LP. Wall stress and patterns of hypertrophy in the human left ventricle. J Clin Invest 1975; 56: 56–64.

35. Spinale FG, Ishihra K, Zile M et al. Structural basis for changes in left ventricular function and geometry because of chronic mitral regurgitation and after correction of volume overload. J Thorac Cardiovasc Surg 1993; 106: 1147–57.

36. Borer JS, Truter S, Herrold EM et al. Myocardial fibrosis in chronic aortic regurgitation: molecular and cellular responses to volume overload. Circulation 2002; 105: 1837–42.

37. Li-Saw-Hee FL, Edmunds E, Blann AD et al. Matrix metalloproteinase-9 and tissue inhibitor metalloproteinase-1 levels in essential hypertension. Relationship to left ventricular mass and anti-hypertensive therapy. Int J Cardiol 2000; 75: 43–7.

38. Hirono O, Fatema K, Nitobe J et al. Long-term effects of benidipine hydrochloride on severe left ventricular hypertrophy and collagen metabolism in patients with essential hypertension. J Cardiol 2002; 39: 195–204.

39. Li H, Simon H, Bocan TM et al. MMP/TIMP expression in spontaneously hypertensive heart failure rats: the effect of ACE- and MMP-inhibition. Cardiovasc Res 2000; 46: 298–306.

40. Mujumdar VS, Tyagi SC. Temporal regulation of extracellular matrix components in transition from compensatory hypertrophy to decompensatory heart failure. J Hypertens 1999; 17: 261–70.

41. Peterson JT, Hallak H, Johnson L et al. Matrix metalloproteinase inhibition attenuates left ventricular remodeling and dysfunction in a rat model of progressive heart failure. Circulation 2001; 103: 2303–9.

42. Iwanaga Y, Aoyama T, Kihara Y et al. Excessive activation of matrix metalloproteinases coincides with left ventricular remodeling during transition from hypertrophy to heart failure in hypertensive rats. J Am Coll Cardiol 2002; 39: 1384–91.

43. Nagatomo Y, Carabello BA, Coker ML et al. Differential effects of pressure or volume overload on myocardial MMP levels and inhibitory control. Am J Physiol Heart Circ Physiol 2000; 278: H151–61.

44. Walther T, Schubert A, Falk V et al. Regression of left ventricular hypertrophy after surgical therapy for aortic stenosis is associated with changes in extracellular matrix gene expression. Circulation 2001; 104: 154–8.

45. Brower GL, Janicki JS. Contribution of ventricular remodeling to pathogenesis of heart failure in rats. Am J Physiol Heart Circ Physiol 2001; 280: H674–83.

46. Dolgilevich SM, Siri FM, Atlas SA et al. Changes in collagenase and collagen gene expression after induction of aortocaval fistula in rats. Am J Physiol Heart Circ Physiol 2001; 281: H207–14.

47. Chancey AL, Brower GL, Peterson JT et al. Effects of matrix metalloproteinase inhibition on ventricular remodeling due to volume overload. Circulation 2002; 105: 1983–8.

48. Tyagi SC, Campbell SE, Reddy HK et al. Matrix metalloproteinase activity expression in infarcted, noninfarcted and dilated cardiomyopathic human hearts. Mol Cell Biochem 1996; 155: 13–21.

49. Thomas CV, Coker ML, Zellner JL et al. Increased matrix metalloproteinase activity and selective upregulation in LV myocardium from patients with end-stage dilated cardiomyopathy. Circulation. 1998; 97: 1708–15.

50. Spinale FG, Coker ML, Heung LJ et al. A matrix metalloproteinase induction/activation system exists in the human left ventricular myocardium and is up-

regulated in heart failure. Circulation 2000; 102: 1944–9.

51. Knauper V, Lopez-Otin C, Smith B et al. Biochemical characterization of human collagenase-3. J Biol Chem 1996; 271: 1544–50.

52. Vu TH, Werb Z. Matrix metalloproteinases: effectors of development and normal physiology. Genes Dev 2000; 14: 2123–33.

53. Li YY, Feldman AM, Sun Y et al. Differential expression of tissue inhibitors of metalloproteinases in the failing human heart. Circulation 1998; 98: 1728–34.

54. Badenhorst D, Maseko M, Tsotetsi OJ et al. Cross-linking influences the impact of quantitative changes in myocardial collagen on cardiac stiffness and remodelling in hypertension in rats. Cardiovasc Res 2003; 57: 632–41.

55. Stroud JD, Baicu CF, Barnes MA et al. Viscoelastic properties of pressure overload hypertrophied myocardium: effect of serine protease treatment. Am J Physiol Heart Circ Physiol 2002; 282: H2324–35.

56. Deswal A, Bozkurt B, Seta Y et al. Safety and efficacy of a soluble P75 tumor necrosis factor receptor (Enbrel, etanercept) in patients with advanced heart failure. Circulation 1999; 99: 3224–6.

CHAPTER 8

Physiological versus pathological cardiac hypertrophy

Julie R McMullen, Junichi Sadoshima, Seigo Izumo

Introduction

Growth of the heart can be induced by physiological stimuli (e.g. postnatal development, chronic exercise training) or pathological stimuli (e.g. pressure or volume overload). Physiological cardiac hypertrophy is characterized by a normal organization of cardiac structure, normal or enhanced cardiac function, and a relatively normal pattern of cardiac gene expression; whereas pathological hypertrophy is associated with an altered pattern of cardiac gene expression, fibrosis, cardiac dysfunction, and increased morbidity and mortality. An unresolved question in cardiac biology has been whether distinct signaling pathways are responsible for the development of pathological and physiological cardiac hypertrophy. Recent studies have suggested that some signaling pathways play unique roles in the regulation of pathological and physiological cardiac hypertrophy. Furthermore, inhibiting pathological cardiac growth while stimulating physiological growth could have therapeutic value in the setting of heart disease.

Cardiac growth

Growth of the heart can broadly be categorized as either physiological or pathological. Physiological growth includes the embryonic and fetal stages of development occurring *in utero*, the rapidly growing phase during postnatal development, aging (or senescence), and compensatory growth of the adult heart in response to stimuli such as exercise (Fig. 8.1). In this context, aging-induced hypertrophy is referring to animals and humans free of heart disease that develop mild left ventricular hypertrophy as a consequence of age-related decreases in the distensibility of the peripheral vasculature.[1] Pathological growth includes compensatory growth in response to pathological stimuli (e.g. overload), cardiomyopathy, decompensated growth, and heart failure (Fig. 8.1). In this review we have concentrated on the molecular mechanisms that contribute to growth of the adult heart in response to pathological or physiological stimuli.

Growth of the adult heart

In the adult, growth of the heart is closely matched to its functional load, and under normal circumstances is mainly constitutive in nature.[2] It is generally believed that shortly after the postnatal period, the majority of cardiac myocytes are unable to re-enter the cell cycle. Thus, growth of the heart is largely due to an increase in myocyte size.[2,3] The inability of postnatal cardiac myocytes to divide has been the subject of some debate.[4-6] However, estimates of DNA labeling indicate that DNA

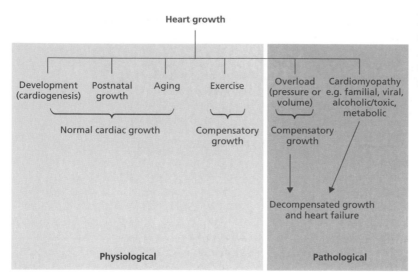

Figure 8.1. *Stages and categories of heart growth. Physiological growth is shown on the left, pathological growth on the right.*

synthesis is taking place in a very small fraction of the total adult cardiac myocyte population.[4,5,7–9]

Cardiac hypertrophy: pathological versus physiological

Pathological cardiac hypertrophy

Pathological cardiac hypertrophy occurs in response to diverse cardiovascular disorders, including hypertension, atherosclerosis, valve disease, myocardial infarction, and cardiomyopathy. When disease causes pressure or volume overload of the heart, wall stress on the left ventricle increases. To counterbalance the chronic increase in wall stress, the heart triggers a hypertrophic response.[10–12] Initially, the enlargement of cardiac myocytes and the formation of new sarcomeres serve to normalize wall stress and permit normal cardiovascular function at rest, i.e. compensated growth. However, function in the hypertrophied heart may eventually decompensate, leading to left ventricle dilation, increased interstitial fibrosis (resulting in increased myocardial stiffness), and heart failure (decompensated or maladaptive growth, Fig. 8.1). Thus, the increase in mass associated with pathological hypertrophy is due in large part to hypertrophy of cardiac myocytes; however hyperplasia of fibroblasts, and the accumulation of

extracellular matrix components, including collagens, also contributes.[13,14] It should be noted that decompensated growth of the heart is commonly considered a progression occurring after compensated growth of the heart. However, there are reports which suggest decompensated growth may also occur in the absence of compensated growth.[15–18]

Pathological hypertrophy caused by chronic pressure overload (e.g. hypertension, left ventricular outflow obstruction, aortic coarction) produces an increase in systolic wall stress and results in concentric ventricular hypertrophy. Concentric hypertrophy is characterized by a parallel pattern of sarcomere addition, leading to an increase in myocyte cell width. Morphologically, this cellular adaptation results in hearts with thick walls and relatively small cavities. By contrast, pathological hypertrophy caused by chronic volume overload (e.g. aortic regurgitation, arteriovenous fistulas) results in eccentric left ventricular hypertrophy. Eccentric hypertrophy is associated with a series pattern of sarcomere addition, leading to an increase in myocyte cell length. Morphologically, this cellular adaptation results in hearts with large dilated cavities and relatively thin walls.[19]

Physiological cardiac hypertrophy

Physiological cardiac hypertrophy occurs in response to increased physical activity or chronic exercise training.[3,20,21] More active animals of the same or related species have heavier hearts e.g. wild hare compared to domestic rabbit, wild rodents compared to laboratory rodents.[20] In humans and animals, intense, prolonged exercise training results in an increase in cardiac mass.[20–24] Isotonic exercise such as running, walking, cycling and swimming, involves movement of large muscle groups and produces eccentric hypertrophy by volume overload.[20,21] In contrast, isometric exercise (e.g. weight lifting, shot put) involves developing muscular tension against resistance without much movement. Such exercise causes a pressure load on the heart rather than a flow load, and results in concentric hypertrophy. Unlike pathological hypertrophy, exercise-induced cardiac hypertrophy does not decompensate into dilated cardiomyopathy or heart failure.

Distinct characteristics of pathological and physiological hypertrophy

Left ventricular hypertrophy is often considered a continuum that transitions from physiological cardiac hypertrophy to pathological cardiac hypertrophy.[25] However, there is reasonable evidence to suggest that physiological and pathological cardiac hypertrophy are two distinct processes that may be mediated by distinct signaling pathways. First, pathological and physiological hypertrophy are associated with distinct phenotypes (Table 8.1). Physiological hypertrophy is characterized by a normal organization of cardiac structure, normal or enhanced cardiac function, and a normal pattern of cardiac gene expression; whereas pathological hypertrophy is associated with an altered pattern of cardiac gene expression, fibrosis, and cardiac dysfunction (Fig. 8.2).[19,22,23,25–30] Second, exercise has been reported to reverse molecular and functional abnormalities in pathological cardiac models.[20,24,31–33] An example is where exercise (swimming-induced) improved cardiac function in rats with pressure overload hypertrophy, even though heart weight was increased.[20,33] Finally, pathological hypertrophy is associated with increased mortality and morbidity, whereas physiological hypertrophy is not.[19,20,26,27]

The mechanistic process which allows the heart to enlarge in response to physiological stimuli while maintaining normal or enhanced function is of clinical relevance, as one potential therapeutic strategy would be to inhibit the pathological growth process while augmenting the physiological growth process.

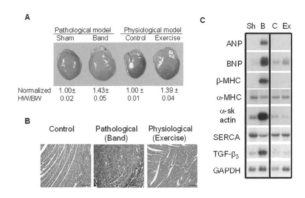

Figure 8.2. *Pathological and physiological cardiac hypertrophy. (A) Representative pictures of hearts from mice subjected to a pathological stimulus (aortic banding (Band) for 1 week, n = 13), a physiological stimulus (chronic exercise: swimming training for 4 weeks (Exercise), n = 6), or no stimulus (sham-operated (Sham), n = 10 or non-exercise trained mice (Control), n = 7). HW/BW: heart weight/body weight ratio, points plotted are mean ± standard error. (B) Histological analysis of heart sections stained with Masson's trichrome. Representative sections from the left venticle wall of control mice (non-exercise trained (= Control)), aortic banded mice (Band) and swimming mice (Exercise). Magnification ×100, bars represent 10 μm. Sections from sham-operated mice were similar to those from non-swim mice (Control). (C) Cardiac gene expression in response to aortic banding or chronic exercise training. Representative Northern blot showing total RNA from ventricles of sham (Sh), band (B), control non-exercise trained (C), and exercise (Ex). Expression of GAPDH was determined to verify equal loading of RNA. See abbreviations list for other abbreviations. See color plate section.*

Table 8.1. *Characteristics of pathological and physiological left ventricular hypertrophy*

	Pathological hypertrophy	Physiological hypertrophy
Stimulus	Chronic pressure load (e.g. hypertension, aortic coarction) or chronic volume load (e.g. valvular disease)	Intermittent pressure load (e.g. strength training: weight lifting) or intermittent volume load (isotonic exercise: running, swimming)
Ventricle morphology	Increased myocyte volume Formation of new sarcomeres Interstitial fibrosis Myocyte necrosis and apoptosis	Increased myocyte volume Formation of new sarcomeres
Fetal gene expression	Upregulated	Relatively normal
Ventricular contractility	Decreased	Normal or enhanced
Ventricular function	Depressed	Normal or enhanced
Coronary flow reserve	Impaired	Normal or enhanced
Completely reversible	Not usually	Yes
Possible progression to heart failure	Yes	No
Associated with increased mortality	Yes	No

Mechanisms responsible for the development of pathological and physiological cardiac hypertrophy

Some investigators have questioned the value of classifying cardiac hypertrophy as 'pathological' or 'physiological' because until recently, it was not clear that distinct molecular mechanisms were responsible for inducing 'pathological' versus 'physiological' hypertrophy.[34] Particular caution should be taken when characterizing transgenic mice, and this is discussed in detail later. However, there is now fairly convincing evidence to suggest that some signaling pathways may play distinct roles in the development of pathological and physiological hypertrophy. These are described below.

Overview of signaling pathways implicated for the development of cardiac hypertrophy

Hypertrophy of ventricular myocytes is commonly associated with stimulation of a hypertrophic pro-

gram of gene expression, an increase in the overall rate of protein synthesis, and organization of contractile proteins into sarcomeric units.[11,28,35,36] Mechanical stimuli, vasoactive substances (e.g. angiotenisn II (Ang II), endothelin-1 (ET-1), prostaglandin $F_{2\alpha}$ ($PGF_{2\alpha}$)), growth factors (insulin-like growth factor-1 (IGF-1), transforming growth factor-beta (TGF-β), fibroblast growth factors (FGFs), and neuregulins), cytokines (interleukin-6 and the cardiotrophin family), hormones (thyroid hormone, growth hormone, adrenergic), and changes in energy metabolism (e.g. fatty acid oxidation, glucose transporters) are all stimuli that activate signal transduction pathways, that have been implicated for the development of cardiac hypertrophy. Hypertrophic stimuli can act directly to induce hypertrophy of cardiac myocytes, or act via paracrine or autocrine mechanisms.

The hypertrophic signals are transduced by signaling pathways including G proteins (heterotrimeric and the small GTP-binding proteins), protein kinase C (PKC), protein kinase A (PKA), mitogen-activated protein kinases (MAPKs), tyrosine kinases, calcineurin, and phosphoinositide 3-

kinase (PI3K) (Fig. 8.3). Many of these pathways activate transcription factors, which interact with DNA and transcriptional co-factors to either promote or suppress transcription. Ultimately, messenger RNA (mRNA) and ribosomal RNA (rRNA) are transported out of the nucleus to the endoplasmic reticulum so that translation may occur on ribosomes. Protein synthesis is the net

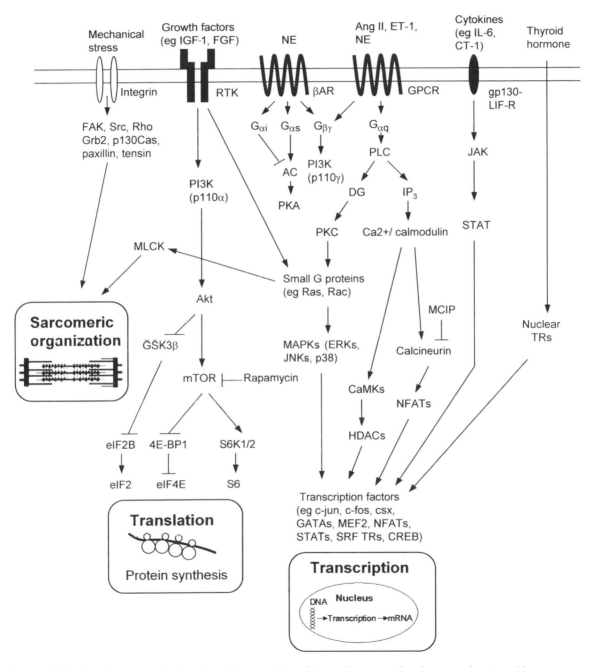

Figure 8.3. *Signaling cascades that have been implicated in mediating cardiac hypertrophy. See abbreviations list for abbreviations.*

result of increased ribosome biogenesis and protein translation.

Pathological versus physiological

Numerous signaling molecules are activated in models of pathological cardiac hypertrophy and failure. However, not all of these molecules or pathways will necessarily contribute to the pathological phenotype. Some signaling molecules may be activated as a protective mechanism against the pathological stimulus. To identify signaling cascades that may play distinct roles for the development of pathological and physiological hypertrophy, it has proven useful to characterize transgenic and knockout mice under basal conditions as well as subjecting them to pathological and physiological stimuli. In mice, models of pressure overload (e.g. aortic banding, minipump infusions of vasoactive substances) are most commonly used to induce pathological hypertrophy. Physiological models include treadmill running, freewheel running, and chronic swimming training. Below, we have described signaling molecules and cascades which appear to play an important role for the development of pathological or physiological hypertrophy.

G proteins

Ang II, ET-1, phenylephrine (PE), and $PGF_{2\alpha}$ all promote a hypertrophic response in cultured cardiac myocytes by binding to G protein-coupled receptors (GPCRs), which are coupled to heterotrimeric G proteins of the G_q family.[37,38] Ligand binding to these receptors activates G_q, which in turn activates a number of signaling pathways including phospholipase C (PLC), PKC, MAPK, and PKA.[37,39]

Cardiac-specific overexpression of $G_{\alpha q}$ in transgenic mice induced cardiac hypertrophy which was associated with altered cardiac gene expression and cardiac dysfunction.[40] Furthermore, transgenic mice expressing a G_{qI} peptide (specific for inhibiting G_q-coupled receptor signaling) in cardiac myocytes,[41] and mice lacking the G proteins $G_{\alpha q}$ and $G_{\alpha 11}$ in cardiac myocytes did not develop cardiac hypertrophy in response to pressure overload, suggesting the $G_{q/11}$ pathway is important for the induction of pathological cardiac hypertrophy.[42]

PLC/PKC pathway

The 'classic' signaling pathway of G_q is the activation of phospholipase C (PLC)β, which hydrolyzes phosphatidyl 4,5-biphosphate (PIP_2) to produce inositol 1,4,5-trisphosphate (IP_3) and diacylglycerol (DG). IP_3 mobilizes intracellular Ca^{2+}, and DG activates PKC.[11,36,43] The increase in cytosolic Ca^{2+} can mediate hypertrophic signals via Ca^{2+}/calmodulin-dependent enzymes and phosphatases (see Calcium signaling, page 123).

Norepinephrine (NE), Ang II, ET-1, PE and mechanical stress have all been reported to increase the activation of PKC in cardiac myocytes, leading to activation of MAPKs.[39,44] PKC appears to play a role in adaptive and maladaptive cardiac responses. Cardiac-specific transgenic mice overexpressing PKCβ developed pathological cardiac hypertrophy associated with cardiac dysfunction, fibrosis and premature death.[45–47] However, PKCβ does not appear essential for hypertrophy because PKCβ knockout mice responded normally to PE or aortic banding.[48] In contrast, mice overexpressing a cardiac-specific constitutively active (ca) PKCϵ or ca PKCδ mutant displayed concentric cardiac hypertrophy with a physiological phenotype, i.e. normal function and no fibrosis.[49] However, in response to ischemia–reperfusion, PKCϵ was protective, whereas PKCδ exacerbated the damage. Recently, it was shown that transgenic mice expressing a PKCδ inhibitor peptide and PKCϵ activator peptide in the heart displayed enhanced postischemic cardioprotection.[50] Cardiac-specific transgenic mice with enhanced PKCα activation did not display cardiac hypertrophy, but showed signs of cardiac dysfunction at one year of age.[51] Interestingly, enhanced PKCα activity caused aggressive heart failure and premature death in a model of pathological hypertrophy (Gαq mice), whereas inhibition of PKCα activity improved systolic and diastolic function.[51]

MAPK pathway

MAPKs are divided into three subfamilies based on the terminal kinase in the pathway: the extracellular signal-regulated kinases (ERKs), the c-jun N-terminal kinases (JNKs), and the p38 MAPKs.[39,52,53] Each of these subfamilies were shown to be activated in response to hypertrophic stimuli including mechanical stress, ET-1, NE and PE in cultured cardiac myocytes, as well as in hearts of animal models of pressure overload and failing human hearts.[11,44,53–63] These studies appeared to suggest that MAPKs were important regulators of pathological cardiac hypertrophy. However, subsequent studies using transgenic and knockout mice have questioned this conclusion:

■ To date, five ERK proteins have been identified (ERK1–ERK5). ERK1 and ERK2 are the most abundantly expressed and have been the best characterized in the heart.[64] ERK1 and ERK2 are regulated by the MAPK kinases: MEK1 and MEK2. Transgenic mice expressing ca MEK1 (specifically activates ERK1/2, but not JNKs or p38) in the heart displayed cardiac hypertrophy with a physiological phenotype.[65] In contrast, transgenic mice expressing a cardiac-specific ca MEK5 mutant (activates ERK5) displayed eccentric cardiac hypertrophy that progressed to dilated cardiomyopathy and resulted in premature death.[66]

■ JNK protein kinases are regulated by MKK4/7. In a study in which JNK activity was inhibited in the heart by a dominant negative (dn) MAPK kinase (MKK4), pressure overload-induced hypertrophy was attenuated,[60] suggesting that JNKs may be necessary regulators of pathological cardiac hypertrophy. However, a more direct assessment of JNKs suggests that they antagonize cardiac growth. Dn JNK1/2 transgenic mice and JNK1/2 gene-targeted mice displayed an enhanced hypertrophic response to pressure overload.[67]

■ P38 is regulated by MKK3/6. MKK3 and MKK6 transgenic mice developed heart failure but this was not associated with hypertrophy of cardiac myocytes.[15] In contrast, transgenic mice expressing a ca TAK1 (upstream of MKK3/6 and p38) mutant specifically in the heart, developed cardiac hypertrophy.[68] However, TAK1 is considerably upstream of p38, and may regulate other signaling effectors such as IκB kinase.[69] Recent work suggests that p38 signaling has an anti-hypertrophic role in myocytes within the adult heart. Transgenic

mice expressing dn mutants of MKK3, MKK6, and p38(α or β) displayed greater hypertrophic responses to pressure overload than non-transgenic controls.[70,71] However, even though p38 does not appear to promote hypertrophy, it may promote dilated cardiomyopathy.[69]

Calcium signaling

Calcium/calmodulin acts as an important second messenger for signals including Ang II, ET-1, α-adrenergic agents and mechanical stretch.[36,72,73] The best described calcium-dependent signaling molecules include calcineurin and calcium/calmodulin-dependent protein kinases (CaMKs). Calcineurin consists of a catalytic A subunit and a regulatory B subunit, and has been implicated as a regulator of the hypertrophic response in conjunction with the nuclear factor of activated T cells (NFAT) family of transcription factors.[74] NFAT translocates to the nucleus, where it associates with other transcription factors such as GATA-binding protein 4 (GATA4) and myocyte enhancer factor 2 (MEF2), to regulate the expression of target genes.[72,75] Cardiac-specific transgenic mice expressing activated forms of calcineurin or NFAT3 developed cardiac hypertrophy and heart failure.[76] Furthermore, calcineurin Aβ-deficient mice displayed an impaired hypertrophic response to pressure overload, Ang II infusion, or isoproterenol infusion.[77] Interestingly, calcineurin signaling has also been implicated for the induction of physiological cardiac hypertrophy. Cardiac-specific overexpression of the calcineurin inhibitory protein myocyte-enriched calcineurin-interacting protein (MCIP) 1 in transgenic mice was shown to inhibit the hypertrophic response in mice with cardiac-restricted overexpression of activated calcineurin, to β-adrenergic receptor stimulation, or exercise training.[78] More recently, it was reported that calcineurin–NFAT coupling participates in pathological, but not physiological, cardiac hypertrophy.[79] NFAT-luciferase reporter mice were subjected to physiological stimuli (exercise training, growth hormone (GH)–IGF-1 infusion) and pathological stimuli (pressure overload, myocardial infarction). NFAT luciferase reporter activity was upregulated in both pathological models, but not the physiological models.[79]

The JAK/STAT pathway

Janus kinases (JAKs) are protein tyrosine kinases which associate with cytokine receptors.[80] Leukemia inhibitory factor (LIF), cardiotrophin-1 (CT-1), and other members of the interleukin-6 (IL-6) cytokine family activate the glycoprotein 130 (gp130) receptor associated with the LIF receptor. Once activated, this receptor interacts with JAK1 causing its activation, which in turn phosphorylates the signal transducer and activator of transcription (STAT) class of transcription factors.[36,81,82] Gp130 appears critical for myocyte survival in response to a pathological stimulus. Mice with ventricular deletion of gp130 have normal cardiac structure and function under basal conditions, but display a rapid onset of dilated cardiomyopathy and massive induction of myocyte apoptosis in response to pressure overload.[83] Transgenic mice with cardiac-specific overexpression of STAT3 displayed cardiac hypertrophy that was protective against doxorubicin-induced cardiomyopathy.[84]

PI3K pathway

PI3Ks are a family of lipid kinases that induce signals by phosphorylating the hydroxyl group at position 3 of membrane lipid phosphoinositides.[85,86] Activation of PI3Ks is coupled to both receptor tyrosine kinases (e.g. insulin and IGF-1R) and GPCRs. There are multiple isoforms of PI3Ks which are divided into three classes (I, II, III) and which have a number of subunits.[85,86] The p110α isoform of PI3K (coupled to receptor tyrosine kinases) has been implicated in playing an important role for developmental, IGF-1R-induced, and exercise-induced cardiac growth.[25,87,88] In cardiac-specific transgenic mice expressing a ca PI3K(p110α) mutant, PI3K activity was elevated 6.5-fold, the heart weight/body weight ratio (HW/BW) was increased by approximately 20%, there was no evidence of fibrosis or myocardial disarray, and cardiac function and lifespan were normal.[13] In mice expressing a dn PI3K(p110α) mutant, PI3K activity was decreased by 77%, and these mice had significantly smaller hearts with normal cardiac function.[13] It was later shown that PI3K(p110α) played a critical role for the induction of exercise and IGF-1R-induced hypertrophy but not pressure overload-induced hyper-

trophy.[24,88] DnPI3K transgenic mice displayed significant hypertrophy in response to aortic banding, but not chronic swimming training or transgenic overexpression of IGF-1R. DnPI3K transgenic mice also showed significant dilation and cardiac dysfunction in response to pressure overload, suggesting that PI3K(p110α) may also be essential for maintaining contractile function in response to pathological stimuli.

PI3K(p110γ) does not regulate heart size under basal conditions,[89] but appears to play a role in the induction of hypertrophy in response to β-adrenergic receptor stimulation. PI3K(p110γ)-deficient mice infused with isoproterenol displayed an attenuated hypertrophic response.[90]

Akt, a serine-threonine kinase (also known as protein kinase B), is the best characterized target of PI3K.[91] Transgenic mice with chronic activation of Akt in the heart have demonstrated a range of phenotypes including massive hypertrophy associated with a pathological phenotype and death, cardiac hypertrophy with preserved systolic function and protection from ischemia–reperfusion injury, hypertrophy associated with enhanced myocardial contractility, and protection from ischemia–reperfusion injury without hypertrophy.[92–95] The range of phenotypes may be related, at least in part, to the degree of Akt expression and subcellular localization. By crossing ca Akt mice with dn PI3K mice it was shown that Akt is genetically downstream of PI3K (the heart size of double transgenics was similar to that of ca Akt mice alone). Furthermore, rapamycin attenuated the increase in heart size in ca Akt transgenics, demonstrating that Akt controlled heart size, at least in part, in a mammalian target of rapamycin (mTOR)-dependent manner.[92]

Glycogen synthase kinase 3 (GSK3), a cellular substrate of Akt, was initially identified for its role in glycogen metabolism. More recently, it has been shown that GSK3 is an important regulatory kinase with a number of cellular targets including cytoskeletal proteins and transcription factors. GSK3 consists of two isoforms: GSK3α and GSK3β.[96–98] Both isoforms are expressed in the heart, however to date most studies have only examined the role of GSK3β.[98] Studies suggest that GSK3β negatively regulates heart growth, and that inhibition of GSK3β by hypertrophic stimuli is an important mechanism for the stimulation of cardiac growth.[99–103]

Overview

A working model of signaling cascades, which may be important for the induction of pathological and physiological cardiac hypertrophy is shown in Fig. 8.4. IGF-1 activating PI3K(p110α) is considered a likely candidate responsible for mediating exercise-induced hypertrophy. Serum levels of IGF-1 were increased in competitive swimmers,[104] and rodents that underwent chronic exercise training.[105,106] Furthermore, cardiac formation of IGF-1, but not ET-1 or Ang II was higher in professional athletes than in control subjects.[107] Not all of the signaling molecules shown on the pathological cascade directly regulate heart size but they are likely to contribute to abnormal ventricular morphology and cardiac dysfunction.

Signaling pathways implicated for the induction of cardiac hypertrophy are considerably more complex than those displayed, and it is becoming evident that there is considerable cross-talk among many pathways, and that parallel signaling pathways are able to compensate for loss of some signaling molecules. For instance, the signaling mechanisms responsible for cardiac hypertrophy induced by pressure overload or cardiac-specific overexpression of $G_{\alpha q}$ do not appear to be identical. Mice deficient in mitogen-activated protein

kinase kinase 1 (MEKK1) displayed an attenuated response to $G_{\alpha q}$ overexpression,[108] but not to aortic banding.[109]

Ongoing questions regarding pathological and physiological hypertrophy will continue until future studies address some of the following issues. The most commonly used animal models of pathological hypertrophy (e.g. aortic banding, hypertension) represent a chronic pressure load that results in concentric hypertrophy. In contrast, models of physiological hypertrophy (e.g. treadmill, voluntary free-wheel, swimming) represent an intermittent volume load that results in eccentric hypertrophy. This raises the following question: can differences in signaling observed in models of pathological and physiological hypertrophy be explained by the duration of the insult (constant versus intermittent) or type of load (volume versus pressure)? Transgenic models with chronic activation of ca PI3K or ca MEK1, at least to some degree, argue that the duration of the stimulus alone is unlikely to account for the differences.[65,87] In both models, the physiological phenotype was not reported to progress to a pathological phenotype.

To correctly characterize cardiac hypertrophy as 'physiological', it has become apparent that such models should be studied in a variety of settings. For instance, PKCδ transgenic mice appeared to display physiological hypertrophy under basal conditions, but activation of PKCδ proved deleterious in

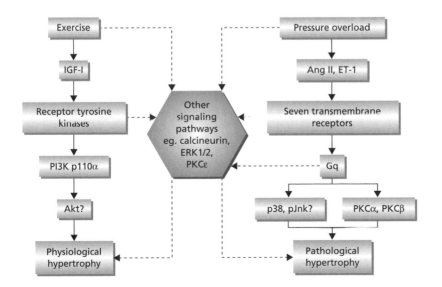

Figure 8.4. *A model illustrating signaling pathways that may be involved in the development of pathological and physiological cardiac hypertrophy. N.B. Not all of the signaling molecules shown on the pathological cascade directly regulate heart size but have been implicated in contributing to abnormal ventricular morphology and cardiac dysfunction.*

response to ischemia–reperfusion.[34,47] By contrast, transgenic mice expressing IGF-1R or PKCε displayed physiological hypertrophy under basal conditions, and were protective in models of pressure overload and ischemia–reperfusion, respectively.[47,88]

Another question that requires further investigation is whether postnatal cardiac growth and exercise-induced hypertrophy are mediated by similar mechanisms.[34] Interestingly, PI3K(p110α) appears essential for developmental heart growth and exercise-induced growth;[24,87] whereas the G_q signaling pathway appears critical for pathological growth,[42] but not developmental growth.[23,31] This suggests that at least some signaling pathways mediate developmental heart growth and exercise-induced growth.

Finally, studies examining the role of Akt in the heart have questioned whether the degree of activation and subcellular localization determine whether the resulting hypertrophy is pathological or physiological.[92–95] These are all important questions which will need further investigation.

Transcription factors

To regulate the long-term alterations in gene expression that are associated with cardiac hypertrophy, intracellular signal transduction pathways are coupled with transcription factors in the nucleus. Numerous transcription factors have been implicated for the activation of cardiac genes in response to hypertrophic stimuli. These include GATA4, GATA6, Csx/Nkx2.5, MEF2, c-jun, c-fos, c-myc, nuclear factor-κB and NFAT,[36,110,111] and have been discussed in detail in an earlier chapter (Chapter 1). Pathological and physiological cardiac hypertrophy appear to be mediated by some distinct signaling pathways.[24,40–42,87,107,112–116] It is therefore likely that the transcriptional regulation of pathological and physiological hypertrophy will be different.

Pathological versus physiological

Re-expression of fetal genes including atrial natriuretic peptide (ANP), brain natriuretic peptide (BNP) and genes for fetal isoforms of contractile proteins, such as skeletal α-actin, atrial myosin light chain (MLC)-1, and β-MHC (myosin heavy chain) have often been associated with models of pathological cardiac hypertrophy. This can be accompanied by downregulation of genes normally expressed at higher levels in the adult than in the embryonic ventricle, such as α-MHC and sarcoplasmic reticulum Ca^{2+} ATPase2A (SERCA2a).[9,28,117] In contrast, re-expression of the fetal gene program does not commonly occur in models of physiological hypertrophy (Fig. 8.2). The significance of changes in the fetal genes with regard to their direct effects on cardiac growth and phenotype is not well understood. However, there is now reasonably convincing evidence to suggest that ANP/ BNP signaling represents an anti-hypertrophic regulatory circuit within cardiac myocytes that antagonizes the growth response.[118–121] It is also noteworthy that a number of studies have demonstrated that cardiac hypertrophy can be dissociated from activation of the fetal gene program.[87,102]

Transcriptional effects of signaling molecules that are reported to result in physiological hypertrophy have been less studied. Microarray studies from ca Akt transgenic mice (hypertrophy with preserved systolic function),[93,122] IGF-1R transgenic mice,[88] ca PI3K transgenic mice,[88] and mice that have undergone exercise training,[123] have provided some insight into the transcriptional profile of physiological cardiac hypertrophy. It is noteworthy that some transgenic models of physiological hypertrophy have been associated with a modest upregulation of fetal genes (e.g. IGF-1R,[88] ca MEK1[65]). Both models were associated with enhanced cardiac function and expression of IGF-1R was protective against the induction of interstitial fibrosis in a pressure-overload model.[88] Consequently, caution must be taken when characterizing transgenic models. Until we have a better understanding of the biological significance of changes in the fetal gene program it would appear more accurate to categorize models of hypertrophy based on functional parameters, histological analysis, and responses to pathological stimuli.

Translational control

The etiologies of some human diseases have been linked with mutations in genes of the translational control machinery, and deregulation of translation has been associated with a range of cancers.[124,125] A defining character of cardiac growth is a net increase in protein synthesis above protein degradation. Under normal circumstances, these two processes are matched. More research is now focusing on studying the signaling pathways that are involved in modulating mRNA, translation and the possible deregulation that may occur during pathological cardiac hypertrophy.

Protein synthesis

Protein synthesis involves three steps: initiation, elongation, and termination.[126,127] The overall rate of protein synthesis is determined by both its efficiency (the rate at which new peptide chains are synthesized per ribosome), and its capacity (the relative abundance of ribosomes, ribosome biogenesis).

Efficiency of protein synthesis

Efficiency is regulated principally at the level of initiation i.e. formation of the methionyl-tRNA initiation complex.[126,128] The molecular mechanisms regulating translational initiation and elongation have not been studied extensively in the heart, but a number of signaling pathways have been shown to alter the activities of initiation and elongation factors in various cell types. These include PKC, cAMP, MAPK, calcium-calmodulin, and PI3K.[126,129,130] In the heart, PI3K-dependent pathways have received the greatest attention. Downstream targets of the PI3K pathway, ribosomal S6 kinases (S6Ks: S6K1 and S6K2) are thought to interact with the translational machinery of the ribosome by the phosphorylation of S6. S6 is a component of 40S ribosomal proteins, positioned at the interface between 40S and 60S ribosomal proteins, and localized to regions involved in mRNA and transfer RNA (tRNA) recognition.[131,132] It has been suggested that S6 phosphorylation maybe a rate-limiting event in protein synthesis, because it is believed to modulate the translation of a family of mRNAs termed 5'TOPs (mRNAs characterized by a terminal oligopyrimidine tract in the 5' untranslated region).[133,134] These mRNAs encode components of the protein synthetic apparatus such as ribosomal subunit pro-

teins and many biosynthetic enzymes which are essential for cell growth.[131,135] S6K1 activity and S6 phosphorylation were elevated in hearts of mice subjected to pressure overload.[136] However, the role of S6Ks regulating heart size *in vivo* remains unclear. Deletion of S6Ks in mice did not attenuate cardiac hypertrophy induced by pressure overload, exercise training or transgenic expression of IGF-1R/ca PI3K.[137] Other protein synthesis regulators include eukaryotic translation initiation factor 4E (eIF4E) and eukaryotic translation initiation factor 2Bε (eIF2Bε).[126,128–130] Phosphorylation of eIF4E (promotes protein translation initiation) was elevated in a canine model of hypertrophy induced by acute pressure overload,[138] whereas phosporylation of eIF2Bε (inhibits protein translation initiation) was decreased in a rat model of myocardial infarction.[139] Ras/ERK signaling was reported to activate protein synthesis via regulation of S6K1 and eIF4E in adult cardiac myocytes in culture.[140] Recently, it was shown that phosphorylation of eIF2Bε by GSK3β regulated β-adrenergic induced increases in cardiac myocyte size *in vitro*.[139]

The mammalian target of rapamycin (mTOR) is thought to control the translational machinery via the activation of S6K1 and via inhibition of 4E binding protein 1 (4E-BP1) (eIF4E inhibitor).[141] Rapamycin, a lipophilic macrolide, inhibits growth by forming a gain-of-function inhibitory complex with FKBP12 (FK506-binding protein, MW of 12 kD).[134,141] This complex binds to mTOR, preventing activation of mTOR targets including S6K1 and the 40S ribosomal S6 protein. Rapamycin inhibited Ang II- and PE-induced increases in protein synthesis in cardiac myocytes *in vitro*.[142,143] More recently, it was reported that rapamycin attenuated and regressed pressure overload-induced cardiac hypertrophy in mice.[136,144]

Capacity of protein synthesis

In the long term, ribosome biogenesis is considered most likely to account for increased protein synthesis during cardiac hypertrophy. This is because at least 80–90% of existing ribosomes in adult hearts (not undergoing hypertrophy) are in the form of polysomes (i.e. they are already engaged in synthesizing proteins).[145,146] In cardiac myocytes it has been proposed that hypertrophic stimuli, by largely undefined pathways, increase the activity of the transcription factor upstream binding factor (UBF), resulting in accelerated rates of the 45S ribosomal gene (rDNA) transcription and rRNA synthesis. The increased pool of rRNA together with an increased translation of ribosomal proteins results in increased translational capacity.[145,147]

Studies with neonatal cardiac myocytes in culture and intact hearts suggest that transcription of the ribosomal genes is a key regulatory event for protein synthesis. Increased rates of rRNA synthesis were observed during beating-induced hypertrophy of cultured neonatal cardiac myocytes. Similar increases in rDNA transcription have been observed in the heart in response to hypertrophic stimuli such as NE, ET-1, adrenergic agonists, and contraction. In some cases, the increase in the rate of rDNA transcription was correlated with the activation of UBF.[146,148–150] More recently, it was shown that cardiac hypertrophy induced by pressure overload in mice was associated with increased expression of UBF.[151]

Sarcomeric organization

The sarcomere is the basic contractile unit of the heart. The formation of sarcomeres is a complex event that involves the synthesis of more than 50 proteins.[2,152] These proteins then become aggregated into filaments, which in turn are organized into specific three-dimensional arrays and aligned with respect to other contractile elements already present in the cell.[2,153] The purpose of cardiac hypertrophy is to accommodate the increased workload (e.g. strain of pressure or volume overload) by increasing the contractile capacity. Therefore, organization of sarcomeres, and thereby an increase in the contractile units, is an essential component for maximizing force generation. The transition from hypertrophy to failure has been associated with loss of contractile filaments in the presence of microtubule densification and desmin disorganization, and loss of proteins of the sarcomeric skeleton (titin, α-actin, myomesin).[154,155] Defects of sarcomeric proteins including cardiac troponin I or T, β-MHC, α-MHC, MLC, α-

tropomyosin, titin and actin have all been associated with familial hypertrophic cardiomyopathy in humans.[156] This is discussed in detail in a later chapter.

Signaling molecules that have been implicated for sarcomere organization include Rac1, RhoA,[157–159] calcium/calmodulin-dependent regulation of myosin light chain kinase (MLCK) and MLC-2v,[160] focal adhesion kinase (FAK) and p130Cas (substrates for the nonreceptor tyrosine kinase Src).[161]

Cell death

Cardiac growth is also regulated by cell death. Cell death can occur in an uncontrolled manner via necrosis or by a highly regulated programmed mechanism termed apoptosis.[162–165] Apoptosis is controlled by a complex interaction of numerous pro-survival and pro-death signals. Low levels of cardiac apoptosis are considered sufficient to induce heart failure,[166] and inhibition of cardiac myocyte apoptosis with a caspase inhibitor improved cardiac function and prevented mortality in pregnant $G_{\alpha q}$ mice.[167]

Consequently, the differential activation of pro-survival and pro-death signals in pathological versus physiological hypertrophy is likely to be an important factor that contributes to the distinct phenotypes of pathological and physiological hypertrophy.

Growth factors (e.g. IGF-1), and cytokines (e.g. CT-1) are reported to have anti-apoptotic effects, in part, via PI3K and/or ERK signaling.[163,164,168,169] In contrast, signaling via $G_{\alpha q}$ or $G_{\alpha s}$ in transgenic mice was shown to promote cardiac myocyte apoptosis.[170,171] Calcineurin has been found to have both pro- and anti-apoptotic effects, depending on the cellular context.[172–174] More recently, it was reported that the mitochondrial death protein Nix was induced by G_q-dependent cardiac hypertrophy and triggered apoptosis.[175] JNK, p38 MAPK, and loss of Gp130 have been associated with increased apoptosis in cardiac myocytes.[63,83,164]

A recent study demonstrated that cardiac myocytes from rats with pathological hypertrophy (Dahl salt-sensitive rats on a high-salt diet) were more sensitive to apoptotic stimulation than myocytes from rats with physiological hypertrophy (treadmill induced). Furthermore, the physiological model was associated with changes in Bcl-2 family members and caspases favoring survival, whereas the pathological model was associated with changes in mitochondrion- and death receptor-mediated pathways (e.g. a decreased Bcl-x_L/Bax ratio, increase in Fas) that have previously been reported in models of pathological hypertrophy and heart failure.[176]

Structural remodeling

Structural remodeling, i.e. changes in the size, shape and function of the heart, is a key determinant of whether cardiac hypertrophy is classified as pathological or physiological. Pathological hypertrophy is associated with structural rearrangement of components of the normal chamber wall that involves cardiac myocyte hypertrophy, cardiac fibroblast proliferation, fibrosis, and cell death (apoptotic and necrotic).[177] Under normal conditions, the fibrillar collagen network ensures the structural integrity of the adjoining myocytes, providing the means by which myocyte shortening is translated into ventricular pump function.[177,178] Fibrosis is an integral feature characteristic of pathological hypertrophy. Accumulation of type 1 collagen, the main fibrillar collagen found in cardiac fibrosis, can stiffen the ventricles and impede contraction and relaxation, impair the electrical coupling of cardiac myocytes with extracellular matrix (ECM) proteins, and reduce capillary density and increase oxygen diffusion distances, leading to hypoxia of myocytes.[177] In contrast, physiological models of cardiac hypertrophy are associated with an increase in myocyte size but not interstitial fibrosis (Fig. 8.2).

A number of receptors, signaling molecules, transcription factors and proteins have been implicated in the development of cardiac fibrosis based on studies from genetically engineered mice. These include β_1-adrenergic receptor, AT_1 receptor, TGF-β, tumor necrosis factor (TNF)-α, G proteins, PKCβ_2, RhoA, calcineurin, calsequestrin, NFATs, Csx/Nkx2.5, and serum response factor.[177,178]

Summary

Pathological and physiological cardiac hypertrophy are both associated with an increase in heart weight due to an increase in myocyte volume; however, pathological hypertrophy is also associated with a complex array of events including re-expression of the fetal gene program, apoptosis, necrosis, alterations in the ECM and remodeling. In the last decade, the generation and characterization of transgenic and knockout mice have greatly accelerated progress made in the field of cardiac hypertrophy and failure. More recently, subjecting these models to pathological and physiological stimuli has helped dissect signaling pathways that appear to play distinct roles for the induction of physiological versus pathological cardiac hypertrophy. This could provide a novel therapeutic strategy for treating patients with heart failure, i.e. inhibiting signaling cascades that cause pathological hypertrophy, while stimulating signaling cascades that promote physiological hypertrophy.

References

1. Rakusan K. Cardiac growth, maturation, and aging. In: Zak R (ed). Growth of the Heart in Health and Disease. New York: Raven Press; 1984: 131–64.
2. Zak R. Overview of the growth process/ factors controlling cardiac growth. In: Zak R (ed). Growth of the Heart in Health and Disease. New York: Raven Press; 1984: 1–24, 165–85.
3. Hudlicka O, Brown MD. Postnatal growth of the heart and its blood vessels. J Vasc Res 1996; 33: 266–87.
4. Pasumarthi KB, Field LJ. Cardiomyocyte cell cycle regulation. Circ Res 2002; 90: 1044–54.
5. Anversa P, Leri A, Kajstura J et al. Myocyte growth and cardiac repair. J Mol Cell Cardiol 2002; 34: 91–105.
6. Anversa P, Nadal-Ginard B. Myocyte renewal and ventricular remodelling. Nature 2002; 415: 240–3.
7. Nakagawa M, Hamaoka K, Hattori T et al. Postnatal DNA synthesis in hearts of mice: autoradiographic and cytofluorometric investigations. Cardiovasc Res 1988; 22: 575–83.
8. Soonpaa MH, Kim KK, Pajak L et al. Cardiomyocyte

DNA synthesis and binucleation during murine development. Am J Physiol Heart Circ Physiol 1996; 271: H2183–9.

9. MacLellan WR, Schneider MD. Genetic dissection of cardiac growth control pathways. Annu Rev Physiol 2000; 62: 289–319.

10. Cooper Gt. Cardiocyte adaptation to chronically altered load. Annu Rev Physiol 1987; 49: 501–18.

11. Sugden PH, Clerk A. Cellular mechanisms of cardiac hypertrophy. J Mol Med 1998; 76: 725–46.

12. Hunter JJ, Chien KR. Signaling pathways for cardiac hypertrophy and failure. N Engl J Med 1999; 341: 1276–83.

13. Weber KT, Brilla CG. Pathological hypertrophy and cardiac interstitium. Fibrosis and renin-angiotensin-aldosterone system. Circulation 1991; 83: 1849–65.

14. Weber KT, Brilla CG, Janicki JS. Myocardial fibrosis: functional significance and regulatory factors. Cardiovasc Res 1993; 27: 341–8.

15. Liao P, Georgakopoulos D, Kovacs A et al. The in vivo role of p38 MAP kinases in cardiac remodeling and restrictive cardiomyopathy. Proc Natl Acad Sci USA 2001; 98: 12283–8.

16. Petrich BG, Gong X, Lerner DL et al. c-Jun N-terminal kinase activation mediates downregulation of connexin43 in cardiomyocytes. Circ Res 2002; 91: 640–7.

17. Petrich BG, Molkentin JD, Wang Y. Temporal activation of c-Jun N-terminal kinase in adult transgenic heart via cre-loxP-mediated DNA recombination. FASEB J 2003; 17: 749–51.

18. Yamamoto S, Yang G, Zablocki D et al. Activation of Mst1 causes dilated cardiomyopathy by stimulating apoptosis without compensatory ventricular myocyte hypertrophy. J Clin Invest 2003; 111: 1463–74.

19. Ferrans VJ. Cardiac hypertrophy: Morphological aspects. In: Zak R (ed). Growth of the Heart in Health and Disease. New York: Raven Press; 1984: 187–239.

20. Schaible TF, Scheuer J. Response of the heart to exercise training. In: Zak R (ed). Growth of the Heart in Health and Disease. New York: Raven Press; 1984: 381–420.

21. Thomas LR, Douglas PS. Echocardiographic findings in athletes. In: Thompson PD (ed). Exercise and Sports Cardiology. New York: McGraw-Hill; 2001: pp 43–70.

22. Kaplan ML, Cheslow Y, Vikstrom K et al. Cardiac adaptations to chronic exercise in mice. Am J Physiol Heart Circ Physiol 1994; 267: H1167–73.

23. Fagard RH. Impact of different sports and training on

24. McMullen JR, Shioi T, Zhang L et al. Phosphoinositide 3-kinase (p110α) plays a critical role for the induction of physiological, but not pathological, cardiac hypertrophy. Proc Natl Acad Sci USA 2003; 100: 12355–60.

25. Hildick-Smith DJ, Shapiro LM. Echocardiographic differentiation of pathological and physiological left ventricular hypertrophy. Heart 2001; 85: 615–19.

26. Cohn JN, Bristow MR, Chien KR et al. Report of the National Heart, Lung, and Blood Institute Special Emphasis Panel on Heart Failure Research. Circulation 1997; 95: 766–70.

27. Levy D, Garrison RJ, Savage DD et al. Prognostic implications of echocardiographically determined left ventricular mass in the Framingham Heart Study. N Engl J Med 1990; 322: 1561–6.

28. Izumo S, Nadal-Ginard B, Mahdavi V. Proto-oncogene induction and reprogramming of cardiac gene expression produced by pressure overload. Proc Natl Acad Sci USA 1988; 85: 339–43.

29. Iemitsu M, Miyauchi T, Maeda S et al. Physiological and pathological cardiac hypertrophy induce different molecular phenotypes in the rat. Am J Physiol Regul Integr Comp Physiol 2001; 281: R2029–36.

30. Richey PA, Brown SP. Pathological versus physiological left ventricular hypertrophy: a review. J Sports Sci 1998; 16: 129–41.

31. Scheuer J, Malhotra A, Hirsch C et al. Physiologic cardiac hypertrophy corrects contractile protein abnormalities associated with pathologic hypertrophy in rats. J Clin Invest 1982; 70: 1300–5.

32. Orenstein TL, Parker TG, Butany JW et al. Favorable left ventricular remodeling following large myocardial infarction by exercise training. Effect on ventricular morphology and gene expression. J Clin Invest 1995; 96: 858–66.

33. Schaible TF, Malhotra A, Ciambrone GJ et al. Chronic swimming reverses cardiac dysfunction and myosin abnormalities in hypertensive rats. J Appl Physiol 1986; 60: 1435–41.

34. Dorn GW 2nd, Robbins J, Sugden PH. Phenotyping hypertrophy: eschew obfuscation. Circ Res 2003; 92: 1171–5.

35. Chien KR, Zhu H, Knowlton KU et al. Transcriptional regulation during cardiac growth and development. Annu Rev Physiol 1993; 55: 77–95.

36. Aoki H, Izumo S. Signal transduction of cardiac myocyte hypertrophy. In: Sperelakis N, Kurachi Y,

Terzic A, Cohen MV (eds). Heart Physiology and Pathology (4e). San Diego: Academic Press; 2001: 1065–86.

37. McKinsey TA, Olson EN. Cardiac hypertrophy: sorting out the circuitry. Curr Opin Genet Dev 1999; 9: 267–74.

38. Clerk A, Sugden PH. Small guanine nucleotide-binding proteins and myocardial hypertrophy. Circ Res 2000; 86: 1019–23.

39. Clerk A, Sugden PH. Activation of protein kinase cascades in the heart by hypertrophic G protein-coupled receptor agonists. Am J Cardiol 1999; 83: 64H–69H.

40. D'Angelo DD, Sakata Y, Lorenz JN et al. Transgenic Galphaq overexpression induces cardiac contractile failure in mice. Proc Natl Acad Sci USA 1997; 94: 8121–6.

41. Akhter SA, Luttrell LM, Rockman HA et al. Targeting the receptor-G_q interface to inhibit in vivo pressure overload myocardial hypertrophy. Science 1998; 280: 574–7.

42. Wettschureck N, Rutten H, Zywietz A et al. Absence of pressure overload induced myocardial hypertrophy after conditional inactivation of Galphaq/Galpha11 in cardiomyocytes. Nat Med 2001; 7: 1236–40.

43. Adams JW, Brown JH. G-proteins in growth and apoptosis: lessons from the heart. Oncogene 2001; 20: 1626–34.

44. Clerk A, Bogoyevitch MA, Anderson MB et al. Differential activation of protein kinase C isoforms by endothelin-1 and phenylephrine and subsequent stimulation of p42 and p44 mitogen-activated protein kinases in ventricular myocytes cultured from neonatal rat hearts. J Biol Chem 1994; 269: 32848–57.

45. Bowman JC, Steinberg SF, Jiang T et al. Expression of protein kinase C beta in the heart causes hypertrophy in adult mice and sudden death in neonates. J Clin Invest 1997; 100: 2189–95.

46. Wakasaki H, Koya D, Schoen FJ et al. Targeted overexpression of protein kinase C beta2 isoform in myocardium causes cardiomyopathy. Proc Natl Acad Sci USA 1997; 94: 9320–5.

47. Chen L, Hahn H, Wu G et al. Opposing cardioprotective actions and parallel hypertrophic effects of delta PKC and epsilon PKC. Proc Natl Acad Sci USA 2001; 98: 11114–19.

48. Roman BB, Geenen DL, Leitges M et al. PKC-beta is not necessary for cardiac hypertrophy. Am J Physiol Heart Circ Physiol 2001; 280: H2264–70.

49. Takeishi Y, Ping P, Bolli R et al. Transgenic over-expression of constitutively active protein kinase C epsilon causes concentric cardiac hypertrophy. Circ Res 2000; 86: 1218–23.

50. Inagaki K, Hahn HS, Dorn GW 2nd et al. Additive protection of the ischemic heart ex vivo by combined treatment with delta-protein kinase C inhibitor and epsilon-protein kinase C activator. Circulation 2003; 108: 869–75.

51. Hahn HS, Marreez Y, Odley A et al. Protein kinase C alpha negatively regulates systolic and diastolic function in pathological hypertrophy. Circ Res 2003; 93: 1111–19.

52. Widmann C, Gibson S, Jarpe MB et al. Mitogen-activated protein kinase: conservation of a three-kinase module from yeast to human. Physiol Rev 1999; 79: 143–80.

53. Pearson G, Robinson F, Beers Gibson T et al. Mitogen-activated protein (MAP) kinase pathways: regulation and physiological functions. Endocr Rev 2001; 22: 153–83.

54. Takeishi Y, Huang Q, Abe J et al. Src and multiple MAP kinase activation in cardiac hypertrophy and congestive heart failure under chronic pressure-overload: comparison with acute mechanical stretch. J Mol Cell Cardiol 2001; 33: 1637–48.

55. Rapacciuolo A, Esposito G, Caron K et al. Important role of endogenous norepinephrine and epinephrine in the development of in vivo pressure-overload cardiac hypertrophy. J Am Coll Cardiol 2001; 38: 876–82.

56. Komuro I, Kudo S, Yamazaki T et al. Mechanical stretch activates the stress-activated protein kinases in cardiac myocytes. FASEB J 1996; 10: 631–6.

57. Choukroun G, Hajjar R, Kyriakis JM et al. Role of the stress-activated protein kinases in endothelin-induced cardiomyocyte hypertrophy. J Clin Invest 1998; 102: 1311–20.

58. Li WG, Zaheer A, Coppey L et al. Activation of JNK in the remote myocardium after large myocardial infarction in rats. Biochem Biophys Res Commun 1998; 246: 816–20.

59. Yano M, Kim S, Izumi Y et al. Differential activation of cardiac c-jun amino-terminal kinase and extracellular signal-regulated kinase in angiotensin II-mediated hypertension. Circ Res 1998; 83: 752–60.

60. Choukroun G, Hajjar R, Fry S et al. Regulation of cardiac hypertrophy in vivo by the stress-activated protein kinases/c-Jun NH(2)-terminal kinases. J Clin Invest 1999; 104: 391–8.

61. Cook SA, Sugden PH, Clerk A. Activation of c-Jun N-terminal kinases and p38-mitogen-activated protein kinases in human heart failure secondary to

ischaemic heart disease. J Mol Cell Cardiol 1999; 31: 1429–34.

62. Clerk A, Michael A, Sugden PH. Stimulation of the p38 mitogen-activated protein kinase pathway in neonatal rat ventricular myocytes by the G protein-coupled receptor agonists, endothelin-1 and phenylephrine: a role in cardiac myocyte hypertrophy? J Cell Biol 1998; 142: 523–35.

63. Wang Y, Huang S, Sah VP et al. Cardiac muscle cell hypertrophy and apoptosis induced by distinct members of the p38 mitogen-activated protein kinase family. J Biol Chem 1998; 273: 2161–8.

64. Bueno OF, Molkentin JD. Involvement of extracellular signal-regulated kinases 1/2 in cardiac hypertrophy and cell death. Circ Res 2002; 91: 776–81.

65. Bueno OF, De Windt LJ, Tymitz KM et al. The MEK1-ERK1/2 signaling pathway promotes compensated cardiac hypertrophy in transgenic mice. EMBO J 2000; 19: 6341–50.

66. Nicol RL, Frey N, Pearson G et al. Activated MEK5 induces serial assembly of sarcomeres and eccentric cardiac hypertrophy. EMBO J 2001; 20: 2757–67.

67. Liang Q, Bueno OF, Wilkins BJ et al. c-Jun N-terminal kinases (JNK) antagonize cardiac growth through cross-talk with calcineurin-NFAT signaling. EMBO J 2003; 22: 5079–89.

68. Zhang D, Gaussin V, Taffet GE et al. TAK1 is activated in the myocardium after pressure overload and is sufficient to provoke heart failure in transgenic mice. Nat Med 2000; 6: 556–63.

69. Liang Q, Molkentin JD. Redefining the roles of p38 and JNK signaling in cardiac hypertrophy: dichotomy between cultured myocytes and animal models. J Mol Cell Cardiol 2003; 35: 1385–94.

70. Braz JC, Bueno OF, Liang Q et al. Targeted inhibition of p38 MAPK promotes hypertrophic cardiomyopathy through upregulation of calcineurin-NFAT signaling. J Clin Invest 2003; 111: 1475–86.

71. Zhang S, Weinheimer C, Courtois M et al. The role of the Grb2-p38 MAPK signaling pathway in cardiac hypertrophy and fibrosis. J Clin Invest 2003; 111(6): 833–41.

72. Frey N, McKinsey TA, Olson EN. Decoding calcium signals involved in cardiac growth and function. Nat Med 2000; 6: 1221–7.

73. Sugden PH. Signalling pathways in cardiac myocyte hypertrophy. Ann Med 2001; 33: 611–22.

74. Olson EN, Williams RS. Calcineurin signaling and muscle remodeling. Cell 2000; 101: 689–92.

75. Wilkins BJ, Molkentin JD. Calcineurin and cardiac hypertrophy: where have we been? Where are we going? J Physiol 2002; 541: 1–8.

76. Molkentin JD, Lu JR, Antos CL et al. A calcineurin-dependent transcriptional pathway for cardiac hypertrophy. Cell 1998; 93: 215–28.

77. Bueno OF, Wilkins BJ, Tymitz KM et al. Impaired cardiac hypertrophic response in calcineurin A beta-deficient mice. Proc Natl Acad Sci USA 2002; 99: 4586–91.

78. Rothermel BA, McKinsey TA, Vega RB et al. Myocyte-enriched calcineurin-interacting protein, MCIP1, inhibits cardiac hypertrophy in vivo. Proc Natl Acad Sci USA 2001; 98: 3328–33.

79. Wilkins BJ, Dai YS, Bueno OF et al. Calcineurin/NFAT coupling participates in pathological, but not physiological, cardiac hypertrophy. Circ Res 2004; 94: 110–18.

80. Pellegrini S, Dusanter-Fourt I. The structure, regulation and function of the Janus kinases (JAKs) and the signal transducers and activators of transcription (STATs). Eur J Biochem 1997; 248: 615–33.

81. Kodama H, Fukuda K, Pan J et al. Leukemia inhibitory factor, a potent cardiac hypertrophic cytokine, activates the JAK/STAT pathway in rat cardiomyocytes. Circ Res 1997; 81: 656–63.

82. Molkentin JD, Dorn IG, 2nd. Cytoplasmic signaling pathways that regulate cardiac hypertrophy. Annu Rev Physiol 2001; 63: 391–426.

83. Hirota H, Chen J, Betz UA et al. Loss of a gp130 cardiac muscle cell survival pathway is a critical event in the onset of heart failure during biomechanical stress. Cell 1999; 97: 189–98.

84. Kunisada K, Negoro S, Tone E et al. Signal transducer and activator of transcription 3 in the heart transduces not only a hypertrophic signal but a protective signal against doxorubicin-induced cardiomyopathy. Proc Natl Acad Sci USA 2000; 97: 315–19.

85. Toker A, Cantley LC. Signalling through the lipid products of phosphoinositide-3-OH kinase. Nature 1997; 387: 673–6.

86. Vanhaesebroeck B, Leevers SJ, Panayotou G et al. Phosphoinositide 3-kinases: a conserved family of signal transducers. Trends Biochem Sci 1997; 22: 267–72.

87. Shioi T, Kang PM, Douglas PS et al. The conserved phosphoinositide 3-kinase pathway determines heart size in mice. EMBO J 2000; 19: 2537–48.

88. McMullen JR, Shioi T, Huang WY et al. The insulin-like growth factor 1 receptor induces physiological heart growth via the phosphoinositide 3-kinase(p110α) pathway. J Biol Chem 2004; 279: 4782–93.

89. Crackower MA, Oudit GY, Kozieradzki I et al.

Regulation of myocardial contractility and cell size by distinct PI3K-PTEN signaling pathways. Cell 2002; 110: 737–49.

90. Oudit GY, Crackower MA, Eriksson U et al. Phosphoinositide 3-kinase gamma-deficient mice are protected from isoproterenol-induced heart failure. Circulation 2003; 108: 2147–52.

91. Chan TO, Rittenhouse SE, Tsichlis PN. AKT/PKB and other D3 phosphoinositide-regulated kinases: kinase activation by phosphoinositide-dependent phosphorylation. Annu Rev Biochem 1999; 68: 965–1014.

92. Shioi T, McMullen JR, Kang PM et al. Akt/protein kinase B promotes organ growth in transgenic mice. Mol Cell Biol 2002; 22: 2799–809.

93. Matsui T, Li L, Wu JC et al. Phenotypic spectrum caused by transgenic overexpression of activated Akt in the heart. J Biol Chem 2002; 277: 22896–901.

94. Condorelli G, Drusco A, Stassi G et al. Akt induces enhanced myocardial contractility and cell size in vivo in transgenic mice. Proc Natl Acad Sci USA 2002; 99: 12333–8.

95. Shiraishi I, Melendez J, Ahn Y et al. Nuclear targeting of Akt enhances kinase activity and survival of cardiomyocytes. Circ Res 2004; 97: 884–91.

96. Ferkey DM, Kimelman D. GSK-3: new thoughts on an old enzyme. Dev Biol 2000; 225: 471–9.

97. Harwood AJ. Regulation of GSK-3: a cellular multiprocessor. Cell 2001; 105: 821–4.

98. Hardt SE, Sadoshima J. Glycogen synthase kinase-3beta: a novel regulator of cardiac hypertrophy and development. Circ Res 2002; 90: 1055–63.

99. Haq S, Choukroun G, Kang ZB et al. Glycogen synthase kinase-3beta is a negative regulator of cardiomyocyte hypertrophy. J Cell Biol 2000; 151: 117–30.

100. Morisco C, Zebrowski D, Condorelli G et al. The Akt-glycogen synthase kinase 3beta pathway regulates transcription of atrial natriuretic factor induced by beta-adrenergic receptor stimulation in cardiac myocytes. J Biol Chem 2000; 275: 14466–75.

101. Morisco C, Seta K, Hardt SE et al. Glycogen synthase kinase 3beta regulates GATA4 in cardiac myocytes. J Biol Chem 2001; 276: 28586–97.

102. Antos CL, McKinsey TA, Frey N et al. Activated glycogen synthase-3 beta suppresses cardiac hypertrophy in vivo. Proc Natl Acad Sci USA 2002; 99: 907–12.

103. Badorff C, Ruetten H, Mueller S et al. Fas receptor signaling inhibits glycogen synthase kinase 3 beta and induces cardiac hypertrophy following pressure overload. J Clin Invest 2002; 109: 373–81.

104. Koziris LP, Hickson RC, Chatterton RT, Jr. et al. Serum levels of total and free IGF-I and IGFBP-3 are increased and maintained in long-term training. J Appl Physiol 1999; 86: 1436–42.

105. Kodama Y, Umemura Y, Nagasawa S et al. Exercise and mechanical loading increase periosteal bone formation and whole bone strength in C57BL/6J mice but not in C3H/Hej mice. Calcif Tissue Int 2000; 66: 298–306.

106. Yeh JK, Aloia JF, Chen M et al. Effect of growth hormone administration and treadmill exercise on serum and skeletal IGF-1 in rats. Am J Physiol Endocrinol Metab 1994; 266: E129–35.

107. Neri Serneri GG, Boddi M, Modesti PA et al. Increased cardiac sympathetic activity and insulin-like growth factor-1 formation are associated with physiological hypertrophy in athletes. Circ Res 2001; 89: 977–82.

108. Minamino T, Yujiri T, Terada N et al. MEKK1 is essential for cardiac hypertrophy and dysfunction induced by G_q. Proc Natl Acad Sci USA 2002; 99: 3866–71.

109. Sadoshima J, Montagne O, Wang Q et al. The MEKK1-JNK pathway plays a protective role in pressure overload but does not mediate cardiac hypertrophy. J Clin Invest 2002; 110: 271–9.

110. Sadoshima J, Izumo S. The cellular and molecular response of cardiac myocytes to mechanical stress. Annu Rev Physiol 1997; 59: 551–71.

111. Akazawa H, Komuro I. Roles of cardiac transcription factors in cardiac hypertrophy. Circ Res 2003; 92: 1079–88.

112. Duerr RL, Huang S, Miraliakbar HR et al. Insulin-like growth factor-1 enhances ventricular hypertrophy and function during the onset of experimental cardiac failure. J Clin Invest 1995; 95: 619–27.

113. Tanaka N, Ryoke T, Hongo M et al. Effects of growth hormone and IGF-1 on cardiac hypertrophy and gene expression in mice. Am J Physiol Heart Circ Physiol 1998; 275: H393–9.

114. Geenen DL, Malhotra A, Buttrick PM. Angiotensin receptor 1 blockade does not prevent physiological cardiac hypertrophy in the adult rat. J Appl Physiol 1996; 81: 816–21.

115. Lembo G, Rockman HA, Hunter JJ et al. Elevated blood pressure and enhanced myocardial contractility in mice with severe IGF-1 deficiency. J Clin Invest 1996; 98: 2648–55.

116. Neri Serneri GG, Boddi M, Poggesi L et al. Activation of cardiac renin-angiotensin system in unstable angina. J Am Coll Cardiol 2001; 38: 49–55.

117. Chien KR, Knowlton KU, Zhu H et al. Regulation of

cardiac gene expression during myocardial growth and hypertrophy: molecular studies of an adaptive physiologic response. FASEB J 1991; 5: 3037–46.

118. Horio T, Nishikimi T, Yoshihara F et al. Inhibitory regulation of hypertrophy by endogenous atrial natriuretic peptide in cultured cardiac myocytes. Hypertension 2000; 35: 19–24.

119. Kishimoto I, Rossi K, Garbers DL. A genetic model provides evidence that the receptor for atrial natriuretic peptide (guanylyl cyclase-A) inhibits cardiac ventricular myocyte hypertrophy. Proc Natl Acad Sci USA 2001; 98: 2703–6.

120. Holtwick R, van Eickels M, Skryabin BV et al. Pressure-independent cardiac hypertrophy in mice with cardiomyocyte-restricted inactivation of the atrial natriuretic peptide receptor guanylyl cyclase-A. J Clin Invest 2003; 111: 1399–407.

121. Molkentin JD. A friend within the heart: natriuretic peptide receptor signaling. J Clin Invest 2003; 111: 1275–7.

122. Cook SA, Matsui T, Li L et al. Transcriptional effects of chronic Akt activation in the heart. J Biol Chem 2002; 277: 22528–33.

123. CardioGenomics. Genomics of Cardiovascular Development, Adaptation, and Remodeling. NHLBI Program for Genomic Applications, Harvard Medical School; 2003. http://www.cardiogenomics.org (accessed 3 May 2005).

124. Calkhoven CF, Muller C, Leutz A. Translational control of gene expression and disease. Trends Mol Med 2002; 8: 577–83.

125. Abbott CM, Proud CG. Translation factors: in sickness and in health. Trends Biochem Sci 2004; 29: 25–31.

126. Rhoads RE. Signal transduction pathways that regulate eukaryotic protein synthesis. J Biol Chem 1999; 274: 30337–40.

127. Orphanides G, Reinberg D. A unified theory of gene expression. Cell 2002; 108: 439–51.

128. Dever TE. Gene-specific regulation by general translation factors. Cell 2002; 108: 545–56.

129. Kleijn M, Scheper GC, Voorma HO et al. Regulation of translation initiation factors by signal transduction. Eur J Biochem 1998; 253: 531–44.

130. Sonenberg N, Gingras AC. The mRNA 5′ cap-binding protein eIF4E and control of cell growth. Curr Opin Cell Biol 1998; 10: 268–75.

131. Brown EJ, Schreiber SL. A signaling pathway to translational control. Cell 1996; 86: 517–20.

132. Fumagalli S, Thomas G. S6 phosphorylation and signal transduction. In: Sonenberg N, Hershey JWB, Mathews MB (eds). Translational Control of Gene

Expression. New York: Cold Spring Harbor Laboratory Press; 2000: 695–717.

133. Chou MM, Blenis J. The 70 kDa S6 kinase: regulation of a kinase with multiple roles in mitogenic signalling. Curr Opin Cell Biol 1995; 7: 806–14.

134. Thomas G, Hall MN. TOR signalling and control of cell growth. Curr Opin Cell Biol 1997; 9: 782–7.

135. Dufner A, Thomas G. Ribosomal S6 kinase signaling and the control of translation. Exp Cell Res 1999; 253: 100–9.

136. Shioi T, McMullen JR, Tarnavski O et al. Rapamycin attenuates load-induced cardiac hypertrophy in mice. Circulation 2003; 107: 1664–70.

137. McMullen JR, Shioi T, Zhang L et al. Deletion of ribosomal S6 kinases does not attenuate pathological, physiological or IGF-1R-PI3K induced-cardiac hypertrophy. Mol Cell Biol 2004; 24: 6231–40.

138. Wada H, Ivester CT, Carabello BA et al. Translational initiation factor eIF-4E. A link between cardiac load and protein synthesis. J Biol Chem 1996; 271: 8359–64.

139. Hardt SE, Tomita H, Katus HA et al. Phosphorylation of eukaryotic translation initiation factor 2Bε by glycogen synthase kinase-3β regulates β-adrenergic cardiac myocyte hypertrophy. Circ Res 2004; 94: 926–35.

140. Wang L, Proud CG. Ras/Erk signaling is essential for activation of protein synthesis by G_q protein-coupled receptor agonists in adult cardiomyocytes. Circ Res 2002; 91: 821–9.

141. Schmelzle T, Hall MN. TOR, a central controller of cell growth. Cell 2000; 103: 253–62.

142. Sadoshima J, Izumo S. Rapamycin selectively inhibits angiotensin II-induced increase in protein synthesis in cardiac myocytes in vitro. Potential role of 70-kD S6 kinase in angiotensin II-induced cardiac hypertrophy. Circ Res 1995; 77: 1040–52.

143. Boluyt MO, Zheng JS, Younes A et al. Rapamycin inhibits alpha 1-adrenergic receptor-stimulated cardiac myocyte hypertrophy but not activation of hypertrophy-associated genes. Evidence for involvement of p70 S6 kinase. Circ Res 1997; 81: 176–86.

144. McMullen JR, Sherwood MC, Tarnavski O et al. Inhibition of mTOR signaling with rapamycin regresses established cardiac hypertrophy induced by pressure overload. Circulation 2004; 109: 3050–5.

145. Hannan RD, Jenkins A, Jenkins AK et al. Cardiac hypertrophy: a matter of translation. Clin Exp Pharmacol Physiol 2003; 30: 517–27.

146. Hannan RD, Rothblum LI. Regulation of ribosomal DNA transcription during neonatal cardiomyocyte hypertrophy. Cardiovasc Res 1995; 30: 501–10.

147. Brandenburger Y, Jenkins A, Autelitano DJ et al. Increased expression of UBF is a critical determinant for rRNA synthesis and hypertrophic growth of cardiac myocytes. FASEB J 2001; 15: 2051–3.

148. Hannan RD, Luyken J, Rothblum LI. Regulation of rDNA transcription factors during cardiomyocyte hypertrophy induced by adrenergic agents. J Biol Chem 1995; 270: 8290–7.

149. Hannan RD, Luyken J, Rothblum LI. Regulation of ribosomal DNA transcription during contraction-induced hypertrophy of neonatal cardiomyocytes. J Biol Chem 1996; 271: 3213–20.

150. Luyken J, Hannan RD, Cheung JY et al. Regulation of rDNA transcription during endothelin-1-induced hypertrophy of neonatal cardiomyocytes. Hyperphosphorylation of upstream binding factor, an rDNA transcription factor. Circ Res 1996; 78: 354–61.

151. Brandenburger Y, Arthur JF, Woodcock EA et al. Cardiac hypertrophy in vivo is associated with increased expression of the ribosomal gene transcription factor UBF. FEBS Lett 2003; 548: 79–84.

152. Vigoreaux JO. The muscle Z band: lessons in stress management. J Muscle Res Cell Motil 1994; 15: 237–55.

153. Sanger JW, Ayoob JC, Chowrashi P et al. Assembly of myofibrils in cardiac muscle cells. Adv Exp Med Biol 2000; 481: 89–102; discussion 103–5.

154. Tsutsui H, Ishihara K, Cooper Gt. Cytoskeletal role in the contractile dysfunction of hypertrophied myocardium. Science 1993; 260: 682–7.

155. Hein S, Kostin S, Heling A et al. The role of the cytoskeleton in heart failure. Cardiovasc Res 2000; 45: 273–8.

156. Margulies KB, Houser SR. Myocyte abnormalities in human heart failure. In: Mann DL (ed). Heart Failure. Philadelphia: Saunders; 2004: 41–56.

157. Aoki H, Izumo S, Sadoshima J. Angiotensin II activates RhoA in cardiac myocytes: a critical role of RhoA in angiotensin II-induced premyofibril formation. Circ Res 1998; 82: 666–76.

158. Hoshijima M, Sah VP, Wang Y et al. The low molecular weight GTPase Rho regulates myofibril formation and organization in neonatal rat ventricular myocytes. Involvement of Rho kinase. J Biol Chem 1998; 273: 7725–30.

159. Pracyk JB, Tanaka K, Hegland DD et al. A requirement for the rac1 GTPase in the signal transduction pathway leading to cardiac myocyte hypertrophy. J Clin Invest 1998; 102: 929–37.

160. Aoki II, Sadoshima J, Izumo S. Myosin light chain kinase mediates sarcomere organization during cardiac hypertrophy in vitro. Nat Med 2000; 6: 183–8.

161. Kovacic-Milivojevic B, Roediger F, Almeida EA et al. Focal adhesion kinase and p130Cas mediate both sarcomeric organization and activation of genes associated with cardiac myocyte hypertrophy. Mol Biol Cell 2001; 12: 2290–307.

162. MacLellan WR, Schneider MD. Death by design. Programmed cell death in cardiovascular biology and disease. Circ Res 1997; 81: 137–44.

163. Haunstetter A, Izumo S. Apoptosis: basic mechanisms and implications for cardiovascular disease. Circ Res 1998; 82: 1111–29.

164. Kang PM, Yue P, Izumo S. New insights into the role of apoptosis in cardiovascular disease. Circ J 2002; 66: 1–9.

165. Kang PM, Izumo S. Apoptosis in heart: basic mechanisms and implications in cardiovascular diseases. Trends Mol Med 2003; 9: 177–82.

166. Wencker D, Chandra M, Nguyen K et al. A mechanistic role for cardiac myocyte apoptosis in heart failure. J Clin Invest 2003; 111: 1497–504.

167. Hayakawa Y, Chandra M, Miao W et al. Inhibition of cardiac myocyte apoptosis improves cardiac function and abolishes mortality in the peripartum cardiomyopathy of Galpha(q) transgenic mice. Circulation 2003; 108: 3036–41.

168. Parrizas M, Saltiel AR, LeRoith D. Insulin-like growth factor 1 inhibits apoptosis using the phosphatidylinositol 3'-kinase and mitogen activated protein kinase pathways. J Biol Chem 1997; 272: 154–61.

169. Sheng Z, Knowlton K, Chen J et al. Cardiotrophin 1 (CT-1) inhibition of cardiac myocyte apoptosis via a mitogen-activated protein kinase-dependent pathway. Divergence from downstream CT-1 signals for myocardial cell hypertrophy. J Biol Chem 1997; 272: 5783–91.

170. Adams JW, Sakata Y, Davis MG et al. Enhanced Galphaq signaling: a common pathway mediates cardiac hypertrophy and apoptotic heart failure. Proc Natl Acad Sci USA 1998; 95: 10140–5.

171. Geng YJ, Ishikawa Y, Vatner DE et al. Apoptosis of cardiac myocytes in Gsalpha transgenic mice. Circ Res 1999; 84: 34–42.

172. Pu WT, Ma Q, Izumo S. NFAT transcription factors are critical survival factors that inhibit cardiomyocyte apoptosis during phenylephrine stimulation in vitro. Circ Res 2003; 92: 725–31.

173. De Windt LJ, Lim HW, Taigen T et al. Calcineurin-mediated hypertrophy protects cardiomyocytes from apoptosis in vitro and in vivo: an apoptosis-independent model of dilated heart failure. Circ Res 2000; 86: 255–63.

135

174. Saito S, Hiroi Y, Zou Y et al. Beta-adrenergic pathway induces apoptosis through calcineurin activation in cardiac myocytes. J Biol Chem 2000; 275: 34528–33.

175. Yussman MG, Toyokawa T, Odley A et al. Mitochondrial death protein Nix is induced in cardiac hypertrophy and triggers apoptotic cardiomyopathy. Nat Med 2002; 8: 725–30.

176. Kang PM, Yue P, Liu Z et al. Alterations in apoptosis regulatory factors during hypertrophy and heart failure. Am J Physiol Heart Circ Physiol 2004; 287: H72–80.

177. Manabe I, Shindo T, Nagai R. Gene expression in fibroblasts and fibrosis: involvement in cardiac hypertrophy. Circ Res 2002; 91:1103–13.

178. Gunasinghe SK, Spinale FG. Myocardial basis for heart failure: role of the cardiac interstitium. In: Mann DL (ed). Heart Failure. Philadelphia: Saunders; 2004: 57–70.

CHAPTER 9

Alterations in signal transduction

Angela Clerk, Peter H Sugden

Introduction

Adaptive and maladaptive growth of the heart involves significant increases in the size and sarcomeric content of the constituent cardiac myocytes, and these morphological responses reflect a genuine cellular hypertrophy (growth in the absence of division). Another facet of the hypertrophic response is that genes which are expressed during fetal development of the heart (e.g. atrial and B-type natriuretic peptides, fetally expressed isoforms of myofibrillar proteins) are upregulated in hypertrophy, and are often used as indices of the response. These changes are elicited by signals operating between and within cells. The principal hypothesis is that neurohumoral factors (e.g. endothelin-1 (ET-1), α_1-adrenergic agonists, angiotensin II (Ang II)) are released locally and act at cell-surface receptors on cardiac myocytes to initiate intracellular signaling events. These modulate cellular functions and lead to the changes in gene and protein expression which evoke the overall response. Much research has focused on identifying the intracellular signaling pathways which are modulated in hypertrophy, and on elucidating the specific roles of each.

When considering potential mechanisms involved, it should be remembered that many of the signaling events associated with hypertrophy are activated acutely and probably act principally as initiators of the response, rather than being involved with chronic maintenance of the hypertrophic state. Furthermore, modulation of intracellular signaling pathways is a normal, functional response of the myocyte to a change in environment (e.g. increased workload). This contrasts with the situation in other disease states such as cancer, in which mutation of signaling components leads to modulation of cellular function. Unlike cancer, there is no evidence for mutation of any signaling intermediate leading to myocardial hypertrophy, probably because any such mutation arising in a single, terminally differentiated cardiac myocyte would never be propagated. A third factor for consideration is that the majority of signaling intermediates exist as members of large families with varying degrees of homology. Much of the evidence implicating specific signaling components in hypertrophy is based on activation of pathways by expression/overexpression of activated, inhibitory or wild-type proteins, and the degree of expression may influence the phenotype significantly. This is potentially associated with loss of signaling specificity with respect to location and substrate selection. Nevertheless, understanding the nature and regulation of the signaling pathways involved in cardiac myocyte hypertrophy is important, since extraneous modulation of these events may lead to the development of rational therapies. Here, we summarize the principal intracellular signaling pathways that are implicated in the hypertrophic response.

G protein-coupled receptors and heterotrimeric G protein signaling

Signaling through G protein-coupled receptors (GPCRs, see Chapter 5) is particularly implicated in cardiac myocyte biology and GPCR agonists regulate functional responses including contractile activity and hypertrophic growth. Following ligand binding to trans-sarcolemmal GPCRs, these receptors interact with membrane-bound heterotrimeric guanine nucleotide-binding (G) proteins. The GPCRs for hypertrophic agonists such as ET-1, α_1-adrenergic agonists and Ang II interact with the $G_{q/11}$ G protein subfamily.[1] As with all heterotrimeric G proteins, $G_{q/11}$ consists of an α subunit (αq, or the related $\alpha 11$), a member of the β subunit family, and a member of the γ subunit family. In the biologically-inactive G_q heterotrimer, αq is ligated (non-covalently) to GDP. Ligand-dependent activation of the G_q protein-coupled receptor (G_qPCR) stimulates exchange of GTP for GDP on αq, and the heterotrimer dissociates into αq.GTP and $\beta\gamma$ dimers. This dissociation leads to activation of phosphoinositide-specific phospholipase Cβ (PLCβ) species (see Phospholipases C and D, next section).[2] Transgenic mouse lines have been derived with cardiospecific overexpression of wild-type or constitutively-activated $G_{\alpha q}$.[1,3] Depending on the extent of overexpression, these mice either show no phenotype, a hypertrophic phenotype or a hypertrophic/heart failure phenotype with premature mortality. Furthermore, interfering with G_qPCR/G_q coupling in transgenic mice diminishes the ability of pressure-overload to induce myocardial hypertrophy. These results and other data from isolated cardiac myocytes[1,3] strongly implicate G_q in cardiac myocyte hypertrophy.

Phospholipases C and D

Phospholipids consist of a glycerol backbone coupled to long-chain fatty acids in the 1 and 2 positions and, via a phosphodiester bond, to a defining head-group on position 3 (Fig. 9.1). Hydrolysis of the glycerol phosphoester bond by phospholipase C (PLC) generates 1,2-diacylglycerol (DAG) (Fig. 9.1A), whereas hydrolysis of the head-group phosphoester bond by phospholipase D (PLD) generates phosphatidic acid (PA) (Fig. 9.1B). DAG kinase phosphorylates DAG in the 3 position to give PA, whereas phosphatidate phosphohydrolase (PPH) removes the phosphate group from PA to generate DAG (Fig. 9.1C). Thus, it was considered

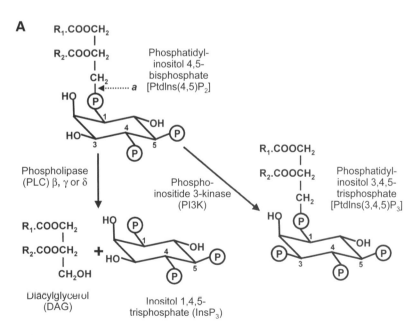

Figure 9.1A. *Phospholipid signaling. (A) PtdIns(4,5)P_2 is hydrolyzed (in position a) by PLC isoforms to generate DAG and InsP$_3$. PtdIns(4,5)P_2 can also be phosphorylated by PI3Ks on the 3' position of the inositol ring to give PtdIns(3,4,5)P_3.*

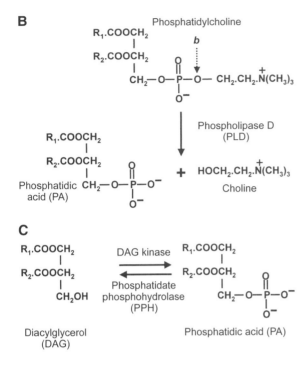

Figure 9.1B, C. *Phospholipid signaling. (B) Phosphatidylcholine is hydrolyzed (in position b) by phospholipase D isoforms to generate PA and choline. (C) Phosphorylation of DAG by DAG kinase generates PA, whereas hydrolysis of PA by PPH produces DAG. The long-chain fatty acids in positions 1 and 2 on the glycerol backbone (R_1 and R_2) differ between inositol phospholipids and phosphatidylcholine such that DAG derived from PLC activity (panel A) and PA derived from PLD activity (panel B) are likely to have independent second messenger functions. Conversion of DAG to PA or PA to DAG (panel C) probably serves to terminate the signal from PLC or PLD, respectively.*

for many years that PLC and PLD could both generate DAG for activation of protein kinase C (PKC, see Protein kinase C, page 140), and termination of the signal was via conversion of DAG to PA. However, the preferred substrates for PLC enzymes are phosphoinositides, particularly phosphatidylinositol 4,5-bisphosphate (PtdIns(4,5)P_2; Fig. 9.1A), whereas PLD is particularly active against phosphatidylcholine (Fig. 9.1B).[4] Phosphoinositides and phosphatidylcholine differ

significantly in the composition of the fatty acyl side chains,[4] suggesting that DAG and PA derived from the different phospholipids define the specificity for interaction with target molecules, and cannot substitute for each other. In this scenario, DAG resulting from PLC hydrolysis of phosphoinositides or PA derived from PLD hydrolysis of phosphatidylcholine are both second messengers, the signals being terminated by DAG kinase to form inactive PA or PPH to form inactive DAG, respectively.

PLC signaling

By hydrolyzing membrane PtdIns(4,5)P_2, phosphoinositide-specific PLC isoforms generate two second messengers, the lipid-phase DAG and soluble inositol 1,4,5-trisphosphate (InsP$_3$) (Fig. 9.1). There are three well-characterized PLC subfamilies: the β-subfamily (β$_1$, β$_2$, β$_3$, β$_4$, ~150 kDa), the γ-subfamily (γ$_1$, γ$_2$, ~145 kDa), and the δ-subfamily (δ$_1$, δ$_2$, δ$_3$, δ$_4$, ~85 kDa).[2] All contain pleckstrin homology (PH) domains which promote membrane association by binding to polyphosphoinositide phospholipids. Of these, expression of PLC β$_1$, β$_3$, γ$_1$, and δ$_1$ has been detected in the heart and in cardiac myocytes.[5,7] A further PLC, PLCε, is highly expressed in the heart, but falls into a separate group since, in addition to possessing PLC activity, this isoform may have a separate function as a guanine nucleotide exchange factor (GEF) for the small G proteins (see Small G proteins, page 141).[8] Classically, G$_q$PCR agonists activate the PLCβ subfamily through αq.[2] Further levels of specificity exist, since α$_1$-adrenergic receptors selectively activate PLC-β$_1$.[9] Other GPCRs, such as β-adrenergic receptors acting through G$_i$, potentially stimulate PLCβ through G$_{βγ}$ subunits.[2] In contrast, receptor protein tyrosine kinases (RPTKs) signal through PLCγ isoforms.

In cardiac myocytes, hypertrophic stimuli acting at G$_q$PCRs (e.g. ET-1, α$_1$-adrenergic agonists[10]), in addition to growth factors acting at RPTKs (e.g. insulin-like growth factor-1, IGF-1[11]), stimulate PLC activity producing DAG and InsP$_3$. The specific role(s) of individual PLC isoforms in cardiac myocyte signaling have not been fully elucidated. The PKC family (see page 140) are a key target for DAG, but a number of other proteins (e.g. chimaerins, Ras.GRPs, Munc13s and DAG kinase γ) are also DAG-responsive.[12] If present in cardiac myocytes, such proteins may also influence the hypertrophic response. The role of InsP$_3$ in the cardiac myocyte is unclear.[13] In other cells, interaction of InsP$_3$ with its receptor, an endoplasmic reticulum Ca^{2+} channel, controls Ca^{2+} fluxes. However, in cardiac myocytes, the ryanodine receptor is the principal Ca^{2+} channel in the sarcoplasmic reticulum (a modified form of the endoplasmic reticulum) and this modulates the Ca^{2+} transient associated with contractile activity. It is unlikely that, in this context, Ca^{2+} release through InsP$_3$ receptors contributes to contractility, but localized release may have signaling implications. While the focus of PLC activity is usually the production of DAG and InsP$_3$, a second consequence is the potential depletion of PtdIns(4,5)P$_2$ from the membrane. Since PtdIns(4,5)P$_2$ is also a substrate for phosphoinositide 3-kinase (PI3K, see The phosphoinositide 3-kinase/protein kinase B/Akt pathway, page 145) which activates cytoprotective pathways, its depletion from the membrane may promote cell death. Indeed, transgenic mice with cardiospecific overexpression of a constitutively-activated G$_{\alpha q}$, which superstimulates PLCβ, have suppressed survival signaling and develop heart failure.[14]

PLD signaling

The two forms of PLD, PLD1 and PLD2, are expressed in the heart.[15,16] PLD1 activity is regulated by PKC and small G proteins of the Rho and ARF families, whereas PLD2 activity appears to be dependent on PtdIns(4,5)P$_2$ levels.[17] Little is known about the role of PLD in the heart, but hydrolysis of phosphatidylcholine by PLD is increased in cardiac myocytes in response to hypertrophic G$_q$PCR agonists including ET-1 or α$_1$-adrenergic agonists.[10] This response is PKC dependent and, therefore, probably reflects PLD1 activity. As discussed above, the PA which is generated directly by PLD activity probably serves as an independent second messenger rather than providing a source of DAG for activation of PKC isoforms. In addition to the PKC-dependent nature of the response, this concept is supported by evidence that phosphatidylcholine-derived DAG species are

generated subsequent to activation of DAG-dependent PKC isoforms.[18,19]

Protein kinase C

The phospholipid-dependent PKC family of Ser/Thr protein kinases is divided into three subfamilies on the basis of their regulation: the 'classical' cPKCs, which require DAG and Ca^{2+}; the 'novel' nPKCs, which require DAG but are Ca^{2+}-independent; and the 'atypical' aPKCs, which are DAG- and Ca^{2+}-independent.[20] Phorbol esters such as phorbol 12-myristate 13-acetate (PMA) mimic DAG and have been extensively used to activate cPKCs and nPKCs. The mode of regulation of aPKCs is currently obscure. Of the cPKCs, cPKCα is clearly expressed in cardiac myocytes, but there is debate about whether the β isoforms (cPKCβ$_I$ and/or cPKCβ$_{II}$) are expressed in nonpathological situations.[21] Some studies suggest that cPKCα and/or cPKCβ are upregulated in human and animal models of heart failure, although this is not a universal view.[22] nPKCδ and nPKCε are readily detected in cardiac myocytes in addition to aPKCs (aPKCζ and/or aPKCλ/ι).[21] Relative to total protein, cPKCα, nPKCδ, nPKCε and aPKCζ are all substantially downregulated in cardiac myocytes during postnatal development although, on a per cell basis, the amount of each isoform is more conserved.[23]

Since DAG is integrated into lipid membranes, activation of cPKCs and nPKCs can be assessed by their translocation from the soluble to the particulate fraction of the cell. Thus, hypertrophic G$_q$PCR agonists such as ET-1 and α$_1$-adrenergic agonists induce rapid (<30 s), though transient (reversed within 5–10 min), translocation of nPKCδ and nPKCε in cardiac myocytes.[24] Although PMA promotes translocation of cPKCα, the consensus is that there is no translocation of cPKCα in response to G$_q$PCR agonists.[24] The question of whether nPKCδ and nPKCε remain active following release from the particulate fraction or are only active while associated with the membrane remains unanswered. However, PKCs are additionally regulated by phosphorylation, and PMA (which induces sustained translocation of cPKCs and nPKCs) promotes phosphorylation of nPKCδ to increase its lipid-independent activity.[25] If this occurs with GqPCR agonists, nPKCδ and nPKCε may well remain active following relocation to the soluble fraction, which would allow them to seek substrates in other compartments of the cell. PKCs are also regulated by their interaction with anchoring proteins which are probably isoform-specific.[26] Such anchoring proteins may serve to bring PKC isoforms to a specific locality either for activation or to phosphorylate specific substrates. The first such protein to be identified was the receptor for activated C kinase 1 (RACK1), which interacts with cPKCβ. Although anchoring proteins have not been unequivocally identified for all PKC isoforms, bioinformatics approaches have identified regions in nPKCδ and nPKCε which may interact with putative anchoring proteins.[21] Peptides designed to interfere with or promote these potential interactions are being used to elucidate PKC isoform-specific cellular responses.[21]

In cardiac myocytes, PKCs probably play fundamental roles in the development of cardiac myocyte hypertrophy and in regulating apoptosis.[21] As mentioned above, GqPCR-linked hypertrophic stimuli activate both nPKCδ and nPKCε, and studies with the activating peptides for these isoforms indicate that either can promote compensated, non-pathological hypertrophy.[27] Consistent with this, cardiospecific overexpression of moderate levels of constitutively-activated nPKCε in transgenic mice results in compensated hypertrophy.[28] Higher expression levels of activated nPKCε produce a heart failure phenotype,[29] but this may reflect some loss of specificity. In these transgenic mice, nPKCε interacts with RACK1,[29] presumably initiating normal cPKCβ signaling but, potentially, highly expressed nPKCε could interact with any PKC anchoring protein or substrate. Since nPKCδ and nPKCε exert specific, potentially antagonistic effects with respect to cardioprotection (nPKCε is probably cardioprotective,[21] whereas nPKCδ, promotes cardiac myocyte apoptosis[30]), interaction of highly-expressed constitutively-activated nPKCε with nPKCδ, substrates could result in a heart failure phenotype. The means by which nPKCδ and nPKCε elicit their responses are not yet known although, in cardiac myocytes, nPKCε signals to

the extracellular signal-regulated kinase 1/2 (ERK1/2) subfamily of the mitogen-activated protein kinases (MAPKs), whereas nPKCδ is implicated in the activation of the two MAPK subfamilies which are particularly responsive to cellular stresses, the c-Jun N-terminal kinases (JNKs) and p38-MAPKs (see Mitogen-activated protein kinases, page 144).[30] Recent functional proteomics studies using overexpressed nPKCε suggest that it may interact with approximately 100 different proteins in potential multiprotein complexes to elicit its effects.[31] If such complexes exist *in vivo*, the unraveling of this network and those of related PKCs will undoubtedly take some time.

The cPKCs, cPKCα and cPKCβ, may contribute to myocardial hypertrophy and/or the development of heart failure. Expression of an activated form of cPKCα in cardiac myocytes can induce a hypertrophic response,[32,33] but antisense suppression of cPKCα expression attenuates only some of the hypertrophic effects of α1-adrenergic agonists.[34] Since hearts from *cPKCα−/−* or wild-type mice respond to pressure-overload with a similar degree of hypertrophy,[35] it is possible that expression of activated cPKCα results in activation of atypical signaling. Cardiospecific overexpression of wild-type cPKCβ or expression of constitutively-activated cPKCβ can induce either myocardial hypertrophy or heart failure in transgenic animals, depending on the level of expression.[22] However, in the context of the debate about cPKCβ expression in cardiac myocytes, the significance of this is unclear. Like *cPKCα−/−* hearts, *cPKCβ−/−* hearts appear to respond normally to hypertrophic stimuli,[36] and, as with cPKCα, the effects of increased cPKCβ signaling may reflect some loss of specificity.

Small G proteins

The small G protein superfamily, of which Ha-Ras was the first to be identified, are the 'molecular switches' of the cell.[37] As with the α subunit of heterotrimeric G proteins, GDP-ligated small G proteins are biologically inactive, and activation is by exchange of GTP for GDP, a reaction which is stimulated by GEFs (Fig. 9.2A). Activation is

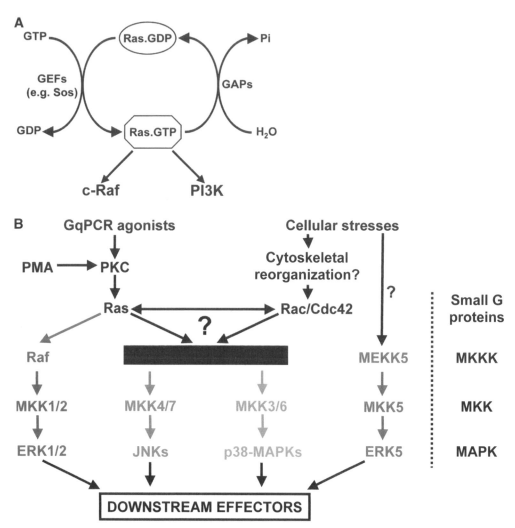

Figure 9.2. *Signaling through Ras and MAPKs. (A) Small G proteins such as Ras act as a 'molecular switch'. Ras is inactive when ligated to GDP and becomes activated by exchange of GTP for GDP. Ras.GTP-loading is increased by GEFs such as Sos. The best-characterized effectors of active Ras.GTP are c-Raf and PI3K, and recruitment of these proteins to the membrane by binding to Ras.GTP leads to their full activation. Small G proteins possess an innate GTPase activity. This is enhanced by GAPs to terminate the signal. (B) Small G proteins signal to the MAPK cascades (right) in which MAPKs are phosphorylated and activated by upstream MKKs, which are in turn phosphorylated and activated by MKKKs. G_qPCR agonists or PMA activate PKC isoforms leading to activation of Ras. Ras is particularly implicated in the activation of the ERK1/2 cascade in which Raf family members phosphorylate and activate MKK1/2, which phosphorylate and activate ERK1/2. However, Ras can signal to other small G proteins such as Rac and Cdc42 and/or may signal directly to the JNK and p38-MAPK cascades. Cellular stresses activate Rac and Cdc42 (possibly by affecting the cytoskeleton), leading to activation of the JNK and p38-MAPK cascades. JNKs and p38-MAPKs are phosphorylated and activated by MKK4/7 and MKK3/6, respectively, but the upstream MKKKs for either cascade are ill-defined. Activated Rac/Cdc42 may promote activation of Ras and lead to stimulation of the ERK cascade. The ERK5 cascade, in which MEKK5 phosphorylates and activates MKK5, which in turn phosphorylates and activates ERK5, is principally activated by cellular stresses, although the precise mechanisms are unknown.*

associated with a conformational change in the small G protein allowing it to interact with its effectors. Small G proteins possess an innate GTPase activity, and hydrolysis of bound GTP to GDP returns the G protein to its inactive state. The innate GTPase activity is low, but it is dramatically enhanced by GTPase-activating proteins (GAPs). Mutations (e.g. V12Ha-Ras) which reduce the innate GTPase activity and interfere with the association between small G proteins and their GAPs produce constitutively-activated forms. In contrast, mutations (e.g. N17Ha-Ras) which increase the association between small G proteins and their GEFs sequester the relevant GEFs and act as inhibitors of small G protein signaling.

There are five small G protein subfamilies: Ras, Rho, Rab, Ran and Sar1/ARF.[37] Generally, the Ras subfamily proteins control cell growth and proliferation, and other facets of cellular biology including cell cycle arrest, senescence, differentiation and survival. These properties are largely dependent on their ability to regulate transcription through the activation of other signaling proteins. Ha-Ras and Ki-Ras are located exclusively at the plasma membrane, although there is evidence of differential localization of Ha-Ras and Ki-Ras with membrane microdomains.[38] Membrane-targeting of Ras proteins is mediated by their post-translational lipidation (C-terminal farnesylation, C-terminal carboxyl methylation and, in the case of Ha-Ras, palmitoylation).[37] Ligand-mediated activation of transmembrane receptors such as RPTKs (e.g. the insulin and IGF-1 receptors, the epidermal growth factor (EGF) receptor) or G_qPCRs activates Ras. The pathways leading from RTPKs to Ras are relatively well understood and involve activation of Ras.GEFs (e.g. Sos),[39] although the pathways leading from G_qPCRs to Ras activation are much less clear.

Ha-Ras, Ki-Ras and N-Ras are all expressed in cardiac myocytes,[40] and there is considerable evidence that Ras promotes myocyte hypertrophy. In these cells, Ras is activated particularly strongly by G_qPCR agonists. In relation to hypertrophy, the two most-relevant functions of Ras.GTP are the activation of the ERK1/2 cascade (see Mitogen-activated protein kinases, page 144) and PI3K (see The phosphoinositide 3-kinase/protein kinase

B/Akt pathway, page 145). However, Ras.GTP can signal through other pathways which have yet to be investigated fully. Agonists such as ET-1 or α_1-adrenergic agonists strongly stimulate Ras.GTP formation in cardiac myocytes.[40] The mechanisms are unclear but probably involve PKC or other DAG-regulated proteins. Transfection of a V12Ha-Ras plasmid or microinjection of V12Ha-Ras protein into cardiac myocytes induces the appearance of the morphological indices of hypertrophy and the fetal gene program.[41] Equally, transfection of N17Ha-Ras blunts particularly the appearance of fetal gene re-expression following exposure of cardiac myocytes to hypertrophic agonists.[41] Transgenic mice with cardiospecific expression of V12Ha-Ras display evidence of myocardial hypertrophy.[41,42] The initial transgenic lines express relatively low levels of V12Ha-Ras and the phenotype is somewhat variable, ranging from no obvious change to severe hypertrophic cardiomyopathy.[41] Subsequent lines (which probably have higher expression levels of V12Ha-Ras) exhibit a more obvious phenotype, albeit one of severe hypertrophic cardiomyopathy rather than adaptive hypertrophy, with diastolic dysfunction, increased fibrosis and premature death.[42] This phenotype is difficult to rationalize in the context of the known growth-promoting functions of Ras, and is clearly more complex than one which simply involves ERK1/2 and/or PI3K activation (see Mitogen-activated protein kinases, page 144, and The phosphoinositide 3-kinase/protein kinase B/Akt pathway, page 145). It should be borne in mind that small G proteins function physiologically as molecular switches. While expression of V12Ha-Ras is clearly relevant in the context of cancer, there is no evidence of constitutive activation of Ras in cardiac myocytes. High-level, constitutive activation of signaling pathways downstream of V12Ha-Ras does not necessarily simulate the normal response in adaptive hypertrophy, and this may account for the detrimental effects seen in V12Ha-Ras transgenic mice.

Fewer studies have focused on the role of non-Ras small G proteins in myocardial hypertrophy. The Rho subfamily (RhoA, Rac1 and Cdc42) regulate the cell shape and migrational properties of non-myocytes by modulating the actin cytoskeleton.[37] The best-characterized effectors of RhoA are protein kinases, namely the two Rho kinases (Rho kinase/ROKα/ROCK-II and p160ROCK/ROCKβ)[43] and the protein kinase C-related kinases (PRKs or PKNs).[44] RhoA is activated in cardiac myocytes by hypertrophic agonists,[45] and a number of approaches, including inactivation of Rho with C3 exonuclease and inhibition of Rho kinases by Y27632, have implicated RhoA.GTP in myofibrillar assembly and fetal gene re-expression.[41] However, cardiospecific over-expression of wild-type RhoA produced a disappointing phenotype associated with severe conduction abnormalities and heart failure leading to premature death.[41] In this context, it is difficult to assess whether RhoA has any role in cardiac myocyte hypertrophy. Rac1 is known to signal through p21-activated kinases (PAKs) which regulate cell shape and activate MAPKs (see Mitogen-activated protein kinases, below).[46] Rac1 has been implicated in cardiac myocyte hypertrophy in isolated myocyte experiments,[41] but cardiospecific expression of constitutively activated Rac1 causes dilated cardiomyopathy and premature lethality.[41] The role of Cdc42 in cardiac myocyte hypertrophy has not been studied. Of the remaining small G proteins, the Rab subfamily (involved in vesicle trafficking to regulate functions such as protein secretion and localization of transmembrane proteins) are implicated in cardiac pathology. Rab protein expression is increased in heart failure, and cardiospecific overexpression of a wild-type Rab1A transgene causes dilated cardiomyopathy in transgenic mice,[47] but a causative role in cardiac myocyte hypertrophy has not been established. The two remaining small G protein subfamilies (Ran and Sar1/ARF) have not been studied in relation to myocardial hypertrophy and failure.

Mitogen-activated protein kinases

The MAPKs are the final components of three-tiered protein kinase cascades. Thus, MAPKs are phosphorylated and activated by upstream MAPK kinases (MKKs) which are themselves phosphorylated and activated by MKK kinases (MKKKs) (Fig. 9.2B).[48] All MAPKs are activated by phos-phorylation of the Thr and Tyr residues in a specific Thr-Xaa-Tyr motif, and three subfamilies have been clearly defined on the basis of this motif. The first MAPKs to be identified, ERK1 (p44-MAPK) and ERK2 (p42-MAPK), possess a Thr-Glu-Tyr motif and are phosphorylated by MKK1/2 (also known as MEK1/2). The MKKKs for this cascade are the Raf family of protein kinases, of which c-Raf and A-Raf are expressed in cardiac myocytes.[24] ERK1/2 are generally activated by anabolic stimuli and, in cardiac myocytes, these MAPKs are potently activated by hypertrophic G_qPCR agonists or PMA,[24] and activation occurs downstream from Ras.GTP.[40] In unstimulated cells, c-Raf is largely cytoplasmic. Although c-Raf does not bind to Ras.GDP, it binds strongly to Ras.GTP, and this interaction translocates c-Raf to the plasma membrane where it becomes fully activated. In addition to Ras.GTP-binding, for c-Raf to be fully active, Ser^{338} and/or Tyr^{341} must be phosphorylated.[49] In cardiac myocytes, activation of Ras and c-Raf is likely to occur downstream from nPKCε, since this isoform is particularly implicated in the activation of ERK1/2.[30] A second signal through Rac1 facilitates the activation of the ERK1/2 cascade by increasing the association between c-Raf and MKK1/2.[45] In cardiac myocytes, an alternative Raf-independent pathway to ERK1/2 activation may exist since there is no evidence of any significant activation of c-Raf by $α_1$-adrenergic agonists, although there is clear activation of MKK1/2 and ERK1/2.[40] Furthermore, activation of ERK1/2 by $α_1$-adrenergic agonists, but not ET-1, appears to be downstream from PI3K.[50]

The JNK and p38-MAPK subfamilies (Fig. 9.2B) are defined on the basis of Thr-Pro-Tyr and Thr-Gly-Tyr phosphorylation motifs, respectively, and both subfamilies are potently activated by cellular stresses.[51] The JNK subfamily derive from three separate genes (JNK1, JNK2 and JNK3), but alternative splicing can generate up to 13 different isoforms of $\sim$ 46 kDa or 54 kDa.[24] The p38-MAPKs comprise six family members deriving from four genes (p38-MAPKα and p38-MAPKβ, which each give rise to two alternatively spliced products, p38-MAPKγ and p38-MAPKδ).[24] JNKs and p38-MAPKs are phosphorylated and activated by upstream kinases, MKK4/7 and MKK3/6, respectively, but the MKKKs for these MAPK cascades are not clearly defined. Since there are a large number of candidates, it is possible that different MKKKs are responsive to specific stimuli. As expected, JNKs and p38-MAPKs are activated in cardiac myocytes by a range of cellular stresses including oxidative stress, hyperosmotic shock and protein synthesis inhibitors.[24,52,53] In most cells, both pathways are activated in parallel but, in the heart, activation of p38-MAPKs and JNKs is not necessarily coupled.[45,54] In the rat, the principal JNK activities which are activated in cardiac myocytes appear to derive from the *jnk1* gene, although multiple isoforms can be separated.[55] Of the p38-MAPKs, p38-MAPKα, p38-MAPKγ and p38-MAPKδ isoforms are expressed,[22] but the principal activity detected is probably p38-MAPKα. ERK5 (also known as 'Big' MAPK 1 (BMK1)) is classed separately from the three principal MAPK subfamilies since it has a Thr-Glu-Tyr phosphorylation motif like ERK1/2, but appears to be particularly stress responsive, and is activated in the heart by oxidative stress and ischemia–reperfusion.[22] ERK5 is phosphorylated and activated by MKK5 which is phosphorylated and activated by MEKK5.

The various MAPKs have all been implicated in the development of cardiac myocyte hypertrophy, although some more clearly than others.[56,57] ERK1/2 are particularly implicated in cardiac myocyte hypertrophy and the development of a compensated hypertrophic phenotype in the heart. Thus, not only are ERK1/2 potently activated in cardiac myocytes by hypertrophic G_qPCR stimuli,[24] but also pharmacological inhibition of the ERK1/2 cascade (with U0126) suppresses morphological and transcriptional changes associated with the response.[58] *In vivo*, cardiospecific overexpression of constitutively activated MKK1 in transgenic mice is one of few interventions that produces compensated hypertrophy rather than heart failure.[59] It is notable that nPKCε is particularly implicated in the activation of the ERK1/2 cascade,[30] and overexpression of nPKCε at moderate levels also promotes a compensated hypertrophy phenotype.[28] The role of JNKs and p38-MAPKs in cardiac myocyte hypertrophy continues to be debated.[57,60] Although G_qPCR agonists activate JNKs and p38-MAPKs in cardiac myocytes, and activation of

these cascades can promote some aspects of hypertrophy, JNKs and p38-MAPKs are also implicated in cardiac myocyte apoptosis and/or cytoprotection.[61] Such mutually exclusive functional roles can only be explained by the context (e.g. subcellular localization, isoform-specific activation, activation of other signaling pathways) of their activation. *In vivo*, cardiospecific activation of the JNK or p38-MAPK cascades in transgenic mice results in heart failure rather than compensated hypertrophy, and inhibition of either cascade enhances hypertrophic growth in response to pressure-overload.[57,60] The overall conclusion is that although these MAPKs may play a significant role in the development of cardiac disorders, they probably do not directly induce cardiac myocyte hypertrophy. Although ERK5 is generally stress responsive, this cascade is also activated by gp130-linked cytokine receptors.[62] The ERK5 cascade may be specifically important in this context, since overexpression of activated MKK5 results in cardiac myocyte hypertrophy with similar features (elongated morphology) to that induced by gp130-linked receptor agonists, and gp130-linked receptor agonist-induced hypertrophy is suppressed by dominant-negative MKK5.[62] A number of studies have examined MAPK activation in animal models of heart failure and in human heart failure, but there is little consensus between the studies. However, since intracellular signaling events can be expected to drive myocytes towards a response rather than necessarily to maintain a phenotype, this is perhaps unsurprising, and studies of end-stage disease may not be particularly useful in delineating causative events.

The phosphoinositide 3-kinase/protein kinase B/Akt pathway

PI3Ks phosphorylate PtdIns(4,5)P_2 and possibly other phosphoinositides in the membrane, leading to an increase in local concentrations of PtdIns(3,4,5)P_3 and other 3-phosphoinositides (Fig. 9.1A). Class IA PI3Ks are heterodimers of a regulatory subunit (p85α, p85β or p55) and a 110 kDa catalytic subunit (α, β or δ), and are activated primarily by RPTKs.[63] Stimulation of RPTKs results in their autophosphorylation on Tyr-residues, and this recruits Class IA PI3K hetero-dimers to the plasma membrane either through direct interaction with the RPTK or via adaptor proteins. The only Class IB PI3K identified consists of a 101 kDa regulatory subunit and a 110 kDa catalytic γ subunit, and is activated by GPCRs. PI3Kγ is brought to the plasma membrane by its interaction with the βγ dimers formed by dissocia-tion of heterotrimeric G proteins. The catalytic subunits of Class I PI3Ks also contain consensus binding sites for Ras.GTP. Interaction of Class I PI3Ks with Ras.GTP, RPTKs or Gβγ brings them to the membrane to phosphorylate phosphoinositide

substrates. It is unclear whether activation of Class I PI3Ks requires receptor-mediated membrane association and interaction with Ras.GTP, or whether a single input is sufficient.

The PI3K-mediated increases in membrane $PtdIns(3,4,5)P_3$ initiate a protein kinase cascade acting through 3-phosphoinositide-dependent kinase 1 (PDK1) and protein kinase B (PKB, also known as Akt) (Fig. 9.3).[64] In unstimulated cells, a proportion of the PDK1 is associated with the plasma membrane through interaction with basal levels of $PtdIns(3,4,5)P_3$. This is mediated through a PH domain with high affinity for $PtdIns(3,4,5)P_3$. When agonist stimulation further increases $PtdIns(3,4,5)P_3$ concentrations, PKB/Akt translo-

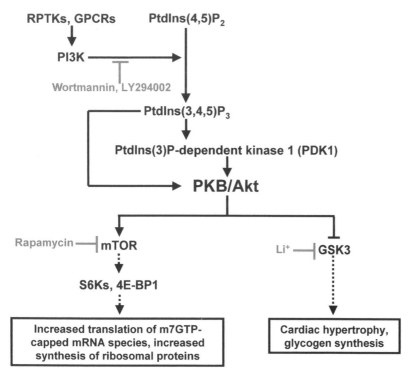

Figure 9.3. *Signaling through PI3K and PKB/Akt. PI3K isoforms are activated by RPTKs or GPCRs and catalyze the phosphorylation of $PtdIns(4,5)P_2$ to give $PtdIns(3,4,5)P_3$. Pharmacological inhibitors, wortmannin or LY294002 inhibit PI3K activity. PDK1 binds to $PtdIns(3,4,5)P_3$ in the sarcolemmal membrane. As $PtdIns(3,4,5)P_3$ levels increase, PKB/Akt is also recruited to the membrane where is becomes activated by phosphorylation of Thr^{308} by PDK1 and Ser^{473} by another (yet to be characterized) protein kinase activity. PKB/Akt phosphorylates and inhibits GSK3 to promote glycogen synthesis and cardiac hypertrophy. GSK3 activity is inhibited by Li^+. Through mTOR (inhibited by rapamycin), PKB/Akt promotes the phosphorylation of p70 small ribosomal subunit S6 kinases (S6Ks) and eukaryotic initiation factor 4E binding protein (4E-BP1) to increase the rate of protein synthesis.*

cates to the membrane through its PH domain-mediated interaction with $PtdIns(3,4,5)P_3$.[64] Translocation to the membrane brings PKB/Akt into the vicinity of active PDK1, and a PH domain-dependent conformational change in PKB/Akt allows phosphorylation of Thr^{308} by PDK1.[64,65] Further phosphorylation of Ser^{473} produces a fully active PKB/Akt, but the identity of the kinase responsible for this second phosphorylation is not established. Although activation of PKB/Akt is the best characterized effect of PI3K activation, $PtdIns(3,4,5)P_3$ influences the activities of other PH domain-dependent signaling pathways, and PDK1 catalyzes facilitating and activating phosphorylations of protein kinases other than PKB/Akt.[64]

PKB/Akt has numerous cellular functions which are generally anabolic and pro-survival. These include increased protein synthesis, increased glycogen synthesis and decreased apoptosis, and is probably the principal pathway through which insulin and IGF-1 exert their effects.[64] The ability of PKB/Akt to stimulate protein synthesis and to inhibit glycogen synthase kinase 3 (GSK3) are most relevant to its roles in hypertrophy. The detailed mechanisms through which it stimulates protein synthesis remain to be fully unraveled, and are beyond the scope of this article. Two known mechanisms involve the stimulation of translational initiation of m7GTP-capped mRNA species and increased synthesis of ribosomal proteins. Both are mediated through PKB/Akt-dependent activation of the protein kinase, mammalian target-of-rapamycin (mTOR).[66] mTOR promotes the phosphorylation of eukaryotic initiation factor 4E (eIF4E)-binding protein. This causes its dissociation from eIF4E itself, which can then bind to the m7GTP mRNA cap and initiate translation. In addition, mTOR participates in the phosphorylation of p70 small ribosomal subunit S6 kinases which can increase translation of mRNAs encoding ribosomal proteins. Phosphorylation of GSK3 by PKB/Akt inhibits its activity and the best characterized effect is the consequent dephosphorylation of glycogen synthase and stimulation of glycogen synthesis.[67] However, it is now recognized GSK3 modifies the biological activity of other protein substrates.

In contrast to the Ras/ERK1/2 cascade, PKB/Akt phosphorylation and activation are strongly stimulated by insulin in cardiac myocytes, but are only weakly stimulated by hypertrophic G_qPCR agonists.[68] A large number of transgenic mouse lines have now been produced in which the activity of the PI3K/PKB/Akt pathway is enhanced in the heart.[69,70] These include mice which express various constructs encoding constitutively activated PI3Kα, constitutively-activated PKB/Akt or PKB/Akt that is predestined for activation by membrane-targeting, and the IGF-1 receptor. Overall, activation of the PI3K/PKB/Akt pathway leads to increased heart and myocyte size, and increased contractile function (as assessed by dP/dt_{max}). Furthermore, inhibition of mTOR by rapamycin suppresses the increase in heart size induced by activated PKB/Akt. Targeted deletion of the 3-phosphoinositide phosphatase PTEN (presumably leading to increased membrane $PtdIns(3,4,5)P_3$ levels) has similar effects on heart size as activation of the PI3K/PKB/Akt pathway, though here contractile function is diminished.[71] Conversely, mice that express inhibitory PI3Kα develop smaller hearts than wild-type controls,[69] and postnatal targeted deletion of PDK1 results in dilated cardiomyopathy and premature death.[72] Generally, activation of the PI3Kα/PKB/Akt pathway does not lead to as pronounced a re-expression of fetal genes as other signaling pathways, suggesting that it has limited involvement in maladaptive hypertrophy. This view is upheld by the observation that inhibitory PI3Kα suppresses heart growth in response to physiological stimuli (e.g. exercise), but not to pathological stimuli (e.g. pressure-overload).[73] In contrast, activation of PI3Kγ may be associated with cardiac growth in response to pathological stimuli, although interfering with $G_{βγ}$-dependent signaling to PI3Kγ does not prevent the development of pressure-overload hypertrophy.[74] It is difficult to understand how signaling from PI3Kα versus PI3Kγ via PKB/Akt can elicit such different effects.

As mentioned earlier, PKB/Akt phosphorylates and inhibits GSK3, which may influence hypertrophy by modulating transcription. Inhibition of GSK3 has been implicated in myocardial hypertrophy in a pathway that is dependent on the dephosphorylation and stabilization of the transcriptional activator, β-catenin.[75,76] Alternatively, the nuclear factor of activated T cells (NFAT) group of transcription factors may be involved in GSK3-mediated myocardial hypertrophy (see Calcineurin, next section). GSK3 phosphorylates NFATc1, decreasing its DNA binding activity and promoting its export from the nucleus.[77] Thus, inhibition of GSK3 should increase NFATc1-dependent transcription. It is not clear whether other NFATs are similarly regulated by GSK3, or whether NFATc1 is directly involved in cardiac myocyte hypertrophy. In transgenic mice, cardio-specific expression of a PKB/Akt-insensitive GSK3 (generated by mutating the PKB/Akt phosphorylation site) reduces the extent of hypertrophy induced by pressure-overload.[78] Anatomical hypertrophy induced by expression of constitutively-activated calcineurin (CaN, see Calcineurin, next section) is also reduced by co-expression of constitutively active GSK3, although the effects on the transcriptional changes associated with hypertrophy are confusing.[78]

Calcineurin

Intracellular Ca^{2+} concentrations ($[Ca^{2+}]_i$) represent an ideal signal to couple contractile function to myocardial mass. Indeed, many hypertrophic agonists are positively inotropic, affecting Ca^{2+} movements and myofibrillar Ca^{2+} sensitivity. Nevertheless, until recent studies implicated the Ca^{2+}-sensitive protein phosphatase CaN (protein phosphatase 2B) in hypertrophy (Fig. 9.4),[79,80] there was no obvious link between $[Ca^{2+}]_i$ and cardiac growth. CaN is a heterodimer consisting of a 59 kDa catalytic subunit (CaNA) tightly associated with a 19 kDa Ca^{2+}-binding regulatory CaNB subunit.[81,82] In striated muscle, CaN is associated with the myofibrillar Z-disc, an interaction that is mediated by the calsarcin proteins.[80] Although CaNB binds Ca^{2+} directly, this only weakly stimulates enzyme activity. However, when $[Ca^{2+}]_i$ increases, it becomes increasingly associated with another Ca^{2+}-binding protein, calmodulin (CaM), and the interaction of $CaM.Ca^{2+}$ with CaNA profoundly increases CaN activity. The role of Ca^{2+} binding to CaNB is somewhat obscure, but it may facilitate the binding of Ca^{2+}-ligated CaM to CaNA. Two other proteins, cyclophilin A which interacts with the immunosuppressant drug cyclosporin A (CsA), and the FK506 (another immunosuppressant drug)-binding protein 12 (FKBP12) interact with CaN. Binding of CsA or FK506 to cyclophilin A or FKBP12, respectively, and the subsequent interaction of the complexes with CaN inhibits its phosphatase activity. This is the basis of the immunosuppressant activities of these drugs.

The role of CaN in cell growth has been best characterized in the immune system. Binding of major histocompatibility complex proteins on antigen presenting cells to T cell receptors, increases T cell $[Ca^{2+}]_i$, thereby activating CaN. The principal substrates for CaN in these cells are the NFAT family of transcription factors. In dormant T cells, NFATs are extensively phosphorylated, resulting in their retention in the cytoplasm. On dephosphorylation of NFATs, a nuclear localization signal is revealed and they migrate to the nucleus. Here, in cooperation with other transcription factors, NFATs promote the transcription of genes associated with T cell proliferation. CsA or FK506 inhibit CaN and nuclear entry of NFATs, thus eliciting immunosuppression.

NFATc4 interacts with the cardiac-restricted GATA4 transcription factor which is of established importance in myocardial development and in hypertrophy, suggesting that CaN may regulate transcription in cardiac myocytes.[83] This implies that an analogous pathway to that in T cells links increases in $[Ca^{2+}]_i$ to myocardial hypertrophy, and a large number of experiments have supported this hypothesis.[79,80] For example, ET-1 causes translocation of endogenous NFATs to the nucleus in cardiac myocytes,[84] and cardiospecific expression of constitutively activated CaNA in transgenic mice can induce morphological myocardial hypertrophy and fetal gene re-expression, but some of these mice die prematurely and the phenotype is somewhat variable.[83] In these transgenic mice, inhibition of CaN by CsA prevents the development of hypertrophy. This suggested a readily available and inexpensive therapeutic approach for the

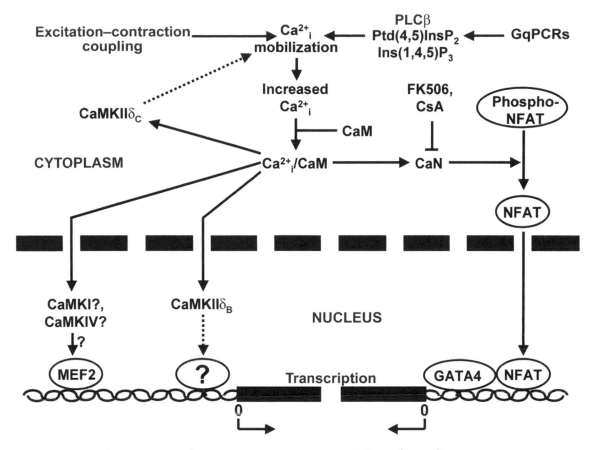

Figure 9.4. *Signaling through Ca^{2+}. In cardiac myocytes, intracellular Ca^{2+} (Ca^{2+})$_i$ concentrations are increased through excitation–contraction coupling and/or signaling from G$_q$PCRs which promote the production of Ins(1,4,5)P$_3$ from Ptd(4,5)InsP$_2$ by increasing PLCβ activity. CaM binds to intracellular Ca$^{2+}_i$ and activates Ca^{2+}-signaling. Interaction of Ca^{2+}/CaM with CaN (which is inhibited by FK506 or CsA) increases the protein phosphatase activity of CaN. Activated CaN dephosphorylates phospho-NFATs which migrate into the nucleus and increase transcription, possibly by interacting with GATA4. The Ca^{2+}/CaM complex also increases the activity of the various CaMKs. Direct binding of Ca^{2+}/CaM to CaMKIIδ$_B$ or CaMKIIδ$_C$ results in its activation. Nuclear-localized CaMKIIδ$_B$ may promote gene expression through an unknown mechanism. CaMKIIδ$_C$ is localized in the cytoplasm. Ca^{2+}/CaM also leads to activation of CaMKI and CaMKIV. This may influence gene expression by stimulating the transcription factor MEF2.*

treatment of myocardial hypertrophy, spawning numerous publications on the ability (or otherwise) of CsA or FK506 to prevent experimentally induced myocardial hypertrophy. In cultured cardiac myocytes, CsA reduces the extent of hypertrophy induced by α$_1$-adrenergic stimulation or Ang II,[83] though whether this reflects the toxic nature of the drug is not clear. Opinion remains divided as to the efficacy of immunosuppressant administration in preventing hypertrophy *in vivo*,[79] but, given the general toxicity of immunosuppressants, it appears unlikely that such a therapy would ever be desirable in humans. Other genetic evidence implicating CaN in myocardial hypertrophy includes transgenic mice with cardiospecific over-expression of endogenous protein inhibitors of CaN, in which the induction of hypertrophy by pressure-overload or by expression of constitutively-activated CaN is reduced.[79] In addition, targeted deletion of *CaNAβ* reduces the extent of anatomical hypertrophy in response to pressure-overload or neurohormonal factors, although the pattern of response of the transcriptional indices of hypertrophy is much less affected.[85] Furthermore, cardiospecific expression of an N-terminally-deleted *NFATc4* transgene (which removes the regulatory phosphorylation sites resulting in constitutive nuclear localization) induces myocardial hypertrophy *in vivo*,[83] which illustrates that the pathway has the potential to promote hypertrophy.

A key problem with the hypothesis that CaN mediates hypertrophic growth is that activation of CaN in non-myocytic cells is normally brought about by relatively slow but sustained increases in $[Ca^{2+}]_i$,[79] whereas rapid oscillations in $[Ca^{2+}]_i$ occur during the contractile cycle in cardiac myocytes. It is difficult to envisage how CaN is activated under these circumstances. However, some forms of heart failure are associated with increased diastolic $[Ca^{2+}]_i$[86] and the increase in baseline $[Ca^{2+}]_i$ may be sufficient for CaN activation. Another problem is that any demonstration of CaN activation *in situ* is difficult because it involves non-covalent interactions that may not survive tissue extraction. Nevertheless, association between CaM and CaN has been shown to occur in hearts in response to pressure-overload.[87] An increase in CsA-sensitive NFAT transactivating activity is also used as a surrogate index of CaN activation. Using an 'NFAT reporter' mouse, in which expression of a luciferase transgene is under the control of a NFAT-sensitive promoter, NFAT-dependent transcription is increased following pressure-overload or myocardial infarction in a CsA-sensitive manner.[88] However, NFAT-dependent transcription does not increase in exercise-induced hypertrophy or following administration of IGF-1 which both increase myocardial mass.[88] This suggests that any activation of CaN/NFATs in the heart is associated with 'maladaptive' hypertrophy rather than physiological 'adaptive' hypertrophy.

Ca²⁺/CaM-dependent protein kinases

Ca^{2+}/CaM-dependent protein kinases (CaMKs) represent a second class of Ca^{2+}-regulated signaling proteins and comprise three subfamilies (I, II and IV).[89] The CaMKII subfamily has several members with the δ_B isoform exhibiting preferential localization to the nucleus whereas the δ_C isoform is cytoplasmic. Although activation of all CaMKs involves their Ca^{2+}/CaM-dependent phosphorylation, the mechanism of activation of CaMKII (involving extensive autophosphorylation) differs from that of CaMKI and CaMKIV (which involves upstream CaMK kinases) (Fig. 9.4).[89] Cell culture experiments implicate CaMKs in cardiac myocyte hypertrophy,[89] but the results from transgenic mouse lines are confusing. Thus, cardiospecific expression of constitutively activated CaMKIV induces a myocardial hypertrophy that probably involves the MEF2 transcription factor, but this later develops into dilated cardiomyopathy.[90] In addition, the CaMKIV-expressing mouse shows increased endogenous CaMKII which may be pro-arrhythmic.[91] However, whether CaMKIV is expressed endogenously in cardiac myocytes is debatable,[89,92] and hypertrophy in response to pressure-overload in *CaMK4*⁻/⁻ mice is indistinguishable from the wild-type response.[92] In another series of experiments, cardiospecific expression of wild-type CaMKIIδ_B or CaMKIIδ_C results ultimately in dilated cardiomyopathy with clear disturbances in Ca^{2+} fluxes.[93–95] Whether these changes in Ca^{2+} fluxes are causative or simply reflect a response to hypertrophy/failure is unclear. Overall, any role of CaMK in myocardial hypertrophy remains to be established.

Cyclic nucleotide-dependent protein kinases

Cyclic AMP- and cyclic GMP-dependent protein kinases (PKA and PKG, respectively) were among the first protein kinases to be discovered. In the cardiac myocyte, cyclic AMP/PKA signaling is important in the regulation of cardiac contractility, being positively inotropic through its effects on

Ca^{2+} movements.[96] Cyclic AMP concentrations are increased by β-adrenergic receptors signaling through G_sPCRs to adenylyl cyclase. Increased cyclic AMP activates the heterotetrameric PKA holoenzyme, by promoting the dissociation of the two catalytic subunits from the inhibitory regulatory subunit dimer. Infusion of β-adrenergic agonists leads to myocardial hypertrophy and, ultimately, to heart failure, but it is not clear whether this is a direct effect of receptor signaling or reflects an indirect response to, for example, the increase in contractility. Transgenic mice that cardiospecifically overexpress the catalytic subunit of PKA develop dilated cardiomyopathy and succumb to early sudden death,[97] but, given the potentially profound effects on contractility, the physiological relevance is unclear. Cyclic GMP concentrations are increased through guanylyl cyclases and activate PKG. Activation of PKG may be anti-hypertrophic, though the molecular mechanism is not established.[98] Importantly, the transmembrane receptors for natriuretic peptides possess an intracellular domain with guanylyl cyclase activity. Thus one role of the expression of natriuretic peptides during maladaptive hypertrophy may be to exert negative feedback control.[20]

Signaling through cytokine receptors: JAKs and STATs

Cardiotrophin-1, interleukin-6 and leukaemia inhibitory factor activate gp130-linked receptors, increasing receptor tyrosine phosphorylation through receptor-associated Janus kinases (JAKs).[99] Classically, the phosphorylated receptors recruit members of the signal transducer and activator of transcription (STAT) family of transcription factors, which are then tyrosine-phosphorylated by JAKs. Phosphorylated STATs are released from the receptor, dimerize and migrate to the nucleus to promote gene expression. However, there is concomitant activation of the ERK1/2 cascade and PKB/Akt via gp130-linked receptors.

There is evidence that signaling through the gp130-linked family of cytokine receptors may have some effects on myocyte hypertrophy, specifically to induce an 'elongated' phenotype which (it is claimed) simulates the volume-overload response *in vivo*.[100] In support of a role for STATs in myocardial hypertrophy, stimulation of gp130-linked receptors in cardiac myocytes or pressure-overload *in vivo* activates STATs.[101] In addition, transgenic mice expressing an inhibitory form of gp130 exhibit a reduced hypertrophic response to pressure-overload compared with wild-type mice, and STAT activation is suppressed. Furthermore, cardiospecific overexpression of STAT3 promotes myocardial hypertrophy in transgenic mice. However, ERK5 (see Mitogen-activated protein kinases, pages 144–5) is also implicated in the hypertrophic response induced by gp130-linked receptors.[62] Apart from potential effects on hypertrophy, gp130-linked receptors, acting principally through ERK1/2 and/or PKB/Akt, have clearly defined anti-apoptotic effects in cardiac myocytes, which may reflect their principal role in the heart.[101]

Concluding remarks: signal transduction

Considerable progress has been made over the last 10 years in identifying the intracellular signaling pathways that are activated by hypertrophic stimuli and that lead to the response. However, there is a tendency to believe that a single signaling pathway is responsible for all the different facets of cardiac myocyte hypertrophy, a view which could be considered somewhat naive. *In vivo*, multiple factors will trigger signaling events in a single cardiac myocyte, activating many or all of the pathways discussed in this chapter. The concentrations of neurohumoral factors in the heart may be only sufficient to elicit submaximal responses, and the degree of activation of any single pathway may vary according to the combinations of factors that are present. Furthermore, it is probable that different routes through the signaling networks can elicit the overall functional response of cardiac myocyte hypertrophy. There is clearly a requirement for signal integration both with respect to initiation and propagation of signaling events, and with respect to downstream consequences (e.g. gene and protein expression), which lead to the hypertrophic response. It is well established that expression of any particular gene is regulated by multiple, interacting factors, but additional post-transcriptional signal integration is required. A clear example in cardiac myocytes involves the upregulation of the immediate early gene *c-jun* (which encodes the c-Jun transcription factor), which is thought to represent an early hallmark of cardiac myocyte or myocardial hypertrophy.[24] Although the ERK1/2 cascade is particularly required for upregulation of *c-jun* mRNA, for the protein to be efficiently expressed, activation of JNKs is also required.[102] This type of signal integration is only just beginning to be explored and represents a challenge for the future.

References

1. Adams JW, Brown JH. G proteins in growth and apoptosis: lessons from the heart. Oncogene 2001; 20: 1626–34.

2. Rhee SG. Regulation of phosphoinositide-specific phospholipase C. Annu Rev Biochem 2001; 70: 281–312.

3. Dorn GW, 2nd, Brown JH. Gq signaling in cardiac adaptation and maladaptation. Trends Cardiovasc Med 1999; 9: 26–34.

4. Hodgkin MN, Pettitt TR, Martin A et al. Diacylglycerols and phosphatidates: which molecular species are intracellular messengers? Trends Biochem Sci 1998; 23: 200–4.

5. Pucéat M, Vassort G. Purinergic stimulation of rat cardiomyocytes induces tyrosine phosphorylation and membrane association of phospholipase Cγ: a major mechanism for InsP$_3$ generation. Biochem J 1996; 318: 723–8.

6. Jalili T, Takeishi Y, Walsh RA. Signal transduction during cardiac hypertrophy: the role of G$_{\alpha q}$, PLC βI, and PKC. Cardiovasc Res 1999; 44: 5–9.

7. Lee WK, Kim JK, Seo MS et al. Molecular cloning and expression analysis of a mouse phospholipase C-δ1. Biochem Biophys Res Commun 1999; 261: 393–9.

8. Lopez I, Mak EC, Ding J et al. A novel bifunctional phospholipase C that is regulated by G$_{\alpha 12}$ and stimulates the Ras/mitogen-activated protein kinase pathway. J Biol Chem 2001; 276: 2758–65.

9. Arthur JF, Matkovich SJ, Mitchell CJ et al. Evidence for selective coupling of α1-adrenergic receptors to phospholipase C-β1 in rat neonatal cardiomyocytes. J Biol Chem 2001; 276: 37341–6.

10. Clerk A, Sugden PH. Regulation of phospholipases C and D in rat ventricular myocytes: stimulation by endothelin-1, bradykinin and phenylephrine. J Mol Cell Cardiol 1997; 29: 1593–604.

11. Guse AH, Kiess W, Funk B et al. Identification and characterization of insulin-like growth factor receptors on adult rat cardiac myocytes: linkage to inositol 1,4,5-trisphosphate formation. Endocrinology 1992; 130: 145–51.

12. Brose N, Rosenmund C. Move over protein kinase C, you've got company: alternative effectors of diacylglycerol and phorbol esters. J Cell Sci 2002; 115: 4399–411.

13. Marks AR. Cardiac intracellular calcium release channels: role in heart failure. Circ Res 2000; 87: 8–11.

14. Howes AL, Arthur JF, Zhang T et al. Akt-mediated cardiac myocyte survival pathways are compromised by Gα_q-induced phosphoinositide 4,5-bisphosphate depletion. J Biol Chem 2003; 278: 40343–51.

15. Kodaki T, Yamashita S. Cloning, expression, and characterization of a novel phospholipase D complementary DNA from rat brain. J Biol Chem 1997; 272: 11408–13.

16. Katayama K, Kodaki T, Nagamachi Y et al. Cloning, differential regulation and tissue distribution of alternatively spliced isoforms of ADP-ribosylation-factor-dependent phospholipase D from rat liver. Biochem J 1998; 329: 647–52.

17. Rizzo M, Romero G. Pharmacological importance of phospholipase D and phosphatidic acid in the regulation of the mitogen-activated protein kinase cascade. Pharmacol Ther 2002; 94: 35–50.

18. Clerk A, Bogoyevitch MA, Andersson MB et al. Differential activation of protein kinase C isoforms by endothelin-1 and phenylephrine and subsequent stimulation of p42 and p44 mitogen-activated protein kinases in ventricular myocytes cultured from neonatal rat hearts. J Biol Chem 1994; 269: 32848–57.

19. Eskildsen-Helmond YEG, Hahnel D, Reinhardt U et al. Phospholipid source and molecular species composition of 1,2-diacylglycerol in agonist-stimulated rat cardiomyocytes. Cardiovasc Res 1998; 40: 182–90.

20. Newton AC. Regulation of the ABC kinases by phosphorylation: protein kinase C as a paradigm. Biochem J 2003; 370: 361–71.

21. Mackay K, Mochly-Rosen D. Localization, anchoring, and functions of protein kinase C isozymes in the heart. J Mol Cell Cardiol 2001; 33: 1301–7.

22. Vlahos CJ, McDowell SA, Clerk A. Kinases as

therapeutic targets for heart failure. Nat Rev Drug Discov 2003; 2: 99–113.

23. Clerk A, Bogoyevitch MA, Fuller SJ et al. Expression of protein kinase C isoforms during cardiac ventricular development. Am J Physiol Heart Circ Physiol 1995; 269: H1087–97.

24. Sugden PH, Clerk A. Cellular mechanisms of cardiac hypertrophy. J Mol Med 1998; 76: 725–46.

25. Rybin VO, Sabri A, Short J et al. Cross-regulation of novel protein kinase C (PKC) isoform function in cardiomyocytes. Role of PKC ε in activation loop phosphorylations and PKC δ in hydrophobic motif phosphorylations. J Biol Chem 2003; 278: 14555–64.

26. Jaken S, Parker PJ. Protein kinase C binding partners. Bioessays 2000; 22: 245–54.

27. Chen L, Hahn H, Wu G et al. Opposing cardioprotective actions and parallel hypertrophic effects of δPKC and εPKC. Proc Natl Acad Sci USA 2001; 98: 11114–19.

28. Takeishi Y, Ping P, Bolli R et al. Transgenic overexpression of constitutively active protein kinase C ε causes concentric cardiac hypertrophy. Circ Res 2000; 86: 1218–23.

29. Pass JM, Zheng Y, Wead WB et al. PKCε activation induces dichotomous cardiac phenotypes and modulates PKCε–RACK interactions and RACK expression. Am J Physiol Heart Circ Physiol 2001; 280: H946–55.

30. Heidkamp MC, Bayer AL, Martin JL et al. Differential activation of mitogen-activated protein kinase cascades and apoptosis by protein kinase C-ε and δ in neonatal rat ventricular myocytes. Circ Res 2001; 89: 882–90.

31. Ping P. Identification of novel signalling complexes by functional proteomics. Circ Res 2003; 93: 595–603.

32. Shubeita HE, Martinson EA, van Bilsen M et al. Transcriptional activation of the cardiac myosin light chain 2 and atrial natriuretic factor genes by protein kinase C in neonatal rat ventricular myocytes. Proc Natl Acad Sci USA 1992; 89: 1305–9.

33. Braz JC, Bueno OF, De Windt LJ et al. PKCα regulates the hypertrophic growth of cardiomyocytes through extracellular signal-regulated kinase 1/2 (ERK1/2). J Cell Biol 2002; 156: 905–19.

34. Kerkalä R, Ilves M, Pikkarainen S et al. Identification of PKCα isoform-specific effects in cardiac myocytes using phosphorothioate oligonucleotides. Mol Pharmacol 2002; 62: 1482–91.

35. Braz JC, Gregory K, Pathak A et al. PKC-α regulates cardiac contractility and propensity toward heart failure. Nat Med 2004; 10: 248–54.

36. Roman BB, Geenen DL, Leitges PM et al. PKC-β is not necessary for cardiac hypertrophy. Am J Physiol Heart Circ Physiol 2001; 280: H2264–70.

37. Takai Y, Sasaki T, Matozaki T. Small GTP-binding proteins. Physiol Rev 2001; 81: 153–208.

38. Hancock JF. Ras proteins: different signals from different locations. Nat Rev Mol Cell Biol 2003; 4: 373–84.

39. Schlessinger J. Cell signaling by receptor tyrosine kinases. Cell 2000; 103: 211–25.

40. Chiloeches A, Paterson HF, Marais RM et al. Regulation of Ras.GTP loading and Ras-Raf association in neonatal rat ventricular myocytes by G protein-coupled receptor agonists and phorbol esters. Activation of the ERK cascade by phorbol esters is mediated by Ras. J Biol Chem 1999; 274: 19762–70.

41. Clerk A, Sugden PH. Small guanine nucleotide-binding proteins and myocardial hypertrophy. Circ Res 2000; 86: 1019–23.

42. Zheng M, Dilly K, Dos Santos Cruz J et al. Sarcoplasmic reticulum calcium defect in Ras-induced hypertrophic cardiomyopathy heart. Am J Physiol Heart Circ Physiol 2004; 286: H424–33.

43. Riento K, Ridley AJ. ROCKs: multifunctional kinases in cell behaviour. Nat Rev Mol Cell Biol 2003; 4: 446–56.

44. Mukai H. The structure and function of PKN, a protein kinase having a catalytic domain homologous to that of PKC. J Biochem (Tokyo) 2003; 133: 17–27.

45. Clerk A, Pham FH, Fuller SJ et al. Regulation of mitogen-activated protein kinases in cardiac myocytes through the small G protein, Rac1. Mol Cell Biol 2001; 21: 1173–84.

46. Bokoch GM. Biology of the p21-activated kinases. Annu Rev Biochem 2003; 72: 743–81.

47. Wu G, Yussman MG, Barrett TJ et al. Increased myocardial Rab GTPase expression: a consequence and cause of cardiomyopathy. Circ Res 2001; 89: 1130–7.

48. Chen Z, Gibson TB, Robinson F et al. MAP kinases. Chem Rev 2001; 101: 2449–76.

49. Mason CS, Springer CJ, Cooper RG et al. Serine and tyrosine phosphorylations cooperate in Raf-1, but not B-Raf activation. EMBO J 1999; 18: 2137–48.

50. Clerk A, Kemp TJ, Harrison JG et al. Integration of protein kinase signaling pathways in cardiac myocytes: signaling to and from the extracellular signal-regulated kinases. Adv Enzyme Regul 2004; 44: 233–48.

51. Kyriakis JM, Avruch J. Mammalian mitogen-activated protein kinase signal transduction pathways

activated by stress and inflammation. Physiol Rev 2001; 81: 807–69.

52. Clerk A, Michael A, Sugden PH. Stimulation of the p38 mitogen-activated protein kinase pathway in neonatal rat ventricular myocytes by the G protein-coupled receptor agonists endothelin-1 and phenylephrine: a role in cardiac myocyte hypertrophy? J Cell Biol 1998; 142: 523–35.

53. Clerk A, Michael A, Sugden PH. Stimulation of multiple mitogen-activated protein kinase sub-families by oxidative stress and phosphorylation of the small heat shock protein, HSP25/27, in neonatal ventricular myocytes. Biochem J 1998; 333: 581–9.

54. Bogoyevitch MA, Gillespie-Brown J, Ketterman AJ et al. Stimulation of the stress-activated mitogen-activated protein kinase subfamilies in perfused heart. p38/RK mitogen-activated protein kinases and c-Jun N-terminal kinases are activated by ischemia–reperfusion. Circ Res 1996; 79: 161–72.

55. Clerk A, Sugden PH. The p38-MAPK inhibitor, SB203580, inhibits cardiac stress-activated protein kinases/c-Jun N-terminal kinases (SAPKs/JNKs). FEBS Lett 1998; 426: 93–6.

56. Sugden PH. Signalling pathways in cardiac myocyte hypertrophy. Ann Med 2001; 33: 611–22.

57. Liang Q, Molkentin JD. Redefining the roles of p38 and JNK signaling in cardiac hypertrophy: dichotomy between cultured myocytes and animal models. J Mol Cell Cardiol 2003; 35: 1385–94.

58. Yue TL, Gu J-L, Wang C et al. Extracellular signal-regulated kinase (ERK) plays an essential role in hypertrophic agonists, endothelin-1 and phenylephrine-induced cardiomyocyte hypertrophy. J Biol Chem 2000; 275: 37895–901.

59. Bueno OF, De Windt LJ, Tymitz KM et al. The MEK1-ERK1/2 signalling pathway promotes compensated hypertrophy in transgenic mice. EMBO J 2000; 19: 6341–50.

60. Petrich BG, Wang Y. Stress-activated MAP kinases in cardiac remodeling and heart failure; new insights from transgenic studies. Trends Cardiovasc Med 2004; 14: 50–5.

61. Clerk A, Cole SM, Cullingford TE et al. Regulation of cardiac myocyte cell death. Pharmacol Ther 2003; 97: 223–61.

62. Nicol RL, Frey N, Pearson G et al. Activated MEK5 induces serial assembly of sarcomeres and eccentric cardiac hypertrophy. EMBO J 2001; 20: 2757–67.

63. Vanhaesebroeck B, Leevers SJ, Ahmadi K et al. Synthesis and function of 3-phosphorylated inositol lipids. Annu Rev Biochem 2001; 70: 535–602.

64. Vanhaesebroeck B, Alessi DR. The PI3K-PDK1 connection: more than just a road to PKB. Biochem J 2000; 346: 561–76.

65. Milburn CC, Deak M, Kelly SM et al. Binding of phosphatidylinositol 3,4,5-trisphosphate to the pleckstrin homology domain of protein kinase B induces a conformational change. Biochem J 2003; 375: 531–8.

66. Harris TE, Lawrence JC, Jr. TOR signalling. Sci STKE 2003; www.stke.org/cgi/content/full/sigtrans; 2003; 212/re15. http://stke.sciencemag.org/cgi/content/full/OC_sigtrans;2003;212/re15 (accessed 30 April 2005).

67. Frame S, Cohen P. GSK3 takes centre stage more than 20 years after its discovery. Biochem J 2001; 359: 1–16.

68. Pham FH, Cole SM, Clerk A. Regulation of cardiac myocyte protein synthesis through phosphatidylinositol 3′ kinase and protein kinase B. Adv Enzyme Regul 2001; 41: 73–86.

69. Sugden PH. Ras, Akt, and mechanotransduction in the cardiac myocyte. Circ Res 2003; 93: 1179–92.

70. McMullen JR, Shioi T, Huang WY et al. The insulin-like growth factor 1 receptor induces physiological heart growth via the phosphoinositide 3-kinase(p110α) pathway. J Biol Chem 2004; 279: 4782–93.

71. Crackower MA, Oudit GY, Kozieradzki I et al. Regulation of myocardial contractility and cell size by distinct PI3K-PTEN signalling pathways. Cell 2002; 110: 737–49.

72. Mora A, Davies AM, Bertrand L et al. Deficiency of PDK1 in cardiac muscle results in heart failure and increased sensitivity to hypoxia. EMBO J 2003; 22: 4666–76.

73. McMullen JR, Shioi T, Zhang L et al. Phosphoinositide 3-kinase(p110α) plays a critical role for the induction of physiological, but not pathological, cardiac hypertrophy. Proc Natl Acad Sci USA 2003; 100: 12355–60.

74. Naga Prasad SV, Esposito G, Mao L et al. Gβγ-dependent phosphoinositide 3-kinase activation in hearts with in vivo overload. J Biol Chem 2000; 275: 4693–8.

75. Haq S, Choukroun G, Kang ZB et al. Glycogen synthase kinase-3β is a negative regulator of cardiomyocyte hypertrophy. J Cell Biol 2000; 151: 117–29.

76. Haq S, Michael A, Andreucci M et al. Stabilization of β-catenin by a Wnt-independent mechanism regulates cardiac myocyte growth. Proc Natl Acad Sci USA 2003; 100: 4610–15.

77. Beals CR, Sheridan CR, Turck CW et al. Nuclear

export of NF-ATc enhanced by glycogen synthase kinase-3. Science 1997; 275: 1930–3.

78. Antos CL, McKinsey TA, Frey N et al. Activated glycogen synthase-3β suppresses cardiac hypertrophy *in vivo*. Proc Natl Acad Sci USA 2002; 99: 907–12.

79. Bueno OF, van Rooij E, Molkentin JD et al. Calcineurin and hypertrophic heart disease: novel insights and remaining questions. Cardiovasc Res 2002; 53: 806–21.

80. Vega RB, Bassel-Duby R, Olson EN. Control of cardiac growth and function by calcineurin signalling. J Biol Chem 2003; 278: 36981–4.

81. Klee CB, Ren H, Wang X. Regulation of the calmodulin-stimulated protein phosphatase, calcineurin. J Biol Chem 1998; 273: 13367–70.

82. Ke H, Huai Q. Structures of calcineurin and its complexes with immophilins-immunosuppressants. Biochem Biophys Res Commun 2003; 311: 1095–102.

83. Molkentin JD, Lu J-R, Antos C et al. A calcineurin-dependent transcriptional pathway for cardiac hypertrophy. Cell 1998; 93: 215–28.

84. van Rooij E, Doevendans PA, de Theije CC et al. Requirement of nuclear factor of activated T-cells in calcineurin-mediated cardiomyocyte hypertrophy. J Biol Chem 2002; 277: 48617–26.

85. Bueno OF, Wilkins BJ, Tymitz KM et al. Impaired cardiac hypertrophic response in calcineurin Aβ-deficient mice. Proc Natl Acad Sci USA 2002; 99: 4586–91.

86. Mittmann C, Eschenhagen T, Scholz H. Cellular and molecular aspects of contractile dysfunction in heart failure. Cardiovasc Res 1998; 39: 267–75.

87. Lim HW, De Windt LJ, Steinberg L et al. Calcineurin expression, activation, and function in cardiac pressure-overload hypertrophy. Circulation 2000; 101: 2431–7.

88. Wilkins BJ, Dai YS, Bueno OF et al. Calcineurin/NFAT coupling participates in pathological, but not physiological, cardiac hypertrophy. Circ Res 2004; 94: 110–18.

89. Zhang T, Miyamoto S, Brown JH. Cardiomyocyte calcium and calcium/calmodulin-dependent protein kinase II: friends or foes? Recent Prog Horm Res 2004; 59: 141–68.

90. Passier R, Zeng H, Frey N et al. CaM kinase signaling induces cardiac hypertrophy and activates the MEF2 transcription factor *in vivo*. J Clin Invest 2000; 105: 1395–406.

91. Wu Y, Temple J, Zhang R et al. Calmodulin kinase II and arrhythmias in a mouse model of cardiac hypertrophy. Circulation 2002; 106: 1288–93.

92. Colomer JM, Mao L, Rockman HA et al. Pressure overload selectively up-regulates Ca^{2+}/calmodulin-dependent protein kinase II in vivo. Mol Endocrinol 2003; 17: 183–92.

93. Zhang T, Johnson EN, Gu Y et al. The cardiac-specific nuclear δ_B isoform of Ca^{2+}/calmodulin-dependent protein kinase II induces hypertrophy and dilated cardiomyopathy associated with increased protein phosphatase 2A activity. J Biol Chem 2002; 277: 1261–7.

94. Maier LS, Zhang T, Chen L et al. Transgenic CaMKIIδ_C overexpression uniquely alters cardiac myocyte Ca^{2+} handling: reduced SR Ca load and activated SR Ca^{2+} release. Circ Res 2003; 92: 904–9.

95. Zhang T, Maier LS, Dalton ND et al. The δ_C isoform of CaMKII is activated in cardiac hypertrophy and induces dilated cardiomyopathy and heart failure. Circ Res 2003; 92: 912–19.

96. Dorn GW II, Molkentin JD. Manipulating cardiac contractility in heart failure: data from mice and men. Circulation 2004; 109: 150–8.

97. Antos CL, Frey N, Marx SO et al. Dilated cardiomyopathy and sudden death resulting from constitutive activation of protein kinase A. Circ Res 2001; 89: 997–1004.

98. Molkentin JD. A friend within the heart: natriuretic peptide receptor signalling. J Clin Invest 2003; 111: 1275–7.

99. Heinrich PC, Behrmann I, Haan S et al. Principles of interleukin (IL)-6-type cytokine signalling and its regulation. Biochem J 2003; 374: 1–20.

100. Wollert KC, Taga T, Saito M et al. Cardiotrophin-1 activates a distinct form of cardiac hypertrophy. Assembly of sarcomeric units in series via gp130/ leukemia inhibitory factor receptor-dependent pathways. J Biol Chem 1996; 271: 9535–45.

101. Molkentin JD, Dorn GW, 2nd. Cytoplasmic signalling pathways that regulate cardiac hypertrophy. Annu Rev Physiol 2001; 63: 391–426.

102. Clerk A, Kemp TJ, Harrison JG et al. Upregulation of *c-jun* mRNA in cardiac myocytes requires the extracellular signal-regulated kinase cascade, but c-Jun N-terminal kinases are required for efficient upregulation of c-Jun protein. Biochem J 2002; 368: 101–10.

CHAPTER 10

Cell-based therapies for cardiac regeneration and repair

Michael D Schneider

Cardiac scarring and persistent muscle cell loss is the hallmark of myocardial infarction in humans. This biological limitation is found, too, with experimental myocardial infarction in other mammals, of less pressing clinical need – an inbuilt restriction contrasting with the exuberant restorative, regenerative growth that enables the total replacement of a limb, tail, or even heart in certain lower species[1,2] – 'lower' being in this respect a sorely misleading designation. Quite recently, and with uncanny speed, the notion of cell implantation as a means to promote cardiac repair has moved from an esoteric laboratory phenomenon to a dramatic and closely watched series of clinical trials on the world stage, involving skeletal muscle as a cardiac surrogate,[3] and progenitor/stem cells from bone marrow and the circulation as the agents of therapeutic angiogenesis and perhaps even as cardiac precursors.[4–8] While some questions get posed with regularity (what cells to use, and how to deliver them), others are being asked more rarely (why does grafting work, and should a patient's own – autologous – cells always be preferred to allogeneic ones?), and some are asked hardly at all (where and how do the endogenous adult cardiac progenitor cells arise?). The interventions now in play are remarkable for their therapeutic promise, their biological complexity, and the likelihood that defective stem cell function will be among the next therapeutic horizons.

If the logic of cell therapy is to replace absent contractile cells, then one might look to the heart itself as a source. Pioneering studies a decade ago demonstrated the capacity of skeletal myoblasts to form stable differentiated myotubes in ventricular myocardium, and of fetal cardiac myocytes likewise to persist stably, when delivered by direct injection to the hearts of dystrophic mice and dogs.[9] Fetal ventricular myocytes had the additional attraction of forming nascent intercalated discs. However, cardiac myocytes themselves fail one test conspicuously, namely, that of being a readily accessible source of donor cells for clinical application. This limit has driven the search for alternative cells as replacements for cardiac muscle, forerunners of cardiac muscle, and agents working less directly to promote heart muscle repair. For detailed reviews, see references 9–16.

What cells might be applicable to cardiac repair?

Skeletal myoblasts

For the reasons of availability and imprecise but plausible similarity to ventricular myocardium, interest focused historically first on autologous skeletal myoblasts – which generate the 'other' striated muscle. Skeletal myoblast engraftment following experimental infarction was shown to improve regional systolic function, although diastolic function was improved even by non-contractile fibroblasts, and few systematic comparisons of donor cell types have been undertaken even to date.[17] The observed impact of skeletal muscle on systolic function, together with the combined attributes of cogent rationale, autologous origin, and easy expansion of patient tissue using cell culture, provided the impetus to a pioneering case study of grafting in a 72 year-old male with refractory heart failure.[3] Because cell therapy was an adjunct to two-vessel anterior wall revascularization, the interpretation of outcome is complex. But, among the encouraging features of this first report, echocardiography demonstrated improved left ventricular ejection fraction, improved segmental contractility, and improved regional motion by tissue Doppler imaging, including benefits in the non-reperfused inferior wall. Of equal note, positron emission tomography confirmed the appearance of new viable tissue in the former area of scar, as did histological follow-up a year and a half later, at the time of the patient's demise.[18] In addition, the grafted cells' differentiation into typical skeletal muscle 'myotubes' was confirmed, with a partial shift to slow-twitch muscle fibers. This work justifiably captured the imagination of the cardiovascular community, as the first implementation in a human of any cell-based therapy for cardiac repair.

This limit has driven the search for alternative cells as replacements for cardiac muscle, forerunners of cardiac muscle, and agents working less directly to promote heart muscle repair (Fig. 10.1). Questions posed in or by this pathfinding study continue to influence much of the clinical, translational, and basic research agenda for the field.

- **Are the grafted cells retained and viable?** Cell death after implantation can be prevalent using direct intramyocardial injection,[19] and effective delivery must be proven, if less direct means are used. More widely available than positron emission tomography, magnetic resonance imaging can provide analogous information, but requires iron fluorophores or tags yet to be developed.
- **What is the measure of success?** Ultimately, of course, one hopes for improved patient survival, but what surrogate endpoints of regional function and status best guide the clinical investigator en route to that goal?
- **Why and how does grafting work?** The overall logic of skeletal muscle as the replacement for cardiac muscle has been discussed above, along with empirical evidence both that it works in animals, and that it probably works better than some extraneous cell types. Yet,

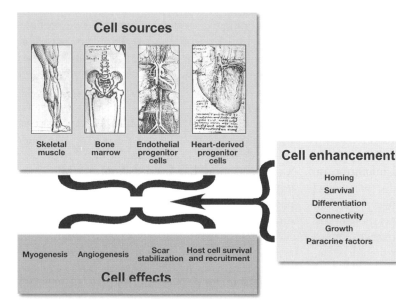

Figure 10.1. *Cell-based therapies for cardiac repair. The principal cell types under study, suggested mechanisms of action, and potential augmentations are shown.*

evolving studies on the cell biology of cardiac repair by skeletal muscle have raised uncertainties. Injected cells were commonly an island of myotubes walled off by scar, or failed to couple electrically to the host myocardium even when contiguous;[20–22] this has been attributed to the absence or paucity of low-resistance gap junctions, proteins essential for normal cardiac electrical connectivity. The failure to connect is striking, in comparison with the physiological coupling of donor cardiomyocytes injected under identical conditions.[22] In addition to their functional isolation, the skeletal myocytes formed after implantation have abnormally short action potentials and heterogeneous calcium transients, when compared with the neighboring host cardiac muscle.[21,23] Consistent with these abnormalities, ventricular arrhythmias have emerged as a potential hazard in the leading clinical studies of skeletal myoblast therapy, requiring implantable defibrillators as an adjunct.[24,25] The added risk of arrhythmia may, though, prove hard to assess conclusively in a population that is prone to arrhythmias already. However, if not part of a seamless syncytium with the host tissue, skeletal muscle grafts must improve contractility in some other way, by some mode of indirect, stretch-induced contraction, or as the instigator of an improved host milieu.

Bone marrow

Bone marrow cells and their derivatives fall conceptually at the opposite end of the spectrum from skeletal myoblasts as a cardiac remedy. Until quite recently, the developmental plasticity of unspecialized bone marrow 'stem' cells was not thought to encompass heart muscle specification; however, this tantalizing possibility was raised in complementary ways. Following myocardial infarction in mice, genetically tagged endogenous bone marrow is recruited to injured myocardium and contributes to new cardiac muscle cells, albeit at a prevalence of 200 per million, 150-fold less than bone marrow's engraftment as endothelial cells.[26] In addition, directly injected bone marrow cells may convert to a cardiac fate,[27] although this finding has been challenged by investigators using either unfractionated bone marrow as the donor population or the Lin⁻c-Kit⁺ fraction that is enriched for hematopoietic stem cells (HSCs).[28–30]

However, to infer that bone marrow is ill-posed for human cardiac repair may be misleading, since cardiac muscle creation (myogenesis) was rarely the cornerstone for clinical trials of these cells. Historically, the suggested use of bone marrow cells and their circulating derivatives – endothelial progenitor cells (EPCs) – arises from an altogether different ideational and technical background: namely, the use of cells to promote therapeutic vessel formation (angiogenesis),[31–33] in lieu of defined angiogenic growth factors or their genes. Successful extrapolation of this concept from peripheral vascular disease to the ischemic myocardium[14,34,35] does not rest on whether the donor cells make a cardiac myocyte or not.

With angiogenesis as the enabling concept, initial human studies of autologous bone marrow cells and of autologous EPCs were reported in 2002, using intracoronary delivery and the clinical setting of acute myocardial infarction.[4,5] In addition, electromechanical mapping has been used, to guide the deployment of cells by endocardial injection, and the clinical context broadened to chronic ischemic cardiomyopathy.[6,8] Collectively, these and other related clinical studies provide substantial evidence that the cells' administration is safe, even in the immediate peri-infarct days, and, although not powered to assert effectiveness, give preliminary yet encouraging indications that perfusion and systolic function are both enhanced. The mechanisms by which systolic function is improved remain open to question: the potential and relative contributions of rescuing the hibernating myocardium, versus allaying apoptosis, versus alternative effects are all areas of active investigation. Importantly, a randomized controlled trial of 60 patients now has been reported, using intracoronary delivery of autologous bone marrow five days after percutaneous coronary intervention for acute myocardial infarction with ST segment elevation (BOOST): the primary endpoint, the increase in LV function six months later by cardiac magnetic resonance imaging (MRI), was improved nearly 10-fold in the bone marrow cell-treated group (6.7%, versus. 0.7% for the controls).[7]

159

Thus, within just two years of the first reported human studies, a highly credible and consistent case has been built, that justifies extending these inquiries to the necessary larger multicenter trials. As an alternative to intracoronary, intramyocardial, transendocardial, and coronary venous instillation, attention also has been directed towards the potential mobilization of bone marrow cells, using chemokines such as stromal cell-derived factor-1 or granulocyte colony-stimulating factor.[36] As with cell administration, there is lively debate concerning the relative importance of new myocyte formation versus angiogenesis and relief from apoptosis.[37]

A formidable complexity to be understood better is the marked heterogeneity of bone marrow. Besides HSCs, which give rise to the blood lineages and whose plasticity in other directions has received the greatest attention, bone marrow also contains mesenchymal stem cells (MSCs), which give rise to the stroma, as well as exceedingly rare subpopulations of each. For HSCs, most if not all long-term self-renewal potential can be tracked to the so-called 'side population' (SP) cell, named for its relative position in flow cytometry measurements of cells' ability to expel the vital dye, Hoechst 33342.[38] For mesenchymal stem cells, which have been studied in both rodent and large-mammal models of infarction,[39–41] an especially intriguing and potentially important constituent is the co-purifying 'multipotential adult progenitor cell' (MAPC).[42] Seemingly equivalent from several differing tissues of origin (bone marrow, muscle, brain),[43] MAPCs can be propagated for many dozens of population doublings and exhibit bona fide multilineage potential. One possible basis for discrepancies among grafting studies thus is not merely their respective content of HSCs versus MSCs, the latter being seldom quantified, but also specific details of these and other HSC and MSC subpopulations. The existence of SP cells and MAPCs outside the bone marrow raises a further question: if one or both of these is important to the benefit provided by grafting bone marrow to the heart, are SP cells and MAPCs equivalent, if taken from other sites such as the heart itself?[44,45] Alternatively, are there specific relevant properties associated with the cardiac niche?

Embryonic stem cells

One cell with indisputable potential to generate cardiac myocytes – both when cultured *in vitro* and when introduced into host myocardium *in vivo* – is the embryonic stem (ES) cell, from which all cells of the organism can be derived.[46–48] Embryonic stem cell lines are created from the inner cell mass of the blastocyst, a few days after conception, when the embryo is just a hollow ball of cells with no structure and no organs, and implantation in the uterus is yet to occur. Nevertheless, their use or, more precisely, their creation has become entangled in a morass of religious concerns as an extension of long-standing objections to abortion.[49] Merely a handful of the Federally-sanctioned human ES cell lines have the requisite properties, but many more have recently been generated, using other resources.[50]

How ES cells become directed to the cardiac lineage after local myocardial delivery might, in principle, correspond exactly or depart in some or all respects from what is know of cardiac specification as it occurs natively, in primitive mesoderm. Paracrine signals have been identified that are required for cardiac specification in ES cells following intramyocardial injection, namely, transforming growth factor-beta (TGF-β) and related growth factors of the bone morphogenetic protein family.[47] Disruption of these pathways, respectively, by a dominant-interfering TGF-β receptor or the extracellular antagonist Noggin resulted in undifferentiated tumor formation. *In vitro*, these growth factors were both sufficient and required to trigger the ES cells' expression of heart-forming transcription factors (Nkx2.5, MEF2-C). Although it is not yet clear if manipulations of these two pathways would be salutary in ES cell-based therapy for cardiac repair, suggestive evidence is now available for a third peptide growth factor, insulin-like growth factor-1 (IGF-1).[51] However, because IGF-1 was administered concomitant with the injected cells, and because IGF-1 was not tested alone, the potential for direct actions of IGF-1 on the host myocardium must be considered.

Heart-derived progenitor cells

Until very recently, no entity residing in the adult heart was known with the capacity to create new cardiac muscle cells. Recent discoveries – embarking from complementary points of departure – have unmasked this tantalizing possibility.[44,45,52–56]

One cardinal property both of pluripotent stem cells and of progenitor cells with a more restricted range of lineage decisions is their growth capacity, i.e. proliferation without replicative senescence. This attribute depends in part on telomerase, the RNA-dependent DNA polymerase (reverse transcriptase) that offsets the erosion of chromosome ends that occurs during DNA replication. Although telomerase reverse transcriptase expression and enzyme activity are markedly downregulated in adult mouse myocardium, compared to embryos or newborns, a small residual telomerase-positive population was identified and isolated, using stem cell antigen-1 (Sca-1) as a surface marker for immunopurification.[44] Cardiac Sca-1 cells are undifferentiated, as assessed by transcripts for the sarcomeric proteins, cytoskeletal proteins, calcium-handling proteins, and enzymes that are characteristic of the cardiac lineage, but do express, even in their basal state, the genes for many cardiogenic transcription factors (Gata4, Mcf2-C, Srf, Tef1). In cell culture, treatment of cardiac Sca-1 cells with the DNA methylation inhibitor 5'-azacytidine triggers the expression of Nkx2.5 – a cardiac-restricted homeodomain transcription factor which is important for cardiac morphogenesis and whose prototype, tinman, in flies is essential for heart formation in that species. Induction of Nkx2.5 was accompanied, in the Sca-1$^+$ cells, by the expression of cardiac structural genes as well (αMHC, βMHC, sarcomeric alpha-actin, cardiac troponin I). Evidence that the inductive effect of azacytidine was not just a promiscuous one, due to abnormal DNA methylation, was provided by showing that cardiac myosin heavy chain (MHC) expression in the cells specifically required an intact signaling pathway for bone morphogenetic proteins, which are triggers of the cardiac lineage in wide-ranging model systems from flies, zebrafish, and avians to mammals.

What about the cells' behavior *in vivo*? The developmental potential of cardiac Sca-1$^+$ cells to become cardiac myocytes was tested, using intravenous delivery immediately after ischemia--reperfusion injury, to obviate the many potential

technical issues arising from direct injection of the mouse ventricular wall, including variance and confounding effects (focal injury, edema, and forced extravasation). Preliminary studies using dye-labeled cells showed selective uptake in the infarct border zone, with little or no homing by Sca-1$^-$ cells, by Sca-1$^+$ cells to undamaged regions of the heart, or by Sca-1$^+$ cells to the hearts of sham-operated control mice. Because this method can be problematic, more conclusive studies of the cells' fate relied instead on genetic tags. A heart muscle-specific transgene was used in the donor cells, αMHC-Cre, enabling induction of the nuclear-localized Cre protein to serve as an unambiguous marker of both donor cell identity and the cells' differentiation *in situ*: αMHC-Cre, like the endogenous gene encoding αMHC, is silent when cardiac Sca-1$^+$ cells are in their fresh, undifferentiated state. The specific choice of Cre protein as the marker – a DNA recombinase – facilitated a second test, namely, using the irreversible Cre-dependent activation of a host cell reporter gene, to ascertain whether or not the appearance of cardiac proteins in donor-derived cells as the index of differentiation was associated with the fusion of donor and host cells. Co-localization of donor-specific markers and differentiation-specific ones in a single cell, in short, can come about by either of two routes: differentiation *per se* (donor cells altering their lineage), or fusion (donor cells, irrespective of their plasticity, fusing with the host to create chimeric cells). Two weeks after intravenous infusion of cardiac Sca-1$^+$ cells, donor-derived myocytes constituted 15–20% of the infarct border zone (the region to which homing occurred), and ~4% of the left ventricular muscle overall,[44] a prevalence 150-fold greater than reported for the recruitment of cardiac myocytes from bone marrow, through the circulation, to the heart.[26] Roughly half the donor-derived myocytes were chimeric cells (donor-host fusion products), and half were exclusively donor derived. Regardless of the presence or absence of fusion, donor-derived myocytes expressed sarcomeric α-actin, cardiac troponin I, and the gap junction protein connexin-43, and were seamlessly interdigitated with the surrounding host myocardium of the border zone. Roughly one in twenty donor-derived myocytes was cycling proliferatively, two weeks after engraftment, using mitotic phosphorylation of histone H3 as an indicator, suggesting both the likelihood of ongoing myocyte generation and the opportunity to enhance or prolong the cells' proliferation further.

Notably, these cells lack certain key markers of HSCs and EPCs (CD45 and CD34), as measured both by surface labeling and by microarray surveys of gene expression. Likewise, there was little or no expression for a triad of HSC transcription factors (Lmo2, GATA2, Tal1/Scl). Hence, the Sca-1 cells resident in myocardium are not merely circulating HSCs and EPCs, captured fortuitously at the time of immunostaining or tissue isolation. Such experiments do not exclude the possibility that cardiac Sca-1 cells derive from a hematopoietic and circulating precursor that loses these distinguishing features after time in the cardiac niche. Also absent by these criteria, or expressed only rarely, was c-Kit, the receptor for stem cell factor, which has been instrumental to the identification of cardiac stem cells by others.[52]

One alternative approach to the successful isolation of progenitor cells from adult myocardium has focused upon the SP phenotype (exploiting either the characteristic dye efflux property, or the responsible ATP-binding cassette transporter, Abcg2).[56,57] A complementary strategy to find primitive cardiac cells with cardiogenic potential has used the persistent expression of Isl-1, a LIM-domain transcription factor that, for the most part, is expressed transiently during cardiac development but remains present in a few hundred cells per adult heart.[54] A cell culture strategy that did not rely on any specific molecular markers obtained cardiogenic cells, instead, by culturing dissociated human or mouse cardiac cells in a low-serum medium supplemented with a serum substitute, defined mitogens, cardiotropin-1, and thrombin.[55]

In each of these instances of heart-derived cells, clinical utilitzation would necessarily be more complex than for more accessible autologous sources like bone marrow, blood, and thigh, but at least three solutions can be envisioned, each meriting detailed inquiries: long-term expansion from biopsy samples (more applicable to chronic states than to acute infarction), using allogeneic cells (as proposed for ES cells and MSCs, with or without immune suppression[51,58]), or activating in situ the cardiac-resident cells' proliferation, migration, and entry into the cardiac fate.

How to empower the existing cardiac self-repair mechanisms to offset more completely the demands of human cardiac disease is one paramount challenge to the field. The remainder of this chapter is devoted to several of the specific outstanding questions.

Cell-free therapy for cardiac repair?

Given the instigation of cell engraftment studies as a route to replacing dead myocytes, it is unsurprising that early experimental attention was devoted to cells that already are, or efficiently convert to, striated muscle – fetal cardiac myocytes, skeletal muscle myoblasts, and embryonic stem cells. For these, and for adult pluripotent cells that generate new myocytes, intrinsic, cell-autonomous effects on regional and global organ-level contractility are easy to envision (although electrical and mechanical integration, as discussed, are not a foregone conclusion). By contrast, the prevalence of myocyte formation by bone marrow cells is in dispute, yet this concern disregards therapeutic angiogenesis as the impetus for human trials of these.

Despite the impeccable logic of using muscle- and vessel-forming cells to salvage or reconstitute the heart, cell-non-autonomous, secretory events must also be taken into consideration as potential mechanisms of benefit. A priori, these would include angiogenic growth factors, most obviously, but also factors for the recruitment of cardiac muscle progenitors from outside or within the heart, factors with beneficial impact on ventricular remodeling, and even myocyte survival factors. It is foreseeable that myocyte survival factors exist, beyond the known candidates like IGF-1,[59,60] and cardiotrophin-1.[61] In one such discovery, thymosin beta-4 was found to be secreted and taken up by cardiac cells, to promote cardiac cell migration and survival in culture, and to reduce infarct size in mice, working, like IGF-1, through activation of the protein kinase Akt.[62] Whether or not defined factors are ultimately found that can supplant cell delivery, one must look closely at the potential array of secretory effects to define how cells are working, an indispensable element of rational therapy.

Self or other? Naïve or optimized?

Simple and safe immunologically, autologous cells as used thus far in all clinical trials of cardiac repair necessitate a production lag – hours or overnight for bone marrow, days for EPCs, weeks for skeletal myoblasts – that stands in contrast to the model of cells-on-demand. For acute myocardial infarction, as seen with thrombolytics, time saved is muscle saved. Especially if paracrine survival factors are instrumental to cells' mechanism of benefit, a move towards autologous cell therapy is likely. However important, speed is but one issue among many, yet other factors might come as well to favor allogeneic cell therapies, including purely logistical ones – superseding the need for local or networked production facilities as required when each patient is his or her own donor, streamlining the task of standardization, or ensuring easier access to therapy with heart-derived cells or rarities like SP cells regardless of source. Biological issues, too, might favor the allogeneic donor. Functional defects have already been seen that constrain the effectiveness of patients' cells in chronic ischemic heart disease,[63] and some obstacles might be surmountable pharmacologically.[64] Hereditary cardiomyopathies in particular might require allogeneic, healthy donors, although cell therapy in this context raises additional challenges: wild-type bone marrow SP cells were recruited to hearts lacking δ-sarcoglycan, but failed to express the missing protein, suggesting impaired differentiation or maturation of the donor-derived cardiomyocytes that formed.[65,66]

The clinical agenda for enhancing the function of donor cells can thus be partitioned into two broad sets of cases: moving beyond normal cells' capacity, or bringing deficient cells up to normal. In either circumstance, use of allogeneic cells would simplify cell engineering, from both the technical standpoint and the attendant question of guaranteeing safety and efficacy. Areas ripe for augmentation include virtually all elements of function discussed earlier in this review – donor cell survival

after grafting,[39] proliferation after grafting,[67–69] homing,[70] differentiation,[71] host cell recruitment,[72,73] and the very survival of jeopardized host myocardium.[62,74]

References

1. Poss KD, Wilson LG, Keating MT. Heart regeneration in zebrafish. Science 2002; 298: 2188–90.
2. Raya A, Koth CM, Buscher D et al. Activation of notch signaling pathway precedes heart regeneration in zebrafish. Proc Natl Acad Sci USA 2003; 100 (Suppl 1): 11889–95.
3. Menasché P, Hagege AA, Scorsin M et al. Myoblast transplantation for heart failure. Lancet 2001; 357: 279–80.
4. Strauer BE, Brehm M, Zeus T et al. Repair of infarcted myocardium by autologous intracoronary mononuclear bone marrow cell transplantation in humans. Circulation 2002; 106: 1913–18.
5. Assmus B, Schachinger V, Teupe C et al. Transplantation of progenitor cells and regeneration enhancement in acute myocardial infarction (top-care-ami). Circulation 2002; 106: 3009–17.
6. Perin EC, Dohmann HF, Borojevic R et al. Transendocardial, autologous bone marrow cell transplantation for severe, chronic ischemic heart failure. Circulation 2003; 107: 2294–302.
7. Wollert KC, Meyer GP, Lotz J et al. Intracoronary autologous bone-marrow cell transfer after myocardial infarction: The boost randomised controlled clinical trial. Lancet 2004; 364: 141–8.
8. Tse HF, Kwong YL, Chan JK et al. Angiogenesis in ischaemic myocardium by intramyocardial autologous bone marrow mononuclear cell implantation. Lancet 2003; 361: 47–9.
9. Murry CE, Whitney ML, Laflamme MA, Reinecke H, Field LJ. Cellular therapies for myocardial infarct repair. Cold Spring Harb Symp Quant Biol 2002; 67: 519–26.
10. Olson EN, Schneider MD. Sizing up the heart: Development redux in disease. Genes Dev 2003; 17: 1937–56.
11. Taylor DA. Cell-based myocardial repair: how should we proceed? Int J Cardiol 2004; 95 (Suppl 1): S8–12.
12. Mathur A, Martin JF. Stem cells and repair of the heart. Lancet 2004; 364: 183–92.
13. Menasche P. Cellular transplantation: hurdles remaining before widespread clinical use. Curr Opin Cardiol 2004; 19: 154–61.
14. Losordo DW, Dimmeler S. Therapeutic angiogenesis

and vasculogenesis for ischemic disease: Part ii: cell-based therapies. Circulation 2004; 109: 2692–7.

15. Dimmeler S, Zeiher AM, Schneider MD. Unchain my heart: the scientific foundations of cardiac repair. J Clin Invest 2005: 115: 572–83.

16. Couzin J, Vogel G. Cell therapy. Renovating the heart. Science 2004; 304: 192–4.

17. Hutcheson KA, Atkins BZ, Hueman MT, Hopkins MB, Glower DD, Taylor DA. Comparison of benefits on myocardial performance of cellular cardio-myoplasty with skeletal myoblasts and fibroblasts. Cell Transplant 2000; 9: 359–68.

18. Hagege AA, Carrion C, Menasche P et al. Viability and differentiation of autologous skeletal myoblast grafts in ischaemic cardiomyopathy. Lancet 2003; 361: 491–2.

19. Zhang M, Methot D, Poppa V et al. Cardiomyocyte grafting for cardiac repair: graft cell death and anti-death strategies. J Mol Cell Cardiol 2001; 33: 907–21.

20. Reinecke H, MacDonald GH, Hauschka SD, Murry CE. Electromechanical coupling between skeletal and cardiac muscle. Implications for infarct repair. J Cell Biol 2000; 149: 731–40.

21. Leobon B, Garcin I, Menasche P et al. Myoblasts transplanted into rat infarcted myocardium are functionally isolated from their host. Proc Natl Acad Sci USA 2003; 100: 7808–11.

22. Rubart M, Pasumarthi KB, Nakajima H et al. Physiological coupling of donor and host cardio-myocytes after cellular transplantation. Circ Res 2003; 92: 1217–1224.

23. Rubart M, Soonpaa MH, Nakajima H, Field LJ. Spontaneous and evoked intracellular calcium transients in donor-derived myocytes following intracardiac myoblast transplantation. J Clin Invest 2004; 114: 775–83.

24. Menasché P, Hagege AA, Vilquin J-T et al. Autologous skeletal myoblast transplantation for severe postinfarction left ventricular dysfunction. J Am Coll Cardiol 2003; 41: 1078–83.

25. Smits PC, van Geuns RJ, Poldermans D et al. Catheter-based intramyocardial injection of autologous skeletal myoblasts as a primary treatment of ischemic heart failure: Clinical experience with six-month follow-up. J Am Coll Cardiol 2003; 42: 2063–9.

26. Jackson KA, Majka SM, Wang H et al. Regeneration of ischemic cardiac muscle and vascular endothelium by adult stem cells. J Clin Invest 2001; 107: 1395–402.

27. Orlic D, Kajstura J, Chimenti S et al. Bone marrow cells regenerate infarcted myocardium. Nature 2001; 410: 701–5.

28. Murry CE, Soonpaa MH, Reinecke H et al. Haematopoietic stem cells do not transdifferentiate into cardiac myocytes in myocardial infarcts. Nature 2004; 428: 664–8.

29. Balsam LB, Wagers AJ, Christensen JL et al. Haematopoietic stem cells adopt mature haematopoietic fates in ischaemic myocardium. Nature 2004; 428: 668–73.

30. Nygren JM, Jovinge S, Breitbach M et al. Bone marrow-derived hematopoietic cells generate cardiomyocytes at a low frequency through cell fusion, but not transdifferentiation. Nat Med 2004; 10: 494–501.

31. Asahara T, Murohara T, Sullivan A et al. Isolation of putative progenitor endothelial cells for angiogenesis. Science 1997; 275: 964–7.

32. Takahashi T, Kalka C, Masuda H et al. ischemia- and cytokine-induced mobilization of bone marrow-derived endothelial progenitor cells for neovascularization. Nat Med 1999; 5: 434–8.

33. Isner JM, Asahara T. Angiogenesis and vasculogenesis as therapeutic strategies for postnatal neovascularization. J Clin Invest 1999; 103: 1231–6.

34. Kawamoto A, Asahara T, Losordo DW. Transplantation of endothelial progenitor cells for therapeutic neovascularization. Cardiovasc Radiat Med 2002; 3: 221–5.

35. Kocher AA, Schuster MD, Szabolcs MJ et al. Neovascularization of ischemic myocardium by human bone-marrow-derived angioblasts prevents cardiomyocyte apoptosis, reduces remodeling and improves cardiac function. Nat Med 2001; 7: 430–6.

36. Orlic D, Kajstura J, Chimenti S et al. Mobilized bone marrow cells repair the infarcted heart, improving function and survival. Proc Natl Acad Sci USA 2001; 98: 10344–9.

37. Ohtsuka M, Takano H, Zou Y et al. Cytokine therapy prevents left ventricular remodeling and dysfunction after myocardial infarction through neovascularization. FASEB J 2004; 18: 851–3.

38. Goodell MA, Brose K, Paradis G, Conner AS, Mulligan RC. Isolation and functional properties of murine hematopoietic stem cells that are replicating in vivo. J Exp Med 1996; 183: 1797–1806.

39. Mangi AA, Noiseux N, Kong D et al. Mesenchymal stem cells modified with akt prevent remodeling and restore performance of infarcted hearts. Nat Med 2003; 9: 1195–201.

40. Toma C, Pittenger MF, Cahill KS, Byrne BJ, Kessler PD. Human mesenchymal stem cells differentiate to a cardiomyocyte phenotype in the adult murine heart. Circulation 2002; 105: 93–8.

41. Shake JG, Gruber PJ, Baumgartner WA et al.

Mesenchymal stem cell implantation in a swine myocardial infarct model: engraftment and functional effects. Ann Thorac Surg 2002; 73: 1919–25; discussion 1926.

42. Jiang Y, Jahagirdar BN, Reinhardt RL et al. Pluripotency of mesenchymal stem cells derived from adult marrow. Nature 2002; 418: 41–9.

43. Jiang Y, Vaessen B, Lenvik T et al. Multipotent progenitor cells can be isolated from postnatal murine bone marrow, muscle, and brain. Exp Hematol 2002; 30: 896–904.

44. Oh H, Bradfute SB, Gallardo TD et al. Cardiac progenitor cells from adult myocardium: Homing, differentiation, and fusion after infarction. Proc Natl Acad Sci USA 2003; 100: 12313–18.

45. Martin CM, Meeson AP, Robertson S et al. Persistent expression of the atp-cassette transporter, abcg2, identifies cardiac stem cells in the adult heart. In: From Stem Cells to Therapy, March 29–April 3 2003; Steamboat Springs, Colorado: Keystone Symposia; 2003: 75.

46. Doetschman TC, Eistetter H, Katz M, Schmidt W, Kemler R. The in vitro development of blastocyst-derived embryonic stem cell lines: formation of visceral yolk sac, blood islands and myocardium. J Embryol Exp Morphol 1985; 87: 27–45.

47. Behfar A, Zingman LV, Hodgson DM et al. Stem cell differentiation requires a paracrine pathway in the heart. FASEB J 2002; 16: 1558–66.

48. Kehat I, Khimovich L, Caspi O et al. Electromechanical integration of cardiomyocytes derived from human embryonic stem cells. Nat Biotechnol 2004; 22: 1282–9.

49. Blackburn E. Bioethics and the political distortion of biomedical science. N Engl J Med 2004; 350: 1379–80.

50. Cowan CA, Klimanskaya I, McMahon J et al. Derivation of embryonic stem-cell lines from human blastocysts. N Engl J Med 2004; 350: 1353–6.

51. Kofidis T, de Bruin JL, Yamane T et al. Insulin-like growth factor promotes engraftment, differentiation, and functional improvement after transfer of embryonic stem cells for myocardial restoration. Stem Cells 2004; 22: 1239–45.

52. Beltrami AP, Barlucchi L, Torella D et al. Adult cardiac stem cells are multipotent and support myocardial regeneration. Cell 2003; 114: 763–76.

53. Matsuura K, Nagai T, Nishigaki N et al. Adult cardiac sca-1-positive cells differentiate into beating cardiomyocytes. J Biol Chem 2004; 279: 11384–91.

54. Laugwitz K-L, Moretti A, Lam J et al. Post-natal isl1+ cardioblasts enter fully differentiated cardiomyocyte lineages. Nature 2005; 433: 647–53.

55. Messina E, De Angelis L, Frati G et al. Isolation and expansion of adult cardiac stem cells from human and murine heart. Circ Res 2004; 95: 911–21.

56. Hierlihy AM, Seale P, Lobe CG, Rudnicki MA, Megeney LA. The post-natal heart contains a myocardial stem cell population. FEBS Lett 2002; 530: 239–43.

57. Martin CM, Meeson AP, Robertson SM et al. Persistent expression of the atp-binding cassette transporter, abcg2, identifies cardiac sp cells in the developing and adult heart. Dev Biol 2004; 265: 262–75.

58. Pittenger MF, Mosca JD, McIntosh KR. Human mesenchymal stem cells: progenitor cells for cartilage, bone, fat and stroma. Curr Top Microbiol Immunol 2000; 251: 3–11.

59. Fujio Y, Nguyen T, Wencker D, Kitsis RN, Walsh K. Akt promotes survival of cardiomyocytes in vitro and protects against ischemia–reperfusion injury in mouse heart. Circulation 2000; 101: 660–7.

60. Chao W, Matsui T, Novikov MS et al. Strategic advantages of insulin-like growth factor-1 expression for cardioprotection. J Gene Med 2003; 5: 277–86.

61. Sheng ZL, Knowlton K, Chen J, Hoshijima M, Brown JH, Chien KR. Cardiotrophin 1 (ct-1) inhibition of cardiac myocyte apoptosis via a mitogen-activated protein kinase-dependent pathway - divergence from downstream ct-1 signals for myocardial cell hypertrophy. J Biol Chem 1997; 272: 5783–91.

62. Bock-Marquette I, Saxena A, White MD et al. Thymosin beta4 activates integrin-linked kinase and promotes cardiac cell migration, survival, and cardiac repair. Nature 2004: 432: 466–72.

63. Heeschen C, Lehmann R, Honold J et al. Profoundly reduced neovascularization capacity of bone marrow mononuclear cells derived from patients with chronic ischemic heart disease. Circulation 2004; 109: 1615–22.

64. Spyridopoulos I, Haendeler J, Urbich C et al. Statins enhance migratory capacity by upregulation of the telomere repeat-binding factor trf2 in endothelial progenitor cells. Circulation 2004; 110: 3136–42.

65. Lapidos KA, Chen YE, Earley JU et al. Transplanted hematopoietic stem cells demonstrate impaired sarcoglycan expression after engraftment into cardiac and skeletal muscle. J Clin Invest 2004; 114: 1577–85.

66. Cossu G. Fusion of bone marrow-derived stem cells with striated muscle may not be sufficient to activate muscle genes. J Clin Invest 2004; 114: 1540–3.

67. Whitney ML, Otto KG, Blau CA, Reinecke H, Murry CE. Control of myoblast proliferation with a synthetic ligand. J Biol Chem 2001; 276: 41191–6.

68. Takeda Y, Mori T, Imabayashi H et al. Can the life

span of human marrow stromal cells be prolonged by bmi-1, e6, e7, and/or telomerase without affecting cardiomyogenic differentiation? J Gene Med 2004; 6: 833–45.

69. Pasumarthi KB, Nakajima H, Nakajima HO, Soonpaa MH, Field LJ. Targeted expression of cyclin d2 results in cardiomyocyte DNA synthesis and infarct regression in transgenic mice. Circ Res 2005; 96: 110–18.

70. Yamaguchi J, Kusano KF, Masuo O et al. Stromal cell-derived factor-1 effects on ex vivo expanded endothelial progenitor cell recruitment for ischemic neovascularization. Circulation 2003; 107: 1322–8.

71. Wang CH, Ciliberti N, Li SH et al. Rosiglitazone facil-itates angiogenic progenitor cell differentiation toward endothelial lineage: A new paradigm in glitazone pleiotropy. Circulation 2004; 109: 1392–400.

72. Yau TM, Li G, Weisel RD et al. Vascular endothelial growth factor transgene expression in cell-transplanted hearts. J Thorac Cardiovasc Surg 2004; 127: 1180–7.

73. Haider H, Ye L, Jiang S et al. Angiomyogenesis for cardiac repair using human myoblasts as carriers of human vascular endothelial growth factor. J Mol Med 2004; 82: 539–49.

74. Schneider MD. Regenerative medicine: Prometheus unbound. Nature 2004; 432: 451–3.

CHAPTER 11

Oxidative stress in the regulation of myocardial hypertrophy and failure

Douglas B Sawyer, David R Pimentel, Wilson S Colucci

Introduction

Hemodynamic overload results in progressive changes in both the structure and function of the myocardium. This process, globally referred to as myocardial remodeling, is characterized by changes in ventricular size and geometry, and alterations in pump function. While early remodeling may be of little functional consequence, with time, the process generally progresses to heart failure that is associated with clinical symptoms, exercise limitation and death.

Compelling evidence from experimental models of heart failure and human studies now supports the concept that oxidative stress is increased in failing myocardium and contributes to the pathogenesis of heart failure. Recent progress in understanding the mechanisms that mediate myocardial remodeling at the molecular and cellular levels has led to evidence that reactive oxygen species (ROS) and the associated oxidative stress play a central role in regulating the phenotype of cardiac myocytes and fibroblasts. While ROS are traditionally viewed as direct cytotoxic agents that cause oxidative damage to proteins, lipids and nucleic acids, a growing body of *in vitro* experiments now suggests that lower levels of ROS can act as intracellular signaling molecules in cardiac myocytes and fibroblasts. Importantly, these emerging observations provide a mechanism by which ROS may determine the myocardial response to extracellular remodeling stimuli such as mechanical strain, neurohormones and cytokines.

Oxidative stress in the failing myocardium

Several clinical observations suggest that there is increased oxidative stress systemically, and in the myocardium of patients with chronic left ventricular systolic failure.[1,2] Similarly, there is evidence of increased oxidative stress in myocardium obtained from animal models of hemodynamic overload including surgical myocardial infarction (MI),[3] pressure overload,[4] and pacing-induced heart failure.[5] Among the first to call attention to oxidative stress in the myocardium were Singal and colleagues, who showed evidence of increased oxidative stress in myocardium from guinea pigs during the transition from compensated hypertrophy to overt failure in association with decreased levels of antioxidant enzymes.[6,7] More recently, Ide et al. have shown that increased oxidative stress in the rapid-pacing model of heart failure is, at least in part, due to increased mitochondrial generation of ROS.[5,8]

The relevance of these observations has been supported by experimental studies in which the administration of antioxidants has ameliorated pathologic remodeling. For example, Singal and colleagues showed that administration of vitamin E prevented the transition from compensated hypertrophy to failure in guinea pigs with pressure-overload due to aortic constriction.[4] In other studies the ROS scavenger dimethylthiourea (DMTU) prevented chamber dilation and pump dysfunction in mice after myocardial infarction.[9] Taken together, these types of observation have led to the thesis that chronically elevated oxidative stress in the myocardium plays a central role in the process of myocardial remodeling leading to heart failure.

Sources of ROS and oxidative stress in failing myocardium

Increased oxidative stress in failing myocardium may reflect an increase in the production of ROS from multiple sources, as well as an absolute or relative deficiency in antioxidant capacity. Mitochondria are well-recognized as a major source of ROS due to the leakage of electrons from the respiratory chain resulting in the conversion of oxygen to superoxide anion. The mitochondrial production of ROS is increased in myocardium obtained from dogs with rapid-pacing-induced heart failure,[5,8] as well as in mice post-MI.[9] In the rapid-pacing dog model, increased ROS production was attributed to a decrease in the function of complex I of the electron transport chain,[5] which may lead to uncoupling of mitochondria and increased leakage of electrons. Such uncoupling would also be apparent as a decrease in myocardial efficiency, or 'oxygen wastage', a term put forth by Opie to describe an increase in oxygen consumption relative to ATP production in the heart.[10]

Another possible source of increased myocardial ROS levels may be increased activity of intracellular oxidases such as NAD(P)H oxidase,[11] xanthine oxidase (XO),[12] or nitric oxide synthase (NOS).[13] NAD(P)H oxidase is a multi-subunit enzyme associated with the plasmalemmal membrane. In vascular smooth muscle cells, NAD(P)H oxidase has been shown to mediate the ROS-dependent effects of angiotensin.[11] Initially described in the immune system, where it is responsible for a cytotoxic high-amplitude burst of ROS, it is now appreciated that NAD(P)H oxidases are present in many other cell types where they are capable of producing much lower, non-cytotoxic levels of ROS that function as signaling intermediates. As discussed below, studies with pharmacological inhibitors have implicated this system in the hypertrophic response of ventricular myocytes.

Other evidence suggests that there is a functionally important increase in the activity of XO in failing myocardium. Using a XO inhibitor, it has been shown that XO contributes to impaired ventricular function characterized by decreased efficiency in dogs with rapid-pacing-induced failure.[12,13] NOS may be another source of ROS. While nitric oxide (NO) is the usual product of NOS, under appropriate conditions the NOS2 isoform, which may be increased in failing myocardium,[14,15] is also capable of generating superoxide.

Antioxidants play an important role in the myocardium, as in other tissues, by maintaining ROS such as O_2^- and H_2O_2 at low levels (Fig. 11.1). Superoxide dismutase (SOD) converts superoxide anions to H_2O_2. There are three known isoforms of SOD in mammalian tissues. Approximately 70% of the total SOD activity in the heart and approximately 90% of the activity in cardiac myocytes is attributable to MnSOD, which is encoded by the nuclear genome but localizes to mitochondria via a mitochondrial targeting sequence. The remaining SOD activity is primarily due to Cu/ZnSOD, which is localized in the cytosol. Extracellular SOD (ECSOD), a Cu^{2+}- and Zn^{2+}-containing enzyme that localizes to the extracellular and intravascular spaces, is the least prevalent isoform in the myocardium.

MnSOD is the only SOD in the mitochondria, and therefore plays a critical role in the control of O_2^- generation during oxidative phosphorylation. Homozygous knockout mice deficient in MnSOD develop normally, but die soon after birth with a cardiomyopathy,[16] emphasizing the importance of MnSOD. On the other hand, homozygous deficiency of CuZnSOD or ECSOD results in no overt cardiac phenotype. In some animal models of heart failure, SOD activity appears to be inappropriately low,[3,4] suggesting that an impairment in antioxidant capacity that leads to increased oxidative stress may contribute to the pathogenesis of heart failure.

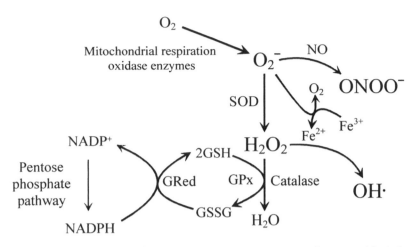

Figure 11.1. *Overview of the regulation of reactive oxygen species. Superoxide anion (O_2^-) formed as a consequence of oxidative respiration or oxidase enzymes is dismutated to hydrogen peroxide (H_2O_2) by superoxide dismutases (SOD). Hydrogen peroxide is metabolized to water by catalase or glutathione peroxidase (GPx). Hydroxyl radicals ($OH^{\bullet}$) can be formed via Fenton chemistry. When nitric oxide is present in sufficient quantity, superoxide may react to form peroxynitrite ($ONOO^-$). The pentose phosphate pathway and glutathione reductase (GRed) help to maintain adequate amounts of reduced glutathione that is required for GPx function and serves as an important regulator of intracellular redox state.*

H_2O_2 is reduced by glutathione peroxidase (GPx) and catalase. GPx is a selenium-containing enzyme that catalyzes the removal of H_2O_2 via the oxidation of reduced glutathione (GSH), which in turn is recycled by glutathione reductase. GPx activity also appears to be decreased in the failing heart,[3] and thus may contribute to increased oxidative stress. The activity of GPx requires stoichiometric quantities of GSH. Therefore, when the levels of GSH are depleted the activity of GPx decreases. Glutathione reductase requires NAD(P)H as a reductant to replenish GSH from GSSG. The generation of NAD(P)H by the pentose phosphate pathway is regulated by glucose 6-phosphate dehydrogenase (G6PD), the rate-limiting enzyme in this pathway, which therefore plays an important role in the cellular antioxidant system.[17,18] Recently, we have shown that G6PD can modulate both redox state and contractile function in cultured cardiac myocytes.[19]

Observations on the antioxidant capacity of failing human myocardium have been inconsistent. Some groups have found no decrease in SOD or GPx activity in samples obtained from explanted human hearts at the time of cardiac transplant as compared to non-failing 'controls'.[20,21] On the other hand, we found reduced MnSOD activity in the failing human heart.[22] The decrease in activity was not associated with decreases in protein or mRNA expression of MnSOD, suggesting that activity is decreased at the post-translational level, perhaps due to oxidative modification of the protein. It is possible that the discrepancy among these studies reflects differences in the handling of 'control' hearts resulting in different degrees of pathological and pharmacological stress. When considered in the context of data showing decreased antioxidant activity in rodent models of heart failure, it seems likely that an absolute or relative reduction in antioxidant capacity is present in the end-stage failing human heart.

Cellular mechanisms of myocardial remodeling

A growing body of literature shows that the cellular events that contribute to myocardial remodeling in heart failure can be regulated by oxidative stress. At the cellular level, the changes in ventricular structure in heart failure include myocyte growth, loss of myocytes via apoptosis or necrosis and/or myocyte slippage, the latter perhaps caused by degradation of fibrillar collagen struts due to activation of matrix metalloproteinases (MMPs). Pump dysfunction may result from loss of functional myocytes and/or a shift in myocyte contractility in part due to re-expression of fetal isoforms involved in contraction (e.g. β-myosin heavy chain (MHC)) and/or calcium homeostasis (e.g. sarcoplasmic reticulum Ca^{2+} ATPase (SERCA)).

Graded effects of ROS on myocyte phonotype *in vitro*

In vitro experiments using ventricular myocytes in primary culture have provided evidence that ROS and oxidative stress can exert important effects on myocyte structure and function similar to those observed in failing myocardium. Furthermore, collectively these studies suggest that the effects of ROS are graded and related to the quantity and/or nature of the ROS.

Initial studies used the copper chelator diethyldithiocarbamic acid (DDC) to inhibit intracellular CuZnSOD.[23] A low concentration of DDC, which caused partial inhibition of SOD resulting in a small increase in myocyte ROS, stimulated myocyte hypertrophy as evidenced by increased protein synthesis and myocyte size, induction of c-fos expression and re-expression of a fetal gene program as reflected by an increase in atrial natriuretic factor *ANF* mRNA and a decrease in *SERCA2* mRNA. Exposure to a higher concentration of DDC, which caused a greater increase in myocyte ROS, led to apoptosis with increased expression of mRNA for the pro-apoptotic protein, bax. The effects of both concentrations of DDC were prevented by antioxidants and mimicked by exogenous O_2^- generation with xanthine/xanthine oxidase. These results indicate that (1) a decrease in antioxidant activity, *per se*, has the potential to cause myocyte remodeling by increasing oxidative stress, and (2) these effects on myocyte phenotype are determined in part by the amount of ROS.

More recent studies in myocytes exposed to exogenous ROS have yielded results consistent with this concept. Adult rat ventricular myocytes in primary culture were exposed to H_2O_2 in concentrations ranging from 10 μM to 1000 μM for 24 hours.[24] Exposure to low concentrations of H_2O_2 (10–30 μM) caused myocyte growth, whereas higher concentrations (100–200 μM) caused apoptosis. Still higher concentrations of H_2O_2 (300–1000 μM) caused both apoptosis and necrosis. A low (hypertrophic) concentration of H_2O_2 (10 μM) stimulated extracellular signal-regulated kinase (ERK), whereas a higher (apoptotic) concentration of H_2O_2 (100 μM) also stimulated c-jun N-terminal kinase (JNK), p38 kinase and protein kinase B (Akt). A MEK1/2 inhibitor (U0126) prevented the hypertrophic effect of low H_2O_2, whereas the apoptotic effect of high H_2O_2 was inhibited by a JNK inhibitor (SP600125) or a dominant-negative JNK adenovirus, and was increased by U0126 or an Akt inhibitor. These studies confirm the concentration-dependent effects of ROS on myocyte phenotype, and further show that these effects are related to the differential activation of kinase signaling pathways that regulate myocyte growth and apoptosis.

Graded, ROS-dependent effects of mechanical strain *in vitro*

Several well recognized stimuli for myocardial remodeling, including mechanical strain, neurohormones and cytokines, also appear to regulate myocyte phenotype via ROS. Mechanical strain of cardiac myocytes results in hypertrophic growth,[25] as well as apoptosis.[26,27] Anversa and colleagues were the first to show that ROS play a role in mediating myocyte apoptosis induced by mechanical stretch of papillary muscles *in vitro*.[26] Using cyclic mechanical strain of cardiac myocytes in primary culture, we likewise found that the strain causes an amplitude-dependent effect on myocyte phenotype that is mediated by ROS.[28] The effects of strain on ROS production were amplitude-dependent, with low levels of ROS in response to low-amplitude strain, and higher levels in response to higher amplitude strain. Low-amplitude strain caused myocyte hypertrophy, sarcomeric reorganization and fetal gene expression, whereas a higher level of strain caused apoptosis. Both strain-induced phenotypes were inhibited by pharmacological antioxidants, thereby implicating a central role for ROS in regulating the graded effect of mechanical strain on phenotype.

ROS and myocyte hypertrophy

Several stimuli known to cause hypertrophic growth of cardiac myocytes *in vitro*, including angiotensin, tumor necrosis factor-α, endothelin and α-adrenergic receptor stimulation, have been shown to act, at least in part, via ROS-dependent mechanisms.[29,30] The role of ROS in mediating hypertrophy remains to be determined, and may well differ for various stimuli. However, a growing body of evidence suggests that an NAD(P)H oxidase is involved. For example, we and others have found in adult rat cardiac myocytes that the NAD(P)H oxidase inhibitor diphenylene iodonium prevents α_1-adrenergic receptor-stimulated activation of ERK, whereas inhibitors of the mitochondrial respiratory chain (e.g. rotenone) have no effect.[31,32]

The neutrophil NAD(P)H oxidase consists of four major subunits, including two membrane-spanning components ($p22^{phox}$ and $gp91^{phox}$), and two cytosolic components ($p67^{phox}$ and $p47^{phox}$).[33] Similar oxidase systems in which the $gp91^{phox}$ subunit may be substituted by one of several related subunits, have been identified in many cell types and appear to provide the ROS that mediates angiotensin-stimulated growth in vascular smooth muscle cells.[34] While the NAD(P)H oxidases present in cardiovascular cell types in general retain an enzyme complex structure similar to neutrophil oxidase, their components and biochemical characteristics may be considerably different.[33] While we have identified the four major subunits of NAD(P)H oxidase (p22phox, gp91phox, p67phox and p47phox) in rat cardiac myocytes,[32] a definitive demonstration of the role of this NAD(P)H oxidase in myocyte hypertrophic signaling will require the development of more specific molecular probes (e.g. antisense RNA, siRNA, knockout mice).

ROS and myocyte death

It is well known that very high levels of ROS, as occur with ischemia–reperfusion injury, can cause myocyte necrosis. More recently it has been appreciated that levels of ROS that are not directly cytotoxic, but perhaps higher than the levels that activate hypertrophic signaling pathways, can cause apoptosis in cardiac myocytes. As noted above, we have shown that both an increase in O_2^- due to inhibition of SOD, and an increase in H_2O_2 due to addition of reagent H_2O_2 can cause apoptosis in cardiac myocytes *in vitro*. von Harsdorf et al. found that apoptosis induced by exogenous O_2^- appears to occur by a pathway distinct from H_2O_2.[35] Whereas H_2O_2 caused activation of the mitochondrial death pathway as reflected by the mitochondrial release of cytochrome c and activation of caspase 3, an increase in O_2^- caused by addition of xanthine and xanthine oxidase (in the presence of catalase) induced apoptosis without apparent activation of the mitochondrial pathway. In contrast, O_2^--induced apoptosis appeared to involve activation of the laminase Mch2α. Since endogenous sources of O_2^- may also lead to an increase in H_2O_2, particularly if there is a decrease in catalase and/or GPx activity, it is possible that both pathways are involved.

We have found that apoptosis stimulated by β-adrenergic receptors (βARs) is mediated by ROS. We and others have found that in adult rat ventricular myocytes *in vitro* βAR stimulation for 24 hours causes apoptosis.[36,37] The norepinephrine causes apoptosis by stimulating β_1ARs that are coupled to Gs-adenylyl cyclase, whereas stimulation of β_2ARs exerts an anti-apoptotic action that is mediated via G_i.[38] β_1AR-stimulated apoptosis was abolished by treatment with an SOD-mimetic or the overexpression of catalase using an adenoviral vector, thus implicating ROS in βAR-stimulated apoptosis.[39] The ability to inhibit βAR-stimulated apoptosis with catalase further suggests that the primary species responsible is H_2O_2 and/or a derivative (e.g. OH-). βAR-stimulated apoptosis was associated with increased expression of bax, and decreased expression of bcl-x_L and bcl-2,[37] thereby implicating a mitochondrial pathway. An early step in myocyte apoptosis is activation of the mitochondrial transition pore, which is regulated by the bcl-2 protein family, leading to the release of cytochrome c from the intermembrane space and the consequent activation of caspase 3. A central role for the mitochondrial death pathway was suggested by our finding that β_1AR stimulation increased the translocation of cytochrome c from the mitochondria to the cytosol and increased caspase-3 activity,[39] the finding by Zaugg et al. that β_1AR stimulation increased the activity of caspase 9, but not caspase 8.[37] More recently, we have found that an inhibitor of the mitochondrial transition pore (bongkrekic acid) prevents βAR-stimulated cytochrome c release, caspase activation and apoptosis.[39]

There is evidence that JNK plays a role in mediating the pro-apoptotic effect of ROS. We further found that βAR-stimulated cytochrome c release and apoptosis, both of which are ROS dependent, were prevented by adenoviral overexpression of a dominant-negative JNK, thus suggesting that JNK plays a key role in mediating the effects of βAR stimulation on the mitochondria.[39] Wei and colleagues have likewise found that H_2O_2 and a superoxide generator (menadione) induce apoptosis that is associated with activation of JNK in the H9C2 cardiac cell line,[40] and that transfection with a dominant negative JNK markedly diminished the extent of DNA fragmentation. Taken together, these observations suggest a model in which ROS, JNK and mitochondria play a central role in mediating βAR-stimulated apoptosis.

Apoptosis induced by inflammatory cytokines may also involve ROS. High concentrations of inflammatory cytokines may cause apoptosis that is dependent on the induction of nitric oxide synthase-2 (NOS2).[41-43] While low concentrations of NO typically produced by NOS3 (endothelial NOS) may be cardioprotective, high levels of NO that are produced by NOS2-inducible NOS appear capable of producing apoptosis which may be mediated through the formation of peroxynitrite,[43] a reaction product of NO and O_2^- that is favored under conditions of increased O_2^- as may occur in the failing heart. We and others have found that NOS2 does contribute to the myocyte apoptosis that occurs in the setting of post-MI remodeling in vivo:[44,45] NOS2 expression is increased in the remote myocardium late after MI in the mouse, and NOS2 knockout mice exhibited less myocyte apoptosis, improved contractile function and increased survival after MI.

ROS regulation of interstitial matrix turnover by cardiac fibroblasts

Collagen and other interstitial matrix proteins connect cardiac myocytes, maintain overall tissue architecture and play an integral role in coordinating the force generated by individual myocytes. The highly dynamic turnover of the interstitial matrix is regulated by the opposing effects of matrix metalloproteinases (MMPs),[46] and the protein synthetic machinery.[47] As the major source of both collagen and MMPs, fibroblasts play a central role in regulating the composition and quantity of cardiac interstitial matrix proteins. Using cardiac fibroblasts in vitro, we have found that ROS can regulate both collagen synthesis and MMP activity.[47] In rat fibroblasts grown in primary culture several sources of ROS (SOD inhibition, xanthine plus xanthine oxidase or exogenous H_2O_2) all decreased fibroblast collagen synthesis as measured by collagenase-sensitive [^{3}H]-proline incorporation. The decrease in collagen synthesis was associated with reduced expression of procollagen mRNA, suggesting that regulation was at least in part at the transcriptional level. All three sources of ROS also increased MMP activity as measured by in-gel zymography, with an ~50% increase in total MMP activity due primarily to increases in the activities of MMP-1, MMP-2 and MMP-9. These effects of ROS would be expected to promote myocardial dilation. Of note, administration of an antioxidant (DMTU) has been shown to reduce the extent of ventricular dilation and suppress myocardial MMP activity in mice post-MI.[9]

Summary

Compelling in vitro and in vivo evidence demonstrates that oxidative stress plays an important role in mediating pathological myocardial remodeling (Fig. 11.2). While the mechanisms responsible for

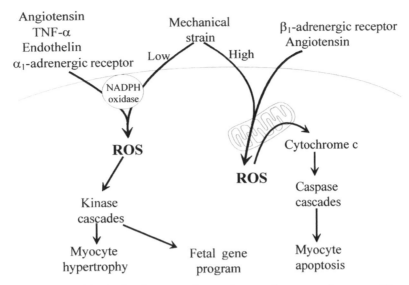

Figure 11.2. *Hypothetical model of role of ROS in the regulation of myocyte phenotype. Hypertrophic stimuli such as angiotensin, norepinephrine and low levels of mechanical strain may work through the activation of specific oxidase enzymes such as the NAD(P)H oxidase, leading to the localized generation of ROS that interact with growth-signaling cascades. The signaling for apoptotic stimuli such as β_1-adrenergic receptor stimulation and higher levels of mechanical strain appears to involve ROS-dependent activation of a mitochondrial pathway. The exact relationships between hypertrophic or apoptotic stimuli, generation of ROS and activation of cellular phenotypes remains to be elucidated.*

increased oxidative stress remain to be elucidated, likely sources of increased ROS production include the mitochondria, NAD(P)H oxidase and xanthine oxidase. Impaired antioxidant enzyme activity may also contribute to increased oxidative stress.

ROS can mimic many of the changes that are characteristic of myocardial remodeling including hypertrophy, apoptosis and fetal gene expression in myocytes, and alterations in the elaboration and metabolism of the interstitial matrix. In cardiac myocytes, ROS can exert a graded effect on phenotype resulting in hypertrophy at low levels and apoptosis at higher levels. ROS mediate these effects in part through the activation of intracellular signaling pathways in the stress-activated protein kinase family. Myocyte hypertrophy involves the ROS-dependent activation of the ras-raf-MEK-ERK cascade. In contrast, myocyte apoptosis may involve the ROS-dependent activation of JNK

and the mitochondrial death pathway. While both hypertrophy and apoptosis may be viewed as 'stress-responses' in the failing heart, these ROS-dependent effects may occur in the absence of overt oxidative damage to intracellular proteins, lipids, and nucleic acids. When viewed in this context, ROS act as intracellular signaling molecules.

While it is now clear that ROS can regulate myocyte biology, relatively little is known about the mechanism by which remodeling stimuli such as mechanical strain, neurohormones or cytokines lead to increased ROS. Even less is known about the precise molecular mechanisms by which ROS initiate changes in protein function or gene expression. A better understanding of these and other aspects of the role of ROS and oxidative stress in myocyte remodeling should lead to the identification of novel therapeutic approaches to the prevention and treatment of heart failure.

References

1. Keith M, Geranmayegan A, Sole MJ et al. Increased oxidative stress in patients with congestive heart failure. J Am Coll Cardiol 1998; 31: 1352–6.

2. Mallat Z, Philip I, Lebret M et al. Elevated levels of 8-iso-prostaglandin F2alpha in pericardial fluid of patients with heart failure: a potential role for in vivo oxidant stress in ventricular dilatation and progression to heart failure. Circulation 1998; 97: 1536–9.

3. Hill MF, Singal PK. Antioxidant and oxidative stress changes during heart failure subsequent to myocardial infarction in rats. Am J Pathol 1996; 148: 291–300.

4. Dhalla AK, Hill MF, Singal PK. Role of oxidative stress in transition of hypertrophy to heart failure. J Am Coll Cardiol 1996; 28: 506–14.

5. Ide T, Tsutsui H, Kinugawa S et al. Mitochondrial electron transport complex I is a potential source of oxygen free radicals in the failing myocardium. Circ Res 1999; 85: 357–63.

6. Singal PK, Dhalla AK, Hill M et al. Endogenous antioxidant changes in the myocardium in response to acute and chronic stress conditions. Mol Cell Biochem 1993; 129: 179–86.

7. Dhalla AK, Singal PK. Antioxidant changes in hypertrophied and failing guinea pig hearts. Am J Physiol 1994; 266: H1280–5.

8. Ide T, Tsutsui H, Kinugawa S et al. Direct evidence for increased hydroxyl radicals originating from superoxide in the failing myocardium. Circ Res 2000; 86: 152–7.

9. Kinugawa S, Tsutsui H, Hayashidani S et al. Treatment with dimethylthiourea prevents left ventricular remodeling and failure after experimental myocardial infarction in mice: role of oxidative stress. Circ Res 2000; 87: 392–8.

10. Opie LH, Thandroyen FT, Muller C et al. Adrenaline-induced 'oxygen-wastage' and enzyme release from working rat heart. Effects of calcium antagonism, beta-blockade, nicotinic acid and coronary artery ligation. J Mol Cell Cardiol 1979; 11: 1073–94.

11. Griendling KK, Ushio-Fukai M. Redox control of vascular smooth muscle proliferation. J Lab Clin Med 1998; 132: 9–15.

12. Ekelund UE, Harrison RW, Shokek O et al. Intravenous allopurinol decreases myocardial oxygen consumption and increases mechanical efficiency in dogs with pacing-induced heart failure. Circ Res 1999; 85: 437–45.

13. Xia Y, Tsai AL, Berka V, Zweier JL. Superoxide generation from endothelial nitric-oxide synthase. A Ca^{2+}/calmodulin-dependent and tetrahydrobiopterin regulatory process. J Biol Chem 1998; 273: 25804–8.

14. Haywood GA, Tsao PS, von der Leyen HE et al. Expression of inducible nitric oxide synthase in human heart failure. Circulation 1996; 93: 1087–94.

15. Habib FM, Springall DR, Davies GJ et al. Tumour necrosis factor and inducible nitric oxide synthase in dilated cardiomyopathy. Lancet 1996; 347: 1151–5.

16. Li Y, Huang T-T, Carlson EJ et al. Dilated cardiomyopathy and neonatal lethality in mutant mice lacking manganese superoxide dismutase. Nat Genet 1995; 11: 376–81.

17. Pandolfi PP, Sonati F, Rivi R et al. Targeted disruption of the housekeeping gene encoding glucose 6-phosphate dehydrogenase (G6PD): G6PD is dispensable for pentose synthesis but essential for defense against oxidative stress. EMBO J 1995; 14: 5209–15.

18. Martini G, Ursini MV. A new lease of life for an old enzyme. Bioessays 1996; 18: 631–7.

19. Jain M, Brenner DA, Cui L et al. Glucose-6-phosphate dehydrogenase modulates cytosolic redox status and contractile phenotype in adult cardiomyocytes. Circ Res 2003; 93: e9–16.

20. Baumer AT, Flesch M, Wang X et al. Antioxidative enzymes in human hearts with idiopathic dilated cardiomyopathy. J Mol Cell Cardiol 2000; 32: 121–30.

21. Dieterich S, Bieligk U, Beulich K et al. Gene expression of antioxidative enzymes in the human heart: increased expression of catalase in the end-stage failing heart. Circulation 2000; 101: 33–9.

22. Sam F, Kerstetter D, Pimentel DR et al. Decreased activity of manganese superoxide dismutase in failing human myocardium. Circulation 2000; 102 (Suppl II) 84.

23. Siwik DA, Tzortzis JD, Pimental DR et al. Inhibition of copper-zinc superoxide dismutase induces cell growth, hypertrophic phenotype, and apoptosis in neonatal rat cardiac myocytes in vitro. Circ Res 1999; 85: 147–53.

24. Kwon SH, Pimentel DR, Remondino A et al. H_2O_2 regulates cardiac myocyte phenotype via concentration-dependent activation of distinct kinase pathways. J Mol Cell Cardiol 2003; 35: 615–21.

25. Sadoshima J, Izumo S. Molecular characterization of angiotensin II-induced hypertrophy of cardiac myocytes and hyperplasia of cardiac fibroblasts. Critical role of the AT1 receptor subtype. Circ Res 1993; 73: 413–23.

26. Cheng W, Li B, Kajstura J et al. Stretch-induced programmed myocyte cell death. J Clin Invest 1995; 96: 2247–59.

27. Leri A, Claudio PP, Li Q et al. Stretch-mediated release of angiotensin II induces myocyte apoptosis by activating p53 that enhances the local renin–angiotensin system and decreases the Bcl-2-to-Bax protein ratio in the cell. J Clin Invest 1998; 101: 1326–42.

28. Pimentel DR, Amin JK, Xiao L et al. Reactive oxygen species mediate amplitude-dependent hypertrophic and apoptotic responses to mechanical stretch in cardiac myocytes. Circ Res 2001; 89: 453–60.

29. Nakamura K, Fushimi K, Kouchi H et al. Inhibitory effects of antioxidants on neonatal rat cardiac myocyte hypertrophy induced by tumor necrosis factor-alpha and angiotensin II. Circulation 1998; 98: 794–9.

30. Amin JK, Xiao L, Pimental DR et al. Reactive oxygen species mediate alpha-adrenergic receptor-stimulated hypertrophy in adult rat ventricular myocytes. J Mol Cell Cardiol 2001; 33: 131–9.

31. Tanaka K, Honda M, Takabatake T. Redox regulation of MAPK pathways and cardiac hypertrophy in adult rat cardiac myocyte. J Am Coll Cardiol 2001; 37: 676–85.

32. Xiao L, Pimentel DR, Wang J et al. Role of reactive oxygen species and NAD(P)H oxidase in alpha(1) adrenoceptor signaling in adult rat cardiac myocytes. Am J Physiol Cell Physiol 2002; 282: C926–34.

33. Griendling KK, Sorescu D, Ushio-Fukai M. NAD(P)H oxidase: role in cardiovascular biology and disease. Circ Res 2000; 86: 494–501.

34. Ushio-Fukai M, Zafari AM, Fukui T et al. p22phox is a critical component of the superoxide-generating NADH/NADPH oxidase system and regulates angiotensin II-induced hypertrophy in vascular smooth muscle cells. J Biol Chem 1996; 271: 23317–21.

35. von Harsdorf R, Li PF, Dietz R. Signaling pathways in reactive oxygen species-induced cardiomyocyte apoptosis. Circulation 1999; 99: 2934–41.

36. Communal C, Singh K, Pimentel DR et al. Norepinephrine stimulates apoptosis in adult rat ventricular myocytes by activation of the β-adrenergic pathway. Circulation 1998; 98: 1329–34.

37. Zaugg M, Xu W, Lucchinetti E et al. Beta-adrenergic receptor subtypes differentially affect apoptosis in adult rat ventricular myocytes. Circulation 2000; 102: 344–50.

38. Communal C, Singh K, Sawyer DB et al. Opposing effects of beta(1)- and beta(2)-adrenergic receptors on cardiac myocyte apoptosis: role of a pertussis toxin-sensitive G protein. Circulation 1999; 100: 2210–12.

39. Remondino A, Kwon SH, Communal C et al. Beta-adrenergic receptor-stimulated apoptosis in cardiac myocytes is mediated by reactive oxygen species/c-Jun NH2-terminal kinase-dependent activation of the mitochondrial pathway. Circ Res 2003; 92: 136–8.

40. Turner NA, Xia F, Azhar G et al. Oxidative stress induces DNA fragmentation and caspase activation via the c-Jun NH2-terminal kinase pathway in H9c2 cardiac muscle cells. J Mol Cell Cardiol 1998; 30: 1789–1801.

41. Pinsky DJ, Cai B, Yang X et al. The lethal effects of cytokine-induced nitric oxide on cardiac myocytes are blocked by nitric oxide synthase antagonism or transforming growth factor beta. J Clin Invest 1995; 95: 677–85.

42. Ing DJ, Zang J, Dzau VJ et al. Modulation of cytokine-induced cardiac myocyte apoptosis by nitric oxide, Bak, and Bcl-x. Circ Res 1999; 84: 21–33.

43. Arstall MA, Sawyer DB, Fukazawa R et al. Cytokine-mediated apoptosis in cardiac myocytes: the role of inducible nitric oxide synthase induction and peroxynitrite generation [see comments]. Circ Res 1999; 85: 829–40.

44. Sam F, Sawyer DB, Xie Z et al. Mice lacking inducible nitric oxide synthase have improved left ventricular contractile function and reduced apoptotic cell death late after myocardial infarction. Circ Res 2001; 89: 351–6.

45. Feng Q, Lu X, Jones DL et al. Increased inducible nitric oxide synthase expression contributes to myocardial dysfunction and higher mortality after myocardial infarction in mice. Circulation 2001; 104: 700–4.

46. Spinale FG, Coker ML, Bond BR et al. Myocardial matrix degradation and metalloproteinase activation in the failing heart: a potential therapeutic target. Cardiovasc Res 2000; 46: 225–38.

47. Siwik DA, Pagano PJ, Colucci WS. Oxidative stress regulates collagen synthesis and matrix metalloproteinase activity in cardiac fibroblasts. Am J Physiol Cell Physiol 2001; 280: C53–60.

Mechanisms for Contractile Depression

CHAPTER 12

Normal and abnormal excitation–contraction coupling

Brian D Hoit

Introduction

Cardiac excitation–contraction (E–C) coupling refers to the cascade of biological processes that begins with the cardiac action potential and ends with myocyte contraction and relaxation. It is intimately related to calcium homeostasis, myofilament sensitivity, and functions of cytoskeletal and sarcomeric proteins. E–C coupling forms the biophysical underpinnings of the inotropic state of the heart and accordingly, plays a central role in the pathogenesis and treatment of heart failure. In addition, physiological processes in the intact heart associated with E–C coupling (e.g. force–frequency relations, post-extrasystolic potentiation) are important because they (1) are fundamental physiological control mechanisms; (2) are frequently utilized as indices of myocardial function; and (3) play a role in the response to exercise.[1–3]

The investigation of E–C coupling highlights the advantages and limitations of a parochial approach to scientific study (i.e. reductionist cellular versus integrative physiological). Thus, whereas E–C coupling involves specific subcellular movements of calcium ions, its ultimate impact on cardiac performance results from a complex interplay between the intrinsic properties of the cardiomyocytes, chamber properties, loading conditions, and the extracellular matrix. Moreover, *in vitro* studies are critically dependent on experimental conditions; for example, the sarcoplasmic reticu-

lum (SR) Ca^{2+} load is sensitive to many experimental variables, including frequency, temperature, mean $[Ca^{2+}]_i$, $[Na^+]_i$, $[K^+]_i$, action potential configuration and duration. Thus clear insight into the physiology of E–C coupling requires complementary studies at the isolated myocyte and intact organism levels. Nevertheless, since E–C coupling in health and disease are, by first principles, manifestations of myocyte calcium handling, a firm grasp of the calcium transient and calcium homeostasis is warranted.

Normal E–C coupling (Fig. 12.1)

The calcium transient
Determinants and characteristics of the calcium transient
The calcium transient is initiated in response to sarcolemmal depolarization by extracellular calcium (Ca^{2+}) influx through voltage-dependent L-type Ca^{2+} channels, which instigates the release of stored Ca^{2+} from the SR via spatially proximate Ca^{2+} release channels (ryanodine receptor, RyR2). This latter amplification step, fittingly termed calcium induced calcium release (CICR), increases the amount of calcium available for myofilament binding and force-generating actin–myosin crossbridges. (A more controversial mechanism for SR Ca^{2+} release is the voltage-sensitive release mechanism (VSRM).[4] Relaxation results from closure of

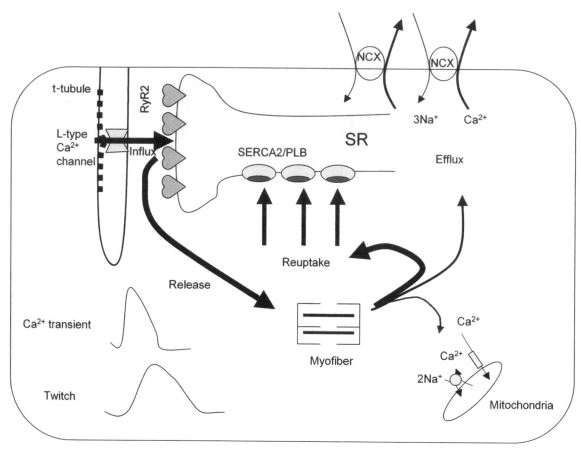

Figure 12.1. *Drawing of a healthy myocyte containing key components of E–C coupling (compare with Fig. 12.2). Influx of calcium is predominantly through the L-type calcium channel. The arrow through channel denotes the amount of activator calcium and is an index of the E–C coupling gain. The relative magnitudes of calcium release, reuptake and efflux are denoted by the arrow widths. The resultant calcium transient and muscle twitch are shown in the lower left of the cell. Other modes of calcium entry and exit from the cell are discussed in the text. Adapted from Scoote M, Poole-Wilson PA, Williams AJ. The therapeutic potential of new insights into myocardial excitation–contraction coupling. Heart 2003; 89: 371–6, with permission.*

the release channels, resequestration of Ca^{2+} by the SR Ca^{2+}-ATPase (SERCA2), and cross-bridge dissolution. To maintain steady state calcium homeostasis, the amount of Ca^{2+} entering the cell with each contraction must be removed before the subsequent contraction. To this end, the sodium–calcium exchanger (NCX) acting in the 'forward' mode competes with SERCA2 for Ca^{2+} and pumps $[Ca^{2+}]_i$ into the extracellular space.

Box 12.1 summarizes the determinants of the cardiac myocyte $[Ca^{2+}]_i$ transient. Factors respon-

sible for the $[Ca^{2+}]_i$ transient amplitude include: (1) calcium current (I_{Ca}), primarily due to Ca^{2+} influx through the L-type Ca^{2+} channel, but in small part due to 'reverse mode' NCX; (2) SR $[Ca^{2+}]_i$ content, which determines the amount of releasable calcium; (3) the efficiency of E–C coupling, or the 'gain' (i.e. the amount of calcium released by the SR for the calcium current, $\Delta[Ca^{2+}]_i/I_{Ca}$); and (4) intracellular Ca^{2+} buffers. The decline of the $[Ca^{2+}]_i$ transient is due to: (1) Ca^{2+} reuptake into SR by SERCA2 (a process

Box 12.1. *Determinants of the calcium ([Ca^{2+}]$_i$) current*

[Ca^{2+}]$_i$ transient magnitude
Calcium current (I_{Ca})
L-type channels
NCX
?Others (T-type channel, tetrodotoxin-sensitive)
SR [Ca^{2+}]$_i$ content
Efficiency of E–C coupling
Intracellular Ca^{2+} buffers

[Ca^{2+}]$_i$ transient decline
SERCA2/phospholamban
NCX
Sarcolemmal Ca^{2+}–ATPase
Mitochondria
Intracellular buffers

modulated by phospholamban); (2) Ca^{2+} extrusion from the cell by the NCX; (3) Ca^{2+} extrusion from the cell by the sarcolemmal Ca^{2+}-ATPase; (4) Ca^{2+} accumulation by mitochondria; and (5) Ca^{2+} binding to intracellular buffers (including fluorescent indicators that are used to measure the transient).[5,6]

Calcium sparks (localized [Ca^{2+}]$_i$ transients) are the elementary SR Ca^{2+} release events that trigger E–C coupling in heart muscle (for a recent detailed review, see reference 7). The basis for the generally accepted 'local control' theory of E–C coupling is that Ca^{2+} sparks are triggered by a local [Ca^{2+}]$_i$ established in the region of RyR2 by the opening of a single L-type Ca^{2+} channel.[8] The amplitude of Ca^{2+} sparks is determined by SR Ca^{2+} load and gating properties of the RyR2.[9,10] Although the exact nature and origin of Ca^{2+} sparks are not completely understood, the prevailing view is that the global [Ca^{2+}]$_i$ transient is produced by the temporal and spatial summation of a large number of Ca^{2+} sparks.[9,11] The mechanisms responsible for terminating sparks are not clear, but proteins accessory to the RyR2 (e.g. sorcin, FKB12) have been suggested to play a key role.[12]

Methods used to study the calcium transient
A major advantage of using isolated cardiomyocytes to assess cardiac mechanics is that the complexities *in vivo* due to loading conditions, chamber geometry, extracellular matrix and neurohormonal influences can be overlooked. Moreover, they are uniquely suited for the study of E–C coupling because they allow application of single-cell voltage clamps used to study cardiac ion channels and transporters, and are suitable for confocal microscopic analysis. Mechanical properties of isolated cardiac myocytes are determined by measurement of force (using microtransducers), analysis of sarcomere length (using laser diffraction microscopy), and video edge detection of cell motion;[13,14] relative simplicity and low instrumentation costs are notable features of the latter. In this method, a myocyte is field-stimulated and focused on a charge-coupled device (CCD) high-speed camera interfaced to an inverted microscope. The voltage along each scan line varies linearly to changes in light intensity on the element, and identifies the lateral cell borders, which are used to derive the rate and extent of shortening (contraction) and the rate of relengthening (relaxation).

The magnitude of the [Ca^{2+}]$_i$ transient modulates the force developed by myofilaments, and factors that modify calcium cycling and/or Ca^{2+} sensitivity of myofilaments may alter significantly the force and extent of myocyte contraction. Quantitative analysis of the characteristics of the [Ca^{2+}]$_i$ transient usually includes the peak systolic, diastolic and amplitude of the [Ca^{2+}]$_i$ transient (the latter is equal to the peak systolic minus diastolic [Ca^{2+}]$_i$), time to the peak, and the time constant of the decline of the [Ca^{2+}]$_i$ transient.[15–17] Calcium homeostasis can be assessed by measurement of the global [Ca^{2+}]$_i$ transient, localized [Ca^{2+}]$_i$ transients (calcium sparks), and kinetics of [Ca^{2+}]$_i$ transients.

The global [Ca^{2+}]$_i$ transient and shortening can be simultaneously measured in isolated cardiomyocytes by adding a [Ca^{2+}]$_i$ measurement system to the edge detector described above.[18] Although [Ca^{2+}]$_i$ transients can be measured in intact myocardial preparations using fluorescence measurements, dye loading into non-myocytes, motion artifacts, filter effects, and unstable background fluorescence may influence the measurement.[19] The bioluminescent Ca^{2+} indicator, aequorin, is used in multicellular preparations to generate $[Ca^{2+}]$–force and force–frequency relations.[20,21]

The most common, commercially available calcium-sensitive fluorescent indicators for measurement of the $[Ca^{2+}]_i$ transient include fura-2, indo-1, and fluo-3. Intracellular Ca^{2+} buffering and dye saturation with the relatively high-affinity Ca^{2+} indicators fura-2 and indo-1 may lead to underestimates of the peak $[Ca^{2+}]_i$.[22] Field stimulation, applications of voltage-clamp pulses, or caffeine exposure trigger fluorescence transients. Fluorescent measurement of calcium transients is influenced by the quality of the isolated cardiomyocytes and technical factors responsible for variable dye loading (e.g. incubation versus diffusion from the patch-clamp pipettes, dye concentration, type of dye, incubation time and temperature, etc).

Examination of Ca^{2+} sparks requires a laser confocal microscope operating in line scan mode (although three-dimensional contour plots can be used for visualization).[7] Fluo-3 is one of the best Ca^{2+} indicators used with line scan devices to visualize Ca^{2+} sparks;[9,11,23] the indicator is introduced into the cell either by using a loading method or by diffusion from a patch pipette. The parameters typically measured when analyzing sparks include: (1) frequency of Ca^{2+} spark occurrence; (2) spark size; (3) spark amplitude; (4) spark time course; (5) time constant of spark decay; and (6) spark location.[24]

L-type Ca^{2+} channel function is assessed by conventional voltage-clamp techniques. Direct measurement of ionic currents is possible by eliminating the capacitive currents during the voltage-clamp step. The passage of ions through channels occurs at a very high throughput rate with high selectivity (i.e. the ability to discriminate a species of ion in a complex milieu) and sensitivity (i.e. the ability to respond to changes in membrane voltages or specific ligands, a function known as the 'gating'). Whole-cell patch-clamp techniques record currents passing through a population of channels, whereas single-channel recording separates the permeation and gating properties of the channel.[5]

Nearly all approaches used to assess SR Ca^{2+} content involve exposure to caffeine, which activates the ryanodine receptor and releases SR Ca^{2+}.

The amplitude of myocyte shortening induced by rapid exposure to caffeine is often used as an index of SR Ca^{2+} content. Although measurement of the peak $[Ca^{2+}]_i$ provides a more accurate measurement of SR Ca^{2+} content, the peak Ca^{2+} level is limited by NCX removal of Ca^{2+}. The most accurate method is to expose myocytes to caffeine with a rapid solution switcher that 'instantaneously' changes the extracellular milieu,[24] and to quantitate the integral of the inward NCX current that results from exchanger-mediated removal of Ca^{2+}.[15,25] Another index of SR Ca^{2+} content is the rapid cooling contracture (RCC); rapid cooling of cardiac muscle simultaneously produces SR Ca^{2+} release and inhibits Ca^{2+} transport mechanisms.[25,26]

The rate of Ca^{2+} sequestration into the SR is often inferred from the rate of myocyte relengthening ($-dL/dt$) or by the exponential time constant (τ) of the $[Ca^{2+}]_i$ transient decline. However, $-dL/dt$ does not precisely track the rate of decline of the $[Ca^{2+}]_i$ transient, and τ is influenced by NCX function and the peak of the $[Ca^{2+}]_i$ transient.[17] These difficulties have been largely overcome by the use of a rapid solution switcher device. Protein and message levels of SERCA2 and phospholamban are frequently used to assess the integrity of SR function.

Ryanodine receptor function is assessed by measurement of Ca^{2+} sparks and the gain of the global $[Ca^{2+}]_i$ transient. E–C coupling gain is quantified as the magnitude and/or the rate of rise of the $[Ca^{2+}]_i$ transient, divided by the amplitude of the L-type Ca^{2+} current;[25] it is a measure of the efficiency by which L-type Ca^{2+} channel current induces Ca^{2+} release from the SR. However, because E–C coupling gain varies directly with SR Ca^{2+} load,[25] comparability of SR Ca^{2+} loading (or accounting for any observed differences) is critical.

The function of the NCX in intact myocytes is typically assessed by measuring the rate of relengthening, and/or the time constant of the $[Ca^{2+}]_i$, during a caffeine-induced contraction or $[Ca^{2+}]_i$ transient, respectively. Because caffeine prevents SR reuptake of Ca^{2+}, the decline of the $[Ca^{2+}]_i$ transient and the associated myocyte relaxation rate are due to cellular extrusion of Ca^{2+} by the NCX. However, the dependence of the NCX on the intracellular $[Na^+]$ concentration and membrane potential complicates this approach. Therefore, precise assessment of exchanger function and density requires that the cell be voltage-clamped and dialyzed with a specific Na^+ concentration solution. An alternative approach is to combine voltage clamps and a rapid solution switcher to measure the outward current produced by the NCX when extracellular Na^+ is abruptly removed.[24] Protein and message levels of the NCX fail to distinguish between the forward and reverse operating modes of the exchanger.

Componenets of E–C coupling

Despite their moniker, t-tubules (invaginations of the sarcolemma and glycocalyx) are both longitudinal and oblique in their orientation, and together form a permeability barrier between the cytosol and the extracellular space.[27] The sarcolemma is the site where calcium enters and leaves the cell through a distribution of ion channels, transporters and pumps. The membranous surface areas are tissue- and species-specific; thus, in the mammalian heart (avian, amphibian, and reptilian hearts lack t-tubules), atrial versus ventricular cells have poorly developed t-tubules.[27] The structural specialization of the sarcolemma includes: (1) SR coupling in the form of dyads by means of the t-tubule; (2) caveolae (invaginations of the sarcolemma which increase surface area and form a scaffold for signaling molecules such as NO synthase and protein kinase C (PKC)); and (3) the intercalated disc (gap junction, intermediate junction, and desmosome).[28]

The SR is an intracellular membrane-bounded compartment consisting of terminal, longitudinal and corbular components. The free walls of the terminal cisternae are apposed to the walls of the t-tubules and form the dyadic cleft; the RyR2 receptors are located in the walls of the terminal cisternae ('feet') and face the dyadic cleft.[29] Longitudinal SR is fairly homogenous and contains primarily the SR proteins, SERCA2 and the associated phosphoprotein, phospholamban. In its dephosphorylated state, phospholamban is an endogenous inhibitor of SERCA2. Phosphorylation by PKA (at Ser16) and CaMKII (at Thr17) lower the K_m of SERCA and result in enhanced calcium uptake.[30] SR calcium is transported from the tubular lumen of the SR to the terminal cisternae where it is stored mostly bound to calsequestrin, a low-affinity high-capacity calcium-binding protein. Calsequestrin forms a complex with the proteins junctin, triadin, and the ryanodine receptor.[31] Junctional SR does not come into contact with the sarcolemma; corbular SR is a form of junctional SR that contains calsequestrin and ryanodine receptors, but is not coupled to the well-recognized calcium cycling events.[28]

Myofilaments comprise the contractile machinery of the cell and occupy 45–60% of the ventricular myocyte volume. The fundamental unit of the myofilament is the sarcomere, bounded by Z-lines on each end, from which the thin actin filaments extend toward the center. At the center of the thick myosin filament is the M-line, where the thick filaments are interconnected by M-protein and myomesin. Titin runs from the M-line to the Z-line; this large structural protein acts as a scaffold for myosin deposition, stabilizes the thick filament, and plays a critical role in determining the passive stiffness of the heart.[32] The Z-lines are the sites of anchor for cytoskeletal intermediate filaments and actin filaments at the intercalated disks and at focal adhesions. The two major structural complexes involved in the connections with the extracellular matrix include the integrin complex and the dystrophin complex, which links actin to laminin and collagen.

The myosin molecule consists of two heavy chains with a globular head, a long α-helical tail, and four myosin light chains (MLCs). The myosin head is the site of ATP hydrolysis and forms cross-bridges with the thin actin filament through an actin-binding domain. Transduction of chemical to mechanical energy and work is the function of myosin ATPase, located in the myosin heads. The most accepted model of energy transduction is the sliding filament theory based on the formation and dissociation of cross-bridges between the myosin head and the thin filament that transition through different energetic states.[6,32] Two MLCs (the alkali or essential light chain, MLC-1 and the phosphorylatable or regulatory light chain, MLC-2) are associated with each myosin head and confer stability to the thick filament. Although phosphorylation of MLC2 by MLC kinase is critical for smooth muscle cell contraction, its physiological significance in cardiac muscle (increased calcium sensitivity and rate of force development) is controversial.[33–36]

The backbone of thin filament is helical double-stranded actin. Tropomyosin is a long flexible double-stranded (largely α-helix) protein which lies in the groove between the actin strands. The troponin complex comprises a calcium-binding subunit, TnC, an inhibitory subunit which binds to actin, TnI, and a tropomyosin-binding subunit, TnT, which is attached to tropomyosin. In the resting state, when $[Ca^{2+}]_i$ is low, calcium-binding sites of TnC are unoccupied and TnI preferentially binds to actin; this favors a configuration in which the troponin–tropomyosin complex sterically hinders myosin–actin interaction. When $[Ca^{2+}]_i$ rises, calcium binds to the calcium-specific sites on TnC and strengthens the interaction of TnC and TnI; TnI dissociates from actin, and a conformational change removes the steric hindrance to myosin–actin interaction.[37,38] Although this is the most accepted model, other potential explanations exist; all models incorporate the concept that myofilaments are dynamically involved in their state of activation and not simply subject to passive changes in $[Ca^{2+}]_i$.[39]

Mitochondria comprise approximately 35% of ventricular myocyte volume, and according to their cellular location are designated as either subsarcolemmal or interfibrillar. Mitochondria are the site of oxidative phosphorylation and ATP generation. Although they have the capacity to buffer large amounts of Ca^{2+} and are a potential source of activator calcium, their contribution to E–C coupling is probably minimal in view of the short time constants involved.[40] Variation in mitochondrial Ca^{2+} during a twitch is imperceptible and thus plays a very minor role in beat-to-beat changes.[40] While mitochondrial Ca^{2+} plays a small role in E–C coupling, slower increases in mitochondrial Ca^{2+} fluxes are important with respect to mitochondrial function and energetics; for example, the matrix enzymes pyruvate dehydrogenase, NAD-dependent isocitrate dehydrogenase, and α-ketoglutarate dehydrogenase, are activated by low $[Ca^{2+}]$.[41,42] In addition, the ability to accumulate large amounts of Ca^{2+} under pathological conditions (e.g. ischemia) may help protect against Ca^{2+} overload; however, Ca^{2+} accumulation by mitochondria ultimately slows ATP production.[43]

E–C coupling in cardiac versus skeletal muscle

Although the details of the process differ in cardiac and skeletal muscle, the general scheme of E–C coupling is similar: depolarization of the sarcolemma (and its branching t-tubules) by an action potential induces calcium release from the SR; calcium binds to the thin filament troponin and initiates interaction of actin and myosin. Thus, in both tissues, the myofilaments are activated in a $[Ca^{2+}]$-dependent manner and transduce chemical energy (ATP) into mechanical force and shortening.

An important difference (based on physiological and confirmed by ultrastructural studies) is that skeletal muscle contraction is dependent primarily on calcium released from the SR, whereas cardiac muscle contraction is also contingent on calcium entry across the sarcolemna. The SR network is more extensive and well organized in skeletal than cardiac muscle; terminal cisternae are large and abut narrow t-tubules in skeletal muscle, whereas in cardiac muscle terminal cisternae are small saccular extensions that abut large t-tubules.

The fraction of myofilaments is greater in skeletal than cardiac muscle, and myofibrils are more defined. In cardiac muscle, myofilaments are not continuous or in series from one cell to the next, but transmit force across cell-to-cell junctions. A critical difference is that skeletal muscle force can be varied by summation of contractions, tetanus, and recruitment of additional fibers. In contrast, increased demands cannot be met in this fashion by cardiac muscle, because the heart is a functional syncytium and as a pump, must relax between contractions to allow filling. In cardiac muscle, the force of contraction is varied largely by changes in the peak systolic $[Ca^{2+}]$ and the sarcomere length.

In skeletal muscle, the stoichiometry of ryanodine receptors to L-type calcium channels is 1:2, whereas the ratio in cardiac muscle is 4–10:1.[44] The stoichiometry of the RyR2 and L-type Ca^{2+} channels suggests that a calcium spark results from activation of a cluster of several RyR2 by activation of one Ca^{2+} channel in a stochastic nature (i.e. a cluster of RyR2 cannot activate another cluster of RyR2).[29] Skeletal muscle ryanodine receptors form arrays in which every other ryanodine receptor is associated with four dihydropyridine receptors; membrane depolarization produces conformational changes of the L-type channel that activate ryanodine receptors, the voltage-activated Ca^{2+} release or VACR.[29] Finally, skeletal muscle differs from cardiac muscle with respect to the number and affinity of calcium- and magnesium-binding sites and contractile protein isoforms of troponin, myosin and MLC.

Properties of contraction

Fundamental to cardiac muscle function are the relations between force and muscle length, velocity of shortening, calcium, and heart rate. The maximal force developed at any sarcomere length is determined by the degree of overlap of thick and thin filaments, and therefore, the number of available cross-bridges.[45] Force increases linearly until the sarcomere length with maximal overlap (~2.2 μm) is achieved, beyond which force and overlap gradually declines to zero ('i.e. the descending limb'). The descending limb of the length–tension relationship is prevented by the strong parallel elastic component in cardiac muscle. The ascending limb of the length–tension relationship (equivalent to the Frank–Starling relationship that relates preload to cardiac performance) is also due to a length-dependent increase in myofilament calcium sensitivity;[46] this has been explained by TnC isoforms, narrower interfilament gaps at long sarcomere length, and increased SR calcium release and uptake at longer sarcomere lengths.[28] The relation between force and velocity of contraction is hyperbolic; at maximum force (isometric force), shortening cannot occur, and at zero force (i.e. 'unloaded' muscle), velocity is at a maximum, V_{max}, reflecting the maximum turnover rate of myosin ATPase. Therefore, alterations in the myosin isoform (i.e. α fast, β slow) such as those seen in response to pressure-overload, have an effect on V_{max}.

Another fundamental property of cardiac muscle is the force–pCa^{2+} relation. In addition to $[Ca^{2+}]$, force–pCa^{2+} measurements are influenced by calcium sensitivity; cooling, acidosis, increased $[PO_4^{2-}]$, $[Mg^{2+}]$, ionic strength and shorter sarcomere lengths decrease Ca^{2+} sensitivity, and caffeine and various inotropic drugs (e.g. levosimendan), are potent calcium sensitizers.[28] Beta-adrenergic stimulation results in a cAMP-dependent phosphorylation of cardiac TnI (at serine 23 and 24) and a resultant decrease in myofilament calcium sensitivity;[47] thus for a positive β-adrenergic receptor (βAR) inotropic effect, the amplitude of the calcium transient must more than compensate for reduced βAR-mediated myofilament sensitivity.

The final property relates heart rate to contraction and relaxation. In rodents, the force–frequency response (FFR) measured *in vitro* (unlike *in vivo*, and in contradistinction to the positive 'treppe' in larger animals) may be negative; this is understood to be related to the Ca^{2+} capacity and load of the SR.[48] By contrast, increases in heart rate accelerate relaxation irrespective of SR characteristics.[49,50] These heart rate-related phenomena are discussed in more detail below.

For all the advantages of studying isolated myocytes and muscle fibers, an integrated and more realistic analysis of cardiovascular function regards the left ventricle as a muscle pump coupled to the systemic and venous circulations. The principal determinants of left ventricular function to be considered include the inotropic (contractile) state, loading conditions (i.e. preload and afterload), and the heart rate. The influence of preload on measures of ventricular performance (e.g. cardiac output) defines the Frank–Starling relationship. Changes in inotropic state independent of loading conditions are defined operationally by shifts of the various ventricular function curves. For example, a drug with positive inotropic activity shifts the Frank–Starling relationship (analogous to the length–shortening curve in papillary muscle preparations) upward and to the left, the stress–shortening relationship (analogous to the force–velocity curve) upward and to the right, and the force–frequency relationship (treppe) upward and to the right.

The rate of left ventricular relaxation may be estimated from the maximal rate of pressure decay ($-dP/dt_{max}$) and indices (e.g. $RT_{\frac{1}{2}}$) that are related to the time necessary for ventricular relaxation, but these measurements are highly dependent on the prevailing load of the intact circulation. In contrast, τ, the time constant of left ventricular relaxation during isovolumic diastole, provides a more accurate, less load-dependent measure of relaxation; τ is shortened by β-adrenergic stimulation (protein kinase A (PKA)-dependent phosphorylation of phospholamban and troponin I) and prolonged with β-adrenergic antagonists.

Non-steady-state E–C coupling

Non-steady state aspects of E–C coupling provide the basis for many physiological phenomena. Mechanical restitution is the relative refractory period that immediately follows a contraction and is usually explained by the recovery of the RyR receptors (since I_{Ca} and SR Ca^{2+} content recover rapidly).[28,51] However, in vivo, we showed that the rate of mechanical restitution is inversely related to the level of phospholamban, reflecting the important influence of phospholamban on the affinity of the SERCA pump for Ca^{2+} and confirming the critical role of the SR Ca^{2+} load on mechanical activity in the intact animal.[52] One possible explanation is that the rate of SR Ca^{2+} uptake is primarily responsible for mechanical restitution and that the resultant SR Ca^{2+} load influences the time constant of the ryanodine channel. In support of this concept, cross signaling between Ca^{2+} channels and the ryanodine receptor have recently been demonstrated with confocal microscopy.[53]

Mechanical restitution is the basis for post-extrasystolic potentiation (PESP), the strong contraction following a weaker extrasystole, since lower $[Ca^{2+}]$ on the extrasystole results in increased I_{Ca} (less Ca^{2+}-induced inactivation of the L-channel), less Ca^{2+} efflux from NCX, and increased SR Ca^{2+} loading on the post-extrasystolic beat.[54,55] The result is a greater amount of released Ca^{2+} and therefore a stronger contraction. In the intact heart, the effect of changing preload (and the impact on Frank–Starling and calcium sensitivity) is an important additional mechanism. PESP contributes to the beat-to-beat variability of the pulse in atrial fibrillation.[56] Mechanical alternans, the alternating contraction amplitude at a constant heart rate that is seen in heart failure, is explained by a similar interplay of RyR2 refractoriness (which is increased in heart failure), I_{Ca} inactivation, NCX competition, and SR Ca^{2+} load.[57]

The relation between pacing rate and force (force–frequency relationship) can be understood similarly by these non-steady-state phenomena. Increased pacing rate overcomes the encroachment on mechanical restitution and produces an increase in force because of rate-dependent increases in I_{Ca}, I_{Na} (which results in less Ca^{2+} efflux by the NCX), diastolic $[Ca^{2+}]_i$ (less time for efflux and greater influx/second), releasable SR Ca^{2+} content, and fractional SR Ca^{2+} release.[58,59] In rat and mouse, the force–frequency response (FFR) (at least in vitro) is usually reported as negative, presumably because of their already high SR Ca^{2+} content and the short duration of their action potentials. In vivo, the FFR (at least the heart rate $-dP/dt_{max}$) in mouse is biphasic.[60] Similar biphasic force–frequency relations are described in primates,[61] and thin muscle strips from human heart failure.[62] In the failing human heart, SR Ca^{2+} only slightly increases with increased frequency (because of reduced SERCA, increased NCX, and decreased filling time), which produces a flat or negative force–frequency relationship.

A phenomenon similar to the force–frequency relationship is observed when the effects of heart rate on the time constant of isovolumic relaxation are examined.[61] Thus, similar to the effect on contraction, relaxation is augmented at higher rates of stimulation. Khoury et al. demonstrated a nonlinear relationship between the heart rate and τ.[61] Thus the time constant decreased monotonically with incremental pacing until a critical heart rate was achieved, after which the relationship reversed and τ increased, suggesting that after exceeding a critical heart rate, the ability of the SR to resequester Ca^{2+} becomes limited, resulting in incomplete relaxation. In these chronically instrumented non-human primates with short duration hyperthyroidism, enhanced frequency-dependent rates of isovolumic left ventricular pressure development and decay were associated with altered regulation of cardiomyocyte myosin heavy chain (MHC) composition and Ca^{2+}-cycling proteins.

Relaxation restitution represents the time for restoration of relaxation in premature beats having progressively longer extrasystolic intervals, and is related to uptake of Ca^{2+} by the SR.[60] In the intact dog, Ca^{2+} handling by the SR was shown to be a primary determinant of early relaxation restitution.[63]

A related phenomenon, frequency-dependent acceleration of relaxation (FDAR) results from CaMKII phosphorylation of phospholamban (or by some other mechanism that increases SR Ca^{2+} transport). CaMKII may be activated by the increased $[Ca^{2+}]_i$ that occurs with increased stimulation rates;[64] however, the precise mechanisms are unresolved.[64,65] The physiological implications for faster relaxation at increased heart rates are obvious.

Role of NO

Nitric oxide (NO) is produced by the myocardium and regulates cardiac function through both vascular-dependent and -independent effects.[66] The influence of NO on E–C coupling was recently reviewed.[67] NO has a modest positive inotropic effect on basal contractility in isolated myocytes and the isolated perfused heart, but a negative inotropic effect in vivo,[68–70] possibly because of nitrosylation ion channels responsible for E–C coupling (e.g. L-type channel, RyR2). The negative inotropic effects on β-adrenergic-stimulated contractility are greater and less controversial, and may comprise a critical component of negative feedback over contractile reserve. NO's positive effects on lusitropy (and in part, for negative inotropic effects) are likely to be due to cGMP-mediated reduction in myofilament Ca^{2+} sensitivity.[71,72] Finally, NO reduces myocardial oxygen consumption ($M\dot{V}O_2$), and increases mechanical efficiency (stroke work/$M\dot{V}O_2$), suggesting that NO regulates energy production as well as influencing consumption.

The effects of NO on E–C coupling are confusing and controversial because of the presence of three NOS isoforms that are spatially localized to highly controlled microdomains and linked to disparate signaling pathways and effectors. For example, NOS3 is compartmentalized to the sarcolemmal and t-tubule caveolae, is associated with the L-type channel, is inactivated by the scaffolding protein caveolin-3 and is activated by Ca^{2+}/calmodulin and Akt phosphorylation. NOS3

produces negative inotropic and positive lusitropic effects via cGMP activation.

Although NOS1 is also activated by Ca^{2+} calmodulin and may be inactivated by caveolin-3, NOS1 is localized to cardiac SR and is involved with calcium homeostasis.[67] NOS1 increases the open probability of the cardiac ryanodine receptor and modulates β-adrenergic mechanics, calcium transients,[73] and the force–frequency relationship,[74] although the mechanisms remain controversial. Inhibition of SR Ca^{2+} uptake by NO may be due to reduced phospholamban phosphorylation, but this mechanism is far from conclusive.[67] Nevertheless, accumulating data suggest that NO plays an important role in E–C coupling vis-à-vis modulation of Ca^{2+} channel activity, myofilament Ca^{2+} sensitivity and mitochondrial respiration.[75]

Excitation–transcription coupling

An emerging concept is that the molecular machinery of E–C coupling is involved in the long-term regulation of gene expression by a process known as excitation–transcription coupling (E–T coupling).[76] Despite periodic oscillations of $[Ca^{2+}]$ from 100 nM to 1 μM during E–C coupling, transcription regulatory proteins (e.g. nuclear factor κB (NFκB), c-jun N-terminal kinase (JNK), nuclear factor of activated T cells (NFAT)) are calcium activated. The amplitude and duration of the calcium signal, the presence of microdomains and anchoring proteins, and linkages through calmodulin, kinases and phosphatases are important mechanisms for discriminating important regulatory cues and resolving this apparent paradox.[77] For example, heart failure is associated with action potential prolongation that may activate calcium-calmodulin kinase (CaMK), which may modulate L-type Ca^{2+} current and regulate uptake and release of Ca^{2+} from the SR.[78]

Calcium-calmodulin regulates proteins involved in calcium transport, ion channels, and cell contraction, metabolism, and proliferation.[79] Phosphorylation substrates for CaMKII that are involved in modulating contraction–relaxation include phospholipase B (PLB), SERCA 2A, L-type Ca^{2+} channels, and the ryanodine receptor.[80,81] CaMKII phosphorylates the transcription factor cAMP response element-binding protein

(CREB) which promotes transcription of c-fos.[82] In addition, CaMKII has autoregulatory properties that are dependent on the frequency of Ca^{2+} spikes,[83] a process thought to have a role in neuronal memory. Little is known about *in vivo* CaMKII activation, but biochemical data suggest that CaMK may be 'primed' to respond to Ca^{2+} spikes.[84] Thus calcium-dependent regulation by CaM and CaMKII has both acute responses affecting E–C coupling, and chronic responses that influence the expression levels of proteins involved in E–C coupling.

The use of transgenic animals

Ablation and overexpression of the proteins involved in calcium homeostasis have provided many insights into the mechanisms that underlie E–C coupling.[57,85–93] Despite their importance, transgenic animal models are not as unambiguous as generally believed, in part because of compensatory mechanisms (at all structural and physiological levels) that complicate interpretation of the phenotype. For example, compensatory changes in the PLBKO mouse include a $\sim$ 25% reduction in the ryanodine receptor and enhanced L-type Ca^{2+} channel inactivation.[86] These changes are potentially important, insofar as SR calcium release influences CaMKII activation.[94] Other limitations of genetically engineered mice include strain-dependent modifying effects, species-dependent differences in protein isoform and phenotypic expression, and confounding due to tissue-specificity and temporal activation of promoters.[95]

Pathological E–C coupling (Fig. 12.2)

Heart failure is characterized by myocyte loss, myocardial hypertrophy and impaired function of viable myocardium, which reflect alterations in signaling pathways that modulate apoptosis, cellular growth and E–C coupling, respectively. Other mechanisms responsible for myocyte contractile dysfunction in heart failure include abnormal protein expression, cytoskeletal abnormalities, and energy starvation. The time-honored concept that hypertrophy (by normalizing excessive load) is initially compensatory and ultimately transitions to a

state of clinical heart failure has recently been challenged.[96,97] Indeed, in one controversial study, the increased wall stress owing to aortic banding was better tolerated (greater shortening) in mice that failed to develop hypertrophy,[98] suggesting a more direct relation between hypertrophy and reduced contractility.

Disparate myocyte and myocardial function is variably reported in hypertrophy and heart failure, in part because of model dependency, the variable techniques and indices used to measure function, diverse conceptual frameworks, and dissimilar experimental conditions. For example, isometric force and unloaded fractional shortening are related to the magnitude of the calcium transient, but the velocity of shortening is additionally related to the myofibrillar ATPase activity, reflecting different biophysical bases for force versus velocity indices of contraction. Other difficulties relate to cellular heterogeneity and variable distribution of wall stresses in viable myocardium. Heterogeneity may have a biophysical basis, as dyssynchronous release of calcium sparks was shown to produce a slower rate of rise of the calcium transient.[99]

Overview of changes in calcium homeostasis in congestive heart failure

In heart failure, there is reduced twitch force and myocyte contraction, reduced relaxation rates, and a flat or negative FFR. The calcium transient amplitude usually mirrors these mechanical changes. The action potential duration is prolonged, particularly at low heart rates. I_{Ca} is usually unchanged, and the density of the current usually keeps pace with cellular hypertrophy in heart failure.[100] SERCA2 expression is downregulated; phospholamban is either unchanged or downregulated, but phosphorylation status of phospholamban is decreased. RyR2 is either unchanged or downregulated, but the RyR2 are dysregulated (increased open probability with less coordinated gating). Changes in RyR2 are not straightforward however, since removal of FK-506-binding protein (FKBP) from RyR2 as occurs in heart failure, enhances fractional SR Ca^{2+} release,[101] and sensitizes the RyR2 to activator Ca^{2+},[102] which would result in increased $[Ca^{2+}]$ transients (a finding not

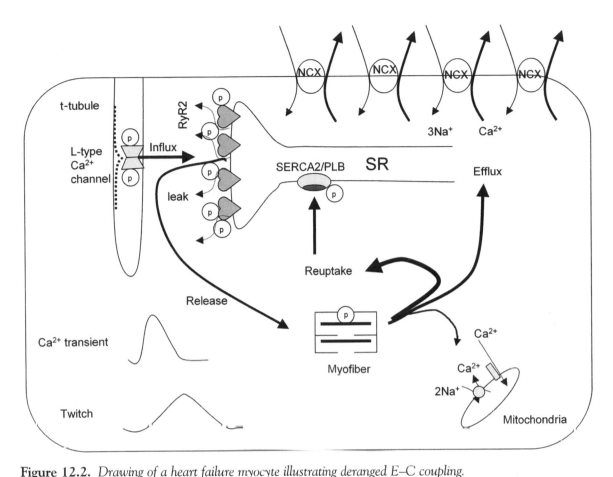

Figure 12.2. *Drawing of a heart failure myocyte illustrating deranged E–C coupling.*
Although most studies fail to show a change in the density of I_{Ca} during failure, results are controversial; therefore the stippled line representing influx is thinner. The efficiency of the inward calcium current (gain) is reduced (smaller activator calcium arrow and increased distance between the calcium channel and RyR2). Adrenergic signaling pathways phosphorylate the L-type channel, RyR2 (resulting in calcium leakage, the small dashed arrows), phospholamban, and TnI. The number of SERCA2 pumps is downregulated (reduced uptake), SR content is reduced (less release) and NCX is upregulated (increased efflux). As a result, the calcium transient and twitch force are reduced. Other changes are discussed in the text. Adapted from Scoote M, Poole-Wilson PA, Williams AJ. The therapeutic potential of new insights into myocardial excitation–contraction coupling. Heart 2003; 89: 371–6, with permission.

typically found in heart failure). Calsequestrin and calreticulin do not change in heart failure. NCX expression is generally higher, and the Na^+–K^+ ATPase is lower (which may produce higher $[Na^+]_i$ and therefore be compensatory). Other ion currents are altered; the most consistent are reductions in the transient outward current (I_{to}) and the inward rectifier potassium current (I_{k1}) The sum of these changes is that there is less SR Ca^{2+} content during steady state. Lower SR Ca^{2+} content reduces SR Ca^{2+} release because there is less Ca^{2+} available for release, and because fractional release for a given I_{Ca} trigger is reduced. The VSRM is depressed in certain models of heart failure.[4] Myofilament calcium sensitivity is generally unchanged.

L-type Ca^{2+} channel

Although inward calcium current is the greatest source of activator calcium, consistent abnormalities in L-type calcium current are not identified in the failing heart. Most studies fail to show a change in the density of I_{Ca} during hypertrophy and failure.[103] Results regarding current density, open channel probability, protein expression and Ca^{2+}-dependent inactivation are controversial;[104–107] reduced protein expression appears compensatory for increased single-channel current. Increased phosphorylation of the channel may explain the blunted response of failing hearts to β-adrenergic stimulation.[108] Other potential triggers of Ca^{2+} release include the NCX, T-type calcium channels, novel sodium and calcium currents, and the VSRM. With the exception of NCX, the importance of these latter channels is debatable and critical review suggests that they are unlikely to have a significant role in the normal or failing heart.[8]

SR Ca^{2+} load

SR Ca^{2+} load can be understood in terms of SERCA2 and NCX activity. Although reduced SR Ca^{2+} load is potentially the biophysical basis of reduced contractility in heart failure as suggested by several studies,[109,110] SR Ca^{2+} content has never been measured *in vivo*. Reduced expression of SERCA2 is generally, but not universally, demonstrated. Phospholamban expression is more variable, but phosphorylation status may be more important.[111,112]

The efficiency of the inward calcium current to induce SR calcium release is the E–C coupling gain.[113] Gain is dependent on the SR Ca^{2+} load and membrane potential, and is reduced in heart failure.[97] Gain may be reduced because of reduced numbers and/or increased phosphorylation of ryanodine receptors, increased ryanodine-L-type channel distance, and structural changes in the t-tubules.[104,114,115] The few studies that have examined changes in the density and shape of t-tubules in hypertrophy and failure have been equivocal.[27]

NCX

Reverse mode NCX current can initiate contraction, but is less efficient than the L-type Ca^{2+} current.[115] NCX expression is increased in hyper-trophied and failing hearts, and this may lead to reduced Ca^{2+} load, and a smaller transient, reduced dependency on L-type Ca^{2+} current for EC coupling, increased susceptibility to Ca^{2+} overload during ischemia–reperfusion and ventricular arrhythmias induced by delayed afterdepolarizations.[109,116–118] However, the elevated [Na$^+$]$_i$ in failing cardiomyocytes, in part due to reduced expression of Na$^+$–K$^+$ pump protein, favors reverse mode NCX function and compensates for the reduced SR Ca^{2+} load. Moreover, it has been shown recently that if NCX function is normalized to cell volume, Ca^{2+} transport via the NCX is unchanged in rat myocytes after infarction.[119]

RyR2

The RyR2 is activated at nanomolar to micromolar concentrations of cytosolic Ca^{2+} and inhibited at millimolar [Ca^{2+}], and is regulated by luminal SR [Ca^{2+}] (i.e. the open probability of the channel is decreased when the [Ca^{2+}] falls).[114] PKA-dependent phosphorylation of RyR2 dissociates the regulatory protein FKBP 12.6 and increases the open probability of the channel. In heart failure, RyR2 levels are reported to be either unchanged or decreased,[114] but regulation of RyR2 is defective. FKBP 12.6 dissociation results in increased Ca^{2+}-dependent activation, subconductance states, and ultimately depletion of calcium stores; diastolic calcium leak through hyperphosphorylated RyR2 receptors is implicated in delayed afterdepolarizations and ventricular arrhythmia. Loss of 'coupled gating' (which ensures that all RyR2 on an SR membrane open and close synchronously) due to dissociation of FKBP 12.6 may also contribute to reduce E–C coupling gain.

Therapeutic implications

Although once the mainstay of therapy, chronic inotropic therapy for heart failure with dobutamine, vesnarinone, and milrinone have been associated with increased mortality.[120,121] Nevertheless, impaired contractility is a cardinal feature of systolic heart failure, and consideration of potential inotropic therapeutic targets is warranted. Moreover, inotropic agents remain essential for the treatment of acute heart failure. Methods for improving inotropy are based on the ability to (1) increase the amount of Ca^{2+} delivered to the myofilaments (upstream mechanism); (2) increase the sensitivity of the myofilaments to the delivered calcium (central mechanism); and (3) increase the efficiency of E–C coupling (downstream mechanism).[122] However, as the sobering experience above indicates, increased contractility does not necessarily translate into improved survival. Important considerations are related to energetics, diastolic function, and ventricular arrhythmia.

Conventional inotropic therapy for acute heart failure includes catecholamines, phosphodiesterase (PDE III) inhibitors, and digitalis. Digitalis inhibits the Na^+–K^+ pump leading to increased intracellular $[Na^+]$, and, via increased reverse mode NCX exchange, increased $[Ca^{2+}]$. It is a weak inotrope that is mortality neutral.[123] Data suggest that the positive inotropic effect is achieved with high energy transfer efficiency and little oxygen wasting.[124] Digitalis also has sympatholytic activity. Preliminary evidence suggests that ouabain-induced increases in protein phosphatase inhibitor may partly restore impaired catecholamine sensitivity in heart failure.[125]

βAR-mediated activation of $G_{\alpha s}$ directly couples to the L-type Ca^{2+} channel,[126] and cAMP-dependent protein kinase A (PKA) phosphorylates the L-type channel, the RyR2 receptor, phospholamban, type I protein phosphatase inhibitor-1 protein, and troponin-I, resulting in an increase in inotropy and lusitropy, and an increase in metabolic activity and myocardial oxygen consumption. Central proteins in this signaling pathway are concentrated at the t-tubule membrane (G_s, adenylate cyclase, A-kinase anchoring protein). On the other hand, compensatory changes in E–C coupling mediated by adrenergic stimulation have the potential to be deleterious. Thus, hyperphosphorylation of the RyR2 increases the open probability of the channel and generates a diastolic leak of calcium. Indeed, beta-blocker therapy, a staple of therapy for chronic heart failure reverses the functional abnormalities in the RyR2 receptor, increases efficiency of E–C coupling, and may be antiarrhythmic.[127]

PKA-dependent TnI phosphorylation by βAR agonists and PDE inhibitors also decreases myofilament Ca^{2+} sensitivity. Although PDE inhibitors produce beneficial vasodilatation and maintain their effectiveness when βAR are downregulated, calcium fluxes are increased (thus increasing the energetic demands as do β agonists), and are less effective when cAMP pools are reduced (as they are in heart failure).

Myofilament Ca^{2+} sensitizers (Box 12.2) directly increase inotropic state by increasing the Ca^{2+} affinity of TnC and increase the amount of force for a given amount of activator calcium; since TnC is a major intracellular Ca^{2+} buffer, they are often associated with a lower peak $[Ca^{2+}]_i$ transient. Reducing the amount of necessary activator calcium lowers oxygen consumption required for Ca^{2+} transport and may blunt Ca^{2+}-dependent transcriptional and translation mechanisms responsible for hypertrophy and failure.[128] Calcium sensitizers usually have mixed effects; many have PDE III activity and increase cAMP and PKA, have a caffeine-like action, opening up SR Ca^{2+} release channels, or increase I_{Ca}. Levosimendan (which is available in Europe) does not have clinically significant PDE III activity, but does have vasodilator activity and cardioprotective effects

---

Box 12.2. *Factors that influence myofilament calcium sensitivity*

Physical factors
Sarcomere length ↑
Cooling ↓
Acidosis ↓

Chemical factors
TnI phosphorylation ↓
(βAR stimulation)
Ionic strength ↓
$[PO_4]$ ↓
$[Mg^{2+}]$ ↓

Drugs
Caffeine ↑
Sulmazole ↑
Pimobendan ↑
Levosimendan ↑
EMD 57033 ↑
CGP 48506 ↑
SCH00013 ↑
Butanedione monoxime ↓

because it opens ATP-dependent potassium channels.[129,130] A potential limitation of Ca^{2+} sensitizers is that they may impair diastolic relaxation and filling. An advantage of pure sensitizers is that transsarcolemmal calcium fluxes are not altered and therefore cells are not likely to become calcium overloaded, reducing the likelihood of arrhythmias.

Dihydropyridine receptor (DHPR) agonists (e.g. Bay K8644) can increase I_{Ca} and produce inotropic effects by increasing SR Ca^{2+} release, Ca^{2+} available to myofilaments, and SR Ca^{2+} load while allowing sufficient steady-state Ca^{2+} extrusion via the NCX, therefore limiting Ca^{2+} overload. The major limitation of I_{Ca} modulators is that they produce vasoconstriction.[131]

Direct NCX inhibitors should have effects similar to drugs (e.g. digoxin) which increase $[Ca^{2+}]_i$ by elevation of $[Na^+]_i$ and reduce the effectiveness of the NCX in extruding calcium. However, calcium overload is a serious limitation with most of the agents; preferential inhibition of NCX influx versus efflux may limit calcium overload. Despite their potential benefit, selectivity and specificity of the various agents are concerns.[132]

Na^+ channel agonists increase cardiac contractility in a cAMP-independent manner, probably via enhanced Na^+–Ca^{2+} exchange activity. The positive inotropic effects of these agents are limited by a potential increase in the incidence of cardiac arrhythmias.[133]

Although the RyR2 has not been a suitable target, the ability to reduce diastolic Ca^{2+} leak from the SR (and therefore increase SR Ca^{2+} content) and facilitate coupled gating of RyR2 make interventions that increase FKBP potentially useful.[134]

Other targets for therapy include manipulation of SERCA2 and its regulating protein, phospholamban; gene therapy has been particularly successful in this regard (see Table QQ and review in reference 135).

Concluding remarks

This chapter has summarized the sequence of biological processes that defines E–C coupling and serves as a foundation for the in-depth chapters that follow. It should be clear that E–C coupling occupies a fundamental position in normal cardiovascular physiology and in the pathogenesis and treatment of heart disease. E–C coupling can be studied in preparations as simple as the isolated cardiomyocyte to those as complex as the intact organism, but only a synthesis of disparate experimental techniques and preparations will provide the insights that will translate into novel approaches to the management of heart disease.

Acknowlededgment

This work was supported by an American Heart Association Grant-in-Aid 0355198B.

References

1. Prabhu SD, Freeman GL. Effect of tachycardiac heart failure on the restitution of left ventricular function in closed-chest dogs. Circulation 1995; 91: 176–85.

2. Ross J, Jr, Miura T, Kambayashi M et al. Adrenergic control of the force–frequency relation. Circulation 1995; 92: 2327–32.

3. Scognamiglio R, Marin M, Miorelli M et al. Postextrasystolic potentiation echocardiography in predicting reversible myocardial dysfunction by surgical coronary revascularization. Am J Cardiol 1998; 81: 36G–40G.

4. Ferrier GR, Howlett SE. Cardiac excitation–contraction coupling: role of membrane potential in regulation of contraction. Am J Physiol Heart Circ Physiol 2001; 280: H1928–44.

5. Chiamvimonvat N, Lalli JM, Yatani A. Single cell patch-clamp analysis of mouse cardiac myocytes. In: Hoit BD, Walsh RA (eds). Cardiovascular Physiology in the Genetically Engineered Mouse. Norwell, MA: Kluwer Academic Publishers; 2002: 91–112.

6. Brenner B. Mechanical and structural approaches to correlation of cross-bridge action in muscle with actomyosin ATPase in solution. Annu Rev Physiol 1987; 49: 655–72.

7. Wang SQ, Wei C, Zhao G et al. Imaging microdomain Ca^{2+} in muscle cells. Circ Res 2004; 94: 1011–22.

8. Wier WG, Balke CW. Ca^{2+} release mechanisms, Ca^{2+} sparks, and local control of excitation–contraction coupling in normal heart muscle. Circ Res 1999; 85: 770–6.

9. Cannell MB, Cheng H, Lederer WJ. The control of calcium release in heart muscle. Science 1995; 268: 1045–49.

10. Lukyanenko V, Wiesner TF, Gyorke S. Termination of Ca^{2+} release during Ca^{2+} sparks in rat ventricular myocytes. J Physiol 1998; 507: 667–77.

11. Ritter M, Su Z, Spitzer KW et al. Caffeine-induced Ca^{2+} sparks in mouse ventricular myocytes. Am J Physiol Heart Circ Physiol 2000; 278: H666–9.

12. Valdivia HH. Modulation of intracellular Ca^{2+} levels in the heart by sorcin and FKBP12, two accessory proteins of ryanodine receptors. Trends Pharmacol Sci 1998; 19: 479–82.

13. Brady AJ. Mechanical properties of isolated cardiac myocytes. Physiol Rev 1991; 71: 413–28.

14. Palmer RE, Brady AJ, Roos KP. Mechanical measurements from isolated cardiac myocytes using a pipette attachment system. Am J Physiol Cell Physiol 1996; 270: C697–704.

15. Yao A, Su Z, Nonaka A et al. Effects of overexpression of the Na^+–Ca^{2+} exchanger on $[Ca^{2+}]_i$ transients in murine ventricular myocytes. Circ Res 1998; 82: 657–65.

16. Kadambi VJ, Ponniah S, Harrer JM et al. Cardiac-specific overexpression of phospholamban alters calcium kinetics and resultant cardiomyocyte mechanics in transgenic mice. J Clin Invest 1996; 97: 533–9.

17. Bers DM, Berlin JR. Kinetics of [Ca]i decline in cardiac myocytes depend on peak [Ca]i. Am J Physiol Cell Physiol 1995; 268: C271–7.

18. Kao JP, Harootunian AT, Tsien RY. Photochemically generated cytosolic calcium pulses and their detection by fluo-3. J Biol Chem 1989; 264: 8179–84.

19. Field ML, Azzawi A, Styles P et al. Intracellular Ca^{2+} transients in isolated perfused rat heart: measurement using the fluorescent indicator Fura-2/AM. Cell Calcium 1994; 16: 87–100.

20. Pieske B, Schlotthauer K, Schattmann J et al. Ca^{2+}-dependent and Ca^{2+}-independent regulation of contractility in isolated human myocardium. Basic Res Cardiol 1997; 92 (Suppl 1): 75–86.

21. Morgan JP. The effects of digitalis on intracellular calcium transients in mammalian working myocardium as detected with aequorin. J Mol Cell Cardiol 1985; 17: 1065–75.

22. Simpson AW. Fluorescent measurement of $[Ca^{2+}]_c$. Basic practical considerations. Methods Mol Biol 1999; 114: 3–30.

23. Bridge JH, Ershler PR, Cannell MB. Properties of Ca^{2+} sparks evoked by action potentials in mouse ventricular myocytes. J Physiol 1999; 518: 469–78.

24. Su SF, Barry WH. Isolated myocyte mechanics and calcium transients. In: Hoit BD, Walsh RA (eds). Cardiovascular Physiology in the Genetically Engineered Mouse. Norwell, MA: Kluwer Academic Publishers; 2002: 71–89.

25. Santana LF, Kranias EG, Lederer WJ. Calcium sparks and excitation–contraction coupling in phospholamban-deficient mouse ventricular myocytes. J Physiol 1997; 503: 21–9.

26. Bers DM. Ryanodine and the calcium content of cardiac SR assessed by caffeine and rapid cooling contractures. Am J Physiol Cell Physiol 1987; 253: C408–15.

27. Brette F, Orchard C. T-tubule function in mammalian cardiac myocytes. Circ Res 2003; 92: 1182–92.

28. Bers DM. Excitation–contraction Coupling and Cardiac Contractile Force (2e). Norwell, MA: Kluwer Academic Publishers; 2001.

29. Lewartowski B. Excitation–contraction coupling in cardiac muscle revisited. J Physiol Pharmacol 2000; 51: 371–86.

30. Kranias EG. Regulation of Ca^{2+} transport by cyclic 3′,5′-AMP-dependent and calcium-calmodulin-dependent phosphorylation of cardiac sarcoplasmic reticulum. Biochim Biophys Acta 1985; 844: 193–9.

31. Muller FU, Kirchhefer U, Begrow F et al. Junctional sarcoplasmic reticulum transmembrane proteins in the heart. Basic Res Cardiol 2002; 97 (Suppl 1): I52–5.

32. Brady AJ. Length dependence of passive stiffness in single cardiac myocytes. Am J Physiol Heart Circ Physiol 1991; 260: H1062–71.

33. Huxley AF, Simmons RM. Proposed mechanism of force generation in striated muscle. Nature 1971; 233: 533–8.

34. Morano I, Ritter O, Bonz A et al. Myosin light chain-actin interaction regulates cardiac contractility. Circ Res 1995; 76: 720–5.

35. Franks K, Cooke R, Stull JT. Myosin phosphorylation decreases the ATPase activity of cardiac myofibrils. J Mol Cell Cardiol 1984; 16: 597–604.

36. van der Velden J, Papp Z, Boontje NM et al. The effect of myosin light chain 2 dephosphorylation on Ca^{2+}-sensitivity of force is enhanced in failing human hearts. Cardiovasc Res 2003; 57: 505–14.

37. Zot AS, Potter JD. Structural aspects of troponin-tropomyosin regulation of skeletal muscle contraction. Annu Rev Biophys Biophys Chem 1987; 16: 535–59.

38. Solaro RJ, Rarick HM. Troponin and tropomyosin: proteins that switch on and tune in the activity of cardiac myofilaments. Circ Res 1998; 83: 471–80.

39. Swartz DR, Moss RL. Influence of a strong-binding myosin analogue on calcium-sensitive mechanical properties of skinned skeletal muscle fibers. J Biol Chem 1992; 267: 20497–506.

40. Zhou Z, Matlib MA, Bers DM. Cytosolic and mitochondrial Ca^{2+} signals in patch clamped mammalian ventricular myocytes. J Physiol 1998; 507: 379–403.

41. Hansford RG. Relation between mitochondrial calcium transport and control of energy metabolism. Rev Physiol Biochem Pharmacol 1985; 102: 1–72.

42. Hansford RG. Relation between cytosolic free Ca^{2+} concentration and the control of pyruvate dehydrogenase in isolated cardiac myocytes. Biochem J 1987; 241: 145–51.

43. Vercesi A, Reynafarje B, Lehninger AL. Stoichiometry of H^+ ejection and Ca^{2+} uptake coupled to electron transport in rat heart mitochondria. J Biol Chem 1978; 253: 6379–85.

44. Bers DM, Stiffel VM. Ratio of ryanodine to dihydropyridine receptors in cardiac and skeletal muscle and implications for E–C coupling. Am J Physiol Cell Physiol 1993; 264: C1587–93.

45. Gordon AM, Huxley AF, Julian FJ. The variation in isometric tension with sarcomere length in vertebrate muscle fibres. J Physiol 1966; 184: 170–92.

46. Hibberd MG, Jewell BR. Calcium- and length-dependent force production in rat ventricular muscle. J Physiol 1982; 329: 527–40.

47. Zhang R, Zhao J, Mandveno A et al. Cardiac troponin I phosphorylation increases the rate of cardiac muscle relaxation. Circ Res 1995; 76: 1028–35.

48. Bers DM. Cardiac Na/Ca exchange function in rabbit, mouse and man: what's the difference? J Mol Cell Cardiol 2002; 34: 369–73.

49. Pieske B, Maier LS, Bers DM et al. Ca^{2+} handling and sarcoplasmic reticulum Ca^{2+} content in isolated failing and nonfailing human myocardium. Circ Res 1999; 85: 38–46.

50. Maier LS, Bers DM, Pieske B. Differences in Ca^{2+}-handling and sarcoplasmic reticulum Ca^{2+}-content in isolated rat and rabbit myocardium. J Mol Cell Cardiol 2000; 32: 2249–58.

51. Freeman G, Colston J. Evaluation of left ventricular mechanical restitution in closed-chest dogs based on single-beat elastance. Circ Res 1990; 67: 1437–45.

52. Hoit BD, Kadambi VJ, Tramuta DA et al. Influence of sarcoplasmic reticulum calcium loading on mechanical and relaxation restitution. Am J Physiol Heart Circ Physiol 2000; 278: H958–63.

53. Cleemann L, Wang W, Morad M. Two-dimensional confocal images of organization, density, and gating of focal Ca^{2+} release sites in rat cardiac myocytes. Proc Natl Acad Sci USA 1998; 95: 10984–9.

54. Hoit BD, Tramuta DA, Kadambi VJ et al. Influence of transgenic overexpression of phospholamban on postextrasystolic potentiation. J Mol Cell Cardiol 1999; 31: 2007–15.

55. Prabhu SD, Freeman GL. Postextrasystolic mechanical restitution in closed-chest dogs. Circulation 1995; 92: 2652–9.

56. Hardman SM, Noble MI, Seed WA. Postextrasystolic potentiation and its contribution to the beat-to-beat variation of the pulse during atrial fibrillation. Circulation 1992; 86: 1223–32.

57. Schmidt A, Kadambi V, Ball N et al. Cardiac specific overexpression of calsequestrin results in left ventricular hypertrophy, depressed force frequency relation and pulsus alternans in vivo. J Mol Cell Cardiol 2000; 37: 1735–44.

58. Li L, Satoh H, Ginsburg KS et al. The effect of Ca^{2+}-calmodulin-dependent protein kinase II on cardiac excitation–contraction coupling in ferret ventricular myocytes. J Physiol 1997; 501: 17–31.

59. Maier LS, Bers DM, Pieske B. Differences in Ca^{2+}-handling and sarcoplasmic reticulum Ca^{2+}-content in isolated rat and rabbit myocardium. J Mol Cell Cardiol 2000; 32: 2249–58.

60. Kadambi VJ, Ball N, Kranias EG et al. Modulation of force–frequency relation by phospholamban in genetically engineered mice. Am J Physiol Heart Circ Physiol 1999; 276: H2245–50.

61. Khoury SF, Hoit BD, Dave V et al. Effects of thyroid hormone on left ventricular performance and regulation of contractile and Ca^{2+}-cycling proteins in the baboon. Implications for the force–frequency and relaxation–frequency relationships. Circ Res 1996; 79: 727–35.

62. Hasenfuss G, Reinecke H, Studer R et al. Relation between myocardial function and expression of sarcoplasmic reticulum Ca^{2+} ATPase in failing and nonfailing human myocardium. Circ Res 1994; 75: 434–42.

63. Prabhu SD. Ryanodine and the left ventricular force–interval and relaxation–interval relations in closed-chest dogs: insights on calcium handling. Cardiovasc Res 1998; 40: 483–91.

64. DeSantiago J, Maier LS, Bers DM. Frequency-dependent acceleration of relaxation in the heart depends on CaMKII, but not phospholamban. J Mol Cell Cardiol 2002; 34: 975–84.

65. Hoit BD. Relaxation . . . it's not getting any easier. J Mol Cell Cardiol 2002; 34: 1135–9.

66. Massion PB, Feron O, Dessy C et al. Nitric oxide and cardiac function: ten years after, and continuing. Circ Res 2003; 93: 388–98.

67. Hare JM. Nitric oxide and excitation–contraction coupling. J Mol Cell Cardiol 2003; 35: 719–29.

68. Chesnais JM, Fischmeister R, Mery PF. Positive and negative inotropic effects of NO donors in atrial and ventricular fibres of the frog heart. J Physiol 1999; 518: 449–61.

69. Paolocci N, Ekelund UE, Isoda T et al. cGMP-independent inotropic effects of nitric oxide and peroxynitrite donors: potential role for nitrosylation. Am J Physiol Heart Circ Physiol 2000; 279: H1982–8.

70. Hare JM, Lofthouse RA, Juang GJ et al. Contribution of caveolin protein abundance to augmented nitric oxide signaling in conscious dogs with pacing-induced heart failure. Circ Res 2000; 86: 1085–92.

71. Zieman SJ, Gerstenblith G, Lakatta EG et al. Upregulation of the nitric oxide–C GMP pathway in aged myocardium: physiological response to L-arginine. Circ Res 2001; 88: 97–102.

72. Layland J, Li JM, Shah AM. Role of cyclic GMP-dependent protein kinase in the contractile response to exogenous nitric oxide in rat cardiac myocytes. J Physiol 2002; 540: 457–67.

73. Barouch LA, Harrison RW, Skaf MW et al. Nitric oxide regulates the heart by spatial confinement of nitric oxide synthase isoforms. Nature 2002; 416: 337–9.

74. Khan SA, Skaf MW, Harrison RW et al. Nitric oxide regulation of myocardial contractility and calcium cycling: independent impact of neuronal and endothelial nitric oxide synthases. Circ Res 2003; 92: 1322–9.

75. Recchia FA, McConnell PI, Bernstein RD et al. Reduced nitric oxide production and altered myocardial metabolism during the decompensation of pacing-induced heart failure in the conscious dog. Circ Res 1998; 83: 969–79.

76. Atar D, Backx PH, Appel MM et al. Excitation–transcription coupling mediated by zinc influx through voltage-dependent calcium channels. J Biol Chem 1995; 270: 2473–7.

77. Anderson ME. Connections count: excitation–contraction meets excitation–transcription coupling. Circ Res 2000; 86: 717–9.

78. Dzhura I, Wu Y, Colbran RJ et al. Calmodulin kinase determines calcium-dependent facilitation of L-type calcium channels. Nat Cell Biol 2000; 2: 173–7.

79. Maier LS, Bers DM. Calcium, calmodulin, and calcium-calmodulin kinase II: heartbeat to heartbeat and beyond. J Mol Cell Cardiol 2002; 34: 919–39.

80. Zhang T, Johnson EN, Gu Y et al. The cardiac-specific nuclear delta(B) isoform of Ca^{2+}/calmodulin-dependent protein kinase II induces hypertrophy and dilated cardiomyopathy associated with increased protein phosphatase 2A activity. J Biol Chem 2002; 277: 1261–7.

81. Peterson BZ, DeMaria CD, Adelman JP et al. Calmodulin is the Ca^{2+} sensor for Ca^{2+}-dependent inactivation of L-type calcium channels. Neuron 1999; 22: 549–58.

82. Hook SS, Means AR. Ca^{2+}/CaM-dependent kinases: from activation to function. Annu Rev Pharmacol Toxicol 2001; 41: 471–505.

83. De Koninck P, Schulman H. Sensitivity of CaM kinase II to the frequency of Ca^{2+} oscillations. Science 1998; 279: 227–30.

84. Swindells MB, Ikura M. Pre-formation of the semi-open conformation by the apo-calmodulin C-terminal domain and implications binding IQ-motifs. Nat Struct Biol 1996; 3: 501–4.

85. Kiriazis H, Kranias EG. Genetically engineered models with alterations in cardiac membrane calcium-handling proteins. Annu Rev Physiol 2000; 62: 321–51.

86. Luo W, Grupp IL, Harrer J et al. Targeted ablation of the phospholamban gene is associated with markedly enhanced myocardial contractility and loss of beta-agonist stimulation. Circ Res 1994; 75: 401–9.

87. He H, Giordano FJ, Hilal-Dandan R et al. Overexpression of the rat sarcoplasmic reticulum Ca^{2+} ATPase gene in the heart of transgenic mice accelerates calcium transients and cardiac relaxation. J Clin Invest 1997; 100: 380–9.

88. Kadambi VJ, Ponniah S, Harrer JM et al. Cardiac-specific overexpression of phospholamban alters calcium kinetics and resultant cardiomyocyte mechanics in transgenic mice. J Clin Invest 1996; 97: 533–9.

89. Kadambi VJ, Sato Y, Ball N et al. Transgenic overexpression of calsequestrin produces pulsus alternans in vivo. J Mol Cell Cardiol 1998; 30: A47.

90. Carr A, Sato Y, Neirouz Y et al. Gene ablation of protein phosphatase inhibitor-1 is associated with depressed cardiac contractility and decreased phospholamban phosphorylation. Circulation 1999; 100: I763.

91. Carr AN, Schmidt AG, Suzuki Y et al. Type 1 phosphatase, a negative regulator of cardiac function. Mol Cell Biol 2002; 22: 4124–35.

92. Kirchhefer U, Neumann J, Baba HA et al. Cardiac

hypertrophy and impaired relaxation in transgenic mice overexpressing triadin 1. J Biol Chem 2001; 276: 4142–9.

93. Reuter H, Han T, Motter C et al. Mice overexpressing the cardiac sodium-calcium exchanger: defects in excitation–contraction coupling. J Physiol 2004; 554: 779–89.

94. Kuschel M, Karczewski P, Hempel P et al. Ser16 prevails over Thr17 phospholamban phosphorylation in the beta-adrenergic regulation of cardiac relaxation. Am J Physiol Heart Circ Physiol 1999; 276: H1625–33.

95. Takeishi Y, Walsh RA. Cardiac hypertrophy and failure: lessons learned from genetically engineered mice. Acta Physiol Scand 2001; 173: 103–11.

96. Mann DL. Mechanisms and models in heart failure: a combinatorial approach. Circulation 1999; 100: 999–1008.

97. Sjaastad I, Wasserstrom JA, Sejersted OM. Heart failure – a challenge to our current concepts of excitation–contraction coupling. J Physiol 2003; 546: 33–47.

98. Esposito G, Rapacciuolo A, Naga Prasad SV et al. Genetic alterations that inhibit in vivo pressure-overload hypertrophy prevent cardiac dysfunction despite increased wall stress. Circulation 2002; 105: 85–92.

99. Litwin SE, Zhang D, Bridge JH. Dyssynchronous Ca^{2+} sparks in myocytes from infarcted hearts. Circ Res 2000; 87: 1040–7.

100. Pogwizd SM, Qi M, Yuan W et al. Upregulation of $Na^+–Ca^{2+}$ exchanger expression and function in an arrhythmogenic rabbit model of heart failure. Circ Res 1999; 85: 1009–19.

101. McCall E, Li L, Satoh H et al. Effects of FK-506 on contraction and Ca^{2+} transients in rat cardiac myocytes. Circ Res 1996; 79: 1110–21.

102. Marks AR. Ryanodine receptors, FKBP12, and heart failure. Front Biosci 2002; 7: d970–7.

103. Benitah JP, Gomez AM, Fauconnier J et al. Voltage-gated Ca^{2+} currents in the human pathophysiologic heart: a review. Basic Res Cardiol 2002; 97 (Suppl 1): I11–18.

104. He J, Conklin MW, Foell JD et al. Reduction in density of transverse tubules and L-type Ca^{2+} channels in canine tachycardia-induced heart failure. Cardiovasc Res 2001; 49: 298–307.

105. Barrere-Lemaire S, Piot C, Leclercq F et al. Facilitation of L-type calcium currents by diastolic depolarization in cardiac cells: impairment in heart failure. Cardiovasc Res 2000; 47: 336–49.

106. Holt E, Tonnessen T, Lunde PK et al. Mechanisms of cardiomyocyte dysfunction in heart failure following myocardial infarction in rats. J Mol Cell Cardiol 1998; 30: 1581–93.

107. Yang Y, Chen X, Margulies K et al. L-type Ca^{2+} channel alpha 1c subunit isoform switching in failing human ventricular myocardium. J Mol Cell Cardiol 2000; 32: 973–84.

108. Chen X, Piacentino V, 3rd, Furukawa S et al. L-type Ca^{2+} channel density and regulation are altered in failing human ventricular myocytes and recover after support with mechanical assist devices. Circ Res 2002; 91: 517–24.

109. Pogwizd SM, Schlotthauer K, Li L et al. Arrhythmogenesis and contractile dysfunction in heart failure: roles of sodium–calcium exchange, inward rectifier potassium current, and residual beta-adrenergic responsiveness. Circ Res 2001; 88: 1159–67.

110. Maier LS, Braunhalter J, Horn W et al. The role of SR Ca^{2+}-content in blunted inotropic responsiveness of failing human myocardium. J Mol Cell Cardiol 2002; 34: 455–67.

111. Schmidt U, Hajjar RJ, Kim CS et al. Human heart failure: cAMP stimulation of SR Ca^{2+}-ATPase activity and phosphorylation level of phospholamban. Am J Physiol Heart Circ Physiol 1999; 277: H474–80.

112. Sande JB, Sjaastad I, Hoen IB et al. Reduced level of serine(16) phosphorylated phospholamban in the failing rat myocardium: a major contributor to reduced SERCA2 activity. Cardiovasc Res 2002; 53: 382–91.

113. Wier WG, Egan TM, Lopez-Lopez JR et al. Local control of excitation–contraction coupling in rat heart cells. J Physiol 1994; 474: 463–71.

114. Marks AR, Reiken S, Marx SO. Progression of heart failure: is protein kinase a hyperphosphorylation of the ryanodine receptor a contributing factor? Circulation 2002; 105: 272–5.

115. Gomez AM, Guatimosim S, Dilly KW et al. Heart failure after myocardial infarction: altered excitation–contraction coupling. Circulation 2001; 104: 688–93.

116. Sipido KR, Maes M, Van de Werf F. Low efficiency of Ca^{2+} entry through the $Na^+–Ca^{2+}$ exchanger as trigger for Ca^{2+} release from the sarcoplasmic reticulum. A comparison between L-type Ca^{2+} current and reverse-mode $Na^+–Ca^{2+}$ exchange. Circ Res 1997; 81: 1034–44.

117. Barry WH. $Na^+–Ca^{2+}$ exchange in failing myocardium: friend or foe? Circ Res 2000; 87: 529–31.

118. Sjaastad I, Bentzen JG, Semb SO et al. Reduced calcium tolerance in rat cardiomyocytes after myocardial infarction. Acta Physiol Scand 2002; 175: 261–9.

119. Gomez AM, Schwaller B, Porzig H et al. Increased exchange current but normal Ca^{2+} transport via Na^+–Ca^{2+} exchange during cardiac hypertrophy after myocardial infarction. Circ Res 2002; 91: 323–30.

120. Feldman AM, Bristow MR, Parmley WW et al. Effects of vesnarinone on morbidity and mortality in patients with heart failure. Vesnarinone Study Group. N Engl J Med 1993; 329: 149–55.

121. Packer M, Carver JR, Rodeheffer RJ et al. Effect of oral milrinone on mortality in severe chronic heart failure. The PROMISE Study Research Group. N Engl J Med 1991; 325: 1468–75.

122. Endoh M. Mechanisms of action of novel cardiotonic agents. J Cardiovasc Pharmacol 2002; 40: 323–38.

123. Hauptman PJ, Kelly RA. Digitalis. Circulation 1999; 99: 1265–70.

124. Hasenfuss G, Mulieri LA, Allen PD et al. Influence of isoproterenol and ouabain on excitation–contraction coupling, cross-bridge function, and energetics in failing human myocardium. Circulation 1996; 94: 3155–60.

125. El-Armouche A, Jaeckel F, Boheler KR et al. Ouabain treatment is associated with upregulation of phosphatase inhibitor-1 and Na^+–Ca^{2+}-exchanger and beta-adrenergic sensitization in rat hearts. Biochem Biophys Res Commun 2004; 318: 219–26.

126. Yatani A, Brown AM. Rapid beta-adrenergic modulation of cardiac calcium channel currents by a fast G protein pathway. Science 1989; 245: 71–4.

127. Scoote M, Williams AJ. The cardiac ryanodine receptor (calcium release channel): emerging role in heart failure and arrhythmia pathogenesis. Cardiovasc Res 2002; 56: 359–72.

128. Arteaga GM, Kobayashi T, Solaro RJ. Molecular actions of drugs that sensitize cardiac myofilaments to Ca^{2+}. Ann Med 2002; 34: 248–58.

129. Lehmann A, Boldt J, Kirchner J. The role of Ca^{2+}-sensitizers for the treatment of heart failure. Curr Opin Crit Care 2003; 9: 337–44.

130. Slawsky MT, Colucci WS, Gottlieb SS et al. Acute hemodynamic and clinical effects of levosimendan in patients with severe heart failure. Study Investigators. Circulation 2000; 102: 2222–7.

131. Doggrell S, Hoey A, Brown L. Ion channel modulators as potential positive inotropic compound for treatment of heart failure. Clin Exp Pharmacol Physiol 1994; 21: 833–43.

132. Doggrell SA, Hancox JC. Is timing everything? Therapeutic potential of modulators of cardiac $Na(+)$ transporters. Expert Opin Investig Drugs 2003; 12: 1123–42.

133. Flesch M, Erdmann E. Na^+ channel activators as positive inotropic agents for the treatment of chronic heart failure. Cardiovasc Drugs Ther 2001; 15: 379–86.

134. Prestle J, Quinn FR, Smith GL. Ca^{2+}-handling proteins and heart failure: novel molecular targets? Curr Med Chem 2003; 10: 967–81.

135. Dorn GW, 2nd, Molkentin JD. Manipulating cardiac contractility in heart failure: data from mice and men. Circulation 2004; 109: 150–8.

CHAPTER 13

Normal and abnormal calcium homeostasis

Donald M Bers, Evangelia G Kranias

Introduction

Ca^{2+} entry into cardiac myocytes is initiated by the cardiac action potential (AP), where depolarization activates inward Ca^{2+} current (I_{Ca}), which in turn contributes to the AP plateau (Fig. 13.1). This Ca^{2+} entry triggers sarcoplasmic reticulum (SR) Ca^{2+} release. The combination of I_{Ca} and SR Ca^{2+} release raises intracellular free $[Ca^{2+}]$, $([Ca^{2+}]_i)$, allowing Ca^{2+} binding to the myofilament protein troponin C, which activates contraction. For relaxation to occur $[Ca^{2+}]_i$ must decline, allowing Ca^{2+}

to dissociate from troponin. This requires Ca^{2+} transport from the cytosol, and four transport pathways can contribute to this transport process: (1) SR Ca^{2+}-ATPase (SERCA); (2) sarcolemmal Na^+–Ca^{2+} exchange (NCX); (3) sarcolemmal Ca^{2+}-ATPase; and (4) mitochondrial Ca^{2+} uniport. For the myocyte to be in a steady state with respect to Ca^{2+} balance, the amount of Ca^{2+} extruded from the cell during relaxation must be the same as the amount of Ca^{2+} entry at each beat. Likewise, the amount of Ca^{2+} released from the SR must equal that re-accumulated by the action of the SERCA.

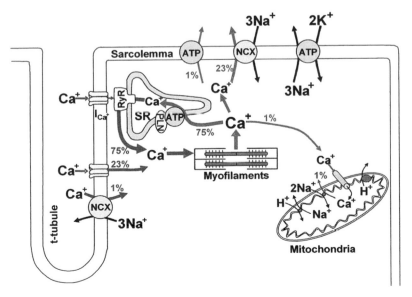

Figure 13.1. Ca^{2+} handling in ventricular myocytes. ATP, ATPase; PLN, phospholamban.

Contraction is graded, depending on the amplitude and kinetics of the $[Ca^{2+}]_i$ transient and other factors. Figure 13.2 shows how much total Ca^{2+} ($[Ca^{2+}]_{Tot} = [Ca^{2+}]_i$ plus bound Ca^{2+}) must be supplied to and removed from the myoplasm during each heart beat. Force development depends non-linearly on $[Ca^{2+}]_i$ and $[Ca^{2+}]_{Tot}$, and is a result of strong myofilament cooperativity with respect to $[Ca^{2+}]_i$.[1–3] Physiological contraction generates both isometric force and shortening (e.g. during ejection of blood). There are two main ways in which the strength of cardiac contraction can be altered: (1) altered Ca^{2+} transient amplitude or duration; and (2) altered myofilament Ca^{2+} sensitivity. Myofilament Ca^{2+} sensitivity is enhanced dynamically by stretching the myofilaments (e.g. during filling), resulting in a stronger contraction. This is a central feature in the classic Frank–Starling response by which the heart adjusts output to altered diastolic filling. Myofilament Ca^{2+} sensitivity is also reduced by β-adrenergic activation and acidosis, but enhanced by certain inotropic drugs. The rapid Ca^{2+} transient kinetics prevent the myofilaments from fully equilibrating with $[Ca^{2+}]_i$ during a normal twitch (especially in the rising phase). Thus, while contraction depends on the Ca^{2+} transient, there is dynamic interplay between Ca^{2+} and myofilaments during excitation–contraction (E–C) coupling. During heart failure (HF), functional expression of different proteins involved in E–C coupling is altered (see below), and these changes contribute to altered Ca^{2+} transients, contractility and arrhythmias in HF.

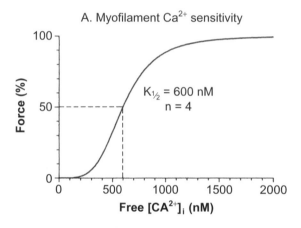

A. Myofilament Ca^{2+} sensitivity

$K_{1/2} = 600$ nM
$n = 4$

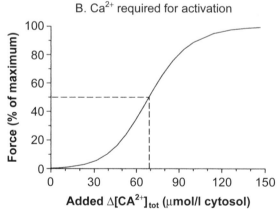

B. Ca^{2+} required for activation

Figure 13.2. *Ca required for contractile activation. (A) Force is shown as a function of $[Ca^{2+}]_i$ (as $100/(1 + (600/[Ca^{2+}]_i)^4)$). (B) Total Ca^{2+} required to activate force, assuming diastolic $[Ca^{2+}]_i = 150$ nM and cytosolic Ca^{2+} buffers which include troponin C (Ca^{2+} and Ca^{2+}/Mg^{2+} sites), myosin, SR Ca^{2+}-ATPase, calmodulin, ATP, creatine phosphate and sarcolemmal sites.[1]*

Myocyte Ca^{2+} cycling

There are both L- and T-type Ca^{2+} channels in heart, but we will focus on L-type I_{Ca}, because it is dominant by far in ventricular myocytes.[1] I_{Ca} is activated by depolarization, but Ca^{2+}-dependent inactivation at the cytosolic side limits the amount of Ca^{2+} entry during the AP. This Ca^{2+}-dependent inactivation occurs very locally and is mediated by calmodulin bound to the carboxyl-terminal of the Ca^{2+} channel.[4,5] L-type Ca^{2+} channels (or DHPRs) are primarily located at junctions between the sarcolemma and SR, where the SR Ca^{2+} release channels (or ryanodine receptors, RyR) exist.[6] During E–C coupling, I_{Ca} activates SR Ca^{2+} release, but this released Ca^{2+} also feeds back on I_{Ca} as a major component of Ca^{2+}-dependent inactivation.[7,8] Indeed, the total Ca^{2+} influx via I_{Ca} is reduced by ~50% when SR Ca^{2+} release occurs.[9] There is only a very short delay between the AP upstroke, I_{Ca} activation and initiation of SR Ca^{2+} release, such that the peak of SR Ca^{2+} release occurs within 5–10 ms of the peak of the AP.[9,10] Thus, SR Ca^{2+} release and I_{Ca} create local negative feedbacks on

Ca^{2+} influx. When there is high Ca^{2+} influx or release, further Ca^{2+} influx is inhibited.

When the SR Ca^{2+} load is elevated, it directly increases the amount of Ca^{2+} available for release, but also greatly enhances the fraction of SR Ca^{2+} that is released for a given I_{Ca} trigger.[11,12] The latter is due to a stimulatory effect of high intra-SR free [Ca^{2+}] ([Ca^{2+}]$_{SR}$) on RyR open probability.[13,14] This increased RyR sensitivity to [Ca^{2+}]$_i$ at high [Ca^{2+}]$_{SR}$ is the basis of aftercontractions, transient inward current and delayed afterdepolarizations that can trigger arrhythmias.[1]

NCX is reversible with a stoichiometry of 3Na$^+$:1Ca^{2+}, so it produces an ionic current (I_{NCX}). NCX can extrude Ca^{2+} (as inward I_{NCX}) or bring Ca^{2+} into the cell (as outward I_{NCX}). High [Ca^{2+}]$_i$ favors Ca^{2+} efflux (inward I_{NCX}), while positive membrane potential (E_m) and high [Na]$_i$ favor more outward I_{NCX} (Fig. 13.3B). The local sub-membrane [Ca^{2+}]$_i$ sensed by NCX molecules ([Ca^{2+}]$_{sm}$) differs from the bulk cytosolic [Ca^{2+}]$_i$ sensed by the myofilaments.[15,16] This is because as the very high local [Ca^{2+}]$_i$ near RyRs in the junctional cleft diffuses toward the myofilaments, it elevates [Ca^{2+}]$_{sm}$ to an intermediate level. Thus [Ca^{2+}]$_{sm}$ during E–C coupling reaches a higher peak value, which occurs earlier than the global [Ca^{2+}]$_i$ (Fig. 13.3A). This also causes Ca^{2+} extrusion via NCX relatively early during the AP (Fig. 13.3C). Overall, the amount of Ca^{2+} extruded via inward I_{NCX} during the cardiac cycle is essentially the same as that which enters via I_{Ca}, since these are the dominant Ca^{2+} efflux and influx pathways in myocytes. While NCX normally works mainly in the Ca^{2+} efflux mode, the amount of Ca^{2+} influx via I_{NCX} can be increased greatly when [Na]$_i$ is elevated, if SR Ca^{2+} release is reduced, or AP duration is prolonged (and all three of these effects are seen in HF; see below).[16,17] In that case, the Ca^{2+} extrusion via NCX (and the sarcolemmal Ca^{2+}-ATPase) must be the same as the total Ca^{2+} influx via I_{Ca} and NCX.

Regulation Of SR Ca^{2+} cycling

Ca^{2+} transport into the SR lumen is mediated by the SERCA, which is under reversible regulation by phospholamban (PLN). Dephosphorylated PLN binds to SERCA2a and inhibits Ca^{2+} pump activity at low [Ca^{2+}]$_i$, resulting in inhibition of SERCA2a apparent Ca^{2+} affinity, without an effect on the maximal velocity (V_{max}). Phosphorylation of PLN alters the PLN–SERCA2a interaction, relieving the Ca^{2+}-ATPase inhibition and enhancing relaxation rates (lusitropic effects) and contractility (inotropic effects).[18,19] PLN can be phosphorylated by cAMP-dependent, Ca^{2+}-calmodulin-dependent protein kinases and protein kinase C (PKC) at distinct sites. *In vivo*, PLN is phosphorylated by both cAMP-dependent and Ca^{2+}-CaM-dependent protein kinases (PKA and CaMK) during β-adrenergic stimulation.[20–25] Relief of PLN inhibitory effects on SERCA2a is the major contributor to the positive inotropic and lusitropic effects of β-adrenergic receptor (β-AR) agonists.[21–25] Dephosphorylation of PLN occurs by a SR-associated type 1 phosphatase, which is regulated by an endogenous inhibitor-1 protein.[26]

RyRs are both SR Ca^{2+} release channels and scaffolding proteins, which localize key regulatory proteins to the junctional complex.[27] These include calmodulin (which can modulate RyR function),[28] FK-506-binding protein (FKBP 12 6; which may stabilize RyR gating and also couple the gating of both individual and adjacent RyR tetramers),[29] PKA and CaMKII (which can alter RyR and I_{Ca} gating),[29,30] phosphatases (PP1 and PP2a),[31] and sorcin (which binds to RyR and DHPR).[32] RyRs are also coupled to other proteins at the luminal SR surface (triadin, junctin and calsequestrin).[33] These proteins participate in both intra-SR Ca^{2+} buffering and modulation of the Ca^{2+}-release process. RyRs are arranged in large organized arrays at the junctions between the SR and sarcolemma under DHPRs (on both the cell surface and in transverse tubules).[34] This constitutes a large functional Ca^{2+} release complex at the junction. This local functional unit concept is supported by observations of Ca^{2+} sparks or spontaneous local Ca^{2+} transients.[35–38] Ca^{2+} sparks are the fundamental units of SR Ca^{2+} release both at rest (where rare, stochastic events occur) and also during E–C coupling (where they are activated in a synchronized manner throughout the cell by I_{Ca}). Ca^{2+} entry via NCX has also been proposed to

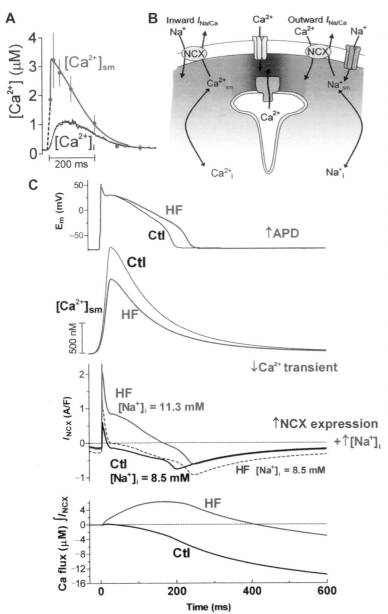

Figure 13.3. Na^+/Ca^{2+} exchanger senses local submembrane $[Ca^{2+}]_i$ ($[Ca^{2+}]_{sm}$). (A) Measured $[Ca^{2+}]_{sm}$ sensed by the NCX (versus $[Ca^{2+}]_i$) during an AP in a rabbit ventricular myocyte. Based on reference 15. (B) Schematic drawing showing that $[Ca^{2+}]_{sm}$ during E–C coupling is lower than $[Ca^{2+}]$ in the junctional cleft (Ca^{2+}_{cleft}), but higher than bulk $[Ca^{2+}]_i$. (C) Simulations to show how I_{NCX} changes during the AP in the rabbit ventricular myocyte, and how outward I_{NCX} is more favored during HF. Graphs indicate longer AP duration (APD), lower Ca^{2+} transients, greater NCX expression ($\pm$ increased $[Na^+]_i$) as measured in HF. Bottom panel shows integrated Ca^{2+} flux via NCX during the AP. Based on Fig. 6 from Despa S, Islam MA, Weber CR et al. Intracellular Na^+ concentration is elevated in heart failure but Na/K pump function is unchanged. Circulation 2002; 105: 2543–8. Ctl, control.

trigger SR Ca^{2+} release in two different ways.[39–41] First, Na^+ current could raise local $[Na^+]_{sm}$ (see Fig. 13.3B), causing Ca^{2+} entry via I_{NCX} to trigger SR Ca^{2+} release. Second, depolarization itself favors Ca^{2+} entry via NCX, which could also affect SR Ca^{2+} release. However, a given Ca^{2+} influx via I_{NCX} is much less effective and slower than I_{Ca} in triggering SR Ca^{2+} release.[47] This is consistent with results that suggest that NCX molecules are rela-

tively excluded from the junctional cleft where RyRs and DHPRs are concentrated.[6]

The mechanism responsible for the termination of SR Ca^{2+} release during Ca^{2+} sparks and E–C coupling is more controversial. Ca^{2+}-induced Ca^{2+}-release (CICR) is inherently a positive feedback mechanism, but it is clear that SR Ca^{2+} release does not go to completion during a twitch.[11,12] The most likely mechanisms include some type of RyR inactivation (or adaptation) and a partial decline in $[Ca^{2+}]_{SR}$, which reduces RyR sensitivity to cytosolic Ca^{2+}. In line with this, the RyR appears to be refractory after a twitch, but increasing $[Ca^{2+}]_{SR}$ can reduce the apparent refractory period.[43]

Function of phospholamban in cardiac control

The physiological role of PLN in the regulation of cardiac function has been elucidated through the generation of genetically altered mouse models. The PLN gene was targeted in embryonic stem cells and mice heterozygous (40% of PLN) and homozygous for PLN deficiency (no PLN) were generated.[44,45] The decreases in PLN levels were associated with a linear increase in the affinity of SERCA2a for Ca^{2+},[45] and with a linear increase in contractile parameters of isolated cardiomyocytes (Fig. 13.4), perfused hearts and intact mice.[45–48] The highly enhanced basal contractile parameters in PLN-deficient hearts could be stimulated only minimally by β-AR agonists.[44,49] The attenuated responses were not associated with alterations in the β-AR agonist signal transduction pathway, such as β-AR density, adenylyl cyclase activity, cAMP levels or the phosphorylation levels of troponin I, C-protein or the 15-kDa sarcolemmal protein.[49] Furthermore, the hyperdynamic cardiac function of PLN-deficient mice was not associated with any alterations in the

levels of SERCA2a, calsequestrin, NCX, myosin, actin, troponin I or troponin T.[50] However, the protein levels of the RyR were decreased by 25%, as a compensatory mechanism to the increased SR Ca^{2+} load by the highly stimulated SERCA2 activity.[50,51] Another compensatory mechanism was the accelerated inactivation kinetics of I_{Ca}, due to increased SR Ca^{2+} load and release.[52] In cardiac energetics, the levels of ATP, creatine kinase activity or creatine kinase reaction velocity were not altered. However, the ADP and AMP levels, as well as the active fraction of mitochondrial pyruvate dehydrogenase were increased.[50] These metabolic adaptations resulted in a new energetic steady state to meet the increased ATP demands in the hyperdynamic PLN-null hearts. The hyperdynamic cardiac function of the PLN-null hearts persisted throughout the aging process, without any alterations in heart-to-body mass ratio, cardiac cell length, or sarcomere length.[53] PLN ablation did not: (1) shorten life span;[53] or (2) compromise the exercise performance,[54] or the ability of the heart to compensate against sustained aortic stenosis.[55] However, PLN ablation increased cardiac susceptibility to ischemic injury,[56] and this was completely restored upon increases in the diminished translocation of PKCε in these hearts.[57]

In a reciprocal set of studies, cardiac overexpression (2-fold) of PLN was associated with significant attenuation of cardiomyocyte contractile parameters and Ca^{2+} kinetics.[58,59] Interestingly, the contractile parameters in the knockout (KO), wild-

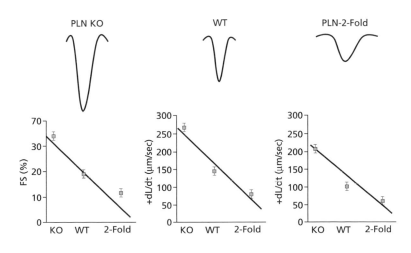

Figure 13.4. *The effects of PLN on cardiac contractility. There is a close linear correlation between the levels of PLN (in PLN-knockouts (KO), wild-types (WT) and 2-fold overexpressors (2-fold)) and fractional shortening (FS), rate of contraction (+dL/dt) and rate of relaxation (–dL/dt) in isolated ventricular myocytes.*

type (WT) and 2-fold PLN overexpression hearts closely correlated with the PLN levels in these hearts (Fig. 13.4), indicating that all the overexpressed PLN was interacting and inhibiting SERCA2. The inhibitory effects of PLN overexpression could be reversed by β-AR agonist stimulation, which resulted in phosphorylation of PLN. Mice with 2-fold PLN overexpression did not exhibit any phenotypic alterations throughout the aging process.[59] However, when the 2-fold overexpressing lines were crossed to generate mice with 4-fold PLN overexpression, the levels of epinephrine and norepinephrine were elevated, as an important compensatory mechanism to phosphorylate PLN and relieve its inhibitory effects.[59] This initial adaptive response became maladaptive in the long-term, since epinephrine induced changes in transcription leading to cardiac remodeling, heart failure and early mortality.

The elevated catecholamine levels to alleviate the PLN inhibitory effects did not allow determination of the functional stoichiometry between PLN and SERCA2 *in vivo*. This was achieved by overexpressing S16A or T17A mutant forms of PLN, that could not be phosphorylated. Maximal inhibition of the apparent affinity of SERCA2a for Ca^{2+} was achieved at a functional stoichiometry of 2.6 PLN:1.0 SERCA2a.[60] Thus, approximately 60% of the SR Ca^{2+} pumps are not inhibited by PLN under basal conditions in mouse hearts.[60]

Collectively, these findings indicate that PLN is a critical regulator of basal cardiac Ca^{2+} cycling and contractile parameters, and PLN is also a key determinant of β-AR agonist responses. Furthermore, only a fraction of the SERCA2 molecules are functionally regulated by PLN *in vivo*. Decreases in PLN levels may result in increased cardiac contractility while increases in PLN levels or activity diminish function (Fig. 13.5).

Further studies, utilizing cardiac overexpression of superinhibitory PLN mutants (N27A, L37A, I40A and V49G) indicated that the increased inhibition of the affinity of SERCA2a for Ca^{2+} resulted in highly depressed cardiomyocyte mechanics and Ca^{2+} kinetics, which were associated with cardiac remodeling.[61–64]

Regulation by β-adrenergic receptor signaling

Sympathetic stimulation of the heart via β-AR increases contractility (inotropy) and relaxation rate (lusitropy), and $[Ca^{2+}]_i$ decline. β-AR stimulation activates a GTP-binding protein (G_S) that stimulates adenylyl cyclase to produce cAMP, which in turn activates PKA. PKA phosphorylates several proteins related to E–C coupling, including PLN, L-type Ca^{2+} channels, RyR, troponin I and myosin binding protein C (Fig. 13.6).

The lusitropic effect of PKA is mediated mainly by phosphorylation of PLN and troponin I, which

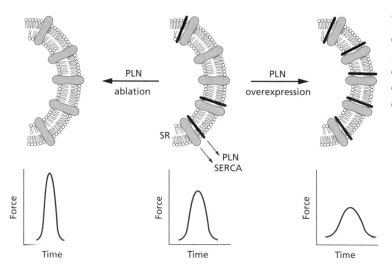

Figure 13.5. *PLN levels and cardiac contractility. Ablation of PLN is associated with enhanced cardiac function while overexpression of PLN results in depressed contractility.*

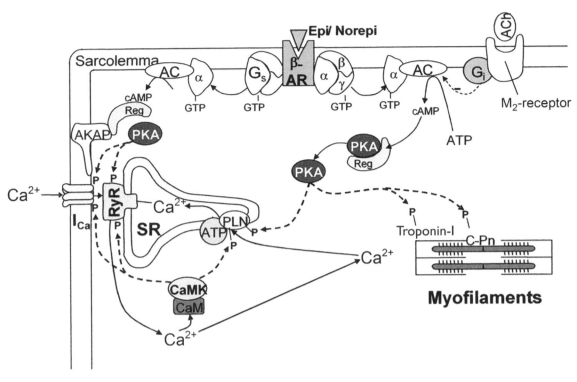

Figure 13.6. *Schematic representation of the effects of β-AR and CaMK activation in cardiac myocytes. AC, adenylyl cyclase; ACh, acetyl choline; AKAP, A kinase-anchoring protein; AR, adrenergic receptor; Epi, epinephrine; Norepi, norepinephrine; P, phosphate; Pn, protein; Reg, regulatory subunit.*

speed up SR Ca^{2+} reuptake and dissociation of Ca^{2+} from the myofilaments, respectively. However, PLN phosphorylation is by far the dominant mechanism for both the lusitropic effect and acceleration of $[Ca^{2+}]_i$ decline.[65] The faster SR Ca^{2+} uptake also contributes to increasing SR Ca^{2+} content. The inotropic effect of PKA activation is mediated by the combination of increased I_{Ca} and greater SR Ca^{2+} availability. This synergistic combination greatly enhances Ca^{2+} transient amplitude, more than offsetting the reduction of myofilament Ca^{2+} sensitivity (caused by troponin I phosphorylation). The myofilament effects of PKA appear to be completely attributable to troponin I phosphorylation (versus myosin binding protein C) because substitution of troponin I with a non-phosphorylatable troponin I abolishes myofilament effects of PKA.[66,67]

Further elucidation of the PLN regulatory effects and the role of dual site PLN phosphoryla-

tion in the heart's responses to β-AR agonists was provided by the generation and characterization of transgenic mice expressing phosphorylation site-specific PLN mutants in the null background. The S16A mutant hearts exhibited a depressed response to isoproterenol and lack of Thr17 phosphorylation. The T17A hearts exhibited Ser16 phosphorylation and a response to isoproterenol similar to that observed with wild-types.[68,69] These findings suggested that during β-AR activation Ser16 phosphorylation may be a prerequisite for Thr17 phosphorylation, and that Ser16 phosphorylation may be sufficient to mediate the β-AR-mediated cardiac inotropy.

CaMKII-dependent Thr17 phosphorylation has been observed to occur independently of Ser16 phosphorylation in hearts perfused with elevated Ca^{2+} (in the presence of okadaic acid to inhibit phosphatase activity),[25] in hearts recovering from an ischemic insult,[70] and in isolated cardiomyocytes subjected to frequency-dependent increases of contractile parameters.[71] So, phosphorylation of Thr17 in PLN may be important during frequency-dependent inotropy and lusitropy or when the heart is stressed, but Ser16 phosphorylation may mediate the β-AR agonist responses *in vivo*.

PKA can also modulate RyR channel open probability. In isolated single-channel recordings, PKA increased initial RyR opening during an abrupt $[Ca^{2+}]_i$ rise, but decreased the steady state open probability at a given $[Ca^{2+}]_i$.[72] In contrast, Marx et al.[29] found that PKA enhanced steady state open probability of single RyRs in bilayers, and attributed this to PKA-dependent phosphorylation causing displacement of FKBP-12.6 from the RyR. These effects remain controversial because not all investigators have observed FKBP dissociation from RyR upon phosphorylation by PKA,[73,74] and Li et al. found no effect of PKA-dependent RyR phosphorylation on Ca^{2+} spark frequency in intact or permeabilized cells (where SR Ca^{2+} load was not increased).[75] During E–C coupling results are somewhat mixed, perhaps because I_{Ca}, SERCA and SR Ca^{2+} content are enhanced by PKA. In a systematic E–C coupling study where I_{Ca} and SR Ca^{2+} content were controlled, PKA was found to have no effect on the amount of SR Ca^{2+} released, but it did increase the initial and maximal rate of Ca^{2+} release.[76]

CaMKII can also affect RyR function, and the site of phosphorylation may be different from that phosphorylated by PKA and does not result in FKBP dissociation from RyR.[77] Not all results agree, but most data suggest that CaMKII exerts an activating effect on RyR open probability in bilayers, on Ca^{2+} sparks in resting cells and during E–C coupling, where fractional release is increased.[30,77–79] CaMKII also produces a modest activating effect on I_{Ca} (facilitation),[80,81] although this is small compared to the powerful activation of I_{Ca} by PKA. Thus CaMKII can modulate I_{Ca}, SERCA and RyR, some of the same key PKA targets involved in E–C coupling. However, the molecular sites and their detailed effects differ.

Local signaling is also important in the β-AR cascade. The cardiac RyR serves as both a PKA target and a scaffolding protein (where PKA and phosphatases 1 and 2A are all bound to the RyR via anchoring proteins).[27,31] The close physical proximity may be functionally critical.[82] β₁-AR activation in ventricular myocytes produces robust inotropic and lusitropic effects, paralleled by phosphorylation of Ca^{2+} channels, phospholamban and troponin I. However, β₂-AR activation can be more restricted to I_{Ca} enhancement,[83] and β₂-ARs are largely localized in sarcolemmal invaginations called caveolae (compared with β₁-ARs which are largely non-caveolar).[84]

Heart failure (HF)

There are many different HF model studies and some disagree about whether particular factors (e.g. I_{Ca}, SERCA, NCX, myofilament Ca^{2+} sensitivity) are increased, decreased or unchanged.[85–92] Table 13.1 provides a summary of some results in human HF, which are also largely consistent with animal HF models.

Reduced twitch force and myocyte contraction is observed in almost all HF models, although at very low heart rates there is typically less difference.[93] In general, there is a less positive (or even a negative) force–frequency relationship in failing versus non-failing heart. Thus, the lower force in HF becomes increasingly apparent at physiological heart rates. These force changes are usually paralleled by changes in Ca^{2+} transient amplitudes. In addition, there is a slowing of the rate of relaxation and $[Ca^{2+}]_i$ decline in HF in most cases. This is related to changes in Ca^{2+} transport. Another ubiquitous finding in HF is a prolongation of AP duration.[89,94–96] This is particularly prominent at very low heart rates, but as frequency increases, the AP duration shortens and there is a smaller difference between failing and non-failing hearts.[93,97] The longer AP duration at low frequency may enhance Ca^{2+} entry and limit Ca^{2+} efflux, thereby partially compensating for the depression seen at physiological heart rate.

The myofilaments

Myofilament Ca^{2+} sensitivity has usually been reported to be unaltered in HF.[98–103] In two rat HF studies, there was a reduction in maximal force and Ca^{2+} sensitivity.[104,105] It should be noted that rats

Table 13.1. *Alterations of expression and function in human heart failure*

	Change in HF[a]	Prevalence	Data type[b] and reference (counter-reference)
Twitch contraction	Lower/slower	~All	F[93,181–183]
Twitch $\Delta[\mathrm{Ca}^{2+}]_i$	Lower/slower	~All	F[93,94,142,181,184]
Force–frequency	From + to more −	~All	F[93,142,182,183,185–188]
Myofibril Ca^{2+} sensitivity	Unaltered (higher)	~All	F[98,100,103,189,190(102,191)]
I_{Ca}/DHPR	Unaltered (lower)	Most	F[94,106,112,192,193(194,195)]; R[196(197,198)]
SR Ca^{2+} ATPase	Lower (unaltered)	Most	F[93,142,181,188,199]; R[197,200–205]; P[206–209(201,204,210)]
SR Ca^{2+} content	Lower	~All	F[133,190,199]
Phospholamban	Lower (unaltered)	Mixed	R[201,203,204,211]; P[207(201,204,210)]
Calsequestrin	Unaltered	~All	R[197,203]; P[207,210]
Calreticulin	Unaltered	~All	P[207]
RyR	Unaltered (lower)	Mixed	R[203(124,212)]; P[125,126,207]; F[132,213(100,214)]
Na^+/Ca^{2+} exchange	Higher (unaltered)	~All	F[137,199,215(16,142)]; R[136,208]; P[136,137,208,215(145)]
Na^+/K^+-ATPase	Lower (unaltered)	~All	P[145,216–218]
AP duration	Higher	~All	F[94,95,219]
I_{to}	Lower	~All	F[95,220,221]
I_{K1}	Lower	~All	F[95,219]
G_i	Higher	~All	F[222]; R[223]; P[222,224–227]
G_s	Unaltered (higher)	~All	R[223(228)]

[a] Where results differ, we have tried to provide balanced assessment. Of course, there may also be true differences in etiology or direct cause of dysfunction.
[b] Supporting data are from F (functional tests in cells, hearts or transport assays), R (mRNA measurements) or P (protein measurements, Western blot or ligand binding. Extended from Table 26 in: Bers DM. Excitation–Contraction Coupling and Cardiac Contractile Force (2e). Dordrecht: Kluwer Academic Press; 2001.

(and mice) shift from fast to slow myosin heavy chain (α to β-MHC) during hypertrophy and HF, whereas larger adult mammals (rabbit, dog and human) are predominantly β-MHC to begin with and do not change much. There are also reports of changes in other myofilament proteins that could alter force development (TnT, TnI and myosin light chains (MLC), see deTombe for review).[91] However, while myofilament changes may occur in HF, it is probably not the most central factor in explaining reduced function.

L-type Ca²⁺-channel

Peak I_{Ca} density has most often been found to be unaltered in HF, but decreases have also been reported (see Table 13.1 and Mukherjee and Spinale).[87] There was no HF-associated change in I_{Ca} at any voltage in human,[106] rat,[107] canine,[96,108] rabbit,[109] or guinea pig ventricular myocytes,[110]

despite strong depression of both contractions and Ca^{2+} transients. This demonstrates that reduced Ca^{2+} transients occur in many HF models with unchanged I_{Ca}. In some cases there appears to be some reduction in the number of Ca^{2+} channels, but an increase in the activation state so that at physiological E_{m} the overall I_{Ca} may be little altered.[111,112] Thus, while I_{Ca} may be reduced in some HF models, it is not primarily responsible for the reduced Ca^{2+} transient.

SERCA2a and PLN

Many studies have measured cardiac SERCA expression and function, since de la Bastie et al. first reported downregulation of SERCA2 expression in pressure-overloaded rat heart.[113] Feldman et al. showed data suggesting that downregulation of SERCA2a expression may mark the transition from hypertrophy to HF.[114] It seems clear that

SERCA is functionally decreased in almost all HF models (despite a few reports to the contrary; Table 13.1). PLN is not altered in HF, indicating decreased Ca^{2+} affinity of SR Ca^{2+} transport.[115] There are also data to suggest that the phosphorylation state of PLN may be reduced in HF.[115–117] This would further reduce the $[Ca^{2+}]$ sensitivity of SR Ca^{2+} uptake and further slow Ca^{2+} transport at physiological $[Ca^{2+}]_i$ (Fig. 13.7). An example where this SERCA2 : PLN ratio changes dramatically is in response to thyroid hormone state, which may decline in HF.[118–120] Increases in thyroid hormone increase SERCA2 expression and decrease PLN expression. This greatly increases the SERCA2 : PLN ratio, and stimulates SR Ca^{2+} transport. The hypothyroid state results in the converse and the greatly reduced SERCA : PLN ratio profoundly depresses SR Ca^{2+} transport.

Reduced SERCA function fits well with the characteristic slowed relaxation and $[Ca^{2+}]_i$ decline of HF. Moreover, when SERCA2 expression in myocytes or failing hearts is increased, or PLN expression is decreased, by adenoviral gene transfer, relaxation and $[Ca^{2+}]_i$ decline can be accelerated.[121,122] Thus, it seems clear that reduced SR Ca^{2+}-transport function is important in the slowed relaxation and $[Ca^{2+}]_i$ decline characteristic of HF, and correction of this depressed SR Ca^{2+} uptake may hold promise as a therapeutic approach in heart failure. However, recent genetic studies indicated that while PLN ablation is beneficial in mouse hearts, PLN null human subjects develop dilated cardiomyopathy and early death.[123] Therefore, therapies aimed at correcting SR Ca^{2+} uptake may not be suitable for all forms of heart failure.

Ryanodine receptor

The SR Ca^{2+}-release channel (*RyR*) mRNA seems to be reduced in human HF, but Western blots and ryanodine binding indicate that RyR protein levels are unchanged.[124–126] In the pacing-induced dog HF model, there seems to be downregulation of RyR,[127,128] but not in the spontaneous hypertensive HF rat.[107] Thus, results are somewhat mixed for RyR number, but RyR regulation may also be altered in HF. Marx et al. showed that in HF, the RyR2 can be hyperphosphorylated by PKA, causing displacement of FKBP12.6 from the RyR.[29] The hyperphosphorylation may be due to less phosphatase associated with the RyR complex, despite a generalized increase in SR phosphatase expression in HF.[129] Without FKBP12.6, the RyR open probability is higher at rest, but shows less coordinated gating. This could increase diastolic SR Ca^{2+} leak, as seen in intact cells treated with FK-506, which also displaces FKBP12.6 from the RyR.[130] Yano et al. also found reduced FKBP/RyR stoichiometry in HF, and greater Ca^{2+} leak from SR vesicles.[128] Recent data in intact rabbit HF myocytes support the notion that there is an increase in resting leak of Ca^{2+} from the SR in HF at a given SR Ca^{2+} content.[131] This leak may contribute to lower SR Ca^{2+} content in HF, and might also alter how the RyR responds to I_{Ca} during E–C coupling.

On the other hand, several recent studies have reported different findings regarding RyR phosphorylation and channel stability in HF. Jiang et al. found no difference in the phosphorylation level or activity of the RyR channel between failing and normal myocytes.[132] Li et al. reported that PKA

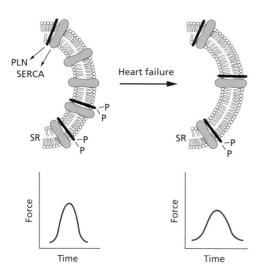

Figure 13.7. *The role of SR function in heart failure. The levels of SERCA and phosphorylated PLN decrease in failing hearts resulting in a decrease in SR Ca^{2+} transport and cardiac contractility. P, phosphate.*

phosphorylation of the RyR has no effect on Ca^{2+} sparks,[75] and two groups found that phosphorylation of the RyR by PKA did not alter FKBP12.6 binding.[73,74] In fact, Xiao et al. reported that FKBP12.6 binds a region of RyR distant from Ser2808.[74] Thus, increased RyR Ca^{2+} leak due to hyperphosphorylation is an attractive hypothesis to explain lower SR Ca^{2+} content and elevated diastolic $[Ca^{2+}]_i$ in HF, but further investigation is required.

Intra-SR Ca^{2+}-buffering capacity is probably unchanged, since calsequestrin (and calreticulin) does not seem to be altered in HF.[85,115] This means that if SR Ca^{2+} content is lower in HF, the free $[Ca^{2+}]_{SR}$ is also lower. There are few measures of SR Ca^{2+} under relatively physiological conditions. Nevertheless, in HF the SR Ca^{2+} content seems to be reduced in the human,[133,134] rabbit,[109,135] and

dog,[136] based on caffeine-induced Ca^{2+} transients. Reduced SR Ca^{2+} content would explain the reduced twitch Ca^{2+} peak and contractile function.

Na^+–Ca^{2+} exchange (NCX)

An increase in NCX in HF appears to be a consistent finding in human[137,138] and most rabbit, guinea pig and dog HF models,[109,110,136,139] as well as in rabbit myocytes from the peri-infarct zone during post-infarct HF.[140] To the extent that NCX is primarily engaged in Ca^{2+} extrusion, higher NCX will be expected to compete better with the SERCA during relaxation (and diastole). This would tend to reduce SR Ca^{2+} content (as above; Fig. 13.8).

The effects of NCX are complicated by its bidirectional nature and its dependence on gradients for $[Na^+]$ and $[Ca^{2+}]$. Indeed, with low Ca^{2+}

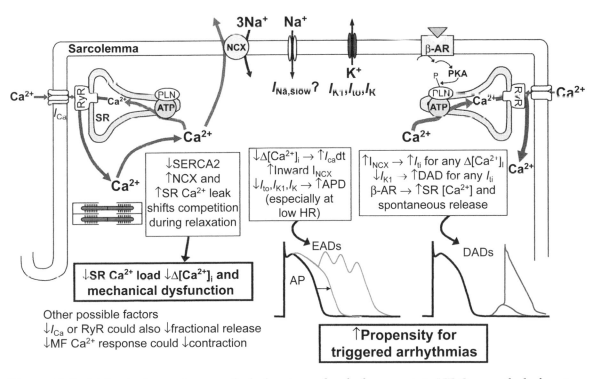

Figure 13.8. *Molecular bases of contractile dysfunction and arrhythmogenesis in HF. Contractile dysfunction probably results from a reduction in SR Ca^{2+} content. Arrhythmogenesis may be triggered by either early or delayed afterdepolarizations (EADs or DADs). Factors which may contribute to EADs (more likely at longer AP duration) are shown. DADs are due to spontaneous SR Ca^{2+} release, and three key factors are indicated which increase propensity for DADs in HF. AR, adrenergic receptor; HR, heart rate; P, phosphate.*

transient amplitudes in HF, greater Ca^{2+} influx via NCX could occur (Fig. 13.3C). This is exacerbated by an increase in $[Na^+]_i$ that is found in HF myocytes.[17,141,142] Also with low $[Ca^{2+}]_i$ and long AP duration in HF, there can be an extended period of Ca^{2+} influx via NCX during the AP,[16,134] which would not be expected for large Ca^{2+} transients and short AP duration. It seems that this Ca^{2+} influx enhancement via NCX would be most likely at low frequency in HF, where AP duration is especially prolonged. This may also explain why the Ca^{2+} transients and contractile force are less depressed compared to control at low heart rates. Thus, it seems likely that NCX upregulation is an important factor in altered Ca^{2+} handling in HF.

Na^+–K^+-ATPASE and $[Na^+]_i$

The way that NCX functions is critically dependent on the level of $[Na^+]_i$, which in turn depends on Na^+–K^+-ATPase. There are several reports which indicate reduced Na^+–K^+-ATPase expression in HF.[143–145] This would be expected to elevate $[Na^+]_i$ and be inotropic (as above). However, in a rabbit HF model two groups found that $[Na^+]_i$ was elevated, but cellular Na^+–K^+-ATPase function was unaltered.[17,146] These groups attributed the rise in $[Na^+]_i$ to either a tetrodotoxin-sensitive pathway or altered Na^+–H^+ exchange. There is also evidence to suggest a very slowly inactivating TTX-sensitive Na^+ current component ($I_{Na,Slow}$) in HF.[147–149] This could also contribute to AP prolongation in HF.

Potassium currents

Other ventricular ion currents may also be altered in HF. The most consistent findings so far are reductions in transient outward and inward rectifier K^+ currents (I_{to} and I_{K1}) in HF (Table 13.1).[88,89] Reduction in I_{to} can reduce the early repolarization during the AP (phase 1). This AP notch may normally serve to enhance the driving force for Ca^{2+} entry once Ca^{2+} channels are activated. Hence, reduced I_{to} could decrease early Ca^{2+} influx and triggering of SR Ca^{2+} release in HF.[150] However, reducing I_{to} may have little effect on overall AP duration in human, dog, rabbit or guinea pig ventricle.[151] Exceptions to this are rat and mouse ventricle, where the very large I_{to} is a predominant

cause of repolarization of the very short AP that is observed in those species. The decrease in I_{K1} may contribute to AP prolongation, but this would be mainly in the very late phases of final repolarization, because of its inward rectification. An even more important aspect of the 40–50% I_{K1} reduction in human, dog and rabbit HF,[95,96,135] is that it destabilizes the diastolic E_m. This may increase the propensity for arrhythmogenesis (Fig. 13.8).

Energetics and heart failure

In hypertrophy and HF there can also be energetic limitations. This may not be apparent under low work conditions, but may manifest as a limited cardiac reserve, when cells, muscles or hearts are challenged with higher work loads.[89,152–154] This may also contribute somewhat to the blunted force–frequency relationship seen in human HF.[93]

Cardiac muscle is a highly oxidative tissue that produces more than 90% of its energy from mitochondrial respiration. The heart uses more than 90% of its oxidative capacity during maximal exercise, indicating that there is no additional capacity of energy production over energy utilization.[155] Oxygen availability, substrate limitation, ATP, ADP, and PCr changes, inorganic phosphate, calcium, redox state, and phosphotransfer systems have all been considered to play a role. One of the candidates for coupling aerobic metabolism and cardiac work is calcium, as it regulates myosin and SR ATPase, the major mitochondrial dehydrogenases, and F0/F1-ATPase.[156] The presence of high-energy phosphotransfer systems, including creatine kinase (CK) and adenylate kinase (AK) is another essential feature of cardiac or striated muscle energy metabolism.[157] Cell architecture and metabolic networks are interrelated to build integrated phosphotransfer systems that improve cellular economy to tightly match cellular functions, and alterations in this fine regulation can compromise cardiac function.[158]

In heart failure, the depression of contractile force is not matched by a concomitant depression of energy consumption, leading to uncoupling of energy consumption.[159,160] Decrease in coronary reserve is one of the important abnormalities, impairing high-energy phosphate synthesis in the failing heart, which may limit nutrient and oxygen delivery to the cardiomyocytes at high workloads. The heart works well when it oxidizes both fatty acid and glucose simultaneously.[161] However, the chief myocardial energy substrates switch from fatty acids to glucose in heart failure, with a downregulation of the enzymes involved in fatty acid oxidation.[162,163] Furthermore, metabolic adaptation becomes insufficient with a lower capacity to oxidize glucose leading to decreased efficiency.[164] Heart failure is not accompanied by overexpression of glycolytic pathways and end-stage heart failure results in decreased glycolytic enzymes.[165,166]

As ventricular myocytes hypertrophy, there is a tendency for the increase in myofilament volume to outstrip the increase in mitochondrial volume.[167,168] This decrease in the mitochondrial volume:myofilament volume may create a supply:demand mismatch and contribute to lower cardiac energetic reserve. In heart failure the situation may become further complicated by altered mitochondrial number, size and structural integrity.[169] Mitochondrial injury is positively correlated with indices of heart failure severity such as plasma norepinephrine, and left ventricle (LV) end-diastolic pressure and ejection fraction.[170] Decreases in the activity of complexes of the respiratory chain or Krebs cycle enzymes have been observed in both human and experimental heart failure. A lower myocardial energy production via oxidative phosphorylation in heart failure is associated with defective oxygen consumption rates and blunted mitochondrial regulation by the phosphate acceptors AMP, ADP and creatine.[165,171-173] A decrease in PCr:ATP ratio is reported in failing human and animal heart, even at moderate workloads. Creatine, creatine transporter, PCr and ATP are significantly reduced,[174,175] and the decrease in the PCr:ATP ratio is a predictor of mortality in congestive heart failure (CHF).[176] An impairment in energy transfer and utilization is also seen in HF. A decrease in total enzyme activity, alteration in the isoenzyme pattern and decreased CK fluxes are hallmarks of cardiac failure.[165,176-180] However, whether these metabolic alterations accompany or even precede the development of HF may depend on the etiology of cardiac diseases.

HF summary

Overall, the contractile dysfunction in HF is probably largely due to a reduction of SR Ca^{2+} content during steady state physiological conditions. Several factors may contribute to the lower SR Ca^{2+} content: (1) reduced SERCA function; (2) increased inhibition by PLN; (3) increased Ca^{2+} efflux via NCX (by competing with the SERCA for Ca^{2+}); and (4) diastolic Ca^{2+} leak from the SR. Furthermore, a lower SR Ca^{2+} content reduces SR Ca^{2+} release both because there is less Ca^{2+} available for release and also because fractional release is reduced for a given I_{Ca} trigger.

Reduced SERCA and increased NCX can both tend to lower SR Ca^{2+} content, but exert opposite effects on relaxation rate and $[Ca^{2+}]_i$ decline. This combination can thus result in unchanged relaxation rate, as observed in a rabbit HF model, where a doubling of NCX appeared to offset ~24% decrease in SR Ca^{2+}-pump function.[109] Hasenfuss et al. found this same situation in 44% of human HF (nearly doubled Na^+–Ca^{2+} exchanger with modest SERCA reduction).[138] These muscles did not show elevated diastolic force at 3 Hz (versus 0.5 Hz). However, another group of HF muscles (34%) that showed slower relaxation and elevated diastolic force at 3 Hz, had a greater reduction in SERCA2 expression, but unaltered NCX. SR Ca^{2+} load may be lower in both cases, but for different main reasons. Piacentino et al. obtained human HF results,[134] similar to the latter group (reduced SERCA function, little NCX alteration), Pogwizd et al. found results more like the former group in non-ischemic rabbit HF (mostly increased NCX function),[135] and O'Rourke et al. found in-between behavior in a rapid-pacing-induced dog HF (both increased NCX and decreased SERCA function).[108] Thus, there may be a real heterogeneity of diastolic dysfunction in HF that depends on the balance of SERCA and NCX function.

References

1. Bers DM. Excitation–Contraction Coupling and Cardiac Contractile Force (2e). Dordrecht: Kluwer Academic Press; 2001.
2. Solaro RJ, Rarick HM. Troponin and tropomyosin:

proteins that switch on and tune in the activity of cardiac myofilaments. Circ Res 1998; 83: 471–80.

3. Moss RL, Buck SH. Regulation of cardiac contraction by Ca^{2+}. In: Page E, Fozzard HA, Solaro RJ (eds). Handbook of Physiology. New York: Oxford University Press; 2001: 420–54.

4. Peterson BZ, DeMaria CD, Adelman JP et al. Calmodulin is the Ca^{2+} sensor for Ca^{2+}-dependent inactivation of L-type calcium channels. Neuron 1999; 22: 549–58.

5. Zuhlke RD, Pitt GS, Deisseroth K et al. Calmodulin supports both inactivation and facilitation of L-type calcium channels. Nature 1999; 399: 159–62.

6. Scriven DR, Dan P, Moore ED. Distribution of proteins implicated in excitation–contraction coupling in rat ventricular myocytes. Biophys J 2000; 79: 2682–91.

7. Sipido KR, Callewaert G, Carmeliet E. Inhibition and rapid recovery of Ca^{2+} current during Ca^{2+} release from sarcoplasmic reticulum in guinea pig ventricular myocytes. Circ Res 1995; 76: 102–9.

8. Sham JS, Song LS, Chen Y et al. Termination of Ca^{2+} release by a local inactivation of ryanodine receptors in cardiac myocytes. Proc Natl Acad Sci USA 1998; 95: 15096–101.

9. Puglisi JL, Yuan W, Bassani JW et al. Ca^{2+} influx through Ca^{2+} channels in rabbit ventricular myocytes during action potential clamp: influence of temperature. Circ Res 1999; 85: e7–16.

10. Zahradnikova A, Zahradnik I, Gyorke I et al. Rapid activation of the cardiac ryanodine receptor by sub-millisecond calcium stimuli. J Gen Physiol 1999; 114: 787–98.

11. Bassani JW, Yuan W, Bers DM. Fractional SR Ca release is regulated by trigger Ca and SR Ca content in cardiac myocytes. Am J Physiol Cell Physiol 1995; 268: C1313–19.

12. Shannon TR, Ginsburg KS, Bers DM. Potentiation of fractional sarcoplasmic reticulum calcium release by total and free intra-sarcoplasmic reticulum calcium concentration. Biophys J 2000; 78: 334–43.

13. Sitsapesan R, Williams AJ. Regulation of the gating of the sheep cardiac sarcoplasmic reticulum Ca^{2+}-release channel by luminal Ca^{2+}. J Membr Biol 1994; 137: 215–26.

14. Lukyanenko V, Gyorke I, Gyorke S. Regulation of calcium release by calcium inside the sarcoplasmic reticulum in ventricular myocytes. Pflugers Arch 1996; 432: 1047–54.

15. Weber CR, Piacentino V, 3rd, Ginsburg KS et al. Na^+–Ca^{2+} exchange current and submembrane

$[Ca^{2+}]$ during the cardiac action potential. Circ Res 2002; 90: 182–9.

16. Weber CR, Piacentino V, 3rd, Houser SR et al. Dynamic regulation of sodium/calcium exchange function in human heart failure. Circulation 2003; 108: 2224–9.

17. Despa S, Islam MA, Weber CR et al. Intracellular Na^+ concentration is elevated in heart failure but Na/K pump function is unchanged. Circulation 2002; 105: 2543–8.

18. Simmerman HK, Jones LR. Phospholamban: protein structure, mechanism of action, and role in cardiac function. Physiol Rev 1998; 78: 921–47.

19. MacLennan DH, Kranias EG. Phospholamban: a crucial regulator of cardiac contractility. Nat Rev Mol Cell Biol 2003; 4: 566–77.

20. Kranias EG, Solaro RJ. Phosphorylation of troponin I and phospholamban during catecholamine stimulation of rabbit heart. Nature 1982; 298: 182–4.

21. Wegener AD, Simmerman HK, Lindemann JP et al. Phospholamban phosphorylation in intact ventricles. Phosphorylation of serine 16 and threonine 17 in response to beta-adrenergic stimulation. J Biol Chem 1989; 264: 11468–74.

22. Talosi L, Edes I, Kranias EG. Intracellular mechanisms mediating reversal of beta-adrenergic stimulation in intact beating hearts. Am J Physiol Heart Circ Physiol 1993; 264: H791–7.

23. Lindemann JP, Jones LR, Hathaway DR et al. Beta-adrenergic stimulation of phospholamban phosphorylation and Ca^{2+}-ATPase activity in guinea pig ventricles. J Biol Chem 1983; 258: 464–71.

24. Garvey JL, Kranias EG, Solaro RJ. Phosphorylation of C-protein, troponin I and phospholamban in isolated rabbit hearts. Biochem J 1988; 249: 709–14.

25. Mundina-Weilenmann C, Vittone L, Ortale M et al. Immunodetection of phosphorylation sites gives new insights into the mechanisms underlying phospholamban phosphorylation in the intact heart. J Biol Chem 1996; 271: 33561–7.

26. Kranias EG, Steenaart NA, Di Salvo J. Purification and characterization of phospholamban phosphatase from cardiac muscle. J Biol Chem 1988; 263: 15681–7.

27. Bers DM. Macromolecular complexes regulating cardiac ryanodine receptor function. J Mol Cell Cardiol 2004; 37: 417–29.

28. Fruen BR, Bardy JM, Byrem TM et al. Differential Ca^{2+} sensitivity of skeletal and cardiac muscle ryanodine receptors in the presence of calmodulin. Am J Physiol Cell Physiol 2000; 279: C724–33.

29. Marx SO, Reiken S, Hisamatsu Y et al. PKA phos-

phorylation dissociates FKBP12.6 from the calcium release channel (ryanodine receptor): defective regulation in failing hearts. Cell 2000; 101: 365–76.

30. Zhang T, Maier LS, Dalton ND et al. The deltaC isoform of CaMKII is activated in cardiac hypertrophy and induces dilated cardiomyopathy and heart failure. Circ Res 2003; 92: 912–19.

31. Marx SO, Reiken S, Hisamatsu Y et al. Phosphorylation-dependent regulation of ryanodine receptors: a novel role for leucine/isoleucine zippers. J Cell Biol 2001; 153: 699–708.

32. Meyers MB, Puri TS, Chien AJ et al. Sorcin associates with the pore-forming subunit of voltage-dependent L-type Ca^{2+} channels. J Biol Chem 1998; 273: 18930–5.

33. Zhang L, Kelley J, Schmeisser G et al. Complex formation between junctin, triadin, calsequestrin, and the ryanodine receptor. Proteins of the cardiac junctional sarcoplasmic reticulum membrane. J Biol Chem 1997; 272: 23389–97.

34. Franzini-Armstrong C, Protasi F, Ramesh V. Shape, size, and distribution of Ca^{2+} release units and couplons in skeletal and cardiac muscles. Biophys J 1999; 77: 1528–39.

35. Cheng H, Lederer WJ, Cannell MB. Calcium sparks: elementary events underlying excitation–contraction coupling in heart muscle. Science 1993; 262: 740–4.

36. Wier WG, Balke CW. Ca^{2+} release mechanisms, Ca^{2+} sparks, and local control of excitation–contraction coupling in normal heart muscle. Circ Res 1999; 85: 770–6.

37. Bridge JH, Ershler PR, Cannell MB. Properties of Ca^{2+} sparks evoked by action potentials in mouse ventricular myocytes. J Physiol 1999; 518: 469–78.

38. Lukyanenko V, Gyorke I, Subramanian S et al. Inhibition of Ca^{2+} sparks by ruthenium red in permeabilized rat ventricular myocytes. Biophys J 2000; 79: 1273–84.

39. Leblanc N, Hume JR. Sodium current-induced release of calcium from cardiac sarcoplasmic reticulum. Science 1990; 248: 372–6.

40. Levesque PC, Leblanc N, Hume JR. Release of calcium from guinea pig cardiac sarcoplasmic reticulum induced by sodium–calcium exchange. Cardiovasc Res 1994; 28: 370–8.

41. Lipp P, Niggli E. Sodium current-induced calcium signals in isolated guinea-pig ventricular myocytes. J Physiol 1994; 474: 439–46.

42. Sipido KR, Maes M, Van de Werf F. Low efficiency of Ca^{2+} entry through the Na^+–Ca^{2+} exchanger as trigger for Ca^{2+} release from the sarcoplasmic reticulum. A comparison between L-type Ca^{2+} current and reverse-mode Na^+–Ca^{2+} exchange. Circ Res 1997; 81: 1034–44.

43. Satoh H, Blatter LA, Bers DM. Effects of $[Ca^{2+}]_i$, SR Ca^{2+} load, and rest on Ca^{2+} spark frequency in ventricular myocytes. Am J Physiol Heart Circ Physiol 1997; 272: H657–68.

44. Luo W, Grupp IL, Harrer J et al. Targeted ablation of the phospholamban gene is associated with markedly enhanced myocardial contractility and loss of beta-agonist stimulation. Circ Res 1994; 75: 401–9.

45. Luo W, Wolska BM, Grupp IL et al. Phospholamban gene dosage effects in the mammalian heart. Circ Res 1996; 78: 839–47.

46. Wolska BM, Stojanovic MO, Luo W et al. Effect of ablation of phospholamban on dynamics of cardiac myocyte contraction and intracellular Ca^{2+}. Am J Physiol Heart Circ Physiol 1996; 271: C391–7.

47. Li L, Chu G, Kranias EG et al. Cardiac myocyte calcium transport in phospholamban knockout mouse: relaxation and endogenous CaMKII effects. Am J Physiol Heart Circ Physiol 1998; 274: H1335–47.

48. Lorenz JN, Kranias EG. Regulatory effects of phospholamban on cardiac function in intact mice. Am J Physiol Heart Circ Physiol 1997; 273: H2826–31.

49. Kiss E, Edes I, Sato Y et al. Beta-adrenergic regulation of cAMP and protein phosphorylation in phospholamban-knockout mouse hearts. Am J Physiol Heart Circ Physiol 1997; 272: H785–90.

50. Chu G, Luo W, Slack JP et al. Compensatory mechanisms associated with the hyperdynamic function of phospholamban-deficient mouse hearts. Circ Res 1996; 79: 1064–76.

51. Santana LF, Gomez AM, Kranias EG et al. Amount of calcium in the sarcoplasmic reticulum: influence on excitation–contraction coupling in heart muscle. Heart Vessels 1997; (Suppl 12): 44–9.

52. Masaki H, Sato Y, Luo W et al. Phospholamban deficiency alters inactivation kinetics of L-type Ca^{2+} channels in mouse ventricular myocytes. Am J Physiol Heart Circ Physiol 1997; 272: H606–12.

53. Slack JP, Grupp IL, Dash R et al. The enhanced contractility of the phospholamban-deficient mouse heart persists with aging. J Mol Cell Cardiol 2001; 33: 1031–40.

54. Desai KH, Schauble E, Luo W et al. Phospholamban deficiency does not compromise exercise capacity. Am J Physiol Heart Circ Physiol 1999; 276: H1172–7.

55. Kiriazis H, Sato Y, Kadambi VJ et al. Hypertrophy and functional alterations in hyperdynamic phospholamban-knockout mouse hearts under chronic aortic stenosis. Cardiovasc Res 2002; 53: 372–81.

56. Cross HR, Kranias EG, Murphy E et al. Ablation of PLB exacerbates ischemic injury to a lesser extent in female than male mice: protective role of NO. Am J Physiol Heart Circ Physiol 2003; 284: H683–90.

57. Gregory KN, Hahn H, Haghighi K et al. Increased particulate partitioning of PKC reverses susceptibility of phospholamban knockout hearts to ischemic injury. J Mol Cell Cardiol 2004; 36: 313–18.

58. Kadambi VJ, Ponniah S, Harrer JM et al. Cardiac-specific overexpression of phospholamban alters calcium kinetics and resultant cardiomyocyte mechanics in transgenic mice. J Clin Invest 1996; 97: 533–9.

59. Dash R, Kadambi V, Schmidt AG et al. Interactions between phospholamban and beta-adrenergic drive may lead to cardiomyopathy and early mortality. Circulation 2001; 103: 889–96.

60. Brittsan AG, Carr AN, Schmidt AG et al. Maximal inhibition of SERCA2 Ca^{2+} affinity by phospholamban in transgenic hearts overexpressing a non-phosphorylatable form of phospholamban. J Biol Chem 2000; 275: 12129–35.

61. Zhai J, Schmidt AG, Hoit BD et al. Cardiac-specific overexpression of a superinhibitory pentameric phospholamban mutant enhances inhibition of cardiac function in vivo. J Biol Chem 2000; 275: 10538–44.

62. Zvaritch E, Backx PH, Jirik F et al. The transgenic expression of highly inhibitory monomeric forms of phospholamban in mouse heart impairs cardiac contractility. J Biol Chem 2000; 275: 14985–91.

63. Haghighi K, Schmidt AG, Hoit BD et al. Superinhibition of sarcoplasmic reticulum function by phospholamban induces cardiac contractile failure. J Biol Chem 2001; 276: 24145–52.

64. Schmidt AG, Zhai J, Carr AN et al. Structural and functional implications of the phospholamban hinge domain: impaired SR Ca^{2+} uptake as a primary cause of heart failure. Cardiovasc Res 2002; 56: 248–59.

65. Li L, Desantiago J, Chu G et al. Phosphorylation of phospholamban and troponin I in beta-adrenergic-induced acceleration of cardiac relaxation. Am J Physiol Heart Circ Physiol 2000; 278: H769–79.

66. Kentish JC, McCloskey DT, Layland J et al. Phosphorylation of troponin I by protein kinase A accelerates relaxation and crossbridge cycle kinetics in mouse ventricular muscle. Circ Res 2001; 88: 1059–65.

67. Pi Y, Kemnitz KR, Zhang D et al. Phosphorylation of troponin I controls cardiac twitch dynamics: evidence from phosphorylation site mutants expressed on a troponin I-null background in mice. Circ Res 2002; 90: 649–56.

68. Luo W, Chu G, Sato Y et al. Transgenic approaches to define the functional role of dual site phospholamban phosphorylation. J Biol Chem 1998; 273: 4734–9.

69. Chu G, Lester JW, Young KB et al. A single site (Ser16) phosphorylation in phospholamban is sufficient in mediating its maximal cardiac responses to beta-agonists. J Biol Chem 2000; 275: 38938–43.

70. Vittone L, Mundina-Weilenmann C, Said M et al. Time course and mechanisms of phosphorylation of phospholamban residues in ischemia–reperfused rat hearts. Dissociation of phospholamban phosphorylation pathways. J Mol Cell Cardiol 2002; 34: 39–50.

71. Hagemann D, Kuschel M, Kuramochi T et al. Frequency-encoding Thr17 phospholamban phosphorylation is independent of Ser16 phosphorylation in cardiac myocytes. J Biol Chem 2000; 275: 22532–6.

72. Valdivia HH, Kaplan JH, Ellis-Davies GC et al. Rapid adaptation of cardiac ryanodine receptors: modulation by Mg^{2+} and phosphorylation. Science 1995; 267: 1997–2000.

73. Stange M, Xu L, Balshaw D et al. Characterization of recombinant skeletal muscle (Ser-2843) and cardiac muscle (Ser-2809) ryanodine receptor phosphorylation mutants. J Biol Chem 2003; 278: 51693–702.

74. Xiao B, Sutherland C, Walsh MP et al. Protein kinase A phosphorylation at serine-2808 of the cardiac Ca^{2+}-release channel (ryanodine receptor) does not dissociate 12.6-kDa FK506-binding protein (FKBP12.6). Circ Res 2004; 94: 487–95.

75. Li Y, Kranias EG, Mignery GA et al. Protein kinase A phosphorylation of the ryanodine receptor does not affect calcium sparks in mouse ventricular myocytes. Circ Res 2002; 90: 309–16.

76. Ginsburg KS, Bers DM. Modulation of excitation–contraction coupling by isoproterenol in cardiomyocytes with controlled SR Ca load and Ca^{2+} current trigger. J Physiol 2004; 556: 463–80.

77. Wehrens XH, Lehnart SE, Reiken SR et al. Ca^{2+}/calmodulin-dependent protein kinase II phosphorylation regulates the cardiac ryanodine receptor. Circ Res 2004; 94: e61–70.

78. Maier LS, Zhang T, Chen L et al. Transgenic CaMKIIdeltaC overexpression uniquely alters cardiac myocyte Ca^{2+} handling: reduced SR Ca^{2+} load and activated SR Ca^{2+} release. Circ Res 2003; 92: 904–11.

79. Li L, Satoh H, Ginsburg KS et al. The effect of Ca^{2+}-calmodulin-dependent protein kinase II on cardiac

excitation–contraction coupling in ferret ventricular myocytes. J Physiol 1997; 501: 17–31.

80. Yuan W, Bers DM. Ca-dependent facilitation of cardiac Ca current is due to Ca-calmodulin-dependent protein kinase. Am J Physiol Heart Circ Physiol 1994; 267: H982–93.

81. Xiao RP, Cheng H, Lederer WJ et al. Dual regulation of Ca^{2+}/calmodulin-dependent kinase II activity by membrane voltage and by calcium influx. Proc Natl Acad Sci USA 1994; 91: 9659–63.

82. Bers DM, Ziolo MT. When is cAMP not cAMP? Effects of compartmentalization. Circ Res 2001; 89: 373–5.

83. Kuschel M, Zhou YY, Spurgeon HA et al. Beta2-adrenergic cAMP signaling is uncoupled from phosphorylation of cytoplasmic proteins in canine heart. Circulation 1999; 99: 2458–65.

84. Rybin VO, Xu X, Steinberg SF. Activated protein kinase C isoforms target to cardiomyocyte caveolae: stimulation of local protein phosphorylation. Circ Res 1999; 84: 980–8.

85. Hasenfuss G. Alterations of calcium-regulatory proteins in heart failure. Cardiovasc Res 1998; 37: 279–89.

86. Richard S, Leclercq F, Lemaire S et al. Ca^{2+} currents in compensated hypertrophy and heart failure. Cardiovasc Res 1998; 37: 300–11.

87. Mukherjee R, Spinale FG. L-type calcium channel abundance and function with cardiac hypertrophy and failure: a review. J Mol Cell Cardiol 1998; 30: 1899–916.

88. Wickenden AD, Kaprielian R, Kassiri Z et al. The role of action potential prolongation and altered intracellular calcium handling in the pathogenesis of heart failure. Cardiovasc Res 1998; 37: 312–23.

89. Nabauer M, Kaab S. Potassium channel down-regulation in heart failure. Cardiovasc Res 1998; 37: 324–34.

90. Phillips RM, Narayan P, Gomez AM et al. Sarcoplasmic reticulum in heart failure: central player or bystander? Cardiovasc Res 1998; 37: 346–51.

91. de Tombe PP. Altered contractile function in heart failure. Cardiovasc Res 1998; 37: 367–80.

92. Houser SR, Piacentino V, 3rd, Mattiello J et al. Functional properties of failing human ventricular myocytes. Trends Cardiovasc Med 2000; 10: 101–7.

93. Pieske B, Kretschmann B, Meyer M et al. Alterations in intracellular calcium handling associated with the inverse force–frequency relation in human dilated cardiomyopathy. Circulation 1995; 92: 1169–78.

94. Beuckelmann DJ, Nabauer M, Erdmann E. Intracellular calcium handling in isolated ventricular myocytes from patients with terminal heart failure. Circulation 1992; 85: 1046–55.

95. Beuckelmann DJ, Nabauer M, Erdmann E. Alterations of K^+ currents in isolated human ventricular myocytes from patients with terminal heart failure. Circ Res 1993; 73: 379–85.

96. Kaab S, Nuss HB, Chiamvimonvat N et al. Ionic mechanism of action potential prolongation in ventricular myocytes from dogs with pacing-induced heart failure. Circ Res 1996; 78: 262–73.

97. Vermeulen JT, McGuire MA, Opthof T et al. Triggered activity and automaticity in ventricular trabeculae of failing human and rabbit hearts. Cardiovasc Res 1994; 28: 1547–54.

98. Gwathmey JK, Hajjar RJ. Relation between steady-state force and intracellular $[Ca^{2+}]$ in intact human myocardium. Index of myofibrillar responsiveness to Ca^{2+}. Circulation 1990; 82: 1266–78.

99. Perreault CL, Bing OH, Brooks WW et al. Differential effects of cardiac hypertrophy and failure on right versus left ventricular calcium activation. Circ Res 1990; 67: 707–12.

100. D'Agnolo A, Luciani GB, Mazzucco A et al. Contractile properties and Ca^{2+} release activity of the sarcoplasmic reticulum in dilated cardiomyopathy. Circulation 1992; 85: 518–25.

101. Wolff MR, McDonald KS, Moss RL. Rate of tension development in cardiac muscle varies with level of activator calcium. Circ Res 1995; 76: 154–60.

102. Wolff MR, Buck SH, Stoker SW et al. Myofibrillar calcium sensitivity of isometric tension is increased in human dilated cardiomyopathies: role of altered beta-adrenergically mediated protein phosphorylation. J Clin Invest 1996; 98: 167–76.

103. Hajjar RJ, Schwinger RH, Schmidt U et al. Myofilament calcium regulation in human myocardium. Circulation 2000; 101: 1679–85.

104. Fan D, Wannenburg T, de Tombe PP. Decreased myocyte tension development and calcium responsiveness in rat right ventricular pressure overload. Circulation 1997; 95: 2312–7.

105. Perez NG, Hashimoto K, McCune S et al. Origin of contractile dysfunction in heart failure: calcium cycling versus myofilaments. Circulation 1999; 99: 1077–83.

106. Beuckelmann DJ, Erdmann E. Ca^{2+}-currents and intracellular $[Ca^{2+}]_i$-transients in single ventricular myocytes isolated from terminally failing human myocardium. Basic Res Cardiol 1992; 87 (Suppl 1): 235–43.

107. Gomez AM, Valdivia HH, Cheng H et al. Defective

excitation–contraction coupling in experimental cardiac hypertrophy and heart failure. Science 1997; 276: 800–6.

108. O'Rourke B, Kass DA, Tomaselli GF et al. Mechanisms of altered excitation–contraction coupling in canine tachycardia-induced heart failure, I: experimental studies. Circ Res 1999; 84: 562–70.

109. Pogwizd SM, Qi M, Yuan W et al. Upregulation of Na^+/Ca^{2+} exchanger expression and function in an arrhythmogenic rabbit model of heart failure. Circ Res 1999; 85: 1009–19.

110. Ahmmed GU, Dong PH, Song G et al. Changes in Ca^{2+} cycling proteins underlie cardiac action potential prolongation in a pressure-overloaded guinea pig model with cardiac hypertrophy and failure. Circ Res 2000; 86: 558–70.

111. Handrock R, Schroder F, Hirt S et al. Single-channel properties of L-type calcium channels from failing human ventricle. Cardiovasc Res 1998; 37: 445–55.

112. Chen X, Piacentino V, 3rd, Furukawa S et al. L-type Ca^{2+} channel density and regulation are altered in failing human ventricular myocytes and recover after support with mechanical assist devices. Circ Res 2002; 91: 517–24.

113. de la Bastie D, Levitsky D, Rappaport L et al. Function of the sarcoplasmic reticulum and expression of its Ca^{2+}-ATPase gene in pressure overload-induced cardiac hypertrophy in the rat. Circ Res 1990; 66: 554–64.

114. Feldman AM, Weinberg EO, Ray PE et al. Selective changes in cardiac gene expression during compensated hypertrophy and the transition to cardiac decompensation in rats with chronic aortic banding. Circ Res 1993; 73: 184–92.

115. Dash R, Frank KF, Carr AN et al. Gender influences on sarcoplasmic reticulum Ca^{2+}-handling in failing human myocardium. J Mol Cell Cardiol 2001; 33: 1345–53.

116. Huang B, Wang S, Qin D et al. Diminished basal phosphorylation level of phospholamban in the postinfarction remodeled rat ventricle: role of beta-adrenergic pathway, G(i) protein, phosphodiesterase, and phosphatases. Circ Res 1999; 85: 848–55.

117. Schwinger RH, Munch G, Bolck B et al. Reduced Ca^{2+}-sensitivity of SERCA 2a in failing human myocardium due to reduced serin-16 phospholamban phosphorylation. J Mol Cell Cardiol 1999; 31: 479–91.

118. Hamilton MA, Stevenson LW, Luu M et al. Altered thyroid hormone metabolism in advanced heart failure. J Am Coll Cardiol 1990; 16: 91–5.

119. Kiss E, Jakab G, Kranias EG et al. Thyroid hormone-induced alterations in phospholamban protein expression. Regulatory effects on sarcoplasmic reticulum Ca^{2+} transport and myocardial relaxation. Circ Res 1994; 75: 245–51.

120. Ojamaa K, Kenessey A, Shenoy R et al. Thyroid hormone metabolism and cardiac gene expression after acute myocardial infarction in the rat. Am J Physiol Endocrinol Metab 2000; 279: E1319–24.

121. del Monte F, Harding SE, Schmidt U et al. Restoration of contractile function in isolated cardiomyocytes from failing human hearts by gene transfer of SERCA2a. Circulation 1999; 100: 2308–11.

122. Miyamoto MI, del Monte F, Schmidt U et al. Adenoviral gene transfer of SERCA2a improves left-ventricular function in aortic-banded rats in transition to heart failure. Proc Natl Acad Sci USA 2000; 97: 793–8.

123. Haghighi K, Kolokathis F, Pater L et al. Human phospholamban null results in lethal dilated cardiomyopathy revealing a critical difference between mouse and human. J Clin Invest 2003; 111: 869–76.

124. Go LO, Moschella MC, Watras J et al. Differential regulation of two types of intracellular calcium release channels during end-stage heart failure. J Clin Invest 1995; 95: 888–94.

125. Schillinger W, Meyer M, Kuwajima G et al. Unaltered ryanodine receptor protein levels in ischemic cardiomyopathy. Mol Cell Biochem 1996; 160–161: 297–302.

126. Sainte Beuve C, Allen PD, Dambrin G et al. Cardiac calcium release channel (ryanodine receptor) in control and cardiomyopathic human hearts: mRNA and protein contents are differentially regulated. J Mol Cell Cardiol 1997; 29: 1237–46.

127. Vatner DE, Sato N, Kiuchi K et al. Decrease in myocardial ryanodine receptors and altered excitation–contraction coupling early in the development of heart failure. Circulation 1994; 90: 1423–30.

128. Yano M, Ono K, Ohkusa T et al. Altered stoichiometry of FKBP12.6 versus ryanodine receptor as a cause of abnormal Ca^{2+} leak through ryanodine receptor in heart failure. Circulation 2000; 102: 2131–6.

129. Neumann J, Eschenhagen T, Jones LR et al. Increased expression of cardiac phosphatases in patients with end-stage heart failure. J Mol Cell Cardiol 1997; 29: 265–72.

130. McCall E, Li L, Satoh H et al. Effects of FK-506 on contraction and Ca^{2+} transients in rat cardiac myocytes. Circ Res 1996; 79: 1110–21.

131. Shannon TR, Pogwizd SM, Bers DM. Elevated sarcoplasmic reticulum Ca^{2+} leak in intact ventricular myocytes from rabbits in heart failure. Circ Res 2003; 93: 592–4.

132. Jiang MT, Lokuta AJ, Farrell EF et al. Abnormal Ca^{2+} release, but normal ryanodine receptors, in canine and human heart failure. Circ Res 2002; 91: 1015–22.

133. Lindner M, Erdmann E, Beuckelmann DJ. Calcium content of the sarcoplasmic reticulum in isolated ventricular myocytes from patients with terminal heart failure. J Mol Cell Cardiol 1998; 30: 743–9.

134. Piacentino V, 3rd, Weber CR, Chen X et al. Cellular basis of abnormal calcium transients of failing human ventricular myocytes. Circ Res 2003; 92: 651–8.

135. Pogwizd SM, Schlotthauer K, Li L et al. Arrhythmogenesis and contractile dysfunction in heart failure: roles of sodium–calcium exchange, inward rectifier potassium current, and residual beta-adrenergic responsiveness. Circ Res 2001; 88: 1159–67.

136. Hobai IA, O'Rourke B. Decreased sarcoplasmic reticulum calcium content is responsible for defective excitation–contraction coupling in canine heart failure. Circulation 2001; 103: 1577–84.

137. Flesch M, Schwinger RH, Schiffer F et al. Evidence for functional relevance of an enhanced expression of the Na^+–Ca^{2+}exchanger in failing human myocardium. Circulation 1996; 94. 992–1002.

138. Hasenfuss G, Schillinger W, Lehnart SE et al. Relationship between Na^+–Ca^{2+}-exchanger protein levels and diastolic function of failing human myocardium. Circulation 1999; 99: 641–8.

139. Sipido KR, Volders PG, de Groot SH et al. Enhanced Ca^{2+} release and Na^+/Ca^{2+} exchange activity in hypertrophied canine ventricular myocytes: potential link between contractile adaptation and arrhythmogenesis. Circulation 2000; 102: 2137–44.

140. Litwin SE, Bridge JH. Enhanced Na^+–Ca^{2+} exchange in the infarcted heart. Implications for excitation–contraction coupling. Circ Res 1997; 81: 1083–93.

141. Pieske B, Maier LS, Piacentino V, 3rd et al. Rate dependence of $[Na^+]_i$ and contractility in nonfailing and failing human myocardium. Circulation 2002; 106: 447–53.

142. Verdonck F, Volders PG, Vos MA et al. Increased Na^+ concentration and altered Na^+/K^+ pump activity in hypertrophied canine ventricular cells. Cardiovasc Res 2003; 57: 1035–43.

143. Dixon IM, Hata T, Dhalla NS. Sarcolemmal Na^+–K^+-ATPase activity in congestive heart failure due to myocardial infarction. Am J Physiol Cell Physiol 1992; 262: C664–71.

144. Semb SO, Lunde PK, Holt E et al. Reduced myocardial Na^+, K^+-pump capacity in congestive heart failure following myocardial infarction in rats. J Mol Cell Cardiol 1998; 30: 1311–28.

145. Schwinger RH, Wang J, Frank K et al. Reduced sodium pump alpha1, alpha3, and beta1-isoform protein levels and Na^+–K^+-ATPase activity but unchanged Na^+–Ca^{2+}exchanger protein levels in human heart failure. Circulation 1999; 99: 2105–12.

146. Baartscheer A, Schumacher CA, van Borren MM et al. Increased Na^+/H^+-exchange activity is the cause of increased $[Na^+]_i$ and underlies disturbed calcium handling in the rabbit pressure and volume overload heart failure model. Cardiovasc Res 2003; 57: 1015–24.

147. Saint DA, Ju YK, Gage PW. A persistent sodium current in rat ventricular myocytes. J Physiol 1992; 453: 219–31.

148. Maltsev VA, Sabbah HN, Higgins RS et al. Novel, ultraslow inactivating sodium current in human ventricular cardiomyocytes. Circulation 1998; 98: 2545–52.

149. Undrovinas AI, Maltsev VA, Sabbah HN. Repolarization abnormalities in cardiomyocytes of dogs with chronic heart failure: role of sustained inward current. Cell Mol Life Sci 1999; 55: 494–505.

150. Sah R, Ramirez RJ, Oudit GY et al. Regulation of cardiac excitation–contraction coupling by action potential repolarization: role of the transient outward potassium current (I_{to}). J Physiol 2003; 546: 5–18.

151. Priebe L, Beuckelmann DJ. Simulation study of cellular electric properties in heart failure. Circ Res 1998; 82: 1206–23.

152. Tian R, Nascimben L, Ingwall JS et al. Failure to maintain a low ADP concentration impairs diastolic function in hypertrophied rat hearts. Circulation 1997; 96: 1313–19.

153. Brandes R, Maier LS, Bers DM. Regulation of mitochondrial [NADH] by cytosolic $[Ca^{2+}]$ and work in trabeculae from hypertrophic and normal rat hearts. Circ Res 1998; 82: 1189–98.

154. Ito K, Yan X, Tajima M et al. Contractile reserve and intracellular calcium regulation in mouse myocytes from normal and hypertrophied failing hearts. Circ Res 2000; 87: 588–95.

155. Mootha VK, Arai AE, Balaban RS. Maximum oxidative phosphorylation capacity of the mammalian heart. Am J Physiol Heart Circ Physiol 1997; 272: H769–75.

156. Balaban RS. Cardiac energy metabolism homeostasis: role of cytosolic calcium. J Mol Cell Cardiol 2002; 34: 1259–71.

157. Bessman SP, Geiger PJ. Transport of energy in muscle: the phosphorylcreatine shuttle. Science 1981; 211: 448–52.

158. Ventura-Clapier R, Garnier A, Veksler V. Energy metabolism in heart failure. J Physiol 2004; 555: 1–13.

159. Schipke JD. Cardiac efficiency. Basic Res Cardiol 1994; 89: 207–40.

160. Saavedra WF, Paolocci N, St John ME et al. Imbalance between xanthine oxidase and nitric oxide synthase signaling pathways underlies mechanoenergetic uncoupling in the failing heart. Circ Res 2002; 90: 297–304.

161. Taegtmeyer H. Metabolism—the lost child of cardiology. J Am Coll Cardiol 2000; 36: 1386–8.

162. Sack MN, Rader TA, Park S et al. Fatty acid oxidation enzyme gene expression is downregulated in the failing heart. Circulation 1996; 94: 2837–42.

163. Razeghi P, Young ME, Alcorn JL et al. Metabolic gene expression in fetal and failing human heart. Circulation 2001; 104: 2923–31.

164. Leong HS, Brownsey RW, Kulpa JE et al. Glycolysis and pyruvate oxidation in cardiac hypertrophy—why so unbalanced? Comp Biochem Physiol A Mol Integr Physiol 2003; 135: 499–513.

165. De Sousa E, Veksler V, Minajeva A et al. Subcellular creatine kinase alterations. Implications in heart failure. Circ Res 1999; 85: 68–76.

166. Dzeja PP, Pucar D, Redfield MM et al. Reduced activity of enzymes coupling ATP-generating with ATP-consuming processes in the failing myocardium. Mol Cell Biochem 1999; 201: 33–40.

167. Lund DD, Tomanek RJ. Myocardial morphology in spontaneously hypertensive and aortic-constricted rats. Am J Anat 1978; 152: 141–51.

168. Anversa P, Olivetti G, Melissari M et al. Morphometric study of myocardial hypertrophy induced by abdominal aortic stenosis. Lab Invest 1979; 40: 341–9.

169. Schaper J, Froede R, Hein S et al. Impairment of the myocardial ultrastructure and changes of the cytoskeleton in dilated cardiomyopathy. Circulation 1991; 83: 504–14.

170. Sabbah HN, Sharov V, Riddle JM et al. Mitochondrial abnormalities in myocardium of dogs with chronic heart failure. J Mol Cell Cardiol 1992; 24: 1333–47.

171. Sanbe A, Tanonaka K, Kobayasi R et al. Effects of long-term therapy with ACE inhibitors, captopril, enalapril and trandolapril, on myocardial energy metabolism in rats with heart failure following myocardial infarction. J Mol Cell Cardiol 1995; 27: 2209–22.

172. Sharov VG, Goussev A, Lesch M et al. Abnormal mitochondrial function in myocardium of dogs with chronic heart failure. J Mol Cell Cardiol 1998; 30: 1757–62.

173. Sharov VG, Todor AV, Silverman N et al. Abnormal mitochondrial respiration in failed human myocardium. J Mol Cell Cardiol 2000; 32: 2361–7.

174. Neubauer S, Remkes H, Spindler M et al. Downregulation of the Na^+-creatine cotransporter in failing human myocardium and in experimental heart failure. Circulation 1999; 100: 1847–50.

175. Beer M, Seyfarth T, Sandstede J et al. Absolute concentrations of high-energy phosphate metabolites in normal, hypertrophied, and failing human myocardium measured noninvasively with (31)P-SLOOP magnetic resonance spectroscopy. J Am Coll Cardiol 2002; 40: 1267–74.

176. Neubauer S, Horn M, Cramer M et al. Myocardial phosphocreatine-to-ATP ratio is a predictor of mortality in patients with dilated cardiomyopathy. Circulation 1997; 96: 2190–6.

177. Nascimben L, Ingwall JS, Pauletto P et al. Creatine kinase system in failing and nonfailing human myocardium. Circulation 1996; 94: 1894–901.

178. Dzeja PP, Redfield MM, Burnett JC et al. Failing energetics in failing hearts. Curr Cardiol Rep 2000; 2: 212–17.

179. Ye Y, Gong G, Ochiai K et al. High-energy phosphate metabolism and creatine kinase in failing hearts: a new porcine model. Circulation 2001; 103: 1570–6.

180. Spindler M, Engelhardt S, Niebler R et al. Alterations in the myocardial creatine kinase system precede the development of contractile dysfunction in beta(1)-adrenergic receptor transgenic mice. J Mol Cell Cardiol 2003; 35: 389–97.

181. Gwathmey JK, Copelas L, MacKinnon R et al. Abnormal intracellular calcium handling in myocardium from patients with end-stage heart failure. Circ Res 1987; 61: 70–6.

182. Davies CH, Davia K, Bennett JG et al. Reduced contraction and altered frequency response of isolated ventricular myocytes from patients with heart failure. Circulation 1995; 92: 2540–9.

183. Maier LS, Braunhalter J, Horn W et al. The role of SR Ca^{2+}-content in blunted inotropic responsiveness of failing human myocardium. J Mol Cell Cardiol 2002; 34: 455–67.

184. Sipido KR, Stankovicova T, Flameng W et al. Frequency dependence of Ca^{2+} release from the sarcoplasmic reticulum in human ventricular myocytes from end-stage heart failure. Cardiovasc Res 1998; 37: 478–88.

185. Mulieri LA, Hasenfuss G, Leavitt B et al. Altered myocardial force–frequency relation in human heart failure. Circulation 1992; 85: 1743–50.

186. Hasenfuss G, Mulieri LA, Leavitt BJ et al. Alteration of contractile function and excitation–contraction coupling in dilated cardiomyopathy. Circ Res 1992; 70: 1225–32.

187. Dipla K, Mattiello JA, Jeevanandam V et al. Myocyte recovery after mechanical circulatory support in humans with end-stage heart failure. Circulation 1998; 97: 2316–22.

188. Schmidt U, Hajjar RJ, Helm PA et al. Contribution of abnormal sarcoplasmic reticulum ATPase activity to systolic and diastolic dysfunction in human heart failure. J Mol Cell Cardiol 1998; 30: 1929–37.

189. Hajjar RJ, Grossman W, Gwathmey JK. Responsiveness of the myofilaments to Ca^{2+} in human heart failure: implications for Ca^{2+} and force regulation. Basic Res Cardiol 1992; 87 Suppl 1: 143–59.

190. Denvir MA, MacFarlane NG, Cobbe SM et al. Sarcoplasmic reticulum and myofilament function in chemically-treated ventricular trabeculae from patients with heart failure. Cardiovasc Res 1995; 30: 377–85.

191. Brixius K, Savvidou-Zaroti P, Mehlhorn U et al. Increased Ca^{2+}-sensitivity of myofibrillar tension in heart failure and its functional implication. Basic Res Cardiol 2002; 97 (Suppl 1): I111–17.

192. Rasmussen RP, Minobe W, Bristow MR. Calcium antagonist binding sites in failing and nonfailing human ventricular myocardium. Biochem Pharmacol 1990; 39: 691–6.

193. Mewes T, Ravens U. L-type calcium currents of human myocytes from ventricle of non-failing and failing hearts and from atrium. J Mol Cell Cardiol 1994; 26: 1307–20.

194. Piot C, Lemaire S, Albat B et al. High frequency-induced upregulation of human cardiac calcium currents. Circulation 1996; 93: 120–8.

195. Ouadid H, Albat B, Nargeot J. Calcium currents in diseased human cardiac cells. J Cardiovasc Pharmacol 1995; 25: 282–91.

196. Schwinger RH, Hoischen S, Reuter H et al. Regional expression and functional characterization of the L-type Ca^{2+}-channel in myocardium from patients with end-stage heart failure and in non-failing human hearts. J Mol Cell Cardiol 1999; 31: 283–96.

197. Takahashi T, Allen PD, Lacro RV et al. Expression of dihydropyridine receptor (Ca^{2+} channel) and calsequestrin genes in the myocardium of patients with end-stage heart failure. J Clin Invest 1992; 90: 927–35.

198. Hullin R, Asmus F, Ludwig A et al. Subunit expression of the cardiac L-type calcium channel is differentially regulated in diastolic heart failure of the cardiac allograft. Circulation 1999; 100: 155–63.

199. Pieske B, Maier LS, Bers DM et al. Ca^{2+} handling and sarcoplasmic reticulum Ca^{2+} content in isolated failing and nonfailing human myocardium. Circ Res 1999; 85: 38–46.

200. Mercadier JJ, Lompre AM, Duc P et al. Altered sarcoplasmic reticulum Ca^{2+}-ATPase gene expression in the human ventricle during end-stage heart failure. J Clin Invest 1990; 85: 305–9.

201. Schwinger RH, Bohm M, Schmidt U et al. Unchanged protein levels of SERCA II and phospholamban but reduced Ca^{2+} uptake and Ca^{2+}-ATPase activity of cardiac sarcoplasmic reticulum from dilated cardiomyopathy patients compared with patients with nonfailing hearts. Circulation 1995; 92: 3220–8.

202. Limas CJ, Olivari MT, Goldenberg IF et al. Calcium uptake by cardiac sarcoplasmic reticulum in human dilated cardiomyopathy. Cardiovasc Res 1987; 21: 601–5.

203. Arai M, Alpert NR, MacLennan DH et al. Alterations in sarcoplasmic reticulum gene expression in human heart failure. A possible mechanism for alterations in systolic and diastolic properties of the failing myocardium. Circ Res 1993; 72: 463–9.

204. Linck B, Boknik P, Eschenhagen T et al. Messenger RNA expression and immunological quantification of phospholamban and SR-Ca^{2+}-ATPase in failing and nonfailing human hearts. Cardiovasc Res 1996; 31: 625–32.

205. Barrans JD, Allen PD, Stamatiou D et al. Global gene expression profiling of end-stage dilated cardiomyopathy using a human cardiovascular-based cDNA microarray. Am J Pathol 2002; 160: 2035–43.

206. Hasenfuss G, Reinecke H, Studer R et al. Relation between myocardial function and expression of sarcoplasmic reticulum Ca^{2+}-ATPase in failing and nonfailing human myocardium. Circ Res 1994; 75: 434–42.

207. Meyer M, Schillinger W, Pieske B et al. Alterations of sarcoplasmic reticulum proteins in failing human dilated cardiomyopathy. Circulation 1995; 92: 778–84.

208. Studer R, Reinecke H, Bilger J et al. Gene expression

of the cardiac Na^+–Ca^{2+}exchanger in end-stage human heart failure. Circ Res 1994; 75: 443–53.

209. Pieske B, Maier LS, Schmidt-Schweda S. Sarcoplasmic reticulum Ca^{2+} load in human heart failure. Basic Res Cardiol 2002; 97 (Suppl 1): I63–71.

210. Movsesian MA, Karimi M, Green K et al. Ca^{2+}-transporting ATPase, phospholamban, and calsequestrin levels in nonfailing and failing human myocardium. Circulation 1994; 90: 653–7.

211. Feldman AM, Ray PE, Silan CM et al. Selective gene expression in failing human heart. Quantification of steady-state levels of messenger RNA in endomyocardial biopsies using the polymerase chain reaction. Circulation 1991; 83: 1866–72.

212. Brillantes AM, Allen P, Takahashi T et al. Differences in cardiac calcium release channel (ryanodine receptor) expression in myocardium from patients with end-stage heart failure caused by ischemic versus dilated cardiomyopathy. Circ Res 1992; 71: 18–26.

213. Holmberg SR, Williams AJ. Single channel recordings from human cardiac sarcoplasmic reticulum. Circ Res 1989; 65: 1445–9.

214. Nimer LR, Needleman DH, Hamilton SL et al. Effect of ryanodine on sarcoplasmic reticulum Ca^{2+} accumulation in nonfailing and failing human myocardium. Circulation 1995; 92: 2504–10.

215. Reinecke H, Studer R, Vetter R et al. Cardiac Na^+/Ca^{2+}exchange activity in patients with end-stage heart failure. Cardiovasc Res 1996; 31: 48–54.

216. Schmidt TA, Allen PD, Colucci WS et al. No adaptation to digitalization as evaluated by digitalis receptor (Na–K-ATPase) quantification in explanted hearts from donors without heart disease and from digitalized recipients with end-stage heart failure. Am J Cardiol 1993; 71: 110–14.

217. Norgaard A, Bagger JP, Bjerregaard P et al. Relation of left ventricular function and Na,K-pump concentration in suspected idiopathic dilated cardiomyopathy. Am J Cardiol 1988; 61: 1312–5.

218. Ishino K, Botker HE, Clausen T et al. Myocardial adenine nucleotides, glycogen, and Na, K-ATPase in patients with idiopathic dilated cardiomyopathy requiring mechanical circulatory support. Am J Cardiol 1999; 83: 396–9.

219. Koumi S, Backer CL, Arentzen CE. Characterization of inwardly rectifying K^+ channel in human cardiac myocytes. Alterations in channel behavior in myocytes isolated from patients with idiopathic dilated cardiomyopathy. Circulation 1995; 92: 164–74.

220. Wettwer E, Amos GJ, Posival H et al. Transient outward current in human ventricular myocytes of subepicardial and subendocardial origin. Circ Res 1994; 75: 473–82.

221. Nabauer M, Beuckelmann DJ, Uberfuhr P et al. Regional differences in current density and rate-dependent properties of the transient outward current in subepicardial and subendocardial myocytes of human left ventricle. Circulation 1996; 93: 168–77.

222. Bohm M, Gierschik P, Jakobs KH et al. Increase of Gi alpha in human hearts with dilated but not ischemic cardiomyopathy. Circulation 1990; 82: 1249–65.

223. Eschenhagen T, Mende U, Nose M et al. Increased messenger RNA level of the inhibitory G protein alpha subunit Gi alpha-2 in human end-stage heart failure. Circ Res 1992; 70: 688–96.

224. Neumann J, Schmitz W, Scholz H et al. Increase in myocardial Gi-proteins in heart failure. Lancet 1988; 2: 936–7.

225. Feldman AM, Cates AE, Veazey WB et al. Increase of the 40 000-mol wt pertussis toxin substrate (G protein) in the failing human heart. J Clin Invest 1988; 82: 189–97.

226. Owen VJ, Burton PB, Mullen AJ et al. Expression of RGS3, RGS4 and Gi alpha 2 in acutely failing donor hearts and end-stage heart failure. Eur Heart J 2001; 22: 1015–20.

227. Bohm M, Gierschik P, Erdmann E. Quantification of Gi alpha-proteins in the failing and nonfailing human myocardium. Basic Res Cardiol 1992; 87 Suppl 1: 37–50.

228. Feldman AM, Cates AE, Bristow MR et al. Altered expression of alpha-subunits of G proteins in failing human hearts. J Mol Cell Cardiol 1989; 21: 359–65.

CHAPTER 14

Mechanisms of reversible ischemic dysfunction

Christophe Depre, Stephen F Vatner

Introduction

Because the cardiac myocyte has a limited capacity for regeneration, most forms of heart disease, including ischemic heart disease, are characterized by a loss of cardiomyocytes.[1-3] Therefore, the protection of ischemic myocardium from cell death has been a major focus for basic and applied research over the past 40 years. Despite the seminal observation by Tennant and Wiggers that the regional dysfunction which accompanies coronary artery occlusion remains reversible if the ischemic period is brief (less than 20 minutes),[4] and despite the morphological characterization of the ischemic myocardium submitted to temporary coronary occlusion,[5] 'it was widely assumed in the 1960s that myocardium perfused by a vessel which became acutely occluded was irreversibly injured'.[6] From the pathological studies, it became clear, however, that reversibility of ischemic damage could be promoted by limiting the time of hypoperfusion, which provided the experimental basis for early reperfusion in patients with acute myocardial infarction.[7,8]

Despite the establishment of the concept of myocardial reperfusion, any evidence of prolonged left ventricular dysfunction following myocardial ischemia was thought at that time to be due to irreversible myocyte damage. In the 1970s, this concept was challenged when the ultrasonic dimension technique permitted measurements of discrete segments of regional myocardial contrac-

tion in the chronically instrumented conscious animal.[9,10] These studies demonstrated that complete occlusion of a major coronary artery induced modest hyperkinesis, i.e. increased contraction, in the remote non-ischemic territory, but either dyskinesis (paradoxical bulging), or akinesis (no systolic contraction) in the central ischemic zone, and hypokinesis (depressed contraction) adjacent to the central ischemic zone (Fig. 14.1) These studies carefully described the time course of changes in regional wall motion compared with changes in myocardial blood flow, regional electrograms and systolic and diastolic function following acute coronary artery occlusion and subsequent reperfusion.[9-12] These studies on coronary occlusion and reperfusion led to the discovery in the dog heart that a relatively short episode of ischemia (up to 15 minutes) is followed by a period of severe postischemic dysfunction, including dyskinesis, despite the full restoration of blood flow and resolution of the electrocardiogram (ECG) changes, complete absence of irreversible damage, and eventually full functional recovery (Fig. 14.2).[9,10] This condition became known as 'myocardial stunning',[13] and was the first evidence that severely dysfunctional myocardium does not necessarily represent irreversibly injured myocardium. In the 1980s, this concept of progressive functional reversibility in a setting of acute ischemia–reperfusion was extended by several observations in patients that long-term myocardial dysfunction due to coronary

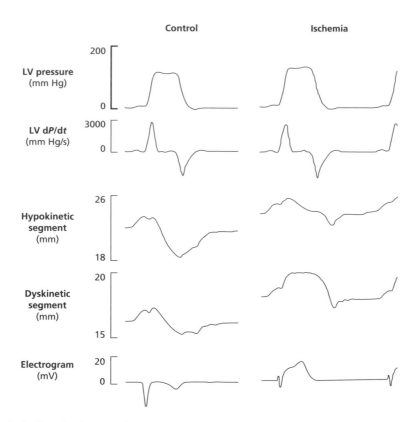

Figure 14.1. *The hallmark of regional myocardial ischemia is regional myocardial dysfunction. This figure shows a recording of phasic waveforms of left ventricular (LV) pressure, LV dP/dt, and two regional segments, the first of which was destined to become hypokinetic and the second of which became dyskinetic, and lastly, the electrogram. Even with complete coronary artery occlusion, it is typical for the central ischemic zone to display dyskinesis, i.e. paradoxical bulging, or akinesis (no motion during systole, not shown), and for regions adjacent to the central ischemic zone to demonstrate hypokinesis, i.e. reduced systolic shortening, while the regional electrogram demonstrates ST segment elevation. Adapted with permission from Vatner SF, Baig H, Manders WT et al. Effects of a cardiac glycoside on regional function, blood flow, and electrograms in conscious dogs with myocardial ischemia. Circ Res 1978; 43: 413–23.*

artery disease does not necessarily result in irreversible damage, but may be reversed upon revascularization,[14] a condition coined 'myocardial hibernation'.[15,16] Its characterization in patients was greatly facilitated by the emergence of imaging techniques of flow and metabolism. Simultaneously, the observation that short episodes of ischemia limit infarct size if the myocardium is submitted to a subsequent severe coronary occlusion was described as 'ischemic preconditioning'.[17] The discovery and characterization of ischemic preconditioning illustrates a mechanism of cardio-

protection that remains the 'gold standard' of cardiac survival mechanisms, but most of all, it was the first demonstration that, all other parameters remaining equal, infarct size can be reduced therapeutically.

Therefore, in four decades, the belief that coronary artery occlusion results in myocardial infarction and irreversible damage evolved into the understanding that viability can be maintained by early reperfusion and through a spectrum of survival and cardioprotective mechanisms described both in animal models and in patients. Most importantly,

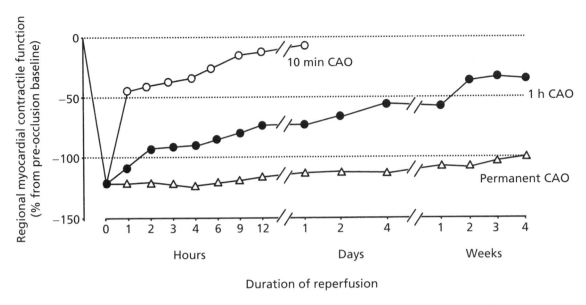

Figure 14.2. *Functional consequences of different durations of coronary artery occlusion. This figure compares the effects of myocardium submitted to 10 min, 1 hour or permanent coronary artery occlusion (CAO) on contractile function. The 10-min episode (open circles) illustrates an example of myocardial stunning with fully reversible post-ischemic dysfunction, 24 h after the brief ischemic insult, whereas the comparison of the two other groups illustrates the importance of coronary artery reperfusion, i.e. with permanent CAO (open triangles) regional function became dyskinetic and never resumed any systolic contraction over the 4-week monitoring period, whereas reperfusion after 1 h (filled circles) demonstrated prolonged stunning, but regional function eventually recovered 70% of systolic contraction. Adapted with permission from Lavallee M, Cox D, Patrick T et al. Salvage of myocardial function by coronary artery reperfusion 1, 2, and 3 hours after occlusion in conscious dogs. Circ Res 1983; 53: 235–47.*

these discoveries fueled remarkable therapeutic improvements that led to a dramatic decrease in mortality accompanying acute myocardial infarction. However, although these mechanisms were characterized meticulously at both the physiological and morphological levels, their molecular basis still remains an area of intense investigation.

Reversibility of acute ischemic damage

Transient coronary artery occlusion results in different syndromes, which include myocardial stunning, coronary reperfusion and ischemic preconditioning. These conditions of reversible ischemia differ from irreversible ischemic damage at both the physiological and molecular levels.

Physiological determinants of myocardial stunning

Stunning is the prolonged dysfunction of the ischemic heart which persists after reperfusion, despite the normalization of blood flow, and which eventually resolves with complete contractile recovery, provided no other ischemic episode intervenes.[9,13]

Due to various experimental approaches using different species, the pathogenesis of myocardial stunning primarily diverges into two hypotheses, i.e. oxygen radical and Ca^{2+} hypothesis.[18,19] The Ca^{2+} hypothesis postulates that myocardial stunning is the result of altered intracellular Ca^{2+} homeostasis, and implies that a transient Ca^{2+} overload occurs during the early phase of reperfusion, which contributes to the pathogenesis of myocardial stunning.[20,21] The oxygen radical hypothesis postulates that myocardial stunning is caused by reactive oxygen species, including superoxide anion, hydrogen peroxide and hydroxyl radical.

The role of impaired intracellular Ca^{2+} handling as a mechanism of contractile dysfunction in stunned myocardium was recently tested by Kim et al. in a pig model of myocardial stunning which was induced regionally by a 90-minute partial stenosis (40% reduction of blood flow) of the left anterior descending artery.[22] In this model, intracellular Ca^{2+} handling in myocytes from stunned and nonischemic regions can be compared in the same heart. The impaired contractile function observed *in vivo* was also present in isolated myocytes *in vitro* in this model. In this study, both Ca^{2+} transients and L-type Ca^{2+} current density were decreased in stunned myocytes, which indicates that the mechanisms of contractile dysfunction in the pig involve impaired Ca^{2+} handling. Of note, there is no degradation of troponin I (TnI) in the pig model of stunning,[22,23] which opposes the data reported below from rodent models. This study also demonstrated that relaxation of stunned myocardium appears to be directionally opposite in rodents and large mammals. Relaxation function is accelerated in stunned rodent myocytes,[24] but is prolonged in isolated myocytes from the pig, which is consistently observed *in vivo* in large mammalian models of stunning including studies in humans. These differences in relaxation may be due to phosphorylation of phospholamban, which is reduced in the swine model of stunning. The species difference in the handling of intracellular Ca^{2+} is not surprising, considering that rodent hearts must contract and relax at physiological frequencies that would be considered pathological tachycardia in large animals. Therefore, in the rodent, approximately 90% of the activator Ca^{2+} for contraction comes from intracellular stores from sarcoplasmic reticulum (SR).[25] In contrast, pigs and most other large animals derive more of their activator Ca^{2+} from extracellular Ca^{2+} entry from the L-type Ca^{2+}

channel. As a result, continued pacing at higher frequencies leads to a negative 'staircase' in the rat ventricle, which is due to a combination of a small decrease in SR Ca^{2+} content and a decrease in fractional release at higher frequencies, i.e. incomplete mechanical restitution.[25] In contrast, large mammalian cardiac muscles show a positive 'staircase', which may be associated with an increase in L-type Ca^{2+} channel current, diastolic Ca^{2+} concentration, and Ca^{2+} availability. In addition, there are well-described differences in sarcomere composition, essentially in the expression of myosin heavy chain (MHC) isoforms (predominant α isoform in rodents versus predominant β isoform in larger mammals).[26] Therefore, it is possible that differences in phospholamban phosphorylation, Ca^{2+} handling, and potential alterations in other proteins (troponin C and troponin T) may explain myocardial stunning in large mammals.

Although the seminal description of stunning was made in a large mammalian model,[9] the Ca^{2+} hypothesis was largely developed in rodent models. Considering the species differences in Ca^{2+} handling summarized above, it is not surprising that divergent data were obtained in rodents. During stunning in rodents, transient Ca^{2+} overload activates the Ca^{2+}-dependent protease calpain I, which degrades, affects the phosphorylation state, and induces covalent modifications of myofilaments, such as TnI.[24,27-30] As a result, there is a decreased responsiveness to Ca^{2+}, manifested by a decrease in the maximal force and a relative insensitivity to extracellular Ca^{2+} concentration. More recently, Murphy et al. described a phenotype of stunned myocardium, characterized by a reduced myofilament Ca^{2+} sensitivity, in transgenic mice expressing the major degradation production of TnI induced by calpain.[31] Consequently, myocardial stunning can be viewed as a short-term alteration of myofilament function. However, as mentioned above, recent studies found no TnI degradation or phosphorylation in large mammalian models.[23,32] Furthermore, it has recently been suggested that proteolysis of TnI could be the result of increased diastolic pressure.[33] There are also supportive studies in human myocardium that TnI proteolysis occurs in myocardium of patients during bypass surgery.[27]

However, those data were collected in conditions where minor amounts of myocardial infarction or other diseases cannot be excluded, as noted by the presence of altered TnI in the myocardium prior to the cross clamp.[27]

The second hypothesis to explain the mechanism of stunning involves free radicals. Many studies from the early 1980s demonstrated that reactive oxygen species (ROS) cause myocardial stunning in both large mammalian and rodent models.[34–40] Evidence supporting this hypothesis is that free radicals are produced at the initial moment of reperfusion in the conscious animal.[41,42] This hypothesis is supported by additional lines of experimental evidence, including the increase of ROS generation in stunned myocardium,[35,43,44] the protection against myocardial stunning by antioxidants,[43] and the contractile dysfunction induced by direct exposure to ROS. Interestingly, myocardial stunning is exacerbated after cardiac denervation, due to an increased release of ROS and the formation of peroxynitrite from the combination of ROS and nitric oxide.[45] The release of ROS in this condition is so severe that irreversible damage and patchy necrosis ensue, whereas the normally innervated heart fully recovers.[45] Although the exact origin of ROS during ischemia–reperfusion and the mechanisms by which ROS cause contractile dysfunction are still unknown, several possibilities have been tested.[18,19,43] The one most possible mechanism, which would bridge the gap between both hypotheses, is that ROS alters intracellular Ca^{2+} handling by the SR Ca^{2+} ATPase pump (SERCA). Other potential targets are the Na^+/Ca^{2+} exchanger and the Na^+/K^+ pump. A deregulation of these pumps would increase Ca^{2+} transport through the sarcolemmal membrane,[46] and, together with the altered SERCA activity, would result in intracellular Ca^{2+} overload.[47] In addition, ROS can directly affect the cross-bridge function of the myofilaments.[48]

In summary, the pathogenesis of myocardial stunning is still not completely defined since it involves multifactorial processes and also requires complex intracellular interactions. However, discrepancies among previous studies could be reconciled on the basis of different experimental designs or species differences.

Physiological determinants of irreversible ischemic damage

Studies on the reversibility of ischemic heart disease began with the development of large mammalian models, especially the dog, reproducing the conditions of ischemic myocardium due to coronary artery disease found in patients. Although the amplitude of ischemia was initially measured by changes in ST elevation on the ECG,[49] the development of piezoelectric crystals rapidly allowed a precise quantification of the loss of myocardial function subsequent to reduction in coronary flow.[9,50–55] The use of this technique showed the beat-to-beat adaptation of both systolic and diastolic function to ischemia,[11] and illustrated the complex interactions between external work, ventricular shortening and velocity of shortening of the ischemic heart.[12] Changes in the volume–pressure relationship were characterized as well, showing that, following coronary occlusion, regional myocardial stiffness first decreases,[56] then increases as irreversible damage extends.[57] The use of microspheres allowed a precise quantification of residual blood flow in the ischemic area.[58,59]

Simultaneously with these physiological contributions, Jennings' group characterized the different pathological stages that lead to irreversible damage and their correlation with the severity of the ischemic episode.[60,61] Remarkably, the physiological and morphological characterization of myocardial ischemia in the canine heart was immediately followed by experimental attempts to salvage myocardium from irreversible damage, through biochemical or pharmacological approaches (such as the administration of glucose-insulin-potassium, corticoids or hyaluronidase).[52,62–64] These endeavors were motivated by the high mortality of patients with myocardial infarction, resulting from acute heart failure and arrhythmias.

One of the first conclusions obtained from the canine model of coronary artery occlusion is that the extent of irreversible damage depends on the imbalance between oxygen supply and oxygen demand. Pharmacological substances (such as digitalis, glucagon, isoproterenol),[52,65,66] or physiological conditions (tachycardia, hyperthermia)[67] that increase the cardiac workload and oxygen demand, will automatically increase the extent of irreversible injury. The opposite can be achieved with pharmacological substances (propranolol, digitalis, nitrates),[52,68] or physiological conditions (increased coronary perfusion pressure, counterpulsation)[69] that decrease oxygen demand or increase oxygen supply.

The direct evidence obtained from histological characterization of myocardial ischemia led to the conclusion that ischemic damage is progressive and goes through different specific phases, from early lesions to late damage, following a wavefront from the subendocardium to subepicardium.[70–73] Cell death occurs in the subendocardium within one hour following occlusion, whereas the entire area-at-risk will be irreversibly damaged after 12–24 hours.[72] Until late after occlusion, the ischemic lesions remain patchy and heterogeneous,[74] due to the fact that cell death is preceded by a full spectrum of sequential events (loss of high-energy phosphates, loss of glycogen, accumulation of protons and lactate, myofilament disorganization . . .) proceeding at a speed that depends on the severity of ischemia and the distribution of collateral flow.[75,76] In addition, neither myocardial metabolic demand nor myocardial perfusion is entirely homogenous throughout the myocardium. These basic observations led to the understanding that early 'intervention' (physiological or pharmacological) could decrease the extent of irreversible lesions and improve the functional recovery of the ischemic area. By far the most beneficial of these potential interventions remains early reperfusion.

Importance of early reperfusion in ischemic reversibility

The work by Ross and colleagues illustrated the importance of early reperfusion in the limitation of contractile dysfunction and irreversible ischemic damage.[51,77,78] At that time, experimental studies on coronary artery reperfusion were further supported by the expanding use of surgical revascularization techniques, such as bypass surgery. In the canine model, in dogs submitted to reperfusion, even up to 6 hours of coronary occlusion resulted in a smaller infarct size, especially in the subepicardium,[74] than dogs submitted to chronic coronary occlusion.[79] Time-course experiments in conscious animals defined the time frame during which myocardial salvage is possible.[80] Although earlier reperfusion elicits a higher risk of tachyarrhythmias,[81,82] it is accompanied, in the acute phase, by an improved cardiac contractile function,[77] and, at a later time point, by decreased infarct size.[78,83] Examples of recovery of regional myocardial function following reperfusion are presented in Fig. 14.2. Whereas a 10-minute episode of coronary occlusion and reperfusion is followed by myocardial stunning for up to one day, and full functional recovery, an episode of one-hour occlusion leads to irreversible damage, which is however significantly less than the damage accompanying a permanent coronary occlusion (Fig. 14.2). Interestingly, ischemic preconditioning paradoxically enhances the rate of tachyarrhythmias, while resulting in reduced necrosis.[84]

At the time of these investigations, the only method of early reperfusion was bypass surgery, a method that was controversial in patients with acute myocardial infarction.[85–87] However, later on, these seminal investigations paved the way for early revascularization of occluded arteries by thrombolysis,[7,88,89] or by angioplasty.[90,91] The TIMI (Thrombolysis In Myocardial Infarction) trial was the first multicenter demonstration in a large series of patients that early reperfusion is the optimal method to salvage myocardial tissue, to preserve ventricular function and to improve survival.[92] These studies also illustrated that cardiac function in an ischemic territory may recover only weeks to months after reperfusion,[51,80,93] similar to the data presented in Fig. 14.2. This finding is of the greatest importance in the assessment of functional recovery after bypass surgery in patients with chronic ventricular dysfunction (hibernating myocardium).[94] This also indicates that myocardial stunning may be elicited following a more severe episode of ischemia–reperfusion, because the dyskinesis that follows reperfusion is not an accurate indication of the ultimate reversibility of function and salvage of ischemic myocardium.[80,95] As shown in Fig. 14.1, temporary ischemia may elicit hypokinesis in the normal, remote myocardium, and dyskinesis in the ischemic territory, yet this abnormality remains reversible upon reperfusion.

Physiological determinants of ischemic preconditioning

Ischemic preconditioning can be defined as the cardioprotective effect conferred by a single or multiple brief episodes of occlusion/reperfusion against the development of irreversible damage following a subsequent, more sustained ischemic episode. Ischemic preconditioning is potentially the most important contribution to the concept of cardiac cell survival, because, although its molecular mechanisms are still the matter of intense investigation, it represents the most striking proof of the principle that infarct size can be manipulated. Although the early experiments characterizing myocardial ischemia in the canine model led rapidly to the experimentation of cardioprotective protocols, none of these has been shown so far to reduce significantly infarct size in patients.[52,62] Reciprocally, ischemic preconditioning is a mechanism of cardioprotection that is both endogenous and remarkably powerful.[18] Interestingly, the seminal observation of ischemic preconditioning was made in the same canine model that had been used to characterize reversible ischemic injury (myocardial stunning) during the previous 30 years,[17] and offered a far greater reduction of infarct size (up to 80%) than any protocol tested before. Importantly, the preconditioning involving either a single or multiple episodes of ischemia, is followed by myocardial stunning. Thus, the first window of preconditioning generally occurs in the presence of myocardial stunning.

Although the first window of ischemic preconditioning confers protection for 1–2 hours after the stimulus is applied,[96] the description of a second window of preconditioning (SWOP),[97,98] which confers protection during the 24–72 hours following the preconditioning stimulus, led to the concept that the ischemic insult can be 'memorized' by the heart, thereby creating a delayed and prolonged mechanism of cardioprotection. This second window could be more relevant than acute preconditioning for patients with recurrent ischemia, because of the prolonged phase of protection that is afforded and because, unlike the acute phase, the SWOP occurs long after the resolution of stunning, but protects against myocardial stunning.[44]

Clinical studies have investigated whether preconditioning occurs *in vivo* in the human heart, although it is more difficult to derive clear conclusions, since results have to be inferred in the absence of a definitive experiment. The most obvious condition would be in patients with chronic coronary artery disease submitted to recurrent episodes of ischemia and at risk for myocardial infarction. Several studies failed to show any improved myocardial function in patients experiencing pre-infarct angina.[99–101] However, pre-infarct angina reduces creatine kinase (CK) release and mortality after acute myocardial infarction,[102] and a recent study showed that a cardioprotective effect can be detected if the ischemic episode intervenes 24–72 hours before infarction, a time frame that is compatible with a SWOP.[103] It should be noted, however, that pre-infarct angina might affect coronary flow reserve or increase the recruitment of collaterals, which would limit irreversible injury independently of any preconditioning. Another potential application of cardioprotection by preconditioning relates to patients undergoing coronary bypass surgery, in whom the preconditioning stimulus can be applied by successive episodes of aortic cross clamping. This procedure was associated with a marked reduction in troponin T release, a better preservation of intracellular ATP concentration and improved post-surgical recovery.[104–107]

Because a *sine qua non* for preconditioning is reperfusion, it can be argued that the conflicting data in patients with myocardial infarction may result from varying degrees of revascularization of the occluded vessels. Therefore, a more controlled method to study the occurrence of preconditioning in the human heart is angioplasty, which requires several episodes of coronary occlusion induced by balloon inflation separated by reperfusion. Studies have shown that, after coronary occlusion secondary to balloon inflation of at least one minute, the physiological, electrographic and biochemical parameters of ischemia are blunted during the subsequent coronary occlusions.[108–110] As mentioned above, these studies remain biased by the potential recruitment of collaterals and coronary flow reserve, although such recruitment cannot totally explain the improvement.[108,109] Also, the endpoint to appreciate the effects of preconditioning is traditionally a decrease in infarct size, which of course is not the endpoint during a procedure of angioplasty. An alternative parameter in clinical practice is to measure the effects of preconditioning on the ST segment elevation on the ECG, but this parameter does not necessarily correlate with reduced infarct size.[111] However, this response is sensitive to agents modulating the activity of the adenosine receptor,[112] and of the K_{ATP} channels,[113,114] which, as discussed below, participate in the mechanisms of ischemic preconditioning.

In another clinical setting, it has been proposed that preconditioning participates in the improved tolerance to exercise in patients with ischemic heart disease. When patients exercise until angina becomes symptomatic, then rest, they can resume exercise without exacerbation of the symptoms, a condition known as the 'warm-up phenomenon'.[115,116] Although this condition seems compatible with preconditioning,[117,118] it may also be due to unrelated factors, such as a recruitment of coronary flow reserve and improved coronary artery vasodilation.[119] In addition, the warm-up phenomenon is not inhibited by K_{ATP} channel blockers.[120]

The molecular mechanisms of ischemic preconditioning are still intensely investigated. The current consensus is that the protection afforded during the acute phase of preconditioning is mediated, upon activation of G_i protein-coupled receptors, by the opening of the mitochondrial K_{ATP} channels, which leads to K^+ entry into the mitochondria.[121–123] K_{ATP} channel opening may improve the resistance of the mitochondria against Ca^{2+} overload during subsequent ischemic episodes.[124] However, it seems that the K_{ATP} channels are not the endpoint of the mechanism of preconditioning, but that their opening leads to the release of free radicals, which in turn activate specific kinases, including PKC and tyrosine kinases, to afford the cardioprotection of preconditioning.[125–127] Although the sequence 'G_i protein–K_{ATP} channels–PKC' represents an attractive sequence of events to explain the phenomenon of preconditioning, the actual effector remains a matter of controversy.[127]

The protective mechanism leading to the SWOP probably involves changes in gene expression.[128]

The same triggers (such as adenosine) and effectors (such as PKC or tyrosine kinases) of the early phase of preconditioning can also activate transcription factors, resulting in the adaptation of transcription of genes involved in the second window. These genes encode cytoprotective molecules, such as the heat-shock proteins,[129] anti-oxidant enzymes (aldose reductase, superoxide dismutase),[130,131] cyclooxygenase-2 (COX-2),[132] and the inducible isoform of nitric oxide (NO) synthase (iNOS).[133–135] More recently, the neuronal isoform of NO synthase (nNOS) has also been suggested to be an important mediator of the SWOP, both in rodents,[136] and in a swine model.[45] Although the role of these specific enzymes in the cardioprotection of delayed preconditioning has been demonstrated in several models, the molecular mechanisms by which protection is conferred remain to be fully elucidated.

Molecular determinants of cell death

The various mechanisms of ischemia–reperfusion injury (including Ca^{2+} overload, ROS, metabolic inhibition and others) lead to cell death by either necrosis or apoptosis. Apoptosis is an energy-dependent process evolving through specific phases, which is followed by the phagocytosis of the apoptotic body by a macrophage, which limits the inflammatory reaction. Cell death by necrosis, reciprocally, occurs as a consequence of severe energy depletion, followed by the incapacity of membrane pumps and carriers to maintain electrochemical gradients, which induces cell swelling and rupture. This process leads to a severe infiltration of the myocardium by polymorphonuclear neutrophils and macrophages, which initiate a cascade of reactions leading to fibroblast recruitment, fibrosis and the replacement of the damaged cardiac tissue by scar tissue. Necrosis is an 'out-of-control' process, which can be limited only by reoxygenation and ATP repletion, whereas apoptosis is a highly regulated mechanism that is controlled by activators (Bad, Bax, Smac/Diablo ...) and inhibitors (Bcl_2, Bcl_{XL}, IAP, ARC, heat-shock proteins ...). Whereas necrosis predominates in the central zone of the ischemic territory, apoptosis predominates in the peripheral zone and thereby delineates the extent of irreversible damage by limiting the inflammatory reaction.

These pro-death mechanisms can be balanced by pro-survival mechanisms. The most classically described mechanisms of cell survival in the cardiac myocyte include the hypoxia-induced factor-1α (HIF-1α), multiple heat-shock proteins and the Akt pathway.[137–153]

HIF-1α is part of a remarkable mechanism of oxygen sensing, which, as a transcription factor, provides a coordinated response to hypoxia. This response includes the increased expression of genes participating in vasculogenesis (vascular endothelial growth factor (VEGF) and its receptor), anaerobic metabolism (glycolytic enzymes) and growth (insulin-like growth factor (IGF), transforming growth factor-β (TGF-β)).[137] As a general stress signal, hypoxia-inducible factor-1α (HIF-1α) can promote both a survival response against ischemia,[154] and a growth response during increased workload.[142]

Heat-shock proteins (HSP) represent about 10% of the total protein content of cardiac myocytes, and the heart expresses almost all of them. Several reports have shown that HSP70 overexpression dramatically reduces infarct size both *in vivo* and *in vitro*.[155–157] The small HSP27,[146] and the recently characterized cardiac-specific co-chaperone mmDJA4 also exert cytoprotective effects.[151] While their function of chaperone preventing the denaturation of binding proteins can explain their cytoprotective role during ischemic stress, more recent evidence suggests that heat-shock proteins are directly involved in the protection against apoptosis. HSP70, HSP10, HSP60, HSP90, HSP27 and αB-crystallin all participate in the inhibition of the mitochondria-mediated activation of caspase-3,[148,150,158–161] whereas HSP27 can also block the receptor-mediated apoptotic cascade.[162] The recently characterized crystalline H11 kinase also promotes both cardiac survival and growth.[163,164]

As reviewed recently,[165] Akt is a major regulator of cell growth and survival, both through a translational adaptation of the heart,[166–168] and through an activation of cytoprotective mechanisms that limit the cell loss induced by ischemia–reperfusion injury.[143,163,169–175] Akt is directly involved in survival mechanisms by blocking glycogen synthase kinase-3β, Bad and caspase-9,[176–178]

as well as by activating the endothelial isoform of NO synthase.[152]

These pathways present multiple interactions, and also offer a tight coupling between survival mechanisms and growth pathways. Glycogen synthase kinase-3β, a downstream target inhibited by Akt, is itself an inhibitor of the expression of heat-shock proteins.[179] Reciprocally, survival molecules, such as heat-shock proteins, modulate the growth activity of both extracellular signal-regulated kinase (ERK) and Akt pathways.[180,181] The same has been shown about the activation of the epidermal growth factor (EGF) pathway by HIF-1α,[182] and, reciprocally, the activation of HIF-1α by the Akt pathway[183], including the growth-promoting kinase mTOR (mammalian target of rapamycin).[184,185]

Reversibility of chronic ischemic damage

The 'ubiquity' of myocardial stunning presented above for acute coronary syndromes can be extended to the chronic setting as part of the mechanism for myocardial hibernation. The reversibility of this condition can be differentiated both at the physiological and molecular levels from ventricular remodeling, which follows a prolonged period of irreversible ischemic damage.

Cardiac remodeling after permanent occlusion

Cardiac remodeling initially is an attempt to compensate for the loss of cardiac muscle, but eventually represents in the long run one of the most common etiologies of heart failure. At the physiological level, cardiac remodeling can be defined as the process modifying ventricular size, shape and function as a consequence of increased load, decreased contractility and neurohormonal regulation.[186] At the molecular level, ventricular remodeling is defined as a combination of cellular hypertrophy,[187,188] apoptosis,[189] and extracellular fibrosis.[190] Irreversible ischemic damage and the fibrotic scar that ensues impair the homogeneity of contraction in the ventricle, which results in a change of its geometry to a more spherical shape.

229

This anatomical modification increases end-diastolic ventricular pressure and volume, which results in a right shift of the Starling curve, decreased output, and increased wall stress that further impairs the external work of the ventricle. The only mechanism to counteract this increased wall stress is an adjustment in wall thickness. This cannot be achieved in the peri-infarct territory (adjacent zone), which is subject to stretching and volume overload, resulting in an increased rate of apoptosis,[191,192] and wall thinning. At the remote non-infarcted territory, which is submitted to both volume and pressure overload, a pattern of cellular hypertrophy develops,[193] to increase the wall thickness and to compensate for the loss of myocardium due to infarction. The neurohormonal trigger of this hypertrophic response consists essentially in a hyperactivation of both the renin–angiotensin and the sympathetic systems, which results from the decreased cardiac output that inevitably accompanies the formation of scar tissue and the process of remodeling. This process will continue until the heart reaches a new equilibrium in wall stress, i.e. a new balance between ventricular pressure and wall thickness. However, if untreated, this equilibrium is fragile in the chronic setting by exposing the cardiac cells to higher energetic and contractile demand, which will inevitably lead to more cell death, wall stress and neurohormonal response, resulting in a spiraling progression into clinical heart failure. Therefore, the contractile function of the remote myocardium initially increases after infarction by hypercompensation, then subsequently deteriorates by cell loss.[194]

In the clinical setting, it is difficult and not necessarily relevant to differentiate the adjacent and the remote territory, therefore the cardiac function is considered as a whole. The best way to limit the damage of ventricular remodeling in patients consists of reducing the infarct size and reducing the wall stress that increases following the changes in shape and loading conditions of the ventricular chamber.[195] A reduction in infarct size is best achieved, as described above, by early reperfusion after coronary artery occlusion. The medical approach to decreasing wall stress mainly involves the inhibitors of the angiotensin converting enzyme (ACE) and β-blockers. Multicenter studies have shown repeatedly that ACE inhibitors improve the physiological parameters of the post-ischemic ventricle and reduce mortality.[196–198] The protective mechanisms following ACE inhibition involve a decrease in afterload, a decrease in the extent of scar tissue, an inhibition of apoptosis, and a decrease in the hypertrophic response of the ventricle.[199–202] Beta-blockers are a classic category of drugs for patients with ischemic heart disease and previous myocardial infarction, which prevent the sympathetic overdrive resulting from the decreased inotropic performance of the remodeled ventricle. As with ACE inhibitors, β-blockers improve both the contractile performance and survival of patients with ischemic heart disease.[203]

Persistent hypoperfusion versus repetitive stunning in hibernating myocardium

Myocardial hibernation represents a condition of chronic regional ventricular dysfunction in patients with coronary artery disease, which is progressively reversible after revascularization. The term 'hibernation' referring to this condition was first used by Diamond et al. in 1978.[15] Rahimtoola expanded on this concept and proposed that the hibernating heart downgrades its contractile function to reach a new equilibrium with reduced blood flow.[14,204] An alternative hypothesis is that baseline blood flow is maintained in the absence of stress but, because of impaired coronary flow reserve, stress results in an imbalance between myocardial supply and demand,[16,205–207] and eventually multiple episodes of demand ischemia and recovery lead to sustained, chronic ischemic dysfunction. The mechanisms underlying these two hypotheses are summarized in Fig. 14.3.

Hibernating myocardium is submitted to repetitive bouts of ischemia due to normally occurring increases in myocardial metabolic demand in the face of significant coronary stenosis and limited coronary reserve, yet it does not develop irreversible damage.[206,208] The diagnosis of this condition is of the highest priority in patients with chronic coronary artery disease, because successful coronary revascularization and restoration of coronary flow reserve in hibernating myocardium is fol-

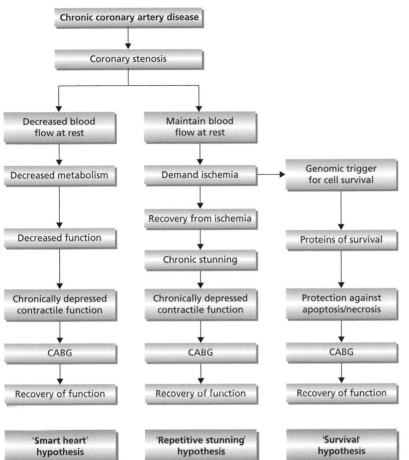

Figure 14.3. *Mechanisms of myocardial hibernation. This figure compares the two main hypotheses of hibernation ('smart heart' versus 'repetitive stunning') and includes a new concept, the role of the survival program which affords protection during chronic ischemia. CABG, coronary artery bypass graft.*

lowed by an improvement of contractile performance,[94,209] whereas, if left untreated, this condition leads to progressive exacerbation of heart failure by the mechanism of remodeling described above.

The original hypothesis generated to explain hibernating myocardium was that chronic hypoperfusion resulted in a self-protecting downregulation in myocardial function and metabolism to match the decreased supply, thereby minimizing energy requirements and preventing irreversible myocardial damage.[14,204] Consequently, myocardial function is recovered upon reflow, a condition referred to as perfusion–contraction matching.[210] This hypothesis is based essentially on clinical reports of post-surgical functional improvement after bypass surgery on the one hand, and measurements of depressed contractile function, accom-

panied by reduced relative regional blood flow in patients. Because the blood flow in the ischemic territory is measured relative to the normal, remote area, it is not necessarily reduced in absolute terms.[211]

The alternative hypothesis suggests that the myocardium is subject to repetitive episodes of ischemia–reperfusion resulting from an imbalance between myocardial metabolic demand, which can increase during exertion, and myocardial oxygen supply, which cannot increase appropriately due to the coronary artery disease and limited coronary reserve. This imbalance results in repetitive episodes of ischemic dysfunction followed by stunning that eventually create a sustained depression of contractile function.[205,207,212,213] This hypothesis is largely based on the measurement of normal

231

absolute blood flow (usually by positron emission tomography) in the hibernating myocardium between the episodes of increased demand.[211]

Because ischemic heart disease in humans can take decades to develop, there is no animal model that reproduces entirely the long-term condition of myocardial hibernation found in patients. However, several laboratories have developed models of either chronic low-flow ischemia or repetitive stunning to reproduce the hallmarks of human hibernating myocardium,[214] including chronic dysfunction, loss of myofibrils, accumulation of glycogen, alteration of Ca^{2+} metabolism, genomic response and some extent of cell degeneration.[32,215–218]

Models supporting the concept of chronic low-flow ischemia

Induction of 1–5 hours' coronary stenosis (20–70% of baseline coronary blood flow) was used as a model of 'short-term hibernation' in the pig.[219–223] This model demonstrated the possibility of short-term perfusion–contraction matching, but little information is available to demonstrate whether this matching can persist for the longer term (more than a week). The effects of 24-hour partial stenosis (40–50% reduction of blood flow) were also examined,[224,225] but this model induces subendocardial necrosis, which is contradictory to the concept of hibernating myocardium. Later, two additional models were introduced to simulate chronic contractile dysfunction as a result of chronic hypoperfusion. One attempt was a fixed stenosis in growing swine for three months,[215,216] and the other was progressive coronary stenosis induced by an ameroid constrictor in mini pig for one month.[218] The results from both models were divergent, because the chronic contractile dysfunction induced in both models was associated with hypoperfusion in the former condition, but with normal flow in the latter. Fallavollita et al.[23] found that, 2 months after a fixed stenosis, the hypokinesis was associated with normal resting flow, but myocardial blood flow reserve, assessed by adenosine, was reduced. However, after 3 months, hibernating myocardium was characterized by a reduced resting blood flow. Therefore, they proposed a revised hypothesis in which contractile dysfunction with normal flow ('chronic stunning'), precedes for a limited time the development of true hibernating myocardium. However, in another study,[218] the regional blood flow remained normal despite a chronic decrease in contractile function, which suggests that repetitive stunning may induce chronic contractile dysfunction.

Models supporting the concept of repetitive stunning

The induction of multiple episodes of myocardial stunning by excitement in conscious pigs leads to a stable chronic contractile dysfunction,[218] and the pathognomonic histologic features of hibernating myocardium.[213] This suggests that myocardial stunning and myocardial hibernation may coexist, or that hibernating myocardium may actually reflect chronic myocardial stunning. More recently, it was demonstrated in the pig that persistent stunning can lead to hibernation.[217] In that model, swine hearts were submitted to repetitive episodes of 90 min coronary stenosis followed by 12 hours' reperfusion, and this cycle was repeated up to six times. This cycle of ischemia–reperfusion is repeated to reproduce the conditions of transient and repeated ischemic episodes found in patients with ischemic heart disease. After the first ischemic episode, the regional function was depressed despite a normalization of blood flow, reflecting persistent stunning, but there was no further decrease in ventricular function after the following episodes of coronary stenosis, reflecting the mismatch between blood flow and contractile function that characterizes myocardial hibernation.[217] This model also showed metabolic characteristics of hibernating myocardium, such as a higher dependence upon glucose and glycogen accumulation.[217]

Genomic determinant of cardiac cell survival

Despite the abundant descriptions of altered myocardial blood flow, contraction and metabolism,[205,211,226–229] mechanistic insights into why hibernating myocardium survives during chronic ischemia have been limited. We present below an overview of the genomic changes that participate in mechanisms of survival, in both the acute and chronic conditions of myocardial ischemia. As it turns out, an investigation of survival mechanisms triggered at the gene level also offers a new molecular link between myocardial stunning and hibernation.

Acute changes in gene expression

The discovery that gene expression can be affected by changes in workload[230–232] rapidly led to the investigation of the adaptation of myocardial gene expression in the ischemic heart. Schaper's laboratory provided pioneering studies on alterations in gene expression in large mammalian models of ischemia–reperfusion. These studies were mainly designed to follow the time course of changes in genes coding for proto-oncogenes and Ca^{2+}-handling proteins.[233–235] The proto-oncogenes were chosen because they represent a rapid mechanism of transcriptional adaptation to stress, whereas Ca^{2+}-handling genes were studied to correlate their expression to the prolonged dysfunction that follows ischemia. Another goal was to investigate whether repetitive episodes of ischemia–reperfusion inducing preconditioning are accompanied by a change in gene expression. Several proto-oncogenes (such as c-fos or junB) showed an increase either during ischemia or reperfusion. Similarly, the Ca^{2+}-ATPase SERCA and calsequestrin were found to increase during post-ischemic dysfunction. These studies were the first to show that even short episodes of ischemia–reperfusion could affect myocardial gene expression. Also, they demonstrated the feasibility of investigating myocardial gene expression in large mammalian models of ischemic heart disease.

The absence of irreversible cellular damage in stunned myocardium may correspond either to an increased resistance of the heart to ischemia, or to the absence of stimuli triggering the pathways of cell death. The latter possibility is less likely, because even mild episodes of ischemia–reperfusion can activate different intracellular stress pathways leading to cell death.[39,236–238] One hypothesis that was tested by gene profiling is whether myocardial stunning triggers the coordinated expression of different sets of genes acting to protect the myocardium against irreversible damage. These experiments were performed in a swine model of myocardial stunning in which there was blood flow reduction through the left anterior descending

coronary artery by about 40% during 90 minutes, followed by full reperfusion.[239] Despite the normalization of blood flow after reperfusion, the contractile function remains depressed, which reflects myocardial stunning. A full functional recovery is typically observed after 48–72 hours, and pathological examination of this myocardium does not show any necrosis or apoptosis. We used this model of reversible ischemia to investigate whether the protection of the myocardium against irreversible damage correlates with a specific gene profile.[163] Interestingly, we found that more than 30% of the genes that were upregulated in stunned myocardium are involved in different mechanisms of cell survival, including: resistance to apoptosis, cytoprotection ('stress response') and cell growth. Many of them had been implicated previously in the survival of other cell types, but had not been described before in the heart. In particular, we found an induction in stunned myocardium of several genes not expressed in the normal myocardium and which participate in the development and growth of different forms of tumors.[163] An example is shown in Fig. 14.4 for H11 kinase, a small heat-shock protein with anti-apoptotic properties.[164,240] Measurement by quantitative polymerase chain reaction (PCR) shows that the expression of this gene starts to increase during ischemia, but peaks during post-ischemic stunning and returns back to normal after 12 hours, when the contractile function recovers. The sensitivity of the quantitative PCR also allows separate measurement of the sub-endocardium from the sub-epicardium, to show that the gene response is of higher amplitude in the sub-endocardium, where the flow reduction is the most important (Fig. 14.4). There is, therefore, a gradient of gene response that matches the gradient of flow reduction, showing that the nuclear response is not an 'all or nothing' phenomenon, but is proportional to the intensity of the initial stimulus. Therefore, this study demonstrates that non-lethal ischemia is accompanied by the expression of a genomic program of cell survival, which potentially counteracts apoptosis.

Prolonged changes in gene expression

Molecular studies of hibernating myocardium have focused mainly on the regulation of genes and

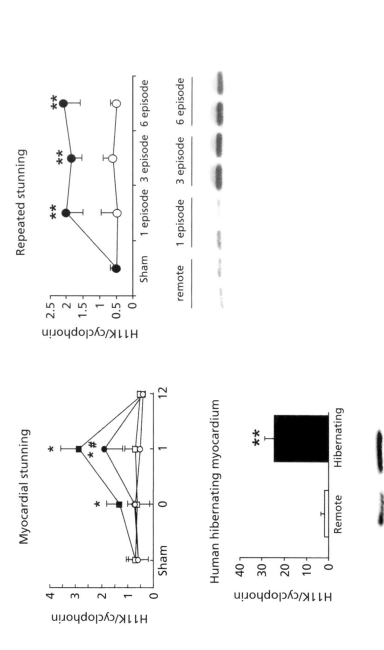

Figure 14.4. *Program of cardiac cell survival in myocardial stunning and hibernation. This figure shows the example of H11 kinase as a survival gene. Top left panel: The mRNA encoding H11 kinase is increased during myocardial stunning after recovery from one episode of a 90-min coronary stenosis in the pig in ischemic territory (filled symbols) but not in the normal area (open symbols), and progressively normalizes during reperfusion as function returns to normal. The gene response is more important in subendocardium (filled squares) than subepicardium (filled circles). Top right panel: When ischemia is repeated, i.e. six episodes of coronary stenosis and reperfusion every 12 hours, the mRNA (graph) increases again in the ischemic territory (filled circles) versus remote area (open circles), and the corresponding protein (immunoblot) progressively builds up. Bottom panel: In human hibernating myocardium compared to the remote territory from the same hearts, a similar regulation of the transcript (graph) and corresponding protein (immunoblot) is observed. *P < 0.05; **P < 0.01 versus remote; #P < 0.05 versus subendocardium. Error bars are standard error of the mean. Top left panel: reproduced with permission from Depre C, Tomlinson JE, Kudej RK et al. Gene program for cardiac cell survival induced by transient ischemia in conscious pig. Proc Natl Acad Sci USA 2001; 98: 9336–41; top right and bottom panels: modified with permission from Depre C, Kim S-J, John AS et al. Program of cell survival underlying human and experimental hibernating myocardium. Circ Res 2004; 95: 433–40.*

proteins which might explain the reduced cardiac contraction in animal models or in the human heart,[32,216,241–243] since these studies focused on proteins regulating Ca^{2+} flux,[32,216,241] inflammatory response,[242] or β-adrenergic receptor density.[243] No study specifically addressed the mechanisms related to prolonged survival. As shown in Fig. 14.3, this survival program may directly participate in the mechanisms of sustained viability of hibernating myocardium. Because it has been proposed that the human hibernating myocardium may result from repetitive episodes of ischemia–reperfusion, we determined whether the recurrence of ischemia plays a role in the expression of a cardioprotective mechanism. Measurements were performed in the swine model described above, in which repeated episodes of ischemia-reperfusion reproduce the pattern of myocardial hibernation described in patients.[217] Myocardial samples were taken from both the hibernating area and the normal territory from the same hearts during the reperfusion following one, three or six episodes of ischemia. H11 kinase was chosen as one of several examples because, as mentioned above, this gene is already regulated by myocardial stunning. As shown in Figure 14.4, the H11 kinase transcript was upregulated after the first episode of ischemia, consistently with the observation described in the acute setting of stunning. However, this gene was re-induced by subsequent ischemic episodes. To determine whether this transcriptional activation described in swine myocardium is followed by an increase of the corresponding protein, myocardial expression of H11 kinase was measured by Western blot in samples submitted to one, three or six cycles of ischemia-reperfusion, and the expression was compared between remote and ischemic areas (Fig. 14.4). There was no significant change in expression in the remote zone of the myocardium after one, three or six episodes. The expression of the protein was not increased after a single episode of ischemia in the ischemic territory compared to the non-ischemic zone, but increased progressively and significantly ($P < 0.05$) by the third episode and remained elevated after six episodes. Therefore, the repetition of ischemic stress induces a reactivation of the survival gene program and the progressive build-up of the corresponding proteins, resulting in increased resistance to additional episodes of ischemic stress. To determine whether a similar adaptation occurs in human hibernating myocardium, tissue samples were obtained from both the hibernating and remote territory of patients at the time of bypass surgery. The mRNA encoding H11 kinase and the corresponding protein were significantly upregulated in hibernating compared to remote myocardium. Therefore, these findings illustrate a novel mechanism underlying hibernating myocardium in the form of an endogenous genomic mechanism of cytoprotection that could subtend cell survival under conditions of prolonged ischemia.[244] The corollary in the animal model shows that a repetitive pattern of ischemia–reperfusion progressively results in a program of cytoprotection and survival.

Summary

Reversible ventricular dysfunction represents a spectrum of conditions that are all characterized by a mismatch between myocardial blood flow, function and morphology. In myocardial stunning, the myocardium is dysfunctional despite normal myocardial blood flow and the absence of necrosis. In myocardial hibernation, the myocardium is dysfunctional despite essentially normal resting blood flow. Following preconditioning, coronary artery occlusion induces markedly reduced infarction, despite the severity of flow reduction. These syndromes are intertwined, as exemplified by the episode of stunning following ischemic preconditioning, or by the prevalence of repetitive stunning in hibernating myocardium. In all cases, myocardial viability is maintained, both by early post-ischemic reperfusion and by the absence of remodeling that complicates irreversible injury. The mechanisms sustaining viability include an acute response (as occurs during stunning and ischemic preconditioning) involving Ca^{2+} homeostasis, free radicals and K^+-ATP channels. In other cases, long-term mechanisms are involved (such as during the second window of preconditioning and myocardial hibernation), and imply a change in gene expression that promotes a cell survival program. The seminal description of these syndromes of survival and their further characterization at both the physiological and molecular level were made possible by relevant chronic animal models, and in the clinical setting, by the development of novel diagnostic techniques (such as positron emission tomography, cardiac magnetic resonance, isotopic scanning) and, most importantly, led to proactive therapeutic strategies (such as early bypass surgery, thrombolysis and angioplasty) that dramatically reduced the mortality of one of the most common forms of disease worldwide, and also delayed or prevented the progression of more severe ischemic dysfunction to the development of heart failure. Whereas 'it was widely assumed in the 1960s that myocardium perfused by a vessel which became acutely occluded was irreversibly injured',[6] it is now clear, four decades later, that the early reperfusion of dysfunctional myocardium ultimately improves cardiac function, but that the improvement in function is markedly delayed because of myocardial stunning. In that sense, the study of the mechanisms of post-ischemic reversibility represents a remarkable example of translational research from the bench to the bedside.

Acknowledgements

This work was supported by National Institutes of Health grants HL33065, HL69020, HL62442, HL59139, AG14121, RR16592 and HL 33107 (SFV), HL 072863 and AHA 0230017N (CD).

References

1. Narula J, Haider N, Virmani R et al. Apoptosis in myocytes in end-stage heart failure. N Engl J Med 1996; 335: 1182–9.
2. Olivetti G, Abbi R, Quaini F et al. Apoptosis in human failing heart. New Engl J Med 1997; 336: 1131–41.
3. Palojoki E, Saraste A, Eriksson A et al. Cardiomyocyte apoptosis and ventricular remodeling after myocardial infarction in rats. Am J Physiol Heart Circ Physiol 2001; 280: H2726–31.
4. Tennant T, Wiggers CJ. Effect of coronary occlusion on myocardial contraction. Am J Physiol 1935; 112: 351–61.
5. Blumgart H, Gilligan D, Schlessinger M. Experimental studies on the effect of temporary occlusion of coronary arteries. Am Heart J 1941; 22: 374–89.
6. Braunwald E. Personnal reflections on efforts to reduce ischemic myocardial damage. Cardiovasc Res 2002; 56: 332–6.
7. Rentrop K, Blanke H, Karsch K. Effects of nonsurgical coronary reperfusion on the left ventricle in human subjects compared with conventional treatment. Study of 18 patients with acute myocardial infarction treated with intracoronary infusion of streptokinase. Am J Cardiol 1982; 49: 1–8.
8. Topol E. Current status and future prospects for acute myocardial infarction therapy. Circulation 2003; 108: III6–13.
9. Heyndrickx GR, Millard RW, McRitchie RJ et al. Regional myocardial functional and electrophysiological alterations after brief coronary artery occlusion in conscious dogs. J Clin Invest 1975; 56: 978–85.
10. Heyndrickx G, Baig H, Nellens P et al. Depression of regional blood flow and wall thickness after brief coronary occlusions. Am J Physiol Heart Circ Physiol 1978; 234: H653–9.
11. Ihara T, Komamura K, Shen Y et al. Left ventricular systolic dysfunction precedes diastolic dysfunction during myocardial ischemia in conscious dogs. Am J Physiol Heart Circ Physiol 1994; 267: H333–43.
12. Pagani M, Vatner S, Baig H et al. Initial myocardial adjustments to brief periods of ischemia and reperfusion in the conscious dog. Circ Res 1978; 43: 83–92.
13. Braunwald E, Kloner RA. The stunned myocardium: prolonged, postischemic ventricular dysfunction. Circulation 1982; 66: 1146–9.
14. Rahimtoola SH. A perspective on the three large multicenter randomized clinical trials of coronary bypass surgery for chronic stable angina. Circulation 1985; 72: V123–35.
15. Diamond G, Forrester J, deLuz P et al. Post-extrasystolic potentiation of ischemic myocardium by atrial stimulation. Am Heart J 1978; 95: 204–9.
16. Braunwald E, Rutherford JD. Reversible ischemic left ventricular dysfunction: evidence for hibernating myocardium. J Am Coll Cardiol 1986; 8: 1467–70.
17. Murry CE, Jennings RB, Reimer KA. Preconditioning with ischemia: a delay of lethal cell injury in ischemic myocardium. Circulation 1986; 74: 1124–36.
18. Kloner RA, Bolli R, Marban E et al. Medical and cellular implications of stunning, hibernation, and preconditioning. An NHLBI workshop. Circulation 1998; 97: 1848–1867.

19. Bolli R. Mechanism of myocardial 'stunning'. Circulation 1990; 82: 723–38.

20. Kusuoka H, Porterfield J, Weisman H et al. Pathophysiology and pathogenesis of stunned myocardium. Depressed Ca activation of contraction as a consequence of reperfusion-induced cellular calcium overload in ferret hearts. J Clin Invest 1987; 79: 950–61.

21. Marban E. Myocardial stunning and hibernation. The physiology behind the colloquialisms. Circulation 1991; 83: 681–8.

22. Kim SJ, Kudej RK, Yatani A et al. A novel mechanism for myocardial stunning involving impaired Ca^{2+} handling. Circ Res 2001; 89: 821–37.

23. Thomas S, Fallavollita J, Lee T et al. Absence of troponin I degradation or altered sarcoplasmic reticulum uptake protein expression after reversible ischemia in swine. Circ Res 1999; 85: 446–56.

24. Gao W, Atar D, Backx P et al. Relationship between intracellular calcium and contractile force in stunned myocardium. Direct evidence for decreased myofilament responsiveness and altered diastolic function in intact ventricular muscle. Circ Res 1995; 76: 1036–48.

25. Bers D. Calcium fluxes involved in control of cardiac myocyte contraction. Circ Res 2000; 87: 275–81.

26. Schwartz K, Lompre AM, Bouveret P et al. Comparison of rat cardiac myosins at fetal stages, in young animals and in hypothyroid adults. J Biol Chem 1982; 257: 14412–17.

27. McDonough J, Arrell D, Van Eyk J. Troponin I degradation and covalent complex formation accompanies myocardial ischemia/reperfusion injury. Circ Res 1999; 84: 9–20.

28. Gao WD, Liu Y, Mellgren R et al. Intrinsic myofilament alterations underlying the decreased contractility of stunned myocardium. A consequence of Ca^{2+}-dependent proteolysis? Circ Res 1996; 78: 455–65.

29. Gao W, Atar D, Liu Y et al. Role of troponin I proteolysis in the pathogenesis of stunned myocardium. Circ Res 1997; 80: 393–9.

30. Van Eyk J, Powers F, Law W et al. Breakdown and release of myofilament proteins during ischemia and ischemia/reperfusion in rat hearts: identification of degradation products and effects on the pCa-force relation. Circ Res 1998; 82: 261–71.

31. Murphy A, Kogler H, Georgakopoulos D et al. Transgenic mouse model of stunned myocardium. Science 2000; 287: 488–91.

32. Luss H, Boknik P, Heusch G et al. Expression of calcium regulatory proteins in short-term hibernation and stunning in the in-situ porcine heart. Cardiovasc Res 1998; 37: 606–17.

33. Feng J, Schaus BJ, Fallavollita JA et al. Preload induces troponin I degradation independently of myocardial ischemia. Circulation 2001; 103: 2035–7.

34. Myers M, Bolli R, Lekich R et al. Enhancement of recovery of myocardial function by oxygen free-radical scavengers after reversible regional ischemia. Circulation 1985; 72: 915–21.

35. Bolli R, Jeroudi M, Patel B et al. Marked reduction of free radical generation and contractile dysfunction by antioxidant therapy begun at the time of reperfusion. Evidence that myocardial 'stunning' is a manifestation of reperfusion injury. Circ Res 1989; 65: 607–22.

36. Bolli R, Patel B, Jeroudi M et al. Demonstration of free radical generation in a 'stunned' myocardium of intact dogs with the use of spin trap alpha-phenyl N-tertbutyl nitrone. J Clin Invest 1988; 82: 476–85.

37. Gross G, Farber N, Haroman H et al. Beneficial actoins of superoxide dismutase and catalase in stunned myocardium of dogs. Am J Physiol Heart Circ Physiol 1986; 250: H372–7.

38. Koerner J, Anderson B, Dage R. Protection against postischemic myocardial dysfunction in anesthetized rabbits with scavengers of oxygen-derived free radicals. J Cardiovasc Pharmacol 1991; 17: 185–91.

39. Przyklenk K, Kloner R. Reperfusion-injury by oxygen-derived free radicals. Circ Res 1989; 64: 86–96.

40. Przyklenk K, Kloner R. Superoxide dismutase plus catalase improves contractile function in the canine model of the 'stunned' myocardium. Circ Res 1986; 58: 148–56.

41. Sekili S, McCay P, Li X et al. Direct evidence that the hydroxyl radical plays a pathogenic role in myocardial 'stunning' in the conscious dog and that stunning can be markedly attenuated without subsequent adverse effects. Circ Res 1993; 73: 705–23.

42. Li X, McCay P, Zughaib M et al. Demonstration of free radical generation in the 'stunned' myocardium in the conscious dog and identification of major differences between conscious and open-chest dogs. J Clin Invest 1993; 92: 1025–41.

43. Zughaib M, Abd-Eltfattah A, Jeroudi M et al. Augmentation of endogenous adenosine attenutates myocardial stunning independently of coronary flow or hemodynamic effects. Circulation 1993; 88: 2359–69.

44. Sun JZ, Tang XL, Knowlton AA et al. Late preconditioning against myocardial stunning. An

endogenous protective mechanism that confers resistance to postischemic dysfunction 24 h after brief ischemia in conscious pigs. J Clin Invest 1995; 95: 388–403.

45. Huang C, Vatner S, Peppas A et al. Cardiac nerves affect myocardial stunning through reactive oxygen and nitric oxide mechanisms. Circ Res 2003; 93: 866–73.

46. Kaneko M, Elimban V, Dhalla NS. Mechanism for depression of heart sarcolemmal Ca^{++} pump by oxygen free radicals. Am J Physiol Heart Circ Physiol 1989; 257: H804–11.

47. Corretti MC, Koretsune Y, Kusuoka H et al. Glycolytic inhibition and calcium overload as consequences of exogenously generated free radicals in rabbit hearts. J Clin Invest 1991; 88: 1014–25.

48. Macfarlane N, Miller D. Depression of peak force without altering calcium sensitivity by the superoxide anion in chemically skinned cardiac muscle of rat. Circ Res 1992; 70: 1217–24.

49. Maroko P, Libby P, Covell J et al. Precordial S-T segment elevation mapping: an atraumatic method for assessing alterations in the extent of myocardial ischemic injury. The effects of pharmacologic and hemodynamic interventions. Am J Cardiol 1972; 29: 223–30.

50. Theroux P, Franklin D, Ross JJ et al. Regional myocardial function during acute coronary artery occlusion and its modification by pharmacologic agents in the dog. Circ Res 1974; 35: 896–908.

51. Theroux P, Ross JJ, Franklin D et al. Coronary artery reperfusion. III. Early and late effects on regional myocardial function and dimensions in conscious dogs. Am J Cardiol 1976; 38: 599–606.

52. Maroko P, Kjekshus J, Sobel B et al. Factors influencing infarct size following experimental coronary artery occlusions. Circulation 1971; 43: 67–82.

53. Hood WJ, McCarthy B, Lown B. Myocardial infarction following coronary ligation in dogs. Hemodynamic effects of isoproterenol and acetylstrophanthidin. Circ Res 1967; 21: 191–9.

54. Hood WJ, Joison J, Kumar R et al. Experimental myocardial infarction. I. Production of left ventricular failure by gradual coronary occlusion in intact conscious dogs. Cardiovasc Res 1970; 4: 73–83.

55. McLaurin L, Rolett E, Grossman W. Impaired left ventricular relaxation during pacing-induced ischemia. Am J Cardiol 1973; 32: 751–7.

56. Forrester J, Diamond G, Parmley W et al. Early increase in left ventricular compliance after myocardial infarction. J Clin Invest 1972; 51: 598–603.

57. Vokonas P, Pirzada F, Hood WJ. Experimental myocardial infarction. XII. Dynamic changes in segmental mechanical behavior of infarcted and non-infarcted myocardium. Am J Cardiol 1976; 37: 853–9.

58. Becker L, Fortuin N, Pitt B. Effect of ischemia and antianginal drugs on the distribution of radioactive microspheres in the canine left ventricle. Circ Res 1971; 28: 263–9.

59. Becker L, Ferreira R, Thomas M. Mapping of left ventricular blood flow with radioactive microspheres in experimental coronary artery occlusion. Cardiovasc Res 1973; 7: 391–400.

60. Jennings R, Sommers H, Smyth G et al. Myocardial necrosis induced by temporary occlusion of a coronary artery in the dog. Arch Pathol 1960; 70: 68–78.

61. Jennings R, Sommers H, Herdson P et al. Ischemic injury of myocardium. Ann NY Acad Sci 1969; 156: 61–78.

62. Maroko PR, Libby P, Sobel BE et al. Effect of glucose-insulin-potassium infusion on myocardial infarction following experimental coronary artery occlusion. Circulation 1972; 45: 1160–75.

63. Watanabe T, Covell J, Maroko P et al. Effects of increased arterial pressure and positive inotropic agents on the severity of myocardial ischemia in the acutely depressed heart. Am J Cardiol 1972; 30: 371–7.

64. Braunwald E, Maroko P, Libby P. Reduction of infarct size following coronary occlusion. Circ Res 1974; 35: 192–201.

65. Lekven J, Kjekshus J, Mjos O. Effects of glucagon and isoproterenol on severity of acute myocardial ischemic injury. Scand J Clin Lab Invest 1973; 32: 129–37.

66. Vatner S, Millard R, Patrick T et al. Effects of isoproterenol on regional myocardial function, electrogram, and blood flow in conscious dogs with myocardial ischemia. J Clin Invest 1976; 57: 1261–71.

67. Redwood D, Smith E, Epstein S. Coronary artery occlusion in the conscious dog. Effects of alteration in heart rate and arterial pressure on the degree of myocardial ischemia. Circulation 1972; 46: 323–32.

68. Vatner S, McRitchie R, Maroko P et al. Effects of catecholamines, exercise, and nitroglycerin on the normal and ischemic myocardium in conscious dogs. J Clin Invest 1974; 54: 563–75.

69. Willerson J, Watson J, Platt M. Effect of hypertonic mannitol and intraaortic counterpulsation on regional myocardial blood flow and ventricular performance in dogs during myocardial ischemia. Am J Cardiol 1976; 37: 514–19.

70. Herdson P, Sommers H, Jennings R. A comparative study of the fine structure of normal and ischemic dog myocardium with special reference to early changes following temporary occlusion of a coronary artery. Am J Pathol 1965; 46: 367–86.

71. Jennings R, Reimer K. Lethal myocardial ischemic injury. Am J Pathol 1981; 102: 241–55.

72. Reimer K, Lowe J, Rasmussen M et al. The wave form phenomenon of ischemic cell death. I. Myocardial infarct size vs. duration of coronary occlusion in dogs. Circulation 1977; 56: 786–94.

73. Jarmakani J, Limbird L, Graham T et al. Effects of reperfusion on myocardial infarct, and the accuracy of estimating infarct size from serum creatine phosphokinase in the dog. Cardiovasc Res 1976; 10: 245–53.

74. Kloner R, Ganote C, Whalen D et al. Effects of a transient period of ischemia on myocardial cells. II. Fine structure during the first few minutes of reflow. Am J Pathol 1974; 74: 399–413.

75. Jennings R. Early phase of myocardial ischemic injury and infarction. Am J Cardiol 1969; 24: 753–65.

76. Haft J, Damato A. Measurement of collateral blood flow after myocardial infarction in the closed-chest dog. Am Heart J 1969; 77: 641–8.

77. Maroko P, Libby P, Ginks W et al. Coronary artery reperfusion. I. Early effects on myocardial contractility and myocardial necrosis. J Clin Invest 1972; 51: 2710–16.

78. Ginks W, Sybers H, Maroko P et al. Coronary artery reperfusion. II. Reduction of myocardial infarct size at one week after the coronary occlusion. J Clin Invest 1972; 51: 2717–23.

79. Sommers H, Jennings R. Experimental acute myocardial infarction. Histologic and histochemical studies of early myocardial infarcts induced by temporary or permanent occlusion of a coronary artery. Lab Invest 1964; 13: 1491–1503.

80. Lavallee M, Cox D, Patrick T et al. Salvage of myocardial function by coronary artery reperfusion 1, 2, and 3 hours after occlusion in conscious dogs. Circ Res 1983; 53: 235–47.

81. Mathur V, Guinn G, Burris W. Maximal revascularization (reperfusion) in intact conscious dogs after 2 to 5 hours of coronary occlusion. Am J Cardiol 1975; 36: 252–61.

82. Huang C-H, Kim S-J, Ghaleh B et al. An adenosine agonist and preconditioning shift the distribution of myocardial blood flow in conscious pigs. Am J Physiol Heart Circ Physiol 1999; 276: H368–75.

83. Constantini C, Corday E, Lang T et al. Revacsularization after 3 hours of coronary arterial occlusion. Effects on regional cardiac metabolic function and infarct size. Am J Cardiol 1975; 36: 368–84.

84. Huang C, Kim S, Ghaleh B et al. An adenosine agonist and preconditioning shift the distribution of myocardial blood flow in conscious pigs. Am J Physiol Heart Circ Physiol 1999; 276: H368–75.

85. Cohn L, Gorlin R, Herman N et al. Aorto-coronary bypass for acute coronary occlusion. J Thorac Cardiovasc Surg 1972; 64: 503–13.

86. McIntosh H, Bucino R. Emergency coronary artery revascularization of patients with acute myocardial infarction. Circulation 1979; 60: 247–50.

87. Bolooki H. Myocardial revascularization after acute infarction. Am J Cardiol 1975; 36: 395–406.

88. Ganz W, Buchbinder N, Marcus H et al. Intracoronary thrombolysis in evolving myocardial infarction. Am Heart J 1981; 101: 4–13.

89. Ganz W, Ninomiya K, Hashida J et al. Intracoronary thrombolysis in acute myocardial infarction: experimental background and clinical experience. Am Heart J 1981; 102: 1145–9.

90. Markis J, Malagold M, Parker J et al. Myocardial salvage after intracoronary thrombolysis with streptokinase in acute myocardial infarction. N Engl J Med 1981; 305: 777–82.

91. Holmes DJ, Vlietstra R, Reeder G et al. Angioplasty in total coronary artery occlusion. J Am Coll Cardiol 1984; 3: 845–9.

92. Sheehan F, Braunwald E, Canner P et al. The effect of intravenous thrombolytic therapy on left ventricular function: a report on tissue-type plasminogen activator and streptokinase from the Thrombolysis in Myocardial Infarction (TIMI Phase I) trial. Circulation 1987; 75: 817–29.

93. Puri P. Contractile and biochemical effects of coronary reperfusion after extended periods of coronary occlusion. Am J Cardiol 1975; 36: 244–51.

94. Vanoverschelde J, Depre C, Gerber B et al. Time course of functional recovery after coronary artery bypass graft surgery in patients with chronic left ventricular ischemic dysfunction. Am J Cardiol 2000; 85: 1432–9.

95. Taylor A, Kieso R, Melton J et al. Echocardiographically detected dyskinesis, myocardial infarct size, and coronary risk region relationships in reperfused canine myocardium. Circulation 1985; 71: 1292–300.

96. Murry CE, Richard VJ, Jennings RB et al. Myocardial protection is lost before contractile function recovers from ischemic preconditioning. Am J Physiol 1991; 260: 1796–804.

97. Marber MS, Latchman DS, Walker JM et al. Cardiac stress protein elevation 24 hours after brief ischemia or heat stress is associated with resistance to myocardial infarction. Circulation 1993; 88: 1264–72.

98. Kuzuya T, Hoshida S, Yamashita N et al. Delayed effects of sublethal ischemia on the acquisition of tolerance to ischemia. Circ Res 1993; 72: 1293–9.

99. Behar S, Reicher-Reiss H, Abinader E et al. The prognostic significance of angina pectoris preceding the occurrence of a first acute myocardial infarction in 4166 consecutive hospitalized patients. Am Heart J 1992; 123: 1481–6.

100. Cupples L, Gagnon D, Wong N et al. Pre-existing cardiovascular conditions and long term prognosis after initial myocardial infarction. The Framingham study. Am Heart J 1993; 125: 863–72.

101. Zahn R, Schiele R, Schneider S et al. Effect of pre-infarction angina pectoris on outcome in patients with acute myocardial infarction treated with primary angioplasty (results from the Myocardial Infarction Registry [MIR]). Am J Cardiol 2001; 87: 1–6.

102. Kloner RA, Shook T, Antman EM et al. Prospective temporal analysis of the onset of preinfarction angina versus outcome: an ancillary study in TIMI-9B. Circulation 1998; 97: 1042–5.

103. Yamagishi H, Akioka K, Hirata K et al. Effects of preinfarction angina on myocardial injury in patients with acute myocardial infarction. J Nucl Med 2000; 41: 830–6.

104. Teoh LKK, Grant R, Hulf JA et al. The effect of preconditioning (ischemic and pharmacological) on myocardial necrosis following coronary artery bypass graft surgery. Cardiovasc Res 2002; 53: 175–80.

105. Yellon D, Alkhulaifi A, Pugsley W. Preconditioning the human myocardium. Lancet 1993; 342: 276–7.

106. Jenkins J, Pugsley W, Alkhulaifi A et al. Ischaemic preconditioning reduces troponin T release in patients undergoing coronary artery bypass surgery. Heart 1997; 77: 314–18.

107. Illes R, Sowyer K. Prospective randomized clinical study of ischemic preconditioning as an adjunct to intermittent cold blood cardioplegia. Ann Thor Surg 1998; 65: 748–53.

108. Cribier A, Korsatz L, Koning R et al. Improved myocardial ischemic response and enhanced collateral circulation with long repetitive coronary occlusion during angioplasty: a prospective study. J Am Coll Cardiol 1992; 20: 578–86.

109. Deutsch E, Berger M, Kussmaul W et al. Adaptation to ischemia during percutaneous transluminal coronary angioplasty. Clinical, hemodynamic, and metabolic features. Circulation 1990; 82: 2044–51.

110. Eltchaninoff H, Cribier A, Tron C et al. Adaptation to myocardial ischemia during coronary angioplasty demonstrated by clinical, electrocardiographic, echocardiographic, and metabolic parameters. Am Heart J 1997; 133: 490–6.

111. Birincioglu M, Yang X-M, Critz SD et al. S-T segment voltage during sequential coronary occlusions is an unreliable marker of preconditioning. Am J Physiol Heart Circ Physiol 1999; 277: H2435–41.

112. Leesar MA, Stoddard M, Ahmed M et al. Preconditioning of human myocardium with adenosine during coronary angioplasty. Circulation 1997; 95: 2500–7.

113. Tomai F, Crea F, Gaspardone A et al. Ischemic preconditioning during coronary angioplasty is prevented by glibenclamide, a selective ATP-sensitive K$^+$ channel blocker. Circulation 1994; 90: 700–5.

114. Saito S, Mizumura T, Takayama T et al. Antiischemic effects of nicorandil during coronary angioplasty in humans. Cardiovasc Drugs Ther 1995; 9: 257–63.

115. Jaffe M, Quinn N. Warm-up phenomenon in angina pectoris. Lancet 1980; 8201: 934–6.

116. Joy M, Cairns A, Spriging D. Observations on the warm up phenomenon in angina pectoris. Br Heart J 1987; 58: 116–21.

117. Bogaty P, Kingma JG, Jr., Robitaille NM et al. Attenuation of myocardial ischemia with repeated exercise in subjects with chronic stable angina: relation to myocardial contractility, intensity of exercise and the adenosine triphosphate-sensitive potassium channel. J Am Coll Cardiol 1998; 32: 1665–71.

118. Okazaki Y, Kodama K, Sato H et al. Attenuation of increased regional myocardial oxygen consumption during exercise as a major cause of warm-up phenomenon. J Am Coll Cardiol 1993; 21: 1597–604.

119. Kay IP, Kittelson J, Stewart RAH. Collateral recruitment and 'warm-up' after first exercise in ischemic heart disease. Am Heart J 2000; 140: 121–5.

120. Correa S, Schlaefer S. Blockade of KATP channels with glibenclamide does not abolish preconditioning during demand ischemia. Am J Cardiol 1997; 79: 75–8.

121. Sasaki N, Murata M, Guo Y et al. MCC-134, a single pharmacophore, opens surface ATP-sensitive potassium channels, blocks mitochondrial ATP-sensitive potassium channels, and suppresses preconditioning. Circulation 2003; 107: 1183–8.

122. Liu Y, Sato T, O'Rourke B et al. Mitochondrial ATP-dependent potassium channels : novel effectors of cardioprotection? Circulation 1998; 97: 2463–9.

123. Garlid KD, Paucek P, Yarov-Yarovoy V et al.

Cardioprotective effect of diazoxide and its interaction with mitochondrial ATP-sensitive K^+ channels: possible mechanism of cardioprotection. Circ Res 1997; 81: 1072–82.

124. Holmuhamedov EL, Wang L, Terzic A. ATP-sensitive K^+ channel openers prevent Ca^{2+} overload in rat cardiac mitochondria. J Physiol 1999; 519: 347–60.

125. Pain T, Yang XM, Critz SD et al. Opening of mitochondrial K(ATP) channels triggers the preconditioned state by generating free radicals. Circ Res 2000; 87: 460–6.

126. Forbes RA, Steenbergen C, Murphy E. Diazoxide-induced cardioprotection requires signaling through a redox-sensitive mechanism. Circ Res 2001; 88: 802–9.

127. Yellon D, Downey J. Preconditioning the myocardium: from cellular physiology to clinical cardiology. Physiol. Rev 2003; 83: 1113–51.

128. Bolli R. The late phase of preconditioning. Circ Res 2000; 87: 972–83.

129. Plumier J, Ross B, Currie R et al. Transgenic mice expressing the human heat-shock protein 70 have improved post-ischemic myocardial recovery. J Clin Invest 1995; 95: 1854–60.

130. Shinmura K, Bolli R, Liu S et al. Aldose reductase is an obligatory mediator of the late phase of ischemic preconditioning. Circ Res 2002; 91: 240–6.

131. Dana A, Jonassen A, Yamashita N et al. Adenosine A1 receptor activation induces delayed preconditioning in rats mediated by manganese superoxide dismutase. Circulation 2000; 101: 2841–8.

132. Shinmura K, Tang X, Wang Y et al. Cyclooxygenase-2 mediates the cardioprotective effects of the late phase of ischemic preconditioning in conscious rabbits. Proc Natl Acad Sci USA 2000; 97: 10197–202.

133. Bolli R, Manchikalapudi S, Tang X et al. The protective effect of the late phase of preconditioning against myocardial stunning in conscious rabbits is mediated by nitric oxide synthase. Circ Res 1997; 81: 1094–107.

134. Takano H, Manchikalapudi S, Tang X et al. Nitric oxide synthase is the mediator of late preconditioning against myocardial infarction in conscious rabbits. Circulation 1998; 98: 441–9.

135. Guo Y, Jones WK, Xuan Y-T et al. The late phase of ischemic preconditioning is abrogated by targeted disruption of the inducible NO synthase gene. Proc Natl Acad Sci USA 1999; 96: 11507–12.

136. Wang Y, Kodani E, Wang J et al. Cardioprotection during the final stage (72 h) of the late phase of ischemic preconditioning is mediated by neuronal NO systhase in concert with cyclooxygenase-2. Circ Res 2004; 95: 84–91.

137. Chi NC, Karliner JS. Molecular determinants of responses to myocardial ischemia/reperfusion injury: focus on hypoxia-inducible and heat shock factors. Cardiovasc Res 2004; 61: 437–47.

138. Benjamin I, McMillan D. Stress (heat shock) proteins. Circ Res 1998; 83: 117–32.

139. Williams R, Benjamin I. Protective responses in the ischemic myocardium. J Clin Invest 2000; 106: 813–18.

140. Huang LE, Bunn HF. Hypoxia-inducible factor and its biomedical relevance. J Biol Chem 2003; 278: 19575–8.

141. Wright G, Higgin JJ, Raines RT et al. Activation of the prolyl hydroxylase oxygen-sensor results in induction of GLUT1, heme oxygenase-1, and nitric-oxide synthase proteins and confers protection from metabolic inhibition to cardiomyocytes. J Biol Chem 2003; 278: 20235–9.

142. Kim C-H, Cho Y-S, Chun Y-S et al. Early expression of myocardial HIF-1α in response to mechanical stresses: regulation by stretch-activated channels and the phosphatidylinositol 3-kinase signaling pathway. Circ Res 2002; 90: 25e–33e.

143. Fujio Y, Nguyen T, Wencker D et al. Akt promotes survival of cardiomyocytes *in vitro* and protects against ischemia–reperfusion injury in mouse heart. Circulation 2000; 101: 660–7.

144. Fujita N, Sato S, Ishida A et al. Involvement of Hsp90 in signaling and stability of 3-phosphoinositide-dependent kinase-1. J Biol Chem 2002; 277: 10346–53.

145. Gabai V, Meriin A, Yaglom J et al. Suppression of stress kinase JNK is involved in HSP27-mediated protection of myogenic cells from transient energy deprivation. J Biol Chem 2000; 275: 38088–94.

146. Vander Heide RS. Increased expression of HSP27 protects canine myocytes from simulated ischemia–reperfusion injury. Am J Physiol Heart Circ Physiol 2002; 282: H935–41.

147. Latchman D. Heat shock proteins and cardiac protection. Cardiovasc Res 2001; 51: 637–46.

148. Lin K, Lin B, Lian I et al. Combined and individual mitochondrial HSP60 and HSP10 expression in cardiac myocytes protects mitochondrial function and prevents apoptotic cell deaths induced by simulated ischemia–reperfusion. Circulation 2001; 103: 1787–92.

149. Ray P, Martin J, Swanson E et al. Transgene over-expression of αB crystallin confers simultaneous protection against cardiomyocyte apoptosis and necrosis

during myocardial ischemia and reperfusion. FASEB J 2001; 15: 393–402.

150. Beere H, Wolf B, Cain K et al. Heat-shock protein 70 inhibits apoptosis by preventing the recruitment of procaspase-9 to the Apaf-1 apoptosome. Nature Cell Biol 2000; 2: 469–75.

151. Depre C, Wang L, Tomlinson J et al. Characterization of pDJA1, a cardiac-specific chaperone found by genomic profiling of the post-ischemic swine heart. Cardiovasc Res 2003; 58: 126–35.

152. Datta SR, Brunet A, Greenberg ME. Cellular survival: a play in three Akts. Genes Dev 1999; 13: 2905–27.

153. Aikawa R, Nawano M, Gu Y et al. Insulin prevents cardiomyocytes from oxidative stress-induced apoptosis through activation of PI3 kinase/Akt. Circulation 2000; 102: 2873–9.

154. Semenza G. HIF-1, O2 and the 3 PHDs. How animal cells signal hypoxia to the nucleus. Cell 2001; 107: 1–4.

155. Radford N, Fina M, Benjamin I et al. Cardioprotective effects of 70-kDa heat shock protein in transgenic mice. Proc Natl Acad Sci USA 1996; 93: 2339–42.

156. Marber M, Mestril R, Chi S et al. Overexpression of the rat inducible 70-kD heat stress protein in a transgenic mouse increases the resistance of the heart to ischemic injury. J Clin Invest 1995; 95: 1446–56.

157. Mestril R, Chi S, Sayen R et al. Expression of inducible stress protein 70 in rat heart myogenic cells confer protection against simulated ischemia–induced injury. J Clin Invest 1994; 93: 759–67.

158. Kamradt MC, Chen F, Sam S et al. The small heat shock protein alpha B-crystallin negatively regulates apoptosis during myogenic differentiation by inhibiting caspase-3 activation. J Biol Chem 2002; 277: 38731–6.

159. Kamradt M, Chen F, Cryns V. The small heat-shock protein αB-crystallin negatively regulates cytochrome c- and caspase-8-dependent activation of caspase-3 by inhibiting its autoproteolytic maturation. J Biol Chem 2001; 276: 16059–63.

160. Li C, Lee J, Ko Y et al. Heat shock protein 70 inhibits apoptosis downstream of cytochrome c release and upstream of caspase-3 activation. J Biol Chem 2000; 275: 25665–71.

161. Pandey P, Saleh A, Nakazawa A et al. Negative regulation of cytochrome c-mediated oligomerization of Apaf-1 and activation of procaspase-9 by heat-shock protein 90. EMBO J 2000; 19: 4310–22.

162. Mehlen P, Schulze-Osthoff K, Arrigo A. Small stress proteins as novel regulators of apoptosis. Heat shock protein 27 blocks Fas/APO-1- and staurosporine-induced cell death. J Biol Chem 1996; 271: 16510–17.

163. Depre C, Tomlinson JE, Kudej RK et al. Gene program for cardiac cell survival induced by transient ischemia in conscious pig. Proc Natl Acad Sci USA 2001; 98: 9336–41.

164. Depre C, Hase M, Gaussin V et al. H11 kinase is a novel mediator of myocardial hypertrophy in vivo. Circ Res 2002; 91: 1007–14.

165. Cantley L. The phosphoinositide 3-kinase pathway. Science 2002; 296: 1655–7.

166. Morisco C, Zebrowski D, Condorelli G et al. The Akt-glycogen synthase kinase 3beta pathway regulates transcription of atrial natriuretic factor induced by beta-adrenergic receptor stimulation in cardiac myocytes. J Biol Chem 2000; 275: 14466–75.

167. Sugden P, Clerk A. Cellular mechanisms of cardiac hypertrophy. J Mol Med 1998; 76: 725–46.

168. Sugden P. Signaling in myocardial hypertrophy. Circ Res 1999; 84: 633–46.

169. Brar B, Stephanou A, Knight R et al. Activation of protein kinase B/Akt by urocortin is essential for its ability to protect cardiac cells against hypoxia/reoxygenation-induced cell death. J Mol Cell Cardiol 2002; 34: 483–92.

170. Cook SA, Matsui T, Li L et al. Transcriptional effects of chronic Akt activation in the heart. J Biol Chem 2002; 277: 22528–33.

171. Jonassen AK, Sack MN, Mjos OD et al. Myocardial protection by insulin at reperfusion requires early administration and is mediated via Akt and p70s6 kinase cell-survival signaling. Circ Res 2001; 89: 1191–8.

172. Mehrhof FB, Muller F-U, Bergmann MW et al. In cardiomyocyte hypoxia, insulin-like growth factor-i-induced antiapoptotic signaling requires phosphatidylinositol-3-OH-kinase-dependent and mitogen-activated protein kinase-dependent activation of the transcription factor cAMP response element-binding protein. Circulation 2001; 104: 2088–94.

173. Matsui T, Tao J, del Monte F et al. Akt activation preserves cardiac function and prevents injury after transient cardiac ischemia in vivo. Circulation 2001; 104: 330–5.

174. Gao F, Gao E, Yue T-L et al. Nitric oxide mediates the antiapoptotic effect of insulin in myocardial ischemia–reperfusion: the roles of PI3-kinase, Akt, and endothelial nitric oxide synthase phosphorylation. Circulation 2002; 105: 1497–502.

175. Negoro S, Oh H, Tone E et al. Glycoprotein 130 regulates cardiac myocyte survival in doxorubicin-induced apoptosis through phosphatidylinositol 3-kinase/Akt phosphorylation and Bcl-x$_L$/caspase-3 interaction. Circulation 2001; 103: 555–61.

176. Datta S, Dudek H, Tao X et al. Akt phosphorylation of Bad couples survival signals to the cell-intrinsic death machinery. Cell 1997; 91: 231–41.

177. Nicholson K, Anderson N. The protein kinase B/Akt signaling pathway in human malignancy. Cell Signal 2002; 14: 381–95.

178. Cardone M, Roy N, Stennicke H et al. Regulation of cell death protease caspase-9 by phosphorylation. Science 1998; 282: 1318–21.

179. Xavier IJ, Mercier PA, McLoughlin CM et al. Glycogen synthase kinase 3beta negatively regulates both DNA-binding and transcriptional activities of heat shock factor 1. J Biol Chem 2000; 275: 29147–52.

180. Rane MJ, Pan Y, Singh S et al. Heat shock protein 27 controls apoptosis by regulating Akt activation. J Biol Chem 2003; 278: 27828–35.

181. Song J, Takeda M, Morimoto RI. Bag1-HSP70 mediates a physiological stress signaling pathway that regulates Raf-1/ERK and cell growth. Nat Cell Biol 2001; 3: 276–82.

182. Gunaratnam L, Morley M, Franovic A et al. Hypoxia inducible factor activates the transforming growth factor-alpha/epidermal growth factor receptor growth stimulatory pathway in VHL(–/–) renal cell carcinoma cells. J Biol Chem 2003; 278: 44966–74.

183. Zundel W, Schindler C, Haas-Kogan D et al. Loss of PTEN facilitates HIF-1-mediated gene expression. Genes Dev 2000; 14: 391–6.

184. Hudson CC, Liu M, Chiang GG et al. Regulation of hypoxia-inducible factor 1α expression and function by the mammalian target of rapamycin. Mol Cell Biol 2002; 22: 7004–14.

185. Humar R, Kiefer F, Berns H et al. Hypoxia enhances vascular cell proliferation and angiogenesis in vitro via rapamycin (mTOR)-dependent signaling. FASEB J 2002; 16: 771–80.

186. Pfeffer MA, Braunwald E. Ventricular remodeling after myocardial infarction. Experimental observations and clinical implications. Circulation 1990; 81: 1161–72.

187. Buja L, Muntz K, Lipscomb K et al. Cardiac hypertrophy in chronic ischemic heart disease. Perspect Cardiovasc Res 1983; 8: 287–94.

188. Anversa P, Beghi C, Kikkawa Y et al. Myocardial infarction in rats. Infarct size, myocyte hypertrophy, and capillary growth. Circ Res 1986; 58: 26–37.

189. Narula J, Pandey P, Arbustini E et al. Apoptosis in heart failure: release of cytochrome c from mitochondria and activation of caspase-3 in human cardiomyopathy. Proc Natl Acad Sci USA 1999; 96: 8144–9.

190. Boluyt M, O'Neill L, Meredith AL et al. Alterations in cardiac gene expression during the transition from stable hypertrophy to heart failure. Circ Res 1994; 75: 23–32.

191. Nadal-Ginard B, Kajstura J, Leri A et al. Myocyte death, growth, and regeneration in cardiac hypertrophy and failure. Circ Res 2003; 92: 139–50.

192. Cheng W, Li B, Kajstura J et al. Stretch-induced programmed myocyte cell death. J Clin Invest 1995; 96: 2247–59.

193. Kajstura J, Zhang X, Reiss K et al. Myocyte cellular hyperplasia and myocyte cellular hypertrophy contribute to chronic ventricular remodeling in coronary artery narrowing-induced cardiomyopathy in rats. Circ Res 1994; 74: 383–400.

194. Bing O. Hypothesis: apoptosis may be a mechanism for the transition to heart failure with chronic pressure overload. J Mol Cell Cardiol 1994; 26: 943–8.

195. Pfeffer M, Braunwald E. Ventricular remodeling after myocardial infarction. Circulation 1990; 81: 1161–72.

196. Pfeffer M, Lamas G, Vaughan D et al. Effect of captopril on progressive ventricular dilatation after anterior myocardial infarction. N Engl J Med 1988; 319: 80–6.

197. Ball S, Hall A, Murray G. Angiotensin-converting enzyme inhibitors after myocardial infarction: indications and timing. J Am Coll Cardiol 1995; 25: 42S–46S.

198. GISSI-3. Effects of lisinopril and transdermal glyceryl trinitrate singly and together on 6-week mortality and ventricular function after acute myocardial infarction. Lancet 1994; 343: 1115–22.

199. Weber K. Extracellular matrix remodeling in heart failure. Circulation 1997; 96: 4065–82.

200. Sadoshima S, Izumo S. The cellular and molecular response of cardiac myocytes to mechanical stress. Annu Rev Physiol 1997; 59: 551–71.

201. Sadoshima J, Xu Y, Slayter II et al. Autocrine release of angiotensin II mediates stretch-induced hypertrophy of cardiac myocytes in vitro. Cell 1993; 75: 977–84.

202. Lindpaintner K, Lu W, Niedermajer N et al. Selective activation of cardiac angiotensinogen gene expression in post-infarction ventricular remodeling in the rat. J Mol Cell Cardiol 1993; 25: 133–43.

203. Foody J, Farrell M, Krumholz H. Beta-blocker therapy in heart failure. JAMA 2002; 287: 883–9.

204. Rahimtoola S. The hibernating myocardium. Am Heart J 1989; 117: 211–21.

205. Wijns W, Vatner SF, Camici PG. Hibernating myocardium. New Engl J Med 1998; 339: 173–81.

206. Vanoverschelde JL, Wijns W, Borgers M et al. Chronic myocardial hibernation in humans. From bedside to bench. Circulation 1997; 95: 1961–71.

207. Camici P, Wijns W, Borgers M et al. Pathophysiological mechansims of chronic reversible left ventricular dysfunction due to coronary artery disease. Circulation 1997; 96: 3205–14.

208. Heusch G. Hibernating myocardium. Physiol Rev 1998; 78: 1055–85.

209. Kalra D, Zoghbi W. Myocardial hibernation in coronary artery disease. Curr Atheroscler Rep 2002; 4: 149–55.

210. Ross JJ. Myocardial perfusion–contraction matching. Implications for coronary heart disease and hibernation. Circulation 1991; 83: 1076–83.

211. Camici PG, Rimoldi OE. Myocardial blood flow in patients with hibernating myocardium. Cardiovasc Res 2003; 57: 302–11.

212. Vanoverschelde J, Wijns W, Depre C et al. Mechanisms of chronic regional postischemic dysfunction in humans. New insights from the study of noninfarcted collateral-dependent myocardium. Circulation 1993; 87: 1513–23.

213. Shen Y, Kudej R, Bishop S et al. Inotropic reserve and histological appearance of hibernating myocardium in conscious pigs with ameroid-induced coronary stenosis. Basic Res Cardiol 1996; 91: 479–85.

214. Borgers M, Ausma J. Structural aspects of the chronic hibernating myocardium in man. Basic Res Cardiol 1995; 90: 44–6.

215. Fallavolita JA, Perry BJ, Canty JM, Jr. 18F-2-Deoxyglucose deposition and regional flow in pigs with chronically dysfunctional myocardium. Evidence for transmural variations in chronic hibernating myocardium. Circulation 1997; 95: 1900–9.

216. Fallavollita JA, Jacob S, Young RF et al. Regional alterations in SR Ca^{2+}-ATPase, phospholamban, and HSP-70 expression in chronic hibernating myocardium. Am J Physiol Heart Circ Physiol 1999; 277: H1418–28.

217. Kim S, Peppas A, Hong S et al. Persistent stunning induces myocardial hibernation and protection. Circ Res 2003; 92: 1233–9.

218. Shen YT, Vatner SF. Mechanism of impaired myocardial function during progressive coronary stenosis in conscious pigs: hibernation versus stunning? Circ Res 1995; 76: 479–88.

219. Arai A, Pantely G, Anselone C et al. Active down-regulation of myocardial energy requirements during prolonged moderate ischemia in swine. Circ Res 1991; 69: 1458–69.

220. Schulz R, Rose J, Martin C et al. Development of short-term myocardial hibernation. Its limitation by the severity of ischemia and inotropic stimulation. Circulation 1993; 88: 684–95.

221. Schulz R, Guth B, Pieper K et al. Recruitment of an inotropic reserve in moderately ischemic myocardium at the expense of metabolic recovery. A model of short-term hibernation. Circ Res 1992; 70: 1282–95.

222. Pantely G, Malone S, Rhen W et al. Regeneration of myocardial phosphocreatine in pigs despite continued moderate ischemia. Circ Res 1990; 67: 1481–93.

223. Matsuzaki M, Gallagher K, Kemper W et al. Sustained regional dysfunction produced by prolonged coronary stenosis: gradual recovery after reperfusion. Circulation 1983; 68: 170–82.

224. Kudej R, Ghaleh B, Sato N et al. Ineffective perfusion–contraction matching in conscious, chronically instrumented pigs with an extended period of coronary stenosis. Circ Res 1998; 82: 1199–205.

225. Chen C, Chen L, Fallon JT et al. Functional and structural alterations with 24-hour myocardial hibernation and recovery after reperfusion: a pig model of myocardial hibernation. Circulation 1996; 94: 507–16.

226. Flameng W, Vanhaecke J, Van Belle H et al. Relation between coronary artery stenosis and myocardial purine metabolism, histology and regional function in humans. J Am Coll Cardiol 1987; 9: 1235–42.

227. Ausma J, Thone F, Dispersyn GD et al. Dedifferentiated cardiomyocytes from chronic hibernating myocardium are ischemia–tolerant. Mol Cell Biochem 1998; 186: 159–68.

228. Depre C, Taegtmeyer H. Metabolic aspects of programmed cell survival and cell death in the heart. Cardiovasc Res 2000; 45: 538–48.

229. Camici PG, Wijns W, Borgers M et al. Pathophysiological mechanisms of chronic reversible left ventricular dysfunction due to coronary artery disease (hibernating myocardium). Circulation 1997; 96: 3205–14.

230. Mercadier J, Lompre A, Wisnewsky C et al. Myosin isoenzymic changes in several models of rat cardiac hypertrophy. Circ Res 1981; 49: 525–32.

231. Schwartz K, Boheler KR, de la Bastie D et al.

Switches in cardiac muscle gene expression as a result of pressure and volume overload. Am J Physiol Regul Integr Comp Physiol 1992; 262: R364–9.

232. Depre C, Shipley G, Chen W et al. Unloaded heart in vivo replicates fetal gene expression of cardiac hypertrophy. Nat Med 1998; 4: 1269–75.

233. Knoll R, Arras M, Zimmermann R et al. Changes in gene expression following short coronary occlusions studied in porcine hearts with run-on assays. Cardiovasc Res 1994; 28: 1062–9.

234. Frass O, Sharma H, Knoll R et al. Enhanced gene expression of calcium regulatory proteins in stunned porcine myocardium. Cardiovasc Res 1993; 27: 2037–43.

235. Brand T, Sharma H, Fleischmann K et al. Proto-oncogene expression in porcine myocardium subjected to ischemia and reperfusion. Circ Res 1992; 71: 1351–60.

236. Currie R, Tangay R, Kingma J. Heat-shock response and limitation of tissue necrosis during occlusion/reperfusion in rabbit hearts. Circulation 1992; 87: 963–71.

237. Sugden P, Clerk A. 'Stress-responsive' mitogen-activated protein kinases (c-Jun N-terminal kinases and p38 mitogen-activated protein kinases) in the myocardium. Circ Res 1998; 83: 345–52.

238. Yin T, Sandhu G, Wolfgang C et al. Tissue-specific pattern of stress kinase activation in ischemic/reperfused heart and kidney. J Biol Chem 1997; 272: 19943–50.

239. Kudej R, Kim S, Shen Y et al. Nitric oxide, an important regulator of perfusion–contraction matching in conscious pigs. Am J Physiol Heart Circ Physiol 2000; 279: H451–6.

240. Silverman G, Bird P, Carrell R et al. The serpins are an expanding superfamily of structurally similar but functionally diverse proteins. J Biol Chem 2001; 276: 33293–6.

241. Luss H, Schafers M, Neumann J et al. Biochemical mechanisms of hibernation and stunning in the human heart. Cardiovasc Res 2002; 56: 411–21.

242. Kalra DK, Zhu X, Ramchandani MK et al. Increased myocardial gene expression of tumor necrosis factor-α and nitric oxide synthase-2: a potential mechanism for depressed myocardial function in hibernating myocardium in humans. Circulation 2002; 105: 1537–40.

243. Shan K, Bick RJ, Poindexter BJ et al. Altered adrenergic receptor density in myocardial hibernation in humans : a possible mechanism of depressed myocardial function. Circulation 2000; 102: 2599–606.

244. Depre C, Kim S-J, John AS et al. Program of cell survival underlying human and experimental hibernating myocardium. Circ Res 2004; 95: 433–40.

CHAPTER 15

Cardiac remodeling

Barry Greenberg

Introduction

Heart failure is pandemic throughout the industrialized world. In the United States alone it is estimated that there are nearly 5 million individuals with this condition. Moreover, the prevalence of heart failure is likely to increase substantially over the next several decades. There are numerous reasons for this, including the aging of the population and increased prevalence of risk factors for heart failure such as obesity and diabetes. Greater awareness of the signs and symptoms of heart failure, availability of diagnostic tests that facilitate early detection of this condition and longer survival of existing patients will also contribute to increased prevalence. Ironically, another major factor is the improved survival of patients who have suffered myocardial injury of one type or another. It is now recognized that injury to the heart, whether due to a myocardial infarction (MI) or some other cause, initiates a complex series of changes in myocardial structure and function that has been termed cardiac remodeling. Although remodeling initially helps to maintain cardiac performance, the process is largely maladaptive, and if allowed to continue it causes progressive deterioration in cardiac function and results in heart failure. The central role of remodeling in the development of cardiac dysfunction is substantiated by evidence from clinical trials showing that therapies that inhibit or reverse the remodeling process are highly effective in preventing and treating heart failure.

This chapter will review the causes, consequences and treatments of cardiac remodeling. Although remodeling is a fundamental pathophysiological response to myocardial injury, the chapter will focus on post-MI cardiac remodeling. This approach was chosen, based on clinical importance as well as the fact that many mechanistic insights and therapeutic advances have evolved from the study of changes in the post-MI heart in relevant experimental animal models and human patients. Moreover, the fundamental cellular and molecular mechanisms involved in post-MI cardiac remodeling are representative of those that are activated during other acute and chronic conditions (outlined in Box 15.1) that cause heart failure.

Box 15.1. *Causes of cardiac remodeling*

- Myocardial infarction
- Hypertension
- Exposure to myocardial toxins (e.g. alcohol, adriamicin)
- Damage due to viral (or other infectious) myocarditis
- Valvular abnormalities resulting in pressure and/or volume overload
- Congenital heart lesions
- Familial cardiomyopathy
- Peripartum cardiomyopathy

Although the chapter will focus on the left ventricle (LV), remodeling of the left atrium (LA) and the right sided chambers of the heart also makes important contributions to the development of heart failure. However, since the mechanisms and pathways that produce remodeling of the various chambers are relatively consistent, the chapter will describe the process as it occurs in the LV.

Description of post-MI cardiac remodeling

Cardiac remodeling has been defined as 'genome expression, molecular, cellular and interstitial changes that are manifested clinically as changes in size, shape and function of the heart after cardiac injury'.[1] In the post-MI heart there is cardiac myocyte hypertrophy and lengthening, ongoing myocyte loss due to apoptosis and restructuring of the extracellular matrix (ECM) with an overall increase in the amount of fibrous tissue.[2–6] As the heart remodels, cardiac myocytes demonstrate altered expression of genes that encode structural and functional proteins including changes in the relative abundance of sarcomeric proteins involved in contractile function such as the myosin heavy chain (MHC) isoforms, proteins that help regulate calcium flux such as sarcoplasmic reticulum Ca^{2+} ATPase (SERCA) and phospholamban, and markers of hypertrophy such as atrial natriuretic factor (ANF).[7–10] Fibroblasts within the heart are stimulated to replicate, migrate and to produce ECM proteins.[11–15] They also release 'secondary' growth factors that further advance the remodeling process through autocrine/paracrine effects.

These changes in cellular structure and function result in the development of eccentric hypertrophy in which there are increases in LV chamber volume, muscle mass and fibrous tissue contents.[1] The normally elliptical LV also changes towards a more spherical shape. As remodeling progresses, abnormalities in the contractile and filling functions of the heart develop. Alterations in electrical impulse conduction that cause inhomogeneous and dys-synchronous LV contraction and an increased propensity towards cardiac arrhythmias further impair cardiac function. Secondary mitral

and tricuspid insufficiency due to abnormalities in the subvalvular apparatus induced by the remodeling process add to the problem.

The structural changes that occur during post-MI remodeling have important consequences on bioenergetic events.[2,10,16–18] Progressive LV dilatation increases myocardial oxygen demands due to increased stress on the chamber wall. Concurrently, oxygen delivery is compromised by a reduction in coronary flow reserve and a relative deficiency in capillaries.[2,17] Within the myocyte there is an imbalance between excessive growth of myofibrils relative to the density of energy producing mitochondria. Furthermore, LV remodeling has been correlated with alterations in the levels of high-energy phosphates in the myocardium.[18] Overall, the imbalance between energy supply and demand that develops adversely affects myocardial function and predisposes towards myocardial ischemia, particularly in the more vulnerable subendocardial region of the LV.

In the post-MI heart, cardiac remodeling has been arbitrarily divided into early and late phases.[1,19] The early phase involves expansion of the infarct zone, while the late phase describes global changes throughout the LV. Wound healing and development of the replacement scar at the infarct site overlaps these phases of post-MI remodeling. Early infarct expansion results from thinning and bulging of devitalized myocardium.[20,21] There is also degradation of the ECM and breakdown of the collagen struts that connect individual myocytes to the ECM due to activation of a family of proteolytic enzymes termed matrix metalloproteinases (MMPs).[4,12–14] This causes myocyte slippage that results in further chamber dilatation and increases in wall stress.

Formation of the replacement scar at the infarct site is an extremely important component of the remodeling process. Rapid deposition of firm scar tissue protects against ventricular rupture and also helps control increases in wall stress by limiting chamber dilatation. Cardiac fibroblasts play a critical role in this process.[22] Phenotypically altered fibroblasts, termed myofibroblasts, migrate to the region surrounding the infarct, where they are involved in the generation of the replacement scar. Myofibroblasts are characterized by increased expression of alpha-smooth muscle actin, a contractile protein that can enhance scar retraction and, thus, limit increases in LV wall stress. Myofibroblasts also exhibit an increase in the density of the type 1 angiotensin (Ang) II (AT_1) receptor.[23,24] As shown in Fig. 15.1, there is evidence that AT_1 receptor upregulation may be caused by tumor necrosis factor-alpha (TNF-α) or interleukin-1 beta (IL-1β).[25] These pro-inflammatory cytokines are known to be elevated in the post-MI heart,[26] and their appearance is temporally and spatially correlated with increased AT_1 receptor on non-myocytes. As indicated in Fig. 15.2, cytokine-stimulated increases in AT_1 receptor density appears to be an important factor in the remodeling process since it increases Ang II-stimulated fibroblast functions involved in the deposition of fibrous tissue.[27]

Late remodeling describes the structural changes that occur in non-infarcted segments of the heart over the months to years after a MI. Over time these regions develop eccentric hypertrophy with increases in chamber volume, muscle mass and fibrosis. Individual myocyte hypertrophy increases cell volume by up to 70–80 %.[2] There is also continued loss of myocytes in the remodeling heart due to apoptosis.[5] Although the extent of myocyte loss due to ongoing apoptotic cell death in the remodeling post-MI is uncertain, it appears to be an important factor in the progression to heart failure.[28] In addition to changes in myocytes, extensive fibrosis in non-infarcted segments of myocardium may develop. In patients with end-stage ischemic cardiomyopathy Beltrami et al. found that two-thirds of the fibrous tissue in the heart was located in regions distant from previous MI sites.[6] Since interstitial fibrosis adversely affects both systolic and diastolic function, the authors concluded that these changes were an important cause of heart failure in MI survivors.

Cardiac remodeling adversely affects outcomes and increases risk of future cardiovascular events. The Survival and Ventricular Enlargement (SAVE) trial included patients with a recent large MI who were without evidence of heart failure. In this study, echocardiograms were obtained in 512 patients at a mean of 11.1 ± 3.2 days post-MI, and again after 1 year in 420 survivors.[29] Over a follow-up period of 3.0 ± 0.6 years, change in LV size

during the first year post-MI was shown to be a potent predictor of survival. When the population was divided into quartiles according to change in LV end-diastolic area, mortality increased in a stepwise fashion from 16.7% to 45.5% from the first to the fourth quartile. Additional evidence of the adverse effects of remodeling comes from the studies of left ventricular dysfunction (SOLVD). As shown in Fig. 15.3, echocardiographic analysis of a cohort of patients from the SOLVD clinical trial and registry populations demonstrated that LV mass was an independent predictor of survival.[30] Since patients included in the SOLVD registry were not required to have a reduced ejection fraction (EF), the population included a substantial number of patients with relatively well-preserved systolic function. Analysis of the results showed that increased mass was associated with an increase in mortality whether the EF was above or below 0.35. These findings demonstrate the independence of mass from EF as a prognostic factor and they emphasize the deleterious effects of hypertrophy on survival.

Causes of cardiac remodeling

Remodeling of the heart is initiated by an event (either acute or chronic) that causes myocardial damage. Although remodeling is usually considered as a response to a discrete event such as a MI or viral infection of the heart, the injurious process can be subtle and/or sustained (Box 15.1). For example, damage to the myocardium induced by toxic agents such as adriamicin or alcohol, or caused by prolonged pressure or volume overload due to a valvular lesion or long-standing hypertension, can initiate remodeling. When myocardial damage is limited, the extent of remodeling may be sufficiently small so as not to adversely affect cardiac function. Thus, patients who experience a 'small' MI demonstrate little or no remodeling over time. In contrast, patients who experience a 'large' MI are likely to experience significant LV remodeling.[31] A characteristic of remodeling is that it progresses after resolution of the initial acute event. Results from clinical trials in MI survivors have shown that each year post-MI, a relatively constant

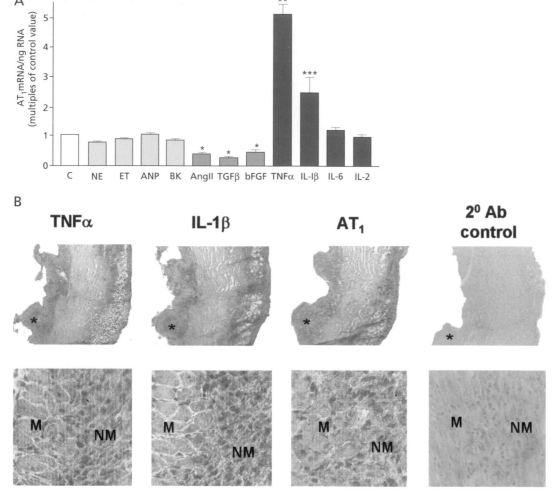

Figure 15.1. *Association between pro-inflammatory cytokines and AT$_1$ receptor upregulation. (A) Effects of various growth factors identified on AT$_1$ mRNA levels. Cells were incubated for 24 hours with each agent, and AT$_1$ mRNA levels, determined by quantitative competitive RT-PCR, are presented as multiples of the control (C) levels. NE (norepinephrine, 10 μmol/l), ET (endothelin, 100 nmol/l), ANP (atrial natriuretic peptide, 1 μmol/l), and BK (bradykinin, 5 nmol/l) had no significant effect on AT$_1$ mRNA levels. Ang II (angiotensin II, 1 μmol/l), TGF-β (transforming growth factor-β,10 ng/ml), and bFGF (b-fibroblast growth factor, 20 nm/ml) significantly reduced AT$_1$ mRNA levels (*P < 0.02). TNF-α (tumor necrosis factor-α,10 ng/ml) had the strongest (~5-fold) effect on enhancement of AT$_1$ mRNA levels (**P < 0.01). IL-1β (interleukin-1, 10ng/ml) enhanced AT$_1$ mRNA levels 2.4-fold (**P < 0.04), whereas IL-6 (10 ng/ml) and IL-2 (10 ng/ml) had no significant effect (n = 3 except for BK (n = 2) and IL-1β [n = 6]). Adapted from Gurantz D, Cowling RT, Villarreal FJ, Greenberg BH. Tumor necrosis factor-alpha upregulates angiotensin II type 1 receptors on cardiac fibroblasts. Circ Res 1999; 85: 272–9. (B) Association between the appearance of pro-inflammatory cytokines and AT$_1$ upregulation in the post-MI heart. Sections were obtained from an adult male rat heart 7 days following coronary artery ligation. A large infarct in the anterior wall of the left ventricle is noted. Immunohistochemical staining demonstrates the appearance of both TNF-α and IL-1β in the peri-infarction zone. There is also evidence of increased AT$_1$ receptor density on non-myocytes (but not on myocytes) that have infiltrated into this region. M, myocytes; NM, non-myocytes. See color plate section.*

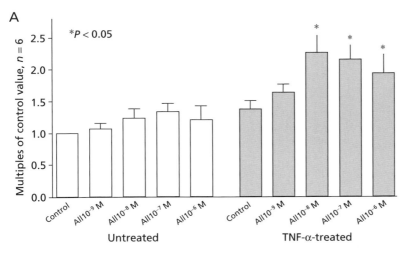

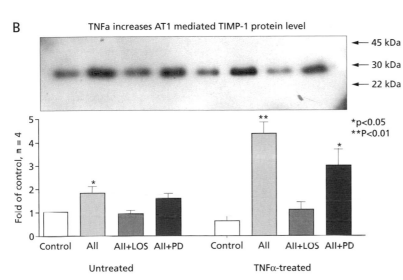

Figure 15.2. *Cytokine induction of the AT$_1$ receptor enhances fibroblast responsiveness to Ang II. (A) Ang II-stimulated [^{3}H]proline incorporation is enhanced by TNF-α pretreatment. In non-pretreated neonatal rat cardiac fibroblasts (open bars), Ang II stimulated [^{3}H]proline incorporation over a concentration range from 10^{-10} to 10^{-6} mol/l. Pretreatment with TNF-α (10 mg/ml) for 48 hours enhanced [^{3}H]proline incorporation by Ang II (filled bars). (B) TNF-α pretreatment enhances Ang II-induced TIMP-1 protein production. Results from Western blot analysis in neonatal rat cardiac fibroblasts non-pretreated (open bars) or pretreated (filled bars) with TNF-α (10 mg/ml). Results from a representative experiment are shown above. Data are presented as mean ± standard error-times control values. *P < 0.05, **P < 0.01; error bars are standard errors. Both panels are adapted from Peng J, Gurantz D, Tran V, Cowling RT, Greenberg BH. Tumor necrosis factor-alpha-induced AT1 receptor upregulation enhances angiotensin II-mediated cardiac fibroblast responses that favor fibrosis. Circ Res 2002; 91: 1119–26.*

percentage of patients develop heart failure,[32,33] and in most cases this is due to the remodeling process.

The major stimuli for cardiac remodeling are increases in LV wall stress, caused by increases in intracardiac pressures and/or volume and neurohormonal agents. These stimuli do not, however, operate through entirely separate pathways. Increased LV wall stress stimulates remodeling not only by activating mechanoreceptors on cardiac cells, but also by increasing neurohormonal activation within the heart. Conversely, systemic neurohormonal systems activated by impaired cardiac performance stimulate cardiac remodeling both by direct effects of effects of molecules like angiotensin II on cells in the heart, and by indirect effects associated with increases in wall stress due to salt/water retention and peripheral vasoconstriction.

The main neurohormonal systems that have been implicated in cardiac remodeling are outlined

A

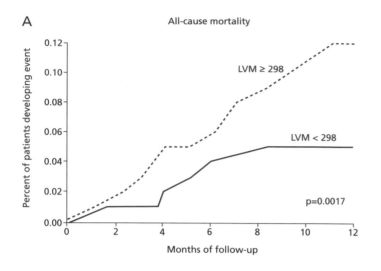

B

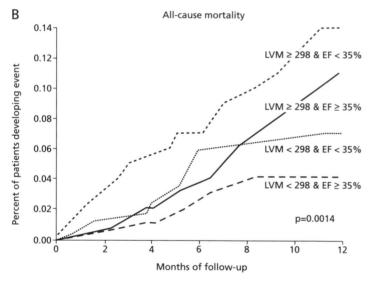

Figure 15.3. *LV mass predicts survival in heart failure patients. (A) Effect of LV mass on survival in heart failure patients. Kaplan–Meier unadjusted survival curves in patients from the SOLVD clinical trial and registry. The population is divided according to whether LV mass was above or below the mean value of 298 g. (B) Kaplan–Meier unadjusted survival curves in patients grouped according to LV mass and ejection fraction (EF). The log-rank statistic P value (0.0014) compares nonparametrically the event rate of the four groups. Both are adapted from Quinones MA, Greenberg BH, Kopelen HA et al. Echocardiographic predictors of clinical outcome in patients with left ventricular dysfunction enrolled in the SOLVD registry and trials: significance of left ventricular hypertrophy. Studies of left ventricular dysfunction. J Am Coll Cardiol 2000; 35: 1237–44.*

in Box 15.2. Results from the SOLVD study indicate that neurohormonal activation precedes the onset of heart failure since asymptomatic patients with LV systolic dysfunction (defined as an EF ≤ 0.35) demonstrated significant increases in circulating neurohormonal levels (compared to an age- and sex-matched control population), despite the absence of the signs and symptoms of heart failure.[34] These findings are consistent with neurohormonal activation playing a role in the initiation and progression of disease.

The renin–angiotensin system (RAS) in post-MI cardiac remodeling

Numerous laboratories around the world have studied the role of the RAS in the pathogenesis of cardiac remodeling. What has made this area of research so interesting and dynamic is that despite intensive study over the past several decades, previously unrecognized or underappreciated aspects of the RAS continue to be discovered, and our understanding of this system continues to evolve.

Box 15.2. *Neurohormones implicated in cardiac remodeling*

Agents that promote growth and remodeling
- Angiotensin II
- Catecholamines
- Endothelin
- Fibroblast growth factor
- Platelet-derived growth factor
- Transforming growth factor-β
- Connective tissue growth factor
- Tumor necrosis factor-α
- Interleukin-1β
- Interleukin-6
- Leukemia inhibitory factor
- Cardiotrophin-1

Agents believed to have anti-growth properties
- Angiotensin-(1–7)
- Bradykinin
- Natriuretic peptides (e.g. ANP, BNP, CNP)

Convincing evidence for a key role of the RAS in post-MI cardiac remodeling comes from a variety of sources. Ang II, the main effector molecule of the RAS, induces cardiac myocyte hypertrophy, and it stimulates cardiac fibroblasts to produce ECM proteins.[35-39] Although the circulatory RAS may be activated transiently post-MI, this system becomes quiescent during the critical asymptomatic phase of post-MI remodeling.[34,40] At the time that patients emerge from this period with signs and symptoms of heart failure, activation of the circulatory RAS is again noted, often in association with the use of diuretics to treat volume overload.[34] Thus, the contribution of the circulatory RAS to post-MI cardiac remodeling (as least during the early phases) appears to be modest.

Evidence that the genes for all components of the RAS are expressed in the heart,[41-45] that renin activity is taken up from the coronary circulation,[46] and that alternative pathways for converting Ang I to Ang II exist in the heart,[47] provide evidence of a localized cardiac RAS. This cardiac rather than the systemic RAS appears to be the source of the majority of Ang II that is found in the heart.[48,49] The cardiac RAS system is regulated independently of the circulatory system and it functions in an autocrine/paracrine manner to influence cardiac structure and function.[50,51] Post-MI, the cardiac RAS is upregulated, and Ang II levels in the heart are increased.[52–56] The importance of the cardiac RAS in remodeling is suggested by the close correlation between the extent of structural changes in the heart and cardiac RAS activation,[52,57] and by evidence that Ang II levels in the interstitial fluid of the heart are much higher than plasma levels.[58]

The use of cardiac-specific transgenic approaches has helped confirm the critical role of the cardiac RAS in the remodeling process.[59–61] Since most effects of Ang II in promoting cardiac remodeling are mediated through its AT_1 receptor, the role of the RAS in post-MI cardiac remodeling has been studied in knockout mice that are null for the AT_1 receptor. When these mice undergo coronary artery ligation to induce a large MI, they develop significantly less LV dilatation, less fibrosis in non-infarcted segments of myocardium, and better LV systolic function over time than wild-type controls, despite the fact that infarct sizes in the two study groups were equal.[60] Moreover, post-MI survival is significantly improved in AT_1 knockout mice compared to wild type controls.

The importance of RAS activation in post-MI remodeling, and the progression to heart failure, has been clearly demonstrated by studies in experimental animal models and in human patients showing that both angiotensin-converting enzyme (ACE) inhibitors and angiotensin receptor blockers (ARBs) inhibit and/or reverse cardiac remodeling, improve survival, and reduce the rate of progression to heart failure.[32,33,62–65]

The role of Ang II stimulation of fibroblasts in cardiac remodeling

Cardiac fibroblasts play a crucial role in post-MI remodeling. They produce ECM proteins such as fibronectin and various collagens, and they regulate the breakdown of fibrous tissue through production of MMPs and tissue inhibitors of metalloproteinases (TIMPs).[22,66–69] They also produce growth factors that act in a paracrine manner to stimulate cardiac myocyte hypertrophy.[26,70,71]

Ang II stimulates fibroblast functions involved in post-MI remodeling including replication, migration and production of ECM proteins and secondary growth factors. These effects are mediated mainly through the AT_1 receptor.[38,72,73] The AT_1 receptor is much more abundant on cardiac fibroblasts than on cardiac myocytes,[74] and in the post-MI heart AT_1 receptor density on cardiac fibroblasts is further increased.[56,75,76] There is evidence that pro-inflammatory cytokines that are present in the post-MI heart induce AT_1 receptor upregulation on cardiac fibroblasts, and that this phenotypic change enhances the responsiveness of these cells to the pro-fibrotic effects of Ang II.[25,27]

In contrast to cardiac fibroblasts, cardiac myocytes have few AT_1 receptors.[74,77] These findings raise the question of whether Ang II directly causes cardiac myocyte hypertrophy. Stimulation of cardiac fibroblasts by Ang II, however, leads to the production of a variety of 'secondary' growth factors including interleukin-6 (IL-6), leukemia inhibitory factor-1 (LIF-1), transforming growth factor (TGF)-β and endothelin-1 (ET-1), all of which are believed to play a role in promoting cardiac remodeling.[71,77–84] Evidence that Ang II induces hypertrophy of cardiac myocytes only when these cells are co-cultured with cardiac fibroblasts, or when conditioned medium from fibroblasts stimulated by Ang II is added to myocyte culture, supports the notion that production of growth factors from fibroblasts is a critical component of the remodeling process,[71,77] and that Ang II-mediated cardiac myocyte hypertrophy may not be a direct effect of the peptide.[70,71,77,85]

Alternative pathways of the RAS involved in cardiac remodeling (Fig. 15.4)

Although Ang II activation of its AT_1 receptor plays an unquestionably important role in post-MI cardiac remodeling, other components of the RAS may be involved. In addition to binding to the AT_1 receptor, Ang II can bind to the Type 2 (AT_2) receptor. Activation of this receptor has been associated with reduction of blood pressure in some experimental models. However, whether signals

from this seven-transmembrane domain G protein-coupled receptor affect cell functions involved in cardiac remodeling is not well defined at this time. Recent evidence suggests that an alternative pathway of the RAS involving ACE2 and Ang-(1–7) may be an important factor in determining cardiac structure and function. ACE2 is a homologue of ACE.[86,87] Human ACE2 cDNA predicts an 805 amino acid protein that has a 42% homology with the N-terminal catalytic domain of ACE. Although first identified in the heart, kidney and testes,[86,87] ACE2 now appears to be considerably more widely distributed throughout the body.[88] As shown in Fig. 15.4, ACE2 functions as a carboxypeptidase that cleaves a single peptide from either the decapeptide Ang I to form Ang-(1–9) or the octapeptide Ang II to form Ang-(1–7). Evidence that the catalytic activity of ACE2 for Ang II is substantially greater than for Ang I suggests that its primary role is to convert Ang II to Ang-(1–7).[89] In contrast to ACE, ACE2 does not convert Ang I to Ang II, it is not affected by ACE inhibitors,[86,87] and it does it not break down bradykinin.[87,89] ACE2 can reduce Ang II levels by several mechanisms including direct degradation, reducing availability of the precursor Ang I, or through Ang-(1–7) inhibition of the C-terminal active site of ACE.[90]

ACE2 has been implicated in regulating cardiac function. Knockout mice null for the ACE2 gene develop LV dilatation in association with severe cardiac contractile abnormalities.[91] These abnormalities are both gender- and time-dependent, since they are more severe in males and they progress with age. Evidence that cardiac dysfunction can be prevented in ACE2 null mice by the ablation of ACE expression suggests that the balance between ACE and ACE2 plays a critical role in regulating cardiac function.[91]

Angiotensin-(1–7) can be generated from either Ang II (by ACE2 activity) or Ang I (in a single step via neutral endopeptidase activity, or in two steps via successive ACE2 and ACE activity). This heptapeptide is believed to have beneficial effects on cardiac structure and function by inhibiting the pressor, proliferative and cell growth-promoting effects of Ang II.[92,93] In vascular smooth muscle cells (VSMCs) Ang-(1–7) inhibits Ang II-

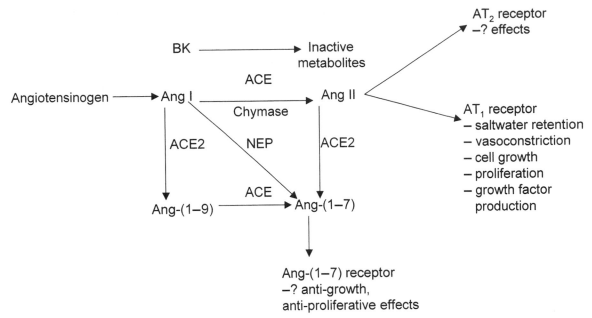

Figure 15.4. *The renin–angiotension system. Ang, angiotensin; AT$_1$, angiotensin II type 1 receptor; AT$_2$, angiotension II type 2 receptor; ACE, angiotensin-converting enzyme; NEP, neutral endopeptidase; BK, bradykinin*

mediated cell growth, and infusion of Ang-(1–7) after vascular injury inhibits neointimal growth through a mechanism independent of effects on heart rate or blood pressure.[94] In the rat coronary ligation model of MI, Ang-(1–7) administration helps preserve cardiac function in association with improved coronary perfusion and endothelial function.[95] Recent evidence that ARB inhibition of hypertrophy in the post-MI rat heart is associated with increases in plasma Ang-(1–7) is consistent with the notion that this peptide may protect against post-MI cardiac remodeling.

The pathways through which Ang-(1–7) favorably affects cellular structure and function have also been studied. In cultured rat aortic VSMCs, the anti-proliferative effects of Ang-(1–7) involve release of prostacyclin and prostacyclin-mediated production of cAMP, activation of cAMP-dependent protein kinase, and attenuation of extracellular signal-regulated kinase1 (ERK1) and ERK2 phosphorylation.[96] In addition, Ang-(1–7) can inhibit Ang II effects by stimulating production of nitric oxide and vasodilatory prostaglandins.[90,96–98] Ang-(1–7) might also counter Ang II effects by virtue of

its ability to inhibit ACE activity,[90] an effect that would reduce the production of Ang II as well as augment levels of anti-growth factors such as bradykinin (BK) by blocking their breakdown. Although Ang-(1–7) might interfere with Ang II signaling through the AT$_1$ receptor, by competing for receptor binding sites or by inducing AT$_1$ receptor internalization,[99] the concentrations required for these effects are substantially higher than are required for the effects on cell functions, making it unlikely that interference with Ang II binding is the major pathway involved.

Angiotensin-(1–7) is generated locally in the myocardium of various species.[100,101] It is formed from Ang I or Ang II in the interstitium of the canine LV, and the concentration of Ang-(1–7) immunoreactivity is increased in the rat heart following induction of a large MI due to coronary artery ligation.[102] In the failing human heart, Ang-(1–7)-forming activity related to both neutral endopeptidase (NEP) and ACE2 is increased.[103] In this setting NEP has a preference for Ang I, while ACE2 appears to have substrate preference for Ang II.[101] The major pathway for the generation of

Ang-(1–7) in the human heart depends on availability of Ang II as a substrate suggesting the importance of ACE2 in this process[101]. In the post-MI rat heart, Ang-(1–7) immunoreactivity is associated with cardiac myocytes but not with interstitial cells or blood vessels, and the most intense signal is noted at 4 weeks in the zone surrounding the replacement scar. Although the increase in Ang-(1–7) in the post-MI heart suggests that ACE2 is likely to be increased in this setting in a manner analogous to that which has been reported with ACE,[52,104] whether or not this actually occurs is uncertain. In one report neither *ACE* nor *ACE2* mRNA levels were increased in non-infarcted segments of LV in rats 4 weeks post-MI.[105]

Treatment of cardiac remodeling

Regardless of whether cardiac remodeling is initiated by myocardial damage due to an MI or one of the other causes outlined in Box 15.1, therapeutic approaches that are effective in treating this condition are relatively consistent. The first (and perhaps most effective) is to prevent the myocardial damage that triggers the remodeling process. The second is to treat hemodynamic abnormalities in order to reduce LV wall stress. The third involves inhibiting the systemic and cellular effects of neurohormonal activation. Additional approaches such as the use of MMP inhibitors counteract processes that are activated by either hemodynamic or neurohormonal stimuli. The treatments that are either currently being used or are being investigated to treat remodeling are listed in Box 15.3.

Rapid intervention designed to achieve patency of the infarct artery during the course of acute coronary occlusion reduces infarct size, improves survival, and limits subsequent cardiac remodeling.[106–108] However, even rapid and complete restitution of flow in an occluded coronary artery does not always prevent cardiac remodeling. In a survey of 284 patients who had successful percutaneous transluminal coronary angioplasty (PTCA), 30% had evidence of LV dilatation (defined as an increase in LV volume of greater than 20%) by 6

Box 15.3. *Approaches to treating cardiac remodeling*

Reduction of myocardial damage
- Treatment of hypertension
- MI prevention
- Emergent revascularization after coronary occlusion to limit damage
- Late patency of the infarct artery
- Correction of valvular or congenital heart lesions
- Minimizing exposure to myocardial toxins (e.g. adriamicin, alcohol)
- Myocardial cell regeneration (e.g. using stem cells or skeletal myoblasts)

Treatment of hemodynamic abnormalities
- Blood pressure control
- Reduction of pressure/volume load in valvular/congenital heart lesions
- Pharmacological unloading of the heart in heart failure patients (e.g. vasodilator drugs)
- Mechanical unloading of the failing heart (e.g. cardiac restraint devices, left ventricular assist devices)
- Cardiac resynchronization therapy (CRT)

Neurohormonal inhibition
- ACEIs and ARBs
- Aldosterone receptor antagonists
- Beta-blockers
- Other neurohormonal blocking approaches (e.g. arginine vasopressin inhibitors, adenosine receptor blockers, cytokine inhibitors, immunomodulatory therapy)

Other
- MMP inhibitors

months post-MI, despite the fact that Thrombolysis In Myocardial Infarction (TIMI) grade 3 flow had been restored within 6 hours of the onset of chest pain.[109] Interestingly, the presence of angina preceding an MI appears to protect against subsequent remodeling. The protective effects of recurrent ischemic episodes are probably related to recruitment of collaterals and/or to ischemic preconditioning of the myocardium. The latter raises the possibility that identification of the chemical mediators of preconditioning could be utilized to define new treatment approaches to help prevent damage that initiates post-MI cardiac remodeling.

The role of unloading the heart in the prevention of remodeling has been somewhat underemphasized over the past several years. However, in both animal models and in human patients, unloading the heart with nitroglycerin in the post-MI period reduces subsequent cardiac remodeling.[110,111] Cardiac remodeling in response to either pressure or volume load is a hallmark of valvular heart disease, and correction of the valve abnormality often results in substantial reversal of structural and functional abnormalities. This is perhaps best exemplified by the reduction in left ventricular hypertrophy that occurs after aortic valve replacement (AVR) for aortic stenosis. The regression in LV hypertrophy post-AVR is due to both a reduction in myocyte hypertrophy and regression of interstitial fibrosis.[112] Patients with aortic insufficiency (AI) often develop remarkable increases in cardiac volume and mass due to the combined pressure and volume load imposed by this valve lesion. Successful AVR can completely reverse these structural changes, provided that sustained LV systolic dysfunction is not present.[113] Furthermore, in patients with chronic severe AI, administration of arterial dilating drugs significantly reverses the remodeling process and delays progression to AVR.[114,115]

Neurohormonal blocking agents prevent or reverse remodeling, reduce progression to heart failure, and improve survival in post-MI patients. The first evidence that neurohormonal blocking agents could favorably affect cardiac structure in this setting came from seminal studies performed by Pfeffer and colleagues in rats that were treated with the ACE inhibitor captopril post-MI induced by coronary artery ligation.[63] The favorable effects of captopril on post-MI cardiac remodeling were then duplicated in patients with reduced LV ejection fraction following an anterior wall MI.[62] The clinical importance of this approach has now been confirmed in several large scale clinical trials in which ACE inhibitors (ACEIs) were shown to reduce both progression to heart failure and all-cause mortality.[32,33] The use of ARBs has been shown to have similar effects to the ACE inhibitors in preventing post-MI remodeling in the rat infarct model.[116] The clinical efficacy of this approach has now been shown in the VALIANT (VALsartan In Acute myocardial iNfarcTion) study in which valsartan, an ARB, was found to have comparable effects on mortality and morbidity to the ACEI captopril in MI survivors with LV dysfunction.[117]

More recently, carvedilol, a non-specific beta-blocker with alpha-1 and antioxidant properties, has been shown to inhibit increases in LV volumes and to reduce mortality in MI survivors with LV dysfunction.[118] The beneficial effects of carvedilol are additive to those of RAS blockade, since the vast majority of patients treated with carvedilol were already receiving an ACEI. The concept that broad-based neurohormonal blockade provides added protection in patients with an MI and LV dysfunction is supported by results of the EPHESUS (Eplerenone Post-AMI Heart Failure Efficacy and Survival) study in which the addition of eplerenone, an aldosterone receptor antagonist, to medical therapy that included both ACEI and beta-blocker resulted in further reduction of mortality.[119] Although it is not known whether this effect was due to inhibition of the remodeling process, there is evidence that addition of an aldosterone receptor antagonist to an ACEI can prevent post-MI remodeling better than an ACEI alone.[120]

Once heart failure develops, the remodeling process by no means becomes quiescent. This was demonstrated in the echocardiographic substudy of SOLVD (Studies of Left Ventricular Dysfunction) which evaluated the effects of the ACEI enalapril on cardiac structure in a representative 301-patient subgroup of the SOLVD study population.[121] As shown in Fig. 15.5, despite the fact that LV volumes and mass were already larger than an age- and sex-matched control population, patients randomized to placebo continued to increase these variables over 12 months of follow-up. Patients randomized to enalapril, however, demonstrated significant attenuation of these adverse structural changes. In addition, carvedilol and other beta-blockers have been shown to have favorable effects on cardiac remodeling, progression to heart failure, and survival in patients with heart failure and systolic dysfunction.

In addition to other neurohormonal blocking agents, a variety of novel approaches to inhibit

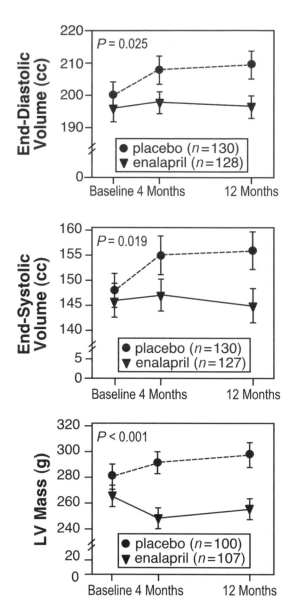

cardiac remodeling (outlined in Box 15.3) are being evaluated. One of the most interesting of these is the use of cardiac resynchronization therapy (CRT). The rationale for this therapy is that dys-synchronous contraction of the LV, associated (usually but not always) with a wide QRS complex on the ECG (which occurs in about 20–30% of heart failure patients) reduces the mechanical efficiency of ventricular contraction. The use of biventricular pacing to deliver CRT has been shown to improve exercise tolerance and symptoms in heart failure patients.[122] A reduction in wall stress and in severity of secondary mitral regurgitation with CRT has also been shown to be associated with a reduction in LV volumes, and an improvement in ejection fraction over time.[123,124] This reversal of the remodeling process would be expected to improve the natural history of heart failure. At this time the latter issue has not been fully resolved. However, a meta-analysis of published clinical trials showed that deaths due to progressive heart failure may be significantly reduced by CRT.[125]

Conclusions

Cardiac remodeling is the result of a prototypic series of events that are initiated by injury to the heart. Damage to myocardium increases stress and load on spared myocytes, and it initiates activation of neurohormonal systems both systemically and within the heart. The net effect is an alteration in cardiac structure that is meant to compensate for decreased function by offsetting the increase in load on the heart. However, remodeling has now been recognized as being a maladaptive response that is associated with progressive deterioration in cardiac function over time. Moreover, remodeling is intrinsic to the development and progression of heart failure. Recognition of the critical role played by remodeling in the pathogenesis of heart failure has helped focus therapies towards prevention and/or reversal of this process. Studies designed to elucidate basic mechanisms involved in remodeling have helped increase our understanding of key pathways involved in regulating cardiac structure and function. This information should serve us well in devising new strategies aimed at inhibiting the

Figure 15.5. *Effects of enalapril on LV remodeling in a representative subgroup of patients in the SOLVD study. Enalapril significantly inhibited increases in LV end-diastolic and end-systolic volumes and in LV mass that were seen in placebo-treated patients. Adapted from Greenberg B, Quinones MA, Koilpillai C et al. Effects of long-term enalapril therapy on cardiac structure and function in patients with left ventricular dysfunction. Results of the SOLVD echocardiography substudy. Circulation 1995; 91: 2573–81.*

remodeling process. Based on the marked impact of treatments (mainly neurohormonal blocking agents) that inhibit remodeling on the clinical course of patients with LV dysfunction, it is likely that new approaches that are successful in treating remodeling will further improve the outlook in post-MI and other patients with LV dysfunction.

References

1. Cohn JN, Ferrari R, Sharpe N. Cardiac remodeling – concepts and clinical implications: a consensus paper from an international forum on cardiac remodeling. On behalf of an International Forum on Cardiac Remodeling. J Am Coll Cardiol 2000; 35: 569–82.
2. Anversa P, Beghi C, Kikkawa Y, Olivetti G. Myocardial infarction in rats. Infarct size, myocyte hypertrophy, and capillary growth. Circ Res 1986; 58: 26–37.
3. Kajstura J, Zhang X, Reiss K et al. Myocyte cellular hyperplasia and myocyte cellular hypertrophy contribute to chronic ventricular remodeling in coronary artery narrowing-induced cardiomyopathy in rats. Circ Res 1994; 74: 383–400.
4. Mukherjee R, Brinsa TA, Dowdy KB et al. Myocardial infarct expansion and matrix metalloproteinase inhibition. Circulation 2003; 107: 618–25.
5. Abbate A, Biondi-Zoccai GG, Bussani R et al. Increased myocardial apoptosis in patients with unfavorable left ventricular remodeling and early symptomatic post-infarction heart failure. J Am Coll Cardiol 2003; 41: 753–60.
6. Beltrami CA, Finato N, Rocco M et al. Structural basis of end-stage failure in ischemic cardiomyopathy in humans. Circulation 1994; 89: 151–63.
7. Mercadier JJ, Lompre AM, Wisnewsky C et al. Myosin isoenzyme changes in several models of rat cardiac hypertrophy. Circ Res 1981; 49: 525–32.
8. Hunter JJ, Chien KR. Signaling pathways for cardiac hypertrophy and failure. N Engl J Med 1999; 341: 1276–83.
9. Braunwald E, Bristow MR. Congestive heart failure: fifty years of progress. Circulation 2000; 10220 (Suppl 4): IV14–23.
10. Razeghi P, Young ME, Alcorn JL et al. Metabolic gene expression in fetal and failing human heart. Circulation 2001; 104: 2923–31.
11. van Krimpen C, Smits JF, Cleutjens JP et al. DNA synthesis in the non-infarcted cardiac interstitium after left coronary artery ligation in the rat: effects of captopril. J Mol Cell Cardiol 1991; 23: 1245–53.
12. Cleutjens JP, Kandala JC, Guarda E, Guntaka RV, Weber KT. Regulation of collagen degradation in the rat myocardium after infarction. J Mol Cell Cardiol 1995; 27: 1281–92.
13. Sun Y, Weber KT. Infarct scar: a dynamic tissue. Cardiovasc Res 2000; 46: 250–6.
14. Woodiwiss AJ, Tsotetsi OJ, Sprott S et al. Reduction in myocardial collagen cross-linking parallels left ventricular dilatation in rat models of systolic chamber dysfunction. Circulation 2001; 103: 155–60.
15. Manabe I, Shindo T, Nagai R. Gene expression in fibroblasts and fibrosis: involvement in cardiac hypertrophy. Circ Res 2002; 91: 1103–13.
16. Katz AM. Cardiomyopathy of overload. A major determinant of prognosis in congestive heart failure. N Engl J Med 1990; 322: 100–10.
17. Karam R, Healy BP, Wicker P. Coronary reserve is depressed in postmyocardial infarction reactive cardiac hypertrophy. Circulation 1990; 81: 238–46.
18. Zhang J, McDonald KM. Bioenergetic consequences of left ventricular remodeling. Circulation 1995; 92: 1011–19.
19. Sutton MG, Sharpe N. Left ventricular remodeling after myocardial infarction: pathophysiology and therapy. Circulation 2000; 101: 2981–8.
20. Pfeffer MA, Braunwald E. Ventricular remodeling after myocardial infarction. Experimental observations and clinical implications. Circulation 1990; 81: 1161–72.
21. Eaton LW, Weiss JL, Bulkley BH, Garrison JB, Weisfeldt ML. Regional cardiac dilatation after acute myocardial infarction: recognition by two-dimensional echocardiography. N Engl J Med 1979; 300: 57–62.
22. Weber KT. Extracellular matrix remodeling in heart failure: a role for de novo angiotensin II generation. Circulation 1997; 96: 4065–82.
23. Lefroy DC, Wharton J, Crake T et al. Regional changes in angiotensin II receptor density after experimental myocardial infarction. J Mol Cell Cardiol 1996; 28: 429–40.
24. Sun Y, Weber KT. Angiotensin II receptor binding following myocardial infarction in the rat. Cardiovasc Res 1994; 28: 1623–8.
25. Gurantz D, Cowling RT, Villarreal FJ, Greenberg BH. Tumor necrosis factor-alpha upregulates angiotensin II type 1 receptors on cardiac fibroblasts. Circ Res 1999; 85: 272–9.
26. Yue P, Massie BM, Simpson PC, Long CS. Cytokine expression increases in nonmyocytes from rats with

postinfarction heart failure. Am J Physiol Heart Circ Physiol 1998; 275: H250–8.

27. Peng J, Gurantz D, Tran V, Cowling RT, Greenberg BH. Tumor necrosis factor-alpha-induced AT1 receptor upregulation enhances angiotensin II-mediated cardiac fibroblast responses that favor fibrosis. Circ Res 2002; 91: 1119–26.

28. Olivetti G, Abbi R, Quaini F et al. Apoptosis in the failing human heart. N Engl J Med 1997; 336: 1131–41.

29. St John SM, Pfeffer MA, Plappert T et al. Quantitative two-dimensional echocardiographic measurements are major predictors of adverse cardiovascular events after acute myocardial infarction. The protective effects of captopril. Circulation 1994; 89: 68–75.

30. Quinones MA, Greenberg BH, Kopelen HA et al. Echocardiographic predictors of clinical outcome in patients with left ventricular dysfunction enrolled in the SOLVD registry and trials: significance of left ventricular hypertrophy. Studies of left ventricular dysfunction. J Am Coll Cardiol 2000; 35: 1237–44.

31. Gaudron P, Eilles C, Kugler I, Ertl G. Progressive left ventricular dysfunction and remodeling after myocardial infarction. Potential mechanisms and early predictors. Circulation 1993; 87: 755–63.

32. Pfeffer MA, Braunwald E, Moye LA et al. Effect of captopril on mortality and morbidity in patients with left ventricular dysfunction after myocardial infarction. Results of the survival and ventricular enlargement trial. The SAVE Investigators [see comments]. N Engl J Med 1992; 327: 669–77.

33. Kober L, Torp-Pedersen C, Carlsen JE et al. A clinical trial of the angiotensin-converting-enzyme inhibitor trandolapril in patients with left ventricular dysfunction after myocardial infarction. Trandolapril Cardiac Evaluation (TRACE) Study Group. N Engl J Med 1995; 333: 1670–6.

34. Francis GS, Benedict C, Johnstone DE et al. Comparison of neuroendocrine activation in patients with left ventricular dysfunction with and without congestive heart failure. A substudy of the Studies of Left Ventricular Dysfunction (SOLVD). Circulation 1990; 82: 1724–9.

35. Sadoshima J, Izumo S. Signal transduction pathways of angiotensin II-induced c-fos gene expression in cardiac myocytes in vitro. Roles of phospholipid-derived second messengers. Circ Res 1993; 73: 424–38.

36. Kim NN, Villarreal FJ, Printz MP, Lee AA, Dillmann WH. Trophic effects of angiotensin II on neonatal rat cardiac myocytes are mediated by cardiac fibroblasts.

Am J Physiol Endocrinol Metab 1995; 269: E426–37.

37. Brilla CG, Zhou G, Matsubara L, Weber KT. Collagen metabolism in cultured adult rat cardiac fibroblasts: response to angiotensin II and aldosterone. J Mol Cell Cardiol 1994; 26: 809–20.

38. Sun Y, Ramires FJ, Zhou G, Ganjam VK, Weber KT. Fibrous tissue and angiotensin II. J Mol Cell Cardiol 1997; 29: 2001–12.

39. Schorb W, Booz GW, Dostal DE et al. Angiotensin II is mitogenic in neonatal rat cardiac fibroblasts. Circ Res 1993; 72: 1245–54.

40. Hodsman GP, Kohzuki M, Howes LG et al. Neurohumoral responses to chronic myocardial infarction in rats. Circulation 1988; 78: 376–81.

41. Paul M, Wagner J, Dzau VJ. Gene expression of the renin–angiotensin system in human tissues. Quantitative analysis by the polymerase chain reaction. J Clin Invest 1993; 91: 2058–64.

42. Dostal DE, Baker KM. Angiotensin II stimulation of left ventricular hypertrophy in adult rat heart. Mediation by the AT1 receptor. Am J Hypertens 1992; 55: 276–80.

43. Katwa LC, Ratajska A, Cleutjens JP et al. Angiotensin converting enzyme and kinase-II-like activities in cultured valvular interstitial cells of the rat heart. Cardiovasc Res 1995; 29: 57–64.

44. Endo-Mochizuki Y, Mochizuki N, Sawa H et al. Expression of renin and angiotensin-converting enzyme in human hearts. Heart Vessels 1995; 10: 285–93.

45. Zhang X, Dostal DE, Reiss K et al. Identification and activation of autocrine renin–angiotensin system in adult ventricular myocytes. Am J Physiol Heart Circ Physiol 1995; 269: H1791–802.

46. Muller DN, Fischli W, Clozel JP et al. Local angiotensin II generation in the rat heart: role of renin uptake. Circ Res 1998; 82: 13–20.

47. Urata H, Boehm KD, Philip A et al. Cellular localization and regional distribution of an angiotensin II-forming chymase in the heart. J Clin Invest 1993; 91: 1269–81.

48. Dostal DE, Baker KM. The cardiac renin–angiotensin system: conceptual, or a regulator of cardiac function? Circ Res 1999; 85: 643–50.

49. van Kats JP, Danser AH, van Meegen JR et al. Angiotensin production by the heart: a quantitative study in pigs with the use of radiolabeled angiotensin infusions. Circulation 1998; 98: 73–81.

50. Dostal DE, Baker KM. The cardiac renin–angiotensin system: conceptual, or a regulator of cardiac function? Circ Res 1999; 85: 643–50.

51. De Mello WC, Danser AH. Angiotensin II and the heart: on the intracrine renin–angiotensin system. Hypertension 2000; 35: 1183–8.

52. Hirsch AT, Talsness CE, Schunkert H, Paul M, Dzau VJ. Tissue-specific activation of cardiac angiotensin converting enzyme in experimental heart failure. Circ Res 1991; 69: 475–82.

53. Lindpaintner K, Lu W, Neidermajer N et al. Selective activation of cardiac angiotensinogen gene expression in post- infarction ventricular remodeling in the rat. J Mol Cell Cardiol 1993; 25: 133–43.

54. Nio Y, Matsubara H, Murasawa S, Kanasaki M, Inada M. Regulation of gene transcription of angiotensin II receptor subtypes in myocardial infarction. J Clin Invest 1995; 95: 46–54.

55. Sun Y, Cleutjens JP, Diaz-Arias AA, Weber KT. Cardiac angiotensin converting enzyme and myocardial fibrosis in the rat. Cardiovasc Res 1994; 28: 1423–32.

56. Lefroy DC, Wharton J, Crake T et al. Regional changes in angiotensin II receptor density after experimental myocardial infarction. J Mol Cell Cardiol 1996; 28: 429–40.

57. Serneri GG, Boddi M, Cecioni I et al. Cardiac angiotensin II formation in the clinical course of heart failure and its relationship with left ventricular function. Circ Res 2001; 88: 961–8.

58. Wei CC, Ferrario CM, Brosnihan KB et al. Angiotensin peptides modulate bradykinin levels in the interstitium of the dog heart in vivo. J Pharmacol Exp Ther 2002; 300: 324–9.

59. Paradis P, Dali-Youcef N, Paradis FW, Thibault G, Nemer M. Overexpression of angiotensin II type 1 receptor in cardiomyocytes induces cardiac hypertrophy and remodeling. Proc Natl Acad Sci USA 2000; 97: 931–6.

60. Harada K, Sugaya T, Murakami K, Yazaki Y, Komuro I. Angiotensin II type 1A receptor knockout mice display less left ventricular remodeling and improved survival after myocardial infarction. Circulation 1999; 100: 2093–9.

61. Mazzolai L, Nussberger J, Aubert JF et al. Blood pressure-independent cardiac hypertrophy induced by locally activated renin–angiotensin system. Hypertension 1998; 31: 1324–30.

62. Pfeffer MA, Lamas GA, Vaughan DE, Parisi AF, Braunwald E. Effect of captopril on progressive ventricular dilatation after anterior myocardial infarction. N Engl J Med 1988; 319: 80–6.

63. Pfeffer JM, Pfeffer MA, Braunwald E. Influence of chronic captopril therapy on the infarcted left ventricle of the rat. Circ Res 1985; 57: 84–95.

64. Michel JB, Lattion AL, Salzmann JL et al. Hormonal and cardiac effects of converting enzyme inhibition in rat myocardial infarction. Circ Res 1988; 62: 641–50.

65. Mankad S, d'Amato TA, Reichek N et al. Combined angiotensin II receptor antagonism and angiotensin-converting enzyme inhibition further attenuates postinfarction left ventricular remodeling. Circulation 2001; 103: 2845–50.

66. Spinale FG, Coker ML, Krombach SR et al. Matrix metalloproteinase inhibition during the development of congestive heart failure: effects on left ventricular dimensions and function. Circ Res 1999; 85: 364–76.

67. Siwik DA, Pagano PJ, Colucci WS. Oxidative stress regulates collagen synthesis and matrix metalloproteinase activity in cardiac fibroblasts. Am J Physiol Cell Physiol 2001; 280: C53–60.

68. Brilla CG, Zhou G, Matsubara L, Weber KT. Collagen metabolism in cultured adult rat cardiac fibroblasts: response to angiotensin II and aldosterone. J Mol Cell Cardiol 1994; 26: 809–20.

69. Carver W, Nagpal ML, Nachtigal M, Borg TK, Terracio L. Collagen expression in mechanically stimulated cardiac fibroblasts. Circ Res 1991; 69: 116–22.

70. Kim NN, Villarreal FJ, Printz MP, Lee AA, Dillmann WH. Trophic effects of angiotensin II on neonatal rat cardiac myocytes are mediated by cardiac fibroblasts. Am J Physiol Endocrinol Metab 1995; 269: E426–37.

71. Harada M, Itoh H, Nakagawa O et al. Significance of ventricular myocytes and nonmyocytes interaction during cardiocyte hypertrophy: evidence for endothelin-1 as a paracrine hypertrophic factor from cardiac nonmyocytes. Circulation 1997; 96: 3737–44.

72. Brilla CG, Zhou G, Matsubara L, Weber KT. Collagen metabolism in cultured adult rat cardiac fibroblasts: response to angiotensin II and aldosterone. J Mol Cell Cardiol 1994; 26: 809–20.

73. Schorb W, Booz GW, Dostal DE et al. Angiotensin II is mitogenic in neonatal rat cardiac fibroblasts. Circ Res 1993; 72: 1245–54.

74. Villarreal FJ, Kim NN, Ungab GD, Printz MP, Dillmann WH. Identification of functional angiotensin II receptors on rat cardiac fibroblasts. Circulation 1993; 88: 2849–61.

75. Nio Y, Matsubara H, Murasawa S, Kanasaki M, Inada M. Regulation of gene transcription of angiotensin II receptor subtypes in myocardial infarction. J Clin Invest 1995; 95: 46–54.

76. Sun Y, Weber KT. Angiotensin II receptor binding following myocardial infarction in the rat. Cardiovasc Res 1994; 28: 1623–8.

77. Gray MO, Long CS, Kalinyak JE, Li HT, Karliner JS. Angiotensin II stimulates cardiac myocyte hypertrophy via paracrine release of TGF-beta 1 and endothelin-1 from fibroblasts. Cardiovasc Res 1998; 40: 352–63.

78. Campbell SE, Katwa LC. Angiotensin II stimulated expression of transforming growth factor- beta1 in cardiac fibroblasts and myofibroblasts. J Mol Cell Cardiol 1997; 29: 1947–58.

79. Fujisaki H, Ito H, Hirata Y et al. Natriuretic peptides inhibit angiotensin II-induced proliferation of rat cardiac fibroblasts by blocking endothelin-1 gene expression. J Clin Invest 1995; 96: 1059–65.

80. Sano M, Fukuda K, Kodama H et al. Interleukin-6 family of cytokines mediate angiotensin II-induced cardiac hypertrophy in rodent cardiomyocytes. J Biol Chem 2000; 275: 29717–23.

81. Piacentini L, Gray M, Honbo NY, Chentoufi J, Bergman M, Karliner JS. Endothelin-1 stimulates cardiac fibroblast proliferation through activation of protein kinase C. J Mol Cell Cardiol 2000; 32: 565–76.

82. Wang F, Trial J, Diwan A et al. Regulation of cardiac fibroblast cellular function by leukemia inhibitory factor. J Mol Cell Cardiol 2002; 34: 1309–16.

83. Kodama H, Fukuda K, Pan J et al. Leukemia inhibitory factor, a potent cardiac hypertrophic cytokine, activates the JAK/STAT pathway in rat cardiomyocytes. Circ Res 1997; 81: 656–63.

84. Murata M, Fukuda K, Ishida H et al. Leukemia inhibitory factor, a potent cardiac hypertrophic cytokine, enhances L-type Ca^{2+} current and $[Ca^{2+}]_i$ transient in cardiomyocytes. J Mol Cell Cardiol 1999; 31: 237–45.

85. Manabe I, Shindo T, Nagai R. Gene expression in fibroblasts and fibrosis: involvement in cardiac hypertrophy. Circ Res 2002; 91: 1103–13.

86. Tipnis SR, Hooper NM, Hyde R et al. A human homolog of angiotensin-converting enzyme. Cloning and functional expression as a captopril-insensitive carboxypeptidase. J Biol Chem 2000; 275: 33238–43.

87. Donoghue M, Hsieh F, Baronas E, Godbout K, Gosselin M, Stagliano N et al. A novel angiotensin-converting enzyme-related carboxypeptidase (ACE2) converts angiotensin I to angiotensin 1–9. Circ Res 2000; 87: E1–9.

88. Harmer D, Gilbert M, Borman R, Clark KL. Quantitative mRNA expression profiling of ACE 2, a novel homologue of angiotensin converting enzyme. FEBS Lett 2002; 532: 107–10.

89. Vickers C, Hales P, Kaushik V et al. Hydrolysis of biological peptides by human angiotensin-converting enzyme-related carboxypeptidase. J Biol Chem 2002; 277: 14838–43.

90. Li P, Chappell MC, Ferrario CM, Brosnihan KB. Angiotensin-(1–7) augments bradykinin-induced vasodilation by competing with ACE and releasing nitric oxide. Hypertension 1997; 29: 394–400.

91. Crackower MA, Sarao R, Oudit GY et al. Angiotensin-converting enzyme 2 is an essential regulator of heart function. Nature 2002; 417: 822–8.

92. Freeman EJ, Chisolm GM, Ferrario CM, Tallant EA. Angiotensin-(1–7) inhibits vascular smooth muscle cell growth. Hypertension 1996; 28: 104–8.

93. Zhu Z, Zhong J, Zhu S et al. Angiotensin-(1–7) inhibits angiotensin II-induced signal transduction. J Cardiovasc Pharmacol 2002; 40: 693–700.

94. Strawn WB, Ferrario CM, Tallant EA. Angiotensin-(1–7) reduces smooth muscle growth after vascular injury. Hypertension 1999; 33: 207–11.

95. Loot AE, Roks AJ, Henning RH et al. Angiotensin-(1–7) attenuates the development of heart failure after myocardial infarction in rats. Circulation 2002; 105: 1548–50.

96. Tallant EA, Clark MA. Molecular mechanisms of inhibition of vascular growth by angiotensin-(1–7). Hypertension 2003; 42: 574–9.

97. Brosnihan KB, Li P, Ferrario CM. Angiotensin-(1–7) dilates canine coronary arteries through kinins and nitric oxide. Hypertension 1996; 27: 523–8.

98. Muthalif MM, Benter IF, Uddin MR, Harper JL, Malik KU. Signal transduction mechanisms involved in angiotensin-(1–7)-stimulated arachidonic acid release and prostanoid synthesis in rabbit aortic smooth muscle cells. J Pharmacol Exp Ther 1998; 284: 388–98.

99. Clark MA, Diz DI, Tallant EA. Angiotensin-(1–7) downregulates the angiotensin II type 1 receptor in vascular smooth muscle cells. Hypertension 2001; 37: 1141–6.

100. Wei CC, Ferrario CM, Brosnihan KB et al. Angiotensin peptides modulate bradykinin levels in the interstitium of the dog heart in vivo. J Pharmacol Exp Ther 2002; 300: 324–9.

101. Zisman LS, Meixell GE, Bristow MR, Canver CC. Angiotensin-(1–7) formation in the intact human heart: in vivo dependence on angiotensin II as substrate. Circulation 2003; 108: 1679–81.

102. Averill DB, Ishiyama Y, Chappell MC, Ferrario CM. Cardiac angiotensin-(1–7) in ischemic cardiomyopathy. Circulation 2003; 108: 2141–6.

103. Zisman LS, Keller RS, Weaver B et al. Increased angiotensin-(1–7)-forming activity in failing human heart ventricles: evidence for upregulation of the angiotensin-converting enzyme homologue ACE2. Circulation 2003; 108: 1707–12.

104. Duncan AM, Burrell LM, Kladis A, Campbell DJ. Effects of angiotensin-converting enzyme inhibition on angiotensin and bradykinin peptides in rats with myocardial infarction. J Cardiovasc Pharmacol 1996; 28: 746–54.

105. Ishiyama Y, Gallagher PE, Averill DB et al. Upregulation of angiotensin-converting enzyme 2 after myocardial infarction by blockade of angiotensin II receptors. Hypertension 2004; 43: 970–6.

106. Randomized trial of intravenous streptokinase, oral aspirin, both, or neither among 17 187 cases of suspected acute myocardial infarction: ISIS-2. ISIS-2 (Second International Study of Infarct Survival) Collaborative Group. Lancet 1988; 2: 349–60.

107. Jeremy RW, Hackworthy RA, Bautovich G, Hutton BF, Harris PJ. Infarct artery perfusion and changes in left ventricular volume in the month after acute myocardial infarction. J Am Coll Cardiol 1987; 9: 989–95.

108. Touchstone DA, Beller GA, Nygaard TW, Tedesco C, Kaul S. Effects of successful intravenous reperfusion therapy on regional myocardial function and geometry in humans: a tomographic assessment using two-dimensional echocardiography. J Am Coll Cardiol 1989; 13: 1506–13.

109. Bolognese L, Neskovic AN, Parodi G et al. Left ventricular remodeling after primary coronary angioplasty: patterns of left ventricular dilation and long-term prognostic implications. Circulation 2002; 106: 2351–7.

110. McDonald KM, Francis GS, Matthews J, Hunter D, Cohn JN. Long-term oral nitrate therapy prevents chronic ventricular remodeling in the dog. J Am Coll Cardiol 1993; 21: 514–22.

111. Mahmarian JJ, Moye LA, Chinoy DA et al. Transdermal nitroglycerin patch therapy improves left ventricular function and prevents remodeling after acute myocardial infarction: results of a multicenter prospective randomized, double-blind, placebo-controlled trial. Circulation 1998; 97: 2017–24.

112. Krayenbuehl HP, Hess OM, Monrad ES, Schneider J, Mall G, Turina M. Left ventricular myocardial structure in aortic valve disease before, intermediate, and late after aortic valve replacement. Circulation 1989; 79: 744–55.

113. Bonow RO, Rosing DR, Maron BJ et al. Reversal of left ventricular dysfunction after aortic valve replacement for chronic aortic regurgitation: influence of duration of preoperative left ventricular dysfunction. Circulation 1984; 70: 570–9.

114. Greenberg B, Massie B, Bristow JD et al. Long-term vasodilator therapy of chronic aortic insufficiency. A randomized double-blinded, placebo-controlled clinical trial. Circulation 1988; 78: 92–103.

115. Scognamiglio R, Rahimtoola SH, Fasoli G, Nistri S, Dalla VS. Nifedipine in asymptomatic patients with severe aortic regurgitation and normal left ventricular function. N Engl J Med 1994; 331: 689–94.

116. Schieffer B, Wirger A, Meybrunn M et al. Comparative effects of chronic angiotensin-converting enzyme-inhibition and angiotensin-II type-1 receptor blockade on cardiac remodeling after myocardial-infarction in the rat. Circulation 1994; 89: 2273–82.

117. Pfeffer MA, McMurray JJ, Velazquez EJ et al. Valsartan, captopril, or both in myocardial infarction complicated by heart failure, left ventricular dysfunction, or both. N Engl J Med 2003; 349: 1893–906.

118. Doughty RN, Whalley GA, Walsh HA et al. Effects of carvedilol on left ventricular remodeling after acute myocardial infarction: the CAPRICORN Echo Substudy. Circulation 2004; 109: 201–6.

119. Pitt B, Remme W, Zannad F et al. Eplerenone, a selective aldosterone blocker, in patients with left ventricular dysfunction after myocardial infarction. N Engl J Med 2003; 348: 1309–21.

120. Hayashi M, Tsutamoto T, Wada A et al. Immediate administration of mineralocorticoid receptor antagonist spironolactone prevents post-infarct left ventricular remodeling associated with suppression of a marker of myocardial collagen synthesis in patients with first anterior acute myocardial infarction. Circulation 2003; 107: 2559–65.

121. Greenberg B, Quinones MA, Koilpillai C et al. Effects of long-term enalapril therapy on cardiac structure and function in patients with left ventricular dysfunction. Results of the SOLVD echocardiography substudy. Circulation 1995; 91: 2573–81.

122. Abraham WT, Fisher WG, Smith AL et al. Cardiac resynchronization in chronic heart failure. N Engl J Med 2002; 346: 1845–53.

123. John Sutton MG, Plappert T, Abraham WT et al. Effect of cardiac resynchronization therapy on left ventricular size and function in chronic heart failure. Circulation 2003; 107: 1985–90.

124. Yu CM, Chau E, Sanderson JE et al. Tissue Doppler echocardiographic evidence of reverse remodeling and improved synchronicity by simultaneously delaying regional contraction after biventricular pacing therapy in heart failure. Circulation 2002; 105: 438–45.

125. Bradley DJ, Bradley EA, Baughman KL et al. Cardiac resynchronization and death from progressive heart failure: a meta-analysis of randomized controlled trials. JAMA 2003; 289:730–40.

CHAPTER 16

The role of inflammatory mediators in cardiac hypertrophy, cardiac remodeling and myocardial dysfunction

Jana Burchfield, RS Ramabadran, Douglas L Mann

Introduction

Although clinicians recognized the pathophysiological importance of inflammatory mediators in the heart as early as 1669,[1] the formal recognition that inflammatory mediators were activated in the setting of hypertrophy and/or heart failure did not occur for another three centuries. Beginning with the original description of inflammatory cytokines in patients with heart failure in 1990,[2] there has been an enduring interest in delineating the role that these molecules play in regulating cardiac structure and function in a variety of different pathophysiological contexts. Accordingly, the intent of the present chapter will be to summarize the current understanding of the role that these molecules play in regulating cardiac hypertrophy, cardiac remodeling and myocardial dysfunction.

Expression of proinflammatory cytokines in the heart

The portfolio of cytokines that comprise the focus of this review includes tumor necrosis factor (TNF), as well as the interleukin-1 (IL-1) and interleukin-6 (IL-6) family of cytokines. These molecules have been referred to as 'proinflammatory cytokines', insofar as they were traditionally thought to be derived exclusively from the immune system and were therefore considered to be primarily responsible for mediating inflammatory responses in tissues. However, inflammatory mediators are now known to be expressed by all nucleated cell types residing in the myocardium, including the cardiac myocyte,[3,4] thus suggesting that these molecules may do more than simply orchestrate inflammatory responses in the heart. The observation that proinflammatory cytokines are consistently and rapidly expressed in response to a variety of different forms of myocardial injury (Box 16.1), suggests that they comprise part of an intrinsic or 'innate' stress response system in the

---

Box 16.1. *Cardiac pathophysiological conditions associated with proinflammatory cytokines*

Acute viral myocarditis
Cardiac allograft rejection
Myocardial infarction
Unstable angina
Myocardial reperfusion injury
Hypertrophic cardiomyopathy[a]
Heart failure[a]
Cardiopulmonary bypass[a]
Magnesium deficiency[a]
Pressure-overload[a]

---

[a]Conditions not traditionally associated with immunologically mediated inflammation.

heart. Thus, analogous to the role that proinflammatory cytokines play as effector molecules in the innate immune system, which is intended to act as an 'early warning system' that allows the host to rapidly discriminate self from non-self,[5] the expression of proinflammatory cytokines in the heart may permit the myocardium to rapidly respond to tissue injury as part of an early warning system that coordinates and integrates a panoply of homeostatic responses within the heart following tissue injury.[6] However, it bears emphasis that the family of proinflammatory molecules that comprise this innate stress response system are phylogenetically ancient, and thus probably evolved in organisms with relatively short life spans (weeks to months). Accordingly, activation of the 'innate' stress response system was never intended to provide long-term adaptive responses to the host organism. And indeed, as will be discussed below, there is now substantial evidence that sustained and/or dysregulated expression of proinflammatory cytokines leads to a variety of maladaptive effects in the heart, including cardiac hypertrophy, remodeling, and decompensation.

Effects of proinflammatory cytokines on hypertrophic growth of the heart

Although all proinflammatory cytokines are capable of provoking hypertrophic responses in cardiac myocytes, the proteins that have received the most attention are related to the IL-6 family of cytokines, which includes interleukin-6 (IL-6), leukemia inhibitory factor (LIF), cardiotrophin-1 (CT-1). Very early studies by Kishimoto and colleagues showed that mice with cardiac restricted overexpression of IL-6 and the IL-6 receptor (IL-6R) in the heart developed a concentric hypertrophic phenotype by 5 months of age (Fig. 16.1).[7] In the same study, cultured neonatal heart muscle cells from normal mice enlarged in response to a combination of IL-6 and a soluble form of IL-6R, thus confirming the specificity of the observed IL-6 effects *in vivo* (Fig. 16.1). CT-1, which is a relatively new member of the IL-6 family of cytokines, was originally identified by screening a cDNA

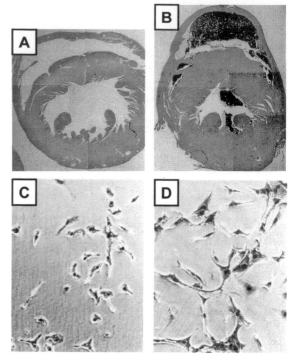

Figure 16.1. *Hypertrophic effects of IL-6. Coronal section of (A) littermate control and (B) double transgenic mice overexpressing IL-6 and IL-6R, showing the development of cardiac hypertrophy in the double transgenic mice. Morphologic appearance of cultured neonatal cardiac myocytes in the absence (C) and presence (D) of IL-6 and IL-6R, showing increased cell size in the cells that were treated with IL-6 and IL-6R. Reproduced with permission from Hirota H, Yoshida K, Kishimoto T et al. Continuous activation of gp130, a signal-transducing receptor component for interleukin 6-related cytokines, causes myocardial hypertrophy in mice. Proc Natl Acad Sci USA 1995; 92: 4862–6. See color plate section.*

expression library for clones that would induce an increase in cell size in cultured cardiomyocytes.[8] Subsequent studies in cultured cardiac myocytes showed that CT-1 signaled through the gp130 and LIF receptor (LIFR), and induced a hypertrophic response in cardiomyocytes that was distinct from the phenotype observed after α-adrenergic stimulation. That is, stimulation with CT-1 resulted in an increase in cardiac cell size that was characterized by an increase in cell length and assembly of sarcomeres in series, without a demonstrable increase in cell width. CT-1 induced a distinct pattern of immediate early genes and upregulated the expression of atrial natriuretic factor (*ANF*) gene.[9] Ensuing studies *in vivo* established that administration of CT-1 to mice for 14 days resulted in a dose-dependent increase in both the heart weight and ventricular weight to body ratios in the treated groups. Interestingly, chronic CT-1 administration stimulated growth of the liver, kidney, and spleen, as well as causing atrophy of the thymus,[10] indicating that the effects of CT-1 were not specific for the heart. Not surprisingly, LIF has also been shown to provoke hypertrophic growth of neonatal and adult cardiac myocytes, analogous to the findings observed with CT-1.[11,12]

IL-1β and TNF have also been shown to provoke hypertrophic growth in cardiac myocytes. Using a cell culture model, Patten and colleagues reported that IL-1β both induced cardiac myocyte hypertrophy, and reinitiated myocyte DNA synthesis.[13] Later studies confirmed these early observations with respect to protein synthesis, but did not confirm the early observations regarding DNA synthesis in neonatal cardiac myocytes.[14] These latter studies demonstrated that treatment with IL-1β resulted in increased expression of ANF and beta-myosin heavy chain (β-*MHC*) mRNA, with decreased expression of mRNA for sarcoplasmic reticulum Ca^{+2}-ATPase (*SERCA*), the calcium-release channel and the voltage-dependent calcium channel.[14] More recently, Isoda et al. showed that targeted overexpression of the human *IL-1* gene resulted in a 1.4- to 2.2-fold increase in the heart weight:body weight ratio when compared to wild-type mice.[15] Lung weight:body weight ratio also increased in the transgenic IL-1 transgenic mice, all of which died within 14 days of birth. Examination of the hearts of these animals revealed concentric hypertrophy with cardiomyocyte hypertrophy in all transgenic lines, and pulmonary edema noted in some animals. Northern blot analysis disclosed re-expression of the fetal gene program, including increased expression of ANF and β-*MHC*, and decreased expression of voltage-dependent calcium channel mRNA

expression, thus recapitulating the previous findings *in vitro*.[14]

Yokoyama et al. were the first to demonstrate that treatment with TNF triggered an increase in protein synthesis in adult cardiac myocytes.[16] Subsequent studies have shown that TNF-induced protein synthesis is sensitive to antioxidants,[17] and that dominant negative Akt (protein kinase B) and c-jun N-terminal kinase (JNK) constructs are sufficient to inhibit TNF-induced increase in cardiomyocyte cell size,[17] suggesting that redox induced activation of these signal transduction pathways plays an important role in hypertrophic growth. The *in vitro* studies in isolated adult cardiac myocytes have been confirmed by subsequent studies in transgenic mouse models, wherein cardiac restricted overexpression of TNF resulted in a concentric hypertrophic phenotype (Fig. 16.2) that was accompanied by reactivation of the fetal gene program.[18–20] Interestingly, recent studies have suggested that part of the hypertrophic phenotype in the TNF transgenic mice (MHCsTNF) is secondary to activation of the renin–angiotensin system in the TNF transgenic mice.[21] This study demonstrated that there was a significant increase in angiotensin-converting enzyme (ACE) mRNA levels (Fig. 16.3A and B) and ACE activity (Fig. 16.3C), as well as increased angiotensin II (Ang II) peptide levels (Fig. 16.3D) in the hearts of the MHCsTNF mice relative to littermate controls. Importantly, the expression of renin and angiotensinogen was not increased in MHCsTNF mice compared with littermate controls. Thus, this study suggested that the increased levels of Ang II peptide levels in the MHCsTNF mice were principally the result of increased ACE activity, as opposed to increased activation of the more proximal components of the renin–angiotensin system, namely renin and angiotensinogen. This study also suggested that treatment of the MHCsTNF mice from 4–8 weeks of age with losartan, an angiotensin type I receptor antagonist, significantly attenuated the cardiac hypertrophy phenotype in the MHCsTNF mice without any significant effect on peripheral hemodynamics, thus suggesting that activation of the renin–angiotensin system was functionally important in the TNF transgenic mice.[21]

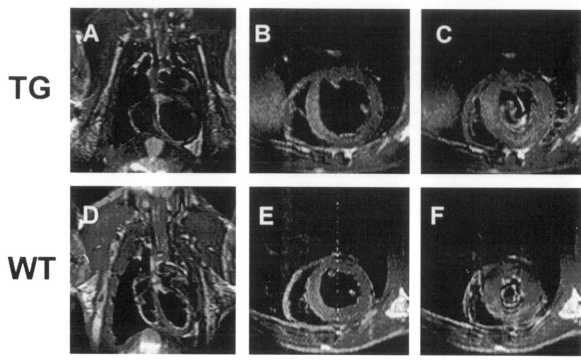

Figure 16.2. *LV remodeling in a transgenic mouse model of TNF overexpression. TNF overexpression was targeted to the cardiac compartment of transgenic mice. Magnetic resonance images of the heart were obtained from 24-week-old transgenic mice (TG; A, B, and C) and an age-matched control mouse (WR; D, E, and F). As shown, there was significant LV dilation in the animal harboring the TNF transgene in the cardiac compartment. Kubota T, McTiernan CF, Frye CS et al. Dilated cardiomyopathy in transgenic mice with cardiac specific overexpression of tumor necrosis factor-alpha. Circ Res 1997; 81: 627–35.*

Effects of proinflammatory cytokines on cardiac remodeling

In addition to provoking cardiac myocyte hypertrophy, inflammatory mediators exert a number of biological effects that are believed to play an important role in the process of left venticular (LV) remodeling, including important changes in the extracellular matrix (Box 16.2).[18,22] The effects of inflammatory mediators on the extracellular matrix appear to be bimodal in that short-term expression of inflammatory mediators leads to fibrillar collagen degradation, which would be expected to favor LV dilation, whereas sustained expression of inflammatory mediators leads to excessive fibrillar collagen deposition, which would be expected to promote increased LV stiffness. With respect to the short-term effects of inflammatory mediators, a

Box 16.2. *Effects of inflammatory mediators on left ventricular remodeling*
Alterations in the biology of the myocyte
myocyte hypertrophy
contractile abnormalities
fetal gene expression
Alteration in the extracellular matrix
MMP activation
degradation of the matrix
fibrosis
Progressive myocyte loss
necrosis
apoptosis

systemic infusion of pathophysiologically relevant concentrations of TNF was shown to lead to a time-dependent change in LV dimension that was accompanied by progressive degradation of the

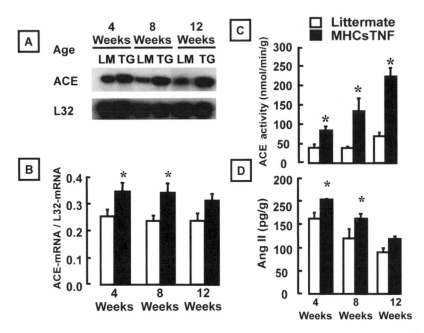

Figure 16.3. ACE mRNA, ACE activity and Ang II peptide levels in TNF transgenic mice (MHCsTNF) and littermate control mice. (A) Ribonuclease protection assay for ACE mRNA in the hearts of the 4-, 8- and 12-week-old MHCsTNF (TG) and littermate control mice (LM) mice. (B) Group data in hearts from 4-, 8- and 12-week-old MHCsTNF (n = 7 hearts for each age) and 4-, 8- and 12-week littermate control mice (n = 7 hearts for each age). (C) ACE activity in the hearts from 4-, 8- and 12-week-old MHCsTNF and the 4-, 8- and 12-week-old littermate control mice. (D) Group data for Ang II peptide levels in the hearts of the MHCsTNF and littermate control mice at 4, 8, and 12 weeks of age. LM, littermate control; TG, transgenic. *P < 0.05 versus age-matched control group by Tukey's test; error bars, standard errors. Reproduced with permission from the American Heart Association: Flesch M, Hoper A, Dell'Italia L et al. Activation and functional significance of the renin–angiotensin system in mice with cardiac restricted overexpression of tumor necrosis factor. Circulation 2003; 108: 598–604.

extracelluar matrix in rats. Subsequent studies in transgenic mice with targeted overexpression of TNF have shown that these mice initially develop degradation of myocardial fibrillar collagen and LV dilation, followed by progressive myocardial fibrosis.[18–20] For example, Kubota et al. showed that a transgenic mouse line that overexpressed TNF in the cardiac compartment developed progressive LV dilation over a 24-week period of observation (Fig. 16.2).[18] Similar findings have also been reported by Bryant et al.[19] and Sivasubramanian et al.,[20] who observed identical findings in terms of LV dilation in transgenic mice with targeted overexpression of TNF in the heart. With respect to the mechanisms that are involved in TNF-induced LV dilation, it has been suggested that TNF-induced

activation of matrix metalloproteinases (MMPs) is responsible for this effect.[20,23] As shown in Figs 16.4 and 16.5, respectively, there was progressive loss of fibrillar collagen and increased MMP activation in the hearts of the transgenic mice overexpressing TNF in the cardiac compartment. The dissolution of the fibrillar collagen weave that surrounds the individual cardiac myocytes and links the myocytes together would be expected to allow for rearrangement ('slippage') of myofibrillar bundles within the ventricular wall.[24] However, Fig. 16.4 shows that long-term stimulation (i.e. 8–12 weeks) with TNF resulted in an increase in fibrillar collagen content that was accompanied by decreased MMP activity (Fig. 16.5B), and increased expression of the tissue inhibitors of matrix metalloproteinases (TIMPs,

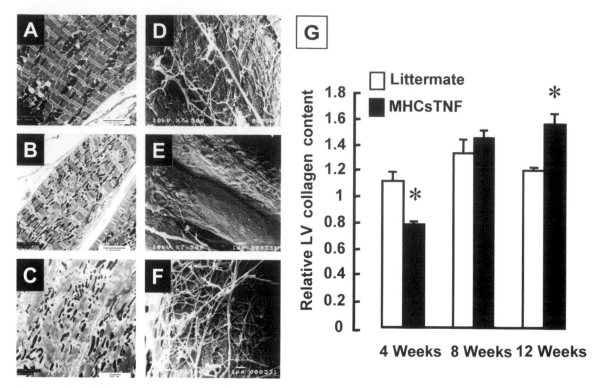

Figure 16.4. *Effects of sustained proinflammatory cytokine expression on myocardial ultrastructure and collagen content. (A)–(C), representative transmission electron micrographs in littermate controls (A) and the TNF transgenic mice at 4 (B) and 8 weeks of age (C). The transmission electron micrographs from the littermate control mice at 4 weeks (A) revealed a characteristic linear array of sarcomeres and myofibrils. In contrast, the myofibril in the 4-week-old TNF transgenic mice were less organized, with loss of sarcomeric registration observed in many of the sections (Fig. 16.2B). The ultrastructural abnormalities in the TNF transgenic mice were further exaggerated in the 12-week-old TNF transgenic mice, which showed a significant loss of sarcomere registration and myofibril disarray (C). (D)–(F) Representative scanning electron micrographs in littermate controls (D) and the TNF transgenic mice a 4 (E) and 8 weeks of age (F). (E) shows that there was a significant loss of fibrillar collagen in the TNF transgenic mice at 4 weeks of age when compared to age-matched littermate controls (D). However, as the TNF transgenic mice aged (12 weeks), there was an obvious increase in myocardial fibrillar collagen content. (G) The myocardial collagen content as determined by picrosirius red staining. There was a loss of myocardial collagen content at 4 weeks of age in the TNF transgenic mice, that was later followed by a progressive increase in myocardial collagen content at 8 and 12 weeks of age. Error bars, standard errors; *P < 0.05 compared to littermate controls. Reproduced with permission from the American Heart Association: Sivasubramanian N, Coker ML, Kurrelmeyer K et al. Left ventricular remodeling in transgenic mice with cardiac restricted overexpression of tumor necrosis factor. Circulation 2001; 104: 826–31.*

Fig. 16.5C). Taken together, these observations suggest that sustained myocardial inflammation provokes time-dependent changes in the balance between MMP activity and TIMP activity. That is, during the early stages of inflammation there is an increase in the ratio of MMP activity to TIMP levels that fosters LV dilation. However, with chronic inflammatory signaling, there is a time-dependent increase in TIMP levels, with a resultant decrease in the ratio of MMP activity to TIMP

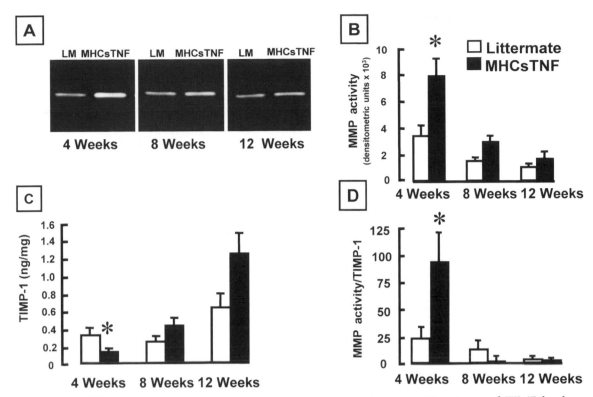

Figure 16.5. *Effects of sustained pro-inflammatory cytokine expression on MMP activity and TIMP levels. (A) A zymogram of total MMP activity in the TNF transgenic mice (MHCsTNF) and littermate (LM) control mice at 4, 8 and 12 weeks of age. (B) Summary of the results of group data for total MMP zymographic activity. MMP activity was significantly (*P < 0.001) greater in the TNF transgenic mice at 4 weeks of age; however, MMP activity was no different from littermate control mice at 8 and 12 weeks of age. (C) The time-dependent changes in TIMP levels at 4, 8 and 12 weeks in the TNF transgenic and littermate control mice. At 4 weeks of age TIMP-1 levels were significantly less in the TNF transgenic mice; however, TIMP-1 levels increased progressively in the TNF transgenic mice from 8–12 weeks of age. (D) The time-dependent changes in the ratio of MMP activity:TIMP levels in the TNF transgenic and littermate control mice. As shown, at 4 weeks of age the ratio of MMP activity:TIMP-1 levels was significantly greater in the TNF transgenic mice, thus favoring collagen degradation (Fig. 16.4G); however, the ratio of MMP activity:TIMP-1 decreased progressively from 8 to 12 weeks of age, thus favoring collagen accumulation (Fig. 16.4G). Error bars, standard errors. Reproduced with permission from the American Heart Association: Sivasubramanian N, Coker ML, Kurrelmeyer K et al. Left ventricular remodeling in transgenic mice with cardiac restricted overexpression of tumor necrosis factor. Circulation 2001; 104: 826–31.*

activity, and a subsequent increase in myocardial fibrillar collagen content. Although the molecular mechanisms that are responsible for the transition between excessive degradation and excessive synthesis of the extracellular matrix are not known, studies in experimental models of chronic injury/inflammation in an array of different organs, including the liver, lung and kidney, have shown an initial increase in MMP expression that is superseded by increased TIMP expression and increased expression of a number of fibrogenic cytokines, most notably tissue growth factor-β (TGF-β).[25,26]

As noted above, sustained myocardial inflammation leads to myocardial fibrosis in transgenic mouse models with cardiac-restricted over-expression of TNF.[18–20] Given that TNF inhibits collagen gene expression and/or collagen synthesis in cardiac fibroblasts *in vitro*,[27,28] it is likely that the increased myocardial fibrosis in transgenic mice with targeted overexpression of TNF is mediated by one or more indirect effects of TNF. For example, stimulation with TNF has been shown to increase the density of angiotensin type I receptors (AT$_1$) on cardiac fibroblasts,[29] as well as increasing the sensitivity of these cells to the profibrotic actions of endogenous angiotensin II.[28] A second potential explanation for the observed fibrosis in the TNF transgenic mice is that TNF increases the expression of TGF-β. Sivasubramanian et al. showed that *TGF-β$_1$* and *TGF-β$_2$* mRNA and protein levels were significantly increased in the hearts of the TNF transgenic mice (MHCsTNF) relative to littermate controls.[20] Furthermore, it bears emphasis that many of the profibrotic actions of Ang II *in vitro*,[30] and *in* vivo,[31] are mediated indirectly through TGF-β. IL-1β has also been shown to exert a potent anti-proliferative effect on cardiac fibroblasts through upregulation of the transcriptional repressor yin yang-1 (YY1).[32] Although the effects of IL-6 have not been directly studied in cardiac fibroblasts, LIF and CT-1 have both been shown to stimulate fibroblast growth.[33,34] Of note, stimulation with LIF also inhibited the differentiation of cardiac fibroblasts into cardiac myofibroblasts, and blunted the effects of TGF-β) on collagen synthesis,[34] suggesting that the LIF may serve as an autocrine/paracrine factor that dampens and negatively modulates ongoing remodeling of the extracellular matrix by preventing excessive fibrosis.

Effects of proinflammatory cytokines on myocardial function

Perhaps the most salient aspect of proinflammatory cytokines is their ability to depress myocardial function, which was first reported in a series of important experimental studies which showed that direct injections of TNF would produce hypotension, metabolic acidosis, hemoconcentration and death within minutes, thus mimicking the cardiac/hemodynamic response seen during endotoxin-induced septic shock.[35] Injections of antibodies raised against TNF were subsequently shown to attenuate the hemodynamic collapse seen in endotoxin shock. Since these original descriptions, a number of studies have demonstrated that TNF, the IL-1 family members (including the recently described member IL-18[36]), as well as the IL-6 family of cytokines are capable of modulating myocardial function.[37] The extant literature suggests that proinflammatory cytokines modulate myocardial function through at least two different pathways: that is, an immediate pathway that is manifest within minutes and is mediated by activation of the neutral sphingomyelinase pathway,[38] and a delayed pathway that requires hours to days to develop, and is mediated by nitric oxide (NO).[39,40] Several studies have shown that treatment with IL-1β will lead to a depression in cardiac myocyte contractility.[41–43] Although the exact signal transduction pathways have not yet been identified for IL-1, the delayed nature of these IL-1β -mediated effects[42,43] suggests that they are mediated by nitric oxide synthase (NOS). Recently, it has been suggested that TNF and IL-1 may produce negative inotropic effects indirectly through activation and/or release of IL-18, which is a recently described member of the IL-1 family of cytokines.[44] Relevant to the present discussion is the observation that specific blockade of IL-18 using neutralizing IL-18-binding protein leads to an improvement in myocardial contractility in atrial tissue that was subjected to ischemia–reperfusion injury.[36] Although the signaling pathways that are responsible for the IL-18-induced negative inotropic effects have not been delineated thus far, it is likely that they will overlap those for IL-1, given that the IL-18 receptor complex utilizes components of the IL-1 signaling chain, including IL-1R-activating kinase and TNFR-associated factor-6. IL-6 has been shown to decrease cardiac contractility via a NO-dependent pathway that is secondary to IL-6-induced phosphorylation of signal transducer and activator of transcription 3 (STAT3).[45] In this study, IL-6 enhanced *de novo* synthesis of the inducible isoform of NO synthase (iNOS) protein, increased NO production, and decreased rat cardiac myocyte contractility after 2 h of incubation. The effects of IL-6 on iNOS production and myocyte contractility were blocked by genistein at concentrations that were sufficient to block IL-6-induced activation of STAT3. Taken together these observations suggest that IL-6 is sufficient to produce negative inotropic effects through STAT3-mediated activation of iNOS.[45]

The negative inotropic effects of TNF have been observed *in vitro*, as well as *in vivo*. Studies in isolated contracting cardiac myocytes have shown that TNF is sufficient to suppress myocyte contractility[46] through an immediate pathway that is manifest within minutes, and is mediated by activation of the neutral sphingomyelinase pathway,[38] as well as a delayed pathway that requires hours to days to develop, and is mediated by NO- mediated blunting of β-adrenergic signaling.[39,40] With respect to the cellular mechanism for the immediate negative inotropic effects of TNF, a study in isolated adult feline cardiac myocytes showed that the negative inotropic effects of TNF were the direct result of alterations in intracellular calcium homeostasis.[46] In this study, treatment with TNF produced a 20–30% decrease in the extent of cell shortening (Fig. 16.6A) and a 40% decrease in peak levels of intracellular calcium (Fig. 16.6B). Moreover, whole-cell patch-clamp studies suggested that the decrease in intracellular calcium was not the result of changes in the voltage-sensitive inward calcium current, thus suggesting that TNF-induced changes in intracellular calcium homeostasis were secondary to alterations in sarcoplasmic reticular handling of calcium.[46] Experimental studies in rats

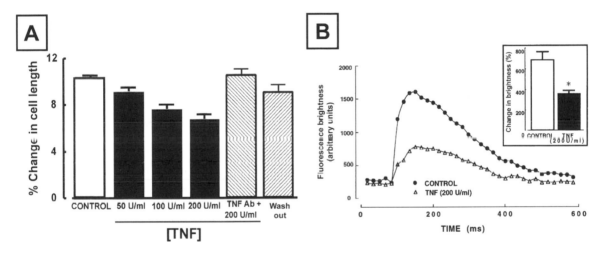

Figure 16.6. *Effect of TNF on contractility of adult cardiac myocytes. (A) In comparison with control cells(open bars), cardiac myocytes treated with TNF (closed bars) developed a concentration-dependent decrease in cell shortening. Pretreatment with a neutralizing anti-TNF antibody (hatched bar) completely attenuated the effects of 200 U/ml TNF on cell shortening. When the cells were washed free of 200 U/ml TNF, and allowed to recover for 45 minutes, the effects of TNF were completely reversible (hatched bar). (B) A typical time–intensity curve for fluo-3 fluorescence brightness (indicative of intracellular calcium concentration) in isolated cardiac myocytes treated with diluent (circles) and 200 U/mlTNF-treated (triangles). As shown, the peak level of intracellular fluorescence brightness was reduced strikingly for the cells treated with 200 U/ml TNF. The inset of this figure, which depicts values obtained for group data, shows that there was ~ 40% decrease in the percentage change in peak intensity of fluorescence brightness for the TNF-treated cells. Error bars, standard errors; *P < 0.05 compared to diluent treated control. Taken together, (A) and (B) suggest that TNF produces negative inotropic effects in isolated cardiac myocytes by producing an alteration in intracellular calcium homeostasis. Reproduced with permission from Yokoyama T, Vaca L, Rossen RD et al. Cellular basis for the negative inotropic effects of tumor necrosis factor-alpha in the adult mammalian heart. J Clin Invest 1993; 92: 2303–12.*

have shown that circulating concentrations of TNF that overlap those observed in patients with heart failure are sufficient to produce persistent negative inotropic effects that are detectable at the level of the cardiac myocyte; moreover, the negative inotropic effects of TNF were completely reversible when the TNF infusion was stopped.[47] Franco and colleagues used cine-magnetic

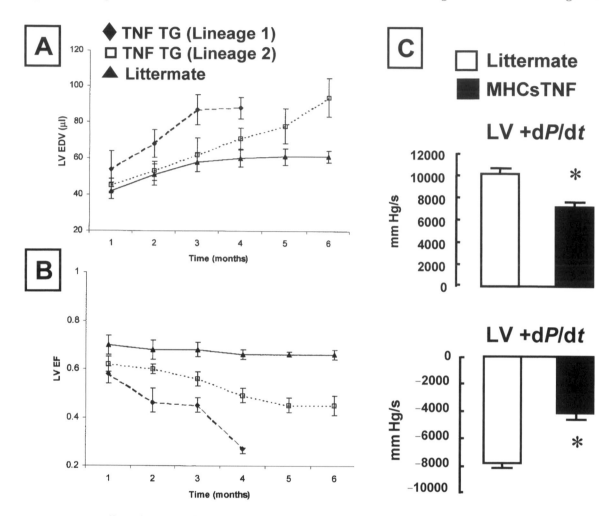

Figure 16.7. *Effect of TNF on left ventricular function. LV volume and LV ejection fraction were serially examined by magnetic resonance imaging in two lines of transgenic mice (TNF TG) with high (Lineage 1) and low (Lineage 2) levels of myocardial TNF expression in comparison to age-matched littermate control mice.[48] (A) Serial changes in LV end-diastolic volume (EDV) in transgenic mice and littermate control mice. (B) Serial changes in LV ejection fraction (EF) in transgenic mice and littermate control mice. (C) LV contractility in mice with targeted overexpression of TNF (MHCsTNF).[20] For these studies, the animals were paced via the atrium to a heart rate at which positive dP/dt was maximal, as defined by examination of the force–frequency curves for each animal, and peak positive and negative dP/dt assessed for MHCsTNF mice and littermate controls. Error bars, standard errors; *P < 0.05 compared to littermate controls. (A) and (B) reproduced with permission from the American Heart Association: Franco F, Thomas GD, Giroir BP et al. Magnetic resonance imaging and invasive evaluation of development of heart failure in transgenic mice with myocardial expression of tumor necrosis factor-alpha. Circulation 1999; 99: 448–54.*

resonance imaging to demonstrate that there was a significant time-dependent increase in LV volume and a significant decrease in LV ejection fraction in transgenic mice with targeted overexpression of TNF.[48] Importantly, these effects were shown to be dependent upon gene dosage: when the line of transgenic mice with high TNF expression (lineage 1) was compared to a transgenic line with lower myocardial TNF expression (lineage 2), there was a significantly greater increase in LV volume (Fig. 16.7A) and a significantly greater decrease in LV ejection fraction (Fig. 16.7B) in the transgenic mouse lines with higher TNF expression.[48] Studies from our laboratory in mice with targeted overexpression of TNF have demonstrated a decrease in peak positive and negative dP/dt when compared to littermate controls (Fig. 16.7C), consistent with the findings reported by Franco et al.[48]

Inflammatory mediators in the heart: cardiac hypertrophy, remodeling and failure

In the present review we have summarized experimental material which suggests that inflammatory mediators play an important role in regulating cardiac structure and function. Although short-term expression of inflammatory mediators in the heart is probably beneficial by upregulating the expression of families of cytoprotective proteins in the heart, the short-term beneficial effects of inflammatory mediators may be contravened by the deleterious effects of these proteins if proinflammatory cytokine expression is sustained and/or dysregulated. That is, as shown in Fig. 16.8, the sustained expression of proinflammatory cytokines in the heart can lead to cardiac hypertrophy and cardiac remodeling, as well as myocardial dysfunction. Accordingly, one of the challenges that faces investigators in this field will be to delineate the signaling pathways that are responsible for the adaptive and maladaptive aspects of cytokine signaling, in order to develop effective strategies that maximize the potential spectrum of beneficial responses conferred by inflammatory mediators in the heart, without simultaneously activating their known potential deleterious effects.

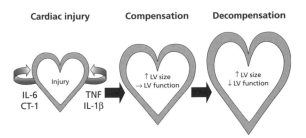

Figure 16.8. *Inflammatory mediators in cardiac hypertrophy, remodeling and failure. Proinflammatory cytokines, including TNF, IL-1, IL-6 and CT-1 are expressed in the heart in response to various forms of tissue injury. Initially, the expression of these molecules may lead to compensatory changes in the heart including (but not limited to) cardiac hypertrophy and cytoprotection. However, sustained expression of these molecules can lead to cardiac remodeling and myocardial dysfunction, resulting in overt cardiac decompensation.*

Acknowledgements
The author gratefully acknowledges Ms Mary Helen Soliz for secretarial assistance. This research was supported by research funds from the Veterans Administration and the NIH (P50 HL-06H and RO1 HL58081 01, RO1 HL61543 01, HL-42250–10/10).

References

1. Lower R. Tractatus de corde: De Motu et Colore Sanguinus et Chyli in eum Tranfitu (1e) London: Jacobi Alleftry; 1669.
2. Levine B, Kalman J, Mayer L et al. Elevated circulating levels of tumor necrosis factor in severe chronic heart failure. N Engl J Med 1990; 223: 236–41.
3. Kapadia S, Lee JR, Torre-Amione G et al. Tumor necrosis factor gene and protein expression in adult feline myocardium after endotoxin administration. J Clin Invest 1995; 96: 1042–52.
4. Kapadia S, Oral H, Lee J et al. Hemodynamic regulation of tumor necrosis factor-α gene and protein expression in adult feline myocardium. Circ Res 1997; 81: 187–95.
5. Hoffmann JA, Kafatos FC, Janeway CA, Jr. et al. Phylogenetic perspectives in innate immunity. Science 1999; 284: 1313–18.

6. Mann DL. Stress-activated cytokines and the heart: from adaptation to maladaptation. Annu Rev Physiol 2003; 65: 81–101.

7. Hirota H, Yoshida K, Kishimoto T et al. Continuous activation of gp130, a signal-transducing receptor component for interleukin 6-related cytokines, causes myocardial hypertrophy in mice. Proc Natl Acad Sci USA 1995; 92: 4862–6.

8. Pennica D, Wood WI, Chien KR. Cardiotrophin-1: a multifunctional cytokine that signals via LIF receptor-gp 130 dependent pathways. Cytokine Growth Factor Rev 1996; 7: 81–91.

9. Wollert KC, Taga T, Saito M et al. Cardiotrophin-1 activates a distinct form of cardiac muscle cell hypertrophy. J Biol Chem 1996; 271: 9535–45.

10. Jin H, Yang R, Keller GA et al. In vivo effects of cardiotrophin-1. Cytokine 1996; 8: 920–6.

11. Matsui H, Fujio Y, Kunisada K et al. Leukemia inhibitory factor induces a hypertrophic response mediated by gp130 in murine cardiac myocytes. Res Commun Mol Pathol Pharm 1996; 93: 149–62.

12. Wang F, Seta Y, Baumgarten G et al. Functional significance of hemodynamic overload-induced expression of leukemia-inhibitory factor in the adult mammalian heart. Circulation 2001; 103: 1296–302.

13. Palmer JN, Hartogensis WE, Patten M et al. Interleukin-1 beta induces cardiac myocyte growth but inhibits cardiac fibroblast proliferation in culture. J Clin Invest 1995; 95: 2555–64.

14. Thaik CM, Calderone A, Takahashi N et al. Interleukin-1 beta modulates the growth and phenotype of neonatal rat cardiac myocytes. J Clin Invest 1995; 96: 1093–119.

15. Isoda K, Kamezawa Y, Tada N et al. Myocardial hypertrophy in transgenic mice overexpressing human interleukin 1alpha. J Card Fail 2001; 7: 355–64.

16. Yokoyama T, Nakano M, Bednarczyk JL et al. Tumor necrosis factor-α provokes a hypertrophic growth response in adult cardiac myocytes. Circulation 1997; 95: 1247–52.

17. Condorelli G, Morisco C, Latronico MV et al. TNF-alpha signal transduction in rat neonatal cardiac myocytes: definition of pathways generating from the TNF-alpha receptor. FASEB J 2002; 16: 1732–7.

18. Kubota T, McTiernan CF, Frye CS et al. Dilated cardiomyopathy in transgenic mice with cardiac specific overexpression of tumor necrosis factor-alpha. Circ Res 1997; 81: 627–35.

19. Bryant D, Becker L, Richardson J et al. Cardiac failure in transgenic mice with myocardial expression of tumor necrosis factor-α (TNF). Circulation 1998; 97: 1375–81.

20. Sivasubramanian N, Coker ML, Kurrelmeyer K et al. Left ventricular remodeling in transgenic mice with cardiac restricted overexpression of tumor necrosis factor. Circulation 2001; 104: 826–31.

21. Flesch M, Hoper A, Dell'Italia L et al. Activation and functional significance of the renin–angiotensin system in mice with cardiac restricted overexpression of tumor necrosis factor. Circulation 2003; 108: 598–604.

22. Thaik CM, Calderone A, Takahashi N et al. Interleukin-1β modulates the growth and phenotype of neonatal rat cardiac myocytes. J Clin Invest 1995; 96: 1093–9.

23. Li YY, Feng YQ, Kadokami T et al. Myocardial extracellular matrix remodeling in transgenic mice overexpressing tumor necrosis factor alpha can be modulated by anti- tumor necrosis factor alpha therapy. Proc Natl Acad Sci USA 2000; 97: 12746–51.

24. Weber KT. Cardiac interstitium in health and disease: the fibrillar collagen network. J Am Coll Cardiol 1989; 13, No. 7, June: 1637–52.

25. Knittel T, Mehde M, Grundmann A et al. Expression of matrix metalloproteinases and their inhibitors during hepatic tissue repair in the rat. Histochem Cell Biol 2000; 113: 443–53.

26. Sime PJ, Marr RA, Gauldie D et al. Transfer of tumor necrosis factor-alpha to rat lung induces severe pulmonary inflammation and patchy interstitial fibrogenesis with induction of transforming growth factor-beta1 and myofibroblasts. Am J Pathol 1998; 153: 825–32.

27. Siwik DA, Chang DL, Colucci WS. Interleukin-1beta and tumor necrosis factor-alpha decrease collagen synthesis and increase matrix metalloproteinase activity in cardiac fibroblasts in vitro. Circ Res 2000; 86: 1259–65.

28. Peng J, Gurantz D, Tran V et al. Tumor necrosis factor-alpha-induced AT1 receptor upregulation enhances angiotensin II-mediated cardiac fibroblast responses that favor fibrosis. Circ Res 2002; 91: 1119–26.

29. Gurantz D, Cowling RT, Villarreal FJ et al. Tumor necrosis factor-alpha upregulates angiotensin II type 1 receptors on cardiac fibroblasts. Circ Res 1999; 85: 272–9.

30. Gray MO, Long CS, Kalinyak JE et al. Angiotensin II stimulates cardiac myocyte hypertrophy via paracrine release of TGF-beta 1 and endothelin-1 from fibroblasts. Cardiovasc Res 1998; 40: 352–63.

31. Schultz JJ, Witt SA, Glascock BJ et al. TGF-beta1 mediates the hypertrophic cardiomyocyte growth induced by angiotensin II. J Clin Invest 2002; 109: 787–96.

32. Felgner PL, Ringold GM. Cationic liposome-mediated transfection. Nature 1989; 337: 387–8.

33. Tsuruda T, Jougasaki M, Boerrigter G et al. Cardiotrophin-1 stimulation of cardiac fibroblast growth: roles for glycoprotein 130/leukemia inhibitory factor receptor and the endothelin type A receptor. Circ Res 2002; 90: 128–34.

34. Wang F, Trial J, Diwan A et al. Regulation of cardiac fibroblast cellular function by leukemia inhibitory factor. J Mol Cell Cardiol 2002; 34: 1309.

35. Tracey KJ, Beutler B, Lowry SF et al. Shock and tissue injury induced by recombinant human cachectin. Science 1986; 234: 470–4.

36. Pomerantz BJ, Reznikov LL, Harken AH et al. Inhibition of caspase 1 reduces human myocardial ischemic dysfunction via inhibition of IL-18 and IL-1beta. Proc Natl Acad Sci USA 2001; 98: 2871–6.

37. Mann DL. Cytokines as mediators of disease progression in the failing heart. In: Hosenpud JD, Greenberg BH (eds). Congestive Heart Failure. Philadelphia: Lippincott Williams and Wilkins; 1999: 213–32.

38. Oral H, Dorn GW, 2nd, Mann DL. Sphingosine mediates the immediate negative inotropic effects of tumor necrosis factor-α in the adult mammalian cardiac myocyte. J Biol Chem 1997; 272: 4836–42.

39. Gulick TS, Chung MK, Pieper SJ et al. Interleukin 1 and tumor necrosis factor inhibit cardiac myocyte β-adrenergic responsiveness. Proc Natl Acad Sci USA 1989; 86: 6753–7.

40. Balligand JL, Ungureanu D, Kelly RA et al. Abnormal contractile function due to induction of nitric oxide synthesis in rat cardiac myocytes follows exposure to activated macrophage-conditioned medium. J Clin Invest 1993; 91: 2314–19.

41. Combes A, Frye CS, Lemster BH et al. Chronic exposure to interleukin 1beta induces a delayed and reversible alteration in excitation–contraction coupling of cultured cardiomyocytes. Pflugers Arch 2002; 445: 246–56.

42. Hosenpud JD. The effects of interleukin-1 on myocardial function and metabolism. Clin Immunol Immunopathol 1993; 68: 175–80.

43. Schulz R, Panas DL, Catena R et al. The role of nitric oxide in cardiac depression induced by interleukin-1β and tumour necrosis factor-α. Br J Pharmacol 1995; 114: 27–34.

44. Dinarello CA. Interleukin-18. Methods 1999; 19: 121–32.

45. Yu X, Kennedy RH, Liu SJ. JAK2/STAT3, not ERK1/2, mediates interleukin-6-induced activation of inducible nitric-oxide synthase and decrease in contractility of adult ventricular myocytes. J Biol Chem 2003; 278: 16304–9.

46. Yokoyama T, Vaca L, Rossen RD et al. Cellular basis for the negative inotropic effects of tumor necrosis factor-alpha in the adult mammalian heart. J Clin Invest 1993; 92: 2303–12.

47. Bozkurt B, Kribbs S, Clubb FJ, Jr. et al. Pathophysiologically relevant concentrations of tumor necrosis factor-α promote progressive left ventricular dysfunction and remodeling in rats. Circulation 1998; 97: 1382–91.

48. Franco F, Thomas GD, Giroir BP et al. Magnetic resonance imaging and invasive evaluation of development of heart failure in transgenic mice with myocardial expression of tumor necrosis factor-alpha. Circulation 1999; 99: 448–54.

CHAPTER 17

Control of cardiac function in health and disease by mechanisms at the level of the sarcomere

R John Solaro, Gerd Hasenfuss

Introduction: control of cardiac function by mechanisms altering sarcomeric structure, function, and response to Ca²⁺

In this chapter, we focus on regulation of the intensity of activity and dynamics of the myocardium by control mechanisms operating at the level of sarcomeric proteins. This mode of regulation includes mechanisms that control cardiac function downstream of alterations in membrane-related Ca^{2+} release, transport, and exchange mechanisms. Experimental evidence for this type of regulation is largely couched in terms of the relation between Ca^{2+} and steady-state tension developed by either single or bundled cells in which the membrane has been removed or made permeable by detergent treatment (skinned or peeled fibers). An altered Ca^{2+} sensitivity of the myofilaments is reflected in a shift of the half-maximally activating Ca^{2+} concentration, the maximum tension, or the steepness of the relation between Ca^{2+} and force. Another perhaps more relevant measure of Ca^{2+} sensitivity is the determination of the temporal relation between Ca^{2+} and force or shortening. In this case, changes in mechanical activity independent of changes in the Ca^{2+} provide strong evidence for an altered myofilament response to Ca^{2+}.

The two main mechanisms for this type of regulation are: (1) regulation of thin filament activation by modifications in Ca^{2+} signaling and in cross-bridge-dependent activation, and (2) regulation of cross-bridge dynamics. By modifications of sarcomeric Ca^{2+} signaling, we mean control by alterations of the rate constants of binding of Ca^{2+} to the thin filament receptor, cardiac troponin C (cTnC), which triggers contraction, and control by alterations in the protein–protein interactions that follow triggering of activation by Ca^{2+}. Cross-bridge-dependent regulation means feedback control of thin filament activation by strongly bound, force-generating interactions between the head of myosin and the thin filament. By regulation of cross-bridge dynamics, we mean alterations in the rate constants of steps in the cross-bridge cycle imposed by alterations at the level of the thick and/or thin filament, or by alterations in other sarcomeric proteins. In addition to modifications in velocity of shortening, these alterations in cross-bridge kinetics modify force by variations in the duty cycle (the time a cross-bridge remains in the strong binding state), and/or the unitary force generated by each cross-bridge.

Fig. 17.1 illustrates the complexity of the sarcomeric protein network that potentially participates in the control devices that modify Ca^{2+} signaling, feedback control by strong cross-bridges, and cross-bridge dynamics. Relatively recent identification of a third filament consisting of titin together with nebulette,[1] along with a better appreciation of the complexity of the Z-disc,[2] has been an exciting advancement in our concepts and

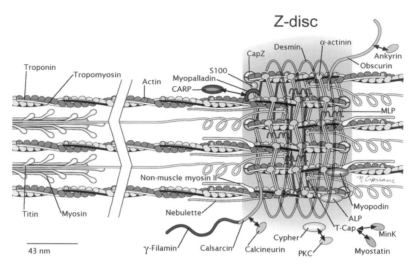

Figure 17.1. *The protein network of the sarcomere surrounding the Z-disc of heart muscle. The proteins and their potential sites of interaction are indicated. For clarity, stoichiometric relationships are not shown. Titin is shown with an extensible region illustrated as a spring, and capped by telethonin (T-cap). CARP, cardiac-restricted ankyrin repeated protein, MLP, muscle specific LIM protein, ALP, actinin associated LIM protein. See text for details. Reproduced with premission from Pyle WG, Solaro RJ. At the crossroads of myocardial signaling: The role of Z-discs in intracellular signaling and cardiac function. Circ Res 2004; 94: 296–305.*

understanding of regulation at the level of the sarcomere. Even though our understanding of how all these proteins participate in modifications in sarcomeric function is still unfolding, there is enough information available to provide a picture of the relative significance of many of these elements in control of cardiac function in physiological and pathophysiological states. A theme of our chapter is the relevance of maladaptive alterations in the sarcomeric response to Ca^{2+} that occur in response to mechanical and biochemical stressors that lead to heart failure.

In our discussions, we consider evidence in support of the following hypotheses:

- altered myofilament response to Ca^{2+} is an essential element in the control of contractility and dynamics
- Signaling and functional changes at the level of the sarcomere are integral elements in the cascade of reactions associated with compensated hypertrophy and the transition to failure
- Sarcomeric Z-discs function not only as a physical anchor for myofilament, cytoskeletal, and membrane proteins, but also as a pivot point for signal transduction and signal processing.

Regulation of thin filament activation at the level of the sarcomere

Control of myofilament response to Ca^{2+}

Substantial recent evidence has provided a detailed picture of the protein–protein interactions that follow Ca^{2+} binding to cTnC, and indicates a rich array of potential points for regulation of the response to Ca^{2+}.[3–11] Figure 17.2 illustrates two states in the process – the relaxed state, and the Ca^{2+}-bound/cross-bridge-bound state. Figure 17.3 depicts a more detailed view of the activation process. The illustrations in Figs 17.2 and 17.3 are derived from extensive analysis aided by *in vitro* determination of protein–protein interactions, mutagenesis, and structures determined by fluorescence and nuclear magnetic resonance (NMR) determinations.[3–10] A major advance in our under-

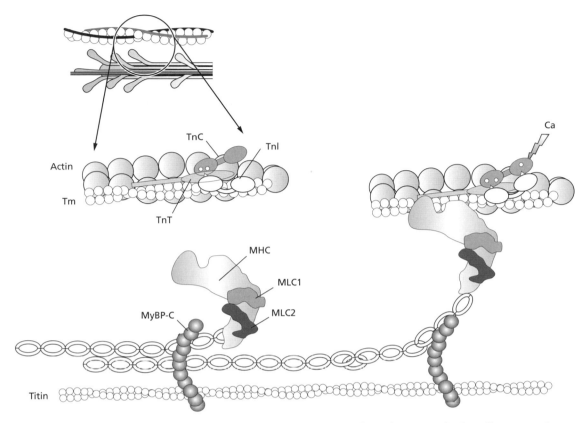

Figure 17.2. *Schematic illustration of a section of the sarcomere with working cross-bridges illustrating the diastolic state on the left, and the systolic state on the right. A structural unit of the thin filament is shown together with the edge of the thick filament indicating the interaction of myosin-binding protein C (MyBP-C) with titin, and with the head-neck region of the myosin heavy chain (MHC) near the myosin light chains (MLC). In diastole, Tm (tropomyosin) blocks the reaction of cross-bridges with actin. In the absence of bound Ca^{2+} at the N-lobe of TnC (troponin C), Tm is held in this position by protein–protein interaction involving TnI (troponin I), which tethers Tn to actin, and by TnT (troponin T), which interacts with actin–Tm and is also responsible for inhibition. In diastole, the N-lobe of TnC does not interact with C-terminal regions of TnI, whereas with Ca^{2+} binding to cTnC, the C-region of TnI binds to the N-lobe of TnC, triggering the movement of Tm through the action of TnT. See text for details.*

standing of thin filament activation occurred with publication of the crystal structure of a complex containing cTnC and large peptides derived from cTnI and the C-terminal region of cTnT.[9]

Maintenance of the myofilament diastolic state and the transition to the systolic state involves complex allosteric, steric, and cooperative mechanisms at the level of the sarcomere.[3–11] In the diastolic state shown in Fig. 17.2, the head of the myosin heavy chain (MHC) or S-1 is impeded

from interacting with the thin filament by the action of tropomyosin and the cardiac troponin (cTn) heterotrimeric complex, consisting of cTnC, cTnI, and cTnT. In this state, a lack of force generating interactions between the thick and thin filaments ensures that the sarcomeres are freely extensible, so that the ventricle may fill with blood with minimal upstream pressure gradients. ATP splitting by the sarcomere is minimal, and most likely reflects the ATPase rate of myosin

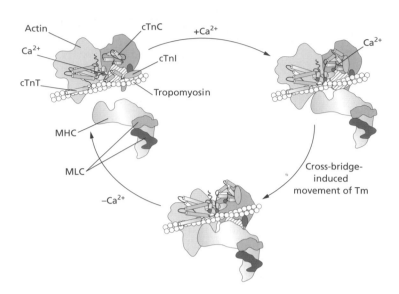

Figure 17.3. *Molecular mechanism of Ca^{2+}-dependent and cross-bridge-dependent activation of the actin–cross-bridge reaction based on the crystal structure of the Tn complex. Three states of the actin–cross-bridge reaction are shown: the diastolic state (upper left), the Ca^{2+}-bound state (upper right), and the fully activated state in which cross-bridge binding has moved Tm further away from the myosin-binding sites on actin, which are hatched. See text for details.*

alone. In diastole it is apparent that a substantial fraction of cross-bridges are sterically blocked from any interaction with the thin filament, in which case the tropomyosin (Tm) is said to be in the 'B' state. A fraction of Tm said to be in the 'C' state permits reaction of cross-bridges with the thin filament, but in a weak, non-force-generating interaction.[12] The right panel of Fig. 17.2 depicts the 'M' state or strong force-generating state of the actin-cross-bridge reaction. Figure 17.2 also indicates that myosin is connected to the cytoskeletal network through myosin-binding protein C (MyBP-C), which binds to the head/neck region of myosin and appears to determine the radial movement of the cross-bridge away from the thick filament. MyBP-C binds to titin. Titin is a giant filamentous protein ($>3 \times 10^6$ kDa) extending from the Z-disc to the center of the sarcomere.[1,2] Titin serves as a template in assembly of the thick filament by acting as a molecular ruler. Titin is a bidirectional spring, responsible for most of the passive tension of cardiac myocytes above slack length, and for the restoration of length below slack length (Fig. 17.1). There is evidence that the stress strain relations of titin may be controlled by protein phosphorylation.[1]

The allosteric nature of the regulation of activation of the thin filaments is evident in that Ca^{2+} occupancy of a binding site at cTnC controls dis-

tant sites in the structural unit consisting of seven actins, one Tm, and one cTn. Steric regulation occurs by physical blocking of the cross-bridge-binding site on the thin filament by the position of Tm (B-state), which has seven quasi repeats that interact generally at the outer domains of actin.[5] In addition to steric hindrance, there is also evidence that Tm binding modifies the actin–myosin interface, thereby altering stereospecific interactions involved in force generation.[13] As discussed below, activation is highly cooperative. Current evidence indicates that Tm is rather immobile on the thin filament in diastole.[14,15] This is shown in both Figs 17.2 and 17.3, which also indicates that Tn itself may block the actin–cross-bridge reaction. The diastolic state of Tm occurs as a result of the following: (1) the tethering action of the inhibitory protein, cTnI, which binds to actin and the C-terminal lobe of cTnC in diastole, and (2) the binding of the N-terminal tail of cTnT to actin–Tm.[16] Regions of cTnI that bind to actin are shown more clearly in Fig. 17.3, and include a highly basic inhibitory peptide (Ip), and a downstream region located in the C-terminal end of cTnI. cTnI and cTnT are also maintained in the Tn complex by interactions with the Ca^{2+}-binding protein, TnC. Two slowly exchanging Ca^{2+} ions bound at the C-terminal lobe of cTnC are important in maintaining the integrity of the Tn–Tm complex.[17,18] A single regulatory site with fast exchange properties and located at the N-terminal lobe of cTnC does not contain Ca^{2+} in diastole.[17,18]

Transition from the B state to a Ca^{2+}-bound state, which promotes transition of Tm to the C state of the thin filament, occurs with systolic release of Ca^{2+}. Ca^{2+} binding to cTnC triggers a series of protein–protein interactions that culminate in movement of Tm away from its blocking position. A key interaction that results from Ca^{2+} binding is induction of strong binding of the Ip and C-terminal domains of cTnI to cTnC, and release of cTnI from actin. The Ip probably binds to a central region of cTnC. C-terminal regions of cTnI outside the Ip bind to a hydrophobic pocket that opens as a result of Ca^{2+} binding to the N-lobe of cTnC (Fig. 17.3). The molecular mechanism for the exposure of this hydrophobic surface is unique to cTnC. In fast skeletal TnC, Ca^{2+} binding alone is able to open the hydrophobic patch in the isolated protein.[19] However, exposure of the cTnC hydrophobic patch requires that cTnC exists in a binary complex with cTnI,[20] or a peptide consisting of residue 147–163.[21] This is an important distinction in as much as phosphorylation of cTnI modifies myofilament activation. Activation of the thin filament also involves a release of cTnT from strong binding to Tm. Tm now becomes mobile or rather weakly bound to the thin filament, and sites on actin with which a cross-bridge may react are accessible. A current view of the thin filament pictures Tm as a continuous flexible chain. Kinks in the chain are produced upon the initial binding of cross-bridges and thus cross-bridges may then bind without a further distortion in the Tm chain.[14,22] Cross-bridges, which contain bound nucleotide in the form of ADP and inorganic phosphate (Pi) in the diastolic state, enter into a state of strong binding that promotes product release and generates force and, depending on the load, promotes sliding of the thin filaments toward the center of the sarcomere. The terminal nucleotide free state (rigor) of a cross-bridge is shown at the right in Fig. 17.2, and in Fig. 17.3. Comparison of the cross-bridges in diastole and systole illustrates the position of the cross-bridge at the beginning and at the end of a power stroke following movement of the lever arm.

Thin filament activation by strong cross-bridges

Activation of the thin filament also occurs by cross-bridge binding itself.[23,24] This is illustrated in Fig. 17.3 by comparison of the position of Tm with only Ca^{2+} bound to the thin filament, and with Ca^{2+} bound together with a strong cross-bridge interaction. The bound cross-bridge causes Tm to move further away from the myosin-binding site. This cross-bridge-dependent activation of the thin filament has important implications with regard to the steep relation between Ca^{2+} and tension that occurs even though there is a single regulatory Ca^{2+}-binding site. The cooperativity apparent in the increased Hill coefficients of the relation between Ca^{2+} and steady-state tension development is likely to be due to both enhanced Ca^{2+} binding to cTnC,[17] and the promotion of the reaction of force-generating cross-bridges in functional units that are near-neighbors to a strongly bound cross-bridge.[23–25] The influence of cross-bridge binding on Ca^{2+} binding is to be expected in as much as Ca^{2+} sets into motion a series of protein interactions that expose a site on actin for reaction with the cross-bridge. The energies of interaction in this 'forward' direction also occur in the 'reverse' direction, and thus it is expected, by what is referred to as detailed balance, that the bound cross-bridge would influence Ca^{2+} binding to TnC. The state of a Tn complex on one side of a thin filament strand may also influence the radial near neighbor Tn complex in the same functional unit.[26] Apart from these radial cooperative interactions, there is a lateral spread of activation by cooperative interactions among functional units.[4,6,8,27] There is evidence that this spread may be extensive with the binding of one Ca^{2+} activating as many as 14 actins.[8,23–25] The lateral spread of activation occurs by end-to-end interactions between contiguous Tm molecules, which wrap around the thin filament, and by interactions among actins. Cross-bridge-dependent activation appears to be critical in explaining dependence of myofilament activation on sarcomere length, which appears to be the molecular basis of the Frank–Starling mechanism.[28] Enhanced binding to TnC induced by strong cross-bridges is amplified in cardiac compared to fast skeletal myofilaments, and thus this mechanism appears especially significant in heart muscle.[28] Moreover, cross-bridge binding to cardiac thin filaments has a bigger effect on submaximal tension than is the case with fast skeletal muscle thin filaments.[24,29] Fitzsimons et al. demonstrated that addition of the same amount of NEM-S1, a form of myosin head that binds to actin but does not cycle, induced a bigger increase in submaximal tension in cardiac versus skeletal myofilaments.[29]

An important feature of Ca^{2+} regulation of heart myofilaments is that insufficient Ca^{2+} is released in the basal state to fully saturate the cTnC molecules on a thin filament. In a basal state of contractility, the combined activation due to strong cross-bridges and Ca^{2+} activates the myofilaments only to about 20–25% of maximum activation. This leaves a reserve of thin filament activation, which in large part is responsible for the 'cardiac reserve' defined by hemodynamic measurements. There is no doubt that the reserve of cross-bridges is called into action by increases in Ca^{2+} delivery and Ca^{2+} binding to the myofilaments. A second kind of reserve is a 'relaxation reserve.' At the relatively fast heart rates associated with exercise, cycle time of the beat must be reduced to match the increased frequency. An important aspect of this tuning of cycle time to the heart rate is an enhanced relaxation rate. Again there is substantial evidence that the enhanced relaxation is due in part to an enhanced removal of Ca^{2+} from the myofilaments by the sarcoplasmic reticulum. However, we present evidence here that modifications at the level of the sarcomere also participate in deploying contraction and relaxation reserve. As with membrane-related mechanisms, we shall see that phosphorylation of sarcomeric proteins affects both contraction and relaxation reserve. Moreover, we shall see that a dominant mechanism for the dependence of stroke volume on filling pressure in the ventricle (Starling's law of the heart) involves the dependence of myofilament activation on sarcomere length.

Control of cross-bridge dynamics

Control of contraction and relaxation reserve at the level of the sarcomere may occur with alterations in the coupled biochemical and mechanical transitions that occur in the cross-bridge cycle.

These alterations could result from isoform switches from V1 to V3 MHC or mutant myosins, alterations in the isoform population of myosin light chains (MLCs), and phosphorylation of sarcomeric proteins, e.g. MLCs, MyBP-C, cTnI, cTnT, or titin. Measurements of the cycle using laser trap techniques to determine mechanics of single cross-bridges have provided insights into possible mechanisms.[30] These techniques permit determination of displacement (step) and the time a cross-bridge spends in the strong binding state. The kinetics of the cross-bridge cycle can be considered in terms of the unitary step displacement of the cross-bridge in its reaction with the thin filament as illustrated in Fig. 17.4. The model in Figure 17.4 depicts the unitary step displacement (d), and the duration of the step (time on; t_{on}). Maximum shortening velocity (V_{max}) is essentially equivalent to d/t_{on}. The duty cycle or fraction of time the cross-bridge spends in the strong binding state (f), and the unitary force of myosin (F) define the average force (F_{ave}). The view of the cross-bridge cycle in Fig. 17.4 indicates that F_{ave} generated by the cross-bridges may be modified by changes in F or f. Moreover, V_{max} may be changed by alterations in d or t_{on}. Palmiter et al. have also modeled two discrete steps in the cycle termed t_{-ADP} and T_{+ATP}. T_{-ADP} is related to the release of ADP from the cross-bridge,[30] which initiates entry into the 'on' state and force generation as the rigor state (AM) approaches. T_{+ATP} is

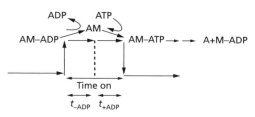

Figure 17.4. *Kinetic model of the duration (time on) of strong binding of cross-bridges (M) to the thin filament (A) based on the work of Palmiter et al.[30,33] Strong binding is indicated as the AM state. The model includes steps (t_{-ADP}) in which ADP is released from the cross-bridge leading to force generation and culmination in the rigor state (AM), and in which ATP binds to AM (t_{+ATP}) leading to dissociation of the cross-bridge from actin. See text for details.*

related to the time the AM state must wait for ATP to bind and end the strong binding state. Using these concepts, Palmiter et al.[30] concluded from their measurements that the slower kinetics of V_{max} and ATPase rate of acto-V3 myosin compared to acto-V1 myosin are due to a 2-fold difference in t_{on}, associated with a 2-fold difference in the rate of ADP release. They found no differences in unitary displacement or force between the two myosin isoforms. Although Noguchi et al. found differences in the relative amounts of V1 in failing hearts (~3%) compared to normals (~6%),[31] they could find no differences in actin sliding velocity or force generation in the motility assay between preparations containing these different populations of myosin isoforms. On the other hand, rat heart myofilaments containing 12% V1 myosin exhibited about 50% greater power than myofilaments containing no V1.[32] The R403Q and L908V mutant myosin, linked to familial hypertrophic cardiomyopathy (FHC), demonstrated no differences in force or motion, but the t_{on} of these mutants was about 25% lower than that of wild-type myosins.[33] The mutant myosins also induced an increased V_{max} of actin sliding in the motility assay.

The influence of sarcomeric protein phosphorylation on single molecule mechanics is an area that remains largely unexplored. Data showing that phosphorylation of myofilament proteins is able to significantly alter V_{max} of skinned fiber preparations, and actin sliding over myosins in the motility assay, provide evidence for the potential that modifications other than myosin isoform switching may affect d, t_{on}, f, and F_{ave}.

Phosphorylation of sarcomeric proteins and alterations in thin filament activation

A detailed discussion of the myofilament protein phosphorylation is beyond the scope of this chapter and this topic has been recently reviewed.[6,10] Thin filament proteins that are substrates for kinases and demonstrated to be phosphorylated in the myocardium are cTnI and cTnT. Phosphorylation of cTnI occurs at sites that are either unique to the cardiac variant, or apparently not substrates for the same kinase in the skeletal variants. A well-studied set of phosphorylation sites occurs at S23, S24 in a unique N-terminal extension of cTnI. S23 and S24 have the requisite sequence homology to be a substrate for cAMP-dependent protein kinase (PKA), but these sites can also be phosphorylated by protein kinase C (PKC). Protein phosphatase 2A is the predominant enzyme dephosphorylating the sites. The N-terminal extension is highly flexible and, when unphosphorylated, is bound to the N-lobe of cTnC, which contains the regulatory binding site (Fig. 17.3). Phosphorylation releases this peptide from the N-lobe, and decreases the Ca^{2+}-binding affinity of cTnC by enhancing the off rate constant, inducing a decrease in Ca^{2+} sensitivity of myofilament activation. Both sites must be phosphorylated for this effect. There is also evidence that phosphorylation of S23, S24 induces an increase in cross-bridge cycling. These effects of S23, S24 phosphorylation would be expected to be important in regulating relaxation kinetics of the myofilaments and, as discussed below, there is evidence that this is indeed the case.

There are also sites on cTnI that are substrates for PKC. They are located at S43, S45 of cTnI in a region that interacts with the C-lobe of cTnC. There is also a site at T144, which is centrally located in the cTnI inhibitory peptide. Phosphorylation of the PKC sites is associated with a depression of maximum ATPase rate and tension, as well as a depression in Ca^{2+} sensitivity. Unloaded velocity of shortening of thin filaments sliding over myosin heads in the motility assay is also depressed by phosphorylation of the PKC sites.[34] There is evidence that the phosphorylation of T144 may be more prominent in this effect. S149 of cTnI has been demonstrated to be an *in vitro* substrate for p21-activated protein kinase (PAK), and to induce an increase in myofilament sensitivity to Ca^{2+}. S149 is located in a C-terminal region of the Ca^{2+} switch region of cTnI, but the physiological significance of phosphorylation of S149 remains to be elucidated. Moreover, recent evidence indicates that activation of the PAK1 isoform in heart cells induces dephosphorylation of cTnI by activation of PP2A.

There are a number of phosphorylation sites on

cTnT that may be of significance in regulation of myocardial function. Early evidence indicated that these sites induce a depression in ATPase rate of heavy meromyosin reacting with reconstituted thin filaments. More recent experiments have extended these studies to determination of force and ATPase rate in the myofilament lattice.[35] A systematic analysis of the functional role of the multiple PKC phosphorylation sites on cTnT localized a region at Thr206, which controls maximum tension, ATPase activity, and Ca^{2+} sensitivity of the myofilaments. *In vitro*, this site was preferentially phosphorylated by the alpha-PKC isoform. However, the relative significance of this phosphorylation *in vivo* awaits further experimentation. Whether these sites are phosphorylated by other kinases requires further investigation. Depression of myofilament force and ATPase rate also occurs, at least *in vitro*, by Rho-dependent kinase phosphorylation of S278 and T287 of cTnT.[36]

The major thick filament phosphorylation sites are myosin light chain 2 (MLC2) and myosin binding protein C (MyBP-C). The N-terminal region of MLC2 is highly homologous with calmodulin and contains Ca^{2+}/Mg^{2+} binding sites as well as a phosphorylatable residue at Ser-19. Ser-19 in the human ventricular MLC2 is a site of phosphorylation by the Ca^{2+}-calmodulin-dependent kinase designated myosin light chain kinase (MLCK) and by PKC.[6,7] An important control mechanism regulating the level of light chain phosphorylation is a regulated myosin light chain phosphatase (MLCP). The targeting domain (MBS) was the first substrate identified for ROK. Phosphorylation of MBS by ROK inhibits MLCP, and thereby is able to increase MLC2 phosphorylation independently of Ca^{2+}.[37] The region around S-19 contains hydrophobic residues that cluster and form an interface with hydrophobic residues on myosin S-1 toward the C-terminus of the lever arm of myosin (Figs 17.2 and 17.3). In view of this position of the MLC2, it is not surprising that modifications in metal binding and/or state of phosphorylation affect the rate of cross-bridge cycling and/or the position of the lever arm relative to the thick filament back bone. In addition to regulation via Rho signaling, the level of light chain phosphorylation is likely to be altered by long-term changes in heart rate, in which Ca^{2+} binding to calmodulin increases, resulting in increased activity of MLCK.

Although Ca^{2+} binding to MLC2 cannot trigger myofilament activation, metal binding on MLC2 appears important in the maintenance of maximum tension, in the rate of transition of cross-bridges into the force-generating state, and in the rate of cross-bridge detachment. This may be an important mechanism as long-term changes in frequency of contraction could alter the amount of Ca^{2+} and/or Mg^{2+} bound to MLC2. A reasonable speculation on the mechanism of these effects is that the light chain determines the flexibility of myosin heads and their relative tendency to move away from the thick filament, thereby altering the probability of attachment to actin at submaximal levels of activation.

There is general agreement that increases in MLC2 phosphorylation also alter the probability of attachment of cross-bridges, thereby inducing an increase in the force level at submaximum Ca^{2+}, and an associated leftward shift of the pCa–force relation in permeabilized cardiac muscle fibers.[6] A reversible increase in myosin head mobility and/or changed conformation, induced by MLC2 phosphorylation, provides a structural basis for the increases in submaximal force developed by skinned fiber preparations in striated muscle. This change in mobility is likely to be due to charge changes at the N-terminus of MLC2 induced by the phosphorylation. A plausible mechanism for this increase in Ca^{2+} sensitivity is that phosphorylation also increases the rate of transition of cross-bridges from weak to strong, force-generating attachments. There is also evidence that MLC2 phosphorylation slows the transition to the weak force-generating state. Interestingly, this sensitization of the myofilament to Ca^{2+} involves cooperative, cross-bridge-dependent activation of the thin filament.

An interesting and provocative finding from the laboratory of Epstein and colleagues is the report of a spatial gradient of light chain phosphorylation across the myocardium.[38] The functional significance of these regional differences in light chain phosphorylation are theorized to be important in the torque of the ventricular tissue, which adds to the compression, and aids in efficient pumping of blood. The demonstration of mutations increasing MLCK activity, genetically linked to cardiac hypertrophy in the human, supports the general importance of this regulation. Changes in myosin light phosphorylation have also been determined to be important in animal models of heart failure in which Rho activation was associated with pacing-induced heart failure. Here it seems likely that alterations in myosin phosphatase may be significantly involved. It is also potentially significant that there are reports of a reduction in MLC2 phosphorylation of atrial MLC2 associated with heart failure. Threats to this mode of economy of energy use include missense mutations in MLC2 in patients with familial hypertrophic cardiomyopathy.

The thick filament-associated protein, MyBP-C, has both structural and regulatory roles.[39] The cardiac isoform (cMyBP-C) is expressed exclusively in the heart throughout development, and has unique features. These include 11 immunoglobulin (Ig) (C1–C11) modules of 90–100 amino acids (versus 10 in skeletal muscle), three phosphorylation sites between modules C1 and C2, and 28 additional amino acids rich in proline in C5. MyBP-C is visualized immunohistochemically as 10 nm transverse stripes with 43-nm intervals along the thick filament in two 200-nm zones separated by 400 nm on either side of the M-line. It appears to wrap transversely around the thick filament, and may also bind to actin in an adjacent thin filament. The binding sites on myosin are just distal to the head neck region of the cross-bridge (Fig. 17.2). MyBP-C also interacts with titin. Phosphorylation of cMyBP-C is thought to interfere with its binding to the neck region of myosin, resulting in radial movement of cross-bridges toward the thin filament, and increased sensitivity to Ca²⁺. However, phosphorylation of MyBP-C alone is not sufficient to increase Ca²⁺ sensitivity. Myofilaments containing cTnI lacking PKA phosphorylation sites show no PKA-dependent alteration in Ca²⁺ sensitivity, despite robust incorporation of phosphate into MyBP-C.[6] However, in isolated heart preparations from mutant mice lacking cMyBP-C, there is an increase in loaded shortening, power output, and rate of force redevelopment, indicating that cMyBP-C limits power output. An important func-

tional and structural role is indicated by the severity of the cardiomyopathies linked to mutations in MyBP-C.

Testing the hypothesis that altered Ca²⁺ is an essential element in the control of cardiac function

This has been greatly aided by the ability to generate transgenic mouse models with specific modifications in the myofilaments. A particularly useful model has been one in which the adult form, cTnI, was replaced with slow skeletal TnI (ssTnI), the embryonic, neonatal isoform in all species studied to date. Data from investigations of cardiac function, under conditions in which it is apparent that the altered myofilament response to Ca²⁺ is a dominant mechanism, provide strong evidence in support of the hypothesis. It is generally accepted that the fall in tension and pressure that occurs during acidosis in the myocardium is associated with an increase in the peak amplitude of the Ca²⁺ transient. Thus the most straightforward interpretation of the mechanism for this depression in cardiac activity is a desensitization of the myofilaments to Ca²⁺. Early studies indicated that the desensitization by acidic pH of Ca²⁺ activation of isolated myofilaments in detergent-extracted fiber bundles was most pronounced in cardiac versus fast and slow skeletal preparations. Compared to adults, Ca²⁺ activation of myofilaments from neonatal dog hearts also demonstrated a greatly reduced deactivation by acidic pH that could be attributed to a thin filament-related process.[3,6] With the demonstration that embryonic and neonatal hearts express ssTnI, together with data indicating that isoform switching of TnI may be the most important difference between adult and immature hearts with respect to response to acidosis, it was natural to generate a transgenic (TG) mouse in which ssTnI replaced cTnI. Compared to non-TG controls, the Ca²⁺ sensitivity is enhanced in myofilaments from ssTnI-TG mouse hearts, and deactivation by acidic pH is significantly blunted. Moreover, papillary muscles from ssTnI-TG hearts generate the same tension when superfused with buffer at physiological pH and buffer at a reduced pH induced by hypercapnia.[3,6] Under these same acidotic conditions, there was the expected significant reduction in force developed by non-TG papillary muscle. In view of evidence that expression of ssTnI in the transgenic mouse heart has very minor effects on cellular Ca^{2+} fluxes, these results support the hypothesis that altered response to Ca^{2+} may dominate regulation of cardiac activity and dynamics under certain conditions.

Another mode of regulation apparently dominated by myofilament response to Ca^{2+} is the dependence of ventricular systolic pressure on ventricular volume, more commonly known as the Frank–Starling relation or Starling's law of the heart.[28] The relation between end-systolic pressure (ESP) and ventricular volume is a measure of Starling's law, which is believed to be rooted in a dependence of cellular tension on sarcomere length (SL). In addition to the increase in maximum tension developed by the myofilaments with increases in SL, there is also an increase in myofilament sensitivity to Ca^{2+}. The mechanism for these changes has been attributed to a decrease in interfilament spacing with increasing SL, but explicit measurements indicate no correlation between lattice spacing and length-dependent activation. More complicated mechanisms involving titin interactions with the thick filament or thin filament may account for length-dependent activation.[1] Whatever the case, myofilaments from ssTnI-TG hearts show a significant blunting of length-dependent activation that is not correlated with a change in interfilament spacing.[39] The reduced length dependence of activation predicts a reduction in the steepness of the ESP–ventricular volume relation. This reduction in slope has, in fact, been demonstrated in preliminary studies with ssTG hearts beating *in situ*, during stimulation with isoproterenol. It has also been reported that PKA-dependent phosphorylation of the myofilaments enhances length-dependent activation. Thus, under β-adrenergic stimulation, differences between ssTnI-TG hearts, in which there is no PKA-dependent phosphorylation of TnI, and non-TG hearts would be expected to be amplified. These results provide additional and compelling

evidence that regulation by alterations in the myofilament response to Ca^{2+} dominates in a fundamental regulatory device in the heart.

Another prominent role of myofilament response to Ca^{2+} in a fundamental regulatory device in the heart is demonstrated from studies of hearts expressing ssTnI in transgenic mouse models with normal amounts of phospholamban, and in hearts of mutant mice with no phospholamban.[40,41] Compared to controls, work-performing ssTnI-TG hearts studied *in vitro* or *in situ* have a reduced inotropic and lusitropic response to β-adrenergic stimulation.[40,41] This effect persists even in the absence of phospholamban, which has been suggested to completely dominate the response of heart to β-adrenergic stimulation.[41] These data thus indicate that, in addition to effects on the efficacy of Starling's law, alterations in the myofilament response to Ca^{2+} potentially influence both the contractile reserve and relaxation reserve of the myocardium. These findings underscore the importance of understanding the contribution of alterations at the level of the sarcomere in cardiac pathology.

Alterations at the level of the cardiac sarcomere in cardiac pathology

In response to a variety of biochemical and mechanical signals, heart cells grow to offset the functional consequences of the stress.[42–44] For example, the initial growth of heart cells that occurs in response to hemodynamic stress such as hypertension provides compensation by permitting control of cardiac output at a physiological end-diastolic volume. However, the growth is not without consequences, and with sustained growth signaling there is a transition to decompensation, failure, and often sudden death. There is strong evidence, reviewed elsewhere in this volume, that altered Ca^{2+} fluxes in the myocytes contribute to systolic and diastolic abnormalities of the failing hearts. Yet, a critical question is whether altered structure and function of sarcomeric proteins is altered in the transition from compensated hypertrophy to heart failure. Linkage of sarcomeric mutations to dilated and hypertrophic myopathies provides a clue.[42] In this case, it is impossible to think about signaling and signal transduction in cell growth and ventricular remodeling without consideration of the intrinsic structural and functional changes in the sarcomeres that apparently precipitate and promote these events. However, it is unclear how the maladaptation to the mutation leads to the clinical syndrome. Nevertheless, this linkage provides strong evidence that sarcomeric alterations such as isoform switching and protein phosphorylation may contribute to the syndrome. Interestingly, both protein phosphorylation and isoform switching involve charge changes,[45] and it is charge change that constitutes many missense mutations in the sarcomeric proteins. Moreover, some of the functional changes associated with phosphorylation and isoform switching are similar to those occurring with mutations.[7,43,45] For example, a reduced level of cTnI phosphorylation at the PKA sites, associated with downregulation of β-adrenergic signaling, induces an enhanced sensitivity of the myofilaments to Ca^{2+}. Enhanced

sensitivity to Ca^{2+} also occurs in a variety of mutations of cTnT and cTnI. In addition to altered protein phosphorylation, there is evidence for proteolytic clipping of sarcomeric proteins, and isoform switching that may also be of significance in the maladaption to hypertrophic signaling.[45] For example, we reported a functionally significant proteolytic clip of cTnT associated with caspase activation in the myocardium.[46] However, it seems more likely that the transition to failure involves maladaptive protein phosphorylation.

In Fig. 17.5 we show a scheme of current concepts of parallel and interrelated pathways signaling cell growth, and indicate potential cross-talk of these pathways with post-translational modifications of the myofilament proteins. Studies carried out up to now have emphasized signaling by adrenergic, angiotensin II (Ang II), and endothelin receptors, through G_s, G_i, and G_q, and PKA and PKC.[7,42-44] However, more recent evidence indicates a potential significance of the Rho subfamily of GTP-binding proteins, RhoA, Rac1, and Cdc42, and potential post-translational modifications of

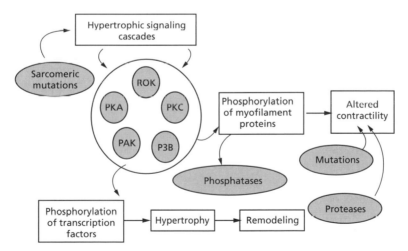

Figure 17.5. *Integration of excitation–contraction coupling, hypertrophic signaling, and myofilament activity and protein phosphorylation. Signaling arising from hemodynamic stress is shown engaging downstream phosphorylation cascades involving various protein kinases including Rho kinase (ROK), protein kinase A (PKA) and protein kinase C (PKC), p38 MAP kinase, and p21-activated kinase (PAK). These kinases influence activity of transcription factors as well as the activity of the myofilaments by altering their state of phosphorylation. This scheme summarizes a hypothesis discussed in the text in which hypertrophic growth of the heart is associated with a reduced contractility, thus leading to heart failure. Also indicated in the scheme is that sarcomeric mutations also may lead to these events, as may the action of proteases.*

myofilament proteins by their downstream effectors, Rho kinase (ROK), p38 MAP kinase, and p21-activated kinase (PAK1). The model of cardiac hypertrophy and failure in Fig. 17.5 integrates the signals that promote cardiac growth and myofilament mass with the signals that modify cardiac myofilament activity. We think that processes that promote cell growth as well as modulation of myofilament activity occur in concert, and for a time provide adequate compensation to the stress; however, at some point in the process, compensation associated with the increase in myofilament mass may be offset by depressed myofilament activity leading to decompensation. We have summarized here and elsewhere,[6,7,42,43] evidence that phosphorylation of cTnI and cTnT at critical sites may lead to decreased tension and shortening velocity. This decrease in intensity and dynamics of myofilament contraction means that heart cells, which have hypertrophied in compensation for an extrinsic or intrinsic stressor, attempt to grow in the face of declining power of contraction. This vicious cycle leads to heart failure. Data in animals support this general idea,[47–50] but much work needs to be done, especially in view of strong evidence that the response of cardiac myofilaments to Ca^{2+} is altered in humans at end-stage heart failure.[51,52]

Although there is upregulation of specific isoforms of PKC in human heart failure,[53] the correlation of this increased PKC activity and sarcomeric protein phosphorylation with stages of human heart failure is difficult inasmuch as samples are obtained at end-stage disease. Normal control samples are difficult to obtain and many times are not 'normal' as they are rejects for transplantation. Moreover despite the general consensus that end-stage heart failure is associated with a depression in cTnI protein phosphorylation at the PKA sites, direct analysis of the cTn from hearts at end-stage failure also indicates that cTnI is phosphorylated at sites other than the N-terminal PKC/PKC sites. Complex relations are also likely to exist among phosphorylation states of other sarcomeric proteins such as titin, MLCs, and MyBP-C.[51] Detailed knowledge of the sites of phosphorylation is critical to correlating the molecular and functional changes. With the emergence of mass spectroscopy

and modern proteomics approaches, new information should be forthcoming soon.

A leading cause of heart failure is ischemic heart disease, which is also associated with pathology at the level of the sarcomeres. Proteolysis, redox state, and generation of free oxygen radicals may all lead to altered response of the myofilaments to Ca^{2+}. Detailed mechanisms are not well defined, but there have been reports of proteolysis of a number of the proteins shown in Fig. 17.1. Removal of a C-terminal region of cTnI appears to be of significance, as does proteolysis of α-actin. In some cases, changes at the level of the sarcomere appear to be responsible for reversible depression cardiac function (stunning or hibernation) in the myocardium. Early studies reported potentially reversible changes in sarcomeric structure occurring during hibernation of dog ventricular myocardium. An interesting recent observation is that, compared to controls, ventricular myocytes from hibernating pig myocardium demonstrate reduced shortening with no significant change in the amplitude of the Ca^{2+} transient.[54] Breakdown of cTnI was not evident in these studies, and the mechanism remains unclear. Ischemia–induced cardiomyopathy in human hearts is also associated with an isoform switch of titin from the N2BA to the N2B form leading to reduced passive tension at the level of the sarcomere.[1] It appears certain that proteomic analysis will also offer new insights into the relation between ischemic injury of varying degrees, and altered structure and function of the sarcomeres.

Z-discs as a pivot point for sarcomeric hypertrophic signal transduction and signal processing

Increases in cell strain imposed by stretching the myocardium during elevations in end-diastolic volume are well recognized as a trigger for hypertrophy. There is emerging and exciting evidence that proteins intrinsically or transiently associated with the Z-discs may be directly involved in transducing this mechanical signal into a biochemical signal.[6,44,54 56] This evidence indicates that when considering control of cardiac functions at the level of the sarcomere, it is important to include control mechanisms at a region of the sarcomere specialized not for tension generation, but for tension transmission, structural stability, and signaling. Z-discs near the periphery of the cardiac myocytes link to the sarcolemma at points containing complex clusters of proteins termed costameres. Costameres link to integrins, which in turn link with the extracellular matrix. The peripheral Z-discs form linkages with more centrally located Z-discs via desmin, and possibly the newly discovered protein, obscurin (Fig. 17.1) In turn, there is evidence that desmin links to the nuclear membrane. These interconnections among the extracellular matrix, the sarcolemma, the Z-disc, and the nucleus provide an exquisite mechanism for cellular components to sense the strain they are under, and to engage appropriate signaling pathways in response to the strain.[57] The major proteins in the network that forms the Z-disc include α-actinin, which cross links neighboring thin filaments at their barbed end terminus, and Z repeats at the N-terminus of titin that link to α-actinin (Fig. 17.1). Components important to the integrity of the Z-disc structure and function include proteins that cap the barbed end of actin in the Z disc (CapZ), and the ends of titin (telethonin or T-Cap). There are numerous signaling molecules that appear to anchor at the Z-disc. PKC-binding proteins, and PKC binding to the sarcomere in general, appear to be influenced by Z-disc proteins. The PKC-binding protein, cypher-1 (also known as ZASP and oracle) has domains that bind to α-actinin and desmin. In addition, the actin capping protein, CapZ, also appears important in PKC signaling. Hearts of transgenic mice with reduced amounts of CapZ also demonstrate hypertrophy and altered response to activation of the PKC pathway.[6] Mutations in cypher-1 are linked to dilated cardiomyopathies and left ventricular compaction.[58] The Z-disc also acts to anchor catalytic subunits of protein phosphatases such as calcineurin, as well as kinases (p21-activated kinase) that appear to activate protein phosphatase 2A.[6] In addition to these kinases and phosphatases, there are signaling proteins (MLP, myopodin, and nuclear factor of activated T cells (NFAT) are examples) that move from the Z-disc to the nucleus and back again to the Z-disc.

Not surprisingly, disturbances in the integrity of the Z-disc may be catastrophic.[56] Mutations in titin are genetically linked to dilated cardiomyopathies, and mutant mice, in which genes expressing muscle-specific LIM protein (MLP) or actinin-associated LIM protein (ALP) are deleted, also exhibit severe dilated cardiomyopathies.[1,6]

Challenges of future research: targeting of sarcomeric proteins by inotropic agents

A major challenge to investigators in the field of sarcomeric control of contractility is translation of their findings to the patient, in the form of new and effective diagnostic and prognostic tools and therapies. Diagnostic and prognostic use of serum levels of cTnI and cTnT as markers of myocardial injury is well-known among clinicians, and remains under investigation. Another line of experimentation involves investigations aimed at discovery of drugs that modify myofilament response to Ca^{2+}.[59,60] These agents are generically referred to as Ca^{2+} sensitizers. The rationale for this approach is couched in data summarized above, supporting the hypothesis that endogenous mechanisms to increase cardiac performance are mediated by altered sarcomeric Ca^{2+} sensitivity. These mechanisms fall into four general categories: (1) covalent mechanisms involving protein phosphorylation; (2) non-covalent mechanisms involving alterations in the chemical environment of the sarcomeres; (3) sarcomere length- and load-dependent mechanisms; and (4) remodeling of the myofilaments by isoform switching. Endogenous mechanisms employing these modes of regulation include length-dependent activation (Frank–Starling mechanism), and adrenergic signaling as described above. Hypothermia, altered pH, ADP and inorganic phosphate levels may also alter the myofilament response to Ca^{2+} and/or the various parameters described in Fig. 17.4. In addition to agents acting directly at the sarcomere, it is known that stimulation of G_q-coupled receptors by phenylephrine, endothelin and angiotensin results in PKC activation and Ca^{2+} sensitization that may result from altered phosphorylation through the

PKC pathway or from PKC-dependent effects on Na^+/H^+-exchanger activation, with subsequent increase of intracellular pH.[61] Increased endogenous availability or exogenous administration of pyruvate also increases contractile force, partially by increased Ca^{2+}-sensitization following intracellular alkalinization.[62]

One objective of drug discovery has been development of a generation of Ca^{2+} sensitizers, which increase inotropy with little effect on diastolic function. If an agent is developed or discovered that increases force by prolonging the time the cross-bridge remains in the force-generating state, as illustrated in Fig. 17.4, then velocity of shortening would decrease. This is equivalent to reducing the cross-bridge off-rate constant 'g'. Since power output is the product of force and velocity, power output may decrease if Ca^{2+} sensitization is due to prolonged cross-bridge attachment. Diastolic function may also be impaired by drugs that increase Ca^{2+} sensitivity by increasing the affinity of cTnC for Ca^{2+}. If the increased affinity of cTnC for Ca^{2+} makes it more difficult for the sarcoplasmic reticulum to remove Ca^{2+} from the myofilaments, there may be a reduction in rate of relaxation and prolonged relaxation time.

All Ca^{2+} sensitizers have beneficial effects on myocardial energetics, however, the quantity of this beneficial effect depends on the mode of action. This can be understood assuming that about 70% of total energy consumption during a contraction–relaxation cycle of a myocyte results from Ca^{2+} cycling and cross-bridge cycling. Those agents that increase Ca^{2+} affinity of cTnC and/or cross-bridge-dependent activation are energetically favorable compared to agents that exert their inotropic effect by increased Ca^{2+} release, because no additional energy for Ca^{2+} transport is needed. However, with these mechanisms, inotropy is still associated with increased energy consumption because more cycling cross-bridges are recruited. Only those agents that increase calcium sensitivity by prolonging cross-bridge attachment time or unitary force may increase inotropy without any increase in energy consumption for both Ca^{2+} cycling and cross-bridge cycling and, therefore, improve economy or efficiency of contraction considerably.

In the case of diastolic function and energetics, one must be aware that most Ca^{2+} sensitizers have additional effects. In particular, most of those drugs increase cyclic AMP by phosphodiesterase inhibition (PDE III inhibition). Ca^{2+} sensitizers such as sulmazole, EMD 57033, ORG 30029, CGP 48056, MCI 154, and pimobendan are able to elicit pronounced phosphodiesterase-inhibiting effects depending on the concentration.[59,60] The Ca^{2+} sensitizers levosimendan, OR 1896, and UDCG-212 may also increase cyclic AMP to various degrees. Levosimendan is of particular interest because it was shown that the molecule dissociates from its binding site of cTnC at low Ca^{2+} levels, and therefore the Ca^{2+} sensitizing effect was suggested to be only present during high systolic calcium levels.[59,60] This may be of particular relevance regarding diastolic function. Accordingly, it has been shown that the positive inotopic effect of levosimendan is associated with an increased instead of a reduced rate of relaxation.[62]

Regarding the three different mechanisms by which calcium sensitizers may increase Ca^{2+} sensitivity, pimobendan and MCI 154 increase Ca^{2+} binding affinity of cTnC, levosimendan and EMD 57033 increase Ca^{2+}-independent activation of cross-bridges (cross-bridge-dependent activation), and OrG 30029, CGP 48503 and also EMD 57033 have direct effects on the cross-bridge cycle with prolonged cross-bridge attachment time.[59] Clinical data are available for the Ca^{2+} sensitizers pimobendan and levosimendan. A low dose of oral pimobendan had favorable short-term effects in patients with congestive heart failure.[63] Pimobendan improved exercise capacity in patients with chronic heart failure. It was well tolerated, however, there was an insignificant trend towards increased mortality during long-term treatment. Levosimendan is a unique Ca^{2+}-sensitizing agent that binds cTnC with high affinity. It does not increase the affinity or Ca^{2+} binding to cTnC, but stabilizes the conformation of the Ca^{2+}–cTnC complex. In addition, levosimendan has some PDE III-inhibiting properties and activates ATP-dependent potassium channels.[64] Levosimendan has a complex pharmacokinetic profile with a long-acting metabolite that has hemodynamic effects persisting for approximately one week.[59] It has been developed for intravenous use thus far. Levosimendan given intravenously exhibited favorable hemodynamic effects by reducing pulmonary capillary wedge pressure and increasing cardiac output. Its effects are more pronounced than that of the selective β_1-adrenoceptor-stimulating agent dobutamine. Furthermore, in two studies using mortality as a secondary endpoint, one-month and six-month mortality were significantly lower with levosimendan, compared to dobutamine.[64–66] Thus, levosimedan may be the most favorable inotopic agent available to treat acute heart failure. Whether it is suited for pulsed intermittent therapy or even for a long-term therapy, has to be proven in clinical trials. Given the fact that, although they may improve short-term hemodynamics in heart failure patients, catecholamines and phosphodiesterase inhibitors have been associated with increased mortality, Ca^{2+} sensitizers may reflect a new group of inotopic agents with favorable effects on hemodynamics, symptoms of heart failure patients, and prognosis. More fundamental studies are required to identify the precise mechanisms of action of these agents. In addition, clinical studies are needed to definitely prove their beneficial effect on morbidity and mortality in patients with heart failure.

Concluding remarks

We set out in this review to present evidence in support of the three hypotheses summarized at the outset. Space limitations did not permit a complete documentation of the evidence, but we think data highlighted here provide a strong case for including mechanisms downstream from Ca^{2+} delivery and removal for the sarcomeres, when considering the mechanism of a broad spectrum of pathologies of the heart. Deeper understanding of the sarcomere has revealed that it is a remarkable organelle with multiple actions, apart from force generation and shortening. The hope that this deeper understanding will translate to better therapeutics and diagnostics does not seem far off.

Acknowledgements
The authors wish to thank their many colleagues for excellent collaborations on the work described in this chapter. The authors also acknowledge support by the National Institutes of Health, Heart, Lung, and Blood Institute (NHLBI), and the German Research Foundation (DFG).

References

1. Granzier HL, Labeit S. The giant protein titin: a major player in myocardial mechanics, signaling, and disease. Circ Res 2004; 94: 284–95.
2. Pyle WG, Solaro RJ. At the crossroads of myocardial signaling: The role of Z-discs in intracellular signaling and cardiac function. Circ Res 2004; 94: 296–305.
3. Kobayashi T, Solaro RJ. Calcium, thin filaments, and integrative biology of cardiac contractility. Annu Rev Physiol 2005; 67: 39–67.
4. Gordon AM, Homsher E, Regnier M. Regulation of contraction in striated muscle. Physiol Rev 2000; 80: 853–924.
5. Lehman W, Hatch V, Korman VL et al. Binding Tm to produce the B state. J Mol Biol 2000; 302: 593–606.
6. Solaro RJ. Modulation of cardiac myofilament activity by protein phosphorylation. In: Page E, Fozzard H, Solaro RJ (eds). Handbook of Physiology: Section 2: The Cardiovascular System. Volume 1 The Heart. New York: Oxford University Press; 2002: 264–300.
7. Solaro RJ, Wolska, BM, Arteaga G et al. Modulation of thin filament activity in long and short term regulation of cardiac function. In: Solaro RJ, Moss RL (eds). Molecular Control Mechanisms in Striated Muscle Contraction. Dordrecht, Netherlands: Kluwer Academic Publishers; 2002: 291–327.
8. Moss RL, Buck SH. Regulation of cardiac contraction by calcium. In: Page E, Fozzard H, Solaro RJ (eds). Handbook of Physiology: Section 2: The Cardiovascular System. Volume 1 The Heart. New York: Oxford University Press, 2002: 420–54.
9. Takeda S, Yamashita A, Maeda K et al. Structure of the core domain of human cardiac troponin in the Ca^{2+}-saturated form. Crystal structure of troponin ternary complex. Nature 2003; 424: 35–41.
10. Metzger JM, Westfall MV. Covalent and noncovalent modification of thin filament action: the essential role of troponin in cardiac muscle regulation. Circ Res 2004; 94: 146–58.
11. Gomes AV, Potter JD, Szczesna-Cordary D. The role of troponins in muscle contraction. IUBMB Life 2002; 54: 323–33.
12. Maytum R, Lehrer SS, Geeves MA. Cooperativity and switching within the three-state model of muscle regulation. Biochemistry. 1999; 38: 1102–10.

13. Lu X, Tobacman LS, Kawai M. Effects of tropomyosin internal deletion Delta23Tm on isometric tension and the cross-bridge kinetics in bovine myocardium. J Physiol 2003; 553: 457–71.

14. Squire JM, Morris EP. A new look at thin filament regulation in vertebrate skeletal muscle. FASEB J 1998; 12: 761–71.

15. Smith DA, Geeves MA. Cooperative regulation of myosin-actin interactions by a continuous flexible chain II: actin-tropomyosin-troponin and regulation by calcium. Biophys J 2003; 84: 3168–80.

16. Tobacman LS, Nihli M, Butters C et al. The troponin tail domain promotes a conformational state of the thin filament that suppresses myosin activity. J Biol Chem 2002; 277: 27636–42.

17. Pan B-S, Solaro RJ. Calcium binding properties of troponin C in detergent skinned heart muscle fibers. J Biol Chem 1987; 262: 7339–49.

18. Robertson SP, Johnson JD, Holroyde MJ et al. The effect of troponin I phosphorylation on the Ca^{2+}-binding properties of the Ca^{2+}-regulatory site of bovine cardiac troponin. J Biol Chem 1982; 257: 260–3.

19. Spyracopoulos L, Li MX, Sia SK et al. Calcium-induced structural transition in the regulatory domain of human cardiac troponin C. Biochemistry 1997; 36: 12138–46.

20. Dong WJ, Xing J, Villain M et al. Conformation of the regulatory domain of cardiac muscle troponin C in its complex with cardiac troponin I. J Biol Chem 1999; 274: 31382–90.

21. Li MX, Spyracopoulos L, Sykes BD. Binding of cardiac troponin-I147–163 induces a structural opening in human cardiac troponin-C. Biochemistry 1999; 38: 8289–98.

22. Smith DA, Maytum R, Geeves MA. Cooperative regulation of myosin–actin interactions by a continuous flexible chain I: actin-tropomyosin systems. Biophys J 2003; 84: 3155–67.

23. Swartz DR, Moss RL, Greaser ML. Calcium alone does not fully activate the thin filament for S1 binding to rigor myofibrils. Biophys J 1996; 71: 1891–904.

24. Moss RL, Razumova M, Fitzsimons DP. Myosin cross-bridge activation of cardiac thin filaments: implications for myocardial function in health and disease. Circ Res 2004; 94: 1290–300.

25. Geeves MA, Lehrer SS. Dynamics of the muscle thin filament regulatory switch: the size of the cooperative unit. Biophys J 1994; 67: 273–82.

26. Tobacman LS. Thin filament-mediated regulation of cardiac contraction. Annu Rev Physiol 1996; 58: 447–81.

27. Rice JJ, Stolovitzky G, Tu Y et al. Ising model of cardiac thin filament activation with nearest-neighbor cooperative interactions. Biophys J 2003; 84: 897–909.

28. Fuchs F. The Frank–Starling relationship: cellular and molecular mechanisms. In: Solaro RJ, Moss RL (eds). Molecular Control Mechanisms in Striated Muscle Contraction. Dordrecht, Netherlands: Kluwer Academic Publishers; 2002; 291–327.

29. Fitzsimons DP, Patel JR, Moss RL. Cross-bridge interaction kinetics in rat myocardium are accelerated by strong binding of myosin to the thin filament. J Physiol 2001; 530: 263–72.

30. Palmiter KA, Tyska M, Dupuis DE et al. Kinetic differences at the single molecule level account for the functional diversity of rabbit cardiac myosin isoforms. J Physiol 1999; 519: 669–78.

31. Noguchi T, Camp P, Jr., Alix SL et al. Myosin from failing and non-failing human ventricles exhibit similar contractile properties. J Mol Cell Cardiol 2003; 35: 91–7.

32. Herron TJ, McDonald KS. Small amounts of alpha-myosin heavy chain isoform expression significantly increase power output of rat cardiac myocyte fragments. Circ Res 2002; 90: 1150–2.

33. Palmiter KA, Tyska MJ, Haeberle JR et al. R403Q and L908V mutant beta-cardiac myosin from patients with familial hypertrophic cardiomyopathy exhibit enhanced mechanical performance at the single molecule level. J Muscle Res Cell Motil 2000; 21: 609–20.

34. Burkart EM, Sumandea, MP, Kobayashi T et al. Phosphorylation or glutamic acid substitution at protein kinase C sites on cardiac troponin I differentially depress myofilament tension and shortening velocity. J Biol Chem 2003; 278: 11265–72.

35. Sumandea MP, Pyle WG, Kobayashi T et al. Identification of a functionally critical PKC phosphorylation residue of cardiac troponin. J Biol Chem 2003; 278: 35135–44.

36. Vahebi S, Kobayashi T, Warren CM et al. Functional effects of rhos-kinase-dependent phosphorylation of specific sites on cardiac troponin. Circ Res 2005; 96:740–7.

37. Somlyo AP, Somlyo AV. Signal transduction by G-proteins, rho-kinase and protein phosphatase to smooth muscle and non-muscle myosin II. J Physiol 2000; 522: 177–85.

38. Davis JS, Hassanzadeh S, Winitsky S et al. A gradient of myosin regulatory light-chain phosphorylation across the ventricular wall supports cardiac torsion. Cold Spring Harb Symp Quant Biol 2002; 67: 345–52.

39. Konhilas JP, Irving TC, Wolska BM et al. Troponin I

in the heart: Influence on length-dependent activation and inter-filament spacing. J Physiol 2003; 547.3: 951–61.

40. Layland J, Grieve DJ, Cave AC et al. Essential role of troponin I in the positive inotropic response to isoprenaline in mouse hearts contracting auxotonically. J Physiol 2004; 556: 835–47.

41. Pena JR, Wolska BM. Troponin I phosphorylation plays an important role in the relaxant effect of beta-adrenergic stimulation in mouse hearts. Cardiovasc Res 2004; 61: 756–63.

42. Solaro RJ, Burkart EM. Functional defects in troponin and the systems biology of heart failure. Invited editorial. J Mol Cell Cardiol 2002; 34: 689–93.

43. Solaro RJ, Montgomery DE, Wang Lynn et al. Integration of pathways that signal cardiac growth with modulation of myofilament activity. J Nuc Cardiol 2002; 9: 523–33.

44. Frey N, Olson EN. Cardiac hypertrophy: the good, the bad, and the ugly. Annu Rev Physiol 2003; 65: 45–79.

45. Marston SB, Redwood CS. Modulation of thin filament activation by breakdown or isoform switching of thin filament proteins: physiological and pathological implications. Circ Res 2003 Dec 12; 93: 1170–8.

46. Communal C, Sumandea M, de Tombe P et al. Functional consequences of caspase activation in cardiac myocytes. Proc Natl Acad Sci USA 2002; 99: 6252–6.

47. Bowman JC, Steinberg SF, Jiang T et al. Expression of protein kinase C beta in the heart causes hypertrophy in adult mice and sudden death in neonates. J Clin Invest 1997; 100: 2189–95.

48. Huang L, Wolska BM, Montgomery DE et al. Increased contractility and altered Ca^{2+} transients of mouse heart myocytes conditionally expressing PKCbeta. Am J Physiol Cell Physiol 2001; 280: C1114–20.

49. Takeishi Y, Chu G, Kirkpatrick DM et al. In vivo phosphorylation of cardiac troponin I by protein kinase C beta2 decreases cardiomyocyte calcium responsiveness and contractility in transgenic mouse hearts. J Clin Invest 1998; 102: 72–8.

50. Takeishi Y, Ping P, Bolli R et al. Transgenic overexpression of constitutively active protein kinase C epsilon causes concentric cardiac hypertrophy. Circ Res 2000; 86: 1218–23.

51. de Tombe PP. Altered contractile function in heart failure. Cardiovasc Res 1998; 37: 367–80.

52. Van der Velden J, Boontje NM, Papp Z et al. Calcium sensitivity of force in human ventricular cardiomyocytes from donor and failing hearts. Basic Res Cardiol 2002; 97 (Suppl 1): I118–26.

53. Bowling N, Walsh RA, Song G et al. Increased protein kinase C activity and expression of Ca^{2+}-sensitive isoforms in the failing human heart. Circulation 1999; 99: 384–91.

54. Bito V, Heinzel FR, Weidemann F et al. Cellular mechanisms of contractile dysfunction in hibernating myocardium. Circ Res 2004; 94: e117.

55. Chien KR. Stress pathways and heart failure. Cell 1999; 98: 555–8.

56. Clark KA, McElhinny AS, Beckerle MC et al. Striated muscle cytoarchitecture: an intricate web of form and function. Annu Rev Cell Dev Biol 2002; 30: 637–706.

57. Knoll R, Hoshijima M, Hoffman HM et al. The cardiac stretch sensor machinery involves a Z-disk complex that is defective in a subset of human dilated cardiomyopathy. Cell 2002; 111: 943–55.

58. Vatta M, Mohapatra B, Jiminez S et al. Mutation in cypher/ZASP in patients with dilated cardiomyopathy and left ventricular non-compaction. J Am Coll Cardiol 2003; 42: 2014–27.

59. Endoh M. Mechanism of action of Ca^{2+} sensitizers – update 2001. Cardiovasc Drugs Ther 2001; 15: 397–403.

60. Arteaga, GA, Kobayashi T, Solaro RJ. Molecular actions of drugs that sensitize cardiac myofilaments to Ca^{2+}. Ann Med 2002; 34: 248–58.

61. Pieske B, Beyermann B, Breu V et al. Functional effects of endothelin and regulation of endothelin receptors in isolated human non-failing and failing myocardium. Circulation 1999; 99: 1802–9.

62. Hasenfuss G, Maier LS, Hermann HP et al. Influence of pyruvate on contractile performance and Ca cycling in isolated failing human myocardium. Circulation 2002; 105: 194–9.

63. Ishiki R, Ishihara T, Izawa H et al. Acute effects of a single low oral dose of pimobendan on left ventricular systolic and diastolic function in patients with congestive heart failure. J Cardiovasc Pharmacol 2000; 35: 897–905.

64. Moiseyiv VS, Poder P, Andrejevs N et al. RUSSLAN Study Investgators. Safety and efficacy of a novel calcium sensitizer, levosimendan, in patients with left ventricular failure due to an acute myocardial infarction. A randomized, placebo-controlled, double-blind study (RUSSLAN). Eur Heart J 2002; 23: 1422–32.

65. Nieminen MS, Akkila J, Hasenfuss G et al. Hemodynamic and neurohumoral effects of continuous infusion of levosimendan in patients with congestive heart failure. J Am Coll Cardiol 2000; 36: 1903–12.

66. Follath F, Cleland JG, Just H et al. Steering

Committee and Investigators of the Levosimendan Infusion versus Dobutamine (LIDO) Study. Efficacy and safety of intravenous levosimendan compared with dobutan in severe low-output heart failure (the LIDO study): a randomized double blind trial. Lancet 2002; 360: 196–202.

CHAPTER **18**

Cardiac protein folding and degradation

Cam Patterson

Introduction

Cells of the adult heart have, at the very least, limited potential for replication, yet these same cells must endure for an individual's lifespan and respond to the range of physiological and pathological stresses that can occur during that time. The adaptations cardiomyocytes undergo in response to stress are well described, and include changes in cell size and shape, alterations in the composition of the contractile apparatus, remodeling of electrical components, and upregulation of protective mechanisms. All of these adaptations must occur while individual cardiomyocytes continue to participate in the mechanical events of cardiac contraction and relaxation. Although signaling mechanisms, transcriptional events, and electrical activities that regulate cardiomyocyte function are well known, it must be kept in mind that cardiomyocytes respond to more mechanical wear and tear than any other organ, and must also be able to 'change out' proteins during the course of phenotypic adaptation during normal development, and in response to hemodynamic changes. Cardiomyocytes must fold large, structurally complicated molecules appropriately to form the contractile apparatus, and must also degrade these proteins when necessary to maintain appropriate homeostasis. Disorder of these events can have devastating consequences for cardiac contractile function. In this chapter, the events that regulate protein folding and degrada-

tion are reviewed, bearing in mind the relationship these events have to regulation of cardiac structure, cell size and the response to injury.

Mechanisms of protein folding and degradation

The central dogma of molecular biology holds that genomic DNA sequence determines protein structure. However, it was not immediately evident from the cracking of the genetic code how an unstructured polypeptide attains its active three-dimensional structure. Careful studies by Christian Anfinsen (who won the Nobel Prize based on this line of inquiry) and Edgar Haber (a founding father of modern experimental cardiology) demonstrated that, in most cases, all the information necessary for a protein to fold is contained in its primary amino acid sequence.[1]

In spite of this, it turns out that the environment within a cell is particularly unfavorable for protein folding, due to constraints that occur during protein synthesis and the likelihood of intermolecular interactions that impede folding in a crowded cellular environment.[2] As such, an evolutionarily ancient system developed to prevent damaged or newly synthesized peptides from aggregating, and to provide a microenvironment that facilitates their proper folding to a thermodynamically favorable active conformation. This system comprises the molecular chaperones, which are present in every cellular compartment to buffer and repair damaged proteins, but which are particularly evolved in number and function in the cytoplasm and endoplasmic reticulum, where folding of newly synthesized proteins occurs (Fig. 18.1). Mechanisms for removing proteins are equally critical to cellular function, and the machinery of protein degradation is similarly archaic. Several pathways exist for protein destruction in eukaryotes. The ubiquitin-proteasome pathway is responsible for the majority of protein degradation, and is the most tightly regulated pathway,[3] but lysosomal degradation and calpain-dependent proteolysis also assist in removing proteins from cells.

Molecular chaperones: the units of folding

Molecular chaperones are defined as 'a functionally related collection of highly conserved and ubiquitous proteins that specifically recognize non-native proteins'.[4] Although they are an amazingly diverse group of proteins, the most abundant cytoplasmic chaperones are also the most closely associated with cardiovascular events. In general, there are three broad families of cytoplasmic chaperones: the Hsp90 family, the Hsp70 family, and the small heat shock proteins.

Folding and aggregation

The classical function of chaperones is to facilitate protein folding, inhibit misfolding, and prevent aggregation. These folding events are regulated by interactions between chaperones and ancillary proteins, the co-chaperones, which in general assist in cycling unfolded substrate proteins on and off the active chaperone complex.[5] The Hsp70 proteins bind to misfolded proteins promiscuously during translation or after stress-mediated protein damage, and provide a hydrophobic microenvironment that favors peptide folding. In contrast, the Hsp90 family members have a more limited range of client proteins; most of these are signaling molecules or transcription factors.[4] The small Hsps (Hsp27 and αB-crystallin) have limited roles in folding, and primarily inhibit aggregation of misfolded proteins so that they can either be disposed of or refolded by other members of the chaperone machinery. Aggregation of misfolded proteins is now implicated as a pathogenic mechanism in chronic degenerative diseases of the nervous system; a potential role for protein aggregates in cardiovascular diseases has not been carefully considered, although it is certainly plausible based on the amount of chronic stress imposed on cardiovascular tissues. Interestingly, several of the small Hsps are expressed in very high levels in the myocardium, and have been implicated in a genetic form of cardiomyopathy,[6] although the extent to which this reflects accumulation of aggregated proteins remains unclear.

Signaling and activation

It is evident from evolutionary considerations that the first function of molecular chaperones has been

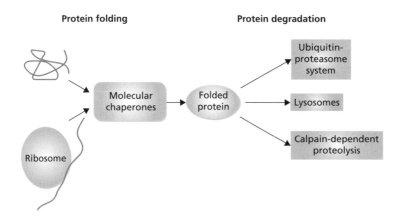

Figure 18.1. *Pathways of protein folding and degradation. Ribosome-bound newly synthesized proteins and damaged (or otherwise misfolded) proteins undergo iterative cycles of refolding through interactions with members of the molecular chaperone family to achieve a stable, functional conformation. Three major pathways exist for degradation of proteins: the ubiquitin-proteasome system, lysosome-dependent protein degradation, and calpain-dependent proteolysis.*

to solve the folding problem. However, the ability of Hsp90 to form multiprotein complexes with a prolific number of co-chaperones has allowed it to assume an additional responsibility as a regulator of key signaling molecules and transcription factors. Many Hsp90-dependent signaling events bear special importance in cardiovascular physiology; for example, steroid hormone receptors such as the estrogen receptors require a complex series of interactions with Hsp90 complexes in order to attain a hormone-binding conformation.[7] The cell cycle-regulatory transcription factor p53 is another protein that requires Hsp90; interestingly, gain-of-functions mutants of p53 that promote growth are preferentially dependent on Hsp90,[8] suggesting that Hsp90 is required for escape from cell cycle arrest induced by mutant p53 in pathological circumstances.

Recently, another story has emerged of a critical circulatory event that is regulated by Hsp90. Hsp90 is a major binding partner of endothelial nitric oxide synthase (eNOS), and generation of nitric oxide (NO) by eNOS is blocked by the Hsp90 inhibitor geldanamycin, indicating that eNOS function is dependent on interactions with Hsp90.[9] Interactions of Hsp90 and eNOS favor generation of NO, and suppress NOS-dependent superoxide generation.[10] Rather than acting as a folding factor in this context, Hsp90 apparently serves as a molecular scaffold to facilitate the interaction between Hsp90 and AKT, which is the kinase required for NO generation by eNOS. Signals such as vascular endothelial growth factor (VEGF) and estrogen, which stimulate NO, do so in part by recruiting AKT and eNOS to a common domain within Hsp90.[11,12] Hsp90 also interacts with the thrombin receptor protease-activated receptor-1 and specifically enhances thrombin-dependent RhoA activation and cytoskeletal changes, without affecting calcium release.[13] In addition, Hsp90 and its co-chaperone Cdc37 are components of the IκB kinase complex, and are required for cytokine-dependent nuclear factor-κB (NF-κB) activation.[14] Conversely, IκB degradation and NF-κB activation can be suppressed by over-expression of Hsp70.[15,16] Taken together, these examples of chaperone interactions with signaling pathways indicate a central and rather complicated role for chaperones in cellular pathways that are relevant to cardiovascular function.

Apoptosis

The interaction of Hsp90 with multiple signaling molecules suggests that recruitment of chaperones in intracellular communication is a common event. However, the extensive involvement of chaperones in apoptotic pathways raises the possibility that the chaperone system and apoptotic events have co-evolved in metazoan species to serve a specific function. Chaperones are powerful modulators of the cell death program, and almost exclusively serve as negative regulators of apoptosis. Anti-apoptotic activity seems to be a common function of cytoplasmic chaperones – Hsp70, Hsp90, and Hsp27 all block cell death under appropriate circumstances.[17,18] In addition, co-chaperone regulators of cytoplasmic chaperones such as Bag-1 and carboxyl-terminus of Hsp70-interacting protein (CHIP) also exert antiapoptotic effects.[19,20] A variety of cell death events are inhibited by the chaperones, including: (1) cell death phosphorylation cascades (especially jun N-terminal kinase[21]); (2) inhibition of caspase 9-APAF apoptosome assembly;[18] (3) suppression of caspase-independent apoptotic events.[22] Interestingly, the effects of the heat shock proteins are, in some cases, independent of their chaperone functions,[21] again suggesting that regulation of apoptosis is a later adaptation of the chaperone machinery to thwart cell death pathways, in coordination with their folding and anti-aggregation functions. Although the anti-apoptotic properties of chaperones have by and large escaped intensive scrutiny by vascular biologists, the expression of these proteins in pathological conditions suggests that they are critical regulators of cell death in cardiovascular pathophysiology.

Protein degradation via the ubiquitin-dependent proteasome system

Ubiquitin is a 76-amino acid molecule that is covalently tagged to proteins, usually in homopolymeric chains, in a reaction catalyzed by three sets of enzymes.[23] Ubiquitin-activating enzyme (E1), which is ubiquitous, attaches to ubiquitin via a thiolester linkage in a charging reaction. The ubiquitin moiety is then transferred to a ubiquitin-conjugating enzyme (E2), of which there are a dozen or so in each cell. Finally, the ubiquitin molecule is attached to a substrate with the assistance of one of the hundreds of ubiquitin ligases (E3s), which serve the primary substrate recognition function of the ubiquitination machinery. There are at least three families of ubiquitin ligases – HECTs, RING fingers, and U-box proteins – and within each family there is tremendous substrate selectivity, which is mediated in part by additional molecules, such as the F-box proteins, which contain substrate-specific interaction motifs. For the most part, ubiquitination provides a recognition signal that targets proteins to the proteasome, although ubiquitination can trigger other events, such as endocytosis, signaling, and membrane trafficking.[24]

Protein degradation

The classical function of ubiquitination is to render substrates susceptible to degradation by the proteasome. The proteasome contains a cylindrical core comprising 28 proteins, and two regulatory caps on either end of the core.[3] Several events are necessary for degradation of ubiquitinated proteins, and specific activities within the proteasome machinery are necessary for each of them. Ubiquitinated proteins must first be recognized by the proteasome, and this requires a recently described ubiquitin chain recognition particle, Rpt5, within the regulatory cap.[25] These proteins must then be de-ubiquitinated (released ubiquitin residues are recycled) and unfolded by an antichaperone activity. These events occur in the regulatory cap, and the unfolded substrate is translocated to the core particle, which contains at least three distinct proteolytic activities that degrade substrates into short peptide fragments.

There are two major reasons for a protein to be degraded: when it is damaged and no longer functional, and when it is no longer needed. The degradation of damaged proteins falls under the auspices of quality control, a function shared by chaperones and the ubiquitin-proteasome system. Indeed, it is now clear that protein quality control mechanisms require close cooperativity between folding and degradation machinery. This link is provided in

part through the function of CHIP, a 35-kDa co-chaperone for Hsp70 and Hsp90, which also contains ubiquitin ligase activity.[20,26–28] CHIP seems to assist the chaperones in recognizing terminally misfolded proteins, and partitions these damaged proteins to the proteasome by tagging them with ubiquitin. The removal of damaged proteins by factors such as CHIP is expected to be of enormous consequence to cells subjected to chronic stress, as otherwise damaged proteins that can aggregate or gain toxic functions would accumulate. However, damage to a protein is not the only reason for it to be degraded. Many proteins serve functions that are temporally restrained; degradation of these proteins provides a means to rapidly turn off a cellular event. Although not widely appreciated until recently, the active degradation of proteins provides a means to control expression of proteins that is as highly regulated (and as complicated) as is gene transcription. The involvement of the ubiquitin-proteasome system in regulation of signaling and cell cycle events provides an example of the importance of these events to cardiovascular processes.

Signaling

By their nature, intracellular signaling events must be activated and terminated with precision and speed. Ubiquitination can terminate signaling by triggering proteasome-dependent degradation of a signaling intermediate in an activated fashion. This relatively simple method is used by AP-1 family members, which are degraded to arrest mitogenic signaling downstream of mitogen-activated protein (MAP) kinase activation.[29] This simple mechanism is used frequently in signaling pathways, but it is by no means the only way that ubiquitination regulates signaling. In fact, the events leading to activation of the pro-hypertrophic transcription factor NF-κB provide a lesson in how many ways ubiquitination can affect signaling. NF-κB consists of dimers of Rel family proteins that exist in the cytoplasm bound to an inhibitor, IκB (α or β). Activation of the IκB kinases IκKα or IκKβ by upstream signals results in phosphorylation, ubiquitination, and degradation of IκB by the proteasome. IκB degradation allows NF-κB to translocate to the nucleus, where it transactivates the expression of genes, many of which are involved in inflammation, stress responses, and hypertrophy.[30] IκB is therefore the 'master switch' for NF-κB activation. The ubiquitin-proteasome system is linked to NF-κB activation via at least three steps. First, the p50 subunit of NF-κB undergoes ubiquitin-dependent limited processing from its precursor, p105.[31] Second, the degradation of IκB is ubiquitin dependent. Lastly, cytokine-stimulated IκB kinase activation occurs in a ubiquitin-dependent, but proteasome-independent, fashion by an activation complex that contains the RING finger ubiquitin ligase TRAF6, which decorates IκB kinase with atypical ubiquitin chains that are required for its activation.[32]

Cell cycle

Proliferation of cells requires a succession of events that require the rapid and sequential synthesis and degradation of proteins required for DNA replication and cell division. By virtue of its rapidity, ubiquitin-dependent proteolysis is the major mechanism for removal of effector proteins as they are no longer needed. Interestingly, the transcription of effector proteins that participate in cell proliferation is itself determined by a master group of proteins that guard the cell cycle, the cyclins and their counterpart cyclin-dependent kinase inhibitors. The up- or downregulation of these proteins determines the place of a cell within the cell cycle or, conversely, a cell's arrest in or exit from the proliferative cycle. For example, cyclin A is degraded by the anaphase-promoting complex as cells enter S phase,[33] whereas the cyclin-dependent kinase inhibitor p27Kip1 is ubiquitinated and degraded to allow passage through the G1 phase of the cell cycle.[34] These events have obvious consequences for cardiovascular events that have a proliferative component; for example, proteolysis of cyclin A is a required event in the withdrawal of cardiomyocytes from the cell cycle during the process of terminal differentiation.[35]

Other pathways for protein degradation

Two other pathways are important for degradation of proteins and have relevance for cardiac pathophysiology. Lysosomes are acidic membrane-bound organelles that participate in the degradation of many cellular components (carbohydrates, proteins, lipids, and nucleic acids) through the activity of hydrolytic enzymes. The non-specificity of lysosomal substrates reflects their critical role in regulating cellular economy, in part through participation in the process of autophagy, and abnormalities in lysosomal function can result in genetic cardiomyopathies.[36,37] In addition, lysosomes serve a specific role in degradation of endocytosed plasma membrane proteins such as epidermal growth factor receptors.

The calpain family represents another pathway for proteolysis. Calpains are cysteine proteases that can be selectively activated by phosphorylation and increased intracellular calcium concentrations. Calpains can either facilitate the removal of proteins or cause their activation via partial proteolysis. For example, calpains induce cleavage of the pro-apoptotic protein Bax to an 18-kDa fragment that may be required for mitochondria-dependent apoptosis.[38] Conversely, specific degradation of troponins by μ-calpain may have special relevance in the regulation of cardiac function.[39]

Myocardial protection and molecular chaperones

It is now clear that molecular chaperones exert multiple effects that confer protection to the heart in the setting of ischemia–reperfusion injury. This is logical, since heat stress and ischemia have many common features, and both result in substantial damage to proteins and induce cellular dysfunction. Members of both the small Hsp family and the Hsp70 family of chaperones are upregulated by ischemia, and overexpression of Hsp70 and the small Hsps αB-crystallin and Hsp27 protect cultured myocytes from ischemic damage.[40] Over the past several years, these studies have led several groups to consider the role of chaperones in protection against ischemia–reperfusion injury to the heart *in vivo*.

Several groups have now demonstrated that overexpression of Hsp70 improves myocardial function and protects against infarction in the setting of induced myocardial ischemia.[41,42] These overexpression studies indicate that Hsp70 family members *can* confer cardioprotection, but it does not automatically follow that they are *required* to protect the heart in the setting of ischemia. Nor is it entirely clear whether the protective effects of chaperones in the setting of ischemia are due entirely to direct effects on myocytes, or whether their presence in cells within the cardiac vasculature also plays a critical role. Defining a requisite role for Hsp70 proteins is made difficult because there are two cytoplasmic isoforms, Hsp70.1 and Hsp70.3, which are functionally redundant. To address this question critically, mice made deficient in both Hsp70.1 and Hsp70.3 by homologous recombination have been created, and their responses to *in vivo* ischemia–reperfusion injury with or without preconditioning have been tested.[43] Surprisingly, infarct sizes after ischemia–reperfusion were no different in Hsp70.1$^{-/-}$/Hsp70.3$^{-/-}$ mice compared with controls in the absence of preconditioning; however, the Hsp70.1$^{-/-}$/Hsp70.3$^{-/-}$ mice had an abnormal response to preconditioning during the late phase. These observations are important but somewhat surprising, insofar as overexpression studies have pointed toward a much more potent role for the cytoplasmic chaperones in protection against ischemia–reperfusion injury. One possible interpretation of these results is that the chaperone system plays a less important role in cardioprotection than has previously been suspected. However, it is also possible that, because the protein refolding systems are exquisitely redundant and undergo autoregulation, it may be difficult to appreciate the importance of this system when single proteins (or, as in this case, even multiple members of the same protein family) are not present.

The abundant constitutive expression of small Hsps in cardiac myocytes is notable, and they may also serve as an important early buffer against ischemia–induced damage in these cells. The redundancy issue is particularly notable for this class of heat shock proteins, as there are at least eight family members with overlapping functions, most of which are expressed in the heart. The most abundant of these is αB-crystallin, and recently mice have been developed that lack both αB-crystallin and the closely linked small Hsp HSPB2. In comparison to wild-type mice, these mice have impaired functional recovery after *in vitro* ischemia–reperfusion injury.[44] The responses of these mice in *in vivo* settings is not yet known, nor are the mechanisms that underlie protective effects mediated by small Hsps. It is interesting to note that mice lacking αB-crystallin and HSPB2 have less reduced glutathione, indicating that they may somehow confer protection against oxidative stress.[44]

If we accept that the existing data generally support a role for chaperone-dependent cardioprotection, but acknowledge that redundancies within this system have made it difficult to assign specific functions to individual chaperone family members, then it is plausible that deletion of regulatory proteins within the chaperone system might be a more informative approach. One such protein is heat shock factor-1 (HSF1), a transcription factor that undergoes trimerization and nuclear localization after stress to activate transcription of many heat shock-responsive chaperones. Interestingly, HSF1 was recently identified in a screen for protection against cell death in cardiomyocytes, and its overexpression in the mouse heart led to reduced infarct size and less cell death after ischemia–reperfusion injury.[45] Whether deletion of HSF1 has the opposite phenotype is not yet known, nor is the mechanism for cardioprotection clear, other than that it suggests that HSF1 transcriptional targets are involved in cardioprotection. In this context, it is interesting to note that hearts of mice lacking HSF1 have impaired antioxidant defenses and altered mitochondrial function, both of which are crucial for cardioprotection in the setting of ischemic injury to the heart.[46] Studies of these mice, or mice lacking other central regulators of chaperone function such as CHIP,[20] may provide answers to the role of molecular chaperones in protection of the myocardium.

Protection of donor hearts after explantation is a promising practical application of molecular chaperones in the context of ischemia–reperfusion injury. The damage that occurs during the explant period is a critical determinant of myocardial engraftment and function after transplantation, and this damage occurs predominantly due to changes in temperature, pH, and oxygen delivery that are toxic to heart cells and their resident proteins. Delivery of Hsp70 by gene therapy techniques has been successful in protection of the myocardium in animal models of heart transplantation.[47] Temporal and delivery considerations make explant chaperone therapy highly practical; extending myocardial preservation during explantation would have enormous repercussions with respect to increasing the donor pool and improving donor:recipient matches, two of the major limitations to heart transplantation at the present time.

Myocardial stunning and calpain-dependent proteolysis

The response to ischemia is regulated by proteolysis as well as by factors affecting protein folding. To date, little is known about ubiquitin-dependent mechanisms regulating cardioprotection. However, calpain-dependent proteolysis appears to be a major determinant of cardiac function after ischemic episodes. Both troponin T and troponin I have been demonstrated to undergo limited proteolysis by μ-calpain.[39] This calpain-dependent proteolysis results in a major cleavage product of troponin I of 193 amino acids (Fig. 18.2), and generated truncated troponin I accumulates after ischemia–reperfusion injury.[48]

Although one might suspect that calpain-dependent degradation of troponins would have predictable effects on cardiac contractility and relaxation, studies have shown that the 193-amino

acid troponin I degradation product actually has a gain-of-function activity that contributes to myocardial stunning and the consequences of ischemia–reperfusion injury. Targeted overexpression of calpastatin, a calpain inhibitor that prevents accumulation of the troponin I degradation product, improves cardiac function in animal models of ischemia–reperfusion injury.[49] Even more compellingly, transgenic overexpression of the 193-amino acid troponin I degradation fragment results in cardiac dysfunction mimicking myocardial stunning in the absence of ischemia,[50] proving that this fragment contains a gain-of-function activity that impairs myocardial function. These studies raise the interesting possibility that calpain inhibition may be a worthy clinical goal to elicit a cardioprotective response.

Dilated cardiomyopathy

Several molecular chaperones, including Hsp70 and Hsp90, are upregulated in human dilated cardiomyopathy, which suggests that misfolded proteins accumulate (and potentially have a toxic role) in this disease.[51] Consistent with this notion, at least one hereditary form of cardiomyopathy is caused by a dysfunctional chaperone. αB-crystallin is a chaperone that is abundantly expressed in the heart and skeletal muscle. Genetic studies have determined that a substitution of glycine for arginine at position 120 in αB-crystallin causes an autosomal dominant disorder that results in

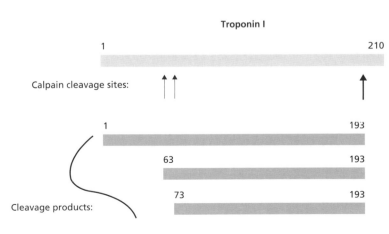

Figure 18.2. *Generation of troponin I cleavage products by calpain-dependent cleavage. The 210-amino acid mature troponin I molecular is susceptible to cleavage by μ-calpain at one major and two minor sites. Cleavage of troponin I by calpain results in a major cleavage product of 193 amino acids and two minor cleavage products. These cleavage products contribute to the pathophysiology of myocardial stunning after ischemia–reperfusion injury to the myocardium.*

cataracts, distal and proximal myopathy, and premature death from intractable congestive heart failure and lethal arrhythmias.[6] Pathologically, the cardiomyopathy is characterized by granulofilamentous deposits of αB-crystallin and phosphorylated desmin, indicating that the consequences of this mutation are due to altered chaperone function of the mutant crystallin. Although to date this is the only example of a chaperone mutation resulting in cardiomyopathy, this observation provides proof of principle that defective chaperone activity leads to cardiac dysfunction, and raises the possibility that subtle defects in chaperone expression or activity may contribute to congestive heart failure in other circumstances.

Defective protein removal has also been linked to dilated cardiomyopathy. Activity of the ubiquitin-proteasome system and accumulation of ubiquitinated proteins (including mitochondrial and sarcomeric proteins) are increased in hearts of patients with dilated cardiomyopathy,[52] although to date no direct link between ubiquitination pathways and dilated cardiomyopathy has been clearly established.

In contrast, an association between dilated cardiomyopathy and defective protein degradation due to lysosomal dysfunction is now well established. A hereditary cardiomyopathy syndrome – Danon disease or X-linked vacuolar cardiomyopathy and myopathy – is caused by mutations in lysosomal membrane-associated protein 2 (LAMP-2).[37] LAMP-2 is a membrane-associated glycoprotein that acts as a receptor for long-lived proteins that are selectively imported into the lysosome for degradation.[53] Consistent with this genetic linkage, mice deficient in LAMP-2 have ultrastructurally abnormal cardiomyocytes and severely impaired cardiac contractility.[36] These studies clearly link abnormal protein turnover and development of dilated cardiomyopathy, although it is not clear whether specific proteins cause this phenotype when inefficiently degraded.

Cardiac hypertrophy

Heart and skeletal muscle cells are subject to changes in cell size – either atrophy or hypertrophy – depending on loading conditions, neurohumoral stimulation, genetic factors, and other influences. Changes in myocyte size are dependent foremost on protein synthesis or degradation, particularly of sarcomeric proteins. There is little formal knowledge about the basis for protein synthesis and degradation in cardiomyocytes, so we are forced to make inferences based on what is understood about skeletal muscle. Atrophy of skeletal muscle primarily occurs by proteasome-dependent protein degradation through pathways that are activated in response to caloric restriction, unloading, and other factors. This effect is mediated at least in part through activation of the N-end rule ubiquitination pathway, which is activated by unmasking of specific amino-terminus degradation sequences within proteins.[54] Myofibrillar proteins are major targets of the proteasome in muscle,[55] which may account for the exaggerated loss of contractile function in the setting of atrophy. Although much less is known about proteasome pathways in cardiac myocytes, it is logical to expect the general mechanisms to be similar.

Recently, two ubiquitination factors have been identified that are restricted to heart and skeletal muscle and are upregulated in response to atrophic stimuli.[56,57] One of these, MAFbx/atrogin-1, is an F-box protein; F-box proteins are substrate-recognition modules of some multicomponent ubiquitin ligases. MAFbx/atrogin-1 participates in a Skp1/Cul1/Roc1 ubiquitin ligase complex, and when deleted by homologous recombination in mice, atrophic responses in skeletal muscle are impaired; cardiac phenotypes due to up- or downregulation of this protein have not yet been reported, but this is obviously a promising area of investigation. The other protein, MuRF1 (muscle RING finger protein 1), is one of a family of three muscle RING finger proteins that contain a tripartite RING, B-Box, coil–coil domain, and deletion of MuRF1 in mice inhibits skeletal muscle atrophy.[57] MuRF1 interacts with the myofibrillar giant spring protein titin at the M line,[58] resulting in disruption of the subdomain that binds MuRF1, suggesting that MuRF1 regulates the stability of this large structural protein, although the functional consequences of this interaction remain to be explored and titin does not appear to be a substrate for the ubiquitin ligase activity of MuRF1. The muscle-specific substrates of these ubiquitination factors are not known, but studies suggest that they are essential components of the muscle atrophic response. As these proteins are present in both skeletal and cardiac muscle, there is a strong possibility that they may have a role in cardiomyopathic changes.

Although not immediately obvious, links also exist between the cardiac hypertrophic response and a requirement for molecular chaperones. Although the precise mechanism for this association remains undetermined, it is possible that either chaperone is upregulated as a reaction to the stress response that elicits hypertrophy. It is equally plausible that chaperones are required to increase the protein-folding capacity within cardiomyocytes as new sarcomeric proteins – which are mostly large, structurally complicated molecules that may burden the protein-folding machinery. Consistent with this model, the potent hypertrophic agonist cardiotrophin-1 (which activates the gp130 transmembrane receptor and intracellular Janus kinase-signal transducer and activator of transcription pathways) causes an increase in several heat shock proteins in cardiomyocytes, including Hsp90, Hsp70, and FK-binding protein 52 (FKBP52, also known as Hsp56).[59] Surprisingly, overexpression of FKBP52, a co-chaperone for Hsp90, by itself can induce hypertrophy in cultured cardiomyocytes, and antisense-dependent knockdown of FKBP52 conversely prevents cardiotrophin-1-induced hypertrophy, indicating an integral role for this co-chaperone in the hypertrophic response.[60]

Whether other chaperones and/or co-chaperones regulate the hypertrophic process is, at present, an open question. How might members of the chaperone system participate in hypertrophy? Certainly, interactions between chaperones and a variety of signaling pathways are now well described, so one likely possibility is that folding factors are regulating signaling events required for hypertrophy. However, one cannot discount the possibility that chaperone function itself is required for cardiomyocytes to increase in size. It is logical to speculate that sarcomeric proteins, which are typically large and structurally complex, would demand significant assistance from chaperones for proper maturation to their active tertiary conformations. Indeed, T-complex polypeptide 1 (TCP-1, also known as type II chaperonin containing T-complex or CCT) is a chaperone that mediates folding of nascent sarcomere components, including actin and myosin heavy chain, in an energy-dependent process.[61,62] In the case of actin, TCP-1 stabilizes specific folding intermediates,[62] indicating that other folding factors are also likely to be involved in assembly of these proteins and therefore in sarcomere biosynthesis. It is possible that chaperone function serves as a rate-limiting step for sarcomere assembly during cardiomyocyte hypertrophy.

Conclusions

In many ways, cells in the heart represent the perfect challenge for cellular quality control mechanisms. Cardiomyocytes undergo constant and brutal stresses that imperil pathways of protein folding and refolding. Many key structural proteins within cardiomyocytes are large and highly ordered, presenting inherent problems for the protein-folding machinery. Cardiomyocytes are prone to large-scale changes in gene regulation in response to changes in hemodynamic and metabolic factors, necessitating removal of unneeded proteins, and accumulation of new proteins. All of this occurs in the context of an organ that cannot fall back on new cell division to repopulate cells that may be lost when these processes go awry. The extent to which protein folding and degradation events regulate key pathophysiological events within the heart is significantly underappreciated, and much remains to be learned about how these processes participate in cardiomyocyte biology. Far from being simply housekeeping events within a cell, understanding these pathways of protein folding and degradation may indicate new paradigms for the understanding and treatment of cardiovascular disease.

Acknowledgements

Work in the author's laboratory is supported by NIH grants GM61728, HL65619, AG021096, and HL61656. CP is an Established Investigator of the

American Heart Association and a Burroughs Wellcome Fund Clinical Scientist in Translational Research.

References

1. Haber E, Anfinsen CB. Regeneration of enzyme activity by air oxidation of reduced subtilisin-modified ribonuclease. J Biol Chem 1961; 236: 422–4.

2. Agashe VR, Hartl FU. Roles of molecular chaperones in cytoplasmic protein folding. Semin Cell Dev Biol 2000; 11: 15–25.

3. Pickart CM. Targeting of substrates to the 26S proteasome. FASEB J 1997; 11: 1055–66.

4. Buchner J. Hsp90 & Co. – a holding for folding. Trends Biochem Sci 1999; 24: 136–41.

5. Höhfeld J. Regulation of the heat shock cognate Hsc70 in the mammalian cell: the characterization of the anti-apoptotic protein BAG-1 provides novel insights. Biol Chem 1998; 379: 269–74.

6. Vicart P, Caron A, Guicheney P et al. A missense mutation in the alphaB-crystallin chaperone gene causes a desmin-related myopathy. Nat Genet 1998; 20: 92–5.

7. Sabbah M, Radanyi C, Redeuilh G et al. The 90 kDa heat-shock protein (hsp90) modulates the binding of the oestrogen receptor to its cognate DNA. Biochem J 1996; 314: 205–13.

8. Blagosklonny MV, Toretsky J, Bohen S et al. Mutant conformation of p53 translated in vitro or in vivo requires functional HSP90. Proc Natl Acad Sci USA 1996; 93: 8379–83.

9. Garcia-Cardena G, Fan R, Shah V et al. Dynamic activation of endothelial nitric oxide synthase by Hsp90. Nature 1998; 392: 821–4.

10. Pritchard KA, Jr., Ackerman AW, Gross ER et al. Heat shock protein 90 mediates the balance of nitric oxide and superoxide anion from endothelial nitric-oxide synthase. J Biol Chem 2001; 276: 17621–4.

11. Fontana J, Fulton D, Chen Y et al. Domain mapping studies reveal that the M domain of hsp90 serves as a molecular scaffold to regulate Akt-dependent phosphorylation of endothelial nitric oxide synthase and NO release. Circ Res 2002; 90: 866–73.

12. Russell KS, Haynes MP, Caulin-Glaser T et al. Estrogen stimulates heat shock protein 90 binding to endothelial nitric oxide synthase in human vascular endothelial cells. J Biol Chem 2000; 275: 5026–30.

13. Pai KS, Mahajan VB, Lau A et al. Thrombin receptor signaling to cytoskeleton requires Hsp90. J Biol Chem 2001; 276: 32642–7.

14. Chen G, Cao P, Goeddel DV. TNF-induced recruitment and activation of the IKK complex require Cdc37 and Hsp90. Molecular Cell 2002; 9: 401–10.

15. Feinstein DL, Galea E, Aquino DA et al. Heat shock protein 70 suppresses astroglial-inducible nitric-oxide synthase expression by decreasing NFκB activation. J Biol Chem 1996; 271: 17724–32.

16. Feinstein DL, Galea E, Reis DJ. Suppression of glial nitric oxide synthase induction by heat shock: effects on proteolytic degradation of IkappaB-alpha. Nitric Oxide 1997; 1: 167–76.

17. Beere HM, Green DR. Stress management: heat shock protein-70 and the regulation of apoptosis. Trends Cell Biol 2001; 11: 6–10.

18. Beere HM, Wolf BB, Cain K et al. Heat-shock protein 70 inhibits apoptosis by preventing recruitment of procaspase-9 to the Apaf-1 apoptosome. Nat Cell Biol 2000; 2: 469–75.

19. Takayama S, Sato T, Krajewski S et al. Cloning and functional analysis of BAG-1: a novel Bcl-2-binding protein with anti-cell death activity. Cell 1995; 80: 279–84.

20. Dai Q, Zhang C, Wu Y et al. CHIP activates HSF1 and confers protection against apoptosis and cellular stress. EMBO J 2003; 22: 5446–58.

21. Yaglom JA, Gabai VL, Meriin AB et al. The function of HSP72 in suppression of c-Jun N-terminal kinase activation can be dissociated from its role in prevention of protein damage. J Biol Chem 1999; 274: 20223–8.

22. Ravagnan L, Gurbuxani S, Susin SA et al. Heat-shock protein 70 antagonizes apoptosis-inducing factor. Nat Cell Biol 2001; 3: 839–43.

23. Cyr DM, Höhfeld J, Patterson C. Protein quality control: U-box E3 ubiquitin ligases join the fold. Trends Biochem Sci 2002; 27: 368–75.

24. Pickart CM. Ubiquitin enters the new millennium. Mol Cell 2001; 8: 499–504.

25. Lam YA, Lawson TG, Velayutham M et al. A proteasomal ATPase subunit recognizes the polyubiquitin degradation signal. Nature 2002; 416: 763–7.

26. Ballinger CA, Connell P, Wu Y et al. Identification of CHIP, a novel tetratricopeptide repeat-containing protein that interacts with heat shock proteins and negatively regulates chaperone functions. Mol Cell Biol 1999; 19: 4535–45.

27. Connell P, Ballinger CA, Jiang J et al. Regulation of heat shock protein-mediated protein triage decisions by the co-chaperone CHIP. Nat Cell Biol 2001; 3: 93–6.

28. Jiang J, Ballinger C, Wu Y et al. CHIP is a U-box-dependent E3 ubiquitin ligase: identification of Hsc70

as a target for ubiquitylation. J Biol Chem 2001; 276: 42938–44.

29. Treier M, Staszewski LM, Bohmann D. Ubiquitin-dependent c-Jun degradation *in vivo* is mediated by the δ domain. Cell 1994; 78: 787–98.

30. Mercurio F, Manning AM. Multiple signals converging on NF-kappaB. Curr Opin Cell Biol 1999; 11: 226–32.

31. Ciechanover A, Gonen H, Bercovich B et al. Mechanisms of ubiquitin-mediated, limited processing of the NF-kappaB1 precursor protein p105. Biochimie 2001; 83: 341–9.

32. Deng L, Wang C, Spencer E et al. Activation of the IkappaB kinase complex by TRAF6 requires a dimeric ubiquitin-conjugating enzyme complex and a unique polyubiquitin chain. Cell 2000; 103: 351–61.

33. Geley S, Kramer E, Gieffers C et al. Anaphase-promoting complex/cyclosome-dependent proteolysis of human cyclin A starts at the beginning of mitosis and is not subject to the spindle assembly checkpoint. J Cell Biol 2001; 153: 137–48.

34. Tomoda K, Kubota Y, Kato J. Degradation of the cyclin-dependent-kinase inhibitor p27Kip1 is instigated by Jab1. Nature 1999; 398: 160–5.

35. Yoshizumi M, Lee W-S, Hsieh C-M et al. Disappearance of cyclin A correlates with permanent withdrawal of cardiomyocytes from the cell cycle in human and rat hearts. J Clin Invest 1995; 95: 2275–80.

36. Tanaka Y, Guhde G, Suter A et al. Accumulation of autophagic vacuoles and cardiomyopathy in LAMP-2-deficient mice. Nature 2000; 406: 902–6.

37. Nishino I, Fu J, Tanji K et al. Primary LAMP-2 deficiency causes X-linked vacuolar cardiomyopathy and myopathy (Danon disease). Nature 2000; 406: 906–10.

38. Wood DE, Thomas A, Devi LA et al. Bax cleavage is mediated by calpain during drug-induced apoptosis. Oncogene 1998; 17: 1069–78.

39. Di Lisa F, De Tullio R, Salamino F et al. Specific degradation of troponin T and I by mu-calpain and its modulation by substrate phosphorylation. Biochem J 1995; 308: 57–61.

40. Mestril R, Hilal-Dandan R, Brunton LL et al. Small heat shock proteins and protection against ischemic injury in cardiac myocytes. Circulation 1997; 96: 4343–8.

41. Mestril R, Chi SH, Sayen MR et al. Overexpression of the rat inducible 70-kD heat stress protein in a transgenic mouse increases the resistance of the heart to ischemic injury. J Clin Invest 1995; 95: 1446–56.

42. Radford NB, Fina M, Benjamin IJ et al. Cardioprotective effects of 70-kDa heat shock protein in

transgenic mice. Proc Natl Acad Sci USA. 1996; 93: 2339–42.

43. Hampton CR, Shimamoto A, Rothnie CL et al. HSP70.1 and -70.3 are required for late-phase protection induced by ischemic preconditioning of mouse hearts. Am J Physiol heart Circ Physiol 2003; 285: H866–74.

44. Morrison LE, Whittaker RJ, Klepper RE et al. Roles for αB-crystallin and HSPB2 in protecting the myocardium from ischemia–reperfusion-induced damage in a KO mouse model. Am J Physiol Heart Circ Physiol 2004; 286: H847–55.

45. Zou Y, Zhu W, Sakamoto M et al. Heat shock transcription factor 1 protects cardiomyocytes from ischemia/reperfusion injury. Circulation 2003; 108: 3024–30.

46. Yan LJ, Christians ES, Liu L et al. Mouse heat shock transcription factor 1 deficiency alters cardiac redox homeostasis and increases mitochondrial oxidative damage. EMBO J 2002; 21: 5164–72.

47. Jayakumar J, Suzuki K, Sammut IA et al. Gene therapy for myocardial protection: transfection of donor hearts with heat shock protein 70 gene protects cardiac function against ischemia–reperfusion injury. Circulation 2001; 104: 303–7.

48. McDonough JL, Arrell DK, Van Eyk JE. Troponin I degradation and covalent complex formation accompanies myocardial ischemia/reperfusion injury. Circ Res 1999; 84: 9–20.

49. Maekawa A, Lee J-K, Nagaya T et al. Overexpression of calpastatin by gene transfer prevents troponin I degradation and ameliorates contractile dysfunction in rat hearts subjected to ischemia/reperfusion. J Mol Cell Cardiol 2003; 35: 1277–84.

50. Murphy AM, Kogler H, Georgakopoulos D et al. Transgenic mouse model of stunned myocardium. Science 2000; 287: 488–91.

51. Barrans JD, Allen PD, Stamatiou D et al. Global gene expression profiling of end-stage dilated cardiomyopathy using a human cardiovascular-based cDNA microarray. Am J Pathol 2002; 160: 2035–43.

52. Weekes J, Morrison K, Mullen A et al. Hyperubiquitination of proteins in dilated cardiomyopathy. Proteomics 2003; 3: 208–16.

53. Eskelinen EL, Illert AL, Tanaka Y et al. Role of LAMP-2 in lysosome biogenesis and autophagy. Mol Biol Cell 2002; 13: 3355–68.

54. Lecker SH, Solomon V, Price SR et al. Ubiquitin conjugation by the N-end rule pathway and mRNAs for its components increase in muscles of diabetic rats. J Clin Invest 1999; 104: 1411–20.

55. Solomon V, Goldberg AL. Importance of the ATP-

ubiquitin-proteasome pathway in the degradation of soluble and myofibrillar proteins in rabbit muscle extracts. J Biol Chem 1996; 271: 26690–7.

56. Gomes MD, Lecker SH, Jagoe RT et al. Atrogin-1, a muscle-specific F-box protein highly expressed during muscle atrophy. Proc Natl Acad Sci USA 2001; 98: 14440–5.

57. Bodine SC, Latres E, Baumhueter S et al. Identification of ubiquitin ligases required for skeletal muscle atrophy. Science 2001; 294: 1704–8.

58. McElhinny AS, Kakinuma K, Sorimachi H et al. Muscle-specific RING finger-1 interacts with titin to regulate sarcomeric M-line and thick filament structure and may have nuclear functions via its interaction with glucocorticoid modulatory element binding protein-1. J Cell Biol 2002; 157: 125–36.

59. Stephanou A, Brar B, Heads R et al. Cardiotrophin-1 induces heat shock protein accumulation in cultured cardiac cells and protects them from stressful stimuli. J Mol Cell Cardiol 1998; 30: 84955.

60. Railson JE, Lawrence K, Buddle JC et al. Heat shock protein-56 is induced by cardiotrophin-1 and mediates its hypertrophic effect. J Mol Cell Cardiol 2001; 33: 1209–21.

61. Srikakulam R, Winkelmann DA. Myosin II folding is mediated by a molecular chaperonin. J Biol Chem 1999; 274: 27265–73.

62. Llorca O, Martin-Benito J, Ritco-Vonsovici M et al. Eukaryotic chaperonin CCT stabilizes actin and tubulin folding intermediates in open quasi-native conformations. EMBO J 2000; 19: 5971–9.

CHAPTER 19

The cytoskeleton

Elizabeth McNally

Introduction

The cytoskeleton provides architecture to a cell and, equally important, transmits signals from the cytoplasm to the nucleus. This signaling is important for the cell's ability to respond to environmental changes. For those cells present in tissues or organs, the cytoskeleton responds to stimuli from the extracellular matrix or to changes in cell–cell contact. Signaling is bidirectional in that information from the matrix, adjacent cells or the cytoplasm can be delivered to the nucleus, and signals from the nucleus can be transmitted back to the cytoplasm, matrix or neighboring cells. The cytoskeleton plays an important role in scaffolding signaling proteins to allow for cell shape alteration and other responses to environmental cues.

There are three major types of proteins that form the backbone of the cytoskeleton structure in eukaryotic cells by forming filamentous networks. These three networks are distinguished on their diameter as viewed using electron microscopy. The smallest filaments are actin-based microfilaments with a 7 nm diameter. Actin filaments are concentrated in the periphery of most cells where they form the cortical cytoskeleton. In striated muscle cells, including skeletal myofibers and cardiomyocytes, actin filaments are highly organized within the cytoplasm forming the sarcomere. In cardiomyocytes, cytoplasmic actin interfaces with the Z line, or Z band, to link the cortical actin

cytoskeleton to the plasma membrane, or sarcolemma, and to the sarcomere. The largest-diameter filaments found in all cell types are microtubules and are 25 nm. Microtubule networks are necessary for chromosome segregation, and provide a network for intracellular transport. Intermediate filaments are 10 nm in diameter and contain filamentous structures of diverse but related protein composition. Intermediate filaments have great tensile strength and, like the cortical actin cytoskeleton, are critical for a cell to withstand deformation. Intermediate filaments are composed of an amino-terminal globular head region and a central elongated rod domain. Carboxy-terminal globular tails may also be present. In cardiomyocytes, there are intermediate filaments found exclusively in nuclei as well as intermediate filaments found in the cytoplasm.

Molecular motors

Molecular motors are responsible for intracellular and cellular movement. There are molecular motors that translocate directly along two of the three filament types.[1] Myosin motors translocate along actin filaments, generally in a single direction, although at least one exception to the directionality of motor movement along actin filaments has been noted. Myosin motors can be classified as conventional or unconventional. Conventional myosins contain two heads and a long rod-like tail that can mediate assembly of multiple myosin

309

molecules into a filamentous structure. Myosin that is found in thick filaments of cardiac and skeletal muscle is conventional myosin, and, in the heart, there are two major forms of conventional myosin in the mature sarcomere. Non-sarcomeric forms of myosin are also present in cardiomyocytes.

Within the cytoskeleton, the second major form of filamentous structure that supports the translocation of molecular motors is the microtubule network. There are two broad classes of microtubule-based motors, dynein and kinesin. Dyneins move toward one end of the microtubules (the minus end) while kinesins move towards the opposite end (the plus end). These motors, like unconventional myosins, often carry cargo attached to the tail region, and this cargo is translocated from one region of a cell to another. Chromosome movement during karyokinesis is accomplished with microtubule-based motors. The near-crystalline arrangement within the cytoplasm of a myocyte places some constraints on intracellular transport, and less is known about unconventional myosins and microtubule-based motors in striated myocytes. Motor proteins do not translocate directly along intermediate filaments, but intermediate filaments may exhibit movement through microtubule- and actin-based filamentous systems.[2] This chapter will focus mainly on the actin cytoskeleton since there is a unique form of the actin cytoskeleton, the sarcomere, in cardiomyocytes. The role of intermediate filament proteins will be reviewed, since these proteins also contribute significantly to normal cardiomyocyte architecture and function.

The actin-based cytoskeleton

In muscle, there is a highly specialized form of microfilament called the myofilament. Myofilaments contain muscle-specific forms of actin and myosin in a highly ordered structure that allows for the rapid deformation or contraction of muscle fibers or cardiomyocytes. At the cellular level, skeletal myofibers are elongated single cells that contain many nuclei. The length of striated myofibers varies greatly, but can extend the length of a muscle. Specialized regions of the myofiber are found at fiber termini in the form of the myotendi-

nous junction and also where nerve terminals end at the neuromuscular junction. In skeletal myofibers, nuclei are generally found in the periphery of myofibers, as the cytoplasm contains sarcomeres. Cardiomyocytes are much smaller and generally only contain one or two nuclei. Specialized cell–cell junctions are found along the short axis of cardiomyocytes and referred to as intercalated discs. Intercalated discs are highly enriched in gap junction subunits to allow for the rapid movement of ions between cells. Intermediate filament proteins provide support to intercalated discs. Like skeletal myofibers, cardiomyocytes contain a similar periodicity of actin and myosin filaments organized into sarcomeres.

Actin filaments are composed of monomers of 43 KDa (G-actin) that assemble into filaments (F-actin) structures. Actin filaments are thinner and more flexible than intermediate filaments, but the addition of actin-binding proteins can stabilize filamentous actin altering its flexibility. Actin filaments are crosslinked by proteins that contain one or two actin-binding sites, to organize actin filaments into parallel or angulated arrays. Actin filaments have a pointed and barbed end, so named for the appearance once myosin heads have been bound to actin. Non-muscle and muscle forms of actin are highly related, reflecting the conserved sequence required for filament formation. Actin sequences are highly conserved in all eukaryotes with little divergence, emphasizing the importance of actin's primary sequence for tertiary and quaternary protein structure. In striated muscle, actin and myosin are organized into sarcomeres where actin-based thin filaments and myosin-based thick filaments are interdigitated. Actin filaments are anchored at their barbed end to Z lines or the Z band, a region of the striated myocyte that stains darkly on electron microscopy. This arrangement of microfilaments is uniquely found in striated muscle cells (Fig. 19.1).

The sarcomere thick filaments

The sarcomere forms the unit of contraction. Myosin is the molecular motor that drives movement and force production in cardiac and skeletal muscle. Muscle myosin is a hexamer composed of two heavy chains and two each of two different

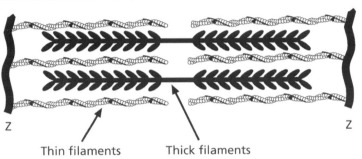

Z   Z

Thin filaments   Thick filaments

Figure 19.1. *The sarcomere. Shown is an electron micrograph of a sarcomere, and the average length of a sarcomere is around 2 μm. The darkly staining regions that flank the sarcomere are Z lines. Myosin-containing thick filaments are in the center of the sarcomere and interact with actin-containing thin filaments by way of myosin heads that protrude from the thick filaments. Thin filament regulatory proteins, the troponins and tropomyosin, provide calcium regulation of the actin–myosin interface. Thin filaments are anchored to the Z line, and the Z line is enriched in proteins such as α-actinin and Cap Z.*

types of light chains. The myosin heavy chain (MHC) is 220 KDa in length, and can be proteolyzed into domains. The amino-terminal domain is approximately 90 KDa in length, and forms a globular head region of myosin referred to as the S1 fragment. The remainder of the myosin heavy chain is dominated by an α helical coiled-coil structure that is responsible for bringing together two MHCs. The globular head region of myosin can be biochemically isolated, and has the property of hydrolyzing ATP. It is this ATP hydrolysis that provides the energy for movement and force production. In the presence of actin, myosin's ATPase activity is stimulated. A crystal structure has been solved for the myosin head region, and in this structure the ATP-binding cleft and the region that forms the interface with actin are physically close.[3,4] The neck region of myosin is a partially elongated structure that provides a bridge between the globular head and the rod-like tail of myosin. The two light chains bind to this neck region, and it is this region that is thought to be the lever arm responsible for movement. Each myosin head can attach and detach from actin about 300 times per minute, allowing for a shortening velocity of up to 15 μm per second. It takes a fully lengthened 3 μm sarcomere less than 0.1 s to fully contract to 2 μm. Muscle myosin heads can move a discrete distance along actin filaments per each unit of ATP that is hydrolyzed. This 'step' can be calculated, and for muscle myosin the step size is considerably smaller than it is for non-muscle forms of myosin. The coordination that allows myosin and actin to shorten in concert is achieved through calcium responsiveness of the actomyosin interface.

Myosin assembles into thick filaments, and each thick filament contains approximately 300 heads, indicating that 150 individual myosin molecules coalesce to form a thick filament. Thick filaments containing multiple myosin molecules can be as long as 1 μm. It is the rod region of myosin that directs the assembly into thick filaments with a characteristic periodicity of 143 Å. In the heart, there are two major forms of sarcomeric MHC.[5] αMHC and βMHC are encoded by two adjacent genes on human chromosome 14. These two myosin genes are highly related to each other and probably arose from a gene duplication event.

The predicted primary sequence of αMHC and βMHC are highly homologous, and yet these two forms of MHC produce myosin hexamers with different *in vitro* and *in vivo* properties. Myosin containing the α heavy chain displays a faster rate of actin-activated ATP hydrolysis, and also displays a faster rate of *in vitro* sliding of actin filaments. *In vivo*, hearts with a greater content of α myosin heavy chain have faster shortening velocity. Mammals differ in their cardiac expression of αMHC versus βMHC. Smaller mammals have predominantly αMHC in the mature heart, with a propensity to express more βMHC in states of heart failure. Larger mammals, such as humans, have primarily βMHC expressed in their hearts, and can increase αMHC expression in states such as hyperthyroidism. Heart failure results in a further increase in βMHC expression.[6] There are two myosin light chains (MLC), a regulatory and essential light chain. The essential light chain (MLC1–3) binds more tightly to the myosin head. Biochemical approaches to remove this light chain from the heavy chain tend to disrupt enzymatic activity of the intact myosin molecule. The regulatory light chain (MLC2) binds to the distal neck of the globular region. The light chains can bind calcium directly, and are phosphorylated. Mutations in *MHC* and *MLC* genes lead to hypertrophic cardiomyopathy in humans.[7] The majority of mutations are missense mutations that have proved useful for structure–function mapping to determine residues critical for myosin function *in vitro* and *in vivo*.

The sarcomere thin filaments

The assembly of actin into filaments requires salt, and is dependent on the concentration of actin monomers. Actin monomers nucleate prior to forming long filaments, and within the cell, the nucleation of actin into filaments is highly regulated by Arp/WASp proteins.[8] Actin grows faster from its plus end (barbed end), and capping proteins present at the minus end (pointed end) can limit growth. Profilin is a small-molecular mass protein that binds to filamentous actin in the ATP-binding site and promotes filament formation.[9] A number of toxins have been useful in understanding actin filament formation. For example, cytochalasin D, a fungus-derived alkaloid, binds to the plus end of actin filaments. By limiting growth, the filaments ultimately become unstable. Latrunculin, a second toxin from sponges, can bind to G-actin, and thereby limit the pool of free G actin available for filament formation. Phalloidin, derived from mushrooms, binds tightly to actin filaments and blocks depolymerization. Actin-binding proteins play an essential role in maintaining higher-order structures of filamentous actin, and can be characterized based on their interaction with actin filaments. Actin-bundling proteins facilitate the formation of parallel or anti-parallel actin filaments. Gelation proteins support angulated actin filaments. Proteins that bind actin monomers regulate the pool of available G-actin, and thereby can regulate the length and distribution of F-actin.

Many actin-binding proteins sequester actin to regions within the cell, and this mechanism is probably most important for the cardiomyocyte in the form of the Z line.[10] The distance from Z line to Z line is approximately 2.2 μm, but this varies with the state of muscle contraction and sarcomere shortening. The Z line contains α-actinin, a small actin-binding protein with an amino-terminal calponin-like, actin-binding site, and four spectrin repeats. α-Actinin-2 is most highly expressed in cardiomyocytes, and anchors actin filaments, or thin filaments, directly to the Z line. In the Z line, α-actinin forms an anti-parallel homodimer, thus linking the barbed ends of actin filaments in adjacent sarcomeres. Cap Z is a small actin-binding protein that caps the barbed end of actin filaments at the Z line of striated myocytes. Cap Z is a heterodimer composed of an α and β subunit.[11] The β1 subunit is preferentially found in the Z line while the β2 subunit is found at the sarcolemma and the intercalated disc, the region of cardiomyocytes specialized for end–end interactions. FATZ is 32-KDa protein that also binds to other proteins with the Z line. FATZ binds to filamin C (γ-filamin), α-actinin and telethonin.[12] Calsarcins are a family of small-molecular mass proteins that are found at the Z line where they interact with α-actinin and filamin C.[13] Interestingly, calsarcins also bind directly to calcineurin, a key regulator in the hypertrophic pathway. The Z line integrates signals from the sarcomere and the plasma membrane, and has the ability to translate these signals to changes in gene expression. The positioning of the Z line is essential for force transmission parallel to the direction of the sarcomere. Additionally, forces perpendicular to the long axis of the sarcomere are transmitted to the Z line through its specialized attachments to the plasma membrane.

Thin filament function is regulated by a series of actin-binding proteins that include α-tropomyosin, troponin I, troponin T and troponin C.[14] α-Tropomyosin is an elongated protein that binds along the groove of actin filaments. The troponin complex provides calcium regulation for the actin–myosin interaction. Troponin C binds calcium directly and, in response, the troponin complex undergoes a conformational change that permits actin and myosin to interact and produce force. Therefore, the thin filament regulatory proteins provide a means to ensure that calcium influx from the sarcoplasmic reticulum is swiftly translated to an efficient powerstroke of actin and myosin. Thin filament regulation is altered in cardiovascular genetic disease. Autosomal dominant mutations in the genes encoding the troponin subunits and cardiac α-tropomyosin can cause hypertrophic cardiomyopathy or dilated cardiomyopathy.

At the sarcolemma, there are additional actin-binding complexes that regulate force transmission through the membrane. Considerable deformation occurs with contraction, and the attachment of the Z line to the membrane is critical to prevent a myocyte from tearing itself apart throughout myofiber or cardiomyocyte contraction. The dystrophin–glycoprotein complex (DGC) forms a complex that anchors the actin cytoskeleton to the membrane and ultimately to the extracellular matrix.[15] Dystrophin is a 427-kDa protein that is the protein product of the Duchenne muscular dystrophy (DMD) locus on chromosome Xp21.[16] DMD is the most common X-linked genetic disorder and, in its most severe form, causes severe muscle degeneration that results in loss of ambulation by the second decade and death by the third decade. Cardiomyopathy accompanies DMD, and highlights that dystrophin function is as important for maintaining cardiomyocyte plasma membrane integrity as it is for skeletal muscle integrity. The dystrophin gene is composed of 79 exons spanning

over 2.5 million base pairs. Two-thirds of DMD cases are associated with large-scale deletions in the gene, that result in the absence of dystrophin protein. Female carriers of dystrophin gene deletions are at risk for developing cardiomyopathy later in life.

The cortical actin cytoskeleton

Dystrophin binds to the actin cytoskeleton by interacting with γ-actin, the major component of cytoplasmic actin underlying the plasma membrane.[17] Dystrophin's amino terminus has a classic actin-binding site similar to that found in α-actinin or in β-spectrin, an important structural protein of the erythrocyte membrane. Dystrophin has 24 spectrin repeats that are interrupted by four hinge regions that provide flexibility to the rod. Unlike β-spectrin and α-actinin, dystrophin exists as a monomer. The spectrin repeats in the mid-portion of dystrophin's rod region constitute a second actin-binding site that is not found in the related protein utrophin. Utrophin is expressed in all cell types and is expressed highly in cardiomyocytes and in regenerating skeletal muscle. Utrophin is thought to partially protect cardiomyocytes from contraction-induced damage when dystrophin is absent. In skeletal muscle, the absence of dystrophin produces results in myofibers that are abnormally susceptible to contraction-induced damage.[18] In cardiac muscle, the absence of dystrophin leads to enhanced cardiomyocyte damage after aortic banding.[19] Therefore, intact dystrophin is essential to protect the plasma membrane from damage during contraction.

The carboxy-terminus of dystrophin binds directly to β-dystroglycan, an integral membrane protein that binds to α-dystroglycan in the extracellular matrix. α- and β-dystroglycan are produced from a single gene and mature dystroglycan is produced from proteolytic processing. α-Dystroglycan is heavily glycosylated, and this glycosylation differs with tissue type. Dystroglycan is expressed in nearly all cell types and plays a role in organizing the extracellular matrix by binding the α2 chain of laminin.[20] Dystroglycan binds the G domains of laminin, perlecan and agrin. In cardiac and skeletal muscle, the sarcoglycan complex is found with the dystrophin–glycoprotein complex. There are six sarcoglycans (α, β, γ, δ, ε and ζ) and the major sarcoglycan complex of the striated myocyte membrane is α, β, γ and δ.[21] Autosomal recessive mutations in sarcoglycan genes lead to muscular dystrophy and cardiomyopathy.[22] Rare instances of autosomal dominant gene mutations have been noted in δ-sarcoglycan.[23]

Recessive mutations in a single sarcoglycan gene produce instability of the protein encoded by the mutated gene, as well as secondary instability of the remainder of the sarcoglycan complex. The sarcoglycan complex is unstable at the plasma membrane when full-length dystrophin is absent. Therefore, sarcoglycan loss is a common molecular feature in these genetically distinct disorders. The cytoplasmic regions of sarcoglycans contain amino acid residues that are phosphorylated, although the kinases responsible for this are unknown. The cytoplasmic regions of sarcoglycans bind directly to filamin C.[24] Filamin is an actin-binding protein with an elongated rod region. The major isoform of filamin expressed in striated muscle is filamin C (or γ-filamin). Filamin C is found at the Z line and at the sarcolemma, and can bind directly to the cytoplasmic regions of γ-sarcoglycan and δ-sarcoglycan. In the absence of sarcoglycan subunits, such as in muscular dystrophy and cardiomyopathy, filamin C is redistributed so that there is an increase in γ-filamin at the sarcolemma. Dystrophin instability is also seen in non-genetic forms of cardiomyopathy. For example, viral infection with Coxsackie virus leads to proteolysis of dystrophin though the E2A protease produced from the virus.[25] In human subjects with decompensated cardiomyopathy, regions of the amino-terminus of dystrophin are selectively lost from the plasma membrane of cardiomyocytes.[26] Moreover, with support from a ventricular assist device, dystrophin loss can be reversed with remodeling.

Microfilaments in non-muscle cells are dispersed throughout the cytoplasm, concentrated into the periphery of the cytoplasm or into extension of cells specialized for attachment. These attachments are referred to as focal adhesions and have a characteristic appearance on electron microscopy. Focal adhesions are composed of a series of proteins that mediate attachment to the plasma membrane and to the cytoskeleton.

Integrins are one group of major cell surface proteins important for linking to the extracellular matrix. Integrins are heterodimers of α and β subunits. Integrins can bind to fibronectin, collagen or laminin in the extracellular complex, and this binding property is largely related to sequence specificity in the α subunit. Both subunits relay signals from the matrix to the cytoskeleton, signaling cytoskeletal rearrangements and activating intracellular signaling pathways. One of the unique properties of integrin signaling is the ability to signal from the outside of the cell to the inside of the cell as well as inside-out signaling. Therefore, through bidirectional signaling, integrins provide an essential role for a cell surrounded by matrix and neighboring cells.

The cytoplasmic regions of integrins bind to the cytoskeletal proteins talin, α-actinin, and vinculin. In non-muscle cells, this is a focal adhesion complex, while in striated muscle, the analogous complex is the membrane attachment to the cytoplasm that overlies the Z line. Integrins are found along the entire plasma membrane but, like the DGC, are enriched at costameres. The laminin-binding form of integrin in striated myocytes is integrin α7β1.[27] Integrin α5β1 is also highly expressed in cardiomyocytes, but binds to fibronectin in the extracellular matrix. Mice generated to lack integrin subunits often do not survive beyond early development, therefore conditional targeting approaches have been taken. With conditional targeting, genes can be deleted specifically in a cell type of interest. Using this approach, the β1 integrin subunit was ablated only in cardiomyocytes and produced cardiomyopathy, indicating that this cell surface complex is essential for normal functioning of a cardiomyocyte.[28]

Integrins, along with the DGC are concentrated over the Z line of cardiomyocytes, although a direct interaction between these complexes has not been demonstrated. The DGC is present at the plasma membrane overlying the M line, but is more prevalent over the Z line.[29] In striated muscle, the DGC binds to γ-actin where it forms a mechanically strong link from the plasma membrane to the cytoskeleton that underlies the plasma membrane. The DGC is found with a rib-like pattern along the sarcolemma with the similar periodicity, and therefore, is preferentially localized over the Z line. Dystrophin can be detected overlying the M line, but there is comparatively less dystrophin overlying the M line compared to that over the Z line. When skeletal muscle membranes are peeled from the underlying sarcomeres, the costameric pattern is also seen along the undersurface of the peeled membrane.[17] Dystrophin and γ-actin co-localize in a costameric pattern. Peeled membranes from *mdx* mice that express no dystrophin, as expected, no longer maintain any dystrophin staining, but importantly no longer maintain the γ-actin pattern on the underside of the membrane. Instead γ-actin remains attached to the sarcomere structure from which the membrane was peeled. Therefore, dystrophin forms a link directly to γ-actin that is overlying sarcomeres, but is distinct from sarcomeric actin (Fig. 19.2).

Anchoring the sarcomere

Titin is a giant protein of striated myocytes, both cardiac and skeletal, that is thought to be important for passive stretch properties and elastic recoil properties of muscle.[30] Titin is over 3 MDa in size, and its amino terminus inserts into the Z line, while the carboxy terminus binds to M line proteins such as myosin-binding protein C. The interaction of titin with myosin-binding protein C may regulate length-dependent activation of myofilaments. The primary structure of titin is dominated by proline, glutamic acid, valine and lysine residues (PEVK region), as well as a repetitive immunoglobulin-like region. These regions are thought to be expandable to adapt to the lengthening myofiber. Titin is found at the sarcolemma as well as along sarcomeres. Titin's vast length spans half the length of the sarcomere, and anchors to the Z line. Titin is a scaffold for many different proteins including Z line-associated proteins ZASP/Cypher and the muscle LIM protein (MLP). These proteins, along with telethonin (T-Cap), form a complex at the Z line that helps myocytes respond to accommodate length (Fig. 19.3).[31]

As titin lengthens, it is thought to relay information to Z line proteins. The absence of MLP leads to dilated cardiomyopathy in the mouse, and this is thought to relate to defective passive force transmission.[32] Mutations in MLP have been found in human dilated cardiomyopathy patients, where it is assumed that defects exist that are similar to those seen in the mouse model. In this case, it is hypothesized that passive stretch leads to loss of

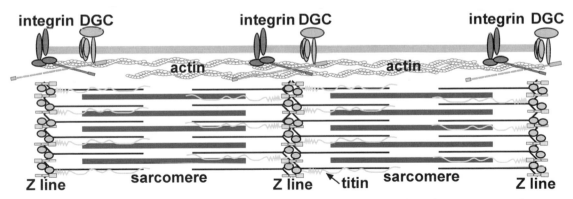

Figure 19.2. *Membrane complexes that concentrate over Z lines. The dystrophin–glycoprotein complex (DGC) is composed of dystrophin and a series of transmembrane proteins including the sarcoglycans and dystroglycan. Dystrophin binds to cortical actin filaments that form a network under the plasma membrane. Integrins bind to vinculin, α-actinin and talin, and also participate in binding to the Z line proteins. Sarcomeres, composed of myosin and actin, anchor to Z lines that contain α-actinin, a spectrin-repeat-containing protein with an actin-binding site. MLP, Cap Z and telethonin are found in the Z line where they interact with titin to help mediate passive force features of the contracting cardiomyocyte.*

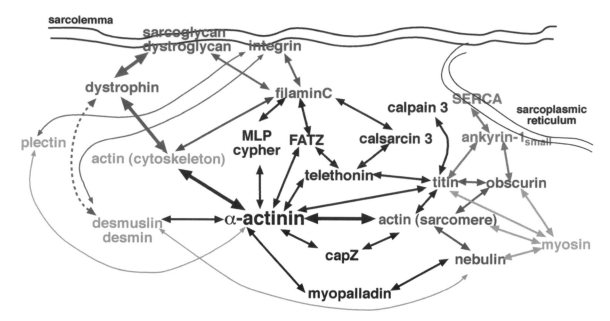

Figure 19.3. *The Z line. The Z line coordinates the interaction of a number of proteins of the sarcolemma and the sarcomere. Within the Z line are proteins that can also translocate to the nucleus as a means of providing signaling to and from the sarcomere and the membrane. Shown in blue are proteins associated with the sarcolemma. Shown in black are those proteins that directly bind to α-actinin, the major constituent of the Z line. The proteins whose names are in red mediate an interaction with the sarcoplasmic reticulum to help mediate calcium release. In dark green are those proteins that in dark green mediate interactions from and within the sarcomere. In blue green are cytoskeleton-associated proteins that provide interactions among different filament types. The broken line indicates indirect interactions. SERCA, sarcoplasmic reticulum Ca²⁺ ATPase. See color plate section.*

myocytes and dilated myopathy. Mutations in ZASP/Cypher have also been described in human dilated cardiomyopathy, although the null mutation in ZASP/Cypher in the mouse leads mainly to early-onset skeletal muscle dysfunction.[33] Titin has two different splice forms that alter the PEVK region, and these splice forms vary their composition during heart failure.[34] Mutations in titin have also been associated with both cardiac and skeletal muscle myopathy in humans and mouse models.

Nebulin is a large protein thought to function as a molecular ruler to 'measure' sarcomere length. While titin is more closely affiliated with thick filaments, nebulin is intimately associated with thin filaments.[35] The carboxy-terminal region of nebulin is affixed to the Z line where it has a binding site for the small protein myopalladin.[36] Nebulin can also interact with desmin to bridge actin filaments to intermediate filaments. Other thin filament proteins, such as the troponins and tropomyosin also bind to nebulin. Cardiomyocytes have a special short form of nebulin, called nebulette. Unlike nebulin, nebulette does not extend fully to the Z line. Overall, nebulin has a highly repetitive structure thought to register with actin thin filaments. Myopallidin can bind directly to cardiac ankyrin repeat protein (CARP), a small protein that can translocate from the nucleus to the Z line. In the nucleus, CARP is a transcriptional regulator.

The nuclear cytoskeleton

In recent years, considerable attention has been focused on the nuclear cytoskeleton (also referred to as the nucleoskeleton). The nuclear membrane is composed of two lipid bilayers, the inner and outer nuclear membrane. The lumen between these bilayers is contiguous with the endoplasmic reticulum. In the cardiomyocyte, the endoplasmic reticulum is the sarcoplasmic reticulum, that includes among its many functions the regulation of calcium release into the cytoplasm for tight control of sarcomere function. The double membrane structure of the nuclear membrane is interrupted by nuclear pores that permit transit of RNA and proteins between the nuclear membrane and the cytoplasm. Underlying the inner nuclear membrane is a nuclear lamina that provides structural support to the nucleus (Fig. 19.4).

The major proteins of the nuclear lamina complex are intermediate filament proteins, mainly the lamins.[37] B type lamins are generally found in dividing cells while the A and C type lamins are mainly found in terminally differentiated cells such as cardiomyocytes and skeletal muscle cells.

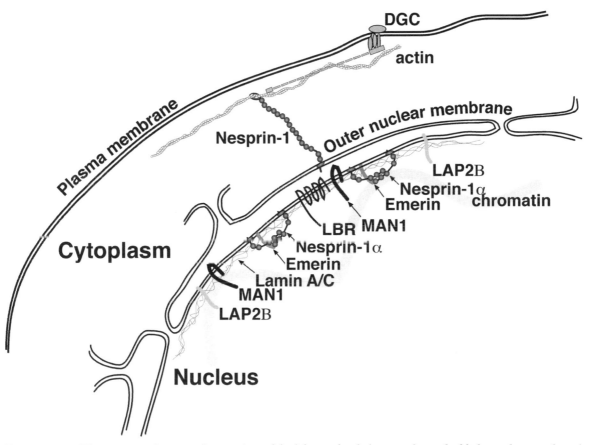

Figure 19.4. *The inner nuclear membrane. A model of the nucleoskeleton is shown highlighting the complex of proteins found at the inner nuclear membrane. Lamin A/C is an intermediate filament protein found exclusively in the nucleus; lamin A/C provides structure and flexibility to the nuclear membrane. Lamin A associates with the membrane through a post-translational modification. Lamin C directly forms filaments with lamin A, and thereby is found at the membrane. Emerin is a membrane-associated protein that binds to lamin A/C. Lamin-associated proteins, LAP 2β, LBR and MAN1 are all found at the inner nuclear membrane. A splice form of nesprin, a nuclear envelope spectrin repeat protein, is found at the inner nuclear membrane where it associates with Lamin A/C. A larger form of nesprin, nesprin-1, has an actin binding site. Genetic models from lower eukaryotes suggest that nesprin-1 is important for localizing nuclei, through interactions with the actin cytoskeleton. The dystrophin–glycoprotein complex (DGC) also interacts with the cortical cytoskeleton.*

Lamins are type V intermediate filament proteins and have both rod-like and globular domains. Like other intermediate filament proteins, lamins dimerize through their coiled coil rod domains. Lamin dimers non-covalently form higher-order tetramers end to end and side to side. Lamins A and C are produced from a single gene that undergoes alternative splicing at the 3′ end to produce a short carboxy-terminus on lamin C and a slightly longer unique sequence at the carboxy-terminus of lamin A. Lamin A is farnesylated and cleaved at its carboxy-terminus to yield the mature form of lamin A, and Zmpste24 is a metalloprotease that is responsible for cleavage of mature lamin A. Autosomal dominant mutations in the gene encoding lamins A and C, *LMNA*, cause dilated cardiomyopathy in humans, although the precise mechanism by which *LMNA* gene mutations lead to dilated cardiomyopathy remains obscure.[38,39] In the mouse, a null allele of *Zmpste24* is associated with a similar phenotype that includes cardiomyopathy and, in this case, defective processing of lamin A was shown.[40]

Lamins A and C interact with a series of transmembrane and non-transmembrane proteins at the inner nuclear membrane.[41] Emerin is a 34-kDa protein encoded by a gene on the X chromosome. Mutations in emerin lead to muscular dystrophy and cardiomyopathy. Emerin associates with the membrane and contains a LEM domain, a region of homology found in lamin-associated protein 2β (LAP 2β), emerin and MAN1. The LEM domain binds to a small molecular mass protein called BAF (barrier to autointegration). When oligomerized, BAF can directly interact with chromatin. Through these connections, the lamina-based network under the inner nuclear membrane can directly interact with heterochromatin. The scaffold structure under the inner nuclear membrane may regulate DNA replication or transcription. It has been hypothesized the lamin-associated structure sequesters heterochromatin and may indirectly or directly repress genes.[42] Mutations in LMNA lead to abnormal heterochromatin structure and nuclear morphology.[43]

As the DGC is important for stabilizing the plasma membrane against mechanical deformation associated with contraction in skeletal and cardio-myocytes, it has been suggested that the scaffold at the inner nuclear membrane similarly provides support the nuclear membrane.[44] The loss of lamin A/C produces nuclei with a distinctly abnormal morphology associated with blebbing, and lacking the normal round shape. Furthermore, these nuclei do not withstand mechanical deformation as well as normal nuclei. A pattern of gene expression associated with stress is activated in LMNA null nuclei.

Anchoring the nucleus

Recently, a very large protein was identified in *C. elegans* that is essential for the proper localization of nuclei within cells. In the worm, syncytial cells are present, in which the nuclei normally align within the syncytium. Mutations in the gene encoding the giant protein ANC-1 cause nuclear disorder within the syncytium.[45] ANC-1 is a large protein (1 MDa) that contains spectrin repeats in its rod domain, and a typical actin-binding site at its amino terminus. The carboxy terminus anchors into the nuclear membrane while the amino terminus anchors to the actin cytoskeleton. There are mammalian homologs of ANC-1 referred to as nesprin (nuclear envelope spectrin repeat protein, and also know as syne or myne).[46] The short form of nesprin is found at the inner nuclear membrane where it associates with lamin A/C and lamin A/C-associated proteins.[47] The nuclear membranes, both inner and outer, are contiguous with the endoplasmic reticulum (sarcoplasmic reticulum) and, ultimately, with the plasma membrane. It is likely that nesprin or nesprin-like proteins help anchor nuclei in mammalian cells. Since heart and skeletal muscle undergo considerable deformation, there may be additional anchor points with the nuclear membrane provided by the short forms of nesprin. In mammalian skeletal muscle, nuclei are normally found in the periphery of myofibers in a similar syncytial arrangement. Cardiomyocytes are generally bi- or uninucleate in mammals, but the placement of nuclei within the cardiomyocyte must accommodate the sarcomeric structure normally found in the cytoplasm.

An additional spectrin repeat protein has been

found at the nuclear membrane of myocytes and this protein is called mAKAP. AKAP proteins anchor protein kinase A and mAKAP is a protein of 250 kDa that is highly expressed in cardiac and skeletal muscle myocytes. Interestingly, protein kinase A binds to the ryanodine receptor that is highly enriched on the sarcoplasmic reticulum.[48] However, as noted above, the sarcoplasmic reticulum is contiguous with the nuclear membranes, where the lumen between the inner and outer nuclear membranes is contiguous with the lumen of the sarcoplasmic reticulum and with the extracellular space. A full understanding of the scaffolding anchors for these complexes and their role in cardiac signaling is likely to be developed in the coming years.

Intermediate filaments and heterofilamentous crosslinkers

Intermediate filament proteins are also present in the cytoplasm and include keratins, vimentin and desmin. Intermediate filaments have more flexibility than other forms of filaments within the cell, and can therefore withstand stress. Intermediate filaments have a central rod domain with globular tail regions that are variable. Intermediate filaments form head-to-tail dimers that assemble into tetramers. Several tetramers assemble to form a protofilament, and protofilaments assemble to form filaments. Keratins are highly expressed in skin, but cytokeratins are also found within cardiomyocytes, where they interact with the subcortical actin filament network. Vimentin is also expressed in cardiomyocytes and can be crosslinked by proteins such as plectin (see below). Desmin is a major intermediate filament protein in the cardiomyocyte, and serves as one of the earliest markers of myocyte development. Desmin is found at the Z line and at the intercalated disc. Desmin is also enriched under the sarcolemma and in the inner nuclear membrane where it is associated with the intermediate filament protein lamin. Desmin mutations lead to myopathy that can be associated with restrictive cardiomyopathy. αB crystallin is a small chaperone protein that helps mediate the formation of desmin filaments. Mutations in the gene encoding αB crystallin can also lead to myopathy. Desmuslin is an intermediate filament protein that binds to α-dystrobrevin, a component of the dystrophin-glycoprotein complex, and to desmin.[49]

Plectin is a large, intermediate filament-binding protein that binds to desmin. Plectin is found in the intercalated disc and associates with vimentin and integrin. A targeted deletion of the plectin gene leads to cardiomyopathy in the mouse, associated with a disruption of intermediate filament organization.[50] Syndromes with skin and muscle defects are associated with plectin mutations in humans, highlighting the importance of the intermediate filament network for cellular stability during cellular deformation.[51] A number of large proteins have recently been described that can crosslink filaments of different types and are considered heterofilamentous crosslinkers.[52] The large protein, ACF7, contains a calponin-like actin-binding protein at its amino terminus, with a long rod region followed by a carboxy terminus that binds microtubules.[53] These large proteins have only recently been discovered, and their role in specialized cells like cardiomyocytes is only now being studied, but it is likely that these cytoskeletal crosslinkers are critical for cardiomyocyte structure and function.

Conclusions

The cytoskeleton in nearly all cell types is highly dynamic and responds to extracellular stimuli as well as nuclear signals. The cytoskeleton of the cardiomyocyte is highly complex, and the Z line is a central location for cell signaling from the sarcomere to the plasma membrane and to the nucleus. The complexity of the sarcoplasmic reticulum and its cytoskeletal links are only now beginning to be understood, with the identification of large proteins such as obscurin that link the Z sarcomere to the sarcoplasmic reticulum.[54] Genetic studies have now implicated many of the genes encoding cytoskeletal proteins in human forms of inherited cardiomyopathy and/or skeletal muscle disease, shedding new light on the importance of the cytoskeleton. The ability to study these proteins through targeted gene disruptions has also been instrumental in identifying proteins critical and non-redundant for cytoskeletal and cellular integrity. Future studies should elucidate a better understanding of the signaling pathways regulated by the cytoskeleton, as many of these pathways are used as adaptations in hypertrophy, both physiological and pathological. As such, the cytoskeleton is a prime target for developing new strategies for the treatment of heart failure and cardiomyocyte dysfunction.

Acknowledgements

The author is supported by the NIH, the American Heart Association, the Muscular Dystrophy Association and the Burroughs Wellcome Fund.

References

1. Vale RD. The molecular motor toolbox for intracellular transport. Cell 2003; 112: 467–80.
2. Helfand BT, Chang L, Goldman RD. The dynamic and motile properties of intermediate filaments. Annu Rev Cell Dev Biol 2003; 19: 445–67.
3. Cope MJ, Whisstock J, Rayment I, et al. Conservation within the myosin motor domain: implications for structure and function. Structure 1996; 4: 969–87.
4. Rayment I. The structural basis of the myosin ATPase activity. J Biol Chem 1996; 271: 15850–3.
5. Morkin E. Control of cardiac myosin heavy chain gene expression. Microsc Res Tech 2000; 50: 522–31.
6. Nakao K, Minobe W, Roden R et al. Myosin heavy chain gene expression in human heart failure. J Clin Invest 1997; 100: 2362–70.
7. Seidman JG, Seidman C. The genetic basis for cardiomyopathy: from mutation identification to mechanistic paradigms. Cell 2001; 104: 557–67.
8. Higgs HN, Pollard TD. Regulation of actin filament network formation through ARP2/3 complex: activation by a diverse array of proteins. Annu Rev Biochem 2001; 70: 649–76.
9. Carlier MF, Wiesner S, Le Clainche C et al. Actin-based motility as a self-organized system: mechanism and reconstitution in vitro. C R Biol 2003; 326: 161–70.
10. Pyle WG, Solaro RJ. At the crossroads of myocardial signaling: the role of Z-discs in intracellular signaling and cardiac function. Circ Res 2004; 94: 296–305.
11. Hart MC, Cooper JA. Vertebrate isoforms of actin capping protein beta have distinct functions in vivo. J Cell Biol 1999; 147: 1287–98.
12. Faulkner G, Pallavicini A, Comelli A et al. FATZ, a filamin-, actinin-, and telethonin-binding protein of the Z-disc of skeletal muscle. J Biol Chem 2000; 275: 41234–42.
13. Frey N, Richardson JA, Olson EN. Calsarcins, a novel family of sarcomeric calcineurin-binding proteins. Proc Natl Acad Sci USA 2000; 97: 14632–7.
14. Metzger JM, Westfall MV. Covalent and noncovalent modification of thin filament action: the essential role of troponin in cardiac muscle regulation. Circ Res 2004; 94: 146–58.
15. Michele DE, Campbell KP. Dystrophin-glycoprotein complex: post-translational processing and dystroglycan function. J Biol Chem 2003; 278: 15457–60.
16. Dalkilic I, Kunkel LM. Muscular dystrophies: genes to pathogenesis. Curr Opin Genet Dev 2003; 13: 231–8.
17. Rybakova IN, Patel JR, Ervasti JM. The dystrophin complex forms a mechanically strong link between the sarcolemma and costameric actin. J Cell Biol 2000; 150: 1209–14.
18. Petrof BJ, Shrager JB, Stedman HH et al. Dystrophin protects the sarcolemma from stresses developed during muscle contraction. Proc Natl Acad Sci USA 1993; 90: 3710–14.
19. Danialou G, Comtois AS, Dudley R et al. Dystrophin-deficient cardiomyocytes are abnormally vulnerable to mechanical stress-induced contractile failure and injury. FASEB J 2001; 15: 1655–7.
20. Henry MD, Campbell KP. Dystroglycan inside and out. Curr Opin Cell Biol 1999; 11: 602–7.
21. Wheeler MT, McNally EM. Sarcoglycans in vascular smooth and striated muscle. Trends Cardiovasc Med 2003; 13: 238–43.
22. Hack AA, Groh ME, McNally EM. Sarcoglycans in muscular dystrophy. Microsc Res Tech 2000; 48: 167–80.
23. Tsubata S, Bowles KR, Vatta M et al. Mutations in the human delta-sarcoglycan gene in familial and sporadic dilated cardiomyopathy. J Clin Invest 2000; 106: 655–62.
24. Thompson TG, Chan YM, Hack AA et al. Filamin 2 (FLN2): a muscle-specific sarcoglycan interacting protein. J Cell Biol 2000; 148: 115–26.
25. Badorff C, Lee GH, Lamphear BJ et al. Enteroviral protease 2A cleaves dystrophin: evidence of cytoskeletal disruption in an acquired cardiomyopathy. Nat Med 1999; 5: 320–6.
26. Vatta M, Stetson SJ, Perez-Verdia A et al. Molecular remodelling of dystrophin in patients with end-stage cardiomyopathies and reversal in patients on assistance-device therapy. Lancet 2002; 359: 936–41.
27. Mayer U. Integrins: redundant or important players in skeletal muscle? J Biol Chem 2003; 278: 14587–90.

28. Ross RS. The extracellular connections: the role of integrins in myocardial remodeling. J Card Fail 2002; 8: S326–31.

29. Williams MW, Bloch RJ. Extensive but coordinated reorganization of the membrane skeleton in myofibers of dystrophic (mdx) mice. J Cell Biol 1999; 144: 1259–70.

30. Maruyama K. Connectin/titin, giant elastic protein of muscle. FASEB J 1997; 11: 341–5.

31. Knoll R, Hoshijima M, Hoffman HM et al. The cardiac mechanical stretch sensor machinery involves a Z disc complex that is defective in a subset of human dilated cardiomyopathy. Cell 2002; 111: 943–55.

32. Arber S, Hunter JJ, Ross J, Jr. et al. MLP-deficient mice exhibit a disruption of cardiac cytoarchitectural organization, dilated cardiomyopathy, and heart failure. Cell 1997; 88: 393–403.

33. Vatta M, Mohapatra B, Jimenez S, et al. Mutations in Cypher/ZASP in patients with dilated cardiomyopathy and left ventricular non-compaction. J Am Coll Cardiol 2003; 42: 2014–27.

34. Neagoe C, Kulke M, del Monte F et al. Titin isoform switch in ischemic human heart disease. Circulation 2002; 106: 1333–41.

35. McElhinny AS, Kazmierski ST, Labeit S et al. Nebulin: the nebulous, multifunctional giant of striated muscle. Trends Cardiovasc Med 2003; 13: 195–201.

36. Bang ML, Mudry RE, McElhinny AS et al. Myopalladin, a novel 145-kilodalton sarcomeric protein with multiple roles in Z-disc and I-band protein assemblies. J Cell Biol 2001; 153: 413–27.

37. Helfand BT, Chang L, Goldman RD. Intermediate filaments are dynamic and motile elements of cellular architecture. J Cell Sci 2004; 117: 133–41.

38. Fatkin D, MacRae C, Sasaki T et al. Missense mutations in the rod domain of the lamin A/C gene as causes of dilated cardiomyopathy and conduction-system disease. N Engl J Med 1999; 341: 1715–24.

39. Burke B, Mounkes LC, Stewart CL. The nuclear envelope in muscular dystrophy and cardiovascular diseases. Traffic 2001; 2: 675–83.

40. Pendas AM, Zhou Z, Cadinanos J et al. Defective prelamin A processing and muscular and adipocyte alterations in Zmpste24 metalloproteinase-deficient mice. Nat Genet 2002; 31: 94–9.

41. Zastrow MS, Vlcek S, Wilson KL. Proteins that bind A-type lamins: integrating isolated clues. J Cell Sci 2004; 117: 979–87.

42. Wilson KL. The nuclear envelope, muscular dystrophy and gene expression. Trends Cell Biol 2000; 10: 125–9.

43. Ostlund C, Worman HJ. Nuclear envelope proteins and neuromuscular diseases. Muscle Nerve 2003; 27: 393–406.

44. Lammerding J, Schulze PC, Takahashi T et al. Lamin A/C deficiency causes defective nuclear mechanics and mechanotransduction. J Clin Invest 2004; 113: 370–8.

45. Starr DA, Han M. Role of ANC-1 in tethering nuclei to the actin cytoskeleton. Science 2002; 298: 406–9.

46. Zhang Q, Ragnauth C, Greener MJ et al. The nesprins are giant actin-binding proteins, orthologous to Drosophila melanogaster muscle protein MSP-300. Genomics 2002; 80: 473–81.

47. Mislow JM, Kim MS, Davis DB et al. Myne-1, a spectrin repeat transmembrane protein of the myocyte inner nuclear membrane, interacts with lamin A/C. J Cell Sci 2002; 115: 61–70.

48. Kapiloff MS, Jackson N, Airhart N. mAKAP and the ryanodine receptor are part of a multi-component signaling complex on the cardiomyocyte nuclear envelope. J Cell Sci 2001; 114: 3167–76.

49. Mizuno Y, Thompson TG, Guyon JR et al. Desmuslin, an intermediate filament protein that interacts with alpha-dystrobrevin and desmin. Proc Natl Acad Sci USA 2001; 98: 6156–61.

50. Andra K, Lassmann H, Bittner R et al. Targeted inactivation of plectin reveals essential function in maintaining the integrity of skin, muscle, and heart cytoarchitecture. Genes Dev 1997; 11: 3143–56.

51. Banwell BL. Intermediate filament-related myopathies. Pediatr Neurol 2001; 24: 257–63.

52. Roper K, Gregory SL, Brown NH. The 'spectraplakins': cytoskeletal giants with characteristics of both spectrin and plakin families. J Cell Sci 2002; 115: 4215–25.

53. Bernier G, Pool M, Kilcup M et al. Acf7 (MACF) is an actin and microtubule linker protein whose expression predominates in neural, muscle, and lung development. Dev Dyn 2000; 219: 216–25.

54. Bagnato P, Barone V, Giacomello E et al. Binding of an ankyrin-1 isoform to obscurin suggests a molecular link between the sarcoplasmic reticulum and myofibrils in striated muscles. J Cell Biol 2003; 160: 245–53.

CHAPTER 20

Adrenergic receptor coupling and uncoupling in heart failure

J David Port, Michael R Bristow

Introduction

Myocardial function is regulated on a beat-by-beat basis, with heart rate and degree of contractility responding dynamically to a variety of physiological and metabolic demands. Primary effectors of this dynamic regulation are the beta-adrenergic receptor (β-AR) signaling pathways. Although they have been investigated extensively, remarkable details regarding the complexities of adrenergic receptor signaling and gene regulation continue to be revealed at an astonishing rate.

In individuals with chronic congestive heart failure, a number of changes in the β-AR signaling pathways occur in part as a result of sustained elevations in concentration of cardiac norepinephrine,[1] the most well recognized of these being downregulation and functional uncoupling of these receptors. In the acute setting, these changes are appropriately adaptive; however, the same changes, both directly and indirectly, can become maladaptive in the chronic setting. The primary focus of this chapter is to explore the effects of chronic congestive heart failure on adrenergic receptor biology, primarily β-ARs, and the pathophysiological consequences thereof. Additional details regarding β-ARs and other G protein-coupled receptors, in general, can be found in Chapter 5. Substantially more detail regarding polymorphisms of adrenergic receptors and their influence on the progression of cardiac disease and its therapeutics are found in recent reviews,[2,3] as well as in Chapter 35. Lastly, descriptions of transgenic mouse models of heart failure, including those wherein adrenergic receptor abundance or activity has been altered, are detailed in Chapter 39.

Adrenergic drive and control of cardiac function in heart failure

In the acute setting, increased physiological demand, secondary to stimuli such as life preservation (fight or flight) or physical exertion, results in rapid activation of adrenergic drive (Fig. 20.1). This response is manifested primarily by increased release of the endogenous catecholamine, norepinephrine, from adrenergic nerve termini within the heart. By definition, acute stimulation of the adrenergic nervous system is of limited duration. In addition to normal temporal constraints, a number of rapidly activated compensatory feedback control mechanisms exist to further dampen the response to catecholamines. This is in distinct contrast to the situation in individuals with chronic congestive heart failure wherein the same counter-regulatory compensatory mechanisms are invoked; however, their capacity to limit the pathophysiological consequences of sustained activation is distinctly limited.

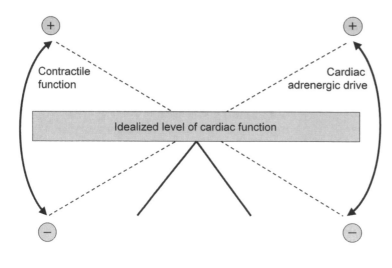

Figure 20.1. *Homeostatic regulation of cardiac contractile function. In the normal heart, contractile function and adrenergic drive are balanced to adapt to normal physiological levels of metabolic demand. In a feedback-controlled manner, when physiological stress or metabolic demands are not adequately met, the activity of the adrenergic nervous system is increased with a resultant increase in cardiac function Adapted from Port JD, Bristow, MR. Altered beta-adrenergic receptor gene regulation and signaling in chronic heart failure. J Mol Cell Cardiol 2001: 33: 887–905, with permission from Elsevier Science.*

This is a two-edged result. The beneficial aspect of a partial desensitization effect is an attenuation of harmful adrenergic signaling without a complete loss of adrenergic support, the latter being potentially devastating to physiological homeostasis.[4,5] Conversely, partial desensitization of the β-AR signaling pathways from sustained stimulation still permits a not inconsequential degree of cellular damage. Moreover, as discussed in greater depth below, adrenergic pathway activation following desensitization has the potential for greater pathological effects than does *de novo* adrenergic signaling.

The extent of deleterious effects of adrenergic signaling on cardiac myocyte biology are well established both *in vitro* and *in vivo*, and date back to the turn of the century.[6] These long-standing observations and the mechanisms underlying them are clearly multifactorial and involve changes in gene expression patterns,[7,8] myocyte viability, as well as producing inordinate increases in intracellular calcium.[9,10] The consequences of increased adrenergic drive on adrenergic signaling pathways, both mediated by and affecting β-ARs, are now discussed in the specific contexts of adrenergic receptor pharmacology, signaling, and gene expression.

Fundamentals of adrenergic receptor pharmacology

It is long established that adrenergic receptors represent a heterogeneous population of ligand binding sites. As early as the 1940s, Ahlquist, utilizing a variety of tissue types and a limited number of selective agonists, pharmacologically distinguished alpha- from beta-adrenergic receptors.[11] The notion that β-ARs could be further subdivided was recognized in classical experiments in heart and lung tissues performed by Lands et al.[12] Significant refinement to pharmacologically characterize adrenergic receptor subtypes followed. However, the singular scientific breakthrough that set in motion the definitive molecular categorization of adrenergic receptor subtypes was their molecular cloning. Undoubtedly, the leading force in this arena was the work of Lefkowitz and colleagues, who were the first to clone many of the adrenergic receptors genes.[13–17] This information not only resolved a number of long-standing discrepancies in pharmacologically based subtype definition, it has also permitted the rigorous biochemical analysis of receptor signaling, and trafficking, as well as the discovery of new and exciting concepts such as receptor homo- and hetero-dimerization.[18] Molecular identification of adrenergic receptors is also the underpinning of what is now a major topic of both basic science and clinical interest, that of adrenergic receptor polymorphisms and their bearing on cardiovascular disease progression and pharmacotherapy.[2,19] In this context, it is almost certain that our current understanding of adrenergic pharmacology, a major component of which is the utilization of β-blocking agents for the treatment of hypertension and heart failure, will be modified by further characterization of adrenergic receptor polymorphisms. Indeed, in recently developed guidelines, the Food and Drug Administration (FDA) is fully anticipating that genetics, i.e. single nucleotide polymorphism or 'SNP' analysis, will be a major determinant of consideration in its assessment of therapeutic efficacy.

Both α- and β-AR subtypes are high-affinity ligand-binding sites for the endogenous catecholamine agonists norepinephrine and epinephrine. A basis of selectivity of response in tissues to endogenous catecholamines is often related to the level of receptor subtype expression within a given cell type or tissue, a classical example of which is the dominance of the $β_1$-AR subtype in cardiac muscle tissues and the $β_2$-AR subtype in the lung and peripheral vasculature. $β_1$- and $β_2$-AR receptors are further distinguished by their relative affinity for endogenous catecholamines with the relative potency of β-agonists at $β_1$-ARs being isoproterenol (ISO) > epinephrine (E) = norepinephrine (NE) and at $β_2$-ARs, ISO > E > NE.[20]

The family of α-ARs is comprised of two major subtypes, $α_1$-ARs and $α_2$-ARs, each with multiple receptor subtypes.[21] Similar in character to the peptidergic angiotensin II and endothelin receptors, the three members of the $α_1$-AR subtype family are coupled to the Gq/G_{11}-family of G proteins. The predominant signaling pathway for $α_1$-ARs is the stimulation of phospholipase C (PLC), which is responsible for the conversion of the phosphoinositide, PIP_2, into two second messenger products, diacylglycerol (DAG), and inositol triphosphate (IP_3).[22] DAG acts primarily to stimulate multiple isoforms of protein kinase C (PKC), serine-threonine kinases, which activate a number of signaling cascades including the induction of pro-growth/hypertrophic transcription factors.[23–26] In a tissue-dependent manner, i.e. cardiac muscle or vascular smooth muscle, the PLC-generated IP_3 binds to either the ryanodine receptor,[27–29] or the IP_3 receptor. In either case, IP_3 binding causes the release of large amounts of calcium from intracellular stores. In turn, calcium's pleiotropic effects cause increased muscle contraction, activation of calcium-sensitive kinases (CaMKII) and phosphatases (PP2B), and produces a pro-growth hypertrophic effect, primarily by activation of transcription factors (e.g. nuclear factor of activated T cells 3 (NFAT3)). Although there is no doubt that $α_1$-ARs are expressed in human myocardium, and that their abundance increases in the failing human heart,[30] their overall abundance, at least in comparison to β-ARs, is relatively low. Further, evidence specifically linking chronic activation of $α_1$-ARs in human heart to a major physiological or pathophyiological role in the context of heart failure is currently limited.[31]

$α_2$-AR subtype receptors, of which there are also three, are primarily coupled to the inhibitory G protein, G_i, which, as its name implies, inhibits adenylyl cyclase-mediated production of cAMP. However, an important function of $α_2$-ARs in a cardiovascular context is the action of the $α_{2c}$-AR subtypes as a presynaptic inhibitory auto-receptor. In situations when adrenergic drive is elevated, such as in patients with congestive heart failure, large amounts of norepinephrine are released from adrenergic nerve termini. Concentrations of norepinephrine in excess of what can be cleared by the re-uptake system (uptake-1 (NET) and uptake-2), spill out of the synaptic cleft and bind to presynaptic $α_2$-ARs which, in turn, attenuate the release of additional norepinephrine. In most instances, this negative feedback loop functions effectively; however, in individuals that express a 'loss of function' deletion mutation of the $α_{2c}$-AR, the ability of norepinephrine to suppress its release is severely attenuated, leading to a higher risk of heart failure.[32] Interestingly, this polymorphism exhibits a higher prevalence (allele frequency) in US African American populations in comparison to US Caucasian populations.[19] Pharmacologically, $α_2$-ARs are not a major target in individuals with heart failure. This being said, it is again possible that our future utilization of β-blocking agents may be tempered by this and other polymorphic variants.

As alluded to above, the dominant population of adrenergic receptors in human heart, as well as most mammalian species, is β-ARs. β-ARs are subdivided into three genetically and pharmacologically distinct subtypes, $β_1$-, $β_2$-, and $β_3$-ARs. $β_1$- and $β_2$-ARs are relatively unique among G protein-coupled receptors (GPCRs), in that their genes are intronless,[15] whereas the $β_3$-AR is not. The $β_3$-AR is quite distinct from other β-ARs in terms of function, localization, and pharmacology.[33] $β_3$-ARs are generally associated with regulation of metabolic rate, however, there is limited evidence that they can also modulate cardiac contractility.[34,35] Given their lack of a prominent role in modulating cardiac drive, $β_3$-ARs will not be discussed further in this chapter.

$β_1$- and $β_2$-ARs have in common at least one major signaling pathway. Both receptor subtypes are positively coupled to adenylyl cyclase stimulation, and to the generation of the second messenger, cAMP. This, in turn, activates protein kinase A (PKA), which subsequently phosphorylates a number of downstream targets including those affecting calcium handling, two major components of which are L-type calcium channels and phospholamban. Both of these proteins are key regulators of myocardial contractility. In addition, β-AR stimulation affects the rate of cellular apoptosis as well as affecting the level of expression of a number of gene products, primarily via transcriptional mechanisms through regulation of cAMP response element-binding protein (CREB) activity.

β-ARs also activate several signaling pathways through cAMP-independent means. Evidence exists for $β_1$-AR, cAMP-independent activation of L-type Ca^{2+} channels,[36] CaM kinase-mediated apoptosis,[37] and activation of extracellular signal-regulated kinase 1/2 (ERK1/2)-mediated signaling.[38] All of these signaling pathways have major significance related to the progression and, potentially, the therapeutic treatment of heart failure. For $β_2$-ARs, there is also evidence of cAMP-independent signaling to the PI3 kinase and ERK1/2 pathways.[39–45]

Pharmacologically, $β$-ARs are major therapeutic targets for the treatment of hypertension and heart failure.[20,46,47] Two major properties of agents acting at these receptors are their relative subtype selectivity and the presence or absence of agonist activity, or intrinsic sympathomimetic activity (ISA). At the most basic level, the prototypical non-selective β-AR agents are the agonist, isoproterenol, and the antagonist, propranolol. In general, there is limited clinical availability of β-AR-selective agonists, with the relative lack of selectivity extending to the endogenous catecholamine, norepinephrine, which has an affinity that is only ~10–30-fold greater for the $β_1$-AR than for the $β_2$-AR.[20,47] In contrast to agonists, far greater subtype selectivity is typically achieved by β-AR antagonists, examples of which are the compounds CGP20712A for the $β_1$-AR, and ICI188511 for the $β_2$-AR. In the clinical setting of chronic heart failure, the most widely evaluated β-AR subtype-selective agents are metoprolol and bisoprolol. Although there are potentially important considerations to the utilization of a selective versus a non-selective β-blocker, such as lipid effects, or the presence of vasospastic angina or asthma, there is no evidence to date that would support major differences in the efficacy of $β_1$-AR-selective agents such as metoprolol, and relatively non-selective β-blockers such as carvedilol[48] in the context of chronic congestive heart failure. Clinical trials with each agent support consistent efficacy of each agent. Nonetheless, the relative superiority of one agent over another remains an active point of discussion within the existing heart failure therapeutics literature.[48,49] Detailed discussions of the use of β-blocking agents and their use in the treatment of congestive heart failure can be found elsewhere.[20,46,47]

β-Adrenergic receptor signaling

As stated above, both $β_1$- and $β_2$-ARs share the ability to activate the heterotrimeric stimulatory G protein, G_s, in an agonist-dependent manner. When the β-AR is occupied by an agonist, the proportion of receptors in the 'R*' or activated state increases, leading to changes in conformation state such that when the β-AR interacts with G_s, there is a dissociation of the α-subunit and the βγ heterodimer. (A number of adrenergic receptors, including β_1- and β_2-ARs, exhibit spontaneous activity i.e. agonist-like activity in the absence of agonist, the degree to which it occurs setting basal adrenergic tone.[50–55] To a large extent, the degree of intrinsic activity is affected by the presence of polymorphic variants of the receptor.[2] Each of these independent signaling entities is important for a broad spectrum of downstream β-AR-signaling events. The free $G_{\alpha s}$ subunit activates adenylyl cyclases (ACs), the predominant isoforms in myocardial tissues being AC Types V and VI, leading to increased formation of second messenger, cAMP. This in turn binds to the regulatory subunits of heterotetrameric PKA, causing dissociation of its catalytic subunits which phosphorylate a number of intracellular targets at consensus serine and threonine sites. PKA and its substrates are coordinately arranged in close proximity through their association with scaffolding A-kinase-anchoring proteins, or AKAPs, which will be discussed in greater detail below. In parallel to the $G_{\alpha s}$ subunit, βγ subunits (5 β subtypes and 12 γ subtypes described[56]) affect a diverse spectrum of signaling pathways including MAPKs and PI3 kinase.[57,58] Importantly, lipid-modified, membrane-associated βγ units, facilitate agonist-dependent β-AR phosphorylation by β-adrenergic receptor kinases (βARKs or GRKs) through their pleckstrin homology (PH) domains. (βARK1 also acts to facilitate crosstalk between agonist-mediated β-AR signaling and the PI3 kinase pathway, with PI3K-generated phophoinositides affecting β-AR endocytosis.)[58,59] This latter effect is of central importance to β-AR desensitization, internalization, and downregulation, and to signaling specificity. β-ARs can also be phosphorylated by PKA and PKC in an agonist-independent manner.[60] Additional downstream targets of βγ subunits include: adenylyl cyclases, PLC, PLA_2, potassium channels, and MAPK cascades (Fig. 20.2).[57,61–66]

The first level of divergence in receptor signaling is at the level of the individual receptor subtype. It is now abundantly clear that there are major differences in the abundance, location, trafficking, and protein–protein interaction between the subtypes, with all of these factors being of potential relevance to the pathophysiological origins and therapeutics of heart failure.

In comparison to the β_2-AR, β_1-AR signaling, at least superficially, appears relatively straightforward. As mentioned above, the primary signaling pathway for the β_1-AR is direct stimulation of the $G_{\alpha s}$, adenylyl cyclase, cAMP, PKA pathway (Fig. 20.2). Activation of PKA leads to the phosphorylation of several critical regulators of intracellular calcium concentration, and in doing so, has a dramatic effect on myocardial contractility. The key targets in the context are the L-type calcium channel, which, when phosphorylated, permits increased transit of extracellular calcium in the vicinity of the junctional triad, inducing calcium-induced calcium release. Another key target of PKA is phospholamban (PLB), a small pentameric molecule which acts to suppress the reuptake of Ca^{2+} by the ATP-dependent calcium pump, SERCA2, into stores within the sarcoplasmic reticulum. PKA-mediated phosphorylation suppresses PLB activity, thereby increasing SR Ca^{2+} uptake. The state of PLB phosphorylation is dynamically modulated by several kinases (PKA, PKC, and CaMKII), and the phosphatase (PP1), which in turn is regulated by a protein phosphatase inhibitor (I-1), in a complex manner.[67] Interestingly, the relative degree of PKA-mediated phosphorylation of PLB does not appear to change substantively during heart failure.[68,69] This is in contrast to the overall degree of PLB phosphorylation which does appear to decrease, presumably due to the activity of specific phosphatases.[70] This results in a decreased ability of the heart to reduce diastolic calcium thereby affecting the rate of relaxation.[71,72] Other important targets of PKA, secondary to β-AR stimulation, are troponin I (TnI),[73,74] and the ryanodine receptor (RyR2).[75] Like PLB, calcium handling by RyR2 is dramatically affected by its state of phosphorylation, with RyR2 being hyperphosphorylated in heart failure,[75] in spite of the fact that PKA activity is generally decreased. Like PLB, net phosphorylation of RyR2 appears to be the result of a balance between kinase and phosphatase activity.[27,29,76]

Underlying differences in β_1- versus β_2-AR signaling are their association with proteins regulating their localization and trafficking and internalization. Depending on the cell type and species, β_2-ARs are known to associate with either AKAP79

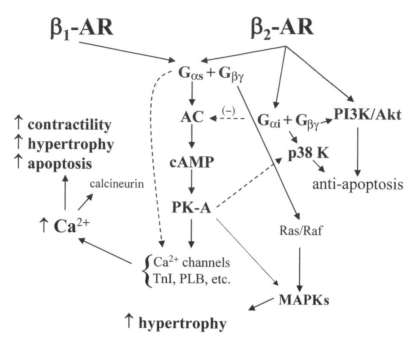

Figure 20.2. *β-Adrenergic receptor signaling pathways. β_1- and β_2-ARs have in common activation of adenylyl cyclase (AC) via the stimulator G-protein, G_s, and formation of the second messenger, cAMP. Through this pathway, both β_1- and β_2-ARs increase intracellular calcium and have positive inotropic effects. However, chronic stimulation of the G_s pathway is known to be pro-apoptotic. When phosphorylated, the β_2-AR has the additional capability of signaling via the inhibitor G-protein, G_i, with both α and $\beta\gamma$ subunits affecting downstream pathways including ones that are anti-apoptotic. Adapted from Port JD, Bristow, MR. Altered beta-adrenergic receptor gene regulation and signaling in chronic heart failure. J Mol Cell Cardiol 2001: 33: 887–905, with permission from Elsevier Science.*

or AKAP250 (also known as AKAP12 and gravin);[77–81] in contrast, limited information exists for cognate AKAP-binding partners for β_1-ARs. The macromolecular complexes scaffolded by AKAPs orchestrate the interaction of the PKA RII catalytic subunit with the β-AR, the L-type calcium channel, PKC, and the calcium-activated phosphatase, calcineurin (PP2B), as well as with specific phosphodiesterases.[82,83] Various permutations of effector molecules and relative stoiciometry can constitute a 'microdomain', leading to increased diversity and specificity of cell signaling within intracellular compartments,[84] particularly in the vicinity of t-tubules and the junctional sarcoplasmic reticulum (SR).[85] This recent biochemical evidence is supported by the long-standing observation that although the β_2-AR is more efficiently

coupled to the production of cAMP ($\sim$5×) on a molar basis, the relative ability of β_1- and β_2-ARs to produce a contractile effect are roughly equivalent, at least in isolated human ventricular trabeculae.[86–88] A necessary conclusion, therefore, is that not all cAMP generated has equivalent access to effecting a contractile response.

Beyond the differential association of β_1 and β_2-ARs with specific AKAPs are their association with differing proteins via their PSD-95/Dlg/ZO-1 (PDZ) recognition motifs, small peptide sequences on the carboxy terminus of the receptor protein. These interactions are of importance not only to internalization and intracellular trafficking but also to their differential coupling to signaling pathways. In this context, the β_1-AR interacts via its carboxy terminus with the PDZ domain-containing protein, PSD-95,[89] which in certain tissues links the β_1-AR to the NMDA-type glutamate receptor channels.[90] PSD-95 appears to attenuate internalization of the β_1-AR, with their interaction being regulated by the kinase, GRK5.[91] Increased GRK5 activity decreases the β_1-AR/PSD-95 association thereby permitting the 'unscaffolded' β_1-AR to internalize. Interestingly, the route(s) of β_1-AR internalization appears to be differentially controlled by phosphorylation with PKA-mediated phosphorylation causing β_1-AR internalization via a caveolae-associated pathway, and GRK2-mediated phosphorylation causing β_1-AR internalization via clathrin-coated pits.[92] These differing pathways of internalization have potential ramifications related to specificity of intracellular signaling. There are also significant differences in the paths of internalization, recycling and protein co-associations between β_1- and β_2-ARs, and other signaling molecules including G proteins and adenylyl cyclases, a topic reviewed in detail elsewhere.[62,84,93,94]

At least three other proteins are able to interact with β_1-ARs via their PDZ domains, endophilins, CNrasGEF (PDZ-GEF1), and MAGI-2. β_1-AR interactions with endophilins (SH3p4/p8/p13) are essentially reciprocal to those of PSD-95 in that endophilins appear to promote β_1-AR internalization as well as inhibiting its coupling to $G_{\alpha s}$.[95] Interaction of the β_1-AR with MAGI-2 also appears to promote internalization.[96] The interaction of β_1-AR with CNrasGEF, which occurs through $G_{\alpha s}$, is linked to the induction of myocardial hypertrophy via the pro-growth Ras pathway (Fig. 20.2).[97]

What is currently unclear at any level of detail is which PDZ-containing proteins are actually critical regulators of β_1-AR signaling in the context of the myocardium, whether normal or failing. Nonetheless, it is quite clear that these interactions are important. In experiments by Xiang et al.,[98,99] disruption of the β_1-AR PDZ recognition motif causes it to become more sensitive to agonist-mediated downregulation, as well as to cause it to internalize and traffic in a manner much like the β_2-AR. Disruption of the β_1-AR PDZ domain also allows the β_1-AR to sequentially couple in a phosphorylation-dependent manner to G_s and G_i,[98]

approximating the well-established effect of phosphorylation on β_2-AR G protein coupling.[100–102]

In distinction to the β_1-AR, the β_2-AR couples to at least one PDZ-containing protein, the Na^+–H^+ exchanger (NHERF/EBP50),[103,104] and the non-PDZ-containing protein, N-ethyl-maleimide-sensitive factor (NSF),[105] both of which are important modulators of agonist-mediated receptor recycling. The β_2-AR also interacts with the PI3K-PTEN pathway, which is relevant to myocardial function and remodeling,[106] and to the $G_{\alpha q}$ and PLA_2 pathways.[101]

Another point of distinction between the β_1- and β_2-AR is the relatively greater pro-apoptotic potential of the β_1-AR,[107–116] an effect which is presumed to accelerate the rate of decline of cardiac function. The ability of the β_1-AR to activate apoptosis is at least partially mediated by increased intracellular calcium secondary to PKA/cAMP-independent, CaMKII-dependent phosphorylation of L-type calcium channels, with a resultant increase in calcineurin activity.[117] In turn, calcineurin dephosphorylates the protein, Bad, allowing it to dimerize and sequester the anti-apoptotic proteins, Bcl-2 and Bcl-x_L. As mentioned above, the β_2-AR signals to an increased degree via the inhibitory G_i protein when phosphorylated (Fig. 20.2).[102,118] β_2-AR signaling via G_i, which is regulated in turn by PDE4,[119] activates the PI3 kinase and Akt/PK-B pathways, the result of which is inhibition of apoptosis.[41,120]

To summarize, stimulation of the β_1-AR pathway produces a generally unopposed pro-apoptotic effect. In contrast, the β_2-AR can produce both pro- and anti-apoptotic effects depending on the relative contribution of the signaling pathways (G_s versus G_i) utilized.

Regulation of β-AR gene expression

Given the major role of adrenergic receptors in dynamically regulating cardiac function, having profound effects on metabolic rate and significant potential to affect cardiac growth and remodeling including programmed cell death, it is not surprising that genes encoding adrenergic receptors are modulated at virtually every level possible.

Regulatory mechanisms documented for modulation of β-AR gene expression include transcriptional, post-transcriptional (mRNA stability), translational, and post-translational (e.g. phosphorylation) ones. The fact that multiple independent levels of regulation exist affords the potential of synergistic mechanisms for modulating adrenergic signaling in a rapid and dynamic manner.

At the transcriptional level, β-AR genes possess a number of well-described promoter, enhancer and suppressor elements, the most obvious of which is the cAMP-responsive element or CRE. In the acute setting, increases in cAMP concentration cause an upregulation of β-AR mRNA.[121,122] However, it is well recognized that chronic exposure of cells to β-AR agonists that increase cAMP, or direct activators of PKA, cause a downregulation of β-AR mRNA and protein.[123–127] Thus, in the context of heart failure, where β-ARs are well established to be downregulated, it is unlikely that CRE-mediated effects are relevant to β-AR gene expression. Remaining at the level of transcription, one potential mechanism that fits with the observed result is activation of the negative regulator of transcription, inducible cyclic AMP early repressor (ICER).[128] In fact, expression of ICER protein is strongly induced in cardiac myocytes by β-AR agonist-mediated stimulation.[129] However, its potential role in human heart and in heart failure remains to be explored.

In addition to cAMP/CRE-mediated effects, β-AR gene expression is transcriptionally regulated by glucocorticoids, retinoic acid, and thyroid hormone,[130–132] via their cognate DNA-binding sites. More specifically, glucocorticoids have a reciprocal effect on β-AR expression causing the upregulation of β_2-ARs and downregulation of β_1-ARs.[131,133–135] The clinical effects of both hypo- and hyperthyroid states on cardiac and β-AR function are also well established (for a review of this subject, see reference 136).

Although less well recognized, post-transcription regulation of β-AR gene expression at the level of mRNA stability is probably a major effector. Similar in nature to the regulatory paradigms of proto-oncogenes and cytokines, the stability of β-AR mRNAs is controlled by signaling pathways affecting the activity of numerous mRNA-binding

proteins with known high affinity for A+U-rich nucleotide sequences or elements (AREs) located within the gene's 3′ untranslated region (3′UTR). Regulation of β-AR expression at the level of mRNA stability is functionally important given that there is a high degree of correlation between β-AR mRNA and protein expression in the human heart (Fig. 20.3).[137]

Previous investigations have demonstrated that increases in cAMP or activation of PKA causes decreases in β-AR mRNA stability,[125–127,138–143] to a degree equivalent to the observed downregulation of β-AR mRNA and protein in the failing human heart. However, this signaling pathway is probably not the dominant effector in modulating β-AR gene expression post-transcriptionally. This role, much as it is for proto-oncogene and cytokine mRNAs, is likely filled by mitogen-activated protein kinases (MAPKs). Specifically, activation of MAPK pathways causes marked stabilization of β-AR mRNA.[143] Reciprocally, inhibition of MAPKs markedly destabilizes β-AR mRNA.[143] Thus, changes in MAPK activity can dynamically and bidirectionally modulate the expression of a large number of genes, including β-ARs, that affect cardiac biology.

Given that in the failing human heart cAMP content is actually diminished, secondary to β-AR downregulation and desensitization,[144] and MAPK (p38 and c-jun N-terminal kinase (JNK)) activities have been shown to be decreased,[145,146] it is entirely possible that the observed downregulation of β-AR mRNA is due to decreased MAPK activity and decreased β-AR mRNA stability, rather than to what has been previously assumed to be the case, i.e. increased β-AR stimulation and the actions of PKA.

A number of mRNA-binding proteins with putative stabilizing or destabilizing effects on β-AR mRNA are likely targets of MAPK phosphorylation, the effects of which presumably include changes in the affinity of the mRNA-binding protein complex to interact with A+U-rich elements. In the context of β-AR mRNAs, the proteins that bind with high affinity to AREs include AUF1/hnRNP D, HuR and hnRNP A1.[147–149] AUF1 has been associated with mRNA destabilization whereas HuR is associated with mRNA

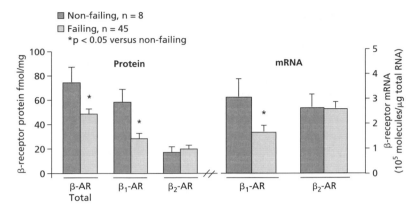

Figure 20.3. *Changes in gene expression associated with dilated cardiomyopathy. β-Adrenergic receptor protein (y-axis, left) and mRNA (y-axis, right) expression versus non-failing controls. Adapted with permission from Lowes BD, Gilbert EM, Abraham WT et al. Myocardial gene expression in dilated cardiomyopathy treated with beta-blocking agents. N Engl J Med 2002; 346: 1357–65. Note, in particular, the close correlation between the downregulation of β₁-AR mRNA and protein abundance. Error bars are standard errors.*

stabilization.[150–156] (150–156). Two other proteins, tristetraprolin (TTP) and butyrate response factor 1 (BRF1), that are associated with increased cytokine mRNA turnover,[156–159] also bind to β-AR AREs and presumably regulate their stability (JD Port and R Tanveer, unpublished data).

The exact role of mRNA stability in regulating adrenergic receptor expression in the context of heart failure remains to be determined. However, given the observation that changes in mRNA turnover and the expression of proteins that affect the rate of mRNA turnover are consistently observed whenever there is a growth, hypertrophic, or inflammatory phenotype,[149,160] it would be very surprising if this mechanism were not of significant overall importance to the heart failure phenotype.

Changes in β-AR signaling as a consequence of heart failure

Two well-established hallmarks of β-AR signaling in heart failure are the subtype-selective downregulation of the β₁-AR and the desensitization of both β₁- and β₂-ARs.[86,87,137,161–176] Undoubtedly, a significant contributor to both of these effects is persistent elevation of NE. However, the effects of NE are by no means limited to β-ARs, as many other downstream components of the β-AR-signaling pathway are also affected. These include changes in the expression of G proteins, most notably increased expression of the inhibitor G protein, $G_{\alpha i}$,[177–181] changes in the abundance of the cardiac isoforms of adenylyl cyclase,[182–185] and increases in expression of the β-adrenergic receptor kinase (βARK1 or GRK2),[186–188] the kinase primarily responsible for agonist-dependent β-AR desensitization and downregulation.

In the normal, non-aged human heart, β₁-ARs represent ~70–80% of the total β-ARs. However, the ratio between β₁- and β₂-ARs approaches a 50/50 distribution in the failing human heart due to the selective downregulation of the β₁-AR subtype (Fig 20.3).[86,165,169,170] There are several potential reasons why the β₁-AR undergoes the observed subtype-selective response. First, as mentioned above, the affinity of NE is greater for the β₁-AR than it is for the β₂-AR; second, the β₁-AR appears to be more closely associated with noradrenergic nerve termini (PSD-95 protein),[89,91,96] presumably exposing them to higher concentrations of NE; lastly, the β₁- and β₂-AR are unique gene products that are regulated differentially at multiple levels.

Downregulation of β-AR in the failing human heart, in conjunction with receptor desensitization, is responsible for a reduced ability of the heart to respond to exogenous stimulation via β-AR pathways.[166,189,190] This is in distinction to a relatively preserved ability of the heart to respond non-selectively to exogenously administered calcium.[88] However, there are other important changes in signaling capacity and/or pathway in addition to simple desensitization. As mentioned above, agonist stimulation of β-ARs leads to their phosphorylation, most prominently by βARKs. For the β_2-AR, this leads not only to desensitization, but also to changes in its signaling pathways. To reiterate, like the β_1-AR, the β_2-AR remains coupled to the $G_{\alpha s}$/AC stimulatory pathway, but unlike the β_1-AR, the β_2-AR, when phosphorylated, increases its ability to couple to the $G_{\alpha i}$/AC inhibitory pathway. β-AR coupling to G_i has consequences beyond inhibition of AC, potentially the most important being linkage to anti-apoptotic signaling pathways and to growth/remodeling-associated pathways.

β_1-Adrenergic mediation of changes in gene expression in the failing heart

One of the remarkable consequences of treatment of chronic heart failure in dilated cardiomyopathy phenotype patients with β-blockers is the 'biological' response of the treatment.[191] That is, in the failing heart the favorable effects of β-blockers on myocardial function and structure are time dependent, and are not obviously manifest until after several months of treatment.[191] After ~3 months of treatment, the therapeutic effects of improved systolic function and reversal of remodeling continue to develop over additional months or even years.[192] The favorable biological response to β-blockers in chronic heart failure is probably due to a partial reversal in the molecular gene profile that characterizes failing, pathological hypertrophied human and rodent ventricles. In the failing human heart, this profile, termed the contractile protein 'fetal gene program', is changed back towards a more normal or adult, non-failing profile by β_1-receptor blockade.[7,193] This implies that the β_1-receptor pathway directly activates fetal gene induction.[194] However, β_1-adrenergic-medi-

ated induction of the fetal gene program, at least in model systems, appears to be cAMP independent,[194] occurring via mechanisms that are as yet not fully understood, but may include ERK1,2 CaMKII, and calcineurin.[194,195]

Summary

During the past few years, significant progress has been made to further our understanding of the complexities of adrenergic receptor signaling and gene regulation as part of both normal homeostasis, and as a consequence of and as a contributor to the pathophysiology of heart failure. Although by no means comprehensive, this chapter has attempted to highlight several key aspects of adrenergic receptor biology in this specific context. Almost certainly, our knowledge of this area will continue to expand at a rapid rate, given the substantial progress being made in the fields of genomics (gene expression and gene variants) and proteomics. Undoubtedly, this new information will help to establish the importance of new regulatory paradigms, ultimately leading to the increased recognition of risk factors and to potential therapeutic targets for the treatment of heart failure.

With advanced (NYHA Class III/IV) heart failure still exhibiting an unacceptably high mortality rate, even in the presence of the best therapeutic agents available (angiotensin-converting enzyme, angiotensin receptor blockers, ARBs, beta-blockers, diuretics, digitalis, etc), it is clear that there is substantial room for improvement. As detailed above, there are many potential novel targets worthy of consideration, not least of which may be pharmacological targets, tailored to individual polymorphic variants of components of the adrenergic signaling cascades.

References

1. Cohn JN, Levine TB, Olivari MT. Plasma norepinephrine as a guide to prognosis in patients with chronic congestive heart failure. N Engl J Med 1984; 311: 819–23.

2. Small KM, McGraw DW, Liggett SB. Pharmacology and physiology of human adrenergic receptor polymorphisms. Annu Rev Pharmacol Toxicol 2003; 43: 381–411.

3. Kirstein SL, Insel PA. Autonomic nervous system pharmacogenomics: a progress report. Pharmacol Rev 2004; 56: 31–52.

4. Bristow MR, Krause-Steinrauf H, Nuzzo R et al. Effect of baseline or changes in adrenergic activity on clinical outcomes in the beta-blocker evaluation trail (BEST). Circulation 2004; 110: 1437–42.

5. Cohn JN, Pfeffer MA, Rouleau J et al. Adverse mortality effect of central sympathetic inhibition with sustained-release moxonidine in patients with heart failure (MOXCON). Eur J Heart Fail 2003; 5: 659–67.

6. Ziegler K. Uber die Wirkung intravenoser adrenalininjektion auf das Gefafssytem und ihre Beziehung zur Arteriosklerose. Bietrage zur Pathologischen Anatonie 1905; 38: 229–54.

7. Lowes BD, Gilbert EM, Abraham WT et al. Myocardial gene expression in dilated cardiomyopathy treated with beta-blocking agents. N Engl J Med 2002; 346: 1357–65.

8. Tan FL, Moravec CS, Li J, Apperson-Hansen C et al. The gene expression fingerprint of human heart failure. Proc Natl Acad Sci USA 2002; 99: 11387–92.

9. Mann DL, Kent RL, Parsons B, Cooper G, 4th Adrenergic effects on the biology of the adult mammalian cardiocyte. Circulation 1992; 85: 790–804.

10. Mann DL. Basic mechanisms of disease progression in the failing heart: the role of excessive adrenergic drive. Prog Cardiovasc Dis 1998; 41(1 Suppl 1): 1–8.

11. Ahlquist R. A study of the adrenotropic receptors. Am J Physiol 1948; 153: 586–600.

12. Lands AM, Arnold A, McAuliff P, Luduena FP, Brown TG, Jr. Differentiation of receptor systems activated by sympathomimetic amines. Nature 1967; 214: 597–8.

13. Cotecchia S, Schwinn D, Randall R, Lefkowitz R, Caron M, Kobilka B. Molecular cloning and expression of the cDNA for the hamster α_1-adrenergic receptor. Proc Natl Acad Sci USA 1988; 85: 7159–63.

14. Dixon R, Kobilka B, Strader C et al. Cloning of the gene and cDNA for mammalian β-adrenergic receptor and homology with rhodopsin. Nature 1986; 321: 75–9.

15. Kobilka B, Dixon R, Frielle T et al. cDNA for the human β_2-adrenergic receptor: a protein with multiple membrane-spanning domains and encoded by a gene whose chromosomal location is shared with that of the receptor for platelet-derived growth factor. Proc Natl Acad Sci USA 1988; 84: 46–50.

16. Frielle T, Collins S, Daniel K, Caron M, Lefkowitz R, Kobilka B. Cloning of the cDNA for the human β1-adrenergic receptor. Proc Natl Acad Sci USA 1987; 84: 7920–4.

17. Regan JW, Kobilka TS, Yang-Feng TL et al. Cloning and expression of a human kidney cDNA for an alpha 2-adrenergic receptor subtype. Proc Natl Acad Sci USA 1988; 85: 6301–5.

18. Angers S, Salahpour A, Bouvier M. Biochemical and biophysical demonstration of GPCR oligomerization in mammalian cells. Life Sci 2001; 68: 2243–50.

19. Small KM, Wagoner LE, Levin AM, Kardia SLR, Liggett SB. Synergistic polymorphisms of β_1- and α_{2C}-adrenergic receptors and the risk of congestive heart failure. N Engl J Med 2002; 347: 1135–42.

20. Bristow MR, Linas S, Port JD. Drugs used in the treatment of heart failure. In: Braunwald E (ed). Heart Disease: a textbook of cardiovascular medicine (7e). Orlando: WB Saunders; 2005: 569–601.

21. Minneman KP, Esbenshade TA. alpha1-Adrenergic receptor subtypes. Ann Rev Pharmacol Toxicol 1994; 34: 117–33.

22. Rohde S, Sabri A, Kamasamudran R, Steinberg SF. The alpha(1)-adrenoceptor subtype- and protein kinase C isoform-dependence of norepinephrine's actions in cardiomyocytes. J Mol Cell Cardiol 2000; 32: 1193–209.

23. Simpson PC, Kariya K, Karns LR, Long CS, Karliner JS. Adrenergic hormones and control of cardiac myocyte growth. Mol Cell Biochem 1991; 104: 35–43.

24. Simpson P, McGrath A, Savion S. Myocyte hypertrophy in neonatal rat heart cultures and its regulation by serum and by catecholamines. Circ Res 1982; 51: 787–801.

25. Simpson P. Norepinephrine-stimulated hypertrophy of cultured rat myocardial cells is an alpha 1 adrenergic response. J Clin Invest 1983; 72: 732–8.

26. Simpson P. Stimulation of hypertrophy of cultured neonatal rat heart cells through an alpha 1-adrenergic receptor and induction of beating through an alpha 1- and beta 1-adrenergic receptor interaction. Evidence for independent regulation of growth and beating. Circ Res 1985; 56: 884–94.

27. Marks AR. Ryanodine receptors/calcium release channels in heart failure and sudden cardiac death. J Mol Cell Cardiol 2001; 33: 615–24.

28. Cheng H, Lederer MR, Xiao RP et al. Excitation–contraction coupling in heart: new

insights from Ca^{2+} sparks. Cell Calcium 1996; 20: 129–40.

29. Marks AR, Marx SO, Reiken S. Regulation of ryanodine receptors via macromolecular complexes: a novel role for leucine/isoleucine zippers. Trends Cardiovasc Med 2002; 12: 166–70.

30. Bristow MR, Minobe W, Rasmussen R, Hershberger RE, Hoffman BB. Alpha-1 adrenergic receptors in the nonfailing and failing human heart. J Pharmacol Exp Ther 1988; 247: 1039–45.

31. Brodde O-E, Michel MC. Adrenergic and muscarinic receptors in the human heart. Pharmacol Rev 1999; 51: 651–90.

32. Small KM, Forbes SL, Rahman FF, Bridges KM, Liggett SB. A four amino acid deletion polymorphism in the third intracellular loop of the human alpha 2C-adrenergic receptor confers impaired coupling to multiple effectors. J Biol Chem 2000; 275: 23059–64.

33. Emorine LJ, Marullo S, Briend-Sutren M-M et al. Molecular characterization of the human beta-3-adrenerigc receptor. Science 1989; 245: 1118–21.

34. Gauthier C, Tavernier G, Charpentier F, Langin D, Le Marec H. Functional beta3-adrenoceptor in the human heart. J Clin Invest 1996; 98: 556–62.

35. Gauthier C, Leblais V, Kobzik L et al. The negative inotropic effect of beta3-adrenoceptor stimulation is mediated by activation of a nitric oxide synthase pathway in human ventricle. J Clin Invest 1998; 102: 1377–84.

36. Lader AS, Xiao YF, Ishikawa Y et al. Cardiac G$_{salpha}$ overexpression enhances L-type calcium channels through an adenylyl cyclase independent pathway. Proc Natl Acad Sci USA 1998; 95: 9669–74.

37. Zhu W-Z, Wang S-Q, Chakir K et al. Linkage of β_1-adrenergic stimulation to apoptotic heart cell death through protein kinase A-independent activation of Ca^{2+}/calmodulin kinase II. J Clin Invest 2003; 111: 617–25.

38. Hu LA, Chen W, Martin NP et al. GIPC interacts with the β_1-adrenergic receptor and regulates β_1-adrenergic receptor-mediated ERK activation. J Biol Chem 2003; 278: 26295–301.

39. Communal C, Colucci WS, Singh K. p38 Mitogen activated protein kinase pathway protects adult rat ventricular myocytes against β adrenergic receptor stimulated apoptosis: evidence for G$_i$ dependent activation. J Biol Chem 2000; 275: 19395–400.

40. Zheng M, Zhang SJ, Zhu WZ et al. Beta 2-adrenergic receptor-induced p38 MAPK activation is mediated by PKA rather than by Gi or Gbeta gamma in adult mouse cardiomyocytes. J Biol Chem 2000; 275: 40635–40.

41. Chesley A, Lundberg MS, Asai T et al. The β_2-adrenergic receptor delivers an antiapoptotic signal to cardiac myocytes through G$_i$-dependent coupling to phosphatidylinositol 3′-kinase. Circ Res 2000; 87: 1172–9.

42. Zheng M, Zhang SJ, Zhu WZ et al. Beta 2-adrenergic receptor-induced p38 MAPK activation is mediated by protein kinase A rather than by Gi or Gbeta gamma in adult mouse cardiomyocytes. J Biol Chem 2000; 275: 40635–40.

43. Chesley A, Lundberg MS, Asai T et al. The beta(2)-adrenergic receptor delivers an antiapoptotic signal to cardiac myocytes through G(i)-dependent coupling to phosphatidylinositol 3′-kinase. Circ Res 2000; 87: 1172–9.

44. Yamauchi J, Hirasawa A, Miyamoto Y, Itoh H, Tsujimoto G. Beta2-adrenergic receptor/cyclic adenosine monophosphate (cAMP) leads to JNK activation through Rho family small GTPases. Biochem Biophys Res Commun 2001; 284: 1199–203.

45. Lavoie C, Mercier JF, Salahpour A et al. Beta 1/beta 2-adrenergic receptor heterodimerization regulates beta 2-adrenergic receptor internalization and ERK signaling efficacy. J Biol Chem 2002; 277: 35402–10.

46. Bristow MR. Beta-adrenergic receptor blockade in chronic heart failure. Circulation 2000; 101: 558–69.

47. Bristow MR, Port JD, Kelly RA. Treatment of heart failure: pharmacological methods. In: Braunwald E, Zipes DP, Libby P (eds). Heart Disease: a textbook of cardiovascular medicine Seventh ed. Orlando: WB Saunders; 2001.

48. Bristow MR, Feldman AM, Adams KF, Jr. Goldstein S. Selective versus nonselective beta-blockade for heart failure therapy: are there lessons to be learned from the COMET trial? J Card Fail 2003; 9: 444–53.

49. Greenberg B. Nonselective versus selective beta-blockers in the management of chronic heart failure: clinical implications of the carvedilol or Metoprolol European Trial. Rev Cardiovasc Med 2004; 5 (Suppl 1): S10–17.

50. Remmers U, Fink C, von Harsdorf R et al. Intrinsic receptor activity of β1-adrenoceptors overexpressed in rat ventricular cardiac myocytes. Circulation 1998; 98: I-123.

51. Zhang SJ, Cheng H, Zhou YY et al. Inhibition of spontaneous beta 2-adrenergic activation rescues beta 1- adrenergic contractile response in cardiomyocytes overexpressing beta 2-adrenoceptor. J Biol Chem 2000; 275: 21773–9.

52. Zhou YY, Yang D, Zhu WZ et al. Spontaneous activation of beta(2)- but not beta(1)-adrenoceptors

expressed in cardiac myocytes from beta(1)beta(2) double knockout mice. Mol Pharmacol 2000; 58: 887–94.

53. Zhou YY, Song LS, Lakatta EG, Xiao RP, Cheng H. Constitutive beta2-adrenergic signalling enhances sarcoplasmic reticulum Ca^{2+} cycling to augment contraction in mouse heart. J Physiol 1999; 521: 351–61.

54. Zhou YY, Cheng H, Song LS, Wang D, Lakatta EG, Xiao RP. Spontaneous beta(2)-adrenergic signaling fails to modulate L-type Ca^{2+} current in mouse ventricular myocytes. Mol Pharmacol 1999; 56: 485–93.

55. Engelhardt S, Grimmer Y, Fan G-H, Lohse MJ. Constitutive activity of the human beta 1-adrenergic receptor in beta 1-receptor transgenic mice. Mol Pharmacol 2001; 60: 712–17.

56. Robishaw JD, Berlot CH. Translating G protein subunit diversity into functional specificity. Curr Opin Cell Biol 2004; 16: 206–9.

57. Naga Prasad SV, Esposito G, Mao L, Koch WJ, Rockman HA. Gbetagamma-dependent phosphoinositide 3-kinase activation in hearts with in vivo pressure overload hypertrophy. J Biol Chem 2000; 275: 4693–8.

58. Naga Prasad SV, Laporte SA, Chamberlain D et al. Phosphoinositide 3-kinase regulates β2-adrenergic receptor endocytosis by AP-2 recruitment to the receptor/β-arrestin complex. J Cell Biol 2002; 158: 563–75.

59. Naga Prasad SV, Barak LS, Rapacciuolo A, Caron MG, Rockman HA. Agonist-dependent recruitment of phosphoinositide 3-kinase to the membrane by beta-adrenergic receptor kinase 1. A role in receptor sequestration. J Biol Chem 2001; 276: 18953–9.

60. Hausdorff W, Caron M, Lefkowitz R. Turning off the signal: desensitization of β-adrenergic receptor function. FASEB J 1990; 4: 2881–9.

61. Morris AJ, Malbon CC. Physiological regulation of G protein-linked signaling. Physiol Rev 1999; 79: 1373–430.

62. Rybin VO, Xu X, Lisanti MP, Steinberg SF. Differential targeting of beta-adrenergic receptor subtypes and adenylyl cyclase to cardiomyocyte caveolae. A mechanism to functionally regulate the cAMP signaling pathway. J Biol Chem 2000; 275: 41447–57.

63. Koch WJ, Hawes BE, Inglese J, Luttrell LM, Lefkowitz RJ. Cellular expression of the carboxyl terminus of a G protein-coupled receptor kinase attenuates G beta gamma-mediated signaling. J Biol Chem 1994; 269: 6193–7.

64. Koch WJ, Hawes BE, Allen LF, Lefkowitz RJ. Direct evidence that Gi-coupled receptor stimulation of mitogen-activated protein kinase is mediated by G beta gamma activation of p21ras. Proc Natl Acad Sci USA 1994; 91: 12706–10.

65. Luttrell LM, van Biesen T, Hawes BE et al. G beta gamma subunits mediate mitogen-activated protein kinase activation by the tyrosine kinase insulin-like growth factor 1 receptor. J Biol Chem 1995; 270: 16495–8.

66. Luttrell LM, van Biesen T, Hawes BE et al. G-protein-coupled receptors and their regulation: activation of the MAP kinase signaling pathway by G-protein-coupled receptors. Adv Second Messenger Phosphoprotein Res 1997; 31: 263–77.

67. El-Armouche A, Pamminger T, Ditz D, Zolk O, Eschenhagen T. Decreased protein and phosphorylation level of the protein phosphatase inhibitor-1 in failing human hearts. Cardiovascular Research 2004; 61: 87–93.

68. Regitz-Zagrosek V, Hertrampf R, Steffen C, Hildebrandt A, Fleck E. Myocardial cyclic AMP and norepinephrine content in human heart failure. Eur Heart J 1994; 15 (Suppl D): 7–13.

69. Kirchhefer U, Schmitz W, Scholz H, Neumann J. Activity of cAMP-dependent protein kinase and Ca^{2+}/calmodulin-dependent protein kinase in failing and nonfailing human hearts. Cardiovasc Res 1999; 42: 254–61.

70. Schwinger RH, Munch G, Bolck B et al. Reduced Ca^{2+}-sensitivity of SERCA 2a in failing human myocardium due to reduced serine-16 phospholamban phosphorylation. J Mol Cell Cardiol 1999; 31: 479 91.

71. Luo W, Grupp IL, Harrer J et al. Targeted ablation of the phospholamban gene is associated with markedly enhanced myocardial contractility and loss of β-agonist stimulation. Circ Res 1994; 75: 401–9.

72. Koss KL, Kranias EG. Phospholamban: a prominent regulator of myocardial contractility. Circ Res 1996; 79: 1059–63.

73. Ward DG, Brewer SM, Gallon CE et al. NMR and mutagenesis studies on the phosphorylation region of human cardiac troponin I. Biochemistry 2004; 43: 5772–81.

74. Kajiwara H, Morimoto S, Fukuda N, Ohtsuki I, Kurihara S. Effect of troponin I phosphorylation by protein kinase A on length-dependence of tension activation in skinned cardiac muscle fibers. Biochem Biophys Res Commun 2000; 272: 104–10.

75. Marx SO, Reiken S, Hisamatsu Y et al. PKA phosphorylation dissociates FKBP12.6 from the calcium

release channel (ryanodine receptor): defective regulation in failing hearts. Cell 2000; 101: 365–76.

76. Reiken S, Gaburjakova M, Guatimosim S et al. Protein kinase A phosphorylation of the cardiac calcium release channel (ryanodine receptor) in normal and failing hearts. Role of phosphatases and response to isoproterenol. J Biol Chem 2002; 278: 444–53.

77. Fraser ID, Cong M, Kim J et al. Assembly of an A kinase-anchoring protein-beta(2)-adrenergic receptor complex facilitates receptor phosphorylation and signaling. Curr Biol 2000; 10: 409–12.

78. Gao T, Yatani A, Dell'Acqua ML et al. cAMP-dependent regulation of cardiac L-type Ca^{2+} channels requires membrane targeting of PKA and phosphorylation of channel subunits. Neuron 1997; 19: 185–96.

79. Shih M, Lin F, Scott JD, Wang HY, Malbon CC. Dynamic complexes of beta2-adrenergic receptors with protein kinases and phosphatases and the role of gravin. J Biol Chem 1999; 274: 1588–95.

80. Malbon CC, Tao J, Wang HY. AKAPs (A-kinase anchoring proteins) and molecules that compose their G-protein-coupled receptor signalling complexes. Biochem J 2004; 379: 1–9.

81. Tao J, Wang HY, Malbon CC. Protein kinase A regulates AKAP250 (gravin) scaffold binding to the beta2-adrenergic receptor. EMBO J 2003; 22: 6419–29.

82. Mongillo M, McSorley T, Evellin S et al. Fluorescence resonance energy transfer-based analysis of cAMP dynamics in live neonatal rat cardiac myocytes reveals distinct functions of compartmentalized phosphodiesterases. Circ Res 2004; 95: 67–75.

83. Perry SJ, Baillie GS, Kohout TA et al. Targeting of cyclic AMP degradation to beta 2-adrenergic receptors by beta-arrestins. Science 2002; 298: 834–6.

84. Steinberg SF, Brunton LL. Compartmentation of g protein-coupled signaling pathways in cardiac myocytes. Annu Rev Pharmacol Toxicol 2001; 41: 751–73.

85. Zaccolo M, Pozzan T. Discrete microdomains with high concentration of cAMP in stimulated rat neonatal cardiac myocytes. Science 2002; 295: 1711–15.

86. Bristow MR, Ginsburg R, Umans V et al. Beta 1- and beta 2-adrenergic-receptor subpopulations in non-failing and failing human ventricular myocardium: coupling of both receptor subtypes to muscle contraction and selective beta 1-receptor down-regulation in heart failure. Circ Res 1986; 59: 297–309.

87. Bristow MR, Hershberger RE, Port JD, Minobe W,

Rasmussen R. Beta 1- and beta 2-adrenergic receptor-mediated adenylate cyclase stimulation in non-failing and failing human ventricular myocardium. Mol Pharmacol 1989; 35: 295–303.

88. Fowler MB, Laser JA, Hopkins GL, Minobe W, Bristow MR. Assessment of the beta-adrenergic receptor pathway in the intact failing human heart: progressive receptor down-regulation and subsensitivity to agonist response. Circulation 1986; 74: 1290–302.

89. Hu LA, Tang Y, Miller WE et al. Beta 1-adrenergic receptor association with PSD-95. Inhibition of receptor internalization and facilitation of beta 1-adrenergic receptor interaction with N-methyl-D-aspartate receptors. J Biol Chem 2000; 275: 38659–66.

90. Hall RA, Lefkowitz RJ. Regulation of G protein-coupled receptor signaling by scaffold proteins. Circ Res 2002; 91: 672–80.

91. Hu LA, Chen W, Premont RT, Cong M, Lefkowitz RJ. G Protein-coupled receptor kinase 5 regulates beta 1-adrenergic receptor association with PSD-95. J Biol Chem 2002; 277: 1607–13.

92. Rapacciuolo A, Suvarna S, Barki-Harrington L et al. Protein kinase A and G protein-coupled receptor kinase phosphorylation mediates β-1 adrenergic receptor endocytosis through different pathways. J Biol Chem 2003; 278: 35403–11.

93. Xiang Y, Rybin VO, Steinberg SF, Kobilka B. Caveolar localization dictates physiologic signaling of beta 2-adrenoceptors in neonatal cardiac myocytes. J Biol Chem 2002; 277: 34280–6.

94. Ostrom RS, Liu X, Head BP et al. Localization of adenylyl cyclase isoforms and G protein-coupled receptors in vascular smooth muscle cells: expression in caveolin-rich and noncaveolin domains. Mol Pharmacol 2002; 62: 983–92.

95. Tang Y, Hu LA, Miller WE et al. Identification of the endophilins (SH3p4/p8/p13) as novel binding partners for the beta1-adrenergic receptor. Proc Natl Acad Sci USA 1999; 96: 12559–64.

96. Xu J, Paquet M, Lau AG et al. Beta 1-adrenergic receptor association with the synaptic scaffolding protein membrane-associated guanylate kinase inverted-2 (MAGI-2). Differential regulation of receptor internalization by MAGI-2 and PSD-95. J Biol Chem 2001; 276: 41310–17.

97. Pak Y, Pham N, Rotin D. Direct binding of the β1 adrenergic receptor to the cyclic amp-dependent guanine nucleotide exchange factor CNrasGEF leads to Ras activation. Mol Cell Biol 2002; 22: 7942–52.

98. Xiang Y, Devic E, Kobilka B. The PDZ binding motif

of the beta 1 adrenergic receptor modulates receptor trafficking and signaling in cardiac myocytes. J Biol Chem 2002; 277: 33783–90.

99. Xiang Y, Kobilka B. The PDZ-binding motif of the β_2-adrenoceptor is essential for physiologic signaling and trafficking in cardiac myocytes. Proc Natl Acad Sci USA 2003: 16: 10776–81.

100. Xiao RP, Avdonin P, Zhou YY et al. Coupling of beta2-adrenoceptor to Gi proteins and its physiological relevance in murine cardiac myocytes. Circ Res 1999; 84: 43–52.

101. Wenzel-Seifert K, Seifert R. Molecular analysis of beta(2)-adrenoceptor coupling to G(s)-, G(i)-, and G(q)-proteins. Mol Pharmacol 2000; 58: 954–66.

102. Kilts JD, Gerhardt MA, Richardson MD et al. Beta(2)-adrenergic and several other G protein-coupled receptors in human atrial membranes activate both G(s) and G(i). Circ Res 2000; 87: 705–9.

103. Hall RA, Premont RT, Chow CW et al. The beta2-adrenergic receptor interacts with the Na$^+$/H$^+$-exchanger regulatory factor to control Na$^+$/H$^+$ exchange. Nature 1998; 392: 626–30.

104. Cao TT, Deacon HW, Reczek D, Bretscher A, von Zastrow M. A kinase-regulated PDZ-domain interaction controls endocytic sorting of the beta2-adrenergic receptor. Nature 1999; 401: 286–90.

105. Cong M, Perry SJ, Hu LA, Hanson PI, Claing A, Lefkowitz RJ. Binding of the beta2 adrenergic receptor to N-ethylmaleimide-sensitive factor regulates receptor recycling. J Biol Chem 2001; 276: 45145–52.

106. Crackower MA, Oudit GY, Kozieradzki I et al. Regulation of myocardial contractility and cell size by distinct PI3K–PTEN signaling pathways. Cell 2002; 110: 737–49.

107. Bisognano JD, Weinberger HD, Bohlmeyer TJ et al. Myocardial-directed overexpression of the human beta(1)-adrenergic receptor in transgenic mice. J Mol Cell Cardiol 2000; 32: 817–30.

108. Communal C, Singh K, Sawyer DB, Colucci WS. Opposing effects of beta(1)- and beta(2)-adrenergic receptors on cardiac myocyte apoptosis: role of a pertussis toxin-sensitive G protein. Circulation 1999; 100: 2210–12.

109. Communal C, Singh K, Colucci W. Gi protein protects adult rat ventricular myocytes from β-adrenergic receptor-stimulated apoptosis in vitro. Circulation 1998; 98: I-742.

110. Communal C, Singh K, Pimentel DR, Colucci WS. Norepinephrine stimulates apoptosis in adult rat ventricular myocytes by activation of the beta-adrenergic pathway. Circulation 1998; 98: 1329–34.

111. Communal C, Singh K, Pimental DR, Colucci WS. The β-adrenergic pathway mediates norepinephrine-stimulated apoptosis in vitro in adult rat cardiac myocytes. Circulation 1997; 96: I-117.

112. Andre C, Couton D, Gaston J et al. Beta2-adrenergic receptor-selective agonist clenbuterol prevents Fas-induced liver apoptosis and death in mice. Am J Physiol Gastrointest Liver Physiol 1999; 276: G647–54.

113. Geng YJ, Ishikawa Y, Vatner DE et al. Apoptosis of cardiac myocytes in Gsalpha transgenic mice. Circ Res 1999; 84: 34–42.

114. Geng Y-J, Homcey CJ, Kim S-J et al. Persistent stimulation of β-adrenergic signaling triggers degeneration and apoptosis of cardiomyocytes. Circulation 1997; 96: I-116.

115. Gu C, Ma YC, Benjamin J, Littman D, Chao MV, Huang XY. Apoptotic signaling through the beta-adrenergic receptor. A new Gs effector pathway. J Biol Chem 2000; 275: 20726–33.

116. Zaugg M, Xu W, Lucchinetti E et al. Beta-adrenergic receptor subtypes differentially affect apoptosis in adult rat ventricular myocytes. Circulation 2000; 102: 344–50.

117. Saito S, Hiroi Y, Zou Y et al. Beta-adrenergic pathway induces apoptosis through calcineurin activation in cardiac myocytes. J Biol Chem 2000; 275: 34528–33.

118. Zamah AM, Delahunty M, Luttrell LM, Lefkowitz RJ. Protein kinase A-mediated phosphorylation of the beta 2-adrenergic receptor regulates its coupling to Gs and Gi. Demonstration in a reconstituted system. J. Biol. Chem. 2002; 277: 31249–56.

119. Baillie GS, Sood A, McPhee I et al. Beta-arrestin-mediated PDE4 cAMP phosphodiesterase recruitment regulates beta-adrenoceptor switching from Gs to Gi. Proc Natl Acad Sci USA 2003; 100: 940–5.

120. Zhu W-Z, Zheng M, Koch WJ et al. Dual modulation of cell survival and cell death by beta 2-adrenergic signaling in adult mouse cardiac myocytes. Proc Natl Acad Sci USA 2001; 98: 1607–12.

121. Collins S, Bouvier M, Bolanowski MA, Caron MG, Lefkowitz RJ. cAMP stimulates transcription of the beta 2-adrenergic receptor gene in response to short-term agonist exposure. Proc Natl Acad Sci USA 1989; 86: 4853–7.

122. Collins S, Altschmied J, Herbsman O et al. A cAMP response element in the beta 2-adrenergic receptor gene confers transcriptional autoregulation by cAMP. J Biol Chem 1990; 265: 19330–5.

123. Hadcock JR, Malbon CC. Down-regulation of β-adrenergic receptors: agonist-induced reduction in

receptor mRNA levels. Proc Natl Acad Sci USA 1988; 85: 5021–5.

124. Hadcock JR, Wang H-Y, Malbon CC. Agonist-induced destabilization of β-adrenergic receptor mRNA. Attenuation of glucocorticoid-induced up-regulation of β-adrenergic receptors. J Biol Chem 1989; 264: 19928–33.

125. Danner S, Frank M, Lohse MJ. Agonist regulation of human β₂-adrenergic receptor mRNA stability occurs via a specific AU-rich element. J Biol Chem 1998; 273: 3223–9.

126. Danner S, Lohse MJ. Cell type-specific regulation of β2-adrenoceptor mRNA by agonists. Eur J Pharmacol 1997; 331: 73–8.

127. Kirigiti P, Bai Y, Yang YF et al. Agonist-mediated down-regulation of rat beta1-adrenergic receptor transcripts: role of potential post-transcriptional degradation factors. Mol Pharmacol 2001; 60: 1308–24.

128. Fitzgerald LR, Li Z, Machida CA, Fishman PH, Duman RS. Adrenergic regulation of ICER (inducible cyclic AMP early repressor) and β1-adrenergic receptor gene expression in C6 glioma cells. J Neurochem 1996; 67: 490–7.

129. Tomita H, Nazmy M, Kajimoto K et al. Inducible cAMP early repressor (ICER) is a negative-feedback regulator of cardiac hypertrophy and an important mediator of cardiac myocyte apoptosis in response to beta-adrenergic receptor stimulation. Circ Res 2003; 93: 12–22.

130. Hadcock JR, Malbon CC. Regulation of β-adrenergic receptors by 'permissive' hormones: glucocorticoids increase steady-state levels of receptor mRNA. Proc Natl Acad Sci USA 1988; 85: 8415–19.

131. Bahouth SW, Park EA, Beauchamp M, Cui X, Malbon CC. Identification of a glucocorticoid repressor domain in the rat beta 1-adrenergic receptor gene. Recept Signal Transduct 1996; 6: 141–9.

132. Bahouth SW, Cui X, Beauchamp MJ, Park EA. Thyroid hormone induces beta1-adrenergic receptor gene transcription through a direct repeat separated by five nucleotides. J Mol Cell Cardiol 1997; 29: 3223–37.

133. Kiely J, Hadcock JR, Bahouth SW, Malbon CC. Glucocorticoids down-regulate beta 1-adrenergic-receptor expression by suppressing transcription of the receptor gene. Biochem J 1994; 302: 397–403.

134. Norris JS, Brown P, Cohen J et al. Glucocorticoid induction of beta-adrenergic receptors in the DDT1 MF-2 smooth muscle cell line involves synthesis of new receptor. Mol Cell Biochem 1987; 74: 21–7.

135. Cornett LE, Hiller FC, Jacobi SE, Cao W, McGraw DW. Identification of a glucocorticoid response element in the rat beta2- adrenergic receptor gene. Mol Pharmacol 1998; 54: 1016–23.

136. Dillmann WH. Cellular action of thyroid hormone on the heart. Thyroid 2002; 12: 447–52.

137. Bristow M, Minobe W, Raynolds M et al. Reduced β1 receptor messenger RNA abundance in the failing human heart. J Clin Invest 1993; 92: 2737–45.

138. Huang L-Y, Tholanikunnel BG, Vakalopoulou E, Malbon CC. The Mr 35 000 β-adrenergic receptor mRNA-binding protein induced by agonists requires both an AUUUA pentamer and U-rich domains for RNA recognition. J Biol Chem 1993; 268: 25769–75.

139. Mitchusson KD, Blaxall BC, Pende A, Port JD. Agonist-mediated destabilization of human beta1-adrenergic receptor mRNA: role of the 3' untranslated translated region. Biochem Biophys Res Commun 1998; 252: 357–62.

140. Tholanikunnel BG, Granneman JG, Malbon CC. The Mr 35,000 β-adrenergic receptor mRNA-binding protein binds transcripts of G-protein-linked receptors which undergo agonist-induced destabilization. J Biol Chem 1995; 270: 12787–93.

141. Tholanikunnel BG, Malbon CC. A 20-nucleotide (A+U)-rich element of the β2-adrenergic receptor (β2-AR) mRNA mediates binding to the β2-AR-binding protein and is obligate for agonist-induced destabilization of receptor mRNA. J Biol Chem 1997; 272: 11471–8.

142. Tholanikunnel BG, Raymond JR, Malbon CC. Analysis of the AU-rich elements in the 3'-untranslated region of beta 2-adrenergic receptor mRNA by mutagenesis and identification of the homologous AU-rich region from different species. Biochemistry 1999; 38: 15564–72.

143. Headley V, Tanveer R, Greene SM, Zweifach A, Port JD. Reciprocal regulation of beta-adrenergic receptor mRNA stability by mitogen activated protein kinase activation and inhibition. Mol Cell Biochem 2004; 285: 109–19.

144. Bohm M, Reiger B, Schwinger RH, Erdmann E. cAMP concentrations, cAMP dependent protein kinase activity, and phospholamban in non-failing and failing myocardium. Cardiovasc Res 1994; 28: 1713–19.

145. Communal C, Colucci WS, Remondino A et al. Reciprocal modulation of mitogen-activated protein kinases and mitogen-activated protein kinase phosphatase 1 and 2 in failing human myocardium. J Card Fail 2002; 8: 86–92.

146. Lemke LE, Bloem LJ, Fouts R et al. Decreased p38 MAPK activity in end-stage failing human

myocardium: p38 MAPK alpha is the predominant isoform expressed in human heart. J Mol Cell Cardiol 2001; 33: 1527–40.

147. Blaxall BC, Pellett AC, Wu SC, Pende A, Port JD. Purification and characterization of beta-adrenergic receptor mRNA-binding proteins. J Biol Chem 2000; 275: 4290–7.

148. Blaxall BC, Pende A, Wu SC, Port JD. Correlation between intrinsic mRNA stability and the affinity of AUF1 (hnRNP D) and HuR for A+U-rich mRNAs. Mol Cell Biochem 2002; 232: 1–11.

149. Pende A, Tremmel KD, DeMaria CT et al. Regulation of the mRNA-binding protein AUF1 by activation of the β-adrenergic receptor signal transduction pathway. J Biol Chem 1996; 271: 8493–501.

150. Zhang W, Wagner BJ, Ehrenman K et al. Purification, characterization, and cDNA cloning of an AU-rich element RNA-binding protein, AUF1. Mol Cell Biol 1993; 13: 7652–65.

151. Sirenko OI, Lofquist AK, DeMaria CT et al. Adhesion-dependent regulation of an A+U-rich element-binding activity associated with AUF1. Mol Cell Biol 1997; 17: 3898–906.

152. Sarkar B, Xi Q, He C, Schneider RJ. Selective degradation of AU-rich mRNAs promoted by the p37 AUF1 Protein Isoform. Mol. Cell. Biol. 2003; 23(18): 6685–93.

153. Brennan CM, Steitz JA. HuR and mRNA stability Cell Mol Life Sci 2001; 58: 266–77.

154. Fan XC, Steitz JA. Overexpression of HuR, a nuclear-cytoplasmic shuttling protein, increases the in vivo stability of ARE-containing mRNAs. EMBO J 1998; 17: 3448–60.

155. Figueroa A, Cuadrado A, Fan J et al. Role of HuR in skeletal myogenesis through coordinate regulation of muscle differentiation genes. Mol Cell Biol 2003; 23: 4991–5004.

156. Raineri I, Wegmueller D, Gross B, Certa U, Moroni C. Roles of AUF1 isoforms, HuR and BRF1 in ARE-dependent mRNA turnover studied by RNA interference. Nucl Acids Res 2004; 32: 1279–88.

157. Baumgarten G, Knuefermann P, Kalra D et al. Load-dependent and -independent regulation of pro-inflammatory cytokine and cytokine receptor gene expression in the adult mammalian heart. Circulation 2002; 105: 2192–7.

158. Carballo E, Lai WS, Blackshear PJ. Evidence that tristetraprolin is a physiological regulator of granulocyte-macrophage colony-stimulating factor messenger RNA deadenylation and stability. Blood 2000; 95: 1891–9.

159. Stoecklin G, Colombi M, Raineri I, Leuenberger S, Mallaun M, Schmidlin M et al. Functional cloning of BRF1, a regulator of ARE-dependent mRNA turnover. EMBO J 2002; 21: 4709–18.

160. Blaxall BC, Dwyer-Nield LD, Bauer AK et al. Differential expression and localization of the mRNA binding proteins, AU-rich element mRNA binding protein (AUF1) and Hu antigen R (HuR), in neoplastic lung tissue. Mol Carcinog 2000; 28: 76–83.

161. Bohm M, Lohse MJ. Quantification of beta-adrenoceptors and beta-adrenoceptor kinase on protein and mRNA levels in heart failure. Eur Heart J 1994; 15 (Suppl D): 30–4.

162. Bristow M, Durham C, Klien J et al. Down-regulation of β-adrenergic receptors and receptor mRNA in heart cells chronically exposed to norepinephrine. Clin Research 1991; 39: 256A.

163. Bristow MR. Myocardial beta-adrenergic receptor downregulation in heart failure. Int J Cardiol 1984; 5: 648–52.

164. Bristow MR, Anderson FL, Port JD et al. Differences in beta-adrenergic neuroeffector mechanisms in ischemic versus idiopathic dilated cardiomyopathy. Circulation 1991; 84: 1024–39.

165. Bristow MR, Ginsburg R, Gilbert EM, Hershberger RE. Heterogeneous regulatory changes in cell surface membrane receptors coupled to a positive inotropic response in the failing human heart. Basic Res Cardiol 1987; 82 (Suppl 2): 369–76.

166. Bristow MR, Ginsburg R, Minobe W et al. Decreased catecholamine sensitivity and beta-adrenergic-receptor density in failing human hearts. N Engl J Med 1982; 307: 205–11.

167. Bristow MR, Hershberger RE, Port JD et al. Beta-adrenergic pathways in nonfailing and failing human ventricular myocardium. Circulation 1990; 82 (2 Suppl): I12–25.

168. Brodde OE. Beta- and alpha-adrenoceptor-agonists and -antagonists in chronic heart failure. Basic Res Cardiol 1990; 85 (Suppl 1): 57–66.

169. Brodde OE. The functional importance of beta 1 and beta 2 adrenoceptors in the human heart. Am J Cardiol 1988; 62: 24C–29C.

170. Brodde OE. Beta 1-and beta 2-adrenoceptors in the human heart: properties, function, and alterations in chronic heart failure. Pharmacol Rev 1991; 43: 203–42.

171. Brodde OE. Beta-adrenoceptors in cardiac disease. Pharmacol Ther 1993; 60: 405–30.

172. Brodde OE, Kretsch R, Ikezono K, Zerkowski HR, Reidemeister JC. Human beta-adrenoceptors: relation of myocardial and lymphocyte beta-adrenoceptor density. Science 1986; 231: 1584–5.

173. Brodde OE, O'Hara N, Zerkowski HR, Rohm N. Human cardiac beta-adrenoceptors: both beta 1- and beta 2-adrenoceptors are functionally coupled to the adenylate cyclase in right atrium. J Cardiovasc Pharmacol 1984; 6: 1184–91.

174. Brodde OE, Schuler S, Kretsch R et al. Regional distribution of beta-adrenoceptors in the human heart: coexistence of functional beta 1- and beta 2-adrenoceptors in both atria and ventricles in severe congestive cardiomyopathy. J Cardiovasc Pharmacol 1986; 8: 1235–42.

175. Brodde OE, Zerkowski HR, Doetsch N et al. Myocardial beta-adrenoceptor changes in heart failure: concomitant reduction in beta 1- and beta 2-adrenoceptor function related to the degree of heart failure in patients with mitral valve disease. J Am Coll Cardiol 1989; 14: 323–31.

176. Murphree SS, Saffitz JE. Distribution of beta-adrenergic receptors in failing human myocardium. Implications for mechanisms of down-regulation. Circulation 1989; 79: 1214–25.

177. Feldman A, Cates A, Bristow M, Van Dop C. Altered expression of the α-subunit of G proteins in failing human heart. J Mol Cell Cardiol 1989; 21: 359–65.

178. Feldman AM, Cates AE, Veazey WB et al. Increase of the 40,000-mol wt pertussis toxin substrate (G protein) in the failing human heart. J Clin Invest 1988; 82: 189–97.

179. Bohm M, Gierschik P, Erdmann E. Quantification of Gi-proteins in the failing and nonfailing human myocardium. Basic Res Cardiol 1992; 87 (Suppl 1): 37–50.

180. Bohm M, Gierschik P, Jakobs KH et al. Increase of Gi alpha in human hearts with dilated but not ischemic cardiomyopathy. Circulation 1990; 82: 1249–65.

181. Ransnas LA. The role of G-proteins in transduction of the beta-adrenergic response in heart failure. Heart Vessels Suppl 1991; 6: 3–5.

182. Ping P, Anzai T, Gao M, Hammond HK. Adenylyl cyclase and G protein receptor kinase expression during development of heart failure. Am J Physiol Heart Circ Physiol 1997; 273: H707–17.

183. Espinasse I, Iourgenko V, Richer C et al. Decreased type VI adenylyl cyclase mRNA concentration and Mg^{2+}-dependent adenylyl cyclase activities and unchanged type V adenylyl cyclase mRNA concentration and Mn^{2+}-dependent adenylyl cyclase activities in the left ventricle of rats with myocardial infarction and longstanding heart failure. Cardiovasc Res 1999; 42: 87–98.

184. Holmer SR, Eschenhagen T, Nose M, Riegger GA. Expression of adenylyl cyclase and G-protein beta subunit in end-stage human heart failure. J Card Fail 1996; 2: 279–83.

185. Wang X, Sentex E, Chapman D, Dhalla NS. Alterations of adenylyl cyclase and G proteins in aortocaval shunt-induced heart failure. Am J Physiol Heart Circ Physiol 2004; 287: H118–25.

186. Ungerer M, Parutti G, Bohm M et al. Expression of β-arrestins and β-adrenergic receptor kinases in the failing human heart. Circ Res 1994; 74: 206–13.

187. Ungerer M, Bohm M, Elce JS, Erdmann E, Lohse MJ. Altered expression of beta-adrenergic receptor kinase and beta 1-adrenergic receptors in the failing human heart. Circulation 1993; 87: 454–63.

188. Benovic JL, Strasser RH, Caron MG, Lefkowitz RJ. Beta-adrenergic receptor kinase: identification of a novel protein kinase that phosphorylates the agonist-occupied form of the receptor. Proc Natl Acad Sci USA 1986; 83: 2797–801.

189. Colucci WS. Observations on the intracoronary administration of milrinone and dobutamine to patients with congestive heart failure. Am J Cardiol 1989; 63: 17A–22A.

190. Colucci WS, Denniss AR, Leatherman GF et al. Intracoronary infusion of dobutamine to patients with and without severe congestive heart failure. J Clin Invest 1988; 81: 1103–10.

191. Eichhorn EJ, Bristow MR. Medical therapy can improve the biological properties of the chronically failing heart. A new era in the treatment of heart failure. Circulation 1996; 94: 2285–96.

192. Gilbert EM, Anderson JL, Deitchman D et al. Long-term beta-blocker vasodilator therapy improves cardiac function in idiopathic dilated cardiomyopathy: a double-blind, randomized study of bucindolol versus placebo. Am J Med 1990; 88: 223–9.

193. Abraham WT, Gilbert EM, Lowes BD et al. Coordinate changes in myosin heavy chain isoform gene expression are selectively associated with alterations in dilated cardiomyopathy phenotype. Mol Med 2002; 8: 750–60.

194. Sucharov CC, Nunley K, Bristow MR. Repression of alpha myosin heavy chain gene expression and promoter activity by the beta-agonist, isoproterenol – role for Ca^{2+}/calmodulin protein and ERK1/2 pathway. J Card Failure 2004; 10: 156.

195. Wang W, Zhu W, Wang S et al. Sustained β_1-adrenergic stimulation modulates cardiac contractility by Ca^{2+}/calmodulin kinase signaling pathway. Circ Res 2004; 95: 798 806.

CHAPTER 21

Environmental factors in cardiac hypertrophy and failure

William H Frishman, Edmund H Sonnenblick, Piero Anversa

Introduction

There are well-known factors associated with the development of pathological ventricular hypertrophy and congestive heart failure (CHF) which include systemic hypertension, valvular heart disease and ischemic heart disease. Specific neurohumoral influences that affect the development of hypertrophy and heart failure include augmentation of the renin–angiotensin–aldosterone system (RAAS) and sympathetic nervous system.[1] Other neurohumoral influences include vasopressin, endothelin, and calcineurin.[2] Pharmacologic interventions designed to inhibit the formation and/or the effects of angiotensin, aldosterone and catecholamines can reverse ventricular hypertrophy and relieve the symptoms of heart failure.[3] Other interventions that treat hypertension and those that correct valvular heart disease will also reverse the hypertrophic process and protect against the development of symptomatic CHF.

There are other factors that may contribute to the hypertrophy–heart failure process, which are the subjects of this chapter. These include the adverse effects of agents used to treat disease (e.g. anthracyclines), nutritional and enzymatic deficiencies, the effects of mineral excess (e.g. iron), and the potentially cardiotoxic effects of illicit drug use (e.g. alcohol, amphetamines and cocaine).

Cardiotoxic drugs used to treat disease

Anthracyclines

The cancer chemotherapeutic agents that have received the most attention because of their high incidence of cardiac toxicities have been the anthracyclines, daunorubicin and doxorubicin.[1-6] The anthracyclines are antibiotics isolated from the soil microbe streptomyces. They act as topoisomerase II inhibitors and have the widest spectrum of activity, effective against sarcomas, lymphomas, leukemias, carcinomas of the breast, lung and thyroid.[7] Their use is limited by an extensively documented cardiotoxicity which can occur acutely, subacutely, chronically, or as late sequelae, years after completion of therapy. These adverse reactions can be relatively benign, commonly manifesting during or shortly after treatment as transient ECG changes or supraventricular arrhythmias. Rarely, pericarditis or myocarditis, which can sometimes be fatal, has been observed within weeks of drug administration.[8]

The most prominent toxic effect is a cardiomyopathy, resulting from chronic, cumulative, dose-related myocardial injury. Advanced cardiomyopathy may present as CHF in 2 to 20% of patients receiving 500–600 mg/m^2 of doxorubicin (Fig. 21.1), usually months following completion of therapy.[9] Occasionally there have been reports

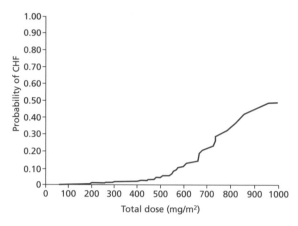

Figure 21.1. *The probability of developing doxorubicin-induced heart failure is related to cumulative dose. There is a continuum of increasing probability of developing congestive heart failure (CHF) rather than an absolute cut-off point that should not be exceeded. At 550 mg/m², however, the slope increases. Reproduced with permission from Von Hoff DD, Layard MW, Basa P et al. Risk factors for doxorubicin-induced congestive heart failure. Ann Intern Med 1979; 91: 710–17.*

of patients developing new onset CHF many years after termination of chemotherapy. These observations were initially reported among older patients in whom the effects of advanced age cannot be excluded. However, there have been reports of late heart failure as a new event in the pediatric population, especially in females.[10] In one recent study, the relation between sex and the cumulative dose of doxorubicin was shown to be interactive; the higher the cumulative dose of the drug, the greater the difference between female and male patients.[10]

Although doxorubicin and daunorubicin have slightly different spectrums of activity, the incidence of cardiomyopathy, the associated histological changes, and proposed risk factors are similar for both. Attempts to limit cardiotoxicity while maximizing antineoplastic efficacy have focused on: (1) assessing myocardial impairment and damage via endocardial biopsy and non-invasive functional testing; (2) altering drug delivery; (3) concomitant use of cardioprotective agents; and (4) development of less toxic analogs, such as epirubicin and idarubicin.

Chronic cardiomyopathy

Cardiomyopathy is a chronic cardiotoxic effect that can manifest as clinical CHF months following termination of anthracycline therapy, with a mean onset 33 days after treatment discontinuation. Anthracycline-induced CHF presents with the typical features of increasing dyspnea, peripheral edema, tachycardia, jugular venous distension, and S_3 gallop. Early reports observed cases of CHF that were refractory to diuretics and digitalis, with a mortality of 30–60%.[11] Other early studies of larger patient cohorts demonstrated a more favorable outcome, with less than 30% dying of fulminant failure immediately after onset of cardiac dysfunction.[12] A majority achieved full recovery of ventricular function, some without further recurrence of CHF.

The cumulative dose of doxorubicin has been repeatedly shown to be the primary determinant of cardiotoxicity. To reduce the risk of CHF, total dosage of doxorubicin was previously empirically limited to less than 550 mg/m² in adults.[8] However, this arbitrary restriction prevented patients without evidence of cardiac dysfunction from receiving maximal antineoplastic therapy. In a large retrospective analysis of 4018 patients who had received doxorubicin, Von Hoff et al. demonstrated the high variability in the dose of doxorubicin that resulted in cardiac failure.[9] None of the five patients who received greater than 1000 mg/m² of doxorubicin developed CHF, whereas 17 patients receiving less than 77 mg/m² of doxorubicin had CHF. It was shown that with increasing dose there was a continuum of increasing cumulative probability of developing CHF, rather than an absolute cut-off point that should not be exceeded. Above 500 mg/m², however, the slope of probability climbs sharply (Fig. 21.1). Risk factors that significantly increased probability of CHF included increasing age, and drug administration every 3 weeks rather than every week. Other suggested risk factors include age less than 15 years, prior mediastinal irradiation, prior daunorubicin therapy, concomitant cyclophosphamide administration, and previous cardiac disease. Performance status, sex and tumor type were not risk factors.[9]

Measurement of left ventricular ejection fraction (LVEF) has been used to assess subclinical cardiac abnormality. Serial echocardiography in children and radionuclide angiocardiography in adults have been performed prior to treatment and after additional cycles of anthracycline, to detect progressive decreases in myocardial contractility.[10] Although cardiac dysfunction is detected earlier on exercise radionuclide testing, angiocardiography at rest is sufficiently effective and is not hampered by effects of decreased exercise tolerance among patients in a debilitated state from cancer, or those with coronary artery disease. Using the radionuclide angiocardiogram to monitor LV function, Schwartz et al developed guidelines for the discontinuation of doxorubicin to minimize the risk of CHF.[13] A baseline radionuclide angiocardiogram was recommended prior to administration of 100 mg/m^2 of doxorubicin. Patients with baseline LVEF below 30% should generally not be started on doxorubicin. Follow-up studies should be performed at least 3 weeks after the last treatment, and before the next proposed treatment, with more frequent monitoring in patients with risk factors such as heart disease, prior irradiation, abnormal ECG, or

cyclophosphamide therapy. Doxorubicin should be discontinued if an absolute decrease in LVEF greater than 10% is associated with a decline to an EF of less than 50%. Among patients with an abnormal baseline LVEF (i.e. below 50%), there is a lower threshold for discontinuation: an absolute decrease in LVEF greater than 10% and/or a final LVEF below 30% (Box 21.1). The Cardiology Committee of the Children's Cancer Study Group has recommended more conservative criteria for the discontinuation of doxorubicin in children.[14]

Despite careful monitoring of patients with non-invasive cardiac testing, patients may still deteriorate after the discontinuation of anthracycline therapy. Methods to increase the sensitivity of detecting early cardiotoxicity are still being sought. These include the measurement of cardiac troponin T levels, measurement of LV diastolic dysfunction, the use of dobutamine stress echocardiography, angiocardiography with radiolabeled antimyosin antibody, angiocardiography with meta$_{10}$-dobenzyl-quanidine, heart rate variability analysis, and measurement of brain natriuretic

---

Box 21.1. *Guidelines for monitoring patients receiving doxorubicin with serial resting radionuclide angiocardiography*

- Perform baseline radionuclide angiocardiography at rest for calculation of left ventricular ejection fraction (LVEF) prior to administration of 100 mg/m^2 doxorubicin. Subsequent studies are performed at least 3 weeks after the indicated total cumulative doses have been given, before consideration of the next dose.

Patients with normal baseline LVEF (>50%)
- Perform the second study after 250–300 mg/m^2
- Repeat study after 400 mg/m^2 in patients with known heart disease, radiation exposure, abnormal electrocardiographic results, or cyclophosphamide therapy; or after 450 mg/m^2 in the absence of any of these risk factors
- Perform sequential studies thereafter prior to each dose
- Discontinue doxorubicin therapy once functional criteria for cardiotoxicity develop, i.e. absolute decrease in LVEF ≥10% (EF units) associated with a decline to a level ≤50% (EF units)

Patients with abnormal baseline LVEF (<50%)
- Doxorubicin therapy should not be initiated with baseline LVEF ≤30%
- In patients with LVEF >30% and <50%, sequential studies should be obtained prior to each dose
- Discontinue doxorubicin with cardiotoxicity; absolute decrease in LVEF ≥10% (EF units) and/or final LVEF ≤30%

Reproduced with permission from Schwartz RG, McKenzie WB, Alexander J et al. Congestive heart failure and left ventricular dysfunction complicating doxorubicin therapy. Seven-year experience using serial radionuclide angiocardiography. Am J Med 1987; 82: 1109–18.

gene expression.[5,15–22] Only an endomyocardial biopsy can permit a definitive evaluation of the risk of developing chronic cardiomyopathy. This procedure provides uniquely valuable information to guide the safety of future therapy.

Mechanisms of anthracycline toxicity

The mechanisms for anthracycline-induced myocardial injury are not clear. However, some form of oxidative damage is suspected. Several theories have been proposed (Box 21.2).

In the past, the leading theory involved the enzymatic reduction of the anthracycline-quinone ring. The anthracyclines contain a quinone structure which, when enzymatically reduced, can transfer an electron to molecular oxygen, which initiates a cascade of oxygen-free radical generation with concomitant O_2 and OH generation, resulting in lipid peroxidation.[7,8] This can result in myocardial cell membrane damage which is seen histologically. The heart, having decreased antioxidant levels, is more susceptible to this type of free-radical injury.[7,8,14,23]

New alternative theories involve the formation of a doxorubicin–iron complex. It is proposed that

---

---

doxorubicin combines with iron to form a complex that undergoes redox cycling, which produces a powerful oxidant that can cause degradation of microsomal and mitochondrial lipids.[8] This anthracycline–iron complex can also catalyze formation of other oxygen radicals, which themselves can cause damage by membrane lipid peroxidation.[24] However, use of free radical scavengers, such as vitamin E and N-acetylcysteine, have not been effective in decreasing anthracycline-induced cardiotoxicity.[24,25] The role of iron is important in both theories, being required in catalytic amounts to initiate hydroxyl radical production in the first, while in the latter, iron forms a complex with the drug.[24]

Another possible mechanism of myocardial toxicity involves a disturbance of calcium and sodium exchange, caused by an interaction with the inner mitochondrial membrane to form a complex that inactivates the electron transport chain, which leads to the production of free oxygen radicals.[26] In addition, Fas-mediated apoptosis has been observed in rats with adriamycin-induced cardiomyopathy,[27] as well as inhibition of pivotal cardiac transcription factors.[28] The presence of myocardial stem cells in humans has recently been described,[29] and repeated cycles of adriamycin therapy may therefore be inhibiting the heart's regenerative capacity. A chemoprotectant, gallopamil, which functions as a calcium antagonist, has been shown to diminish the cardiotoxic effects of anthracyclines and may mediate protection via inhibition of this mechanism of toxicity.[30]

Approaches to minimizing anthracycline cardiotoxicity risk factors and precautions

The risk of developing cardiotoxic effects is significantly increased by the presence of certain factors that reduce cardiac tolerance to anthracyclines. These include pre-existing hypertension or cardiovascular disease, liver disease, prior mediastinal radiation, prior doxorubicin or daunorubicin therapy, concomitant cyclophosphamide therapy, very old or very young age.[5] Before initiating chemotherapy, caution should be used in elderly patients who may have inadequate bone marrow reserves, and in children under 4 years of age, who seem to be especially susceptible to anthracycline cardiotoxicity. It is believed that anthracyclines may induce loss or damage of myocardial cells to a level below that required to generate a normal adult myocardial mass, thus placing young children especially at risk for irreversible cardiac injury.[14,24] Since doxorubicin is excreted primarily by the bile, and daunorubicin by both urinary and biliary pathways, baseline measurements of hepatic and renal function should be performed so that doses may be adjusted accordingly. The clinician should also be aware of potential precipitating factors of clinical heart failure, such as volume-overload (during intravenous chemotherapy), surgical trauma, general anesthesia, and alcohol abuse.

Chemoprotectants and adjunctive therapies to reduce cardiotoxicity

As described earlier, previous attempts to block the cardiotoxicity of anthracyclines in cancer-treated patients with the antioxidant vitamin E (α-tocopherol) and N-acetylcysteine, have been unsuccessful. The prevention of cardiotoxicity has focused on the use of other experimental cardioprotectants to limit free radical injury (Box 21.3). The cardioprotectants ICRF-159 (razoxane) and 187 (dexrazoxane) are enantiomers belonging to the bis- (dioxopiperazine) family. These drugs act as iron chelators and were initially developed as antitumor agents. Their cardioprotective mechanism is unclear, but it is thought that the iron chelation prevents formation of free radicals rather than neutralization.[31] Reduction in available iron possibly blocks the formation of the doxorubicin–iron complex and the generation of oxygen radi-

cals.[32] Chelators like deferoxamine are not useful due to very poor uptake. Multiple animal studies demonstrated the effectiveness of ICRF-187 in preventing acute and long-term cardiotoxicity when administered 30 minutes prior to doxorubicin, and Herman et al. showed a greater tolerance to increased cumulative doses.[33–35]

Human studies have been equally promising. In a randomized prospective trial in women with metastatic breast cancer treated with 5-FU, doxorubicin, and cyclophosphamide (CAF), Speyer et al. demonstrated that the group pretreated with ICRF 187 had significantly fewer patients removed from the study due to cardiotoxicity.[31] In addition, there were significantly fewer incidences of CHF and significantly fewer patients developing reductions in LVEF that required discontinuation of chemotherapy. Pretreated patients had significantly lower biopsy scores among those who consented to biopsy. These patients also tolerated significantly greater cycles of chemotherapy with higher cumulative doses of doxorubicin. Eleven patients received greater than 1000 mg/m^2 without developing CHF, whereas no patient in the control group received as much as 1000 mg/m^2. A number of other studies have further demonstrated the cardioprotective effects of dexrazoxane.[36–38] Randomized trials in patients with metastatic breast cancer, soft tissue sarcoma, and small cell lung cancer have confirmed the cardioprotective effect of ICRF 187 in doxorubicin-based combination chemotherapy.[32,36–39] Swain et al., in two phase III, double-blind, prospectively randomized, placebo-controlled multicenter trials, found the likelihood of developintg a cardiac event to be 2.0 to 2.6 times greater without the addition of dexrazoxane.[39] Despite the increased doxorubicin received in the Speyer study, tumor response rate and patient survival remained equivalent. One study actually demonstrated a non-significant decrease in tumor response among patients with metastatic breast cancer treatment with CAF and ICRF-187,[24] which raises the concern that ICRF-187 may interfere with antineoplastic drug efficacy. Nevertheless, the overwhelming majority of evidence demonstrates that the ICRF-187 given with an anthracycline seems to reduce cardiotoxicity without lessening its antineoplastic action.[24,31,32,36,37]

Box 21.3. *Pharmacological strategies used to reduce anthracycline cardiotoxicity*

- Antioxidants
- Iron chelators (ICRF-159 and -187)
- Coenzyme Q$_{10}$ and carnitine
- Antihistamines, anti-adrenergics, cromolyn sodium
- Glutamine

Adapted from Frishman WH, Sung HM, Yee HCM et al. Cardiovascular toxicity with cancer chemotherapy. Curr Probl Cardiol 1996; 21: 225–88, with permission.

Based on clinical evidence, the Food and Drug Administration has approved dexrazoxane (ICRF-187, Zinecard®) for marketing. The drug is indicated in parenteral form for reducing the incidence and severity of cardiomyopathy associated with doxorubicin administration in women with metastatic breast cancer who have received an accumulative doxorubicin dose of 300 mg/m^2 and who, in their physician's opinion, would benefit from continuing therapy with doxorubicin.

Finally, there is preliminary evidence that lessening of cardiotoxicity with anthracyclines may occur with use of the cardioprotective substances coenzyme Q_{10}, carnitine and glutamine. In addition, there is some evidence that anthracyclines can increase the rates of release of catecholamine and histamine, possible potentiators of cardiotoxicity, effects which can be prevented by pretreatment with antihistamines, antiadrenergics, or mast cell stabilizers, such as cromolyn sodium.[40–42] There is also evidence to suggest that the antioxidant lipid-lowering drug probucol, and glutamine can protect against anthracycline toxicity.[43,44]

Newer anthracyclines

Epirubicin, 4′-epidoxorubicin, is the 4′ epimer of the anthracycline antibiotic doxorubicin, and has shown antitumoral activity against a broad spectrum of human tumors. Clinical studies show that whereas epirubicin's therapeutic activity is comparable to that of doxorubicin,[41,44] its toxicity in animals, as well as in humans, has been found to be lower, particularly with regard to its cardiac toxicity.[45–47] Animal studies investigating the acute and chronic toxicity of epirubicin have demonstrated that, mg for mg, epirubicin has a lower propensity for producing cardiotoxic effects than doxorubicin, and has been used with cumulative dosing approaching 1000mg/m^2.[48] Epirubicin's cardiotoxicity appears after higher cumulative doses than doxorubicin: 935 mg/m^2 for epirubicin versus 550mg/m^2 for doxorubicin.[30] This allows for more treatment cycles than with doxorubicin. Furthermore, studies have indicated the cardiotoxic effects can be further diminished by the use of cardioprotective agents such as ICRF-187 and gallopamil, which functions as a calcium antagonist.[30,49]

In addition, there is evidence from ultrastructural and biochemical studies with the drug that it is better tolerated than doxorubicin.[40] The lower frequency of cardiotoxic effects with epirubicin as compared to doxorubicin may be related to its greater release from the myocardium.[40,48] Additional explanations include: (1) differences in metabolic elimination; (2) less mitochondrial toxicity; (3) lower decrease in oxygen consumption by cardiac myocytes; (4) less superoxide radical production; (5) less lipid peroxidation in myocardial mitochondria; (6) epirubicin accumulates half as much as doxorubicin in the myocardium; and (7) less inhibition of Na$^+$–Ca^{2+} exchange in the heart sarcolemmal vesicles.[46,50–52]

Anthracycline analogues such as idarubicin, hydroxyrubicin, esorubicin and pirarubicin, have been reported to cause less early cardiotoxicity than daunorubicin or doxorubicin, but their long-term effects have not been assessed.[26]

Although cardiotoxic effects are less common with the new anthracyclines (epirubicin), the physician still needs to employ similar precautions in patients to prevent cardiac complications, and the follow-up approach with therapy is similar to that used with other anthracyclines. The risk of cardiac toxicity is increased with factors such as previous cardiovascular disease; previous therapy with anthracyclines at high cumulative doses or with other potentially cardiotoxic agents; with concomitant or previous radiation to the mediastinal-pericardial area; or in patients with anemia, bone marrow suppression, infections, leukemic pericarditis, or myocarditis.[8,48,53] Nevertheless, the low incidence of acute toxicity coupled with oral administration makes idarubicin a useful addition to the anthracyclines.

Nutritional and enzymatic abnormalities

Carnitine

L-Carnitine (the physiologically required form of carnitine) is a nitrogenous constituent of muscle that plays a primary role in the oxidation of fatty acids in mammals (Fig. 21.2). Because the mammalian heart uses fatty acids as its primary source

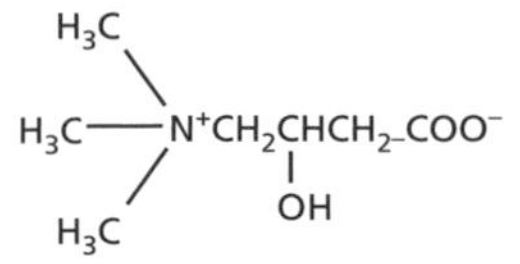

H₃C
    \
H₃C——N⁺CH₂CHCH₂-COO⁻
    / |
H₃C   OH

Figure 21.2. *Chemical formula of L-carnitine.*

of energy, it is understandable that a deficiency in the supply of carnitine can result in various cardiac abnormalities.

In addition, clinical and experimental studies have shown the therapeutic usefulness of L-carnitine and its derivative, propionyl-L-carnitine (PLC), in the treatment of various cardiovascular diseases, including ischemic heart disease, CHF, arrhythmia, and peripheral vascular disease (Box 21.4).

Physiological actions

Carnitine has many physiologically important roles. It is most important for the oxidation of long-chain fatty acids. It also allows for the removal of short- and medium-chain fatty acids from the cell.

Box 21.4. *Potential therapeutic benefits from exogenous treatment with L-carnitine in cardiovascular disease*

■ Correction of carnitine-deficiency state
■ Facilitation of fatty acid and glucose oxidation in situations of limited oxygen availability
■ Removal from mitochondria of harmful fatty acyl groups that accumulate as a result of normal and pathological metabolism
■ Free-radical scavenger following reperfusion of ischemic myocardium

Reproduced with permission from Frishman WH, Retter A, Misailidis J et al. Innovative pharmacologic approaches to the treatment of myocardial ischemia. In: Frishman WH, Sonnenblick EH, Sica DA (eds). Cardiovascular Pharmacotherapeutics (2e). New York: McGraw Hill; 2003: 655–90.

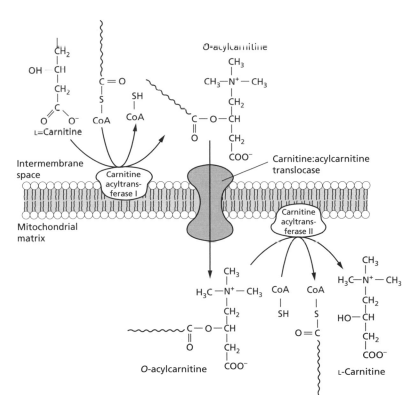

Figure 21.3. *The formation of acylcarnitines and their transport across the inner mitochondrial membrane – the carnitine 'shuttle'. Reproduced with permission from Garrett RH, Grisham CM. Biochemistry. Fort Worth, TX: Saunders College Publications; 1995: 739.*

347

Carnitine also facilitates the aerobic metabolism of carbohydrates,[54,55] and enhances the rate of oxidative phosphorylation.

Carnitine's role in fatty acid oxidation is the following (Fig. 21.3): fatty acids are activated in the outer mitochondrial membrane upon conjugation to a molecule of CoA. They must then enter the mitochondrial matrix to be oxidized. Long-chain fatty acyl-CoA molecules, however, are unable to traverse the inner mitochondrial membrane. Therefore, a special transport system is needed. An activated long-chain fatty acid is able to cross the inner mitochondrial membrane by conjugating itself to a carnitine molecule producing an acylcarnitine. The enzyme for this reaction is carnitine acyltransferase I, which is found on the outer mitochondrial membrane. The acylcarnitine is then shuttled across the inner mitochondrial membrane by a translocase. At this point, the long-chain fatty acid reconjugates with a CoA molecule found in the matrix, and it is now ready to undergo oxidation. This reconjugation is catalyzed by carnitine acyltransferase II. The free carnitine molecule is shuttled back to the mitochondrial cytosol by a translocase in exchange for a new incoming acylcarnitine. This specialized transport system is commonly known as the 'carnitine shuttle'.

Carnitine also facilitates removal from the mitochondria of short- and medium-chain fatty acids that accumulate there as a result of normal and pathological metabolism. These acyl-CoA groups possess detergent-like properties and are therefore harmful to cellular and intracellular membranes. The short- and medium-chain fatty acyl-CoAs are transesterified to carnitine and are then transported out of the mitochondria by the carnitine-acylcarnitine translocase. This allows the 'freeing up' of CoA molecules in the mitochondria under conditions where short- and medium-chain fatty acids are being produced at a rate that is faster than they can be used.

In addition, as carnitine conjugates with these acyl groups, it lowers the mitochondrial acetyl-CoA:CoA ratio, and glucose oxidation is stimulated. This happens because a decrease in the ratio of acetyl-CoA:CoA stimulates the pyruvate dehydrogenase complex, which catalyzes the rate-limiting step of glucose oxidation.[54]

Carnitine is a physiological modulator of the mitochondrial acetyl-CoA:CoA ratio, thus it can influence many biological processes.

Congestive heart failure

Several studies show that in situations of cardiac hypertrophy and CHF, myocardial carnitine levels are significantly decreased, and plasma and urinary free and acyl-carnitine levels are significantly increased.[56–59] It is therefore logical that the restoration of normal carnitine levels through the administration of exogenous carnitine would be of therapeutic value when treating these cardiomyopathies. Interestingly, it has also been shown that the therapeutic value of carnitine treatment extends beyond the restoration of normal carnitine levels, also affecting improvement in cardiac function through other metabolic pathways.[60,61]

Several experimental and clinical studies have demonstrated that carnitine can improve cardiac function in subjects with hypertrophied hearts and those with heart failure.

Schonekess et al. studied the effects of PLC on hypertrophied rat hearts.[54,62] The results of the study showed that those rats treated with PLC had a significant increase in contractile function and cardiac work. The study noted that while those hearts treated with PLC showed a sharp rise in their glucose oxidation rate, the increases in fatty acid oxidation rates were not significant. This allows for the possibility that the primary effect of carnitine administration in hypertrophied hearts is an increase in carbohydrate oxidation. The mechanism for this increase is (as mentioned before in great detail) the conjugation of carnitine with acetyl groups, thereby lowering the acetyl-CoA:CoA ratio. This, in turn, stimulates the enzyme pyruvate dehydrogenase (PDH) complex, the rate-limiting step of glucose oxidation. Additionally, the anaplerotic effect of the propionyl group in PLC feeds intermediates into the tricarboxylic acid cycle. Another proposed explanation of the improvement of cardiac function is that somehow PLC increases the efficiency of translating ATP production into cardiac work.[57] Either way, data from these experiments seem to indicate that the beneficial effect of PLC on the mechanical functioning of hypertrophied hearts does not result merely from normalization of fatty acid oxidation.[63]

Kobayashi et al.[64] studied the myocardial carnitine levels and therapeutic efficacy of L-carnitine in BIO 14.6 hamsters with cardiomyopathies, and in patients with chronic CHF. The results of the study showed that both cardiomyopathic hamsters and patients with heart failure had reduced free carnitine levels and increased levels of acylcarnitine esters. It was also shown that significant myocardial damage in the cardiomyopathic hamsters was prevented by the intraperitoneal administration of L-carnitine in the early stages of the cardiomyopathy. Similarly, 55% of the patients with CHF moved to a lower New York Heart Association (NYHA) class, and 66% showed an overall improvement in their condition.

A large-scale clinical trial was carried out,[65] in which 574 patients with heart failure (NYHA class II and III) and LVEF <40% were randomized to carnitine versus placebo. Exercise tolerance was significantly improved with carnitine in the patients with LVEFs of 30% to 40 %. However, the overall mortality rate was slightly higher (3.0% versus 1.9%), as well as the subsequent follow-up admission rate (6.3% versus 5.3%), in the carnitine group.

Additional clinical investigations included a double-blind, phase II study of PLC versus placebo in a group of 60 patients with mild to moderate (NYHA class II and III) CHF.[66] The group was made up of men and women between the ages of 48 and 73 years who were chronically treated with digitalis and diuretics for at least 3 months, and who still displayed symptoms. Thirty randomly chosen patients were given an oral dosage of 500 mg of PLC, three times daily for 180 days. After 30, 90, and 180 days, the maximum exercise time was evaluated, as was the LVEF. Throughout the period of testing, the patients treated with PLC showed significant increases in both the exercise tolerance and EF tests. On the basis of these results, the authors concluded that PLC was a drug of undoubted therapeutic value for patients with CHF.[66]

Pucciarelli et al. also conducted a randomized, double-blind study of PLC versus placebo.[67] The study included 50 patients of both sexes between the ages of 48 and 69 suffering from mild to moderate CHF. Results of the study demonstrated that those patients in the PLC group had a greater maximum exercise time and an increased LVEF.

Coenzyme Q_{10}

Coenzyme Q_{10} (CoQ_{10}; 2,3-dimethoxy-5-methyl-6-decaprenyl benzoquinone), also known as ubiquinone, is an endogenous cellular membrane constituent that has antioxidant properties that may be useful in preventing cellular damage due to myocardial ischemia–reperfusion injury.[68,69] A clinical experiment has already been developed with this naturally-occurring substance in various cardiovascular and noncardiovascular disorders.

From a biochemical standpoint, CoQ_{10} appears to serve several functions. Endogenous CoQ_{10} plays a pivotal role in oxidative respiration (Fig. 21.4). It has a direct regulatory role on succinyl and NADH dehydrogenases. It serves as a catalyst and has an integral role in regulating the cytochrome b–cl complex. Furthermore, CoQ_{10} may have direct membrane-stabilizing properties separate from its role in oxidative phosphorylation. CoQ_{10}

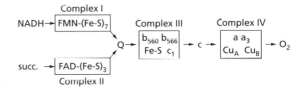

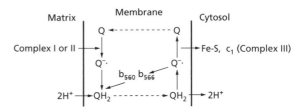

Figure 21.4. *Role of coenzyme Q_{10} in the respiratory chain. Succ., succinyl. Reproduced with permission from Frishman WH, Retter A, Misailidis J et al. Innovative pharmacologic approaches to the treatment of myocardial ischemia. In: Frishman WH, Sonnenblick EH, Sica DA (eds). Cardiovascular Pharmacotherapeutics (2e). New York: McGraw Hill; 2003: 655–90.*

may have mitogenic properties.[70] It may function in membrane secretion,[71] which would explain the occurrence of high concentrations of CoQ_{10} in the Golgi apparatus. Relatively little is known about the biodegradation of ubiquinone. The turnover rate of CoQ_{10} in various tissues ranges from 50 to 125 h.[72]

Congestive heart failure

The results of clinical trials testing the efficacy of CoQ_{10} in the treatment of CHF have been inconsistent. It has been proposed that free-radical injury may promote myocardial decompensation, and that antioxidants may be useful in preventing free-radical damage.[73] Furthermore, CoQ_{10} may help to replenish myocardial ATP stores, and thus improve the heart's energy production.[74] Some studies confirm these beneficial effects. In 1997, Sacher et al.[75] studied the clinical and hemodynamic effects of CoQ_{10} in congestive cardiomyopathy. In this open-label study consisting of 17 patients with NYHA class III and IV heart failure, patients were given 4 months of CoQ_{10} therapy (30 mg three times daily). Significant improvement occurred in a number of areas, including NYHA functional class, LVEF, and cardiac output. In a recent meta-analysis that evaluated eight randomized controlled trials, Soja et al. demonstrated significant improvement in a number of hemodynamic heart parameters, including stroke volume, cardiac output, and ejection fraction.[74] Furthermore, numerous published case studies have touted the benefits of incorporating CoQ_{10} as part of a treatment regimen for heart-failure patients.[76,77] The agent is approved for clinical use to treat heart failure patients in many countries.

However, some recent studies have cast doubts regarding CoQ_{10}'s efficacy in treating CHF. In 1999, Watson et al. conducted a randomized, double-blind study comparing oral CoQ_{10} to placebo for patients with ischemic or idiopathic dilated cardiomyopathy and chronic LV dysfunction.[77] In this study, 30 patients randomly received 33 mg CoQ_{10} three times daily, or placebo for 3 months. Cardiac output, echocardiographic LV volumes, and quality of life indices were measured at baseline and after treatment. There were no improvements in any of the cardiac parameters

with CoQ_{10} at the study's conclusion. More recently, Khatta et al. conducted a randomized, double-blind controlled trial to determine the effect of CoQ_{10} on peak O_2 consumption, exercise duration, and EF in 55 patients with NYHA class III and IV symptoms.[78] Patients were given 200 mg/day of CoQ_{10}, or placebo for 6 months. LVEF (measured by LV radionuclide ventriculography), peak oxygen consumption, and exercise duration were measured at baseline and at the end of the 6-month trial. All three of these parameters remained unchanged at the end of the study period. The authors concluded that there was no clinical benefit in adding CoQ_{10} to standard heart-failure treatment regimens.

Proponents of CoQ_{10} have argued that the two aforementioned studies included patients with late-stage CHF, in whom little benefit would be expected from CoQ_{10}; the sample sizes were too small; the studies were too short; and the dosage of CoQ_{10} was too low.[79,80] Researchers who found little benefit with CoQ_{10} therapy argue that although studies should be done on patients with early-stage CHF, it would be difficult to recruit patients when there is already effective standard therapy, and that previous studies that have demonstrated a benefit of CoQ_{10} were small, anecdotal, and inconsistent.[81,82] It is hoped that more randomized, double-blind, controlled trials will be performed to fully determine whether CoQ_{10} is efficacious in treating CHF.

Trace minerals

Cobalt

Cobaltous chloride is sometimes used to treat iron deficiency and chronic renal failure. Excessive cobalt intake may cause cardiomyopathy and CHF, with pericardial effusions due to deposition of cobalt–lipoic acid complexes in the heart. High cobalt consumption has also been implicated in thyroid enlargement, polycythemia, neurological abnormalities, and interference with pyruvate and fatty acid metabolism.[83]

Iron

Excessive iron levels in the body may come from increased absorption (e.g. hemochromatosis), or chronic toxic ingestions. Increased iron loads can also come from excessive blood transfusion, or as a consequence of ineffective erythropoeisis as in thalassemia major.

Iron accumulates as ferritin and hemosiderin in almost every cell in the body. Cardiac deposition of ferritin and hemosiderin in the myocardium can cause atrial or ventricular arrhythmias and impaired contractility of cardiac muscle.[84]

The most useful laboratory test is the measurement of iron concentration, total iron-binding capacity, and transferring saturation. With marked iron overload, the serum ferritin concentration will exceed 500 μg/l. With iron overload, echocardiography and nuclear imaging studies may demonstrate a dilated cardiomyopathy, and a myocardial biopsy demonstrates increased iron concentration.

Therapy for iron overload includes the elimination of excessive dietary iron and phlebotomy. In addition, the treatment of cardiomyopathy will require use of diuretics, inhibitors of the renin–angiotensin–aldosterone system, beta-blockers, and digoxin. Cardiomyopathy may respond to reducing the iron overload.[84] There appears to be no benefit from iron chelation therapy.

Selenium

Selenium is an antioxidant and essential mineral with immune-enhancing and cancer-fighting properties. The metabolic relationships between the mineral selenium and vitamin E are very close. Selenium is a co-factor of the enzyme glutathione peroxidase, which serves as an antioxidant and is found in the platelets and the arterial walls. In contrast to vitamin E, which prevents formation of lipid hydroperoxides in cell membranes and low density lipoprotein (LDL) by acting as a biological free-radical trap, selenium and glutathione peroxidase help destroy lipid hydroperoxides already formed by peroxidation of polyunsaturated fatty acids. Selenium thereby defends against the free-radical oxidative stress that escapes the protection of vitamin E.

In some areas of the world, soil deficiencies in selenium have produced Keshan disease, a disorder of cardiac muscle characterized by multifocal myocardial necrosis that causes cardiomyopathy, CHF, and cardiac arrhythmias. Men with low levels of serum selenium ($<1.4\,\mu$M) demonstrated increased thickness in the intima and media of the common carotid arteries.

Thiamine

Thiamine (vitamin B_1), one of the earliest described vitamin substances, plays an essential role in the treatment of 'wet beri-beri', a dietary thiamine and protein deficiency state characterized by high-output cardiac failure, tachycardia, extensive edema, and depressed ventricular function.[85] Patients with chronic alcoholism or CHF receiving chronic diuretic therapy may present with thiamine deficiency and experience neurological as well as cardiovascular effects.[86]

Thiamine pyrophosphate is the physiologically active form of thiamine and is involved in a series of biochemical reactions involving carbohydrate metabolism. In the citric acid cycle, it serves as a coenzyme in decarboxylation of α-keto acids such as pyruvate and α-ketoglutarate. Thiamine is also involved as a cofactor for the enzyme transketolase, which is involved in the hexose monophosphate shunt.[87]

Inadequate intake or increased excretion may cause thiamine deficiency. Inadequate intake is commonly due to consumption of non-enriched grains such as rice and wheat, which are stripped of vitamins by the milling process.[87] Ingestion of raw fish containing microbial thiaminases may destroy the vitamin in the gastrointestinal tract. Chronic alcoholism is associated with low thiamine intake, but also with impaired absorption and vitamin storage. Because most chronic alcoholics ingest high-carbohydrate diets, the physiological demand for thiamine is increased. Patients undergoing long-term renal dialysis or intravenous feedings are at high risk of thiamine deficiency.[88] A rare cause of deficiency is thiamine-responsive megaloblastic anemia, a clinical triad of megaloblastic anemia with ringed sideroblasts, diabetes mellitus, and progressive sensorineural deafness. The responsible gene, *SLC19A2*, has recently been mapped and cloned, and its product was determined to normally function as a high-affinity thiamine transmembrane transport protein.[89]

Thiamine deficiency will lead to neurological symptoms and muscle wasting (dry beri-beri) as well as cardiac involvement manifested by tachycardia, augmented peripheral vasodilation, and biventricular myocardial failure with extensive accumulation of edema ('wet beri-beri').[85] In deficiency states, the oxidation of α-ketoacids is dysfunctional, pyruvate will accumulate in the blood, and glucose cannot be utilized anaerobically. Type B lactic acidosis, (or slow lactic acidosis, in contrast to type A or fast lactic acidosis caused by oxygen deprivation) may result from decreased conversion of lactate to glucose. More importantly, however, the central nervous system is deprived of the ATP from the oxidation of ketoacids, and the Wernicke–Korsakoff syndrome results.[90] In peripheral blood vessels, the decrease in available ATP is believed to cause endothelial swelling, smooth muscle relaxation, and arteriovenous shunting. As resistance vessels become paralyzed, a hyperkinetic state ensues, with increased ventricular filling pressures and decreased peripheral vascular resistance. Pulse pressures may widen, and renal and cerebral blood flow decrease in favor of skeletal muscle blood flow, with attendant tachycardia, edema, weakness, dyspnea, and paresthesias of the lower extremities.[85] Cardiac involvement with thiamine deficiency is often related to increased physical exertion and excessive caloric intake, while the polyneuritic form is seen in patients who are relatively inactive or on caloric restriction.[87]

Clinical heart failure may develop explosively in beriberi, with some patients succumbing from their illness within 48 h of symptom onset. Shoshin beriberi is a fulminant form of thiamine deficiency characterized by hypotension, tachycardia, and lactic acidosis, with patients dying of pulmonary edema if not treated.[91]

In cardiac catheterization studies in patients with beriberi heart disease, right atrial and pulmonary wedge pressures are elevated, cardiac index is increased, and the arteriovenous oxygen difference is increased.[92] These abnormalities may return to normal after treatment with thiamine. LVEF may be normal or decreased, and may fall further with exercise.[93]

The diagnosis of thiamine deficiency as a cause of heart failure is often made with a recent history of thiamine deficiency for 3 months or longer, an absence of another cause for heart failure, evidence of polyneuritis, and an improvement in clinical symptoms with thiamine replacement. One author cautions that although chronic malnutrition may cause thiamine-induced dilated cardiomyopathy in alcoholics, alcoholism per se may simply predispose the patient to a viral myocarditis, which may cause overlapping symptoms with normal thiamine stores.[94] Clinical presentation alone may thus be highly suggestive of a deficiency but is insufficient for the diagnosis.

Several laboratory modalities provide quantitative assessments of thiamine status. Thiamine pyrophosphate effect (TPPE) measures transketolase enzyme activity when thiamine diphosphate is added to serum. Normal individuals show a TPPE of between zero and 15%; larger effects signify relative deficiency states; and greater than 25% defines severe deficiency.[85] A direct measurement of erythrocyte transketolase levels may be performed on whole blood, and activities that rise from a depressed value after thiamine replacement confirm the diagnosis of a deficiency state.[85] High-performance liquid chromatography may even be more precise for identifying a thiamine-deficiency state, as direct measurements of each phosphorylation state of thiamine may be made.[95]

Patients with beriberi heart disease will fail to respond adequately to digitalis and diuretics alone. However, the improvement after the administration of thiamine (up to 100 mg intravenously followed by 25 mg/day orally for 1 to 2 weeks) may be dramatic. A marked diuresis, a decrease in heart size and heart rate, and a clearing of pulmonary congestion may occur within 12 to 48 h.[92] However, the acute reversal of vasodilation induced by correction of the thiamine deficiency may cause the unprepared heart to undergo low-output failure. Therefore, patients with heart failure should receive a glycoside, angiotensin-converting enzyme (ACE) inhibitor, and diuretic therapy along with thiamine replacement treatment.

Thiamine supplements can improve LV function, and biochemical evidence of deficiency in some patients with moderate to severe CHF who are receiving long-term furosemide therapy. In a study by Shimon et al,[88] thiamine replacement for 7 weeks in patients receiving furosemide for 3 months or longer was shown to improve LVEF by 22%, decrease heart size and heart rate, and increase blood pressure and rate of diuresis. Patients were administered 1 week of intravenous thiamine (200 mg/day) and then 6 weeks of oral thiamine (200 mg/day).

Cardiotoxicity from illicit drug use

Substance abuse is an immense problem in the United States and around the world. Although the psychological effects of this condition are widely known due to popular exposure in the media, the medical consequences are often unknown to the user. In this chapter, the effects of substance abuse on the cardiovascular system are discussed, and include the agents alcohol, the amphetamines, and cocaine.

Alcoholic cardiomyopathy

In 1884, Bolinger described cardiac enlargement in persons who had habitually consumed excessive amounts of beer. Later, in 1893, Steel recognized a clinical syndrome of heart failure in the same population.[96] Since then, the definition of alcoholic heart muscle disease, has been refined to include cardiomegaly, ventricular chamber dilation, myocardial contractile abnormalities, and pathological alterations in the heart muscle involving both ventricles in patients whose sole causative agent is ethanol consumption of >80 g/day for 10 years or more.[97] Some recent surveys suggest that less, but still significant consumption of alcohol may play a part in some 'idiopathic' cases.[98] Symptoms generally become clinically evident between 30 and 60 years of age. Symptomatic heart failure is typically of sudden onset.[96] Alcoholic heart muscle disease is a leading cause of secondary cardiomyopathy in the United States. In 1987, 4.2% of reported deaths resulting from cardiomyopathy were due to alcohol.[99]

Mechanisms of alcoholic cardiomyopathy

For decades, the nature of alcoholic heart disease was obscured by its confusion with beri-beri. Alcoholic cardiomyopathy, however, is a low output form of heart failure, unlike beri-beri which is characterized by peripheral vasodilation and high output failure. In 1956, alcoholic cardiomyopathy was described in well-nourished alcoholics.[96] Since then, the association of alcoholic cardiomyopathy and excessive alcohol consumption has been well established.[100,101] In fact, a direct correlation of lifetime alcohol dosage with LV mass, and an inverse correlation with EF has also been established.[102] Timmis et al. studied parameters of myocardial contractility in volunteers who were stratified according to their previous alcohol exposure.[103] Their results demonstrated a progressive decline in ventricular function in response to ethanol exposure. The authors suggest that this finding implies some degree of chronic myocardial impairment which is proportional to the degree of ethanol exposure. Also, the low prevalence of clinical nutritional deficiency in patients with alcoholic cardiomyopathy, and the infrequency of heart disease in patients with cirrhosis support the contention that cardiac abnormalities are not dependent on malnutrition.[100] Recently it was shown that the vulnerability to cardiomyopathy among chronic alcohol abusers is partially genetic and related to the presence of the ACE DD genotype.[104]

In early stages, the disorder tends to be subclinical. Cardiac hypertrophy and dilatation are clinically evident by non-invasive means only. LV end-diastolic pressure is increased, and LVEF decreased.[96,105] Actually, up to one-third of all chronic alcoholics have a depressed EF.[102]

In experimental models, it was shown that rats given 30% of calories as ethanol every day for eight months displayed hemodynamic effects similar to those seen in humans with long-term alcohol intake. Systemic arterial pressure, LV peak systolic pressure, and myocardial contractility were decreased with no change in heart rate. The alcoholic rats demonstrated a 5.2-fold elevation in LV end-diastolic pressure. Finally, ventricular chamber volume was increased through myocardial remodeling, with evidence of decreased LV thickness. These alterations resulted in a 571% increase in the diastolic volume.[106]

Patients who succumb to the disease have hearts up to 900 g (normal 300–325 g), dilated and hypertrophied atria and ventricles, and irregular foci of thickened, fibrotic endocardium, that is often overlaid by mural thrombi. Microscopically, diffuse interstitial fibrosis, interstitial chronic inflammation, and hypertrophic, as well as atrophic myocytes are present.[96]

Decreases in glycogen, mitochondrial respiration, and fatty acid oxidation are also produced experimentally after chronic ethanol consumption.[107,108] Changes in key enzyme levels cause a shift to the glycolytic pathway. Also, the rates of mitochondrial protein synthesis in atrial and ventricular myocardial tissue can become significantly reduced, affecting protein turnover.[108,109] These changes in metabolism and mechanics of the ventricle are believed to be pertinent to the early pathogenesis of clinical alcoholic cardiomyopathy.[110]

Although alcoholic cardiomyopathy is more common in men, due to their higher frequency of alcoholism, women are actually more sensitive to the cardiac effects of ethanol.[96] In a large, cross-sectional study by Urbano-Marquez et al., alcoholic women were compared to both alcoholic men and non-alcoholic women, comparing myopathy and cardiomyopathy between the sexes.[111] They found that although female alcoholics had a mean lifetime dose of alcohol which was only 60% that of male counterparts, they suffered myopathy and cardiomyopathy as frequently. The threshold dose for the development of cardiomyopathy and myopathy was considerably less in women than in men. Also, the decline in EF with increasing alcohol dose was significantly steeper. Kupari and Koskinen, however, found that in alcoholic women, indices of LV function were affected to the same extent as in men, after adjustment for body surface area and other indexes of systolic or diastolic dysfunction.[112]

Altura et al found that ethanol caused vasoconstriction of the microscopic resistance and capacitance vessels.[113] Acetaldehyde and acetate conversely caused vasodilatation on the same vessels. They suggested that it is microvascular ischemia over a long period of time that leads to the syndrome of alcoholic myopathy. Microscopic findings similar to those seen in ischemia/anoxia (swelling of the mitochondria and appearance of I-bands of myocardial fibers) support this theory.[107,108]

Guarnieri and Lakatta suggested that the mechanism of myocardial depression in cardiomyopathy results from ethanol's effect on electromechanical coupling through inhibition of the calcium–myofilament interaction.[114] Using ferret papillary muscle cells, these investigators demonstrated no change in transmembrane action potential or cytosolic calcium current. This depression in ferret papillary cells (10% depressed force in concentrations similar to intoxicating alcohol levels in man) was quickly reversible by removal of ethanol from the perfusate (using aequorin).

Although the exact mechanism has not been elucidated, the development of alcoholic heart muscle disease has been attributed to the increased stimulation from the peripheral sympathetic nervous system. Alcohol may augment obstruction in hypertrophic cardiomyopathy by this mechanism. Excess catecholamines and adrenal hypertrophy have been identified in rats receiving excessive alcohol, who subsequently develop cardiac hypertrophy.[115,116] The alpha-1 blocker, prazosin, given concurrently with ethanol in experimental rats, did not prevent, and actually enhanced the cardiomegaly seen in rats given ethanol alone, suggesting that postsynaptic alpha-1 adrenoceptor stimulation is not an important contributor to ethanol-induced cardiomegaly.[115] In contrast, when similar rats were first exposed to metoprolol, a β-adrenergic blocker (in high levels only – 100 mg/kg three times daily), it prevented the previously seen cardiac hypertrophy, suggesting that there is a β-adrenoceptor-mediated link in the cardiac hypertrophy process induced by ethanol.[116]

In early alcoholic cardiomyopathy, the heart may return to normal following discontinuation of alcohol use. Demakis et al. studied the natural course of alcoholic cardiomyopathy from time of diagnosis for an average of 40.5 months.[117] They divided patients into groups based on progression: clinical improvement remained unchanged, symptoms deteriorated. The two most significant factors in improvement of symptoms and prolonged clinical outcome were abstention from alcohol, and a short time before beginning symptomatic treatment (i.e. digitalis, diuretics). Even patients with NYHA class IV heart failure have been demonstrated to have improvement in symptoms as well as objective criteria (LVEF) with abstinence.[118] Previous studies showed some benefit of prolonged bed rest, however only 21% of patients with improvement maintained the benefit upon returning to exercise.[117]

Abstention is the cornerstone to treatment of patients with alcoholic heart muscle disease.[119] As stated above, this is the only true hope for reversal of the process. However, a recent study showed that both abstinence and controlled drinking of up to 60 g of ethanol per day (four standard drinks) were comparably effective in promoting improvement in cardiac function.[120] Patients should also receive treatment for heart failure including diuretics, ACE inhibitors, angiotensin II blockers, β-blockers, and digoxin, as with all forms of dilated cardiomyopathy.[121]

Patients with alcoholic heart muscle disease and signs and symptoms of heart failure have an average life expectancy of less than 3 years.[96] However, in comparison, patients suffering from alcoholic cardiomyopathy have a significantly better long-term prognosis than patients with idiopathic dilated cardiomyopathy (IDC). In one study, the actuarial survival of a patient with IDC was only 30% versus 81% in alcoholic heart muscle disease. The transplant-free survival was 20% and 81% respectively.[122]

There are also data to suggest that there may be a potential benefit from moderate alcohol use in patients with pre-existing LV dysfunction.[123] In a long-term follow-up observational study published by Abramson et al.,[124] there was a correlation between moderate alcohol consumption and a lower heart failure incidence, unrelated to a reduction in myocardial infarction (MI) risk. It was found that in persons without heart failure, the relative risk of developing symptomatic CHF was 1.00, 0.79, and 0.53 for those who consumed no alcohol, 1 to 20 oz, and 21 to 70 oz per month, respectively. Even after adjustments for age, sex, and incidence of MI, this relationship held. These findings were independent of the type of alcohol consumed.[124]

In patients with documented ischemic cardiomyopathy (EF <35%), a similar reduction in the relative risk of both all-cause mortality (85%) and MI (55%) was observed among patients reporting light to moderate alcohol consumption versus those reporting none. There was no difference in the two groups with regard to death from ischemia, arrhythmia, or progressive heart failure.[125]

Box 21.5. *Risks and benefits of alcohol consumption*

Risks
- Alcohol heart muscle disease (alcoholic cardiomyopathy)
- Hypertension
- Arrhythmias

Benefits (1–2 drinks per day)
- Protection from coronary artery disease
- Protection from heart failure
- Hypertension control
- Beneficial effects on lipids
- Beneficial effects on platelet aggregation and adhesion

Updated with permission from Del Vecchio A, Frishman WH, Fadal A, Ismail A. Cardiovascular manifestations of substance abuse. In: Frishman WH, Sonnenblick EH (eds). Cardiovascular Pharmacotherapeutics. New York; McGraw Hill; 1997: 1115–49.

Amphetamines

Amphetamines potentiate the effects of catecholamines. They are thought to act in two ways: first, they cause the direct release of stored catecholamines from the pre-synaptic membrane; second, they function as a weak inhibitor of monoamine oxidase (MAO). MAO is the enzyme which functions to inactivate neurotransmitter molecules such as norepinephrine and dopamine.[126] Through these actions, amphetamines act much like cocaine. Similar to cocaine, they have been shown to elevate blood pressure in a dose-dependent manner.[127] At high doses, they have been shown to cause tachycardia, palpitations, and sweating. At extremely high doses, arrhythmias may occur.[128,129]

Cardiovascular effects

Unlike cocaine, the effects of amphetamines on the cardiovascular system are less well described. The most commonly reported vascular complication of amphetamine abuse is intracerebral hemorrhage.[128] However, reports of pulmonary hypertension, acute aortic dissection, ruptured berry aneurysm, and sudden death have also been described.[123]

Cardiomyopathy

The development of cardiomyopathy in patients has been documented after the use of smoked,[130] oral,[131,132] and intravenous[133,134] forms of amphetamine. Patients are generally young to middle-aged adults, often with some previous use of amphetamine for a condition such as depression,[131] or obesity,[132] and generally present in frank CHF. Patients complain of dyspnea and fatigue. Edema is also found in some cases.[131]

On clinical examination, patients often present with bilateral pulmonary congestion. On echocardiography, findings of cardiac enlargement, dilatation and decreased contractile function, as evidenced by reduced EFs, are consistent with a dilated congestive cardiomyopathy.[130,132] Often these patients are thought to have an idiopathic form of cardiomyopathy.[132]

Pathological examination of this type of cardiomyopathy shows enlarged, dilated ventricles and atria with scattered foci of fibrosis. The coronary vessels are most often without significant disease.[131,132] Microscopically, widespread interstitial edema with scattered, lymphocytic and histiocytic cellular infiltrate can be found. Muscle fiber degeneration and muscle necrosis are also present.[135] Electron microscopy displays mitochondrial abnormalities including intramitochondrial granules and loss of matrix.[132]

As stated, amphetamines potentiate the actions of catecholamines. Experimental evidence supports that catecholamines in excess doses may induce myocardial necrosis.[136] In fact, focal myocarditis was defined using experimental administration of catecholamines almost 100 years ago.[137] Factor and Sonnenblick have suggested that catecholamine-induced transient microspasm is the common denominator in all forms of congestive cardiomyopathy.[138]

Treatment of cardiomyopathy involves abstention from amphetamines. However, the occurence of reversibility is not always certain with amphetamine cardiomyopathy, although it has been documented in other forms of catecholamine-induced cardiomyopathy including pheochromocytoma.[139] Hong et al. presented a patient who, upon abstention and symptomatic therapy, displayed marked clinical improvement, with a modest reduction in ventricular dimension size, however with no change in LV systolic dysfunction.[130] Besides abstention, symptomatic treatment of heart failure helps to decrease dyspnea and fatigue.[132] As mentioned earlier with alcoholic cardiomyopathy, diuretics, β-blockers, digoxin and ACE inhibitors should all be employed appropriately to treat symptoms.

Cocaine

The clinical importance of cocaine abuse rests in the many cardiovascular complications that can occur.[140] These include the more common acute ischemic syndromes and arrhythmias, as well as case reports of complications such as ischemic colitis,[141] dissection, and pneumopericardium.[142] Despite the widespread frequency of cocaine abuse, only a small percentage of users suffer from cardiovascular effects severe enough to seek medical treatment.[143] Long-term use of cocaine in humans has been shown to predispose the user to premature atherosclerosis, left ventricular hypertrophy (LVH), and cardiomyopathy, while acute use causes vasoconstriction, thrombus formation, increased myocardial oxygen demand, and conduction defects.[144] These effects of cocaine are not unique to parenteral use or use of massive doses; do not require underlying heart disease; and seizures are not required, nor are they a typical accompanying feature to cardiac toxicity.[145]

Negative effects (myocardium)

Many investigators have suggested that there is a direct inhibitory effect of cocaine on the myocardium.[146–148] In anesthetized dogs, cocaine infusion caused an immediate decrease in ventricular performance, which was a linear function of cumulative dose; a decrease in ventricular relaxation; increases in LV end-diastolic pressure without changes in cardiac output; and slight decreases in coronary blood flow, most likely related not to increases in vascular resistance but to poor contractile function. Plasma and tissue norepinephrine were initially increased, indicating that cocaine has a direct inhibitory effect on the myocardium, decreasing contractile performance and relaxation.[147] Similarly, within seconds after cocaine administration in anesthetized dogs, LV dP/dt decreased by 18 to 28%, while coronary blood flow increased 45 to 73%, supporting a direct effect on the myocardium, independent of coronary blood flow.[148] Also, studies of rabbit heart septae, while maintaining coronary blood flow and heart rate, demonstrated a negative inotropic effect of cocaine in a dose dependent manner over the range of 10^{-5} M to 10^{-3} M. These are blood levels of cocaine that have been reported in humans.[149]

Researchers believe that the mechanism behind these direct effects of cocaine are related to its local anesthetic properties. Direct inhibition of sarcolemmal, voltage-dependent Na^+ channels decreases Na^+ transport across the sarcolemma. Less intracellular Na^+ is then available to power the Na^+–Ca^{2+} exchanger, thereby decreasing trans-sarcolemmal calcium influx, resulting in less intracellular Ca^{2+}. Inadequate Ca^{2+} loading of the sarcoplasmic reticulum inhibits excitation–contraction coupling.[150–153]

There are contradictory reports whether cocaine has any effect on the ability of the sarcoplasmic reticulum (SR) to release and accumulate calcium. Some studies have stated that it has no effect.[150,152] However, others have demonstrated that, at concentrations >50 μM, cocaine inhibits the ability of the SR to release and accumulate calcium.[154] Yuan and Acosta reported inhibition of calcium release from the SR at only very high cocaine concentrations (10^{-3} M).[153] At the present time, it is still uncertain whether any of cocaine's effects on the SR play a role in its overall cardiovascular actions.

In 'skinned' ferret ventricular myocytes, where sarcolemma and SR are functionally removed, cocaine also demonstrated the potential to alter the direct interaction of myocardial contractile proteins with calcium. The decreased sensitivity to calcium, similar to the effect on vascular smooth muscle (see below), is probably due to a direct action on the proteins because these skinned fibers

are also devoid of functional autonomic nerve endings. This may contribute to the overall effect.[155]

Pilati et al. found that in vivo exposure to high doses of cocaine (10 mg/kg), similarly caused diminished contractile performance.[156] This decrease was persistent even after cocaine-free perfusate was used. However, in vitro exposure caused similar, yet readily reversible changes in performance. The authors state that the decreases in cardiac performance may involve more than just cocaine's direct action on the myocyte. Catecholamine cardiotoxicity did not seem to account for these effects, since the catecholamine levels did not rise in the animals receiving high-dose cocaine (10 mg/kg). They suggested either transient ischemia with subsequent stunning, and side-effects of anesthesia as possible causes for the differential effect. The possibility that cocaine's metabolites are responsible for its in vivo effect has also been explored.[157,158] Cardiovascular effects of cocaine can develop minutes to hours after use. Cholinesterase inhibition in vivo was shown to have cardiovascular-protective effects.[157] Norcocaine, a metabolite of cocaine produced by hepatic cholinesterase activity, demonstrated potent, Na^+ channel-blocking ability.[158] Perhaps norcocaine has longer-lasting effects that were not diminished by a cocaine-free perfusate.

Left ventricular hypertrophy

Studies have shown a relationship between cocaine abuse and left ventricular hypertrophy (LVH). The authors have suggested that this may be a clinical substrate for facilitating myocardial ischemia, arrhythmias and sudden death.[159,160] In a study of two-dimensional echocardiograms in 30 cocaine abusers and 30 matched controls, LV mass index (LVMI) was found to be significantly higher in the former. Also, posterior wall thickness (PWT) was increased (≥ 1.2cm) in 13 (43%) cocaine abusers versus 4 (13%) of the controls.[159] Forty-nine normotensive cocaine abusers and 30 matched controls underwent treadmill exercise stress testing to 85% of maximum predicted heart rate. These patients were then examined by two-dimensional echocardiography. LVH, defined as LVMI ≥ 105 g/m² and PWT ≥ 1.2cm, was seen in 16/49 (33%) of cocaine abusers versus 3/30 (10%) of controls.[160]

Om et al. demonstrated similar findings in patients with no known history of hypertension.[161] They suggest transient elevations in blood pressure after cocaine use as the etiology. In fact, Ciggarroa et al. showed that cocaine abusers with LVH had an exaggerated pressor response to sympathetic stimuli such as exercise.[160] Another possible explanation is long-term α-adrenergic stimulation of the myocardium, which has been shown to cause increased myofibrillar protein synthesis and cellular hypertrophy in animal studies. This is unlikely due to the low density of α receptors in human myocardium. Concomitant alcohol abuse and unknown contaminants in 'street' cocaine have also been suggested as etiological factors.[159,161]

A major study disputes the above findings. The Coronary Artery Risk Development in Young Adults (CARDIA) study examined the association of LVH and self-reported cocaine and alcohol use among 3446 black and white participants with a mean age of 29.9 years.[162] The population was divided among those who (1) denied ever using cocaine; (2) had used cocaine between one and 10 times (experimental users); (3) had used cocaine more than ten times (recurrent users). They found no consistent association between LVH and levels of cocaine use. Unfortunately, although this is a large study, it is not representative of long-term cocaine abuse. Even the so-called recurrent user group includes patients who may have used cocaine as few as 11 times, perhaps many years previously. There is no mention of the chronicity of use, or even the approximate amounts used. A more extensive analysis of these data, as well as future follow-up of the study sample, may help us to better understand the true effects of cocaine on LVH. A recent study by Henning et al. showed that cocaine, like norepinephrine, increased the total protein content and expression of beta myosin heavy chain in cardiac ventricular myocytes, and caused ventricular hypertrophy.[163] This process was not inhibited by α- or β-adrenergic receptor blockade.

Congestive heart failure

Scattered reports have described patients with acute and chronic dilated cardiomyopathy associated with cocaine use, which resolved after two weeks of drug abstinence.[164] Infants born to cocaine-addicted mothers had their cardiomyopathy resolve in two days of life. It is thought that these effects may be similar to the catecholamine-induced cardiomyopathy of pheochromocytoma, a direct toxic effect of cocaine rather than one secondary to ischemia.[165]

References

1. Hachamovitch R, Strom JA, Sonnenblick EH et al. Left ventricular hypertrophy in hypertension, and the effects of antihypertensive drug therapy. Curr Probl Cardiol 1988; 13: 371–421.

2. Frishman WH, Retter A, Mobati D et al. Innovative drug targets for treating cardiovascular disease: adhesion molecules, cytokines, neuropeptide Y, calcineurin, bradykinin, urotensin, and heat shock protein. In: Frishman WH, Sonnenblick EH, Sica DA (eds). Cardiovascular Pharmacotherapeutics (2e). New York: McGraw Hill; 2003: 705–39.

3. Dahlöf B, Devereux RB, Kjeldsen SE et al. for the LIFE Study Group. Cardiovascular morbidity and mortality in the Losartan Intervention For Endpoint reduction in hypertension study (LIFE): a randomized trial against atenolol. Lancet 2002; 359: 995–1003.

4. Shan K, Lincoff AM, Young JB. Anthracycline-induced cardiotoxicity. Ann Intern Med 1996; 125: 47–58.

5. Singal PK, Iliskovic N. Doxorubicin-induced cardiomyopathy. N Engl J Med 1998; 339: 900–5.

6. Nelson MA, Frishman WH, Seiter K et al. Cardiovascular considerations with anthracycline use in patients with cancer. Heart Disease 2001; 3: 157–68.

7. Doroshow JH. Doxorubicin-induced cardiac toxicity. N Engl J Med 1991; 324: 843–5.

8. Frishman WH, Sung HM, Yee HCM et al. Cardiovascular toxicity with cancer chemotherapy. In: Frishman WH, Sonnenblick EH (eds). Cardiovascular Pharmacotherapeutics. New York: McGraw Hill; 1997; 1479–503.

9. Von Hoff DD, Layard MW, Basa P et al. Risk factors for doxorubicin-induced congestive heart failure. Ann Intern Med 1979; 91: 710–17.

10. Lipshultz SE, Lipsitz SR, Mone SM et al. Female sex and drug dose as risk factors for late cardiotoxic effects of doxorubicin therapy for childhood cancer. N Engl J Med 1995; 332: 1738–43.

11. Moreg JS, Oylon DJ. Outcomes of clinical congestive heart failure induced by anthracycline chemotherapy. Cancer 1992; 70: 2637–41.

12. Mushlin PS, Olson RD. Anthracycline cardiotoxicity: new insights. Ration Drug Ther 1988; 22: 1–9.

13. Schwartz RG, McKenzie WB, Alexander J et al. Congestive heart failure and left ventricular dysfunction complicating doxorubicin therapy. Seven-year experience using serial radionuclide angiocardiography. Am J Med 1987; 82: 1109–18.

14. Lipshultz SE, Colan SD, Gelber RD et al. Late cardiac effects of doxorubicin therapy for acute lymphoblastic leukemia in childhood. N Engl J Med 1991; 324: 808–15.

15. Herman EH, Zhang J, Lipshultz SE et al. Correlation between serum levels of cardiac troponin-t and the severity of the chronic cardiomyopathy induced by doxorubicin. J Clin Oncol 1999; 17: 2237–43.

16. Tjeerdsma G, Meinardi MT, van Der Graaf WT et al. Early detection of anthracycline induced cardiotoxicity in asymptomatic patients with normal left ventricular systolic function: autonomic versus echocardiographic variables. Heart 1999; 81: 419–23.

17. Cottin Y, l'Huillier I, Casasnovas O et al. Dobutamine stress echocardiography identifies anthracycline cardio-toxicity. Eur J Echocardiog 2000; 1: 180–3.

18. Hashimoto I, Ichida F, Miura M et al. Automatic border detection identifies subclinical anthracycline cardio-toxicity in children with malignancy. Circulation 1999; 99: 2367–70.

19. Agarwala S, Kumar R, Bhatnagar V et al. High incidence of adriamycin cardiotoxicity in children even at low cumulative doses: role of radionuclide cardiac angiography. J Pediatr Surg 2000; 35: 1786–9.

20. Chen S, Garami M, Gardner DG. Doxorubicin selectively inhibits brain versus atrial natriuretic peptide gene expression in cultured neonatal rat myocytes. Hypertension 1999; 34: 1223–31.

21. Sparano JA, Wolff AC, Brown D. Troponin for predicting cardiotoxicity from cancer therapy. Lancet 2000; 346: 1947–8.

22. Frishman WH, Sung HM, Yee HCM et al. Cardiovascular toxicity with cancer chemotherapy. Curr Probl Cardiol 1996; 21: 225–88.

23. Li T, Singal PK. Adriamycin-induced early changes in myocardial antioxidant enzymes and their modulation by probucol. Circulation 2000; 102: 2105–10.

24. Schimmel KJM, Richel DJ, van den Brink RBA, Guchelaar H-J. Cardiotoxicity of cytotoxic drugs. Cancer Treat Rev 2004; 30: 181–91.

25. Myers C, Bonow R, Palmeri S et al. A randomized controlled trial assessing the prevention of doxorubicin cardiomyopathy by N-acetylcysteine. Semin Oncol 1983; 10: 53–5.

26. Rhoden W, Hasleton P, Brooks N. Anthracyclines and the heart. Br Heart J 1993; 70: 499–502.

27. Nakamura T, Ueda Y, Juan Y et al. Fas-mediated apoptosis in adriamycin-induced cardiomyopathy in rats. In vivo study. Circulation 2000; 102: 572–8.

28. Poizat C, Sartorelli V, Chung G. Inhibition of pivotal cardiac transcription factors in doxorubicin-induced cardiomyopathy. Circulation 1999; 100 (Suppl I): I-213.

29. Beltrami AP, Barlucchi L, Torella D et al. Adult cardiac stem cells are multipotent and support myocardial regeneration. Cell 2003; 114: 763–76.

30. Polverino W, Basso A, Bonelli A et al. 4'-epidoxorubicin: its cardiotoxicity. Possible cardiac protection with gallopamil, a drug with calcium-antagonist action. Minerva Cardioangiol 1992; 40: 23–30.

31. Speyer JL, Green MD, Zeleniuch-Jacquotte A et al. ICRF-187 permits longer treatment with doxorubicin in women with breast cancer. J Clin Oncol 1992; 10: 117–27.

32. Speyer JL, Green MD, Kramer E et al. Protective effect of the bispiperazinedione ICRF-187 against doxorubicin-induced cardiac toxicity in women with advanced breast cancer. N Engl J Med 1988; 319: 745–52.

33. Herman EH, Ferrans VJ, Young RS et al. Effect of pretreatment with ICRF-187 on the total cumulative dose of doxorubicin tolerated by beagle dogs. Cancer Res 1988; 48: 6918–25.

34. Herman EH, Ferrans VJ, Myers CE et al. Comparison of the effectiveness of (+/–)-1,2-bis(3,5-dioxopiperazinyl-1-yl)propane (ICRF-187) and N-acetylcysteine in preventing chronic doxorubicin cardiotoxicity in beagles. Cancer Res 1985; 45: 276–81.

35. Herman EH, Ferrans VJ. Amelioration of chronic anthracycline cardiotoxicity by ICRF-187 and other compounds. Cancer Treat Rev 1987; 14: 225–9.

36. Venturini M, Michelotti A, Del Mastro L et al. Multicenter randomized controlled clinical trial to evaluate cardioprotection of dexrazoxane versus no cardioprotection in women receiving epirubicin chemotherapy for advanced breast cancer. J Clin Oncol 1996; 41: 3112–20.

37. Lopez M, Vici P, Di Lauro L et al. Randomized prospective clinical trial of high dose epirubicin and dexrazoxane in patients with advanced breast cancer and soft tissue sarcomas. J Clin Oncol 1998; 16: 1–7.

38. Wexler LH, Andrich MP, Venzon D et al. Randomized trial of the cardioprotective agent ICRF-187 in pediatric sarcoma patients treated with doxorubicin. J Clin Oncol 1996; 14: 362–72.

39. Swain S, Whaley F, Gerber MC et al. Cardioprotection with dexrazoxane for doxorubicin containing therapy in advanced breast cancer. J Clin Oncol 1997; 15: 1318–32.

40. Decorti G, Klugmann F, Candussio L. Characterization of histamine secretion induced by anthracyclines in rat peritoneal mast cells. Biochem Pharmacol 1986; 35: 1939–42.

41. Klugmann FB, Decorti G, Candussio L et al. Amelioration of 4′-epidoxorubicin-induced cardiotoxicity by sodium cromoglycate. Eur J Cancer Clin Oncol 1989; 25: 361–8.

42. de Jong J, Schoofs PR, Onderwater RC et al. Isolated mouse atrium as a model to study anthracycline cardiotoxicity: the role of the beta-adrenoceptor system and reactive oxygen species. Res Commun Chem Pathol Pharmacol 1990; 68: 275–89.

43. Siveski-Iliskovic N, Hill M, Chow DA et al. Probucol protects against adriamycin cardiomyopathy without interfering with its antitumor effect. Circulation 1995; 91: 10–15.

44. Cao YH, Kennedy R, Klimberg VS. Glutamine protects against doxorubicin-induced cardiotoxicity. J Surg Res 1999; 85: 178–82.

45. Launchbury AP, Habboubi N. Epirubicin and doxorubicin: a comparison of their characteristics, therapeutic activity and toxicity. Cancer Treat Rev 1993; 19: 197–228.

46. Bertazzoli C, Rovero C, Ballerini L et al. Experimental systemic toxicology of 4′-epidoxorubicin, a new, less cardiotoxic anthracycline antitumor agent. Toxicol Appl Pharmacol 1985; 79: 412–22.

47. Pouna P, Bonoron-Adele S, Gouverneur G et al. Evaluation of anthracycline cardiotoxicity with the model of isolated, perfused rat heart: comparison of new analogues versus doxorubicin. Cancer Chemother Pharmacol 1995; 35: 257–61.

48. Plosker GL, Faulds D. Epirubicin. A review of its pharmacodynamic and pharmacokinetic properties, and therapeutic use in cancer chemotherapy. Drugs 1993; 45: 788–856.

49. Seymour, L, Bramwell, V, Moran, LA. Use of dexrazoxane as a cardioprotectant in patients receiving doxorubicin or epirubicin chemotherapy for the treatment of cancer. The Provincial Systemic Treatment Disease Site Group. Cancer Prev Control 1999; 3: 145–59.

50. Pouna P, Bonoron-Adele S, Gouverneur G et al. Development of the model of rat isolated perfused heart for the evaluation of anthracycline cardiotoxicity and its circumvention. Br J Pharmacol 1996; 117: 1593–9.

51. Coukell AJ, Faulds D. Epirubicin: An updated review of its oharmacodynamic and pharmacokinetic properties and therapeutic efficacy in the management of breast cancer. Drugs 1997; 53: 453–82.

52. Neri B, Cini-Neri G, Bandinelli M et al. Doxorubicin and epirubicin cardiotoxicity: experimental and clinical aspects. Int J Clin Pharmacol Ther Toxicol 1989; 27: 217–21.

53. Gennari A, Salvadori B, Donati S et al. Cardiotoxicity of epirubicin/paclitaxel-containing regimens: role of cardiac risk factors. J Clin Oncol 1999; 17: 3596–602.

54. Schonekess BO, Allard MF, Lopaschuk GD. Propionyl L-carnitine improvement of hypertrophied heart function is accompanied by an increase in carbohydrate oxidation. Circ Res 1995; 77: 726–34.

55. Reggiani C, Canepari M, Micheletti R et al. Effect of propionyl-L-carnitine on the kinetic properties of the myofibrillar system in pressure overload cardiac hypertrophy. Ann N Y Acad Sci 1995; 752: 204–6.

56. Regitz V, Shug AL, Fleck E. Defective myocardial carnitine metabolism in congestive heart failure secondary to dilated cardiomyopathy and to coronary, hypertensive and valvular heart diseases. Am J Cardiol 1990; 65: 755–60.

57. Pierpont ME, Judd M, Goldenberg IF et al. Myocardial carnitine in end-stage congestive heart failure. Am J Cardiol 1989; 64: 56–60.

58. Masmura M, Kobayashi A, Yamazaki N. Myocardial free carnitine and fatty acylcarnitine levels in patients with chronic heart failure. Jpn Circ J 1990; 54: 1471–6.

59. Matsui S, Sugita T, Matoba M. Urinary carnitine excretion in patients with heart failure. Clin Cardiol 1994; 17: 301–5.

60. Micheletti R, Giacalone G, Canepari M et al. Propionyl-L-carnitine prevents myocardial mechanical alterations due to pressure overload in rats. Am J Physiol Heart Circ Physiol 1994; 266: H2190–7.

61. Ferrari R, Di Lisa F, de Jong, JW et al. Prolonged propionyl-L-carnitine pre-treatment of rabbit: biochemical, hemodynamic and electrophysiological effect on myocardium. J Mol Cell Cardiol 1992; 24: 219–32.

62. Schonekess BO, Allard MF, Lopaschuk GD. Propionyl L-carnitine improvement of hypertrophied rat heart function is associated with an increase in cardiac efficiency. Eur J Pharm 1995; 286: 155–66.

63. el Alaoui-Talibi Z, Bouhaddioni N, Moravec J. Assessment of the cardiostimulant action of propionyl-L-carnitine on chronically volume-overloaded rat hearts. Cardiovasc Drugs Ther 1993; 7: 357–63.

64. Kobayashi A, Masmura Y, Yamazaki N. L-Carnitine treatment for congestive heart failure. Jpn Circ J 1992; 56: 86–94.

65. De Giuli F, Pasini E, Opasich C et al. Effects of propionyl-L-carnitine in patients with heart failure [abstract]. J Mol Cell Cardiol 1995; 27: V44.

66. Mancini M, Rengo F, Lingetti M et al. Controlled study on the efficacy of propionyl-L-carnitine in patients with congestive heart failure. Arzneimittelforschung 1992; 42: 1101–4.

67. Pucciarelli G, Mastursi M, Latte S et al. The clinical and hemodynamic effects of propionyl-L-carnitine in the treatment of congestive heart failure. Clin Ter 1992; 141: 379–84.

68. Tran MT. Role of coenzyme Q_{10} in chronic heart failure, angina, and hypertension. Pharmacotherapy 2001; 21: 797–806.

69. Mortensen SA. Perspectives on therapy of cardiovascular diseases with coenzyme Q_{10} (ubiquinone). J Clin Invest 1993; 71: S116–23.

70. Crane FL, Sun IL, Sun EE et al. Cell growth stimulation by plasma membrane electron transport. FASEB J 1991; 5: A1624.

71. Rodriguez M, Moreau P, Paulik M et al. NADH activated cell-free transfer between Golgi apparatus and plasma membranes of rat liver. Biochim Biophys Acta 1992; 1107: 131–8.

72. Rengo F, Abete P, Landino P et al. Role of metabolic therapy in cardiovascular disease. J Clin Invest 1993; 71: S124–8.

73. Keith M, Geranmayegan A, Sole MJ et al. Increased oxidative stress in patients with congestive heart failure. J Am Coll Cardiol 1998; 31: 1352–6.

74. Soja AM, Mortensen SA. Treatment of congestive heart failure with coenzyme Q10 illuminated by meta-analyses of clinical trials. Mol Aspects Med 1997; 18 (Suppl): S159–68.

75. Sacher HL, Sacher ML, Landau SW et al. The clinical and hemodynamic effects of coenzyme Q10 in congestive cardiomyopathy. Am J Ther 1997; 4: 66–72.

76. Sinatra ST. Coenzyme Q10: a vital therapeutic nutrient for the heart with special application in congestive heart failure. Conn Med 1997; 61: 707–11.

77. Watson PS, Scalia GM, Galbraith A et al. Lack of effect of coenzyme Q on left ventricular function in patients with congestive heart failure. J Am Coll Cardiol 1999; 33: 1549–52.

78. Khatta M, Alexander BS, Krichten CM et al. The effect of coenzyme Q_{10} in patients with congestive heart failure. Ann Intern Med 2000; 132: 636–40.

79. Sinatra S. Coenzyme Q_{10} and congestive heart failure [letter]. Ann Intern Med 2000; 133: 745.

80. Mortensen SA. Coenzyme Q_{10} as an adjunctive therapy in patients with congestive heart failure [letter]. J Am Coll Cardiol 2000; 36: 304.

81. Gottlieb S, Khatta M, Fisher M. Response to S Sinatra: coenzyme Q_{10} and congestive heart failure. Ann Intern Med 2000; 133: 745.

82. Watson P, Scalia G, Galbraith A et al. Response to S Mortensen: coenzyme Q_{10} as an adjunctive therapy in patients with congestive heart failure. J Am Coll Cardiol 2000; 36: 304.

83. Frishman WH, Sinatra S, Kruger N. Nutriceuticals and cardiovascular illness. In: Frishman WH, Weintraub M, Micozzi M (eds). Complementary and Integrative Therapies for Cardiovascular Disease. St Louis: Elsevier; 2004: 58–85.

84. Bacon BR. Iron overload (hemochromatosis). In: Goldman L, Ausiello D (eds). Cecil Textbook of Medicine (22e). Phildelphia: Saunders; 2004: 1302–6.

85. Leslie D, Gheorghiade M. Is there a role for thiamine supplementation in the management of heart failure? Am Heart J 1996; 131: 1248–50.

86. Seligmann H, Halkin H, Rauchfleisch S et al. Thiamine deficiency in patients with congestive heart failure receiving long-term furosemide therapy: A pilot study. Am J Med 1991; 91: 151–5.

87. Marcus R, Coulston AM. Water soluble vitamins. In: Hardman G, Limbird LE (eds). Goodman Gilman's The Pharmacological Basis of Therapeutics (10e). New York: McGraw Hill; 2001: 1753–71.

88. Shimon I, Almog S, Vered Z et al. Improved left ventricular function after thiamine supplementation in patients with congestive heart failure receiving long-term furosemide therapy. Am J Med 1995; 98: 485–90.

89. Neufeld EJ, Fleming JC, Tartaglini E et al. Thiamine-responsive megaloblastic anemia syndrome: a disorder of high-affinity thiamine transport. Blood Cells Mol Dis 2001; 27: 135–8.

90. Luft FC. Lactic acidosis update for critical care clinicians. J Am Soc Nephrol 2001; 12: S15–19.

91. Blankenhorn MA. Effect of vitamin deficiency on the heart and circulation. Circulation 1995; 11: 288–91.

92. Akbarian M, Yankopoulos NA, Abelmann WH. Hemodynamic studies in beriberi heart disease. Am J Med 1966; 41: 197–212.

93. Webster MWI, Ikram H. Myocardial function in alcoholic cardiac beriberi. Int J Cardiol 1987; 17: 213–15.

94. Constant J. The alcoholic cardiomyopathies – genuine and pseudo. Cardiology 1999; 91: 92–5.

95. Talwar D, Davidson H, Cooney J et al. Vitamin B(1) status assessed by direct measurement of thiamin pyrophosphate in erythrocytes of whole blood by HPLC: comparison with erythrocyte transketolase activation assay. Clin Chem 2000; 46: 704–10.

96. Rubin E, Urbano-Marquez A. Alcoholic cardiomyopathy. Alcoholism. Clin Exper Res 1994; 18: 111–14.

97. Preedy VR, Richardson PJ. Ethanol induced cardiovascular disease. Br Med Bull 1994; 1: 152–63.

98. McKenna CJ, Codd MB, McCann HA et al. Alcohol consumption and idiopathic dilated cardiomyopathy: a case control study. Am Heart J 1998; 135: 833–7.

99. Piano MR, Schwertz DW. Alcoholic heart disease. Heart Lung 1994; 23; 3–17.

100. Regan TJ, Ettinger PO, Halder B et al. The role of ethanol in cardiac disease. Ann Rev Med 1977; 28: 393–409.

101. Mouschmousch B, Abi-Monsour P. Alcohol and the heart: the long term effect of alcohol on the cardiovascular system. Arch Intern Med 1991; 151: 36–41.

102. Urbano-Marquez A, Estruch R, Navarro-Lopez F et al. The effects of alcoholism on skeletal and cardiac muscle. N Engl J Med 1989; 320: 409–15.

103. Timmis GC, Ramos RC, Gordon S et al. The basis for differences in ethanol induced myocardial depression in normal subjects. Circulation 1975; 51: 1144–8.

104. Fernandez-Sola J, Nicolas JM, Oriola J et al. Angiotensin-converting enzyme gene poly-morphism is associated with vulnerability to alcoholic cardiomyopathy. Ann Intern Med 2002; 137: 321–6.

105. Lazarevic AM, Nakatani S, Neskovic AN et al. Early changes in left ventricular function in chronic asymptomatic alcoholics: relation to duration of heavy drinking. J Am Coll Cardiol 2000; 35: 1599–606.

106. Capasso JM, Li P, Guideri G et al. Left ventricular dysfunction induced by chronic alcohol ingestion in rats. Am J Physiol Heart Circ Physiol 1991; 261: H212–19.

107. Moriya N, Nihira M, Sato S et al. Ultrastructural changes of liver, heart, lung and kidney. Jpn J Alcohol Studies Drug Depend 1992; 27: 189–200.

108. Segel LD, Rendig SV, Choquet Y et al. Effects of chronic graded ethanol consumption on the metabolism, ultrastructure and mechanical function of the rat heart. Cardiovasc Res 1975; 9: 649–63.

109. Siddiq T, Salisbury JR, Richardson PJ et al. Synthesis of ventricular mitochondrial proteins in vivo : effect of acute ethanol toxicity. Alcoholism: Clin Exper Res 1993; 17: 894–9.

110. Preedy VR, Patel VB, Why HJ et al. Alcohol and the heart: biochemical alterations. Cardiovasc Res 1996; 139–47.

111. Urbano-Marquez A, Estruch R, Fernandez-Sola J et al. The greater risk of alcoholic cardiomyopathy and myopathy in women. JAMA 1995; 274: 149–54.

112. Kupari M, Koskinen P. Comparison of the cardiotoxicity of ethanol in women versus men. Am J Cardiol 1992; 70: 645–9.

113. Altura BM, Altura BT, Gebrewold A. Comparative effects of ethanol, acetaldehyde and acetate on arterioles and venules in skeletal muscle: direct in situ studies on the microcirculation and their possible relationship to alcoholic myopathy. Microcirc Endothelium Lymphatics 1990; 6: 107–26.

114. Guarnieri T, Lakatta EG. Mechanism of myocardial contractile depression by clinical concentration of ethanol. A study of ferret papillary muscles. J Clin Invest 1990; 85: 1462–7.

115. Adams MA, Hirst M. Lack of cardiac alpha 1 adrenoceptor involvement in ethanol induced cardiac hypertrophy. Can J Physiol Pharmacol 1989; 67: 240–5.

116. Adams MA, Hirst M. Metoprolol suppresses the development of ethanol induced cardiac hypertrophy in the rat. Can J Physiol Pharmacol 1990; 68: 562–7.

117. Demakis JG, Proskey A, Rahimtoola SH et al. The natural course of cardiomyopathy. Ann Intern Med 1974; 80: 293–7.

118. Guillo P, Mansourati J, Maheu B et al. Long-term prognosis in patients with alcoholic cardiomyopathy and severe heart failure after total abstinence. Am J Cardiol 1997; 79: 1276–8.

119. Gavazzi A, De Maria R, Parolini M et al. on behalf of the Italian Multicenter Cardiomyopathy Study Group (SPIC). Am J Cardiol 2000; 85: 1114–18.

120. Nicolas JM, Fernandez-Sola J, Estruch R et al. The effect of controlled drinking in alcoholic cardiomyopathy. Ann Intern Med 2002; 136: 192–200.

121. Fauchier L, Babuty D, Poret P et al. Comparison of long term outcome of alcoholic and idiopathic dilated cardiomyopathy. Eur Heart J 2000; 21: 267–9.

122. Prazak P, Pfisterer M, Osswald S et al. Differences of disease progression in congestive heart failure due to alcoholic as compared to idiopathic dilated cardiomyopathy. Eur Heart J 1996; 17: 251–7.

123. Frishman WH, Del Vecchio A, Sanal S et al. Cardiovascular manifestations of substance abuse. Part 2: Alcohol, amphetamines, heroin, cannabis and caffeine. Heart Dis 2003; 5: 253–71.

124. Abramson JL, Williams SA, Krumholz HM et al. Moderate alcohol consumption and risk of heart failure among older persons. JAMA 2001; 285: 1971–7.

125. Cooper HA, Exner DV, Domanski MJ. Light to moderate alcohol consumption and prognosis in patients with left ventricular systolic dysfunction. J Am Coll Cardiol 2000; 35: 1753–9.

126. Kosten TR, Hollister LE. Drugs of abuse. In: Katzung BG (ed). Basic and Clinical Pharmacology (8e). Norwalk: Appleton and Lange; 2000: 532–48.

127. Lester SJ, Baggott M, Welm S et al. Cardiovascular effects of 3,4-methylenedioxy-methamphetamine. A double-blind, placebo-controlled trial. Ann Intern Med 2000; 133: 969–73.

128. Committee on Alcoholism and Addiction and Council on Mental Health. Dependence on amphetamines and other stimulant drugs. JAMA 1966; 197: 1023.

129. Badon LA, Hicks A, Lord K et al. Changes in cardiovascular responsiveness and cardio-toxicity elicited during binge administration of ecstasy. J Pharmacol Exp Ther 2002; 302: 898–907.

130. Hong R, Matsuyama E, Nur K. Cardiomyopathy associated with the smoking of crystal methamphetamine. JAMA 1991; 265: 1152–4.

131. Smith HJ, Roche AH, Jausch MF et al. Cardiomyopathy associated with amphetamine administration. Am Heart J 1976; 91: 792–7.

132. Jacobs LJ. Reversible dilated cardiomyopathy induced by methamphetamine. Clin Cardiol 1989; 12: 725–7.

133. O'Neill ME, Arnolda LF, Coles DM et al. Acute amphetamine cardiomyopathy in a drug addict. Clin Cardiol 1983; 6: 189–91.

134. Call TD, Hartneck J, Dickinson WA et al. Acute cardiomyopathy secondary to intravenous amphetamine. Ann Intern Med 1982; 97: 559–60.

135. Eagan TS, Watson RR. Methamphetamine in heart disease. Cardiovasc Rev Rep 2002; 23: 320–4.

136. Kahn DS, Rona G, Chappel CI. Isoproterenol induced cardiac necrosis. Ann NY Acad Sci 1969; 156: 285–93.

137. Pearce RM. Experimental myocarditis: a study of the histological changes following intravenous injections of adrenalin. J Exper Med 1906; 8: 400.

138. Factor SM, Sonnenblick EH. The pathogenesis of clinical and experimental congestive cardiomyopathies: recent concepts. Prog Cardiovasc Dis 1985; 27: 395–420.

139. Imperato-McGinley J, Gautier T, Ehlers K et al. Reversibility of catecholamine induced dilated cardiomyopathy in a child with pheochromocytoma. N Engl J Med 1987; 316: 793–7.

140. Frishman WH, Del Vecchio A, Sanal S et al. Cardiovascular manifestations of substance abuse. Part 1: Cocaine. Heart Dis 2003; 5: 187–201.

141. Linder JD, Monkemuller KE, Raijman I et al. Cocaine associated ischemic colitis. Southern Med J 2000; 93: 909–13.

142. Albrecht CA, Jafri A, Linville L et al. Cocaine-induced pneumopericardium. Circulation 2000; 102: 2792–4.

143. Pentel PR, Thompson T, Hatsukami DK et al. 12 lead and continuous ECG recording of subjects during inpatient administration of smoked cocaine. Drug Alcohol Depend 1994; 35: 107–16.

144. Hollander JE, Burstein JL, Hoffman RS et al. Cocaine associated myocardial infarction. Clinical safety of thrombolytic therapy. Chest 1995; 107: 1237–41.

145. Isner JM, Estes NA, Thompson PD et al. Acute cardiac events temporally related to cocaine abuse. New Engl J Med 1986; 315: 1438–43.

146. Gardin JM, Wong N, Alker K et al. Acute cocaine administration induces ventricular regional wall motion abnormalities in an anesthetized rabbit model. Am Heart J 1994; 128: 1117–29.

147. Abel FL, Wilson SP, Zhao RR et al. Cocaine depresses the canine myocardium. Circulatory Shock 1989; 28: 309–19.

148. Hale SL, Alker KJ, Rezkella SH et al. Nifedipine protects the heart from the acute deleterious effects of cocaine if administered before but not after cocaine. Circulation 1991; 83: 1437–43.

149. Morcos NC, Fairhurst A, Henry WL. Direct myocardial effects of cocaine. Cardiovasc Res 1993; 27: 269–73.

150. Renard DC, Delaville FJ, Thomas AP. Inhibitory effects of cocaine on Ca^{2+} transients and contraction in single cardiomyocytes. Am J Physiol Heart Circ Physiol 1994; 266: H555–67.

151. Perrault CL, Hauge NL, Morgan KG et al. Negative inotropic and relaxant effects of cocaine on myopathic human ventricular myocardium and epicardial coronary arteries in vitro. Cardiovasc Res 1993; 27: 262–8.

152. Stewart G, Rubin E, Thomas AP. Inhibition by cocaine of excitation contraction coupling in isolated cardiomyocytes. Am J Physiol Heart Circ Physiol 1991; 260: H50–7.

153. Yuan C, Acosta D. Inhibitory effect of cocaine on calcium mobilization in cultured rat myocardial cells. J Mol Cell Cardiol 1994; 26: 1415–19.

154. Tomita F, Basset AL, Myerberg RJ et al. Effects of cocaine on sarcoplasmic reticulum in skinned rat heart muscle. Am J Physiol Heart Circ Physiol 1993; 264: H845–50.

155. Perrault CL, Hauge BJ, Morgan JP. The effects of cocaine on intracellular Ca^{2+} handling and myofilament Ca^{2+} responsiveness of ferret ventricular myocardium. Br J Pharmacol 1990; 101: 679–85.

156. Pilati CF, Espinal AR, Pukys TF. Persistent left ventricular dysfunction after cocaine treatment in rabbits. Proc Soc Exper Biol Med 1993; 203: 100–7.

157. Kambam JR, Franks JJ, Mets B et al. The effects of hepatectomy and plasma cholinesterase inhibition on cocaine metabolism and cardiovascular responses in pigs. J Lab Clin Med 1994; 124: 715–22.

158. Crumb WJ, Clarkson CW. Characteristics of the sodium channel blocking properties of the major metabolites of cocaine in single cardiac myocytes. J Pharmacol Exper Therap 1992; 261: 910–17.

159. Brickner ME, Willard JE, Eichhorn EJ et al. Left ventricular hypertrophy associated with chronic cocaine abuse. Circulation 1991; 84: 1130–5.

160. Ciggarroa CG, Boehrer JD, Brickner ME et al. Exaggerated pressor response to treadmill exercise in chronic cocaine abusers with left ventricular hypertrophy. Circulation 1992; 86: 226–31.

161. Om A, Ellahham S, Vetrovec GW et al. Left ventricular hypertrophy in normotensive cocaine users. Am Heart J 1993; 125: 1441–3.

162. Hoegerman GS, Lewis CE, Flack J et al. Lack of association of recreational cocaine and alcohol use with left ventricular mass in young adults. J Am Coll Cardiol 1995; 25: 895–900.

163. Henning RJ, Silva J, Reddy V et al. Cocaine increases beta-myosin heavy-chain protein expression in cardiac myocytes. J Cardiovasc Pharmacol Ther 2000; 5: 313–22.

164. Missouris CG, Swift PA, Singer DRJ. Cocaine use and acute left ventricular dysfunction [case report]. Lancet 2001; 357: 1586.

165. Goldfrank LR, Hoffman RS. The cardiovascular effects of cocaine. Ann Emerg Med 1991; 20: 165–75.

Arrhythmogenesis

CHAPTER 22

Cellular and molecular basis of electrogenesis of normal cardiac tissue

Wen Dun, Penelope A Boyden

Introduction

Basic impulse propagation mechanisms

Passive properties

Excitability of a myocyte is dictated by the make-up of its active and passive membrane properties. The passive electrical behavior of a cell membrane is similar to a fixed resistor and capacitor in parallel, and determines the behavior of flow of the subthreshold currents of the excitatory impulse. The membrane length constant is the distance at which potential has decayed to $1/e$ (37%) of the value at the source of the current application. Because atrial cells are smaller in diameter than ventricular cells,[1] r_i (internal resistance of the fiber per unit length) is higher in atrial cells assuming that the cytosol of atrial and ventricular myocytes has a similar conductivity. This effect makes the membrane length constant shorter in atrial versus ventricular cells.[1]

Impulse propagation

By replacing a fixed resistor with a set of variable resistors, one can modify the linear cable model to simulate impulse propagation in a strand of cardiac 'cells' where depolarizing current flow and extracellular/intracellular resistances, r_o/r_i, have independent effects (continuous medium) on the velocity of propagation. Under these conditions maximal upstroke velocity, dV/dt_{max} (V_{max}), an experimental measure of maximal depolarizing current during the cardiac action potential, is related to the propagation velocity by $\theta = \sqrt{dV/dt_{max}}$. In arrhythmic substrates it is more likely that due to macro- and microstructural discontinuities, impulse propagation is discontinuous.[2]

To understand propagation in a two- or three-dimensional structure, one must consider the curvature of the wavefront. From wave theory, a wave with a flat wave front has a propagation velocity equal to that of the wave in a one-dimensional strand (a cable). For wave fronts curving outwards (convex), actual propagation velocity is reduced compared to that of the flat wave. Here the gradient of local current flow at the 'head' of the wave front is a small 'source' relative to the large 'sink' of tissue not yet excited. Thus propagation is inefficient and slow. For wave fronts curving inwards (concave), the opposite holds.[3]

Safety factor

Critical to the understanding of propagation in both homogeneous and inhomogeneous tissue is the concept of safety factor of propagation. Safety factor has been described as the ratio of electrical charge generated by the excitatory impulse, divided by the charge needed to bring the cell (tissue) to threshold.[4] A change in the safety factor for propagation can occur for example with decreased electrical coupling (an increase) or decreased excitability (e.g. sodium current, I_{Na}, changes – a decrease).

Ion channels between cells are formed by connexin proteins. Their physiology and molecular composition are discussed in Chapter 25.

Ion channels and transporters

Voltage-gated channels

Fundamental to most types of voltage-gated ion channels is that the opening of the pore of the channel protein is energetically coupled to the movement of a part of the channel protein in response to a changed electric field. Diverse structural domains within the specific protein then modulate this gated channel to occlude or inactivate the channel. Recent crystal structures of several ion channel proteins (e.g. K^+ channel[5]) have allowed for a more precise detailing of the atomic structures that make up the voltage-gated ion channel.

Cardiac sodium channels

Voltage-gated sodium channels are responsible for the current (fast I_{Na}) that initiates most types of cardiac action potentials. Sustained currents through these channels can also contribute to the action potential (AP) plateau. The cardiac sodium channel is composed of an α subunit (Nav1.5) and two to four β subunits. Other α subunits, thought to encode neuronal type Na^+ channels (Nav1.1, Nav1.3, and Nav1.6, see reference 6 for nomenclature) have been identified in ventricular cells of some species,[7] and may also contribute to Na^+-dependent potential depolarization. Within the same species (guinea pig) I_{Na} differs between atrial and ventricular cells.[8] Atrial I_{Na} has a greater density, and different gating kinetics than its ventricular counterpart. This may be due to a different subset of β subunits in atria versus those in ventricle.[9] In cell expression systems, the Na^+ channel β subunits are multifunctional since they can modulate the channel's gating, and expression, as well as functioning as cell adhesion molecules.[10]

The Na^+ channel α subunit consists of four homologous domains, each of which contains six α-helical transmembrane segments. Voltage sensors of the folded protein are in the four S4 regions, and upon sensing voltage cause a change in conformation leading to the active or open state of the channel. Then channels can inactivate even though depolarization is maintained. A small number of Na^+ channels can fail to inactivate and reopen or show a bursting pattern.[11] Recovery from 'fast' inactivation of the Na^+ channel occurs over tens of milliseconds, while that from slow inactivation is longer (tens of seconds). It is this rate of recovery from inactivation of the Na^+ channel that contributes greatly to the refractoriness of a cardiac cell's excitability. Hence, refractoriness of cardiac tissue is due to the recovery of Na^+ channels from inactivation, which is time and voltage dependent. Thus during the course of normal AP repolarization the membrane potential is directly related to the degree of Na^+ channel availability.

Fast inactivation of open channels develops rapidly, as the intracellular peptide between domains II and IV occludes the channel by binding to a receptor site of residues of the III–IV linker, similar to the 'ball and chain' inactivation of K^+ channels.[12] Inactivation can be modified further, and the highly charged and structured regions of the C terminus of the Na^+ channel protein have been shown to modulate channel inactivation.[13] Post-translational modifications of the protein include glycosylation and phosphorylation. Both protein kinase C (PKC)- and protein kinase A (PKA)-dependent phosphorylation alter Na^+ channel density and its gating kinetics.[14,15] A PKA-mediated effect on the channel may also promote trafficking of the channel to the cell surface. In fact PKA-dependent enhancement of Na^+ channel bursting behavior has been reported, and these findings implicate domains of the C terminus in this effect.[11] PKC stimulation inhibits channel bursting via phosphorylation of the α subunit.[16] Both effects become particularly important in some Na^+ channel disease-induced mutants.[11,16] Whether the same holds for dysfunction of Na^+ channels in acquired disease is unknown.

Cardiac calcium channels

The L-type (L for long-lasting, I_{CaL}) and T type (T for transient, I_{CaT}) Ca^{2+} currents were initially described in neuronal tissues but subsequently recognized to exist in the various tissues of the heart. Bean first described multiple cardiac Ca^{2+} channels in canine atrial cells.[17] At that time, two types of Ca^{2+} currents carried by Ba^{2+} ions were recognized. Subsequently, I_{CaL} and I_{CaT} have been recorded in tissues of most species under various conditions. However, within the same species, the density of I_{CaL} and I_{CaT} currents varies depending on the location of the myocyte within the heart. There is a large density of both the L- and the T-type channels in rabbit sino-atrial (SA) node cells,[18] and large T-type currents in latent atrial pacemaker cells.[19] In Purkinje cells dispersed both from free-running fiber bundles and the subendocardium of the left ventricle (LV),[20–22] there is a large peak T/L current:density ratio. Notably, T currents have not been observed in human atrial,[23,24] human ventricular,[24,25] or human Purkinje cells (unpublished data, Boyden laboratory).

Cardiac L and T type Ca^{2+} channels are different in the following biophysical properties:

- **voltage range of activation**: the T channel activation occurs at more negative voltages than the L channel, e.g. in 5mM $[Ca^{2+}]_o$ the threshold for activation is –50 and –30mV for T and L respectively[20]
- **voltage range of inactivation**: in 5mM $[Ca^{2+}]_o$, the T channel can be inactivated by membrane depolarization positive to –70mV. The L channel remains fully available for activation at potentials positive to –40 mV
- **mechanism of inactivation**: T channel inactivates solely by membrane depolarization. For the L channel, both membrane depolarization and Ca^{2+} ions participate in the inactivation process.

Voltage-dependent inactivation of the L-type Ca^{2+} current is clearly evident as channels incorporated into lipid bilayers inactivate even when Ca^{2+} is buffered,[26] and as noted from the dependence of the time course of Ba^{2+} current decay on voltage.[27,28]

The molecular determinants of voltage-dependent inactivation of Ca^{2+} channels are less well understood than those of K^+ or Na^+ channels. Several critical locations throughout the channel protein have been implicated in the voltage-dependent inactivation process. They are: the I–II linker, the proximal C terminus, the EF hand area in IC, the S6s of domain III and IV.[29–36] One model proposed suggests that a domain of the I–II linker docks to one or all of the S6 segments at the cytoplasmic end.[37] Importantly, this mechanism is not involved in 'recovery from inactivation', only the channel's response to depolarization.

The molecular basis of the cardiac L-type Ca^{2+} channel structure is due to the combination of the α_{1C} subunit (Cav1.2; see reference 38 for nomenclature) with obligatory β_2, α_2/δ and γ subunits. The γ subunit (33 kDa) is exclusively expressed in skeletal muscle. The voltage sensor for activation is the highly charged S4 segment. Ca^{2+}-binding sites formed by glutamates in the pore loop of each repeat not only distinguish a Ca^{2+} channel from a Na^+ channel, but are critical for selectivity of the channel.[39] Sections of the α_1 subunit pore-forming segments, intracellular loops, and C terminus all contribute to channel inactivation. Mutating single amino acids in IIIS6 and IVS6 domains have an effect on current decay, suggesting that an area in the inner channel mouth is a key player in channel inactivation. Point mutations in the intracellular I–II loop and the IVS5–IVS6 linker both affect Ca^{2+} channel inactivation. These sections of the protein are also critical sites of β subunit and/or G protein interactions.

Voltage-dependent inactivation of the L-type Ca^{2+} channel is slow relative to the Ca^{2+}-induced inactivation of the channel. The concentration of Ca^{2+} that controls this feedback inhibition of the channel is very high, occurring in the microdomain between the sarcolemma (or the t-tubular membrane) and the Ca^{2+} release channel/sarcoplasmic reticulum (SR) complex. For L-type Ca^{2+} channels, calmodulin (CaM) is prebound to the residues of the C terminus of the protein.[40] When cytosol Ca^{2+} increases, it binds to the prebound CaM, which then interacts with the specific IQ motif on the C terminus, moving it towards the inner pore and thus blocking the channel's opening. Recovery from Ca^{2+}-induced inactivation is voltage independent *per se*, but it requires a fall in cytosolic Ca^{2+} which is related to voltage-dependent Ca^{2+} efflux of the Na^+–Ca^{2+} exchanger.

369

Currently, four potential β subunits (β1–β4) have been recognized, but in cardiac cells the β_2 subunit predominates, providing a rate-limiting step in the expression of Ca^{2+} channel proteins.[41] The α_{1C} subunit (Cav1.2) when expressed alone is sufficient for L-type channel activity, but when expressed with a cytosolic β subunit (55kDa),[42] which binds to the α subunit at the intracellular I–II loop, peak currents increase 4-fold, apparently by accelerating opening of the pore and reducing the rate of channel closure.[43–47] Furthermore, there is a shift in the activation curve, a slowing of activation, and an enhancement of the inactivation process.[45]

The two-component α_2–δ subunit (170 kDa) remains linked *in vitro*, with the α_2 subunit being an extracellular highly glycosylated protein. δ is the short membrane-spanning protein with a fully glycosylated extracellular portion. This subunit complex affects both ionic and gating currents of the expressed Ca^{2+} channels by increasing the number of functional channels at the cell surface.[48–51] In one series of experiments, coexpression of α_2–δ with α_{1C} β_3 subunits prevented voltage-dependent facilitation.[52] In these latter studies, this voltage-dependent facilitation was due to an increase in the number of channels that were able to produce gating current, as well as the number of channels that opened in response to voltage.[52] The mostly extracellular α_2–δ subunit causes these changes by a direct interaction with the pore unit of the α subunit.[53]

In addition to Cav1.2 (α_{1C}), mRNA and protein from α_{1D} (Cav1.3) subunits have been measured in heart tissues.[54,55] Cav1.3 Ca^{2+} channel proteins are sparse, but this type of Ca^{2+} protein may serve a specific functional role. *Cav1.3* mRNA has been detected in the mouse sino-atrial node (SAN),[56] and atria. Currents mediated by Cav1.3 proteins (plus β_{2a} subunits) activate more quickly, and decay more slowly than those of Cav1.2 proteins.[57] Furthermore, steady-state inactivation and activation relations of Ba^{2+} currents through Cav1.3 channels are shifted in the hyperpolarizing direction. These specific kinetic differences between Cav1.2 and Cav1.3 currents account in part for the decreased sensitivity of the expressed Cav1.3 currents to dihydropyridine block.[58]

Finally, mice lacking α_{1D} subunit proteins are deaf, exhibit sinoatrial dysfunction and bradycardia, suggesting a role for α_{1D} proteins in normal pacemaker activity.[58,59]

The molecular basis of neuronal and cardiac T-type Ca^{2+} channels has been defined.[60,61] In both cases, the low-voltage T-type Ca^{2+} channel protein (α_{1H}, α_{1G}; Cav3.3 and Cav3.2) has high sequence identity with the α_{1C} subunit, particularly in the membrane-spanning regions.[60] Charged residues of the S4 regions are conserved between α_{1C} and α_{1H}, α_{1G}, while a ring of glutamates important in α_{1C} channel selectivity have been partially replaced by aspartates. T-type Ca^{2+} currents inactivate with voltage but not by Ca^{2+}.[62,63] Some have suggested that a 'ball and chain' type mechanism involving the amino side of the C terminus contributes to T-type channel inactivation.[62,64] Intracellular loop motifs involved in β subunit binding,[65,66] or Ca^{2+} binding[67] of the L-type α_{1C} protein are missing in both the α_{1G} and α_{1H} proteins.

Potassium conductance

I_{K1}-inward rectifier: In heart, inward rectification was first described for currents in Purkinje fibers.[68] Numerous studies that followed suggested that the inward rectifier potassium current (I_{K1}) of a cardiac cell plays an important role in setting the resting membrane potential and shaping the late repolarization phase of the action potential.[69–72] Mammalian atrial cells have been reported to have less I_{K1} than ventricular cells, and this seems to hold true for human myocytes. Studies in the canine cells reveal that I_{K1} in ventricular cells is substantially (and significantly) larger than that in atrial cells. In human hearts, ventricular myocytes' I_{K1} is three times larger than I_{K1} in atrial cells, and ventricular I_{K1} inactivation kinetics are half those in atrial cells.[73,74] These differences contribute to functionally important and differing action potential time courses in ventricular and atrial cells (Fig. 22.1).

I_{K1} channels show voltage-dependent activation and inactivation like the Kv channels. The increase of current following hyperpolarization is referred to as activation, and the reduction of channel current at more positive potentials is due to a property of the channel, inward rectification.

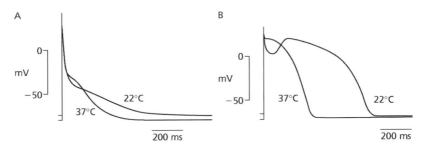

Figure 22.1. *Transmembrane action potentials elicited from human atrial (A) and human ventricular (B) myocytes paced at 0.2Hz. The effects of temperature on each waveform are illustrated. Modified with permission from Amos GJ, Wettwer E, Metzger F et al. Differences between outward currents of human atrial and subepicardial ventricular myocytes. J Physiol 1996; 491.1: 31–50.*

I_{K1} channels are sensitive to both the potassium equilibrium potential and the transmembrane potential. Little outward current flows positive to the potassium equilibrium potential, due to the inward rectifying properties of the channel. However, inspection of the whole-cell current–voltage relations in atrial and ventricular cells illustrates that the marked N-shape of the ventricular current–voltage relation is less prominent in atrial cells (Fig. 22.2, Table 22.1). This suggests that the mechanism of inward rectification differs between the atria and ventricles. Detailed single-channel data have shown that single-channel characteristics of I_{K1} in atria and ventricles do not differ.[75] What differed was the percentage of patches containing single I_{K1} channels. The reason for this heterogeneity between myocytes from different chambers may be explained by differing post-

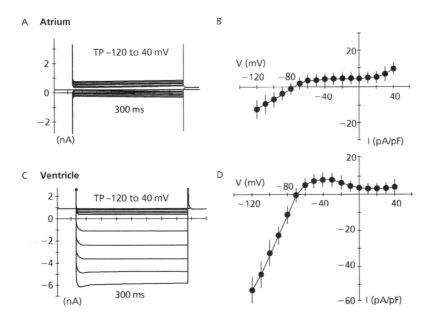

Figure 22.2. *Typical current tracings from a human atrial (A) and ventricular (C) cell during voltage-clamp protocols to elicit inwardly rectifying currents (I_{K1}). Average current density relations are shown in (B) and (D), respectively. TP, holding or test voltages. Electrophysiological characteristics of single I_{K1} channels are shown in Table 22.1. Modified with permission from Koumi S, Backer CL, Arentzen CE. Characterization of inwardly rectifying K^+ channel in human cardiac myocytes: alteration in channel behavior in myocytes isolated from patients with idiopathic dilated cardiomyopathy. Circulation 1995; 92: 164–74.*

Table 22.1. *Electrophysiological characteristics of single I_{k1} channels from human atrium and ventricle*

	n	γ (pS)	$τ_o$ (ms)	P_o	%
Atrium	18	27 ± 2	30.2 ± 4.0	0.56 ± 0.06	34.7
Ventricle	23	28 ± 3	31.5 ± 4.6	0.59 ± 0.08	88.6

γ, slope conductance; $τ_o$, time constant for open time distributions; P_o, channel open probability. Values are presented as mean ± standard deviation for values from six patients who provided both atrial and ventricular specimens. Values were not statistically different between the two groups. Modified with permission from Koumi S, Backer CL, Arentzen CE. Characterization of inwardly rectifying K⁺ channel in human cardiac myocytes: alteration in channel behavior in myocytes isolated from patients with idiopathic dilated cardiomyopathy. Circulation 1995; 92: 164–74.

translational modifications of the α subunit underlying I_{K1} conductance.

The inward rectification of the I_{K1} channel is due to internal block of channel by Mg^{2+},[76] and cytoplasmic polyamines.[77] A study of the structure and function of IRK1 (inward rectifier K⁺ channel) showed that a negatively charged aspartate in the putative second transmembrane domain is essential for time-dependent block by the cytoplasmic polyamines, and a negatively charged residue contributes to the formation of the binding pocket for Mg^{2+} and polyamines.[78]

On the basis of sequence homology, inward rectifier potassium channels (I_{K1}) are now classified into seven subfamilies (Kir1 to Kir7).[79,80] Kir subunits are made up of two transmembrane domains (TM1 and TM2) flanking a well-conserved pore-loop (P-loop) and large hydrophilic N and C termini located on the cytoplasmic side of the membrane. Functional Kir channels (I_{K1}) are assembled from four subunits that are symmetrically arranged around a central pore-lined TM2 and the P-loop. Transcription of cloned Kir channels suggests that the Kir2 subunits (Kir2.1, Kir2.2 and Kir2.3) underlie cardiac I_{K1}. Three distinct Kir2 subfamily members have been cloned to date, and all encode strong inward rectifiers. The expressed channels differ in single-channel conductance (Kir2.1: 20pS, Kir2.2: 35pS, Kir2.3: 10pS) and in sensitivity to phosphorylation and other second messengers.[81–83]

Human atrial I_{K1} can be inhibited by $α_1$-adrenergic stimulation via a PKC-dependent pathway.[84] Expressed Kir2.1 inward rectifier potassium currents could be downregulated by a specific stimulator of PKC, however currents can also be increased by cAMP-dependent protein kinase (PKA).[85] This suggests that PKA and PKC mediate opposing effects on Kir2.1 channels. Human cardiac Kir2.1 is termed as Kir2.1a and Kir2.1b, respectively, according to the homology to the mouse tissue (Kir2.1a). Kir2.1b is 70% homologous to that of the mouse tissue,[86] and has functional PKC phosphorylation sites. Thus, Kir2.1b can be inhibited by direct PKC-dependent regulation, whereas Kir2a cannot.[83] In rabbit cardiomyocytes, macroscopic I_{K1} appears to be heteromeric channels assembled from both Kir2.1 and Kir2.2.[87] Therefore, the regulation of whole-cell I_{K1} channel by the second messengers and phosphorylation depends on the dominant Kir subunit, which may differ in different cell types and species.

I_{to} transient outward current: In studies using human atrial myocytes, it is clear that various types of outward currents can be differentiated according to rate of inactivation and steady-state density.[1,88] Most cells possess a large transient outward current (I_{to}) that is activated by voltage, is 4-aminopyridine sensitive, and intracellular calcium independent. A close evaluation of the availability of the atrial cell current after various prepotentials has shown that voltage can inactivate I_{to}, but a large sustained component remains. In contrast, in similar protocols using human ventricular myocytes, most of the transient outward current is inactivated with prepulses to various voltages, thus little sustained component remains (Fig. 22.3). The presence of a

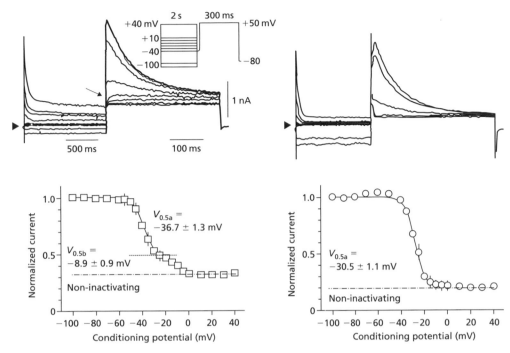

Figure 22.3. *Typical voltage-clamp protocol and current tracings obtained in a human atrial (left panel) and ventricular cell (right panel) during pulses that inactivate the voltage-dependent transient outward current (I_{to1}). Resulting graphs depicting voltage dependence of inactivation are shown below. Note that prepulses inactivate both the peak ($V_{0.5a}$) and sustained ($V_{0.5b}$) component of outward currents in the atrial cell (left panels) and only the transient component in the ventricular cell (right panel). Modified with permission from Amos GJ, Wettwer E, Metzger F et al. Differences between outward currents of human atrial and subepicardial ventricular myocytes. J Physiol 1996; 491.1: 31–50.*

large-density sustained outward current during depolarizing steps (I_{KUR}, see below) is a clear difference in ionic make-up of these two cell types. Undoubtedly its existence in atrial myocytes contributes significantly to the marked accelerated repolarization of the action potential. Furthermore, differences in the kinetics of recovery of the transient outward current in the two cell types exist, with slower recovery kinetics existing in atrial cells (Fig. 22.4). Such differences would underlie the relative differences in changes in action potential duration (APD) occurring in response to rate in the two different cell types. The properties of heterologously expressed Kv4.2 and/or Kv4.3 channels closely but not precisely resemble native I_{to} ($I_{to,fast}$) currents.[89] Kv1.4 underlies the cardiac rapidly activating and slowly inactivating I_{to} ($I_{to,s}$). In human, Kv4.3 and Kv1.4 heterologously encode

the major pore-forming α subunits of functional I_{to}.[90,91] While there is a large gradient of the expression of I_{to} across the ventricular wall, in higher mammals this gradient is not due to a transmural gradient of KV4 mRNA, but probably to the transmural gradient of an interacting protein KChIP2 (see below).[92]

Upon depolarization, Kv4 K channels undergo structural changes in response to voltage. In particular, the S4 regions sense the voltage change and the channel gate opens, but during maintained depolarization, the channel enters an inactivated state. For Kv4 channels this inactivation mechanism was originally thought to be due to a 'ball and chain'-type mechanism, but studies deleting regions of both the N and C termini and mutating regions within the pore have not removed inactivation.[93–95] Accordingly, current hypotheses sug-

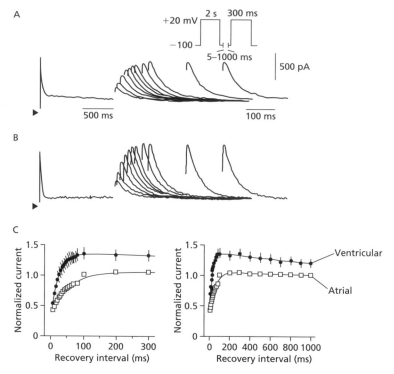

Figure 22.4. *Typical voltage-clamp protocol and resultant current tracings in a human atrial (A) and ventricular myocyte (B), illustrating the time course of recovery of the transient outward current (I_{to1}). Note that in both cell types the transient component recovers reasonably quickly yet there are significant differences between the two cell types (C). Modified with permission from Amos GJ, Wettwer E, Metzger F et al. Differences between outward currents of human atrial and subepicardial ventricular myocytes. J Physiol 1996; 491.1: 31–50.*

gest that inactivation from the Kv4 closed state is the mechanism,[96] perhaps due to a change at the inner mouth of the channel pore.

KChIPs are Ca^{2+}-binding accessory subunit proteins with EF hand domains. KChIP2 family members are highly expressed in the heart.[97,98] When expressed with Kv4 α subunits, KChiPs increase cell surface expression, slow inactivation kinetics and accelerate recovery kinetics of I_{to}.[99,100] KChIP2b's and 2d's ability to alter Kv4.3 inactivation is due to Ca^{2+}-independent as well as Ca^{2+}-dependent steps.[101] Presumably the crevice of the KChIP2 structure binds to residues of the Kv4.3 N terminus, and provides a network structure for the effects on inactivation of the α subunit. Ca^{2+} binding to the EF hand domains of the KChIP cause local conformational changes, again to affect inactivation.[101]

In human myocytes and cells expressing human Kv4.3, α-adrenergic stimulation down-regulates I_{to}, shifts the voltage dependence of inactivation to more negative potentials, and slows the recovery from inactivation through activation of PKC.[102] The PKC-binding site is in the

C terminus of hKv4.3. PKA phosphorylation of the Kv4.2 subunit requires the presence of a KChIP (KChIP3), and results in a slowing of inactivation kinetics with no effect on recovery kinetics.[103] What role this plays in native I_{to} remains unknown.

I_{KUR}: The sustained outward current component that is prominent in atrial but not ventricular myocytes (Fig. 22.5) appears to have a unique molecular composition and pharmacology,[104–106] and has been called I_{KUR} or $I_{h,KUR}$. This current is large in atrial cells and is insensitive to TEA, DTX, CTX, barium, extracellular sodium, dofetilide, and E4031, but is sensitive to low concentrations of 4-aminopyridine and quinidine. Interestingly, the density of this sustained component of outward current increases if these adult atrial cells are maintained in cell culture (Fig. 22.6). The reason for this increase is not known, but it is not mediated by PKC-dependent pathways,[107] and may be related to loss of cell-to-cell contact.[108]

The expressed Kv1.5 currents are similar to those of I_{Kur} in rat, human and canine atrial

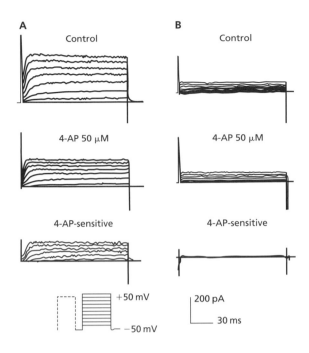

Figure 22.5 (left). *Voltage-clamp protocols in a human atrial (A) and ventricular (B) myocyte, illustrating the differences in the presence of a rapidly activating outward current (I_{KUR}) and its pharmacology. 4-AP, 4-aminopyridine. Modified with permission from Li G-R, Feng J, Yue L, Carrier M, Nattel S. Evidence of two components of delayed rectifier K current in human ventricular myocytes. Circ Res 1996; 78: 689–96.*

Figure 22.6 (below). *The effects of culturing for 8 days on the timecourse and density of outward currents (transient I_{to}, and sustained I_{sus}) in adult human atrial myocytes. When atrial myocytes are maintained in culture (B), there is a large increase in the size of the sustained outward current compared to those of freshly isolated cells (A). Modified with permission from Hatem SN, Benardeau A, Rucker-Martin C et al. Differential regulation of voltage-activated potassium currents in cultured human atrial myocytes. Am J Physiol Heart Circ Physiol 1996; 271: H1609–19.*

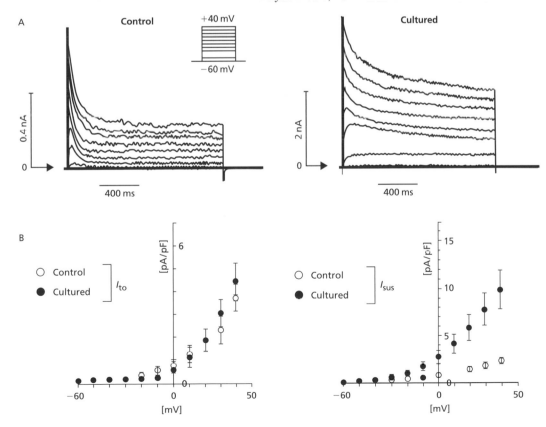

cells.[105,109,110] The molecular basis of I_{KUR} was thought to be due to Kv3.1 in canine cells,[111] but to Kv1.5 in other species including humans. Recently it was shown that canine I_{KUR} is due to Kv1.5 protein.[110] Like the Kv4 channels, Kv1.5 channels activate with depolarization but, importantly, do not immediately inactivate, thus contributing a sustained current during depolarization. Diversity of the Kv1.5 current phenotype comes through interaction with a set of β subunits (called Kv β1.2 and Kv β1.3 subunits) which assemble with Kv1.5 to produce currents with partial fast inactivation, a shift in voltage dependence and a slowing of deactivation.[112] Importantly, coassembly of Kvβs with Kv1.5 enhances the response of this channel to both PKC and PKA stimulation.[113,114].

Calcium-dependent transient outward current: The calcium-dependent transient outward current has been difficult to study because it is dependent on Ca^{2+} influx via the L-type Ca^{2+} channel and release of Ca^{2+} from the SR. It is inhibited by the removal of Ca^{2+}, by blockade of SR release and the presence of a SITs- or DIDS-type compound. Studies in rabbit and canine atrial cells have described this current as being $I_{Cl(Ca)}$.[115,116] While initial reports implied the existence of a Ca^{2+}-dependent outward current in human atrial myocytes,[117] recent studies suggest that calcium-dependent chloride currents do not exist in either human atrial,[118] or human ventricular myocytes. What may gain importance in human cells is the Ca^{2+}-activated K^+ channel which is an apamin-sensitive small conductance K_{Ca} channel (SK). Its activation is also dependent on Ca^{2+} influx via the L-type Ca^{2+} channel, and release of Ca^{2+} from the SR. It is encoded by three genes, *SK1*, *SK2*, and *SK3*.[119] However, only the SK2 isoform is present in heart myocytes.[120] This channel was first identified in human and mouse cardiac cells, and plays a crucial role in cardiac AP profile. Importantly, there is differential expression of the SK2 channel with more abundant SK2 channel in the atria compared with the ventricles. A study on rabbit atrial and ventricular cells also showed that $I_{K,Ca}$ is larger in the atrium than that in the ventricle.[121]

I_{ks} delayed rectifiers: The delayed rectifier potassium current (I_k) was first described by Noble and Tsien,[122] using the two-microelectrode voltage-clamp technique in the multicellular Purkinje fiber preparation. I_K can be separated into rapid and slow components (I_{kr} and I_{ks}, respectively).[123,124] Native slow delayed rectifier, I_{ks}, is a major outward current responsible for ventricular muscle action potential. Because I_{ks} activation occurs at about 0 mV, and this voltage is more positive than the normal Purkinje fiber action potential plateau voltage, I_{ks} may not play an important role in Purkinje fiber APD. Conversely, in ventricular muscle, action potential plateau voltage is more positive (about 20 mV) allowing I_{ks} to be substantially more activated. Therefore, I_{ks} significantly contributes to the repolarization of the action potential in ventricular myocytes. When the repolarization is delayed by certain factors, the prolonged APD favors greater I_{ks}, activation which serves to limit further APD prolongation. The density of I_{ks} in different myocardial regions is also variable. In canine heart, I_{ks} density is higher in epicardial and endocardial cells than in midmyocardial cells.[125] This is thought to underlie the long action potential of the isolated midmyocardial cell.

Upon depolarization, I_{ks} is a slowly activating current with only modest rectification, and can accumulate at high stimulation rates.[126] This is not seen in rabbit I_{ks} and human I_{ks}. Some have reported that I_{ks} is absent in human ventricular myocytes, however, Li et al. observed that a functional I_{ks} is present in human ventricular myocytes.[127]

Native I_{ks} is composed of a pore-forming α subunit, KCNQ1, and a function-altering β subunit, KCNE1. KCNQ1 is a 676-amino acid protein that, when expressed alone in *Xenopus* oocytes or cultured mammalian cells, induces a K^+ current with biophysical properties unlike any known native cardiac current from normal cells. However, when coexpressed with KCNE1, depolarization induces a slowly activating current similar to native cardiac I_{ks}.[128,129] While the stoichiometry for assembly of KCNQ1 and KCNE1 subunits remains unknown, it has been suggested that functional I_{ks} channels contain two KCNE1 subunits.[130] In *Xenopus* oocytes' expression system, KCNE2, another β subunit, has no effect on KCNQ1 gating or current amplitude.[131] But, in mammalian cells, KCNE2 coexpression drastically modifies KCNQ1 current characteristics, decreasing its current density.[132] KCNE3, a third β subunit with 35% identity to KCNE1, causes a marked change in 'lone' KCNQ1 current properties, as well as colocalizing with KCNQ1 in the membrane.[133] KCNQ1/KCNE3 currents have a linear I–V relationship, are nearly instantaneous, and exhibit some time-dependent gating at positive voltages.

Native I_{ks} is enhanced by β-adrenergic stimulation via an elevation of intracellular cAMP and an activation of PKA. Cytosolic cAMP increases I_{ks} current amplitude, slows deactivation kinetics and shifts the activation curve to more negative potentials.[134] The cAMP regulation of recombinant KCNQ1/KCNE1 complex requires PKA anchoring by A-kinase-anchoring proteins (AKAPs). Therefore, the PKA-mediated response of native cardiac I_{ks} depends on the coexpression of AKAP and KCNQ1/KCNE1 in the cell membrane.[135] AKAPs comprise a group of proteins with a proposed role in mediating the attachment of PKA to subcellular structures. Neuronal AKAP is AKAP79. Cardiac AKAP is AKAP15/18. Marx et al. have demonstrated that β-adrenoceptor modulation of KCNQ1 is mediated by both PKA and protein phosphatase 1, via another targeting protein, yotiao.[136] In their expression system, they did not detect AKAP79 or AKAP15/18 with KCNQ1.

I_{kr}: The rapidly-activating delayed rectifier K$^+$ current, I_{kr}, plays an important role in action potential repolarization of cardiac cells in many mammalian species, including humans (Fig. 22.7). I_{kr} has also been shown to contribute to repolarization and thus pacemaking in the SAN cells. The density of I_{kr} in different myocardial cell types is variable. In guinea pig, for example, I_{kr} is 2-fold higher in atrial than in ventricular myocytes.[126] There are regional differences in I_{kr} expression in different species. For example, cells of the epicardial left ventricular free wall have a higher density of I_{kr} than those of mid-myocardial or subendocardial wall.[137,138] In dog and rabbit ventricle, I_{kr} densities are similar among the three layers.[125,139]

Upon depolarization, I_{kr} activates and inactivates very rapidly, thus accounting for its marked inward rectification.[140] Upon hyperpolarization, inactivated channels go to the open state prior to their slower deactivation to the close state. Rectification of the current results from an intrinsic gating process, since it is not mediated by channel block by Mg^{2+}, polyamines, or other intracellular particles as is the case for the I_{kr} channels.[77] Rapid and strong voltage-dependent I_{kr} channel inactivation occurs about 100 times faster than channel opening at potentials near 0 mV.[141] The time constant of inactivation has a classical 'bell'-shaped voltage dependence that is distinct from C-type inactivation mechanisms described for voltage-gated K$^+$ channels.[142] Thus, it seems likely

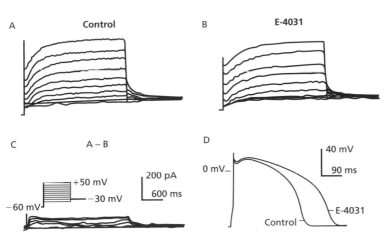

Figure 22.7. *E-4031-sensitive and -insensitive currents in a typical human ventricular myocyte (A–C). (D) The effects of E-4031 on the action potential of this myocyte. Modified with permission from Li G-R, Feng J, Yue L, Carrier M, Nattel S. Evidence of two components of delayed rectifier K current in human ventricular myocytes. Circ Res 1996; 78: 689–96.*

that inactivation of I_{kr} has its own intrinsic voltage sensitivity. The K^+-selective channel encoded by KCNH2 (ERG) (see Table 1 in reference 143) shows currents similar to native I_{kr} in terms of the characteristic inward rectification and high sensitivity to La^{3+}.[144,145] KCNH2 has the usual voltage-gated K^+ channel topology, with six transmembrane-spanning regions, a pore region, and long cytoplasmic N- and C-terminal regions. However, there are unexplained differences between native I_{kr} and the first cloned KCNH2 studied: the KCNH2 current expressed in *Xenopus* oocytes or mammalian cell lines exhibits 4–10 times slower activation and deactivation kinetics than those of native I_{kr}. Additional channel β subunits, KCNE1 or KCNE2 (MiRP-1)[146] have been identified to physically associate with the pore-forming α subunit, KCNH2. When KCNE1 is coexpressed with KCNH2, an I_{kr}-like K^+ current similar to KCNH2 alone, but larger in amplitude was seen.[147] Single-channel recording and immunoprecipitation studies reveal that KCNE1 increases the number of functional KCNH2 channels, but does not alter the amount of KCNH2 protein at the cell surface. The mixed complex of KCNH2 and KCNE2 proteins forms channels to produce currents similar to native I_{kr} in terms of gating kinetics, unitary conductance, regulation by potassium, and distinctive biphasic inhibition by methanesulfonalinide drugs.[131] Regulation of KCNH2 by KCNE2 has been challenged, since KCNE2 and KCNH2 expressed in mammalian cells did not necessarily recapitulate the native I_{kr} current.[148] Thus, additional studies are needed to confirm the existence of KCNH2/KCNE2 complexes in native cells. KCNE3, another β subunit which is weakly expressed in the normal heart strongly inhibits KCNH2 currents.[133]

Activation of cAMP-dependent PKA results in KCNH2 phosphorylation, leading to a rapid reduction of current amplitude, acceleration of deactivation, and a depolarizing shift of the activation curve.[149] cAMP can also directly bind to the KCNH2 protein to produce a hyperpolarizing shift of the activation curve. Notably, this stimulatory effect of cAMP on KCNH2 is enhanced by co-expression of β subunit proteins (MiRP1 or KCNE1).[149] PKC activated by β-adrenoceptor

stimulation enhances native I_{kr}, through a reduction in channel inactivation which underlies the rectification of the channel.[150] The intracellular mechanisms of this modulating effect remain unknown.

HCN channel

The hyperpolarization-activated cyclic nucleotide-modulated (HCN) channel family encodes the pacemaker channel, I_f, and contributes to depolarization of phase 4 of the transmembrane potential. Unlike most voltage-gated channels, which open when the membrane potential is depolarized, HCN and I_f channels open upon membrane hyperpolarization. While the topology of the HCN channel is similar to voltage-gated K^+ channels, the inverted voltage dependence of HCN gating appears to reside in the important interactions between specific residues of the S4–S5 linker and the intracellular activation gates.[151] I_f is carried by both K^+ and Na^+ under normal physiological conditions, is present not only in the SAN, atrioventricular node (AVN) and Purkinje fibers, but also in atrial and ventricular myocytes. However, the I_f of ventricular and atrial cells tends to activate with slower kinetics and have lower peak densities than SAN I_f.[152–154] In addition, $V_{0.5}$ of ventricular I_f is shifted to more hyperpolarizing voltages (range of –95 to –135 mV) compared with I_f of the SAN (range of –65 to –90 mV). Since the voltage of activation of ventricular I_f is far from the normal resting membrane potential, the presence of this current should be without functional consequence under normal physiological conditions.

Four HCN isoforms (1–4) have been identified, and each has a distinct pattern of gene expression and tissue distribution.[155–157] In the heart, three HCN channel types (HCN1, 2 and 4) exist. In the SAN and Purkinje fibers, HCN4 is the predominant HCN channel type, while in atrial and ventricular myocytes, HCN2 is the major isoform. Therefore, HCN4 expression is correlated with spontaneously active cells of the conduction tissue, whereas HCN2 expression is essentially correlated with quiescent cells.[152,158]

The α subunit of the HCN channels contains six membrane-spanning helices (S1–S6), including a charged voltage-sensing S4 segment and an ion-conducting pore between S5 and S6. The tetrameric channel has the property of being gated by membrane hyperpolarization and cAMP.[159] The C terminus of the channel carries a cyclic nucleotide-binding domain (CNBD). cAMP binding to the β-roll domain of the C terminus enhances channel opening by relieving a 'tonic' inhibition on gating,[160] and C linker residues regulate the tonic inhibition.

I_f channels are regulated by various neurotransmitters as well as metabolic stimuli. For example, cAMP shifts the HCN channel activation curve to more positive membrane potentials through binding to the CNBD. Thus, the channel activity is enhanced by neurotransmitters such as norepinephrine that activate adenylyl cyclase, whereas it is decreased by substances such as acetylcholine that inhibit cAMP synthesis. A modulatory auxiliary subunit, KCNE2 (MiRP1), can enhance HCN4-generated currents, slow its activation kinetics and shift the voltage for half-maximal activation of currents to more negative voltages.[161]

Na+–Ca2+ exchanger

The cardiac Na^+–Ca^{2+} exchanger protein transports Ca^{2+} ions across the sarcolemma in exchange for Na^+ ions, and is important in maintaining Ca^{2+} homeostasis in the myocyte. Normally, Na^+–Ca^{2+} exchange works in the so-called forward mode, i.e. extruding Ca^{2+} ions in exchange for extracellular Na^+ ions. Reverse-mode operation of the Na^+–Ca^{2+} exchanger could provide additional Ca^{2+} influx into the cell. In the forward mode, Ca^{2+} ions are being transported out against their electrochemical gradient, therefore energy is needed. It is generally accepted that it is the Na^+ ion distribution which provides the energy. Stoichiometric determinations have shown that three Na^+ ions are transported for one Ca^{2+} ion, and thus the exchanger is electrogenic. Under normal conditions, the reversal potential of the Na^+–Ca^{2+} exchanger has been shown to be -30mV.[162] Accordingly, negative to this potential, Na^+ is moving in and Ca^{2+} out, and inward current is generated; positive to this potential, the Na^+–Ca^{2+} exchanger works in reverse mode, and outward current is generated. For NCX1 transporter protein, it has been estimated that the turnover rate is up to 5000/s,[163,164] where the K_m of $[Ca^{2+}]_i$ is approximately 6 µmol/l.[165] Recent data derived from steady-state voltage and Ca^{2+} dependence of the Na^+–Ca^{2+} exchanger protein have suggested that within <32 ms of an action potential upstroke, peak Ca^{2+} in a submembrane space is >3.2 µmol/l.[166] Thus, Na^+–Ca^{2+} exchanger current greatly influences both the atrial and ventricular action potential.[167] Further, a component of observed transmural electrical heterogeneity of the LV has been ascribed to basal differences in I_{NaCa} currents across the wall.[168]

The Na^+–Ca^{2+} exchanger protein is now considered to consist of nine transmembrane segments with a large ($\sim$550 amino acids) intracellular loop (loop f) between segments 5 and 6.[169] Distinct regions of the protein have been shown to be involved in the Na^+–Ca^{2+} translocation process,[170] while other regions, particularly loop f, are involved in the intrinsic regulation of the Na^+–Ca^{2+} exchanger by both Na^+ and Ca^{2+} ions.[171] Ca^{2+}-dependent regulation of exchanger activity is via a high-affinity binding site that is distinct from the Ca^{2+} transport site,[165] is about 130 amino acids in length, and located in the center of loop f.[172] Ca^{2+}-dependent regulation of Na^+–Ca^{2+} exchanger activity is allosteric, such that when $[Ca^{2+}]_i$ levels are reduced (approximately <150 nM) Na^+–Ca^{2+} current deactivates.[173] A corollary is that when $[Ca^{2+}]_i$ is elevated, steady-state activation of Na^+–Ca^{2+} exchanger current increases, by as much as 67% for a doubling of $[Ca^{2+}]_i$. Such activation in normal cells promotes Ca^{2+} efflux with concomitant production of inward currents. If Ca^{2+} transients occur as traveling Ca^{2+} waves, activated Na^+–Ca^{2+} exchanger current would contribute to occurrence of Ca^{2+}-activated membrane currents.

Possible causes of regional/chamber specific heterogeneity

As indicated in a recent review,[174] several clones have been identified as being expressed in all chambers of the heart, yet the underlying native currents may not be (Fig. 22.8). There are several reasons why this may occur, three of which will be generally outlined here.

Current	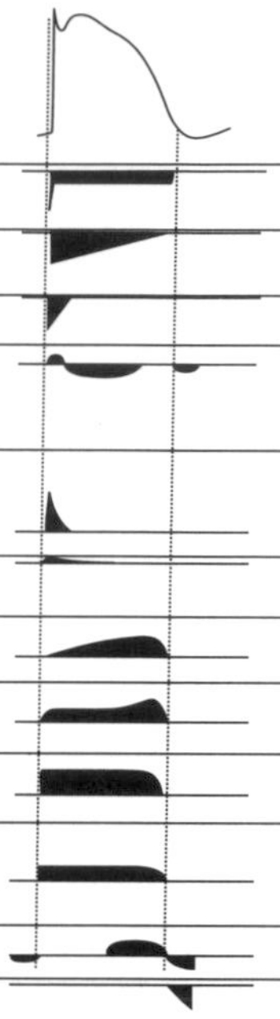	Regional expression				Probable clone
		Sinus, AV node	Atrium	Ventricle	His-purkinje system	
I_{Na}			✓	✓	✓	hH1; Nav1.5
I_{Ca-L}		✓	✓	✓	✓	Ca^{2+} channel; Cav1.2
I_{Ca-T}		✓	✓		✓	Cav1.3
I_{Na-Ca}		✓	✓	✓	✓	Na^+–Ca^{2+} exchanger
I_{TO1}		✓	✓	✓	✓	Kv4.2/3, Kv1.4, Kv1.5, Kv2.1 +/or Kv1.2
I_{TO2}			✓	✓	✓	
I_{Ks}		✓	✓	✓	✓	KvLQT1 + minK; KCNQ1 + KCNE1
I_{Kr}		✓	✓	✓	✓	HERG
I_{Kur}			✓	✓		Kv1.5
I_{Kp}/I_{cl}			✓	✓		CFTR (I_{Cl}); ?TWIK/ORK family (I_K), CTBAK-1
$I_{Kl}/I_{K-ATP}/I_{k-ACh}$			✓	✓		Kir family
I_f		✓	✓	✓	✓	HCN1, 2, 4

Figure 22.8. *Relationships among the various cardiac ion currents and probable clones for each with respect to action potential shown (2nd column). Regional expression of current is also depicted. Genes whose products are thought to be responsible for each current are indicated in far column. Modified with permission from Roden DM, George AL, Jr. Structure and function of cardiac sodium and potassium channels. Am J Physiol Heart Circ Physiol 1997; 273: H511–25.*

First, regional/chamber heterogeneity may result from differences in transcriptional changes at the level of the DNA. Here alternative splicing of the same gene may result in many proteins from one gene. If alternative splicing occurs, the protein products (ion channels) may differ in their properties. A recent example of this type of phenomenon has been described for the *HERG* gene, which encodes a protein for the native I_{Kr} delayed rectifying current. In these latter studies, three cDNA isoforms of the mouse *ERG* homologue were described.[175] Each form is unique in its base pairs, and dictates the expression of a unique protein. When each is expressed under appropriate conditions, a specific current with a differing phenotype occurs. The largest difference is in the deactivation of I_{Kr} tail currents, suggesting that if expressed in myocytes, currents flowing via these proteins would accumulate differently during rapid pacing rates.

Second, ionic current regional/chamber heterogeneity may be due to the expression of the same protein (e.g. ion channel α subunit), but with differences in the post-translational processing and/or coassembly with other proteins, which would subsequently alter the phenotypic expression of the original protein. An example of such may be in the various combinations of α and β subunits of some K+ channels. However, perhaps more interestingly is the linking of an α subunit gene product with the protein minK (KCNE1). An example is from experiments where the HERG protein and minK gene product when expressed together, form channels whose current is double the size of that of HERG expressed alone (Fig. 22.9).[147]

Third, differences in phenotypic expression of an ion channel protein between atrial and ventricular cells may be due to differences in the cell sur-

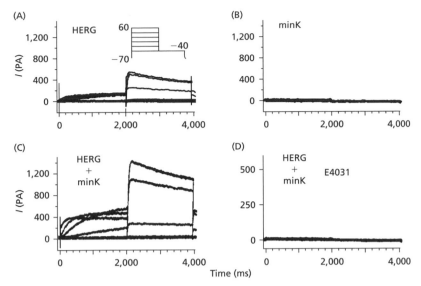

Figure 22.9. *Functional expression of HERG (KCNH2) and mink (KCNE1) in Chinese hamster ovary cells. When HERG is expressed alone, currents typical of I_{Kr} are expressed (A). When minK is expressed alone in CHO cells, no voltage-gated K^+ currents exist (B). However, when minK and HERG are co-transiently expressed, K^+ currents doubled in amplitude (C) yet remain sensitive to E4031 (D). Modified with permission from McDonald TV, Yu Z, Ming Z et al. A minK–HERG complex regulates the cardiac potassium current I_{Kr}. Nature 1997; 388: 289–92.*

face distribution of the ion channel protein. Clearly, when the α subunit of an inwardly rectifying K^+ current (I_{K1}) is expressed with an anchoring protein called PSD-95, large whole-cell currents occur, and ion channel proteins become clustered on the cell surface.[176] There is at least one example of an ion channel protein altering its cell surface pattern in human atrial cells. This is in the developmental change in cell surface localization of the Kv1.5 K^+ current protein.[177] Whether this anatomical change in protein redistribution is linked to functional phenotypic changes in developing human atrial cells is not known at this time.

Cellular/multicellular basis of arrhythmias

During normal sinus rhythm, the cardiac impulse for each heart beat originates in the SAN, and spreads in an orderly fashion throughout the atria, through the atrioventricular node and His–Purkinje system, and throughout the ventricles. An arrhythmia is present if the sinus rate is abnormally rapid or slow, if the sinus rhythm is occasionally or frequently interrupted by impulses originating from some other site, or if all electrical impulses originate at a site other than the sinus node at a frequency greater than that of the node. Electrophysiological mechanisms causing arrhythmias fall under two general categories: abnormal impulse generation, and abnormal impulse propagation.

Abnormal impulse initiation

Impulses responsible for these types of arrhythmias are caused by spontaneous diastolic depolarizations during phase 4 of the action potential. If these depolarizations reach threshold, then automatic firing of the cardiac cell exists. Some cells have the capacity for phase 4 depolarization and automaticity as part of their normal electrical properties (e.g. latent atrial pacemaker cells, Purkinje fibers). Working atrial and ventricular cells do not become automatic when their electrical profile is normal. However if resting potential is reduced, then these cell types can elicit non-driven electrical activity due to abnormal automaticity. Usually, action potentials associated with abnormal automaticity depend heavily on Ca^{2+} current for activation, and delayed rectifier currents for repolarization.

381

It is more likely that abnormal initiation results from a triggered arrhythmia. Here an action potential must always precede the initiation of the arrhythmic beat, hence a triggered beat. There are two cellular forms of triggered arrhythmias. The first is an arrhythmia caused by early afterdepolarizations (EADs). EADs occur during the repolarization phase of the preceding action potential which had been initiated from a normal resting potential. EADs are voltage changes occurring in the positive direction relative to the membrane potential expected during repolarization. If these potential changes are large enough, they can elicit extra beats and propagate throughout the myocardium. EADs require that the APD is first prolonged, allowing the L-type Ca^{2+} current to inactivate and then recover to provide the depolarizing current of EADs.[178] Intracellular Ca^{2+} then changes as a consequence of these voltage changes.

Delayed afterdepolarizations (DADs) occur just near the completion of or after the end of the AP repolarization phase.[179] An impulse is triggered from a DAD when the DAD has reached the threshold of excitability of the cell. The Ca^{2+}-dependent I_{ti} causes a DAD, but a traveling intracellular Ca^{2+} wave underlies the Ca^{2+} changes that lead to the I_{ti} that initiates the DAD.

Abnormal impulse conduction

The first example of this type of arrhythmia is block. If the resting potential of a cardiac cell is altered such that less Na^+ current is available, then the safety factor for propagation is reduced and conduction can fail. Cell-to-cell uncoupling would be expected to lead to conduction block or slowing, although from computer simulations we know that the margin of safety for conduction remains large even in the presence of ongoing uncoupling.[180] Under these conditions, Ca^{2+} currents may maintain propagation. Sites at fiber branching, whether normal or resulting from marked fibrosis, can also alter cardiac cell propagation. At these sites, current-to-load mismatches occur, and result in conduction slowing and block.[181] Unidirectional block, a prerequisite for re-entrant excitation, can result at several points of heterogeneities, for instance in areas of excitability heterogeneity (local distribution of refractoriness), at cell-to-cell uncoupling areas, and microscopic anatomic heterogeneities related to tissue geometry.

Re-entrant excitation of the heart occurs when the propagating impulse does not die out but continues to propagate and re-excite the heart at tissue sites where refractoriness has recovered. As stated above, to begin re-entry unidirectional block and slowed conduction as well as an appropriate path length are required. Re-entry has been further categorized according to the nature of the area of block around which the re-entering impulse propagates. Two types of re-entry are anatomic re-entry and functional re-entry. Functional re-entry can be further subdivided into leading circle,[182] or anisotropic[183] re-entry. For both cases of functional re-entry, concepts have been developed based on two-dimensional cardiac tissues. In anatomic re-entry the location of the circuit remains fixed, while in functional re-entry the excitatory wavefront may follow a complex path. In the three dimensions of the myocardial wall this can lead to spiral wave re-entry where the tip of the wave meanders depending on the excitatory properties ahead of it. A scroll wave is a spiral wave in three dimensions.

References

1. Amos GJ, Wettwer E, Metzger F et al. Differences between outward currents of human atrial and subepicardial ventricular myocytes. J Physiol 1996; 491.1: 31–50.
2. Spach MS, Miller WTI, 3rd, Gesrlowitz DB et al. The discontinuous nature of propagation in normal canine cardiac muscle. Evidence for recurrent discontinuities of intracellular resistance that affect the membrane currents. Circ Res 1981; 48: 39–54.
3. Cabo C, Pertsov AM, Baxter WT et al. Wave-front curvature as a cause of slow conduction and block in isolated cardiac muscle. Circ Res 1994; 75: 1014–28.
4. Kleber AG, Rudy Y. Basic mechanisms of cardiac impulse propagation and associated arrhythmias. Physiol Rev 2004; 84: 431–88.
5. MacKinnon R. Potassium channels. FEBS Lett 2003; 555: 62–5.
6. Catterall WA, Goldin AL, Waxman SG. International union of pharmacology. XXXIX.

Compendium of voltage-gated ion channels: sodium channels. Pharmacol Rev 2003; 55: 575–8.

7. Maier SKG, Westenbroek RE, Schenkman KA et al. An unexpected effect for brain type sodium channels in coupling of cell surface depolarization to contraction in the heart. Proc Natl Acad Sci USA 2002; 99: 4073–8.

8. Li G-R, Lau C-P, Shrier A. Heterogeneity of sodium current in atrial versus epicardial ventricular myocytes of adult guinea pig hearts. J Mol Cell Cardiol 2002; 34: 1185–94.

9. Fahmi A, Patel M, Stevens EB et al. The sodium channel beta subunit SCN3b modulates the kinetics of SCN5a and is expressed heterogeneously in sheep heart. J Physiol 2001; 537: 693–700.

10. Isom LL, DeJongh KS, Catterall WA. Auxiliary subunits of voltage gated ion channels. Neuron 1994; 12: 1183–94.

11. Tateyama M, Rivolta I, Clancy CE et al. Modulation of cardiac sodium channel gating by protein kinase A can be altered by disease linked mutation. J Biol Chem 2003; 278: 46718–26.

12. Zagotta WN, Aldrich RW. Voltage-dependent gating of Shaker A-type potassium channels in Drosophila muscle. J Gen Physiol 1990; 95: 29–60.

13. Cormier JW, Rivolta I, Tateyama M et al. Secondary structure of the human cardiac Na channel C terminus. J Biol Chem 2002; 277: 9233–41.

14. Murray KT, Hu N, Daw JR et al. Functional effects of protein kinase C activation on the human cardiac Na+ channel. Circ Res 1997; 80: 370–6.

15. Zhou J, Yi J, Hu N et al. Activation of protein kinase A modulates trafficking of the human cardiac sodium channel in Xenopus oocytes. Circ Res 2000; 87: 33–8.

16. Tateyama M, Kurokawa J, Terrenoire C et al. Stimulation of protein kinase C inhibits bursting in disease induced mutant human cardiac sodium channels. Circ 2003; 107: 3216–22.

17. Bean BP. Two kinds of calcium channels in canine atrial cells: differences in kinetics, selectivity, and pharmacology. J Gen Physiol 1985; 86: 1–30.

18. Hagiwara N, Irisawa H, Kameyama M. Contribution of two types of calcium currents to the pacemaker potentials of rabbit sino-atrial node cells. J Physiol 1988; 395: 233–53.

19. Zhou Z, Lipsius SL. T-type calcium currents in latent pacemaker cells isolated from cat right atrium. J Mol Cell Cardiol 1994; 26: 1211–19.

20. Tseng G-N, Boyden PA. Multiple types of Ca currents in single canine Purkinje myocytes. Circ Res 1989; 65: 1735–50.

21. Hirano Y, Fozzard HA, January CT. Characteristics of L- and T-type Ca^{2+} currents in canine cardiac Purkinje cells. Am J Physiol Heart Circ Physiol 1989; 256: H1478–92.

22. Hirano Y, Fozzard HA, January CT. Inactivation properties of T-type calcium current in canine cardiac Purkinje cells. Biophys J 1989; 56: 1007–16.

23. Van Wagoner DR, Pond AL, McCarthy PM et al. Atrial L-type Ca^{2+} currents and human atrial fibrillation. Circ Res 2000; 85: 428–36.

24. Mewes T, Ravens U. L-type calcium currents of human myocytes from ventricle of non-failing and failing hearts and from atrium. J Mol Cell Cardiol 1994; 26: 1307–20.

25. Bosch R, Zeng X, Grammer JB et al. Ionic mechanisms of electrical remodeling in human atrial fibrillation. Cardiovascular Res 1999; 44: 121–31.

26. Rosenberg RL, Hess P, Reeves JP et al. Calcium channels in planar lipid bilayers: insights into mechanism of ion permeation and gating. Science 1986; 231: 1564–6.

27. Hadley RW, Lederer WJ. Ca^{2+} and voltage inactivate Ca^{2+} channels in guinea pig myocytes through independent mechanisms. J Physiol 1991; 444: 257–68.

28. Argibay JA, Fischmeister R, Hartzell HC. Inactivation, reactivation and pacing dependence of calcium current in frog cardiocytes: correlation with current density. J Physiol 1988; 401: 201–26.

29. Zhang J-F, Ellinor PT, Aldrich RW et al. Molecular determinants of voltage dependent inactivation in calcium channels. Nature 1994; 372: 97–100.

30. Bourinet E, Zamponi GW, Stea A et al. The alpha 1E calcium channel exhibits permeation properties similar to low-voltage-activated calcium channels. J Neurosci 1996; 16: 4983–93.

31. Bernatchez G, Talwar D, Parent L. Mutations in the EF-hand motif impair the inactivation of barium currents of the cardiac alpha1C channel. Biophys J 1998; 75: 1727–39.

32. Zamponi GW, Soong TW, Bourinet E et al. Beta subunit coexpression and the alpha1 subunit domain I-II linker affect piperidine block of neuronal calcium channels. J Neurosci 1996; 16: 2430–43.

33. Hering S, Aczel S, Grabner M et al. Transfer of high sensitivity for benzothiazepines from L-type to class A (BI) calcium channels. J Biol Chem 1996; 271: 24471–5.

34. Hering S, Aczel S, Kraus RL et al. Molecular mechanism of use-dependent calcium channel block by phenylalkylamines: role of inactivation. Proc Natl Acad Sci USA 1997; 94: 13323–8.

35. Soldatov NM, Zuhlke RD, Bouron A et al. Molecular structures involved in L-type calcium channel inactivation. Role of the carboxyl-terminal region encoded by exons 40–42 in alpha1C subunit in the kinetics and Ca^{2+} dependence of inactivation. J Biol Chem 1997; 272: 3560–6.

36. Berrou L, Bernatchez G, Parent L. Molecular determinants of inactivation within the I-II linker of alpha1E (CaV2.3) calcium channels. Biophys J 2001; 80: 215–28.

37. Stotz SC, Zamponi GW. Identification of inactivation determinants in the domain IIS6 region of high voltage-activated calcium channels. J Biol Chem 2001; 276, No.35: 33001–10.

38. Ertel EA, Campbell KP, Harpold MM et al. Nomenclature of voltage gated calcium channels. Neuron 2000; 25: 533–5.

39. Ellinor PT, Yang J, Sather WA et al. Ca^{2+} channel selectivity at a single locus for high affinity Ca^{2+} interactions. Neuron 1995; 15: 1121–32.

40. Pitt GS, Zuhlke RD, Hudmon A et al. Molecular basis of calmodulin tethering and Ca^{2+}-dependent inactivation of l-type Ca^{2+} channels. J Biol Chem 2001; 276: 30794–802.

41. Colecraft HM, Alseikhan B, Takahashi SX et al. Novel functional properties of Ca^{2+} channel beta subunits revealed by their expression in adult rat heat cells. J Physiol 2002; 541; 435–52.

42. Birnbaumer L, Qin N, Olcese R et al. Structure and functions of calcium channel B subunits. J Bioenerg Biomembr 2000; 30: 357–75.

43. Neely A, Olcese R, Baldelli P et al. Dual activation of the cardiac Ca^{2+} channel alpha 1C subunit and its modulation by the beta subunit. Am J Physiol Cell Physiol 1995; 268: C732–40.

44. Neely A, Wei X, Olcese R et al. Potentiation by the beta subunit of the ratio of the ionic current to the charge movement in the cardiac calcium channel. Science 1993; 262: 575–8.

45. Perez Garcia MT, Kamp TJ, Marban E. Functional properties of cardiac L type calcium channels transiently expressed in HEK293 cells. Role of alpha 1 and beta subunit. J Gen Physiol 1995; 105: 289–305.

46. Lacerda AE, Kim HS, Ruth P et al. Normalization of current kinetics by interaction between alpha 1 and beta subunits of the skeletal muscle dihydropyridine sensitive Ca^{2+} channel. Nature 1991; 352: 527–30.

47. Mitterdorfer J, Froschmayr M, Grabner M et al. Calcium channels: the beta subunit increases the affinity of dihydropyridine and Ca^{2+} binding sites of the alpha 1-subunit. FEBS Lett 1994; 352: 141–5.

48. Singer D, Biel M, Lotan I et al. The roles of the sub-

49. Klugbauer N, Lacinova L, Marais E et al. Molecular diversity of the calcium channel Alpha2Delta subunit. J Neurosci 1999; 19: 684–91.

50. Bangalore R, Mehrke G, Gingrich K et al. Influence of L-type Ca channel Alpha2/Delta subunit on ionic and gating current in transiently transfected HEK293 cells. Am J Physiol Heart Circ Physiol 1996; 39: H1521–28.

51. Cens T, Restituito S, Vallentin A et al. Promotion and inhibition of L-type Ca^{2+} channel facilitation by distinct domains of the subunit. J Biol Chem 1998; 273: 18308–15.

52. Platano D, Qin N, Noceti F et al. Expression of the $\alpha2\delta$ subunit interferes with prepulse facilitation in cardiac l-type calcium channels. Biophys J 2000; 78: 2959–72.

53. Gurnett CA, De Waard M, Campbell KP. Dual function of the voltage-dependent Ca^{2+} channel alpha 2 delta subunit in current stimulation and subunit interaction. Neuron 1996; 16: 431–40.

54. Wyatt CN, Campbell V, Brodbeck J et al. Voltage dependent binding and calcium channel current inhibition by an anti-alpha1D subunit antibody in rat dorsal root ganglion neurons and guinea pig myocytes. J Physiol 1997; 502: 307–19.

55. Takimoto K, Li D, Nerbonne JM, Levitan ES. Distribution, splicing and glucocorticoid-induced expression of cardiac alpha 1C and alpha 1D voltage-gated Ca^{2+} channel mRNAs. J Mol Cell Cardiol 1997; 29: 3035–42.

56. Bohn G, Moosmang S, Conrad H et al. Expression of T and L type calcium channel mRNA in murine sinoatrial node. FEBS Lett 2000; 481: 73–6.

57. Koschak A, Reimer D, Huber I et al. Alpha 1D (Cav1.3) subunits can form l-type Ca^{2+} channels activating at negative voltages. J Biol Chem 2001; 276: 22100–6.

58. Platzer J, Engel J, Schrott-Fischer A et al. Congenital deafness and sinoatrial node dysfunction in mice lacking class D L type Ca^{2+} channels. Cell 2000; 102: 89–97.

59. Zhang Z, Xu Y, Song H et al. Functional roles of Cav1.3 (alpha1D) calcium channel in sinoatrial nodes. Insight gained using gene-targeted null mutant mice. Circ Res 2002; 90; 981–7.

60. Perez Reyes E, Cribbs LL, Daud A et al. Molecular characterization of a neuronal low voltage activated T type calcium channel. Nature 1998; 391: 896–900.

61. Cribbs LL, Lee JH, Yang J et al. Cloning and characterization of alpha1H from human heart, a member

of the T-type Ca^{2+} channel gene family. Circ Res 1998; 83: 103–9.

62. Staes M, Talavera K, Klugbauer N et al. The amino side of the C-terminus determines fast inactivation of the T-type calcium channel Alpha1G. J Physiol 2001; 530: 35–45.

63. Droogmans G, Nilius B. Kinetic properties of the cardiac T-type calcium channel in the guinea-pig. J Physiol 1989; 419: 627–50.

64. Burgess DE, Crawford O, Delisle BP et al. Mechanism of inactivation gating of human T-type (low-voltage activated) calcium channels. Biophys J 2002; 82: 1894–906.

65. Pragnell M, De Waard M, Mori Y et al. Calcium channel beta-subunit binds to a conserved motif in the I-II cytoplasmic linker of the alpha 1-subunit. Nature 1994; 368: 67–70.

66. Lambert RC, Maulet Y, Mouton J et al. T-type Ca^{2+} current properties are not modified by Ca^{2+} channel beta subunit depletion in nodosus ganglion neurons. J Neurosci 1997; 17: 6621–8.

67. de Leon M, Wang Y, Jones L et al. Essential Ca^{2+}-binding motif for Ca^{2+}-sensitive inactivation of L-type Ca^{2+} channels. Science 1995; 270: 1502–6.

68. Weidmann S. Rectifier properties of Purkinje fibers. Am J Physiol 1955; 183: 671.

69. Hume JR, Uehara A. Ionic basis of the different action potential configurations of single guinea pig atrial and ventricular myocytes. J Physiol 1985; 368: 525–44.

70. Giles WR, Imaizumi Y. Comparison of potassium currents in rabbit atrial and ventricular cells. J Physiol 1988; 405: 123–45.

71. Shimoni Y, Clark RB, Giles WR. Role of an inwardly rectifying potassium current in rabbit ventricular action potential. J Physiol 1992; 448: 709–27.

72. Nichols CG, Makhina EN, Pearson WL et al. Inward rectification and implications for cardiac excitability. Circ Res 1996; 78: 1–7.

73. Varro A, Namasi PP, Lathrop DA. Potassium currents in isolated human atrial and ventricular cardiocytes. Acta Physiol Scand 1993; 149: 133–42.

74. Wang A, Yue L et al. Differential distribution of inward rectifier potassium channel transcripts in human atrium versus ventricle. Circulation 1998; 98: 2422–8.

75. Koumi S, Backer CL, Arentzen CE. Characterization of inwardly rectifying K^+ channel in human cardiac myocytes: alteration in channel behavior in myocytes isolated from patients with idiopathic dilated cardiomyopathy. Circulation 1995; 92: 164–74.

76. Vandenberg CA. Inward rectification of a potassium channel in cardiac ventricular cells depends on internal magnesium ions. Proc Natl Acad Sci USA 1987; 84: 2560–4.

77. Lopatin AN, Makhina EN, Nichols CG. Potassium channel block by cytoplasmic polyamines as the mechanism of intrinsic rectification. Nature 1994; 372: 366–9.

78. Taglialatela M, Ficker E, Wible BA et al. C-terminus determinants for Mg^{2+} and polyamine block of the inward rectifier K^+ channel IRK1. EMBO J 2004; 14: 5532–41.

79. Doupnik CA, Davidson N, Lester HA. The inward rectifier potassium channel family. Curr Opin Neurobiol 1995; 5: 268–77.

80. Nichols CG, Lopatin AN. Inward rectifier potassium channels. Annu Rev Physiol 1997; 59: 171–91.

81. Cohen NA, Sha Q, Makhina EN et al. Inhibition of an inward rectifier potassium channel (Kir2.3) by G-protein $\beta\gamma$ subunits. J Biol Chem 1996; 271: 32301–5.

82. Zhu G, Qu Z, Cui N et al. Suppression of Kir2.3 activity by protein kinase C phosphorylation of the channel protein at threonine 53. J Biol Chem 1999; 274: 11643–6.

83. Karle CA, Zitron E, Zang W et al. Human cardiac inwardly-rectifying K^+ channel Kir2.1b is inhibited by direct protein kinase C-dependent regulation in human isolated cardiomyocytes and in an expression system. Circulation 2002; 106: 1493–9.

84. Koumi S, Sato R. Modulation of the inwardly rectifying K^+ channel in isolated human atrial myocytes by alpha 1-adrenergic stimulation. J Membr Biol 1995; 148: 185–91.

85. Fakler B, Brandle U, Glowatzki E et al. Kir2.1 inward rectifier K^+ channels are regulated independently by protein kinases and ATP hydrolysis. Neuron 1994; 13: 1413–20.

86. Wible BA, De Biasi MD, Majumder K et al. Cloning and functional expression of an inwardly rectifying K^+ channel from human atrium. Circ Res 1995; 76: 343–50.

87. Zobel C, Cho HC, Nguyen TT et al. Molecular dissection of the inward rectifier potassium current (I_{k1}) in rabbit cardiomyocytes: evidence for heteromeric co-assembly of $K_{ir2.1}$ and $K_{ir2.2}$. J Physiol 2003; 550: 365–72.

88. Firek L, Giles WR. Outward currents underlying repolarization in human atrial myocytes. Cardiovasc Res 1995; 30: 31–8.

89. Nerbonne JM. Molecular basis of functional voltage-gated K^+ channel diversity in the mammalian myocardium. J Physiol 2000; 525: 285–98.

90. Dixon JE, Shi W, Wang HS et al. Role of the Kv4.3 K+ channel in ventricular muscle. A molecular correlate for the transient outward current. Circ Res 1996; 79: 659–668.

91. Kong W, Po S, Yamagishi T et al. Isolation and characterization of the human gene encoding Ito: further diversity by alternative mRNA splicing. Am J Physiol Heart Circ Physiol 1998; 275: H1963–70.

92. Rosati B, Pan Z, Lypen S et al. Regulation of KChIP2 potassium channel Beta subunit gene expression underlies the gradient of transient outward current in canine and human ventricle. J Physiol 2001; 533.1: 119–25.

93. Bahring R, Boland LM, Varghese A et al. Kinetic analysis of open and closed state inactivation transitions in human Kv4.2 A type potassium channels. J Physiol 2001; 535: 65–81.

94. Jerng HH, Covarrubias M. K+ channel inactivation mediated by the concerted action of the cytoplasm. Biophys J 1997; 72: 163–74.

95. Jerng HH, Shaidulla M, Covarrubias M. Inactivation gating of Kv4 potassium channels: molecular interaction involving the inner vestibule of the pore. J Gen Physiol 1999; 113: 641–60.

96. Beck EJ, Bowlby M, An WF et al. Remodeling inactivation gating of Kv4 channels by KChIP1, a small-molecular-weight calcium-binding protein. J Physiol 2002; 538: 691–706.

97. Rosati B, Grau F, Rodriguez S et al. Concordant expression of KChIP2 mRNA, protein and transient outward current throughout the canine ventricle. J Physiol 2003; 548: 815–22.

98. Patel SP, Campbell DL, Strauss HC. Elucidating KChIP effects on Kv4.3 inactivation and recovery kinetics with a minimal KChIP2 isoform. J Physiol 2002; 545: 5–11.

99. Shibata R, Misonou H, Campomanes CR et al. A fundamental role for KChIPs in determining the molecular properties and trafficking of Kv4.2 potassium channels. J Biol Chem 2003; 278: 36445–54.

100. An WF, Bowlby MR, Betty M et al. Modulation of A type potassium channels by a family of calcium sensors. Nature 2000; 403: 553–6.

101. Patel SP, Parai R, Parai R, Campbell DL. Regulation of Kv4.3 voltage-dependent gating kinetics by KChIP2 isoforms. J Physiol 2004; 557: 19–41.

102. Po SS, Wu RC, Juang GJ et al. Mechanism of alpha-adrenergic regulation of expressed hKv4.3 currents. Am J Physiol Heart Circ Physiol 2001; 281: H2518–27.

103. Schrader LA, Anderson AE, Mayne A et al. PKA modulation of Kv4.2-encoded A-type potassium channels requires formation of a supramolecular complex. J Neurosci 2002; 22: 10123–33.

104. Wang Z, Fermini B, Nattel S. Delayed rectifier outward current and repolarization in human atrial myocytes. Circ Res 1993; 73: 276–85.

105. Wang Z, Fermini B, Nattel S. Sustained depolarization induced outward current in human atrial myocytes. Evidence for a novel delayed rectifier K current similar to Kv1.5 cloned channel currents. Circ Res 1993; 73: 1061–76.

106. Wang Z, Fermini B, Nattel S. Rapid and slow components of delayed rectifier current in human atrial myocytes. Cardiovasc Res 1994; 28: 1540–6.

107. Hatem SN, Benardeau A, Rucker-Martin C et al. Differential regulation of voltage-activated potassium currents in cultured human atrial myocytes. Am J Physiol Heart Circ Physiol 1996; 271: H1609–19.

108. Hershman KM, Levitan ES. Cell to cell contact between adult rat myocytes regulates Kv1.5 and Kv4.2 K+ channel mRNA expression. Am J Physiol Cell Physiol 1998; 275: C1473.

109. Yue L, Feng J, Li G-R, Nattel S. Characterization of an ultrarapid delayed rectifier potassium channel involved in canine atrial repolarization. J Physiol 1996; 496: 647–62.

110. Fedida D, Eldstrom J, Hesketh JC et al. Kv1.5 is an important component of repolarizing K current in canine atrial myocytes. Circ Res 2003; 93; 744–5.

111. Yue L, Wang Z, Rindt H et al. Molecular evidence for a role of Shaw (Kv3) potassium channel subunits in potassium currents of dog atrium. J Physiol 2000; 527.3: 467–78.

112. Deal KK, England SK, Tamkun MM. Molecular physiology of cardiac potassium channels. Physiol Rev 1996; 76: 49–67.

113. Kwak YG, Hu N, Wei J et al. Protein kinase A phosphorylation alters Kvbeta 1.3 subunit mediated inactivation of the Kv1.5 potassium channel. J Biol Chem 1999; 274: 13928–32.

114. Williams CP, Hu N, Shen W et al. Modulation of the human Kv1.5 by protein kinase C activation: role of the Kv β1.2 subunit. J Pharmacol Exp Ther 2002; 302: 545–550.

115. Zygmunt AC, Gibbons WR. Properties of the calcium-activated chloride current in heart. J Gen Physiol 1992; 99: 391–414.

116. Zygmunt AC. Intracellular calcium activates a chloride current in canine ventricular myocytes. Am J Physiol Heart Circ Physiol 1994; 267: H1984–95.

117. Escande D, Coulombe A, Faivre JF et al. Two types of transient outward currents in adult human atrial

cells. Am J Physiol Heart Circ Physiol 1987; 252: H142–8.

118. Li GR, Feng J, Wang Z et al. Comparative mechanisms of 4 aminopyridine resistant Ito in human and rabbit atrial myocytes. Am J Physiol Heart Circ Physiol 1995; 269: H463–72.

119. Stocker M, Pedarzani P. Differential distribution of three Ca^{2+}-activated K^+ channel subunit, SK1, SK2, and SK3, in the adult rat central nervous system. Mol Cell Neurosci 2000; 15: 476–93.

120. Xu Y, Tuteja D, Wang Z et al. Molecular identification and functional roles of a Ca^{2+} activated K channel in human and mouse hearts. J Biol Chem 2003; 278: 49085–94.

121. Clark RB, Giles WR, Imaizumi Y. Properties of the transient outward current in rabbit atrial cells. J Physiol 1988; 405: 147–68.

122. Noble D, Tsien RW. Outward membrane currents activated in the plateau range of potentials of cardiac Purkinje fibers. J Physiol 1969; 200: 205–31.

123. Sanguinetti MC, Jurkiewicz NK. Two components of cardiac delayed rectifier K current. Differential sensitivity to block by Class III antiarrhythmic agents. J Gen Physiol 1990; 96: 195–215.

124. Carmeliet EE. Voltage and time dependent block of the delayed K current in cardiac myocytes by dofetilide. J Pharmacol Exp Ther 1992; 262: 809–17.

125. Liu DW, Antzelevitch C. Characteristics of the delayed rectifier current (IKr and IKs) in canine ventricular epicardial, midmyocardial, and endocardial myocytes. A weaker IKs contributes to the loner action potential of the M cell. Circ Res 1995; 76: 351–65.

126. Sanguinetti MC, Jurkiewicz NK. Delayed rectifier outward K^+ current is composed of two currents in guinea pig atrial cells. Am J Physiol Heart Circ Physiol 1991; 260: H393–9.

127. Li G-R, Feng J, Yue L, Carrier M, Nattel S. Evidence of two components of delayed rectifier K current in human ventricular myocytes. Circ Res 1996; 78: 689–96.

128. Sanguinetti MC, Curran ME, Zou A et al. Coassembly of KvLQT1 and minK (Isk) proteins to form IKs cardiac potassium channels. Nature 1996; 384: 80–3.

129. Barhanin J, Lesage F, Guillemare E et al. KvLQT1 and Isk(minK) proteins associate to form the IKs cardiac potassium current. Nature 1996; 384: 78–80.

130. Wang KW, Goldstein SA. Subunit composition of minK potassium channels. Neuron 1995; 14: 1303–9.

131. Abbott GW, Sesti F, Splawski I et al. MiRP1 forms IKr potassium channels with HERG and is associated with cardiac arrhythmia. Cell 1999; 16: 175–87.

132. Tinel N, Diochot S, Borsotto M et al. KCNE2 confers background current characteristics to the cardiac KCNQ1 potassium channel. EMBO J 2000; 19: 6326–30.

133. Schroeder BC, Waldegger S, Fehr S et al. A constitutively open potassium channel formed by KCNQ1 and KCNE3. Nature 2000; 403: 196–9.

134. Volders PG, Stengl M, van Opstal JM et al. Probing the contribution of Iks to canine repolarization: key role for beta adrenergic receptor stimulation. Circ 2003; 107; 2753–60.

135. Potet F, Scott JD, Mohammad-Panah R et al. AKAP proteins anchor cAMP-dependent protein kinase to KvLQT1/IsK channel complex. Am J Physiol Heart Circ Physiol 2001; 280: H2038–45.

136. Marx SO, Kurokawa J, Reiken S et al. Requirement of a macromolecular signaling complex for beta adrenergic receptor modulation of the KCNQ1-KCNE1 potassium channel. Science 2002; 295: 496–9.

137. Bryant SM, Wan X, Shipsey SJ et al. Regional differences in the delayed rectifier current (I_{Kr} and I_{Ks}) contribute to the differences in action potential duration in basal left ventricular myocytes in guinea-pig. Cardiovasc Res 1998; 40: 322–31.

138. Main MC, Bryant SM, Hart G. Regional differences in action potential characteristics and membrane currents of guinea-pig left ventricular myocytes. Exp Physiol 1998; 83: 747–61.

139. Xu X, Rials SJ, Wu Y et al. Left ventricular hypertrophy decreases slowly but not rapidly activating delayed rectifier potassium currents of epicardial and endocardial myocytes in rabbits. Circulation 2001; 103: 1585–90.

140. Yang T, Snyders DJ, Roden DM. Rapid inactivation determines the rectification and $[K^+]o$ dependence of the rapid component of the delayed rectifier K^+ current in cardiac cells. Circ Res 1997; 80: 782–9.

141. Sanguinetti MC. Dysfunction of delayed rectifier potassium channels in an inherited cardiac arrhythmia. Anal NY Acad Sci 1999; 868: 406–13.

142. Wang S, Morales MJ, Lui S et al. Time, voltage and ionic concentration dependence of rectification of h-erg expressed in Xenopus oocytes. FEBS Lett 1996; 389: 167–73.

143. Gutman GA, Chandy G, Adelman JP et al. International Union of Pharmacology. XLI. Compendium of voltage-gated ion channels: potassium channels. Pharmacological Reviews 2003; 55: 583–6.

144. Sanguinetti MC, Jiang C, Curran ME et al. A

mechanistic link between an inherited and an acquired cardiac arrhythmia: HERG encodes the IKr potassium channel. Cell 1995; 81: 299–307.

145. Trudeau MC, Warmke JW, Ganetzky B et al. HERG, a human inward rectifier in the voltage-gated potassium channel family. Science 1995; 269: 92–5.

146. Abbott GW, Goldstein SA. Potassium channel subunits: the MiRP family. Mol Intervent 2001; 1: 95–107.

147. McDonald TV, Yu Z, Ming Z et al. A minK–HERG complex regulates the cardiac potassium current I_{Kr}. Nature 1997; 388: 289–92.

148. Weerapura M, Nattel S, Chartier D et al. A comparison of currents carried by HERG, with and without coexpression of MiRP1, and the native rapid delayed rectifier current. Is MiRP1 the missing link? J Physiol 2002; 540.1: 15–27.

149. Cui J, Melman Y, Palma E et al. Cyclic AMP regulates the HERG K^+ channel by dual pathways. Curr Biol 2000; 10: 671–4.

150. Heath BM, Terrar DA. Protein kinase C enhances the rapidly activating delayed rectifier potassium current, Ikr, through a reduction in C-type inactivation in guinea-pig ventricular myocytes. J Physiol 2000; 522.3: 391–402.

151. Decher N, Chen J, Sanguinetti MC. Voltage-dependent gating of hyperpolarization-activated, cyclic nucleotide-gated pacemaker channels: molecular coupling between the S4-S5 and C-linkers. J Biol Chem 2004; 279: 13859–66.

152. Shi W, Wymore R, Yu H et al. Distribution and prevalence of hyperpolarization-activated cation channel (HCN) mRNA expression in cardiac tissues. Circ Res 1999; 85: e1–6.

153. Thuringer D, Lauribe P, Escande D. A hyperpolarization-activated inward current in human myocardial cells. J Mol Cell Cardiol 1992; 24: 451–5.

154. Yu H, Chang F, Cohen IS. Pacemaker current I_f in adult canine cardiac ventricular myocytes. J Physiol 1995; 485: 469–83.

155. Ludwig A, Zong X, Jeglitsch M et al. A family of hyperpolarization-activated mammalian cation channels. Nature 1998; 393: 587–91.

156. Santoro B, Liu DT, Yao H et al. Identification of a gene encoding a hyperpolarization-activated pacemaker channel of brain. Cell 1998; 93: 717–29.

157. Santoro B, Chen S, Luthi A et al. Molecular and functional heterogeneity of hyperpolarization-activated pacemaker channels in the mouse CNS. J Neurosci 2000; 20: 5264–75.

158. Han W, Bao W, Wang Z et al. Comparison of ion channel subunit expression in canine cardiac Purkinje fibers and ventricular muscle. Circ Res 2002; 91: 790–7.

159. DiFrancesco D, Tortora P. Direct activation of cardiac pacemaker channels by intracellular cAMP. Nature 1991; 351: 145–7.

160. Ulens C, Siegelbaum S. Regulation of hyperpolarization-activated HCN channels by cAMP through a gating switch in binding domain symmetry. Neuron 2003; 40: 959–70.

161. Decher N, Bundis F, Vajna R et al. KCNE2 modulates current amplitudes and activation kinetics of HCN4: influence of KCNE family members on HCN4 currents. Pflugers Arch 2003; 446: 633–40.

162. Kimura J, Miyamae S, Noma A. Identification of sodium–calcium exchange current in single ventricular cells of guinea pig. J Physiol 1987; 384: 199–222.

163. Niggli E, Lederer WJ. Molecular operations of the sodium–calcium exchanger revealed by conformation currents. Nature 1991; 349: 621–4.

164. Hilgemann DW, Nicoll DA, Philipson KD. Charge movement during Na^+ translocation by native and cloned cardiac Na^+/Ca^{2+} exchanger. Nature 1991; 352: 715–18.

165. Matsuoka S, Nicoll DA, Hryshko LV et al. Regulation of the cardiac Na^+–Ca^{2+} exchanger by Ca^{2+}. Mutational analysis of the Ca^{2+} binding domain. J Gen Physiol 1995; 105: 403–20.

166. Weber CR, Piacentino V, 3rd, Ginsburg KS et al. Na^+–Ca^{2+} exchange current and submembrane $[Ca^{2+}]$ during the cardiac action potential. Circ Res 2002; 90: 182–9.

167. Janvier NC, Boyett MR. The role of Na^+–Ca^{2+} exchange current in the cardiac action potential. Cardiovasc Res 1996; 32: 69–84.

168. Zygmunt AC, Goodrow RJ, Antzelevitch C. INaCa contributes to electrical heterogeneity within the canine ventricle. Am J Physiol Heart Circ Physiol 2000; 278: H1671–8.

169. Nicoll DA, Ottolia M, Lu L et al. A new topological model of the cardiac sarcolemmal Na^+–Ca^{2+} exchanger. J Biol Chem 1999; 274: 910–17.

170. Doering AE, Nicoll DA, Lu Y et al. Topology of a functionally important region of the cardiac Na^+–Ca^{2+} exchanger. J Biol Chem 1998; 273: 778–83.

171. Matuoska S, Nicoll DA, He Z et al. Regulation of cardiac Na^+–Ca^{2+} exchanger by endogenous XIP region. J Gen Physiol 1997; 109: 273–86.

172. Levitsky DO, Nicoll DA, Philipson KD. Identification of the high affinity Ca^{2+}-binding domain of the cardiac Na^+–Ca^{2+} exchanger. J Biol Chem 1994; 269: 22847–52.

173. Weber CR, Ginsburg KS, Philipson KD et al.

Allosteric regulation of Na/Ca exchange current by cytosolic Ca in intact cardiac myocytes. J Gen Physiol 2001; 117: 119–31.

174. Roden DM, George AL, Jr. Structure and function of cardiac sodium and potassium channels. Am J Physiol Heart Circ Physiol 1997; 273: H511–25.

175. Lees-Miller JP, Kondo C, Wang L et al. Electrophysiological characterization of an alternatively processed ERG K^+ channel in mouse and human hearts. Circ Res 1997; 81: 719–26.

176. Horio Y, Hibino H, Inanobe A et al. Clustering and enhanced activity of an inwardly rectifying potassium channel, Kir4.1, by an anchoring protein, PSD-95/SAP90. J Biol Chem 1997; 272.20: 12885–8.

177. Mays DJ, Foose JM, Philipson LH et al. Localization of the Kv1.5 K^+ channel protein in explanted cardiac tissue. J Clin Invest 1995; 96: 282–92.

178. January C, Riddle JM. Early afterdepolarizations: mechanisms of induction and block. A role for the L type Ca current. Circ Res 1989; 64: 977–90.

179. Cranefield PF, Aronson RS. Cardiac Arrhythmias: the role of triggered activity. Mount Kisco: Futura Publishing Co. Inc.; 1988.

180. Rudy Y, Quan W. A model study of the effects of the discrete cellular structure on electrical propagation in cardiac tissue. Circ Res 1987; 61: 815–23.

181. Rohr S, Kleber AG, Kucera JP. Optical recording of impulse propagation in designer cultures. Cardiac tissue architectures inducing ultra low conduction. Trends Cardiovasc Med 1999; 9: 173–9.

182. Allessie MA, Bonke FIM, Schopman FJG. Circus movement in rabbit atrial muscle as a mechanism of tachycardia. III. The 'leading circle' concept: A new model of circus movement in cardiac tissue without the involvement of an anatomical obstacle. Circ Res 1977; 41: 9–18.

183. Wit AL, Dillon S, Coromilas J et al. Anisotropic reentry in the epicardial borderzone of myocardial infarcts. Ann NY Acad Sci 1990; 591: 86–108.

CHAPTER 23

Abnormalities of inward currents underlying cardiac depolarization: role in acquired and heritable arrhythmias

Jonathan C Makielski

Inward currents in the heart and arrhythmia

Inward currents carried by either sodium or calcium ions through specific membrane channel proteins depolarize the cell during systole, while outward currents carried by potassium through other channels repolarize the cell to a negative resting potential during diastole. Five major inward currents of the heart are listed in Table 23.1, along with the associated human gene for the α subunit that forms the channel pore, and the tissues where they play a dominant role. In normal electrogene-

sis (see Chapter 22) these inward currents carry the depolarizing wave front in the heart, from the atria, through the atrio-ventricular node and specialized conduction tissue (His–Purkinje system), and the ventricle. They also underlie pacemaker depolarization in the sinus node and other automatic tissue. Outward potassium currents are major determinants of repolarization (see Chapter 24), but the persistence and slow decay of inward currents also affect timing of repolarization, duration of the action potential, and refractoriness. Cardiac arrhythmia comes in two basic types – too slow and too fast. Slow rhythms or bradycardia are caused by failure of impulse generation (pacemaking) and/or conduction. Rapid rhythm (tachycardia) is caused by three general mechanisms: enhanced normal pacemaking, triggered automaticity, and re-entry.[1]

Of the inward currents, the Na$^+$ channel (Fig. 23.1) plays an important role in both acquired and inherited arrhythmia and will be a major focus of this chapter. Inherited arrhythmias, such as the long QT syndrome 3 (LQT3) and Brugada syndrome are caused by mutations in the Na$^+$ channel that lead either to 'gain of function' or to 'loss of function' (Figs 23.1 and 23.2). Although uncommon, inherited arrhythmias are important clinical entities and may provide a model for the more common acquired arrhythmias,[2] such as those found in hypertrophy and failure. Moreover, subtle inherited channel abnormality may contribute to acquired arrhythmia, and genetic considerations

Table 23.1. *Inward depolarizing currents in the human heart*

Name	Current	Gene	Tissue
Na$^+$ current	I_{Na}	*SCN5A*	A, HP, V
L-type Ca^{2+} current	I_{Ca-L}	*CAV1*	N, HP
T-type Ca^{2+} current	I_{Ca-T}	*CAV3*	A
Pacemaker current	I_f, I_h	*HCN*	N
Na$^+$/Ca^{2+} exchange	I_{NaCaEx}	*NCX*	A, V, N?

The letters under tissue indicate where the current plays a major role where A is atria, V is ventricle, HP is His–Purkinje network, N is nodal tissue (sinus and atrioventricular node). For references see general reviews,[93] and more recent studies for calcium channels in human heart.[120]

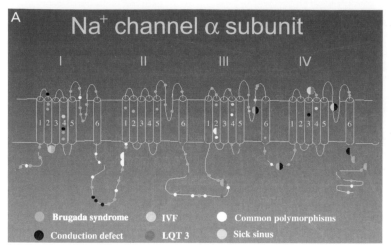

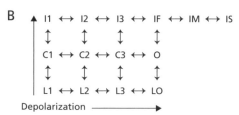

Figure 23.1. *The voltage dependent Na⁺ channel (SCN5A) structure and gating kinetics. (A) Topological diagram of the transmembrane structure of SCN5A, with the extracellular side at the top. The locations of arrhythmia-causing mutations and common (>0.5%) polymorphisms are shown. (B) Kinetic state diagram for SCN5A. See text for further explanation. IVF, idiopathic ventricular fibrillation. See color plate section for Fig. 23.1(A).*

could therefore have practical clinical importance for arrhythmia in acquired heart disease. This chapter will first focus on the Na⁺ channel and the mechanisms by which it causes arrhythmia, then discuss inherited arrhythmias, acquired arrhythmias, and antiarrhythmic drugs. Finally, the role abnormalities of other inward currents such as calcium channels will be considered.[3]

The cardiac Na⁺ channel SCN5A

The pore-forming cardiac Na⁺ channel α subunit protein (Fig. 23.1A), originally called hH1 when it was cloned,[4] is encoded by the gene SCN5A;[5] the channel protein is now referred to as SCN5A or Na$_V$1.5.[6] The channel has four repeat homologous domains each with six membrane-spanning segments as depicted in Fig. 23.1A and as previously reviewed.[7,8] In humans the channel exists as two splice variants with either 2015 or 2016 amino acids due to alternative splicing at position 1077.[9] Mutations in the channel that cause a change at the amino acid level are mainly 'missense' muta-

tions and result from a mutation at a single base pair in the gene, to allow for the substitution of one amino acid for another. Such mutations are indicated by using the single letter amino acid code and giving the usual amino acid, followed by the position number, and then the substituted amino acid. For example, the common polymorphism H558R denotes that the normal histidine is substituted by an arginine at position 558. Other mutations cause a deletion in amino acids and are designated by 'del' or 'Δ' with the deleted amino acids given; sometimes but not always the position in the sequence is also given. For example one of the first LQT3 mutations discovered was called ΔKPQ,[10] also more precisely Δ1505–1507KPQ, where three amino acids were deleted. Sometimes base pair mutations cause a non-sense, frameshift, or stop codon that fails to generate a protein. SCN5A is distinct from brain and skeletal muscle Na⁺ channels, and is relatively specific to atrial, ventricular, and Purkinje cells in the heart, although SCN5A exists in other tissues such as the brain and gastrointestinal tract.[11] In addition, brain isoforms of Na⁺ channels exist in the heart.[12,13] Auxiliary sub-

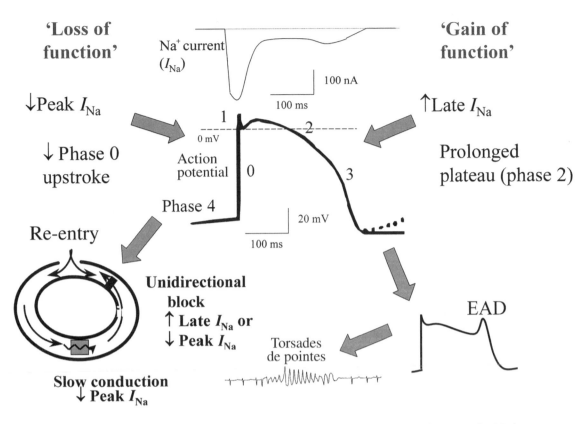

Figure 23.2. *Mechanisms of arrhythmia with alterations in the Na$^+$ current. A diagram of a Na$^+$ current trace (I_{Na}) measured in nanoamperes (nA) and extending over several hundred milliseconds (ms) is shown at the top; a cellular action potential measured in millivolts (mV) with the inside of the cell negative (downward) at rest in phase 4, and depolarized to positive levels (upward) during phases 0 through 3, with the 0 mV line shown as a broken line in the middle; and arrhythmia mechanisms are shown at the bottom. On the left side of the diagram, abnormalities in the Na$^+$ current causing 'loss of function' or a decreased peak I_{Na} lead to a decrease in the upstroke of the action potential (phase 0) causing slow conduction or unidirectional block, which then may cause arrhythmia by re-entry mechanisms shown diagrammatically at the lower left. On the right side, 'gain of function' or an increase in the late I_{Na} may cause prolongation of the action potential plateau (phase 2), which may lead to triggered arrhythmia by the cellular mechanism of early afterdepolarization (EAD) shown as a hump on the action potential in the insert at the bottom right, which causes torsades des pointes arrhythmia shown in the insert at bottom.*

units (β1, β2, β3, β4) may also play a role in the heart,[12] by increasing expression levels and by altering kinetics. In addition to the ubiquitous splice variant insert at Q1077, eight common polymorphisms (>0.5% in at least one ethnic group) and numerous rare variants exist in the heart.[14] With the exceptions noted below, functional implications of these variants are not known.

Time and voltage-dependent kinetics of the cardiac Na$^+$ current I_{Na}

During the cardiac cycle the Na$^+$ channel opens and closes in response to the membrane voltage; these kinetic transitions of the channel are considered as a kinetic model (Fig. 23.1B) involving three basic states, closed but ready to open (C), open and conducting current (O), and inactivated (I). During diastole when the membrane is resting at a negative potential (–80 mV), most channels are in the C1 state. When the membrane is depolarized to the threshold level of about –75 mV, the channel activates over less than 1 ms, transitioning through C2 and C3 to the open state O. At the same time channels undergo a slower transition over 2–5 ms to the inactivated states in the top row (I1, I2, I3, IF). Depolarization generally favors transitions to the right and upward as noted on Fig. 23.1B. When the membrane is repolarized back to resting potentials, the transitions are reversed and the channel recovers back to the closed states over about 10–50 ms.

Two important kinetic phenomena – slow recovery from inactivation and late I_{Na} – require the addition of the IM and IS states at the top, and the 'L' states at the bottom (Fig. 23.1B). The literature on this issue can be quite confusing because both phenomena are referred to as 'slow inactivation',[15] but they are best considered separately. Slow recovery from inactivation (also called intermediate inactivation[16]) is a component of I_{Na} that recovers slowly, over hundreds of milliseconds or longer.[17] In terms of kinetics (Fig. 23.1B) the channel enters into different inactivated states, a fast inactivated state (IF) from which recovery is fast (<10 ms), intermediate states (IM) from which recovery is intermediate (10–500 ms), and slow or ultra-slow inactivated states (IS) from which recovery is even slower (>500 ms). The distinction between intermediate versus slow versus ultra-slow is not consistently defined; the ranges given here are arbitrary. It is clear, however, that the longer the channel is depolarized, the longer the recovery time.[17,18] Inactivation is often measured by a 'steady-state inactivation' protocol, where the membrane is first conditioned by holding it at a specific potential, then depolarized to see how much current can be elicited to give the amount of current that is not inactivated. Slow recovery from inactivation, however, means that the inactivation process is not at steady-state for very long times, if ever, so that each measurement of 'steady-state inactivation' is dependent upon the details of the protocol used to measure it, specifically the duration of the conditioning pulse. Slow recovery from

inactivation can be important for cardiac excitability if recovery is not complete in time for the next heartbeat. In that case less I_{Na} is available as channels accumulate in the inactivated states in a rate-dependent manner, with repetitive depolarization leading to a 'loss of function'.

Late I_{Na} is the current that continues to flow through the Na^+ channel long after the majority of the current has inactivated (>10 ms). In heart, late I_{Na} has been variously called steady-state Na^+ current,[19] window current,[20] slow inactivation,[21,22] persistent current,[23] and late current.[24,25] The term 'late' I_{Na} is preferred over 'steady-state' and 'persistent' because those terms suggest that late I_{Na} is not subject at all to inactivation, but in many cases late I_{Na} does slowly inactivate.[26] Late I_{Na} has been reported in native heart Na^+ channels,[27] as a rarely occurring bursting mode of I_{Na}, as if the inactivation gate is non-functional. Late I_{Na} has been modeled as a mode with separate gating (depicted as L for 'late' mode in Fig. 23.1B),[28] although the exact kinetic mechanisms and models for late I_{Na} require further study.[15] Although small in magnitude (generally less than 0.5% of the peak I_{Na}), this late I_{Na} as a depolarizing influence during the cardiac action potential plateau was recognized nearly 40 years ago when it was shown that tetrodotoxin, a relatively specific blocker of I_{Na}, shortened cardiac action potential duration.[29]

SCN5A and mechanisms of arrhythmia

The mechanisms by which abnormalities of I_{Na} cause arrhythmia can be considered from molecular defect or genotype through biophysical and cellular phenotypes, to the clinical syndrome or clinical phenotype. In this paradigm a channel protein abnormality (either acquired or a mutation) causes an altered biophysical phenotype at the channel level, which interacts with other ion channels and causes an altered cellular action potential phenotype, which at the tissue level causes an arrhythmia resulting in a clinical syndrome. For the biophysical phenotype, the mechanisms of arrhythmia caused by abnormalities in SCN5A can be classified into 'loss of function' and 'gain of function'.[10] Figure 23.2 shows a diagram of an I_{Na} trace with the late I_{Na} greatly accentuated relative to peak I_{Na} for illustrative purposes. 'Loss of function' refers to a decrease in peak I_{Na}; this acts at the cellular level to decrease the upstroke of the action potential (middle of Fig. 23.2). At the tissue level this causes slow conduction or block, providing the substrate for re-entry arrhythmias depicted diagrammatically at bottom left. Loss of function of I_{Na} in the conduction system leads to atrioventricular conduction abnormalities or heart block, and loss of I_{Na} may contribute to dysfunction in pacemaker tissue.[30] 'Gain of function' is the biophysical phenotype that refers to an increase in late I_{Na} that acts at the cellular level to prolong the action potential duration or phase 2. This could promote unidirectional block and re-entry at the tissue level by prolonging refractoriness. At the cellular level it could lead to a type of triggered activity called early afterdepolarizations (EAD) depicted diagrammatically at lower right (Fig. 23.2), which in turn may trigger the ventricular arrhythmia called torsade de pointes.[31] A concomitant finding is a prolonged QT interval on the electrocardiogram (ECG) that reflects the underlying prolongation of the action potentials in the ventricle. These considerations apply mainly to ventricular arrhythmias. Atrial fibrillation is a special case where the abnormality at the cellular level is a shortening of the action potential to allow for a decreased refractory period and rapid re-entry.[32] In theory, a loss of late inward current function (either Na^+ or Ca^{2+} channel) could cause this, but no mutations in inward currents have been implicated in familial atrial fibrillation, and structural remodeling is thought to be more important for atrial fibrillation in acquired heart disease than ionic current remodeling.[33] Nevertheless, abnormalities in inward currents may contribute to the mechanisms of atrial fibrillation in acquired heart disease,[34] and Na^+ and Ca^{2+} channel pore proteins and auxilary/regulatory proteins remain candidate genes for familial atrial fibrillations.

Electrophysiological heterogeneity is important for the generation of arrhythmia at the tissue level by forming the substrate for re-entry. Cardiac tissue has intrinsic structural heterogeneity as it is organized into the heart as an organ. Another aspect of heterogeneity is the differential expression of ion channels in different parts of the heart. The transient outward current particularly has a strong transmural gradient, being more prominent in the epicardium.[35] The Na^+ channel may also have a heterogeneous distribution across the ventricular wall (epicardium to endocardium) and regionally (apex to base, right ventricle and left ventricle and atrium).[36] Alterations in Na^+ channel function might interact with the intrinsic electrical heterogeneity of the heart to cause arrhythmia.[37] It has not yet been shown whether or not acquired or inherited transmural or regional distribution abnormalities in the Na^+ channel itself may contribute to arrhythmia. Mechanisms of arrhythmia in general have been recently reviewed,[1] this chapter will now focus on the biophysical phenotypes of inward currents that may contribute to arrhythmia.

Heritable disorders of SCN5A causing arrhythmia

Mutations in *SCN5A* cause as many as five distinct and overlapping arrhythmia syndromes including LQT3, Brugada syndrome, idiopathic ventricular fibrillation, conduction disorder, and sick sinus syndrome (Table 23.2). The trafficking and biophysical function of these mutations are studied by making the mutation in a wild-type *SCN5A* back-

Table 23.2. *Arrhythmogenic inherited disorders involving mutations resulting in altered inward currents in the heart*

Syndrome	Gene, current affected
LQT3	*SCN5A, I_{Na}*
Brugada	*SCN5A, I_{Na}*
Idiopathic ventricular fibrillation	*SCN5A, I_{Na}*
Conduction disorder	*SCN5A, I_{Na}*
Sick sinus syndrome	*SCN5A, I_{Na}*
	HCN4, I_f
LQT4 (Ankyrin)	*ANK4, I_{Na}*
Timothy syndrome	*CaV2.1, I_{Ca}*

ground then expressing them in an expression system such as oocytes or in mammalian cell culture such as HEK cells. They can then be studied by voltage clamp, or by immunocytochemistry or other biochemical assays. It is important to recognize that the results obtained may depend upon the choice of expression system, choice of background wild-type channel sequence (such as the presence of common polymorphisms,[38,39] rare variants, splice variants[9]), conditions of study such as temperature,[40] and specifics of the voltage-clamp protocols.[41] Finally, these channels are studied under artificial conditions in surrogate non-muscle cells, therefore any function at this level may or may not reflect what occurs in human cardiac muscle under a potentially very different regulatory environment. Computer simulations, or *in silico* experiments, are useful for predicting effects on the action potential,[41] but they are even farther removed from what occurs *in vivo* than the *in vitro* experiments, and depend upon assumptions made by the modelers. In an *in vivo* experiment, the *LQT3* gene was inserted into a transgenic mouse and produced arrhythmia,[42] but it must be noted that mouse cardiac electrophysiology is different from human electrophysiology. These caveats should be kept in mind as the functional consequences of these mutations are described in the literature and as they are discussed below.

Heritable disorders of inward current causing arrhythmia – 'gain of function' mutations in *SCN5A*

LQT3 is a distinct type of inherited congenital LQT without deafness, and is less common than syndromes LQT1 and LQT2 associated with potassium channel mutations.[43] Like the other congenital LQT syndromes, patients have a prolonged QT interval on the ECG with syncope and sudden death from a characteristic ventricular arrhythmia known as torsade de pointes.[31] Patients with LQT3 tend to have an extremely long QT interval, it occurs classically later in life, and they tend to have an enhanced rate dependence of QT interval shortening.[43] LQT3 was associated with mutations in *SCN5A*,[44] that when tested in expression sys-

tems showed a 'gain of function' as an increase in late I_{Na} for ΔKPQ.[10] Mathematical modeling of the action potential[45] with the altered Na$^+$ channel kinetics supported the mechanism of action potential prolongation.

At the time of writing, more than 31 different mutations in *SCN5A* have been described in LQT3 patients; the electrophysiological function has been characterized for 20 of them, and this number is certain to grow. The usual way to measure late I_{Na} is to apply a prolonged depolarization and determine if late I_{Na} is increased at plateau level clamp steps; the majority of those characterized (16/20) show increased late I_{Na} compared to wild-type channels. Late I_{Na} may be increased by several biophysical mechanisms such as inactivation failure,[46] and increased window current.[47,48] For four mutations (E1295K,[48] A1330P,[49] I1768V,[50] and D1790G[51]), no increased late I_{Na} was found, although a later report did find an increased late I_{Na} for D1790G.[52] For I1768V, however, a subsequent study found late I_{Na} when the protocol included a ramp,[41] emphasizing the protocol dependence of the biophysical phenotype. If some *SCN5A* mutations truly have no late I_{Na}, what may be the mechanism for prolonged QT in these mutations? For E1295K it was suggested that positive shifts in activation and inactivation midpoints altered later repolarizing K$^+$ currents.[48] For D1790G it was shown by a computer action potential model that the negative shift in midpoint could prolong action potential duration (APD) and generate EADs through effects on calcium sensitive exchange and ion currents.[53] All *LQT3* mutations have, to a greater or lesser degree, these other kinetic effects including alterations in the midpoints of activation and inactivation, decay rates, development, and recovery rates for both fast inactivation and slow inactivation. Perhaps these indirect mechanisms also play a role in generating the clinical phenotype. In any event, it is probably better to not think of LQT3 exclusively as 'gain of function'.

The general scheme of *SCN5A* mutations causing 'gain of function' and increased late I_{Na} with prolonged QT interval and arrhythmia in LQT3 applies broadly, but details of the particular biophysical phenotype generated by specific mutations may correlate with particulars of the clinical phenotype. As an example, classically LQT3 reportedly occurred in older children and adults, and in comparison to LQT1 and LQT2, events were more likely to occur at rest or during sleep.[43] Later it was apparent that LQT3 could also occur in very young children,[49,54] and a molecular/cellular phenotype of LQT3 was described for mutations found in documented cases of sudden infant death syndrome or SIDS.[55] Patients with classical LQT3 had longer resting QT intervals and enhanced rate-dependent QT interval shortening compared with other LQT syndromes, and the rate-dependent QT shortening was greater than that found in normal individuals.[56] Late I_{Na} for the LQT3 mutation ΔKPQ was noted to be not absolutely persistent, but rather decayed over time,[26] as if still subject to slow inactivation. Recovery from this slow inactivation of late I_{Na} was even slower than recovery of peak I_{Na},[26] and at faster stimulation rates inactivation accumulated, leading to decreased late I_{Na} current,[57] perhaps accounting for the enhanced rate-dependent QT shortening on the ECG. Functional characterization of the mutations associated with SIDS showed that the late I_{Na} did not decay,[55] precluding rate-dependent mechanisms for these mutations. Perhaps this relative lack of rate dependence accounts for the early lethality of these mutations. This example shows that details of the genotype defect may affect the molecular/cellular phenotype, and ultimately account for characteristics of the clinical phenotype.

In addition to the mutation itself, the particular genetic background within SCN5A may play a role in expression of the clinical phenotype. A patient with a long QT was found to have a mutation at position 1766 substituting a methionine for a leucine or M1766L.[58] In the normal wild-type background, this channel did not express, suggesting a 'loss of function' mutation. When expressed in the presence of the common polymorphism H558R, a common polymorphism with a frequency ~30% that the patient also had, it exhibited a normal peak I_{Na} but increased late I_{Na}.[38] This is an example of a 'double-hit' mutation where one of the 'hits' is the most common polymorphism in the human population.

Finally, it is important to point out that the genetic abnormality need not be in the channel itself to produce late I_{Na}. Many patients with LQT cannot be mapped to a known LQT genetic abnormality, implying that other genes are affected. These other genes could produce LQT by causing late I_{Na} by indirect mechanisms. Based upon in vitro studies, the loss of the β_1 subunit might be expected to both decrease peak I_{Na} and increase late I_{Na} because late I_{Na} is increased in the absence of the β_1 subunit in expression studies.[59] So far, however, no defects in Na$^+$ channel subunit genes have been detected with arrhythmia syndromes. Other candidate genes may include the cytoskeleton. Acute disruption of cytoskeleton attachments causes both a decrease in peak I_{Na} and an increase in late I_{Na},[60] invoking both 'loss of function' and 'gain of function' mechanisms. A transgenic mouse model with the cytoskeleton element ankyrin disrupted increased late I_{Na},[61] and subsequently mutations in the Ankyrn B gene were found in the syndrome that had been earlier mapped to a distinct locus and called LQT4.[62] Additional candidate genes for LQT syndromes acting through late I_{Na} may lie in the cytoskeletal system, or other systems regulating Na$^+$ channel activity.

Heritable disorders of inward current causing arrhythmia – 'loss of function' mutations

'Loss of function' leads to at least five, sometimes overlapping, arrhythmia syndromes: Brugada syndrome,[63] idiopathic ventricular fibrillation,[64] conduction disorder,[65,66] sick sinus syndrome,[67] and atrial standstill.[68] Some mutations cause more than one of these syndromes in the same or different individuals,[69] as might be expected from their common pathogenetic mechanism, but it is not yet clear what particular biophysical phenotype for a particular mutation tends to give it a distinct clinical phenotype. Brugada syndrome is an autosomal dominant disorder with a male predominance characterized by sudden death from ventricular fibrillation in patients with right bundle branch block (RBBB) and ST segment elevation V1–V3.[70] Many, but not even the majority, of these patients are found to have mutations of SCN5A that, when investigated, cause a decrease in peak I_{Na}.[63] At the time of writing, over 60 distinct mutations in *SCN5A* have been found in these patients (Fig. 23.1A), but only about 24 of these mutations have been made and studied *in vitro*. Mutations in *SCN5A* have also been described in patients with idiopathic ventricular fibrillation and an inherited tendency for sudden death, but lack resting or drug-induced ECG criteria for Brugada syndrome.[64] As the ECG can be variable in Brugada syndrome, and the underlying mechanism of loss of function of I_{Na} appears to be in common, it is likely that this may be within the spectrum of the same syndrome. 'Loss of function' of I_{Na} because of mutation in *SCN5A* has also been reported for familial conduction disorder that includes syncope, atrioventricular block, and intraventricular conduction delay,[65] and for families with sick sinus syndrome.[67] The presence of SCN5A in sinus node tissue and a role in pacemaking[30] may account for sinus node dysfunction in familial sick sinus syndrome with *SCN5A* mutations. A rarer condition called atrial standstill has been associated in a family to a mutation in *SCN5A* cosegregating with a gap junction mutation,[68] and also co-existing with Brugada syndrome.[71]

Mutations cause loss of function by a number of mechanisms. A mutation that causes a stop or nonsense codon would cause a failure in transcription.[63] Some mutations cause a trafficking defect; the protein is made but does not make it to the cell surface.[72,73] In other cases, the channel is expressed and makes it to the cell surface, but an alteration in kinetic properties makes I_{Na} less available. One such change is an enhancement of slow or intermediate inactivation leading to a rate-dependent decrease in the availability of peak I_{Na}.[16] Another mechanism is a negative or hyperpolarizing shift in the voltage-dependent availability relationship (also called steady-state inactivation) as described for R1512W,[74,75] Y1795H,[76] A1924T,[74] S1710L,[64] and 1795 Ins D.[77] A positive shift in activation may decrease I_{Na} by increasing the gap between resting and threshold potential as suggested for G514C,[66] L567Q,[78] A735V,[79] S1710L,[64] and 1795 Ins D.[77] A common kinetic alteration found in these mutations is an enhanced entry into 'intermediate inactivation' from which channels recover slowly, as

described for G298S,[80] T521I(with H558R),[39] D1595N,[80] R1512W,[74,75] T1620M (with temperature and β_1 dependence),[40,63] Y1795H,[76] S1710L,[64] and 1795 Ins D.[81] Note that many of these mutations, most notably 1795 Ins D, have more than one of these defects. In contrast to these mutations with kinetic effects, in some cases no or little current was seen. Failure to express the channel at all may occur when a non-sense or stop codon is inserted,[63] representing a failure in transcription. Four mutations have been shown to have a defect in trafficking with retention in the endoplasmic reticulum (ER) including R1232W,[72] R1432G,[82] M1766L,[58] and G1743R.[73] One mutation G1406R had no I_{Na} but normal trafficking, as protein was demonstrated at the surface,[69] a possible example of a non-conducting channel.

Double mutations may also play a role in 'loss of function' arrhythmia mutations. In the case of the conduction disorder mutation *T512I*, the presence of H558R actually ameliorated the increased slow inactivation leading to loss of function in T512I alone.[39] H558R is not the only example of the importance of double mutations. In one example, two mutations (*T1620M* and *R1232W* were required to account for a Brugada syndrome phenotype.[72] These findings stress the importance of studying a mutation in the correct genetic background, including the patient's own particular genetic make-up.

SCN5A and arrhythmia in acquired heart disease

The heritable arrhythmia syndromes described above generally occur in the absence of structural heart disease. They are rare compared to arrhythmias occurring in such common conditions as ischemic injury, hypertrophy and failure. For the electrophysiological mechanisms underlying arrhythmia in acquired heart disease, it is useful to distinguish between an intrinsic alteration in the channel density or function, also called electrical remodeling, and a normal or physiological response of the unaltered channel to the abnormal stimuli found in the disease state, what we might call extrinsic alterations. In acute ischemia, for example, the resulting acute membrane depolarization will inactivate Na$^+$ channels causing a 'loss of function' mechanism for arrhythmia.[83] Metabolites released in ischemia such as lysophospholipids are known to be arrhythmogenic, and may act through Na$^+$ channels to cause increased bursting and arrhythmia by a 'gain of function' mechanism.[84]

Potassium channels are known to be extensively involved in the electrical remodeling found in chronic ischemia, hypertrophy, and failure (see Chapter 24). Alterations in calcium homeostasis (see Chapter 13) are also well known, but alterations of I_{Na} have been less prominently considered in the electrical remodeling in hypertrophy and failure.[85] In general, peak I_{Na} has been found to be unaltered in models of acquired heart disease such as hypertrophy in rats,[86,87] and human atrium,[88] and in tachycardia pacing-induced canine heart failure.[89] Human heart Na$^+$ channel mRNA and protein has been reported to be unchanged in heart failure.[90] Thus, in hypertrophy and heart failure no 'loss of function' mechanism caused by remodeling would appear to be operative. In a canine model of atrial fibrillation, however, peak I_{Na},[91] and message for SCN5A are both decreased.[92] Peak I_{Na} was also decreased in the border zone of a canine infarction.[93] This would tend to shorten action potentials and refractory periods, facilitating re-entry. What about 'gain of function' or an increase in late I_{Na} as an arrhythmogenic mechanism in hypertrophy and failure? This would tend to lengthen action potentials, a nearly universal finding in heart failure and hypertrophy models.[94] The first report to suggest an increase in late I_{Na} was in hamsters,[95] but a subsequent study showed no such increase.[96] A prominent late I_{Na} was seen in isolated ventricular cells both for a canine ischemic model of heart failure,[97] and for the rapid pacing canine model of heart failure.[98] Ventricular cells isolated from failing human hearts explanted from transplant patients showed a prominent late I_{Na},[99] that was later shown to be significantly greater than the late I_{Na} from non-human failing hearts.[100] Block of late I_{Na} by tetrodotoxin shortened action potential duration and eliminated EADs, suggesting a role for late I_{Na} in arrhythmogenesis in heart failure.[97] Thus it appears that the ionic basis for prolonged action potentials found in heart failure may include increased late I_{Na}, as well

as downregulation of potassium channels. The biophysical mechanism for late I_{Na} in heart failure may be caused by a subpopulation of channels with a different gating mode.[101] This kinetic change might be caused indirectly by structural remodeling in other proteins such as cytoskeleton (Chapter 19), or perhaps changes in phosporylation because of altered surface receptor regulation (Chapter 5). Isoform or subunit switching is another possibility. Late I_{Na} has been shown to be increased in the absence of the β_1 subunit, and it is greater in the SCN4A isoform,[18] but neither downregulation of the β_1 subunit nor isoform switching to SCN4A was found in a dog pacing model of heart failure.[18]

SCN5A mutations and acquired arrhythmia

The distinction between acquired and inherited abnormalities in inward currents for causing arrhythmia may not be clear-cut, as discussed in Chapter 29. Inherited allelic SCN5A variants in presumably normal subjects might predispose to drug-acquired long QT arrhythmia.[102] The mutation S1103Y is common (>5%) in the black population, and may predispose to acquired ventricular arrhythmia.[103,104] The common (>25%) polymorphism H558R showed a marked decrease in current density with the ubiquitous splice variant containing an insert Q1077,[9] but the implications of this for clinical arrhythmia are not known.

Drugs, Na$^+$ current and arrhythmia

Drugs that interact with the Na$^+$ channel (Vaughan Williams Class I) such as lidocaine (and the analogue mexiletine), disopyramide and flecainide,[105] could be considered a type of acquired abnormality of the channel. Amiodarone, a mainstay of antiarrhythmic therapy, also affects Na$^+$ channels. These drugs generally block the Na$^+$ channel and are thought to be antiarrhythmic for the re-entrant mechanism, by converting the slow conduction or the unidirectional block in the re-entrant circuit to complete block. Clearly they can be proarrhythmic by the 'loss of function' mechanism, and this or other mechanisms may account for the proarrhythmia seen most dramatically in the Cardiac Arrhythmia Suppression Trial.[106] This study in post-myocardial infarction (MI) patients showed excess mortality in those treated to suppress ventricular ectopy, and probably represents an interaction of the acquired disease state, the Na^+ channel, and the drug. Na^+ channel-blocking drugs, however, may have unique roles in the case of inheritable SCN5A arrhythmia syndromes. In the LQT3 syndrome, these drugs have a preferential block for late I_{Na}.[107–109] In a Brugada syndrome mutation where the channel shows a trafficking defect, mexiletine was shown to rescue the channel function *in vitro*.[73] It may seem paradoxical that a drug that generally causes a loss of function might ultimately result in a correction of a loss of function defect. Mexiletine is regarded as a 'weak' Na^+ channel-blocking drug because it recovers from normal channels very rapidly, and does not affect the Na^+ channel activity under normal conditions. By upregulating the defective Na^+ channel density, the net effect could be an increase in I_{Na} without block under physiological conditions. This potentially novel pharmacological mechanism has not yet been proven to work *in vivo*.

Abnormalities of other inward currents and arrhythmia

Table 23.3 summarizes reported abnormalities of all inward currents in the heart. Calcium channels in the heart are crucial for excitation–contraction coupling.[110,111] They also maintain the action potential plateau, and play an excitatory role in the sinus and atrioventricular node, and may underlie excitability in myocardial tissue where the Na^+ current is suppressed by depolarization.[83] Only recently, the first mutation in the cardiac L-type calcium channel was described that causes arrhythmia in combination with autism (Timothy Syndrome).[112] Because of the importance of calcium channels for so many intracellular functions including cell signaling as well as contractility, such mutations might be expected to cause severe and early lethal abnormalities. Perhaps this accounts for the rarity of mutations in this gene presenting as arrhythmia.

Table 23.3. *Abnormalities of inward currents in acquired heart disease*

	I_{Na}	Late I_{Na}	I_{CA-L}	I_{CA-T}	I_F	Na^+–Ca^{2+} exchange
LVH	NC–	↑↓	↑	↑	↑, NC	
HF	NC	↑	↑↓	↑	NC	↑
MI	↓	–	↓	–	–	–
AF	↓	–	↓	NC	–	–

NC, no change; ↓, decreased; ↑, increased; – not known. Where results differ both are shown. For references see an original study,[93] and additional recent references.[91,121,122]

The Ca^{2+} channel has been extensively studied in acquired heart disease because of its role in contractility. Current through the L-type calcium channel in heart in acquired heart disease has been reported to be unchanged, increased, or decreased as previously reviewed.[94,113] It is important to note that phosphorylation may be an important factor in acquired changes in L-type calcium current.[114,115] The T-type calcium channel is a distinct calcium current and channel that has been shown to be altered in acquired heart disease (Table 23.3), but the role of this channel in arrhythmia is uncertain, and no arrhythmogenic mutations have been found to date. The Na^+–Ca^{2+} exchanger is electrogenic and extrudes one Ca^{2+} molecule for every three Na^+ molecules, generating a net inward depolarizing current when running in the normal direction, with a reversal potential of about –20 mV.[116] This exchanger has been reported to be altered in acquired heart disease (Table 23.3), and may contribute to triggered automaticity through delayed afterdepolarizations.[117] The calcium channels and Na^+–Ca^{2+} exchanger are key players in calcium homeostasis. The role of altered calcium homeostasis in acquired heart disease and arrhythmia is a complex topic,[118] and it is addressed further in Chapter 13.

The pacemaker current (I_f) is a depolarizing current activated by hyperpolarization, and contributes to phase 4 depolarization. In humans the pacemaker current flows through a channel encoded by members of the cyclic nucleotide-gated channels.[119] A mutation in a gene *HCN4* encoding one of these channels causes a familial sick sinus syndrome.[120] As for a role in acquired arrhythmia, I_f was increased in a rat hypertrophy model,[121] and a trend toward greater I_f was found in human heart failure.[122]

Conclusion

Inherited abnormalities in the cardiac sodium channel underlie multiple arrhythmia syndromes, and the mechanisms for the arrhythmia can be traced from the genotype through the molecular/biophysical phenotype and cellular phenotype to the clinical syndrome. These examples of inherited abnormalities suggest pathogenetic mechanisms for acquired arrhythmia in hypertrophy and heart failure, and indeed acquired abnormalities of I_{Na} have been described. Abnormalities in I_{Na} are clearly important in arrhythmias generated in acute ischemia, and antiarrhythmic drugs that act on this channel also play an important role in both antiarrhythmia and proarrhythmia. Inherited abnormalities of the cardiac calcium channels that cause arrhythmia syndromes are less common, perhaps because any such mutations would be more critical to development and be lethal to the embryo. Acquired abnormalities in calcium channels may play a role through altered calcium homeostasis. The importance of the interaction of inherited channel abnormalities with acquired heart disease may become increasingly recognized in the future.

Acknowledgements

Supported by NIHNHLBI grants HL71092 and HL-66378.

References

1. Zipes DP. Mechanisms of clinical arrhythmias. J Cardiovasc Electrophysiol 2003; 14: 902–12.
2. Balser JR. Inherited sodium channelopathies: models for acquired arrhythmias? Am J Physiol Heart Circ Physiol 2002; 282: H1175–80.
3. Foell JD, Balijepalli RC, Delisle BP et al. Molecular heterogeneity of calcium channel β subunits in canine and human heart: evidence for differential subcellular localization. Physiol Genomics 2004; 17: 183–200.
4. Gellens ME, George AL, Jr., Chen LQ et al. Primary structure and functional expression of the human cardiac tetrodotoxin-insensitive voltage-dependent sodium channel. Proc Natl Acad Sci USA 1992; 89: 554–8.
5. George AL, Jr. Varkony TA, Drabkin HA et al. Assignment of the human heart tetrodotoxin-resistant voltage-gated Na⁺ channel α-subunit gene (SCN5A) to band 3p21. Cytogenet Cell Genet 1995; 68: 67–70.
6. Goldin AL, Barchi RL, Caldwell JH et al. Nomenclature of voltage-gated sodium channels. Neuron 2000; 28: 365–8.
7. Fozzard HA, Hanck DA. Structure and function of voltage-dependent sodium channels: comparison of brain II and cardiac isoforms. [Review]. Physiol Rev 1996; 76: 887–926.
8. Catterall WA. From ionic currents to molecular mechanisms: the structure and function of voltage-gated sodium channels. Neuron 2000; 26: 13–25.
9. Makielski JC, Ye B, Valdivia CR et al. A ubiquitous splice variant and a common polymorphism affect heterologous expression of recombinant human SCN5A heart sodium channels. Circ Res 2003; 93: 821–8.
10. Bennett PB, Yazawa K, Makita N et al. Molecular mechanism for an inherited cardiac arrhythmia. Nature 1995; 376: 683–5.
11. Ou Y, Gibbons SJ, Miller SM et al. SCN5A is expressed in human jejunal circular smooth muscle cells. Neurogastroenterol Motil 2002; 14: 477–86.
12. Malhotra JD, Chen C, Rivolta I et al. Characterization of sodium channel α- and β-subunits in rat and mouse cardiac myocytes. Circulation 2001; 103: 1303–10.
13. Maier SK, Westenbroek RE, Schenkman KA et al. An unexpected role for brain-type sodium channels in coupling of cell surface depolarization to contraction in the heart. Proc Natl Acad Sci USA 2002; 99: 4073–8.
14. Ackerman MJ, Splawski I, Makielski, JC et al. Spectrum and prevalence of cardiac sodium channel variants among black, white, Asian, and Hispanic individuals: implications for arrhythmogenic susceptibility and Brugada/long QT syndrome genetic testing. Heart Rhythm 2004; 1: 600–7.
15. Goldin AL. Mechanisms of sodium channel inactivation. Curr Opin Neurobiol 2003; 13: 284–90.

16. Wang DW, Makita N, Kitabatake A et al. Enhanced Na$^+$ channel intermediate inactivation in Brugada syndrome. Circ Res 2000; 87: E37–43.

17. Shander GS, Fan Z, Makielski JC. Slowly recovering cardiac sodium current in rat ventricular myocytes: effects of conditioning duration and recovery potential. J Cardiovasc Electrophysiol 1995; 6: 786–95.

18. Valdivia CR, Nagatomo T, Makielski JC. Late currents affect kinetics for heart and skeletal Na$^+$ channel α and β_1 subunits expressed in HEK293 cells. J Mol Cell Cardiol 34: 1029–39.

19. Colatsky TJ. Another layer of ventricular heterogeneity? Alpha 1 agonists prolong repolarization in Purkinje fibers but not M-cells [editorial; comment]. Cardiovasc Res 1999; 43: 827–9.

20. Attwell D, Cohen I, Eisner D et al. The steady state TTX-sensitive ('window') sodium current in cardiac Purkinje fibres. Pflugers Arch 1979; 379: 137–42.

21. Gintant GA, Datyner NB, Cohen IS. Slow inactivation of a tetrodotoxin-sensitive current in canine cardiac Purkinje fibers. Biophys J 1984; 45: 509–12.

22. Carmeliet E. Slow inactivation of the sodium current in rabbit cardiac Purkinje fibres. Pflugers Arch 1987; 408: 18–26.

23. Saint DA, Ju YK, Gage PW. A persistent sodium current in rat ventricular myocytes. J Physiol 1992; 453: 219–31.

24. Salata JJ, Wasserstrom JA. Effects of quinidine on action potentials and ionic currents in isolated canine ventricular myocytes. Circ Res 1988; 62: 324–37.

25. Ono K, Kiyosue T, Arita M. Isoproterenol, DBcAMP, and forskolin inhibit cardiac sodium current. Am J Physiol Cell Physiol 1989; 256: C1131–7.

26. Nagatomo T, Fan Z, Ye B et al. Temperature dependence of early and late currents in human cardiac wild-type and long QT Δ KPQ Na$^+$ channels. Am J Physiol Heart Circ Physiol 1998; 275: H2016–24.

27. Patlak JB, Ortiz M. Slow currents through single sodium channels of the adult rat heart. J Gen Physiol 1985; 86: 89–104.

28. Clancy CE, Rudy Y. Na$^+$ channel mutation that causes both Brugada and long-QT syndrome phenotypes: a simulation study of mechanism. Circulation 2002; 105: 1208–13.

29. Dudel J, Peper K, Rudel R et al. Effect of tetrodotoxin on membrane currents in mammalian cardiac fibres. Nature 1967; 213: 296–7.

30. Ju Y, Gage PW, Saint DA. Tetrodotoxin-sensitive inactivation-resistant sodium channels in pacemaker cells influence heart rate. Pflugers Arch 1996; 431: 868–75.

31. El Sherif N, Turitto G. Torsade de pointes. Curr Opin Cardiol 2003; 18: 6–13.

32. Nattel S. Atrial electrophysiology and mechanisms of atrial fibrillation. J Cardiovasc Pharmacol Ther 2003; 8 (Suppl 1): S5–11.

33. Cha TJ, Ehrlich JR, Zhang L et al. Dissociation between ionic remodeling and ability to sustain atrial fibrillation during recovery from experimental congestive heart failure. Circulation 2004; 109: 412–18.

34. Yue L, Melnyk P, Gaspo R et al. Molecular mechanisms underlying ionic remodeling in a dog model of atrial fibrillation. Circ Res 1999; 84: 776–84.

35. Antzelevitch C, Sicouri S, Litovsky SH et al. Heterogeneity within the ventricular wall. Electrophysiology and pharmacology of epicardial, endocardial, and M cells [Review]. Circ Res 1991; 69: 1427–49.

36. Antzelevitch C. Electrical heterogeneity, cardiac arrhythmias, and the sodium channel. Circ Res 2000; 87: 964–5.

37. Antzelevitch C. The Brugada syndrome: ionic basis and arrhythmia mechanisms. J Cardiovasc Electrophysiol 2001; 12: 268–72.

38. Ye B, Valdivia CR, Ackerman MJ et al. A common human SCN5A polymorphism modifies expression of an arrhythmia causing mutation. Physiol Genomics 2003; 12: 187–93.

39. Viswanathan PC, Benson DW, Balser JR. A common SCN5A polymorphism modulates the biophysical effects of an SCN5A mutation. J Clin Invest 2003; 111: 341–6.

40. Dumaine R, Towbin JA, Brugada P et al. Ionic mechanisms responsible for the electrocardiographic phenotype of the Brugada syndrome are temperature dependent. Circ Res 1999; 85: 803–9.

41. Clancy CE, Tateyama M, Liu HJ et al. Non-equilibrium gating in cardiac Na$^+$ channels – an original mechanism of arrhythmia. Circulation 2003; 107: 2233–7.

42. Nuyens D, Stengl M, Dugarmaa S et al. Abrupt rate accelerations or premature beats cause life-threatening arrhythmias in mice with long-QT3 syndrome. Nat Med 2001; 7: 1021–7.

43. Schwartz PJ, Priori SG, Spazzolini C et al. Genotype-phenotype correlation in the long-QT syndrome: gene-specific triggers for life-threatening arrhythmias. Circulation 2001; 103: 89–95.

44. Wang Q, Shen J, Splawski I et al. SCN5A mutations associated with an inherited cardiac arrhythmia, long QT syndrome. Cell 1995; 80: 805–11.

45. Clancy CE, Rudy Y. Linking a genetic defect to its

cellular phenotype in a cardiac arrhythmia. Nature 1999; 400: 566–9.

46. Chandra R, Starmer CF, Grant AO. Multiple effects of KPQ deletion mutation on gating of human cardiac Na$^+$ channels expressed in mammalian cells. Am J Physiol Heart Circ Physiol 1998; 274: H1643–54.

47. Wang DW, Yazawa K, George AL, Jr et al. Characterization of human cardiac Na$^+$ channel mutations in the congenital long QT syndrome. Proc Natl Acad Sci USA 1996; 93: 13200–5.

48. Abriel H, Cabo C, Wehrens XH et al. Novel arrhythmogenic mechanism revealed by a long-QT syndrome mutation in the cardiac Na$^+$ channel. Circ Res 2001; 88: 740–5.

49. Wedekind H, Smits JP, Schulze-Bahr E et al. De novo mutation in the SCN5A gene associated with early onset of sudden infant death. Circulation 2001; 104: 1158–64.

50. Rivolta I, Clancy CE, Tateyama M et al. A novel SCN5A mutation associated with long QT-3: altered inactivation kinetics and channel dysfunction. Physiol Genomics 2002; 10: 191–7.

51. An RH, Wang XL, Kerem B et al. Novel LQT-3 mutation affects Na$^+$ channel activity through interactions between α- and β$_1$-subunits. Circ Res 1998; 83: 141–6.

52. Baroudi G, Chahine M. Biophysical phenotypes of SCN5A mutations causing long QT and Brugada syndromes. FEBS Lett 2000; 487: 224–8.

53. Wehrens XH, Abriel H, Cabo C et al. Arrhythmogenic mechanism of an LQT-3 mutation of the human heart Na$^+$ channel alpha-subunit: a computational analysis. Circulation 2000; 102: 584–90.

54. Schwartz PJ, Priori SG, Dumaine R et al. A molecular link between the sudden infant death syndrome and the long-QT syndrome. N Engl J Med 2000; 343: 262–7.

55. Ackerman MJ, Siu BL, Sturner WQ et al. Postmortem molecular analysis of SCN5A defects in sudden infant death syndrome. JAMA 2001; 286: 2264–9.

56. Schwartz PJ, Locati EH, Napolitano C, Priori SG. The long QT syndrome. In: Zipes DP, Jalife J (eds). Cardiac Electrophysiology. Philadelphia: WB Saunders Co, 1995: 788–811.

57. Nagatomo T, January CT, Ye B et al. Rate-dependent QT shortening mechanism for the LQT3 DeltaKPQ mutant. Cardiovasc Res 2002; 54: 624–9.

58. Valdivia CR, Ackerman MJ, Tester DA et al. A novel SCN5A arrhythmia mutation, M1766L, with expression defect rescued by mexiletine. Cardiovasc Res 2002; 55: 279–89.

59. Valdivia CR, Nagatomo T, Makielski JC. Late Na$^+$ currents affected by alpha subunit isoform and beta1 subunit co-expression in HEK293 cells. J Mol Cell Cardiol 2002; 34: 1029–39.

60. Undrovinas AI, Shander GS, Makielski JC. Cytoskeleton modulates gating of voltage-dependent sodium channel in heart. Am J Physiol Heart Circ Physiol 1995; 269: H203–14.

61. Chauhan VS, Tuvia S, Buhusi M et al. Abnormal cardiac Na$^+$ channel properties and QT heart rate adaptation in neonatal ankyrin(B) knockout mice. Circ Res 2000; 86: 441–7.

62. Mohler PJ, Schott JJ, Gramolini AO et al. Ankyrin-B mutation causes type 4 long-QT cardiac arrhythmia and sudden cardiac death. Nature 2003; 421: 634–9.

63. Chen Q, Kirsch GE, Zhang D et al. Genetic basis and molecular mechanism for idiopathic ventricular fibrillation. Nature 1998; 392: 293–6.

64. Akai J, Makita N, Sakurada H et al. A novel SCN5A mutation associated with idiopathic ventricular fibrillation without typical ECG findings of Brugada syndrome. FEBS Lett 2000; 479: 29–34.

65. Schott JJ, Alshinawi C, Kyndt F et al. Cardiac conduction defects associate with mutations in SCN5A. Nat Genet 1999; 23: 20–1.

66. Tan HL, Bink-Boelkens MT, Bezzina CR et al. A sodium-channel mutation causes isolated cardiac conduction disease. Nature 2001; 409: 1043–7.

67. Benson DW, Wang DW, Dyment M et al. Congenital sick sinus syndrome caused by recessive mutations in the cardiac sodium channel gene (SCN5A). J Clin Invest 2003; 112: 1019–28.

68. Groenewegen WA, Firouzi M, Bezzina CR et al. A cardiac sodium channel mutation cosegregates with a rare connexin40 genotype in familial atrial standstill. Circ Res 2003; 92: 14–22.

69. Kyndt F, Probst V, Potet F et al. Novel SCN5A mutation leading either to isolated cardiac conduction defect or Brugada syndrome in a large French family. Circulation 2001; 104: 3081–6.

70. Brugada P, Brugada J. Right bundle branch block, persistent ST segment elevation and sudden cardiac death: a distinct clinical and electrocardiographic syndrome. A multicenter report [see comments]. J Am Coll Cardiol 1992; 20: 1391–6.

71. Takehara N, Makita N, Kawabe J et al. A cardiac sodium channel mutation identified in Brugada syndrome associated with atrial standstill. J Intern Med 2004; 255: 137–42.

72. Baroudi G, Acharfi S, Larouche C et al. Expression and intracellular localization of an SCN5A double mutant R1232W/T1620M implicated in Brugada Syndrome. Circ Res 2002; 90: E11–16.

73. Valdivia CR, Tester DJ, Rok BA et al. A trafficking defective, Brugada syndrome-causing SCN5A mutation rescued by drugs. Cardiovasc Res 2004; 62: 53–62.

74. Rook MB, Alshinawi CB, Groenewegen WA et al. Human SCN5A gene mutations alter cardiac sodium channel kinetics and are associated with the Brugada syndrome. Cardiovasc Res 1999; 44: 507–17.

75. Deschenes I, Baroudi G, Berthet M et al. Electrophysiological characterization of SCN5A mutations causing long QT (E1784K) and Brugada (R1512W and R1432G) syndromes. Cardiovasc Res 2000; 46: 55–65.

76. Rivolta I, Abriel H, Tateyama M et al. Inherited Brugada and long QT-3 syndrome mutations of a single residue of the cardiac sodium channel confer distinct channel and clinical phenotypes. J Biol Chem 2001; 276: 30623–30.

77. Bezzina C, Veldkamp MW, van Den Berg MP et al. A single Na$^+$ channel mutation causing both long-QT and Brugada syndromes. Circ Res 1999; 85: 1206–13.

78. Wan X, Chen S, Sadeghpour A et al. Accelerated inactivation in a mutant Na$^+$ channel associated with idiopathic ventricular fibrillation. Am J Physiol Heart Circ Physiol 2001; 280: H354–60.

79. Vatta M, Dumaine R, Varghese G et al. Genetic and biophysical basis of sudden unexplained nocturnal death syndrome (SUNDS), a disease allelic to Brugada syndrome. Hum Mol Genet 2002; 11: 337–45.

80. Wang DW, Viswanathan PC, Balser JR et al. Clinical, genetic, and biophysical characterization of SCN5A mutations associated with atrioventricular conduction block. Circulation 2002; 105: 341–6.

81. Veldkamp MW, Viswanathan PC, Bezzina C et al. Two distinct congenital arrhythmias evoked by a multidysfunctional Na$^+$ channel. Circ Res 2000; 86: E91–7.

82. Baroudi G, Pouliot V, Denjoy I et al. Novel mechanism for Brugada syndrome: defective surface localization of an SCN5A mutant (R1432G). Circ Res 2001; 88: E78–83.

83. Gilmour RF, Jr, Heger JJ, Prystowsky EN et al. Cellular electrophysiologic abnormalities of diseased human ventricular myocardium. Am J Cardiol 1983; 51: 137–44.

84. Shander GS, Undrovinas AI, Makielski JC. Rapid onset of lysophosphatidylcholine-induced modification of whole cell cardiac sodium current kinetics. J Mol Cell Cardiol 1996; 28: 743–53.

85. Janse MJ. Electrophysiological changes in heart failure and their relationship to arrhythmogenesis. Cardiovasc Res 2004; 61: 208–17.

86. Gulch RW, Baumann R, Jacob R. Analysis of myocardial action potential in left ventricular hypertrophy of Goldblatt rats. Basic Res Cardiol 1979; 74: 69–82.

87. Gaughan JP, Hefner CA, Houser SR. Electrophysiological properties of neonatal rat ventricular myocytes with alpha1-adrenergic-induced hypertrophy. Am J Physiol Heart Circ Physiol 1998; 275: H577–90.

88. Sakakibara Y, Wasserstrom JA, Furukawa T et al. Characterization of the sodium current in single human atrial myocytes. Circ Res 1992; 71: 535–46.

89. Kaab S, Nuss HB, Chiamvimonvat N et al. Ionic mechanism of action potential prolongation in ventricular myocytes from dogs with pacing-induced heart failure. Circ Res 1996; 78: 262–73.

90. Kaab S, Dixon J, Duc J et al. Molecular basis of transient outward potassium current downregulation in human heart failure – a decrease in KV4.3 mRNA correlates with a reduction in current density. Circulation 1998; 98: 1383–93.

91. Yue L, Feng J, Gaspo R et al. Ionic remodeling underlying action potential changes in a canine model of atrial fibrillation. Circ Res 1997; 81: 512–25.

92. Gaspo R, Sun H, Fareh S et al. Dihydropyridine and beta adrenergic receptor binding in dogs with tachycardia-induced atrial fibrillation. Cardiovasc Res 1999; 42: 434–42.

93. Pu J, Boyden PA. Alterations of Na$^+$ currents in myocytes from epicardial border zone of the infarcted heart. A possible ionic mechanism for reduced excitability and postrepolarization refractoriness. Circ Res 1997; 81: 110–19.

94. Makielski JC, Fozzard HA. Ion channels and cardiac arrhythmia in heart disease. In: Page E, Fozzard HA, Solaro RJ (eds). Handbook of Physiology. New York: Oxford University Press, 2002: 709–40.

95. Jacques D, Bkaily G, Jasmin G et al. Early fetal like slow Na$^+$ current in heart cells of cardiomyopathic hamster. Mol Cell Biochem 1997; 176: 249–56.

96. Deroubaix E, Thuringer D, Coulombe A et al. Dilation and action potential lengthening in cardiomyopathic Syrian hamster heart. Basic Res Cardiol 1999; 94: 274–83.

97. Undrovinas AI, Maltsev VA, Sabbah HN. Repolarization abnormalities in cardiomyocytes of

dogs with chronic heart failure: role of sustained inward current. Cell Mol Life Sci 1999; 55: 494–505.

98. Valdivia CR, Chu WW, Pu JL et al. Increased late sodium current in myocytes from a canine heart failure model and from failing human heart. J Mol Cell Cardiol 2005; 38: 475–83.

99. Maltsev VA, Sabbah HN, Higgins RSD et al. Novel, ultraslow inactivating sodium current in human ventricular cardiomyocytes. Circulation 1998; 98: 2545–52.

100. Makielski JC, Kamp TJ, Valdivia CR. Late sodium current is increased in ventricular myocytes from failing human hearts compared with nonfailing hearts. Jpn J Electrocardiol 2000; 20 (Suppl): 99–100.

101. Undrovinas AI, Maltsev VA, Kyle JW et al. Gating of the late Na$^+$ channel in normal and failing human myocardium. J Mol Cell Cardiol 2002; 34: 1477–89.

102. Yang P, Kanki H, Drolet B et al. Allelic variants in long-QT disease genes in patients with drug-associated torsades de pointes. Circulation 2002; 105: 1943–8.

103. Chen S, Chung MK, Martin D et al. SNP S1103Y in the cardiac sodium channel gene SCN5A is associated with cardiac arrhythmias and sudden death in a white family. J Med Genet 2002; 39: 913–15.

104. Splawski I, Timothy KW, Tateyama M et al. Variant of SCN5A sodium channel implicated in risk of cardiac arrhythmia. Science 2002; 297: 1333–6.

105. Grant AO. Mechanisms of action of antiarrhythmic drugs: from ion channel blockage to arrhythmia termination. Pacing Clin Electrophysiol 1997; 20: 432–44.

106. The Cardiac Arrhythmia Suppression Trial (CAST) investigators. Preliminary report: effect of encainide and flecainide on mortality in a randomized trial of arrhythmia suppression after myocardial infarction. N Engl J Med 1989; 321: 406–12.

107. Wang DW, Yazawa K, Makita N et al. Pharmacological targeting of long QT mutant sodium channels. J Clin Invest 1997; 99: 1714–20.

108. Nagatomo T, January CT, Makielski JC. Preferential block of late sodium current in the LQT3 ΔKPQ mutant by the class I(C) antiarrhythmic flecainide. Mol Pharmacol 2000; 57: 101–7.

109. Ono K, Kaku T, Makita N et al. Selective block of late currents in the DeltaKPQ Na$^+$ channel mutant by pilsicainide and lidocaine with distinct mechanisms. Mol Pharmacol 2000; 57: 392–400.

110. Huang B, El Sherif T, Gidh-Jain M et al. Alterations of sodium channel kinetics and gene expression in the postinfarction remodeled myocardium. J Cardiovasc Electrophysiol 2001; 12: 218–25.

111. He J, Conklin MW, Foell JD et al. Reduction in density of transverse tubules and L-type Ca^{2+} channels in canine tachycardia-induced heart failure. Cardiovasc Res 2001; 49: 298–307.

112. Splawski I, Timothy KW, Sharpe LM et al. Ca(v)1.2 calcium channel dysfunction causes a multisystem disorder including arrhythmia and autism. Cell 2004; 119: 19–31.

113. Tomaselli GF, Marban E. Electrophysiological remodeling in hypertrophy and heart failure. Cardiovasc Res 1999; 42: 270–83.

114. Ouadid H, Albat B, Nargeot J. Calcium currents in diseased human cardiac cells. J Cardiovasc Pharmacol 1995; 25: 282–91.

115. Kamp TJ, He JQ. L-type Ca^{2+} channels gaining respect in heart failure. Circ Res 2002; 91: 451–3.

116. Janvier NC, Boyett MR. The role of Na–Ca exchange current in the cardiac action potential [Review]. Cardiovasc Res 1996; 32: 69–84.

117. Pogwizd SM, Bers DM. Na/Ca exchange in heart failure: contractile dysfunction and arrhythmogenesis. Ann N Y Acad Sci 2002; 976: 454–65.

118. Pogwizd SM, Bers DM. Cellular basis of triggered arrhythmias in heart failure. Trends Cardiovasc Med 2004; 14: 61–6.

119. Robinson RB, Siegelbaum SA. Hyperpolarization-activated cation currents: from molecules to physiological function. Annu Rev Physiol 2003; 65: 453–80.

120. Schulze-Bahr E, Neu A, Friederich P et al. Pacemaker channel dysfunction in a patient with sinus node disease. J Clin Invest 2003; 111: 1537–45.

121. Cerbai E, Barbieri M, Mugelli A. Occurrence and properties of the hyperpolarization-activated current I_f in ventricular myocytes from normotensive and hypertensive rats during aging. Circulation 1996; 94: 1674–81.

122. Hoppe UC, Jansen E, Sudkamp M et al. Hyperpolarization-activated inward current in ventricular myocytes from normal and failing human hearts. Circulation 1998; 97: 55–65.

CHAPTER 24

Repolarization abnormalities

Stefan Kääb, Michael Näbauer

Introduction

Heart failure is the most common discharge diagnosis in the United States with a prognosis worse than many cancers.[1] Whatever the initiating factors are (such as ischemic heart disease, hypertension, viral infection), the final common phenotype is one of a dilated, poorly contracting heart. The name 'heart failure' indicates the major symptoms of patients suffering from this disease, namely decreased ability to exercise and shortness of breath. However, with respect to cause of death, clinical trials and basic research demonstrated that patients with heart failure die for two reasons: progressing circulatory failure or cardiac arrhythmias leading to sudden cardiac death.[2] With annual mortality rates exceeding 10% in patients with heart failure, 35–50% of the deaths are sudden and unexpected. Importantly, relative rates of sudden cardiac death are higher in patients with mild to moderate degrees of cardiac failure.[3,4]

In the majority of cases, sudden death in heart failure is caused by sustained ventricular tachyarrhythmias.[5] When describing mechanisms of arrhythmogenesis in the context of hypertrophy and heart failure, one should discriminate between the mechanism underlying the initiating beat ('trigger') and that of favorable pre-existing conditions ('substrate') facilitating the onset and maintenance of the arrhythmia. These principle mechanisms may be altered by various modulating factors (e.g.

electrolytes, catecholamines, ischemia, antiarrhythmic medication).[6] One such trigger of ventricular arrhythmias is intracellular calcium overload, giving rise to delayed afterdepolarizations in the context of heart failure-related upregulation of the sodium–calcium exchanger, in the setting of downregulation of the inward rectifier and maintained beta-adrenergic responsiveness.[7] This chapter will focus on a second trigger of arrhythmias in hypertrophy and heart failure, namely abnormal repolarization. In heart failure, the arrhythmic tendency is aggravated by the combination of both major mechanisms.

Altered repolarization as a major arrhythmogenic substrate and trigger in cardiac hypertrophy and heart failure

Prolonged action potentials in cardiac myocytes, indicating abnormal repolarization, have been a hallmark phenotype in a multitude of animal models of hypertrophy and heart failure for more than two decades. These findings have been transferred and extended to the human pathophysiology of cardiac hypertrophy and failure on a molecular, cellular and clinical level where the QT interval of the surface electrocardiogram (ECG) is the key marker of repolarization (Fig. 24.1).[8–10] As a con-

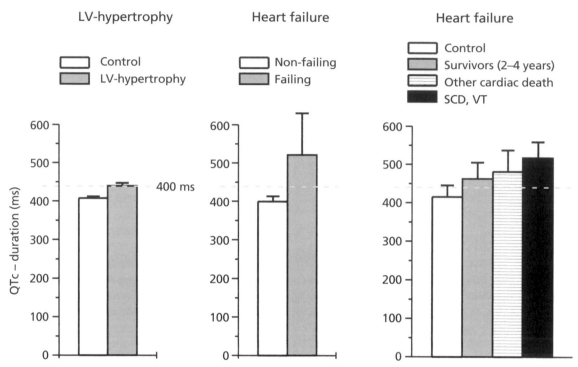

Figure 24.1. *Prolonged cardiac repolarization in human LV-hypertrophy and heart failure: QTc, as a clinical surrogate marker, is prolonged in human LV-hypertrophy and heart failure. LV, left ventricular; SCD, sudden cardiac death; VT, ventricular tachycardia. The data are summarized from references 8–10.*

sequence, human heart failure has recently been defined as a common, acquired form of the long QT syndrome.[11] Myocytes from failing hearts show prolongation of action potentials due to downregulation of K^+-currents (Fig. 24.2).[12–14] As a result, repolarization *in vivo* is abnormally labile,[15] predisposing the individual to early afterdepolarizations, inhomogeneous repolarization, and ventricular tachyarrhythmias.

Numerical models of electrical activity have begun to shed light on the mechanism of cardiac arrhythmias. The initial insight came in simulations carried out on a cellular level, which rationalized the mechanisms of long QT-related action potential prolongation and afterdepolarizations. More recently, massively parallel network simulations of whole-heart electrical activity have successfully reproduced polymorphic ventricular tachycardia – an arrhythmia commonly seen in heart failure (video animations are available

online).[16,17] The biological hypothesis of repolarization-related arrhythmias has thus been validated numerically from first principles.[18]

Repolarization abnormalities: altered heterogeneity of repolarization is the key substrate for arrhythmogenesis

The heart failure-associated prolongation of the action potential alone would not necessarily suffice to produce re-entrant ventricular arrhythmias, particularly if the prolongation were homogeneous. Regional variations in action potential duration, however, create dispersion of repolarization and refractoriness that provides an arrhythmogenic substrate.[19,20] Recent studies point to the fact that heterogeneity of repolarization is not restricted to

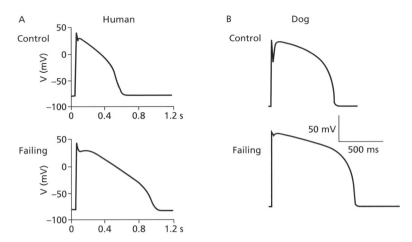

Figure 24.2. *Action potential prolongation in heart failure. Prolongation of the cardiac action potential is a hallmark phenotype seen in heart failure both in humans (A) and in animal studies (B). Adapted with permission from Beuckelmann DJ, Näbauer M, Erdmann E. Alterations of K^+ currents in isolated human ventricular myocytes from patients with terminal heart failure. Circ Res 1993; 73: 379–85; and Kääb S, Nuss HB, Chiamvimonvat N et al. Ionic mechanism of action potential prolongation in ventricular myocytes from dogs with pacing-induced heart failure. Circ Res 1996; 78: 262–73.*

adjacent sites located in the same cell layer, but rather extends to sites across the ventricular wall.[21,22]

Mechanisms of repolarization abnormalities

With human heart failure being viewed as a common, acquired form of the long QT syndrome,[11] it is more than appropriate to look at the rare genetic forms of cardiac repolarization abnormalities as model diseases for arrhythmias in the context of hypertrophy and heart failure.

Genetic repolarization abnormalities: a model for arrhythmogenesis in cardiac hypertrophy and heart failure

In 1995, Mark Keating and colleagues identified two genes responsible for congenital long QT syndrome, a prevalent cause of sudden cardiac death in individuals without any structural heart disease.[23,24] Functional changes in cardiac ion channels that orchestrate the regular heart beat were central to the disorder. This revelation initiated a plethora of experimental and clinical studies, was followed by the discovery of novel disease genes, and eventually culminated in attempts at genotype–phenotype correlation, genotype-based risk stratification, and first attempts of genotype-specific therapy.

Rare mutations in disease genes: channelopathies and more?

The genetically transmitted long QT syndrome is caused by discrete mutations in genes that encode mostly ion channels. Either the Na^+ channel gene *SCN5A* or one of four genes that make up repolarizing potassium channels (I_{Kr} and I_{Ks}) can be affected.[25] Long QT-associated mutations in *SCN5A* produce channels with increased Na^+ flux,[24] whereas mutations in affected genes encoding K^+ channels lead to loss of function. Hundreds of mutations in three long QT syndrome-causing genes, namely *KCNQ1* (I_{Ks}, α subunit, LQT1), *KCNH2* (I_{Kr}, α subunit, LQT2), and *SCN5A* (I_{Na}, α subunit, LQT3), are now known to account for approximately two-thirds of cases of congenital long QT syndrome.[26] In addition, mutations in two auxillary subunits of potassium channels (KCNE1, MIRP, β subunit of I_{Kr} – LQT5; KCNE2, minK, β subunit of I_{Ks} – LQT6) cause rare forms of long QT syndrome.[27,28] Very recently, genetic studies have broadened the repertoire of genes that cause long QT syndrome as the first non-channelopathy form of long QT syndrome was revealed to be linked to the gene encoding ankyrin-B (LQT4).[29]

Interestingly, the dynamic and quantitative genetic effects on cardiac repolarization go both ways: mutations in some of the same genes that cause enhanced cell repolarization may result in converse disorders, such as short QT syndrome, also linked to enhanced risk for cardiac arrhythmias and sudden cardiac death.[30,31]

While we have learned a lot in the past decade about the primary defects and general aspects of arrhythmogenesis underlying congenital long QT syndromes, advances in a more specific genotype–phenotype correlation are slow. Due to variable penetrance and expressivity, and the still-growing genetic heterogeneity, it is impossible to establish the diagnosis of long QT syndrome or predict a specific phenotype simply by identifying a single mutation in one of the known disease genes.[32]

Several factors such as gender,[33] hormonal changes, heart failure, electrolyte abnormalities, concomitant drugs, and food intake have been implicated to account for triggering arrhythmic events in previously asymptomatic patients.[32] None of these factors, however, has been conclusively linked to the severity of the clinical phenotype. The term 'modifiers' is used to refer to all those factors that may be responsible for modulating the clinical manifestation.

Common gene variants modifying cardiac repolarization

In the context of monogenic diseases, the term 'modifiers' is also, and importantly, used for genetic variants that account for the inter-individual variability among patients harboring the same mutation. A few studies have tried to gain a deeper insight in the consequence of mutations by incorporating in their models at least another factor that may influence the phenotype.[32] Baroudi et al. were the first to introduce the concept that the interaction of common polymorphisms and rare mutations may exert profound effects on functional

consequences. The combination of a mutation in the SCN5A gene (T1620M) known to affect channel gating, resulted in impaired protein trafficking when expressed in tsA201 cells.[34]

Moreover, recent evidence demonstrates that, even in the absence of a disease-causing mutation, polymorphisms may influence QT duration in the general population,[35] and, specifically, susceptibility to arrhythmias in the general population.[36] Splawski et al. suggested that the polymorphism SCN5A (Y1102) that is present in 13.2% of the African-American population is strongly associated with the development of arrhythmias, namely drug-induced long QT syndrome, in the absence of a clinical phenotype of LQT3 or any other heritable arrhythmia that has been linked to mutations in the Na$^+$ channel gene.[36] The idea that polymorphisms in genes encoding cardiac ion channels or other genes that alter cardiac de- or repolarization may also contribute to enhanced arrhythmia susceptibility in more common forms of cardiac disease (e.g. hypertrophy and heart failure) represents a probable, but as yet unproven hypothesis.[37] Preliminary data indicate that multiple, small, individually insignificant genetic contributions add up to a detectable quantitative trait.[38] Whether these polymorphisms will combine to enhance overall susceptibility to arrhythmias in the context of e.g. heart failure remains to be seen.

Secondary repolarization abnormalities

Heart failure presents with a similar phenotype to congenital long QT syndromes, namely prolonged QT interval indicating altered repolarization. While in congenital long QT syndrome mutations in genes encoding cardiac ion channels cause functional changes leading to altered repolarization, in hypertrophy and heart failure repolarization abnormalities are induced primarily by altered gene expression involving ion channels and regulatory subunits mediated by signaling pathways involving transmitters such as catecholamines, acetylcholine, and angiotensin II, to name a few (Fig. 24.3).[11]

In addition to changes in expression of genes that directly control cardiac repolarization,

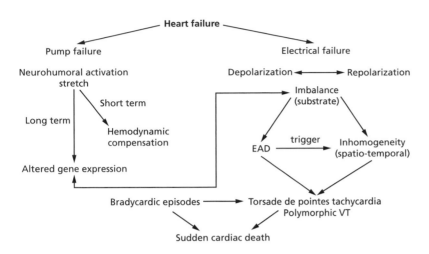

Figure 24.3. *Sudden cardiac death in heart failure. EAD: early afterdepolarization; VT: ventricular tachycardia. Modified from Marban E. Heart failure: the electrophysiologic connection. J Cardiovasc Electrophysiol 1999; 10: 1425–8.*

increased heterogeneity of sympathetic innervation is well described in cardiomyopathy patients,[39] and has been correlated with heterogeneity of recovery of excitability.[40]

Altered gene expression in hypertrophy and heart failure

Regulation of gene expression may reflect cardiac disease (e.g. heart failure, atrial fibrillation or hypertrophy) or may be related to common gene variants that may explain congenital inter-individual gene expression patterns (so-called genomic convergence.[41] In a similar way, gene expression could be modified with inter-individual variations by regulatory intronic or intergenic DNA elements responsive to hormone status, neurohormonal messengers and other signaling pathways providing a potential explanation for inter-individual variation of the responsiveness to the induction of disease-related altered gene expression. However, at present, the hypothesis that could link certain single nucleotide polymorphisms (SNPs) or haplotype blocks to a specific gene expression pattern and, subsequently, to arrhythmia susceptibility is not proven.

On the other hand, a multitude of studies demonstrate that direct regulation of gene expression is a major adaptive or maladaptive process in probably all cardiovascular diseases. Sometimes, findings of different expression patterns cannot be judged as being a causative or secondary adaptive process (Fig. 24.3).

Electrophysiological changes in heart failure are not confined to the ventricles. Yet, the most consistent electrophysiological changes in the ventricles are prolongation of the action potential (Fig. 24.2), especially at slow rates, a reduction in the transient outward current I_{to}, the rapid and slow components of the delayed rectifier I_{Kr} and I_{Ks} (predominantly in animal models) (Fig. 24.4, Table 24.1),[22,42–44] and the inward rectifier I_{K1}.[45,46] Abnormalities in intracellular calcium handling play a major role in the genesis of delayed afterdepolarizations. Triggered activity based on delayed afterdepolarizations has been demonstrated in failing myocardium, and is caused by spontaneous release of calcium from the sarcoplasmic reticulum. The propensity for the occurrence of delayed afterdepolarizations is enhanced by increased activity of Na^+–Ca^{2+} exchanger, reduced inward rectifier, and residual β-adrenergic responsiveness required to raise the reduced sarcoplasmic calcium content to a level where spontaneous calcium release occurs. Early afterdepolarizations and automaticity have also been demonstrated as potential mechanisms for arrhythmogenesis in the context of failing myocardium.[7,46,47]

Changes in gene expression on the ventricular level probably represent secondary changes in the course of hypertrophy and heart failure. The primary factors that directly trigger changes in electrical properties and gene expression (neurohumoral and transcription factors) are still widely unknown. At present little is known about the

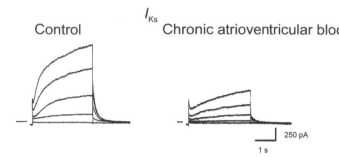

Figure 24.4. *Downregulation of delayed rectifier currents in cardiac hypertrophy and failure. Downregulation of* I_{Ks} *in chronic atrioventricular block hypertrophy in dog ventricle. Modified with permission from Volders PG, Sipido KR, Vos MA et al. Downregulation of delayed rectifier* K^+ *currents in dogs with chronic complete atrioventricular block and acquired torsade de pointes. Circulation 1999; 100: 2455–61.*

Table 24.1. *Experimental studies demonstrating downregulation of delayed rectifier currents in cardiac hypertrophy or failure*

	I_{Ks}	I_{Kr}
Rabbit pacing heart failure[43]	−48%	−36%
Pig pacing heart failure[22]	Delayed rectifier current	−50%
Dog chronic AV block hypertrophy[42]	−50%	No change
Rabbit 1 kidney 1 clip hypertrophy[44]	−40%	No change
Functional similarity to congenital repolarization disorders	LQT1	LQT2

These acquired ion channel disorders have the same ion channel substrate and thus functional similarity to the congenital repolarization disorders LQT1 and LQT2.

mechanisms and factors of transcriptional regulation and the time course of adaptive or maladaptive changes, but a great number of studies have examined altered gene expression in cardiac tissue as a way to understand basic mechanisms of arrhythmogenesis in defined disease states. Studies attempting a genome-wide analysis of expression profiles of failing and non-failing human hearts could demonstrate both regional and disease-specific differences.[48–51] So far, gene array studies are helpful in defining comprehensive molecular portraits of gene expression in defined regions and defined disease states. Further studies and refined bioinformatic analysis will help to identify common pathways of heart failure and transcriptional changes underlying arrhythmogenesis.

Sex differences in cardiac repolarization

There is an increased awareness of the extent to which cardiac function is influenced by sex. One of the most dramatic and potentially lethal differ-ences is that seen in repolarization of the heart.[52] Age- and sex-related differences in QTc duration have been noted in the normal population and in patients with long QT syndrome. Sex differences in QT and QTc intervals have been observed to change during lifetime in the general population. While in children the QT interval is not significantly different between boys and girls, boys' QT times decline significantly by up to 20 ms at the onset of puberty, while those of girls remain unchanged. The difference between both sexes then progressively declines until they become similar again around age 50 years. So far these sex differences in QT interval in the general population have not been conclusively linked to disease and arrhythmia manifestation or mortality.

On the other hand, the greater risk in women for torsade de pointes induced by drugs that prolong repolarization is generally accepted to be one of the most dramatic and important differences between men and women regarding sex and arrhythmia.[53] More specifically, female sex is an independent risk factor for syncope and sudden cardiac death in familial long QT syndrome.[54]

Mechanisms underlying these differences are incompletely defined but are believed to involve gonadal steroids. In orchiectomized men the JT interval increases compared to that of non-orchiectomized men. Correspondingly in virilized women the JT interval is shorter compared to that of normal women. All these observations underscore the effects of sex hormones on QT interval, and point towards a prominent role of male androgens such as testosterone or 5-dehydrotestosterone (DHT) in the sex bias of repolarization, as their levels make up for the most prominent changes during puberty, orchiectomy and virilization.[52]

In addition, even though the electrophysiological phenotype appears to be similar (i.e. the baseline repolarization duration is equal), there are distinct repolarization mechanisms in male and female hearts that are not influenced by sex hormones.[55]

In female patients, there is evidence that estrogen or progesterone, or more specifically, the progesterone to estrogen ratio, may increase susceptibility to ventricular arrhythmias in the context of QT prolonging drugs.[56,57] In summary, experimental and clinical evidence demonstrate that hormones (testosterone and estrogen) are important modulators of ventricular repolarization, in part through their regulation of ionic currents (i.e. I_{Ca-L}, I_{K1}, I_{Kr}) that control repolarization. Recent studies on common SNPs in *KCNH2* (encoding I_{Kr}, a major repolarizing current in the human heart and major substrate of most drugs inducing QT prolongation) reveal sex differences and associated effects on QTc.[35,38] A current hypothesis postulating SNPs that modulate gene expression in response to sex hormones is not proven.

Despite a longer corrected QT interval and higher incidence of torsade de pointes, there are no conclusive data connecting these sex-related QT differences to a higher incidence of cardiac arrhythmias in the context of hypertrophy and heart failure in women as compared to men. Prospective studies will be required to help better understand the basic mechanisms involved in sex differences in electrophysiology and arrhythmias in heart failure.

Ischemia and repolarization

Hypertrophy and heart failure may be accompanied by chronic and acute states of ischemia. Mechanisms of repolarization abnormalities in acute ischemia differ in a substantial way from the previously discussed pathophysiology.

In the setting of acute ischemia and concomittant intracellular ATP depletion, opening of the K_{ATP} channels, which are quiescent under physiological conditions, results in action potential shortening,[58,59] which may facilitate re-entry tachycardias, and serves as a proarrhythmic substrate.[60] Particularly in zones next to infarcted myocardium where ischemia and ATP-depletion poses a constant threat, 'border zone' arrhythmias have been linked to activation of K_{ATP} channels and thus an increased dispersion of repolarization between ischemic and non-ischemic areas.[61] The clinical use of IK_{ATP} blockers as antiarrhythmic agents has been limited by the widespread expression of K_{ATP} channels in various tissues, leading to significant extracardiac side-effects of established IK_{ATP} blockers such as glibenclamide (e.g. hypoglycemia and hyperinsulinemia by inhibition of the pancreatic K_{ATP} channel, or interference with coronary vasodilatation during ischemia by blocking vascular K_{ATP}-channels).

Increasing knowledge of the pharmacology and molecular biology of ATP-dependent potassium currents will hopefully allow for the development of a specific blocker of the myocardial sarcolemmal channel which is thought to be composed of hetero-octamers consisting of tetramers of the Kir6.2 pore-forming subunit, and tetramers of the regulatory sulfonylurea receptor SUR2A.

Potassium and repolarization

Potassium has been linked to cardiac arrhythmias for almost a century, and its effects on the QT interval are not new. It is recognized that potassium supplementation could have arrhythmia-suppressing properties,[62,63] and that the potential mechanism is the shortening of cardiac action potentials.[64] Today it is widely accepted that these effects are mediated mostly by the rapidly activating delayed rectifier potassium current (I_{Kr}) that is paradoxically increased when extracellular potassium is increased.[65] In the context of both congestive heart failure and QT-prolonging medications, hypokalemia may further increase labile repolarization (Fig. 24.5) both by reducing intrinsic potassium outward current (I_{Kr}) and by increasing drug binding to the channel, resulting in excessive prolongation of repolarization.[66]

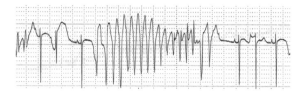

Figure 24.5. *Typical torsade de pointes tachycardia. A 67-year-old female patient with congestive heart failure (NYHA class III) presented with multiple runs of torsade de pointes tachycardia when diuretic therapy had caused serum K^+ to decrease to 3.3 mmol/l. QTc 510 ms. Note initial short-long-short sequence, and inverted and distorted T wave preceding torsade de pointes tachycardia.*

Clinical implications

Summarizing the crucial role of an orderly cardiac repolarization process, it becomes evident that any factor excessively prolonging cardiac repolarization or specifically increasing heterogeneity of refractoriness by site-specific prolongation or shortening of cardiac action potentials is increasing the propensity to cardiac arrhythmias.

Repolarization reserve

Besides mutations that directly affect cardiac repolarization, a number of additional factors such as female sex, left ventricular hypertrophy, heart failure, drugs, and hypokalemia have been linked to QT prolongation and risk of arrhythmias (Fig. 24.6). Evidence for an increased susceptibility to QT prolongation upon challenge of repolarization led to the concept of altered repolarization reserve as a proarrhythmic substrate.[67] The concept of reduced repolarization reserve as the basis for indi-

vidual susceptibility to arrhythmias in the context of QT prolongation[67] has been confirmed in several studies,[68,69] and applies for congenital long QT syndrome as well as for other causes of QT prolongation, such as hypertrophy and heart failure, and QT-prolonging drugs. Current evidence suggests that primary, genetic factors (major and minor gene variants) reduce repolarization reserve, as well as the effect of secondary changes as seen in e.g. downregulation of repolarizing potassium currents in hypertrophy and heart failure. Both of these changes may be subclinical until unmasked by extra triggers such as QT-prolonging drugs or hypokalemia.[53]

Modifying repolarization abnormalities

Potassium supplementation

When *KCNH2*, the gene coding for the α subunit of I_{Kr}, was first linked to congenital long QT syndrome, the ground for gene specific therapy was laid.[70] Compton et al. were the first to systematically study the effects of potassium supplementation in patients with congenital long QT syndrome. They showed that acute intravenous potassium supplementation (increase ≥1.5 mmol/l), an intervention known to increase repolarizing K^+ currents, led to a 24% reduction of QTc in seven patients carrying mutations in *KCNH2* (*LQT2*), compared with a 4% reduction in five control probands.[71] This investigation was followed by a recent study testing a 4-week oral potassium administration (accompanied by high dose spironolactone) in eight patients with mutations in *KCNH2* that demonstrated an increase in serum potassium from 4.0 ± 0.3 to 5.2 ± 0.3 mmol/l, resulting in a decrease in QTc from 526 ± 94 to 423 ± 36 ms.[72]

In another small clinical study, Choy et al. demonstrated that increasing serum potassium concentrations could partially reverse altered repolarization, namly QTc interval, in the context of heart failure or antiarrhythmic drugs associated with QT prolongation.[9]

At present, clinical and experimental evidence demonstrate effects of serum potassium on QTc. In

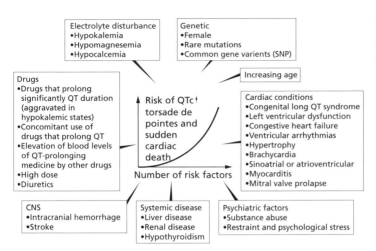

Figure 24.6. *A multitude of factors determine individual risk of arrhythmia in the context of QT-prolongation. Additive effects act exponentially.*

states of altered repolarization such as in heart failure and congenital long QT syndrome, prolonged QTc could be reduced substantially by increasing serum potassium levels. In subjects with normal renal function, attempts to achieve a long-lasting rise of serum potassium to correct prolonged QTc interval have not been satisfactory.[73] Potassium as a chronic therapy for long QT syndromes has obvious potential risks. It has a relatively narrow therapeutic margin of safety, and as long-term therapy it will require careful patient follow-up. It requires that I_{Kr} channels be intact and responsive. However, even a reduced number of functional channels (less than 50% in mutations exhibiting a dominant negative effect) suffices to achieve a QTc shortening in LQT2 patients during K^+ supplementation.[72] Despite missing conclusive evidence, one can assume that prolonged QTc interval in all forms of long QT syndrome (including hypertrophy, heart failure etc.) can be corrected acutely in part by increasing serum potassium levels. Prospective multicenter studies are needed to demonstrate potential benefits of such an approach with respect to cardiac events and mortality. Potassium supplementation, with the goal to increase serum potassium levels to above 1mmol/l is not suited for long-term therapy at present.

While increased serum potassium levels may be beneficial in long QT syndromes, hypokalemic states certainly are hazardous, and may trigger arrhythmias and sudden cardiac death due to excess QT prolongation. Potassium must be con-

sidered as a vital parameter in controlling cardiac repolarization. Regular but moderate and controlled potassium supplementation may prevent occurrence and severity of hypokalemia.

Gene therapy

Given our current knowledge of repolarization abnormalities in hypertrophy and heart failure, cardiac arrhythmias in this context present a promising but less well-developed target for gene-based therapeutic strategies.[74] Our understanding of the basic principles of cellular electrophysiology can be used to target specific genes that will modify active membrane properties of the cardiac myocyte. Proof of concept has been demonstrated in a number of *in vitro* and *in vivo* models in animals, with a variety of delivered genes. Studies in isolated cardiac ventricular myocytes demonstrated the ability to alter the action potential profile by infection with a channel gene-containing virus,[75,76] thus potentially antagonizing acquired long QT syndrome. Although such work is conceptually attractive, widespread and homogeneous gene delivery with long-term expression will be required before human trials can be anticipated.

The future goal is to improve our understanding of both extrinsic and genetic modifiers and their interplay with primary and secondary structural defects as seen in congestive heart failure and hypertrophy. This will provide some hope for defining an individual's risk or threshold for suffering ventricular arrhythmias in the context of repolarization abnormalities and, potentially, for individualized therapy.

References

1. Stewart S, MacIntyre K, Hole DJ et al. More 'malignant' than cancer? Five-year survival following a first admission for heart failure. Eur J Heart Fail 2001; 3: 315–22.

2. Stevenson WG, Stevenson LW, Middlekauff HR et al. Improving survival for patients with advanced heart failure: a study of 737 consecutive patients. J Am Coll Cardiol 1995; 26: 1417–23.

3. Gradman A, Deedwania P, Cody R et al. Predictors of total mortality and sudden death in mild to moderate heart failure. Captopril-Digoxin Study Group. J Am Coll Cardiol 1989; 14: 564–70.

4. Kjekshus J. Arrhythmias and mortality in congestive heart failure. Am J Cardiol 1990; 65: 42I–48I.

5. Stevenson WG, Stevenson LW, Middlekauff HR et al. Sudden death prevention in patients with advanced ventricular dysfunction. Circulation 1993; 88: 2953–61.

6. Coronel R. Repolarization abnormalities in heart failure. Cardiovasc Res 2002; 54: 11–12.

7. Pogwizd SM, Bers DM. Cellular basis of triggered arrhythmias in heart failure. Trends Cardiovasc Med. 2004; 14: 61–6.

8. Singh JP, Johnston J, Sleight P et al. Left ventricular hypertrophy in hypertensive patients is associated with abnormal rate adaptation of QT interval. J Am Coll Cardiol 1997; 29: 778–84.

9. Choy AM, Lang CC, Chomsky DM et al. Normalization of acquired QT prolongation in humans by intravenous potassium. Circulation 1997; 96: 2149–54.

10. Fu GS, Meissner A, Simon R. Repolarization dispersion and sudden cardiac death in patients with impaired left ventricular function. Eur Heart J 1997; 18: 281–9.

11. Marban E. Heart failure: the electrophysiologic connection. J Cardiovasc Electrophysiol. 1999; 10: 1425–8.

12. Beuckelmann DJ, Näbauer M, Erdmann E. Alterations of K^+ currents in isolated human ventricular myocytes from patients with terminal heart failure. Circ Res 1993; 73: 379–85.

13. Kääb S, Nuss HB, Chiamvimonvat N et al. Ionic mechanism of action potential prolongation in ventricular myocytes from dogs with pacing-induced heart failure. Circ Res 1996; 78: 262–73.

14. Kääb S, Dixon J, Duc J et al. Molecular basis of transient outward potassium current downregulation in human heart failure: a decrease in Kv4.3 mRNA correlates with a reduction in current density. Circulation 1998; 98: 1383–93.

15. Berger RD, Kasper EK, Baughman KL et al. Beat-to-beat QT interval variability: novel evidence for repolarization lability in ischemic and nonischemic dilated cardiomyopathy. Circulation 1997; 96: 1557–65.

16. www.cmbl.jhu.edu/movies/normal.mpg (accessed 4 May 2005)

17. www.cmbl.jhu.edu/movies/ead.mpg (accessed 4 May 2005)

18. Marban E. Cardiac channelopathies. Nature 2002; 415: 213–18.

19. Brugada P, Wellens HJ. Early afterdepolarizations: role in conduction block, 'prolonged repolarization-dependent reexcitation,' and tachyarrhythmias in the human heart. Pacing Clin Electrophysiol 1985; 8: 889–96.

20. Tomaselli GF, Beuckelmann DJ, Calkins HG et al. Sudden cardiac death in heart failure. The role of abnormal repolarization. Circulation 1994; 90: 2534–9.

21. Akar FG, Rosenbaum DS. Transmural electrophysiological heterogeneities underlying arrhythmogenesis in heart failure. Circ Res 2003; 93: 638–45.

22. Lacroix D, Gluais P, Marquie C et al. Repolarization abnormalities and their arrhythmogenic consequences in porcine tachycardia-induced cardiomyopathy. Cardiovasc Res 2002; 54: 42–50.

23. Curran ME, Splawski I, Timothy KW et al. A molecular basis for cardiac arrhythmia: HERG mutations cause long QT syndrome. Cell 1995; 80: 795–803.

24. Wang Q, Shen J, Splawski I et al. SCN5A mutations associated with an inherited cardiac arrhythmia, long QT syndrome. Cell 1995; 80: 805–11.

25. Keating MT, Sanguinetti MC. Molecular and cellular mechanisms of cardiac arrhythmias. Cell 2001; 104: 569–80.

26. Ackerman MJ. Cardiac channelopathies: it's in the genes. Nat Med 2004; 10: 463–4.

27. Splawski I, Tristani-Firouzi M, Lehmann MH et al. Mutations in the hminK gene cause long QT syndrome and suppress IKs function. Nat Genet 1997; 17: 338–40.

28. Abbott GW, Sesti F, Splawski I et al. MiRP1 forms IKr potassium channels with HERG and is associated with cardiac arrhythmia. Cell 1999; 97: 175–87.

29. Mohler PJ, Schott JJ, Gramolini AO et al. Ankyrin-B mutation causes type 4 long-QT cardiac arrhythmia and sudden cardiac death. Nature 2003; 421: 634–9.

30. Gaita F, Giustetto C, Bianchi F et al. Short QT syndrome: a familial cause of sudden death. Circulation 2003; 108: 965–70.

31. Brugada R, Hong K, Dumaine R et al. Sudden death associated with short-QT syndrome linked to mutations in HERG. Circulation 2004; 109: 30–5.

32. Priori SG. Inherited arrhythmogenic diseases: the complexity beyond monogenic disorders. Circ Res 2004; 94: 140–5.

33. Priori SG, Schwartz PJ, Napolitano C et al. Risk stratification in the long-QT syndrome. N Engl J Med 2003; 348: 1866–74.

34. Baroudi G, Acharfi S, Larouche C et al. Expression and intracellular localization of an SCN5A double mutant R1232W/T1620M implicated in Brugada syndrome. Circ Res 2002; 90: E11–16.

35. Bezzina CR, Verkerk AO, Busjahn A et al. A common polymorphism in KCNH2 (HERG) hastens cardiac repolarization. Cardiovasc Res 2003; 59: 27–36.

36. Splawski I, Timothy KW, Tateyama M et al. Variant of SCN5A sodium channel implicated in risk of cardiac arrhythmia. Science 2002; 297: 1333–6.

37. Arking DE, Chugh SS, Chakravarti A et al. Genomics in sudden cardiac death. Circ Res 2004; 94: 712–23.

38. Pfeufer A, Jalilzadeh S, Perz S et al. Common variants in myocardial ion channel genes modify the QT interval in the general population: results from the KORA study. Circ Res 2005; 96: 693–701.

39. Henderson EB, Kahn JK, Corbett JR et al. Abnormal I-123 metaiodobenzylguanidine myocardial washout and distribution may reflect myocardial adrenergic derangement in patients with congestive cardiomyopathy. Circulation 1988; 78: 1192–9.

40. Calkins H, Allman K, Bolling S et al. Correlation between scintigraphic evidence of regional sympathetic neuronal dysfunction and ventricular refractoriness in the human heart. Circulation 1993; 88: 172–9.

41. Morley M, Molony CM, Weber TM et al. Genetic analysis of genome-wide variation in human gene expression. Nature 2004 ;430: 743–7.

42. Volders PG, Sipido KR, Vos MA et al. Downregulation of delayed rectifier K^+ currents in dogs with chronic complete atrioventricular block and acquired torsades de pointes. Circulation 1999; 100: 2455–61.

43. Tsuji Y, Opthof T, Kamiya K et al. Pacing-induced heart failure causes a reduction of delayed rectifier potassium currents along with decreases in calcium and transient outward currents in rabbit ventricle. Cardiovasc Res 2000; 48: 300–9.

44. Xu X, Rials SJ, Wu Y et al. Left ventricular hypertrophy decreases slowly but not rapidly activating delayed rectifier potassium currents of epicardial and endocardial myocytes in rabbits. Circulation 2001; 103: 1585–90.

45. Näbauer M, Kääb S. Potassium channel down-regulation in heart failure. Cardiovasc Res 1998; 37: 324–34.

46. Tomaselli GF, Marban E. Electrophysiological remodeling in hypertrophy and heart failure. Cardiovasc Res 1999; 42: 270–83.

47. Nuss HB, Kääb S, Kass DA et al. Cellular basis of ventricular arrhythmias and abnormal automaticity in heart failure. Am J Physiol Heart Circ Physiol 1999; 277: H80–91.

48. Grzeskowiak R, Witt H, Drungowski M et al. Expression profiling of human idiopathic dilated cardiomyopathy. Cardiovasc Res 2003; 59: 400–11.

49. Steenman M, Chen YW, Le Cunff M et al. Transcriptomal analysis of failing and nonfailing human hearts. Physiol Genomics 2003; 12: 97–112.

50. Kääb S, Barth AS, Margerie D et al. Global gene expression in human myocardium-oligonucleotide microarray analysis of regional diversity and transcriptional regulation in heart failure. J Mol Med 2004; 82: 308–16.

51. Yung CK, Halperin VL, Tomaselli GF et al. Gene expression profiles in end-stage human idiopathic dilated cardiomyopathy: altered expression of apoptotic and cytoskeletal genes. Genomics 2004; 83: 281–97.

52. Pham TV, Rosen MR. Sex, hormones, and repolarization. Cardiovasc Res 2002; 53: 740–51.

53. Roden DM. Drug-induced prolongation of the QT interval. N Engl J Med 2004; 350: 1013–22.

54. Zareba W, Moss AJ, Locati EH et al. International Long QT Syndrome Registry. Modulating effects of age and gender on the clinical course of long QT syndrome by genotype. J Am Coll Cardiol 2003; 42: 103–9.

55. Pham TV. Gender differences in cardiac development: are hormones at the heart of the matter? Cardiovasc Res 2003; 57: 591–3.

56. Rodriguez I, Kilborn MJ, Liu XK et al. Drug-induced QT prolongation in women during the menstrual cycle. JAMA 2001; 285: 1322–6.

57. Romhilt DW, Chaffin C, Choi SC et al. Arrhythmias on ambulatory electrocardiographic monitoring in women without apparent heart disease. Am J Cardiol 1984; 54: 582–6.

58. Nakaya H, Takeda Y, Tohse N et al. Effects of ATP-sensitive K^+ channel blockers on the action potential shortening in hypoxic and ischaemic myocardium. Br J Pharmacol 1991; 103: 1019–26.

59. Venkatesh N, Lamp ST, Weiss JN. Sulfonylureas, ATP-sensitive K^+ channels, and cellular K^+ loss during hypoxia, ischemia, and metabolic inhibition in mammalian ventricle. Circ Res 1991; 69: 623–37.

60. Wilde AA, Janse MJ. Electrophysiological effects of ATP sensitive potassium channel modulation: implications for arrhythmogenesis. Cardiovasc Res 1994; 28: 16–24.

61. Picard S, Rouet R, Ducouret P et al. KATP channels and 'border zone' arrhythmias: role of the repolarization dispersion between normal and ischaemic ventricular regions. Br J Pharmacol 1999; 127: 1687–95.

62. Sampson JJ, Anderson EM. Treatment of certain cardiac arrhythmias with potassium salts. JAMA 1932; 99: 2257–61.

63. Anderson BN, Jr., Bellet S, Bettinger JC et al. The effect of intravenous administration of potassium chloride on ectopic rhythms, ectopic beats and disturbances in A-V conduction. Am J Med 1956; 21521–33.

64. Weidmann S. Shortening of the cardiac action potential due to a brief injection of KCl following the onset of activity. J Physiol 1956; 132: 157–63.

65. Sanguinetti MC, Jurkiewicz NK. Role of external Ca^{2+} and K^+ in gating of cardiac delayed rectifier K^+ currents. Pflugers Arch 1992; 420: 180–6.

66. Yang T, Roden DM. Extracellular potassium modulation of drug block of IKr. Implications for torsade de pointes and reverse use-dependence. Circulation 1996; 93: 407–11.

67. Roden DM. Taking the 'idio' out of 'idiosyncratic': predicting torsades de pointes. Pacing Clin Electrophysiol 1998; 21: 1029–34.

68. Pak PH, Nuss HB, Tunin RS et al. Repolarization abnormalities, arrhythmia and sudden death in canine tachycardia-induced cardiomyopathy. J Am Coll Cardiol 1997; 30: 576–84.

69. Kääb S, Hinterseer M, Näbauer M et al. Sotalol testing unmasks altered repolarization in patients with suspected acquired long-QT-syndrome-a case-control pilot study using i.v. sotalol. Eur Heart J 2003; 24: 649–57.

70. Sanguinetti MC, Jiang C, Curran ME et al. A mechanistic link between an inherited and an acquired cardiac arrhythmia: HERG encodes the IKr potassium channel. Cell 1995; 81: 299–307.

71. Compton SJ, Lux RL, Ramsey MR et al. Genetically defined therapy of inherited long-QT syndrome. Correction of abnormal repolarization by potassium. Circulation 1996; 94: 1018–22.

72. Etheridge SP, Compton SJ, Tristani-Firouzi M et al. A new oral therapy for long QT syndrome: long-term oral potassium improves repolarization in patients with HERG mutations. J Am Coll Cardiol 2003; 42: 1777–82.

73. Tan HL, Alings M, Van Olden RW, Wilde AA. Long-term (subacute) potassium treatment in congenital HERG-related long QT syndrome (LQTS2). J Cardiovasc Electrophysiol 1999; 10: 229–33.

74. Tomaselli GF, Donahue JK. Somatic gene transfer and cardiac arrhythmias: problems and prospects. J Cardiovasc Electrophysiol 2003; 14: 547–50.

75. Nuss HB, Johns DC, Kääb S et al. Reversal of potassium channel deficiency in cells from failing hearts by adenoviral gene transfer: a prototype for gene therapy for disorders of cardiac excitability and contractility. Gene Ther 1996; 3: 900–12.

76. Nuss HB, Marban E, Johns DC. Overexpression of a human potassium channel suppresses cardiac hyperexcitability in rabbit ventricular myocytes. J Clin Invest 1999; 103: 889–96.

CHAPTER 25

Cell–cell communication abnormalities

Jeffrey E Saffitz, Madison S Spach

Introduction

This chapter reviews the fundamental process of intercellular communication via gap junctions in the heart. It is divided into two main sections. The first considers basic features of gap junction biology in the heart and cell–cell propagation of the cardiac action potential. The second addresses electrophysiological effects of remodeling of gap junctions during normal development and aging, and in selected cardiac disease states. Particular attention is focused on changes in the expression, distribution and/or function of connexins and gap junction channels in heart disease, and the role of abnormalities in cell–cell communication in arrhythmogenesis.

Basis of cell–cell propagation of the cardiac action

Connexins, connexons and gap junctions

Cardiac muscle is composed of individual cells each invested with an insulating lipid bilayer that would effectively prevent intercellular current flux were there not specialized cell–cell electrical junctions to serve this purpose. Electrical activation of the heart involves the flow of ions from one cell to another via gap junctions, specialized regions of the sarcolemma containing transmembrane protein

assemblies that adjoin in the extracellular space to create aqueous pores that directly link the cytoplasmic compartments of neighboring cells.[1] A gap junction consists of a plaque of tens to thousands of closely packed channels that permit intercellular passage of ions and small molecules up to ~1 kDa in molecular weight (Fig. 25.1).[2]

Gap junction channels are composed of members of a multi-gene family of proteins called con-

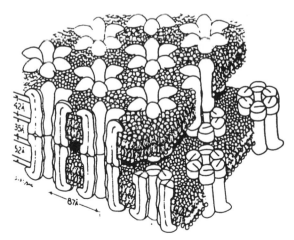

Figure 25.1. *A model of the structure of a gap junction originally proposed by Makowski et al.[2] Intercellular channels are created by pairing of hexameric assemblies in the plasma membranes of adjacent cells. Gap junctions may contain tens to thousands of individual channels.*

nexins. The human genome contains 21 genes encoding distinct connexins.[3] These proteins are named by the abbreviation Cx followed by the molecular weight of the specific protein. The general topology of a connexin in the sarcolemma and a comparison of the structures of several connexins are shown in Fig. 25.2.[4] All connexins contain four transmembrane domains and two extracellular loops which are highly conserved among connexin isoforms (Fig. 25.2). In contrast, the intracellular loop between the second and third transmembrane domains, and the carboxy terminus have unique amino acid sequences that largely account for the differences in molecular weights, and which confer specific biophysical properties on channels composed of different connexins.[1]

An individual intercellular channel is created by stable, non-covalent interactions of two hemi-channels referred to as connexons, each located in the plasma membranes of adjacent cells (see Fig. 25.1). Electron cyrocrystallography studies have confirmed that each connexon is formed by six connexin subunits, each having four transmembrane α-helices (Fig. 25.3).[5] A complete intercellular channel, therefore, consists of a dodecamer containing 48 closely packed transmembrane α-helices (Fig. 25.3).[5] This arrangement confers considerable structural stability on the macromolecular assembly.[5]

Expression of multiple connexins in the heart

Like most differentiated cells, individual cardiac myocytes express multiple connexins. Messenger RNAs encoding Cx37, Cx40, Cx43, Cx45, Cx46 and Cx50 have been detected in homogenates of mammalian heart muscle,[6–11] but not all of these proteins are expressed by cardiac myocytes. Expression of Cx37 is confined to the endothelium of the coronary vasculature.[9] Cx50, originally localized to the atrio-ventricular junction in the early postnatal rat heart, may be expressed by an interstitial cell.[10] Other connexins, such as Cx43 and Cx40, are expressed by both cardiac myocytes and vessel wall cells (endothelium and/or smooth muscle).

Mammalian cardiac myocytes express Cx43, Cx45, and Cx40, but different tissues of the heart exhibit different connexin expression phenotypes. The major cardiac gap junction protein, Cx43, is

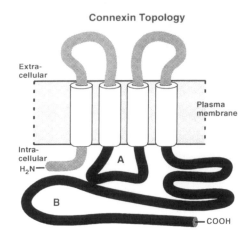

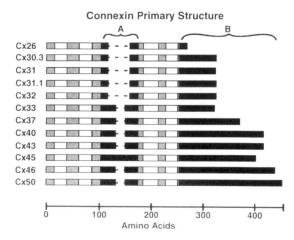

Figure 25.2. *Upper panel: a model of the topology of a connexin molecule. The transmembrane domains are shown in white, the N-terminus and extracellular domains are shown in grey (hatched), and the intracellular hydrophilic loop connecting the second and third transmembrane domains (A) and the C-terminal tail (B) are shown in black. Lower panel: comparison of the sequences of multiple members of the connexin family. The amino acid number begins at the N-terminus. The grey (hatched) and white segments correspond to the regions of the molecules shown in the upper panel. Divergent sequences, shown in black, are located mainly in the intracellular loop (A) and the C-terminal tail (B) regions. Reprinted with permission from Kanno S, Saffitz JE. The role of myocardial gap junctions in electrical conduction and arrhythmogenesis. Cardiovasc Pathol 2001; 10: 169–77.*

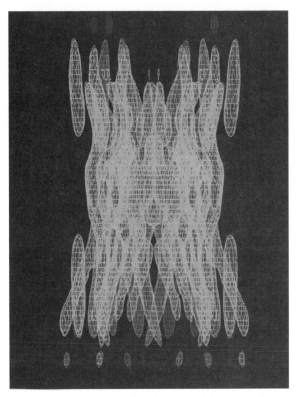

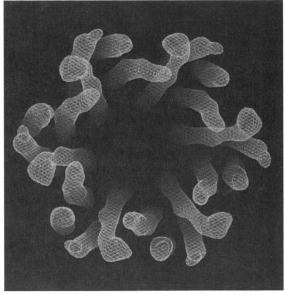

Figure 25.3. Left panel: side view of the three-dimensional density map of a recombinant cardiac gap junction channel. The outer boundary of the MAP shows that the diameter of the channel is ~80 Å within the two membrane regions and then narrows in the extracellular gap to a diameter of ~50 Å. The rod-like features in the MAP corresponding to membrane α-helices. Right panel: a top view looking toward the extracellular gap of the three-dimensional map of recombinant gap junction channel. For clarity, only the cytoplasmic and most of the membrane-spanning regions of one connexon are shown. The rod-shaped features reveal the packing of 24 transmembrane α-helices within each connexin. Reprinted from Unger VM, Kumar NM, Gilula NB et al. Three-dimensional structure of a recombinant gap junction membrane channel. Science 1999; 283: 1176–80. See color plate section.

expressed in atrial and ventricular myocytes and in selected regions of the atrioventricular conduction system.[12–14] Recent work, for example, has implicated differential expression of Cx43 in dual-pathway electrophysiology of the atrioventricular node.[15] Cx43 has also been identified in heterogeneous regions of the rabbit and canine sinus nodes.[16,17] Cx40 is expressed by atrial myocytes and by His–Purkinje fibers, but not by adult ventricular myocytes.[13,14,18–21] Studies in Cx40-null mice have revealed differential effects on conduction within the right and left bundle branches,[22] suggesting a high degree of specific spatial heterogeneity in Cx40 expression and function. Cx45 is expressed in low levels in atrial and ventricular myocytes, but is concentrated in the conduction system.[23]

Channels formed by individual connexins exhibit distinct unitary conductances, pH and voltage dependence, and permeabilities to ions and small molecules such as fluorescent dyes. Cx43 channels have main conductance states of 90–115 pS, and are relatively insensitive to changes in transjunctional voltage,[24–27] compared with channels composed of Cx40 or Cx45. Cx40 channels have larger conductances (150–160 pS),[26,28–30] whereas Cx45 channels exhibit a much lower main conductance state of ~25 pS, and are highly sensitive to transjunctional voltage.[26,31]

Because individual myocytes generally express more than a single species of connexin, the possibility exists for formation of individual channels composed of more than one connexin isoform. Theoretical combinations of connexins in hybrid channels and a currently used system of nomenclature[32] are illustrated in Fig. 25.4. Individual connexons can be homomeric or heteromeric, and individual channels can be homotypic or heterotypic. The potential for structurally and functionally diverse forms of hybrid gap junction channels of the heart is great. In expression systems, cardiac connexins form various types of hybrid channels that exhibit distinct channel properties. For example, Cx43 and Cx45 form heterotypic channels with lower unitary conductances than homotypic Cx43 channels, and Cx43/Cx45 channels do not pass the fluorescent dye Lucifer yellow,[33,34] which flows freely across Cx43 channels. Evidence is mounting that hybrid channels also occur in the heart where they may play important roles in modulating cell–cell communication in both physiological and disease states. Hybrid gap junction channels may be especially important at interfaces between cardiac tissues with disparate connexin expression phenotypes, such as the Purkinje–myocyte junction. Because ventricular myocytes express mainly Cx43, and Purkinje fibers express abundant Cx40, it is possible that heteromultimeric Cx43/Cx40 channels with unique properties facilitate preferential conduction from Purkinje fibers to ventricular myocytes.

Use of genetically engineered mouse models to characterize the role of individual cardiac connexins

The biological significance of different connexin expression phenotypes in cardiac physiology has been investigated most directly in genetically engineered mice. The first such model was reported in 1995 by Reaume et al.[35] who produced a germ-line Cx43 knockout mouse. Cx43-null animals develop to term, but die shortly after birth, apparently as a result of a conotruncal malformation which causes right ventricular outflow tract obstruction.[35] Heterozygous (Cx43[+/−]) mice have a normal life span and show no apparent abnormalities in cardiac structure or contractile function.[36] Their hearts contain ~50% of the wild-type level of Cx43, and ventricular myocytes are connected by fewer, but not smaller, gap junctions.[37] The effects of reduced Cx43 expression in Cx43[+/−] mice on Cx43 ventricular conduction appear to be modest under normal basal conditions. Initial studies in whole hearts[38,39] reported a decrease in conduction velocity, but other studies showed little or no conduction slowing.[40] High resolution optical mapping of conduction in microspecific strands of neonatal mouse myocytes *in vitro*[41] has revealed only a very small change in conduction velocity in Cx43[+/−] cells. However, the maximal upstroke velocity of the transmembrane action potential, dV/dT_{max}, is increased and action potential duration is reduced in heterozygous compared with wild-type strands.[41] These observations suggest that heterozygous null mutation of *Cx43* produces a complex electrical phenotype in neonatal cells *in vitro*, characterized by changes in both ion channel function and cell–cell coupling.

More recently, Cre-loxP strategies have been used to knockout Cx43 in a cardiac-restricted manner.[42,43] Cardiac specific excision of *Cx43* alleles leads to a dramatic phenotype characterized by malignant ventricular tachyarrhythmia and sudden cardiac death.[42,43] One potentially interesting

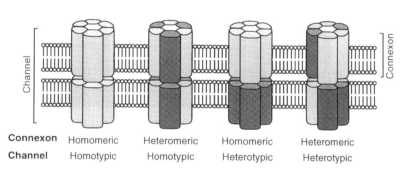

Figure 25.4. *Nomenclature for connexons and gap junction channels formed by one or more connexins. Modified with permission from Kumar NM, Gilula NB. The gap junction communication channel. Cell 1996; 84: 381–8.*

aspect of this model is that the Cre-loxP approach results in 5–10% of ventricular cells with normal connexin expression. This raises the possibility that heterogeneous coupling plays a significant role in arrhythmogenesis in these mouse models. This point is supported by additional studies of Cx43-wild-type/Cx43-null chimerics which also demonstrate arrhythmias.[44]

Cell–cell coupling has been studied directly in ventricular myocyte pairs isolated from Cx43-null murine hearts.[45] Transjunctional conductance in myocyte pairs from knockout mice are dramatically reduced to 1–2% of controls.[45] These findings indicate that residual coupling in ventricular myocytes in the absence of Cx43 is very low. Presumably, this is provided by Cx45 but this has not been directly proven.

Germ-line knockout of Cx40 does not affect ventricular conduction velocity, but does cause atrial conduction disturbances and atrial arrhythmias.[46,47] Cx40 knockout mice also exhibit a right bundle branch block pattern on the surface electrocardiogram (ECG),[46,47] which appears to be related to disparate effects of Cx40 knockout on the His–Purkinje system. Tamaddon et al.[22] mapped activation of the mouse bundle branches and observed ~40% reduction in propagation velocity in the right bundle branch, but no apparent delay in the left bundle branch. Thus, altered patterns of epicardial activation in Cx40-null mice seem to be directly attributable to conduction slowing in the right bundle branch.

Germ-line deletion of Cx45 results in early embryonic lethality.[48] Using an alternative strategy to delineate the role of Cx45 in cardiac conduction, Alcoléa et al. created mice with bi-allelic replacement of Cx40 by Cx45.[49] These animals demonstrate increased P-wave duration, and a prolonged and fractionated QRS complex. Conduction velocities in the right atrium and the ventricular myocardium are normal, and conduction through the AV node is unaffected.[49] Electrical mapping of the His bundle branches revealed slow conduction in the right bundle branch only. Thus, replacement of Cx40 by Cx45 results in a phenotype similar, but not identical, to that seen in Cx40-null mice.

Intracellular pathways of connexin assembly, trafficking and degradation

Evidence from a variety of experimental systems has indicated that connexins have very short half-lives (~1–3 h),[50–52] and traffic into and out of gap junctions at an astonishingly rapid rate (Fig. 25.5).[54,55] Newly synthesized connexins are co-translationally incorporated into the endoplasmic reticulum membrane, where they assemble into hexameric hemichannels (connexons) before passing through the Golgi apparatus and moving in vesicles to non-junctional regions of the plasma membrane.[56,57] Then, they migrate to the periphery of gap junction plaques where they dock with hemichannels in a neighboring cell, form functional channels, and become incorporated into the plaque. Time lapse photography of Cx43 labeled with green fluorescent protein in living mammalian cells has demonstrated dynamic flow of fluorescent vesicles to and from the cell surface.[54] Elegant studies in HeLa cells using tetracysteine genetic tags and multicolor imaging of connexin trafficking have shown that new connexons become incorporated at the periphery of a gap junction and move continuously from the periphery to the center of the plaque where protein is eventually removed.[55]

The determinants of connexin turnover in the heart have not been fully elucidated. Studies of connexin degradation in cultured cells,[58,59] including cardiac myocytes,[60] and in isolated adult rat hearts,[52] have shown that Cx43 is degraded by both lysosomal and proteasomal pathways. Although much of the connexin in a cell is apparently located in gap junctions, a considerable amount resides in the Golgi apparatus, non-junctional plasma membrane and intracellular vesicles. Clearly, the highly dynamic flow of connexins through gap junctions must be tightly controlled to maintain normal levels of functional channels at the cell surface, and prevent inappropriate accumulation or loss of connexins in the cell. Even relatively minor adjustments in rates of connexin degradation could rapidly lead to large changes in intercellular communication.

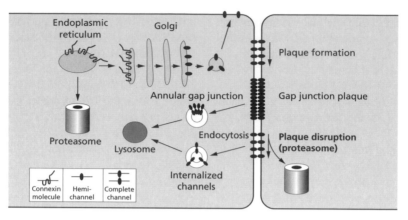

Figure 25.5. *Intracellular pathways of connexin synthesis, assembly and degradation. Connexins are synthesized in the endoplasmic reticulum where they fold and assemble into hexameric hemichannels (connexons) before passing through the Golgi apparatus to non-junctional regions of the plasma membrane. Connexins migrate to the periphery of gap junction channels where they dock with hemichannels in a neighboring cell, and become incorporated into the plaque. Connexons are removed from the plaque and brought into the cell for degradation via lysosomal or proteasomal pathways. Reprinted with permission from Beardslee MA, Laing JG, Beyer EC et al. Rapid turnover of connexin43 in the adult rat heart. Circ Res 1998; 83: 629–35.*

Regulation of connexin expression and gap junction channel function

Regulation of the extent to which cardiac myocytes are electrically coupled by gap junction channels is complex and not well understood. In general, it appears that cells can rapidly (within minutes) change the number of functional channels at the cell surface, through multiple mechanisms involving mobilization of intracellular connexin to junctional plaques, internalization of junctional channels, changes in channel open probability and, potentially, changes in rates of connexin synthesis and degradation. It is likely that the major mechanisms regulating connexin trafficking, degradation and channel function involve changes in protein phosphorylation occurring principally in the intracellular C-terminal domains.[61] Phosphorylation of multiple serine and tyrosine residues in this region has been shown to be important in Cx43 and Cx45 assembly into gap junction channels, and in changes in channel function, intracellular translocation and degradation.[61]

Unlike typical regulatory proteins that normally reside in an inactive unphosphorylated state and become activated transiently by phosphorylation, Cx43 in cardiac myocytes exists in multiple phosphorylated isoforms under basal conditions.[61] At least five phosphorylated isoforms of Cx43 can be identified by phosphopeptide mapping, but the exact patterns of phosphorylated residues in these isoforms are still unknown.[61] Much of the Cx43 in normal cardiac myocytes is phosphorylated at multiple sites, resulting in a preponderance of isoforms that migrate on polyacrylamide gels with apparent molecular weights of 43–46 kDa. These various isoforms presumably reflect protein in different pools such as newly synthesized Cx43 trafficking to the cell surface, Cx43 in functional or non-functional channels in gap junction plaques, and Cx43 that has left junctions en route to degradation. Little is currently known about the endogeneous, presumably constituitively active kinases responsible for maintaining high basal levels of phosphorylated connexins.

Despite the fact that Cx43 is normally phosphorylated under basal conditions, transient changes in phosphorylation mediated by activation of kinases and/or phosphatases in response to physiological or pathophysiological perturbations can lead to marked changes in channel function and connexin distribution.[61] Numerous kinases have been implicated. For example, Cx43 can be phosphorylated in specific serine residues by mitogen-activated protein (MAP) kinases,[61] protein kinases C (PKC)[61] and A (PKA),[62] and casein kinase I,[63] and on tyrosines by v-src and c-src.[61] However, the precise signaling mechanisms responsible for changing intercellular coupling in cardiac myocytes are incompletely understood.

Multiple mechanisms have been implicated in regulation of connexin distribution and gap junction channel function. For example, elegant work has elucidated intramolecular interactions involved in pH gating of gap junction channels.[64–66] Specific regions of the C-terminal tail and intracellular cytoplasmic loop can undergo conformational changes in response to intracellular acidification resulting in channel closure.[64–66] Another emerging subject of great interest concerns potential regulatory roles of connexin-interacting proteins. The best characterized interaction is between Cx43 and the PDZ-domain protein zonula occludens-1 (ZO-1) which has been implicated in formation and maintenance of gap junction plaques and internalization of Cx43 under conditions in which cells have uncoupled.[66–68] Cx45 has also been shown to interact with ZO-1.[69] The full

repertoire of connexin-interacting proteins and their biological roles in the healthy and diseased heart has only barely been explored.

Three-dimensional topography of gap junctions at intercalated discs

Gap junctions are invariably located at intercalated discs,[70–72] complex structures containing cell–cell adhesive junctions responsible for mechanically coupling cardiac myocytes (Fig. 25.6). The intimate spatial juxtaposition of gap junctions and adhesion junctions presumably reflects the need for mechanical stabilization of sarcolemmal regions in neighboring cells in order to form and maintain large arrays of intercellular channels. Human cardiomyopathies caused by mutations in cell–cell junction proteins exhibit marked reductions in gap junctions, despite the presence of normal levels of connexin protein in the tissue.[73,74] This observation is consistent with the hypothesis that normal electrical coupling depends on normal mechanical coupling, and that defects in mechanical coupling may lead to remodeling of gap junctions.

Ventricular myocytes are interconnected by gap junctions that vary widely in size.[70] The largest ventricular gap junctions have a ribbon-like shape, and are oriented perpendicular to the long axis of the cell (Fig. 25.7).[70] When the ends of individual myocytes are viewed in cross-sections, these large

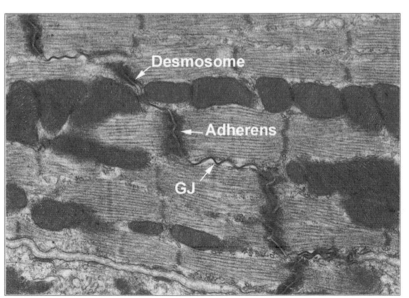

Figure 25.6. *A transmission electron micrograph of an intercalated disc connecting to ventricular myocytes sectioned in a plane parallel to the long axis of the cells. There is an intimate spatial relationship between gap junctions (GJ) and adhesive junctions (fascia adherens junctions and desmosomes).*

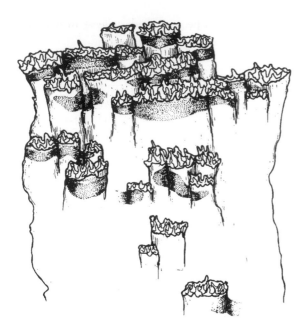

Figure 25.7. *A diagram showing the relationship between large gap junctions (shown at stippled patches on the cell surface membrane) and fascia adherens junctions (shown as finger-like projections) at the intercalated disc at the end of a typical ventricular myocyte. The largest gap junctions have a rectangular or ribbon-like shape. They are located in membrane regions between offset shelves of adhesion junctions. The long axis of the large gap junctions is oriented perpendicular to the long axis of the myocyte. Reprinted with permission from Hoyt RH, Cohen ML, Saffitz JE. Distribution and three-dimensional structure of intercellular junctions in canine myocardium. Circ Res 1989; 64: 565–74.*

ribbon-like gap junctions can be seen forming a well-defined ring of junctions outlining the periphery of the cell.[70,75] This ring of large gap junctions at the ends of ventricular myocytes presumably facilitates efficient intercellular current transfer, provides favorable local current-to-load conditions and enhances the safety of conduction.

Tissue-specific distributions of gap junctions in the heart

The heart consists of multiple tissues with distinctly different structures and conduction proper-

ties. Although tissue-specific differences in cell size and shape, and differential expression of ion channel proteins account for differences in conduction, specific spatial patterns of intercellular connections probably play an important part as well. For example, ventricular myocytes are extensively interconnected to one another in varying degrees of end-to-end and side-to-side apposition.[70,76] This pattern of connections provides numerous points for wavefronts to propagate across cell borders in both longitudinal and transverse directions. Because of the elongated shape of the cells, however, wavefronts traveling in the transverse direction must cross more intercellular junctions per unit distance traveled and, thus, encounter greater resistance and propagate more slowly than wavefronts traveling an equal distance in the longitudinal direction. Anisotropy of conduction velocities in the ventricle, therefore, appears to be determined to a large extent by the shape of the myocytes. The crista terminalis, a discrete atrial bundle that conducts impulses from the sinus node to the atrioventricular (AV) junction, has conduction properties and exhibits a pattern of myocyte interconnections distinctly different from those in the left ventricle.[21] Conduction in the crista is very rapid (~1 m/s), and the ratio of longitudinal to transverse conduction velocity is ~10:1 compared with only ~3:1 in the ventricle.[77] Whereas ventricular myocytes are extensively connected in both end-to-end and side-to-side orientations, the great majority of interconnections in the crista terminalis occur between cells oriented in end-to-end apposition.[21] This arrangement undoubtedly contributes to the highly anisotropic conduction in the crista. Transverse propagation in this tissue would probably be impeded by the relative paucity of connections between cells in side-to-side apposition.

Remodeling of gap junctions and electrical uncoupling in response to acute ischemia

In the setting of acute severe cellular injury, cardiac myocytes uncouple from one another. Although this response is presumably adaptive, in that it limits the spread of chemical mediators of injury, it also contributes to transient conduction abnormalities that may promote arrhythmogenesis. For example, the incidence, frequency and duration of arrhythmias is much greater in isolated hearts of Cx43[+/-] mice than wild-type mice subjected to acute occlusion of the left anterior descending coronary artery.[78] The onset of ventricular arrhythmias also occurs significantly earlier in Cx43-deficient hearts.[78] Thus, a background level of reduced coupling accelerates the onset and increases the incidence, frequency and duration of ventricular tachyarrhythmias induced by acute ischemia.

Electrical uncoupling during ischemia is a complex phenomenon that involves closure of gap junction channels and changes in connexin phosphorylation and distribution. Gap junction channels close under conditions of intracellular acidosis and elevated intracellular Ca^{2+} levels.[79,80] Fatty acids and amphiphilic lipid metabolites such as acylcarnitines can also close intercellular channels.[81,82] However, these effects apparently occur under conditions achieved only after severe, prolonged ischemic injury. Mechanisms responsible for initiating uncoupling during earlier stages of ischemic injury probably involve stress-activated signaling pathways causing changes in connexin phosphorylation. For example, the onset of uncoupling in no-flow ischemia in isolated perfused rat hearts is associated with progressive dephosphorylation of Cx43, accumulation of non-phosphorylated Cx43 in junctions, and translocation of Cx43 from junctions to an intracellular pool.[83] The precise mechanistic relationships between channel closure, changes in phosphorylation, and intercellular translocation are not known. It is likely, however, that specific changes in Cx43 phosphorylation result first in channel closure. Other phosphorylation and/or dephosphorylation events may facilitate subsequent intracellular movement, possibly by promoting binding to ZO-1 or other connexin-interacting proteins.

Brief intervals of ischemic preconditioning have been shown to decrease the rate of uncoupling, and delay the time to complete uncoupling in isolated rat hearts subjected to no-flow ischemia.[84] Dephosphorylation of Cx43, and translocation of Cx43 from gap junctions to the cytosol are also dramatically diminished by preconditioning.[84] These effects appear to be regulated, at least in part, by activation of K^+_{ATP} channels and PKC.[84]

Despite the fact that cells uncouple in response to injury, there is considerable evidence that gap junction communication persists for some time after the onset of myocardial ischemia.[85] For example, healed infarcts are considerably smaller in Cx43[+/-] mice than in wild-type mice subjected to permanent occlusion of the left anterior descending coronary artery.[86] This suggests that biochemical mediators of injury can spread from areas of severe injury to less injured regions, and lead to a larger area of infarction than would occur under conditions of reduced coupling. These observations raise important concerns about potential therapeutic strategies designed to reduce arrhythmias in patients with ischemic heart disease by maintaining open gap junction channels. While this might limit conduction abnormalities that promote arrhythmias, it could also lead to more extensive necrosis in patients who develop acute myocardial infarction.

Enhanced cell–cell communcation during the acute hypertrophic response

Early compensatory hypertrophic growth of cardiac myocytes appears to be associated with enhanced cell–cell communication. This has been demonstrated and characterized mainly in studies *in vitro*. For example, neonatal rat ventricular myocytes incubated with a membrane-permeable form of cAMP for 24 h show increased Cx43 expression, increased numbers of gap junctions and greater conduction velocity.[87] Incubation of cultured ventricular myocytes with angiotensin II for 6–24 h induces similar changes, which have been shown to be related to an increase in Cx43 protein synthesis.[88] More recent studies have shown that application of linear pulsatile stretch of cultured neonatal ventricular myocytes for only 1 h causes a 2-fold increase in Cx43 expression, associated with a 30% increase in conduction velocity.[89,90] Signaling pathways involving transforming growth factor β, vascular endothelial growth factor and angiotensin II have been directly implicated in stretch-induced upregulation of Cx43 expression in cultured myocytes.[90,91]

Diminished connexin expression in chronic heart disease

Reduced Cx43 expression in gap junctions occurs in the hearts of patients with end-stage heart disease of diverse etiologies, including ischemia, hypertension, valvular abnormalities and primary cardiomyopathies.[92–94] Reduced Cx43 expression has generally been observed in viable but structurally altered cells that exhibit degenerative features typical of hibernating myocardium.[94] Diminished Cx43 expression in end-stage heart disease is also associated with selective loss of larger gap junctions at the ends of cells.[94]

Recently, it has been reported that, whereas heart failure is associated with reduced Cx43 expression, Cx45 expression is enhanced in the failing human ventricle (Fig. 25.8).[95] Previous studies have consistently demonstrated that coexpression of Cx45 and Cx43 results in an overall reduction in cell–cell coupling, presumably related to properties of Cx43/Cx45 hybrid channels.[33,34] Thus, concomitant reduction in expression of Cx43 and increased expression of Cx45 could diminish cell–cell communication and contribute to the pathophysiology of heart failure and sudden cardiac death.

Little is known about mechanisms responsible for changes in connexin expression and gap junction remodeling in chronic forms of heart failure. There may be a role for c-jun N-terminal kinase (JNK), a stress-related protein kinase activated in response to injury caused by ischemia–reperfusion or hemodynamic overload. Activation of JNK pathways leads to rapid, marked downregulation of Cx43 expression in neonatal ventricular myocytes.[96] Recent studies have also demonstrated that conditional activation of JNK *in vivo* diminishes ventricular Cx43 expression and causes slow conduction.[97]

Electrophysiological effects of remodeling of gap junctions

Remodeling of gap junction distribution during normal postnatal growth

As shown in the first section of this chapter, the multiple factors that produce remodeling of cardiac gap junctions produce different distributions and amounts of gap junction connexins. These changes are now considered to be an important factor in the origin of cardiac arrhythmias.[98] In addition to remodeling the arrangement and the number of gap junctions secondary to ischemic heart disease,[76,99] changes in the distribution of gap junctions occur normally during growth hypertrophy after birth,[75,100,101] and during aging.[102,103] Postnatal growth remodeling of cell–cell electrical coupling is relevant to arrhythmias because pathological reiteration of several developmental features of the gap junctions occur when heart disease and aging alter normal adult myocardium.[104] Prior to considering abnormalities of cell–cell communication, it is therefore necessary to examine changes in conduction events produced by the

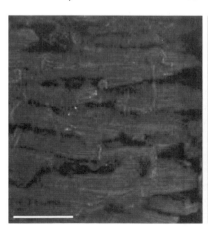 

Figure 25.8. *Confocal images of control human myocardium (left) and failing human myocardium (right) stained with an anti-Cx45 antibody showing upregulation of Cx45 in the failing heart. Bar = 20 μm. Reprinted with permission from Yamada KA, Rogers JW, Sundset R et al. Upregulation of connexin45 in heart failure. J Cardiovasc Electrophysiol 2003; 14: 1205–12. See color plate section.*

manner in which cell size and the distribution of gap junctions are remodeled during development, to achieve the cell–cell coupling pattern of the normal mature heart (Fig. 25.9). Such information should become increasingly useful in extrapolating cell–cell communication results obtained in synthetic layers of neonatal cells[105–107] to naturally occurring cardiac tissues, as well as in evaluating the effects on conduction of transplantation of small fetal and neonatal cells into diseased adult hearts.[108,109]

In the neonatal ventricle, gap junctions are distributed in a relatively uniform manner with a periodicity of 4 to 11 μm along the perimeter of cardiomyocytes (Fig. 25.9).[75,100,101] This pattern also occurs in the neonatal atrium[110] and in cultured monolayers of neonatal cardiomyclytes.[105–107] As growth enlargement of the cells occurs, the gap junctions and cell adhesion molecules (e.g. N-cadherins) become concentrated at the intercalated discs located at the ends of the cells (Fig. 25.9).[75,100,101] In the human ventricle, these postnatal changes in the spatiotemporal pattern of gap junction and cell adhesion molecules continue to about 6 years of age.[75] The associated remodeling of myocyte size and shape from the neonatal to the adult state is illustrated for typical canine ventricular myocytes in Fig. 25.10A.

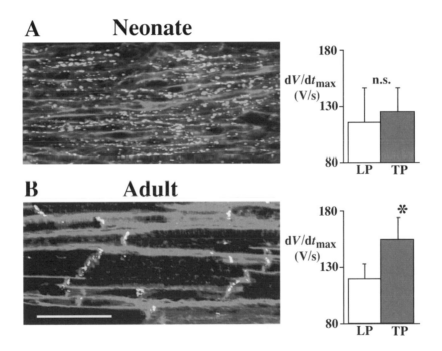

Figure 25.9. *Connexin43 distribution in neonatal and adult canine ventricular muscle (left) and the associated anisotropic changes in dV/dt$_{max}$ (right). Gap junctions (green or yellow) are labeled with antibodies to Cx43. The sarcolemma (red) is labeled with wheat germ agglutinin.[111] (A) Neonatal distribution of gap junctions typical of that in canine hearts from birth to 2 months of age. In the neonatal ventricular muscle, there is no significant difference in mean LP dV/dt$_{max}$ and mean TP dV/dt$_{max}$, which also occurs in synthetic neonatal cellular monolayers.[112] (B) Mature canine ventricular muscle. Bar = 50 μm. In adult ventricular muscle, mean TP dV/dt$_{max}$ is significantly greater than mean LP dV/dt$_{max}$. Data presented as mean ±1 SD; *P < 0.001 (n = 24). Modified with permission from Spach MS, Heidlage JF, Dolber PC et al. Electrophysiological effects of remodeling cardiac gap junctions and cell size: experimental and model studies of normal cardiac growth. Circ Res 2000; 86: 302–11. See color plate section.*

Effects of postnatal remodeling of ventricular microstructure on anisotropic conduction

In cardiac tissue preparations, conduction velocity is measured at a macroscopic size scale (1 mm or more). In neonatal canine ventricular muscle, the average velocity is 0.33 m/s during propagation along the long axis of the fibers (LP) and 0.12 m/s during propagation transverse to the long axis of the fibers (TP).[113] These directionally different velocities are greater in the adult ventricle; e.g. 0.50 m/s during LP and 0.17 m/s during TP. In both neonatal and adult cardiac bundles, dV/dt_{max} varies from site to site in the direction of propagation, and dV/dt_{max} changes at each impalement site when the direction of conduction is shifted from LP to TP.[113,114] When dV/dt_{max} is measured at many sites, however, mean TP and mean LP dV/dt_{max} are not significantly different in the neonatal ventricle (Fig. 25.9A), but with growth to maturity mean TP dV/dt_{max} in the adult ventricle becomes significantly greater than mean LP dV/dt_{max} (Fig. 25.9B).

These propagation events provide insight into the dual nature of impulse transmission in cardiac muscle when viewed at different size scales. At the larger macroscopic size scale, conduction appears to be a continuous process, whereas at a microscopic or cellular level conduction is discontinuous.[115,116] At a microscopic level, the action potential propagates through fluctuating values of resistance produced by the low resistance of the cytoplasm throughout the cells, and by the relatively higher resistance of the gap junctions localized at the intercalated discs. These repetitive resistive discontinuities cause propagation events to have a stochastic nature at a microscopic level; i.e. the excitatory events during propagation are constantly changing with varying intracellular events and delays between cells.[117] These microscopic stochastic variations become averaged at the larger macroscopic size scale, at which the conduction velocity appears to be a smooth process like that which occurs in an electrically continuous medium. In a continuous medium, the velocity and resistance are inversely related according to the following equation:[118]

$$\theta = k(1/\sqrt{r_a}) \text{ (eqn 25.1)}$$

where θ is the conduction velocity, k represents a constant for the same sarcolemmal membrane properties throughout, and r_a is the effective axial resistance that is produced by the cumulative resistance of the gap junctions and the cytoplasm of the cells in the direction of propagation. The resistance of interstitial space also contributes to the effective axial resistance,[119] although it is seldom included in models of cardiac conduction. Due to the elongated shape of cardiomyocytes and the spatial arrangement of their gap junctions, the axial resistance is relatively high in a direction across cardiac fibers, whereas along the long axis of the fibers the axial resistance is relatively low.[116–118] Therefore, due to the inverse relation between conduction velocity and axial resistance (eqn 25.1), at a macroscopic size scale conduction is faster along the long axis of cardiac fibers than across the fibers.

According to continuous medium theory, however, the time course of the transmembrane potential (e.g. dV/dt_{max}) should not be different at different sites, nor should dV/dt_{max} change at a given site when a directional change in conduction results in a change in axial resistance.[118] The variations in dV/dt_{max} are accounted for at a microscopic level, because cardiac propagation is discontinuous due to the cells and their interconnections creating microscopic discontinuities of axial resistance.[116] In turn, these discontinuities produce electrical load variations that alter the rate of rise of the action potential from site to site. Because electrical load is difficult to quantitate or measure, variations of dV/dt_{max} within areas that have similar membrane ionic properties serve as a useful index of variations in electrical load. For example, the greater the electrical load on the membrane at a given site, the lower the value of dV/dt_{max} which, in turn, decreases the safety factor of conduction. The discontinuous nature of propagation is therefore important because the loading variations produced by the discrete nature of the cardiomyocytes and their electrical connections can markedly alter the safety factor of conduction.

Two-dimensional conduction models based on cardiac microstructure

In situ anisotropic coupling has been estimated over multiple cells lengths and widths from respective measurements of longitudinal and transverse space constants.[120] However, due to the complex electrical circuit produced by the arrangement of gap junctions in cardiac tissues,[117,121] it has not been possible to measure the conductance (or resistance) of individual gap junctions nor the conductance between two cells in multicellular preparations. Consequently, conductance measurements of side–side and/or end–end coupling between two cells have required that the measurements be made in isolated cell pairs extracted from the tissue of interest.[122,123] Experimental limitations also have prevented measurement of impulse transmission from cell to cell in cardiac tissue

preparations. In the absence of experimental data in multicellular preparations, two-dimensional (2D) cellular models have been developed to gain electrophysiological insights at this small size scale.[113,117]

Variations in cell shape and size, and the distribution of gap junctions of any one cellular model, however, represent only one of an infinite number of possible arrangements. Rather than focus on a specific structure, we wish to use the results of the 2D cellular models shown in Fig. 25.10 to illustrate how the combined changes in cell size and in the distribution of gap junctions can account for the above-described postnatal growth changes in

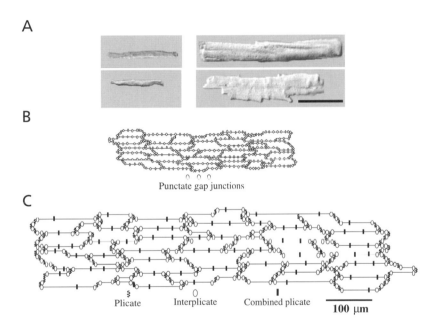

Figure 25.10. *Isolated neonatal and mature ventricular myocytes and diagrams of the basic units of 2D cellular models representing neonatal and adult canine ventricular muscle. (A) Representative neonatal (left) and mature (right) canine ventricular myocytes. The cells demonstrate the considerable difference in size of neonatal and adult cardiomyocytes. Bar = 50 μm. (B) Neonatal 33-cell unit. To represent the distribution of neonatal gap junctions shown in Fig. 25.9A, punctate gap junctions (ovals) were located 5–10 μm apart along the sides and at the ends of the myocytes. Each punctate gap junction was assigned a conductance of 0.16 μS. (C) Adult 33-cell unit. The distribution of the adult gap junctions at the intercalated discs is based on the work of Hoyt et al.[70] The different types of adult gap junctions were assigned the following conductance values: plicate 0.50 μS, interplicate 0.33 μS, and combined plicate 0.16 μS. To form 2D sheets or bundles of different sizes, the 33-cell units were replicated longitudinally and vertically by fitting the ends and sides together. Reproduced with permission from Spach MS, Heidlage JF, Dolber PC et al. Electrophysiological effects of remodeling cardiac gap junctions and cell size: experimental and model studies of normal cardiac growth. Circ Res 2000; 86: 302–11.*

anisotropic propagation. Construction of the models was based on the measured geometry of isolated neonatal and mature cardiomyocytes. The distribution of neonatal gap junctions was obtained from photomicrographs (Fig. 25.9A), and the distribution of adult gap junctions at the intercalated discs was based on data from Saffitz's laboratory.[70,76] Details of the models can be found in prior papers, including assignment of conductance values to the gap junctions, numerical analysis techniques to approximate the fast sodium current, and verification of the models.[114,117,124]

Macroscopic conduction related to dV/dt_{max} and cell delays

In the neonatal model, the macroscopic LP velocity was 0.36 m/s, and the TP velocity was 0.13 m/s. In the adult cellular model, the longitudinal and transverse velocities were 0.48 m/s and 0.17 m/s, respectively.[113] These conduction velocities show good agreement with the experimental data in neonatal and adult preparations. The model results also illustrate that the effects of the irregularly shaped cells of different sizes and the different gap junction distributions are averaged at the larger macroscopic size scale, to produce conduction velocities as measured experimentally.

In both the adult and the neonatal cellular networks, dV/dt_{max} varied throughout each cell, with lower values near the ends than in the middle region of the myocytes (not shown).[113] These intracellular dV/dt_{max} fluctuations accounted for the experimental variation in dV/dt_{max} from site to site. When averaged over many cells, however, in the adult cellular model mean TP dV/dt_{max} was significantly greater than mean LP dV/dt_{max}, whereas in the neonatal model there was no significant difference in the mean values of LP dV/dt_{max} and TP dV/dt_{max} (Fig. 25.11A). Thus, in the adult and neonatal cellular models, the TP–LP mean dV/dt_{max} relationships shown in Fig. 25.11A were in good agreement with the experimental data (Fig. 25.9). As shown in Fig. 25.11B, the mean cell–cell delays during LP and TP had the same relationships as the mean dV/dt_{max} values in the neonatal and adult cellular networks. In the adult, the mean cell–cell delay during TP (178 µs) was much greater than during LP (90 µs), whereas in the

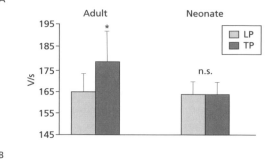

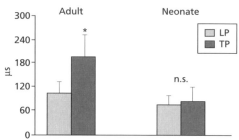

Figure 25.11. *Mean dV/dt_{max} and cell-to-cell delays during anisotropic propagation in the adult and neonatal 2D cellular models. (A) Mean dV/dt_{max} during LP and TP in the adult and neonatal 2D models. (B) Mean cell–cell delays (time difference in dV/dt_{max}) across adjoining end–end cell borders during LP, and across lateral cell borders during TP. The cell–cell delay results were acquired in groups of 22 adjoining cells. Data are mean ±1 SD; *P < 0.001; n.s., not significant. Modified with permission from Spach MS, Heidlage JF, Dolber PC et al. Electrophysiological effects of remodeling cardiac gap junctions and cell size: experimental and model studies of normal cardiac growth. Circ Res 2000; 86: 302–11.*

neonatal cellular network there was no significant difference in mean cell–cell delay during TP (75 µs) and LP (68 µs).

One interpretation of the greater mean TP dV/dt_{max} values and mean cell–cell delays in adult compared to neonatal cardiac muscle is that there are prominent differences in the cellular distribution of gap junctions. However, the size of neonatal and adult cardiomyocytes is also quite different. To separate the effects of gap junction distribution from those of cell size during TP, consider the following 2D cellular model experiments. Two networks were the adult and neonatal 2D models of Fig. 25.10B and C. To reverse the relationship between cell size and gap junction distribution in these models, two hypothetical cellular networks were created: one that approximated the adult large cell geometry but with the neonatal gap junction distribution and conductance values (Fig. 25.12, cell type 2), and another network that mimicked the neonatal small cell geometry but with the adult gap junction distribution and conductance values (Fig. 25.12, cell type 3). These four networks represent the four possible combinations of differ-

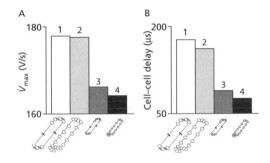

Figure 25.12. *Mean dV/dt_{max} (A) and cell–cell delays (B) of action potential transmission during transverse propagation in cellular networks that differ one from the other with respect to the distribution of gap junctions and cell size. (A) Mean values of dV/dt_{max} in four networks that represent the four possible combinations of different adult and neonatal cell sizes and gap junction distributions. The cell drawings below each graph represent each cellular network. Cells 1 and 4 represent the adult and neonatal cellular networks of Fig. 25.10B and C; cell 2 represents a hypothetical network with adult cell size scaling but with neonatal gap junction distribution and conductance values; and cell 3 represents a hypothetical network with neonatal cell size scaling but with adult gap junction distribution and conductance values. Modified with permission from Spach MS, Heidlage JF, Dolber PC et al. Electrophysiological effects of remodeling cardiac gap junctions and cell size: experimental and model studies of normal cardiac growth. Circ Res 2000; 86: 302–11.*

ences in adult and neonatal cell size and gap junction distribution.

During TP in all of the cellular networks, the time required to excite the membrane of each cell (cytoplasmic time) was only 21 to 29 μs, but the lateral cell–cell delays were considerably longer (Fig. 25.12B). Across the different networks, the mean cell–cell delay correlated strongly with mean dV/dt_{max} ($r = 0.99$; $P < 0.01$). The fact that the values of mean dV/dt_{max} and of mean cell–cell delay have the same relationships with respect to cell size and gap junction distribution, suggests that a link exists between these two parameters – the greater the cell–cell delay, the greater dV/dt_{max}, and vice versa. The results of Fig. 25.12 illustrate that cell size can be as important as gap junction distribution in accounting for different TP–LP dV/dt_{max} relationships. For example, in the 2D models consisting of small cells, a change from the neonatal diffuse gap junction distribution to the adult pattern (while maintaining the adult gap junction conductances) produced little change during TP in mean dV/dt_{max} or lateral cell–cell delays. Correspondingly, reversal of the neonatal–adult gap junction distributions in 2D networks of adult cells produced little change in these parameters during TP. On the other hand, with either type of gap junction distribution, a change from the small cell network to the large cell network produced a large increase in both mean dV/dt_{max} and mean cell–cell delay. Calculations of cell–cell currents (not shown) demonstrated that dV/dt_{max} increases because increases in cell–cell transmission time reduce the electrical current flowing downstream up to the time of dV/dt_{max}, the decrease in downstream load thus enhancing dV/dt_{max}.[113]

These features of cell–cell transmission and dV/dt_{max} provide the following general principles concerning the relative roles of cell size and gap junction distribution: as the degree of electrical coupling between cells (number of functional gap junction channels per unit area of sarcolemma) decreases in relation to cell size, conduction becomes more discontinuous with an associated increase in cell–cell delay. Conversely, for a given cell size, an increase in the number of gap junction channels along the sides (TP) or at the ends of the cells (LP), as well as a decrease in cell size for a given number of channels, decreases the discontinuous nature of conduction with an associated decrease in cell–cell delay.

Effects of cell–cell transmission on the spatial distribution of depolarization

The discrete cellular structure of normal canine adult ventricular muscle is illustrated in the light micrograph of Fig. 25.13A. To show the effects of this structure on the microscopic distribution of the depolarization phase of the action potential in adult ventricular muscle, Fig. 25.13B presents a plot of the transmembrane potential, V_m, along a line during LP and TP in the 2D cellular model of this tissue (Fig. 25.10C). During both LP and TP, depolarization extended over 6–10 cells for a distance of approximately one space constant; i.e. 1.3 mm during LP and 0.4 mm during TP.[124] During LP, the spatial pattern of depolarization approximated a smooth curve consisting of large changes of V_m within each cell, and small V_m discontinuities at the connections between cells (Fig. 25.13B, LP). However, the TP pattern was just the opposite with large discontinuities of V_m at the connections between cells, whereas V_m showed little change across each cell. During both LP and TP, the effect of the irregularities of V_m as a function of distance was well described by a single exponential ($r = 0.99$) over many cells in the foot of the spatial action potential, as shown for TP in Fig. 25.13B. The foot of the action potential is produced by the passive spread of currents from 'upstream' cells in which the sodium current has become active. Thus, the discontinuous changes in the V_m foot produced by the discrete cellular structure of cardiac bundles are averaged at the larger macroscopic size scale, and appear consistent with the continuous (exponential) change of V_m produced by the passive spread of currents in cardiac bundles.[118,125]

Abnormal remodeling of gap junctions and conduction abnormalities

Remodeling of the normal adult distribution and/or amount of gap junctional membrane (number of connexons) has been described in most cardiac diseases associated with arrhythmias. A commonly noted remodeling feature is cellular lateralization of gap junctions. For example, lateralization occurs in ventricular myocytes during healing after myocardial infarction,[126] with right ventricular hypertrophy secondary to pulmonary hypertension,[127] in hypertrophic cardiomyopathy,[128] and in human atria during aging[103] and after prolonged atrial fibrillation.[129] Cell lateralization also has been related to re-entrant ventricular tachycardia circuits that develop following premature stimuli during the healing stage of canine infarcts.[130] Within the epicardial border zone of healing canine infarcts, Peters et al.[131] found cell lateralization of Cx43 to occur within a narrow region corresponding to the critical common pathway of these 'figure of 8' re-entrant circuits. However, cellular lateralization of Cx43 has not been a feature in other areas of the epicardial border zone outside the central common pathway.[123]

Although there is an association between inducibility of arrhythmias and cell lateralization of connexins, it is not clear what effect the cellular redistribution of gap junctions has on cell–cell transmission, nor what role the remodeled lateral distribution of gap junctions plays in altering conduction velocity at the macroscopic size scale. Taken at face value, one might expect an increase in gap junction plaques along the sides of the myocytes to enhance lateral coupling between cells and, thus, increase the transverse conduction velocity at the macroscopic size scale. However, just the opposite occurs in two conditions associated with cell lateralization of Cx43: transverse conduction velocity decreases markedly in the central common pathway of re-entrant circuits located within the epicardial border zone of healing infarcts,[130,131] and in aging human atrial bundles.[103,132] Thus, knowledge of the functional consequences of cell lateralization of the gap junction is quite incomplete. We highlight this remodeling feature to illustrate some of the accompanying microstructural factors that can change and thereby alter impulse transmission between individual cells and groups of cells.

Altered coupling between individual myocytes

In rat ventricular cells bordering healing infarcts, after day 3 the lateralized gap junctions are

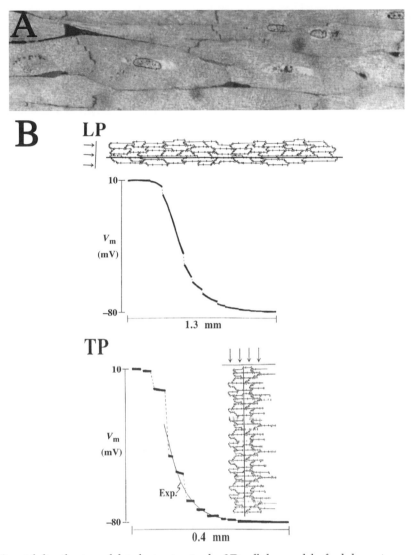

Figure 25.13. *Spatial distribution of depolarization in the 2D cellular model of adult canine ventricular muscle. (A) Light micrograph of a representative section of normal adult canine left ventricle. Potassium ferrocyanide was used to enhance sarcolemmal staining. (B) Spatial distribution of depolarization during LP and TP in the 2D cellular model of adult canine ventricle. LP: the spatial distribution curve during LP is a plot of the transmembrane potential (V_m) along the horizontal line overlying the cellular network shown at the top. TP: the spatial distribution curve during TP is a similar plot of V_m along the vertical line overlying the associated network of cells. In each depolarization curve, the bold solid portion represents V_m within the myocytes, and the broken lines represent the step changes in V_m at the gap junctions and cell borders. In the TP spatial depolarization curve, the superimposed solid line (Exp) is the best fit of single exponential to the V_m values during the foot of the action potential ($r = 0.99$). Panel A modified with permission from Luke RA, Saffitz JE. Remodeling of ventricular conduction pathways in healed canine infarct border zones. J Clin Invest 1991; 87: 1594–602; panel B reproduced with permission from Spach MS, Heidlage JF. The stochastic nature of cardiac propagation at a microscopic level: an electrical description of myocardial architecture and its application to conduction. Circ Res 1995; 76: 366–80.*

associated with adherens junctions (cadherin) and/or desmosomes (desmoplakin),[133] a feature that would provide favorable conditions for maintaining gap junctions by making them more resistant to shear forces.[76] However, variable amounts of remodeled post-infarct lateral gap junctions are located in invaginations of the sarcolemma in the cell interior, where they cannot contribute to cell–cell transmission.[133,134] Due primarily to experimental difficulties in tissue preparations, no experimental data are available to provide conductance values that exist between remodeled individual myocytes with lateralized gap junctions. Interestingly, in cell pairs isolated from the healing epicardial border zone of canine infarcts, Yao et al. demonstrated no reduction in total gap junction conductance of end–end coupled cells;[123] however, a 90% decrease was observed in side–side electrical coupling. Cellular lateralization, however, was not present in a majority of the cell pairs they measured. Although the pairs were removed from the epicardial border zone, they were not taken from a documented central common re-entrant path in which the myocytes demonstrate marked cellular lateralization.[123]

Because the size of gap junction plaques changes during remodeling, an important functional component is the value of the effective gap junctional conductance for a given gap junction size. As shown in the model developed by Wilders and Jongsma,[135] electrical field effects caused by cytoplasmic access resistance produce a decrease in the effective gap junctional conductance of the total plaque. This effect becomes more prominent with increasing gap junction size. For example, for gap junctions with an area of 0.5 μm^2, the effective conductance is <50% of the value obtained by simply adding up the unitary conductances of all channels in a given plaque (uncorrected conductance), and for gap junctions >4 μm^2, the effective gap junction conductance is <20%.[136] Their analysis also indicates that changes in cell geometry, with accompanying changes in the resistance produced by differences in cytoplasmic volume, are more important contributors to anisotropic conduction velocities than commonly realized.[136]

Altered side–side coupling between groups of cardiomyocytes

In addition to altered connections between individual cells, long-term remodeling following ventricular infarction,[76] and with increasing age of human atrial bundles,[132] involves remodeling of the interstitium with the deposition of collagen that assumes a prominent role in uncoupling side-to-side fiber connections.[76,102] Fig. 25.14A illustrates that 3–10 weeks after canine ventricular infarction, in the border zone of the infarct there is deposition of collagen bundles running parallel to the long axis of the fibers. Ultrastructural studies have shown that the peri-infarct myocytes separated by interspersed bundles of collagen are connected to one another by fewer and shorter gap junctions than those of normal cells;[76] the lateral cell–cell connections are reduced by 75%, whereas primarily end-to-end connections are reduced by only 22%. Accordingly, the average number of cells connected to individual myocytes is reduced in an anisotropic manner; i.e. side-to-side coupling between myocytes is disrupted, while little change occurs in end-to-end cellular coupling.

We wish to emphasize that collagenous bundles encircling small groups of cells mark sites where it is not possible to have side-to-side electrical coupling between the cell groups. Thus, deposition of collagen bundles (microfibrosis) is associated with disruption of connections between cells with side-to-side apposition. Figure 25.14B illustrates the proliferation of connective tissue septa in a typical human atrial pectinate bundle older than 60 years. In older human atrial bundles, these septa are generally long, and completely surround groups of cells for variable lengths (sometimes over 1 mm).[102] The atrial cell groups are separated side-to-side by collagenous septa in Fig. 25.14B, and range in diameter from approximately 20 μm to 200 μm (2–6 cells wide). In normal atrial bundles of mature adults aged 20–40 years, the collagenous septa are short and only incompletely surround small groups of cells (not shown).[102] In these bundles, the gap junctions are located primarily at the ends of the myocytes, a feature in common with mature atrial myocytes.[21,110]

As noted previously, an associated gap junction remodeling feature of aging human atrial bundles

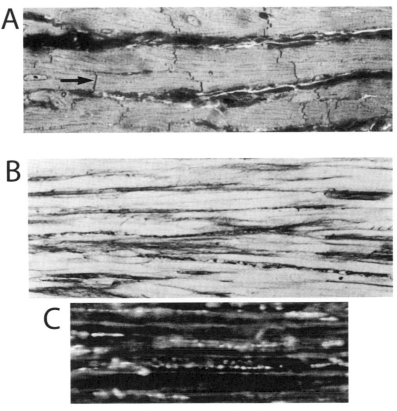

Figure 25.14. *Altered side–side connections between groups of cardiomyocytes secondary to collagen deposition. (A) Representative light micrograph of epicardial border zone adjacent to healed canine myocardial infarct. Myocytes in the border zone of the infarct are separated by interspersed collagen bundles (dark interstitial material) running parallel to the long axis of the myocytes with disruption of connections between cells in a side-to-side apposition. The use of potassium ferrocyanide enhanced sarcolemmal staining, thereby facilitating identification of cell borders.[76] The horizontal arrow marks an intercalated disc connecting two cells with end-to-end apposition. (B) Collagenous septa in an atrial bundle from a 64-year-old patient. The picrosirius technique used stains collagenous septa gray to black and leaves the background colorless.[137] (C) Cellular distribution of connexin43 in a representative section from a 76-year-old patient. Panel A modified with permission from Luke RA, Saffitz JE. Remodeling of ventricular conduction pathways in healed canine infarct border zones. J Clin Invest 1991; 87: 1594–602; Panel B modified with permission from Spach MS, Dolber PC, Heidlage JF. Influence of the passive anisotropic properties on directional differences in propagation following modification of the sodium conductance in human atrial muscle. A model of reentry based on anisotropic discontinuous propagation. Circ Res 1988; 62: 811–32; Panel C modified with permission from Spach MS, Heidlage JF, Dolber PC et al. Changes in anisotropic conduction caused by remodeling cell size and the cellular distribution of gap junctions and Na^+ channels. J Electrocardiol 2001; 34: 69–76.*

is that Cx43 becomes distributed over the entire surface of individual myocytes (Fig. 25.14C). However, the functional consequences of gap junction lateralization in these aging atrial myocytes are not clear. For example, whether or not some of the lateralized gap junctions are located in invaginations of the sarcolemma (where they cannot contribute to cell-to-cell transmission) is not known, and no data are available on the conductance values between any two myocytes within the discrete cell groups formed by collagenous septa.

Conduction changes associated with deposition of collagenous bundles

Transverse conduction at a macroscopic size scale in normal mature cardiac bundles appears as a uniform process with smooth extracellular potential waveforms, which provides a way to depict excitation spread across the fibers in the form of isochrones (Fig. 25.15A). Following the deposition of collagen bundles in the epicardial border zone of healed ventricular infarcts and in aging atrial bundles, transverse conduction becomes markedly discontinuous and non-uniform. The non-uniform nature of transverse propagation in these tissues is evidenced by the complex multiphasic[102,132] or fractionated[138,139] extracellular waveforms that are produced during TP (Fig. 25.15B, positions 1 and 2). Additionally, although the membrane resting potential remains normal with fast action potential upstrokes, the effective transverse conduction velocity decreases to values as low as the very slow conduction that occurs in the AV node. For example, the average TP velocity of the young atrial bundles is 0.12 m/s, whereas in the aging bundles with microfibrosis the average TP velocity decreases to 0.08 m/s.[132]

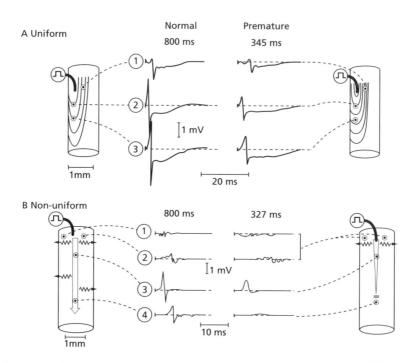

Figure 25.15. *Representative propagation responses to premature action potentials in (A) young mature atrial bundles with uniform anisotropic properties, and (B) older atrial bundles with nonuniform properties. In each panel, normal excitation sequence and a few extracellular waveforms are shown on the left, and those of an early premature beat are shown on the right. A box above each set of waveforms shows the interstimulus interval (m). (A) Isochrones are drawn with a 1 ms difference. (B) The transverse excitation spread was so complex that isochrones could not be constructed. 'Sawtooth' curves represent the irregular course of transverse excitation spread. The elongated arrow in the normal sequence represents the narrow region of fast longitudinal conduction. On the right, the elongated triangle represents decremental conduction to block. Modified with permission from Spach MS, Dolber PC, Heidlage JF. Influence of the passive anisotropic properties on directional differences in propagation following modification of the sodium conductance in human atrial muscle. A model of reentry based on anisotropic discontinuous propagation. Circ Res 1988; 62: 811–32.*

Another major functional consequence of the deposition of collagenous septa is that the anisotropic propagation response of premature action potentials is altered (Fig. 25.15). When the premature stimulus interval is shortened in mature adult uniform anisotropic bundles, the extracellular potential waveforms decrease in peak-to-peak amplitude, but they maintain the same general shape and smooth contour (Fig. 25.15A). The even contours of the extracellular waveforms and the isochrones of premature beats indicate that propagation at a microscopic size scale continues in all directions as a relatively smooth process. In older non-uniform anisotropic atrial bundles, however, progressive shortening of the premature interval results in abnormalities of longitudinal conduction, while stable but slower transverse conduction continues. As illustrated in Fig. 25.15B, progressive shortening of the premature interval in atrial bundles with prominent collagenous septa produced unidirectional conduction block along the long axis of the fibers, as evidenced by the progressive disappearance of extracellular deflections in that direction (Fig. 25.15, positions 3 and 4). Simultaneously, however, slow but markedly discontinuous conduction continued across the fibers (Fig. 25.15B, positions 1 and 2). That the unidirectional block was due to anisotropic differences in coupling between cells, or groups of cells, is illustrated by the fact that the same unidirectional longitudinal block occurs when the stimulus site is shifted to other locations on the bundle. This feature would not occur if the unidirectional block was caused by spatial differences of membrane ionic properties within the bundles.

The conduction abnormalities depicted in Fig. 25.15 illustrate only a few of the conduction disturbances now known to occur following changes from the normal state of cell–cell electrical coupling.[130,131,140] A hallmark of tissues in which there is an increasing propensity to develop re-entrant circuits is the ability to initiate unidirectional propagation failure in one area, with slowing of conduction in other areas; e.g. the progressive slowing of conduction decreases the size of the structure needed to contain a re-entrant circuit. Figure 25.16

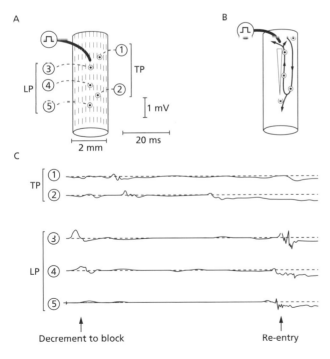

Figure 25.16. *Re-entry in a small area in response to premature action potentials in an older, non-uniform anisotropic atrial bundle. The basic stimulus rate was 171/min. (A) The premature stimulus (symbol) was delivered at the same site as that used for the basic rate. The numbers indicate the locations where each waveform was recorded. (B) Solid lines with arrows indicate the perimeter of the re-entrant circuit. The elongated triangle indicates initial decremental longitudinal conduction to the point of conduction failure. (C) Waveforms at corresponding recording sites that demonstrate unidirectional longitudinal block in the presence of continued very slow transverse discontinuous conduction, followed by re-entry. LP, longitudinal propagation; TP, transverse propagation. Modified with permission from Spach MS, Dolber PC, Heidlage JF. Influence of the passive anisotropic properties on directional differences in propagation following modification of the sodium conductance in human atrial muscle. A model of reentry based on anisotropic discontinuous propagation. Circ Res 1988; 62: 811–32.*

illustrates that these conduction disturbances related to the abnormal anisotropic distribution of side-to-side connections can lead to re-entry within areas less than 2 mm^2 in aging atrial bundles. In mature atrial bundles with normal side-to-side cellular coupling, however, premature action potentials did not produce unidirectional longitudinal block to set the stage for re-entry.

Conclusion

This chapter illustrates that the multiple factors that produce remodeling of cardiac gap junctions are important in the origin of cardiac arrhythmias. We have described the basis of cell–cell communication, beginning with the molecular structure of cardiac connexins, and going on to factors that control their expression and turnover, as well as noting new insights provided by the selective genetic manipulation of cardiac connexins. We also have presented some of the information available as to how cell-to-cell communication is altered by remodeling the cellular distribution of gap junctions during normal postnatal development to the normal mature adult state, followed by changes in gap junction distribution that occur with disease and aging. An important feature of remodeling of the cellular distribution of gap junctions is that there can be accompanying changes in cell geometry and deposition of collagenous septa in the interstitium. These accompanying changes can have considerable effects on anisotropic conduction, making it possible for conduction abnormalities that lead to re-entrant arrhythmias to occur. The results imply future chemical and genetic therapeutic strategies that directly alter the expression and turnover of cardiac connexins. Additionally, regulation of the deposition and turnover of interstitial collagen provides an important target for enhancement of side-to-side electrical coupling between small groups of cells when microfibrosis occurs in abnormal states.

Acknowledgements

This work was supported by the National Heart, Lung, and Blood Institute of the NIH (Grants HL 50537 (MSS), and HL50598 and HL58507 (JES)).

References

1. Saffitz JE, Yeager M. Intracardiac cell communication and gap junctions. In Spooner PM, Rosen M (eds). Foundations of Cardiac Arrhythmias. New York: Marcel Dekker; 2000: 171–204.
2. Makowski L, Caspar DL, Phillips WC et al. Gap junction structures. II. Analysis of the x-ray diffraction data. J Cell Biol 1977; 74: 629–45.
3. Sohl G, Willecke K. An update on connexin genes and their nomenclature in mouse and man. Cell Commun Adhes 2003; 10: 173–80.
4. Kanno S, Saffitz JE. The role of myocardial gap junctions in electrical conduction and arrhythmogenesis. Cardiovasc Pathol 2001; 10: 169–77.
5. Unger VM, Kumar NM, Gilula NB et al. Three-dimensional structure of a recombinant gap junction membrane channel. Science 1999; 283: 1176–80.
6. EC Beyer. Molecular cloning and developmental expression of two chick embryo gap junction proteins. J Biol Chem 1990; 265; 14439–43.
7. Kanter HL, Saffitz JE, Beyer EC. Cardiac myocytes express multiple gap junction proteins. Circ Res 1992; 70; 438–44.
8. Kanter HL, Laing JG, Beyer EC et al. Multiple connexins colocalize in canine ventricular myocyte gap junctions. Circ Res 1993; 73: 344–50.
9. Reed EK, Westphale EM, Larson DM et al. Molecular cloning and functional expression of human connexin37, an endothelial cell gap junction protein. J Clin Invest 1993; 91: 997–1004.
10. Gourdie RG, Green CR, Severs NJ et al. Immunolabeling patterns of gap junction connexins in the developing and mature rat heart. Anat Embryol 1992; 185: 363–78.
11. Paul DL, Ebihara L, Takemoto LJ et al. Connexin46, a novel lens gap junction protein, induces voltage-gated currents in nonjunctional plasma membrane of *Xenopus* oocytes. J Cell Biol 1991; 115: 1077–89.
12. van Kempen MJ, Fromaget C, Gross D et al. Spatial distribution of connexin43, the major cardiac gap junction protein, in the developing and adult rat heart. Circ Res 1991; 68: 1638–51.
13. Kanter HL, Laing JG, Beau SL et al. Distinct patterns of connexin expression in canine Purkinje fibers and ventricular muscle. Circ Res 1993; 72: 1124–31.
14. Davis LM, Kanter HL, Beyer EC et al. Distinct gap junction protein phenotypes in cardiac tissues with

disparate conduction properties. J Am Coll Cardiol 1994; 24: 1124–32.

15. Nikolski VP, Jones SA, Lancaster MK et al. Cx43 and dual-pathway electrophysiology of the atrioventricular node and atrioventricular nodal reentry. Circ Res 2003; 92: 469–75.

16. Anumonwo JMB, Wang H-Z, Trabka-Janik E et al. Gap junctional channels in adult mammalian sinus nodal cells: Immunolocalization and electrophysiology. Circ Res 1992; 71: 229–39.

17. Kwong KF, Schuessler RB, Green KG et al. Differential expression of gap junction proteins in the canine sinus node. Circ Res 1998; 82: 604–12.

18. Bastide B, Neyses L, Ganten D et al. Gap junction protein connexin40 is preferentially expressed in vascular endothelium and conductive bundles of rat myocardium and is increased under hypertensive conditions. Circ Res 1993; 73: 1138–49.

19. Gros D, Jarry-Guichard T, Ten Velde I et al. Restricted distribution of connexin40, a gap junctional protein, in mammalian heart. Circ Res 1994; 74: 839–51.

20. Gourdie RG, Severs NG, Green CR et al. The spatial distribution and relative abundance of gap-junctional connexin40 and connexin43 correlate to functional properties of components of the cardiac atrioventricular conduction system. J Cell Sci 1993; 105: 985–91.

21. Saffitz JE, Kanter HL, Green KG et al. Tissue-specific determinants of anisotropic conduction velocity in canine atrial and ventricular myocardium. Circ Res 1994; 74: 1065–70.

22. Tamaddon HS, Vaidya D, Simon AM et al. High-resolution optical mapping of the right bundle branch in connexin40 knockout mice reveals slow conduction in the specialized conduction system. Circ Res 2000; 87: 929–36.

23. Coppen SR, Dupont E, Rothery S et al. Connexin45 expression is preferentially associated with the ventricular conduction system in mouse and rat heart. Circ Res 1998; 82: 232–43.

24. Moreno AP, Sáez JC, Fishman GI et al. Human connexin43 gap junction channels - Regulation of unitary conductances by phosphorylation. Circ Res 1994; 74: 1050–7.

25. Brink PR, Ramanan SV, Christ GJ. Human connexin43 gap junction channel gating: evidence for mode shifts and/or heterogeneity. Am J Physiol Cell Physiol 1996; 271: C321–31.

26. Veenstra RD. Size and selectivity of gap junction channels formed from different connexins. J Bioenerg Biomembr 1996; 28: 317–37.

27. Moreno AP, Fishman GI, Spray DC. Phosphorylation shifts unitary conductance and modifies voltage dependent kinetics of human connexin43 gap junction channels. Biophys J 1992; 62: 51–3.

28. Beblo DA, Wang H-Z, Beyer EC et al. Unique conductance, gating, and selective permeability properties of gap junction channels formed by connexin40. Circ Res 1995; 77: 813–22.

29. Bukauskas FF, Elfgang C, Willecke K et al. Biophysical properties of gap junction channels formed by mouse connexin40 in induced pairs of transfected human HeLa cells. Biophys J 1995; 68: 2289–98.

30. Beblo DA, Veenstra RD. Monovalent cation permeation through the connexin40 gap junction channel. Cs, Rb, K, Na, Li, TEA, TNA, TBA and effects of anions Br, Cl, F, acetate, aspartate, glutamate, and NO$_3$. J Gen Physiol 1997; 104: 509–22.

31. Veenstra RD, Wang H-Z, Beyer EC et al. Selective dye and ionic permeability of gap junction channels formed by connexin45. Circ Res 1994; 75: 483–90.

32. Kumar NM, Gilula NB. The gap junction communication channel. Cell 1996; 84: 381–8.

33. Steiner E, Ebihara L. Functional characterization of canine connexin45. J Membr Biol 1996; 150: 153–61.

34. Moreno AP, Fishman GI, Beyer EC et al. Voltage dependent gating and single channel analysis of heterotypic channels formed by Cx45 and Cx43. Prog Cell Res 1995; 4: 405–8.

35. Reaume AG, de Sousa PA, Kulkarni S et al. Cardiac malformation in neonatal mice lacking connexin43. Science 1995; 267: 1831–4.

36. Betsuyaku T, Kovacs A, Saffitz JE et al. Structure and function in young and senescent mice heterozygous for a connexin43 null mutation. J Molec Cell Cardiol 2002; 34: 175–84.

37. Saffitz JE, Green KG, Kraft WJ et al. Effects of diminished expression of connexin43 on gap junction number and size in ventricular myocardium. Am J Physiol Heart Circ Physiol 2000; 278: H1662–70.

38. Guerrero PA, Schuessler RB, Davis LM et al. Slow ventricular conduction in mice heterozygous for a Cx43 null mutation. J Clin Invest 1997; 99: 1991–8.

39. Thomas SA, Schuessler RB, Berul CI et al. Disparate effects of deficient expression of connexin43 on atrial and ventricular conduction: Evidence for chamber-specific molecular determinants of conduction. Circulation 1998; 97: 686–91.

40. Morley GE, Vaidya D, Samie FH et al. Characterization of conduction in the ventricles of normal and heterozygous Cx43 knockout mice using

optical mapping. J Cardiovasc Electrophysiol 1999; 10: 1361–75.

41. Thomas SP, Kucera JP, Bircher-Lehmann L et al. Impulse propagation in synthetic strands of neonatal cardiac myocytes with genetically reduced levels of connexin43. Circ Res 2003; 92: 1209–16.

42. Gutstein DE, Morley GE, Tamaddon H et al. Conduction slowing and sudden arrhythmic death in mice with cardiac-restricted inactivation of connexin43. Circ Res 2001; 88: 333–9.

43. Eckardt D, Theis M, Degen J et al. Functional role of connexin43 gap junction channels in adult mouse heart assessed by inducible gene deletion. J Mol Cell Cardiol 2004; 36: 101–10.

44. Gutstein DE, Morley GE, Vaidya D et al. Heterogeneous expression of gap junction channels in the heart leads to conduction defects and ventricular dysfunction. Circulation 2001; 104: 1194–9.

45. Yao JA, Gutstein DE, Liu F et al. Cell coupling between ventricular myocyte pairs from connexin43-deficient murine hearts. Circ Res 2003; 93: 736–43.

46. Simon AM, Goodenough DA, Paul DL. Mice lacking connexin40 have cardiac conduction abnormalities characteristic of atrioventricular block and bundle branch block. Curr Biol 1998; 8: 295–8.

47. Hagendorff A, Schumacher B, Kirchoff S et al. Conduction disturbances and increased atrial vulnerability in connexin40-deficient mice analyzed by transesophageal stimulation. Circulation 1999; 99: 1508–15.

48. Kumai M, Nishii K, Nakamura K et al. Loss of connexin45 causes a cushion defect in early cardiogenesis. Development 2000; 127: 3501–12.

49. Alcoléa S, Jarry-Guichard T, de Bakker J et al. Replacement of connexin40 by connexin45 in the mouse: impact on cardiac electrical conduction. Circ Res 2004; 94: 100–9.

50. Laird DW, Puranam KL, Revel JP. Turnover and phosphorylation dynamics of connexin43 gap junction protein in cultured cardiac myocytes. Biochem J 1991; 273: 67–72.

51. Darrow BJ, Laing JG, Lampe PD et al. Expression of multiple connexins in cultured neonatal rat ventricular myocytes. Circ Res 1995; 76: 381–7.

52. Beardslee MA, Laing JG, Beyer EC et al. Rapid turnover of connexin43 in the adult rat heart. Circ Res 1998; 83: 629–35.

53. Saffitz JE, Laing JG, Yamada KA. Connexin expression and turnover. Implications for cardiac excitability. Circ Res 2000; 86: 723–8.

54. Jordan K, Solan JL, Dominguez M et al. Trafficking, assembly and function of a connexin43-green fluo-rescent protein chimera in live mammalian cells. Mol Biol Cell 1999; 10: 2033–50.

55. Gaietta G, Deerinck TJ, Adams SR et al. Multicolor and electron microscopic imaging of connexin trafficking. Science 2002; 296: 503–7.

56. Lauf U, Giepmans BNG, Lopez P et al. Dynamic trafficking and delivery of connexins to the plasma membrane and accretion of gap junctions in living cells. Proc Natl Acad Sci USA 2002; 99: 10446–51.

57. Martin PEM, Blundell G, Ahmad S et al. Multiple pathways in the trafficking and assembly of connexin 26, 32 and 43 into gap junction intercellular communication channels. J Cell Sci 2001; 114: 3845–55.

58. Laing JG, Beyer EC. The gap junction protein connexin43 is degraded via the ubiquitin proteasome pathway. J Biol Chem 1995; 270: 26399–403.

59. Laing JG, Tadros P, Westphale EM et al. Degradation of connexin43 gap junctions involves both the proteasome and the lysosome. Exp Cell Res 1997; 236: 483–92.

60. Laing JG, Tadros PN, Green K et al. Proteolysis of connexin43-containing gap junctions in normal and heat-stressed cardiac myocytes. Cardiovasc Res 1998; 38: 711–8.

61. Lampe PD, Lau AF. Regulation of gap junctions by phosphorylation of connexins. Arch Biochem Biophys 2000; 384: 5–15.

62. TenBroek E, Lampe PD, Solan JL et al. Ser364 of connexin43 and the upregulation of gap junction assembly by cAMP. J Cell Biol 2001; 155: 1307–18.

63. Cooper CD, Lampe PD. Casein kinase 1 regulates connexin-43 gap junction assembly. J Biol Chem 2002; 277: 44962–8.

64. Morley GE, Taffet SM, Delmar M. Intramolecular interactions mediate pH regulation of connexin43 channels. Biophys J 1996; 70: 1294–1302.

65. Duffy HS, Sorgen PL, Girvin ME et al. pH-dependent intramolecular binding and structure involving Cx43 cytoplasmic domains. J Biol Chem 2002; 277: 39706–14.

66. Duffy HS, Ashton AW, O'Donnell P et al. Regulation of connexin43 protein complexes by intracellular acidification. Circ Res 2004; 94: 215–22.

67. Toyofuku T, Akamatsu Y, Zhang H et al. c-Src regulates the interaction between connexin-43 and ZO-1 in cardiac myocytes. J Biol Chem 2001; 276: 1780–8.

68. Singh D, Lampe PD. Identification of connexin-43 interacting proteins. Cell Commun Adhes 2003; 10: 215–20.

69. Laing JG, Manley-Markowski RN, Koval M et al.

Connexin45 interacts with zonula occludens-1 in osteoblastic cells. Cell Commun Adhes 2001; 8: 209–12.

70. Hoyt RH, Cohen ML, Saffitz JE. Distribution and three-dimensional structure of intercellular junctions in canine myocardium. Circ Res 1989; 64: 565–74.

71. Severs NJ. The cardiac gap junction and intercalated disc. Int J Cardiol 1990; 26: 137–73.

72. Page E. Cardiac gap junctions: In: Fozzard HA, Haber E, Jenings RB, Katz AM, Morgan HE (eds). The Heart and Cardiovascular System. New York: Raven Press; 1992: 1003–47.

73. Kaplan SR, Gard JJ, Carvajal-Huerta L et al. Structural and molecular pathology of the heart in Carvajal syndrome. Cardiovasc Pathol 2004; 13: 26–32.

74. Kaplan SR, Gard JJ, Protonotarios N et al. Remodeling of gap junctions in arrhythmogenic right ventricular cardiomyopathy due to a deletion in plakoglobin (Naxos disease). Heart Rhythm 2004; 1: 3–11.

75. Peters NS, Severs NJ, Rothery SM et al. Spatiotemporal relation between gap junctions and fascia adherens junctions during postnatal development of human ventricular myocardium. Circulation 1994; 90: 713–25.

76. Luke RA, Saffitz JE. Remodeling of ventricular conduction pathways in healed canine infarct border zones. J Clin Invest 1991; 87: 1594–602.

77. Spach MS, Miller WT, Geselowitz DB et al. The discontinuous nature of propagation in normal canine cardiac muscle. Evidence for recurrent discontinuities of intracellular resistance that affect the membrane currents. Circ Res 1981; 48: 39–54.

78. Lerner DL, Yamada KA, Schuessler RB et al. Accelerated onset and increased incidence of ventricular arrhythmias induced by ischemia in Cx43-deficient mice. Circulation 2000; 101: 547–52.

79. Burt JM. Block of intercellular communication: interaction of intracellular H+ and Ca2+. Am J Physiol Cell Physiol 1987, 352: C607–12.

80. White RL, Doeller JE, Verselis VK et al. Gap junctional conductance between pairs of ventricular myocytes is modulated synergistically by H+ and Ca2+. J Gen Physiol 1990; 95: 1061–75.

81. Burt JM, Massey KD, Minnich BN. Uncoupling of cardiac cells by fatty acids: Structure-activity relationship. Am J Physiol Cell Physiol 1991; 260: C439–48.

82. Wu J, McHowat J, Saffitz JE et al. Inhibition of gap junctional conductance by long-chain acylcarnitines and their preferential accumulation in junctional sarcolemma during hypoxia. Circ Res 1993; 72: 879–89.

83. Beardslee MA, Lerner DL, Tadros PN et al. Dephosphorylation and intracellular redistribution of ventricular Cx43 during electrical uncoupling induced by ischemia. Circ Res 2000; 87: 656–62.

84. Jain SK, Schuessler RB, Saffitz JE. Mechanisms of delayed electrical uncoupling induced by ischemic preconditioning. Circ Res 2003; 92: 1138–44.

85. García-Dorado D, Rodríguez-Sinovas A, Ruiz-Meana M. Gap junction-mediated spread of cell injury and death during myocardial ischemia–reperfusion. Cardiovasc Res 2004; 61: 386–401.

86. Kanno S, Kovacs A, Yamada KA et al. Connexin43 as a determinant of infarct size following coronary occlusion in mice. J Am Coll Cardiol 2003; 41: 681–6.

87. BJ Darrow, VG Fast, Kléber AG et al. Functional and structural assessment of intercellular communication: increased conduction velocity and enhanced connexin expression in dibutyryl cAMP-treated cultured cardiac myocytes. Circ Res 1996; 79: 174–83.

88. Dodge SM, Beardslee MA, Darrow BJ et al. Effects of angiotensin II on expression of the gap junction channel protein connexin43 in neonatal rat ventricular myocytes. J Am Coll Cardiol 1998; 32: 800–7.

89. Zhuang J, Yamada KA, Saffitz JE et al. Pulsatile stretch remodels cell-to-cell communication in cultured myocytes. Circ Res 2000; 87: 316–22.

90. Pimentel RC, Yamada KA, Kléber AG et al. Autocrine regulation of myocyte Cx43 expression by VEGF. Circ Res 2002; 90: 671–7.

91. Shyu KG, Chen CC, Wang BW et al. Angiotensin II receptor antagonist blocks the expression of connexin43 induced by cyclical mechanical stretch in cultured neonatal rat cardiac myocytes. J Mol Cell Cardiol 2001; 33: 691–8.

92. Peters NS, Green CR, Poole-Wilson PA et al. Reduced content of connexin43 gap junctions in ventricular myocardium from hypertrophied and ischemic human hearts. Circulation 1993; 88: 864–75.

93. Peters NS. New insights into myocardial arrhythmogenesis: distribution of gap junctional coupling in normal, ischaemic and hypertrophied human hearts. Clin Sci 1996; 90: 447–52.

94. Kaprielian RR, Gunning M, Dupont E et al. Downregulation of immunodetectable connexin43 and decreased gap junction size in the pathogenesis of chronic hibernation in the human left ventricle. Circulation 1998; 97: 651–60.

95. Yamada KA, Rogers JW, Sundset R et al. Upregulation of connexin45 in heart failure. J Cardiovasc Electrophysiol 2003; 14: 1205–12.

96. Petrich BG, Gong X, Lerner D et al. c-Jun N-terminal kinase activation mediates downregulation of connexin43 in cardiomyocytes. Circ Res 2002; 91: 640–7.

97. Petrich BG, Eloff BC, Lerner DL et al. Targeted activation of c-Jun N-terminal kinase in vivo induces cardiomyopathy and conduction slowing. J Biol Chem 2004; 279: 15330–8.

98. Spooner PM, Joyner RW, Jalife J (eds). Discontinuous Conduction in the Heart. Armonk, NY: Futura Publishing Co, Inc.:1997.

99. Severs NJ. Pathophysiology of gap junctions in heart disease. J Cardiovasc Electrophysiol 1994; 5: 462–75.

100. Gourdie RG, Green CR, Severs NJ et al. Immuno-labeling patterns of gap junction connexins in the developing and mature rat heart. Anat Embryol 1992; 185: 363–78.

101. Angst BD, Khan LUR, Severs NJ et al. Dissociated spatial patterning of gap junctions and cell adhesion junctions during postnatal differentiation of ventricular myocardium. Circ Res 1997; 80: 88–94.

102. Spach MS, Dolber PC. Relating extracellular potentials and their derivatives to anisotropic propagation at a microscopic level in human cardiac muscle: evidence for electrical uncoupling of side-to-side fiber connections with increasing age. Circ Res 1986; 58: 356–71.

103. Spach MS, Heidlage JF, Dolber PC et al. Changes in anisotropic conduction caused by remodeling cell size and the cellular distribution of gap junctions and Na$^+$ channels. J Electrocardiol 2001; 34: 69–76.

104. Gourdie RG, Litchenberg WH, Eisenberg LM. Gap junctions and heart development. In: De Mello WC, Janse MJ (eds). Heart Cell Communication in Health and Disease. Norwell, MA: Kluwer Academic Publishers Group; 1998: 19–43.

105. Rohr S, Scholly DM, Kléber AG. Patterned growth of neonatal rat heart cells in culture: morphological and electrophysiological characterization. Circ Res 1991; 68: 114–30.

106. Fast VG, Kléber AG. Microscopic conduction in cultured strands of neonatal rat heart cells measured with voltage-sensitive dyes. Circ Res 1993; 73: 914–25.

107. Fast VG, Darrow BJ, Saffitz JE et al. Anisotropic activation spread in heart cell monolayers assessed by high-resolution optical mapping. Role of tissue discontinuities. Circ Res 1996; 79: 115–27.

108. Soonpaa MII, Koh GY, Klug MG et al. Formation of nascent intercalated discs between grafted fetal cardiomyocytes and host myocardium. Science 1994; 262: 98–101.

109. Rubart M, Pasumarthi KBS, Nakajima H et al. Physiological coupling of donor and host cardiomyocytes after cellular transplantation. Circ Res 2003; 92: 1217–24.

110. Litchenberg WH, Norman LW, Holwell AK et al. The rate and anisotropy of impulse propagation in the postnatal terminal crest are correlated with the remodeling of Cx43 gap junction pattern. Cardiovasc Res 2000; 45: 379–87.

111. Dolber PC, Beyer EC, Junker JL et al. Distribution of gap junctions in dog and rat ventricle studied with a double-label technique. J Mol Cell Cardiol 1992; 24: 1443–57.

112. Fast VG, Kléber AG. Anisotropic conduction in monolayers of neonatal rat heart cells cultured on collagen substrate. Circ Res 1994; 75: 591–5.

113. Spach MS, Heidlage JF, Dolber PC et al. Electrophysiological effects of remodeling cardiac gap junctions and cell size: experimental and model studies of normal cardiac growth. Circ Res 2000; 86: 302–11.

114. Spach MS, Heidlage JF, Darken ER et al. Cellular dV/dt$_{max}$ reflects both membrane properties and the load presented by adjoining cells. Am J Physiol Heart Circ Physiol 1992; 263: H1855–63.

115. Spach MS, Heidlage JF, Dolber PC. The dual nature of anisotropic discontinuous conduction in the heart. In: Zipes DP, Jalife J (eds). Cardiac Electrophysiology: From Cell to Bedside (3e). Philadelphia, PA, WB Saunders Company; 2000: 213–22.

116. Spach MS. Discontinuous cardiac conduction: its origin in cellular connectivity with long- term adaptive changes that cause arrhythmias. In: Spooner P, Joyner RW, Jalife J (eds). Discontinuous Conduction in the Heart. Armonk, NY: Futura Publishing Co. Inc.; 1997: 5–51.

117. Spach MS, Heidlage JF. The stochastic nature of cardiac propagation at a microscopic level: an electrical description of myocardial architecture and its application to conduction. Circ Res 1995; 76: 366–80.

118. Clerc L. Directional differences of impulse spread in trabecular muscle from mammalian heart. J Physiol 1976; 255: 335–46.

119. Spach MS, Heidlage JF, Dolber PC et al. Extracellular discontinuities in cardiac muscle: Evidence for capillary effects on the action potential foot. Circ Res 1998; 83: 1144–64.

120. Akar FG, Roth BJ, Rosembaum DS. Optical meas-

urement of cell–cell coupling in intact heart using subtreshold electrical stimulation. Am J Physiol Heart Circ Physiol 2001; 281: H533–42.

121. Socolar SJ: The coupling coefficient as an index of junctional conductance. J Membr Biol 1977; 34: 29–37.

122. Kieval RS, Spear JF, Moore EN. Gap junctional conductance in ventricular myocyte pairs isolated from postischemic rabbit myocardium. Circ Res 1992; 71: 127–36.

123. Yao J-A, Hussain W, Patel P et al. Remodeling of gap junctional channel function in epicardial border zone of healing canine infarcts. Circ Res 2003; 92: 437–43.

124. Spach MS and Heidlage JF: A multidimensional model of cellular effects on the spread of electrotonic currents and on propagating action potentials. In: Pilkington TC, Loftis B, Thompson JF et al. (eds). High-Performance Computing in Biomedical Research. Boca Raton: CRC Press; 1993: 289–318.

125. Weidmann S: Electrical constants of trabecular muscle from mammalian heart. J Physiol 1970; 210: 1041–54.

126. Severs NJ. Pathophysiology of gap junctions in heart disease. J Cardiovasc Electrophysiol 1994; 5: 462–75.

127. Uzzaman M, Honjo H, Takagishi Y et al. Remodeling of gap junctional coupling in hypertrophied right ventricles of rat with monocrotaline-induced pulmonary hypertension. Circ Res 2000; 86: 871–8.

128. Sepp R, Severs N, Gourdie RG. Altered patterns of cardiac intercellular junction distribution in hypertrophic cardiomyopathy. Heart 1996; 76: 412–7.

129. Kostin S, Klein G, Szalay Z et al. Structural correlate of atrial fibrillation in human patients. Cardiovasc Res 2002; 54: 361–79.

130. Wit AL. Discontinuous conduction in the epicardial border zone of infarcted hearts and its role in anisotropic reentry. In: Spooner P, Joyner RW, Jalife J (eds). Discontinuous Conduction in the Heart. Armonk, NY: Futura Publishing Co. Inc.; 1997: 453–69.

131. Peters NS, Coromilas J, Severs NJ et al. Disturbed connexin43 gap junction distribution correlates with the location of reentrant circuits in the epicardial border zone of healing canine infarcts that cause ventricular tachycardia. Circulation 1997; 95: 988–96.

132. Spach MS, Dolber PC, Heidlage JF. Influence of the passive anisotropic properties on directional differences in propagation following modification of the sodium conductance in human atrial muscle. A model of reentry based on anisotropic discontinuous propagation. Circ Res 1988; 62: 811–32.

133. Matsushita T, Oyamada M, Fujimoto K et al. Remodeling of cell–cell and cell- extracellular matrix interactions at the border zone of rat myocardial infarcts. Circ Res 1999; 85: 1046–55.

134. Smith JH, Green CR, Peters NS et al. Altered patterns of gap junction distribution in ischemic heart disease: an immunohistochemical study of human myocardium using laser scanning confocal microscopy. Am J Pathol 1991; 139: 801–21.

135. Wilders R, Jongsma HJ. Limitations of the dual voltage clamp method in assaying conductance and kinetics of gap junction channels. Biophys J 1992; 63: 942–1053.

136. Jongsma HJ, Wilders R. Gap junctions in cardiovascular disease. Circ Res 2000; 86: 1193–7

137. Dolber PC, Spach MS. Picrosirius red staining of cardiac muscle following phosphomolybdic acid treatment. Stain Tech 1987; 62: 23–6.

138. Gardner PI, Ursell PC, Fenoglio JJ et al. Electrophysiologic and anatomic basis for fractionated electrograms recorded from healed myocardial infarcts. Circulation 1985: 72: 596–611.

139. Ciaccio EJ. Localization of the slow conduction zone during reentrant ventricular tachycardia. Circulation 2000; 102: 464–9.

140. Delmar M, Michaels DC, Johnsom T et al. Effects of increasing intercellular resistance on transverse and longitudinal propagation in sheep epicardial muscle. Circ Res 1987; 60: 780–5.

CHAPTER 26

Novel targets for arrhythmia prevention or treatment

Harry A Fozzard

Introduction

The huge growth in our understanding of the biology of the heart and the circulation has opened a vast number of ways to intervene in the pathophysiological processes related to arrhythmias at the cellular, molecular, and genetic level. Only careful study will show which of these many new approaches will be safe and successful. But I do not propose to offer here an encyclopedic tabulation of all possible ideas for intervention in the natural history of arrhythmogenic diseases. The chapters in this arrhythmia section and in the entire book already eloquently present many choices, and I do not have a crystal ball for divining which is the antiarrhythmic magic bullet. The unique contribution I may be able to bring to the discussion is a perspective or viewpoint, derived from a lifetime of struggle against the forces of sudden cardiac death. This has led me to a series of conclusions or lessons, derived from mistakes that we in the field have made. After defining the nature of the lethal arrhythmia problem, I propose to draw on the past to explain these conclusions, and then provide an example that illustrates a provocative and promising approach to arrhythmia prevention in heart failure.

I am enthusiastic that we are making great progress, but I am cautious when considering any specific approach. In the 50 years since I began my commitment to the study of arrhythmias, I have seen too many treatments advocated with great enthusiasm, but eventually shown to be useless or, more dramatically, even harmful. However understandable our haste to control lethal arrhythmias, for most of the last half-century it is important to realize that the control untreated group in trials of arrhythmia management was usually the safest place to be. Our most successful treatments have been only indirectly focused on arrhythmias.

Background

Arrhythmia is a disorder of the electrical system of the heart, resulting usually from intrinsic cardiac disease, but also from other physiological processes that disturb cardiac function. The cardiac electrical system is complicated, and must be described in multiple layers. Firstly, the cardiac physical geometry is complex, with clusters of primary and secondary pacemaker cells, atrial chambers with trabeculation and entry and exit openings, and ventricular chambers with papillary muscles and radically different shape between the right and the left sides, and the atrioventricular conduction system that connects them and coordinates ventricular excitation. This anatomy plays a large role in how the electrical signals are propagated. Secondly, at the cell level, the orchestration of a host of ion channels and pumps in the myocyte membrane generates characteristic action potentials for each muscle type. All elements of this membrane process are subject to modulation non-uniformly, by nervous, hormonal and metabolic mediators. Finally, the organ, cellular and membrane protein landscape is controlled by a genetic system that sets the stage for electrical function and modifies it continuously. It is a wonder that the cardiac electrical system is so amazingly stable in the face of the multitude of physiological and pathological challenges, rather than that it malfunctions occasionally. Clearly, there must be powerful compensatory mechanisms to keep all of these factors in balance. However, it only needs to malfunction once to result in lethal arrhythmia and death. To consider ways to maintain or restore system stability when it has been disturbed, we must consider all the organ, cellular and molecular complexity of the cardiac electrical system.

History

In this, as in any field with a long history, it is useful to understand some of the roots of our present challenge in the management of arrhythmias. In the period after the Second World War (WWII), the US experienced an epidemic of coronary heart disease. It was most dramatically manifested as sudden cardiac death, typically unexpected.[1] Myocardial infarction was increasing in frequency, and 25–35% of those admitted to the hospital with that diagnosis died during their stay. About half of these deaths were the result of ventricular arrhythmia, usually during the first 2–3 days of hospitalization. Perhaps an equal number died of some type of ventricular arrhythmia before arriving at the hospital. Cumulatively, this implied that 25–30% of deaths in the US were the result of lethal ventricular arrhythmia, a massive public health problem. To put this in perspective, this was about the same as the number of deaths from cancer. Although it was initially assumed that out-of-hospital sudden deaths were the result of myocardial infarction because of the high incidence of coronary disease, the methods used to detect pathological changes of infarction at autopsy were not sensitive, requiring many hours to develop. Consequently, examination of the hearts of these

out-of-hospital arrhythmia deaths was of limited value in determining the presence of a myocardial infarction of a few hours' duration. When enough of these patients were resuscitated outside the hospital and brought to the hospital, where more detailed *in vivo* studies could be done, we found that about two-thirds of them did not have acute myocardial infarction.[2] Almost all of them did have some cardiac disease, usually but not always coronary artery disease.[1]

The arrhythmias known to be associated with myocardial ischemia/infarction captured our attention because of their frequency, but that hid a substantial number of serious/lethal arrhythmias not resulting from myocardial ischemia. In those dying of arrhythmia outside of the hospital, or in those successfully resuscitated, about 20% had little or no coronary disease, but most of those had some other types of cardiac disease – left ventricular hypertrophy from hypertension, valvular disease, or cardiomyopathy. Only a small number had no signs of cardiac disease. Although I start discussing the ischemic arrhythmia problem, many of the concepts are appropriate to the other causes.

There were surprisingly few effective therapies for ventricular arrhythmias in the 1950s, but attention to this epidemic of cardiac death resulted in a variety of new approaches. Special nursing units were established (coronary care units), where patients' rhythms could be monitored during the critical first three days. This was first by visual observation of oscilloscope tracings, a very boring process, and then with computerized recognition programs. Bioengineering approaches were the first to have a significant impact on arrhythmias. They first yielded transthoracic, then transvenous, pacemakers, which addressed the problems of those with heart block or sinus failure. The direct current (DC) defibrillator replaced a flawed alternating current (AC) defibrillator system, which surgeons had been using for some years at the operating table. By delivering shocks timed to parts of the cardiac electrical cycle, it was useful for initial conversion of both ventricular and atrial arrhythmias. Oral quinidine and procainamide, and even intravenous procainamide, were used with little benefit, and in retrospect probably harm. Eventually, lidocaine and other ion channel blockers became available in the 1970s, but also with surprisingly little long-term benefit.[3]

Trying to see the wood among all the trees

The clustering of patients with myocardial ischemia/infarction in categorical nursing units not only focused our attention on the arrhythmias, but also on subtle features of hemodynamic compromise. There was initially considerable resistance to separating these patients with myocardial infarction into special nursing units, and especially to the idea of making invasive hemodynamic studies during the acute phase of myocardial infarction. The National Heart Lung and Blood Institute (NHLBI) initiated a huge increase in our understanding by funding ten special myocardial infarction research units, where intensive study of patients with acute myocardial infarction could be made. Use of simple blood pressure monitoring, pulmonary arterial and left atrial pressures via the Swan–Gans balloon catheter, and finally acute ventricular and coronary angiography showed that myocardial death did not occur at once, but progressed over hours or days.[4] Arrhythmias accelerated the hemodynamic deterioration, and hemodynamic deterioration caused arrhythmias. Circulatory support turned out to be our most effective antiarrhythmic tool, as well as beneficial in preservation of ischemic heart muscle. At the same time, control of heart rate and rhythm was hemodynamically beneficial. The separation of cardiac management into hemodynamic and electrical components turned out to be completely artificial. There was no sudden fall in death rate in coronary care units. Gradually, the death rate in hospitalized patients with myocardial infarction began to fall from about 30% (higher in acute care hospitals, where patients arrived earlier) to 10–15% by the 1980s.

A long-standing dispute for the origin of myocardial infarction was whether it resulted from progressive arteriosclerotic occlusion of vessels rather than formation of a clot in an otherwise fully patent vessel. After abortive efforts for many years because of inadequate drugs, thrombolytic therapy was widely used and showed conclusively that the occlusion was from a clot, with sometimes as little as 30–40% fixed occlusion. This resolved the dispute, but posed a serious problem for clinicians concerned with long-term management. Angioplasty or bypass surgery was available for high-grade fixed obstructions, and they were very useful for managing angina. But their effect on subsequent myocardial infarction was less impressive. Indeed, bypass surgery is not useful for those with 30–40% coronary obstruction, since the grafts tend to occlude postoperatively from inadequate flow. Nevertheless, two lessons for us were obvious: (1) if there is an acute loss of blood flow to a region of the heart, the best treatment is to restore the flow as quickly as possible; and (2) if there is loss of an effective cardiac rhythm, the best treatment is to restore the rhythm as quickly as possible.

Subsequent studies, finally bringing cardiology into the era of large-scale multicenter controlled trials, have focused on whether the best approach to rapid restoration of flow is thrombolytic therapy or angioplasty. Gradually, all of these approaches have produced a decline in the in-hospital mortality for acute myocardial infarction from ~35% in 1964 to ~5% in 2004. Although dramatic, this success has resulted in exaggeration of another problem. What happened to those patients who otherwise would have died during the acute phase of myocardial infarction? However desirable their survival, it is not reasonable to consider them cured. Rather, they often had damaged hearts and eventually developed cardiac failure. We soon discovered that patients who had sudden cardiac arrest, but who did not progress to transmural infarction, had a worse post-discharge prognosis over the next year.[5] It still seems paradoxical that transmural myocardial infarction, if survived, protected against lethal arrhythmias in the short term, but it fitted with the well-known observation that transmural infarction could alleviate chronic angina. Again, it became obvious that ongoing cardiac ischemia was the highly arrhythmogenic factor, and that some intervention to improve blood flow was necessary. What is the present situation, after 40 years of intense effort to reduce lethal arhythmias? The Center for Disease Control now estimates that sudden cardiac death represents 18–20% of total deaths in the US,[6] although the average age of death has shifted from the 50s to the 70s.

'Out of the blue' lethal ventricular arrhythmia

Although we made great progress in the care of hospitalized patients, the problem remained that a large number of patients died unexpectedly outside the hospital, where our antiarrhythmic therapies were useless.[7] Three strategies to address this problem were (1) to bring treatment to the victim when the arrhythmia occurs; (2) to identify individuals at risk and pretreat them with some effective agent; and (3) to determine risk factors for the disease itself and reduce them. All of these strategies have proved useful to varying degrees. One of the first approaches was to increase the chances that an observed sudden death could be treated on the spot through teaching of cardiopulmonary resuscitation and rapid response of emergency medical personnel. The community emergency medical teams continue to play an important community role, but the success rate has been too low to consider this the answer to sudden cardiac death.[8] Following the lesson described above, we have begun placing portable defibrillators in places where large numbers of people are gathered. The important lesson from this entire effort is that after cardiac arrest has occurred, restoration of rhythm quickly is the only effective treatment, and for the large number of those at risk, this is very difficult to achieve outside the hospital.

A huge amount of effort has been invested in identifying individuals at risk of arrhythmia, so that preventive therapies could be used. With the exception of rare genetic causes, adequate prediction has been limited to patients who have survived sudden death or myocardial infarction. This effort has been frustrating, but with some important successes. In the mid-1970s the NHLBI sponsored a 'thinktank' meeting of some 100 cardiac scientists to consider if any drugs available at that time would be useful. This led directly in 1977 to institution of the β-adrenergic blocker (propranalol) trial in patients who survived acute myocardial infarction.[9] It and several similar trials demonstrated that β-blockers were safe for long-term use in most patients and reduced the rate of death substantially. The mechanism of β-blocker benefit is not entirely clear, even now,[10,11] but its success means that understanding of the mechanism would be of great value in planning new interventions. Subsequently, a study of drug side-effects serendipitously and unexpectedly suggested that aspirin would reduce myocardial infarction and death rate in patients after acute myocardial infarction.[12] The mechanism of action of aspirin is probably to stabilize platelets, but because inflammation play a part in plaque rupture, alternative mechanisms are possible. Both treatments are now considered standard therapy, but this fact reveals another major problem. Although considered standard, a surprisingly large number of eligible patients are not prescribed or do not take these drugs. The implantable defibrillator was initially used only for the small number of survivors of sudden death, but gradually with careful controlled trials the criteria for its use have been expanded.[13] Nevertheless, this is not a satisfactory approach for the large number of individuals with modest risk. The beneficial use of ablation to cure certain atrial arrhythmias has not yet been successful for potentially lethal ventricular arrhythmias. To our great disappointment, antiarrhythmic drugs that act on ion channels have so far proved to be either useless or even harmful,[14] and there are a host of drugs used for other medical problems that turn out to have a harmful effect; demonstrated again by the recent withdrawal of Vioxx from the market because of increased cardiovascular risk. The lesson from these huge efforts is that most individuals in the population who are at risk of sudden cardiac death cannot be reliably identified, and that large-scale drug treatment can be risky, expensive and/or difficult to implement.

Finally, the third strategy is to reduce the disease or its manifestations that set the stage for the lethal arrhythmia. This has unquestionably been the most successful approach, although it was very slow in coming. Early efforts to reduce coronary arteriosclerosis by lowering serum cholesterol were failures. One dramatic one during the 1950s was a blocker of cholesterol metabolism, which reduced levels of cholesterol. However, the atheromatous plaques simply filled with the cholesterol precursor. The Coronary Drug Project used other methods, such as a metabolically inactive thyroid hormone and estrogen, only to find that they increased the death rate in survivors of myocardial infarction.[15] This was followed by the more successful Multiple

Risk Factor Intervention Trial (MRFIT) program, which sought to reduce smoking, control blood pressure, and lower cholesterol in a similar population. Lowering blood pressure and cholesterol was clearly beneficial, although those whose serum potassium fell were at greater risk. But it also was clear that our methods to halt smoking or lower cholesterol were simply too primitive.[16] In other early studies, lowering cholesterol gave the confounding result that cardiovascular deaths were reduced, but death from accidents and violence was increased, resulting in no differences in total mortality.[17] Finally, the statins became available, and then we could truly, and fairly safely, lower cholesterol in the broad population. The death rate from cardiovascular disease fell, although there may be beneficial effects of statin therapy not correlated with the change in cholesterol.[18] As already noted, lowering blood pressure so that left ventricular hypertrophy did not occur had an excellent benefit. It is sobering to realize that when I began my residency at Yale, there were no antihypertensive drugs in the Yale Medical Center pharmacy, because the prevailing view was that treating hypertension was harmful. In spite of the present clear evidence of huge benefit from blood pressure control, we again find that most patients who could benefit do not have adequate blood pressure control.[19] The lesson to be learned from these efforts is that the best way to reduce sudden cardiac death is reduction of the diseases and the pathological processes that lead to ischemia, hypertrophy, and heart failure.

Although the long-term benefit to attacking the disease itself is evident, no one can deny the need also to find satisfactory ways to benefit patients toward the end of their disease. And once a clearly beneficial standard of care has been established, we badly need some way to ensure that eligible patients receive that care. This would have an impact greater than most of the therapies currently under study. Under the present healthcare system, exhortation from academia will probably fail; perhaps there is a way to couple physicians' compensation to proper care. When dealing with diseases rooted in modern lifestyle, we cannot expect to abolish them. But we can delay them and ameliorate their manifestations. Unfortunately, the recent

increase in the incidence of diabetes may reverse some of the progress we have made.

Targeting the heart failure arrhythmia problem

Exploitation of basic science insight, coupled with an important clinical fact, is the great opportunity now open to us. Heart failure now represents a large risk factor for lethal arrhythmia, and basic insight into the pathophysiology of failure has been substantial (see also Chapters 12 and 13). We had found in the coronary care unit (CCU) that arrhythmia and acute heart failure are closely related. This also seems to be true for arrhythmia and chronic heart failure, regardless of its etiology.[20] Much of our progress in the last several decades has been to delay death and extend survival of those with heart failure, and this has greatly increased the number of patients living with heart failure. Arrhythmia remains a major cause of disability and death for these patients. The exact mechanism of arrhythmia in heart failure, other than that due to ischemia, has been elusive, and probably several mechanisms will be found. Two definite abnormalities occur in the electromechanical system. First, the ventricular action potential is prolonged, and this is probably related to downregulation of one or more types of K^+ channels.[21] Some promising approaches are being developed to use cellular and genetic methods to normalize the repolarization abnormality. Second, calcium cycling is deranged.[22] One promising approach for arrhythmia management involves attempts to ameliorate the consequences of this Ca^{2+} regulation defect in heart failure.

A major purpose of the cardiac electrical system is to activate contraction in a specific pattern. The most obvious activation sequence of importance determines the interval between atrial contraction and ventricular contraction until it has filled the ventricles optimally, utilizing delay in the atrioventricular (AV) node and ventricular conducting (Purkinje) system. AV node slow conduction depends on Ca^{2+} channel currents, which are smaller and activate more slowly than Na^+ currents, resulting in a slow but fairly reliable conduction velocity. Also important is the coordination of ventricular contraction by rapid conduction via Na^+ channels in the Purkinje system. First, the ventricular septum is activated and stiffens so that increased pressure on the left side does not distend the septum into the right ventricle, then contraction begins at the left ventricular apex and progresses to the base. This sequence ejects blood from the ventricle with optimal efficiency. Although the endocardial surface is activated first, the timing difference between the endocardial and epicardial parts of the ventricle is probably too small to be of importance during normal conduction. Ventricular coordination can be improved through biventricular pacing, and there may be a long-term effect on lethal arrhythmia.[23] Another intriguing approach is to find ways to use genetically altered stem cells to create a pacemaker of the sinus node type in the ventricle, which would relieve the patient of the need for an implanted electronic pacemaker, and perhaps even allow the heart rate to respond to normal physiological stimuli.[24]

Cardiac contraction is activated by opening of Ca^{2+} channels during the action potential plateau. Ca^{2+} enters the cell and triggers release of Ca^{2+} from the terminal cisternae of the sarcoplasmic reticulum (SR) through the ryanodine receptor (RyR), a process called calcium-triggered calcium release. Under normal conditions less than one-fifth of the Ca^{2+} needed for contraction comes from the transmembrane current and about four-fifths from the SR.[25] In the steady state about four-fifths of the released Ca^{2+} is pumped back into the SR by its Ca^{2+} pump, and the remainder is extruded from the cell, mainly by the surface membrane Na^+/Ca^{2+} exchanger. An important regulator of SR uptake in cardiac myocytes is phospholamban, an accessory protein to the SR Ca^{2+} pump. Phospholamban is itself modulated by protein kinase A (PKA)-induced phosphorylation in response to sympathetic stimulation. Unphosphorylated phospholamban slows Ca^{2+} uptake by the SR, by interaction with the SR Ca^{2+} pump, but PKA-dependent phosphorylation relieves that slowing,[26] allowing the normal heart to respond with increased contraction strength when sympathetic nerve activity increases the heart rate. The released Ca^{2+} itself binds to

troponin C, activating a cascade of protein interactions that result in myosin cross-bridge cycling and sliding of the myosin filaments relative to actin filaments, producing shortening or force development by the myocytes.

The RyR is critical to normal activation of contraction.[27] Its pore is a huge tetrameric protein complex (2.2 MDa). Several of these RyR molecules are clustered in the terminal cisternal membrane of the SR that is adjacent to the sarcolemma (both the t-tubule and surface membranes). Also part of the complex are calmodulin and FK-binding protein (FKBP12.6), and their roles in RyR function is more speculative. The current model is that FKBP protein stabilizes the gating function of the RyR (Fig. 26.1). In heart muscle, Ca^{2+} entering the cytoplasm through Ca^{2+} channels located in the surface membrane adjacent to the RyR-containing SR binds to the complex and causes opening of the RyR pore. This Ca^{2+} is mainly responsible for switching the contraction on. Multiple regulatory processes have been suggested for the RyR, which would provide fine modulation of contraction.

In heart failure there are several changes to this Ca^{2+} cycling system.[28] The most obvious change is a slower rise in intracellular Ca^{2+} to a lower peak and a slower return to normal diastolic levels.[29] This does not seem to be related to direct changes in the L-type Ca^{2+} currents.[30] However, this topic deserves continued study, because regulation of Ca^{2+} channels by adrenergic stimulation and other intracellular mechanisms is complex, and the conditions for Ca^{2+} channel study are sufficiently unphysiological that reproducing the heart failure conditions *in vitro* is difficult. Alteration in the Ca^{2+} transient is at least partly related to slower release of Ca^{2+} from the SR, and subsequent slower uptake, and this is accompanied by slower muscle relaxation.[31] There are also changes in the Na^+/Ca^{2+} exchange pump level, but the role of this transporter in heart failure is not yet entirely resolved.[32] As failure progresses, Ca^{2+} fails to fall back to normal during diastole, and tends to remain above normal for some time. The elevated levels of diastolic Ca^{2+} result from some combination of reduced uptake and slow leak from the SR through the RyR, the SR Ca^{2+} release channel, but

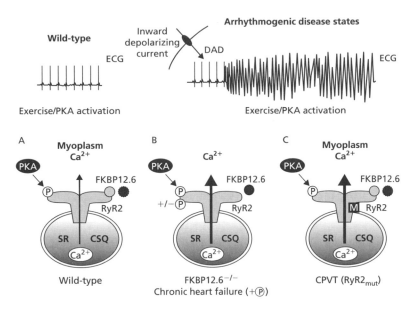

Figure 26.1. *Calcium cycling in heart failure and in catecholamine-induced polymorphic ventricular tachycardia (CPVT). (A) During exercise- or catecholamine-induced stress, wtRyR2 releases more Ca^{2+}, increasing myocardial contractility with a normal electrocardiogram (ECG). The proposed mechanism for this increase in Ca^{2+} release is partial dissociation of the FKBP12.6 from RyR2 as a result of phosphorylation of RyR2. In heart failure and CPVT (B and C) the amount of Ca^{2+} release is greater, and this can lead to an elevated diastolic myoplasmic Ca^{2+}. This elevation in myoplasmic Ca^{2+} can in turn lead to diastolic afterdepolarizations (DADs) that can initiate fatal ventricular tachyarrhythmias and cause sudden death. The increase can be mediated through several mechanisms. (B) In FKP12.6 knockout animals or chronic heart failure, where RyR2 is hyperphosphorylated, FKBP12.6 is either absent or completely dissociated from RyR2. (C) In CPVT hearts the amount of dissociation is normal, but some other factor associated with the causative mutation in RyR increases SR Ca^{2+} release. CSQ, calsequestrin, which also buffers intracellular Ca^{2+}. Modified from Allen PD. Not all sudden death is the same. Circ Res 93; 484–6, with permission.*

the relative importance of these mechanisms is not yet clear.[33] The leak of the RyR clearly is responsible for some rare genetic causes of exercise-induced ventricular tachycardia,[34] and it has been proposed as a problem for heart failure in general.[35] Some investigators have reported that increased phosphorylation of the RyR occurs in heart failure, and that this causes a basal Ca^{2+} leak.[36,37]

Leak of Ca^{2+} during diastole has a long history as a mechanism for ventricular arrhythmia.[38] The first clue was overdrive tachycardia in the presence of high levels of digitalis,[39] which triggered runs of ventricular tachycardia. Recording of the transmembrane events showed that rapid stimulation caused transient depolarizations during the inter-

val after action potential repolarization (DADs),[40] which were due to transient inward currents.[41] The depolarizations could reach threshold for excitation, and began a run of spontaneous action potentials. The depolarization follows a transient rise in intracellular Ca^{2+}. The exact mechanism of this depolarization is not entirely resolved, but it is either stimulation of the electrogenic Na^+/Ca^{2+} exchange or Ca^{2+} activation of a membrane channel that permits inward current. The rise in Ca^{2+}, however, clearly results from a spontaneous release of Ca^{2+} from the SR, presumably through the RyR channel. Therefore, a logical target for prevention of arrhythmia is release of Ca^{2+} through the RyR channel during diastole.

The RyR-stabilizing protein FKBP12.6 has been reported to be reduced in various animal models of heart failure.[42,43] This also occurs in inherited exercise-induced ventricular tachycardia syndrome,[44] where it is related to missense mutations in the RYR.[45] At least one possible cause of this dissociation may be heavy phosphorylation of the RyR,[41,46] and β-adrenergic blockers can restore the FKBP12.6 binding ratio and reduce Ca^{2+} leak *in vitro*.[36] The mechanism of heavy phosphorylation of RyR in experimentally induced heart failure is not clear, but several models of heart failure, including ischemia, hypertrophy, and tachycardia-induced cardiomyopathy have all shown this pattern. The logical sequence appears to be first some cardiac process that initiates heart failure, including ischemia, hypertrophy, and cardiomyopathy. Then adrenergic stimulation leads to excess phosphorylation of the RyRA, which results in dissociation of FKBP 12.6. This dissociation results in the leak of Ca^{2+} from the SR, which then initiates an inward current that precipitates a ventricular arrhythmia. A plausible therapeutic approach would be to prevent the Ca^{2+} leak.

There are two recent reports that indeed a derivative of 1,4 benzothiazepine, JTV519, may reverse the dissociation of FKBP12.6 from the RyR, and prevent both arrhythmia and the progression of the heart failure itself. The RyR is a Ca^{2+} channel, albeit a very different one from those in the surface membrane, but we have seen that nature is conservative in its use of ion channel protein motifs. Firstly, Kohno et al. induced heart failure in dogs by the pacemaker-induced tachycardia method and reported that the BTZ derivative restored normal *in vitro* SR function.[47] The drug did not influence uptake of Ca^{2+} into the SR, but did prevent Ca^{2+} leak, and also improved Ca^{2+} release. Yano et al.,[48] from the same group of investigators, then showed that administration of the BTZ derivative during the overdrive protocol not only prevented the dissociation of FKBP12.6 from the RyR, but surprisingly also prevented the manifestations of heart failure. Admittedly, this tachycardia-induced cardiomyopathy may not exactly mimic human heart failure, but this result implies that the RyR defect may play a key role in left ventricular (LV) remodeling and deterioration of ventricular function.

If indeed this RyR disregulation by dissociation of FKBP12.6 is a cause of heart failure, rather than a consequence, then suppression of the expression of FKBP12.6 should produce heart failure. However, Wehrens et al.[49] reported that FKBP12.6-deficient mice have normal cardiac mechanical function, no remodeling, and normal baseline electrophysiological parameters. Nevertheless, these FKBP12.6-deficient mice showed polymorphic ventricular tachycardia and sudden death during exercise, after modest doses of epinephrine. Those with haploinsufficiency had about 40% reduction in FKBP12.6 levels, and pretreatment with the BTZ derivative prevented arrhythmias during the exercise-epinephrine challenge. However, those with complete loss of FKBP12.6, although showing no sign of cardiac dysfunction without challenge, all developed lethal arrhythmia upon challenge, in spite of pretreatment with the BTZ derivative. Programmed electrical stimulation in these three groups of mice yielded similar results.

The studies from both groups make a strong argument that Ca^{2+} leak from the SR can lead to lethal arrhythmia, that the leak is probably the result of dissociation of the RyR regulatory accessory protein FKBP12.6 that occurs in heart failure (probably because of heavy phosphorylation of RyR), and that a BTZ derivative can reverse the dissociation and prevent arrhythmias. They do not agree on a very important question – whether the FKBP12.6 dissociation and consequent SR dysfunction is a cause or a result of heart failure. Furthermore, the extent and nature of RyR phosphorylation remains controversial.[50] No doubt part of the problem is their need to use animal models of heart failure, which perhaps simulate different aspects of this very complex pathophysiological process. This is an exciting new opportunity to understand the cellular and molecular mechanisms at the very basic level, and it is also an exciting new idea for therapy to prevent the mechanical and electrical consequences of heart failure.

Summary

Sudden cardiac death is very common in heart disease, and especially in the increasing population of patients with heart failure, and this constitutes both personal tragedy and a public health crisis. During the last half-century we have learned that these sudden deaths cannot be treated effectively by resuscitation outside the hospital. Furthermore, they cannot be prevented by any of the presently available ion channel-directed antiarrhythmic drugs, which seem to be too disruptive of cardiac function. We have found some very effective drugs to reduce sudden death, but they either modulate cardiac properties or interfere with disease mechanisms. They have delayed cardiac death to more advanced ages, and changed the clinical picture, but they are unevenly used in the relevant population. Large numbers of individuals at risk and who could benefit from the drugs, cannot be identified by present methods.

During the same time of this clinical progress, there has been a huge increase in our understanding of normal and pathological physiology of the heart and the circulation. There is now a plethora of opportunities for intervention at the cellular, molecular, and genetic levels. As an example of such an opportunity, I have discussed a possible way to intervene in heart failure.

Past experience encourages us to be cautious in applying any treatment to the large population at risk for sudden cardiac death, especially any treatment that directly disturbs the cardiac electrical system. Rather, we should use this basic scientific understanding either to modulate the circulation gently, or to interfere with the disease process itself.

References

1. Kullar LH. Sudden death: definition and epidemiologic considerations. Prog Cardiovasc Dis 1980; 23: 1–12.
2. Cobb LA, Werner JA, Trobaugh GB. Sudden cardiac death. I. A decade's experience with out-of-hospital resuscitation. Mod Concepts Cardiovasc Dis 1980; 49: 31–6.
3. MacMahan S, Collins R, Peto R et al. Effects of prophylactic lidocaine in suspected myocardial infarction. JAMA 1988; 260: 1910–15.
4. Maroko PR, Braunwald E. Modification of myocardial infarct size after coronary occlusion. Ann Int Med 1973; 79: 720–34.
5. Schaffer WA, Cobb LA. Recurrent ventricular fibrillation and modes of death in survivors of out-of-hospital ventricular fibrillation. N Engl J Med 1975; 293: 259–62.
6. Zheng Z, Croft JB, Giles WH et al. State-specific mortality from sudden cardiac death: United Stat, 1999. MMWR Morb Mortal Wkly Rep 2002; 51: 123–6.
7. Eisenberg MS, Pantridge JF, Cobb LA et al. The revolution and evolution of prehospital cardiac care. Arch Intern Med 1996; 156: 1611–19.
8. Becker LB, Han BH, Mayer PM. Racial differences in the incidence of cardiac arrest and subsequent survival. N Engl J Med 1993; 329: 600–3
9. β-Blocker heart attack trial research group. A randomized trial of propranalol in patients with acute myocardial infarction. I. Mortality results. JAMA 1982; 247: 1707–14.
10. Friedman LM, Byington RP, Capone RJ et al. Effect of propranalol in patients with myocardial infarction and ventricular arrhythmia. J Am Coll Cardiol 1986; 7: 1–8.
11. Packer M. Do beta-blockers prolong survival in heart failure only by inhibiting the beta1 receptor? J Card Fail 2003; 9: 429–43.
12. Antiplatelet Trialist Collaboration. Collaborative overview of randomized trials of antiplatelet therapy. BMJ 1994; 308: 81–95.
13. Moss AJ. MADIT-I and MADIT-II J Cardiovasc Electrophysiol 2003; 14: S96–8.
14. Echt DS, Liebson PR, Mitchell B et al. Mortality and morbidity in patients receiving encainide, flecainide, and placebo. N Engl J Med 1991; 324: 781–8.
15. Coronary Drug Project. JAMA 1978; 240: 1483–4.
16. Cohen JD, Grimm RH, Smith WM. Multiple risk factor intervention trial. Prev Med 1981; 10: 501–18.
17. Wysowski DK, Gross TP. Deaths due to accidents and violence in two recent trials of cholesterol-lowering drugs. Arch Intern Med 1990; 150: 2169–83.
18. Scandinavian Simvastatin Survival Study. Randomized trial of cholesterol lowering. Lancet 1994; 344: 633–9.
19. Hyman DJ, Pavlik VN. Characteristics of patients with uncontrolled hypertension in the United States. N Engl J Med 2001; 345: 479–86.
20. Tomaselli GF, Zipes DP. What causes sudden death in heart failure? Circulation Res 2004; 95: 754–63.
21. Beuckelmann DJ, Nabauer M, Erdmann E. Alterations of K currents in isolated human ventricular myocytes from patients with terminal heart failure. Circ Res 1993; 73: 379–85.
22. Bers DM, Eisner DA, Valdivia HH. Sacroplasmic reticulum Ca and heart failure: roles of diastolic leak

and Ca transport. Circulation Res 2003; 93: 487–90.

23. Bristow MR, Saxon LA, Boehmer J et al. Cardiac-resynchronization therapy with or without an implanted defibrillator in advanced chronic heart failure. N Engl J Med 2004; 350: 2140–50.

24. Potapova I, Plotnikov A, Lu Z et al. Human mesenchymal stem cells as a gene delivery system to create cardiac pacemakers. Circulation Res 2004; 94: 952–959.

25. Bers DM. Excitation-Contraction Coupling and Cardiac Contractile Force. Boston: Kluver Academic Publishers; 1991.

26. Simmerman, HK, Jones LR. Phospholamban: protein structure, mechanism of action, and role in cardiac function. Physiol Rev 1998; 78: 921–47.

27. Bers DM. Macromolecular complexes regulating cardiac ryanodine receptor function. J Mol Cell Cardiol 2004; 37: 417–30.

28. Sjaastad I, Wasserstrom JA, Sejersted OM. Heart failure - a challenge to our current concepts of excitation-contraction coupling. J Physiol 2003; 546: 33–47.

29. Beuckelmann DJ and Erdmann E. Ca currents and intracellular Ca transients in single ventricular myocytes isolated from terminally failing human myocardium. Basic Res Cardiol 1992; 87: 235–43.

30. Chen X, Piacentino V, Furukawa S et al. L-type Ca channel density and regulation are altered in failing human ventricular myocytes and recover after support with mechanical devices. Circ Res 2002; 91: 517–24.

31. Hasenfuss G and Pieske B. Calcium cycling in congestive heart failure. J Mol Cell Cardiol 2002; 34: 951–69.

32. Schillinger W, Fiolet JWT, Schlotthauer H et al. Relevance of Na/Ca exchange in heart failure. Cardiovasc Res 2003; 57: 921–933.

33. Eisner DA, Trafford AW. Heart failure and the ryanodine receptor. Circ Res 2002; 91: 979–81.

34. George CH, Higgs GV, Lai FA, Ryanodine receptor mutations associated with stress-induced ventricular tachycardia mediate increased Ca release in stimulated cardiomyocytes. Circ Res 2003; 93: 531–40.

35. Allen PD. Not all sudden death is the same. Circ Res 93; 484–6.

36. Ono K, Yano M, Ohkusa T et al. Altered interaction of FKBP12.6 with ryanodine receptor as a cause of abnormal Ca release in heart failure. Cardiovasc Res 2000; 48: 323–31.

37. Reiken S, Gaburjakova M, Gaburjakova J et al. β-adrenergic receptor blockers restore cardiac calcium release channel (ryanodine receptor) structure and function in heart failure. Circulation 2001; 104: 2843–28.

38. January CT, Fozzard HA. Delayed afterdepolarizations in heart muscle; mechanisms and relevance. Pharmacol Rev 1989; 40: 219–27.

39. Wittenberg SM, Streuli F, Klocke FJ. Accelleration of ventricular pacemakers by transient increases in heart rate in dogs during ouabain administration. Circ Res 1970; 26: 705–16.

40. Ferrier GR. Digitalis arrhythmias, role of oscillatroy afterpotentials. Prog Cardiovasc Dis 1977; 19: 459–74.

41. Lederer WJ, Tsien RW. Transient inward current underlying arrythmogenic effects of cardiotonic steroids in Purkinje fibers. J Physiol 1976; 263: 73–100.

42. Marx SO, Reiken S, Hisamatsu Y et al. PKA phosphorylation dissociates FKBP12.6 from the calcium release channel (ryanodine receptor): defective regulation in failing hearts. Cell 2000; 101: 365–76.

43. Yano M, Ono K, Ohkusa T et al. Altered stoichiometry of FKBP12.6 versus ryanodine receptor as a cause of abnormal Ca leak through the ryanodine receptor in heart failure. Circulation 2000; 102: 2131–6.

44. Wehrens XH, Lehnert SE, Huang F et al. FKBP12.6 deficiency and defective calcium release channel (ryanodine receptor) function linked to exercise-induced sudden cardiac death. Cell 2003; 113: 829–40.

45. Lehnart SE, Wehrens XHT, Laitman PJ et al. Sudden death in familial polymorphic ventricular tachycardia associated with calcium release (ruanodine receptor) leak. Circulation 2004; 109: 3208–14.

46. Reiken S, Gaburjakova M, Guatimosim S et al. Protein kinase A phosphorylation of the cardiac calcium release channel (ryanodine receptor) in normal and failing hearts. J Biol Chem 2003; 278: 444–53.

47. Kohno M, Yano M, Kobayashi S et al. A new cardioprotective agent, JTV519, improves defective channel gating of ryanodine receptor in heart failure. Am J Physiol Heart Circ Physiol 2003; 284: H1035–42.

48. Yano, M, Kobayashi S, Kohno M, et al. FKBP12.6-mediated stabilization of calcium release channel (ryanodine receptor) as a novel therapeutic strategy against heart failure. Circulation 2003; 107: 477–84.

49. Wehrens XHT, Lehnart SE, Reiken SR et al. Protection from cardiac arrhythmia through ryanodine receptor-stabilizing protein calstabin2. Science 2004; 304: 292–6.

50. Xiao B, Jiang MT, Zhao M et al. Characterization of a novel PKA phosphorylation site, serine-2030, reveals no PKA hyperphosphorylation of the cardiac ryanodine receptor in canine heart failure. Circ Res 2005; 96: 847–55.

CHAPTER 27

Genetic factors underlying susceptibility to common arrhythmias

Peter M Spooner

Introduction

The transition between the last and the present millennium of scientific progress will surely be most noted for the successful definition of the complete DNA sequences of several unique individuals, including even some who pioneered this remarkable accomplishment. To place the event in historical perspective, recall that its origins go back only a bit more than a century to Mendel's discoveries concerning the inheritance of simple traits affecting his harvest of garden peas. Discovery of the paired double-stranded structure of DNA, now 50 years ago, subsequently provided the basis for explaining how the phenotypes observed by Mendel were transmitted from one generation to another. Since that time, much has been learned about how these same principles help explain human traits such as eye or hair color, and how the ability to read DNA sequences of genes at specific chromosomal loci can provide information on the incidence of many congenital diseases, their segregation within different generations, and how such changes convey a just-emerging dimension of health susceptibility not previously envisioned. Current understanding of how mutations result in functional physiological differences and how single-gene, that is 'monogenic', syndromes are characterized by well-defined disease phenotypes that occur in concert with Mendel's laws, is well advanced. Indeed single gene mutations with high

'penetrance', that is they occur in most individuals who inherit the mutation, are now well recognized as the cause of more than 1000 different syndromes,[1] and the stories of their discovery represent some of the most exciting pages in medical history. Today, as the details in the newly completed human genome projects are being digested, the task for the next period of discovery would appear to be to understand not just mutations that result in frank 'congenital' defects and 'Mendelian' diseases, but the full complement of DNA variation, from the less frequent 'high impact' events that result in extreme phenotypes, to the more complicated 'polygenic' conditions, involving multiple genes and 'lower impact' common single base polymorphisms (SNPs), that define more subtle functional differences between a healthy and a diseased individual.

Although there remains much to learn about how development and progression of common disorders are influenced by 'high' and 'low' impact variants, by changes in coding and non-coding transcription 'regulatory' sequences, by factors that contribute to variations in 'penetrance' and variability in phenotypic expression in different individuals, that is 'expressivity', much progress has in fact been made. The results suggest that even though inheritance patterns and the environmental milieu in which these influences exert their effects are highly complex, clearly outside Mendel's garden, there are many new ways these problems can be approached. Today this includes not just identifying which of the various biological influences might be most influential in a specific cardiovascular disorder, but also addressing how gene expression is regulated, how 'modifying' influences work and how these influences are transmitted, by either classical or more recently revealed epigenetic mechanisms, from one generation to another. A second major issue is how genetic expression is influenced by confounding environmental and behavioral factors, such that expression of disease phenotypes can appear almost unique in each individual.

Understanding monogenic disorders was the first step, while dissecting the different genetic influences that play a role in polygenic syndromes, the stage we are at today, represents the 'second major wave' in understanding how genes influence specific pathologies. The next challenge, the 'third wave', represents the transition from 'genetics' to 'genomics', from considering variation in one versus multiple genes, and discovering how relatively unique constellations of subtle sequence changes influence incidence and outcomes of common diseases. Fortunately, as is often the case in science, new concepts and technical advances in expanding fields such as informatics, and the population sciences, especially genetic epidemiology, have emerged equal to the challenge. Another especially promising development lies in the beginnings of an understanding and use of patterns of genetic architecture, and the ways ethnic, regional and familial groups express and inherit large-scale patterns of genetic variation. With new information not just on isolated mutations but also with blocks of adjoining, biologically unrelated sequences or 'haplotypes' transmitted together in larger 'architectures', now being defined in the Haplotype Map Project, it may be possible to infer much regarding disease susceptibility or resistance by more limited screening, and knowledge of which variants are most like to be inherited together. Limiting issues then become those of statistical inference and analysis: while Mendel and his peas provided the start, today it seems it will more likely be the output of mega- and tera-pixel chips, high-throughput scanners, and association informatics able to take advantage of the new technologies

that will provide the clues to this next level of complexity.

Arrhythmia mechanisms in complex diseases

This chapter addresses the issue of how genetic variation affecting the heart's pathways of electrical function may contribute to development of lethal arrhythmias in common (that is not-rare) forms of heart disease, especially in patients with heart failure (HF). The topic is important because lethal arrhythmias appear the precipitating cause of death in about half of all patients who succumb to both diastolic and systolic forms of failure. Before reviewing what has been accomplished, it is important to note that the goal of applying genetic approaches to this problem is not so much to supplant conventional approaches, but more to improve diagnosis and therapy selection by adding dimensions of personal information not otherwise available. Other, important points to keep in mind are that cardiac arrhythmias present a number of unique genetic issues: the first is to recognize that arrhythmias are a highly complex phenotype which, as illustrated in Table 27.1, span a range of disorders encompassing most cardiac pathologies.[2–4] A second is that patterns of genomic variation involved in common arrhythmias are qualitatively different from disease associations made with more typical disease genes, where phenotype–genotype relationships are more structured and less variable. Atrial and ventricular arrhythmias in many cases do not represent a primary phenotype or hallmark manifestation of the disease gene(s) with which a particular susceptibility may segregate: variations in many different genes may all result in the same, almost identical arrhythmia. An example may help: whereas variations in several different genes have been observed to affect susceptibilities for common forms of hypertension, the basis of many can ultimately be shown to be related to pathways of Na^+ homeostasis.[5] In contrast, genes predisposing to arrhythmias may be linked to a broad array of different biological pathways, ranging from mutations in electrogenic ion channels to alterations in transcription

Table 27.1. *Genetic susceptibilities in cardiac disease*

Syndrome	Gene(s) affected	Proteins identified	Cell and tissue phenotypes
Disorders with primary defects in electrogenesis			
Long QT syndromes (LQTS)			
LQT-1	*KCNQ1 (KVLQT1)*	K$^+$ channel α subunit	Delayed I_{Ks} repolarization
LQT-2	*KCNH2 (HERG)*	K$^+$ channel α subunit	Delayed I_{Kr} repolarization
LQT-3	*SCN5A*	Na$^+$ channel α subunit	Defective depolarization
LQT-4	*ANK2*	Membrane anchor	Altered [Ca^{2+}]$_i$, EADs
LQT-5	*KCNE1 (minK)*	K$^+$ channel β subunit	Defects in I_{Ks} repolarization
LQT-6	*KCNE2 (MiRP1)*	K$^+$ channel β subunit	Defects in I_{Kr} repolarization
Lenegre–Lev conduction disease	*SCN5A*	Na$^+$ channel α subunit	Defects in depolarization
Brugada's syndrome (BrS)	*SCN5A*	Na$^+$ channel α subunit	Decrease in depolarization
Short QT Syndrome (SQTS)	*KCNH2/KCNQ1*	K$^+$ channel α subunit	Enhanced repolarization
Catecholaminergic polymorphic ventricular tachycardia (CPVT)	*RYR2/CA5Q2*	SR Ca^{2+} release channel	Disordered SR Ca^{2+} release
Familial 'lone' atrial fibrillation (FAF)	*KCNQ1/KCNE1*	K$^+$ channel α/β subunit	Delayed I_{Ks} repolarization
Sinus node dysfunction	*HCN4*	Pacing channel α subunit	Defective I_f regulation
'Timothy syndrome'	*CACNA1C*	L-Ca^{2+} channel α subunit	Defects in Ca^{2+} cycling(?)
Disorders with structural defects affecting conduction and electrogenesis			
Arrhythmogenic right ventricula cardiomyopathy (ARVC)	*RYR2/JUP/PKP2PH*	SR Ca^{2+} release channel/ plakoglobin and plakofilin-2	Decreased SR Ca^{2+} release/cell necrosis and fatty cell infiltration
Hypertrophic cardiomyopathies Sarcomeric protein defects	*MYH7, MYBPC3, TNNC1, TNNT2, TNNI3, TPM1, MYL3, MYL2,ACTC, TTN,* etc	β myosin heavy chain, myosin-binding protein C, troponin C, troponin T, troponin I, α tropomyosin, essential light chain, regulatory light chain, cardiac actin, titin, etc.	Defects in myocardial conduction, defects in actin–myosin interactions, defects in myocardial energetics, defects in Ca^{2+} cycling, fiber disarray, fibrosis, replacement cell infiltration
Kinases	*PRKAG2*	AMP kinase subunit	
Dilated cardiomyopathies Sarcomeric proteins	*MYHt, MYBPC3, TNNT2, TTN, ACTC,* etc.	β myosin heavy chain, myosin-binding protein C, troponin T. titin, actin, AMP kinase subunit, etc	Defects in myocardial conduction, defects in actin–myosin interaction, defects in energetics, defects in Ca^{2+} cycling, fibrosis, disarray, cell infiltration, muscular dystrophies, etc
Cytoskeletal proteins	*DES, SGCD, VCL, DMD,* etc	Desmin, β sarcoglycan, metavinculin, dystrophin, etc	
Nuclear membrane proteins	*LMNA*	Lamins A and C	DCM and conduction disease
Ca^{2+} regulatory proteins	*PLB*	Phospholamban	DCM and defects in Ca^{2+} cycling

factors that affect collagen synthesis. A third point is that it should be appreciated that arrhythmias in complex common diseases represent a fleeting disease outcome, a highly transitory phenotype. Rather than an established, readily observable change in a static biological parameter such as enzyme synthesis levels, precipitating molecular events in arrhythmogenesis can be a millisecond aberration, occurring perhaps only once every 10 000 or 20 000 heartbeats. Such events may occur in single or multiple cardiac cell types, in cells that are just in specific regions of the heart such as the endocardium, or just in electrically specialized tissue, such as the Purkinje network. Resulting electrical phenotypes are variable and often not diagnostically unique. For example, in one case the same familial Na^+ channel mutation was reported to result in symptoms characteristic of two different syndromes, long QT-3 (LQT-3) and the Brugada syndrome, in the same individuals.[6] In another case a different mutation in this same Na^+ channel resulted in symptoms of both LQT-3 and Brugada's syndromes and, in addition, a form of conduction system disease, all in the same pedigree.[7] Not only does this make gene sleuthing difficult, but it is potentially deceptive in that presumably characteristic electrophysiological signatures may derive from different sources. Unless molecular or cellular electrophysiology studies are done, an infrequent occurrence, it may be difficult to establish underlying causation on the basis of *in vivo* electrocardiographic (ECG) or other phenotypic diagnoses alone. Effective pharmaceutical targeting based only on ECG phenotypes would obviously be difficult indeed, although this has in fact had been pursued for many years.

Contemporary views on mechanisms of arrhythmogenesis suggests two principal sources of arrhythmia susceptibility may be important: (1) factors which result in initiation of an aberrant electrical impulse, often at sites external to the sinus node; and (2) elements involved in abnormal propagation and transcellular conduction within the conducting network and working myocardium.[8] Abnormal automaticity, that is spontaneous depolarizations able to initiate a self-perpetuating impulse, is a principal mechanism in the origin of many arrhythmias, as are two forms of 'triggered' afterdepolarizations which may follow a normal wave of excitation.[9] Early afterdepolarizations (EADs) have been associated with alterations in intracellular Ca^{2+} cycling between the cytoplasm and the sarcoplasmic reticulum (SR), and appear a frequent source of aberrant impulses in the ventricle.[10] EADs secondary to mutations in sarcoplasmic Na^+ and K^+ channel subunits have been seen in patients with monogenic arrhythmias including some forms of the long QT syndrome (LQTS),[11,12] as well as patients with non-ischemic heart failure.[13] Delayed after-depolarizations (DADs) have been implicated as an important contributor to abnormal impulses developing in the pulmonary veins,[14] a common source of ectopy in atrial fibrillation (AF).[15] Genetic changes predisposing to EADs and DADs, such as those resulting in compartmental imbalances in cell Ca^{2+}, are a possible unifying focus in the search for common elements of both acquired and inherited arrhythmia susceptibility.

In addition to alterations in electrogenesis, genetic variation affecting macroscopic pathways of electrical propagation are the second critical element in most clinically important arrhythmias. Structural defects affecting conduction velocity, and directional heterogeneities occur prominently in patients with ventricular hypertrophy, stage II–IV congestive failure, and individuals whose myocardium has undergone significant remodeling, for example as a result of hypertension, ischemia and infarction. Tissue and cellular pathologies contributing to this dimension of susceptibility include apoptosis, necrosis, cell replacement, altered collagen deposition and fibrosis, and disturbances in myocardial fiber, bundle and tissue architecture. Such changes result in spatial and temporal electrical heterogeneities, leading to conduction delays, conduction block, and re-entrant patterns of excitation that are the frank and frequent precursors to fibrillation. Replacement of viable contracting cells with infiltrating non-conductive cell types (e.g. myofibroblasts or fatty cells) also occurs in a number of common and inherited conditions. Congenital arrhythmogenic right ventricular dysplasias (ARVD) are a notable example of the latter,[16] while ischemic post-infarction patients with areas of necrosis and infarction may be illustrative of the former.[17] The role of progressive structural changes in cardiac electrical heterogeneities is one that has traditionally been underappreciated, and there is still the possibility that genetic influences which act at this level also contribute to inherited arrhythmias. These views are beginning to change, in large part as a result of detailed pathological studies on sudden death victims published over the past few years.[18] Such studies have made it clear that both specific (e.g. atrioventricular (AV) bundle malformations and/or non-specific structural anomalies (e.g. diffuse fibrosis) occur in most individuals who suffer acute life-threatening ventricular arrhythmias.

Disturbances in global patterns of electrical performance in contrast, may derive from more specific molecular and cellular events, and it is at this level that changes in function can be most directly related to gene variation. Two primary types of pathology appear. The first is disturbances in membrane transport and electrogenesis – direct alterations in sarcolemmal ion channels and transport proteins involved in establishing atrial and ventricular action potentials. Specific proteins include some of the same Na^+, Ca^{2+} and K^+ channel α and β subunits, discovered to underlie many of the rare inherited ventricular arrhythmic syndromes like LQTS and Brugada's syndrome.[19,20] Various lines of evidence indicate that other, as yet unimplicated, membrane and perhaps mitochondrial proteins, for example the Na^+/Ca^{2+} exchanger and additional Ca^{2+}-binding proteins may also play a role. The second source of potential genetic influences are those that may affect proteins and pathways involved in cell-to-cell transcellular conduction.[21] Specific targets here include variants that affect function and expression of gap junction proteins, such as the cardiac connexins 43, 40 and 47, which establish pathways for electrical and small molecule communication between opposing membranes of neighboring cells.[22] Proteins in phosphorylation pathways regulating gap junction gating, and those influencing the very rapid turnover of connexin subunits could also be critical, especially in hearts undergoing extensive remodeling,[23] one of the characteristic features of structural forms of HF. Transcellular conduction pathways in the heart are strongly affected by alter-

ations in the extracellular matrix. Factors affecting cardiac fibrosis include alterations in the synthesis and turnover of various cardiac collagens, and there is some evidence that genetic influences at this level are important in HF patients. Angiotensin II (Ang II) and aldosterone, both modulate cardiac collagen turnover and appear critical endocrine mediators of cardiac fibrosis.[17,24] Significantly, inhibitors of the aldosterone mineralocorticoid receptor and the Ang II receptor, therapies commonly used in HF patients, are effective in reducing arrhythmic sudden deaths, while also reducing collagen metabolism.[25,26] There is a broad literature suggesting a high degree of genetic variation is found in renin–angiotensin–aldosterone (RAAS) synthetic pathways. While findings from these studies have received an increasing degree of attention in work on other complex cardiovascular disorders, notably hypertension, they have only begun to be explored as a source and possible solution for rhythm disturbances.[5,27]

Figure 27.1 provides a theoretical scheme by which genetic variants, can be ascribed to specific pathways considered potentially causative in arrhythmogenesis. Variation in elements in 'final common pathways' of electrogenesis, that is within those Na^+ and K^+ channel genes directly contributing to the cardiac action potential, are by this view more likely to result in abnormalities at the 'proximal' end of this spectrum, while others, for example mutations in pathways mechanistically distant, such as those leading to atherosclerotic plaque formation are viewed as mechanistically more 'distal'. Others intermediate in 'impact', for example polymorphisms affecting clotting factor proteins conveying risk for coronary thrombosis and infarction, are placed somewhere between these two extremes. It should be noted too that levels of influence of different factors would be expected to vary in individuals with different predisposing forms of cardiac disease. Thus while the scheme in Fig. 27.1 may be useful in thinking tactically about different genetic risks that might occur in various cardiac pathologies, as noted above, it is still unrealistic to believe electrogenic and conductive elements can be separated in real patients with real forms of disease. A final point regarding Fig. 27.1 is that although

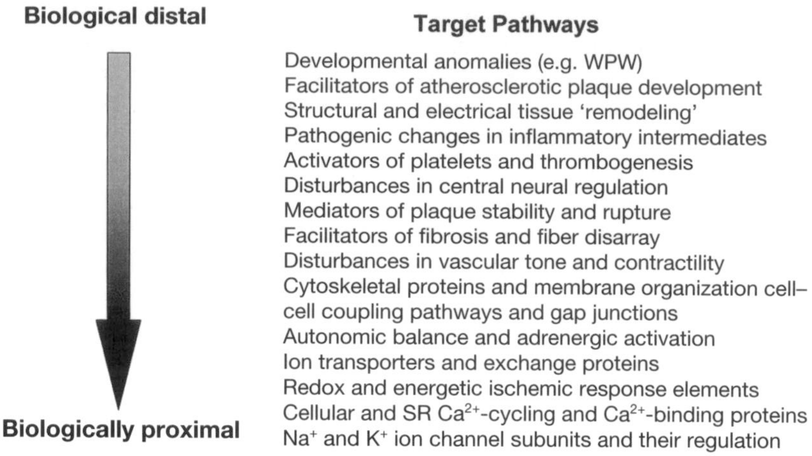

Biological distal

Target Pathways

Developmental anomalies (e.g. WPW)
Facilitators of atherosclerotic plaque development
Structural and electrical tissue 'remodeling'
Pathogenic changes in inflammatory intermediates
Activators of platelets and thrombogenesis
Disturbances in central neural regulation
Mediators of plaque stability and rupture
Facilitators of fibrosis and fiber disarray
Disturbances in vascular tone and contractility
Cytoskeletal proteins and membrane organization cell–
cell coupling pathways and gap junctions
Autonomic balance and adrenergic activation
Ion transporters and exchange proteins
Redox and energetic ischemic response elements
Cellular and SR Ca^{2+}-cycling and Ca^{2+}-binding proteins
Na^+ and K^+ ion channel subunits and their regulation

Biologically proximal

Figure 27.1. *Potential pathways in which genetic variation may contribute to enhanced susceptibility to atrial or ventricular arrhythmias. Target pathways and individual elements of risk depicted as more 'biologically distal', are those furthest removed in time and/or biological impact on the propensity to initiate an acute arrhythmia. These include both 'conditioning' and conductive risks which impact on structural as well as electrogenic factors that predispose to acute re-entrant forms of electrical instability, and are in many cases disease specific. Pathways and elements suggested to be more 'proximal' are in general those more associated with acute triggering of an aberrant impulse and with disturbances in electrogenic balance, such as delays in membrane repolarization, and are less disease specific. While a general order from proximal to distal is depicted in the figure, the relative importance will clearly vary in different individuals with different predisposing cardiac pathologies. Progressive, common cardiac diseases most often involve disturbances in several of the various pathways illustrated, and genetically inherited influences in these individuals probably involve multiple 'low impact' sources of susceptibility that evolve over time. Arrhythmia prevention is most effective in addressing pathological factors depicted as 'distal', whereas acute strategies are targeted primarily on those factors in final common pathways of electrical instability. WPW, Wolfe–Parkinson–White syndrome. Modified with permission from Spooner PM, Albert C, Benjamin EJ et al. Sudden cardiac death, genes and arrhythmogenesis: consideration of new population and mechanistic approaches from an NHBLI workshop, part II. Circulation 2001; 103:2447–52.*

commonly occurring genetic variants are indeed likely to play a specific role in different common disease syndromes, their relative effects may be different, as a function of environmental and behavioral factors, along with influences such as ethnic origins. Underlying population structure is a critical, often overlooked factor in variation – susceptibility associations. As emphasized in a recent review, 'Different subsets of genes will influence phenotypic variation in different subgroups in the same population. Because multigene genotypes will have a multinomial distribution, different combinations of susceptibility genes will

be involved in defining disease risks in specific, but different individuals'.[28]

Atrial fibrillation and genomics

Atrial fibrillation is the most prevalent of any cardiac rhythm disturbance, with the number of affected individuals in the US estimated at ~2.2 million in 1999, and projected to increase to ~5.5 million by 2050.[29,30] Clinical phenotypes include high rates of atrial activation (~250–500/min), incomplete atrial emptying, increased atrial thrombosis and stroke, and left ventricular remodeling secondary to increased ventricular pacing. Overall mortality is increased by ~1.5–1.9-fold on an age-adjusted risk basis.[31] Both structural and electrogenic etiologies have been noted,[32–34] with strong evidence that increased ectopy, especially in junctional tissues near the pulmonary veins, is a major mechanism in early, that is 'paroxysmal', and subsequent, that is 'persistent', and more 'permanent' manifestations of the disease.[35] Physiological and genomic adaptations, including extensive electrical and structural remodeling, are a significant part of this process, as evidenced by recent studies indicating changes in more than 100 unique protein species occur during pacing-induced atrial fibrillation (AF) progression.[36] Discovering mechanisms involved in altered protein expression in AF represents a significant challenge, with one of the major questions being whether the broad changes observed might reflect changes in transcriptional regulation. While AF does not usually present an immediately lethal threat, new approaches able to predict and thwart upstream events in its progression would be highly useful clinical additions to today's palliative drug- and ablation-based therapies.

Associations between HF and AF are particularly interesting, in that patients with HF have a notably higher incidence of AF. Data from clinical trials suggest the proportion of HF patients with AF increases from ~4% in New York Heart Association (NYHA) class I patients, to ~50% in class IV.[37] As with HF, aging is a major population risk factor for AF, an association which raises interest in overlap in causative mechanisms, especially in light of commonalities in pathologies including neurohumoral and sympathetic deficiencies, enhancements in interstitial fibrosis, and abnormalities in Ca^{2+} cycling reported frequently in both.[35] Epidemiological data indicate that of patients who develop both conditions, AF and HF each appear first in about the same percentage of individuals (38% AF first; 41% HF first), with the remainder (21%) presenting concurrently.[38]

'Lone AF', that is AF occurring in the absence of apparent structural or electrical pathology, is rare, with an estimated prevalence of ~3–15% of presenting cases, depending on the population sampled and method of ascertainment.[34] In cases where it does occur, familial patterns of incidence are common. Phenotypes vary in the probands from more than 100 different families evaluated in laboratories in both the US and Europe.[39] In addition to cases of 'lone' AF, an increasing number of pedigrees are being reported where AF segregates with familial ventricular diseases, including several in which enhanced susceptibility to sudden cardiac death (SCD) appears commonly. These include some of the familial hypertrophic cardiomyopathies (FHC),[40,41] patients with familial dilated cardiomyopathies (DCM),[42] and cases of LQTS. Also of interest is the single family with the recently elucidated LQT-4 syndrome found associated with mutations in the cell membrane-anchoring protein ankyrin.[43] Frequent episodes of AF were reported to segregate closely with delays in ventricular repolarization in affected individuals with the variant ankyrin allele.

Work done to identify genetic AF susceptibilities in 'lone AF' has implicated surprisingly few sites of potential variance. Linkage studies on three unrelated European families identified a broad locus near 10q22–24 some years ago, but, as yet, specific gene identities have not been published,[44] Another study on a single large family with apparently Mendelian inheritance implicated a different broad region, 6q14–16, but again affected genes remain to be identified.[45] Interestingly, this same region on chromosome 6 was linked to cases of familial DCM in earlier work.[46] Commonalities in AF linkage to the 10q22–24 region mentioned above have also been reported in patients presenting with familial DCM, again suggesting common mechanisms, but as before, clues to the biology await better definition of the genes and loci involved.[45]

As is detailed in the first chapter in this section 'Cellular and molecular basis of electrogenesis of normal cardiac tissue' (Chapter 22), there is considerable overlap in the distribution of various ion channel subunit proteins between atrial and ventricular muscle. Mutations in ion channel proteins are, as detailed below, a principal source of arrhythmogenesis in the ventricle,[19] resulting in interest that mutational consequences of some of the same channel proteins might have functional implications for both atrial and ventricular electrophysiology. Channel subunits of interest include the α and β subunits of the KVLQT1 channel responsible for the repolarizing I_{ks} current. A single missense mutation (S140G) in the *KCNQ1* (*KVLQT1*) gene was reported to be linked to a bradycardic form of familial AF, and some of these same individuals also displayed a prolonged QTc interval,[47] suggestive of phenotypes seen with LQT-1 patients.[48] Interestingly, a polymorphism in mink, the β subunit of this same channel, was identified as a risk allele for familial AF in a larger case control study.[49] AF incidence in individuals with the rarer 38G minK polymorphism was higher than in individuals homozygous for the more frequent 38S, while heterozygotes displayed an intermediate phenotype. Lastly, it is worth noting the relatively unique finding of 'idiopathic' sinus node dysfunction in a family in which AF mapped to a dominant negative mutation in the atrial pacemaking current (I_f) channel HCN.[50] Reductions in sinus rate, secondary to a decrease in I_f have been reported in HF patients, as has enhanced sensitively to acetylcholine, responses which would result in similarly decreased rates of sinus activation.[13] Direct evidence that genetic variants identified in patients with 'lone AF' either contribute to common forms of 'acquired' AF, including individuals with HF, is presently lacking, but speculation that there may be some overlap in underlying mechanisms runs high as further studies continue.

Genomics and ventricular arrhythmias

Ventricular fibrillation (VF) and tachycardias (VT), the precursors of most SCDs, are the major cause of acute ischemic arrhythmic mortality. VF/VT are also the principal direct cause of death in 20–50% of patients who die of HF.[51,52] As lifetime risk of HF in the US appears in about one in five in the population at large,[53] this implies that up to 10% of all such deaths may follow directly from this pathology. Unfortunately, other than nonspecific clinical stratifiers, such as reduced left ventricular function,[54] there are few sensitive

indicators able to identify individuals at highest risk, nor is there information on which groups or individuals would benefit most from those preventative strategies that are available, principally implantable defibrillators, antiarrhythmic pacers, and resynchronization devices, an increasingly costly clinical dilemma.

Causes of sudden arrhythmic deaths in heart failure are multiple, involving both electrogenic and structural factors and effects on Ca cycling.[13,55] Approximately half of all infarction survivors eventually develop symptoms of HF. In this group, ischemia is an especially potent arrhythmogenic trigger, contributing to a large proportion of mortalities.[56,57] Although conventional risk factors (e.g. abnormal lipid levels, diabetes, obesity, smoking, etc) and other subclinical influences are common in the HF population, they have not proven useful in defining risks for SCD or arrhythmias in individuals within these groups.[58] The hypothesis that there may be identifiable genetic susceptibilities which predispose to arrhythmic death in heart failure and infarction patients is thus one being pursued with increasing intensity, although it too remains unproven. Also, despite increasing evidence that new markers, such as the inflammatory indicator c-reactive protein,[59] may be helpful in this regard, there is little evidence that these have the power required to identify susceptibilities in individuals as opposed to larger groups and populations. Unfortunately, as was pointed out by Myerburg and colleagues,[60] it is precisely the large groups of asymptomatic individuals with subclinical disease, and those with early stage cardiac diseases who contribute the most to the high prevalence of SCD in the population at large (Fig. 27.2). As the figure shows however, HF patients enrolled in these various studies are among those with the highest overall SCD incidence, which suggests arrhythmia 'prevention' *per se* should be one of the highest priorities in reducing mortality from these increasingly epidemic conditions.

Three lines of investigation support the idea that gene variation may be an important contributor to arrhythmic death in common 'acquired' forms of heart disease. First, a large series of studies on rare 'monogenic' syndromes suggests a causal relationship between lethal ventricular arrhyth-

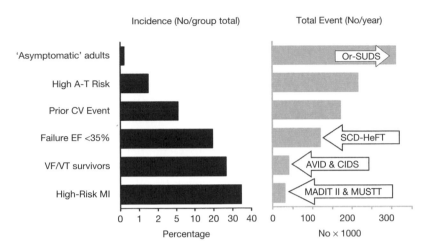

Figure 27.2. *Relationship between incidence and prevalence of sudden cardiac deaths in the US, suggesting the largest number of sudden deaths related to ventricular arrhythmias occurs in largely 'asymptomatic' individuals with subclinical forms of common cardiac diseases. The figure illustrates the important paradoxical observations of Myerburg et al.*[60] *that although specific risks in patients with most advanced forms of cardiovascular disease are the highest, these cohorts contribute to a much lesser extent to overall prevalence than individuals with lesser forms of pathology. Different patient cohorts with common forms of pathology have been evaluated in various clinical trials as indicated in the arrows on the right side of the figure. For example the MADIT II study was a clinical trial of implantable cardio defibrillator (ICD) therapy in post-myocardial infarction patients with high incidence of sudden arrhythmic death.*[61] *Individual risk was less in the clinical subgroups addressed in the AVID,*[62] *CIDS,*[63] *and recent SCD-Heft*[64] *trials. The Oregon Sudden Unexpected Death (Or-SUDS) study is a study being conducted in Portland by Dr. Sumeet Chugh, to assess clinical and genetic risk for SCD and lethal ventricular arrhythmias, in a broad-based case control design in the community. A-T, athero-thrombotic; CV, cardiovascular; EF, ejection fraction; MI, myocardial infarction.*

mias and mutationally derived dysfunction in electrogenic Na^+ and K^+ ion channel proteins of the cardiocyte membrane, the RyR2 Ca^{2+} release channel of the sarcoplamic reticulum, the L-type Ca channel,[65] and related pathways that modulate their activity, intracellular processing, insertion and membrane anchoring.[48] Variations in at least eight such proteins result in what have been termed 'ion channelopathies'.[66] These include (Table 27.1) at least six forms of the LQTS, short QT syndrome,[67] Andersen's syndrome,[68] Lenègre–Lev conduction disease,[69] Brugada's syndrome,[70] catecholamine-induced ventricular tachycardia,[71] arrhythmogenic right cardiomyopathy,[72] and most probably other, as yet unrecognized linkages. Hundreds of mutations in channel proteins have been identified and studied in enormous detail. Much has been learned regarding channel function and dysfunction, but unifying principals remain obscure. As opposed to observations on other 'high impact' monogenic ion channel diseases (e.g. cystic fibrosis), arrhythmogenic channel mutations appear to be due to multiple types of variation (deletion, early termination, substitution, etc) spread throughout relevant coding sequences. 'Hot spots', that is high-frequency sites of variation, and upstream sites of regulatory variation have only rarely, been encountered .[73] This has given rise to the concept that many of these diseases involve highly unique, almost personal, patterns of DNA variation, a finding which would make DNA screening across larger groups more difficult. As noted recently,[74] genotype–phenotype associations in monogenic ventricular channelopathies are quite variable, complicated by a high degree of genetic heterogeneity, expressivity and penetrance, probably reflecting modifying influences, including other genes, not yet identified.[75]

A second line of evidence concerning ventricular susceptibilities, one which highlights the importance of environmental interactions, concerns the so-called 'acquired' LQTS (aLQTS). This disorder derives from arrhythmogenic interactions between drugs and related circulating small molecules and the same protein underlying the LQT-2 syndrome, the HERG (KCNE2) K^+ channel α subunit.[76] Normally silent variations in this channel are believed to enhance susceptibility for binding of these exogenous ligands to HERG, resulting in episodes of life-threatening 'torsade-de-pointes' ventricular arrhythmias.[77] Interacting species include a number of antibiotics (e.g. erythromycin), antihistamines (e.g. terfenadine), gastrointestinal drugs (e.g. cisapride) and other drugs, including paradoxically, multiple antiarrhythmics (e.g. sotalol, dofetilide, quinidine, ibutilide etc), all of which result in delayed repolarization.[78,79] Arrhythmogenicity following exposure to aLQTS-inducing ligands may be magnified in some individuals who also have metabolically compromising genetic variants in hepatic pathways involved in the clearance of the interacting species. Defects, for example, in hepatic P450 cytochromes induced either by (1) independent genetic variation; or (2) agents that block P450 cytochrome activities (e.g. naturally occurring inhibitors in grapefruit or licorice extracts) can alter circulating levels of HERG-interacting drugs, further compromising channel function, even when offending compounds are present at levels believed safe in 'normal', as opposed to 'hypo-' or 'hyperactive' drug metabolizers. Although the prevalence of aLQTS-like susceptibilities are believed to be low in the overall population,[80] they could be involved in more cases of idiopathic SCD than generally acknowledged. It is worth noting that a scan for genetic variation in β-adrenergic receptor proteins showed no correlation between common polymorphisms at this level and arrhythmic events in aLQTS patients, alleviating to some extent, suspi-

cion they play significant role in arrhythmia triggering in these individuals.[81]

A final line of evidence relating to enhanced ventricular susceptibility comes from two epidemiological studies supporting the long-established belief that 'family history' is a predisposing factor in many common heart diseases. The Paris Prospective Study was a longitudinal follow-up survey of >7000 male civil servants followed an average of 23 years. Cardiac deaths in this population were retrospectively coded as either SCDs or myocardial infarctions depending on whether individuals succumbed rapidly after first appearance of cardiac symptoms. A majority of deaths in both categories were likely caused by infarction, with the principal difference believed to be the extent of underlying electrical susceptibility to arrhythmia following acute ischemia. Statistical analyses revealed that a parental history of SCD conveyed a 1.6–1.8-fold increase in SCD in offspring, independent of risks conveyed by conventional risk factors (lipid levels, diabetes, smoking, blood pressure etc). In a very limited number of cases where there was a history of both patients experiencing a sudden death event, relative risk of SCD in offspring was increased ~9-fold.[82] A similar study of 'primary cardiac arrest' (PCA) events in a 200-patient case-controlled study in Seattle resulted in the conclusion that risk of PCA is increased ~50% in first-degree relatives of index cases, as compared to controls.[83] Again, multivariate analysis indicated risk for sudden arrhythmic deaths was distinguishable from risk for myocardial infarction and statistically independent of conventional risk factors for atherothrombotic coronary disease. Subsequent reports have provided little further clarification regarding the particular genetic variants that might be involved in either population. There are some data that the effect seen in the Seattle study appears more significant in early-, but not late- (>age 65 years) onset PCAs,[84] and that there may be a correlation of SCDs with increased plasma fatty acid levels, suggestive of enhanced sympathetic activity in the Paris study.[85] Additional evidence consistent with this interpretation was provided by a recent report documenting destabilizing changes in heart rate responses (i.e. sympathetic balance) of those suffering SCD[86]. There is much evidence indicating that hyper-sympathetic activation-enhanced triggering of arrhythmogenesis occurs in common heart diseases,[8] and although this extends to the heart failure population where variants in β-receptors have been linked to HF susceptibility,[87] genetic variants in adrenergic pathways have not yet been associated with enhanced incidence of arrhythmias in this population.

Translating from rare syndromes to common heart diseases

Since the initial discoveries concerning the origins of the monogenic channelopathies in the mid-1990s, there has been much interest in the idea that 'high impact' mutations in channels contributing to action potential currents might identify key points of 'low impact' SCD susceptibility. While this seems a logical premise, it remains an unproven one. There are multiple explanations why this may be the case. Several that have been discussed are: (1) today's technologies are insufficiently sensitive to detect the potentially high number and low level of sequence variants causative in different diseases or individuals; (2) individual susceptibility loci may be so diverse or sufficiently conditional that they might only be detectable in the context of special environmental or behavioral influences; (3) there may be additional undiscovered 'protective' influences that segregate independently from disease-conferring alleles; and (4) very rare 'private' mutations, rather than more frequent 'public' polymorphisms may predominate. Although evidence is sparse, there are multiple biological clues which bear on this question. One comes from the finding of enhanced arrhythmia susceptibility in individuals of African descent, who have a common sequence variation in the SCN5A cardiac Na^+ channel gene.[88] A single C>A transversion suggested present in up to 4.6 million African Americans was associated with a 'low impact' enhancement in arrhythmia risk. The effect was of such reduced magnitude that it was only considered potentially significant in the context of other influences, such as amiodarone or perhaps aLQTS-inducing drugs. Another possible

clue has come from studies such as one recent one involving a common polymorphism in the HERG channel gene.[89] In this work, a single A>C base substitution was associated with a prolongation in cardiac repolarization more prominent in women as compared to men. Biophysical properties of the cloned variant channel were, as in the studies of the SCN5A polymorphism, consonant with a 'low impact' shift in channel electrical properties. Overall however, the significance of these types of findings, and many more likely to emerge in studies of the rare syndromes, remains mostly speculative.

Information on other sources of susceptibility remains relatively speculative, but intriguing observations have been made in patients with inherited hypertrophic and dilated cardiomyopathies (HCM and DCM respectively).[90,91] These conditions receive extensive coverage in Section IV 'Genetic basis for cardiomyopathy' later in this volume, and thus are not dealt with in any detail here. It is, however, worth noting that lethal arrhythmias are a prominent phenotype in many of these conditions. Primary defects in both HCM and DCM have been linked, as highlighted in Table 27.1 to genetic variation in more than a dozen different proteins, including many structural components of the sarcomere (e.g. cardiac actin, myosin, troponin, myosin-binding proteins etc.),[90,91] as well as some of the same Ca^{2+}-handling proteins identified as important sites of molecular dysfunction and perhaps arrhythmogenesis in HF models.[92] Highly disordered patterns of muscle cell orientation, fiber disarray, increased collagen synthesis and interstitial fibrosis are frequent anatomical findings in patients with familial HCMs,[93] where alterations in myocardial conduction may be a significant element in arrhythmia propagation.[94] Increases in collagen expression and fibrosis in heart failure patients have been ascribed in part to alterations in RAAS pathways, and specifically to elevations in circulating Ang II and aldosterone. Genetic variation in RAAS pathways is quite common, and polymorphisms in genes for angiotensin-converting enzyme, the Ang II receptor type I, the aldosterone synthase gene, and angiotensinogen have been well studied.[95] SNPs in RAAS pathway enzymes have for example been implicated as susceptibility sites in studies determining risks for hypertension,[96] and one study of a single family with HCM secondary to a mutation in myosin-binding protein C, reported that allele frequencies of five specific RAAS polymorphisms had a determining effect on cardiac structural phenotype in affected individuals.[27] Inhibitors of Ang II reduce electrical stability in HF patients, and inhibitors of the aldosterone mineralocorticoid receptor, spironolactone and eplerenone, drugs now used extensively in the HF population, result in a very significant reduction in arrhythmic deaths, of the same magnitude as that achieved independently with β-blockers. In studies where changes in collagen metabolism were also assessed, it was noted this effect occurred in parallel with a reduction in collagen metabolism.[24-26] Further, both HF and HCM patients have been reported to harbor defects in Ca^{2+} pathways conducive to arrhythmogenesis,[91-93] and genetic variation in RyR2 is, as noted earlier, important in inherited cases of arrhythmogenic right ventricular dysplasia (ARVD) and familial catecholaminergic polymorphic tachycardia.[70] Mutations in the SR Ca^{2+}-handling proteins phospholamban[100] and calsequestrin[101] are also associated with adrenergically mediated arrhythmias in experimental models and patients.[102] Taken together these results suggest that genetic variation in Ca^{2+}-related pathways, especially in the context of structural disease, may be critical to arrhythmia risks in HF patients, an idea highlighted in Fig. 27.3 which presents this author's current interpretation of the evidence regarding how disturbances in major arrhythmogenic pathways lead to SCD.

Future directions and improved approaches

Answers to the questions of whether, how and when genetic approaches to identify increased risks for atrial or ventricular arrhythmias will become clinically useful will depend on further progress in identifying new molecular susceptibilities, and in the development of new technologies for their assessment. Pioneering studies over the past decade have made much progress in identifying electrogenic sources of genetic variation in rare predisposing syndromes, and some clues relating to

variants of this nature may be relevant to much larger populations as well. Most recently, the focus of investigation is changing to the discovery of variants that may affect arrhythmogenesis in ways previously unanticipated while encompassing new technologies able to identify 'lower impact" variants, widely distributed in the population at large. Clinical applications of these new approaches are initially likely to include susceptibility analyses in high-risk families where there is an unexpectedly high history of arrhythmic events in addition to their continued utility in for example familial HCM and DCM. As progress is made, this approach seems likely to have important implications also for therapy selection. More widespread scanning in asymptomatic populations will probably remain unrealized until much more is learned about how specific variants underlie specific risks in high-incidence disease. This transition from understanding the genetic basis of rare inherited diseases, to deciphering genetic etiologies in patients and populations with common forms of diseases will require transition from the primary pursuit of family-based studies, to studies focused primarily on population genetics.[3] While the popular rare disease paradigm has been valuable in identifying molecular susceptibilities underlying ion channel diseases, extension of the analogy to patients with conditions such as HF remains unrealized and new concepts need to be proposed and evaluated. Understanding the many more dimensions of biological risk, those associated with as yet undetected electrogenic, structural and triggering abnormalities, for example in pathways of neural remodeling, in regulation of cytokine, inflammatory and immunological influences, and in the acute effects of ischemia; alterations in intracellular Ca^{2+} cycling, all represent important challenges at the frontiers of this quest. While it is a difficult problem, it is not insurmountable, reflecting instead one of the field's primary intellectual challenges. While new approaches like whole genome association studies, and technologies offering 'order of magnitude' improvements in SNP and polymorphism screening will need to be perfected and validated, we need new ideas able to translate basic laboratory findings into new approaches to therapy in the same way that expansion of

ELECTROGENESIS

CONDUCTION

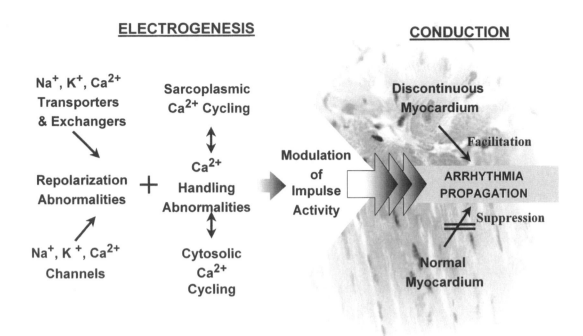

Figure 27.3. *Electrogenic and conductive risks and mechanisms believed to be important in arrhythmias in patients with common cardiac diseases. Disturbances in Na$^+$ and K$^+$ ion channel pathways, for example those elicited by acute ischemia, and in pathways of intracellular Ca^{2+} metabolism, are depicted as commonly occurring mechanisms impacting myocardial electrical stability and arrhythmia propagation. Alterations in electrogenesis may be facilitated in structurally disordered hearts, whereas in normal continuous myocardium such impulses may be suppressed. Genetic risks influencing both aberrant impulse formation and conduction are believed to alter susceptibilities in accord with how 'proximal' or 'distal' such risks may be with respect to final common pathways of arrhythmia propagation (see Fig. 27.1). Different elements in this figure may apply differentially to patients with different predisposing cardiac diseases, and the figure should not be interpreted as indicating each such event occurs in every case, or is necessary in each and all cases. Thus different risks and mechanisms would be expected to be more important in some pathologies than others. For example derangements in Ca^{2+} metabolism might be expected to play a more prominent role in arrhythmias occurring with heart failure, than perhaps in individuals with a structurally normal myocardium who experience an acute ischemic infarction.*

Mendel's ideas on peas led to the discovery of DNA and now delineation of the genome and the beginnings of an understanding the complexities of its function.

References

1. www.genetests.org (accessed 5 May 2005)

2. Roberts R, Brugada R. Genetics and arrhythmias. Annu Rev Med 2003; 54: 257–67.

3. Risch NJ. Searching for genetic determinants in the new millennium. Nature 2000; 405: 847–56.

4. Arking DE, Chugh SS, Chakravarti A, Spooner PM. Genomics in sudden cardiac death. Circ Res 2004; 94.

5. Turner ST, Boerwinkle E. Genetics of blood pressure, hypertensive complications, and antihyper-

tensive drug responses. Pharmacogenomics 2003; 4: 53–65.

6. Bezzina C, Veldkamp MW, van Den Berg MP et al. A single Na$^+$ channel mutation causing both long-QT and Brugada syndromes. Circ Res 1999; 85: 1206–13.

7. Grant AO, Carboni MP, Neplioueva V et al. Long QT syndrome, Brugada syndrome, and conduction system disease are linked to a single sodium channel mutation. J Clin Invest 2002; 110: 1201–9.

8. Zipes DP. Mechanisms of clinical arrhythmias. J Cardiovasc Electrophysiol 2003; 14: 902–12.

9. Cranefield P. Action potentials, afterpotentials and arrhythmias. Circ Res; 41: 415–23.

10. Wit AL, Rosen MR. Pathophysiologic mechanisms of cardiac arrhythmias. Am Heart J 1983; 106: 798–811.

11. El-Sherif N. Mechanism of ventricular arrhythmias in the long QT syndrome: on hermeneutics. J Cardiovasc Electrophysiol 2001; 12: 973–6.

12. Damiano BRM. Effects of pacing on triggered activity induced by early afterdepolarizations. Circulation; 69: 1013–25.

13. Janse MJ. Electrophysiological changes in heart failure and their relationship to arrhythmogenesis. Cardiovasc Res 2004; 61: 208–17.

14. Chen YJ, Chen SA, Chen YC et al. Effects of rapid atrial pacing on the arrhythmogenic activity of single cardiomyocytes from pulmonary veins: implication in initiation of atrial fibrillation. Circulation 2001; 104: 2849–54.

15. Shah DC, Haissaguerre M, Jais P. Toward a mechanism-based understanding of atrial fibrillation. J Cardiovasc Electrophysiol 2001; 12: 600–1.

16. Marcus F, Towbin JA, Zareba W et al. Arrhythmogenic right ventricular dysplasia/cardiomyopathy (ARVD/C): a multidisciplinary study: design and protocol. Circulation 2003; 107: 2975–8.

17. Sun Y, Kiani MF, Postlethwaite AE et al. Infarct scar as living tissue. Basic Res Cardiol 2002; 97: 343–7.

18. Chugh SS, Kelly KL, Titus JL. Sudden cardiac death with apparently normal heart. Circulation 2000; 102: 649–54.

19. Keating MT, Sanguinetti MC. Molecular and cellular mechanisms of cardiac arrhythmias. Cell 2001; 104: 569–80.

20. Antzelevitch C, Brugada P, Brugada J et al. Brugada syndrome: 1992–2002: a historical perspective. J Am Coll Cardiol 2003; 41: 1665–71.

21. Kucera JP, Kleber AG, Rohr S. Slow conduction in cardiac tissue: insights from optical mapping at the cellular level. J Electrocardiol 2001; 34 (Suppl): 57–64.

22. Kanno S, Saffitz JE. The role of myocardial gap junctions in electrical conduction and arrhythmogenesis. Cardiovasc Pathol 2001; 10: 169–77.

23. Peters NS, Wit AL. Myocardial architecture and ventricular arrhythmogenesis. Circulation 1998; 97: 1746–54.

24. Zannad F, Alla F, Dousset B et al. Limitation of excessive extracellular matrix turnover may contribute to survival benefit of spironolactone therapy in patients with congestive heart failure: insights from the randomized aldactone evaluation study (RALES). Rales Investigators. Circulation 2000; 102: 2700–6.

25. Pitt B, Zannad F, Remme WJ et al. The effect of spironolactone on morbidity and mortality in patients with severe heart failure. Randomized Aldactone Evaluation Study Investigators. N Engl J Med 1999; 341: 709–17.

26. Pitt B, Remme W, Zannad F et al. Eplerenone, a selective aldosterone blocker, in patients with left ventricular dysfunction after myocardial infarction. N Engl J Med 2003; 348: 1309–21.

27. Ortlepp JR, Vosberg HP, Reith S et al. Genetic polymorphisms in the renin-angiotensin-aldosterone system associated with expression of left ventricular hypertrophy in hypertrophic cardiomyopathy: a study of five polymorphic genes in a family with a disease causing mutation in the myosin binding protein C gene. Heart 2002; 87: 270–5.

28. Sing CF, Stengard JH, Kardia SL. Genes, environment, and cardiovascular disease. Arterioscler Thromb Vasc Biol 2003; 23: 1190–6.

29. Ayala C WW, Croft JB, Hyduk A et al. Public Health and Aging: Atrial Fibrillation as a Contributing Cause of Death and Medicare Hospitalization – United States, 1999. Morb Mortal Wkly Rep 2003; 52: 128–31.

30. Go AS, Hylek EM, Phillips KA et al. Prevalence of diagnosed atrial fibrillation in adults: national implications for rhythm management and stroke prevention: the Anticoagulation and Risk Factors in Atrial Fibrillation (ATRIA) Study. JAMA 2001; 285: 2370–5.

31. Benjamin EJ, Wolf PA, D'Agostino RB et al. Impact of atrial fibrillation on the risk of death: the Framingham Heart Study. Circulation 1998; 98: 946–52.

32. Falk RH. Atrial fibrillation. N Engl J Med 2001; 344: 1067–78.

33. Nattel S, Li D. Ionic remodeling in the heart: pathophysiological significance and new therapeutic opportunities for atrial fibrillation. Circ Res 2000; 87: 440–7.

34. Chugh SS, Blackshear JL, Shen WK et al. Epidemiology and natural history of atrial fibrillation: clinical implications. J Am Coll Cardiol 2001; 37: 371–8.

35. Allessie MA, Boyden PA, Camm AJ et al. Pathophysiology and prevention of atrial fibrillation. Circulation 2001; 103: 769–77.

36. Thijssen VL, van der Velden HM, van Ankeren EP et al. Analysis of altered gene expression during sustained atrial fibrillation in the goat. Cardiovasc Res 2002; 54: 427–37.

37. Maisel WH, Stevenson LW. Atrial fibrillation in heart failure: epidemiology, pathophysiology, and rationale for therapy. Am J Cardiol 2003; 91: 2D–8D.

38. Wang TJ, Larson MG, Levy D et al. Temporal relations of atrial fibrillation and congestive heart failure and their joint influence on mortality: the Framingham Heart Study. Circulation 2003; 107: 2920–5.

39. Darbar D, Herron KJ, Ballew JD et al Familial atrial fibrillation is a genetically heterogeneous disorder. J Am Coll Cardiol 2003; 41: 2185–92.

40. Morita H, DePalma SR, Arad M et al. Molecular epidemiology of hypertrophic cardiomyopathy. Cold Spring Harb Symp Quant Biol 2002; 67: 383–8.

41. Arad M, Seidman JG, Seidman CE. Phenotypic diversity in hypertrophic cardiomyopathy. Hum Mol Genet 2002; 11: 2499–506.

42. Sebillon P, Bouchier C, Bidot LD et al. Expanding the phenotype of LMNA mutations in dilated cardiomyopathy and functional consequences of these mutations. J Med Genet 2003; 40: 560–7.

43. Mohler PJ, Schott JJ, Gramolini AO et al. Ankyrin-B mutation causes type 4 long-QT cardiac arrhythmia and sudden cardiac death. Nature 2003; 421: 634–9.

44. Brugada R, Tapscott T, Czernuszewicz GZ et al. Identification of a genetic locus for familial atrial fibrillation. N Engl J Med 1997; 336: 905–11.

45. Ellinor PT, Shin JT, Moore RK et al. Locus for atrial fibrillation maps to chromosome 6q14–16. Circulation 2003; 107: 2880–3.

46. Sylvius N, Tesson F, Gayet C et al. A new locus for autosomal dominant dilated cardiomyopathy identified on chromosome 6q12-q16. Am J Hum Genet 2001; 68: 241–6.

47. Chen YH, Xu SJ, Bendahhou S et al. KCNQ1 gain-of-function mutation in familial atrial fibrillation. Science 2003; 299: 251–4.

48. Keating MT, Sanguinetti MC. Molecular genetic insights into cardiovascular disease. Science 1996; 272: 681–5.

49. Lai LP, Su MJ, Yeh HM et al. Association of the human minK gene 38G allele with atrial fibrillation: evidence of possible genetic control on the pathogenesis of atrial fibrillation. Am Heart J 2002; 144: 485–90.

50. Schulze-Bahr E, Neu A, Friederich P et al. Pacemaker channel dysfunction in a patient with sinus node disease. J Clin Invest 2003; 111: 1537–45.

51. Uretsky BF, Sheahan RG. Primary prevention of sudden cardiac death in heart failure: will the solution be shocking? J Am Coll Cardiol 1997; 30: 1589–97.

52. National Heart lBI. Fact Book Fiscal Year 2002. In: Washington, DC; 2003.

53. Lloyd-Jones DM, Larson MG, Leip EP et al. Lifetime risk for developing congestive heart failure: the Framingham Heart Study. Circulation 2002; 106: 3068–72.

54. Poole-Wilson PA, Uretsky BF, Thygesen K et al. Mode of death in heart failure: findings from the ATLAS trial. Heart 2003; 89: 42–8.

55. Tomaselli GF, Beuckelmann DJ, Calkins HG et al. Sudden cardiac death in heart failure. The role of abnormal repolarization. Circulation 1994; 90: 2534–9.

56. Uretsky BF, Thygesen K, Armstrong PW et al. Acute coronary findings at autopsy in heart failure patients with sudden death: results from the assessment of treatment with lisinopril and survival (ATLAS) trial. Circulation 2000; 102. 611–6.

57. Every N, Hallstrom A, McDonald KM et al. Risk of sudden versus nonsudden cardiac death in patients with coronary artery disease. Am Heart J 2002; 144: 390–6.

58. Rifai N, Ridker PM. Inflammatory markers and coronary heart disease. Curr Opin Lipidol 2002; 13: 383–9.

59. Albert CM, Ma J, Rifai N et al. Prospective study of C-reactive protein, homocysteine, and plasma lipid levels as predictors of sudden cardiac death. Circulation 2002; 105: 2595–9.

60. Myerburg RJK, Castellanos A. Epidemiology of sudden cardiac death: population characteristics, conditioning risk factors, and dyndamic risk factors. In: Spooner PM, Brown AM (eds). Ion Channels in the Cardiovascular System: function and dysfunction. New York: Futura Publishing Company; 1994: 15–35.

61. Moss AJ, Zareba W, Hall WJ et al. Prophylactic implantation of a defibrillator in patients with myocardial infarction and reduced ejection fraction. N Engl J Med 2002; 346: 877–83.

62. The Antiarrhythmics versus Implantable Defibrillators (AVID) Investigators. A comparison of antiarrhythmic-drug therapy with implantable defibrillators in patients resuscitated from near-fatal ventricular arrhythmias. N Engl J Med 1997; 337: 1576–84.

63. Connolly SJ, Gent M, Roberts RS et al. Canadian Implantable defibrillator Study (CIDS). A randomized trial of the implantable cardioverter defibrillator against amiodarone. Circulation. 2000; 101:1297-1302.

64. Bardy GH, Lee KL, Mark DB et al. Amiodarone or an implantable cardioverter–defibrillator for congestive heart failure. N Engl J Med 2005; 352: 225–37.

65. Splawski I, Timothy KW, Decher N, et al. Severe arrhythmia disorder caused by cardiac L-type calcium channel mutations. Proc Natl Acad Sci USA 2005 Apr 29 [Epub ahead of print].

66. Marban E. Cardiac channelopathies. Nature 2002; 415: 213–8.

67. Brugada R, Hong K, Dumaine R et al. Sudden death associated with short-QT syndrome linked to mutations in HERG. Circulation 2004; 109: 30–5.

68. Lange PS, Er F, Gassanov N et al. Andersen mutations of KCNJ2 suppress the native inward rectifier current IK1 in a dominant-negative fashion. Cardiovasc Res 2003; 59: 321–7.

69. Probst V, Kyndt F, Potet F et al. Haploinsufficiency in combination with aging causes SCN5A-linked hereditary Lenegre disease. J Am Coll Cardiol 2003; 41: 643–52.

70. Splawski I, Shen J, Timothy KW et al. Spectrum of mutations in long-QT syndrome genes. KVLQT1, HERG, SCN5A, KCNE1, and KCNE2. Circulation 2000; 102: 1178–85.

71. Priori SG, Napolitano C, Tiso N et al. Mutations in the cardiac ryanodine receptor gene (hryr2) underlie catecholaminergic polymorphic ventricular tachycardia. Circulation 2001; 103: 196–200.

72. Tiso N, Stephan DA, Nava A et al. Identification of mutations in the cardiac ryanodine receptor gene in families affected with arrhythmogenic right ventricular cardiomyopathy type 2 (ARVD2). Hum Mol Genet 2001; 10: 189–94.

73. Zhang L, Vincent GM, Baralle M et al. An intronic mutation causes long QT syndrome. J Am Coll Cardiol 2004; 44: 1283–91.

74. Moss AJ, Zareba W, Kaufman ES et al. Increased risk of arrhythmic events in long-QT syndrome with mutations in the pore region of the human ether-a-go-go-related gene potassium channel. Circulation 2002; 105: 794–9.

75. Priori SG. Inherited arrhythmogenic diseases: the complexity beyond monogenic disorders. Circ Res 2004; 94: 140–5.

76. Sesti F, Abbott GW, Wei J et al. A common polymorphism associated with antibiotic-induced cardiac arrhythmia. Proc Natl Acad Sci USA 2000; 97: 10613–8.

77. Roden DM. Cardiovascular pharmacogenomics. Circulation 2003; 108: 3071–4.

78. Roden DM. Pharmacogenetics and drug-induced arrhythmias. Cardiovasc Res 2001; 50: 224–31.

79. Roden DM. Drug Therapy: Drug induced prolongation of the QT interval. New Eng J Med 2004; 350: 1013–22.

80. Yang P, Kanki H, Drolet B et al. Allelic variants in long-QT disease genes in patients with drug-associated torsades de pointes. Circulation 2002; 105: 1943–8.

81. Kanki H, Yang P, Xie HG et al. Polymorphisms in beta-adrenergic receptor genes in the acquired long QT syndrome. J Cardiovasc Electrophysiol 2002; 13: 252–6.

82. Jouven X, Desnos M, Guerot C et al. Predicting sudden death in the population: the Paris Prospective Study I. Circulation 1999; 99: 1978–83.

83. Friedlander Y, Siscovick DS, Weinmann S et al. Family history as a risk factor for primary cardiac arrest. Circulation 1998; 97: 155–60.

84. Friedlander Y, Siscovick DS, Arbogast P et al. Sudden death and myocardial infarction in first degree relatives as predictors of primary cardiac arrest. Atherosclerosis 2002; 162: 211–6.

85. Jouven X, Charles MA, Desnos M et al. Circulating nonesterified fatty acid level as a predictive risk factor for sudden death in the population. Circulation 2001; 104: 756–61.

86 Jouven X, Empana JP, Schwartz PJ et al. Heart-rate profile during exercise as a predictor of sudden death. N Engl J Med. 2005;352:1951-8.

87. Mialet Perez J, Rathz DA, Petrashevskaya NN et al. Beta 1-adrenergic receptor polymorphisms confer differential function and predisposition to heart failure. Nat Med 2003; 9: 1300–5.

88. Splawski I, Timothy KW, Tateyama M et al. Variant of SCN5A sodium channel implicated in risk of cardiac arrhythmia. Science 2002; 297: 1333–6.

89. Bezzina CR, Verkerk AO, Busjahn A et al. A common polymorphism in KCNH2 (HERG) hastens cardiac repolarization. Cardiovasc Res 2003; 59: 27–36.

90. Seidman JG, Seidman C. The genetic basis for cardiomyopathy: from mutation identification to mechanistic paradigms. Cell 2001; 104: 557–67.

91. Fatkin D, Graham RM. Molecular mechanisms of inherited cardiomyopathies. Physiol Rev 2002; 82: 945–80.

92. Shannon TR, Pogwizd SM, Bers DM. Elevated sarcoplasmic reticulum Ca^{2+} leak in intact ventricular myocytes from rabbits in heart failure. Circ Res 2003; 93: 592–4.

93. Marian AJ, Salek L, Lutucuta S. Molecular genetics and pathogenesis of hypertrophic cardiomyopathy. Minerva Med 2001; 92: 435–51.

94. Wu TJ, Ong JJ, Hwang C et al. Characteristics of wave fronts during ventricular fibrillation in human hearts with dilated cardiomyopathy: role of increased fibrosis in the generation of reentry. J Am Coll Cardiol 1998; 32: 187–96.

95. Johnson JA, Humma LM. Pharmacogenetics of Cardiovascular Drugs. Briefings in Fundamental Genomics and Proteomics 2002; 1: 1–14.

96. Zhu X, Chang YP, Yan D et al. Associations between hypertension and genes in the renin-angiotensin system. Hypertension 2003; 41: 1027–34.

97. Marks AR. Ryanodine receptors/calcium release channels in heart failure and sudden cardiac death. J Mol Cell Cardiol 2001; 33: 615–24.

98. Fatkin D, McConnell BK, Mudd JO et al. An abnormal Ca^{2+} response in mutant sarcomere protein-mediated familial hypertrophic cardiomyopathy. J Clin Invest 2000; 106: 1351–9.

99. Semsarian C, Ahmad I, Giewat M et al. The L-type calcium channel inhibitor diltiazem prevents cardiomyopathy in a mouse model. J Clin Invest 2002; 109: 1013–20.

100. Schmitt JP, Kamisago M, Asahi M et al. Dilated cardiomyopathy and heart failure caused by a mutation in phospholamban. Science 2003; 299: 1410–3.

101. Viatchenko-Karpinski S, Terentyev D, Gyorke I, Terentyeva R et al. Abnormal calcium signaling and sudden cardiac death associated with mutation of calsequestrin. Circ Res 2004; 94: 471–477.

102. Wehrens XH, Marks AR. Altered function and regulation of cardiac ryanodine receptors in cardiac disease. Trends Biochem Sci 2003; 28: 671–8.

Genetic basis for cardiomyopathy

CHAPTER 28

Epidemiology of left ventricular hypertrophy and dilated cardiomyopathy

Jonathan N Bella, Richard B Devereux, Craig T Basson

Introduction

Heart disease is a major cause of morbidity and mortality worldwide, and is of major public health importance. In 2001, heart disease affected more than 64 million Americans, and accounted for 1 in every 2.6 deaths in the United States.[1] Of these, five million Americans were diagnosed with heart failure. Heart failure had an overall death rate of 18.7 per 1000 in 2001, and cost an estimated $28.8 billion in direct and indirect costs in 2004.

Heart failure is a syndrome arising when myocardial performance is insufficient to meet the metabolic demand of vital tissues and organs. This complex pathophysiological state represents the final common pathway of a variety of cardiovascular conditions that include hypertension, coronary heart disease, valvular heart disease, cardiomyopathy and congenital heart disease.[2] In response to these cardiovascular insults, the left ventricle compensates by either hypertrophy or dilatation to preserve and maintain cardiac pump performance. Conditions that cause pressure overload, such as hypertension and aortic stenosis, induce concentric left ventricular (LV) hypertrophy, while conditions that promote volume overload, such as valvular regurgitation, lead to LV dilatation and eccentric hypertrophy.

Left ventricular hypertrophy

Prognosis

While compensatory LV hypertrophy or dilatation may initially preserve cardiac pump function, they ultimately become maladaptive. Indeed, LV hypertrophy has been shown to be an independent predictor of increased cardiovascular morbidity and mortality in clinical and population based samples.[3–15] As may be seen in Table 28.1, individuals with LV hypertrophy consistently have ≥2-fold increases in rates of adverse events, as indicated by the odds ratios.

In the 1960s, the association of electrocardiographic LV hypertrophy with cardiovascular events was first described in clinical and population-based studies.[3–5] In the first study to relate direct measurements of echocardiographic LV mass to prognosis, 140 men with uncomplicated essential hypertension were followed for five years to determine the incidence of 'hard' cardiovascular events, i.e. cardiac death, myocardial infarction, stroke or angina pectoris requiring coronary bypass surgery.[6] The 20% ($n = 29$) of patients with LV mass exceeding a predefined partition value had approximately 4-fold higher rate of morbid events (24%) than the patients without LV hypertrophy (6%).

Other studies have subsequently extended these findings by demonstrating that increased LV mass strongly predicts cardiac and cerebrovascular

Table 28.1. *Prognosis associated with left ventricular hypertrophy by various methods*

Reference	Method of diagnoisis	Number of patients analyzed	Endpoint	Odds ratio
Sokolow and Perloff, 1961[3]	ECG	439	Death	8.0
Breslin et al., 1966[4]	ECG	631	Death	4.8
Kannel et al., 1969[5]	ECG	5055	CV events	2.5
Casale et al., 1986[6]	Echo	140	Death or CV events	4.0
Silberberg et al., 1989[7]	Echo	119	Death	3.7
Aronow et al., 1988[8]	Echo	554	Ventricular fibrillation or sudden death	4.7
Levy et al., 1990[9]	Echo	3220	All-cause mortality and All CV events	2.4 and 2.5
Koren et al., 1991[10]	Echo	280	CV death and all CV events	14.2 and 3.0
Ghali et al., 1992[11]	Echo	785	All-cause and cardiac mortality	2.1
Bikkina et al., 1994[12]	Echo	447	Stroke, TIA	2.7
Gardin et al., 2001[13]	Echo	5888	Incident CHD, CHF and stroke	3.4
Quinones et al., 2000[15]	Echo	1172	Death	1.4

CHD, coronary heart disease; CHF, congestive heart failure; CV, cardiovascular; TIA, transient ischemic attack.

morbidity and mortality, independently of traditional risk factors.[7–15] A report from the Framingham Heart Study showed that increased LV mass strongly predicted all-cause and cardiac mortality and coronary heart disease events in adults over 40 years, independently of conventional risk factors.[9] Furthermore, they found that age-adjusted incidence of stroke or transient ischemic attack was substantially higher in the highest quartile of LV mass (18.4% in men and 12.2% in women) than in the lowest quartile (5.2% and 2.2%, respectively).[12] Similarly, among older adults (age >65 years) in the Cardiovascular Heath Study, the highest quartile of LV mass conferred a hazards ratio of 3.36 for incident congestive failure, compared with the lowest quartile.[13] In adults with coronary artery disease, Liao et al. found that the attributable risk of death from LV hypertrophy was greater (2.4%) than that of multivessel coronary artery disease (1.6%) or low ejection fraction at catheterization (2.0%).[14] Quinones et al.[15] extended this result to the patients enrolled in the Studies of Left Ventricular Dysfunction (SOLVD) trials, finding higher LV mass to be independently associated with higher age-adjusted risks of death

(risk ratio = 2.75) and cardiovascular hospitalization (risk ratio = 1.81).[15]

Prevalence of and methods to detect left ventricular hypertrophy

The prevalence of LV hypertrophy is dependent upon the method used and the population under study. The electrocardiogram (ECG) has been the traditional method to detect ventricular hypertrophy in epidemiological studies. Electrocardiographic diagnosis of LV hypertrophy is based mainly on the magnitude of electrical potential generated by high LV mass and the abnormalities of ventricular conduction and repolarization and of the left atrium that parallel changes in ventricular anatomy.[16,17] However, specificity of electrocardiographic criteria for LV hypertrophy is better than sensitivity (Table 28.2).[16,18–21] In epidemiological studies, LV hypertrophy prevalence by various electrocardiographic criteria ranges from 5% to 7%.[5,9,22]

Echocardiography has provided a safe non-invasive method to evaluate cardiac anatomy and function. The development of necropsy-validated echocardiographic formulae has permitted accurate assessment of LV mass.[23,24] Echocardiographic

Table 28.2. *Sensitivity and specificity of common electrocardiographic criteria for left ventricular hypertrophy*

Criterion	Sensitivity (%)	Specificity (%)
Sokolow–Lyon voltage SV1 + RV5/V6 >35 mm	22	100
RaVL >11 mm	11	100
Romhilt–Estes Score >5 points	33	94
Voltage = 3 points (any one of the following: R or S in limb lead >20 mm, SV1/V2/V3 >25 mm or RV4/V5/V6 >25 mm)		
Strain pattern		
without digitalis = 3 points		
with digitalis = 1 point		
Left axis deviation >−15° = 2 points		
QRS duration >0.09 s = 1 point		
Left atrial enlargement = 3 points		
Intrinsicoid defelection V5/V6 >0.05 s = 1 point		
Cornell voltage RAVL + SV3 >28 mm in men and >25 mm in women	42	96
Framingham	7	98
Strain pattern (>1 mm ST segment depression in V2–V6) and at least one of the following: RI + SIII >25 mm, SV1/V2 + RV5/V6 >35 mm, S in right precordial lead >25 mm or R in left precordial lead >25 mm		
Cornell voltage–duration product SV3 + RaVL([+6 mm in women]) QRS>2440 mm × ms	51	95

LV mass has been shown to have excellent inter-study reproducibility,[25] and be a more sensitive tool for detection of ventricular hypertrophy than electrocardiography. In population-based samples, the prevalence of echocardiographic LV hypertrophy is 15–18% compared to 5–7% detected by electrocardiography.[9,20,22]

Cardiovascular magnetic resonance has emerged as a potential gold standard for non-invasive detection of LV mass.[26] Studies have shown good agreement between magnetic resonance-derived and actual LV mass in humans, with good inter-study reproducibility.[27,28] However, the high cost of immobile laboratories, and patient aversion to claustrophobic imaging milieus limit its widespread use.

Stimuli to left ventricular hypertrophy

The left ventricle normally grows continuously from infancy to adulthood, with cardiomyocyte enlargement or hypertrophy accounting for most of the increase in size.[29] In apparently normal children and adults, LV mass is closely correlated with body size.[30,31] Traditionally, body size has been taken into account, by adjusting LV mass for body surface area. However, this method of indexing ventricular mass may misclassify obesity-induced ventricular hypertrophy as normal. Height-based indexations of LV mass, which identify both blood pressure and obesity-associated increases in myocardial mass, have been shown to maintain and perhaps enhance prediction of cardiac risk.[32]

After puberty, men have higher ventricular mass than women in relation to body size. LV mass/body surface area is 10–20% greater in men than in women, which parallels the sex difference in fat-free mass, and may reflect genetic, hormonal or exercise effects that influence both skeletal and cardiac muscle.[29] Thus, age and sex need to be taken into account when establishing upper normal limits for LV mass.

In addition to demographic factors, hemodynamic variables play an important role in determining LV mass. Our understanding of the full impact of blood pressure on the heart has been enhanced by the use of 24-hour ambulatory blood pressure monitoring. LV mass or wall thicknesses are more closely related to 24-hour than casual blood pressures.[33] In a study of normotensive and hypertensive adults, patients with concentric LV hypertrophy had the highest ambulatory systolic and diastolic blood pressures, while those with eccentric LV hypertrophy had lower ambulatory than clinic blood pressures.[33] Exaggerated blood pressure increase during exercise may also contribute to the development of LV hypertrophy.[34,35]

Numerous studies have shown that obesity is associated with increased LV mass.[30,36] High-salt diets have also been linked to hypertensive LV hypertrophy.[37] Both these factors may increase stroke volume, thereby increasing chamber volume and predisposing to eccentric LV hypertrophy.[38] The important role of volume load in the pathogenesis of LV hypertrophy is underscored by the fact that chamber size and stroke volume are more closely related to LV mass than systolic blood pressure in normotensive and hypertensive adults, and in population-based samples.[38–42] In addition to hemodynamic pressure and volume load, LV mass is also affected by a negative relation between LV contractility and myocardial mass.[38,42] In a recent population-based report, almost half of the variability in LV mass was associated with inter-individual differences in stroke volume, contractility, systolic blood pressure, body mass indices and sex.[42]

Another stimulus to LV hypertrophy is abnormal glucose metabolism. Several epidemiological studies have shown that adults with diabetes have higher LV mass, independently of other stimuli to LV hypertrophy.[43–45] This relationship may be important in view of the increasing prevalence of diabetes in the United States.[46]

These stimuli to LV hypertrophy induce not only an increase in cardiac mass and wall thickness, but also a fundamental reconfiguration of the protein, molecular and genetic components of the myocardium, which will be discussed in detail in subsequent chapters.

Genetic epidemiology of LV hypertrophy

Only one-half to two-thirds of the inter-individual variability of LV mass can be explained by its clinical and hemodynamic correlates.[22,30–45] Recent evidence indicates that LV mass is influenced by genetic factors. Monozygotic twins have substantially more similar LV mass than dizygotic twins.[47–49] Adams et al. evaluated within-pair differences in LV wall thicknesses and dimensions in 31 monozygotic twin pairs, 10 dizygotic twin pairs, 6 siblings and 30 unrelated individuals.[47] They found that they were lower within twin-pair differences for LV internal diameter and posterior wall thickness, but not for interventricular septal wall thickness, suggesting that familial influences, including both genetic and environmental factors, are important determinants of cardiac size. In contrast, Bielen et al. found significant heritability (h^2) for LV wall thickness (0.29 and 0.28 for interventricular septal wall thickness and posterior wall thickness, respectively), but not for LV internal diameter in 32 monozygotic and 21 dizygotic twin pairs, after adjusting for age, weight, blood pressure and skinfold thicknesses.[49] In 22 African-American normotensive twin pairs, LV mass/body surface area, adjusted for sex and systolic blood pressure, had a h^2 of 0.58.[48] However, twins share environmental factors to a unique degree,[50] and h^2 estimates, which in twins include this shared environmental component, are usually higher than estimates derived from other relatives.

Epidemiological studies have confirmed that LV mass is heritable, independently of co-variates.[51–54] The Framingham Heart Study, using intraclass correlation methods, assessed the h^2 of LV mass in their adult, overwhelmingly Caucasian population.[51] The estimated h^2 of adjusted LV mass was between 0.24 (from aunt/uncle–niece/nephew correlation) and 0.32 (sibling–sibling correlation), with an intermediate estimate of 0.30 from parent–child correlation. The Tecumseh Offspring Study has reported similar parent–child correlation (0.28, $P = 0.006$) for LV mass.[52] A recent report from the Hypertension Genetic Epidemiology Network (HyperGEN) study, in a population-based sample of Caucasian and African-American hypertensives, indicated that sibling correlations for LV mass among African-Americans ranged from 0.22 (brother–sister) to 0.44 (brother–brother), compared to 0.05 (brother–sister) to 0.22 (sister–sister) among Caucasians, while sibling correlations for relative wall thickness, a measure of ventricular concentricity, were lower in African-American siblings (0.04–0.12) than their Caucasian counterparts (0.19–0.28),[53] suggesting ethnic heterogeneity among genes influencing ventricular geometry. Among adult American Indians, the h^2 of LV mass and relative wall thickness were both 0.17 after adjusting for a comprehensive set of covariates that included age, sex, body size, blood pressure, heart rate, diabetes and medications.[54] The substantial h^2 for LV mass in population-based samples of varying ethnicity indicates the robust genetic influence on common forms of LV hypertrophy.

The search for candidate genes that influence LV mass in population-based samples has intensified recently. Potential candidate genes include ones encoding proteins regulating cardiac structure, hemodynamic load, calcium homeostasis, hormones, substrate metabolism, growth factors, energy metabolism and cell signaling.[55] Genome-wide linkage analyses in extended families, and association studies have been performed in population-based samples to identify genes influencing LV mass. Such linkage analyses have been highly fruitful.[56,57] Genome scans from three population-based studies have shown evidence for linkage of LV mass to chromosome 7,[56] chromosome 12,[57] and chromosome 22.[58] However, the specific genes responsible for the observed linkage results have yet to be identified in these epidemiological studies.

Multiple studies have tried to link single-nucleotide polymorphisms (SNPs) in regulatory and pathway genes with common forms of LV hypertrophy. Considerable attention has been devoted to polymorphisms in enzymes and hormones involved in the renin–angotensin–aldosterone system. The angiotensin-converting enzyme (ACE) insertion deletion polymorphism,[59,60] angiotensin II type I receptor gene A1166C polymorphism,[60] and angiotensinogen gene M235T polymorphism,[60,61] have been implicated in exercise-induced LV hypertrophy. The angiotensin II type 2 receptor gene (+1675 G/A) polymorphism has been associated with abnormal LV geometry in young men with mild hypertension.[62] A SNP in the aldosterone synthase gene ([CYP11B2] –344 C/T) has also been found to be associated with eccentric LV hypertrophy in essential hypertension.[63–65] A recent study suggests that a polymorphism of the β_1 adrenergic receptor gene (glycine for arginine at position 389) affects LV mass in patients with renal failure.[66] Moreover, the G protein β_3 subunit (C825T) polymorphism has been associated with LV mass in hypertension.[67,68]

Other candidate genes have been selected based on their role in myocardial fatty acid oxidation. Recent studies have shown that high LV mass and dilated cardiomyopathy are associated with abnormal fatty acid metabolism.[69,70] Jamshidi et al. reported that a SNP within an intron of the peroxisome proliferator-activated receptor alpha (PPARα) influenced LV growth in response to exercise and hypertension.[71] In a recent report, the Framingham investigators found statistically significant associations of the angiotensin receptor type 1 (AGTR2) gene with LV mass ($P = 0.05$) and LV chamber size ($P = 0.007$), the β_2 adrenergic receptor (ADRB2) gene with LV mass ($P = 0.02$) and LV wall thickness ($P = 0.005$), and the cardiac troponin T (TNNT2) gene with LV chamber diameter ($P = 0.0005$).[72]

A note of caution must be introduced to the analysis of these SNP association studies. There is little evidence that these genetic variations are causal. In fact, they may be in linkage dysequilibirum with unidentified key variations in specific populations. Thus, these studies have limited potential for translation and/or therapeutics.

Hypertrophic cardiomyopathy

Hypertrophic cardiomyopathy is a Mendelian autosomal dominant form of LV hypertrophy.[73] It is a relatively common heritable disorder (1 in 500 individuals), and affects men and women equally.[74] The clinical diagnosis of hypertrophic cardiomyopathy is easily and reliably established by echocardiography (Fig. 28.1). The diagnostic criterion for hypertrophic cardiomyopathy is a maximal LV wall thickness of 15 mm or greater, in the absence of

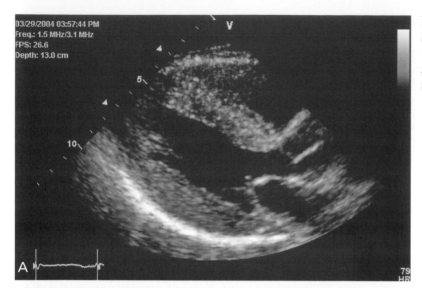

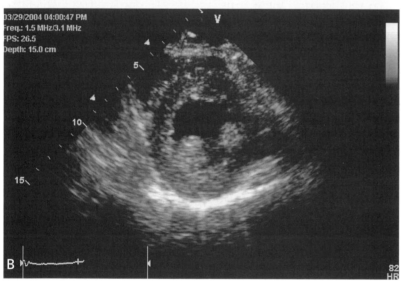

Figure 28.1. *Symmetrical hypertrophic cardiomyopathy. (A) Parasternal long- and (B) short-axis echocardiographic images.*

another cardiac or systemic condition capable of producing the magnitude of hypertrophy present, with or without dynamic LV outflow tract obstruction.[75] Interventricular septal wall thickness/free wall thickness ≥1.3 or interventricular septal wall thickness/posterior wall thickness ≥1.3 can be used as alternative echocardiographic criteria, as these have been shown to be present in 77% and 55% of the genetically affected hypertrophic cardiomyopathy adult population.[76] The presence of LV outflow tract obstruction, however, has important poten-

tial clinical implications.[77,78] Indeed, LV outflow tract obstruction has been shown to be an independent predictor of the progression to heart failure or death from cardiac causes.[75,77,78]

Genetic epidemiology of hypertrophic cardiomyopathy

Hypertrophic cardiomyopathy is primarily a disorder of the cardiac sarcomere, the functional contractile unit in myocytes (Fig. 28.2), and is caused by any of at least 10 genes that encode protein

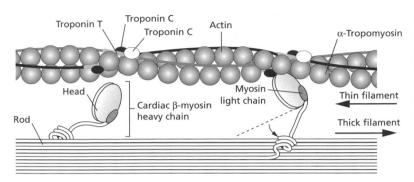

Figure 28.2. *The cardiac sarcomere. Adapted with permission from Kamisago M, Sharma SD, DePalma SR et al. Mutations in sarcomere protein genes as a cause of dilated cardiomyopathy. N Engl J Med 2000; 343: 1688–96.*

components of the sarcomere composed of thick or thin filaments with contractile, structural or regulatory functions (Table 28.3).[73,79–91] These are myosin-binding protein C (MYBPC3), β-myosin heavy chain (MYH7), cardiac troponin-I (TTNI3), cardiac troponin-T (TNNT2), myosin ventricular regulatory light chain (MYL2), myosin ventricular essential light chain (MYL3), titin (TNN), α-tropomyosin (TPMI), α-actin (ACTC) and α-myosin heavy chain (MYH6). Most mutations are missense mutations, with a single amino acid residue substituted for another. Intragenic heterogeneity compounds the diversity of genes responsible for hypertrophic cardiomyopathy, with over 200 mutations now identified.[90] Indeed, most individuals/families have private mutations that are unique. Recently, missense mutations in the gene that encodes the γ-2 regulatory subunit of adenosine monophosphate-activated protein kinase (PRKAG2) have been reported to cause familial Wolff–Parkinson–White syndrome associated with

conduction abnormalities and LV hypertrophy simulating hypertrophic cardiomyopathy.[92,93] Interestingly, recent data suggest that although PRKAG2 mutations do increase LV mass, they do so by producing a glycogen storage disease with a consequent increase in intracellular cardiomyocyte deposition that increases cell size, similar effects being seen in other glycogen storage diseases such as Pompe's disease.[94] PRKAG2 mutations are not important causes of isolated true hypertrophic cardiomyopathy or isolated Wolff–Parkinson–White disease.[95,96] Variants in muscle LIM protein (CRP3/MLP) gene, which acts as an essential promoter of myogenesis, may be associated with hypertrophic cardiomyopathy.[97,98] Causal genetic diversity accounts in part for the phenotypic variability of affected individuals, but environmental effects and other potential modifier genes influence phenotypic expression in individuals carrying identical disease mutations.[99,100]

The advent of molecular genetic studies has

Table 28.3. *Chromosomal loci and genes causing hypertrophic cardiomyopathy*

Chromosomal locus	Reference number	Gene
1q32	79, 103	Cardiac troponin T
2q31	88	Titin
3p21	87	Essential myosin light chain
11p11	81, 83	Cardiac myosin binding protein-C
12q23-p21	87	Regulating myosin light chain
14q11-q12	89	Cardiac α-myosin heavy chain
14q12	80, 84	Cardiac β-myosin heavy chain
15q14	82	Cardiac actin
15q22	103	α-tropomyosin
19q13	85	Cardiac troponin I

provided important insights into the clinical and genetic heterogeneity of hypertrophic cardiomyopathy, including preclinical diagnosis of individuals with genetic mutations who show no evidence of LV hypertrophy.[79–90] Furthermore, genotype–phenotype studies have indicated both variable expressivity and incomplete penetrance. Mutations involving *MYBPC3* have been reported to have age-related penetrance and late-onset phenotype, with a more favorable outcome in that group.[80,83,101,102] Mutations involving *TTNT2* involve incomplete penetrance and poor prognosis.[103]

While DNA-based identification of mutant genes is the definitive method for establishing the diagnosis of hypertrophic cardiomyopathy, this is not currently a routine strategy because it is technically complex, time consuming, and expensive. Furthermore, there is no single predominant mutation in hypertrophic cardiomyopathy.[73,79–100] In a recent study, Richards et al. analyzed the entire coding sequences of nine genes (*MYH7*, *MYBPC3*, *TTNI3*, *TTNT2*, *MYL2*, *MYL3*, *TPMI*, *ACTC*, and *TNN*) in 197 index cases of hypertrophic cardiomyopathy.[104] They identified disease-causing mutations in 124 patients (63%). Ninety-seven different mutations were identified, of which 60 mutations were novel. *MYBPC3* mutations accounted for 42%, followed by *MYH7* mutations (40%), *TTNT2* (6.5%), *TNNI3* (6.5%), *MYL2* (4%) and *MYL3* (<1%). Furthermore, in those patients with sporadic (i.e. negative family history) hypertrophic cardiomyopathy, 15 of 25 (60%) had mutations increasing the likelihood of future disease transmission. However, a more recent report by Van Driest et al. found in a large tertiary referral population that thin filament mutations are even less prevalent than previously estimated.[105] In 389 unrelated patients with hypertrophic cardiomyopathy, only 18 patients (4.6%) had thin filament mutations: 8 (2%) had *TTNT2* mutations, 6 (1.5%) had *TNNI3* mutations, 3 (0.7%) had *TPMI* mutations and 1 (0.3%) had an *ACTC* mutation. Moreover, of the 12 unique missense mutations identified, 9 (75%) were novel mutations. They also reported that the 1% of patients who had mutations that confer higher risk were <25 years of age at the time of presentation.[106]

Thus, a patient with the diagnosis of hypertrophic cardiomyopathy should be informed of the familial nature of the disease, and first-degree relatives should be screened. In the absence of and perhaps in addition to (in view of the prognostic significance of some clinical findings) DNA-based testing, the recommended clinical screening employs history and physical examination, 12-lead ECG, and two-dimensional echocardiography at annual evaluations during adolescence (12 to 18 years).[107] Relatives with normal echocardiograms during adulthood should have subsequent clinical re-evaluations at least every five years, because of the risk of delayed-onset LV hypertrophy.[107] Screening may begin in childhood, and is particularly important if there is a high-risk family history or involvement in intense competitive sports programs.[73,107,108] Affected individuals should be evaluated routinely with annual echocardiography.[105]

Recently, novel echocardiographic tissue Doppler imaging parameters (Fig. 28.3) to assess diastolic function were used to predict development of hypertrophic cardiomyopathy.[109–111] In patients with causal genetic mutations associated with hypertrophic cardiomyopathy, Nagueh et al. found that patients who developed LV hypertrophy during the ensuing two years had reduced tissue Doppler early and late diastolic velocities (Fig. 28.4), with increased left atrial volume and abnormal pulmonary venous flow patterns compared to age- and sex-matched unaffected relative controls.[110] Furthermore, McMahon et al. found that children with hypertrophic cardiomyopathy and abnormal diastolic function by tissue Doppler imaging are at higher risk of adverse events including sudden death and significant cardiac symptoms.[111] Thus, tissue Doppler assessment of diastolic function may identify not only affected individuals with preclinical disease, but also those at higher risk who require more frequent follow-up and early intervention.

Natural history of hypertrophic cardiomyopathy

Since hypertrophic cardiomyopathy may present clinically at any stage of life from infancy to old age, clinical manifestations are variable, and patients may remain stable over long periods of time, with

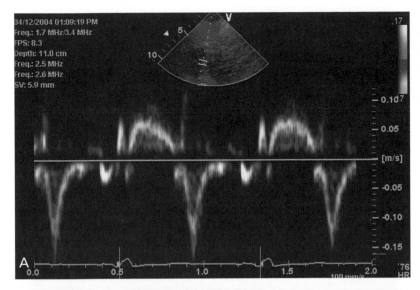

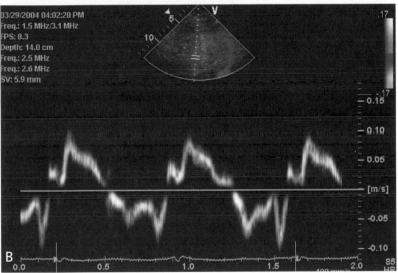

Figure 28.3. *Tissue Doppler echocardiographic images of the interventricular septum in (A) a normal individual, and (B) a hypertrophic cardiomyopathy patient. See color plate section.*

25% achieving normal longevity [112] However, the course of many patients may be punctuated by adverse clinical events.[113–114] The clinical course of hypertrophic cardiomyopathy commonly follows one or more pathways (Fig. 28.5) that include premature sudden death, progressive symptoms of exertional dyspnea, chest pain or syncope in the presence of preserved LV systolic function, progression to advanced heart failure ('burnt-out phase') or embolic stroke.[107,112–115] In contrast to mortality rates of 3–6% per year derived from ter-

tiary referral centers,[107,112–115] more recent data from community-based cohorts cite annual mortality rates of about 1%.[78,112,113,116]

Dilated cardiomyopathy

Dilated cardiomyopathy is characterized by ventricular dilatation and low contractility and affects up to 4/10000 individuals.[117,118] The diagnosis of dilated cardiomyopathy is often made through

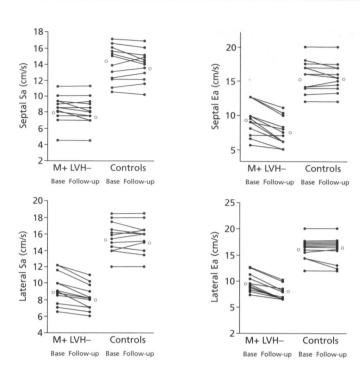

Figure 28.4. *Individual data points showing interventricular septal and lateral wall systolic annular (Sa) and early diastolic filling annular (Ea) velocities at baseline (Base), and follow-up of subjects with genetic mutations for hypertrophic cardiomyopathy (M+) and no left ventricular hypertrophy (LVH−) and controls. Adapted with permission from Nagueh S, McFalls J, Myer D et al. Tissue Doppler imaging predicts the development of hypertrophic cardiomyopathy in subjects with subclinical disease. Circulation 2003; 108: 395–8.*

echocardiography (Fig. 28.6), showing reduced LV systolic function (LV ejection fraction of <45% and/or fractional shortening <25%), and LV

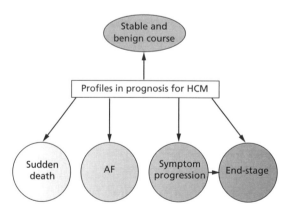

Figure 28.5. *Pathways of disease progression in hypertrophic cardiomyopathy. AF, atrial fibrillation; HCM, hypertrophic cardiomyopathy. Adapted with permission from Maron BJ, McKenna WJ, Danielson GK et al. American College of Cardiology/European Society of Cardiology Clinical Expert Consensus Document on Hypertrophic Cardiomyopathy. J Am Coll Cardiol 2003; 42: 1687–713.*

chamber dilatation (LV end-diastolic diameter >117% of the value predicted for age and body surface area).[118] Detection of dilated cardiomyopathy usually prompts a search for the pathogenesis of LV dilatation and decreased systolic function, e.g. ischemia, infection or autoimmune disease. In many instances, however, an underlying cause is not discovered, and the diagnosis of idiopathic dilated cardiomyopathy is made.[118,119]

Prevalence and correlates of left ventricular systolic dysfunction

Recent studies indicate that the prevalence of LV systolic dysfunction ranges from 1.8 to 14.0%, depending on the setting and study design (Table 28.4).[120–129] Of these, 2.9–14% had mild LV systolic dysfunction, and 1.8–7.7% had moderate to severe LV systolic dysfunction.[120–129] The prevalence of asymptomatic LV systolic dysfunction, i.e. LV systolic dysfunction without overt congestive heart failure, was 0.9–12.9%. Reduced LV systolic function was more common in men than in women, in older than in younger individuals, and in those with than without coronary heart disease.[120,121,128] Men had lower LV ejection fractions

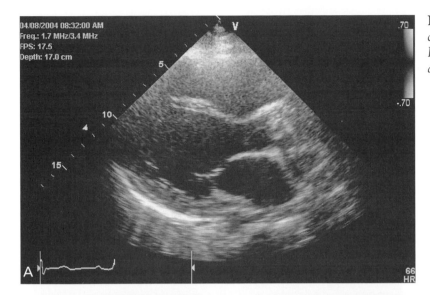

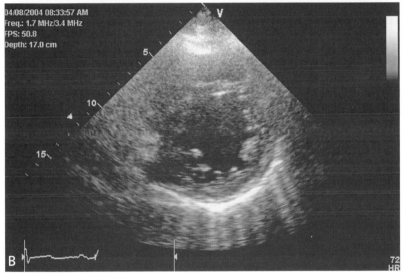

Figure 28.6. *Dilated cardiomyopathy. (A) Parasternal long- and (B) short-axis echocardiographic images.*

by 3–5% than women.[120,121,130] Few studies address the issue of ethnicity and LV systolic dysfunction. In a biracial population-based sample of hypertensive adults, the HyperGEN study reported similar prevalences of mild (10.7% and 8.6%) and moderate–severe (3.9% and 4.1%) LV systolic dysfunction in black and white participants, respectively.[121] However, there was a statistically significant association between black ethnicity and reduced LV ejection fraction after adjusting for covariates.[119] Among adult American Indian participants in the Strong Heart Study, the preva-

lences of mild and moderate–severe LV systolic dysfunction were 11.1% and 2.9%.[120]

Comprehensive assessment of correlates of LV systolic dysfunction in population-based samples has been fruitful. First, epidemiological studies have shown that while older age is associated with lower LV systolic function in univariate analyses, there was no independent association between age and LV systolic function, in multivariate analyses that considered arterial stiffness and renal dysfunction as covariates.[120,121] This suggests that much of the association between older age and lower

Table 28.4. *Prevalence of left ventricular systolic dysfunction*

Reference	Number of patients dysfunction	Mild left ventricular systolic dysfunction		Moderate to severe left ventricular systolic	
		Total (%)	Asymptomatic (%)	Total (%)	Asymptomatic (%)
Devereux et al., 2001[120]	3184	14.0	12.5	2.9	2.1
Devereux et al., 2001[121]	2086	14.0	12.9	4.0	3.4
Davies et al., 2001[122]	3960	5.3	3.3	1.8	0.9
Hedberg et al., 2001[123]	412	6.8	3.2	–	–
Nielsen et al., 2001[124]	126	2.9	1.0	–	–
Mosterd et al., 1999[125]	2267	3.7	2.9	–	–
Kupari M et al., 1997[126]	501	10.8	8.6	–	–
Schunkert et al., 1998[127]	1566	2.7	1.1	–	–
McDonagh et al., 1997[128]	1467	–	–	7.7	5.9
Wang et al., 2003[129]	4257	–	1.8	–	1.2

ventricular systolic dysfunction is mediated by age-related impairment in arterial and arteriolar function, as manifested by higher pulse pressure and serum creatinine levels. Second, body mass index and measures of body adiposity decreased from participants with normal LV systolic function to those with severe systolic dysfunction.[120,121] Indeed, lower body mass index was an independent predictor of mortality in patients with congestive heart failure.[131] Individuals with poor cardiovascular health may have suffered parallel reductions in LV systolic function and body weight. Third, diabetes mellitus is an independent predictor of reduced LV systolic function.[45,120,121,132,133] In view of the increasing prevalence of diabetes in the United States, this is a particularly relevant association.[46] Fourth, there is a stepwise increase in measures of arterial stiffening (pulse pressure/stroke index), atherosclerosis (ankle/arm index), and microvascular dysfunction (serum creatinine and albuminuria) from individuals with normal LV function to those with severe LV systolic dysfunction.[120] These results extend the well-described association between LV systolic dysfunction and hypertension.[120,121]

Genetic epidemiology of dilated cardiomyopathy

In recent years, there has been increasing recognition that many cases of idiopathic dilated cardio-myopathy are familial. In a report by Michels et al., using stringent criteria to diagnose idiopathic dilated cardiomyopathy, >20% of individuals with dilated cardiomyopathy had an affected first-degree relative.[134] More recent population-based studies indicate that 30–40% of cases of dilated cardiomyopathy are familial, and can have autosomal (dominant or recessive), X-linked or mitochondrial transmission.[135–137] These observations have not only fostered molecular genetic studies to identify causal mutation, but also prompted screening of first-degree family members of probands since they are at risk for inherited dilated cardiomyopathy. Such strategies may allow early intervention to prevent symptoms and complications of heart failure and reduce cardiovascular morbidity and mortality.

The most common mode of inheritance of familial dilated cardiomyopathy is autosomal dominant. As indicated in Table 28.5, in addition to genetic heterogeneity of dilated cardiomyopathy, there is also diversity in conditions that accompany this disorder.[138–172] Recognition of these associated manifestations is important, as these phenotypes are expressed in advance of dilated cardiomyopathy, and thus may be used as surrogate markers of cardiac status. Identification of causal genetic mutations associated with dilated cardiomyopathy has identified several potential mechanisms by which cardiac muscle may fail, including abnormalities in force generation, force transmission or energy production.[90,119] Other mutant genes have been shown to be involved in cytoskeletal and intracellular pathways linking the extracellular matrix to the sarcomere, sarcolemma and nuclear membrane.[173]

Due to the high morbidity and mortality associated with congestive heart failure,[2,174,175] and the available treatment strategies to prevent complications and prolong survival,[176–184] identification of relatives with familial dilated cardiomyopathy is warranted. While a DNA-based approach may

provide definitive diagnoses of familial dilated cardiomyopathy, this strategy is even more daunting than that for hypertrophic cardiomyopathy, because of the marked genetic heterogeneity (Table 28.5) and age-related penetrance of the disease.[183] In a series of European families with dilated cardiomyopathy, detectable expression of dilated cardiomyopathy was 10% for age <20 years, 34% for 20–30 years, 60% for 30–40 years and 90% for age >40 years.[185] Thus, a clinical strategy to diagnose familial dilated cardiomyopathy has been proposed (Fig. 28.7).[186]

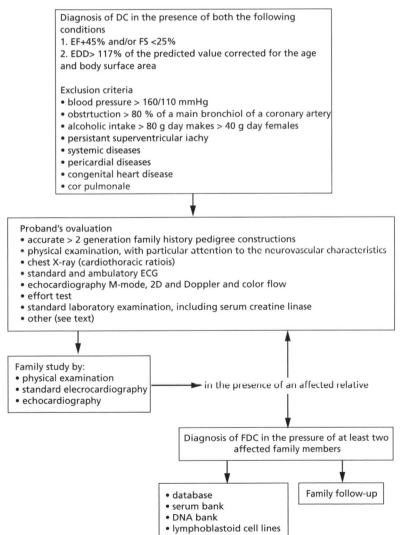

Figure 28.7. *Proposed screening algorithm for familial dilated cardiomyopathy. EF, ejection fraction; FS, fractional shortening; DC, dilated cardiomyopathy; EDD, LV end-diastolic diameter; FDC, familial dilated cardiomyopathy. Adapted with permission from Mestroni L, Maisch B, McKenna WJ et al. Guidelines for the study of familial dilated cardiomyopathy. Eur Heart J 1999; 20: 93–102.*

Table 28.5. *Chromosomal loci, inheritance pattern and genes causing dilated cardiomyopathy and associated phenotypes*

Chromosome locus	Reference number	Inheritance pattern	Gene	Associated phenotype
1p1–q21	138	Autosomal dominant	Lamin A/C	Conduction disease, skeletal myopathy
1q32	139	Autosomal dominant	Cardiac troponin T	None
1q42–q43	158	Autosomal dominant	Cardiac ryanodine receptor	Arrhythmogenic right ventricular dysplasia, exercise-induced ventricular tachhycardia
2q14–q22	152	Autosomal dominant	Unknown	Conduction disease
2q31	140, 141	Autosomal dominant	Titin	None
2q32.1–q32.3	159	Autosomal dominant	Unknown	Arrhythmogenic right ventricular dysplasia
2q35	154	Autosomal dominant	Desmin	Skeletal myopathy
3p23	160	Autosomal dominant	Unknown	Arrhythmogenic right ventricular dysplasia
3p22–p25	153	Autosomal dominant	Unknown	Conduction disease
4q12	155	Autosomal recessive	β-sarcoglycan	Skeletal myopathy
5q33–q34	142	Autosomal dominant	δ-sarcoglycan	None
6p24	167	Autosomal recessive	Desmoplakin	Wooly hair, keratoderma
6q12–q16	143, 144	Autosomal dominant	Phospholamban	None
6q23–q24	156, 157	Autosomal dominant	Unknown	Sensorineural hearing loss, muscular dystrophy
9q13–q22	145	Autosomal dominant	Unknown	None
10p12–p14	151	Autosomal dominant	Unknown	Arrhythmogenic right ventricular dysplasia
10q21–q23	149	Autosomal dominant	Unknown	Mitral valve prolapse
10q22.1–q23	150	Autosomal dominant	Vinculin	None
10q22.2–q23	161	Autosomal dominant	Cypher/ZASP[a]	Ventricular non-compaction
11p15	162	Autosomal dominant	Unknown	Ventricular non-compaction
11p15.1	163	Autosomal dominant	Cysteine-rich protein, 3/muscle LIM protein	None
12p11	173	Autosomal dominant	Plakophilin-2	Arrhythmogenic right ventricular cardiomyopathy
12p12.1	164	Autosomal dominant?	ABCC9[b]	Ventricular tachycardia
14q11	146	Autosomal dominant	Cardiac β-myosin heavy chain	None
14q12–q22	165	Autosomal dominant	Unknown	Arrhythmogenic right ventricular dysplasia
14q23–q24	165	Autosomal dominant	Unknown	Arrhythmogenic right ventricular dysplasia
15q14	146	Autosomal dominant	Cardiac actin	None
15q22.1	147	Autosomal dominant	α-tropomyosin	None
17q21	167	Autosomal recessive	Plakoglobin	Arrhythmogenic right ventricular dysplasia, wooly hair, keratoderma
18q12.1–q12.2	168	Autosomal dominant	α-dystrobrevin	Ventricular non-compaction
Xp21	170	X-linked	Dystrophin	Skeletal myopathy
Xq28	169	X-linked	Tafazzin	Short stature, neutropenia, ventricular non-compaction

[a] ZASP, Z band alternatively spliced PDZ-motif protein; [b] ABCC9, ATP-binding cassette, subfamily C, member 9.

Natural history of left ventricular systolic dysfunction and dilated cardiomyopathy

Only half of patients with dilated cardiomyopathy survive 5 years,[2] with most premature death occurring from complications of congestive heart failure.[174,175] Only recently have the frequency and natural history of asymptomatic LV systolic dysfunction been evaluated in population-based samples. The Framingham investigators reported that, the hazards ratio for developing congestive heart failure and death in adults with asymptomatic LV systolic dysfunction were 4.7 (95% confidence interval = 2.7–8.1), and 1.6 (95% confidence interval = 1.1–2.4), after adjusting for cardiovascular risk factors.[129] The median survival of individuals with asymptomatic LV systolic dysfunction was 7.1 years.

Recent studies have provided insight into the natural history of familial dilated cardiomyopathy. In a prospective study of 408 asymptomatic relatives of 110 patients with dilated cardiomyopathy, 45 (20%) had LV dilatation, 13 (6%) had asymptomatic LV dysfunction, and 7 (3%) had dilated cardiomyopathy.[187] Over a mean follow-up period of 39 months, 3 (2%) family members who were initially unaffected developed LV enlargement, and 12 (27%) with LV enlargement progressed to dilated cardiomyopathy.[187] In a 10-year follow-up of 127 previously healthy first-degree relatives of individuals with dilated cardiomyopathy with follow-up LV ejection fraction measurements, 9 (7%) developed dilated cardiomyopathy.[188] The likelihood of developing idiopathic dilated cardiomyopathy was greater in family members with LV dilatation at baseline (hazard ratio = 8.3, 95% confidence interval: 3–26, $P < 0.001$).

In a study of 240 patients with dilated cardiomyopathy, of whom 32 were familial and 209 were sporadic, the six-year survival was 6% in the familial groups and 23% in the sporadic group.[189] These results suggest that familial dilated cardiomyopathy is a rapidly progressing disease. However, in a recent longitudinal study of 99 patients with dilated cardiomyopathy, of whom 30 were familial and 69 were sporadic cases, the five-year survival was similar in both groups (51 and 55%, respectively).[190] Whether differences in the inheritance patterns or causal genetic mutations account for the differences in survival among patients with familial dilated cardiomyopathy remains uncertain.

Shared etiologies of hypertrophic and dilated cardiomyopathy

The observation that both hypertrophic and dilated cardiomyopathy can be caused by mutations in sarcomeric proteins (Tables 28.3 and 28.5) suggests that these phenotypes may represent either mutation-specific effects on these genes, or potentially different stages in a final common pathway.[90,91,119,173,191] Genetically engineered mice heterozygous for sarcomeric protein mutations develop hypertrophic cardiomyopathy,[192] while homozygous mice develop dilated cardiomyopathy.[193,194] In humans, the deterioration of some cases of hypertrophic cardiomyopathy to end-stage heart failure, resembling dilated cardiomyopathy, suggests shared pathogenesis.[90, 91, 102,107,115] Indeed, a novel missense mutation of myosin-binding protein C (*Arg820Gln* mutation) was associated with 'burnt-out phase' hypertrophic cardiomyopathy.[195] Recently, it was shown that missense mutations in muscle LIM protein may be rare causes of both hypertrophic and dilated cardiomyopathy in humans.[98,163] These observations in animal models and humans support shared elements of pathways in cardiac remodeling to hypertrophy and dilatation. However, there are many instances of hypertrophic cardiomyopathy that never progress to dilated cardiomyopathy, and many documented instances of mildly dilated left ventricles with low normal ejection fraction and LV mass within the normal range that progress to dilated cardiomyopathy. These exceptions to the general pattern of types of cardiomyopathies 'breeding true' may provide novel insights into the underlying biology of the hypertrophy–dysfunction connection, as well as into specific gene–gene or gene–environment interactions.

Conclusions

Advances in molecular genetics have expanded our knowledge of the epidemiology of LV hypertrophy. Rigorous quantitative genetic epidemiology is identifying the causal mutations responsible for LV hypertrophy and providing insight into its pathophysiology. Many genes have already been identified that may act independently or synergistically to increase the risk of developing LV hypertrophy. Clinical outcome studies will characterize the implications of these genetic variations for the natural history of LV hypertrophy and identify the impact of genetic variations on progression or regression of LV hypertrophy. As novel genes responsible for LV hypertrophy, hypertrophic cardiomyopathy, and dilated cardiomyopathy are discovered, our understanding of the inciting events triggering cardiac remodeling and failure will expand. Although the identification and characterization of genes contributing to LV hypertrophy and dysfunction are challenging, they offer much promise for future prevention, early intervention and treatment of these major public health issues.

References

1. Heart Disease and Stoke Statistics – 2004 Update. American Heart Association Website www.myamericanheart.org (accessed 25 May 2005).
2. Cohn JN, Bristow MR, Chien KR et al. Report of the National Heart, Lung and Blood Institute Special Emphasis Panel on Heart Failure Research. Circulation 1997; 95: 766–70.
3. Sokolow M, Perloff D. The prognosis of essential hypertension treated conservatively. Circulation 1961; 23: 697–706.
4. Breslin DJ, Gifford RW, Fairburn JF. Essential hypertension: a twenty-year follow-up study. Circulation 1966; 33: 87–97.
5. Kannel WB, Gordon T, Offutt D. Left ventricular hypertrophy by electrocardiogram: prevalence, incidence and mortality in the Framingham Heart Study. Ann Intern Med 1969; 71: 89–105.
6. Casale PN, Devereux RB, Milner M et al. Value of echocardiographic assessment of LV mass in predicting cardiovascular morbid events in hypertensive men. Ann Intern Med 1986; 105: 173–8.
7. Silberberg JS, Barre PE, Prichard SS et al. Impact of left ventricular hypertrophy on survival in end-stage renal disease. Kidney Int 1989; 36: 286–90.
8. Aronow WS, Epstein S, Koenigsberg M et al. Usefulness of echocardiographic left ventricular mass, ventricular tachycardia and complex ventricular arrhythmias in predicting ventricular fibrillation and sudden death in elderly patients. Am J Cardiol 1988; 62: 1124–5.
9. Levy D, Garrison RJ, Savage DD et al. Prognostic implication of echocardiographically determined left ventricular mass in the Framingham Heart Study. N Engl J Med 1990; 322: 1561–6.
10. Koren MJ, Devereux RB, Casale PN et al. Relation of left ventricular mass and geometry to morbidity and mortality in uncomplicated essential hypertension. Ann Intern Med 1991; 114: 345–52.
11. Ghali JK, Liao Y, Simmons B et al. The prognostic role of left ventricular hypertrophy in patients with or without coronary artery disease. Ann Intern Med 1992; 117: 831–6.
12. Bikkina M, Levy D, Evans JC et al. Left ventricular mass and risk of stroke in an elderly cohort. JAMA 1994; 272: 33–6.
13. Gardin JM, McClelland R, Kitzman D et al. M-mode echocardiographic predictors of six- to seven-year incidence of coronary heart disease, stroke, congestive heart failure and mortality in an elderly cohort (The Cardiovascular Health Study). Am J Cardiol 2001; 87: 1051–7.
14. Liao Y, Cooper RS, McGee DL et al. The relative effects of left ventricular hypertrophy, coronary artery disease and ventricular dysfunction on survival among black adults. JAMA 1995; 273: 1592–7.
15. Quinones MA, Greenberg BH, Kopelen HA et al. Echocardiographic predictors of clinical outcome in patients with left ventricular dysfunction enrolled in the SOLVD registry and trials: significance of left ventricular hypertrophy. J Am Coll Cardiol 2000; 35: 1237–44.
16. Sokolow M, Lyon TP. The ventricular complex in left ventricular hypertrophy as obtained by unipolar precordial and limb leads. Am Heart J 1949; 37: 161–8.
17. Roman MJ, Kligfield PD, Devereux RB et al. Geometric and functional correlates of electrocardiographic repolarization abnormalities in aortic regurgitation. J Am Coll Cardiol 1987: 9: 500–8.
18. Romhilt DW, Estes EH. A point-score system for the diagnosis of ECG left ventricular hypertrophy. Am Heart J 1968; 75: 752–8.
19. Casale PN, Devereux RB, Kligfield P et al.

Electrocardiographic detection of left ventricular hypertrophy: development and prospective validation of improved criteria. J Am Coll Cardiol 1985; 6: 572–80.

20. Levy D, Labib SB, Anderson KM et al. Determinants of sensitivity and specificity of electrocardiographis criteria for left ventricular hypertrophy. Circulation 1990; 81: 815–20.

21. Okin PM, Roman MJ, Devereus RB et al. Electrocardiographic identification of increased left ventricular mass by simple voltage-duration products. J Am Coll Cardiol 1995; 23: 412–23.

22. Bella JN, Devereux RB, Roman MJ et al. Relations of left ventricular mass to fat-free and adipose body mass: The Strong Heart Study. Circulation 1998; 98: 2538–44.

23. Devereux RB, Reichek N. Echocardiographic determination of left ventricular mass in man: anatomic validation of the method. Circulation 1987; 55: 613–18.

24. Devereux RB, Alonso DR, Lutas EM et al. Echocardiographic assessment of left ventricular hypertrophy: comparison to necropsy studies. Am J Cardiol 1986; 57: 250–8.

25. Palmieri V, Dahlof B, DeQuattro V et al. Reliability of echocardiographic assessment of left ventricular structure and function: The PRESERVE Study. J Am Coll Cardiol 1999; 34: 1625–32.

26. Myerson S, Bellenger NG, Pennell DJ. Assessment of left ventricular mass by cardiovascular magnetic resonance. Hypertension 2002; 39: 750–5.

27. Katz J, Milliken M, Stray-Gundersen J et al. Estimation of human myocardial mass with MR imaging. Radiology 1988; 169: 495–8.

28. Bellenger NG, Davies LC, Francis JM et al. Reduction in sample size for studies of remodeling in heart failure by the use of cardiovascular magnetic resonance. J Cardiovasc Magn Reson 2000; 2: 271–8.

29. de Simone G, Devereux RB, Daniels SR et al. Gender differences in left ventricular growth. Hypertension 1995; 26: 979–83.

30. Hammond IW, Devereux RB, Alderman MH et al. Relation of blood pressure and body build to left ventricular mass in normotensive and hypertensive adults. J Am Coll Cardiol 1988; 12: 996–1004.

31. de Simone G, Devereux RB, Roman MJ et al. Relation of obesity and gender to left ventricular hypertrophy in normotensive and hypertensive adults. Hypertension 1994; 23: 600–6.

32. de Simone G, Devereux RB, Daniels SR et al. Effect of growth on variability of left ventricular mass: assessment of allometric signals in adults and children and their capacity to predict cardiovascular risk. J Am Coll Cardiol 1995; 25: 1056–62.

33. Devereux RB, James GD, Pickering TG. What is normal blood pressure?: comparison of ambulatory pressure level and variability in patients with normal or abnormal left ventricular geometry. Am J Hypertens 1993; 6: 211s–215s.

34. Gonzales IJ, Bella JN, Blackshear J et al. Abnormal left ventricular mass and geometry in normotensive patients with hypertensive response to exercise. J Am Soc Echocardiogr 2002; 15: 509.

35. Castello R, Gonzales IJ, Bella JN et al. Importance of diastolic and systolic blood pressure response to exercise on left ventricular mass and geometry. J Am Coll Cardiol 2003; 41: 217A.

36. Gottdiener J, Reda D, Materson B et al. Importance of obesity, race and age to cardiac structural and functional effects of hypertension. J Am Coll Cardiol 1994; 24: 1492–8.

37. Schmieder RE, Messerli FH, Garavaglia GE et al. Dietary salt intake: a determinant of cardiac involvement in essential hypertension. Circulation 1988; 108: 7–13.

38. Ganau A, Devereux RB, Pickering TG, Roman MJ, Schnall PL, Santucci S et al. Relation of left ventricular hemodynamic load and contractile performance to left ventricular mass in hypertension. Circulation 1990; 81: 25–36.

39. Jones EJ, Devereux RB, O'Grady MJ et al. Hemodynamic volume load stimulates arterial and cardiac enlargement. J Am Coll Cardiol 1997; 29: 1303–10.

40. Bella JN, Wachtell K, Palmieri V et al. Relation of left ventricular geometry and function to systemic hemodynamics in hypertension: The LIFE Study. J Hypertens 2001; 19: 127–34.

41. Chen CH, Ting CT, Lin SJ et al. Which arterial and cardiac parameters best predict left ventricular mass? Circulation 1998; 98: 422–8.

42. Devereux RB, Roman MJ, de Simone G et al. Relations of left ventricular mass to demographic and hemodynamic variables in American Indians: The Strong Heart Study. Circulation 1997; 96: 1416–23.

43. Devereux RB, Roman MJ, Paranicas M et al. Impact of diabetes on cardiac structure and function: The Strong Heart Study. Circulation 2000; 101: 2271–6.

44. Palmieri V, Bella JN, Arnett DK et al. Effect of type 2 diabetes mellitus on left ventricular geometry and systolic function in hypertensive subjects: Hypertension Genetic Epidemiology Network (HyperGEN) Study. Circulation 2001; 103: 102–7.

45. Bella JN, Devereux RB, Roman MJ et al. Separate and joint effects of systemic hypertension and diabetes mellitus on left ventricular structure and function in American Indians (The Strong Heart Study). Am J Cardiol 2001; 87: 1260–5.

46. Mokdad AH, Bowman BA, Ford ES et al. The continuing epidemics of obesity and diabetes in the United States. JAMA 2001; 286: 1195–200.

47. Adams T, Yanowitz F, Fisher AG et al. Heritability of cardiac size: an echocardiographic and electrocardiographic study of monozygotic and dizygotic twins. Circulation 1985; 71: 39–44.

48. Harshfield GA, Grim C, Hwang C et al. Genetic and environmental influences on echocardiographically determined left ventricular mass in black twins. Am J Hypertens 1990; 3: 538–43.

49. Bielen E, Fagard R, Amery A. The inheritance of left ventricular structure and function assessed by imaging and Doppler echocardiography. Am Heart J 1991; 21: 1743–9.

50. Christian JC, Kang KW, Norton JA. Choice of an estimate of genetic variance from twin data. Am J Hum Genet 1974; 26: 154–61.

51. Post WS, Larson MG, Myers RH et al. Heritability of left ventricular mass. The Framingham Heart Study. Hypertension 1997; 30: 1025–8.

52. Palatini P, Krause L, Amerena J et al. Genetic contribution to the variance in left ventricular mass: The Tecumseh Offspring Study. J Hypertens 2001; 19: 1217–22.

53. Arnett DK, Hong Y, Bella JN et al. Sibling correlations of left ventricular mass and geometry in hypertensive African Americans and whites: The HyperGEN Study. Am J Hypertens 2001; 14: 1226–30.

54. Bella JN, MacCluer JW, Roman MJ et al. Heritability of left ventricular dimensions and mass in American Indians: The Strong Heart Study. J Hypertens 2004; 22: 281–6.

55. Arnett DK, de las Fuentes L, Broeckel U. Genes for left ventricular hypertrophy. Curr Hypertens Rep 2004; 6: 36–41.

56. Arnett DK, Lynch A, Kitzman D et al. Linkage of left ventricular (LV) structural phenotypes to chromosome 7. Circulation 2002; 105: e102.

57. Göring HHH, Diego VP, Cole SA et al. A QTL for left ventricular mass on chromosome 12p in American Indians. Am J Hum Genet 2004; 75: 376.

58. Benjamin EJ, DeStefano A, Larson MG et al. Genetic linkage analyses for left ventricular mass phenotypes in the Framingham Heart Study. Circulation 2000; 102: 4131.

59. Montgomery HE, Clarkson P, Dollery CM et al. Association of angiotensin-converting enzyme I/D polymorphism with change in left ventricular mass in response to physical training. Circulation 1997; 96: 741–7.

60. Karjalainen J, Kujala UM, Stolt A et al. Angiotensinogen gene M235T polymorphism predicts left ventricular hypertrophy in endurance athletes. J Am Coll Cardiol 1999; 34: 494–9.

61. Diet F, Graf C, Mahnke N et al. ACE and angiotensinogen gene phenotypes and left ventricular mass in athletes. Eur J Clin Invest 2001; 10: 836–42.

62. Schmieder RE, Erdmann J, Delles C et al. Effect of angiotensin II type 2 receptor (+1675 G/A) on left ventricular structure in humans. J Am Coll Cardiol 2001; 37: 175–82.

63. Kupari M, Hautanen A, Lankinen L et al. Associations between human aldosterone (CYP11B2) gene polymorphisms and left ventricular size, mass and function. Circulation 1998; 97: 569–75.

64. Delles C, Erdmann J, Jacobi J et al. Aldosterone synthase (CYP11B2) −344 C/T polymorphism is associated with left ventricular structure in human arterial hypertension. J Am Coll Cardiol 2001 37: 878–84.

65. Stella P, Bigatti G, Tizzoni L et al. Association between aldosterone synthase (CYP11B2) polymorphism and left ventricular mass in essential hypertension. J Am Coll Cardiol 2004; 43: 265–70.

66. Stanton T, Inglis GC, Padmanabhan S et al. Variation at the beta-1 adrenoceptor gene locus affects left ventricular mass in renal failure. J Nephrol 2002; 15: 512–18.

67. Poch E, Gonzalez D, Gomez-Angelats E et al. G-protein beta (3) subunit gene variant and left ventricular hypertrophy in essential hypertension. Hypertension 2000; 35: 214–18.

68. Semplicini A, Siffert W, Sartori M et al. G-protein beta3 subunit gene 825T allele is associated with increased left ventricular mass in young subjects with mild hypertension. Am J Hypertens 2001; 14: 1191–5.

69. de las Fuentes L, Herrero P, Peterson LR et al. Myocardial fatty acid metabolism: independent predictor of left ventricular hypertrophy in hypertensive heart disease. Hypertension 2003; 41: 83–7.

70. Davila-Roman VG, Vedala G, Herrero P et al. Altered myocardial fatty acid and glucose metabolism in idiopathic dilated cardiomyopathy. J Am Coll Cardiol 2002; 40: 271–7.

71. Jamshidi Y, Montgomery HE, Hense HW et al. Peroxisome proliferator-activated receptor alpha

gene regulates left ventricular growth in response to exercise and hypertension. Circulation 2002; 105: 950–5.

72. O'Donnell CJ for the Writing Group of the Cardiogenomics Program. A systematic search for genetic variants associated with left ventricular structure and function phenotypes in the NHLBI Cardiogenomics Program for Genomics Applications. Circulation 2004; 109: e77–e78.

73. Spirito P, Seidman CE, McKenna WJ et al. The management of hypertrophic cardiomyopathy. N Engl J Med 1997; 775–85.

74. Maron BJ, Gardin JM, Flack JM et al. Prevalence of hypertrophic cardiomyopathy in a general population of young adults. Echocardiographic analysis of 4111 subjects in the CARDIA Study. Coronary Artery Risk Development In (Young) Adults. Circulation 1995; 92: 785–9.

75. Klues HG, Schiffers A, Maron BJ. Phenotypic spectrum and patterns of left ventricular hypertrophy in hypertrophic cardiomyopathy: morphologic observations and significance as assessed by two-dimensional echocardiography in 600 patients. J Am Coll Cardiol 1995; 26: 1699–708.

76. Solomon SD, Wolff S, Watkins H et al. Left ventricular hypertrophy and morphology in familial hypertrophic cardiomyopathy associated with mutations of the beta-myosin heavy chain. J Am Coll Cardiol 1993; 22: 306–7.

77. Maron MS, Olivotto I, Betocchi S et al. Effect of left ventricular outflow tract obstruction on clinical outcome in hypertrophic cardiomyopathy. N Engl J Med 2003; 348: 295–303.

78. Kloffard MJ, Ten Cate FJ, Van der Lee C et al. Hypertrophic cardiomyopathy in a large community-based population: clinical outcome and identification of risk factors for sudden cardiac death and clinical deterioration. J Am Coll Cardiol 2003; 41: 987–83.

79. Thierfelder L, Watkins H, MacRae C et al. Alpha-tropomyosin and cardiac troponin T mutations cause familial hypertrophic cardiomyopathy: a disease of the sarcomere. Cell 1994; 77: 701–12.

80. Anan R, Greve G, Thierfelder L et al. Prognostic implications of novel beta cardiac myosin heavy chain gene mutations that cause familial hypertrophic cardiomyopathy. J Clin Invest 1994; 93: 280–5.

81. Niimura H, Bachincki LL, Sangwatanaroj S et al. Mutations in the gene for cardiac myosin-binding protein C and late-onset familial hypertrophic cardiomyopathy. N Engl J Med 1998; 338: 1248–57.

82. Mogensen J, Klausen IC, Pedersen AK et al. Alpha-cardiac actin is a novel disease gene in familial hypetrophic cardiomyopathy. J Clin Invest 1999; 103: R39–43.

83. Erdmann J, Raible J, Maki-Abadi J et al. Spectrum of clinical phenotypes and gene variants in cardiac-myosin binding protein C mutation carriers with hypertrophic cardiomyopathy. J Am Coll Cardiol 2001; 38: 322–30.

84. Gruver EJ, Fatkin D, Dodds GA et al. Familial hypertrophic cardiomyopathy and atrial fibrillation caused by Arg663His beta-cardiac myosin heavy chain mutation. Am J Cardiol 1999; 83: 13H–18H.

85. Kimura A, Harada H, Park JE et al. Mutations in the cardiac troponin I gene associated with hypertrophic cardiomyopathy. Nat Genet 1997; 16: 379–82.

86. Niimura H, Patton KK, McKenna WJ et al. Sarcomere protein gene mutations in hypertrophic cardiomyopathy of the elderly. Circulation 2002; 105: 446–51.

87. Poetter K, Jiang H, Hassanzadeh S et al. Mutations in either the essential or regulatory light chains of myosin are associated with rare myopathy in human heart and skeletal muscle. Nature Genet 1996; 13: 63–9.

88. Satoh M, Takahashi M, Sakamoto T et al. Structural analysis of the titin gene in hypertrophic cardiomyopathy: identification of a novel disease gene. Biochem Biophys Res Commun 1999; 262: 411–17.

89. Solomon SD, Geisterfer-Lowrance AA, Vosberg HP. A locus for familial hypertrophic cardiomyopathy is closely linked to the cardiac myosin heavy chain genes, CRI-L436, and CRI-L329 on chromosome 14 at q11-q12. Am J Hum Genet 1990; 47: 389–94

90. Seidman JG, Seidman CE. The genetic basis of cardiomyopathy: from mutation identification to mechanistic paradigms. Cell 2001; 104: 557–67.

91. Watkins H. Genetic clues to disease pathways in hypertrophic and dilated cardiomyopathies. Circulation 2003; 107: 1344–6.

92. Blair E, Redwood C, Ashrafian H et al. Mutations in the gamma (2) subunit of AMP-activated protein kinase cause familial hypertrophic cardiomyopathy: evidence for the central role of energy compromise in disease pathogenesis. Hum Mol Genet 2001; 10: 1215–20.

93. Gollob MH, Green MS, Tang AS et al. Identification of a gene responsible for familial Wolff-Parkinson-White syndrome. N Engl J Med 2001; 344: 1823–31.

94. Arad M, Benson DW, Perez-Atayde AR et al. Constitutively active AMP kinase mutations cause

glycogen storage disease mimicking hypertrophic cardiomyopathy. J Clin Invest 2002; 109: 357–62.

95. Vaughan CJ, Hom Y, Okin DA et al. Molecular genetic analysis of PRKAG2 in sporadic Wolff-Parkinson-White syndrome. J Cardiovasc Electrophysiol 2003; 14: 263–8.

97. Arber S, Hunter JJ, Ross J Jr et al. MLP-deficient mice exhibit a disruption of cardiac cytoarchitectural organization, dilated cardiomyopathy and heart failure. Cell 1997; 88: 393–403.

98. Geier C, Perrot A, Ozcelik C et al. Mutations in human muscle LIM protein gene in families with hypertrophic cardiomyopathy. Circulation 2003; 107: 1390–5.

99. Osterop AP, Kofflard MJ, Sandkujl LA et al. AT1 receptor A/C1166 polymorphism contributes to cardiac hypertrophy in subjects with hypertrophic cardiomyopathy. Hypertension 1998; 32: 825–30.

100. Lechin M, Quinones MA, Omran A et al. Angiotensin-I converting enzyme genotypes and left ventricular hypertrophy in patients with hypertrophic cardiomyopathy. Circulation 1995; 92: 1808–12.

101. Charron P, Dubourg O, Desnos N et al. Clinical features and prognostic implications of familial hypertrophic cardiomyopathy related to the cardiac myosin-binding protein C. Circulation 1998; 97: 2230–6.

102. Maron BJ, Niimura H, Casey SA et al. Development of left ventricular hypertrophy in adults with hypertrophic cardiomyopathy caused by cardiac myosin-binding protein C mutations. J Am Coll Cardiol 2001; 38: 315–21.

103. Watkins H, McKenna WJ, Thierfelder L et al. Mutations in genes for cardiac troponin T and α-tropomyosin in hypertrophic cardiomyopathy. N Engl J Med 1995; 332: 1058–64.

104. Richard P, Charron P, Carrier L et al. Hypertrophic cardiomyopathy: distribution of genes, spectrum of mutations and implications for a molecular diagnostic strategy. Circulation 2003; 107: 2227–32.

105. Van Driest SL, Ellsworth EG, Ommen SR et al. Prevalence and spectrum of thin filament mutations in an outpatient referral population with hypertrophic cardiomyopathy. Circulation 2003; 108: 445–51.

106. Ackerman MJ, Van Driest SL, Ommen SR et al. Prevalence and age-dependence of malignant mutations in hypertrophic cardiomyopathy: a comprehensive outpatient perspective. J Am Coll Cardiol 2002; 19: 2049–51.

107. Maron BJ, McKenna WJ, Danielson GK et al. American College of Cardiology/European Society of Cardiology Clinical Expert Consensus Document on Hypertrophic Cardiomyopathy. J Am Coll Cardiol 2003; 42: 1687–713.

108. Maron BJ, Pellicia A, Spirito P. Cardiac disease in young trained athletes. Insights into methods for distinguishing athlete's heart from structural heart disease, with particular emphasis on hypertrophic cardiomyopathy. Circulation 1995; 91: 1596–601.

109. Ho CY, Sweitzer NK, McDonough B et al. Assessment of diastolic function with Doppler tissue imaging to predict phenotype in preclinical hypertrophic cardiomyopathy. Circulation 2002; 105: 2992–7.

110. Nagueh S, McFalls J, Myer D et al. Tissue Doppler imaging predicts the development of hypertrophic cardiomyopathy in subjects with subclinical disease. Circulation 2003; 108: 395–8.

111. McMahon CJ, Nagueh SF, Pignatelli RH et al. Characterization of left ventricular diastolic function by tissue Doppler imaging and clinical status in children with hypertrophic cardiomyopathy. Circulation 2004; 109: 1756–62.

112. Maron BJ, Casey SA, Hauser RG et al. Clinical course of hypertrophic cardiomyopathy with survival to advanced age. J Am Coll Cardiol 2003; 42: 882–8.

113. Maki S, Ikeda H, Muro A et al. Predictors of sudden death in hypertrophic cardiomyopathy. Am J Cardiol 1998; 82: 774–8.

114. Maron BJ, Olivotto I, Bellone P, Conte MR, Cecchi F, Flygenring BP et al. Clinical profile of stroke in 900 patients with hypertrophic cardiomyopathy. J Am Coll Cardiol 2002; 39: 301–7.

115. Maron BJ, Casey SA, Poliac LC et al. Clinical course of hypertrophic cardiomyopathy in a regional United States cohort. JAMA 1999; 281: 650–5.

116. Maron BJ, Spirito P. Impact of patient selection biases on the perception of hypertrophic cardiomyopathy and its natural history. Am J Cardiol 1993; 72: 970–2.

117. Grogan M, Redfield MM, Bailey KR et al. Long-term outcome of patients with biopsy-proven myocarditis: comparison with idiopathic dilated cardiomyopathy. J Am Coll Cardiol 1995; 26: 80–4.

118. Manolio TA, Baughman KL, Rodeheffer R et al. Prevalence and etiology of idiopathic dilated cardiomyopathy (summary of a National Heart, Lung and Blood Institute Workshop). Am J Cardiol 1992; 69: 1459–66.

119. Schönberger J, Seidman CE. Many roads to a broken heart: the genetics of dilated cardiomyopathy. Am J Med Genet 2001; 69: 249–60.

120. Devereux RB, Roman MJ, Paranicas M et al. A population-based assessment of left ventricular systolic dysfunction in middle-aged and older adults: The Strong Heart Study. Am Heart J 2001; 141: 439–46.

121. Devereux RB, Bella JN, Palmieri V et al. Left ventricular systolic dysfunction in a biracial sample of hypertensive adults. The HyperGEN Study. Hypertension 2001; 38: 417–23.

122. Davies M, Hobbs F, Davis R et al. Prevalence of left ventricular systolic dysfunction in the Echocardiographic Heart of England Screening Study. Lancet 2001; 358: 439–44.

123. Hedberg P, Lonnberg I, Jonason T et al. Left ventricular systolic dysfunction in 75 year old men and women: a population-based study. Eur Heart J 2001; 22: 676–83.

124. Nielsen OW, Hilden J, Larsen CT et al. Cross-sectional study estimating the prevalence of heart failure and left ventricular systolic dysfunction in community patients at risk. Heart 2001; 86: 172–8.

125. Mosterd A, Hoes AW, de Bruyne MC et al. Prevalence of heart failure and left ventricular dysfunction in the general population. The Rotterdam Study. Eur Heart J 1999; 20: 447–55.

126. Kupari M, Lindroos M, Iivanainen AM et al. Congestive heart failure in old age: prevalence, mechanisms and 4-year prognosis in the Helsinki Aging Study. J Intern Med 1997; 241: 387–94.

127. Schunkert H, Broeckel U, Hense HW et al. Left ventricular dysfunction. Lancet 1998; 351: 372.

128. McDonagh TA, Morrison CE, Lawrence A et al. Symptomatic and asymptomatic left ventricular systolic dysfunction in an urban population. Lancet 1997; 350: 829–33.

129. Wang TJ, Evans JC, Benjamin EJ et al. Natural history of asymptomatic left ventricular systolic dysfunction in the community. Circulation 2003; 108: 977–82.

130. Bella JN, Palmieri V, Roman MJ et al. Gender differences in left ventricular function in a population-based sample: The Strong Heart Study. J Am Soc Echocardiogr 1999; 12: 382.

131. Brophy JM, Dagenais GR, McSherry F et al. A multivariate model for predicting mortality in patients with heart failure and systolic dysfunction. Am J Med 2004; 116: 300–4.

132. Devereux RB, Roman MJ, Paranicas M et al. Impact of diabetes on cardiac structure and function: The Strong Heart Study. Circulation 2000; 101: 2271–6.

133. Palmieri V, Bella JN, Arnett DK et al. Effect of type 2 diabetes mellitus on left ventricular geometry and systolic function in hypertensive subjects: Hypertension Genetic Epidemiology Network (HyperGEN) Study. Circulation 2001; 103: 102–7.

134. Michels VV, Moll PP, Miller FA et al. The frequency of familial dilated cardiomyopathy in a series of patients with idiopathic dilated cardiomyopathy. N Engl J Med 1992; 326: 77–82.

135. Keeling PJ, Gang Y, Smith G et al. Familial dilated cardiomyopathy in the United Kingdom. Br Heart J 1995; 73: 417–21.

136. Grunig E, Tasman JA, Kucherer H et al. Frequency and phenotypes of dilated cardiomyopathy. J Am Coll Cardiol 1998; 31: 186–94.

137. Mestroni L, Rocco C, Gregori D et al. Familial dilated cardiomyopathy: evidence for genetic and phenotypic heterogeneity: Heart Muscle Disease Study Group. J Am Coll Cardiol 1999; 34: 181–90.

138. Fatkin D, MacRae C, Sasaki T et al. Missense mutations in the rod domain of the lamin A/C gene as a cause of dilated cardiomyopathy and conduction-system disease. N Engl J Med 1999; 341: 1759–62.

139. Durand JB, Bachinski LL, Bieling LC et al. Localization of a gene responsible for familial dilated cardiomyopathy to chromosome 1q32. Circulation 1995; 92: 3387–9.

140. Siu BL, Niimura II, Osborne JA et al. Familial dilated cardiomyopathy locus maps to chromosome 2q31. Circulation 1999; 99: 1022–6.

141. Gerull B, Gramlich M, Atherton J et al. Mutations of TTN, encoding the giant muscle filament titin, cause familial dilated cardiomyopathy. Nature Genet 2002; 30: 201–4.

142. Tsubata S, Bowles KR, Vatta M et al. Mutations in human delta-sarcoglycan gene in familial and sporadic dilated cardiomyopathy. J Clin Invest 2000; 106: 655–62.

143. Sylvius N, Tesson F, Gayet C et al. A new locus for autosomal dilated cardiomyopathy identified on chromosome 6q12-q16. Am J Med Genet 2001; 68: 241–6.

144. Schmitt JP, Kamisago M, Asahi M et al. Dilated cardiomyopathy and heart failure caused by mutation in phospholamban. Science 2003; 299: 1410–13.

145. Kranijovic M, Pinamonti B, Sinagra G et al. Linkage of familial dilated cardiomyopathy to chromosome 9. Am J Hum Genet 1995; 57: 846–52.

146. Kamisago M, Sharma SD, DePalma SR et al. Mutations in sarcomere protein genes as a cause of dilated cardiomyopathy. N Engl J Med 2000; 343: 1688–96.

147. Olson TM, Michels VV, Thibodeau SN et al. Actin mutations in dilated cardiomyopathy, a heritable form of heart failure. Science 1998; 280: 750–2.

148. Olson TM, Kishimoto NY, Whitby FG et al. Mutations that alter the surface charge of alpha-tropomyosin are associated with dilated cardiomyopathy. J Moll Cell Cardiol 2001; 33: 723–32.

149. Bowles KR, Gajarski R, Porter P et al. Gene mapping an autosomal dominant familial dilated cardiomyopathy to chromosome 10q21–23. J Clin Invest 1996; 98: 1355–60.

150. Olson TM, Illenberger S, Nishimoto NY et al. Metavinculin mutations alter actin interactions in dilated cardiomyopathy. Circulation 2002; 105: 431–7.

151. Li D, Ahmad F, Gardner MJ et al. The locus of a novel gene responsible for arrhythmogenic right ventricular dysplasia characterized by early onset and high penetrance maps to chromosome 10p12-p14. Am J Hum Genet 2000; 66: 148–56.

152. Jung M, Poepping I, Perrot A et al. Investigation of a family with dilated cardiomyopathy defines a novel locus on chromosome 2q14-q22. Am J Med Genet 1999; 65: 1068–77.

153. Olson TM, Keating MT. Mapping a cardiomyopathy locus to chromosome 3p22–25. J Clin Invest 1996; 97: 528–32.

154. Li D, Tapscoft T, Gonzales O et al. Desmin mutation responsible for idiopathic dilated cardiomyopathy. Circulation 1999; 100: 461–4.

155. Barresi R, Di Blasi C, Negri T et al. Disruption of heart sarcoglycan complex and severe cardiomyopathy caused by beta-sarcoglycan mutations. J Med Genet 2000; 37: 102–7.

156. Messina DN, Speer MC, Pericak-Vance MA et al. Linkage of familial dilated cardiomyopathy with conduction defect and muscular dystrophy to chromosome 6q23. Am J Hum Genet 1997; 61: 909–17.

157. Schönberger J. Levy H, Grunig E et al. Dilated cardiomyopathy and sensorineural hearing loss: a heritable syndrome that maps to 6q23–24. Circulation 2000; 101: 1812–18.

158. Rampazo A, Nava A, Erne P et al. A new locus for arrhythmogenic right ventricular cardiomyopathy (ARVD2) maps to chromosome 1q42-q43. Hum Mol Genet 1995; 4: 2151–4.

159. Rampazo A, Nava A, Miorin M et al. ARVD4, a new locus for arrhythmogenic right ventricular cardiomyopathy maps to chromosome 2 long arm. Genomics 1995; 45: 259–63.

160. Ahmad F, Li D, Karibe A et al. Localization of a gene responsible for arrhythmogenic right ventricular dysplasia to chromosome 3p23. Circulation 1998; 98: 2791–5.

161. Vatta M, Mohapatra B, Jimenez S et al. Mutations in Cypher/ZASP in patients with dilated cardiomyopathy and left ventricular non-compaction. J Am Coll Cardiol 2003; 42: 2014–27.

162. Sasse-Klaassen S, Probst S, Gerull B et al. Novel gene locus for autosomal dominant left ventricular noncompaction maps to chromosome 11p15. Circulation 2004; 109: 2720–3.

163. Knoll R, Hoshijima M, Hoffman HM et al. The cardiac mechanical stretch sensor machinery involves a Z disc complex that is defective in a subset of human dilated cardiomyopathy. Cell 2002; 111: 943–55.

164. Binengraeber M, Olson TM, Selivanov VA et al. ABCC9 mutations identified in human dilated cardiomyopathy disrupt catalytic K_{ATP} channel gating. Nature Genet 2004; 36: 382–7.

165. Severini GM, Krajinovic M, Pinamonti B et al. A new locus for arrhythmogenic right ventricular dysplasia on the long arm of chromosome 14. Genomics 1996; 31: 193–200.

166. Rampazo A, Nava A, Danieli GA et al. The gene for arrhythmogenic right ventricular cardiomyopathy maps to chromosome 14q23-q24. Hum Mol Genet 1994; 3: 959–62.

167. Norgett EE, Hatsell SJ, Carvajal-Huerta L et al. Recessive mutation in desmoplakin disrupts desmoplakin-intermediate filament interactions and causes dilated cardiomyopathy, wooly hair and keratoderma. Hum Mol Genet 2000; 9: 2761–6.

168. Coonar AS, Protonatorios N, Tsatsopulou A et al. Gene for arrhythmogenic right ventricular cardiomyopathy with diffuse nonepidermolytic palmoplantar keratoderma and wooly hair (Naxos disease) maps to chromosome 17q21. Circulation 1998; 97: 2049–58.

169. Ichida F, Tsubata S, Bowles KR et al. Novel gene mutations in patients wit left ventricular noncompaction or Barth syndrome. Circulation 2001; 103: 1256–63.

170. Bolhuis PA, Hensels GW, Hulsebos TJ et al. Mapping of the locus for X-linked cardioskeletal myopathy with neutropenia and abnormal mitochondria (Barth syndrome) to Xq28. Am J Hum Genet 1991; 48: 481–5.

171. Towbin JA, Hejtmancik JF, Brink P et al. X-linked dilated cardiomyopathy: molecular genetic evidence of linkage to the Duchenne muscular dystrophy (dystrophin) gene at the Xp21 locus. Circulation 1993; 87: 1854–65.

172. Gerull B, Hauser A, Wichter T et al. Mutations in desmosomal protein plakophilin-2 are common in arrhythmogenic right ventricular cardiomyopathy. Nature Genet 2004; 36: 1162–4.

173. Bowles NE, Bowles KR, Towbin JA. The 'final

common pathway' hypothesis and inherited cardio-vascular disease: the role of cytoskeletal proteins in dilated cardiomyopathy. Herz 2000; 25: 168–75.

174. Schocken DD, Arrieta MI, Leaverton PE et al. Prevalence and mortality rate of congestive heart failure in the United States. J Am Coll Cardiol 1992; 20: 301–6.

175. Ho KKL, Anderson KM, Kannel WB et al. Survival after onset of congestive heart failure in Framingham Heart Study subjects. Circulation 1993; 88: 107–15.

176. Pffefer MA, Braunwald E, Moye LA et al. Effect of captopril on mortality and morbidity in patients with left ventricular dysfunction after myocardial infarction. Results of the Survival and Ventricular Enlargement trial. The SAVE Investigators. N Engl J Med 1992; 327: 669–77.

177. Effect of enalapril on mortality and the development of heart failure in asymptomatic patients with reduced left ventricular ejection fractions. The SOLVD Investigators. N Engl J Med 1992; 327: 685–91.

178. CIBIS Investigators and Committees. A randomized trial of beta-blockers in heart failure. The Cardiac Insufficiency Bisoprolol Study (CIBIS). Circulation 1994; 1765–73.

179. Effect of metoprolol CR/XL in chronic heart failure: Metoprolol CR/XL Randomised Intrevention Trial in Congestive Heart Failure (MERIT-HF). Lancet 1999, 353. 2001–7.

180. Pitt B, Zannad F, Remme WJ et al. The effect of spironolactone on morbidity and mortality in patients with severe heart failure. Randomized Aldactone Evaluation Study Investigators. N Engl J Med 1999; 341: 709–17.

181. Dries DL, Strong MH, Cooper RS et al. Efficacy of angiotensin-converting enzyme inhibition in reducing progression from asymptomatic left ventricular dysfunction to symptomatic heart failure in black and white patients. J Am Coll Cardiol 2002; 40: 311–17.

182. Poole-Wilson PA, Swedberg K, Cleland JG et al. Comparison of carvedilol and metoprolol on clinical outcomes in patients with chronic heart failure in the Carvedilol or Metoprolol European Trial (COMET): randomized controlled trial. Lancet 2003; 362: 7–13.

183. Cohn JN, Tognoni G. Valsartan Heart Failure Trial Investigators. A randomized trial of angiotensin-receptor blocker valsartan in chronic heart failure. N Engl J Med 2001; 345: 1667–75.

184. Granger CB, McMurray JJ, Yusuf S et al. Effect of candesartan in patients with chronic heart failure and reduced left ventricular systolic function intolerant to angiotensin converting enzyme inhibitors: the CHARM-Alternative Trial. Lancet 2003; 362: 772–6.

185. Mestroni L, Krajinovic M, Severini JM et al. Familial dilated cardiomyopathy. Br Heart J 1994; 72: 35–41.

186. Mestroni L, Maisch B, McKenna WJ et al. Guidelines for the study of familial dilated cardiomyopathy. Eur Heart J 1999; 20: 93–102.

187. Baig MK, Goldman JH, Caforio AL et al. Familial dilated cardiomyopathy: cardiac abnormalities are common in asymptomatic relatives and may represent early disease. J Am Coll Cardiol 1998; 195–201.

188. Michels VV, Olson TM, Miller FA et al. Frequency of development of idiopathic dilated cardiomyopathy among relatives of patients with idiopathic dilated cardiomyopathy. Am J Cardiol 2003; 91: 1389–92.

189. Csanady M, Hogye M, Kallai A et al. Familial dilated cardiomyopathy: a worse prognosis compared with sporadic forms. Br Heart J 1995; 74: 171–3.

190. Michels VV, Driscoll DJ, Miller FA et al. Progression of familial and non-familial dilated cardiomyopathy: long term follow-up. Heart 2003; 89: 757–761.

191. Franz WM, Muller OJ, Katus HA. Cardiomyopathies: from genetics to prospect of treatment. Lancet 2001; 258: 1627–37.

192. Geisterfer-Lowrance AAT, Christe M, Connor DA et al. A mouse model for familial hypertrophic cardiomyopathy. Science 1996; 272: 731–4.

193. Fatkin D, Christe ME, Aristizabal O et al. Neonatal cardiomyopathy in mice homozygous for the Arg403Gln mutation in α-cardiac myosin heavy chain gene. J Clin Invest 103; 147–53.

194. McConnell BK, Jones KA, Fatkin D et al. Dilated cardiomyopathy in homozygous myosin-binding protein-C mice. J Clin Invest 1999; 104: 1235–44.

195. Konno T, Shimizu M, Ino II et al. A novel missense mutation in the myosin-binding protein C gene is responsible for hypertrophic cardiomyopathy with left ventricular dysfunction and dilatation in elderly patients. J Am Coll Cardiol 2003; 41: 781–6.

CHAPTER 29

Left ventricular hypertrophy in special populations

Mark H Drazner

Introduction

Left ventricular hypertrophy (LVH) is associated with numerous adverse cardiovascular outcomes including the development of a depressed left ventricular ejection fraction,[1] heart failure,[2] and overall mortality.[3] Deciphering the genetic risk factors for LVH may translate into improved prevention, detection, risk stratification, and ultimately treatment of this important disease. Nevertheless, despite the availability of high-throughput sequencing and genotyping, there has been little progress in identifying those polymorphisms in genes which lead to LVH, outside of mutations in sarcomeric proteins which cause familial hypertrophic cardiomyopathy. This lack of progress in elucidating the genetic risk factors of LVH may be understandable when one considers LVH as a complex trait, i.e. impacted by two or more interacting genes, each contributing some small amount of risk to the development of the phenotype, as well as being influenced by environmental factors.

One potential strategy to dissect the genetic basis of a complex trait such as LVH is to focus on populations that have an increased prevalence of LVH, reasoning that such populations may have an increased frequency of LVH-susceptibility genetic variants. African-Americans have emerged as a group that may be attractive for such targeted studies.[4,5] In this chapter, we will explore the available data that suggest that African-Americans as compared to whites are at an increased risk of LVH. Next, we will review the contentious debate about whether the increased risk of LVH in African-Americans should be attributed to genetic or nongenetic risk factors.

Are African-Americans at increased risk of LVH?

Electrocardiographic studies

Although the electrocardiogram (ECG) has limited sensitivity for the diagnosis of LVH,[6] it was the initial diagnostic test available to compare the prevalence of LVH in African-Americans and whites. An early study found ~2-fold increased prevalence of electrocardiographic LVH in African-American as compared to white men, and in African-American as compared to white women,[7] while another found nearly 10-fold increased prevalence of LVH in African-American men compared to white men.[8] More recent studies have confirmed that electrocardiographic LVH is more prevalent in African-Americans than in whites,[9] even after adjusting for other risk factors in multivariable modeling.[10–12]

Despite the fairly consistent finding of ethnic disparities in prevalent electrocardiographic LVH, the question as to whether LVH is more common in African-Americans than whites remained open because the accuracy of the ECG as a diagnostic tool to assess LVH in African-Americans was challenged. The initial study which raised this concern found that traditional ECG criteria had lower specificity in African-Americans than in whites.[13] Such a difference in operating characteristics would lead to a higher rate of false positive cases of LVH in African-Americans, and thereby overestimate ethnic disparities in prevalence of this condition.[13] Subsequent studies that have addressed this question likewise have demonstrated decreased specificity of classical LVH criteria including the Sokoloff–Lyon criteria in African-Americans,[14–17] largely attributed to increased QRS voltages in African-Americans, and have led to the suggestion that ethnic-specific ECG criteria for LVH are needed.

Echocardiographic studies

Given the limitations of electrocardiography in assessing LVH, it is not surprising that echocardiography was used next to assess whether ethnic disparities in left ventricular (LV) mass and LVH existed. The initial echocardiographic studies, performed ~20–25 years ago, comprised small numbers of patients and yielded conflicting results with some reporting increased LV mass and/or LVH in African-Americans as compared to whites,[18,19] and others finding no differences in cardiac structure between the two ethnic groups.[20,21] In 1998, Devereux et al. published a meta-analysis of the echocardiographic studies available at that point (n = 9, including those cited above) and concluded that there was no consistent trend for an increased left ventricular mass among African-Americans as compared to whites, but there was for an increased left ventricular wall thickness.[22]

Since the Devereux meta-analysis, there has been continued interest in this subject and at least 14 subsequent studies have addressed this question (Table 29.1). Many, but not all, of the studies found higher left ventricular mass and/or prevalent LVH in African-Americans as compared to whites. Several of the larger studies warrant special comment.

The Cardiovascular Health Study

The Cardiovascular Health Study (CHS) is a National Heart, Lung, and Blood Institute prospec-

tive population-based multicenter study of non-institutionalized elderly ($\geq$ 65 years of age) individuals.[35] The initial cohort recruited between 1989 and 1990 was predominantly white with a second cohort of African-Americans subsequently recruited between 1992 and 1993. Echocardiograms were obtained in the initial cohort at baseline, and repeated after five years of follow-up. The African-American cohort had their initial echocardiogram during the time interval when the initial cohort had their year 5 follow-up echocardiogram. There was no significant difference in echocardiographic LV mass between African-American (n = 162) and white (n = 1170) men (171 ± 50 g versus 173 ± 51 g, respectively), or between African-American (n = 342) and white (n = 1953) women (138 ± 42 g versus 137 ± 41 g, respectively). Furthermore, the relationship of LV mass to body weight was steeper in both white men and women as compared to their African-American counterparts.

Although these data suggest that LV mass is not increased in African-Americans as compared to whites, there are limitations to this study. First, the white and African American cohorts were recruited into the CHS at different time points, introducing the possibility of selection bias. For example, LV mass in the African-American cohort was measured predominantly after two years of follow-up in the CHS, while the LV mass of the white participants was measured either at baseline or after 5 years of follow-up. Whether this difference in length of follow-up prior to assessment of LV mass introduced a bias is uncertain, though plausible. Another consideration is survivor bias. If African-Americans have higher rates of LVH at younger ages than white individuals, and LVH is associated with increased mortality,[3] then African-Americans with LVH may not live long enough ($\geq$ 65 years of age) to be enrolled in the CHS. In essence, then, it is possible that the elderly population of CHS is not well suited to address this question.

The CARDIA study

The Coronary Artery Disease Risk Development in Young Adults (CARDIA) study is a National Heart, Lung, and Blood Institute prospective study

Table 29.1. *Recent studies of left ventricular mass (LVM) and LV hypertrophy (LVH) in blacks and whites*

Author	Year	Study name or type[d]	Number of subjects		Age (approximate years)	Comparison between blacks and whites[a]				
						LV mass			LVH	
			Black	White		Crude	BSA	Height[2.7]	BSA	Height[2.7]
Zabalgoitia et al.[23]	1998	HOT	112	332	60	=	=	B > W		B > W
Olutade et al.[24]	1998	University-based	34	39	53	=	=	B >W		=
Mayet et al.[25]	1998	Hypertension clinic	46	46	44		B > W			
Chapman et al.[14]	1999	Hypertension clinic	137	271	47		B > W		B > W	
Rautaharju et al.[16]	2000	Cardiovascular Health Study	504	123	75	=				
El-Garbawy et al.[26]	2001	University based	82	63	46		B > W			
Dekkers et al.[27]	2002	Longitudinal study of children	328	359	23	B > W				
Stanton et al.[28]	2002	Hypertension clinic	19	19	46				B > W	
East et al.[29]	2003	Cardiac catheterization lab	462	1999	66		B > W		B > W	
Lorber et al.[30]	2003	CARDIA 10-year follow-up	566	792	35			B > W[c]		
Fox et al.[31]	2004	Jackson cohort of ARIC	1729	–	59					~40%[b]
Hinderliter et al.[32]	2004	Duke Biobehavorial Investigation of Hypertension	88	83	33			B > W[c]		
Rodriguez et al.[33]	2004	Northern Manhattan Study	417	377	71			B > W		
Kizer et al.[34]	2004	HyperGen	1060	580	53	B > W		B > W	B > W	B > W

[a]B, black, W, white, =, comparable in blacks and whites; BSA, body surface area.
[b]Prevalence of LVH in black men and women was 37% and 41%, respectively. No white subjects were included in the Jackson cohort with which to compare.
[c]No *P* value reported in Lorber et al.,[30] and *P = 0.09* in Hinderliter et al.[32]
[d]HOT, Hypertension Optimal Treatment; CARDIA, Coronary Artery Disease Risk Development in Young Adults; ARIC, Atherosclerosis Risk in Communities Study; HyperGEN, Hypertension Genetic Epidemiology Network.

of young adults who were African-American or white and free of long-term disease or disability.[36] At the baseline echocardiographic study, African-American men and women as compared to their white counterparts had higher LV mass after multivariable adjustment for other risk factors.[37] At 5 years of follow-up, at which time the subjects were ~35 years of age, only African-American women among the four ethnic and gender subgroups had an increase in LV mass over time.[38] Additionally, the prevalence of LVH in the overall cohort at this point remained low (2%, 30 cases in 1358 subjects), although it was higher in African-American versus white women (3% versus 1%), and in African-American men versus white men (5% versus 1%).[30]

The Jackson cohort of ARIC

The Atherosclerosis Risk in Communities Study (ARIC) is a prospective, epidemiological study sponsored by the National Institutes of Health and National Heart, Lung and Blood Institute in which subjects were recruited from four communities in the United States. In one of these centers (Jackson, Mississippi), 1729 enrolled subjects, all of whom were African-American, underwent echocardiography at baseline. The echocardiographic cohort represented 21% of those subjects originally approached for study enrollment.[39] The prevalence of LVH as defined by indexation to allometric height (51 $g/m^{2.7}$) in this middle-aged cohort was very high, ranging from 37% in men to 41% in women.[31] The prevalence of LVH was further increased in those with higher blood pressure and increased body mass index.

It is worthwhile to pause and reflect on these findings. First, the prevalence of LVH in the middle-aged Jackson cohort (~ 40%) was 8–10-fold greater than that found in the younger African-Americans (3–5 %) of CARDIA. Though one must be cautious in drawing conclusions based on two studies with different methods of recruitment in disparate geographic locations in the United States, the juxtaposition of these findings nevertheless suggests an alarming increase in the prevalence of LVH among African-Americans between the ages of 35 and ~60 years. Second, the prevalence of LVH in the African-American Jackson cohort appears to be ~2–3-fold higher when compared to similarly aged whites in the Framingham Heart Study.[40] Third, given the important adverse prognosis associated with LVH,[3,41] these data suggest an urgent need for therapies targeted at preventing and reversing LVH in African-Americans, in order to reduce the exaggerated cardiovascular mortality they endure as compared to whites.[42]

HyperGEN

The Hypertension Genetic Epidemiology Network (HyperGEN) is a component of the Family Blood Pressure Program sponsored by the National Heart, Lung, and Blood Institute. Recruitment into HyperGEN was based on a sibling-pair design such that subjects were eligible if ≥2 family members developed essential hypertension prior to the age of 60 years. A recent analysis from this study focused on whether African-American ethnicity was an independent risk factor for LVH.[34] This study is important due to its large size (1060 African-Americans and 580 whites) and adjustment for potential confounders such as length of college education in multivariable models. The principal finding from this study was that African-American ethnicity was associated with an odds ratio of 1.8 (95% confidence intervals 1.3–2.5) for LVH, defined by indexation to height[2.7] and an odds ratio of 2.3 (95% confidence intervals 1.2–4.3) for LVH defined by indexation to body surface area in multivariable models. The LV diastolic dimension (adjusted in multivariable model) was not increased in African-Americans as compared to whites, compatible with concentric LVH as occurs in response to hypertension.

There are limitations to this study.[43] First, due to the sibling-pair design, only subjects who had a hypertensive sibling were enrolled. Whether the conclusions of the study apply to patients with hypertension who do not have a hypertensive sibling is uncertain. Another concern is whether confounding secondary to socioeconomic status was adequately accounted for in the multivariable models. Although the level of college education was adjusted for in the multivariable models, this may not be sufficient to account for the impact of lower levels of socioeconomic status. Completion of high school education, for example, may have been a more appropriate surrogate for the impact of low socioeconomic status.

NOMAS

The Northern Manhattan Study (NOMAS) is a prospective, population-based multi-ethnic cohort in which subjects were recruited from northern Manhattan, New York, via random telephone digit dialing. Among the 1916 subjects, there were 377 whites and 417 African-Americans. The importance of considering socioeconomic status when assessing inter-individual differences in LV mass is demonstrated in this study.[33] LV mass/height$^{2.7}$ was inversely associated with socioeconomic status as assessed by four levels of education (<high school; completed high school; some college; ≥ college graduate), with mean LV mass/height$^{2.7}$ values of 48.4, 48.6, 47.1, and 45.3 g/m$^{2.7}$, respectively ($P <$ 0.001). Given the known differences of socioeconomic status between African-Americans and whites, it is not surprising then that socioeconomic status interacted with the association of ethnicity and LV mass. Whereas LV mass/height$^{2.7}$ adjusted for age, sex, systolic blood pressure, diabetes, body mass index, and physical activity was comparable in African-Americans (44.7 g/m$^{2.7}$) and whites (45.3 g/m$^{2.7}$) among those who had graduated from college, similarly adjusted LV mass/height$^{2.7}$ values appeared higher in African Americans (50 g/m$^{2.7}$) than whites (45.3 g/m$^{2.7}$) among those who did not complete high school.

Summary: LVH and LV mass in African-Americans as compared to whites

Although there is consistent evidence of an increased prevalence of electrocardiographic LVH in African-Americans, concerns over the diagnostic accuracy of that modality to address this question persist. Why African-Americans appear to have increased QRS voltage as compared to whites in the absence of cardiac hypertrophy is an area that also needs further investigation.

Fortunately, there has been considerable ongoing interest in using echocardiography to determine whether African-Americans do have increased LV mass and prevalence of LVH as com-

pared to whites (Table 29.1). One study (the Jackson cohort[31]) found near epidemic levels of LVH (~ 40%) in the African-American community in Mississippi, although there were no white subjects enrolled in this study with which to compare. However, this prevalence of LVH is considerably higher than that found in the predominantly white cohort of the Framingham Heart Study,[40] a finding which would undoubtedly explain part of the increased cardiovascular mortality endured by African-Americans as compared to whites.[42] Whether differences in age are sufficient to explain the 8–10-fold higher prevalence of LVH in the older subjects (~60 years of age) of the Jackson cohort of ARIC, as compared to the younger (~35 years of age) African-Americans in CARDIA[38] is not yet clear, but does suggest an opportunity for therapeutic interventions aimed at preventing the development of LVH in young to middle-aged African-Americans. It is somewhat more difficult to reconcile the findings from the Jackson cohort with the CHS. The latter is a study of even older African-Americans (75 years) and yet there was no association of ethnicity with increased LV mass assessed by echocardiography,[16] leaving one to speculate as to the possibility of selection or survivor bias. In contrast to CHS, a recently published analysis of the HyperGEN study[34] found that middle-aged African-Americans, as compared to white hypertensives had ~2-fold increased odds of having LVH, although it is possible that inadequate adjustment for lower socioeconomic status may have confounded this conclusion.[33,43] Finally, the Dallas Heart Study,[44] and the Multi-Ethnic Study of Atherosclerosis,[45] are two recently completed studies in which large subjects of African-Americans and whites underwent cardiac magnetic resonance imaging, arguably the gold-standard modality to measure LV mass. These studies should provide further information to the emerging database which compares the risk of LVH in African-Americans to that in whites.

In conclusion, the majority of, but not all, echocardiographic studies suggest that African-Americans are at increased risk of LVH as compared to whites. If so, then it is of considerable interest to consider the basis of this apparent ethnic disparity in the hypertrophic response.

Is the increased risk of LVH in African-Americans due to genetic or non-genetic (environmental) risk factors?

Determining the etiology of the increased prevalence of LVH in African-Americans is complicated and touches upon a raging debate in the medical community: specifically, what role does genetics play in explaining racial classification and the differences in health outcomes between races (e.g. African-Americans and whites).[46–50] This debate at its core addresses the respective roles of nature versus nurture in human disease states. As schematized in Fig. 29.1A, it is tempting to speculate that ethnic-specific genetic differences explain the increased prevalence of LV hypertrophy in African-Americans. However, the schema in

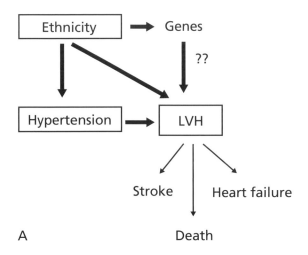

A

Figure 29.1. *Pathways linking African-American ethnicity with left ventricular hypertrophy (LVH). (A) The association of African-American ethnicity with hypertension and LVH is schematized. It is tempting to speculate that ethnic-specific genetic risk factors explain the association of African-American ethnicity with LVH. (B) Highlights additional complexities when considering the association between African-American ethnicity and LVH. Specifically, an alternative hypothesis now shown is that environmental factors such as low socioeconomic status (SES) are the basis for the increased prevalence of LVH in African-Americans. Socioeconomic status is a complex construct and only certain of its components, and potential pathways that link low SES to LVH and hypertension, are depicted.*

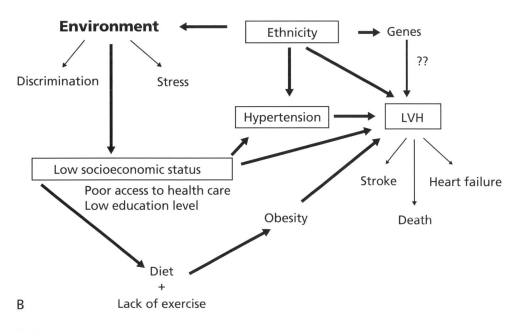

B

Fig. 29.1A is deliberately simplified, and a more complete diagram is shown in Fig. 29.1B, where some of the potential contributions of the environment have now been incorporated. Due to the known disparities in socioeconomic status between African-Americans and whites, and the link between socioeconomic status and a number of health outcomes including overall mortality,[51] Fig. 29.1B emphasizes its potential role in contributing to the exaggerated hypertrophic response in African-Americans. It should be recognized, however, that other environmental exposures including perceived racism may also contribute to the increased prevalence of hypertension and its complications in African-Americans.[52,53] Differences in quality of care provided to different ethnic groups are also likely to be important.[54]

Low socioeconomic status may lead to ethnic disparities in LV mass, through a variety of pathways in addition to impaired access to care. For example, low socioeconomic status may be associated with obesity via an unhealthy diet or lack of exercise, and obesity may then contribute to the development of LVH.[55] At the least, an increased burden of obesity in African-Americans may represent an environmental factor which can interact and modify an underlying genetic risk factor for LVH, even if those genetic factors are present in both ethnic groups.

Probably, most complex diseases in the end will be attributed to a combination of numerous interacting genetic and environmental factors,[56] and a distinction between the two may be even counterproductive if it minimizes the importance of either factor. Nevertheless, there are a number of arguments that can be proposed to support the role of a genetic or non-genetic (i.e. environmental) cause of the disproportionate LVH in African-Americans. These are summarized in Table 29.2, and discussed below.

Ethnic-specific polymorphisms exist

The frequencies of polymorphisms can vary dramatically between ethnic groups.[49] It is now known that some polymorphisms are unique or privileged to a specific ethnic group, and this is more often the case in individuals of African ancestry than in white individuals.[47,57] In theory, if such ethnic-specific

polymorphisms predisposed to LVH, they could contribute to the increased prevalence of LVH in African-Americans. However, before accepting this possibility, it is important to recognize that to date no such ethnic-specific polymorphism has been shown to be meaningfully associated with cardiac hypertrophy. Additionally, it has been noted that the overall genetic differences between different ethnic groups (inter-race) are less than the overall genetic differences within an ethnic group (intra-race).[50] This observation suggests that ethnic-specific polymorphisms would be unlikely to be important contributors to the inter-individual differences in LV mass and hypertrophy.

African-Americans have an increased risk of other hypertensive end-organ damage

African-Americans are known to have an increased prevalence of other hypertensive complications including stroke and end-stage renal disease.[58] Those in favor of an underlying genetic predisposition to hypertensive end-organ damage believe this to support their hypothesis, while those in favor of a non-genetic cause would argue that this observation merely reflects the greater burden of hypertension (i.e. severity or duration as above) endured by African-Americans.

LVH is disproportionate to the increased burden of hypertension in African-Americans

African-American ethnicity remained associated with increased LV mass and LVH, independently of blood pressure in multivariable models,[33,34] supporting the possibility that African-Americans are uniquely susceptible to complications of hypertension, possibly on a genetic basis. However, blood pressure in studies to date has been measured by isolated, daytime readings, and not by 24-hour blood pressure monitoring. Because 24-hour blood pressure monitoring may allow a better assessment of the hypertension burden,[59] it is possible that the true blood pressure burden in African-Americans has been underestimated in studies to date. This would lead to residual confounding in multivariable

Table 29.2. *Arguments for a genetic or non-genetic (environmental) etiology of increased prevalence of LVH in African-Americans*

Argument	Genetic	Non-genetic (environment)
Ethnic-specific genetic variants exist	African Americans have greater genetic diversity; supports plausibility of ethnic-specific genetic risk factors	No clear association of ethnic-specific genetic variant with LVH to date; intra-race genetic variation much greater than inter-race genetic variation.
African-Americans have increased risk of other hypertensive end-organ damage	Ethnic susceptibility to effects of hypertension	Reflects increased severity or duration of hypertension.
LVH is disproportionate to hypertension	Increased risk of LVH persists despite adjustment for blood pressure in multivariable models	Inadequate adjustment for severity of HTN (not 24-hour blood pressure); inadequate adjustment for duration of HTN.
African-Americans are at increased risk of other complex diseases	Multiple underlying genetic differences	Unlikely for one ethnic group to have a large number of disadvantageous genes; a more parsimonious explanation is that one underlying common exposure (low socioeconomic status) is the root cause.
Low socioeconomic status is associated with LVH	Increased risk of LVH persists despite adjustment for socioeconomic status in multivariable models	Inadequate capture of socioeconomic status as a construct leads to residual confounding.

models, suggesting an independent association of African-American ethnicity with LVH, when in fact this association was mediated via differences in the blood pressure burden. A similar argument can be made for the duration of hypertension. African-Americans are more likely than whites to be uninsured,[60] with poorer access to preventive healthcare that can delay the diagnosis of hypertension. A delay in the diagnosis of hypertension would lead to an underestimation of the true duration of hypertension. Thus, when duration of hypertension is adjusted for in multivariable models, it will be disproportionately shorter in African-Americans as compared to whites, again leading to residual confounding.

African-Americans are at increased risk of other complex diseases

When comparing African-Americans to whites, striking differences exist in the prevalence and outcome of a broad range of disease states beyond cardiovascular illnesses, including diabetes mellitus,[61] cancer,[62] and HIV.[63] Putative genetic causes have begun to be identified for a number of these, including at least one case where there are important ethnic differences in the frequency of a polymorphism which impacts outcome.[47,64] Nevertheless, there are strong arguments to suggest that the inter-ethnic differences in prevalence and outcome across broad disease states support a non-genetic cause as the basis for these ethnic disparities. First, because it is unlikely that genetic risk factors for such a broad array of diseases would overlap to any substantial degree, one would have to postulate that the genetic makeup of African-Americans includes a large number of ethnic-specific polymorphisms that are associated with adverse health outcomes. Clearly, this is an unlikely scenario. A far more parsimonious explanation is that a common exposure (e.g. low socioeconomic status) shared by many African-Americans is the root cause of their adverse health outcomes including prevalent LVH. The plausibility of this hypothesis is supported by the powerful impact that socioeconomic status has on health outcomes, including overall premature mortality.[51]

Low socioeconomic status is associated with LVH

Low socioeconomic status is a risk factor for increased LV mass.[33] In two studies to date, African-American ethnicity remained associated with an increased risk of LVH, despite adjustment in multivariable models for socioeconomic status as reflected by educational status,[33,34] supporting the belief that differences in socioeconomic status do not fully account for ethnic disparities in LVH. However, it has been noted that socioeconomic status is a complex construct with many domains, and the traditional variables used to reflect socioeconomic status in multivariable modeling (e.g. educational status) may not fully capture its impact, thereby leading to residual confounding of the association of ethnicity and adverse health outcomes.[65]

Summary

Attempts to dissect the genetic basis of complex diseases are under way in full force, even as it remains uncertain how important such genetic factors are in comparison to the influence of environmental exposures. Recognizing that the relative risk conferred by a contributing polymorphism may well be below the limits of detection with linkage analysis,[66] case-control association studies are likely to be important in this quest. As has been highlighted recently,[66–69] such studies will need to be carefully designed and have sufficient power to detect relative risk ratios of 1.5 to 2. Replication will be necessary to minimize false-positive associations, a substantial risk given that the large number (possibly 15 million[66]) of polymorphisms in the human genome rapidly overwhelms conventional thresholds of statistical significance.

It is not yet certain whether the apparent ethnic disparities in LVH reflect a unique (i.e. genetic) susceptibility to end-organ damage from hypertension, or whether it represents residual confounding from an underestimation of the hypertension burden or environmental influences acting upon African-Americans. Nevertheless, as data emerge which show that African-Americans have both an increased prevalence of LVH and a greater degree of genetic diversity, it seems reasonable that association studies in this population may offer unique advantages to identify those genetic factors which contribute to inter-individual differences in LV mass and hypertrophy. Simultaneously, intensive efforts should be focused on identifying environmental factors that contribute to LVH. Undoubtedly, a greater understanding of both genetic and environmental influences will afford the best opportunity to minimize disparities in health outcomes between ethnic groups in the United States, a major goal of modern health initiatives.[70]

References

1. Drazner MH, Rame JE, Marino EK et al. Increased left ventricular mass is a risk factor for the development of a depressed left ventricular ejection fraction within five years: the Cardiovascular Health Study. J Am Coll Cardiol 2004; 43: 2207–15.

2. Gottdiener JS, Arnold AM, Aurigemma GP et al. Predictors of congestive heart failure in the elderly: the Cardiovascular Health Study. J Am Coll Cardiol 2000; 35: 1628–37.

3. Levy D, Garrison RJ, Savage DD et al. Prognostic implications of echocardiographically determined left ventricular mass in the Framingham Heart Study. N Engl J Med 1990; 322: 1561–6.

4. Arnett DK, Hong Y, Bella JN et al. Sibling correlation of left ventricular mass and geometry in hypertensive African Americans and whites: the HyperGEN study. Am J Hypertens 2001; 14: 1226–30.

5. Harshfield GA, Grim CE, Hwang C et al. Genetic and environmental influences on echocardiographically determined left ventricular mass in black twins. Am J Hypertens 1990; 3: 538–43.

6. Levy D, Labib SB, Anderson KM et al. Determinants of sensitivity and specificity of electrocardiographic criteria for left ventricular hypertrophy. Circulation 1990; 81: 815–20.

7. Bartel A, Heyden S, Tyroler HA et al. Electrocardiographic predictors of coronary heart disease. Arch Intern Med 1971; 128: 929–37.

8. Sutherland SE, Gazes PC, Keil JE et al. Electrocardiographic abnormalities and 30-year mortality among white and black men of the Charleston Heart Study. Circulation 1993; 88: 2685–92.

9. Rautaharju PM, LaCroix AZ, Savage DD et al. Electrocardiographic estimate of left ventricular mass versus radiographic cardiac size and the risk of cardiovascular disease mortality in the epidemiologic follow-up study of the First National Health and Nutrition Examination Survey. Am J Cardiol 1988; 62: 59–66.

10. Antikainen R, Grodzicki T, Palmer AJ et al. The determinants of left ventricular hypertrophy defined by Sokolow-Lyon criteria in untreated hypertensive patients. J Hum Hypertens 2003; 17: 159–64.

11. Arnett DK, Rautaharju P, Crow R et al. Black-white differences in electrocardiographic left ventricular mass and its association with blood pressure (the ARIC study). Am J Cardiol 1994; 74: 247–52.

12. Xie X, Liu K, Stamler J, Stamler R. Ethnic differences in electrocardiographic left ventricular hypertrophy in young and middle-aged employed American men. Am J Cardiol 1994; 73: 564–7.

13. Lee DK, Marantz PR, Devereux RB et al. Left ventricular hypertrophy in black and white hypertensives. Standard electrocardiographic criteria overestimate racial differences in prevalence. JAMA 1992; 267: 3294–9.

14. Chapman JN, Mayet J, Chang CL et al. Ethnic differences in the identification of left ventricular hypertrophy in the hypertensive patient. Am J Hypertens 1999; 12: 437–42.

15. Okin PM, Wright JT, Nieminen MS et al. Ethnic differences in electrocardiographic criteria for left ventricular hypertrophy: the LIFE study. Losartan Intervention For Endpoint. Am J Hypertens 2002; 15: 663–71.

16. Rautaharju PM, Park LP, Gottdiener JS et al. Race- and sex-specific ECG models for left ventricular mass in older populations. Factors influencing overestimation of left ventricular hypertrophy prevalence by ECG criteria in African-Americans. J Electrocardiol 2000; 33: 205–18.

17. Crow RS, Prineas RJ, Rautaharju P et al. Relation between electrocardiography and echocardiography for left ventricular mass in mild systemic hypertension (results from Treatment of Mild Hypertension Study). Am J Cardiol 1995; 75: 1233–8.

18. Dunn FG, Oigman W, Sungaard-Riise K et al. Racial differences in cardiac adaptation to essential hypertension determined by echocardiographic indexes. J Am Coll Cardiol 1983; 1: 1348–51.

19. Hammond IW, Devereux RB, Alderman MH et al. The prevalence and correlates of echocardiographic left ventricular hypertrophy among employed patients with uncomplicated hypertension. J Am Coll Cardiol 1986; 7: 639–50.

20. Hammond IW, Alderman MH, Devereux RB et al. Contrast in cardiac anatomy and function between black and white patients with hypertension. J Natl Med Assoc 1984; 76: 247–55.

21. Savage DD, Henry WL, Mitchell JR et al. Echocardiographic comparison of black and white hypertensive subjects. J Natl Med Assoc 1979; 71: 709–12.

22. Devereux RB, Okin PM, Roman MJ. Pre-clinical cardiovascular disease and surrogate endpoints in hypertension: does race influence target organ damage independent of blood pressure? Ethn Dis 1998; 8: 138–48.

23. Zabalgoitia M, Ur Rahman SN, Haley WE et al. Impact of ethnicity on left ventricular mass and relative wall thickness in essential hypertension. Am J Cardiol 1998; 81: 412–7.

24. Olutade BO, Gbadebo TD, Porter VD et al. Racial differences in ambulatory blood pressure and echocardiographic left ventricular geometry. Am J Med Sci 1998; 315: 101–9.

25. Mayet J, Chapman N, Li CK et al. Ethnic differences in the hypertensive heart and 24-hour blood pressure profile. Hypertension 1998; 31: 1190–4.

26. El-Gharbawy AH, Kotchen JM, Grim CE et al. Predictors of target organ damage in hypertensive blacks and whites. Hypertension 2001; 38: 761–6.

27. Dekkers C, Treiber FA, Kapuku G et al. Growth of left ventricular mass in African American and European American youth. Hypertension 2002; 39: 943–51.

28. Stanton AV, Mayet J, Chapman N et al. Ethnic differences in carotid and left ventricular hypertrophy. J Hypertens 2002; 20: 539–43.

29. East MA, Jollis JG, Nelson CL et al. The influence of left ventricular hypertrophy on survival in patients with coronary artery disease: do race and gender matter? J Am Coll Cardiol 2003; 41: 949–54.

30. Lorber R, Gidding SS, Daviglus ML et al. Influence of systolic blood pressure and body mass index on left ventricular structure in healthy African-American and white young adults: the CARDIA study. J Am Coll Cardiol 2003; 41: 955–60.

31. Fox E, Taylor H, Andrew M et al. Body mass index and blood pressure influences on left ventricular mass and geometry in African Americans: The Atherosclerotic Risk In Communities (ARIC) Study. Hypertension 2004; 44: 55–60.

32. Hinderliter AL, Blumenthal JA, Waugh R et al. Ethnic differences in left ventricular structure: relations to hemodynamics and diurnal blood pressure variation. Am J Hypertens 2004; 17: 43–9.

33. Rodriguez CJ, Sciacca RR, Diez-Roux AV et al. Relation between socioeconomic status, race-ethnicity, and left ventricular mass: the Northern Manhattan study. Hypertension 2004; 43: 775–9.

34. Kizer JR, Arnett DK, Bella JN et al. Differences in left ventricular structure between black and white hypertensive adults: the Hypertension Genetic Epidemiology Network study. Hypertension 2004; 43: 1182–8.

35. Fried LP, Borhani NO, Enright P et al. The Cardiovascular Health Study: design and rationale. Ann Epidemiol 1991; 1: 263–76.

36. Friedman GD, Cutter GR, Donahue RP et al. CARDIA: study design, recruitment, and some characteristics of the examined subjects. J Clin Epidemiol 1988; 41: 1105–16.

37. Gardin JM, Wagenknecht LE, Anton-Culver H et al. Relationship of cardiovascular risk factors to echocardiographic left ventricular mass in healthy young black and white adult men and women. Circulation 1995; 92: 380–7.

38. Gardin JM, Brunner D, Schreiner PJ et al. Demographics and correlates of five-year change in echocardiographic left ventricular mass in young black and white adult men and women: the Coronary Artery Risk Development in Young Adults (CARDIA) study. J Am Coll Cardiol 2002; 40: 529–35.

39. Skelton TN, Andrew ME, Arnett DK et al. Echocardiographic left ventricular mass in African-Americans: the Jackson cohort of the Atherosclerosis Risk in Communities Study. Echocardiography 2003; 20: 111–20.

40. Levy D, Anderson KM, Savage DD et al. Echocardiographically detected left ventricular hypertrophy: prevalence and risk factors. The Framingham Heart Study. Ann Intern Med 1988; 108: 7–13.

41. Liao Y, Cooper RS, McGee DL et al. The relative effects of left ventricular hypertrophy, coronary artery disease, and ventricular dysfunction on survival among black adults. JAMA 1995; 273: 1592–7.

42. National Center for Health Statistics. Health, United States, 2003 with Chartbook on Trends in the Health of Americans (Table 36,) Hyattsville, Maryland: National Center for Health Statistics; 2003.

43. Drazner MH. Left ventricular hypertrophy is more common in black than white hypertensives: is this news? Hypertension 2004; 43: 1160–1.

44. Victor RG, Haley RW, Willett DL et al. The Dallas Heart Study: a population-based probability sample for the multidisciplinary study of ethnic differences in cardiovascular health. Am J Cardiol 2004; 93: 1473–80.

45. Bild DE, Bluemke DA, Burke GL et al. Multi-ethnic study of atherosclerosis: objectives and design. Am J Epidemiol 2002; 156: 871–81.

46. Cooper RS, Kaufman JS, Ward R. Race and genomics. N Engl J Med 2003; 348: 1166–70.

47. Burchard EG, Ziv E, Coyle N et al. The importance of race and ethnic background in biomedical research and clinical practice. N Engl J Med 2003; 348: 1170–5.

48. Sankar P, Cho MK, Condit CM et al. Genetic research and health disparities. JAMA 2004; 291: 2985–9.

49. Risch N, Burchard E, Ziv E et al. Categorization of humans in biomedical research: genes, race and disease. Genome Biol 2002; 3(comment2007): 1–12.

50. Kaplan JB, Bennett T. Use of race and ethnicity in biomedical publication. JAMA 2003; 289: 2709–16.

51. Cooper RS, Kennelly JF, Durazo-Arvizu R et al. Relationship between premature mortality and socioeconomic factors in black and white populations of US metropolitan areas. Public Health Rep 2001; 116: 464–73.

52. Knox SS, Hausdorff J, Markovitz JH. Reactivity as a predictor of subsequent blood pressure: racial differences in the Coronary Artery Risk Development in Young Adults (CARDIA) Study. Hypertension 2002; 40: 914–9.

53. Steffen PR, McNeilly M, Anderson N et al. Effects of perceived racism and anger inhibition on ambulatory blood pressure in African Americans. Psychosom Med 2003; 65: 746–50.

54. Epstein AM, Ayanian JZ. Racial disparities in medical care. N Engl J Med 2001; 344: 1471–3.

55. Gottdiener JS, Reda DJ, Materson BJ et al. Importance of obesity, race and age to the cardiac structural and functional effects of hypertension. The Department of Veterans Affairs Cooperative Study

Group on Antihypertensive Agents. J Am Coll Cardiol 1994; 24: 1492–8.

56. Chakravarti A, Little P. Nature, nurture and human disease. Nature 2003; 421: 412–4.

57. Stephens JC, Schneider JA, Tanguay DA et al. Haplotype variation and linkage disequilibrium in 313 human genes. Science 2001; 293: 489–93.

58. American Heart Association. Statistical Fact Sheet-Populations. African Americans and cardiovascular diseases. 2003: 1–8.

59. Staessen JA, Thijs L, Fagard R et al. Predicting cardiovascular risk using conventional vs ambulatory blood pressure in older patients with systolic hypertension. Systolic Hypertension in Europe Trial Investigators. JAMA 1999; 282: 539–46.

60. Institute of Medicine. Coverage Matters: Insurance and Health Care. Washington DC: National Academy Press; 2001.

61. Wong MD, Shapiro MF, Boscardin WJ et al. Contribution of major diseases to disparities in mortality. N Engl J Med 2002; 347: 1585–92.

62. Kiefe CI. Race/ethnicity and cancer survival: the elusive target of biological differences. JAMA 2002; 287: 2138–9.

63. McGinnis KA, Fine MJ, Sharma RK et al. Understanding racial disparities in HIV using data from the veterans aging cohort 3-site study and VA administrative data. Am J Public Health 2003; 93: 1728–33.

64. Stephens JC, Reich DE, Goldstein DB et al. Dating the origin of the CCR5-Delta32 AIDS-resistance allele by the coalescence of haplotypes. Am J Hum Genet 1998; 62: 1507–15.

65. Kaufman JS, Cooper RS, McGee DL. Socioeconomic status and health in blacks and whites: the problem of residual confounding and the resiliency of race. Epidemiology 1997; 8: 621–8.

66. Botstein D, Risch N. Discovering genotypes underlying human phenotypes: past successes for mendelian disease, future approaches for complex disease. Nat Genet 2003; 33 (Suppl):228–37.

67. Cardon LR, Bell JI. Association study designs for complex diseases. Nat Rev Genet 2001; 2: 91–9.

68. Tabor HK, Risch NJ, Myers RM. Opinion: Candidate-gene approaches for studying complex genetic traits: practical considerations. Nat Rev Genet 2002; 3: 391–7.

69. Lohmueller KE, Pearce CL, Pike M et al. Meta-analysis of genetic association studies supports a contribution of common variants to susceptibility to common disease. Nat Genet 2003; 33: 177–82.

70. Department of Health and Human Services. Healthy People 2010: understanding and improving health. Washington DC: Government Printing Office; 2000.

CHAPTER 30

Monogenic causes of cardiac hypertrophy: hypertrophic cardiomyopathy and related phenotypes

Charles Redwood, Hugh Watkins

Introduction

Left ventricular hypertrophy (LVH), reflecting a growth in individual myocyte size rather than an increase in cell number (hyperplasia), is one of the most powerful predictors of cardiac death. Most commonly, LVH is produced as a secondary consequence of increased cardiac load; in that setting the extent of hypertrophy is influenced by multiple susceptibility genes, and LVH functions as a classical polygenic quantitative trait. However, cardiac hypertrophy can also be triggered by inherited single gene disorders in the absence of other cardiac or systemic disease. Genetic analyses have the power to identify previously unknown pathogenic pathways causing LVH, because such analyses can be applied systematically without reliance on existing assumptions regarding etiology. Accordingly, monogenic disorders that result in cardiac hypertrophy have been an important area of study, with ramifications that extend beyond the direct clinical impact of these relatively uncommon conditions.

Inherited syndromes that include cardiac hypertrophy

In many monogenic forms of cardiac hypertrophy, the cardiac phenotype is a component of a more extensive syndrome. In this category there are a number of metabolic disorders inherited as autoso-mal recessive traits, such as those affecting cardiac fatty acid uptake (CD36 deficiency)[1] or metabolism (very-long-chain acyl-coenzyme A dehydrogenase deficiency).[2] Similarly, defects in nuclear-encoded genes involved in mitochondrial function produce recessive cardiac hypertrophy syndromes (for example Friedreich's ataxia),[3] whereas inherited mutations of mitochondrial tRNA genes produce maternally transmitted syndromes that include ventricular hypertrophy.[4] Noonan syndrome is an important example of an autosomal dominant syndrome which affects multiple organ systems with prominent features of cardiac hypertrophy.[5] Additionally, a number of storage disorders affect the heart, sometimes leading to massive hypertrophy; these include disorders of glycogen storage (for example, Pompe disease), and lysosomal storage (such as Danon disease and Fabry disease). All of the above syndromes share in common a tendency for the cardiac hypertrophy to manifest in infancy or childhood and, in part because of the severity of the associated syndromes, most tend to be rare in adult populations.

Hypertrophic cardiomyopathy

By far the most common cause of inherited left ventricular hypertrophy in the general population is the autosomal dominant disorder hypertrophic cardiomyopathy (HCM) which affects up to 1 in

500 of the population.[6] HCM is characterized by unexplained asymmetric left ventricular (LV) hypertrophy which, at the whole heart level, is associated with systolic hypercontractility and a small ventricular cavity. The hypertrophy typically manifests in late childhood or early adult life and, in the main, is not progressive thereafter. The defining histological feature of HCM is myocyte disarray, and the principle clinical issue is that, in some families, HCM is associated with a high incidence of sudden death. In contrast with the previous groups of disorders, the clinical features of HCM are confined to the heart. Thus the designation HCM, which in the past has sometimes been used simply as a descriptor of the morphological features of asymmetric LVH, has come to be used specifically for the familial trait with isolated cardiac involvement. Open questions, which have come to the fore with the demonstration that HCM is genetically heterogeneous, are whether it should be considered as one disease or many, and whether it should be defined by a molecular taxonomy or by the disease phenotype.

In recent years HCM has been studied intensively as a genetic model of primary cardiac hypertrophy. Other chapters in this section concentrate on describing the clinical implications of molecular analyses in HCM and the relevant genetically engineered mouse models; here we describe the genetic basis of HCM, how the mutations affect the properties of the proteins encoded by the known disease genes, and how these alterations may bring about pathological hypertrophy and ventricular remodeling. Insights from these analyses suggest there may be shared mechanisms with other monogenic cardiac hypertrophy syndromes, and that these mechanisms may, in turn, shed light on susceptibility to LVH as a quantitative trait.

Mutations in β-myosin heavy chain cause HCM

A seminal paper from the Seidman laboratory published in 1990 was the first to identify a specific mutant allele that causes HCM.[7] The affected gene was identified by linkage mapping of a large kindred as MYH7 (which encodes β-myosin heavy

chain, βMHC), and the reported missense mutation in exon 13 resulted in an arginine to glutamine amino acid substitution at residue 403. Since this first report, at least a further 100 HCM mutations in this gene have been reported, accounting for approximately one-quarter of all affected individuals. MHC molecules assemble in vivo to form bipolar thick filaments, which align in a parallel arrangement with the thin actin filaments (see Fig. 30.1). The ATP-driven sliding of these two sets of filaments with respect to one another is responsible for the contraction and relaxation of cardiac muscle. The disease-causing mutations in MYH7 are missense mutations causing single amino acid substitutions chiefly in the head portion of the protein which contains the actin- and ATP-binding sites, although more recently mutations within the light meromyosin domain, responsible for thick filament formation, have also been reported.[8]

Mutations in other contractile protein genes cause HCM: a disease of the sarcomere?

Linkage analysis in other affected families revealed that mutations in the genes encoding cardiac troponin T,[9,10] α-tropomyosin,[9,10] and cardiac myosin binding protein-C (MyBP-C)[11,12] (also components of the contractile apparatus; Fig. 30.1) can also cause HCM. This gave rise to the proposal that HCM is a 'disease of the sarcomere',[9] and suggested that genes encoding other cardiac contractile proteins were good candidates as additional disease genes. Candidate gene screening has indeed shown that mutations in a further five genes encoding sarcomeric proteins cause the disease. These include components of both the thick and thin filaments: both the regulatory and essential myosin light chains (MLC),[13] cardiac actin,[14] cardiac troponin I,[15] and the giant protein titin,[16] which spans the entire half-sarcomere from the Z-disc to the M line (see Fig. 30.1). A single mutation in the gene encoding α-myosin heavy chain,[17] a minor component of ventricular myosin, has been reported, and similarly a single cardiac troponin C mutation has also been found.[18] However, these mutations were only identified in single cases with no evidence of

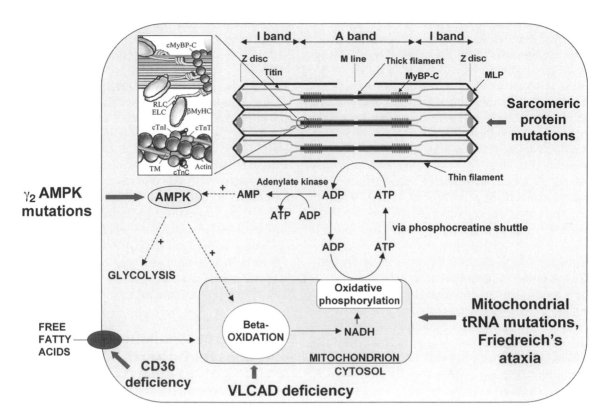

Figure 30.1. A schematic showing the contractile and metabolic processes in the cardiac myocyte which are affected by mutations in disease genes for inherited monogenic forms of cardiac hypertrophy. The disease hypertrophic cardiomyopathy (HCM) is caused by mutations in at least nine genes encoding sarcomeric proteins. These include the thick filament proteins β-myosin heavy chain (βMHC), the regulatory and essential myosin light chains (RLC and ELC) and cardiac myosin binding protein-C (cMyBP-C); the thin filament proteins cardiac actin, cardiac troponin I (cTnI), cardiac troponin T (cTnT) and α-tropomyosin (TM); and the giant protein titin. Mutations in the gene encoding muscle LIM protein (MLP), which is located at the periphery of the Z-disc, also cause HCM. A similar phenotype, along with pre-excitation and conduction disease, results from mutations in the γ_2 subunit of AMP-activated protein kinase (AMPK). Cardiac hypertrophy is also caused by other inherited defects, such as those affecting mitochondrial function (mitochondrial tRNA mutations and Friedreich's ataxia), and those causing deficiencies in proteins involved in fatty acid uptake (CD36) and metabolism (very long chain acyl coenzyme A dehydrogenase, VLCAD). The inset showing the detail of the contractile apparatus is adapted from Spirito P, Seidman CE, McKenna WJ, Maron BJ. The management of hypertrophic cardiomyopathy. N Engl J Med 1997; 336: 775–85.

segregation with disease, and no further mutations have been found to substantiate them; thus the status of these two genes as disease causing remains unproven.

Most large cohort studies have found that disease-causing mutations can be identified in 50–70% of those screened, and have suggested that mutations in MYH7 and MYBPC3 are the most common, each accounting for approximately one-quarter to one-third of all cases, with the remaining genes each contributing 5% or less. For example, a recent study of 197 unrelated patients found disease-causing mutations in 124 cases (~63% of the total), 40% of these being in MYH7

(25% of the total) and 42% in *MYBPC3* (26% of the total) with troponin T and troponin I mutations each found in 8% (5% of the total).[19] These studies are prone to ascertainment bias due to different referral practices. For example, it is likely that the contribution of troponin T mutations was overestimated in early reports, because a large proportion of the families studied had a high incidence of sudden death, a phenpotype associated with troponin T mutations.[10] The vast majority of HCM-causing alleles encode missense mutations; the one exception to this is mutations of *MYBPC3* which are often truncation mutations (see below). The missense mutations tend to be scattered through much of each disease-gene, and the total number of disease alleles is very high, such that many families are found to have 'private' mutations.

A mutation in a specific gene may cause disease in the heterozygous state through haploinsufficiency, by diminishing the transcription or translation of the mutant allele, or by reducing the stability of the gene product. Alternatively, the mutation may act in a dominant-negative manner, in which the mutant protein is expressed and interferes with the normal protein function. These two mechanisms may be discriminated by measuring the amount of mutant protein with respect to wild type. Such studies on HCM mutant proteins have proved difficult due to limited availability of myocardial samples from affected individuals, and, moreover, the difficulty in separation of wild-type and mutant protein differing by only a single amino acid. However, in nearly all cases in which detection of mutant protein has been technically possible, its presence has been confirmed, often in approximately equal abundance as the wild type protein. For example, the presence of Arg403Gln βMHC[20] (which has a missing Arg-C protease site) and Asp175Asn α-tropomyosin[21] (which migrates faster than wild type α-tropomyosin upon SDS-polyacrylamide gel electrophoresis (PAGE)) has been demonstrated in the skeletal muscles of affected patients. Using a combination of high-pressure liquid chromatography (HPLC), mass spectrometry and capillary zone electrophoresis, the levels of the βMHC mutants Val606Met (12% of total βMHC) and Gly584Arg (23%) have been measured in myocardial samples.[22] It is thus widely

accepted that HCM mutant proteins are stably incorporated into the sarcomere and act via a dominant-negative mechanism as 'poison polypeptides'. The possible exception to this are some of the cardiac MyBP-C mutants; over half of the reported *MYBPC3* mutations encode truncated proteins (chiefly due to aberrant splicing), and most of these are missing the C-terminal domains responsible for localizing the protein to the thick filament backbone.[23] Western blot analysis of the myocardium from a patient with one of these truncation mutations failed to detect the expected mutant protein,[24] thus suggesting haploinsufficiency as a possible disease mechanism in this case. However, it is still entirely possible that a small amount of mutant protein is expressed and acts via a 'poison polypeptide' mechanism, and further study of this class of mutation is required.

Functional consequences of HCM mutations in contractile proteins

Numerous biophysical, biochemical and physiological studies have sought to determine the fundamental changes in contractility caused by HCM disease gene mutations.[25] *In vitro* determinations of contractile function (such as actomyosin ATPase measurements, *in vitro* sliding filament assays, skinned muscle fiber studies, and optical trap techniques) have been employed using protein derived from biopsy material or a suitable heterologous expression system. In addition, a number of mouse models of HCM have been created, and are the focus of Chapter 37. The initial *in vitro* work focused on the βMHC mutant proteins which, in a number of early studies, were found to generate less force than wild-type myosin, and to translocate actin filaments at lower velocities.[26] These findings led to the proposal that the hypertrophy produced by these mutations was compensatory in response to the lower force generated by the mutant sarcomeres.[27] More recent work however has contradicted this, and has suggested that these βMHC mutants are able to translocate actin filaments at higher (not lower) velocities.[28] Furthermore, functional analysis of mutant alleles of other HCM disease genes has suggested that these are likely to act via a 'hypercontractile' mechanism. Most HCM mutations in the thin filament regulatory proteins troponin T, troponin I, and α-tropomyosin, appear to cause an increase in the Ca^{2+} sensitivity of the regulation of contraction,[29] thus causing increased force at any submaximal Ca^{2+} concentration, and probably relaxation abnormalities. Increased Ca^{2+} sensitivity has also been observed with myosin regulatory light chain mutants.[30]

Mutations in non-sarcomeric protein genes can also cause HCM

The tenet that HCM is only caused by mutations that directly affect contractility has had to be revised in recent years as disease-causing mutations have been reported in the genes encoding muscle LIM protein (MLP) and the γ_2 subunit of AMP-activated protein kinase (AMPK). MLP is a positive regulator of myogenesis, and promotes myocyte differentiation. It is colocalizes with α-actinin and titin at the periphery of the Z-disc and, along with telethonin, has been suggested to form a stretch-dependent signaling complex.[31] MLP was investigated as a candidate gene for inherited cardiomyopathies based on the dilated cardiomyopathy (DCM) phenotype of a MLP-knockout mouse. Three different missense mutations were found in three unrelated HCM patients; each mutation affects an amino acid involved in the binding of α-actinin, and the Cys58Gly mutant showed a reduced interaction with this protein.[32]

A clinical phenotype of HCM in combination with Wolff–Parkinson–White syndrome is caused by mutations in PRKAG2 which encodes the γ_2 subunit of the AMPK.[33–35] Affected individuals have marked cardiac hypertrophy, which may be asymmetric, and a tendency to later ventricular dilation and contractile dysfunction. The electrophysiological features of pre-excitation are present from birth, whereas progressive conduction disease occurs later in life. Histological analysis shows that this disorder is distinct from classical 'sarcomeric' HCM, as disarray is not present and, instead, glycogen deposition is prominent.[35] It is a matter of semantics as to whether this condition is classified as HCM (which it clinically resembles), or as a

glycogen storage disorder (based on histological grounds); the importance lies in what these genetic findings reveal in terms of disease pathways. In this context, it is of note that the glycogen content appears insufficient to account for the cardiac hypertrophy by a direct mass effect. The AMPK kinase, an αβγ trimer of which α is the catalytic subunit, is activated upon ATP depletion, turning on energy-producing pathways (such as glycolysis and β-oxidation of fatty acids), and inhibiting ATP-consuming processes (for example, fatty acid synthesis), and has been termed the fuel gauge of the cell.[36] The effects of the PRKAG2 mutations on kinase activity are currently contentious. Analysis of the equivalent amino acid substitutions in the γ_1 subunit,[37] and in the yeast homolog Snf1/Snf4 kinase,[35] have suggested they may cause increased kinase activity and/or constitutive activation. However cellular co-expression of α and β AMPK subunits with either wild-type or mutant γ_2 has shown the mutations result in reduced maximal activity and/or AMP dependence.[38]

Possible pathways that may be activated to trigger pathological hypertrophy in HCM

Altered Ca^{2+} homeostasis

The heart responds to a wide variety of hypertrophic stimuli (such as pressure overload, peptide growth factors and vasoactive peptides) by the activation of a number of interrelated signaling pathways, which result in changes in gene expression, increases in protein synthesis, and altered Ca^{2+} homeostasis. These pathways may involve receptors for both α- and β-adrenergic agents and growth factors leading to the activation the extracellular signal-regulated kinase (ERK), p38 and c-jun N-terminal kinase (JNK) mitogen-activated protein kinase (MAPK) cascades, protein kinase C (PKC), calmodulin-dependent protein kinase II, and calcineurin (Fig. 30.2).[39] A major question is which, if any, of these mechanisms are triggered by the primary changes brought about by HCM mutations in contractile and non-contractile proteins. As the cytoplasmic Ca^{2+} concentration, $[Ca^{2+}]_i$, is intimately linked to the regulation of contractility,

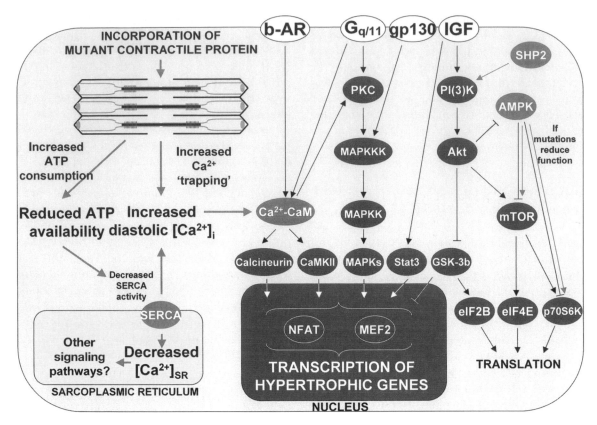

Figure 30.2. *A diagram showing some of the known cardiac hypertrophic signaling pathways and possible mechanisms by which mutations in known disease genes may interact with them. Activation of a number of different receptor classes leads to hypertrophic signaling; these include the angiotensin II, endothelin-1 and α-adrenergic receptors (all of which act via the GTP-binding proteins G_q/G_{11}) along with β-adrenergic receptors (β-AR), gp130 tyrosine kinase and the insulin-like growth factor-1 receptor (IGF). The signals are transduced by a variety of routes including the mitogen-activated protein kinase cascade (MAPK/MAPKK/MAPKKK), protein kinase C (PKC), the phosphatidyl inositol kinase (PI(3)K)/Akt/glycogen synthase kinase 3β (GSK3β) pathway, calmodulin-dependent protein kinase II (CaMKII), and the phosphatase calcineurin. These pathways result in the modulation of transcription via modification of nuclear factors of activated T cells (NFATs), myocyte enhancer factor-2 (MEF2) and other intermediates. Effects on translation are mediated via mammalian target of rapamycin (mTOR) and GSK3β. The mechanisms by which HCM mutant contractile proteins are postulated to increase diastolic $[Ca^{2+}]_i$, and hence activate Ca^{2+}-dependent pathways involving calcineurin, CaMKII and Ca^{2+}-sensitive PKC isoforms, are indicated with red arrows, as are possible mechanisms by which mutations in the $γ_2$ subunit of AMPK and in SHP2 (which cause Noonan syndrome) may affect signaling.*

it has been hypothesized that one of the possible pathogenic mechanisms of the HCM mutations may involve Ca^{2+} dysregulation,[40] possibly leading to increased $[Ca^{2+}]_i$ and activation of the Ca^{2+}/calmodulin-sensitive calcineurin, CaMKII and/or Ca^{2+}-sensitive protein kinase C isoforms.

There is some experimental evidence for this in mouse models of HCM; for example, ventricular myocytes paced at physiological rates from Ile79Asn mutant troponin T hearts showed elevated diastolic $[Ca^{2+}]_i$ compared with those from non-transgenic animals.[41] In most animal models, diastolic $[Ca^{2+}]_i$ increases have not been reported, although significant changes in the resting level are difficult to detect using the common spectroscopic indicators (for example fura-2). Furthermore, the $[Ca^{2+}]_i$ surrounding the contractile apparatus varies over a 10-fold range approximately once a second, due to the cardiac contraction/relaxation cycle, suggesting that sustained small increases in $[Ca^{2+}]_i$ may be sensed in a distinct subcellular pool.

It has been suggested that increases in $[Ca^{2+}]_i$ may be caused by 'calcium trapping' by the mutant sarcomere,[42] mediated by changes in the Ca^{2+} affinity of troponin C, widely believed to be the major Ca^{2+} buffer during the Ca^{2+} transient.[43] It is well established that the HCM mutants in troponin T, troponin I and α-tropomyosin in general result in increased myofilament calcium sensitivity, likely to be mediated by an increase in Ca^{2+} binding at the low-affinity, regulatory site of troponin C. In addition, 'hypercontractile' βMHC or MyBP-C mutants give an increase in strong myosin head binding, which in turn would increase Ca^{2+} binding by troponin C. A slower release of the 'trapped' calcium at the end of the transient might reduce the efficiency of reuptake by the sarcoplasmic reticulum (SR), and if this was sufficiently marked it could result in a heightened diastolic $[Ca^{2+}]_i$ as well as reduced SR Ca^{2+} load. It is important to note however that increases in thin filament Ca^{2+} sensitivity do not necessarily result in hypertrophy; for example, cardiomyocytes from mice expressing the slow skeletal isoform of troponin I show increased Ca^{2+} sensitivity, but the hearts appear to have normal morphology.[44]

Energetic compromise

A further postulated mechanism leading to the pathological hypertrophy of HCM, and which may in part also act through calcium dysregulation, involves energetic compromise.[45] The myosin ATPase accounts for at least 70% of ATP hydrolysis in the cardiac myocyte, and so any perturbation of either the motor itself, or its regulation, which would alter the efficiency of ATP usage by the sarcomere, would have major consequences for the cell. Biophysical studies have predicted an inefficiency in chemomechanical coupling for a number of different classes of HCM-mutant contractile proteins,[25,46] and this may identify a unifying feature of

the diverse sarcomeric mutations. These proposals have been supported by empirical data showing increased tension cost in mutant mouse fibers.[47] Chronic increase in the energy cost of maintaining normal power output will give rise to ATP depletion, which is predicted to affect other highly ATP-dependent processes in the cell (Fig. 30.2). In particular, the SR calcium pump (SERCA), the ATPase with the highest minimal energy requirement in muscle cells,[48] may be compromised with a resultant increase in cytoplasmic Ca^{2+}. There is strong evidence that energy compromise occurs in the hearts of HCM patients; a recent paper described similar (~30%) reductions in phophocreatine/ATP ratios in patients with either βMHC, MyBP-C or troponin T mutations, and this change was apparent prior to the development of hypertrophy.[49] In addition, in a mouse model of the Arg403Gln myosin mutation, a comparable reduction in phosphocreatine/ATP was measured, and it was suggested that the free energy change for ATP hydrolysis was decreased to close to the minimal requirement for the SERCA.[50] This finding has been reiterated in a recent study of a troponin T mutant mouse with profoundly altered cardiac energetics.[51]

Abnormalities of energetics in other inherited forms of cardiac hypertrophy

The inherited metabolic syndromes that cause cardiac hypertophy (see above) have in common that they result in impaired ATP production in the cardiomyocyte, for example by impairing substrate utilization, the mitochondrial electron transport chain or the export of ATP from the mitochondrion (Fig. 30.1). These observations lend further weight to the hypothesis that energetic compromise can initiate a signaling pathway leading to pathological cardiac hypertrophy. Similarly, the finding that the AMPK γ_2 mutations cause HCM in combination with Wolff–Parkinson–White syndrome[33] further implicates abnormalities of energy homeostasis. A reduction in the ability to sense low ATP levels (as predicted by reductions in AMP sensitivity) might directly lead to energetic compromise. However, it should be noted that the γ_2 isoform is a minority component (with γ_1 accounting for most γ regulatory activity), and this may imply that the γ_2 mutations perturb some unique role of this subunit (rather than an overall loss of activity). Through mechanisms that are currently unclear, the γ_2 mutations lead to marked glycogen accumulation; because glycogen inhibits AMPK activity this will result in global suppression of AMPK in the heart (affecting both mutant and wild-type complexes). Recent studies have also suggested that AMPK directly interacts with components of hypertrophic signaling pathways; in particular, AMPK activation (for example through cellular energy starvation) activates TSC1/TSC2, and thereby inhibits mTOR to inhibit cell growth, and enhance cell survival, by inhibiting protein translation.[52] These results imply that reduction in AMPK activity (secondary to glycogen accumulation) may increase translation and act to promote hypertrophy (Fig. 30.2).[53] Likewise, the Noonan syndrome gene product, SHP2, is a tyrosine phosphatase which has a role in the regulation of the PI3 kinase and ERK/MAPK cascades;[54] the mutations which cause this syndrome appear to be activating and may cause hypertrophy by altered signaling via these pathways (Fig. 30.2).

In the glycogen storage disorders, an inability to break down glycogen (for example due to alpha-1,4-glucosidase deficiency in Pompe disease) may render the heart relatively substrate deficient, because the abundant glycogen is metabolically inaccessible. The accumulated glycogen will, however, also lead to reduction in AMPK activity (either by direct inhibition or by sequestration, and hence prevention of its activation by upstream kinases[55]), and this would be expected to have growth-stimulating consequences. This may help explain the disproportionate myocardial hypertrophy (several-fold increase in heart weight), despite the fact that glycogen remains a minority constituent.

Unanswered genetic questions: further disease genes and modifiers for HCM

Systematic screens of the known HCM disease genes typically only identify pathogenic mutations in about two-thirds of probands or families screened. Some mutations may be missed by current techniques, but this is unlikely to explain the shortfall. At the same time, screening of the remaining candidate contractile protein genes has not revealed any more numerically important causes of HCM. It is therefore anticipated that further HCM disease genes will encode non-sarcomeric proteins and, as such, will probably reflect downstream components of the disease pathway. Based on our model of HCM pathogenesis we predict that these will probably include proteins involved in energy metabolism or calcium handling in the myocyte.

More subtle variants in the same set of genes will be prime candidates for modifier genes for HCM and, by extrapolation, other forms of cardiac hypertrophy. Despite carrying an identical disease-causing allele, affected relatives in families with HCM show a wide range of hypertrophy response. No formal segregation data exist, but inspection of pedigrees suggests clustering of the extent and pattern of hypertrophy in close relatives, suggesting significant effects of shared genetic background. Some data on phenotypically discordant monozygotic twins have been cited in support of non-genetic modifiers in HCM, but monozygotic twins can differ in their mitochondrial genomes, with obvious implications for their ATP-generating capacity. Indeed, co-inheritance of deleterious mitochondrial variants with sarcomeric missense mutations is one of the few convincingly documented modifier effects in HCM to date.[56] The other clear illustration of additive genetic effects is the surprisingly high frequency of compound heterozygosity in HCM families;[19] presumably, many individuals with single HCM mutations escape clinical detection, whereas those who inherit two different alleles manifest obvious disease and so constitute a significant proportion (~5%) of patient series. Finally, where two deleterious alleles exist in the same copy of the same gene, then there is consistent distortion of the genotype–phenotype relationship throughout the pedigree.[57] This situation, and the existence of compound heterozygosity more generally, illustrates the hazards of screening patients simply by typing for known mutations; instead, systematic screening is mandated.

Conclusion

The genetic complexity underlying inherited forms of cardiac hypertrophy has slowed the direct clinical application of molecular analyses. However, the corollary is that the same complexity has been informative in revealing key steps involved in hypertrophy signaling pathways in the heart. Despite the diverse array of initiating genetic defects, functional analyses suggest that these may converge on a limited repertoire of perturbations to cellular homeostasis; in particular, abnormalities of energy homeostasis and calcium handling appear to be central. On the basis of these mechanistic insights, new approaches to therapy will be explored that may ultimately be relevant to common, acquired, forms of cardiac hypertrophy.

References

1. Tanaka T, Sohmiya K, Kawamura K. Is CD36 deficiency an etiology of hereditary hypertrophic cardiomyopathy? J Mol Cell Cardiol 1997; 29: 121–7.
2. Cox GF, Souri M, Aoyama T et al. Reversal of severe hypertrophic cardiomyopathy and excellent neuropsychologic outcome in very-long-chain acyl-coenzyme A dehydrogenase deficiency. J Paediatr 1998; 133: 247–53.
3. Lodi R, Cooper JM, Bradley JL et al. Deficit of in vivo mitochondrial ATP production in patients with Friedreich ataxia. Proc Natl Acad Sci USA 1999; 96: 11492–5.
4. Merante F, Tein I, Benson L et al. Maternally inherited hypertrophic cardiomyopathy due to a novel T-to-C transition at nucleotide 9997 in the mitochondrial tRNA(glycine) gene. Am J Hum Genet 1994; 55: 437–46.
5. Burch M, Mann JM, Sharland M et al. Myocardial disarray in Noonan syndrome. Br Heart J 1992; 68: 586–8.
6. Maron BJ, Gardin JM, Flack JM et al. Prevalence of hypertrophic cardiomyopathy in a general population of young adults. Echocardiographic analysis of 4111 subjects in the CARDIA Study. Coronary Artery Risk Development in (Young) Adults. Circulation 1995; 92: 785–9.
7. Geisterfer-Lowrance AA, Kass S, Tanigawa G et al. A molecular basis for familial hypertrophic cardiomyopathy: a beta cardiac myosin heavy chain gene missense mutation. Cell 1990; 62: 999–1006.
8. Blair E, Redwood C, de Jesus Oliveira M et al. Mutations of the light meromyosin domain of the beta-myosin heavy chain rod in hypertrophic cardiomyopathy. Circ Res 2002; 90: 263–9.
9. Thierfelder L, Watkins H, MacRae C et al. Alpha-tropomyosin and cardiac troponin T mutations cause familial hypertrophic cardiomyopathy: a disease of the sarcomere. Cell 1994; 77: 701–12.
10. Watkins H, McKenna WJ, Thierfelder L et al. Mutations in the genes for cardiac troponin T and alpha-tropomyosin in hypertrophic cardiomyopathy. N Engl J Med 1995; 332: 1058–64.
11. Watkins H, Conner D, Thierfelder L et al. Mutations in the cardiac myosin binding protein-C gene on chromosome 11 cause familial hypertrophic cardiomyopathy. Nat Genet 1995; 11: 434–7.
12. Bonne G, Carrier L, Bercovici J et al. Cardiac myosin binding protein-C gene splice acceptor site mutation is associated with familial hypertrophic cardiomyopathy. Nat Genet 1995; 11: 438–40.
13. Poetter K, Jiang H, Hassanzadeh S et al. Mutations in either the essential or regulatory light chains of myosin are associated with a rare myopathy in human heart and skeletal muscle. Nat Genet 1996; 13: 63–9.
14. Mogensen J, Klausen IC, Pedersen AK et al. Alpha-cardiac actin is a novel disease gene in familial hypertrophic cardiomyopathy. J Clin Invest 1999; 103: 39–43.
15. Kimura A, Harada H, Park JE et al. Mutations in the cardiac troponin I gene associated with hypertrophic cardiomyopathy. Nat Genet 1997; 16: 379–82.
16. Satoh M, Takahashi M, Sakamoto T et al. Structural analysis of the titin gene in hypertrophic cardiomyopathy: identification of a novel disease gene. Biochem Biophys Res Commun 1999; 262: 411–17.
17. Niimura H, Patton KK, McKenna WJ et al. Sarcomere protein gene mutations in hypertrophic cardiomyopathy of the elderly. Circulation 2002; 105: 446–51.
18. Hoffmann B, Schmidt-Traub H, Perrot A et al. First mutation in cardiac troponin C, L29Q, in a patient with hypertrophic cardiomyopathy. Hum Mutat 2001; 17: 524.
19. Richard P, Charron P, Carrier L et al. Hypertrophic cardiomyopathy: distribution of disease genes, spectrum of mutations, and implications for a molecular diagnosis strategy. Circulation 2003; 107: 2227–32.
20. Cuda G, Fananapazir L, Zhu W-S et al. Skeletal muscle expression and abnormal function of β-myosin in hypertrophic cardiomyopathy. J Clin Invest 1993; 91: 2861–5.

21. Bottinelli R, Coviello DA, Redwood CS et al. A mutant tropomyosin that causes hypertrophic cardiomyopathy is expressed in vivo and associated with an increased calcium sensitivity. Circ Res 1998; 82: 106–15.

22. Nier V, Schultz I, Brenner B et al. Variability in the ratio of mutant to wildtype myosin heavy chain present in the soleus muscle of patients with familial hypertrophic cardiomyopathy. A new approach for the quantification of mutant to wildtype protein. FEBS Lett 1999; 461: 246–52.

23. Flashman E, Redwood C, Moolman-Smook J et al. Cardiac myosin binding protein C: its role in physiology and disease. Circ Res 2004; 94: 1279–89.

24. Rottbauer W, Gautel M, Zehelein J et al. Novel splice donor mutation in the cardiac myosin-binding protein-C gene in familial hypertrophic cardiomyopathy. J Clin Invest. 1997; 100: 475–82.

25. Redwood CS, Moolman-Smook JC, Watkins H. Properties of mutant contractile proteins that cause hypertrophic cardiomyopathy. Cardiovasc Res 1999; 44: 20–36.

26. Lowey S. Functional consequences of mutations in the myosin heavy chain at sites implicated in familial hypertrophic cardiomyopathy. Trends Cardiovasc Med 2002; 12: 348–54.

27. Bonne G, Carrier L, Richard P et al. Familial hypertrophic cardiomyopathy: from mutations to functional defects. Circ Res 1998; 83: 580–93.

28. Palmiter KA, Tyska MJ, Haeberle JR et al. R403Q and L908V mutant beta-cardiac myosin from patients with familial hypertrophic cardiomyopathy exhibit enhanced mechanical performance at the single molecule level. J Muscle Res Cell Motil 2000; 21: 609–20.

29. Hernandez OM, Housmans PR, Potter JD. Invited Review: pathophysiology of cardiac muscle contraction and relaxation as a result of alterations in thin filament regulation. J Appl Physiol 2001; 90: 1125–36.

30. Levine RJC, Yang Z, Epstein ND et al. Structural and functional responses to mammalian thick filaments to alterations in myosin regulatory light chains. J Struct Biol 1998; 122: 149–61.

31. Knoll R, Hoshijima M, Hoffman HM et al. The cardiac mechanical stretch sensor machinery involves a Z disc complex that is defective in a subset of human dilated cardiomyopathy. Cell 2002; 111: 943–55.

32. Geier C, Perrot A, Ozcelik C et al. Mutations in the human muscle LIM protein gene in families with hypertrophic cardiomyopathy. Circulation 2003; 107: 1390–5.

33. Blair E, Redwood C, Ashrafian H et al. Mutations in the gamma 2 subunit of AMP-activated protein kinase

cause familial hypertrophic cardiomyopathy: evidence for the central role of energy compromise in disease pathogenesis. Hum Mol Genet 2001; 10: 1215–20.

34. Gollob MH, Green MS, Tang AS et al. Identification of a gene responsible for familial Wolff–Parkinson–White syndrome. N Engl J Med 2001; 344: 1823–31.

35. Arad M, Benson DW, Perez-Atayde AR et al. Constitutively active AMP kinase mutations cause glycogen storage disease mimicking hypertrophic cardiomyopathy. J Clin Invest 2002; 109: 357–62.

36. Hardie DG, Carling D. The AMP-activated protein kinase – fuel gauge of the mammalian cell? Eur J Biochem 1997; 246: 259–73.

37. Hamilton SR, Stapleton D, O'Donnell JB Jr et al. An activating mutation in the gamma1 subunit of the AMP-activated protein kinase. FEBS Lett 2001; 500: 163–8.

38. Daniel T, Carling D. Functional analysis of mutations in the gamma 2 subunit of AMP-activated protein kinase associated with cardiac hypertrophy and Wolff–Parkinson–White syndrome. J Biol Chem 2002; 277: 51017–24.

39. Nicol RL, Frey N, Olson EN. From the sarcomere to the nucleus: role of genetics and signaling in structural heart disease. Annu Rev Genomics Hum Genet 2000; 1: 179–223.

40. Seidman JG, Seidman C. The genetic basis for cardiomyopathy: from mutation identification to mechanistic paradigms. Cell 2001; 104: 557–67.

41. Knollmann BC, Kirchhof P, Sirenko SG et al. Familial hypertrophic cardiomyopathy-linked mutant troponin T causes stress-induced ventricular tachycardia and Ca^{2+}-dependent action potential remodeling. Circ Res 2003; 92: 428–36.

42. Fatkin D, McConnell BK, Mudd JO et al. An abnormal Ca^{2+} response in mutant sarcomere protein-mediated familial hypertrophic cardiomyopathy. J Clin Invest 2000; 106: 1351–9.

43. Smith GA, Dixon HB, Kirschenlohr HL et al. Ca^{2+} buffering in the heart: Ca^{2+} binding to and activation of cardiac myofibrils. Biochem J 2000; 346: 393–402.

44. Fentzke RC, Buck SH, Patel JR et al. Impaired cardiomyocyte relaxation and diastolic function in transgenic mice expressing slow skeletal troponin I in the heart. J Physiol 1999; 517: 143–57.

45. Ashrafian H, Redwood C, Blair E et al. Hypertrophic cardiomyopathy:a paradigm for myocardial energy depletion. Trends Genet 2003; 19: 263–8.

46. Sweeney HL, Feng HS, Yang Z et al. Functional analyses of troponin T mutations that cause hypertrophic cardiomyopathy: insights into disease pathogenesis

and troponin function. Proc Natl Acad Sci USA 1998; 95: 14406–10.

47. Montgomery DE, Tardiff JC, Chandra M. Cardiac troponin T mutations: correlation between the type of mutation and the nature of myofilament dysfunction in transgenic mice. J Physiol 2001; 536: 583–92.

48. Kammermeier H. High energy phosphate of the myocardium: concentration versus free energy change. Basic Res Cardiol 1987; 82 (Suppl 2): 31–6.

49. Crilley JG, Boehm EA, Blair E et al. Hypertrophic cardiomyopathy due to sarcomeric gene mutations is characterized by impaired energy metabolism irrespective of the degree of hypertrophy. J Am Coll Cardiol 2003; 41: 1776–82.

50. Spindler M, Saupe KW, Christe ME et al. Diastolic dysfunction and altered energetics in the $\alpha MHC^{403/+}$ mouse model of familial hypertrophic cardiomyopathy. J Clin Invest 1998; 101: 1775–83.

51. Javadpour MM, Tardiff JC, Pinz I et al. Decreased energetics in murine hearts bearing the R92Q mutation in cardiac troponin T. J Clin Invest 2003; 112: 768–75.

52. Inoki K, Zhu T, Guan KL. TSC2 mediates cellular energy response to control cell growth and survival. Cell 2003; 115: 577–90.

53. Bolster DR, Crozier SJ, Kimball SR et al. AMP-activated protein kinase suppresses protein synthesis in rat skeletal muscle through down-regulated mammalian target of rapamycin (mTOR) signaling. J Biol Chem 2002; 277: 23977–80.

54. Neel BG, Gu H, Pao L. The 'Shp'ing news: SH2 domain-containing tyrosine phosphatases in cell signaling. Trends Biochem Sci 2003; 28: 284–93.

55. Hudson ER, Pan DA, James J et al. A novel domain in AMP-activated protein kinase causes glycogen storage bodies similar to those seen in hereditary cardiac arrhythmias. Curr Biol 2003; 13: 861–6.

56. Arbustini E, Fasani R, Morbini P et al. Coexistence of mitochondrial DNA and beta myosin heavy chain mutations in hypertrophic cardiomyopathy with late congestive heart failure. Heart 1998; 80: 548–58.

57. Blair E, Price SJ, Baty CJ et al. Mutations in cis can confound genotype-phenotype correlations in hypertrophic cardiomyopathy. J Med Genet 2001; 38: 385–8.

CHAPTER 31

Monogenic dilated cardiomyopathy

Timothy M Olson

Introduction

Prior to 1992, the importance of hereditary factors in the pathogenesis of idiopathic dilated cardiomyopathy (DCM) was not fully recognized. Even a focused family history identified a hereditary form of DCM in only 6–8% of cases.[1] Moreover, when familial and non-familial cases of idiopathic DCM were compared, no differences in baseline clinical, serological, or histopathological characteristics or in long-term outcome were observed to help distinguish familial cases.[1,2] The concept of DCM as a monogenic disorder received major impetus from a 1992 study in which first-degree relatives of index patients were screened by echocardiography.[1] DCM, defined as left ventricular ejection fraction <50%, and left ventricular dimensions >95th percentile for body surface area and age, was identified in presymptomatic members of several families, accounting for a 20% frequency of familial disease in this patient cohort. These findings were reproduced in a similar study, which identified familial disease in 25% of cases.[3] In both studies, the average age at diagnosis in probands and their relatives was in the fourth to fifth decade, yet children in their first decade of life were also diagnosed with either symptomatic or clinically silent DCM. In studies that have performed screening echocardiograms on relatives of the index cases, 9–18% of asymptomatic individuals had left ventricular (LV) enlargement with normal ejection fraction, sug-

gesting that cardiac dilation is a precursor of DCM.[1,3] Indeed, progression to DCM was confirmed in subsequent longitudinal follow-up studies.[4,5] If less stringent criteria, such as isolated LV enlargement or sudden unexplained death, are used to diagnosis DCM in relatives, the frequency of familial disease may be as high as 35–48%.[4,6] The importance of genetics in the pathogenesis of DCM, in fact, may be even greater than suggested by these studies. For example, sporadic cases can be caused by de novo mutations, inheritance of two copies of a recessive mutation due to parental consanguinity, or the combined effects of two or more distinct mutations which are clinically silent in isolation.

Familial DCM is most commonly inherited as an autosomal dominant trait, conferring a 50% risk of DCM on offspring of an individual with DCM.[1,3,4,6,7] Less commonly, DCM is an X-linked disorder in males who have inherited a mutation from their mothers who exhibit mild or no cardiac disease. Barth syndrome, characterized by cardiac and skeletal myopathy, short stature, 3-methylglutaconic aciduria and neutropenia, is an X-linked disorder caused by mutations in the G4.5 gene that encodes tafazzin.[8] Left ventricular non-compaction can also be caused by defects in this gene. Males with Barth syndrome usually develop fatal infantile DCM. DCM alone, however, in the absence of non-cardiac features of Barth syndrome, has not been clearly documented. Autosomal recessive and maternally inherited cardiomyopathy due to defects in fatty acid oxidation and mitochondrial oxidative phosphorylation, respectively, may present as DCM.[9] Patients with disorders of cardiac energy metabolism, however, usually have hypertrophic cardiomyopathy, neuromuscular disease, and metabolic derangements. Traits such as subtle skeletal myopathy, cardiac conduction system disease, and atrial arrhythmia segregate with DCM in certain families with autosomal dominant or X-linked disease.[6,7,10,11] These phenotypic subtypes may suggest specific gene defects and predict progression of DCM.[12] Most reports of familial DCM as an isolated disorder or as part of a syndrome, however, have shown age-dependent penetrance, and variable expression of disease among members of the same family. In other words, some carriers of a gene mutation may not have DCM or may have a partial cardiomyopathy phenotype, such as isolated LV enlargement or conduction system disease.

Identifying genes for dilated cardiomyopathy

Dramatic technological advances in DNA and genomic analysis have evolved over the last half century, beginning with Watson and Crick's discovery of the double helical structure of DNA, and culminating in the decoding of the entire human genome through the Human Genome Project. In 1990, the year the Human Genome Project officially began, the first gene for hypertrophic cardiomyopathy (HCM) was identified by linkage analysis in a large family. Over the next 5 years, the application of human molecular genetics led to discovery of several additional genes for familial HCM, providing new insight into the molecular pathogenesis of maladaptive cardiac hypertrophy. Coincident with the era of HCM gene discovery, recognition of DCM as potentially a monogenic disorder was emerging. Clearly, the same strategies used to identify HCM genes could be employed in DCM gene discovery.[13]

In the field of DCM genetics, two general approaches have been used to discover disease-causing gene defects – gene localization by genetic linkage analysis, and mutation identification by DNA sequence analysis of candidate genes. Genetic linkage analysis is the strategy used to identify familial disease-causing genes, based on chromosomal position. Linkage is established by identifying a polymorphic DNA marker of known location within the genome that cosegregates with the disease phenotype. The power of linkage analysis is its ability to narrow the list of potential candidate genes to a defined region of the genome, without prior knowledge of the disease mechanism. Indeed, several familial DCM genes identified by linkage analysis have provided new, even unexpected, insights into the molecular basis of disease.

In its application to DCM, however, genome-wide linkage analysis has important limitations. Large multigenerational families with many living, unambiguously affected individuals, critical for detecting linkage, are relatively rare. Children and young adults with a mutation may have normal cardiac size and function due to incomplete, age-dependent penetrance. At the other extreme, high mortality rates limit the number of living family members with DCM and DNA samples available for analysis. Even in families suitably large to identify a chromosome locus for DCM, family size may nevertheless impede refined mapping to demarcate a region harboring a relatively small number of positional candidate genes. As a result, several studies that have mapped DCM loci have not yet led to identification of the disease-causing gene (Table 31.1).

An alternative strategy to linkage analysis in large families is direct mutation analysis of candidate genes in small families and sporadic cases in which chromosome position information is unobtainable.[13] Rationales for this approach include: (1) the rarity of large families suitable for gene mapping; (2) the possibility that DCM genes in such families do not represent the full spectrum of genetic causes for DCM; and (3) the potential to discover de novo or recessive mutations in genes unique to sporadic DCM. A non-positional candidate gene approach necessitates a mechanistic hypothesis for DCM and previous identification of genes for specific cellular structures or pathways in the heart. This strategy may appear daunting since a very large number of genes are expressed in the

Table 31.1. *Dilated cardiomyopathy loci identified by genetic linkage analysis*

Locus	Locus size (cM)	Associated phenotypes	Age of affected individuals (years)	Reference
2q14–q22	11	Conduction defects	17–55	Jung et al., 1999[14]
3p22-p25	30	Automaticity and conduction defects, atrial arrhythmia	6–88	Olsen and Keating, 1996[10]
6q12–q16	16	None	17–66	Sylvius et al., 2001[15]
6q23	3	Conduction defects, skeletal myopathy	N/A	Messina et al., 1997[16]
6q23–q24	3	Hearing loss	27-61	Schonberger et al., 2000[17]
9q13–q22	15	None	N/A	Krajinovic et al., 1995[18]
10q21–q23	4	Mitral valve prolapse	14–78	Bowles et al., 1996[19]

Genetic distance of 1 cM corresponds to physical distance of approximately 1000 kilobases. cM, centimorgans; N/A, data not available.

heart and, by definition, the molecular bases for idiopathic DCM are unknown. Moreover, once a putative mutation in a candidate gene is discovered co-segregation with DCM cannot be statistically 'proven' without a large, extended family. Distinguishing a disease-causing or disease-modifying mutation from a benign DNA polymorphism thus requires a high level of scrutiny. Data in support of a mutation include: (1) its absence in a large number of control individuals; (2) its disruption of a highly conserved domain critical to protein structure and function; (3) demonstration that the encoded mutant protein is dysfunctional *in vitro* or *in vivo*; and (4) identification of additional mutations in the same gene in other patients with DCM. Despite its limitations, the candidate gene approach has proven complementary to positional cloning in defining genetic defects in DCM. Genetic, biochemical and physiological studies of human cardiomyopathy, together with genetic models of cardiomyopathy in mice, have defined molecular pathways for myocardial failure,[20–22] facilitating focused candidate gene selection in DCM. Moreover, a non-positional candidate gene approach has become increasingly feasible as the identity, structure and tissue expression pattern of most human genes are now known, and as high-throughput systems for mutation scanning have been developed.

In other heritable cardiac disorders, like hypertrophic cardiomyopathy, or long QT syndrome, a spectrum of mutations in relatively few functionally related genes appear to account for a large fraction of prototypical phenotypes. By contrast, DCM has proven to be markedly heterogeneous in both phenotype and genotype. Over the past decade, a substantial number of DCM genes have been discovered, yet no common gene (genetic heterogeneity) or mutation (allelic heterogeneity) for isolated DCM has emerged. In the following sections, key discoveries will be presented to emphasize conceptual advances in human DCM genetics. A comprehensive catalog of genes for monogenic, non-syndromic DCM, identified by linkage analysis and/or candidate gene approaches, is displayed in Table 31.2.

X-linked DCM and Duchenne muscular dystrophy: a common etiology

Dystrophin is a large structural protein localized to the inner cell membrane in myocytes.[39] Its amino terminus binds to actin, whereas its carboxy terminus binds to the dystrophin–glycoprotein complex, a cluster of proteins spanning the cell membrane. Dystrophin, therefore, links the sarcomere to the

Table 31.2. *Dilated cardiomyopathy genes*

Locus	Gene symbol, protein	Inheritance	Age of affected individuals (years)	Variably associated phenotypes	Allelic disorders	Primary strategy	Protein class	References
1q32	*TNNT2*, cardiac troponin T2	Autosomal dominant	1–84[23], 1–53[24]	None	HCM	Both	Thin filament of sarcomere	Durand et al., 1995[23], Kamisago et al., 2001[24]
1q21.2–q21.3	*LMNA*, lamin A/C	Autosomal dominant	19–53	Conduction defects, atrial arrhythmia, myopathy, increased CK	Six distinct disorders (see text)	Linkage analysis	Nuclear membrane	Kass et al., 1994[25], Fatkin et al., 1999[11]
1q42–q43	*ACTN2*, actinin, alpha 2	Autosomal dominant	7–42	None	None	Candidate gene	Z-disc, intercalated disc	Mohapatra et al., 2003[26]
2q31	*TTN*, titin	Autosomal dominant	9–53	None	HCM, primary myopathy	Linkage analysis	Sarcomeric cytoskeleton	Siu et al., 1999[27], Gerull et al., 2002[28]
2q35	*DES*, desmin	Autosomal dominant	N/A	None	Primary myopathy	Candidate gene	'True' cytoskeleton	Li et al., 1999[29]
5q33–q34	*SGCD*, δ sarcoglycan	Autosomal dominant, sporadic	0–38	None	Primary myopathy	Candidate gene	Membrane-associated cytoskeleton	Tsubata et al., 2000[30]
6q22.1	*PLN*, phospholamban	Autosomal dominant	20–30	None	HCM	Candidate gene	Sarcoplasmic reticulum	Schmitt et al., 2003[31]
10q22.1–q23	*VCL*, vinculin	Autosomal dominant	29–70	None	None	Candidate gene	Intercalated disc	Olson et al., 2002[32]
10q22.3–q23.2	*LDB3*, LIM domain binding 3 (ZASP)	Autosomal Dominant, sporadic	0–68	LV non-compaction, LVH, conduction defects	LV non-compaction	Candidate gene	Z-disc	Vatta et al., 2003[33]
11p15.1	*CSRP3*, cardiac LIM protein (MLP)	Autosomal dominant	32–70	None	HCM	Candidate gene	Z-disc	Knoll et al., 2002[34]
14q12	*MYH7*, cardiac β myosin heavy chain	Autosomal dominant	0–57	None	HCM	Linkage analysis	Thick filament of sarcomere	Kamisago et al., 2000[24]
15q11–q14	*ACTC*, cardiac actin	Autosomal dominant	2–41	None	HCM	Candidate gene	Thin filament of sarcomere	Olson et al., 1998[35]
15q22.1	*TPM1*, alpha-tropomyosin 1	Autosomal dominant	0–33	None	HCM	Candidate gene	Thin filament of sarcomere	Olson et al., 2001[36]
Xp21.2	*DMD*, dystrophin	X-linked	15–23 (M),[37] 45–53 (F),[37] 13–36 (M)[38]	Subclinical myopathy, Increased CK	Primary myopathy	Linkage analysis	Membrane-associated cytoskeleton	Towbin et al., 1993[37], Muntoni et al., 1993[38]

Genes for which DCM is associated with primary skeletal myopathies, multi-system syndromes, or metabolic disorders are excluded. References are for initial reports of DCM-associated mutations. HCM, hypertrophic cardiomyopathy; CK, creatine kinase; LVH, left ventricular hypertrophy; M, male; F, female.

extracellular basement membrane. Its function has been viewed as both passive and active, conferring stability to the sarcolemma during myocyte contraction, and transducing force from the intracellular contractile apparatus to non-contractile proteins of the extracellular matrix, respectively.

Mutations in the genes encoding dystrophin and other proteins of the dystrophin–glycoprotein complex cause muscular dystrophies.[40] While skeletal myopathy is the primary feature of these disorders, later-onset cardiac myopathy often develops. Patients with mutations in the dystrophin gene are clinically classified into two groups. In Duchenne muscular dystrophy, skeletal manifestations begin at 3–6 years of age, and cardiomyopathy is invariably present by 18 years of age, although cardiac symptoms may be masked by physical inactivity.[39] Conversely, in Becker muscular dystrophy, skeletal myopathy is milder and cardiac myopathy may be the primary reason for morbidity and mortality. On the most extreme end of this phenotypic spectrum are families with X-linked dominant DCM, where males develop a rapidly progressive dilated cardiomyopathy in adolescence or early adulthood, and females have a milder, more indolent form of disease. Skeletal muscle involvement is subclinical, evident only by elevated serum muscle creatine kinase (CK). A positive family history may be less obvious than in autosomal dominant DCM, or even absent in the case of a spontaneous mutation.

In 1993, the first gene for DCM was identified in families with X-linked DCM.[37,38] Focusing genotype and linkage analysis on the X chromosome, a gene defect was mapped to the proximal region of the dystrophin gene.[37] Consistent with the clinical phenotype, dystrophin was significantly decreased or absent in cardiac muscle, but normally abundant in skeletal muscle. Later that year, a deletion that removed the muscle-specific promoter and first exon of the dystrophin gene was identified in another family with X-linked DCM.[38] It was postulated that compensatory upregulation of gene expression in skeletal muscle mediated by the 'brain' promoter, but not cardiac muscle, accounted for the cardiac-specific phenotype. Indeed, later identification of a point mutation affecting proximal gene splicing suggested a similar

mechanism.[41] Alternative mechanisms for cardioselective effects of dystrophin gene mutations in X-linked DCM, however, are implicated by mutations discovered in non-regulatory domains.[42]

Identification of dystrophin as a DCM gene had several implications. It demonstrated the value of linkage analysis as a tool for DCM gene discovery. Without positional information implicating a mutation in dystrophin, mutation scanning of genes on the X chromosome would have been a formidable task, given the large number of potential candidate genes expressed in the heart, and the very large size of dystrophin itself. Mutations in the dystrophin gene highlighted the importance of 'gene dosage' in the pathogenesis of DCM, namely age at onset and rate of progression. Males with a dystrophin mutation, carrying only a single X chromosome, are hemizygous, i.e. a defect is present in their one and only gene. Female mutation-carriers, with two copies of the dystrophin gene, are heterozygous, i.e. only one gene is mutated. In this way, females in X-linked families are similar to both females and males in families with autosomal dominant DCM caused by heterozygous mutations, in whom onset of symptomatic disease may be delayed until 30–50 years of age.[1,7] Identification of dystrophin mutations in X-linked DCM (Fig. 31.1A) suggested that other genes for skeletal myopathies were candidates for DCM. Moreover, discovering that genetic defects in dystrophin cause DCM, led to further investigations to determine whether dystrophin had a broader role in the pathobiology of heart failure. Indeed, disruption of dystrophin by enteroviral proteases was identified in myocarditis, establishing a common mechanism for both inherited and acquired forms of heart failure.[43]

Autosomal dominant DCM: a genetically heterogeneous disorder

In 1994, the first locus for autosomal dominant DCM was mapped to chromosome 1.[25] A very large, multigenerational family was studied, consisting of 23 individuals classified as 'affected'. The cardiac phenotype segregating in the family included both early-onset atrial arrhythmia, and conduction system disease with later-onset DCM.

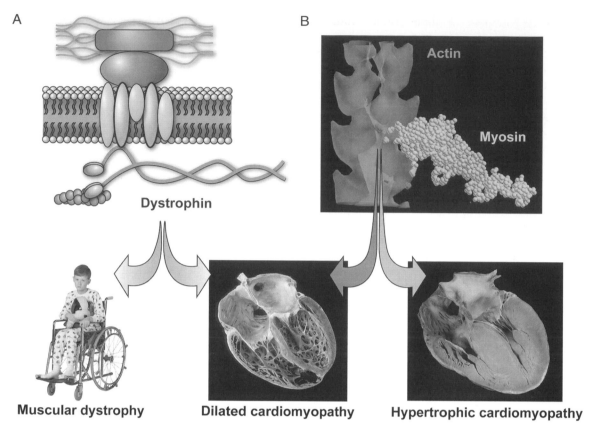

Figure 31.1. *Heritable defects in dystrophin (A) cause muscular dystrophy and/or dilated cardiomyopathy. Similarly, amino acid substitutions in cardiac actin (B) cause either dilated cardiomyopathy or hypertrophic cardiomyopathy.*

Consequently, rhythm disturbances, observed in several family members without diagnostic criteria for DCM, provided an early marker for DCM. While the disease-causing gene was not identified for another 5 years,[11] a critical first step toward gaining new insights into the pathogenesis of heart failure had been achieved by linkage analysis.

Over the next seven years, 10 additional unique loci for autosomal DCM were identified by genome-wide linkage analysis, establishing DCM as a genetically heterogeneous disorder. These studies also emphasized the phenotypic heterogeneity of DCM among and within families due to incomplete penetrance, variable expression, and associated non-cardiac phenotypes. Indeed, the ability of some studies to localize and refine mapping of DCM loci was based on use of less stringent

diagnostic criteria for DCM. For example, LV dilation in systole or diastole with normal ejection fraction,[27] reduced ejection fraction with normal cardiac dimensions,[14] and sudden death without an echocardiographic or pathological diagnosis of DCM,[16] were sufficient to classify individuals as 'affected' in some studies. Early-onset arrhythmias and/or conduction system disease preceding DCM were observed in four families.[10,14,16,25] In isolation, rhythm disturbances were considered either diagnostic of the familial cardiac syndrome in dichotomous classification schemes,[16,25] or suggestive of the inherited disease in schemes that assigned point values to specific phenotypic traits.[10,14] Similarly, individuals with non-cardiac traits variably expressed with DCM, such as skeletal myopathy,[16] and sensorineural hearing loss,[27] were classified as 'affected' in some studies. Linkage studies that capitalized on early-onset cardiac and non-cardiac phenotypes, co-inherited with DCM, increased the number of family members classified as 'affected', and the power to detect linkage.

The disease-causing genes at four mapped DCM loci have ultimately been discovered.[11,23,24,28] Linkage studies led to identification of DCM genes that may not have been considered prime candidates without positional information placing them within a mapped DCM locus. Among the loci at which DCM genes have not yet been identified, only three are relatively small.[16,17,19] Because essentially all human genes are now characterized and mapped to specific regions of the genome, it is conceivable that DCM genes at these loci will be identified in the near future. Four mapped loci, however, are very large with >100 potential candidate genes.[10,14,15,18] Refined mapping will require identification of genetic recombination events within these loci, either in family members with DCM who were not previously studied, or in those who have now developed DCM or a phenotypic trait that portends DCM. Alternatively, a previously unsuspected positional candidate gene may come to light, based on its functional similarity to a newly discovered DCM gene. In addition, new technologies for high throughput mutation scanning and DNA sequencing have made analysis of a large number of candidate genes increasingly feasible.

Collectively, the many DCM loci mapped by linkage analysis have brought to light the extraordinary genetic heterogeneity of familial DCM. If the genes at these loci are representative of the spectrum of DCM genes in smaller families and sporadic cases, it appears that a common genetic basis for DCM may not exist. In this regard, future prospects for routine, comprehensive genetic testing appear challenging. Nevertheless, clinically relevant observations have emerged from the detailed phenotypic evaluation of extended families. Age at onset for familial DCM may be highly variable among related family members who carry the same mutation. Moreover, cardiac and non-cardiac phenotypes may be associated with DCM, and these traits may serve as early markers for disease. Consequently, serial evaluation of at-risk family members by complete physical examination, echocardiography and electrocardiography is warranted.

Inherited defects in actin: DCM gene discovery using a candidate gene approach

In 1998, the first gene for autosomal dominant DCM was identified by a non-positional candidate gene strategy.[35] Previous advances in cardiomyopathy genetics provided a conceptual framework supporting the hypothesis that defects in cardiac actin could lead to DCM. First, it had been established that mutations in genes encoding sarcomeric proteins could cause hypertrophic cardiomyopathy (HCM). Since these proteins were involved in myocellular contraction, the proposed mechanism for maladaptive hypertrophy was chronic reduction of force generation. Second, discovery of dystrophin as a gene for X-linked DCM showed that dysfunction of a protein transmitting force from the contractile apparatus to the extracellular matrix caused DCM. In this context, cardiac actin was a plausible candidate gene for DCM because: (1) it is highly, almost exclusively, expressed in the heart; and (2) it forms sarcomeric thin filaments whose ends opposite to myosin cross-bridge domains are anchored to Z-discs and intercalated discs, sites of contractile force transmission between sarcomeres and adjacent myocytes, respectively. Conceivably, mutations that disrupted force transmission within and between cardiac myocytes could cause DCM, rather than HCM. Mutation scanning of the cardiac actin gene identified two missense mutations in small DCM families with 3-4 affected individuals, altering highly conserved amino acids. Age at diagnosis in these families was highly variable, ranging from 1 to 41 years, and two mutation carriers had dilated left ventricles with preserved systolic function. Haploinsufficiency of cardiac actin in mice created by heterozygous disruption of cardiac actin does not cause DCM. Consequently, the heterozygous mutations identified in humans are likely to alter actin function by incorporation of both normal and mutant proteins into actin filaments, i.e. a dominant-negative effect. This is the proposed mechanism for most mutations subsequently identified in other DCM genes, although some may cause loss of protein function. Consistent with the mechanistic hypothesis of defective force transmission, atomic modeling placed these amino acid substitutions in domains that do not interface with myosin to generate contractile force.[44] Subsequent *in vitro* studies confirmed that these mutations compromised actin function.[45] One of the mutations did not affect *in vitro* actomyosin motility, yet it caused a 3-fold reduction in α-actinin binding affinity. Notably, α-actinin is localized to Z-discs and intercalated discs, principal sites of actin filament anchoring and contractile force transmission.

The families with identified mutations in the cardiac actin gene were too small for genome-wide linkage analysis, and mapping studies in larger DCM families have not implicated the cardiac actin locus on chromosome 15q14. Thus, the candidate gene approach has complemented traditional linkage analysis in the discovery of DCM genes. Indeed, this approach has become increasingly attractive, as the Human Genome Project comes to its completion. Like linkage studies, candidate gene studies have demonstrated the profound genetic heterogeneity of DCM. In fact, 10 DCM genes have been identified by a hypothesis-based candidate gene approach, more than the number of genes identified by linkage analysis. Most of these investigations have focused on genes encoding cytoskeletal and sarcomeric proteins that confer structural integrity and/or mediate contractile force dynamics in the myocardium.

Mutations in genes encoding cytoskeletal proteins: a unifying etiology for DCM

The cytoskeleton is an intracellular scaffold consisting of many proteins that link, anchor or tether other cellular components.[46] In cardiac myocytes, the cytoskeleton plays a critical role both in maintaining cellular integrity in the face of ongoing mechanical stress and in mechanical force transduction. In this regard, both contractile proteins of the sarcomere, such as actin and myosin, and proteins of the cytoskeleton are important in contractile force dynamics. Cytoskeletal alterations occur in idiopathic, ischemic, and tachycardia-induced cardiomyopathic remodeling,[47,48] yet molecular genetic studies were needed to establish a cause–effect relationship. For improved clarity, the following classification scheme for cytoskeletal proteins has been proposed, based on their structural and functional properties: membrane-associated proteins, e.g. dystrophin; 'true' cytoskeletal proteins, e.g. desmin; intercalated disc proteins, e.g. vinculin; and proteins of the sarcomeric skeleton, e.g. titin.[46] Mutations in genes for each of these components have been identified in DCM, establishing heritable defects in the cytoskeleton as a primary mechanism for heart failure.

In 1999, the first mutation in a 'true' cytoskeletal protein gene, desmin, was reported in a small family with DCM.[29] Like the discovery of cardiac actin as a DCM gene, a candidate-gene approach was employed. The rationale for selecting desmin as a candidate included high cardiac expression, reports that desmin mutations in mice and humans cause cardioskeletal myopathy, and knowledge that certain mutations in the same gene could cause either skeletal myopathy with cardiac involvement, or isolated DCM with no clinically apparent skeletal muscle disease.[37,38] An incompletely penetrant missense mutation was identified in four family members, two of whom had DCM. It was speculated that location of the defect in the tail domain of desmin accounted for its heart-specific effects, since previously reported mutations in skeletal myopathy occurred in the central rod domain. However, the identical mutation was subsequently reported in a family with skeletal myopathy but no cardiac involvement.[49] These findings demonstrated that the same mutation could have different phenotypic manifestations, presumably determined by other genetic and/or environmental factors. The mechanism by which the mutation causes desmin dysfunction remains unknown, since *in vitro* studies were uninformative.[49]

By 2003, six additional DCM genes that encode cytoskeletal proteins were identified by either linkage analysis or candidate gene strategies – δ-sarcoglycan (*SCCD*), vinculin (*VCL*), titin (*TTN*), cardiac LIM protein (*CSRP3*), α-2 actinin (*ACTN2*) and LIM domain binding 3 (*LBD3*).

Sarcoglycan, like dystrophin, is a component of the dystrophin-glycoprotein complex expressed in both cardiac and skeletal muscle. Mutations in the δ-sarcoglycan gene cause autosomal recessive limb-girdle muscular dystrophy, i.e. affected individuals have two copies of the same mutation (homozygosity), or two different mutations in the same gene (compound heterozygosity). By contrast, mutation scans of δ-sarcoglycan in patients with DCM identified two dominant mutations, i.e. a mutation in one of two genes (heterozygosity) was sufficient to cause DCM.[30] A missense mutation was found to segregate with DCM in a small family, and the same *de novo* 3-base pair (bp) deletion was found in two unrelated individuals with sporadic DCM. The majority of mutation carriers presented as children with congestive heart failure and rapidly progressive DCM; one was only 9 months of age. Muscle CK was mildly elevated in one patient, but none had clinically apparent skeletal muscle disease.

Vinculin and its isoform metavinculin are protein components of intercalated discs, structures that anchor actin filaments and transmit contractile force between adjacent cardiac myocytes. Based on a mechanistic hypothesis of altered force transmission,[35] and human and animal studies potentially implicating vinculin in the pathogenesis of DCM, vinculin was investigated as a candidate gene.[32] Vinculin (*VCL*) encodes a smaller, ubiquitously expressed protein and a larger protein isoform expressed exclusively in cardiac and smooth muscle. Mutation analyses were confined to a single alternatively spliced exon that is expressed only in the muscle-specific isoform, called metavinculin. An incompletely penetrant missense mutation was identified in a small family and a 3-bp deletion was found in a sporadic case, both associated with adult-onset DCM. A rare polymorphism was also reported, postulated to confer risk for DCM. Mutant proteins significantly altered metavinculin-mediated cross-linking of actin filaments in an established *in vitro* assay. Ultrastructural examination of cardiac tissue, performed in one patient, revealed disruption of intercalated discs.

Titin is a huge muscle protein that spans sarcomeres from M lines, the myosin-binding midportions of thick filaments, to Z-discs, the anchoring sites for actin-containing thin filaments.[50] Thus it is a cytoskeletal protein with intimate apposition to contractile proteins, critical for muscle assembly and for conferring structural stability and resting tension to the sarcomere. An initial link between titin defects and myocardial disease was made when a missense mutation, altering α-actinin binding, was identified in a patient with hypertrophic cardiomyopathy.[51] Titin (*TTN*) was first investigated as a candidate gene for DCM based on its co-localization with a DCM locus mapped to chromosome 2 by linkage analysis.[27] Exons encoding a cardiac-specific region of titin were initially targeted, but no mutations were identified. Subsequent comprehensive scans of 313 *TTN* exons expressed in the heart revealed a missense mutation.[28] In another family with autosomal dominant DCM, a 2-bp insertion/frameshift mutation, predicted to truncate the translated protein, was also identified. Since each mutation in *TTN* involved exons expressed in both cardiac and non-cardiac isoforms, it remained unclear why there was lack of skeletal muscle disease. Notably, a recessive *TTN* mutation in zebrafish with a DCM phenotype was reported at the same time.[52] Consistent with the emerging paradigm for genes encoding cytoskeletal proteins, *TTN* mutations were reported in patients with tibial muscular dystrophy soon after *TTN* was identified as a DCM gene.

A more recent discovery of a cytoskeletal gene for DCM followed insights derived from a mouse model.[22] Targeted disruption of the murine gene encoding the Z-disc protein MLP (muscle LIM protein) caused a DCM phenotype.[53] Like humans with DCM, *MLP* knockout mice have age-dependent onset of DCM and heart failure. Consequently, biochemical and biophysical analyses of cardiac muscle before DCM developed identified defective sensing of passive mechanical stretch as the primary defect.[34] *MLP* knockout mice were rescued from DCM by concomitant knockout of the phospholamban gene (*PLN*), postulated to remove the mechanical stress stimulus. Remarkably, later it would be discovered that *PLN* mutations *cause* DCM in humans, opposite to the effect observed in *MLP* knockout mice.

Phenotypic characterization of the *MLP* knockout mouse prompted investigation of *CSRP3*, the gene encoding MLP in humans, as a candidate for human DCM.[34] The same missense mutation in *CSRP3* was identified in 10 patients with DCM from a cohort of 536 European patients, but this mutation was not found in a cohort of 285 Japanese patients nor in >500 normal controls from both populations. Haplotype analyses suggested a common founder as the most likely explanation for these findings. Family segregation data were limited, yet mutant MLP was shown to cause an *in vitro* defect in interaction and localization with TCAP, another Z-disc protein that binds to titin. Genetic and functional studies of α-actinin and ZASP have further implicated defects in the Z-disc as a mechanism for DCM.[26,33] In a subsequent study, distinct mutations in *CSRP3* were identified in patients with hypertrophic cardiomyopathy.[54] This extended the paradigm, previously established for contractile proteins of the sarcomere, that defective proteins encoded by the same gene can cause either DCM or HCM (discussed below).

DCM with conduction disease: novel disease gene discovery by linkage analysis

In 1999, the first gene for autosomal dominant DCM to be identified by genetic linkage analysis, lamin A/C (*LMNA*), was reported.[11] Lamin A/C encodes an intermediate filament protein of the inner nuclear membrane.[55] The specific functions of the alternatively spliced gene products, lamin A and lamin C, are unknown, but they are thought to confer structural integrity to the nuclear envelope and may influence gene expression. Notwithstanding, discovery of *LMNA* as a DCM gene expanded the list of myocellular structures implicated in the pathogenesis of myopathic heart failure beyond the cytoskeleton and sarcomere. Like other DCM genes that encode cytoskeletal proteins, *LMNA* was previously identified as a gene for a skeletal myopathy – autosomal dominant Emery–Dreifuss muscular dystrophy. As discussed above, the DCM locus previously mapped to the centromeric region of chromosome 1 was identified in a family whose phenotype included atrial arrhythmias and conduction system disease with absence of overt skeletal

muscle disease.[25] Accordingly, 11 families with similar phenotypes were scanned for mutations in *LMNA*, a positional candidate gene on chromosome 1.[11] Distinct missense mutations that segregated with DCM, atrial fibrillation, and/or conduction system disease were identified in five of the families. Most mutation carriers younger than 30 years of age did not have cardiac disease, yet sudden death occurred frequently, typically in the fourth to sixth decades of life.

Remarkably, *LMNA* mutations have been identified in five other, very distinct human diseases: limb girdle muscular dystrophy type 1B, Dunnigan partial lipodystrophy, mandibuloacral dysplasia, Charcot–Marie–Tooth neuropathy type 2 B1, and Hutchinson–Gilford progeria.[55] The pleiotropic manifestations of *LMNA* mutations emphasize both the critical importance of the nuclear membrane and its diverse but poorly understood functions. Unlike mutations in the cardiac actin gene, which appear to cause DCM or HCM depending on the functional domain they disrupt,[35,44] genotype–phenotype correlations in the laminopathies are generally inapparent. Indeed, a frameshift mutation in *LMNA* was reported to cause isolated DCM with conduction defects, DCM with Emery–Dreifuss-like skeletal muscle disease, and DCM with limb girdle muscular dystrophy-like myopathy among members of the same family.[56] Findings in a recent study and review of *LMNA* mutations in DCM provide a rationale for targeted mutation screening in a subset of patients.[12] The presence of skeletal muscle involvement, atrial arrhythmia, conduction defects and DCM with mild LV dilation were predictive of lamin A/C mutations. Moreover, patients with *LMNA* mutations had more progressive disease and worse outcome than patients without *LMNA* mutations, albeit penetrance and expression were quite variable.

DCM and HCM: shared disease genes, divergent cardiac remodeling pathways

Coordinated, synchronous contraction of the heart requires mechanical coupling of individual myocytes to the rest of the myocardium. Contractile force is generated by actin–myosin interaction within cardiac sarcomeres, and transmitted to adjacent sarcomeres via Z-discs, to neighboring myocytes via intercalated discs, and to the extracellular matrix via costameres and the dystrophin–glycoprotein complex. Actin, the main protein of thin filaments, plays a dual role in force dynamics by regulating contractile force generation through thick filament (myosin) interaction, and transmitting force through filament ends anchored to Z-discs and intercalated discs. At the time mutations in the cardiac actin gene were identified and hypothesized to cause DCM by defective transmission of contractile force,[35] it had been well established that molecular defects in force-generating proteins of the sarcomere caused HCM. Subsequently, missense mutations in myosin-binding domains of the cardiac actin gene were identified in familial and sporadic HCM.[44,57] Thus, distinct defects in the same sarcomeric protein were shown to cause either congestive heart failure or maladaptive hypertrophy. Mapping of the identified cardiac actin mutations in an atomical model of the actin–myosin complex,[44] coupled with *in vitro* analysis,[45] indicated that HCM-associated mutations were in myosin binding domains. By contrast, DCM-associated mutations were in domains important for anchoring the thin filament (Fig. 31.1B).

A candidate gene strategy further validated the hypothesis that defects in a sarcomeric protein could cause either HCM or DCM. Mutational analyses were carried out in the gene encoding α-tropomyosin,[36] previously implicated in the pathogenesis of HCM. The thin filament consists of both actin and tropomyosin. Tropomyosin lies within a groove along the surface of the actin filament, providing stability to the thin filament and regulating actin–myosin interaction. The electrostatic surface charge of tropomyosin is predominantly negative, whereas domains in the actin groove interfacing with tropomyosin are positively charged, supporting an electrostatic basis for protein interaction. Missense mutations in the α-tropomyosin gene (TPM1) were identified in two small families with DCM. Protein modeling predicted these mutations would cause electrostatic charge reversal on the surface of tropomyosin, compromising thin filament stability. In fact, electron microscopy demonstrated an irregular, fragmented appearance of sarcomeres in cardiac tissue from one of the patients.

Genetic studies of the thin filament proteins actin and tropomyosin suggest that differential effects of mutations on contractile force generation and transmission determine the cardiac remodeling pathway. Yet, a specific mutation is unlikely to exclusively alter the generation or transmission of force in myocytes, and alone may not completely account for the resultant phenotype. In a hamster model of cardiomyopathy caused by mutation of the δ-sarcoglycan gene, either HCM or DCM developed, depending on the strain of animal.[57] The modifying effects of other genes are clearly important in cardiac remodeling pathways,[20] even in cardiomyopathies attributable to single gene defects.

Evidence that DCM and HCM are allelic disorders also came from a linkage study in familial DCM.[24] Initially, a DCM locus was identified on chromosome 14 where the gene encoding cardiac β-myosin heavy chain (an HCM gene) is located, prompting its investigation as a positional candidate. A missense mutation was identified that segregated with relatively early-onset DCM, yet variable penetrance was observed in the family. A different missense mutation was found in another family that also had early onset of DCM. In fact, LV dilation was recognized prenatally in one case. Mutation analyses of other genes encoding sarcomeric proteins identified the same 3-bp deletion in the troponin T gene in two small, unrelated families in whom sudden death was prominent, even in infancy.

Mutations in β-myosin heavy chain and troponin T genes were predicted to cause a deficit in force generation, the same effect on force dynamics implicated in most HCM-associated mutations. In a genetically engineered mouse model, heterozygosity for a myosin mutation causes HCM, whereas homozygosity causes neonatal DCM with myocardial necrosis,[59] suggesting that the amount of defective protein determines the cardiomyopathy phenotype. By analogy, human mutations that have a more extreme effect on sarcomere function may cause DCM. Such mutations could compromise structural integrity of sarcomeres and render the cardiac myocyte more vulnerable to injury and death under physiological hemodynamic stress. Because myocytes lack the capacity to regenerate, hypertrophy of viable myocytes may be inadequate compensation for cumulative cell loss, leading to dilation and pump failure rather than cardiac hypertrophy.

A new mechanism for familial DCM: defective myocellular calcium regulation

High-fidelity regulation of intracellular calcium flux, orchestrated by the coordinated actions of many proteins, is central to excitation–contraction coupling in the heart. Indeed, defective myocellu-

lar calcium cycling is central to the pathobiology of heart failure.[60] Phospholamban (PLN) is a small protein of the sarcoplasmic reticulum (SR) that regulates the calcium ATPase SERCA2a pump, critical for calcium reuptake by the SR and for cardiac relaxation. Accordingly, the phospholamban gene (*PLN*) was investigated as a candidate gene, and a missense mutation that segregated with DCM was identified in a familial case.[31] Affected family members developed cardiomyopathy at age 20 to 30 years, which rapidly progressed to clinical heart failure. The DCM phenotype was replicated in a transgenic mouse model, while cellular and biochemical studies determined that mutant phospholamban exerted a dominant-negative effect on wild-type phospholamban by sequestering protein

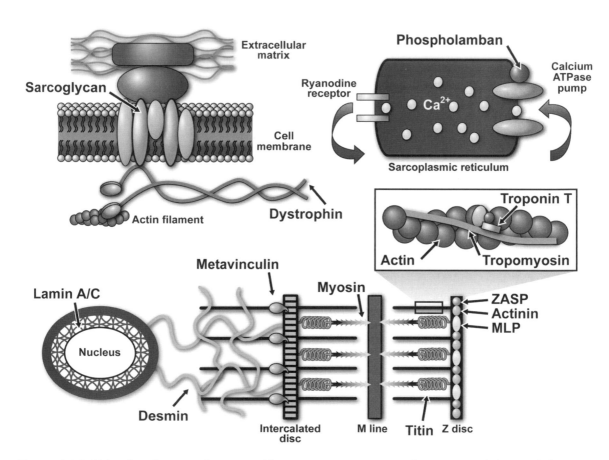

Figure 31.2. *Dilated cardiomyopathy is caused by mutations in genes encoding proteins of the cytoskeleton, sarcoplasmic reticulum, nuclear membrane, and sarcomere. MLP, muscle LIM protein; ZASP, Z-band alternatively spliced PDZ motif-containing protein.*

kinase A. In a separate study, heterozygosity and homozygosity for a truncation mutation, causing functional deletion of *PLN*, was identified in two families.[61] Heterozygotes developed either LV hypertrophy or DCM, while homozygotes developed a malignant form of DCM necessitating cardiac transplantation at a young age. Collectively, these studies defined intracellular calcium dysregulation as a new mechanism for DCM.

Conclusions

Little more than 10 years ago, the importance of genetics in the pathogenesis of DCM was unrecognized. We now know that DCM is familial in 20–30% of cases, and attributable to single gene defects. Given the profound genetic heterogeneity of DCM and lack of evidence for a common DCM gene, routine genetic testing is not currently feasible. Targeted lamin A/C screening in patients with associated skeletal muscle or conduction system disease, however, is a possible exception. Because DCM is often clinically silent until advanced myocardial disease has developed, the importance of screening echocardiography in family members cannot be over-emphasized. Penetrance may be highly variable, and a family member at risk for DCM is probably never too young or too old for clinical screening. Early diagnosis and treatment of asymptomatic DCM may clearly prevent or attenuate development of heart failure.

The Human Genome Project, together with technological advances in DNA and whole genome analysis, have provided new and powerful tools with which to uncover the molecular basis of heart failure in DCM. Through the application of

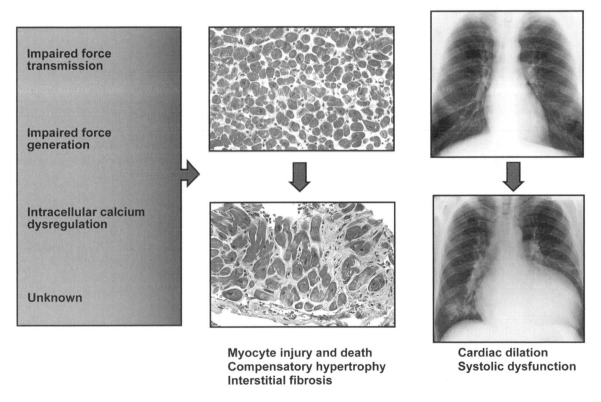

Figure 31.3. *Defects in myocellular proteins governing contractile force dynamics, intracellular calcium regulation, and unknown cellular processes lead to cardiac remodeling and the phenotype of dilated cardiomyopathy.*

complementary genetic linkage and candidate gene approaches, seven DCM loci have been mapped, and mutations in 14 genes causing DCM have been discovered. The diverse structure and function of proteins these genes encode match the genetic heterogeneity of DCM. Heritable defects in proteins of the cytoskeleton (sarcomeric, extra-sarcomeric, submembranous, intercalated disc), sarcomere (thin and thick filament, Z-disc), nuclear membrane and sarcoplasmic reticulum have been identified (Fig. 31.2). The implicated mechanisms for heart failure include defective transmission of contractile force, decreased generation of contractile force, and dysregulation of intracellular calcium (Fig. 31.3). Biochemical, cellular and animal studies of mutant proteins promise to provide additional mechanistic insight. Discovery of novel genes for DCM is on-going and will continue to advance our understanding of the pathobiology of heart failure.

References

1. Michels VV, Moll PP, Miller FA et al. The frequency of familial dilated cardiomyopathy in a series of patients with idiopathic dilated cardiomyopathy. N Engl J Med 1992; 326: 77–82.

2. Michels VV, Driscoll DJ, Miller FA et al. Progression of familial and non-familial dilated cardiomyopathy: long term follow up. Heart 2003; 89: 757–61.

3. Keeling PJ, Gang Y, Smith G et al. Familial dilated cardiomyopathy in the United Kingdom. Br Heart J 1995; 73: 417–21.

4. Baig MK, Goldman JH, Caforio ALP et al. Familial dilated cardiomyopathy: cardiac abnormalities are common in asymptomatic relatives and may represent early disease. J Am Coll Cardiol 1998; 31: 195–201.

5. Michels VV, Olson TM, Miller FA et al. Frequency of development of idiopathic dilated cardiomyopathy among relatives of patients with idiopathic dilated cardiomyopathy. Am J Cardiol 2003; 91: 1389–92.

6. Grunig E, Tasman JA, Kucherer H et al. Frequency and phenotypes of familial dilated cardiomyopathy. J Am Coll Cardiol 1998; 31: 186–94.

7. Mestroni L, Rocco C, Gergori D et al. Familial dilated cardiomyopathy: evidence for genetic and phenotypic heterogeneity. J Am Coll Cardiol 1999; 34: 181–90.

8. Bione S, D'Adamo P, Maestrini E et al. A novel X-linked gene, G4.5. is responsible for Barth syndrome. Nat Genet 1996; 12: 385–9.

9. Kelly DP, Strauss AW. Inherited cardiomyopathies. N Engl J Med 1994; 330: 913–9.

10. Olson TM, Keating MT. Mapping a cardiomyopathy locus to chromosome 3p22-p25. J Clin Invest 1996; 97: 528–32.

11. Fatkin D, MacRae C, Sasaki T et al. Missense mutations in the rod domain of the lamin A/C gene as causes of dilated cardiomyopathy and conduction-system disease. N Engl J Med 1999; 341: 1715–24.

12. Taylor MRG, Fain PR, Sinagra G et al. Natural history of dilated cardiomyopathy due to lamin A/C gene mutations. J Am Coll Cardiol 2003; 41: 771–80.

13. Olson TM, Keating MT. Defining the molecular genetic basis of idiopathic dilated cardiomyopathy. Trends Cardiovasc Med 1997; 7: 60–3.

14. Jung M, Poepping I, Perrot A et al. Investigation of a family with autosomal dominant dilated cardiomyopathy defines a novel locus on chromosome 2q14-q22. Am J Hum Genet 1999; 65: 1068–77.

15. Sylvius N, Tesson F, Gayet C et al. A new locus for autosomal dominant dilated cardiomyopathy identified on chromosome 6q12-q16. Am J Hum Genet 2001; 68: 241–6.

16. Messina DN, Speer MC, Pericak-Vance MA et al. Linkage of familial dilated cardiomyopathy with conduction defect and muscular dystrophy to chromosome 6q23. Am J Hum Genet 1997; 61: 909–17.

17. Schonberger J, Levy H, Grunig E et al. Dilated cardiomyopathy and sensorineural hearing loss: a heritable syndrome that maps to 6q23–24. Circulation 2000; 101: 1812–8.

18. Krajinovic M, Pinamonti B, Sinagra G et al. Linkage of familial dilated cardiomyopathy to chromosome 9. Am J Hum Genet 1995; 57: 846–52.

19. Bowles KR, Gajarski R, Porter P et al. Gene mapping of familial autosomal dominant dilated cardiomyopathy to chromosome 10q21–23. J Clin Invest 1996; 98: 1355–60.

20. Seidman JG, Seidman CE. The genetic basis for cardiomyopathy: from mutation identification to mechanistic paradigms. Cell 2001; 104: 557–67.

21. Towbin JA, Bowles NE. The failing heart. Nature 2002; 415: 227–33.

22. Ross J. Dilated cardiomyopathy. Concepts derived from gene deficient and transgenic animal models. Circ J 2002; 66: 219–24.

23. Durand J-B, Bachinski LL, Bieling LC et al. Localization of a gene responsible for familial dilated cardiomyopathy to chromosome 1q32. Circulation 1995; 92: 3387–9.

24. Kamisago M, Sharma SD, DePalma SR et al. Mutations in sarcomere protein genes as a cause of

dilated cardiomyopathy. N Engl J Med 2000; 343: 1688–96.

25. Kass S, MacRae C, Graber HL et al. A gene defect that causes conduction system disease and dilated cardiomyopathy maps to chromosome 1p1–1q1. Nat Genet 1994; 7: 546–51.

26. Mohapatra B, Jimenez S, Lin JH et al. Mutations in the muscle LIM protein and alpha-actinin-2 genes in dilated cardiomyopathy and endocardial fibroelastosis. Mol Genet Metab 2003; 80: 207–15.

27. Siu BL, Niimura H, Osborne JA et al. Familial dilated cardiomyopathy locus maps to chromosome 2q31. Circulation 1999; 99: 1022–6.

28. Gerull B, Gramlich M, Atherton J et al. Mutations of TTN, encoding the giant muscle filament titin, cause familial dilated cardiomyopathy. Nat Genet 2002; 30: 201–4.

29. Li D, Tapscoft T, Gonzalez O et al. Desmin mutation responsible for idiopathic dilated cardiomyopathy. Circulation 1999; 100: 461–4.

30. Tsubata S, Bowles KR, Vatta M et al. Mutations in the human δ-sarcoglycan gene in familial and sporadic dilated cardiomyopathy. J Clin Invest 2000; 106: 655–62.

31. Schmitt JP, Kamisago M, Asahi M et al. Dilated cardiomyopathy and heart failure caused by a mutation in phospholamban. Science 2003; 299: 1410–3.

32. Olson TM, Illenberger S, Kishimoto NY et al. Metavinculin mutations alter actin interaction in dilated cardiomyopathy. Circulation 2002; 105: 431–7.

33. Vatta M, Mohapatra B, Jimenez S et al. Mutations in Cypher/ZASP in patients with dilated cardiomyopathy and left ventricular non-compaction. J Am Coll Cardiol 2003; 42: 2014–27.

34. Knoll R, Hoshijima M, Hoffman HM et al. The cardiac mechanical stretch sensor machinery involves a Z disc complex that is defective in a subset of human dilated cardiomyopathy. Cell 2002; 111: 943–55.

35. Olson TM, Michels VV, Thibodeau SN et al. Actin mutations in dilated cardiomyopathy, a heritable form of heart failure. Science 1998; 280: 750–2.

36. Olson TM, Kishimoto NY, Whitby FG et al. Mutations that alter the surface charge of alpha-tropomyosin are associated with dilated cardiomyopathy. J Mol Cell Cardiol 2001; 33: 723–32.

37. Towbin JA, Hejtmancik JF, Brink P et al. X-linked dilated cardiomyopathy: molecular genetic evidence of linkage to the Duchenne muscular dystrophy (dystrophin) gene at the Xp21 locus. Circulation 1993; 87: 1854–65.

38. Muntoni F, Cau M, Ganau A et al. Deletion of the dystrophin muscle-promoter region associated with x-linked dilated cardiomyopathy. New Engl J Med 1993; 329: 921–5.

39. Beggs AH. Dystrophinopathy, the expanding phenotype. Dystrophin abnormalities in x-linked dilated cardiomyopathy. Circulation 1997; 95: 2344–7.

40. Campbell KP. Three muscular dystrophies: loss of cytoskeleton-extracellular matrix linkage. Cell 1995; 80: 675–9.

41. Milasin J, Muntoni F, Severini GM et al. A point mutation in the 5? splice site of the dystrophin gene first intron responsible for X-linked dilated cardiomyopathy. Hum Mol Genet 1996; 5: 73–9.

42. Ortiz-Lopez R, Li H, Su J et al. Evidence for a dystrophin missense mutation as a cause of X-linked dilated cardiomyopathy. Circulation 1997; 95: 2434–40.

43. Badorff C, Lee GH, Lamphear BJ. Enteroviral protease 2A cleaves dystrophin: evidence of cytoskeletal disruption in an acquired cardiomyopathy. Nat Med 1999; 5: 320–6.

44. Olson TM, Doan TP, Kishimoto NY et al. Inherited and de novo mutations in cardiac actin cause hypertrophic cardiomyopathy. J Mol Cell Cardiol 2000; 32: 1687–94.

45. Wong WW, Doyle TC, Cheung P et al. Functional studies of yeast actin mutants corresponding to human cardiomyopathy mutations. J Muscle Res Cell Motil 2001; 22: 665–74.

46. Hein S, Kostin S, Heling A et al. The role of the cytoskeleton in heart failure. Cardiovasc Res 2000; 45: 273–8.

47. Ganote C, Armstrong S. Ischaemia and the myocyte cytoskeleton: review and speculation. Cardiovasc Res 1993; 27: 1387–1403.

48. Eble DM, Spinale FG. Contractile and cytoskeletal content, structure, and mRNA levels with tachycardia-induced cardiomyopathy. Am J Physiol 1995; 268:H2426-H2439.

49. Dalakas MC, Dagvadorj A, Goudeau B et al. Progressive skeletal myopathy, a phenotypic variant of desmin myopathy associated with desmin mutations. Neuromuscul Disord 2003; 13: 252–8.

50. Gregorio CC, Granzier H, Sorimachi H et al. Muscle assembly: a titanic achievement? Curr Opin Cell Biol 1999; 11: 18–25.

51. Satoh M, Takahashi M, Sakamoto T et al. Structural analysis of the titin gene in hypertrophic cardiomyopathy: identification of a novel disease gene. Biochem Biophys Res Commun 1999; 262: 411–17.

52. Xu X, Meiler SE, Zhong TP et al. Cardiomyopathy in zebrafish due to mutation in an alternatively spliced exon of titin. Nat Genet 2002; 30: 205–9.

53. Arber S, Hunter JJ, Ross J et al. MLP-deficient mice exhibit a disruption of cardiac cytoarchitectural organization, dilated cardiomyopathy, and heart failure. Cell 1997; 88: 393–403.

54. Geier C, Perrot A, Ozcelik C et al. Mutations in the human muscle LIM protein gene in families with hypertrophic cardiomyopathy. Circulation 2003; 107: 1390–5.

55. Mounkes L, Kozlov S, Burke B et al. The laminopathies: nuclear structure meets disease. Curr Opin Genet Dev 2003; 13: 223–30.

56. Brodsky GL, Muntoni F, Miocic S et al. Lamin A/C gene mutation associated with dilated cardiomyopathy with variable skeletal muscle involvement. Circulation 2000; 101: 473–6.

57. Mogensen J, Klausen IC, Pedersen AK et al. Alpha-cardiac actin is a novel disease gene in familial hypertrophic cardiomyopathy. J Clin Invest 1999; 103: R39–R43.

58. Sakamoto A, Ono K, Abe M et al. Both hypertrophic and dilated cardiomyopathies are caused by mutation of the same gene, delta-sarcoglycan, in hamster: an animal model of disrupted dystrophin-associated glycoprotein complex. Proc Natl Acad Sci USA 1997; 94: 13873–8.

59. Fatkin D, Christe ME, Aristizabal O et al. Neonatal cardiomyopathy in mice homozygous for the Arg403Gln mutation in the alpha cardiac myosin heavy chain gene. J Clin Invest 1999; 103: 147–53.

60. Chien KR, Ross J, Hoshijima M. Calcium and heart failure: the cycle game. Nat Med 2003; 9: 508–9.

61. Haghighi K, Kolokathis F, Pater L et al. Human phospholamban null results in lethal dilated cardiomyopathy revealing a critical difference between mouse and human. J Clin Invest 2003; 111: 869–76.

CHAPTER 32

Cardiomyopathy in muscular dystrophies

Daniel E Michele, Kevin P Campbell

Introduction

Muscular dystrophies encompass a group of distinct genetic disorders, sharing a common feature of skeletal muscle weakness and wasting that is believed to be due to a disease process that occurs within skeletal muscle cells. The pathological hallmarks of muscular dystrophy patients include high serum levels of cytosolic muscle enzymes and, upon muscle biopsy, evidence of ongoing muscle necrosis, muscle regeneration, intersitial fibrosis and fatty replacement. The vast genetic heterogeneity of the more than 30 forms of muscular dystrophy has been realized. It has long been recognized that many muscular dystrophy patients have significant risk of developing cardiovascular disease, with a description of a patient dying from an enlarged heart in one of the earliest known papers describing muscular dystrophy patients.[1] Given that there are significant similarities, both structural and functional, between cardiac and skeletal muscle, similar disease mechanisms may be involved in both skeletal muscle fibers and cardiac muscle cells. Elevations in cardiac-specific cytoplasmic markers such as cardiac troponin I in some muscular dystrophy patients may be analogous to release of creatine kinase from skeletal muscle. However, many proteins mutated in muscular dystrophy patients are also expressed in the vasculature or the cardiac conduction system, and have the potential to modify cardiovascular function in addition to

any primary defect in cardiac muscle cells themselves.

In this chapter, we summarize our current understanding of the genetic diversity and molecular mechanisms that lead to the development of cardiomyopathy in muscular dystrophy patients. Much of our current knowledge has been derived from smaller case studies, some natural history studies in human patients, and study of genetic mouse models. Together, these studies are beginning to form a framework for important recommendations regarding clinical care of human patients. Hopefully, with additional basic research, and much needed multicenter clinical assessments and trials, pharmacological and/or genetic therapies can be rigorously tested and applied to prevent the important clinical problem of heart disease in the muscular dystrophy patient population. More broadly, the work on genetically defined forms of cardiomyopathy will hopefully provide important insight into the critical molecular mechanisms behind the more common acquired forms of heart disease.

Overview

The proteins affected by genetic mutations that cause muscular dystrophy are localized to a number of key functional systems within muscle cells, including the extracellular matrix, the muscle cell membrane, the intracellular cytoskeleton, the secretory pathway, the sarcomere and even the nuclear envelope (Fig. 32.1). Although, in general, cardiomyopathy is a recognized clinical problem in many forms of muscular dystrophy, in some cases it appears that cardiac muscle is spared (Table 32.1). This may be due to differential expression of the mutated gene in skeletal muscle, compared to cardiac muscle and smooth muscle, but also may be due to differences in the significance of the functional processes affected by the mutation in cardiac or skeletal muscle. There is not an extensive amount of large-scale natural history data determining the relative risk of cardiomyopathy in

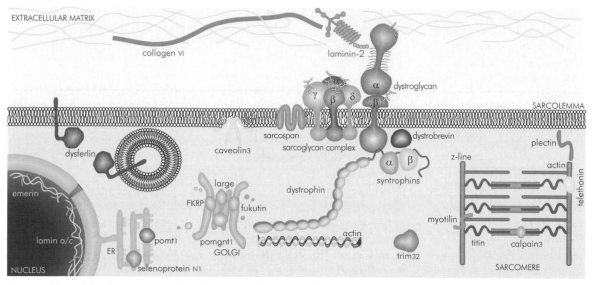

Figure 32.1. *Muscle proteins and molecular pathways involved in the pathogenesis of human muscular dystrophies.*

Table 32.1. *Cardiac involvement in human muscular dystrophies.*

Gene mutated	Muscular dystrophy	Inheritance	Cardiac involvement
Membrane-associated proteins			
dystrophin (*DMD*)	Duchenne muscular dystrophy	X-linked	Very common
	Becker muscular dystrophy	X-linked	Very common
α-sarcoglycan (*SGCA*)	Limb girdle muscular dystrophy 2D	AR	Rare
β-sarcoglcyan (*SGCB*)	Limb girdle muscular dystrophy 2E	AR	Common
γ-sarcoglycan (*SGCG*)	Limb girdle muscular dystrophy 2C	AR	Common
δ-sarcoglycan (*SGCD*)	Limb girdle muscular dystrophy 2F	AR	Common
caveolin 3 (*CAV3*)	Limb girdle muscular dystrophy 1C	AD	None reported
	Rippling muscle disease	AD	Rare
	HyperCKemia	AD	None reported
dysferlin (*DYSF*)	Limb girdle muscular dystrophy 2B	AR	Rare
	Miyoshi myopathy	AR	None reported
plectin (*PLEC*)	Epidermolysis bullosa with muscular dystrophy	AR	None reported

Table 32.1. *Cardiac involvement in human muscular dystrophies – continued*

Gene mutated	Muscular dystrophy	Inheritance	Cardiac involvement
Enzymes involved in glycosylation			
fukutin (*FCMD*)	Fukuyama congenital muscular dystrophy	AR	Common
fukutin-related protein (*FKRP*)	Limb girdle muscular dystrophy 2I	AR	Common
	Congenital muscular dystrophy 1C	AR	Common
protein O-mannosyltransferase 1 (*POMT1*)	Walker–Warburg syndrome	AR	None reported
protein O-mannosyl-*N*-acetylglucosaminyl transferase (*POMGNT1*)	muscle-eye-brain disease	AR	None reported
LARGE	Congenital muscular dystrophy 1D	AR	None reported
Extracellular matrix proteins			
laminin α_2 (merosin) (*LAMA2*)	Merosin-deficient congenital muscular dystrophy	AR	Rare
collagen VI (*COL6*)	Ullrich congenital muscular dystrophy	AR	None reported
	Bethlem myopathy	AD	Rare
Sarcomeric proteins			
Titin (*TTN*)	Tibial muscular dystrophy/LGMD2J	AR	None reported*
calpain-3 (*CAPN3*)	Limb girdle muscular dystrophy 2A	AR	None reported
telethonin (*TCAP*)	Limb girdle muscular dystrophy 2G	AR	Common
myotilin (*TTID*)	Limb girdle muscular dystrophy 1A	AD	None reported
Nuclear membrane proteins			
emerin (*EMD*)	Emery–Dreifuss muscular dystrophy	X-linked	Very common
lamin A/C (*LMNA*)	Emery–Dreifuss muscular dystrophy	AD	Very common
	limb girdle muscular dystrophy 1B	AD	Very common
Cytosolic proteins			
TRIM32	Limb girdle muscular dystrophy 2H	AR	None reported
selenoprotein N1 (*SEPN1*)	Rigid spine congenital muscular dystrophy	AR	Common?
Nucleotide repeats			
D4Z4 region of chromosome 4	Facioscapulohumeral muscular dystrophy	AD	Rare
DMPK(CUG repeat)	Myotonic dystrophy 1	AD	Very common
ZNF9 (CCUG repeat)	Myotonic dystrophy 2	AD	Very common
PABPN1 (poly-alanine)	Oculopharyngeal muscular dystrophy	AD/AR	None reported

Muscular dystrophies are classified by the proposed function of the gene (mRNA). Muscular dystrophy nomenclature is by the convention of the World Muscle Society and cardiac involvement is indicated. *Mutations in the *TTN* can cause DCM, but mutations in *TTN* causing muscular dystrophy do not seem to cause heart disease. AD, autosomal dominant; AR, autosomal recessive.

muscular dystrophies, but small clinical reports and anecdotal evidence indicate that in many cases, the risk of cardiovascular disease is probably very significant. There are a number of excellent recent reviews that describe the cardiovascular complications of some of the muscular dystrophies in greater detail.[2–8] In some forms of muscular dystrophy, the European Neuromuscular Centre has proposed some guidelines for clinical cardiovascular care.[9] In addition, research in mouse models of muscular dystrophies has strongly supported the possibility that pharmacological or genetic therapies will be able to correct or improve cardiovascular disease in muscular dystrophy patients.

Because of the vast genetic heterogeneity of genes causing muscular dystrophy, and because current classification systems of muscular dystrophies are based in part on the historical order of clinical and genetic identification, we will instead focus on the functional compartments of the muscle cell affected by these mutations (Table 32.1).

Membrane-associated proteins

Mutations in several proteins localized to the plasma membrane of muscle cells have been implicated in the pathogenesis of various forms of muscular dystrophy. The proteins of the dystrophin–glycoprotein complex (Fig. 32.1) appear to play a critical role in linking the submembrane cytoskeleton to the extracellular matrix. In additional, several other molecules that regulate membrane structure or membrane-bound signaling proteins appear to be important in human muscular dystrophies. Many of these muscular dystrophies associated with defects in the muscle cell plasma membrane present with clinically significant cardiovascular disease.

Dystrophinopathies

The first mutation identified to cause muscular dystrophy was in a protein called dystrophin which causes two X-linked forms, Duchenne's (DMD) and Becker's muscular dystrophy (BMD).[10,11] The dystrophin gene is one of the largest genes in nature with 79 exons, and alterations in promoter usage and splicing can generate several dystrophin isoforms.[12] The major form expressed in skeletal and cardiac muscle is the full-length 427 kDa isoform. Dystrophin is anchored to the sarcolemma and the muscle cell plasma membrane, through its association with the transmembrane dystrophin–glycoprotein complex (DGC), which includes dystrophin, dystroglycan, sarcoglycans, sarcospan, syntrophins and dystrobrevin.[13] The N terminus of dystrophin binds to the actin cytoskeleton, a long rod domain containing 24 spectrin-like repeats, and regions near the C terminus bind to dystroglycan (Fig. 32.1). Therefore, dystrophin, by anchoring the DGC to the actin cytoskeleton, is thought to provide structural support to the muscle cell membrane and prevent membrane damage during cycles of contraction and relaxation.[14] Mutations in dystrophin cause loss of dystrophin from the muscle cell membrane and a concomitant reduction in the expression of the DGC.[15,16] Although dystrophin is also expressed in smooth muscle, there is also significant expression of a dystrophin homolog called utrophin in smooth muscle, and utrophin appears to be able to maintain DGC expression in smooth muscle in the absence of dystrophin.[17]

DMD was first identified in the mid 1800s, and predominantly affects boys (X-linked) with a frequency of 1 in 3500 male births, with clinical symptoms usually appearing between the ages of 3 and 6.[18] By the early teens, DMD patients are often wheelchair bound, require respiratory therapy, and nearly always die in their late teens or early twenties. Although the exact percentages are uncertain, it has been estimated that roughly 15–20% of DMD patients die from heart failure or cardiomyopathy, and the rest from predominantly respiratory failure (40%) or a combination of pneoumonia, respiratory failure and cardiovascular complications.[4,19,20] At least 90% of older DMD patients have some form of cardiovascular involvement detectable by either electrocardiogram (ECG) or echocardiography.[4]. Sinus tachycardia is almost always present after age 5, and is usually the first sign of cardiac involvement.[4,21] Many patients progress to dilated cardiomyopathy, although sometimes hypertrophy is seen early on.[22] In contrast, BMD patients only develop mild skeletal muscle disease, often do not require a wheelchair, and can live a near normal lifespan. However, despite their mild skeletal muscle disease, heart muscle can be severely affected, with cardiac death due to heart failure occurring in up to 50% of patients.[4] Many BMD patients can present with cardiomyopathy, while having little or no symptoms of skeletal muscle weakness.[23] Cardiovascular involvement occurs in up to 90% of BMD patients, detectable by echocardiography and ECG, with disease often progressing to dilated cardiomyopathy or severe heart failure requiring cardiac transplant.[4,24,25] It is not clear if the high prevalence of cardiomyopathy with mild skeletal muscle disease in BMD patients is solely due to a longer lifespan in which cardiac disease can become apparent, or whether somehow the specific mutations that cause mild BMD differentially severely affect cardiac muscle and while only having mild effects on skeletal muscle.

The importance of dystrophin in the development of heart disease is also supported by the identification of mutations in dystrophin that cause X-linked dilated cardiomyopathy (XLDCM).[26,27] XLDCM is often described as an exclusively cardiovascular disease, a particularly rapidly progressing heart disease that usually affects males in their teenage years. In many cases of XLDCM, patients do have mild skeletal muscle involvement which blurs the distinction between XLDCM and BMD.[4,28] However, certain XLDCM causing mutations in the 5′ end of the dystrophin gene appear to regulate the interaction of the dystrophin promoter with various molecules that regulate gene transcription.[27,29] In many of these patients, dystrophin is completely absent in cardiac muscle, but skeletal muscle shows upregulation of non-muscle dystrophin isoforms.[27,29] Therefore, differences in dystrophin gene expression regulation in cardiac and skeletal muscle may explain the cardiovascular-specific features of certain XLDCM patients.

Female carriers of BMD and DMD mutations would also be predicted to have partial dystrophin deficiency in the myocardium, due to random X-inactivation in cardiac myocytes. Indeed, up to 90% of adult female DMD carriers have significant cardiovascular involvement with echocaordiographic and ECG changes, and up to 10% develop

cardiac failure.[4,30] Because these patients are generally asymptomatic with respect to both cardiac and skeletal muscle disease, this group should be monitored closely for signs of preclinical cardiovascular disease.

The major mechanism for dystrophin-related cardiomyopathy is probably due to disruption of DGC function, leading to destabilization of the cardiac myocyte membrane, and resulting in increased myocardial cell necrosis. Unlike skeletal muscle, cardiac muscle is unable to undergo significant amounts of regeneration, and necrotic tissue is replaced with connective tissue and/or fat. More significant ECG changes appear to reflect scarring of the left ventricular wall. The fibrotic changes in the heart are probably responsible for the changes in electrical conduction seen in human patient ECGs.[31,32] The left ventricle appears more significantly affected in DMD patients, perhaps indicating that increased myocardial wall stress induces contraction-induced damage.[33] The compensation of dystrophin deficiency by utrophin appears to maintain the utrophin–glycoprotein complex expression in smooth muscle.[17] However, there is some evidence in human patients that blood flow, at least to the limb muscles, is altered in DMD patients.[34] Data from mouse models suggest the mechanism of alterations in blood flow might be due to abnormal amounts of nitric oxide produced by neuronal nitric oxide synthase (NOS) in skeletal muscle.[35,36] Neuronal NOS is associated with the DGC in striated muscle cells, and regional nitric oxide normally counterbalances α-adrenergic-stimulated contraction of the microvasculature during the 'fight or flight' response. It is unclear whether this mechanism is involved in cardiomyopathy, as positron emission tomography studies have not consistently shown abnormal or reduced coronary blood flow in human DMD and BMD patients.[37]

Mdx mice have naturally occurring mutations in the dystrophin gene, and display many features of human muscular dystrophy.[38] There is pathological evidence of cardiomyopathy, and histological evidence of necrosis in the hearts of mdx mice beginning around 24 months of age, along with elevated levels of cardiac troponin I.[39] The cardiovascular phenotype in mdx mice appears to worsen under conditions of pressure overload, or isoproterenol infusion, suggesting cardiovascular stress plays a significant modulatory role in disease pathogenesis.[40,41] Mdx/myoD double knockout mice have a more severe cardiomyopathy, even though myoD is expressed in skeletal muscle cells, suggesting that a more severe muscular dystrophy might exacerbate the cardiovascular phenotype.[42] Transgenic expression of the dystrophin gene specifically within cardiac myocytes specifically rescues the cardiovascular phenotype within mdx mice.[43] Mdx mice do not show evidence of coronary vasospasm, nor do they respond to agents which prevent coronary vasospasm in other mouse models. This suggests that the primary defect in mdx mice causing cardiomyopathy is a defect within the cardiac myocyte, and not secondary to a deficiency of dystrophin within smooth muscle of the coronary vasculature.[39]

Sarcoglycanopathies

Sarcoglycans were originally identified by biochemical purification of the DGC in skeletal muscle.[13] There are six known sarcoglycan genes in humans, α, β, γ, δ, ε and ζ. The sarcoglycans are believed to assemble as a five-protein complex of four single pass transmembrane sarcoglycans, and a tetraspan-like protein sarcospan (Fig. 32.1).[44] The major striated muscle sarcoglycan complex in the DGC is composed of α-, β-, γ- and δ- sarcoglycan[13] ε-sarcoglycan shares significant sequence identity with α-sarcoglycan, but is expressed much more broadly in nearly all tissues.[45] ε-sarcoglycan replaces α-sarcoglycan, and assembles with β-, γ-and δ-sarcoglycan in the smooth muscle sarcoglycan–sarcospan complex.[46] ζ-sarcoglycan was recently described and appears to be expressed in skeletal muscle and smooth muscle, but it is still unclear whether it is an essential component of the major sarcoglycan complexes in these tissues.[47]

Subsequent to the identification of sarcoglycans in the DGC, mutations in α-, β-, γ- and δ-sarcoglycan were identified in limb girdle muscular dystrophies 2C–2F, respectively.[48–53] Patients generally develop a muscular dystrophy that is variable between DMD-like and BMD-like severity, and cardiomyopathy is common in patients with mutations in β-, γ- and δ-sarcoglycan mutations. Signs of cardiac involvement (ECG or echocardiographic abnormalities) have been reported in approximately 30–70% of sarcoglycanopathy patients, with nearly 20% showing clear signs of dilated cardiomyopathy.[54–56] Interestingly, mutations in ε-sarcoglycan in humans do not cause muscular dystrophy, but instead are responsible for a hereditary form of myoclonus dystonia.[57] Finally, heterozygous mutations in δ-sarcoglycan have been identified in patients with inherited and idiopathic dilated cardiomyopathy (DCM).[13] Interestingly, these patients do not appear to have any skeletal muscle disease. It is unclear what accounts for the apparent tissue-specific effect of these particular DCM mutations in δ-sarcoglycan.

Interest in the smooth muscle expression of the sarcoglycan complex originated largely from mouse studies where the sarcoglycan genes have been knocked out. In particular, α-sarcoglycan knockout mice do not develop significant evidence of cardiomyopathy, while β- and δ-sarcoglycan knockout mice develop cardiomyopathy.[58–61] Interestingly, while all three lines have the deficiencies of the sarcoglycan complex in cardiac muscle, the latter two lines of mice have the sarcoglycan complex disrupted in smooth muscle in addition to cardiac muscle (Fig. 32.2).[58,59] Morphological studies showed evidence of coronary vasospasm in these mice, as well as vasospasm in skeletal muscle and other non-muscle tissues.[59] Furthermore, the cardiomyopathy could be prevented in β- and δ-sarcoglycan knockout mice by pharmacological agents, nicorandil or verapamil, which prevented vasospasm (these agents could not prevent mdx mouse cardiomyopathy).[39,58] Together these results have suggested that the disruption of the sarcoglycan complex causes a primary defect in smooth muscle function that contributes to the pathogenesis of cardiomyopathy. Similar results of coronary vasospasm have been reported in γ-sarcoglycan knockout mice, which also develop cardiomyopathy.[62] There is some disagreement in the field over the expression distribution of γ-sarcoglycan in animal models, but it appears clear that in human patients with mutations in γ-sarcoglycan the entire smooth muscle sarcoglycan complex is disrupted.[17] Transgenic expression of γ-sarcoglycan or δ-sarcoglycan

Figure 32.2. *Cardiovascular distribution of the sarcoglycan complex in cardiac myocytes and blood vessels contributes to the pathogenesis of vasospasm and cardiomyopathy. In α-sarcoglycan-deficient patients (LGMD2D) the cardiac myocyte sarcoglycan complex is disrupted, but the sarcoglycan complex expression in smooth muscle is maintained by expression of ε-sarcoglycan. In β- and δ-sarcoglycan-deficient patients (LGMD2E and 2F), the smooth muscle sarcoglycan complex is also disrupted and vasospasm and cardiomyopathy occurs.*

specifically in cardiac myocytes of corresponding knockout mice can prevent cardiomyopathy, and suggests that the smooth muscle disruption of the sarcoglycan complex alone is not sufficient to cause cardiomyopathy.[63] These data together support the hypothesis that the disruption of the sarcoglycan complex in cardiac myocytes is critical to rendering the muscle cells more susceptible to ischemic challenges resulting from the coronary vessel dysfunction.[64]

Defects in the sarcoglycan complex are also responsible for inherited cardiomopathy in inbred hamster lines.[65,66] The cardiomyopathic hamster displays evidence of coronary vasospasm that is responsive to verapamil, and is thought to play a significant role in the focal necrosis seen in the heart.[67] Interestingly, two sublines, Bio14.6 and T0–2, were inbred specifically for hypertrophic cardiomyopathy and dilated cardiomyopathy, respectively. However, both lines have the same deletion of the first exon of the gene encoding δ-sarcoglycan.[68] It is still unclear how the same mutations in δ-sarcoglycan result in different forms of cardiomyopathy. Clearly, genetic modifiers between the two hamster strains probably play an important role in the differing pathogenesis of these two strains of animals.

Caveolinopathies

Caveolins comprise a three-member gene family that encode essential components of plasma membrane invaginations called caveolae.[69] Caveolae are specializations of the plasma membrane and are rich in proteins involved in cellular signaling. Caveolin-3 is a muscle-specific caveolin isoform. Caveolin-3 is expressed in skeletal, cardiac and smooth muscle, and is localized to the plasma membrane in striated muscle.[70] However, in smooth muscle, caveolin-1 is the predominant isoform localized to the muscle cell membrane.[71]

Dominant mutations in caveolin-3 have been identified in four different forms of muscle disease including limb girdle muscular dystrophy 1C (LGMD1C), rippling muscle disease (RMD), distal myopathy and hyperCKemia.[72–76] The clinical features of each disease are distinct with, for example, LGMD1C having a typical muscular dystrophy phenotype and RMD having distinctive percussion-induced muscle contractions. Although each disease has differences in clinical presentation, there can clearly be overlap, with the same mutations causing several different disorders, occasionally even within the same family.[77,78] There is one report of patients with homozygous mutations in caveolin-3, and they appear to have a more severe phenotype.[79]

Cardiomyopathy has not been reported in LGMD1C. In fact, some mutations have been shown to cause severely decreased skeletal muscle caveolin-3 membrane localization, while having little effect on cardiac muscle caveolin-3 localization.[80] There have been two cases of RMD where patients were concluded to have died from cardiac arrythmias but were later shown to have CAV3 mutations,[81] and there is a form of recessive RMD where severe cardiac arrythmias are common.[82] Despite, the relative lack of cardiomyopathy in muscular dystrophy patients, mutations in caveolin-3 have been recently described in a family with hypertrophic cardiomyopathy, by a candidate gene approach.[83] The patients in this family did not have muscular dystrophy. Interestingly the identified mutated residue is also a target for an LGMD1C mutation, albeit a different substitution. The molecular basis for the apparent tissue specificity of these mutations at the same codon of caveolin-3 is unclear.

Despite the lack of evidence for cardiomyopathy being common in caveolin-related muscular dystrophy patients, mouse models of caveolinopathies have suggested that caveolin-3 may indeed be important for cardiovascular function, at least in the mouse. The caveolin-3 knockout mouse displays evidence for a mild muscular dystrophy and progressive cardiomyopathy with hypertrophy, dilation and fibrosis.[84] In addition, transgenic overexpression of caveolin-3 in the heart also produces a similar cardiomyopathic phenotype.[85] Finally, transgenic overexpression of a LGMD1C mutant caveolin-3 in skeletal muscle and in heart causes muscular dystrophy and hypertrophic cardiomyopathy.[86] Therefore, the mouse models that increase or eliminate caveolin-3 expression, or increase expression of mutant caveolin-3 expression, have not clarified exactly how dominant mutations in caveolin-3, that generally result in protein mislocalization, cause disease in human patients. However, several mechanisms including alterations in mitogen-activated protein kinase (MAPK) signaling, NOS signaling, or alterations of interactions with either the DGC or dysferlin have been proposed.[77]

Dysferlinopathies
Dysferlin is a transmembrane protein that is expressed in both the sarcolemma and intracellular vesicles.[87] Mutations have been linked to two allelic forms of muscular dystrophy that primarily affect the distal or the limb girdle muscles, Miyoshi myopathy and LGMD2B, respectively.[88] There have been no reports of cardiomyopathy in Miyoshi myopathy, but there are a few individual LGMD2B patients presenting with ECG abnormalities or Holter ECG abnormalities.[89] Recently, a mouse model of complete dysferlin deficiency showed dysferlin is involved in the muscle membrane repair process by regulating the Ca^{2+}-dependent fusion of vesicles to repair holes in the plasma membrane.[90] Interestingly, dysferlin localization is altered in many forms of muscular dystrophy, suggesting that this membrane repair pathway may be activated, and dysferlin may play a modulatory role in other forms of muscular dystrophy.[87,91] Therefore, the potential role of dysferlin in membrane repair in the heart awaits further examination. The SJL inbred mouse strain has a naturally occurring mutation in dysferlin that decreases dysferlin protein expression.[92] In addition, there is a dysferlin homolog called myoferlin, which is expressed in skeletal muscle, and the heart and may compensate for the loss of dysferlin in certain striated muscles.[93]

Epidermolysis bullosa with muscular dystrophy (plectin)
Plectin is localized to the Z-discs in striated muscle and to places where the cytoskeleton interacts with the membrane such as intercalated discs, desmosomes and gap junctions in the heart.[94] Mutations in plectin have been linked to autosomal recessive epidermolysis bullosa with muscular dystrophy.[95,96] Patients with plectin mutations have severe skin blistering and a late-onset mild muscular dystrophy. No cardiovascular involvement has been reported. Plectin-null mice die at 2–3 days of age, and have skin blistering, mild skeletal muscle necrosis, and disruptions of the intercalated discs in the heart.[97]

Rigid spine congenital muscular dystrophy (selenoprotein N1)
Selenoproteins are enzymes involved in oxidation–reduction reactions and have a selenocysteine residue in the catalytic site. Selenoprotein N1 (SEPN1) is localized to the endoplasmic reticulum

and appears to be an integral membrane protein.[98] Mutations in SEPN1 have been linked to rigid spine congenital muscular dystrophy and a myopathy called multiminicore disease.[99,100] There are several early reports of patients with a rigid spine syndrome and cardiac involvement, but it is not completely clear whether any or all of these patients have SEPN1 mutations.[101–104] Selenium defiency in the diet in humans is associated with a cardiomyopathy called Keshan disease.[105] Keshan disease is an interesting disorder of endemic cardiomyopathy resulting from outbreaks of myocarditits in children in rural China. Fortunately, Keshan disease can be completely prevented by oral sodium selenite.[106] Keshan disease appears to be caused by the selenium deficiency resulting in both an increased susceptibility to Coxsackievirus-induced myocarditis, and alterations in virulence of certain coxsackievirus strains.[107,108] Related to muscular dystrophy-associated cardiomyopathy, coxsackievirus protease 2A has been shown to cleave dystrophin in experimental myocarditis (described later in this chapter, page 555).

Enzymes involved in glycosylation

Despite the importance of the DGC in muscular dystrophy and associated cardiomyopathy described above, there have been no mutations in humans described in the central protein of this complex, dystroglycan. Dystroglycan plays a critical role in the DGC as it binds intracellularly to dystrophin and extracellularly to laminin and other extracellular matrix proteins, thereby providing a critical link between the cytoskeleton and the extracellular matrix (Fig. 32.1).[109] Recently, a number of muscular dystrophies have been characterized with mutations in proteins that appear to be involved in glycosylation of glycoproteins.[110,111] These enzymes are critical in the glycosylation pathway of dystroglycan, and the abnormal glycosylation of dystroglycan in these patients disrupts its function as an extracelluar matrix receptor.[112] Therefore, these disorders can be thought of as 'dystroglycanopathies' because of the disruption of the functional link from the cytoskeleton to the extracellular matrix through dystroglycan. Similar to mutations in other components of the DGC, cardiomyopathy seems to be an important feature in 'dystroglycanopathy' patients. In addition, many of these patients have abnormal central nervous system development and eye development, and severe mental retardation. The traditional knockout of the dystroglycan gene in the mouse is embryonic lethal due to a malformation of embryonic basement membranes.[113] However, the specific deletion of the dystroglycan gene in skeletal muscle causes muscular dystrophy.[114,115] The loss of dystroglycan function in the central nervous system in the mouse recapitulates nearly all of the neurological features of these disorders.[116]

Fukuyama congenital muscular dystrophy

Fukuyama congenital muscular dystrophy (FCMD) primarily affects the Japanese population, is the second most common muscular dystrophy in Japan (to Duchenne muscular dystrophy), and the majority of cases are linked to a founder mutation in the gene encoding fukutin (FCMD).[117] FCMD has sequence similarity to the Fringe family of glycosyltransferases and is localized to the Golgi apparatus, but its exact function is unknown.[118,119] FCMD patients have been shown to be deficient in normally glycosylated α-dystroglycan and α-dystroglycan-dependent ligand-binding activity in FCMD skeletal muscle is disrupted.[112,120] There are several reports of severe myocardial fibrosis in autopsied specimens of FCMD cases.[19,121] In a postmortem study of 10 FCMD patients, one died of confirmed heart failure, and eight died of sudden death of undetermined causes.[122] In addition, the fully glycosylated form of α-dystroglycan is not detectable in heart samples from FCMD patients.[120] A chimeric mouse model from fukutin-deficient embryonic stem cells has been produced (the knockout mouse is embryonic lethal) with brain and muscle phenotypes similar to human patients, but no cardiac characterization was reported.[123]

Limb girdle muscular dystrophy 2I and MDC1C (fukutin-related protein)

Mutations in a protein with sequence homology to fukutin, fukutin-related protein (FKRP), have been described in patients with limb girdle muscular dystrophy (LGMD2I) and congenital muscular dystrophy (MDC1C).[124,125] There is a reduction in the fully glyocosylated form of α-dystroglycan in skeletal muscle of patients with FKRP mutations.[126] The exact function of FKRP is unknown, but FKRP is localized to the Golgi body.[123] Originally, mutations in FKRP were thought to spare the brain, but recent evidence indicates that rare patients can have developmental abnormalities similar to FCMD and muscle-eye-brain disease.[127] LGMD2I is milder and of later onset than MDC1C, and all LGMD2I patients share a common founder missense mutation (L276I).[128] The carrier frequency of the missense mutation is as high as 1:400 in the United Kingdom. In general the severity of LGMD2I is milder than Duchenne muscular dystrophy. Similar to Duchenne muscular dystrophy and sarcoglycanopathies, dilated cardiomyopathy is a frequent complication in LGMD2I patients (~45% of patients), and left ventricular functional defects are common in MDC1C even before age 10.[128,129]

Walker–Warburg syndrome and muscle-eye-brain disease (POMT1 and POMGNT1)

Walker–Warburg syndrome (WWS) and muscle-eye-brain disease (MEB) were originally thought to be allelic disorders due to the similarities of their clinical presentation including congenital muscular dystrophy, abnormal brain and eye development, and severe mental retardation.[130,131] The clinical features of WWS and MEB are similar to severe FCMD, and patients with severe FKRP mutations. WWS is generally more severe than MEB, with patients failing to survive past three or four years of age. Mutations in WWS and MEB patients have been described in genes encoding enzymes involved in O-mannosylation within the secretory pathway, in genes encoding an O-mannosyl-transferase (POMT1) and the O-mannosyl-N-acetyl-glucosamine transferase (POMGNT1), respectively.[132,133] Dystroglycan contains relatively unique O-mannosyl linked sugars that are believed to be important for ligand binding.[134] WWS and MEB patient skeletal muscle shows abnormal glycosylation of α-dystroglycan, and a loss of dystroglycan matrix receptor function.[112,135] Because there are no additional homologs for POMGNT1 in humans, and POMT2 is required for POMT1 function, it is likely these mutations also effect glycosylation of dystroglycan in the heart. No cardiac phenotypes have been reported in WWS or MEB. However, due to the rarity and severity of these disorders, and the young age of the patients at the time of diagnosis, it could be that cardiovascular phenotypes have not yet developed or are overlooked.

Congenital muscular dystrophy 1D (LARGE)

LARGE is a protein with two domains that have homology to glycosyltransferases. An association of LARGE in muscular dystrophy was first shown by identification of mutations in LARGE through genetic linkage in the spontaneous mutant myodystrophy mouse (*myd*).[136] *myd* mice have abnormal glycosylation of dystroglycan similar to FCMD and MEB in both muscle and brain resulting in a loss of ligand-binding activity.[112] *myd* mice also have brain abnormalities identical to those of the mice with the dystroglycan gene (*DAG*) knocked out in the brain, and recapitulate features of human lissencephaly.[112,116] Mutations in LARGE have been recently reported in a single patient with muscular dystrophy and severe mental retardation, but no cardiovascular phenotype was reported.[137] However, evidence of late onset focal myocardial necrosis/fibrosis has been shown in *myd* mice suggesting cardiac involvement might be seen in human patients.[138]

Extracellular matrix proteins

Merosin (laminin-2)-deficient muscular dystrophy

Laminin is a heterotrimer of α, β and γ subunits, and resides in the extracellular matrix.[139] Laminins can bind to dystroglycan through their C-terminal G-domains, thereby completing the link from the actin cytoskeleton through dystrophin and dystroglycan to the extracellular matrix. Mutations in the laminin α_2 chain (merosin) cause merosin-deficient congenital muscular dystrophy.[140] Laminin α_2

is expressed in the heart, but there is only one small report where one out of six patients had significant cardiac involvement with a reported congestive cardiomyopathy at 1.4 years of age.[141] The dy/dy mouse is a naturally occurring model of laminin α_2-deficient muscular dystrophy, and has been reported preliminarily to have some cardiac involvement, although the lifespan of this mouse is only around 8 weeks.[142] The lack of significant cardiac involvement in most patients could be due to compensation by other laminin α chains in the heart, or perhaps due to the young age at the time of diagnosis.

Bethlem myopathy and Ullrich congenital muscular dystrophy (collagen VI)

Collagen VI is a component of collagen fibrils in the extracellular matrix. Increased collagen deposition is often a consequence of myocardial remodeling following necrosis. Dominant mutations in the gene encoding collagen VI (COL6) cause Bethlem myopathy, and recessive mutations in COL6 have been linked to Ullrich congenital muscular dystrophy.[143,144] Cardiac involvement with hypertrophic cardiomyopathy has been reported in single patients with Bethlem myopathy (one out of 27 examined), but overall cardiac involvement appears to be rare in these muscular dystrophies.[145] A mouse model of collagen VI deficiency has been recently characterized with a mild myopathy, but no cardiovascular phenotype was reported.[146]

Sarcomeric proteins

The sarcomere is the key functional unit within muscle necessary to produce both force and motion. It is well known that mutations in sarcomere proteins can affect heart function. Mutations in sarcomere proteins are responsible for nearly all cases of familial hypertrophic cardiomyopathy, and mutations in sarcomere proteins have been identified in dilated cardiomyopathy patients.[147] Many hypertrophic cardiomyopathy patients surprisingly do not have skeletal muscle disease, despite expression of the mutant proteins in skeletal muscle.[148] However, there are mutations in sarcomeric pro-

teins that do cause muscular dystrophy and other myopathies.[149] In some cases, it appears as though muscular dystrophy mutations in sarcomeric proteins do not cause significant heart disease. Therefore, with sarcomeric proteins it appears as though different mutations in these proteins may produce markedly different cardiac- or skeletal muscle-specific phenotypes depending on the location of the mutation and the functional or structural property of the protein that is affected.

Tibial muscular dystrophy/LMGD 2J (titin)

Titin is a huge protein that spans and binds to the Z-line and the M-line of the sarcomere in striated muscle. Titin is thought to be important in maintaining sarcomere structure and the elasticity of the sarcomere.[150] Mutations in the gene encoding titin have been linked to dilated cardiomyopathy.[151] However, mutations in titin have also been linked to tibial muscular dystrophy/LGMD2J.[152] As its name implies, tibial muscular dystrophy, primarily affects the tibialis anterior, causing patients to have 'foot drop'. Mutations in titin have also been linked to muscular dystrophy in a spontaneous mutant mouse, the mdm or muscular dystrophy with myositis mouse.[153] There have been neither cases of cardiomyopathy in large studies of tibial muscular dystrophy, nor reports of cardiomyopathy in the mdm mouse.[152] Although the molecular basis for the tissue specificity of titin mutations is unclear, all the mutations that cause muscular dystrophy reside near the calpain-binding site near the M-line on titin, while mutations that cause cardiomyopathy reside primarily near the Z-line region of titin.[154] Therefore, the interactions of titin and calpain may be important in the muscular dystrophy phenotype of titin mutations. In fact, tibial muscular dystrophy patients show a secondary loss of calpain 3 in skeletal muscle.[155]

Limb girdle muscular dystrophy 2A (calpain 3)

Calpain 3 is a member of family of calcium-activated proteases and is expressed in adult skeletal muscle. Although, not typically thought of as a sarcomere protein, as described above calpain 3 is known to bind to sarcomeric titin.[156] Patients with limb girdle muscular dystrophy 2A have recessive mutations in calpain 3 leading to a loss of calpain 3 activity.[157] Although calpain 3 is expressed in fetal cardiac muscle during cardiogenesis, the protein expression declines to non-detectable levels in the adult heart.[158] Therefore, the skeletal muscle-specific expression of calpain 3 may explain the lack of cardiovascular phenotypes in muscular dystrophy patients with mutations both in calpain 3 and in titin.

Limb girdle muscular dystrophy 2G (telethonin)

Telethonin (also called T-cap) is a small protein that binds to titin and other Z-line proteins such as α-actinin, at the Z-line.[159] Recessive mutations in telethonin result in loss of telethonin expression and limb girdle muscular dystrophy 2G, with an interesting feature of rimmed vacuoles within the muscle.[160] Telethonin is expressed in both skeletal and cardiac muscle, and cardiovascular involvement has been reported in one family with half of the members showing cardiovascular symptoms.[160] Telethonin interacts with titin near the sites that are mutated in DCM, and also interacts with muscle LIM protein (MLP).[161] Telethonin has been proposed to be a mechanical stress sensor in cardiac muscle along with MLP, and mutations in MLP also cause both HCM and DCM.[161,162]

Limb girdle muscular dystrophy 1A (myotilin)

Myotilin is expressed in both adult cardiac and skeletal muscle, binds to α-actinin, and is believed to help cross link actin filaments at the Z-line.[163] Myotilin is more broadly expressed during development, and may be involved in actin filament assembly in other tissues. Mutations in the gene encoding myotilin have been linked to autosomal dominant limb girdle muscular dystrophy 1A.[164] Patients typically present with proximal muscle weakness and a characteristic dysarthic pattern of speech.[164] Despite the expression of myotilin in the heart, no cardiac involvement has been reported in LGMD1A, even though large families with this disorder have been characterized.[165]

Nuclear membrane proteins

Like the plasma membrane, the nuclear membrane also has a submembrane network of proteins that are involved in structure and function within the nucleus. A number of mutations in nuclear membrane proteins have been linked to a variety of cardiovascular and muscle diseases. These genetic studies have helped to define the important role of the nuclear membrane in cellular function, both in muscle and in other tissues. Most of the muscular dystrophy patients with genetic defects in nuclear membrane proteins present with cardiovascular phenotypes. However, as described below, in some cases the various disorders associated with the mutation in the same protein appear to only have partial clinical overlap.

X-linked Emery–Dreifuss muscular dystrophy (emerin)

Emerin is expressed underneath the nuclear membrane.[166] Emerin appears to interact directly with the nuclear membrane and a protein called BAF. BAF in turn binds to heterochromatin.[167] In addition, a protein called nesprin appears to cross link emerin and lamin A/C, and help to form a subnuclear membrane cytoskeleton.[168] Therefore, it is still not clear whether or not the primary function of emerin is to regulate gene expression through an interaction with chromatin, or to regulate nuclear membrane structure. Mutations in emerin have been linked to X-linked Emery–Dreifuss muscular dystrophy (XL-EDMD).[169] Mutations in the gene encoding emerin tend to either cause the loss of stable protein expression or mislocalization of emerin from the nucleus.[170,171] XL-EDMD is characterized by early contractures (prior to muscle weakness), a slowly developing progressive muscle weakness that effects the proximal upper arms and distal lower limbs, and significant cardiac conduction defects. The cardiac defects are the most serious clinical aspect of XL-EDMD with sudden death being common from complete heart block or other forms of arrythmia.

Longitudinal studies of XL-EDMD patients have been described, and conduction system defects are very common.[172] Conduction defects usually first present as prolonged PR interval and sinus bradycardia. Complete sinoatrial and atrio-ventricular block, atrial flutter or complete paralysis are common, in some cases without any patient-perceived symptoms. In many cases the atria become dilated, and in some cases patients develop dilated cardiomyopathy, heart failure or sudden death.[172,173] Because conduction defects and sudden death are common in XL-EDMD, 24-hour Holter monitoring is recommended, and pacemaker insertion can be life saving.[172] Female carriers of emerin mutations do not have significant skeletal muscle disease, but can present with cardiac conduction defects.[174,175]

Autosomal dominant Emery–Dreifuss muscular dystrophy and limb girdle muscular dystrophy 1b (lamin A/C)

Lamin A and lamin C are broadly expressed, and are proteins produced from the same gene (Lamin A/C, LMNA) but differ in their C terminus.[176] As mentioned above, lamin A/C are components of intermediate filaments that reside underneath the nuclear membrane and form the submembrane cytoskeleton. Mutations in lamin A/C have been described in autosomal dominant Emery–Dreifuss muscular dystrophy (AD-EDMD),[177] and autosmal dominant limb girdle muscular dystrophy 1B (LGMD1B).[178] AD-EDMD is clinically similar to XL-EDMD, with early contractures, humoral-peroneal weakness and cardiac conduction defects. LGMD1B is clinically defined by the absence of early contractures, and slowly progressive limb girdle weakness, with cardiac conduction defects. The complexities of variable clinical phenotypes from mutations in the same gene are confounded by the fact that mutations in LMNA can also cause DCM without muscular dystrophy,[179] familial partial lipodystrophy,[180,181] autosomal recessive Charcot–Marie–Tooth syndrome type 2B1,[182] mandibuloacral dysplasia (MAD),[183] atypical Werner's syndrome,[184] and Hutchinson–Gilford progeria syndrome.[185,186] Although many of these disorders are considered to be clinically distinct, there can be overlapping clinical features. For example, MAD patients also have partial lipodystrophy,[183] and DCM patients with LMNA mutations also have cardiac conduction defects.[179] Recently a mutation in a single family has been

described that has features of Charcot–Marie–Tooth syndrome, muscular dystrophy, heart conduction defects and DCM.[187]

The molecular basis of the different clinical features of LMNA mutations is currently unclear. Originally, it was believed that perhaps different mutations in LMNA affect separate functions or interactions of lamin A/C in a genotype/phenotype-specific manner. However, experimental evidence and sufficient genetic evidence for this hypothesis is lacking. As described above, certain genetic loci can display features of several disorders. Futhermore, LMNA mutants with mutations representing different disease mutations, appear to have no consistent genotype/phenotype dependent effect on levels of protein expression or localization.[188] The most striking example of experimental evidence for the lack of genotype-specific effects on lamin A/C function is a knock-in mouse where a specific mutation was knocked in to the LMNA locus to represent a model of AD-EDMD.[189] Heterozygous mice do not display any evidence of muscular dystrophy like human patients, but homozygous mice display a phenotype consistent with Hutchinson–Gilford progeria syndrome. Therefore, there appears to be a relative continuum of clinical entities that can be caused by lamin A/C mutations. Depending on the severity of the mutation, whether the mutation is homozygous or heterozygous, and probably the genetic background, certain clinical phenotypes are more prominent or apparent than others.[190]

How LMNA mutations cause disease, particularly in skeletal and cardiac muscle is currently unclear. As with emerin, it is unclear whether or not lamin A/C primarily functions in nuclear structure or regulation of gene expression. Knockout mice for lamin A/C have been generated and display a severe EDMD phenotype and die early.[191] Cells from lamin A/C null mice, and some human EDMD patients show some rare evidence of nuclear membrane damage.[191–193] However, lamin A/C is also localized into speckles that appear to be involved in spatial organization of gene transcription and splicing.[194] Alternatively, it may be that the structural and gene regulation hypotheses are not mutually exclusive. Recently, lamin A/C null cells, when subjected to external mechanical strain, show increased nuclear deformation and decreased nuclear factor-κB (NF-κB) transcriptional response or alterations in other transcription factor localization.[195,196] Therefore, both nuclear membrane structure and transcriptional responses to external mechanical strain may be altered by *LMNA* mutations, and may be involved in some of the disease processes in striated muscle where changes in mechanical strain of the cells is frequent.

Cytosolic proteins

Limb girdle muscular dystrophy 2H (TRIM 32)

TRIM32 is a member of the tripartite motif family of proteins of which there are more than 37 family members.[197] Mutations in TRIM family members have been identified in a number of human diseases. TRIM32 is ubiquitously expressed in the cytosol of cells, and is believed to be involved in the ubiquitin–proteosome pathway.[198] Mutations in *TRIM32* have been identified in autosomal recessive LGMD2H, a mildly progressive limb girdle muscular dystrophy primarily within the Manitoba Hutterite population.[199] There is no cardiovascular involvement reported in LGMD2H.

Nucleotide repeat disorders

Nucelotide repeats with varying length and complexity are quite common within the human genome. In some cases the repeats occur within coding regions of genes, but often they reside in non-coding sequence. Several human diseases have been linked to alterations in nucleotide repeats including fragile X syndrome and Huntington's disease.[200] Alterations in nucleotide repeats have also been linked to forms of muscle disease.

Facioscapulohumeral muscular dystrophy

Facioscapulohumeral muscular dystrophy (FSHD) has been linked to deletions of an integral number of tandem 3.2 kb repeats within a region of chromosome 4 known as D4Z4.[201,202] Normal individuals have between 11 and 150 repeats while <11 repeats is associated with FSHD.[203] FSHD is an autosomal dominant disorder that often produces assymetrical weakness affecting the muscles of the face, upper back and arms. The molecular mechanism of FSHD is unclear, but recent evidence suggests that the D4Z4 region might be a transcriptional repressor of nearby genes on chromosome 4, and the repeats may be a binding site for transcriptional repressor factors.[204] Cardiovascular involvement is uncommon in FSHD, but in rare cases clinically significant ECG abnormalities or induced ECG abnormalities have been detected.[205]

Myotonic dystrophy

Myotonic dystrophy is one of the most common forms of muscular dystrophy in humans with a frequency of about 1 in 8000 live births.[206] Two forms of autosomal myotonic dystrophy have been described. Myotonic dystrophy type 1 (DM1) has been linked to a trinucleotide repeat expansion within the 3′ untranslated region of myotonic dystrophy protein kinase gene (*DMPK*).[207–209] In normal individuals, the (CTG)n repeat length is n = 5–34, but raises to n = 50–2000 in DM1. DM1 has a number of clinical symptoms including active myotonia (which may have both muscle and peripheral nerve components), central nervous system defects, cataracts, endocrine abnormalities including diabetes, gastrointestinal tract dysfunction and cardiac involvement.[7] Myotonic dystrophy type 2 (DM2) shares many clinical features with DM1, and has been linked to an expansion of a (CCTG)n repeat within the first intron of the gene encoding zinc finger protein 9.[210]

The cardiovascular complications in myotonic dystrophy are significant, and sometimes the earliest signs of disease.[7] The most common findings in DM1 are cardiac conduction defects in approximately 65% of patients.[211] The mean age of death is 53 years in DM1 patients, with 30% of fatalities being due to cardiovascular events, and 40% due to respiratory failure. Upon ECG monitoring, patients show long PR intervals, wide QRS and late potentials.[212] Ventricular arrythmias, as well as atrial flutter or fibrillation are also quite common, sometimes resulting in clinical symptoms.[213] Malignant ventricular tachycardia may also be a significant cause of sudden cardiac death in DM1 patients, and require treatment in addition to pacemaker implantation.[214] Direct myocardial dysfunction is not frequently observed, but in some cases of cardiac hypertrophy, diastolic dysfunction, ischemic heart disease and mitral valve prolapse have all been described in DM1 patients.[215] A recent large-scale retrospective study suggests approximately 14% of patients have detectable left ventricular systolic dysfunction, while only 2% received a diagnosis of heart failure.[216] Some studies suggest significant coronary vascular abnormalities may precede the development of cardiomyopathy in DM1 patients.[217] Autopsy analysis indicates that the conduction system and ventricular myocytes are both common targets for pathogenic processes.[218] Although not as extensively studied, DM2 patients also have a high prevalence of cardiac conduction defects.[219]

The exact molecular basis of myotonic dystrophy is currently unclear, but the field is developing rapidly. The first myotonic dystrophy hypothesis was that the nucleotide expansion in DM1 may be causing either abnormal expression of DMPK itself, or a heterochromatin-like silencing effect on neighboring genes. The DMPK gene was therefore disrupted in mice,[220,221] Mice developed a mild myopathy with no typical signs of myotonic dystrophy such as skeletal muscle weakness. However, DMPK homozygous knockout mice do have cardiac conduction defects, and the severity of the defect seems to be related to the gene dosage of DMPK, because heterozygous mice have a less severe cardiac conduction defect.[222] To test the potential effects of the expansion on neighboring gene expression, the Six5 gene was knocked out in mice, and heterozygous and homozygous mice have cataracts,[223,224] so DMPK activity and Six5 may play a role in some aspects of myotonic dystrophy. However, this could not explain how an expansion on another chromosome could result in a similar myotonic dystrophy (DM2) and cardiac conduction defects. Most of the recent data to date indicate that a majority of the myotonic dystrophic phenotype can be ascribed to a toxic gain of function of mutant mRNA containing expanded repeats. These repeat-containing mRNAs tend to

cluster within nuclei and sequester nuclear factors, such as a CUG/CCUG-binding protein called muscleblind, that are required for gene transcription regulation and gene splicing machinery.[225–227] Several muscle genes have been shown to be alternatively spliced in myotonic dystrophy such as the insulin receptor and ClC-1 chloride channels.[228,229] In support of this latter hypothesis, transgenic mice overexpressing expanded CUG repeat mRNAs, and the muscle-blind knockout mouse both have many of the features of myotonic dystrophy.[230,231] Furthermore, a transposon insertion in the ClC-1 gene causes myotonia in mice.[232] Therefore, some of the myotonic dystrophy cardiovascular phenotype may be due to a loss of DMPK function, but it is likely that abnormal splicing also plays a role in disease phenotypes and cardiac conduction defects in both DM1 and DM2 patients.

Oculopharyngeal muscular dystrophy

Oculopharyngeal muscular dystrophy (OPMD) is inherited in both an autosomal dominant and recessive pattern, and patients present with pharyngeal weakness and ptosis of the eyelids.[233] OPMD has been linked to trinucleotide expansion within the polyadenylate-binding protein nuclear 1 gene.[234] In this muscular dystrophy, the nucleotide expansion is in the coding region of the gene, and encodes for an expanded polyalanine track of 10 alanines (normal) to 11 alanines (recessive OPMD) or 12–17 alanines (dominant OPMD). Mutant PABPN1 appears to accumulate in the nucleus as intranuclear inclusions, and the expanded polyalanines may cause oligomerization of this protein into filaments.[235] No cardiovascular phenotypes have been reported in OPMD patients.

Acquired forms of cardiomyopathy

Muscular dystrophy genes in many cases appear to provide a genetic explanation for relatively rare forms of human cardiovascular disease. However, muscular dystrophy proteins have also been identified as targets for pathological processes within more common forms of acquired cardiovascular disease.

Enteroviral infections are a common cause of human myocarditis. Enteroviral infections trigger a number of cascades within cells to allow for viral replication. Enteroviral infection, particularly from coxsackievirus B3, has been shown to result in the expression of enteroviral protease 2A that can cleave dystrophin.[236] This dystrophin cleavage results in disruption of the DGC and sarcolemmal integrity of virus-infected cells. Interestingly, when the same coxsackievirus B3 infection is performed in mdx mice, the myocarditis is much worse.[237] Although, this clearly indicates dystrophin is probably not the only pathogenic player in coxsackievirus infection-induced myocarditis, it does suggest dystrophin cleavage may be involved in reducing sarcolemmal integrity to allow viral dissemination and further infection.

In addition, a number of muscular dystrophy proteins appear to be remodeled following ischemia–reperfusion injury. For instance, dystrophin appears to be lost from the sarcolemma following ischemia, and in randomly sampled human heart failure patients.[238,239] This may be due to some downstream effect of ischemia on the cleavage of β-dystroglycan.[240] Some of the changes in DGC protein expression may be explained by the activation of myocardial calpains.[241] Caveolin 3 also changes its localization from the sarcolemma to the cytosol following myocardial ischemia or experimental myocardial infarction.[242] It is not clear if these changes in muscular dystrophy proteins play a causal role in the development of myocardial dysfunction following ischemia, but it is well known that genetic alterations in these proteins can indeed cause cardiovascular phenotypes

Clinical recommendations and therapies for cardiomyopathy in muscular dystrophy patients

Recently, the European Neuromuscular Center (ENMC) has produced a guide for the clinical management of heart disease in muscular dystrophy patients and can be consulted for details.[9] In specific muscular dystrophies, such as myotonic dystrophy, clinical recommendations from other groups have also been recently published.[7,243] Even though the cardiovascular complications in some forms of muscular dystrophy are rare, many dystrophies do not have large natural history studies, and

therefore cannot precisely define the prevalence or risk of cardiovascular disease. Therefore, it is probably warranted that all patients with muscular dystrophy receive a cardiovascular examination (ECG and echocardiography) at the time of diagnosis, especially if the genetic diagnosis is unclear. In addition, the ENMC recommends routine cardiovascular follow-up in nearly all patients with muscular dystrophy at regular intervals: annually for DMD and sarcoglycanopathy patients over age 10, every two years for DMD and sarcoglycanopathy under age 10, and every 5 years for BMD patients or DMD carriers after the age of 16. In cases where cardiac conduction defects are common (XL-EDMD, AD-EDMD/LGMD1B, myotonic dystrophy) rigorous 12-lead ECGs should be performed at diagnosis, and annually afterwards.

As is the case with skeletal muscle disease in muscular dystrophy patients, there are no direct therapies defined that will completely prevent or cure cardiovascular disease in muscular dystrophy patients. Standard care for heart failure treatment applies in muscular dystrophy cases where there is a progressive trend toward cardiomyopathy or heart failure. Angiotensin-converting enzyme (ACE) inhibitors and sometimes β-blockers, depending on symptoms, appear to be beneficial in improving symptoms in DMD/BMB and sarcoglycanopathies. Although verapamil and nicorandil have been used in the sarcoglycanopathy mouse and hamster to prevent cardiomyopathy, no trials of the efficacy of agents, such as calcium channel blockers, targeting coronary vasospasm have been reported in human patients, so their use should be approached with some caution. In cardiomyopathies with conduction defects, there are no clear indications that anti-arrhythmic drugs are effective. Pacemakers are highly recommended in myotonic dystrophy and XL-EDMD when conduction defects are progressive, even if there are no clinical symptoms. However, in laminopathies (AD-EDMD/LGMD1D) pacemakers may not be the method of choice. Dilated cardiomyopathy can occur in laminopathy patients in concert with ventricular conduction defects. Sudden death is often seen even with pacing.[244] Therefore, implantable defibrillators are currently generally recommended in patients with laminopathies. Cardiac transplantation has been performed and is a viable alternative in some cases of BMB, XLDCM and DMD carriers, where cardiovascular symptoms are severe but skeletal muscle function is less affected or normal.[3]

Primarily due to a lack of understanding of the molecular mechanisms underlying inherited cardiomyopathies in muscular dystrophy patients, there are a number of groups trying to develop genetic therapies to directly restore normal proteins to dystrophic skeletal muscle or the heart. Approaches include gene transfer vectors, oligonucleotide-directed genetic repair, and myoblast- or stem cell-mediated cell therapies. There are a few reports using adenovirus or adeno-associated virus to deliver mini-dystrophin genes or normal δ-

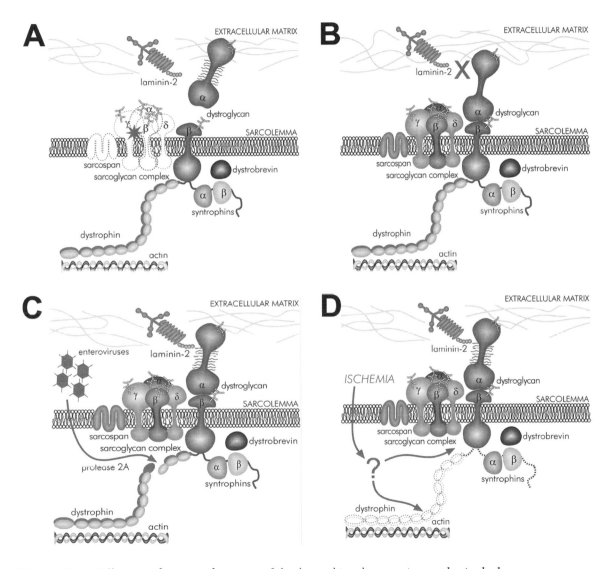

Figure 32.3. *Different pathways to disruption of the dystrophin–glycoprotein complex in the heart. Dystrophin–glycoprotein complex function can be disrupted: (A) by direct mutation of its components, (B) secondary to the loss of enzymes required for dystroglycan function, (C) secondary to enteroviral infection, and (D) secondary to myocardial ischemia–reperfusion injury.*

sarcoglycan to the heart of the mdx mouse or the cardiomyopathic hamster, respectively.[245–247] The preliminary results are encouraging and seem to show some limited or significant effect on restoring sarcolemmal integrity. However, full prevention of functional cardiomyopathy has not been demonstrated. In addition, gene transfer of therapeutic proteins targeting calcium handling or β-adrenergic receptor signaling within cardiac myocytes have been tested in the cardiomyopathic hamster, and seem to be able to improve cardiac function.[248,249] Issues of gene vector delivery, vector serotype, efficiency and efficacy are currently being examined in a number of animal models.

Summary

Clearly, a number of advances in our understanding of the genetic basis of human cardiomyopathy have come from the genetic, molecular and physiological studies on muscular dystrophy patients and mouse models. The complexity of the molecular mechanisms are beginning to be realized as important muscle proteins can be affected by direct genetic mutation, mutations in proteins or mRNAs that bind to or modify proteins (such as glycosylation enzymes), and non-genetic pathways (such as viral infections and changes in extracellular environment due to other systemic diseases) (Fig. 32.3). Hopefully, by understanding these complex molecular pathways, the most critical proteins and pathways that lead to muscular dystrophy and its associated cardiomyopathies will be identified. With advances in patient care, particular respiratory care, muscular dystrophy patients are living much longer today than in the past, and cardiovascular disease will probably become an even more important issue in the clinical care of muscular dystrophy patients.[20] In addition, the molecular understanding of these relatively rare genetic disorders will hopefully define critical pathways that may be involved and targeted in more common acquired forms of muscle and cardiovascular disease.

Acknowledgements

We thank Ms Maria C. Martin for graphic design of all the figures in this chapter. We thank the members of the Campbell laboratory and Dr Robert M Weiss for critical reading of this manuscript. Daniel E Michele was supported by the National Institutes of Health grant F32 HL073601-01. This work was supported in part by the Muscular Dystrophy Association. Kevin P Campbell is an investigator for the Howard Hughes Medical Institute.

References

1. Conte G, Gioja L. Srofola del sistema muscolare. Annali Clinici dell'Ospedale delgi Incurabili di Napoli 1836; 2: 66–79.
2. Kaplan, JC, Fontaine B. Neuromuscular disorders: gene location. Neuromuscular Disorders 2004; 14: 85–106.
3. Finsterer J, Stollberger C. Cardiac involvement in primary myopathies. Cardiology 2000; 94: 1–11.
4. Finsterer J, Stollberger C. The heart in human dystrophinopathies. Cardiology 2003; 99: 1–19.
5. McNally E, Allikian M, Wheeler MT et al. Cytoskeletal defects in cardiomyopathy. J Mol Cell Cardiol 2003; 35: 231–41.
6. Muntoni, F. Cardiomyopathy in muscular dystrophies. Curr Opin Neurol 2003; 16: 577–583.
7. Pelargonio G, Dello Russo A, Sanna T et al. Myotonic dystrophy and the heart. Heart 2002; 88: 665–70.
8. Sachdev B, Elliott PM, McKenna,WJ. Cardiovascular complications of neuromuscular disorders. Curr Treat Options Cardiovasc Med 2002; 4: 171–9.
9. Bushby K, Muntoni F, Bourke JP. 107th ENMC international workshop: the management of cardiac involvement in muscular dystrophy and myotonic dystrophy. 7th–9th June 2002, Naarden, the Netherlands. Neuromuscul Disord 2003; 13: 166–72.
10. Hoffman EP, Brown RH, Jr., Kunkel LM. Dystrophin: the protein product of the Duchenne muscular dystrophy locus. Cell 1987; 51: 919–28.
11. Hoffman EP, Kunkel LM. Dystrophin abnormalities in Duchenne/Becker muscular dystrophy. Neuron 1989; 2: 1019–29.
12. Ahn AH, Kunkel LM. The structural and functional diversity of dystrophin. Nature Genetics 1993; 3: 283–91.
13. Ervasti, JM, Campbell, KP. Membrane organization of the dystrophin-glycoprotein complex. Cell 1991; 66: 1121–31.

14. Campbell KP. Three muscular dystrophies: loss of cytoskeleton-extracellular matrix linkage. Cell 1995; 80: 675–9.

15. Bonilla E, Samitt CE, Miranda AF et al. Duchenne muscular dystrophy: deficiency of dystrophin at the muscle cell surface. Cell 1988; 54: 447–52.

16. Ervasti JM, Ohlendieck K, Kahl SD et al. Deficiency of a glycoprotein component of the dystrophin complex in dystrophic muscle. Nature 1990; 345: 315–19.

17. Barresi R, Moore SA, Stolle CA et al. Expression of gamma-sarcoglycan in smooth muscle and its interaction with the smooth muscle sarcoglycan-sarcospan complex. J Biol Chem 2000; 275: 38554–60.

18. Emery AEH. Duchenne Muscular Dystrophy or Meryon's disease. In: Emery AEH (ed). The Muscular Dystrophies. New York: Oxford University Press; 2001: 55–71.

19. Moriuchi T, Kagawa N, Mukoyama M et al. Autopsy analyses of the muscular dystrophies. Tokushima J Exp Med 1993; 40: 83–93.

20. Eagle M, Baudouin SV, Chandler C et al. Survival in Duchenne muscular dystrophy: improvements in life expectancy since 1967 and the impact of home nocturnal ventilation. Neuromuscul Disord 2002; 12: 926–9.

21. Melacini P, Vianello A, Villanova C et al. Cardiac and respiratory involvement in advanced stage Duchenne muscular dystrophy. Neuromuscul Disord 1996; 6: 367–76.

22. Nigro G, Comi LI, Politano L et al. The incidence and evolution of cardiomyopathy in Duchenne muscular dystrophy. Int J Cardiol 1990; 26: 271–7.

23. Angelini C, Fanin M, Freda MP et al. Prognostic factors in mild dystrophinopathies. J Neurol Sci 1996; 142: 70–8.

24. Hoogerwaard EM, de Voogt WG, Wilde AA et al. Evolution of cardiac abnormalities in Becker muscular dystrophy over a 13-year period. J Neurol 1997; 244: 657–63.

25. Melacini P, Fanin M, Danieli GA et al. Cardiac involvement in Becker muscular dystrophy. J Am Coll Cardiol 1993; 22: 1927–34.

26. Muntoni F, Cau M, Ganau A et al. Brief report: deletion of the dystrophin muscle-promoter region associated with X-linked dilated cardiomyopathy. N Engl J Med 1993; 329: 921–5.

27. Muntoni F, Melis MA, Ganau A et al. Transcription of the dystrophin gene in normal tissues and in skeletal muscle of a family with X-linked dilated cardiomyopathy. Am J Hum Genet 1995; 56: 151–7.

28. Palmucci L, Mongini T, Chiado-Piat L et al. Dystrophinopathy expressing as either cardiomyopathy or Becker dystrophy in the same family. Neurology 2000; 54: 529–30.

29. Milasin J, Muntoni F, Severini GM et al. A point mutation in the 5′ splice site of the dystrophin gene first intron responsible for X-linked dilated cardiomyopathy. Hum Mol Genet 1996; 5: 73–9.

30. Politano L, Nigro V, Nigro G et al. Development of cardiomyopathy in female carriers of Duchenne and Becker muscular dystrophies. JAMA 1996; 275: 1335–8.

31. Perloff JK, Roberts WC, de Leon AC Jr et al. The distinctive electrocardiogram of Duchenne's progressive muscular dystrophy. An electrocardiographic-pathologic correlative study. Am J Med 1967; 42: 179–88.

32. Perloff JK, Henze E, Schelbert HR. Alterations in regional myocardial metabolism, perfusion, and wall motion in Duchenne muscular dystrophy studied by radionuclide imaging. Circulation 1984; 69: 33–42.

33. Miyoshi K. Echocardiographic evaluation of fibrous replacement in the myocardium of patients with Duchenne muscular dystrophy. Br Heart J 1991; 66: 452–5.

34. Sander M, Chavoshan B, Harris SA et al. Functional muscle ischemia in neuronal nitric oxide synthase-deficient skeletal muscle of children with Duchenne muscular dystrophy. Proc Natl Acad Sci USA 2000; 97: 13818–23.

35. Thomas GD, Shaul PW, Yuhanna IS et al. Vasomodulation by skeletal muscle-derived nitric oxide requires alpha-syntrophin-mediated sarcolemmal localization of neuronal nitric oxide synthase. Circ Res 2003; 92: 554–60.

36. Thomas GD, Sander M, Lau KS et al. Impaired metabolic modulation of alpha-adrenergic vasoconstriction in dystrophin-deficient skeletal muscle. Proc Natl Acad Sci USA 1998; 95: 15090–5.

37. Gnecchi-Ruscone T, Taylor J, Mercuri E et al. Cardiomyopathy in Duchenne, Becker, and sarcoglycanopathies: a role for coronary dysfunction? Muscle Nerve 1999; 22: 1549–56.

38. Durbeej M, Campbell KP. Muscular dystrophies involving the dystrophin-glycoprotein complex: an overview of current mouse models. Curr Opin Genet Dev 2002; 12: 349–61.

39. Cohn RD, Durbeej M, Moore SA et al. Prevention of cardiomyopathy in mouse models lacking the smooth muscle sarcoglycan-sarcospan complex. J Clin Invest 2001; 107: R1–7.

40. Danialou G, Comtois AS, Dudley R et al. Dystrophin-deficient cardiomyocytes are abnormally

vulnerable to mechanical stress-induced contractile failure and injury. FASEB J 2001; 15: 1655–7.

41. Kamogawa Y, Biro S, Maeda M et al. Dystrophin-deficient myocardium is vulnerable to pressure over-load in vivo. Cardiovasc Res 2001; 50: 509–15.

42. Megeney LA, Kablar B, Perry RL et al. Severe cardiomyopathy in mice lacking dystrophin and MyoD. Proc Natl Acad Sci USA 1999; 96: 220–5.

43. Hainsey TA, Senapati S, Kuhn DE et al. Cardiomyopathic features associated with muscular dystrophy are independent of dystrophin absence in cardiovasculature. Neuromuscul Disord 2003; 13: 294–302.

44. Crosbie RH, Lim LE, Moore SA et al. Molecular and genetic characterization of sarcospan: insights into sarcoglycan-sarcospan interactions. Hum Mol Genet 2000; 9: 2019–27.

45. Ettinger AJ, Feng G, Sanes JR. Epsilon-sarcoglycan, a broadly expressed homologue of the gene mutated in limb-girdle muscular dystrophy 2D [erratum appears in J Biol Chem 1998; 273: 19922]. J Biol Chem 1997; 272: 32534–8.

46. Straub V, Ettinger AJ, Durbeej M et al. epsilon-sarcoglycan replaces alpha-sarcoglycan in smooth muscle to form a unique dystrophin-glycoprotein complex. J Biol Chem 1999; 274: 27989–96.

47. Wheeler MT, Zarnegar S, McNally EM. Zeta-sarcoglycan, a novel component of the sarcoglycan complex, is reduced in muscular dystrophy. Hum Mol Genet 2002; 11: 2147–54.

48. Piccolo F, Roberds SL, Jeanpierre M et al. Primary adhalinopathy: a common cause of autosomal recessive muscular dystrophy of variable severity [erratum appears in Nat Genet 1995; 11: 104]. Nat Genet 1995; 10: 243–5.

49. Nigro V, de Sa Moreira E, Piluso G et al. Autosomal recessive limb-girdle muscular dystrophy, LGMD2F, is caused by a mutation in the delta-sarcoglycan gene. Nat Genet 1996; 14: 195–8.

50. Roberds SL, Leturcq F, Allamand V et al. Missense mutations in the adhalin gene linked to autosomal recessive muscular dystrophy. Cell 1994; 78: 625–33.

51. Bonnemann CG, Modi R, Noguchi S et al. Beta-sarcoglycan (A3b) mutations cause autosomal recessive muscular dystrophy with loss of the sarcoglycan complex [erratum appears in Nat Genet 1996; 12: 110]. Nat Genet 1995; 11: 266–73.

52. Lim LE, Duclos F, Broux O et al. Beta-sarcoglycan: characterization and role in limb-girdle muscular dystrophy linked to 4q12. Nat Genet 1995; 11: 257–65.

53. Noguchi S, McNally EM, Ben Othmane K et al.

Mutations in the dystrophin-associated protein gamma-sarcoglycan in chromosome 13 muscular dystrophy. Science 1995; 270: 819–22.

54. Fanin M, Melacini P, Boito C et al. LGMD2E patients risk developing dilated cardiomyopathy. Neuromuscul Disord 2003; 13: 303–9.

55. Melacini P, Fanin M, Duggan DJ et al. Heart involvement in muscular dystrophies due to sarcoglycan gene mutations. Muscle Nerve 1999; 22: 473–9.

56. Politano L, Nigro V, Passamano L et al. Evaluation of cardiac and respiratory involvement in sarcoglycanopathies. Neuromuscul Disord 2001; 11: 178–85.

57. Zimprich A, Grabowski M, Asmus F et al. Mutations in the gene encoding epsilon-sarcoglycan cause myoclonus-dystonia syndrome. Nat Genet 2001; 29: 66–9.

58. Coral-Vazquez R, Cohn RD, Moore SA et al. Disruption of the sarcoglycan-sarcospan complex in vascular smooth muscle: a novel mechanism for cardiomyopathy and muscular dystrophy. Cell 1999; 98: 465–74.

59. Durbeej M, Cohn RD, Hrstka RF et al. Disruption of the beta-sarcoglycan gene reveals pathogenetic complexity of limb-girdle muscular dystrophy type 2E. Mol Cell 2000; 5: 141–51.

60. Araishi K, Sasaoka T, Imamura M et al. Loss of the sarcoglycan complex and sarcospan leads to muscular dystrophy in beta-sarcoglycan-deficient mice. Hum Mol Genet 1999; 8: 1589–98.

61. Hack AA, Lam MY, Cordier L et al. Differential requirement for individual sarcoglycans and dystrophin in the assembly and function of the dystrophin-glycoprotein complex. J Cell Sci 2000; 113: 2535–44.

62. Wheeler MT, Korcarz CE, Collins KA et al. Secondary coronary artery vasospasm promotes cardiomyopathy progression. Am J Pathol 2004; 164: 1063–71.

63. Wheeler MT, Allikian MJ, Heydemann A et al. Smooth muscle cell-extrinsic vascular spasm arises from cardiomyocyte degeneration in sarcoglycan-deficient cardiomyopathy. J Clin Invest 2004; 113: 668–75.

64. Cohn RD, Campbell KP. Molecular basis of muscular dystrophies. Muscle Nerve 2000; 23: 1456–71.

65. Mizuno Y, Noguchi S, Yamamoto H et al. Sarcoglycan complex is selectively lost in dystrophic hamster muscle. Am J Pathol 1995; 146: 530–6.

66. Nigro V, Okazaki Y, Belsito A et al. Identification of the Syrian hamster cardiomyopathy gene. Hum Mol Genet 1997; 6: 601–7.

67. Sonnenblick EH, Fein F, Capasso JM et al. Microvascular spasm as a cause of cardiomyopathies and the calcium-blocking agent verapamil as potential primary therapy. Am J Cardiol 1985; 55: 179B–184B.

68. Sakamoto A, Ono K, Abe M et al. Both hypertrophic and dilated cardiomyopathies are caused by mutation of the same gene, delta-sarcoglycan, in hamster: an animal model of disrupted dystrophin-associated glycoprotein complex. Proc Natl Acad Sci USA 1997; 94: 13873–8.

69. Razani B, Woodman SE, Lisanti MP. Caveolae: from cell biology to animal physiology. Pharmacol Rev 2002; 54: 431–67.

70. Song KS, Scherer PE, Tang Z et al. Expression of caveolin-3 in skeletal, cardiac, and smooth muscle cells. Caveolin-3 is a component of the sarcolemma and co-fractionates with dystrophin and dystrophin-associated glycoproteins. J Biol Chem 1996; 271: 15160–5.

71. Hagiwara Y, Nishina Y, Yorifuji H et al. Immunolocalization of caveolin-1 and caveolin-3 in monkey skeletal, cardiac and uterine smooth muscles. Cell Struct Funct 2002; 27: 375–82.

72. Minetti C, Sotgia F, Bruno C et al. Mutations in the caveolin-3 gene cause autosomal dominant limb-girdle muscular dystrophy. Nat Genet 1998; 18: 365–8.

73. Betz RC, Schoser BG, Kasper D et al. Mutations in CAV3 cause mechanical hyperirritability of skeletal muscle in rippling muscle disease. Nat Genet 2001; 28: 218–19.

74. Carbone I, Bruno C, Sotgia F et al. Mutation in the CAV3 gene causes partial caveolin-3 deficiency and hyperCKemia. Neurology 2000; 54: 1373–6.

75. Merlini L, Carbone I, Capanni C et al. Familial isolated hyperCKaemia associated with a new mutation in the caveolin-3 (CAV-3) gene. J Neurol Neurosurg Psychiatry 2002; 73: 65–7.

76. Tateyama M, Aoki M, Nishino I et al. Mutation in the caveolin-3 gene causes a peculiar form of distal myopathy. Neurology 2002; 58: 323–5.

77. Woodman SE, Sotgia F, Galbiati F et al. Caveolinopathies: mutations in caveolin-3 cause four distinct autosomal dominant muscle diseases. Neurology 2004; 62: 538–43.

78. Fischer D, Schroers A, Blumcke I et al. Consequences of a novel caveolin-3 mutation in a large German family. Ann Neurol 2003; 53: 233–41.

79. Kubisch C, Schoser BG, von During M et al. Homozygous mutations in caveolin-3 cause a severe form of rippling muscle disease. Ann Neurol 2003; 53: 512–20.

80. Cagliani R, Bresolin N, Prelle A et al. A CAV3 microdeletion differentially affects skeletal muscle and myocardium. Neurology 2003; 61: 1513–19.

81. Ricker K, Moxley RT, Rohkamm R. Rippling muscle disease. Arch Neurol 1989; 46: 405–8.

82. Koul RL, Chand RP, Chacko A et al. Severe autosomal recessive rippling muscle disease. Muscle Nerve 2001; 24: 1542–7.

83. Hayashi T, Arimura T, Ueda K et al. Identification and functional analysis of a caveolin-3 mutation associated with familial hypertrophic cardiomyopathy. Biochem Biophys Res Commun 2004; 313: 178–84.

84. Woodman SE, Park DS, Cohen AW et al. Caveolin-3 knock-out mice develop a progressive cardiomyopathy and show hyperactivation of the p42/44 MAPK cascade. J Biol Chem 2002; 277: 38988–97.

85. Aravamudan B, Volonte D, Ramani R et al. Transgenic overexpression of caveolin-3 in the heart induces a cardiomyopathic phenotype. Hum Mol Genet 2003; 12: 2777–88.

86. Ohsawa Y, Toko H, Katsura M et al. Overexpression of P104L mutant caveolin-3 in mice develops hypertrophic cardiomyopathy with enhanced contractility in association with increased endothelial nitric oxide synthase activity. Hum Mol Genet 2004; 13: 151–7.

87. Piccolo F, Moore SA, Ford GC et al. Intracellular accumulation and reduced sarcolemmal expression of dysferlin in limb–girdle muscular dystrophies. Ann Neurol 2000; 48: 902–12.

88. Liu, J, Aoki, M, Illa, I et al. Dysferlin, a novel skeletal muscle gene, is mutated in Miyoshi myopathy and limb girdle muscular dystrophy. Nat Genet 1998; 20: 31–6.

89. Cagliani R, Fortunato F, Giorda R et al. Molecular analysis of LGMD-2B and MM patients: identification of novel DYSF mutations and possible founder effect in the Italian population. Neuromuscul Disord 2003; 13: 788–95.

90. Bansal D, Miyake K, Vogel SS et al. Defective membrane repair in dysferlin-deficient muscular dystrophy. Nature 2003; 423: 168–72.

91. Prelle A, Sciacco M, Tancredi L et al. Clinical, morphological and immunological evaluation of six patients with dysferlin deficiency. Acta Neuropathol 2003; 105: 537–42.

92. Bittner RE, Anderson LV, Burkhardt E et al. Dysferlin deletion in SJL mice (SJL-Dysf) defines a natural model for limb girdle muscular dystrophy 2B. Nat Genet 1999; 23: 141–2.

93. Davis DB, Delmonte AJ, Ly CT et al. Myoferlin, a candidate gene and potential modifier of muscular dystrophy. Hum Mol Genet 2000; 9: 217–26.

94. Wiche G, Krepler R, Artlieb U et al. Occurrence and immunolocalization of plectin in tissues. J Cell Biol 1983; 97: 887–901.

95. Pulkkinen L, Smith FJ, Shimizu H et al. Homozygous deletion mutations in the plectin gene (PLEC1) in patients with epidermolysis bullosa simplex associated with late-onset muscular dystrophy. Hum Mol Genet 1996; 5: 1539–46.

96. Smith FJ, Eady RA, Leigh IM et al. Plectin deficiency results in muscular dystrophy with epidermolysis bullosa. Nat Genet 1996; 13: 450–7.

97. Andra K, Lassmann H, Bittner R et al. Targeted inactivation of plectin reveals essential function in maintaining the integrity of skin, muscle, and heart cytoarchitecture. Genes Dev 1997; 11: 3143–56.

98. Petit N, Lescure A, Rederstorff M et al. Selenoprotein N: an endoplasmic reticulum glycoprotein with an early developmental expression pattern. Hum Mol Genet 2003; 12: 1045–53.

99. Moghadaszadeh B, Petit N, Jaillard C et al. Mutations in SEPN1 cause congenital muscular dystrophy with spinal rigidity and restrictive respiratory syndrome. Nat Genet 2001; 29: 17–18.

100. Ferreiro A, Quijano-Roy S, Pichereau C et al. Mutations of the selenoprotein N gene, which is implicated in rigid spine muscular dystrophy, cause the classical phenotype of multiminicore disease: reassessing the nosology of early-onset myopathies. Am J Hum Genet 2002; 71: 739–49.

101. Colver AF, Steer CR, Godman MJ et al. Rigid spine syndrome and fatal cardiomyopathy. Arch Dis Child 1981; 56: 148–51.

102. Mussini JM, Mathe JF, Prost, A et al. [Rigid-spine syndrome in a female patient (author's translation)]. Rev Neurol 1982; 138: 25–37.

103. Niamane R, Birouk N, Benomar A et al. Rigid spine syndrome. Two case-reports. Rev Rhum Engl Ed 1999; 66: 347–50.

104. Spranger M, Spranger S, Ziegan J et al. Three familial cases presenting with an immobile spine. Rigid spine or Emery-Dreifuss syndrome? Clin Genet 1996; 50: 229–31.

105. Beck MA, Levander OA, Handy J. Selenium deficiency and viral infection. J Nutr 2003; 133: 1463S–1467S.

106. Ge K, Xue A, Bai J et al. Keshan disease-an endemic cardiomyopathy in China. Virchows Arch A Pathol Anat Histopathol 1983; 401: 1–15.

107. Beck MA, Kolbeck PC, Rohr LH et al. Benign human enterovirus becomes virulent in selenium-deficient mice. J Med Virol 1994; 43: 166–70.

108. Beck MA. Rapid genomic evolution of a non-virulent coxsackievirus B3 in selenium-deficient mice. Biomed Environ Sci 1997; 10: 307–15.

109. Ervasti JM, Campbell KP. A role for the dystrophin-glycoprotein complex as a transmembrane linker between laminin and actin. J Cell Biol 1993; 122: 809–23.

110. Michele DE, Campbell KP. Dystrophin-glycoprotein complex: post-translational processing and dystroglycan function. Journal of Biological Chemistry 2003; 278: 15457–60.

111. Muntoni F, Brockington M, Torelli S et al. Defective glycosylation in congenital muscular dystrophies. Curr Opin Neurol 2004; 17: 205–9.

112. Michele DE, Barresi R, Kanagawa M et al. Post-translational disruption of dystroglycan-ligand interactions in congenital muscular dystrophies. Nature 2002; 418: 417–22.

113. Williamson RA, Henry MD, Daniels KJ et al. Dystroglycan is essential for early embryonic development: disruption of Reichert's membrane in Dag1-null mice. Hum Mol Genet 1997; 6: 831–41.

114. Cohn RD, Henry MD, Michele DE et al. Disruption of DAG1 in differentiated skeletal muscle reveals a role for dystroglycan in muscle regeneration. Cell 2002; 110: 639–48.

115. Cote PD, Moukhles H, Lindenbaum M et al. Chimaeric mice deficient in dystroglycans develop muscular dystrophy and have disrupted myoneural synapses. Nat Genet 1999; 23: 338–42.

116. Moore SA, Saito F, Chen J et al. Deletion of brain dystroglycan recapitulates aspects of congenital muscular dystrophy. Nature 2002; 418: 422–5.

117. Kobayashi K, Nakahori Y, Miyake M et al. An ancient retrotransposal insertion causes Fukuyama-type congenital muscular dystrophy. Nature 1998; 394: 388–92.

118. Aravind L, Koonin EV. The fukutin protein family—predicted enzymes modifying cell-surface molecules. Curr Biol 1999; 9: R836–7.

119. Esapa CT, Benson MA, Schroder JE et al. Functional requirements for fukutin-related protein in the Golgi apparatus. Hum Mol Genet 2002; 11: 3319–31.

120. Hayashi YK, Ogawa M, Tagawa K et al. Selective deficiency of alpha-dystroglycan in Fukuyama-type congenital muscular dystrophy. Neurology 2001; 57: 115–21.

121. Miura K, Shirasawa H. Congenital muscular dystrophy of the Fukuyama type (FCMD) with severe

myocardial fibrosis. A case report with postmortem angiography. Acta Pathol Jpn 1987; 37: 1823–35.

122. Itoh M, Houdou S, Kawahara H et al. Morphological study of the brainstem in Fukuyama type congenital muscular dystrophy. Pediatr Neurol 1996; 15: 327–31.

123. Takeda S, Kondo M, Sasaki J et al. Fukutin is required for maintenance of muscle integrity, cortical histiogenesis and normal eye development. Hum Mol Genet 2003; 12: 1449–59.

124. Brockington M, Yuva Y, Prandini P et al. Mutations in the fukutin-related protein gene (FKRP) identify limb girdle muscular dystrophy 2I as a milder allelic variant of congenital muscular dystrophy MDC1C. Hum Mol Genet 2001; 10: 2851–9.

125. Brockington M, Blake DJ, Prandini P et al. Mutations in the fukutin-related protein gene (FKRP) cause a form of congenital muscular dystrophy with secondary laminin alpha2 deficiency and abnormal glycosylation of alpha-dystroglycan. Am J Hum Genet 2001; 69: 1198–209.

126. Brown SC, Torelli S, Brockington M et al. Abnormalities in alpha-dystroglycan expression in MDC1C and LGMD2I muscular dystrophies. Am J Pathol 2004; 164: 727–37.

127. Topaloglu H, Brockington M, Yuva Y et al. FKRP gene mutations cause congenital muscular dystrophy, mental retardation, and cerebellar cysts. Neurology 2003; 60: 988–92.

128. Mercuri E, Brockington M, Straub V et al. Phenotypic spectrum associated with mutations in the fukutin-related protein gene. Ann Neurol 2003; 53: 537–42.

129. Poppe M, Cree L, Bourke J et al. The phenotype of limb-girdle muscular dystrophy type 2I. Neurology 2003; 60: 1246–51.

130. Voit T. Congenital muscular dystrophies: 1997 update. Brain Dev 1998; 20: 65–74.

131. Cormand B, Pihko H, Bayes M et al. Clinical and genetic distinction between Walker–Warburg syndrome and muscle-eye-brain disease. Neurology 2001; 56: 1059–69.

132. Beltran-Valero de Bernabe D, Currier S, Steinbrecher A et al. Mutations in the O-mannosyltransferase gene POMT1 give rise to the severe neuronal migration disorder Walker–Warburg syndrome. Am J Hum Genet 2002; 71: 1033–43.

133. Yoshida A, Kobayashi K, Manya H et al. Muscular dystrophy and neuronal migration disorder caused by mutations in a glycosyltransferase, POMGnT1. Dev Cell 2001; 1: 717–24.

134. Chiba A, Matsumura K, Yamada H et al. Structures of sialylated O-linked oligosaccharides of bovine peripheral nerve alpha-dystroglycan. The role of a novel O-mannosyl-type oligosaccharide in the binding of alpha-dystroglycan with laminin. J Biol Chem 1997; 272: 2156–62.

135. Kim D-S, Hayashi YK, Matsumoto H et al. POMT1 mutation results in defective glycosylation and loss of laminin-binding activity in α-DG. Neurology 2004; 62: 1009–11.

136. Grewal PK, Holzfeind PJ, Bittner RE et al. Mutant glycosyltransferase and altered glycosylation of alpha-dystroglycan in the myodystrophy mouse. Nat Genet 2001; 28: 151–54.

137. Longman C, Brockington M, Torelli S et al. Mutations in the human LARGE gene cause MDC1D, a novel form of congenital muscular dystrophy with severe mental retardation and abnormal glycosylation of alpha-dystroglycan. Hum Mol Genet 2003; 12: 2853–61.

138. Holzfeind PJ, Grewal PK, Reitsamer HA et al. Skeletal, cardiac and tongue muscle pathology, defective retinal transmission, and neuronal migration defects in the Large(myd) mouse defines a natural model for glycosylation-deficient muscle-eye-brain disorders. Hum Mol Genet 2002; 11: 2673–87.

139. Colognato H, Yurchenco PD. Form and function: the laminin family of heterotrimers. Dev Dyn 2000; 218: 213–34.

140. Helbling-Leclerc A, Zhang X, Topaloglu H et al. Mutations in the laminin alpha 2-chain gene (LAMA2) cause merosin-deficient congenital muscular dystrophy. Nat Genet 1995; 11: 216–18.

141. Gilhuis HJ, ten Donkelaar HJ, Tanke RB et al. Nonmuscular involvement in merosin-negative congenital muscular dystrophy. Pediatr Neurol 2002; 26: 30–36.

142. Rash SM, Wanitkin S, Shiota T, Sahn DJ, Pillers DM. Congenital muscular dystrophy mouse model dy/dy has hypertrophic cardiomyopathy by echocardiography. J Invest Med 1998; 46: 107A.

143. Jobsis GJ, Keizers H, Vreijling JP et al. Type VI collagen mutations in Bethlem myopathy, an autosomal dominant myopathy with contractures. Nat Genet 1996; 14: 113–15.

144. Camacho Vanegas O, Bertini E, Zhang RZ et al. Ullrich scleroatonic muscular dystrophy is caused by recessive mutations in collagen type VI. Proc Natl Acad Sci USA 2001; 98: 7516–21.

145. de Visser M, de Voogt WG, la Riviere GV. The heart in Becker muscular dystrophy, facioscapulohumeral dystrophy, and Bethlem myopathy. Muscle Nerve 1992; 15: 591–6.

146. Bonaldo P, Braghetta P, Zanetti M et al. Collagen VI deficiency induces early onset myopathy in the mouse: an animal model for Bethlem myopathy. Hum Mol Genet 1998; 7: 2135–40.

147. Chien KR. Genotype, phenotype: upstairs, downstairs in the family of cardiomyopathies. J Clin Invest 2003; 111: 175–8.

148. Michele DE, Metzger JM. Physiological consequences of tropomyosin mutations associated with cardiac and skeletal myopathies. J Mol Med 2000; 78: 543–53.

149. Fananapazir L, Dalakas MC, Cyran F et al. Missense mutations in the beta-myosin heavy-chain gene cause central core disease in hypertrophic cardiomyopathy. Proc Natl Acad Sci USA 1993; 90: 3993–7.

150. Gregorio CC, Granzier H, Sorimachi H et al. Muscle assembly: a titanic achievement? Curr Opin Cell Biol 1999; 11: 18–25.

151. Gerull B, Gramlich M, Atherton J et al. Mutations of TTN, encoding the giant muscle filament titin, cause familial dilated cardiomyopathy. Nat Genet 2002; 30: 201–4.

152. Hackman P, Vihola A, Haravuori H et al. Tibial muscular dystrophy is a titinopathy caused by mutations in TTN, the gene encoding the giant skeletal-muscle protein titin. Am J Hum Genet 2002; 71: 492–500.

153. Garvey SM, Rajan C, Lerner AP et al. The muscular dystrophy with myositis (mdm) mouse mutation disrupts a skeletal muscle-specific domain of titin. Genomics 2002; 79: 146–9.

154. Hackman JP, Vihola AK, Udd AB. The role of titin in muscular disorders. Ann Med 2003; 35: 434–41.

155. Haravuori H, Vihola A, Straub V et al. Secondary calpain3 deficiency in 2q-linked muscular dystrophy: titin is the candidate gene. Neurology 2001; 56: 869–77.

156. Sorimachi H, Kinbara K, Kimura S et al. Muscle-specific calpain, p94, responsible for limb girdle muscular dystrophy type 2A, associates with connectin through IS2, a p94-specific sequence. J Biol Chem 1995; 270: 31158–62.

157. Richard I, Broux O, Allamand V et al. Mutations in the proteolytic enzyme calpain 3 cause limb-girdle muscular dystrophy type 2A. Cell 1995; 81: 27–40.

158. Fougerousse F, Anderson LV, Delezoide AL et al. Calpain3 expression during human cardiogenesis. Neuromuscul Disord 2000; 10: 251–56.

159. Faulkner G, Lanfranchi G, Valle G. Telethonin and other new proteins of the Z-disc of skeletal muscle. IUBMB Life 2001; 51: 275–82.

160. Moreira ES, Wiltshire TJ, Faulkner G et al. Limb-girdle muscular dystrophy type 2G is caused by muta-

tions in the gene encoding the sarcomeric protein telethonin. Nat Genet 2000; 24: 163–6.

161. Knoll R, Hoshijima M, Hoffman HM et al. The cardiac mechanical stretch sensor machinery involves a Z disc complex that is defective in a subset of human dilated cardiomyopathy. Cell 2002; 111: 943–55.

162. Geier C, Perrot A, Ozcelik C et al. Mutations in the human muscle LIM protein gene in families with hypertrophic cardiomyopathy. Circulation 2003; 107: 1390–5.

163. Salmikangas P, van der Ven PF, Lalowski M et al. Myotilin, the limb-girdle muscular dystrophy 1A (LGMD1A) protein, cross-links actin filaments and controls sarcomere assembly. Hum Mol Genet 2003; 12: 189–203.

164. Hauser MA, Horrigan SK, Salmikangas P et al. Myotilin is mutated in limb girdle muscular dystrophy 1A. Hum Mol Genet 2000; 9: 2141–7.

165. Gilchrist JM, Pericak-Vance M, Silverman L et al. Clinical and genetic investigation in autosomal dominant limb-girdle muscular dystrophy. Neurology 1988; 38: 5–9.

166. Bengtsson L, Wilson KL. Multiple and surprising new functions for emerin, a nuclear membrane protein. Curr Opin Cell Biol 2004; 16: 73–9.

167. Haraguchi T, Koujin T, Segura-Totten M et al. BAF is required for emerin assembly into the reforming nuclear envelope. J Cell Sci 2001; 114: 4575–85.

168. Mislow JM, Holaska JM, Kim MS et al. Nesprin-1alpha self-associates and binds directly to emerin and lamin A in vitro. FEBS Lett 2002; 525: 135–40.

169. Bione S, Maestrini E, Rivella S et al. Identification of a novel X-linked gene responsible for Emery-Dreifuss muscular dystrophy. Nat Genet 1994; 8: 323–7.

170. Nagano A, Koga R, Ogawa M et al. Emerin deficiency at the nuclear membrane in patients with Emery-Dreifuss muscular dystrophy. Nat Genet 1996; 12: 254–9.

171. Ellis JA, Craxton M, Yates JR et al. Aberrant intracellular targeting and cell cycle-dependent phosphorylation of emerin contribute to the Emery–Dreifuss muscular dystrophy phenotype. J Cell Sci 1998; 111: 781–92.

172. Boriani G, Gallina M, Merlini L et al. Clinical relevance of atrial fibrillation/flutter, stroke, pacemaker implant, and heart failure in Emery-Dreifuss muscular dystrophy: a long-term longitudinal study. Stroke 2003; 34: 901–8.

173. Merlini L, Granata C, Dominici P et al. Emery-Dreifuss muscular dystrophy: report of five cases in a family and review of the literature. Muscle Nerve 1986; 9: 481–5.

174. Emery AE. X-linked muscular dystrophy with early contractures and cardiomyopathy (Emery–Dreifuss type). Clin Genet 1987; 32: 360–7.

175. Fishbein MC, Siegel RJ, Thompson CE et al. Sudden death of a carrier of X-linked Emery–Dreifuss muscular dystrophy. Ann Intern Med 1993; 119: 900–5.

176. Worman HJ, Courvalin JC. The nuclear lamina and inherited disease. Trends Cell Biol 2002; 12: 591–8.

177. Bonne G, Di Barletta MR, Varnous S et al. Mutations in the gene encoding lamin A/C cause autosomal dominant Emery–Dreifuss muscular dystrophy. Nat Genet 1999; 21: 285–8.

178. Muchir A, Bonne G, van der Kooi AJ et al. Identification of mutations in the gene encoding lamins A/C in autosomal dominant limb girdle muscular dystrophy with atrioventricular conduction disturbances (LGMD1B). Hum Mol Genet 2000; 9: 1453–9.

179. Fatkin D, MacRae C, Sasaki T et al. Missense mutations in the rod domain of the lamin A/C gene as causes of dilated cardiomyopathy and conduction-system disease. N Engl J Med 1999; 341: 1715–24.

180. Cao H, Hegele RA. Nuclear lamin A/C R482Q mutation in canadian kindreds with Dunnigan-type familial partial lipodystrophy. Hum Mol Genet 2000; 9: 109–12.

181. Shackleton S, Lloyd DJ, Jackson SN et al. LMNA, encoding lamin A/C, is mutated in partial lipodystrophy. Nat Genet 2000; 24: 153–6.

182. De Sandre-Giovannoli A, Chaouch M, Kozlov S et al. Homozygous defects in LMNA, encoding lamin A/C nuclear-envelope proteins, cause autosomal recessive axonal neuropathy in human (Charcot-Marie-Tooth disorder type 2) and mouse. Am J Hum Genet 2002; 70: 726–36.

183. Novelli G, Muchir A, Sangiuolo F et al. Mandibuloacral dysplasia is caused by a mutation in LMNA-encoding lamin A/C. Am J Hum Genet 2002; 71: 426–31.

184. Chen L, Lee L, Kudlow BA et al. LMNA mutations in atypical Werner's syndrome. Lancet 2003; 362: 440–5.

185. De Sandre-Giovannoli A, Bernard R, Cau P et al. Lamin a truncation in Hutchinson–Gilford progeria. Science 2003; 300: 2055.

186. Eriksson M, Brown WT, Gordon LB et al. Recurrent de novo point mutations in lamin A cause Hutchinson-Gilford progeria syndrome. Nature 2003; 423: 293–8.

187. Goizet C, Yaou RB, Demay L et al. A new mutation of the lamin A/C gene leading to autosomal dominant axonal neuropathy, muscular dystrophy, cardiac disease, and leuconychia. J Med Genet 2004; 41:e29.

188. Ostlund C, Bonne G, Schwartz K et al. Properties of lamin A mutants found in Emery–Dreifuss muscular dystrophy, cardiomyopathy and Dunnigan-type partial lipodystrophy. J Cell Sci 2001; 114: 4435–45.

189. Mounkes LC, Kozlov S, Hernandez L et al. A progeroid syndrome in mice is caused by defects in A-type lamins. Nature 2003; 423: 298–301.

190. Bonne G, Levy N. LMNA mutations in atypical Werner's syndrome. Lancet 2003; 362: 1585–6; author reply 1586.

191. Sullivan T, Escalante-Alcalde D, Bhatt H et al. Loss of A-type lamin expression compromises nuclear envelope integrity leading to muscular dystrophy. J Cell Biol 1999; 147: 913–20.

192. Favreau C, Dubosclard E, Ostlund C et al. Expression of lamin A mutated in the carboxyl-terminal tail generates an aberrant nuclear phenotype similar to that observed in cells from patients with Dunnigan-type partial lipodystrophy and Emery–Dreifuss muscular dystrophy. Exp Cell Res 2003; 282: 14–23.

193. Vigouroux C, Auclair M, Dubosclard E et al. Nuclear envelope disorganization in fibroblasts from lipodystrophic patients with heterozygous R482Q/W mutations in the lamin A/C gene. J Cell Sci 2001; 114: 4459–68.

194. Kumaran RI, Muralikrishna B, Parnaik VK. Lamin A/C speckles mediate spatial organization of splicing factor compartments and RNA polymerase II transcription. J Cell Biol 2002; 159: 783–93.

195. Lammerding J, Schulze PC, Takahashi T et al. Lamin A/C deficiency causes defective nuclear mechanics and mechanotransduction. J Clin Invest 2004; 113: 370–8.

196. Nikolova V, Leimena C, McMahon AC et al. Defects in nuclear structure and function promote dilated cardiomyopathy in lamin A/C-deficient mice. J Clin Invest 2004; 113: 357–369.

197. Reymond A, Meroni G, Fantozzi A et al. The tripartite motif family identifies cell compartments. Embo J 2001; 20: 2140–51.

198. Freemont PS. RING for destruction? Curr Biol 2000; 10: R84–87.

201. Frosk P, Weiler T, Nylen E et al. Limb-girdle muscular dystrophy type 2H associated with 199ation in TRIM32, a putative E3-ubiquitin-ligase gene. Am J Hum Genet 2002; 70: 663–72.

200. Sutherland GR, Richards RI. Simple tandem DNA repeats and human genetic disease. Proc Natl Acad Sci USA 1995; 92: 3636–41.

201. Wijmenga C, Frants RR, Hewitt JE et al. Molecular genetics of facioscapulohumeral muscular dystrophy. Neuromuscul Disord 1993; 3: 487–91.

202. Wijmenga C, Hewitt JE, Sandkuijl LA et al. Chromosome 4q DNA rearrangements associated with facioscapulohumeral muscular dystrophy. Nat Genet 1992; 2: 26–30.

203. Lunt PW. 44th ENMC International Workshop: Facioscapulohumeral Muscular Dystrophy: Molecular Studies 19–21 July 1996, Naarden, The Netherlands. Neuromuscul Disord 1998; 8: 126–30.

204. Gabellini D, Green MR, Tupler R. Inappropriate gene activation in FSHD: a repressor complex binds a chromosomal repeat deleted in dystrophic muscle. Cell 2002; 110: 339–48.

205. Laforet P, de Toma C, Eymard B et al. Cardiac involvement in genetically confirmed facioscapulohumeral muscular dystrophy. Neurology 1998; 51: 1454–6.

206. Mankodi A, Thornton CA. Myotonic syndromes. Curr Opin Neurol 2002; 15: 545–52.

207. Fu YH, Pizzuti A, Fenwick RG, Jr. et al. An unstable triplet repeat in a gene related to myotonic muscular dystrophy. Science 1992; 255: 1256–8.

208. Brook JD, McCurrach ME, Harley HG et al. Molecular basis of myotonic dystrophy: expansion of a trinucleotide (CTG) repeat at the 3′ end of a transcript encoding a protein kinase family member. Cell 1992; 68: 799–808.

209. Mahadevan M, Tsilfidis C, Sabourin L et al. Myotonic dystrophy mutation: an unstable CTG repeat in the 3′ untranslated region of the gene. Science 1992; 255: 1253–5.

210. Liquori CL, Ricker K, Moseley ML et al. Myotonic dystrophy type 2 caused by a CCTG expansion in intron 1 of ZNF9. Science 2001; 293: 864–7.

211. Mathieu J, Allard P, Potvin L et al. A 10-year study of mortality in a cohort of patients with myotonic dystrophy. Neurology 1999; 52: 1658–62.

212. Olofsson BO, Forsberg H, Andersson S et al. Electrocardiographic findings in myotonic dystrophy. Br Heart J 1988; 59: 47–52.

213. Lazarus A, Varin J, Ounnoughene Z et al. Relationships among electrophysiological findings and clinical status, heart function, and extent of DNA mutation in myotonic dystrophy. Circulation 1999; 99: 1041–6.

214. Merino JL, Carmona JR, Fernandez-Lozano I et al. Mechanisms of sustained ventricular tachycardia in myotonic dystrophy: implications for catheter ablation. Circulation 1998; 98: 541–6.

215. Fragola PV, Calo L, Luzi M et al. Doppler echocardiographic assessment of left ventricular diastolic function in myotonic dystrophy. Cardiology 1997; 88: 498–502.

216. Bhakta D, Lowe MR, Groh WJ. Prevalence of structural cardiac abnormalities in patients with myotonic dystrophy type I. Am Heart J 2004; 147: 224–7.

217. Annane D, Merlet P, Radvanyi H et al. Blunted coronary reserve in myotonic dystrophy. An early and gene-related phenomenon. Circulation 1996; 94: 973–7.

218. Nguyen HH, Wolfe JT, 3rd, Holmes DR, Jr. et al. Pathology of the cardiac conduction system in myotonic dystrophy: a study of 12 cases. J Am Coll Cardiol 1988; 11: 662–71.

219. Flachenecker P, Schneider C, Cursiefen S et al. Assessment of cardiovascular autonomic function in myotonic dystrophy type 2 (DM2/PROMM). Neuromuscul Disord 2003; 13: 289–93.

220. Reddy S, Smith DB, Rich MM et al. Mice lacking the myotonic dystrophy protein kinase develop a late onset progressive myopathy. Nat Genet 1996; 13: 325–35.

221. Jansen G, Groenen PJ, Bachner D et al. Abnormal myotonic dystrophy protein kinase levels produce only mild myopathy in mice. Nat Genet 1996; 13: 316–24.

222. Berul CI, Maguire CT, Aronovitz MJ et al. DMPK dosage alterations result in atrioventricular conduction abnormalities in a mouse myotonic dystrophy model. J Clin Invest 1999; 103: R1–7.

223. Sarkar PS, Appukuttan B, Han J et al. Heterozygous loss of Six5 in mice is sufficient to cause ocular cataracts. Nat Genet 2000; 25: 110–4.

224. Klesert TR, Cho DH, Clark JI et al. Mice deficient in Six5 develop cataracts: implications for myotonic dystrophy. Nat Genet 2000; 25: 105–9.

225. Fardaei M, Rogers MT, Thorpe HM et al. Three proteins, MBNL, MBLL and MBXL, co-localize in vivo with nuclear foci of expanded-repeat transcripts in DM1 and DM2 cells. Hum Mol Genet 2002; 11: 805–14.

226. Mankodi A, Urbinati CR, Yuan QP et al. Muscleblind localizes to nuclear foci of aberrant RNA in myotonic dystrophy types 1 and 2. Hum Mol Genet 2001; 10: 2165–70.

227. Miller JW, Urbinati CR, Teng-Umnuay P et al. Recruitment of human muscleblind proteins to (CUG)(n) expansions associated with myotonic dystrophy. EMBO J 2000; 19: 4439–48.

228. Mankodi A, Takahashi MP, Jiang H et al. Expanded CUG repeats trigger aberrant splicing of ClC-1 chloride channel pre-mRNA and hyperexcitability of

skeletal muscle in myotonic dystrophy. Mol Cell 2002; 10: 35–44.

229. Charlet BN, Savkur RS, Singh G et al. Loss of the muscle-specific chloride channel in type 1 myotonic dystrophy due to misregulated alternative splicing. Mol Cell 2002; 10: 45–53.

230. Mankodi A, Logigian E, Callahan L et al. Myotonic dystrophy in transgenic mice expressing an expanded CUG repeat. Science 2000; 289: 1769–73.

231. Kanadia RN, Johnstone KA, Mankodi A et al. A muscleblind knockout model for myotonic dystrophy. Science 2003; 302: 1978–80.

232. Steinmeyer K, Klocke R, Ortland C et al. Inactivation of muscle chloride channel by transposon insertion in myotonic mice. Nature 1991; 354: 304–8.

233. Brais B. Oculopharyngeal muscular dystrophy: a late-onset polyalanine disease. Cytogenet Genome Res 2003; 100: 252–60.

234. Brais B, Bouchard JP, Xie YG et al. Short GCG expansions in the PABP2 gene cause oculopharyngeal muscular dystrophy. Nat Genet 1998; 18: 164–7.

235. Fan X, Dion P, Laganiere J et al. Oligomerization of polyalanine expanded PABPN1 facilitates nuclear protein aggregation that is associated with cell death. Hum Mol Genet 2001; 10: 2341–51.

236. Badorff C, Lee GH, Lamphear BJ et al. Enteroviral protease 2A cleaves dystrophin: evidence of cytoskeletal disruption in an acquired cardiomyopathy. Nat Med 1999; 5: 320–6.

237. Xiong D, Lee GH, Badorff C et al. Dystrophin deficiency markedly increases enterovirus-induced cardiomyopathy: a genetic predisposition to viral heart disease. Nat Med 2002; 8: 872–7.

238. Armstrong SC, Latham CA, Shivell CL et al. Ischemic loss of sarcolemmal dystrophin and spectrin: correlation with myocardial injury. J Mol Cell Cardiol 2001; 33: 1165–79.

239. Vatta M, Stetson SJ, Perez-Verdia A et al. Molecular remodelling of dystrophin in patients with end-stage cardiomyopathies and reversal in patients on assistance-device therapy. Lancet 2002; 359: 936–41.

240. Armstrong SC, Latham CA, Ganote CE. An ischemic beta-dystroglycan (betaDG) degradation product: correlation with irreversible injury in adult rabbit cardiomyocytes. Mol Cell Biochem 2003; 242: 71–9.

241. Yoshida H, Takahashi M, Koshimizu M et al. Decrease in sarcoglycans and dystrophin in failing heart following acute myocardial infarction. Cardiovasc Res 2003; 59: 419–27.

242. Ratajczak P, Damy T, Heymes C et al. Caveolin-1 and -3 dissociations from caveolae to cytosol in the heart during aging and after myocardial infarction in rat. Cardiovasc Res 2003; 57: 358–69.

243. Lazarus A, Varin J, Babuty D et al. Long-term follow-up of arrhythmias in patients with myotonic dystrophy treated by pacing: a multicenter diagnostic pacemaker study. J Am Coll Cardiol 2002; 40: 1645–52.

244. Becane HM, Bonne G, Varnous S et al. High incidence of sudden death with conduction system and myocardial disease due to lamins A and C gene mutation. Pacing Clin Electrophysiol 2000; 23: 1661–6.

245. Yue Y, Li Z, Harper SQ et al. Microdystrophin gene therapy of cardiomyopathy restores dystrophin-glycoprotein complex and improves sarcolemma integrity in the mdx mouse heart. Circulation 2003; 108: 1626–32.

246. Li J, Wang D, Qian S et al. Efficient and long-term intracardiac gene transfer in delta-sarcoglycan-deficiency hamster by adeno-associated virus-2 vectors. Gene Ther 2003; 10: 1807–13.

247. Ikeda Y, Gu Y, Iwanaga Y et al. Restoration of deficient membrane proteins in the cardiomyopathic hamster by in vivo cardiac gene transfer. Circulation 2002; 105: 502–8.

248. Hoshijima M, Ikeda Y, Iwanaga Y et al. Chronic suppression of heart-failure progression by a pseudophosphorylated mutant of phospholamban via in vivo cardiac rAAV gene delivery. Nat Med 2002; 8: 864–71.

249. Tomiyasu K, Oda Y, Nomura M et al. Direct intra-cardiomuscular transfer of beta2-adrenergic receptor gene augments cardiac output in cardiomyopathic hamsters. Gene Ther 2000; 7: 2087–93.

CHAPTER 33

Complex cardiomyopathies: laminopathies and arrhythmogenic right ventricular dysplasia/ cardiomyopathy

Matthew RG Taylor, Carl V Barnes, Luisa Mestroni

Introduction

Dilated cardiomyopathy (DCM) may present with either a pure and isolated myocardial involvement or with a complex DCM phenotype.[1] In the pure form, the pathological features are dominated by ventricular dilation and systolic dysfunction and are frequently the result of mutations in various cytoskeletal and sarcomeric genes, such as *ACTC*, *DES*, *DMD*, *MYH7*, *SGCD*, *TNNT2*, *TTN* (reviewed further in Chapter 31). Complex forms of DCM, complicated by additional problems including cardiac conduction disease, arrhythmia and skeletal muscular dystrophy, also exist. Many of these have been associated with mutations in the gene encoding lamin A/C (*LMNA*).[2–4] Lamins A and C are key components of the nuclear cytoskeleton, and have important roles in maintaining nuclear envelope architecture. Mutations in the emerin gene (*EMD*) may also lead to DCM and arrhythmia,[5,6] usually in the setting of pedigrees manifesting X-linked patterns of inheritance.

Cardiomyopathy and arrhythmia coexist in another syndrome characterized by predominantly right ventricular involvement and high risks of sudden death in the young,[7] arrhythmogenic right ventricular dysplasia/cardiomyopathy (ARVD/C). This disease has gained attention due to suspicions that it is currently under-diagnosed,[8] and because of the recent identification of causative mutations in an ion channel gene, the cardiac ryanodine

receptor gene (*RYR2*),[9] and in desmosomal genes, desmoplakin, plakoglobin and plakophilin-2 (encoded by *DSP*, *JUP*, and *PKP2*, respectively).[10]

The focus of this chapter is on the discussion of these two principal forms of cardimyopathy complicated by cardiac conduction and/or arrhythmia abnormalities: laminopathies and ARVD/C. The continued molecular genetic research into these and other forms of cardiomyopathies, including the evaluation of animal models, will probably lead to the identification of additional genes responsible for these diseases. From this foundation it is hoped that a more complete understanding of the underlying pathogenic mechanisms leading to these complex phenotypes will be possible. The development of clinical molecular genetic testing for mutations in the genes involved in these diseases is upon us, and the integration of genetic counseling and these new diagnostic tests into clinical cardiology is currently under way.

The laminopathies

Initial identification of lamin A/C gene mutations in DCM

Early studies of the familial cardiomyopathies involved linkage analysis of suitably informative families. From this approach, Kass et al. studied a large seven-generation kindred with autosomal dominant (AD) DCM complicated by conduction system disease.[11] A genome-wide linkage study was undertaken with a high logarithm of the odd (LOD) score (maximum LOD = 13.2) present in the centromeric region of chromosome 1p1–1q1.[11] Subsequent data confirmed that the 1p1–2q1 region contributed to DCM and conduction system disease in additional families.[3] Concurrently, interest in the AD form of Emery–Dreifuss muscular dystrophy (EDMD) [OMIM#181350] led to the discovery that mutations in the gene encoding lamin A/C (*LMNA*) resulted in an AD-EDMD phenotype of humeroperoneal muscular dystrophy, joint contractures, and later-onset cardiac disease.[12] The cardiac disease was of interest, as both conduction system disease and ventricular dilation and dysfunction were observed. Armed with the reproducibility of the cardiac linkage data suggesting a major DCM locus was located at 1p1-1q1, the similarities between the EDMD cardiac disease, and that observed in several DCM conduction-system disease families, and the localization of *LMNA* to the centromeric portion of chromosome 1, Fatkin et al. performed molecular genetic screening of *LMNA* in eleven families with DCM and conduction disease.[3] Five putative disease-causing missense mutations were detected in five unrelated families. The mutations discovered in exons 1 (Arg60Gly and Leu85Arg) and 3 (Asn195Lys and Glu203Gly) affect both of the two principal isoforms of LMNA, lamin A and lamin C (Fig. 33.1A); one mutation was localized to exon 10 (Arg571Ser), and was predicted to only affect the alternatively spliced lamin C isoform.

Clinical data were available for 39 affected members from the five families harboring *LMNA* mutations. The average age of onset of disease was 38 years, and a presentation with asymptomatic electrocardiographic (ECG) abnormalities was common. Sinus node disease or atrioventricular conduction abnormalities were present in 87% of affected subjects, 54% had pacemakers implanted to treat bradyarrhythmias or high-grade atrioventricular blocks, and 64% had DCM. None of the evaluated mutation carriers had clinical signs or symptoms of skeletal muscle disease, emphasizing that this phenotype was distinct from that reported in AD-EDMD complicated by cardiac disease. However, elevations in creatine kinase were reported in four individuals from the family with the Arg571Ser mutation, and a vastus lateralis biopsy in an 'active jogger' carrying the Asn195Lys mutation showed only 'mild, nonspecific changes with minimal variation in fiber size and few internal nuclei'. These findings suggested that even though muscular disease was not clinically obvious in these patients, subclinical abnormalities of muscle may still be encountered in families affected predominantly by DCM due to *LMNA* mutations.

Background on *LMNA* gene products

The proteins encoded by *LMNA* are essential elements of the nuclear lamina, a complex protein-derived network that lies beneath and supports the nuclear envelope (Fig. 33.2A). The lamina is believed to provide integral structural support to the nucleus,[13] as well as having a role in the organization of chromatin and the process of DNA replication.[14] There are several lamin proteins, which are classified as type V intermediate filaments,[15] major constituents of the interphase nuclear lamina; they are subdivided into A-type and B-type lamins. The B-type lamins, lamins B1 and B2, are encoded by separate genes *LMNB1*,[16] and *LMNB2*[17] located at 5q23.3–q31.1 and 19p13.3, respectively, and are constitutively expressed in many tissues.[18,19] The A-type lamins in humans demonstrate a more restrictive expression pattern, being expressed chiefly in differentiated cells.[15] They are represented by two principal isoforms, lamin A and C, which are alternatively spliced products of the single *LMNA* gene located on chromosome 1q21.2. Lamins A and C dimerize and assemble into higher order structures, providing a supporting framework for the nuclear lamina.[20] The lamina structure provides a networking structure below the nuclear envelope where dimerized lamins A and C are found arranged in higher-ordered structures, associating with the nuclear membrane, nuclear pore complexes, other nuclear envelope proteins and chromatin (Fig. 33.2B). Less abundant isoforms lamin AΔ10 and lamin C2 are also produced from *LMNA*. Both are more closely similar to lamin C in sequence, lamin AΔ10 has been reported in a lung tumor cell line,[21] and lamin C2 appears to be largely restricted to testicular tissue,[22] although one study has suggested that

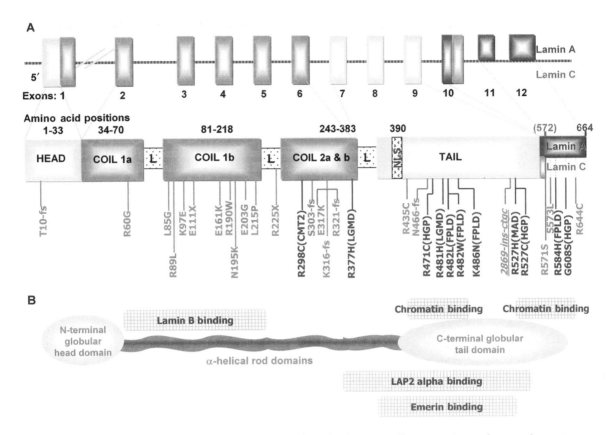

Figure 33.1. *The LMNA gene. (A) The exons encoding the domains of lamin A/C are shown; alternative splicing in exon 10 results in the two principal products. The location of mutations reported in dilated cardiomyopathy (red) and other laminopathies (blue) are indicated by their corresponding amino acid residue position and the abbreviation for the amino acid. CMT2, Charcot–Marie–Tooth type 2; LGMD, limb girdle muscular dystrophy; HGP, Hutchinson–Gilford progeria; MAD, mandibuloacral dysplasia; FPLD, familial partial lipodystrophy; fs, frameshift; NLS, nuclear localization signal region, L-linker segment. All mutation numbers refer to amino acid positions except for the italicized entry in the tail region representing a 4 bp insertion at mRNA position 2869/amino acid residue 525. (B) The general structure of the lamin A or C protein product with the various reported binding sites (to other proteins and chromatin) is depicted. See color plate section.*

it may be expressed in the failing human heart.[23] The *LMNA* products share a common intermediate filament structure of a central rod domain and globular head and tail domains (Fig. 33.1B). The central rod domain is further subdivided into four alpha-helical coiled-coiled domains separated by flexible linker regions.[15] Additional heptad repeats in this central rod, a nuclear localization region in the tail, and a propensity for carboxy isoprenylation and methylation distinguish the nuclear lamins from other cytoplasmic intermediate filaments.[15,24]

Diseases of lamin A/C: laminopathies

In addition to DCM and AD-EDMD, mutations in *LMNA* are associated with several other, seemingly disparate, diseases. The AD limb girdle muscular dystrophy (LGMD) type 1 B [OMIM#159001],[25] and an autosomal recessive (AR) form of EDMD [OMIM#604929][26] represent additional muscular dystrophy phenotypes. AR Charcot–Marie–Tooth type 2 [OMIM#605588],[27] Dunnigan-type familial partial lipodystrophy [OMIM#151660],[28,29] mandibuloacral dysplasia [OMIM#248370],[30] and

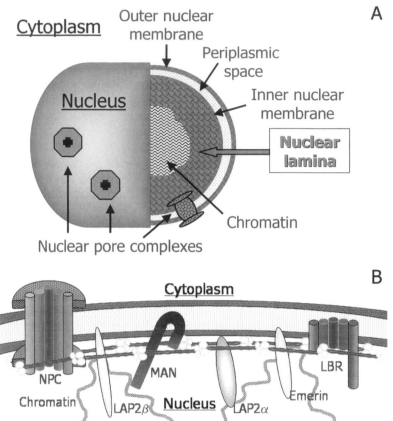

A **Figure 33.2.** (A) *Location of the nuclear lamina is depicted in relationship to nuclear envelope. In addition to roles in supporting nuclear structure and architecture, lamins are believed to interact with nuclear pore complexes and chromatin. (B) Lamin A/C dimers align beneath and in contact with the inner nuclear membrane allowing interactions between lamins A and C and integral inner nuclear membrane protein (MAN), lamin-associated polypeptides alpha and beta (LAP2α, LAP2β), emerin, lamin binding receptor (LBR) and the nuclear pore complexes (NPC).*

B

most recently, Hutchinson–Gilford progeria syndrome [OMIM#176670][31] have all been associated with *LMNA* mutations (Table 33.1). The remarkable diversity of disease phenotypes linked to *LMNA* mutations is probably a reflection of the many important functions assumed by lamin A and C proteins.

Additional mutations in *LMNA*: evidence of variable skeletal muscle involvement

Multiple additional *LMNA* DCM mutations have now been reported (Fig. 33.1A), and the DCM disease phenotype has expanded to include skeletal muscle disease in addition to biventricular dilation and conduction system disease.[2,4] Typifying this is the MDDC1 family where affected family members suffered from DCM, conduction system disease and variable mild skeletal muscle involvement

(Fig. 33.3A).[2] Interestingly, in this family the proband had a pure DCM with no skeletal muscle involvement, while an EDMD phenotype and LGMD-like phenotype were separately observed in two other affected relatives. Endomyocardial and skeletal muscle biopsies from this family were abnormal (Fig. 33.3B and C). Although the majority of mutations described to date predict missense alterations, the MDDC1 family has a base pair deletion in exon 6 that predicted a premature stop codon (del959T). A predicted addition of 158 novel carboxy-terminal amino acids after the lysine residue at position 316 is the result of this mutation; it is unclear whether this aberrant protein is produced, or has any function. In another family affected with DCM and conduction defects, a nucleotide insertion in exon 1 leading to a premature stop codon was detected.[12] The conduction disease present in this family was severe, and pace-

Table 33.1. *Diseases linked to LMNA mutations*

Disease	OMIM[a] name	OMIM #	Cardiac phenotype	Extracardiac involvement
AD DCM +/– conduction defects	CMD1A	115200	Dilated cardiomyopathy, conduction defects, arrhythmia	Muscular dystrophy
AD Emery–Dreifuss muscular dystrophy	EDMD2	181350	Arrhythmia, dilated cardiomyopathy	Muscular dystrophy, joint contractures (elbows, Achilles)
AR Emery–Dreifuss muscular dystrophy	EDMD3	604929	None	Muscular dystrophy, joint contractures
AD familial partial lipodystrophy	FPL	151660	None	Lipodystrophy, insulin resistance, hyperlipidemia
AR mandibuloacral dysplasia	MAD	248370	None	Hypoplastic mandible and clavicles, short stature, alopecia
AR Charcot–Marie–Tooth, type 2B1	CMT2B1	605588	None	Distal limb peripheral neuropathy, foot deformities
AD Hutchinson–Gilford progeria	HGPS	176670	Coronary artery disease/ myocardial infarction/ ischemic cardiomyopathy	Premature aging, short stature, loss of subcutaneous fat, alopecia

[a]OMIM-Online Mendelian Inheritance in Man (www.ncbi.nlm.nih.gov/entrez/query.fcgi?db=OMIM).
AD, autosomal dominant; AR, autosomal recessive.

makers or defibrillators were implanted; three individuals in the pedigree (presumptive mutation carriers) experienced sudden death before age 55 years.[32]

Because of the complex and intriguing phenotypes associated with *LMNA* mutations, investigators have studied their genetic epidemiology. Arbustini et al. estimated the frequency of *LMNA* mutations in families with AD DCM and atrioventricular blocks (AVB).[33] In this study, 73 cases of DCM were screened, resulting in the identification of five novel mutations. These five families represented 33% of all the studied families with both DCM and AVB, indicating that this may be a high-risk group for *LMNA* mutations. An additional study estimated the frequency of *LMNA* mutations in a cohort of 49 consecutively ascertained DCM families with comprehensive phenotypic data.[4] This survey detected four *LMNA* mutations, representing 8% of the families studied. Analysis of genotype–phenotype correlations and of the survival rate (Fig. 33.4) found that *LMNA* mutation carriers had a significantly poorer cumulative survival compared to DCM patients without *LMNA* mutations.[4] The presence of skeletal muscle disease, supraventricular arrhythmia, and/or conduction defects was a significant predictor of carrying a *LMNA* mutation. Importantly, one of the mutations was detected in a DCM patient who had a negative family history, arguing that *LMNA* mutations are also relevant in apparently isolated or sporadic forms of DCM. The study population was unselected for skeletal muscle or conduction system disease, and the authors suggested that the relatively high prevalence of mutations in this cohort (8%), as well as the poor prognosis, argued for the consideration of *LMNA* clinical testing in patients presenting with DCM, especially if skeletal muscle disease, arrhythmia, or conduction defects were present.

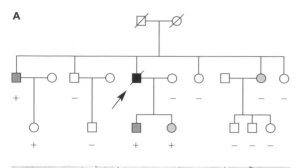

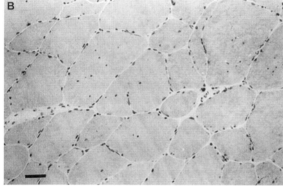

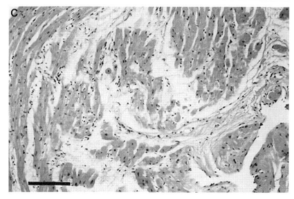

Figure 33.3. *(A) Pedigree of MDDC1 family; proband indicated by arrow; DCM (black), EDMD (Green), LGMD (Blue); LMNA mutation carriers (+) and non-carriers (–) indicated. (B) and (C) Abnormalities of skeletal and cardiac muscle in MDDC1 patients. Abnormalities of cardiac and skeletal muscle in MDDC1 patients. (B) Skeletal muscle biopsy showing variability of fiber size, with both atrophic and hypertrophic fibers, several splittings, an increase in internal nuclei, focal signs of degeneration and regeneration with basophilic fibers, and a mild increase in endomysial connective tissue (hematoxylin-eosin stain, magnification ×25, scale bar = 25 μm). (C) Endomyocardial biopsy (hematoxylin-eosin stain, magnification ×40, scale bar = 0.1 mm) of individual II-1 at disease onset (ejection fraction, 53%) showing interstitial edema, hypertrophy, fibrosis, and cellular infiltrates, suggestive of healing myocarditis. (Reproduced with permission from Brodsky GL, Muntoni F, Miocic S et al. A lamin A/C gene mutation associated with dilated cardiomyopathy with variable skeletal muscle involvement. Circulation 2000; 101: 473–6. See color plate section.*

Overlap with other disease phenotypes

It is apparent that there is overlap between the different disease phenotypes reportedly linked to *LMNA*. For example, many families ascertained on the basis of AD-EDMD manifest substantial cardiac involvement. In one study of 53 EDMD patients from 23 families (6 autosomal dominant and 17 sporadic), *LMNA* mutations were found in 18 families.[34] Substantial phenotypic variability was observed between and within families, including the presence of relatives of EDMD patients who had no signs of EDMD but were affected with cardiac conduction abnormalities and DCM. Cardiomyopathy has also been reported to complicate a partial lipodystrophy phenotype caused by a *LMNA* mutation.[35]

Animal models

Several investigators have explored the functions of products of *LMNA* in animal models. *LMNA* knockout mice have been generated that are viable and appear healthy at birth. However, by 3–4 weeks of age there is evidence of growth retarda-

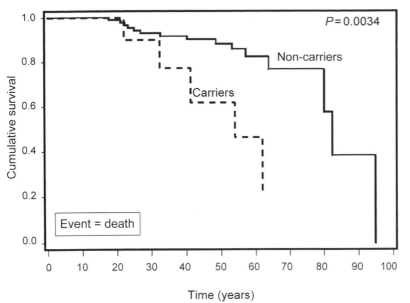

Figure 33.4. *Kaplan–Meier cumulative survival curves for cardiovascular death in DCM patients carriers of LMNA mutations (dotted lines) compared to non-carrier patients (continuous lines).*

tion and an EDMD-like phenotype, characterized by contractures and skeletal muscle wasting.[36] Loss of adipose tissue is also present, but other features of lipodystrophy such as insulin resistance are not reported.[37] Furthermore, the cells harvested from these animals have misshapen nuclei and ultrastructural perturbations of the nuclear envelope. Additional work has shown that these LMNA null cells have increased nuclear deformation and have less viability when exposed to biomechanical strains. Alterations in gene expression were also documented in response to mechanical strain in these cells.[38]

In a different study, the *L530P* mutation was introduced into mice in an effort to produce a mouse with an autosomal EDMD phenotype. Interestingly, the resulting heterozygous mice are phenotypically normal, at least up until and beyond 6 months of age. On the other hand, homozygous mutant mice develop, at a very early age, a phenotype resembling Hutchinson–Gilford progeria with shortened stature, alopecia, osteoporosis and thin skin.[39] Death typically occurs by 4–5 weeks of age. Degeneration and atrophy of cardiac and skeletal muscles was also observed. After the association between *LMNA* mutations and human autosomal recessive Charcot–Marie–Tooth syndrome Type 2 (AR-CMT2) was established, an analysis was per-

formed on *LMNA* knockout mice. Reduced axonal density and demyelination were observed, mirroring the peripheral nerve findings in human AR-CMT2.[27]

Arrhythmogenic right ventricular dysplasia/cardiomyopathy (ARVD/C)

Clinical features and epidemiology

Arrhythmogenic right ventricular dysplasia/cardiomyopathy (ARVD/C) is a disease characterized by fatty or fibrofatty replacement and progressive degeneration of the right ventricular (RV) myocardium with occasional extension to the left ventricle.[40] Clinically, affected individuals may experience palpitations and syncope due to tachyarrhythmias of right ventricular origin, and some will progress to symptomatic heart failure in advanced disease.[41] However, clinical manifestations range from asymptomatic patients to others who present with sudden cardiac death.[7,42,43] Young adults and competitive athletes with ARVD/C appear to be at increased risk for sudden death. ARVD/C remains a leading cause for atraumatic sudden death in this population, accounting for 3–20% of all cases, second only to hypertrophic cardiomyopathy, in one report.[44,45] Onset of

575

symptoms typically occurs in late adolescence and young adulthood, but may present much later in life as well.

Motivated in part by the presumed under-diagnosis of and lack of standardized diagnostic criteria for ARVD/C, an expert panel first published consensus diagnostic criteria in 1994 (Table 33.2).[46] According to the diagnostic criteria, the affected

Table 33.2. *Diagnostic criteria for ARVD*

Diagnostic approach	Major criterion	Minor criterion	Task force[c]
Imaging[a]	• Severe dilation and reduction of RV EF (preserved LV function)	• Mild global RV dilation or reduction of RV EF	Yes
	• Localized RV aneurysms	• Mild segmental dilation of RV	Yes
	• Severe segmental dilation of RV	• Regional RV hypokinesis	Yes
MRI specific criteria[b]	• Fatty infiltration of the right ventricular myocardium with high signal intensity on T1-weighted images	• Regional contraction abnormalities	No
	• Fibrofatty replacement, which leads to diffuse thinning of the right ventricular myocardium	• Global diastolic dysfunction	No
Cardiac biopsy	• Fibrofatty or fatty replacement of myocardium		Yes
Electrocardiogram	• Epsilon waves or localized QRS prolongation (>110 ms: V1–V3)	• Inverted T waves (V1–V3) in absence of LBBB (age >12 years)	Yes
		• Late potential (signal-averaged ECG)	Yes
Arrhythmia		• LBBB-type VT	Yes
		• Frequent ventricular extrasystoles (>1000/24 hours by Holter)	Yes
Family history	• Familial disease confirmed by autopsy or surgery	• Family history of sudden death (<age 35 years) due to suspected ARVD/C	Yes
Molecular genetics	• Confirmed pathologic mutation in known ARVD gene	• Suspected pathological mutation in known ARVD gene	No
	• Confirmatory genetic linkage studies	• Suggestive genetic linkage studies	No

[a] Imaging using echocardiography, angiography, magnetic resonance imagery, or radionucleotide scintigraphy.
[b] MRI specific criteria have been discussed and proposed, but consensus has not been reached.
[c] Criteria from McKenna WJ, Thiene G, Nava A et al. Diagnosis of arrhythmogenic right ventricular dysplasia/cardiomyopathy. Br Heart J 1994; 71: 215–8.
RV, right ventricle; LV, left ventricle; EF, ejection fraction; LBBB, left bundle branch block; VT, ventricular tachycardia. Adapted with permission from Kayser HW, van der Wall EE, Sivananthan MU et al. Diagnosis of arrhythmogenic right ventricular dysplasia: a review. Radiographics 2002; 22: 639–48.

status is defined by the presence of: (1) two major criteria, or (2) one major plus two minor criteria, or (3) four minor criteria. The unknown status is defined by the presence of 1–3 minor criteria, and the unaffected status is defined by the presence of a normal heart or the determination of other causes of myocardial dysfunction. The diagnostic criteria incorporate structural, histological and electrophysiological abnormalities in addition to family history. While such guidelines have improved disease recognition, ARVD/C remains an under-recognized, under-diagnosed entity with disease status often confirmed only at autopsy.[47] Consequently, the exact prevalence of ARVD/C remains unknown, but recent estimates suggest roughly 6 per 10 000 individuals are affected in the general population, with significantly higher rates observed in certain geographic regions, such as Northeastern Italy.[48,49] Diagnosis relies on the use of multiple modalities including echocardiography (Fig. 33.5), ECG, signal-averaged ECG (SAECG), magnetic resonance imaging (MRI), right heart ventriculography and endomyocardial biopsy. Although endomyocardial biopsy may confirm the characteristic histopathology (Fig. 33.6), the sensitivity is limited by the patchy distribution of the affected myocardium, as well as the highly variable fat content of 'normal' myocardium.[49] Similarly, the non-invasive techniques limited diagnostic sensitivity and specificity, necessitating employment of multiple testing strategies. MRI has emerged as another possible diagnostic tool in recent years, and its diagnostic utility continues to be evaluated and discussed,[50] but the current sensitivity and specificity of this approach is not well established, and probably depends on the expertise of the interpreting radiologist.[51] Clinical molecular genetic testing is under development, and is anticipated to provide another diagnostic option for presymptomatic or early-stage patients. Confirming ARVD/C is essential as prognosis, treatment and implications for potentially affected family members differ considerably from similar arrhythmogenic conditions, such as right ventricular outflow tract tachycardia (RVOT).[52]

Two-dimensional echocardiography remains the standard technique for detecting RV dilatation and focal aneurysms. Electrocardiogram findings in ARVD/C include repolarization abnormalities (e.g. T wave inversions in V2–V3), right bundle branch block, and selective prolongation of the QRS complex in leads V1–V3 of more than 25 ms beyond the QRS duration in lead V6.[53] Sustained/non-sustained ventricular tachycardia typically displays a left bundle-branch pattern indicative of its RV origin. Post-excitation epsilon waves are more specific for ARVD/C, but are observed in less than 30% of ECGs. Late potentials on SAECG have been reported in 50–100% of patients,[54] and combining SAECG with measurements of QT dispersion (QTd >40 ms) may improve the sensitivity of this non-invasive approach.[55]

The familial nature of ARVD/C is now well established, with 30–50% of relatives of ARVD/C patients manifesting abnormalities suggestive of the disease when evaluated by non-invasive approaches.[56] The importance of the heritable component of ARVD/C is reflected in the incorporation of family history data as diagnostic criteria. The recognition that subtle signs and symptoms in a first-degree relative of an ARVD/C patient probably represent early disease led Hamid et al. to propose less stringent diagnostic criteria in such instances.[57] Currently, one criterion in first-degree relatives supports the diagnosis of ARVD/C.

Molecular genetics

ARVD/C is a genetically heterogeneous disease with most families exhibiting autosomal dominant inheritance. Genetic studies have identified 9 disease loci (Table 33.3) in families with autosomal dominant disease and isolated cardiac disease. A chromosomal locus (10q22.3) has also been identified in an autosomal dominant form associated with skeletal muscle involvement.[58] Furthermore, a rare autosomal recessive form of ARVD/C called Naxos disease, named after the Greek island from which the majority of cases were identified, is characterized by palmoplantar keratoderma and wooly hair in addition to ARVD/C. Research within the past decade has successfully demonstrated disease-causing mutations in four genes, plakoglobin (*JUP*), desmoplakin (*DSP*), plakophilin-2 (*PKP2*) and the cardiac ryanodine receptor (*hRYR2*) in certain ARVD/C families (Table 33.3).[9,10,59,60]

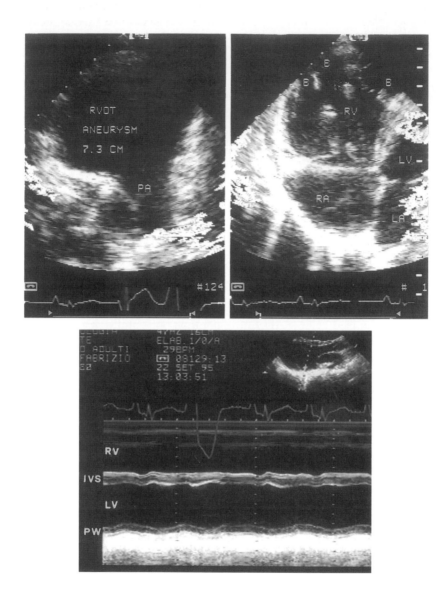

Figure 33.5. *ARVD/C Echocardiography: top, two-dimensional echocardiography. Left, parasternal short-axis view at basal level. Right, modified apical four-chamber view. Both frames in systole. Severe right ventricular (RV) enlargement with aneurysmal dilatation of outflow tract (RVOT) and multiple wall bulges (right) are present. PA, pulmonary artery; RA, right atrium; LV, left ventricle; and LA, left atrium. Bottom, M-mode echocardiographic tracing at ventricular level. RV is severely enlarged; LV is not enlarged, but severe hypokinesis of both septum (IVS) and posterior wall (PW) is evident. Reproduced with permission from: Pinamonti B, Pagnan L, Bussani R et al. Right ventricular dysplasia with biventricular involvement. Circulation 1998; 98: 1943–5.*

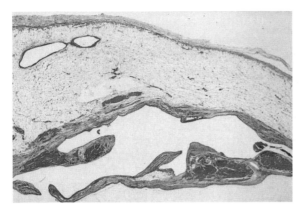

Figure 33.6. *Histological findings in ARVD/C showing fatty tissue replacement of myocardial muscle in right free ventricular wall (×2.5). Reproduced with permission from Severini GM, Krajinovic M, Pinamonti B et al. A new locus for arrhythmogenic right ventricular dysplasia on the long arm of chromosome 14. Genomics 1996; 31: 193–200. See color plate section.*

Naxos disease is caused by a two basepair deletion in *JUP* that introduces a premature stop codon, resulting in a protein truncated by 56 amino acids

(Fig. 33.7C).[59] Mutations in the gene encoding another desmosomal protein, desmoplakin (*DSP*), was described in two distinct ARVD/C phenotypes. A missense mutation (Ser299Arg) in the N-terminal domain of *DSP* resulted in a pure form of ARVD/C in an Italian family with autosomal dominant transmission.[60] Furthermore, a homozygous point mutation (Gly2375Arg) in the C-terminal domain of *DSP* cosegregated with the disease in a family with recessive inheritance, which had associated hair and dermatological abnormalities distinct from Naxos disease. A mutation in the N-terminal domain of *DSP* has also been shown to cause Carvajal syndrome, an autosomal recessive syndrome first described in three Ecuadorian families and characterized by palmoplantar keratoderma, wooly hair and left ventricular dilation.[61] The constellation of skin/hair findings, cardiac arrhythmias and association with *DSP* mutations have led some to classify Carvajal syndrome in the ARVD/C spectrum.[56]

Function of plakoglobin, desmoplakin and plakophilin

Plakoglobin is a highly conserved 744 amino acid cytoskeletal protein localized to desmosomes and

Table 33.3. *ARVD loci and genes*

Disease	OMIM[a] name	OMIM#	Chromosomal locus	Gene
AD ARVD/C	ARVD1	107970	14q24.3	
	ARVD2	600996	1q42–q43	*hRYR2* (cardiac ryanodine receptor II)
	ARVD3	602086	14q11–21	
	ARVD4	602087	2q32.1–q32.3	
	ARVD5		3p23	
	ARVD6	604401	10p12–p14	
	ARVD8	605676	6p24	*DSP* (desmoplakin)
	ARVD9		8p23.2	
			12p11	*PKP2* (plakophilin-2)
AR ARVD/C with wooly hair and keratoderma	ARVD/C	601214	17q21	*JUP* (plakoglobin)
		125647	6p24	*DSP* (desmoplakin)
AD ARVD/C with myofibrillar myopathy	ARVD7		10q22.3	

[a]OMIM-Online Mendelian Inheritance in Man (www.ncbi.nlm.nih.gov/entrez/query.fcgi?db=OMIM).

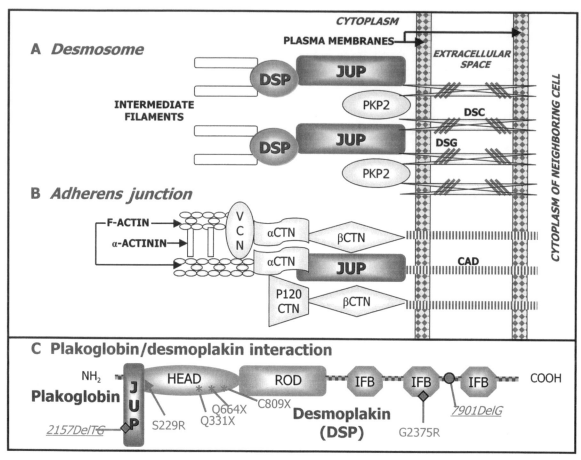

Figure 33.7. *Cell–cell adhesion complexes. Basic desmosomal (A) and adherens junctional (B) structures. Locations and proposed interactions/bindings of desmoplakin (DSP) and Plakoglobin (JUP) are depicted. Plakophilin-2 (PKP2), Desmocollin (DSC) Desmoglein (DSG); alpha, beta, p120 catenin (CTN), and vinculin (VCN). (C) Interaction between plakoglobin and N-terminal head of desmoplakin. Described pathogenic mutations and amino acid positions are indicated (2157DelTG, S229R, Q331X, Q664X, C809X, G2375R, 7901DelG; italicized underlined numbers refer to nucleotide position). Arrow, autosomal dominant ARVD/C mutation with amino acid position. *, DSP mutations causing striate palmoplantar keratoderma. Diamond indicates autosomal recessive ARVD/C mutations in JUP (Naxos disease) and DSP. Circle indicates autosomal recessive mutation causing left ventricular cadiomyopathy, skin and hair disease (Carvajal syndrome). IFB, intermediate filament-binding domain.*

adherens junctions. Adherens junctions form a continuous ring between adjacent cells that connect to the actin cytoskeleton, whereas desmosomes are focally located (likened to 'spot-welds' between connecting cells), and bind to intermediate filament networks (Fig. 33.7A and B). Desmosomes are present in many tissues, with high expression noted in heart and skin tissue. In addition to their role in structural adhesion, these complexes also perform inter- and intracellular signaling functions.[62] Plakoglobin associates with several other desmosomal proteins, including DSP, desmoglein I, and the cadherin–catenin complex responsible for anchoring actin connected to adherens junctions.

The 24-exon *DSP* gene encodes a 2871-amino

acid protein with an N-terminal domain that binds to the desmosome through its association with plakoglobin, a rod domain, and a C-terminal intermediate filament-binding domain (Fig. 33.7C). The first disease-causing mutation in *DSP* was described in the striate subtype of hereditary palmoplantar keratoderma.[63] The heterozygous single base pair mutation in the N-terminal domain creates a null allele and DSP haploinsufficiency. The mutation in Carvajal syndrome, conversely, is in the C-terminal intermediate filament-binding domain. The Ser299Arg mutation in autosomal dominant ARVD/C modifies a putative phosphorylation site in the amino-terminal plakoglobin-binding domain of DSP. The recessive mutation described by Alcali et al.[64] changes a single amino acid in the C-terminal domain, potentially disrupting intermediate filament binding. How such mutations lead to the observed allelic and phenotypic heterogeneity is unclear, and there remains no clear association between individual mutations and tissue-specific effects. The gene for plakophilin-2 on chromosome 12 encodes an 881-amino acid protein. PKP2, a member of the armadillo-repeat protein family, is a desmosomal cadherin that binds to desmoplakin and is essential for proper formation of the desmosomal plaque. A recent report identified various mutations in 32 of 102 unrelated individuals including two small kindreds that exhibited incomplete penetrance.[10] Thus, PKP2 may be the first reported 'common' ARVD/C-causing gene. Together, desmoplakin and plakoglobin represent a potential common pathway through which ARVD/C arises by disruption of cell–cell architecture, with resultant tissue degeneration. This process potentially also leads to apoptosis, which has been observed in ARVD/C for many years.

The ion channel pathway: the ryanodine receptor

Recently the identification of pathogenic mutations in the cardiac ryanodine receptor (*hRYR2*) in autosomal dominant ARVD/C type 2 (ARVD/C2) has introduced a second functional pathway through which ARVD/C may arise. The large, 105 exon *hRYR2* gene encodes a 566-kDa monomer, which associates into a functional tetrameric channel. The hRYR2 protein ion channel couples myocyte excitation–contraction by regulating calcium efflux from the sarcoplasmic reticulum into the cytoplasm (Fig. 33.8).[65] A number of other protein interactions with the ion channel have been described including: spinophilin, protein phosphatase 1 and 2A, FK-506-binding protein, calmodulin, muscle A kinase-anchoring protein and protein kinase A.

Tiso et al.[9] established linkage in four Italian families to chromosome 1q42–q43. An analysis of candidate genes in the region led to the identification of four missense mutations in *hRYR2*. Interestingly, the *hRYR1* gene encoding the ryanodine receptor expressed in skeletal muscle was shown to have mutations in homologous domains that lead to malignant hyperthermia. Additional research has demonstrated distinct missense mutations in *hRYR2* in patients with catecholaminergic polymorphic ventricular tachycardia (CPVT) and familial polymorphic ventricular tachycardia (FPVT), conditions characterized by arrythmogenicity in the absence of structural abnormalities.[66,67]

ARVD: the final common pathway

A link between the desmosome and ion channel pathways is believed to lie in the altered intracellular calcium concentration caused by defects in these structures. Adherens junctions play a role in the activation of stretch-sensitive calcium-permeable channels, via cadherins leading to alterations in mechanical intracellular signaling.[68] Furthermore, stretching of cardiomyocytes modulates the process of calcium release from ryanodine receptor release channels.[69] Therefore, a genetically impaired response to mechanical stress might adversely affect intracellular calcium concentration and the excitation–contraction coupling.[10,61] The calcium imbalance could also induce apoptosis and cellular necrosis, leading in turn to fibro-fatty substitution. The prevalent or selective involvement of the right ventricle could derive from its extensibility and different mechanics compared to the left ventricle.

Genotype–phenotype correlations

Currently, little is known about genotype–phenotype correlations in ARVD/C. Studies examining the *JUP* mutation in Naxos disease have shown

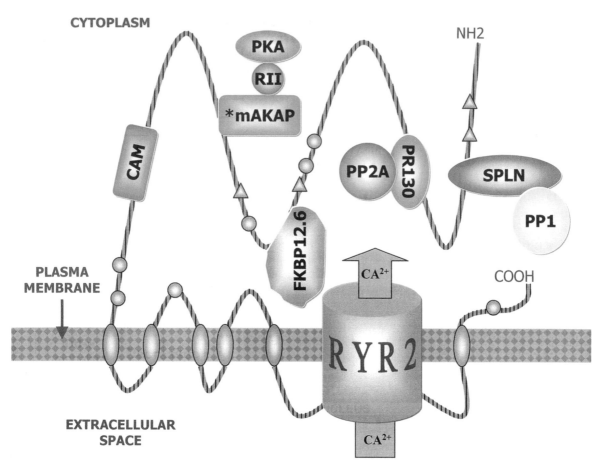

Figure 33.8. *Cardiac ryanodine receptor (RYR2) and associated regulatory proteins with RYR2 amino acid-binding domain position listed in parentheses: spinophilin (SPLN)/protein phosphatase 1 (PP1) (554–588), PR130/protein phosphatase 2A (PP2A) (1603–1631), FK-506-binding protein (FKBP12.6), calmodulin (CAM), and muscle A-kinase anchoring protein (mAKAP)/RII/protein kinase A (PKA) (3003–3039). *mAKAP was screened for mutations in families linked to the ARVD3 locus (14q12). Circles denote known RYR2 mutations in ARVD/C (R176Q, L433P, N2386I, T2504M), Triangles denotes known RYR2 mutations in familial polymorphic ventricular tachycardia (FPVT)/catcholinergic polymorphic ventricular tachycardia (CPVT) (S2246L, P2328S, R2474S, N4104K, Q4201R, R4497C, V4653F).*

that homozygous individuals display high disease penetrance while heterozygotes may exhibit subtle clinical findings but do not develop overt disease. Individual *DSP* mutations do not appear to predict phenotype either, however, the dermatological abnormalities observed in both hereditary palmoplantar keratoderma and Carvajal syndrome disrupt the C-terminal intermediate filament-binding domain, while the mutation in the pure form of *ARVD/C* is localized to the N-terminal plakoglo-

bin-binding domain. Rampazzo et al. postulated that ARVD/C families with the *S299R* mutation lacked dermatological findings due to the sparing of the intermediate filament-binding domain.[60]

In the case of *hRYR2* mutations, ARVD/C2 families had a phenotype characterized by effort-induced ventricular arrhythmias, high disease penetrance and equal male/female ratio, however, phenotypic variance could not be ascribed to particular mutations, partly due to the small number of families. Studies examining the functional effect of ARVD/C mutations compared to mutations found in other stress-induced tachycardias suggest the two types of mutations may oppositely affect the hRYR2 gating protein, FKB12.6, causing differential calcium release from the sarcoplasmic reticulum.[70]

Further investigation is ongoing to identify the mechanisms underlying observed differences in affected male to female ratio, variable expressivity within families, and increased risk for sudden death in younger individuals. Likewise, the observed ARVD/C phenotype associated with skeletal muscle disease implicates an as yet unidentified disease gene, perhaps one that is already known to be involved in DCM with skeletal muscle involvement. In addition to genetic factors, environmental influences are also being investigated since 50–70% of ARVD/C appears sporadic. Bowles et al. have identified the presence of enteroviruses in patients with sporadic ARVD/C, and found a high frequency of enteroviral DNA sequence detection in myocardial tissue.[71] Even though a causal role has yet to be established, the analogy with myocarditis and DCM suggest that viral infection could account for a subset of ARVD/C cases as well.

Animal models

As in humans, a naturally occurring ARVD/C phenotype has been reported in several animal models, including dogs, cats and mink. Recently, Asano et al. identified the gene causing ARVD/C in the KK/rvd mouse strain by positional cloning.[72] The human homologous gene, LAMR1 maps to 3p21.3 and encodes laminin receptor 1. The mutated LAMR1 appears to affect transcription regulation and induce apoptosis, which has been identified in ARVD/C in previous studies. In order to improve clinical and basic science research, and create collaborative efforts in pursuing this relatively rare condition, investigators have established both North American and international registries for ARVD/C.[8,56]

Clinical genetic implications

Advances in the understanding of the genetic nature of the complex cardiomyopathies challenge clinicians to apply this new knowledge to the care of affected patients. Beyond the direct implications of a diagnosis to an individual patient, the involvement of patients' relatives at risk of the disease and the adoption of a family-based approach to these disorders represent new challenges to many cardiologists and general practitioners. Increased attention to family history as well as to pedigree construction and interpretation will help guide the clinician in determining the likely mode of inheritance, and in offering recommendations for clinical and echocardiography-based screening of at-risk relatives. From these approaches, the clinician may need to provide genetic counseling and a genetic explanation of the cardiomyopathy to the patient(s). As family-based models may be unfamiliar to many cardiologists treating adults, the involvement of a clinical geneticist or genetic counselor may be helpful in these situations. The appropriate use of molecular genetic testing in such patients and families is still unclear, and formal guidelines for DNA testing are needed to guide their use. As multiple genes are involved in both simple and complex cardiomyopathies, it is anticipated that molecular diagnostic panels will eventually be developed to simultaneously or sequentially test for mutations in the cardiomyopathy genes.

Conclusions

The human cardiomyopathies are a fascinating collection of diseases, which in more recent years have been defined as genetic diseases in many patients and families. Complex forms of cardiomyopathy, complicated by skeletal muscle disease and/or arrhythmia are now understood to also be genetic diseases, and the identification of specific genes, including LMNA and the ARVD/C genes, represent successes in the molecular characterization of these conditions. Undoubtedly, more remains to be discovered and understood, and the development of therapies that target and correct the underlying genetic defects is still a distant possibility. More immediately, however, the increased recognition of the genetic nature of these conditions, and the growing number of clinical genetic tests is challenging the clinician to integrate genetic counseling, genetic evaluation and genetic testing into the medical evaluation and management of affected patients and families.

References

1. Mestroni L, Maisch B, McKenna WJ et al. Guidelines for the study of familial dilated cardiomyopathies. Eur Heart J 1999; 20: 93–102.

2. Brodsky GL, Muntoni F, Miocic S et al. A lamin A/C gene mutation associated with dilated cardiomyopathy with variable skeletal muscle involvement. Circulation 2000; 101: 473–6.

3. Fatkin D, MacRae C, Sasaki T et al. Missense mutations in the rod domain of the lamin A/C gene as causes of dilated cardiomyopathy and conduction-system disease. N Engl J Med 1999; 341: 1715–24.

4. Taylor MRG, Fain P, Sinagra G et al. Natural history of dilated cardiomyopathy due to lamin A/C gene mutations. J Am Coll Cardiol 2003; 41: 771–80.

5. Buckley AE, Dean J, Mahy IR. Cardiac involvement in Emery Dreifuss muscular dystrophy: a case series. Heart 1999; 82: 105–8.

6. Merchut MP, Zdonczyk D, Gujrati M. Cardiac transplantation in female Emery–Dreifuss muscular dystrophy. J Neurol 1990; 237: 316–19.

7. McRae AT 3rd, Chung MK, Asher CR. Arrhythmogenic right ventricular cardiomyopathy: a cause of sudden death in young people. Cleve Clin J Med 2001; 68: 459–67.

8. Corrado D, Fontaine G, Marcus FI et al. Arrhythmogenic right ventricular dysplasia/cardiomyopathy: need for an international registry. Study Group on Arrhythmogenic Right Ventricular Dysplasia/Cardiomyopathy of the Working Groups on Myocardial and Pericardial Disease and Arrhythmias of the European Society of Cardiology and of the Scientific Council on Cardiomyopathies of the World Heart Federation. Circulation 2000; 101: E101–6.

9. Tiso N, Stephan DA, Nava A et al. Identification of mutations in the cardiac ryanodine receptor gene in families affected with arrhythmogenic right ventricular cardiomyopathy type 2 (ARVD2). Hum Mol Genet 2001; 10: 189–94.

10. Gerull B, Heuser A, Wichter T et al. Mutations in the desmosomal protein plakophilin-2 are common in arrhythmogenic right ventricular cardiomyopathy. Nat Genet. 2004; 11: 1162–4.

11. Kass S, MacRae C, Graber HL et al. A gene defect that causes conduction system disease and dilated cardiomyopathy maps to chromosome 1p1–1q1. Nat Genet 1994; 7: 546–51.

12. Bonne G, Di Barletta MR, Varnous S et al. Mutations in the gene encoding lamin A/C cause autosomal dominant Emery-Dreifuss muscular dystrophy. Nat Genet 1999; 21: 285–8.

13. Hutchison CJ, Alvarez-Reyes M, Vaughan OA. Lamins in disease: why do ubiquitously expressed nuclear envelope proteins give rise to tissue-specific disease phenotypes? J Cell Sci 2001; 114: 9–19.

14. Mounkes L, Kozlov S, Burke B, Stewart CL. The laminopathies: nuclear structure meets disease. Curr Opin 2003; 13: 1–8.

15. Hutchison CJ. Lamins: building blocks or regulators of gene expression? Nat Rev Mol Cell Biol. 2002; 3: 848–58.

16. Lin F, Worman HJ. Structural organization of the human gene (*LMNB1*) encoding nuclear lamin B1. Genomics 1995; 27: 230–6.

17. Hoger TH, Zatloukal K, Waizenegger I et al. Characterization of a second highly conserved B-type lamin present in cells previously thought to contain only a single B-type lamin. Chromosoma 1990; 99: 379–90.

18. Rober RA, Weber K, Osborn M. Differential timing of nuclear lamin A/C expression in the various organs of the mouse embryo and the young animal: a developmental study. Development 1989; 105: 365–78.

19. Lehner CF, Stick R, Eppenberger HM et al. Differential expression of nuclear lamin proteins during chicken development. J Cell Biol. 1987; 105: 577–87.

20. Stuurman N, Heins S, Aebi U. Nuclear lamins: their structure, assembly and interactions. J Struct Biol 1998; 122: 42–66.

21. Machiels BM, Zorenc AH, Endert JM et al. An alternative splicing product of the lamin A/C gene lacks exon 10. J Biol Chem 1996; 271: 9249–53.

22. Alsheimer M, von Glasenapp E, Schnolzer M et al. Meiotic lamin C2: the unique amino-terminal hexapeptide GNAEGR is essential for nuclear envelope association. Proc Natl Acad Sci USA 2000; 97: 13120–5.

23. Berry DA, Keogh A, dos Remedios CG. Nuclear membrane proteins in failing human dilated cardiomyopathy. Proteomics 2001; 1: 1507–12.

24. McKeon FD, Kirschner MW, Caput D. Homologies in both primary and secondary structure between nuclear envelope and intermediate filament proteins. Nature 1986; 319: 463–8.

25. Muchir A, Bonne G, van der Kooi AJ et al. Identification of mutations in the gene encoding

lamins A/C in autosomal dominant limb girdle muscular dystrophy with atrioventricular conduction disturbances (LGMD1B). Hum Mol Genet 2000; 9: 1453–9.

26. Raffaele Di Barletta M, Ricci E, Galluzzi G et al. Different mutations in the LMNA gene cause autosomal dominant and autosomal recessive Emery–Dreifuss muscular dystrophy. Am J Hum Genet 2000; 66: 1407–12.

27. De Sandre-Giovannoli A, Chaouch M, Kozlov S et al. Homozygous defects in LMNA, encoding lamin A/C nuclear-envelope proteins, cause autosomal recessive axonal neuropathy in human (Charcot–Marie–Tooth disorder type 2) and mouse. Am J Hum Genet 2002; 70: 726–36.

28. Speckman RA, Garg A, Du F et al. Mutational and haplotype analyses of families with familial partial lipodystrophy (Dunnigan variety) reveal recurrent missense mutations in the globular C-terminal domain of lamin A/C. Am J Hum Genet 2000; 66: 1192–8.

29. Shackleton S, Lloyd DJ, Jackson SN et al. LMNA, encoding lamin A/C, is mutated in partial lipodystrophy. Nat Genet 2000; 24: 103–4.

30. Novelli G, Muchir A, Sangiuolo F et al. Mandubuloacral dysplasia is caused by a mutation in LMNA-encoding lamin A/C. Am J Hum Genet 2002; 71: 426–31.

31. Eriksson M, Brown WT, Gordon LB et al. Recurrent de novo point mutations in lamin A cause Hutchinson-Gilford progeria syndrome. Nature. 2003; 423: 293–8.

32. Sebillon P, Bouchier C, Bidot LD et al. Expanding the phenotype of LMNA mutations in dilated cardiomyopathy and functional consequences of these mutations. J Med Genet 2003; 40: 560–7.

33. Arbustini E, Pilotto A, Repetto A et al. Autosomal dominant dilated cardiomyopathy with atrioventricular block: a lamin A/C defect-related disease. J Am Coll Cardiol 2002; 39: 981–90.

34. Bonne G, Mercuri E, Muchir A et al. Clinical and molecular genetic spectrum of autosomal dominant Emery–Dreifuss muscular dystrophy due to mutations of the lamin A/C gene. Ann Neurol 2000; 48: 170–80.

35. van der Kooi AJ, Bonne G, Eymard B et al. Lamin A/C mutations with lipodystrophy, cardiac abnormalities, and muscular dystrophy. Neurology 2002; 59: 620–3.

36. Sullivan T, Escalante-Alcalde D, Bhatt H et al. Loss of A-type lamin expression compromises nuclear envelope integrity leading to muscular dystrophy. J Cell Biol 1999; 147: 913–20.

37. Cutler DA, Sullivan T, Marcus-Samuels B et al. Characterization of adiposity and metabolism in LMNA-deficient mice. Biochem Biophys Res Commun 2002; 291: 522–7.

38. Lammerding J, Schulze PC, Takahashi T et al. Lamin A/C deficiency causes defective nuclear mechanics and mechanotransduction. J Clin Invest 2004; 113: 370–8.

39. Mounkes LC, Kozlov S, Hernandez L et al. A progeroid syndrome in mice is caused by defects in A-type lamins. Nature 2003; 423: 298–301.

40. Basso C, Thiene G, Corrado D et al. Arrhythmogenic right ventricular cardiomyopathy. Dysplasia, dystrophy, or myocarditis? Circulation 1996; 94: 983–91.

41. Nava A, Bauce B, Basso C et al. Clinical profile and long-term follow-up of 37 families with arrhythmogenic right ventricular dysplasia. J Am Coll Cardiol 2000; 36: 2226–33.

42. Corrado D, Thiene G, Nava A et al. Sudden death in young competitive athletes: clinicopathologic correlations in 22 cases. Am J Med 1990; 89: 588–96.

43. Lobo FV, Heggtveit HA, Butany J et al. Right ventricular dysplasia: morphological findings in 13 cases. Can J Cardiol 1992; 8: 261–8.

44. Maron BJ, Shirani J, Poliac LC et al. Sudden death in young competitive athletes: clinical, demographic and pathological profiles. JAMA 1996; 276: 199–204.

45. Corrado D, Basso C, Thiene G. Arrhythmogenic right ventricular cardiomyopathy: diagnosis, prognosis, and treatment. Heart 2000; 83: 588–95.

46. McKenna WJ, Thiene G, Nava A et al. Diagnosis of arrhythmogenic right ventricular dysplasia/cardiomyopathy. Br Heart J 1994; 71: 215–18.

47. Danieli GA, Rampazzo A. Genetics of arrhythmogenic right ventricular cardiomyopathy. Curr Opin Cardiol 2002; 17: 218–21.

48. Ahmad F. The molecular genetics of right ventricular dysplasia-cardiomyopathy. J Clin Invest 2003; 26: 167–78.

49. Gemayel C, Pellicia A, Thompson PD. Arrhythmogenic right ventricular dysplasia. J Am Coll Cardiol 2001; 38: 1773–81.

50. Kayser HW, van der Wall EE, Sivananthan MU et al. Diagnosis of arrhythmogenic right ventricular dysplasia: a review. Radiographics 2002; 22: 639–48; discussion 649–50.

51. Tandri H, Calkins H, Marcus FI. Controversial role of magnetic resonance imaging in the diagnosis of arrhythmogenic right ventricular dysplasia. Am J Cardiol 2003; 92: 649.

52. Marcus FI. Update of arrhythmogenic right ventricular dysplasia. Card Electrophysiol Rev 2002; 6: 54–6.

53. Fontaine G, Fontaliran F, Hebert JL et al.

Arrhythmogenic right ventricular dysplasia. Annu Rev Med 1999; 50: 17–35.

54. Nasir K, Rutberg J, Tandri H et al. Utility of SAECG in arrhythmogenic right ventricle dysplasia. Ann Noninvasive Electrocardiol 2003; 8: 112–20.

55. Nasir K, Bomma C, Khan FA et al. Utility of a combined signal-averaged electrocardiogram and QT dispersion algorithm in identifying arrhythmogenic right ventricular dysplasia in patients with tachycardia of right ventricular origin. Am J Cardiol 2003; 92: 105–9.

56. Marcus F, Towbin JA, Zareba W et al. Arrhythmogenic right ventricular dysplasia/cardiomyopathy (ARVD/C): a multidisciplinary study: design and protocol. Circulation 2003; 107: 2975–8.

57. Hamid MS, Norman M, Quraishi A et al. Prospective evaluation of relatives for familial arrhythmogenic right ventricular cardiomyopathy/dysplasia reveals a need to broaden diagnostic criteria. J Am Coll Cardiol 2002; 40: 1445–50.

58. Melberg A, Oldfors A, Blomström-Lundqvist C et al. Autosomal dominant myofibrillar myopathy with arrhythmogenic right ventricular cardiomyopathy linked to chromosome 10q. Ann Neurol 1999; 46: 684–692.

59. McKoy G, Protonotarios N, Crosby A et al. Identification of a deletion in plakoglobin in arrhythmogenic right ventricular cardiomyopathy with palmoplantar keratodera and woolly hair (Naxos disease). Lancet 2000; 355: 2119–24.

60. Rampazzo A, Nava A, Malacrida S et al. Mutation in human desmoplakin domain binding to plakoglobin causes a dominant form of arrhythmogenic right ventricular cardiomyopathy. Am J Hum Genet 2002; 71: 1200–6.

61. Norgett EE, Hatsell SJ, Carvajal-Huerta L et al. Recessive mutations in desmoplakin disrupts desmoplakin-intermediate filament interactions and causes dilated cardiomyopathy, woolly hair and keratoderma. Hum Genet 2000; 9: 2761–6.

62. McMillan JR, Shimizu H. Desmosomes: structure, function in normal and diseased epidermis. J Dermatol 2001; 28: 291–8.

63. Armstrong DK, McKenna KE, Purkis PE et al. Haploinsufficiency of desmoplakin causes a striate subtype of palmoplantar keratoderma. Hum Mol Genet 1999; 8: 143–8.

64. Alcalai R, Metzger S, Rosenheck S et al. A recessive mutation in desmoplakin causes arrhythmogenic right ventricular dysplasia, skin disorder, and woolly hair. J Am Coll Cardiol 2003; 42: 319–27.

65. Wehrens XH, Marks AR. Altered function and regulation of cardiac ryanodine receptors in cardiac disease. Trends Biochem Sci 2003; 28: 671–8.

66. Priori SG, Napolitano C, Tiso N et al. Mutations in the cardiac ryanodine receptor gene (hRyR2) underlie catecholaminergic polymorphic ventricular tachycardia. Circulation 2001; 103: 196–200.

67. Laitinen PJ, Brown KM, Piippo K et al. Mutations of the cardiac ryanodine receptor (RyR2) gene in familial polymorphic ventricular tachycardia. Circulation 2001; 103: 485–90.

68. Ko KS, Arora PD, McCulloch CA. Cadherins mediate intercellular mechanical signaling in fibroblasts by activation of stretch-sensitive calcium-permeable channels. J Biol Chem 2001; 276: 35967–77.

69. Petroff MG, Kim SH, Pepe S et al. Endogenous nitric oxide mechanisms mediate the stretch dependence of Ca^{2+} release in cardiomyocytes. Nat Cell Biol 2001; 3: 867–73.

70. Tiso N, Salamon M, Bagattin A et al. The binding of the RyR2 calcium channel to its gating protein FKBP12.6 is oppositely affected by ARVD2 and VTSIP mutations. Biochem Biophys Res Commun 2002; 299: 594–8.

71. Bowles NE, Ni J, Marcus F et al. The detection of cardiotropic viruses in the myocardium of patients with arrhythmogenic right ventricular dysplasia/cardiomyopathy. J Am Coll Cardiol 2002; 39: 892–5.

72. Asano Y, Takashima S, Asakura M et al. Lamr1 functional retroposon causes right ventricular dysplasia in mice. Nat Genet 2004; 36: 123–30.

CHAPTER 34

Mitochondrial cardiomyopathies

Arthur B Zinn, Charles L Hoppel

Introduction

The heart requires large amounts of energy to maintain its normal function while in the resting or working states, as well as during the fed and fasted states. A small proportion of this energy is derived from anaerobic metabolism, but the myocardium depends primarily on aerobic metabolism within the mitochondria for the majority of its energy production, especially during periods requiring increased cardiac output. Fatty acids are the major fuel source for the myocardium, although it also utilizes glucose (or glycogen) and several amino acids. Hence, genetic and acquired defects that affect aerobic mitochondrial metabolism can be associated with clinically significant myocardial dysfunction. This chapter deals exclusively with genetic disorders, as opposed to acquired disorders, of mitochondrial function that produce cardiomyopathy and/or other forms of cardiac dysfunction.

The initial steps in the pathways of carbohydrate, fatty acid and protein metabolism that ultimately provide substrates for mitochondrial oxidation occur in the cytosol (Fig. 34.1).[1] The major carbohydrate fuel source is glucose. Dietary glucose can be used directly for energy production or stored for future energy utilization as glycogen. The steps in energy production involve the conversion of glucose to pyruvate via the cytosolic pathway called glycolysis, the transport of pyruvate into the mitochondrial matrix, the conversion of pyruvate to acetyl-CoA by the pyruvate dehydrogenase complex, the oxidative degradation of acetyl-CoA with the concomitant production of reducing equivalents (NADH and $FADH_2$) by the tricarboxylic acid cycle, and the utilization of these reducing equivalents to generate ATP by the oxidative phosphorylation (OXPHOS) system. Glucose can be replenished from pyruvate in the fasting state by the gluconeogenic pathway. When present in excess, glucose also can be stored as glycogen in liver, skeletal muscle and cardiac muscle for future energy production. The pathway leading from glucose to glycogen is called glycogen synthesis, while the pathway leading from glycogen to glucose (actually glucose-6-phosphate) is termed glycogenolysis. Glycogen synthesis, glycogenolysis, and glycolysis all occur in the cytosol. Thus, while disorders affecting these pathways can potentially cause cardiomyopathy, these disorders do not represent disorders of mitochondrial function, and are not reviewed in this chapter.

Similarly, amino acids that are derived from protein degradation are catabolized by a combination of cytosolic and mitochondrial pathways (Fig. 34.1). With rare exception, genetic defects affecting these mitochondrial pathways do not lead to cardiomyopathy. The exceptions to the rule are disorders that lead to over-production of specific acyl-CoAs that interfere with the OXPHOS system. For example, propionic acidemia is an

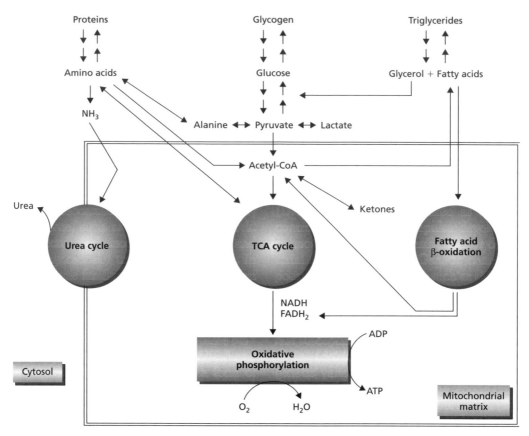

Figure 34.1. *Overview of mitochondrial energy metabolism. Abbreviations used: acetyl-CoA, acetyl-coenzyme A; ADP, adenosine diphosphate; ATP, adenosine triphosphate; FADH$_2$, reduced flavin adenine dinucleotide; NADH, reduced nicotinamide adenine dinucleotide; NH$_3$, ammonia; and TCA cycle, tricarboxylic acid cycle.*

autosomal recessive disorder of isoleucine, valine, threonine, and methionine catabolism that leads to excessive concentrations of propionyl-CoA in the mitochondrion. Propionic acidemia is a rare disorder and produces cardiomyopathy in only a small proportion of affected patients. Thus, the amino acid disorders are uncommon, indirect causes of mitochondrial cardiomyopathy, and will not be discussed further.

As opposed to genetic disorders of glucose and amino acid metabolism, inborn errors of fatty acid oxidation are a significant cause of mitochondrial cardiomyopathy. The pathways, pathophysiology, clinical features, and treatment of fatty acid oxidation disorders are discussed in the next section of this chapter. In brief, fatty acids are converted to

their acyl-CoA esters and then transported by a carnitine-dependent pathway into the mitochondrial matrix, where the fatty acid β-oxidation pathway degrades them, producing acetyl-CoA and reducing equivalents (NADH and FADH$_2$) (Fig. 34.1). Acetyl-CoA, in turn, enters the tricarboxylic acid (TCA) cycle (generating additional reducing equivalents) and the OXPHOS system, both of which are located within the mitochondrion. Thus, the TCA cycle and the OXPHOS system are the final common pathway for amino acid, glucose and fatty acid oxidation. It is not surprising, therefore, that inborn errors of the TCA cycle or the OXPHOS system can lead to mitochondrial cardiomyopathy. Inborn errors of the TCA cycle are not discussed further in this chapter because they are quite rare, and because they produce primarily neurological syndromes rather than cardiomyopathy. OXPHOS defects are discussed in detail because they are a significant cause of cardiomyopathy and other cardiac abnormalities.

These biochemical processes take place within the mitochondria, which are microscopic intracellular organelles that have a unique double membrane system.[1,2] The outer membrane surrounds the mitochondria, whereas the inner (boundary) membrane encloses the matrix or inside of the mitochondria. Thus, mitochondria have two spaces, the intermembrane space between the outer and inner membrane and the matrix space. Depending on the energy status, the inner boundary membrane and the outer membrane can fuse to form structures called contact sites that are involved in mitochondrial protein translocation, in energy coupling by transport of ADP, and in transport of long-chain fatty acids via the carnitine system. Attached to the inner membrane are cristae, which appear to be infoldings of the inner membrane. These cristae are attached to the inner membrane via pedicles called crista junctions. The cristae contain the respiratory complexes of the electron transport chain and the phosphorylation system. The inner membrane and cristae enclose the matrix. The matrix contains the majority of the mitochondrial proteins (all encoded by nuclear genes), the mitochondrial DNA and RNA, and granules containing calcium and phosphate.[2]

Disorders of fatty acid oxidation

Pathways

Dietary fat contains primarily saturated and unsaturated long-chain fatty acids of chain length C16 and C18. In addition, the average person's diet also contains long-chain fatty acids of different length (C12 to C18), as well as medium-chain (C6 to C10) and short-chain (C4 to C6) fatty acids. The fatty acid oxidation pathways vary with the chain length of the fatty acid. Long-chain fatty acids are transported across the plasma membrane by a long-chain specific transporter, require 'activation' to their acyl-CoA derivatives by an enzyme located in the outer mitochondrial membrane, and require a carnitine-dependent pathway to cross the mitochondrial inner and outer membranes into the mitochondrial matrix.[3] In contrast, medium- and short-chain fatty acids cross the plasma membrane by their own transport system and do not require activation to their acyl-CoA esters or a carnitine-dependent pathway to cross the mitochondrial inner and outer membranes into the mitochondrial matrix. Because dietary fat is composed primarily of long-chain fatty acids, and since cardiomyopathy is much more commonly associated with defects of long-chain fatty acid oxidation than with defects of medium- or short-chain fatty acid oxidation, the metabolism of long-chain fatty acids is the focus of this chapter.

The pathways of long-chain fatty acid metabolism and their inter-relationship with carnitine are depicted in Figs 34.2 and 34.3.[3,4] Long-chain fatty acids and carnitine are transported across the plasma membrane by energy-dependent carriers, the long-chain fatty acid transporter and the carnitine transporter (CT). Once inside the cytosolic space, long-chain fatty acids are converted to their corresponding acyl-CoA esters by fatty acyl-CoA synthetase, which is located in the outer mitochondrial membrane (OMM). The fatty acyl-CoA ester then undergoes a series of three carnitine-dependent processes to cross the inner mitochondrial membrane (IMM) to the mitochondrial matrix:

- carnitine palmitoyltransferase I (CPT I)
- carnitine-acylcarnitine translocase (CACT)
- carnitine palmitoyltransferase II (CPT II)

CPT I converts a long-chain fatty acyl-CoA (e.g. palmitoyl-CoA, a saturated C16-fatty acid) and carnitine to palmitoylcarnitine and free CoA; CACT carries palmitoylcarnitine across the inner mitochondrial membrane in exchange for free carnitine; and CPT II converts palmitoylcarnitine and free CoA to palmitoyl-CoA and free carnitine (which is recycled by the translocase). There is no loss or gain of free carnitine or free CoA in this series of reactions.

Once they are in the mitochondrial matrix, fatty acyl-CoA esters are degraded sequentially by a set of four enzymes, which shorten the fatty

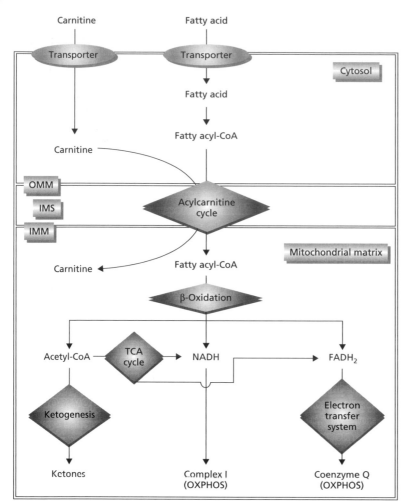

Figure 34.2. *Overview of long-chain fatty acid oxidation. Abbreviations used: acetyl-CoA, acetyl-coenzyme A; coenzyme Q, ubiquinone; FADH$_2$, reduced flavin adenine dinucleotide; IMM, inner mitochondrial membrane; IMS, intermembrane space; OMM, outer mitochondrial membrane space; OXPHOS, oxidative phosphorylation; NADH, reduced nicotinamide adenine dinucleotide; and TCA cycle, tricarboxylic acid cycle.*

acyl-CoA chain length (n) by two carbons, forming acetyl-CoA and a shorter acyl-CoA ($n − 2$), and generate reducing equivalents in the form of NADH and FADH$_2$ (Figs 34.2 and 34.4).[3–5] These four enzymes are:

- acyl-CoA dehydrogenase
- enoyl-CoA hydratase
- 3-hydroxyacyl-CoA dehydrogenase
- 3-ketoacyl-CoA thiolase.

There is more than one form of each of these enzymes, each serving to transform fatty acids of different chain length. Four acyl-CoA dehydrogenases have been described, all flavin-dependent: very

long chain acyl-CoA dehydrogenase (VLCAD), long chain acyl-CoA dehydrogenase (LCAD), medium chain acyl-CoA dehydrogenase (MCAD) and short chain acyl-CoA dehydrogenase (SCAD). VLCAD appears to be primarily responsible for the degradation of the normal dietary long-chain fatty acids, while the physiological role of LCAD in humans is uncertain. Two forms of enoyl-CoA hydratase, 3-hydroxyacyl-CoA hydratase and 3-ketoacyl-CoA thiolase exist, a long-chain form and a medium- and/or short-chain form for each enzyme. The long-chain forms of three of these enzymes – the enoyl-CoA hydratase, the 3-hydroxyacyl-CoA dehydrogenase and the thiolase – are contained within a single functional unit, called

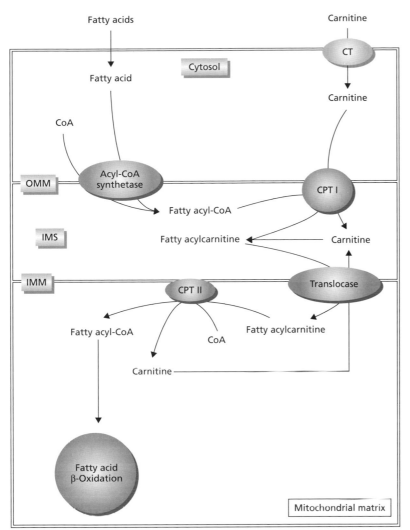

Figure 34.3. *Overview of fatty acid oxidation: the carnitine cycle. Abbreviations used: CoA, free coenzyme A; CPT I, carnitine palmitoyltransferase I; CPT II, carnitine palmitoyltransferase II; IMM, inner mitochondrial membrane; IMS, intermembrane space; OMM, outer mitochondrial membrane space.*

the trifunctional protein (TFP). Thus, VLCAD and the TFP are responsible for the initial stages of long-chain fatty acid β-oxidation.

The ultimate end products of fatty acid β-oxidation are acetyl-CoA and the reducing equivalents, NADH and FADH$_2$. Acetyl-CoA can enter the Krebs cycle (in the liver, kidney, skeletal muscle and heart muscle), or be used by the liver for ketogenesis. NADH and FADH$_2$ can enter the mitochondrial respiratory chain, directly in the case of NADH or indirectly via electron transfer flavoprotein (ETF) and electron transfer flavoprotein dehydrogenase (ETF DH) in the case of FADH$_2$ (Fig. 2).[3,4]

Disorders

A partial listing of inborn errors of fatty oxidation and their associated cardiac findings is provided in Table 34.1. In general, most disorders of long-chain fatty acid oxidation are associated with cardiomyopathy, whereas cardiomyopathy is an uncommon finding in patients with defects of medium- or short-chain fatty acid oxidation.[4–6] The biochemical hallmark of the disorders of fatty acid oxidation that involve the liver is non-ketotic or hypoketotic hypoglycemia.[4,5] To date, all of the disorders that are associated with cardiomyopathy also lead to hypoglycemia associated with impaired ketogenesis.

591

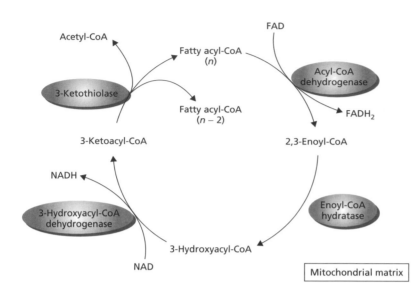

Figure 34.4. *Overview of fatty acid oxidation: the β-oxidation cycle. Abbreviations used: CoA, coenzyme A; FAD, flavin adenine dinucleotide; FADH₂, reduced flavin adenine dinucleotide; NAD, nicotinamide adenine dinucleotide; NADH, reduced nicotinamide adenine dinucleotide; n, the number of carbons in the fatty acid chain; n – 2, the number of carbons in the original fatty acid chain minus 2 carbons.*

Table 34.1. *Inborn errors of mitochondrial fatty acid oxidation that cause cardiomyopathy*

Category of disorder: specific defect (abbreviation)	Frequency of cardiomyopathy
Defects of acylcarnitine formation and transport	
Plasma membrane carnitine transporter (CT)	++++
Carnitine palmitoyltransferase I (CPT I)	–
Carnitine-acylcarnitine translocase (CACT)	++++
Carnitine palmitoyltransferase II (CPT II)	++++ → –
Defects of β-oxidation	
Very-long-chain acyl-CoA dehydrogenase (VLCAD)	++++ → –
Trifunctional protein (TFP)	+++
Long-chain 3-hydroxyacyl-CoA dehydrogenase (LCHAD)	+++
Medium-chain acyl-CoA dehydrogenase (MCAD)	–
Medium/short-chain 3-hydroxyacyl-CoA dehydrogenase (M/SCHAD)	+
Short-chain acyl-CoA dehydrogenase (SCAD)	–
Defects of electron transfer	
Electron transfer flavoprotein (ETF)	++
ETF dehydrogenase (ETF DH)	++

++++, 75–100%; +++, 50–75%; ++, 25–50%; +, 0–25%; –, rarely found.

Defects in long-chain fatty acid transport across the plasma membrane and the mitochondrial membranes

CPT I deficiency is the notable exception to the general rule that defects of long-chain fatty acid oxidation are associated with cardiomyopathy. This exception can be explained by the fact that CPT exists in two tissue-specific isoforms, a liver-specific form and a muscle-specific form, which are encoded by distinct loci. Almost all of the patients with CPT I deficiency who have been evaluated have deficiency of the liver-specific isoform, which is associated exclusively with hepatic disease. None of the few patients with deficiency of the muscle isoform developed cardiomyopathy. The biochemical hallmarks of CPT I deficiency are increased plasma concentrations of total and free carnitine, with a normal plasma concentration of long-chain acylcarnitines and an absence of abnormal accumulation of products of incomplete fatty acid oxidation in blood or urine. These findings are consistent with the physiological role of CPT I (Fig. 34.3).

The other defects of long-chain fatty acid transport, either across the plasma membrane (CT deficiency),[7,8] or the inner mitochondrial membranes (CACT deficiency[9,10] and CPT II deficiency),[11] generally produce cardiomyopathy in the neonatal period or in infancy. CT deficiency can present with failure to thrive, skeletal muscle weakness, cardiomyopathy, or sudden death. Oral carnitine must be provided in pharmacological doses to overcome the renal loss of carnitine and, more importantly, the absence of a functional physiological transporter in skeletal muscle and heart muscle. This treatment will prevent the development of cardiomyopathy and the other clinical features, as well as reversing many of these features in most patients once they have developed. CACT deficiency, generally presents with fulminant hepatic failure, hyperammonemia, skeletal muscle weakness and hypertrophic cardiomyopathy. Patients who present in the neonatal period have a poor prognosis, and often die; milder, later onset cases may have an improved prognosis. CPT II can present in the neonatal period with a multiple malformation syndrome (brain malformations and cystic renal disease) plus hypertrophic cardiomyopathy. Alternatively, the disorder can present later in childhood, adolescence, or adulthood with an abnormal phenotype limited to exercise-induced myoglobinuria. The prognosis for the congenital form of the disorder is poor, whereas later-onset cases are amenable to therapy.

Defects in fatty acid β-oxidation

The defects of long-chain fatty acid β-oxidation, VLCAD,[12,13] TFP deficiency[14] and isolated long-chain 3-hydroxyacyl-CoA dehydrogenase (LCHAD) deficiency,[15] also present with a range of clinical phenotypes. VLCAD deficiency can present in the neonatal period with cardiomyopathy,

hepatic dysfunction or sudden death, or in later life with an isolated skeletal myopathy. Patients with TFP deficiency present differently than do patients with isolated LCHAD deficiency, even though LCHAD is one of the enzymatic activities contained within the TFP.[14] TFP deficiency is primarily a combined skeletal myopathy and cardiomyopathy, with relatively few patients exhibiting significant hepatic disease. However, severe hepatic dysfunction is often the most prominent feature of isolated LCHAD deficiency. Affected neonates can present with severe hepatic failure, whereas older children can present with a more insidious, chronic course of hepatic disease. In additional to hepatic disease, patients with LCHAD deficiency also manifest cardio- and skeletal myopathy. Patients who have neonatal-onset disease often manifest retinal pigmentary degeneration and peripheral neuropathy later in the course of the disease. Although the reasons underlying the discordant presentations of TFP deficiency and isolated LCHAD deficiency are unknown, the commonly held view is that the differences are due to the effects of the long-chain 3-hydroxyacids that accumulate in isolated LCHAD deficiency.[16] Further support for the unique pathogenesis of isolated LCHAD deficiency is the observation that mothers who are heterozygous for isolated LCHAD deficiency are at greatly increased risk of developing HELLP syndrome (hemolysis, elevated liver enzymes, and low platelets) and AFLP (acute fatty liver of pregnancy) – but not cardiomyopathy – during a pregnancy in which they are carrying a fetus with homozygous LCHAD deficiency.[17] This situation appears to be an example of the unusual situation in which the fetus produces metabolites toxic to the pregnant mother, or more specifically, to the pregnant mother's liver.

As stated above, the defects of medium- and short-chain fatty acid β-oxidation, i.e. MCAD deficiency, M/SCAD deficiency and SCAD deficiency, generally do not produce cardiomyopathy, although occasionally they may do so (Table 34.1).[4,5] These disorders are associated primarily with impaired hepatic ketogenesis and/or skeletal myopathy.[18,19] The last group of disorders listed in Table 34.1 include the defects in electron transfer, ETF deficiency and ETF DH deficiency.[4,5] These two disorders affect several pathways that involve flavin-dependent dehydrogenase in addition to fatty acid β-oxidation, and thus produce a complex phenotype. Classically, affected patients present in the neonatal period with metabolic acidosis with or without craniofacial dysmorphism and congenital malformations (primarily cystic brain and renal abnormalities). Patients who survive the neonatal period often go on to develop cardiomyopathy. A small proportion of these patients present in infancy or thereafter with recurrent metabolic acidosis, hypoglycemia, hepatic dysfunction, skeletal myopathy and/or cardiomyopathy. Prognosis and treatment for these disorders is poor, although the subset of patients who respond to riboflavin supplementation can do well.

Diagnosis

The diagnosis of inborn errors of fatty acid β-oxidation depends on recognition of the need to consider the possibility of this category of disorders, followed by a sequential clinical and laboratory evaluation of the patient.[4,5] Clinically, the patient should be evaluated for liver dysfunction and skeletal muscle disease, in addition to performing a comprehensive cardiac evaluation. Sick newborns also should be evaluated for congenital brain malformations and renal cysts by the appropriate imaging study. The initial laboratory studies should include the following: serum glucose, plasma β-hydroxybutyrate and β-acetoacetate, serum ammonia, serum liver function studies, serum triglycerides and cholesterol, plasma free fatty acids, plasma and urine carnitine analysis (total carnitine, free carnitine, total acylcarnitines and an acylcarnitine profile), urine organic acids and urine acylglycines. The finding of non-ketotic or hypo-ketotic hypoglycemia during an acute presentation or following an age-appropriate fast is the hallmark of inborn errors of fatty acid oxidation. Exaggerated lipolysis, with increased serum triglycerides and fatty acid concentrations, also may be seen during fasting. Hyperammonemia is found in some patients who have defects in long-chain fatty acid oxidation, while abnormal liver function studies provide a non-specific clue. The presence and identity of abnormally increased plasma and/or urinary acylcarnitines, urine organic acids, or urine acylglycines often provides an important clue to the nature of the specific defect.

It is important to consider the possibility of an inborn error of fatty acid oxidation when the patient first presents, since that is often the best time to document abnormal laboratory findings. Important diagnostic laboratory clues can be obscured by intravenous or oral glucose therapy. Similarly, important diagnostic clues may not be present in the clinically stable patient. A patient who has a clinical history that is highly suggestive of a fatty acid oxidation disorder, but has normal laboratory findings when clinically stable, should be studied further by performing additional in vivo or in vitro studies.

Specialized in vivo stress tests, such as a controlled fasting study, a fat challenge test, or a carnitine challenge tests will often reveal an underlying inborn error of fatty acid oxidation. Similarly, in vitro tests such as measuring fatty acid oxidation in cultured skin fibroblasts, enzyme analysis of cultured skin fibroblasts, liver, or skeletal muscle, or polarographic analysis of isolated mitochondria from liver, or skeletal muscle, generally will permit a specific diagnosis to be made.[16,20] In some instances, especially where a particular disorder is characterized by a high degree of genetic homogeneity, genetic testing will provide an efficient means of diagnosis. This is the case, for example, for LCHAD deficiency in which one mutation accounts for approximately half the mutant alleles found in affected patients.

Following these steps generally will lead the clinician to a specific diagnosis, which will then permit appropriate guidance regarding treatment, prognosis and genetic counseling.

Treatment

The acute management of a critically ill patient with an inborn error of fatty acid oxidation includes intravenous administration of glucose and, in some disorders, intravenous carnitine supplementation.[5] The mainstay of chronic care for these patients is that they must avoid fasting and restrict dietary fat intake. In the case of some defects of long-chain fatty acid oxidation, supplementation with medium-chain triglycerides may be helpful, but medium-chain triglyceride supplementation should never be given to a patient with a defect in medium- or short-chain fatty acid oxidation. Chronic carnitine treatment can be life saving in the case of the plasma membrane carnitine transporter deficiency, and may be helpful in the other disorders.[7] Care should be exercised, however, in providing carnitine supplementation to patients who have defects in long-chain fatty acid oxidation that lead to accumulation of long-chain acylcarnitines.[5,9,10] The concern is that exogenous carnitine might further increase the intracellular concentration of long-chain acylcarnitines, which are potentially arrhythmogenic due to their effect on the calcium channel. Finally, riboflavin supplementation is helpful for the rare patient who has riboflavin-responsive multiple acyl-CoA dehydrogenase deficiency (ETF or ETF DH deficiency).

The prognosis for the cardiomyopathy and other clinical features associated with inborn errors of fatty acid oxidation varies within and among patients with the various disorders. As a rule, the disorders that have their onset during the neonatal period, i.e. CACT deficiency, CPT II deficiency, VLCAD deficiency, LCHAD/TFP deficiency and ETF/ETF DH deficiency, can be life threatening and often are refractory to treatment. Many of these disorders also have milder infantile, juvenile, or adult-onset forms, which are more amenable to treatment and have a better prognosis.

Disorders of oxidative phosphorylation

Pathways

The mitochondrial oxidative phosphorylation (OXPHOS) pathway is composed of five multimeric protein units, termed complexes I, II, III, IV and V.[1,21,22,23] The common names, composition and genetic origin of these complexes are provided in Table 34.2. Complexes I, III, IV and V are composed of some subunits encoded by the mitochondrial genome (mtDNA) and others encoded by the nuclear genome (nDNA), whereas complex II is composed only of nuclear encoded subunits. The dual genetic origin of the OXPHOS complex has important implications for the pathophysiology of inborn errors of the OXPHOS. We therefore will discuss several of the key differences between nDNA and mtDNA.

The mitochondrial genome is located within the mitochondrion and therefore is a cytoplasmic rather than a nuclear component. The structure of mtDNA differs from that of nDNA in the following ways:[21,23]

- *mtDNA is a circular DNA composed of two non-identical DNA strands, an outer (or heavy) strand and an inner (or light) strand:* mtDNA contains approximately 16 500 base pairs and therefore is several orders of magnitude smaller than the nuclear genome. The heavy and light strands

Table 34.2. *Components, common names and genetic composition of the oxidative phosphorylation (OXPHOS) complexes*

| Complex | Common name | Genetic origin of subunits | | |
		mtDNA	nDNA	Total
I	NADH–ubiquinone oxidoreductase	7	≥39	≥46
II	Succinate–ubiquinone oxidoreductase	0	4	4
III	Ubiquinol–cytochrome c oxidoreductase	1	10	11
IV	Cytochrome c oxidase	3	10	13
V	ATP synthase	2	14	16

ATP, adenosine triphosphate; mtDNA, mitochondrial DNA; nDNA, nuclear DNA; NADH, reduced nicotinamide adenine dinucleotide; ubiquinol, reduced ubiquinone.

encode for different gene products; most genes (28) are located on the heavy strand.[24] Together, the strands encode for 37 genes: 2 ribosomal RNA (rRNA) genes, 22 transfer RNA (tRNA) genes and 13 messenger RNA (mRNA) genes. All 13 mRNAs encode for polypeptides that serve as subunits of the OXPHOS complexes. These 13 subunits make up only 15% of the total number of subunits in the five OXPHOS complexes. The majority of subunits (~80 subunits) are nuclear-encoded

- *the genetic codes of mtDNA and nDNA differ*: several codons are read differently by the mitochondrial and nuclear translational machinery
- *mtDNA has no introns*: the mtDNA genes do not contain introns. This presumably conferred a selective advantage to mtDNA during its evolution – smaller genomes replicate faster. Over 90% of the nucleotides are coding
- *mtDNA has a higher mutation rate than nDNA*: the frequency of mtDNA mutations is approximately 10-fold greater than that of nDNA mutations, because the mitochondrion does not contain histones or an efficient system for DNA repair. In addition, a mutation that occurs in mtDNA is more likely to have deleterious consequences than is a nDNA mutation, because a greater proportion of the nucleotides in mtDNA is part of the coding sequences.

There are also several key biological differences between mtDNA and nDNA, including the following:

- *mtDNA is inherited exclusively from the mother*: although sperm contain mitochondria, sperm mitochondria do not enter the egg (except on very rare occasions). The zygote therefore derives all of its mitochondria, and all the mtDNA contained therein, from the egg. mtDNA inheritance therefore is termed maternal or matrilineal
- *mtDNA multiplicity*: each mitochondrion contains 2–10 copies of mtDNA. Each cell, in turn, contains a large number of mitochondria, ranging from tens to hundreds of mitochondria per cell. Thus, there are tens to thousands of copies of mtDNA per cell. This multiplicity of

mtDNA is quite different than the case for nDNA, in which all cells contain two copies of nDNA (except mature germ cells, which are haploid)

- *mtDNA exhibits heteroplasmy leading to an 'average' phenotype*: a cell may contain more than one type of mtDNA, including a normal ('wild') type and one or more mutant types. The presence of different types of mtDNA within the same cell is termed heteroplasmy; the presence of only one type of mtDNA is termed homoplasmy. A cell can be homoplasmic for a normal or abnormal type of mtDNA
- *mtDNA undergoes replicative segregation*: the ratio of normal to abnormal mtDNA in a cell that is heteroplasmic may shift with cellular (nuclear) division, because the partitioning of mitochondria into daughter cells during cell division is random, the rates of mtDNA replication for normal and mutant mtDNA may be different, and there may be selection for or against the mutant mtDNA. Thus, a cell may experience redistribution toward homoplasmy for either the normal or the mutant type of mtDNA. The process of replicative segregation may occur any time after fertilization, both before and after birth, in any replicating cell. Thus, replicative segregation may lead to different degrees of heteroplasmy in different tissues and organs that may change over time
- *threshold effect*: the phenotypic consequences of mtDNA heteroplasmy may vary for different tissues or organs, depending on each tissue's or organ's dependence on mitochondrial energy production. Thus, a particular mutation may have clinical consequences that are tissue specific or organ specific.

These unique biological properties of mtDNA lead to the following rules for inheritance for a clinical disorder associated with a mtDNA mutation:

- mitochondrial inheritance is matrilineal
- an affected woman passes her mtDNA to all of her children
- an affected man does not pass his mtDNA to any of his children
- an affected individual's phenotype may exhibit

variable expression in different tissue and organs

- an affected individual's phenotype may vary over time.

It also is important to note that a patient with a mitochondrial-encoded disorder may have it as the consequence of a sporadic, i.e. *de novo*, mutation, and will have a negative family history. This is almost always the case for patients who have large structural rearrangements (deletions or duplications) of mtDNA rather than those who have point mutations (see below).

Disorders

The traditional approaches to classifying mitochondrial disorders were based on the pattern of clinical phenotypes, pathological findings and/or enzyme analysis.[21–23,25] Although still useful, the limitations of these approaches to classification became apparent when the underlying genetic causes of the various disorders were discovered. Genetic studies revealed that a particular disease, or pathological finding, or enzyme deficiency could be associated with more than one genetic defect, and conversely, the same genetic defect could produce different diseases. Current efforts to classify OXPHOS disorders based on their genetic basis are imperfect, but are evolving as more is learned about the various causes of impaired mitochondrial respiration. The genetic approach divides the disorders into two broad categories, mtDNA mutations and nDNA mutations, which are then further subdivided into the general outline shown in Box 34.1.

More than 100 mutations of mtDNA-encoded genes have been described to date, but only a relatively small subset of these mutations produce cardiomyopathy or other cardiac abnormalities.[24] Most of the patients who have been identified in this category have tRNA mutations. Similarly, only a few mutations that involve nuclear genes that encode for the OXPHOS subunits, or their transport, assembly or stabilization, or mtDNA replication or maintenance have been described in patients who have cardiomyopathy. It is anticipated that many more nDNA defects will be recognized in the future, since more than 1000

Box 34.1. *Classification of disorders affecting mitochondrial oxidative phosphorylation (OXPHOS)*

Mutations of mtDNA-encoded genes
- Large deletions and/or duplications
- Point mutations:
  - rRNA
  - tRNA
  - mRNA (encoding OXPHOS subunits)

Mutations of nDNA-encoded genes
- Mutations of nuclear genes that encode for OXPHOS subunits
- Mutations of nuclear genes required for transport, assembly or stabilization of OXPHOS complexes
- Mutations of nuclear genes required for mtDNA replication or maintenance:
  - multiple mtDNA deletions
  - mitochondrial depletion syndromes
- Mutations of other nuclear genes required for normal functioning of OXPHOS complexes

---

mRNA, messenger RNA; mtDNA, mitochondrial DNA; nDNA, nuclear DNA; rRNA, ribosomal RNA; and tRNA, transfer RNA.

---

nuclear genes are thought to be involved in mitochondrial synthesis and function. Finally, the number of disorders associated with other nuclear genes that affect mitochondrial OXPHOS is small, but is increasing, as we better understand the pathogenesis of many clinical syndromes that are associated with cardiomyopathy.

In general, almost all patients who have been identified as having mutations of mtDNA-encoded genes, mutations of nuclear genes that encode for the OXPHOS subunits, mutations of nuclear genes required for transport, assembly or stabilization of OXPHOS genes, or mutations of nuclear genes required for mtDNA replication or maintenance have complex systemic disorders rather than isolated cardiomyopathy.[21,23,26,27] This is not surprising when we consider the fact that all cells, tissues and organs, except the mature red cell, are dependent on mitochondrial energy metabolism. Disorders of mitochondrial OXPHOS were recognized initially in patients who had neurological or neuromuscular disease, and the number of neurological manifestations associated with this group of disorders is now quite large, involving all components of the central and peripheral nervous systems. Many of the patients found to have neurological abnormalities associated with an underlying mitochondrial OXPHOS abnormality also were shown to have other systemic findings. With time, mitochondrial abnormalities were found in patients who had a highly diverse range of non-neurological disorders, e.g. failure to thrive, visual impairment (retinal pigmentary degeneration), cardiac disease (cardiomyopathy and conduction abnormalities), hepatic dysfunction, pancreatic disease (exocrine dysfunction), proximal renal tubular dysfunction, rhabdomyolysis, blood cell disorders (sideroblastic anemia, neutropenia, and thrombocytopenia), endocrinopathy (adrenal insufficiency, diabetes mellitus, hypoparathyroidism, and multiple symmetric lipomatosis) and psychiatric disease (depression and schizophrenia-like disorders). Many patients have been identified who have one or more of these systemic manifestations, but no abnormal neurological manifestations.

The cardiac problems that have been identified in patients with disorders of mitochondrial OXPHOS include cardiomyopathy and cardiac conduction abnormalities. Hypertrophic and dilated cardiomyopathy are the most common forms of cardiomyopathy that have been identified in patients who have mitochondrial disease,[26,27] but there have also been reports of families with histiocytoid cardiomyopathy,[28,29] and non-compaction of the left ventricle.[30] Several different cardiac conduction defects have been identified including partial and complete heart block and Wolff–Parkinson–White syndrome. This chapter focuses on defects that produce cardiomyopathy. Most, but not all, of the mtDNA mutations or nDNA mutations that produce cardiomyopathy do so in infancy or early childhood. On the other hand, the disorders caused by mutations of 'other' nuclear genes that affect mitochondrial OXPHOS subunits (last category in Box 34.1) can be associated with adult-onset cardiomyopathy.

A partial listing of inborn errors of mitochondrial OXPHOS is provided in Tables 34.3 and 34.4,

using the classification scheme outlined in Box 34.1. Table 34.3 and Table 34.4 list the disorders associated with mutations of mtDNA-encoded genes and nDNA-encoded genes, respectively. Mitochondrial-encoded disorders that have been identified in only a single laboratory are considered 'provisional', whereas those identified in more than one laboratory are considered 'confirmed'.[24] All of the confirmed mtDNA mutations that can cause cardiomyopathy, but only selected provisional mutations, are discussed below.

Mutations of mitochondrial-encoded genes

The first mitochondrial-encoded disorder listed in Table 34.3 is Kearns–Sayre syndrome, which is caused by large deletions or, less commonly, by large duplications of the mtDNA, affecting several genes.[21,23] The deletions can involve all regions of the mtDNA, but usually spare the origin of replication in the heavy strand. The most common deletion involves a 5 kb segment that encompasses a segment from the ND5 gene to the ATPase 6 gene (see Fig. 34.5). The boundaries of this deletion are defined by repeating DNA sequences ('mutational hot-spots'). Kearns–Sayre syndrome is classically associated with chronic progressive external ophthalmoplegia, blindness, an elevated cerebrospinal fluid protein concentration, and onset before age 20 years. In addition, most patients develop cardiac conduction defects, whereas a minority develops hypertrophic cardiomyopathy in the later stages of the disease.

Cardiomyopathy also is associated with mtDNA point mutations affecting OXPHOS structural genes, rRNAs and tRNAs (Table 34.3 and Fig. 34.5).[27] Two point mutations that may cause cardiomyopathy have been identified provisionally in the cytochrome b subunit of complex III. The G15243A mutation causes severe hypertrophic cardiomyopathy in infancy. The other mutation, the G15498A mutation, causes an unusual disorder called histiocytoid cardiomyopathy.[28] This disorder generally presents with severe arrhythmias before 2 years of age, preferentially affects females, is characterized histologically by accumulation of foamy histiocyte-like cells in the myocardium, and has a poor prognosis, often culminating in sudden death. Recently, another

Table 34.3. *Mitochondrial-encoded OXPHOS defects that may cause cardiomyopathy*

mtDNA gene	Mutation	Clinical features
Large deletion and/or duplication	Common deletion (5 kb)	Kearns–Sayre syndrome: onset before age 20 years, blindness, elevated cerebrospinal fluid protein, cardiac conduction abnormality, cardiomyopathy (rare, late in disease)
Point mutations of OXPHOS genes		
Complex III (cyt b)	G15243A	Hypertrophic cardiomyopathy (neonatal onset)
	G15498A	Histiocytoid cardiomyopathy (infancy onset)
Complex V (ATPase 6)	T8993G	Leigh syndrome
		NARP
		Hypertrophic cardiomyopathy
Point mutations of rRNA genes		
12S	A1555G	Aminoglycoside-induced deafness
		Cardiomyopathy
Point mutations of tRNA gene		
Gly	T9997C	Hypertrophic cardiomyopathy
Ile	A4269G	Dilated cardiomyopathy
	A4295G	Hypertrophic cardiomyopathy
	A4300G	Hypertrophic cardiomyopathy
Leu (UUR)	A3243G	MELAS
		Diabetes mellitus ± deafness
		Hypertrophic cardiomyopathy
		Wolff–Parkinson–White syndrome
	A3260G	Skeletal myopathy
		Hypertrophic cardiomyopathy
	C3303T	Skeletal myopathy
		Hypertrophic cardiomyopathy (infantile onset)
Lys	A8344G	MERRF
		Multiple symmetric lipomatosis
		Hypertrophic cardiomyopathy
	G8363A	Hearing loss
		Hypertrophic cardiomyopathy

ATPase 6, ATP synthase subunit 6; cyt b, cytochrome b; mtDNA, mitochondrial DNA; MELAS, mitochondrial encephalomyopathy, lactic acidosis and stroke-like episodes; MERRF, myoclonic epilepsy and ragged-red fibers; NARP, neuropathy, ataxia, and retinitis pigmentosa.

mtDNA mutation (the tRNA^{Lys} A8344G mutation, see below) was found in a patient with histiocytoid cardiomyopathy, confirming that this phenotype may be associated relatively frequently with mitochondrial defects.[29]

Another mutation affecting an OXPHOS structural gene, the T8993G mutation in the ATPase 6 gene (complex V), is a well-confirmed cause of cardiomyopathy.[23,27] This mutation can cause Leigh syndrome (subacute necrotizing encephalomyopathy) in infancy or NARP (neuropathy, ataxia, and retinitis pigmentosa) in later childhood or adulthood. Leigh syndrome is a neurodegenerative disease that classically affects the pons and thalamus, and leads to optic atrophy, hyotonia, pyramidal signs, developmental delay

Table 34.4. *Nuclear-encoded OXPHOS defects that may cause cardiomyopathy*

OXPHOS complex or protein	Gene	Clinical features
Mutation of nuclear genes that encoded for OXPHOS subunits		
Complex I	NDUFV2	Hypertrophic cardiomyopathy (neonatal onset) Encephalopathy
	NDUFS2	Hypertrophic cardiomyopathy (neonatal onset) Encephalopathy
Mutations of nuclear genes involved in mitochondrial transport, assembly or stabilization of OXPHOS complexes		
Complex IV	COX10	Anemia Hypertrophic cardiomyopathy Leigh syndrome
	COX15	Hypertrophic cardiomyopathy (fatal infantile disease)
	SCO2	Hypertrophic cardiomyopathy Leigh syndrome (infantile onset)
Mutations of nuclear genes required for mtDNA replication or maintenance		
Mitochondrial depletion syndrome	–	Hypertrophic cardiomyopathy Encephalomyopathy
Multiple mtDNA deletions	–	Progressive external ophthalmoplegia Hypertrophic cardiomyopathy
Mutations of other nuclear genes required for normal functioning of OXPHOS complexes		
Frataxin	FRDA	Friedreich ataxia Onset before age 20 years Ataxia Sensorimotor neuropathy Diabetes mellitus Hypertrophic cardiomyopathy
Linoleoyl-specific acyltransferase	Tafazzin	Barth syndrome Onset in infancy Cataracts Hypertrophic cardiomyopathy Neutropenia Skeletal myopathy

COX, cytochrome c oxidase (the functional name for complex IV); NDUFV2 and NDUFS7, subunits of complex I; COX10, COX15 and SCO2, names of genes that encode for proteins involved in complex IV assembly; FRDA, name of the gene that encodes for Freidriech ataxia.

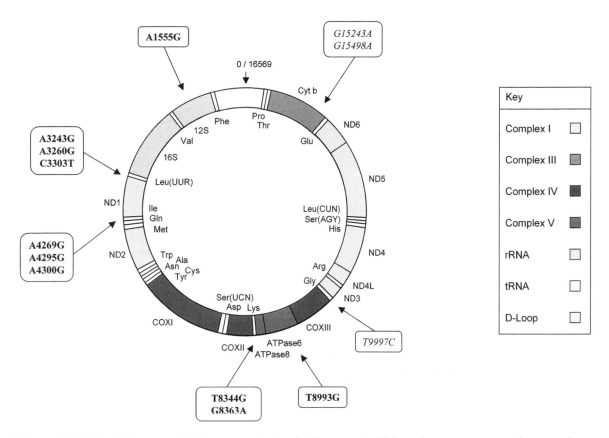

Figure 34.5. *Morbidity map of the human mitochondrial genome for defects that may cause cardiomyopathy. The mitochondrial genome contains two circular strands of DNA, an outer strand and an inner strand. The two strands encode for a total of 37 genes, 2 ribosomal RNAs (rRNAs), 22 transfer RNAs (tRNAs) and 13 messenger RNAs (mRNAs). For simplicity, the two strands have been shown as a single circular DNA. The 2 rRNA genes and the 22 tRNA genes are indicated inside the circle. The rRNA genes are designated by the sedimentation rate of the product they encode (12S and 16S), whereas the tRNA genes are designated by the standard three-letter abbreviation for the amino acids they transfer. The 13 mRNA genes are designated by the standard abbreviations for the OXPHOS subunits they encode: complex I (ND, NADH dehydrogenase), ND1, ND2, ND3, ND4, ND4L, ND5, ND6 and ND7; complex III cytochrome b, (cyt b); complex IV (COX, cytochrome c oxidase), COXI, COXII and COXIII; and complex V (ATPase, ATP synthase), ATPase6 and ATPase8. The origin of replication is located in the displacement loop (D-loop); the nucleotides of the mtDNA sequence also begin and end in the displacement loop. Mutations that have been confirmed to cause cardiomyopathy are indicated in bold type, while mutations that have been implicated provisionally (i.e. only demonstrated by one laboratory) are indicated in italics.[24] See color plate section*

and seizures. However, many patients with Leigh syndrome also have systemic manifestations, including cardiomyopathy.[21,23] Leigh syndrome or Leigh-like syndrome is the consequence of many different genetic defects, and this group of disorders is discussed below (see nuclear-encoded

defects of genes that are required for assembly of OXPHOS complexes). Cardiomyopathy also can be seen in older patients who have NARP.

Cardiomyopathy has been identified in patients with the A1555G mutation in the 12S rRNA, a mutation more commonly associated with sporadic or aminoglycoside-induced deafness (Table 34.3 and Fig. 34.5).[24] This is an interesting mutation that alters the properties of the 12S rRNA so that it more closely resembles its bacterial homolog, which was the evolutionary precursor for the human mitochondrial 12S rRNA molecule. Cardiomyopathy is not commonly seen in patients with the A1555G mutation. Other mutations in the 12S and 16S rRNA genes have been identified in patients with hypertrophic cardiomyopathy, but they have not been confirmed.

Point mutations of several mitochondrial-encoded tRNA genes lead to hypertrophic cardiomyopathy (Table 34.3 and Fig. 34.5). The tRNA^Ile and tRNA^Leu genes appear to be 'hot-spots' for mutations that produce cardiomyopathy,[23,27,31] but pathogenetic mutations also have been identified in the tRNA^Gly and tRNA^Lys genes.[23,27] Mutations in several other tRNA genes have been reported, but require confirmation. Three tRNA^Ile mutations have been confirmed, i.e. A4269G, A4295G and A4300G, and other mutations have been identified provisionally in this gene. The originally described patient with the A4269G mutation was a child who had a systemic disorder (short stature, deafness, glomerulosclerosis and seizures), and then died as a young adult from dilated cardiomyopathy. The A4295G mutation appears to lead to isolated hypertrophic cardiomyopathy in infancy. The A4300G leads to maternally inherited hypertrophic cardiomyopathy but has no other systemic effects. Compared to mutations of other tRNAs described below, the tRNA^Ile gene appears to be particularly susceptible to undergoing mutations that lead to isolated hypertrophic cardiomyopathy.

Efforts to explain the cardioselectivity of the tRNA^Ile mutations have been inconclusive. For example, two of these mutations, A4269G and A4317G, showed only small reductions in the efficiency of aminoacylation of the tRNA they produced.[31] Similarly, the observation that the tRNA^Ile A4300G mutation can be present in either the heteroplasmic or homoplasmic state in patients with cardiomyopathy is a departure from one of the long-standing rules for determining the pathogenicity of mtDNA mutations.[32] Clearly, further studies are needed to understand the pathophysi-

ology of these tRNA mutations and their cardioselectivity.

Three tRNA^Leu mutations have been confirmed: A3243G, A3260G and C3303T.[33,34] The A3243G mutation is the most commonly diagnosed mitochondrial tRNA mutation. It is the most common cause of MELAS (mitochondrial encephalomyopathy, lactic acidosis and stroke-like episodes), which is a highly variable clinical syndrome. Children with MELAS generally present with a cyclic vomiting, migraine, seizures, recurrent stroke-like episodes that can be associated with hemiparesis and lactic acidosis. Adults with MELAS more typically present in the third or fifth decade with hearing loss and/or diabetes mellitus before the onset of stroke-like episodes (i.e. strokes that produce clinical findings, and radiological images that are inconsistent with the normal pattern of vascular anatomy). Patients with the A3243G mutation can also present with a wide spectrum of other clinical findings. Hypertrophic cardiomyopathy and/or Wolff–Parkinson–White syndrome develop in a significant proportion of patients who have systemic forms of MELAS, but neither of these cardiac abnormalities present as isolated findings in patients with the A3243G mutation. Patients with the latter mutation develop the maternally inherited myopathy and cardiomyopathy syndrome. The syndrome has a high degree of clinical variability, which appears consistent with the heteroplasmic nature of the underlying mutation. The C3303T mutation leads to infantile-onset hypertrophic cardiomyopathy with or without skeletal myopathy.[33] Some, but not all, of the tissue and organ variability for these mutations can be explained by heteroplasmy.[34] The bioenergetic consequences of the A3243G mutation have also been confirmed by *in vivo* phosphorus magnetic resonance spectroscopy.[35]

The T9997C mutation in the tRNA^Gly gene leads to maternally inherited hypertrophic cardiomyopathy and systemic manifestations, including intestinal pseudo-obstruction in some patients. Thus far, the importance of this mutation has not been confirmed, so that its inclusion in this chapter is based on its 'provisional' status. Cell fusion studies with enucleated human cultured skin fibroblasts from patients with the T9997C mutation and ρ^0 cells (mitochondria-depleted osteosarcoma cells) showed that the mutation caused complex I and complex IV deficiency, suggesting the pathogenetic basis for this disorder.[36]

Two tRNA^{Lys} mutations have been described: A8344G and G8363A (Table 34.3 and Fig. 34.5). The A8344G mutation is the primary cause of MERRF (myoclonic epilepsy and ragged red fibers), a neuromuscular disorder that is associated with myopathy and ataxia, as well as myoclonic epilepsy. An alternate phenotype for this mutation is the highly distinctive multiple symmetric lipomatosis syndrome, which is a rare disorder characterized by an unusual distribution of brown fat, i.e., a 'horse-collar' pattern. Approximately one-third of patients with the classical form of MERRF develop cardiac muscle dysfunction. A second tRNA^{Lys} mutation, G8363A, leads to maternally inherited hearing loss and hypertrophic cardiomyopathy.

Mutations of nuclear-encoded genes

It is anticipated that mutations affecting the nuclear-encoded OXPHOS genes would be responsible for a large proportion of OXPHOS disorders, but relatively few mutations have been identified (Table 34.4). Mutations affecting two of the nuclear-encoded subunits of complex I, the NDUFV2 and NDUFS2, have been shown to cause neonatal- or infantile-onset cardiomyopathy plus other systemic manifestations.[24,37] No mutations that cause cardiomyopathy have been confirmed in any of the structural OXPHOS genes for complex II, III, IV or V.

Efforts to identify mutations of nuclear genes involved in mitochondrial transport, assembly, or stabilization of OXPHOS complexes have been more successful than the efforts to identify mutations in the structural genes themselves. The genetic evaluation of patients with enzymatically confirmed cytochrome c oxidase (complex IV) deficiency has been particularly successful.[38] Cardiomyopathic mutations have been identified in three genes, COX10,[39] COX15[40] and SCO2 (Table 34.4).[41] COX10 and COX15 encode for proteins that have roles in heme A synthesis. Patients who have COX10 deficiency develop Leigh syndrome, proximal tubular dysfunction,

and/or cardiomyopathy. COX15 deficiency leads to Leigh syndrome, sideroblastic anemia and hypertrophic cardiomyopathy. SCO2 is a copper-transporter located in the inner mitochondrial membrane and its deficiency leads to cytochrome c oxidase (COX) deficiency because COX is a copper-dependent enzyme. Clinically, SCO2 deficiency produces a Leigh syndrome-like picture plus hypertrophic cardiomyopathy. Parenthetically, the most common cause of Leigh syndrome is SURF1 deficiency. The SURF1 gene encodes for another protein that is involved in assembly of cytochrome c oxidase. What is interesting from the point of view of trying to establish genotype/phenotype correlations for various forms of COX deficiency is that patients with SURF1 mutations have purely neurological presentations, while patients with COX10, COX15 and SCO2 deficiency have more complex phenotypes including proximal renal dysfunction, sideroblastic anemia and/or hypertrophic cardiomyopathy.

Progress has been made in the past several years in identifying mutations of nuclear genes involved in mitochondrial DNA replication and maintenance. These mutations produce two striking biochemical phenotypes: multiple mtDNA deletions and mitochondrial depletion syndrome.[23] Multiple mtDNA deletions have been identified in several different clinical syndromes, of which ophthalmoplegia appears to be the most common finding. Both autosomal dominant and autosomal recessive forms of multiple mtDNA deletions exist. One of the autosomal recessive forms leads to a systemic disorder that includes ophthalmoplegia and hypertrophic cardiomyopathy, but the genetic basis of this particular disorder has not been identified. However, the other clinical syndromes associated with multiple mtDNA deletions are either the consequence of an error involving a gene required for mtDNA replication (e.g. polymerase γ deficiency), or a gene required for maintaining intramitochondrial nucleotide pools (e.g. thymidine phosphorylase deficiency).[23] Similarly, cardiomyopathy has been found in a few patients with a mitochondrial depletion syndrome, but the underlying genetic cause of the disorder has not been established.[23]

The final group of nuclear-encoded disorders that is listed in Table 34.4 is those associated with genes that play a role in mitochondrial homeostasis, but do not encode for the OXPHOS structural genes or are not involved in mtDNA replication or maintenance. This is a biochemically diverse group of disorders, which will no doubt grow in number in the future. The first member of this group of disorders is Barth syndrome, an X-linked Mendelian disorder associated with mutations in the G4.5 gene, which encodes for tafazzin.[42] Tafazzin encodes for a linoleoyl-specific acyltransferase.[43] This acyltransferase has a key role in the biosynthesis of cardiolipin, which is a critical component of mitochondrial membranes and is required for normal mitochondrial respiratory chain activity. Clinically, Barth syndrome is associated with cardiomyopathy (either hypertrophic cardiomyopathy or non-compaction of the left ventricle), skeletal myopathy and neutropenia.[42] Other genes have been implicated in the causation of non-compaction of the left ventricle, all of which are nuclear encoded and not directly involved in OXPHOS synthesis or maintenance.

Friedreich ataxia, an autosomal recessive disorder associated with onset in late adolescence or early adulthood, ataxia, sensorimotor neuropathy, diabetes mellitus and hypertrophic cardiomyopathy, is caused by an abnormal triplet repeat (GAA) in the FRDA gene.[44] Frataxin, the protein product of FRDA gene, plays a role in intramitochondrial iron homeostasis. Patients with Friedreich ataxia are thought to have an increased free iron concentration in the mitochondrion, which leads to oxidative damage to the iron-sulfur clusters that are contained in complex I, complex II, complex III and aconitase (one of the tricarboxylic acid cycle enzymes).[45]

Diagnosis

The diagnosis of inborn errors of OXPHOS depends on recognition of the need to consider the possibility of this category of disorders, followed by a sequential clinical and laboratory evaluation of the patient.[21,23,46] The clinical presentations of patients with OXPHOS disorders are incredibly diverse and variable. Historically, genetic disorders of OXPHOS were recognized primarily in patients with neurological abnormalities, especially myopathy or encephalomyopathy. With time, the range of neurological features has grown considerably, as has the range of systemic manifestations (see above).

The initial laboratory evaluation of a patient suspected of having a mitochondrial disorder should include analysis of the following: serum glucose and electrolytes, serum lactate and pyruvate, plasma β-hydroxybutyrate and β-acetoacetate, serum liver function studies, plasma amino acids (especially alanine, the transamination product of pyruvate), plasma and urine carnitine analysis (total carnitine, free carnitine, total acylcarnitines and an acylcarnitine profile), and urine organic acid analysis.

It is important to recognize that many patients who have OXPHOS disorders do not have lactic acidemia, pyruvic acidemia, or hyperalaninemia. The lactate/pyruvate ratio should be calculated because it can be useful in distinguishing patients who have lactic acidemia associated with an OXPHOS defect, from patients who have other genetic defects. Patients who have an OXPHOS defect often have an increased lactate/pyruvate ratio, whereas patients who have a lactic acidemia associated with another genetic cause generally have a normal lactate/pyruvate ratio. Many patients with an OXPHOS defect do not have lactic acidemia because they have organ-specific defects, or for unknown reasons. Some patients will exhibit lactic acidemia following a glucose load, and patients should be tested before and after a meal or a standardized glucose load. Similarly, many patients will exhibit lactic acidemia if they are tested during a period of severely impaired cardiac function or shock. Patients found to have lactic acidemia during such times should be retested after they have clinically stabilized. Most patients do not exhibit either hypoglycemia or impaired ketosis, especially patients who have tissue-specific disorders that do not involve the liver. The plasma carnitine analysis often will show a non-specific decrease in total and free carnitine concentration, with an unremarkable acylcarnitine profile. The urine carnitine analysis might be normal or show excessive carnitine and acylcarnitine excretion, especially in patients who have impaired renal tubular function. The urine organic acid analysis might reveal markedly increased excretion of lactic acid and pyruvic acid, or a non-specific pattern of mildly increased excretion of lactic acid and Krebs cycle metabolites, or a normal pattern.

Specialized biochemical testing including *in vivo* stress tests, e.g. a controlled exercise challenge test with biochemical monitoring and/or *in vitro* tests, e.g. enzyme analysis of cultured skin fibroblasts, liver, or skeletal muscle, or polarographic analysis of isolated mitochondria from liver, or skeletal muscle, will often permit a specific diagnosis to be made.[46,47] In those instances where clinical evaluation suggests a particular clinical syndrome, genetic testing often will provide the most direct and successful approach to diagnosis. Furthermore, many mutations are expressed in readily available cell types, e.g. white blood cells or urinary sediment cells, and therefore mutational analysis often is the least invasive and least expensive approach to obtaining a diagnosis. However, most pathogenetic mutations of mtDNA-encoded genes are heteroplasmic rather than homoplasmic. Heteroplasmic point mutations generally may be identified in blood cells, whereas heteroplasmic large deletions generally cannot be identified. In addition, mutational analysis of a complete 'panel' of mtDNA and nDNA mutations is not clinically available at this time, thereby limiting the sensitivity of genetic testing for patients who have a suspected mitochondrial cardiomyopathy.

Treatment

The management of cardiomyopathy in patients who have OXPHOS disorders does not differ from the management of the cardiac disease in other patients, because the management of the underlying OXPHOS defects has achieved limited success. There have been no carefully controlled double-blind studies of any of the current therapeutic modalities for the OXPHOS disorders.[21,23] Dietary modifications are of limited benefit because the OXPHOS system is the final common pathway for the aerobic metabolism of carbohydrates, proteins and fats. Frequent feedings might be helpful. Carnitine supplementation should be provided for patients who have decreased plasma free carnitine, but the role of carnitine supplementation for patients with normal plasma carnitine concentration is uncertain. A vitamin and cofactor 'cocktail' generally is prescribed for these patients, but it too has had limited success. The cocktail usually contains cofactors for the OXPHOS complexes (riboflavin), precursors for redox metabolites (niacin), and electron acceptors or free radical scavengers (coenzyme Q_{10}, vitamin C and vitamin K).[48] Additionally, dichloroacetate, an activator of the pyruvate dehydrogenase complex, is under investigational use on the rationale that it might increase flux through functionally intact mitochondria in patients with heteroplasmic mutations, but it has not been evaluated in patients who have cardiomyopathy.

More specific therapy has been introduced for a subset of patients who have particular biochemical or clinical phenotypes. For example, the combination of vitamin K and vitamin C was administered to a patient who had a skeletal myopathy associated with a complex III defect, on the rationale that these vitamins would provide an electron shuttle that would bypass the genetic deficiency. Clinical, biochemical and magnetic resonance spectroscopic studies confirmed that this patient experienced remarkable improvement. Unfortunately, and for reasons that are simply unknown, this therapy has not produced comparable improvement for other patients with complex III deficiency. A few patients with complex I deficiency have had significant, sustained improvement after receiving riboflavin supplementation, but this too has not been a generally successful approach to therapy. Finally, organ transplantation is of potential benefit to patients with organ-specific disorders. In particular, heart transplantation appears to have been successful for a patient who had cardiomyopathy (but no neurological abnormalities or skeletal myopathy) associated with a mitochondrial depletion syndrome. Thus, heart transplantation might be a viable therapy for patients who have $tRNA^{Ile}$ or $tRNA^{Leu}$ mutations that produce isolated cardiomyopathy.[49]

The prognosis for patients who have cardiomyopathy as the consequence of an OXPHOS disorder is highly variable.[50] The prognosis appears to depend primarily on the severity of the disease rather than on the treatment program adopted. Recent studies have documented that patients with a mitochondrial disorder associated with cardiomyopathy have a higher mortality than do patients with a mitochondrial disorder who do not have cardiomyopathy.

Conclusion

Genetic disorders of mitochondrial energy metabolism, primarily including fatty acid disorders and OXPHOS disorders, are important causes of cardiomyopathy. Both categories of disorders are well-recognized causes of cardiomyopathy in childhood and increasingly are being identified in adults. We have learned a great deal about these disorders in the past decade.

The fatty acid oxidation disorders that produce cardiomyopathy almost invariably are associated with systemic defects that also affect the liver and skeletal muscle. The biochemical hallmark of these disorders is non-ketotic or hypo-ketotic hypoglycemia, while acylcarnitine profiles and urinary organic acid profiles provide clues that can permit diagnostic and pathophysiological distinctions to be made among the various disorders. Treatment for this group of disorders varies within and among the specific conditions, but can be quite effective if initiated before irreversible damage occurs. The potential benefits of therapy are especially true for the cardiomyopathy that can be associated with these disorders.

The OXPHOS disorders produce a wide spectrum of unusual clinical phenotypes that include isolated cardiomyopathy, cardiomyopathy associated with other organ-specific disorders (e.g. hearing loss or diabetes mellitus), or cardiomyopathy that is part of systemic disorder. The biochemical findings of OXPHOS disorders are less consistent, and even a thorough biochemical evaluation can be negative. Diagnosis requires a combination of biochemical, morphological and genetic studies. Treatment for this group of disorders is poor.

Despite the advances that have been made in our understanding about the identity, diagnosis, treatment and prognosis for these disorders, many questions about them remain unanswered.

What is the frequency of these mitochondrial disorders in patients with 'metabolic' cardiomyopathy?

Efforts have been made to identify the frequency of fatty acid oxidation disorders and OXPHOS defects in patients with 'metabolic' cardiomyopathy, but thus far suffer from two basic limitations: the populations studied have been relatively small and reflect various ascertainment biases; and the methods used to diagnosis the disorders have not been uniform and do not include state of the art biochemical or genetic testing.

What is the frequency of mitochondrial disorders in patients with 'common' cardiomyopathy?

The majority of patients, both adults and children, who have cardiomyopathy do not have obvious clinical evidence of any underlying metabolic disorders, particularly a defect in fatty acid oxidation or OXPHOS. Many of these patients have been shown to have a defect in either the myocardial contractile or cytoskeletal proteins, but it still remains uncertain whether a subset of the undiagnosed patients have a single-gene (either nuclear or mitochondrial-encoded) defect in mitochondrial metabolism, or a complex trait that includes one or more genetic variations in mitochondrial metabolism. For example, recent studies suggest that certain mtDNA polymorphisms are associated with cardiomyopathy, particularly dilated cardiomyopathy.[51,52]

What is the pathophysiology of these disorders?

Our understanding of the pathophysiological basis of the fatty acid oxidation disorders is reasonably well understood, but we still do not understand the variable age of onset and tissue expression of these disorders. A better understanding of the genotype/phenotype correlation is needed, as well as an understanding of the role of epigenetic factors. The situation for the OXPHOS disorders is less satisfactory. The traditional explanations for both tissue- and organ-specific expression, as well as variable age of onset, were the phenomena of heteroplasmy and the threshold effect, but these explanations have not proven totally satisfactory.

The recent observations showing that the variable expression of cytochrome c oxidase deficiency depends at least in part on modifying nuclear genes involved in the assembly or stabilization of the complex is a very important clue.[38] Similarly, the mitochondrial abnormalities that have a role in producing apoptosis may also be involved in the pathogenesis of these disorders.[53] Finally, mouse models of genetic mitochondrial defects promise to provide useful insights into the mechanisms of cardiomyopathy in mitochondrial disorders.[54]

What are more successful approaches to treating patients with these disorders?

Although the treatment for many fatty acid oxidation disorders is successful, alternative dietary approaches are needed, e.g. triheptanoate supplementation offers great promise.[55] The treatment options for the OXPHOS disorders are poor, and a range of approaches needs to be developed as our understanding of the underlying pathophysiology of these disorders improves. For example, combined organ transplantation needs to be explored. At the very least, controlled clinical trials of the 'accepted' therapies should be performed to minimize the misinformation, false hope and expense associated with the current treatment options.

The work and imagination necessary to resolve these unanswered questions are formidable obstacles, but appear feasible given the enormous progress made over the past decade in understanding these disorders.

References

1. Beattie DS. Bioenergetics and oxidative metabolism. In: Devlin TM (ed). Textbook of Biochemistry with Clinical Correlations (4e). New York: Wiley-Liss; 2000: 537–95.
2. Lesnefsky EJ, Moghaddas S, Tandler B et al. Mitochondrial dysfunction in cardiac disease: ischemia – perfusion, aging, and heart failure. J Mol Cell Cardiol 2001; 33: 1065–89.
3. McGarry JD. Lipid metabolism I: utilization and storage of energy in lipid form. In: Devlin TM (ed). Textbook of Biochemistry with Clinical Correlations (5e). New York: Wiley-Liss; 2000: 693–725.
4. Rinaldo P, Matern D, Bennett MJ. Fatty acid oxidation disorders. Annu Rev Physiol 2002; 64: 477–502.
5. Saudubray JM, Martin D, de Lonlay et al. Recognition and management of fatty acid oxidation defects: a series of 107 patients. J Inherit Metab Dis 1999; 22: 488–502.
6. Bergmann SR, Herrero P, Sciacca R et al. Characterization of altered myocardial fatty acid metabolism in patients with inherited cardiomyopathy. J Inherit Metab Dis 2001; 24: 657–74.
7. Pierpont MEM, Breningstall GN, Stanley CA, Singh A. Familial carnitine transporter defect: a treatable cause of cardiomyopathy in children. Am Heart J 2000; 139: S96–S106.
8. Lahjouji K, Mitchell GA, Qureshi A. Carnitine transport by organic cation transporters and systemic carnitine deficiency. Mol Genet Metab 2001; 73: 287–97.
9. Roschinger W, Muntau AC, Duran M et al. Carnitine-acylcarnitine translocase deficiency: metabolic consequences of an impaired mitochondrial carnitine cycle. Clin Chim Acta 2000; 298: 55–68.
10. Rubio-Gozalbo ME, Bakker JA, Waterham HR, Wanders RJ. Carnitine-acylcarnitine translocase deficiency, clinical, biochemical and genetic aspects. Mol Aspects Med 2004; 25: 521–32.
11. Sigauke E, Rakheja D, Kitson K, Bennett MJ. Carnitine palmitoyltransferase II deficiency: a clinical, biochemical, and molecular review. Lab Invest 2003; 83: 1543–54.
12. Pons R, Cavadini P, Baratta S et al. Clinical and molecular heterogeneity in very-long-chain acyl-coenzyme A dehydrogenase deficiency. Pediatr Neurol 2000; 22: 98–105.
13. Vianey-Saban C, Divry P, Brivet M et al. Mitochondrial very-long-chain acyl-coenzyme A dehydrogenase deficiency: clinical characteristics and diagnostic considerations in 30 patients. Clin Chim Acta 1998; 269: 43–62.
14. Spierkerkoetter U, Khuchua Z, Yue Z, Strauss AW. The early-onset phenotype of mitochondrial trifunctional protein deficiency: a lethal disorder with multiple tissue involvement. J Inherit Metab Dis 2004; 27: 294–6.
15. den Boer ME, Wanders RJ, Morris AA et al. Long-chain 3-hydroxyacyl-CoA dehydrogenase deficiency: clinical presentation and follow-up of 50 patients. Pediatrics 2002; 109: 99–104.
16. Shen JJ, Matern D, Millington DS et al. Acylcarnitines in fibroblasts of patients with long-chain 3-hydroxyacyl-CoA dehydrogenase deficiency and other fatty acid oxidation disorders. J Inherit Metab Dis 2000; 23: 27–44.

17. Bellig LL. Maternal acute fatty liver of pregnancy and the associated risk for long-chain 3-hydroxyacyl-coenzyme a dehydrogenase (LCHAD) deficiency in infants. Adv Neonatal Care 2004; 4: 26–32.

18. Iafolla A, Thompson R, Roe C. Medium-chain acyl-coenzyme A dehydrogenase deficiency: clinical course in 120 affected children. J Pediatr 1994: 124: 409–15.

19. Wilson C, Champion M, Collins J et al. Outcome of medium-chain acyl-CoA dehydrogenase deficiency after diagnosis. Arch Dis Child 1999: 80: 459–62.

20. Sim KG, Hammond J, Wilcken B. Strategies for the diagnosis of mitochondrial fatty acid β-oxidation disorders. Clin Chim Acta 2002; 323: 37–58.

21. Leonard JV, Schapira AHV. Mitochondrial respiratory chain disorders I: mitochondrial DNA defects. Lancet 2000; 335: 299–304.

22. Leonard JV, Schapira AHV. Mitochondrial respiratory chain disorders II: neurodegenerative disorders and nuclear gene defects. Lancet 2000; 355: 389–94.

23. DiMauro S, Schon EA. Mitochondrial respiratory–chain diseases. N Engl J Med 2003; 348: 2656–68.

24. MITOMAP: A Human Mitochondrial Genome Database; 2005. www.mitomap.org/ (accessed 12 May 2005).

25. Finsterer J. Mitochondriopathies. Eur J Neurol 2004; 11: 163–86.

26. Antozzi C, Zeviani M. Cardiomyopathies in disorders of oxidative metabolism. Cardiovasc Res 1997; 35: 184–99.

27. Santorelli FM, Tessa A, d'Amati G et al. The emerging concept of mitochondrial cardiomyopathies. Am Heart J 2001; 141: el.

28. Andreu AL, Checcarelli N, Iwata S et al. A missense mutation in the mitochondrial cytochrome b gene in a revisited case with histiocytoid cardiomyopathy. Pediatr Res 2000; 48: 311–4.

29. Vallance HD, Jeven G, Wallace DC, Brown MD. A case of sporadic infantile histiocytoid cardiomyopathy caused by the A8344G (MERFF) mitochondrial DNA mutation. Pediatr Cardiol 2004; 25: 538–40.

30. Pignatelli RH, McMahon CJ, Denfield WJ et al. Clinical characterization of left ventricular noncompaction in children: a relatively common form of cardiomyopathy. Circulation 2003; 108: 2672–8.

31. Degoul F, Brule H, Cepanec C et al. Isoleucylation properties of native human mitochondrial tRNA[Ile] and tRNA[Ile] transcripts. Implications for cardiomyopathy-related point mutations (4269, 4317) in the tRNA[Ile] gene. Hum Mol Genet 1998; 7: 347–54.

32. Taylor RW, Giordano C, Davidson MM et al. A homoplasmic mitochondrial transfer ribonucleic acid mutation as a cause of maternally inherited hypertrophic cardiomyopathy. J Am Coll Cardiol 2003; 41: 1786–96.

33. Bruno C, Kirby DM, Koga Y et al. The mitochondrial DNA C3303T mutation can cause cardiomyopathy and/or skeletal myopathy. J Pediatr 1999; 135: 197–202.

34. Iwanaga R, Koga Y, Aramaki S et al. Inter- and/or intra-organ distribution of mitochondrial C3303T or A3243G mutation in mitochondrial cytopathy. Acta Neuropathol 2001; 101: 179–84.

35. Lodi R, Rajagopalan B, Blamire AM et al. Abnormal cardiac energetics in patients carrying the A3243G mtDNA mutation measured in vivo using phosphorus MR spectroscopy. Biochim Biophys Acta 2004; 1657: 146–50.

36. Raha S, Merante F, Shoubridge E et al. Repopulation of ρ° cells with mitochondria from a patient with a mitochondrial DNA point mutation in tRNA[Gly] results in respiratory chain dysfunction. Hum Mutat 1999; 13: 245–54.

37. Benit P, Beugnot R, Chretien D et al. Mutant NDUFV2 subunit of mitochondrial complex 1 causes early onset hypertrophic cardiomyopathy and encephalopathy. Hum Mutat 2003; 21: 582–6.

38. Shoubridge EA. Cytochrome c oxidase deficiency. Am J Med Genet (Semin Med Genet) 2001; 106: 46–52.

39. Antonicka H, Leary SC, Guercin GH et al. Mutations in COX10 result in a defect in mitochondrial heme A biosynthesis and account for multiple, early-onset clinical phenotypes associated with isolated COX deficiency. Hum Mol Genet 2003; 12: 2693–702.

40. Antonicka H, Mattman A, Carlson CG et al. Mutations in COX15 produce a defect in the mitochondrial heme biosynthetic pathway, causing early-onset fatal hypertrophic cardiomyopathy. Am J Hum Genet 2003; 72: 101–14.

41. Sue CM, Karadimas C, Checcardelli N et al. Differential features of patients with mutations in two COX assembly genes, Surf-1 and SCO2. Ann Neurol 2000; 47: 589–95.

42. Barth PG, Valianpour F, Bowen VM et al. X-linked cardioskeletal myopathy and neutropenia (Barth syndrome): an update. Am J Med Genet 2004; 126A: 349–54.

43. Xu Y, Kelley RI, Blanck TJJ, Schlame M. Remodeling of cardiolipin by phospholipid transacylation. J Biol Chem 2003; 278: 51380–85.

44. Pandolfo M. Iron metabolism and mitochondrial abnormalities in Friedreich ataxia. Blood Cell Mol Dis 2002; 29: 536–47.

45. Seznec H, Simon D, Bouton C et al. Friedreich ataxia, the oxidative stress paradox. Hum Mol Genet 2004; 14: 463–74.

46. Thorburn DR, Smeitink J. Diagnosis of mitochondrial disorders: clinical and biochemical approach. J Inherit Metab Dis 2001; 24: 312–6.

47. Rustin P, Lebidois J, Chretien D et al. Endomyocardial biopsies for early detection of mitochondrial disorders in hypertrophic cardiomyopathies. J Pediatr 1994; 124: 224–8.

48. Lerman-Sagie T, Rustin P, Lev D et al. Dramatic improvement in mitochondrial cardiomyopathy following treatment with idebenone. J Inherit Metab Dis 2001; 24: 28–34.

49. Santorelli FM, Gaglairdi MG, Dionisi-Vici C et al. Hypertrophic cardiomyopathy and mtDNA depletion. Successful treatment with heart transplantation. Neuromuscul Disord 2002; 12: 56–9.

50. Scaglia F, Towbin JA, Craigen WJ et al. Clinical spectrum, morbidity, and mortality in 113 pediatric patients with mitochondrial disease. Pediatrics 2004; 114: 925–31.

51. Davila-Roman VG, Vedala G, Herrero P et al. Altered myocardial fatty acid and glucose metabolism in idiopathic dilated cardiomyopathy. J Am Coll Cardiol 2002; 40: 271–7.

52. Marin-Garcia J, Goldenthal MJ. Understanding the impact of mitochondrial defects in cardiovascular disease: a review. J Card Fail 2002; 8: 347–61.

53. Zhang D, Mott JL, Farrar P et al. Mitochondrial DNA mutations activate the mitochondrial apoptotic pathway and cause dilated cardiomyopathy. Cardiovasc Res 2003; 57: 147–57.

54. Russell LK, Finck BN, Kelly DP. Mouse models of mitochondrial dysfunction and heart failure. J Mol Cell Cardiol 2005; 38: 81–91.

55. Roe CR, Sweetman L, Roe DS et al. Treatment of cardiomyopathy and rhabdomyolysis in long-chain fat oxidation disorders using an anaplerotic odd-chain triglycerides. J Clin Invest 2002; 110: 259–69.

CHAPTER 35

Adrenergic receptor polymorphisms in heart failure: molecular and physiological phenotypes

Kersten M Small, Jeanne Mialet-Perez, Lynne E Wagoner, Stephen B Liggett

Introduction

Acute activation of the sympathetic nervous system acts to increase cardiac output and critical organ perfusion. However, chronic activation, particularly via norepinephrine stimulation of myocyte β_1-adrenergic receptors (β_1AR) ultimately leads to a deterioration of ventricular function. This phenomenon is accompanied by a number of mechanisms which have positive, compensatory effects as well as deleterious features.[1] The seemingly paradoxical improvement of ventricular function by β-blockers used in the treatment of chronic heart failure is due to blunting of this sympathetic drive,[2] and indeed norepinephrine (NE) levels are inversely correlated to survival in heart failure.[3] Given these adaptive and maladaptive events that occur via the sympathetic nervous system, we considered that genetic variants of adrenergic receptors that alter receptor expression or function, may have relevant physiological consequences in heart failure. The incomplete penetrance in familial cardiomyopathies,[4] and extensive interindividual variability of adrenergic regulation in humans,[5] the clinical characteristics of heart failure,[4,6] and the response to therapy in the syndrome,[7–9] also suggest a genetic component to common heart failure syndromes such as ischemic and dilated cardiomyopathies. Concerning the receptors, of the nine adrenergic receptors, six have coding polymorphisms (defined as a genetic variant with an allele

frequency of $\geq 1\%$), which alter receptor function in recombinant cells.[10] Figure 35.1 shows a schematic diagram of adrenergic receptors expressed in the cardiac presynaptic nerve terminal and the myocyte. Sympathetic stimulation is controlled in part by α_2AR located in presynaptic nerve termini which regulate NE release at the synapse. The α_{2A}AR subtype controls NE release

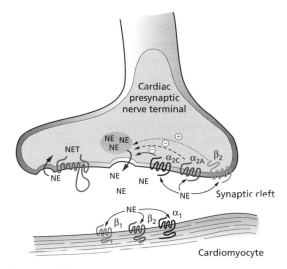

Figure 35.1. *Representation of adrenergic receptors of cardiac presynaptic nerve termini and cardiomyocytes. NET, norepinephrine transporter; NE, norepinephrine. Modified with permission from Gerson et al. J Nucl Cardiol 2003; 16; 583–9 (© American Society of Nuclear Cardiolgy).*

from presynaptic nerves undergoing high-frequency stimulation, while the α_{2C}AR inhibits release from low-frequency (basal) stimulation.[11] Indeed, the α_{2A}/α_{2C} double knockout mouse develops a catecholamine-mediated cardiomyopathy, indicating that even the non-stressed setting regulation of NE release is critical for normal cardiac function and adaptation.[11] Furthermore, the α_{2C}AR knockout, while apparently without pathology in the non-stressed state, develops a marked cardiomyopathy and heart failure after induction of pressure-overload by aortic banding.[12] A presynaptic β_2AR has also been identified, whose activation is thought to increase NE release.[13] At the myocyte, released NE acts primarily at the β_1AR, but NE can activate β_2ARs (albeit at lower affinity), and both subtypes respond to circulating epinephrine. In this review the salient phenotypes of adrenergic receptor polymorphisms relevant to heart failure, as ascertained in transfected cells or transgenic mice, are summarized, and the consequences of certain polymorphisms in the human disease are discussed.

Signaling phenotypes of adrenergic receptor polymorphisms

α_{2A}AR

A single non-synonymous polymorphism has been identified in each of the α_2AR subtypes (Table 35.1). For the α_{2A}AR, a C to G transversion found at nucleotide 753 results in an Asn to Lys change at amino acid 251 within the third intracellular loop of the receptor.[14] This polymorphism is rare, occurring at allele frequencies of 0.4% and 5% in Caucasians and African-Americans, respectively. In Chinese hamster ovary (CHO) cells stably expressing either α_{2A}-Asn251 or α_{2A}-Lys251, ligand-binding studies showed that agonist and antagonist binding were not altered by the presence of this polymorphism. An increase in agonist-promoted [^{35}S]GTPγS binding, however, was observed for α_{2A}-Lys251 compared to α_{2A}-Asn251. This enhanced agonist-promoted G protein coupling was also evident in multiple signaling pathways. Agonist-promoted inhibition of

Table 35.1. *Adrenergic receptor polymorphisms*

| Receptor | Position | | Alleles | | Minor allele frequency (%) | |
	Nucleotide	Amino acid	Major	Minor	Caucasians	African-Americans
α_{1A}AR	1441	492	Cys[a]	Arg	46	70
α_{2A}AR	753	251	Asn	Lys	0.4	5
α_{2B}AR	901–909	301–303	no deletion	delete Glu-Glu-Glu	31	12
α_{2C}AR	964–975	322–325	no deletion	delete Gly-Ala-Gly-Pro	4	38
β_1AR	145	49	Ser	Gly	15	13
	1165	389	Arg	Gly	27	42
β_2AR	46	16	Gly	Arg	39	50
	79	27	Gln	Glu	43	27
	491	164	Thr	Ile	2–5	2–5
β_3AR	190	64	Trp	Arg	10	?

[a]In African-Americans, Arg is the major allele.

forskolin-stimulated adenylyl cyclase activity was increased up to 40% for α_{2A}-Lys251 compared with α_{2A}-Asn251, and agonist-promoted activation of mitogen-activated protein kinase (MAPK) was increased to an even greater extent (>2-fold).

$\alpha_{2B}AR$

For the $\alpha_{2B}AR$, a common in-frame deletion polymorphism was identified that results in loss of Glu-Glu-Glu at amino acids 301–303 in the third intracellular loop (ICL3) of the receptor (Table 35.1).[15] This polymorphism is present at allele frequencies of 31% in Caucasians and 12% in African-Americans. In CHO cells, only minor differences in agonist and antagonist binding were observed between the wild-type receptor or α_{2B}-Del301–303. Agonist-promoted inhibition of forskolin-stimulated cyclase studies showed a modest decrease in receptor G_i-coupling for the deletion-containing receptor, with α_{2B}-Del301–303 showing an 18% decrease in maximal inhibition of adenylyl cyclase and a 2-fold increase in the EC_{50} for this response. Because this polymorphism is located within an acidic region of the third intracellular loop of the receptor essential for short-term agonist-promoted receptor phosphorylation and desensitization by GRK2,[16] studies investigating the effects of the deletion polymorphism on these functions were explored. Whole-cell phosphorylation studies were carried out in COS-7 cells

transiently expressing GRK2 and each $\alpha_{2B}AR$ exposed to vehicle or 10 mM epinephrine for 20 min. Results from these studies showed that the α_{2B}-Del301–303 had ~50% reduced phosphorylation compared to the wild-type receptor. In functional studies (inhibition of adenylyl cyclase), this decrease in receptor phosphorylation correlated with a complete absence of receptor desensitization.[15]

$\alpha_{2C}AR$

Lastly and most pertinent to this review is the $\alpha_{2C}AR$ polymorphism, which consists of an in-frame deletion of four amino acids within the third intracellular loop of the receptor, Gly-Ala-Gly-Pro, at positions 322–325.[17] This polymorphism is common in African-Americans, occurring with an allele frequency of 40% in this population (Table 35.1). In Caucasians, however, α_{2C}-Del322–325 is rare, occurring at a frequency of 4% in this group. When wild-type and deletion-containing $\alpha_{2C}ARs$ were expressed to equal levels in CHO cells, high-affinity agonist binding was moderately decreased, suggesting an altered agonist–receptor–G protein interaction. Indeed, in functional studies of adenylyl cyclase inhibition, a marked decrease in function was observed (~10% inhibition of forskolin-stimulated adenylyl cyclase activity for Del322–325 compared to ~70% for wild-type, Fig. 35.2). Decreased function was also observed

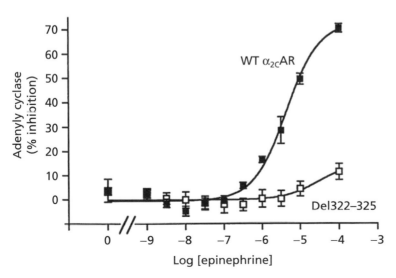

Figure 35.2. α_{2C}-Del322–325 polymorphism confers decreased coupling to inhibition of adenylyl cyclase. Experiments were performed using membranes prepared from CHO cells stably expressing either wild-type or polymorphic $\alpha_{2C}AR$ at equivalent levels. WT, wild type. Error bars are standard errors.

for agonist-mediated stimulation of MAPK and inositol 1,4,5-trisphosphate (IP_3) production.[17] Thus the α_{2C}Del322–325 phenotype is one of marked loss of function. As is discussed below, this polymorphism has been found to be associated with heart failure and intermediate phenotypes in African-Americans.[18]

β_1AR

A single nucleotide polymorphism (SNP) in the amino terminus of the β_1AR was found which results in a substitution of Gly for Ser at amino acid 49 (Table 35.1). This polymorphism occurs at similar allele frequencies in both Caucasians (15%) and African-Americans (13%).[19] To date, two reports have been published describing the functional effects of this polymorphism in transfected cells. Rathz et al. showed that in CHW-1102 and HEK-293 cells recombinantly expressing either the β_1-Gly49 or β_1-Ser49 receptors, no differences in agonist or antagonist binding affinities, or stimulation of adenylyl cyclase were found.[20] Due to the location of this polymorphism in the amino terminus of the receptor, phenotypic differences in receptor trafficking were also explored. Long-term agonist exposure in the context of blocked new receptor synthesis revealed that the β_1-Gly49 receptor underwent enhanced agonist-promoted downregulation compared to β_1-Ser49 (Fig. 35.3A). Agonist-promoted internalization of both receptors was to the same extent, as was the rate of new receptor synthesis and membrane insertion. Additional studies showed that the glycosylation patterns differed between the two receptors. As shown in Fig. 35.3B, some of the β_1-Ser49 receptors exist in a highly glycosylated form at 105 kDa which was never found with the β_1-Gly49 receptor. The high molecular weight band was sensitive to the glycosylation inhibitor tunicamycin (Tunica) and was not found to be due to receptor dimerization. The results from this study indicate that the β_1-Gly49 receptor displays enhanced agonist-promoted receptor downregulation compared to the β_1-Ser49, an effect which appears to occur after receptor internalization and is associated with an alteration in the degradative pathway, possibly due to differences in the glycosylation status. A report by Levin et al. also demonstrates enhanced ago-

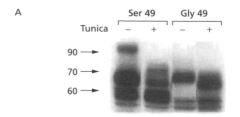

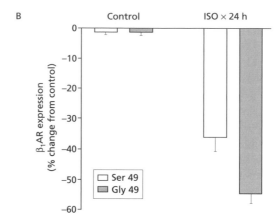

Figure 35.3. *β_1-Gly49 displays altered receptor trafficking and glycosylation. β_1-Gly49 displays altered glycosylation (A) and enhanced long-term agonist-promoted downregulation (B) compared to β_1-Ser49. Receptors were expressed to equivalent levels in HEK-293-cells. Error bars are standard errors.*

nist-promoted downregulation for the β_1-Gly49 receptor in HEK-293 cells.[21] However, in contrast to results described by Rathz et al.,[20] this group showed that the β_1-Gly49 receptor has slightly increased agonist binding affinity, and had increased basal and agonist-stimulated adenylyl cyclase activities, with increased sensitivity to the inverse agonist metoprolol. The discrepancies between these two reports may be the result of differences in receptor expression levels, which were up to 10-fold higher in the second study, and/or use of different isolates of HEK-293 cells which have been noted to display differences in expression of post-receptor signaling proteins.[22] Nonetheless, the β_1-Gly49 polymorphism consistently shows enhanced agonist-promoted downregulation in recombinant cell-based studies. As is discussed below, this phenotype may confer a protective effect in the failing heart that is chronically exposed to elevated levels of NE.

A second non-synonymous polymorphism of the β_1AR was identified whereby Arg or Gly can be present at amino acid 389.[23] This polymorphism lies in a predicted small intracellular α-helix (based on the crystal structure of bovine rhodopsin), which is between transmembrane domain seven and the membrane-anchoring palmitoylation site. The 'wild-type' amino acid has been considered to be Gly, as it was present in the first human receptor to be cloned. However, Arg is typically more common, although the difference between the frequencies of the two alleles in African-Americans is not very different (Gly = 42%, Arg = 58%).[19] In this review, for purposes of consistency, β_1-Gly389

will be considered the reference genotype, since all previous structure/function studies have been carried out with this receptor. In transfected cells, agonist-promoted [^{35}S]GTPγS binding was increased with Arg389, as was agonist high-affinity binding, indicating enhanced agonist–receptor–G_s interaction.[23] In functional studies (stimulation of adenylyl cyclase), basal and particularly agonist stimulation of adenylyl cyclase was higher for β_1-Arg389 compared to β_1-Gly389 (Fig. 35.4A). As discussed below, a similar gain-of-function phenotype has been found physiologically in the hearts of transgenic mice with targeted expression of Arg389 or Gly389 b_1ARs.[24]

Another study showed that the β_1-Gly389 receptor undergoes greater desensitization by

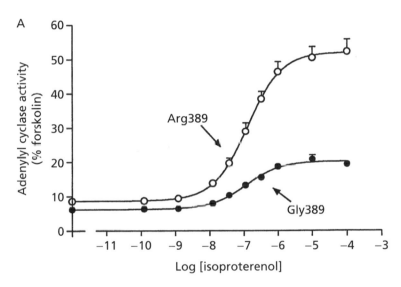

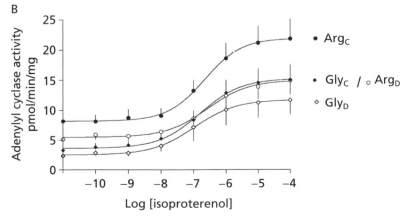

Figure 35.4. β_1-Arg389 displays enhanced agonist-promoted coupling to stimulation of adenylyl cyclase and increased agonist-promoted receptor desensitization compared to β_1-Gly389. (A) adenylyl cyclase assays were performed using membranes from Chinese hamster fibroblasts expressing equivalent levels of the two receptors. (B) cells were pre-exposed to vehicle (designated 'C' for control) or 10 μM norepinephrine (designated 'D' for desensitized) for 20 min, membranes prepared, and adenylyl cyclase activities measured. Error bars are standard errors.

G protein receptor kinases (GRKs).[25] In these studies, cells were exposed to 10 μM NE for 20 minutes, washed with cold phosphate-buffered saline (PBS), membranes prepared, and agonist-promoted adenylyl cyclase activities determined. In doing so, the extent of short-term agonist-promoted desensitization of adenylyl cyclase was found to be increased in β_1-Arg389 (~34%) cells compared to β_1-Gly389 cells (~21%), suggesting that the receptor conformation of β_1-Arg389 is more favorable for GRK-mediated phosphorylation. This is consistent with the notion that the 'active' state is the most preferred conformation for GRKs. As shown in functional studies, the β_1-Arg389 achieves a conformation that provides for enhanced signal transduction (i.e. 'more active') compared to β_1-Gly389. Examination of the absolute levels of adenylyl cyclase activity of both β_1-Arg389 and β_1-Gly389 receptors at basal and agonist-stimulated levels, with or without agonist pre-exposure, revealed that the influence of genetic variation on receptor signaling was of the same magnitude as that of desensitization (Fig. 35.4B). In terms of β_1AR function, the desensitized β_1-Arg389 receptor ('Arg$_D$') signals as well as the non-desensitized (control) β_1-Gly389 receptor ('Gly$_C$'). Similar differences in desensitization were observed in transgenic mice expressing these two β_1AR variants as well (see below). The results from these studies indicate that both desensitization and genetic variation are important determinants of the ultimate level of β_1AR signaling.

The cardiac consequences of polymorphic β_1AR in transgenic mice have also been examined. In these studies, targeted expression to the heart of either β_1-Gly389 or β_1-Arg389 was achieved via the α-myosin heavy chain promoter.[24] Two lines were primarily studied, each expressing equivalent levels of receptor (~1 pmol/mg for each line). At 3 months of age, work-performing heart preparations showed that baseline and agonist-stimulated contractility were higher in β_1-Arg389 versus β_1-Gly389 mice (Fig. 35.5A). Baseline chronotropic responses were also greater in β_1-Arg389 mice. These results confirm the concept that β_1-AR genetic variants have physiological relevance at the level of the intact heart. At 6 months of age, however, a change in the β_1-Arg389 phenotype

was observed. While baseline contractility was still increased, there was no contractile response to agonist (Fig. 35.5C). There was a minor degree of desensitization of the β_1-Gly389 hearts. Interestingly, the chronotropic response to agonist was preserved in β_1-Arg389 hearts (Fig. 35.5B and D). Agonist-promoted adenylyl cyclase activities in cardiac membranes showed a ~45% desensitization of β_1-Arg389 (6-month compared to 3-month), while no statistical difference was noted for β_1-Gly389 hearts. Forskolin-stimulated adenylyl cyclase activities were also depressed in β_1-Arg389 6-month old hearts compared to 3-month old hearts. Such was not observed for β_1-Gly389. The desensitization observed in these proximal signaling events was further explored with Western blots which revealed β_1-Arg389 (but not β_1-Gly389) hearts had decreased G$_{\alpha s}$ and adenylyl cyclase type V. Both displayed increases in G$_{\alpha i}$ and GRK2. The degree of desensitization of β_1-Arg389 hearts due to changes in these proximal signal transduction elements, though, was not felt to be sufficient to explain the complete lack of a physiological response to agonist observed at 6 months in these mice. Additional studies showed that within the same time period, there were significant changes in fetal- and hypertrophy-associated genes and Ca^{2+}-handling proteins (Fig. 35.6), only in the β_1-Arg389 hearts. Ultimately by 9-months, β_1-Arg389 mice displayed a marked decrease in cardiac function and remodeling of the heart. In vivo echo-cardiography showed that β_1-Arg389 mice had reduced fractional shortening at rest compared with β_1-Gly389 mice (26% versus 42%), and histopathology revealed myocyte loss with replacement fibrosis in β_1-Arg389 that was not seen in β_1-Gly389 mice.

Additional studies in young mice revealed differential response to β-blockade based on genetic variability, with acute and chronic responses to the β-antagonist propranolol being greater in the β_1-Arg389 mice compared to β_1-Gly389 mice. As shown in Fig. 35.7A, β_1-Arg389 hearts have a linear response (decrease in contractility) to acute propranolol, while β_1-Gly389 hearts are unresponsive except at the highest dose. In chronic studies, 4-month-old mice were treated for one month with oral propranolol, and heart rate decrease was used

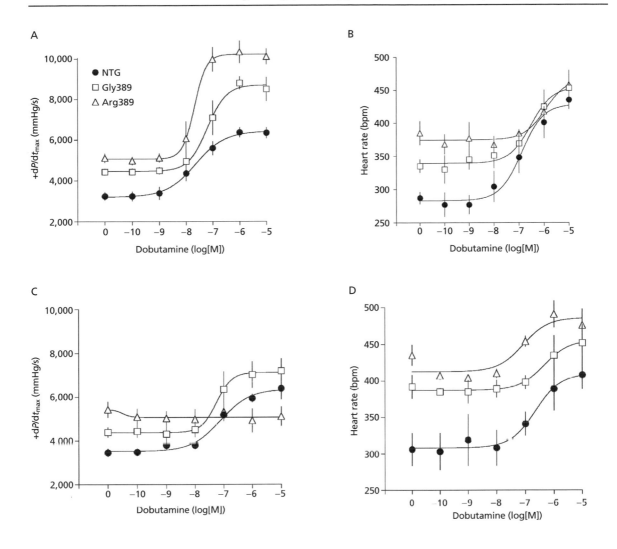

Figure 35.5. *Hemodynamic responses in cardiac-targeted transgenic mice reveal a phenotypic switch in β₁-Arg389 mice. Work-performing heart studies at 3 months of age (A and B) show that β₁-Arg389 transgenic mice display enhanced contractility (+dP/dt) and heart rates at baseline, and enhanced contractile response to the β-agonist dobutamine compared to β₁-Gly389 transgenic mice. At 6 months of age (C and D), β₁-Arg389 mice show a loss of contractile response to agonist. NTG, non-transgenic; bpm, beats per minute. Error bars are standard errors.*

as the physiological endpoint. Only β₁-Arg389 mice displayed such a decrease in heart rate (Fig. 35.7B). Taken together, the data suggest that β₁-Arg389 may be a risk for failure, and that the phenotype is time dependent. This may be important, as the timing of interventions (such as β-blockers) may play a role in the genotype effect of the β₁AR polymorphisms. The data also suggest that β₁-Arg389 is the favorable response allele for β-blockers, and has been the basis for human clinical trials to ascertain the relationship between the β₁AR genotype and response to β-blockers.

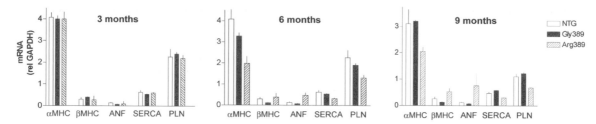

Figure 35.6. β₁-Arg389 and β₁-Gly389 transgenic mice display different cardiac gene expression profiles at 6 and 9 months compared to non-transgenic mice. Expression of fetal, hypertrophy-associated, and calcium-cycling genes at 3, 6 and 9 months are shown. α-and β-MHC, a- and b- myosin heavy chains; ANF, atrial natriuretic factor; NTG, non-transgenic; SERCA, sarcoplasmic reticulum Ca^{2+} ATPase; PLN, phospholamban. Error bars are standard errors.

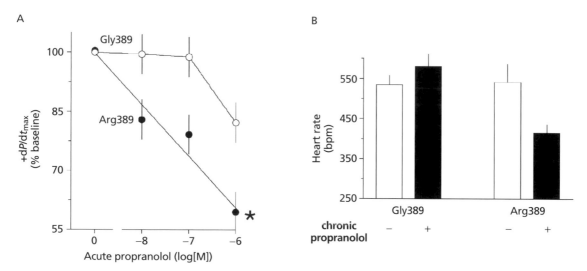

Figure 35.7. Acute and chronic responses to β-blockade are greater in β₁-Arg389 transgenic mice compared to β₁-Gly389 mice. (A) The contractile response of acute administration of the β-antagonist propranolol. (B) The effects of chronic oral administration of propranolol on heart rate. Error bars are standard errors.

β₂AR

There are three non-synonymous polymorphic loci in the coding block of the β₂AR. The most uncommon, but the one with the most significant functional phenotype, is a Thr-Ile substitution at amino acid 164 of the fourth transmembrane domain.[26] The conformation(s) of the β₂-Ile164 receptor are not favorable for $G_{\alpha s}$ coupling, thus basal and agonist-stimulated adenylyl cyclase activities are substantially lower with this receptor compared to wild type. This same coupling phenotype

was also noted in myocytes from transgenic mice expressing the β₂-Thr164 and β₂-Ile164 receptors.[27] No pathological sequelae was noted in either transgenic mouse, although they were not subjected to conditions that might provoke cardiac phenotypes.

The more common polymorphisms of β₂AR are at amino acid positions 16 and 27, which are within the extracellular amino terminus. While ligand binding affinities and G_s coupling are not perturbed by these polymorphisms, long-term agonist-promoted downregulation is altered.[28] The β_2-Gly16 Gln27 receptor has increased downregulation compared to the Arg16 counterpart (41% versus 26%). In contrast, the rare β_2-Arg16 Glu27 receptor undergoes no detectable downregulation in transfected cells, and very little (in comparison to the other haplotypes) in airway smooth muscle cells endogenously expressing the receptors. The β_2-Gly16 Glu27 receptor has enhanced agonist-promoted downregulation. Trafficking studies indicate that these phenotypes are probably due to altered degradation of receptor after the initial internalization step.[28]

β_3AR

There is some controversy as to the function of β_3AR in the heart. In $\beta_1^{-/-}/\beta_2^{-/-}$ mice, isoproterenol causes a very small (~5%) and transient decrease in the rate of contraction of isolated myocytes.[29] In transgenic mice overexpressing the β_3AR, one group has reported that agonists promote increased contractility in intact hearts,[30] while another has reported a negative inotropic effect in isolated right ventricular strips.[31] Several studies in human heart have suggested a negative inotropic effect of β_3AR that involves nitric oxide release and/or cGMP generation. Nevertheless, there is one non-synonymous polymorphism of the human β_3AR at nucleotide 190, where Trp at amino acid position 64 is substituted by Arg.[32] One group has reported that the Arg β_3AR has wild-type coupling.[33] Another report has indicated that the polymorphic β_3AR is ~30% uncoupled from agonist-stimulated adenylyl cyclase activation.[34] No human studies have been carried out to ascertain potential relationships between the β_3AR polymorphism and heart failure phenotypes.

Clinical studies of adrenergic receptor polymorphisms in heart failure

A number of studies have shown associations between polymorphisms of certain adrenergic receptor subtypes and cardiovascular phenotypes. These include associations between β_2AR polymorphisms with exercise response,[35] and vascular response to agonists,[36,37] in normal individuals, various heart failure phenotypes,[38,39] and hypertension.[40] Studies have also implicated the α_{2B}-Del301–303 in altered vascular responses to agonist,[41] and fatal myocardial infarction or sudden death.[42,43] Here we will review in detail studies of polymorphism of the β_1AR and α_{2C}AR genes, as they are both highly polymorphic and form a critical 'circuit' in sympathetic control of cardiac function (Fig. 35.1).

α_2AR STUDIES

As stated above, presynaptic α_{2C}AR regulates NE release during low-frequency stimulation (which can be equated with control of chronic release) via a negative feedback system. The α_{2C}-Del322–325 polymorphism has depressed coupling in transfected cells, so this variant would predispose to higher NE levels in the synapse. (Note that the α_{2A}AR also controls NE release, however, the coding SNP in this receptor causes an increase in function, which in the context of heart failure would be considered protective). Given this, investigation of 189 control individuals and 159 patients with either idiopathic dilated or ischemic cardiomyopathy and left ventricular ejection fractions <35% revealed that the α_{2C}-Del322–325 variant does indeed represent a risk factor for the development of heart failure.[18] In this study, Chi-square tests of independence were used to test for association between heart failure and genotype or allele, and, since previous determinations of allele frequencies revealed that the α_{2C}-Del322–325 polymorphism was much more common in African-Americans than Caucasians, the two ethnic groups were analyzed separately. In African-Americans, the allele frequency of the α_{2C}-Del322–325 polymorphism was found to be higher in patients with heart failure (0.615 versus 0.411). This association was maintained when examining genotypes as well, with 52.6 % of heart failure patients homozygous for the α_{2C} polymorphism compared to 16.6% of controls. The adjusted risk factor for this association was 5.65 (95% confidence interval (CI) = 2.67 to 11.95, $P < 0.001$). Additional analysis in the African-American heart failure cohort revealed several other associations consistent with these findings. The odds ratio for onset of failure before age 40 among α_{2C}-Del322–325 carriers compared to wild-type was 4.07 (95% CI = 1.25-13.3, $P < 0.02$). Using the median left ventricular ejection fraction (LVEF) (22%) of the entire cohort in an analysis of α_{2C} genotype and LVEF, the odds ratio for LVEF of <22% in those homozygous for α_{2C}-Del322–325 compared to those homozygous for wild-type was 3.63 (95% CI = 1.17-11.22, $P = 0.03$). So, these secondary analyses reveal that not only is α_{2C}-Del322–325 a potential risk for failure, it may also affect age of onset and severity. In Caucasians, a population in which the deletion polymorphism is uncommon, no significant association of this variant and heart failure was found. This may be the result of a lack of statistical power due to the small number of homozygous α_{2C}-Del322–325 cases and controls. Another study of primarily Caucasian individuals also showed that the allele frequency of the deletion polymorphism was not significantly different between control individuals and patients with heart failure.[12] In this study, however, patients carrying the α_{2C}-Del322–325 polymorphism had a worse clinical status and significantly reduced cardiac function, characterized by reduced LVEF and contractility, compared with patients without this polymorphism.[12]

Another study has shown an apparent compensatory increase in the activity of the NE uptake-1 transporter in heart failure patients carrying the α_{2C} deletion polymorphism.[44] In this small study of 39 heart failure patients, nine of whom carried the α_{2C}-Del322–325 polymorphism, neuronal imaging with ^{123}I-metaiodobenzylguanidine (MIBG) revealed that the heart-to-mediastinum ratio of this radio-labeled NE analog was significantly higher in patients with the polymorphism than those without (25.1 versus 18.2 counts/pixel at 4 hours, $P = 0.03$). These data suggest that increased reuptake of NE may develop as an adaptive mechanism to counter the deleterious effects of increased synaptic NE imposed by dysfunctional α_{2C}ARs.

β_1AR studies

Multiple clinical studies investigating the role of polymorphic β_1AR in heart failure have been carried out. In relation to the position 49 variants, a study by Borjesson et al. showed that the allele frequency of the β_1-Gly49 variant was not significantly different in patients with idiopathic heart failure ($n = 184$) compared to healthy control individuals ($n = 77$).[45] An improvement in long-term survival in patients carrying the β_1-Gly49 polymorphism, however, was observed, with 62% of patients homozygous for β_1-Ser49 experiencing death or hospitalization, compared with 39% of patients carrying β_1-Gly49 (homozygotes and heterozygotes, $P = 0.005$). When corrected for other variables, a 2-fold risk of the β_1-Ser49 allele for death or transplantation was noted (95% CI = 0.99–4.16, $P = 0.05$). In another study consisting of >1000 normotensive and hypertensive individuals, Ranade et al. found that the position 49 polymorphism was significantly associated with resting heart rate ($P = 0.004$).[46] This association was independent of other variables including age, sex, ethnicity, hypertension status and β-blocker use. This study, as well as the aforementioned study of survival/hospitalizations, suggests a potential protective role for the β_1-Gly49 variant in heart failure. Of note, one study has reported that β_1-Gly49 is associated with idiopathic dilated cardiomyopathy.[47] This conclusion was based on small sample sizes (37 patients and 40 controls), where the β_1-Gly49 allele was found in four patients and none of the controls. The results of this study are probably spurious due to the small sample size.

Because of the enhanced function of the β_1-Arg389 receptor and its potential to chronically stimulate the heart, the role of this polymorphism in heart failure has also been investigated. To date, no association between the β_1-Arg389 polymorphism and the risk of heart failure has been found, including in one study of ~400 controls and a similar number of idiopathic dilated cardiomyopathy patients.[18,48,49] However, as discussed below, the 389 locus may have a synergistic interaction with the α_{2C}AR polymorphism. So, as a single gene the risk may be low, but in combination with other members of the signaling circuit the β_1-389 polymorphism may indeed predispose to failure.

When considering disease modification and drug response, the β_1-Arg389 variant has been shown to be associated with improved exercise capacity in stable patients with heart failure.[50] In a study of 263 patients with ischemic or dilated cardiomyopathy, using the modified Naughton treadmill protocol, maximal oxygen consumption ($\dot{V}O_2$) was greater with patients homozygous for β_1-Arg389 compared to those homozygous for β_1-Gly389 (17.7 ± 0.4 versus 14.5 ± 0.6 ml/kg/min, $P = 0.006$). Heterozygotes had an intermediate $\dot{V}O_2$ that was marginally different ($P = 0.04$) than β_1-Gly389 homozygotes. The baseline clinical characteristics, except for systolic blood pressure, between the three groups were indistinguishable.[50] There was no confounding by heart failure etiology, age, sex, baseline LVEF, or β-blocker usage. Other studies have also noted associations with β_1-Arg389 and blood pressure or the hypotensive response to β-blockers.[51-53] Taken together, it appears that genetic variation of the β_1AR, which is also expressed in the vasculature and the kidney, may play a role in hypertension and β-blocker responsiveness, and thus interplay with heart failure risk. Other studies of β_1AR 389 polymorphisms directed towards cardiac function have revealed additional associations. In human atrial appendage tissue from individuals homozygous for β_1-Arg389, NE was shown to have increased inotropic potency as well as increased agonist-stimulated accumulation of cAMP compared to atria from homozygous β_1-Gly389 individuals.[54] This is in contrast to a study by Molenaar et al. which showed no differences in the potency of NE on atrial contractility in individuals with various β_1 polymorphisms (both 49 and 389 variants).[55] Another report showed that the β_1-Gly389 polymorphism suppresses the occurrence of ventricular tachycardia in patients with dilated cardiomyopathy.[49] Also, somewhat unexpectedly, the β_1-Gly389 polymorphism has been shown to be significantly associated with increased LV mass in patients with renal disease.[56]

As described above, the α_{2C}-Del322–325 polymorphism has been shown to confer a ~5-fold risk for development of heart failure, whereas no risk has been shown for the β_1-Arg389. In addition to these single locus risk analyses, the combination of the β_1-Arg389 and α_{2C}-Del322–325 polymorphisms has also been considered to represent an even greater risk than either polymorphic receptor alone, since there would be increased NE acting on a hyper-responsive cardiac β_1AR.[18] In order to test for interactions between the α_{2C}AR and β_1AR polymorphisms, logistic-regression methods were used to model the effect of each genotype and their interaction on the risk of heart failure. Likelihood ratio tests were used to assess the influence of each locus, and the interaction between them both before and after adjustment for the potential confounding effects of age and sex. The reference group was defined as that which had the most frequent genotypic combinations and it was assigned an odds ratio of 1.00. This group consisted of those who carried at least one wild-type α_{2C} and one β_1-Gly389 allele. Other combinations consisted of maintaining carrier status for one gene and having the other homozygous for α_{2C}-Del322–325 or β_1-Arg389. The final genotypic combination, then, was homozygosity for both the α_{2C}-Del322–325 and β_1-Arg389. The findings were confined to African-Americans due to the low frequency of the α_{2C}AR polymorphism in Caucasians. The double homozygous combination revealed an unadjusted odds ratio of 12.67 (95% CI = 2.70-59.42, $P = 0.001$). When adjusted for age and sex, the odds ratio remained highly significant, at 10.11 (95%) CI = 2.11-48.53, $P = 0.004$). The combination of the two genotypes had a multiplicative association (i.e. more than an additive effect) with the risk of heart failure ($P = 0.05$ by the likelihood-ratio test for interaction). There was no evidence for confounding by diagnostic group (ischemic versus dilated) or other factors such as smoking, obesity, or diabetes. Taken together, the data indicated a two-gene polymorphic combination that represents a non-trivial risk for heart failure in African-Americans.

β_1AR polymorphisms as pharmacogenic loci for the response to β-blockers

The cell- and transgenic-based studies described above suggested that the β_1-Arg389 polymorphism may represent a favorable response allele in the treatment of heart failure by β-blockers. In the most straightforward scenario, the hyperactive β_1-Arg389 could act as a risk factor for development of heart failure, but, perhaps those who have the variant have a greater chance of responding to β-blocker treatment either in terms of LVEF improvement or more importantly, survival. In a small pilot study of 224 individuals, we noted that the LVEF response to carvedilol was significantly higher in Arg389 homozygotes compared to Gly389 homozygotes (8.7 ± 1.1% versus 0.93 ± 1.7% increase, respectively, $P = 0.02$).[24]

Given that the main therapeutic target of all β-blockers is the β_1AR subtype, it is most likely that variations in this receptor also have an influence on therapy with other agents in this class. Future studies will need to specifically test this, as well as the potential for polymorphisms of other genes in the adrenergic system, and related systems such as the renin–angiotensin axis[57] to also influence therapy. In addition, outcomes such as survival and exacerbations need to be considered. As such, in the not too distant future, there will probably be a panel of polymorphic sites in multiple genes that can be utilized to tailor β-blocker therapy.

Conclusions

Given the central role of sympathetic and related neurohormonal pathways in maintaining homeostasis, and their adaptive and maladaptive effects in heart failure, it is not altogether surprising that genetic variability of adrenergic receptors has effects on cardiac phenotype. The unexpected number of functionally relevant polymorphisms in most of the adrenergic receptor subtypes, and their various cellular phenotypes, has made it a challenge to ascertain their clinical impact. To date some of these appear to affect risk, clinical characteristics or the response to therapy. Additional studies are required to place these all into perspective, so as to understand how knowledge of genotypes can be utilized to improve diagnosis, prognosis and treatment.

References

1. Liggett SB. β-adrenergic receptors in the failing heart: the good, the bad, and the unknown. J Clin Invest 2001; 107: 947–8.
2. Insel PA, Motulsky HJ. Physiologic and pharmaco-logic regulation of adrenergic receptors. New York: Marcel Decker, Inc.; 1987: 201–36.
3. Cohn JN, Levine B, Olivari MT et al. Plasma norepi-nephrine as a guide to prognosis in patients with chronic congestive heart failure. N Engl J Med 1984; 311: 819–23.
4. Bouvier M, Leeb-Lundberg LM, Benovic JL et al. Regulation of adrenergic receptor function by phos-phorylation. II. Effects of agonist occupancy on phos-phorylation of α_1 and β_2-adrenergic receptors decreases its coupling to G_s. J Biol Chem 1987; 262: 3106–13.
5. Liggett SB, Shah SD, Cryer PE. Human tissue adren-ergic receptors do not predict in vivo responses to epi-nephrine. Am J Physiol Endocrinol Metab 1989; 256: E600–9.
6. Givertz MM. Underlying causes and survival in patients with heart failure. N Engl J Med 2000; 342: 1120–2.
7. van Campen LC, Visser FC, Visser CA. Ejection frac-tion improvement by beta-blocker treatment in patients with heart failure: an analysis of studies pub-lished in the literature. J Cardiovasc Pharmacol 1998; 32 (Suppl 1): S31–S35.
8. Bolger AP, Al Nasser F. Beta-blockers for chronic heart failure: surviving longer but feeling better? Int J Cardiol 2003; 92: 1–8.
9. Wollert KC, Drexler H. Carvedilol Prospective Randomized Cumulative Survival (COPERNICUS) Trial: Carvedilol as the sun and center of the β-blocker world? Circ 2002; 106: 2164–6.
10. Small KM, McGraw DW, Liggett SB. Pharmacology and physiology of human adrenergic receptor poly-morphisms. Annu Rev Pharmacol Toxicol 2003; 43: 381–411.
11. Hein L, Altman JD, Kobilka BK. Two functionally distinct α_2-adrenergic receptors regulate sympathetic neurotransmission. Nature 1999; 402: 181–4.
12. Brede M, Wiesmann F, Jahns R et al. Feedback inhi-bition of catecholamine release by two different alpha2-adrenoceptor subtypes prevents progression of heart failure. Circ 2002; 106: 2491–6.
13. Nedergaard OA, Abrahamsen J. Modulation of nora-drenaline release by activation of presynaptic beta-adrenoceptors in the cardiovascular system. Ann N Y Acad Sci 1990; 604: 528–44.
14. Small KM, Forbes SL, Brown KM et al. An Asn to Lys polymorphism in the third intracellular loop of the human α_{2A}-adrenergic receptor imparts enhanced agonist-promoted G_i coupling. J Biol Chem 2000; 275: 38518–23.
15. Small KM, Brown KM, Forbes SL et al. Polymorphic

deletion of three intracellular acidic residues of the α_{2B}-adrenergic receptor decreases G protein-coupled receptor kinase-mediated phosphorylation and desensitization. J Biol Chem 2001; 276: 4917–22.

16. Jewell-Motz EA, Liggett SB. An acidic motif within the third intracellular loop of the α_2C2 adrenergic receptor is required for agonist-promoted phosphorylation and desensitization. Biochem 1995; 34: 11946–53.

17. Small KM, Forbes SL, Rahman FF et al. A four amino acid deletion polymorphism in the third intracellular loop of the human α_{2C}-adrenergic receptor confers impaired coupling to multiple effectors. J Biol Chem 2000; 275: 23059–64.

18. Small KM, Wagoner LE, Levin AM et al. Synergistic polymorphisms of β_1- and α_{2C}-adrenergic receptors and the risk of congestive heart failure. N Engl J Med 2002; 347: 1135–42.

19. Moore JD, Mason DA, Green SA et al. Racial differences in the frequencies of cardiac β_1-adrenergic receptor polymorphisms: analysis of c145A>G and c1165G>C. Hum Mutat 1999; 14: 271.

20. Rathz DA, Brown KM, Kramer LA et al. Amino acid 49 polymorphisms of the human Beta 1-adrenergic receptor affect agonist-promoted trafficking. J Cardiovasc Pharmacol 2002; 39: 155–60.

21. Levin MC, Marullo S, Muntaner O et al. The myocardium-protective Gly-49 variant of the beta 1-adrenergic receptor exhibits constitutive activity and increased desensitization and down-regulation. J Biol Chem 2002; 277: 30429–35.

22. Lefkowitz RJ, Pierce KL, Luttrell LM. Dancing with different partners: protein kinase A phosphorylation of seven membrane-spanning receptors regulates their G protein-coupling specificity. Mol Pharmacol 2002; 62: 971–4.

23. Mason DA, Moore JD, Green SA et al. A gain-of-function polymorphism in a G-protein coupling domain of the human β_1-adrenergic receptor. J Biol Chem 1999; 274: 12670–4.

24. Perez JM, Rathz DA, Petrashevskaya NN et al. β_1-adrenergic receptor polymorphisms confer differential function and predisposition to heart failure. Nat Med 2003; 9: 1300–5.

25. Rathz DA, Gregory KN, Fang Y et al. Hierarchy of polymorphic variation and desensitization permutations relative to β_1- and β_2-adrenergic receptor signaling. J Biol Chem 2003; 278: 10784–9.

26. Green SA, Cole G, Jacinto M et al. A polymorphism of the human β_2-adrenergic receptor within the fourth transmembrane domain alters ligand binding and functional properties of the receptor. J Biol Chem 1993; 268: 23116–23121.

27. Turki J, Lorenz JN, Green SA et al. Myocardial signalling defects and impaired cardiac function of a human β_2-adrenergic receptor polymorphism expressed in transgenic mice. Proc Natl Acad Sci USA 1996; 93: 10483–8.

28. Green S, Turki J, Innis M et al. Amino-terminal polymorphisms of the human β_2-adrenergic receptor impart distinct agonist-promoted regulatory properties. Biochem 1994; 33: 9414–19.

29. Devic E, Xiang Y, Gould D et al. Beta-adrenergic receptor subtype-specific signaling in cardiac myocytes from beta(1) and beta(2) adrenoceptor knockout mice. Mol Pharmacol 2001; 60: 577–83.

30. Kohout TA, Takaoka H, McDonald PH et al. Augmentation of cardiac contractility mediated by the human beta(3)-adrenergic receptor overexpressed in the hearts of transgenic mice. Circ 2001; 104: 2485–91.

31. Tavernier G, Toumaniantz G, Erfanian M et al. beta3-Adrenergic stimulation produces a decrease of cardiac contractility ex vivo in mice overexpressing the human beta3-adrenergic receptor. Cardiovasc Res 2003; 59: 288–96.

32. Clement K, Vaisse C, Manning BSJ et al. Genetic variation in the beta-3-adrenergic receptor and an increased capacity to gain weight in patients with morbid obesity. N Engl J Med 1995; 333: 352–4.

33. Candelore MR, Deng L, Tota LM et al. Pharmacological characterization of a recently described human β3-adrenergic receptor mutant. Endocrinology 1996; 137: 2638–41.

34. Pietri-Rouxel F, St John Manning B, Gros J et al. The biochemical effect of the naturally occurring Trp64ÆArg mutation on human b3-adrenoceptor activity. Eur J Biochem 1997; 247: 1174ñ9.

35. Brodde OE, Buscher R, Tellkamp R et al. Blunted cardiac responses to receptor activation in subjects with Thr164Ile beta(2)-adrenoceptors. Circ 2001; 103: 1048–50.

36. Dishy V, Sofowora GG, Xie H-G et al. The effect of common polymorphisms of the β_2-adrenergic receptor on agonist-mediated vascular desensitization. N Engl J Med 2001; 345: 1030–5.

37. Kraus WE, Longabaugh JP, Liggett SB. Electrical pacing induces adenylyl cyclase in skeletal muscle independent of the b-adrenergic receptor. Amer J Physiol Endocrinol Metab 1992; 263: E226–30.

38. Liggett SB, Wagoner LE, Craft LL et al. The Ile164 β_2-adrenergic receptor polymorphism adversely affects the outcome of congestive heart failure. J Clin Invest 1998; 102: 1534–9.

39. Wagoner LE, Craft LL, Singh B et al. Polymorphisms

of the β₂-adrenergic receptor determine exercise capacity in patients with heart failure. Circ Res 2000; 86: 834–40.

40. Bray MS, Krushkal J, Li L et al. Positional genomic analysis identifies the β₂-adrenergic receptor gene as a susceptibility locus for human hypertension. Circ 2000; 101: 2877–82.

41. Snapir A, Koskenvuo J, Toikka J et al. Effects of common polymorphisms in the alpha1A-, alpha2B-, beta1- and beta2-adrenoreceptors on haemodynamic responses to adrenaline. Clin Sci 2003; 104: 509–20.

42. Snapir A, Heinonen P, Tuomainen TP et al. An insertion/deletion polymorphism in the alpha2B-adrenergic receptor gene is a novel genetic risk factor for acute coronary events. J Am Coll Cardiol 2001; 37: 1516–22.

43. Snapir A, Mikkelsson J, Perola M et al. Variation in the alpha2B-adrenoceptor gene as a risk factor for pre-hospital fatal myocardial infarction and sudden cardiac death. J Am Coll Cardiol 2003; 41: 195–6.

44. Gerson MC, Wagoner LE, McGuire N et al. Activity of the uptake-1 norepinephrine transporter as measured by I-123 MIBG in heart failure patients with a loss-of-function polymorphism of the presynaptic α₂C-adrenergic receptor. J Nucl Cardiol 2003; 10: 583–9.

45. Borjesson M, Magnusson Y, Hjalmarson A et al. A novel polymorphism in the gene coding for the beta(1)-adrenergic receptor associated with survival in patients with heart failure. Eur Heart J 2000; 21: 1810–12.

46. Ranade K, Jorgenson E, Sheu WH et al. A polymorphism in the beta1 adrenergic receptor is associated with resting heart rate. Am J Hum Genet 2002; 70: 935–42.

47. Podlowski S, Wenzel K, Luther HP et al. β₁-Adrenoceptor gene variations: a role in idiopathic dilated cardiomyopathy? J Mol Med 2000; 78: 87–93.

48. Tesson F, Charron P, Oeuchmaurd M et al. Characterization of a unique genetic variant in the β₁-adrenoceptor gene and evaluation of its role in idio-

pathic dilated cardiomyopathy. J Mol Cell Cardiol 1999; 31: 1025–32.

49. Iwai C, Akita H, Shiga N et al. Suppressive effect of the Gly389 allele of the β₁-adrenergic receptor gene on the occurrence of ventricular tachycardia in dilated cardiomyopathy. Circ J 2002; 66: 723–8.

50. Wagoner LE, Craft LL, Zengel P et al. Polymorphisms of the β₁-adrenergic receptor predict exercise capacity in heart failure. Am Heart J 2002; 144: 840–6.

51. Sofowora GG, Dishy V, Muszkat M et al. A common β₁-adrenergic receptor polymorphism (Arg389Gly) affects blood pressure response to β-blockade. Clin Pharmacol Ther 2003; 73: 366–71.

52. Johnson JA, Zineh I, Puckett BJ et al. Beta 1-adrenergic receptor polymorphisms and antihypertensive response to metoprolol. Clin Pharmacol Ther 2003; 74: 44–52.

53. Humma LM, Puckett BJ, Richardson HE et al. Effects of β₁-adrenoceptor genetic polymorphisms on resting hemodynamics in patients undergoing diagnostic testing for ischemia. Am J Cardiol 2001; 88: 1034–7.

54. Sandilands AJ, O'Shaughnessy KM, Brown MJ. Greater inotropic and cyclic AMP responses evoked by noradrenaline through Arg389 beta(1)-adrenoceptors versus Gly389 beta(1)-adrenoceptors in isolated human atrial myocardium. Br J Pharmacol 2003; 138: 386–92.

55. Molenaar P, Rabnott G, Yang I et al. Conservation of the cardiostimulant effects of (–)-norepinephrine across Ser49Gly and Gly389Arg beta(1)-adrenergic receptor polymorphisms in human right atrium in vitro. J Am Coll Cardiol 2002; 40: 1275–82.

56. Stanton T, Inglis GC, Padmanabhan S et al. Variation at the β₁-adrenoceptor gene locus affects left ventricular mass in renal failure. J Nephrol 2002; 15: 512–18.

57. McNamara DM, Holubkov R, Janosko K et al. Pharmacogenetic interactions between beta-blocker therapy and the angiotensin-converting enzyme deletion polymorphism in patients with congestive heart failure. Circ 2001; 103: 1644–8.

CHAPTER 36

Transcriptional profiling in heart failure

Kenneth B Margulies, Sunil Matiwala

Introduction

Progress in genome sequencing and annotation combined with technological advances permitting accurate, high-throughput analyses of mRNA abundance enables investigators to elucidate increasingly comprehensive views of transcriptional changes within tissue specimens from varied sources. High-throughput devices using quantitative real-time polymerase chain reaction (PCR) and robotic processing technology can perform hundreds of quantitatively robust assays in parallel, while custom or commercially available microarrays can simultaneously examine the expression of tens of thousands of transcripts from a single biological specimen. This unprecedented degree of analytical power provides exciting opportunities to provide novel insights into cell biology and an enormous variety of pathophysiological processes. At the same time, growth in analytical capacity creates a host of new challenges for virtually every facet of the scientific inquiry particularly study design, choice of appropriate control experiments, data management, analysis and interpretation. Indeed, increasing application of high-throughput technologies such as microarrays has challenged the age old paradigm of hypothesis-driven research and data collection, by providing the technological platforms for parallel queries of massive data sets, thereby providing the alternative construct of data-driven research and hypothesis generation.

In many ways the diversity and the complexity of processes and adaptations occurring during myocardial failure, render transcriptional profiling of the failing heart both a particularly daunting challenge and a fantastic opportunity for elucidating previously undiscovered aspects of disease pathophysiology. Accordingly, after a brief review of the technical aspects of transcriptional profiling, this chapter will focus on four major applications of transcriptional profiling in failing hearts: profiling genetically modified mouse hearts, profiling animal models of acquired heart disease, profiling human heart tissues, and profiling to elucidate myocardial responses to therapeutic interventions. We will also highlight how these distinct applications of transcription profiling in failing hearts are mutually informative, providing new insights that might escape more narrowly focused inquiries.

Techniques for assessing mRNA abundance

Gel-based techniques

With classic northern hybridization techniques, total cellular or tissue-derived RNA is resolved by denaturing electrophoretic separation on either formaldehyde- or glyoxal-based agarose gels. The smaller transcripts move faster than the larger ones through the gel matrix towards the positive electrode, because of the negatively charged phosphate backbone on RNA. The denatured RNA fragments are subsequently transferred onto either nitrocellulose or nylon membranes and fixed by ultraviolet light-mediated cross-linking. The immobilized single stranded RNA fragments on the membrane are allowed to hybridize with a radio-active or chemilumenescent-labeled, single-stranded complementary DNA (cDNA) or RNA. Autoradiography and densitometry allow visualization and quantification of bands corresponding to specific sizes of RNA transcripts on the agarose gel.

Polymerase chain reaction (PCR)

PCR is a rapid and versatile technique of clonal amplification of transcripts of interest that can be applied to detect and quantify very small amounts of RNA. To amplify mRNA, typically a single unique oligonucleotide, or 'primer' is annealed to the specific mRNA transcript. In the presence of a retrovirus-derived enzyme, reverse transcriptase and deoxynucleotidetriphosphates, a complementary DNA strand, or cDNA, is synthesized. It is this cDNA strand, identical to the target gene of interest, that is subjected to PCR amplification To facilitate PCR, specific oligonucleotide primers that are typically 15 to 25 nucleotides in length and complementary to the plus and minus DNA strands in the reverse and forward directions, serve as primers for the thermostable *taq* polymerase. In PCR, a mixture of primers, template cDNA, deoxynucleotides and *taq* polymerase are heated at temperatures of 93°C to 95°C to denature the target, and incubated at temperatures that vary from 50°C to 70°C to facilitate the formation of thermodynamically stable duplexes between each primer and template DNA. The *taq* polymerase synthesizes complementary sequences of the gene of interest by incubation for various times at 72°C. At the end of first cycle, two copies of the target gene sequence are generated. The process is then repeated over subsequent cycles in a logarithmic fashion. Newly synthesized DNA strands serve as templates in subsequent cycles of amplification. In a typical reaction 30 cycles can generate a sufficient quantity of amplified mRNA (as cDNA) to clone or sequence. Quantification of the mRNA of interest in a given sample can be accurately estimated by determining the number of PCR amplification cycles required to increase the mRNA into the linear range of the detection system. Higher abundance transcripts require fewer amplification cycles while less abundant transcripts require more cycles to achieve the linear range. With advances in technology and automation, up to 384 PCR reactions can be performed in tandem with each reaction plate completed in a matter of hours.

DNA microarrays

One drawback of the methods described above is that one can study only limited number of preselected genes at a time. In this context, microarrays provide a scalable alternative technique that permits thousands of known transcripts and expressed sequence tags (ESTs) to be evaluated in single experiment. DNA microarrays can be either cDNA-spotted arrays or an oligonucleotide array where DNA is either spotted onto a solid substrate such as a glass slide, or synthesized directly on the array using PCR or photolithography. The basic principle of microarray technique is illustrated in Fig. 36.1.[1] The RNA is extracted from the source and can be amplified if needed. Subsequently, RNA is reverse transcribed to cDNA and is labeled with either fluorescent or radioactive tags. The labeled cDNA is allowed to hybridize with the microarray. The complementary strands are hybridized to the array while non-specific material is washed away. Expressed genes are identified by the position of specifically bound probe on the array, and quantified by a signal intensity algorithm. Commercially available microarrays providing relatively comprehensive coverage of mouse, rat and human genomes have been available for several years. In addition, custom DNA microarrays, designed to examine a more limited set of transcripts, are a versatile and sometimes cost-effective alternative to commercial microarrays.

Analysis of data from microarrays

Due to the tremendous amount of data that are generated and the potential for manufacturing defects, data analysis is challenging and critical in microarray-based inquiries. Though there are cur-

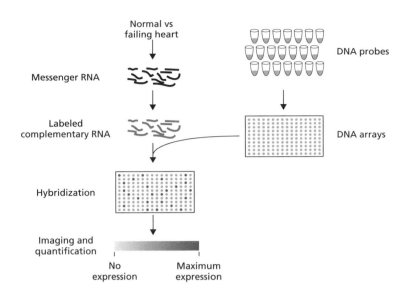

Figure 36.1. *High-throughput analysis of gene expression in heart failure by hybridization with high-density DNA arrays. Reproduced with permission from Schneider MD, Schwartz RJ. Chips ahoy: gene expression in failing hearts surveyed by high-density microarrays. Circulation 2000; 102: 3026–7.*

rently no set guidelines for design and analysis of microarray experiments, there is an emerging consensus that microarray analysis should incorporate at least four key components: data normalization, data filtering, computational analysis and validation.[2–5] In general, normalization is designed to correct for potential differences in the amount of source RNA loaded onto the microarray. Among the approaches described are per gene normalization using regression techniques,[4,6] and 'global normalization'. The latter approach assumes that the composite signal intensity will be constant for each array because upregulated and downregulated transcripts will offset one another, and a scaling factor is calculated to adjust intensities represented for each transcript. Often, multiple normalization approaches are combined.

The objective of filtering is to remove unwanted or misleading data prior to final statistical analysis, or to pre-group biological samples on the basis of common or disparate characteristics. For example, very low signal intensities often represent non-specific background fluorescence, while the highest signal intensities may represent fluorescence saturation that can skew the results. Similarly, biological samples that represent similar or identical etiology, sex, age, etc, can be sorted such that the resulting analyses are more meaningful. Various filtering approaches have been employed, including filtering based on raw signal intensity, fold-change magnitude or other statistical manipulations. Because there is no general consensus about an optimal approach, many investigators use a combination of filtering methods, recognizing that alternative choices may yield differing results.[6]

The third step in microarray data analysis includes use of interrogation of the normalized and filtered data, using conventional statistical tests and clustering algorithms. Here again, there is no consensus as to which statistical method is optimal for microarray analysis. While experiments sampling hundreds or thousands of genes simultaneously clearly require some protection against false positive results, the Student's *t* test with the Bonferroni correction is generally perceived as too stringent. One alternative approach is to use parametric and non-parametric analysis of variance (ANOVA). Clustering tools are helpful in unraveling alternative patterns of gene expression that may provide novel insight into coordinated regulation and the functional biology of transcriptional changes. Commonly used methods for clustering are hierarchical clustering, K-means clustering and self-organizing maps.

One area where there is more of a consensus is the need for validation of microarray experiments with PCR, Northern blots or other approaches. Although changes in mRNA abundance do not necessarily result in changes in protein abundance or functional assays, parallel changes at these levels may provide another type of validation of microarray results.

Transcriptional profiling in transgenic models of cardiomyopathy

The advent of genetically modified mouse models of myocardial hypertrophy and heart failure, achieved through mutation, deletion or overexpression of particular genes, has empowered a variety of new types of inquiries into the pathophysiology of heart failure. Among these new inquiry strategies, the combination of transgenic mouse models with high-throughput techniques for transcription profiling has begun to yield several important insights that are relevant to both disease pathogenesis and transcription profiling in other settings. In general, the new insights from transcription profiling of transgenic models include lessons about molecular integration, the time dependence of pathophysiological adaptations, and phenotypic convergence as heart failure advances.

A broad-based sampling of transcriptional adaptations in response to manipulation of a single molecular target provides a dramatic illustration of the integrated nature of myocardial biology. Indeed, every study that has taken this approach has demonstrated altered transcription of numerous targets other than the transgene itself. Typically, the transcriptional changes in genetically modified mouse hearts involve adaptations in several functionally different pathways that may or may not be intuitively related to the over- or underexpressed molecule. For example, Tang et al. reported that cardiac-specific tumor necrosis factor-α (TNF-α) overexpression induces dysregulation of more than 1000 genes at a stage of model development characterized by compensatory hypertrophy without significant contractile dysfunction or left ventricular (LV) dilation.[7] In response to overexpression of this inflammatory cytokine, it is not surprising that more than 20% of the transcriptional changes involved immune response genes. However, somewhat less expected were alterations in large numbers of genes involved

with metabolism, cell communication and growth. Of course, the number of dysregulated genes varies from model to model based on the microarray platform employed, severity of phenotype produced, stringency of the statistical analysis and other factors. Nevertheless, the consistent observation that a single molecular manipulation induces multiple and varied transcriptional adaptations supports the general concept that myocardial cellular biology is highly integrated, and that even a single molecular manipulation induces multiple response cascades that might either amplify or counterregulate the impact of the original perturbation.

Another lesson emerging from the profiling of genetically manipulated mouse models concerns the time dependence of the transcriptional adaptations to the altered transgene. Several studies comparing the transcriptional profiles of early and later phases of the evolution of transgenic mouse models have observed important and relatively consistent changes over time. For example, in evaluating a limited number of transcripts at five separate time points in four related transgenic models of adrenergically mediated cardiomyopathy, Gaussin et al identified some molecules, (uncoupling protein 2 (UCP2) and four-and-a-half LIM domain protein-1 (FHL1)) that were activated early and persistently during disease progression, others that were activated only early in disease progression (α-myosin heavy chain (α-MHC), M-protein, collagen type III) and still others that were activated only during the latter phases of disease progression.[8] Of these, only the last category included genes that were 'classic' myocardial markers of pathological hypertrophy and failure. Complementary investigations performed by Yun et al. utilized broad transcriptional profiling methods and reported quantitative and qualitative differences in the transcription profiles of young and old transgenic mice overexpressing the α_{1b}-adrenergic receptor (α_{1b}-AR).[9] These investigators found that young mice had a greater number of dysregulated genes including genes related to extracellular matrix constituents, apoptosis and myocyte growth and inflammation. In older mice, fewer genes were dysregulated compared with age-matched controls, and the abnormalities in older mice mainly included genes related to embryonic development and inflammation, with very few genes demonstrating a consistent pattern of dysregulation from 2 to 12 months despite consistent α_{1b}-AR overexpression during this interval.

Further extending these findings, Aronow et al. performed broad transcriptional profiling in four separate mouse models of cardiac hypertrophy at a time when their phenotypes remained relatively well-compensated (no pulmonary edema).[10] In their comparison of 'physiological hypertrophy' induced by protein kinase Cε (PKCε) overexpression, mild hypertrophy and contractile depression due to calsequestrin (CSQ) overexpression, and severe hypertrophy and contractile depression induced by calcineurin (CN) or $G_{\alpha q}$ overexpression, these investigators observed that the more severely abnormal hearts demonstrated the greatest number of dysregulated genes. Moreover, as highlighted in Fig. 36.2, there were no dysregulated genes common to all four hypertrophy models, only one gene (atrial natriuretic factor, ANF) was dysregulated in all three models with pathological hypertrophy, and the greatest concordance was found between the two models with the most severe phenotypes. One composite conclusion from these and other studies is that early transcriptional changes observed prior to the development

of severe hypertrophy or dysfunction are more likely to represent model-specific changes that point to direct molecular effects of the manipulated transgene rather than non-specific responses to the development of heart failure, myocardial stress or injury. Conversely, the more advanced the structural and functional defects observed in a given transgenic model, the greater the likelihood that transcriptional dysregulation will be a consequence, rather than a cause, of heart failure.

A corollary of this tendency for a time-dependent increase in phenotype-driven adaptations as a transgenic model progresses, is that there will be convergence of the transcriptional profiles of alternative models if they are examined after a severely diseased phenotype has developed. Indeed, this convergence has been observed in several studies including similarity between older muscle LIM protein (MLP)$^{-/-}$ and CSQ overexpressing mice,[11] the overlap with $G_{\alpha q}$-overexpressing,[10] and TNF-α overexpressing models,[7] and the similarity between the CSQ and calcineurin-overexpressing mice.[10] A feature of this transcriptional convergence with more advanced disease is that the shared profiles of these diverse models include many of the 'typical' transcriptional adaptations often observed in acquired, as opposed to genetic, models of heart

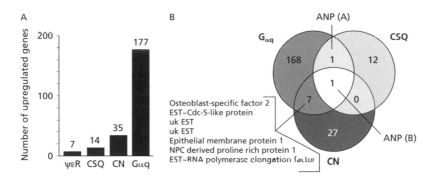

Figure 36.2. *Dynamically regulated genes in three separate transgenic mouse models of pathologic cardiac hypertrophy. (A) The absolute number of upregulated genes in each model corresponds to severity of phenotype ($G_{\alpha q}$ > CN > CSQ > $\psi\varepsilon R$). $G_{\alpha q}$, G protein alpha subunit; CN, calcineurin; CSQ, calsequestrin; $\psi\varepsilon R$ signifies protein kinase C-ε activation peptide. (B) comparison of dynamically upregulated sequences in the three pathological hypertrophy models identifies few dysregulated genes common to more than one model. EST, expressed sequence tag; uk, unknown; NPC, nasopharyngeal carcinoma. Reproduced with permission from Aronow BJ, Toyokawa T, Canning A et al. Divergent transcriptional responses to independent genetic causes of cardiac hypertrophy. Physiol Genomics 2001; 6: 19–28.*

failure, including upregulation of fetal genes such as those encoding ANP, brain natriuretic peptide (BNP), β-MHC, vascular smooth muscle actin and skeletal-α-actin. Such findings support the general contention that fetal genes expressed in failing hearts are useful markers of disease and/or myocardial plasticity,[10,12] but are less likely to represent mediators of failure in response to either genetic or acquired defects.

Transcriptional profiling in animal models of acquired cardiomyopathy

Animal models of acquired human myocardial hypertrophy and failure remain a mainstay of mechanistic research. In general, animal models of acquired cardiomyopathy begin with myocardial insults, such as ischemia/infarction, mechanical overload, neurohormonal excess or microbial infection, that affect many molecular processes and cell types at once. Accordingly, the initial phase of myocardial adaptations in acquired models of cardiomyopathy is expected to be more complex than that produced by temporally and/or spatially controlled, single-gene manipulations in genetically modified mouse models. While this added complexity may cloud interpretation of the myocardial adaptations to the insult, the types of insults explored in animal models of acquired cardiomyopathy are more relevant to the etiologies that account for the great majority of instances of human myocardial hypertrophy and failure (hypertension, ischemic heart disease, valvular disease and myocarditis). At the same time, animal models of acquired hypertrophy and failure permit homogeneity of host and insult, potential for serial tissue sampling, and correlative phenotypic analysis that cannot be achieved in human tissue studies. Thus, carefully executed studies using clinically relevant animal models of acquired myocardial disease combined with high throughput microarray-based transcriptional profiling provide a powerful opportunity to generate novel, clinically relevant insights into myocardial adaptations to external stress.

To date, transcriptional profiling of animal models has been limited to rodent models of hypertrophy and failure, because of the availability of custom and commercial microarrays for mice and rats but not other animal models. For example, well-characterized rat models of hypertension and hypertensive heart disease have been utilized to help characterize the myocardial transcriptional responses to sustained pressure overload. Ueno et al. examined the time course of transcriptional responses following administration of a high salt diet to Dahl salt-sensitive (DSS) rats from 6 to 15 weeks of age.[13] During this interval, these animals have sustained, severe hypertension, and their myocardial response evolves from normal architecture and function (6 weeks), to hypertrophy with limited fibrosis and preserved function (11 weeks), to continued hypertrophy with severe fibrosis and reduced function (15 weeks). In contrast to the relative transcriptional stability in age-matched animals fed a low-salt diet, the transcription profiles of DSS rats fed a high-salt diet exhibited a variety of abnormal patterns, some of which are reminiscent of the temporal responses in many of the transgenic models described above. There was a limited subset of genes activated during the first two weeks of hypertension, and most of these transcripts returned to baseline levels as the model progressed. One exception was ANF which increased progressively and dramatically during the evolution of pressure overload hypertrophy and failure. This pattern is quite similar to patterns observed in transgenic models progressing to decompensation.[7,10,14] A somewhat larger group of genes, not activated during the first two weeks of hypertension, were significantly altered during late compensation (11 weeks). Some of these transcripts, including aldolase and β-actin returned to more normal levels during decompensation, while other transcripts, such as 12-lipoxogenase, remained activated from compensation through decompensation. Though these studies did not identify any genes activated only during decompensation, some genes were selectively decreased during decompensation, leading the authors to speculate about a potential protective role for these molecules, such as D-binding protein. Overall, studies like these help identify candidate molecules that may be triggering transitions from normal to adapted phenotypes, and from compensation to decompensation. Analogous studies have been performed examining renal transcriptional adaptations to during sustained hypertension,[15] but no studies have compared renal and myocardial responses in this or other models.

Another approach to examining the myocardial adaptations to sustained hypertension has been to employ sustained infusions of vasoconstrictors. Using this approach, Friddle et al. used microarray profiling to identify genes that showed expression changes during induction of hypertrophy during a two-week period of isoproterenol or angiotensin infusion.[16] These investigators also identified transcriptional changes associated with the regression of hypertrophy after the vasoconstrictor infusion was discontinued. Although stringent statistical inclusion criteria and a requirement for analogous responses to isoproterenol and angiotensin limited the total number of genes identified, these studies found 32 genes altered only during the induction phase of hypertrophy, eight genes altered only during the regression phase of hypertrophy and 15 genes with significant reciprocal changes in expression during induction and regression of hypertrophy. The transcriptional changes persisting despite regression of hypertrophy included several classical fetal gene markers of hypertrophy (ANP, BNP α-vascular smooth muscle actin) raising the possibility that these fetal genes are markers of myocardial plasticity rather than hypertrophy *per se*. Indeed, this interpretation is consistent with studies that demonstrate considerable transcriptional overlap (both magnitude and direction) between the myocardial responses to hemodynamic overload and hemodynamic unloading.[12] Conversely, identification of genes specifically associated with regression of hypertrophy supports the concept of a distinct transcriptional program for 'reverse remodeling' that is not simply the inverse or deactivation of the hypertrophy induction profile.[16]

Rodent models of myocardial infarction have demonstrated great similarity with their clinical counterpart and have guided major successes in human therapeutics designed to limit adverse remodeling following myocardial infarction.[17] Accordingly, transcriptional profiling of infarcted rodent hearts offers the possibility of identifying new processes for investigation and therapeutics. By time course analysis of rat myocardium over a 16-week period following coronary ligation, Stanton et al. demonstrated both qualitative and quantitative distinctions between the transcriptional adaptations observed in the infarct zone and the non-infarcted myocardium.[18] As shown in Fig. 36.3, there were far more transcriptional changes in the infarct zone than the non-infarcted region. In this and other studies, a variety of temporal patterns were observed in both regions including early changes followed by subsequent normalization, delayed activation with subsequent deactivation and consistent or progressive changes throughout the early and delayed post-infarction period.[18,19] Particularly instructive is the analysis of functional clusters of genes combined with regional and temporal variations in myocardial infarction models. Not surprisingly, two separate studies have observed a robust and multifaceted activation of genes involved with extracellular matrix remodeling in the infarct region, with a temporal pattern that is consistent with structural remodeling and wound healing.[18–21] Perhaps less expected are changes among genes involved with myocyte energy metabolism including downregulation of transcripts involved in fatty acid catabolism, yet these findings are consistent with functional studies indicating that glucose is the favored energy source in the recovering myocardium.[22] Among genes activated throughout the 16-week post-infarction period are known regulators of transcription such as the cardiac ankyrin repeat protein (CARP) and the transforming growth factor β-stimulating clone (TSC-22), suggesting their possible role in regulating the transcriptional response to infarction both within and beyond the infarct zone.

Sehl et al. highlighted an additional important dimension to the transcriptional adaptations observed in these and other studies, namely, clear demonstration that altered transcription is occurring in a variety of different cells within the myocardium in response to tissue injury.[19] For example, in studies of infarcted rat hearts, phospholemman was activated primarily in cardiac myocytes adjacent to the area of necrosis, while activation of vimentin, p41-Arp2/3 protein complex subunit (ARC), and elongation factor 1-α were confined to vascular structures, and cathepsin B was expressed in infiltrating macrophages involved with the repair process. Another important global finding from this

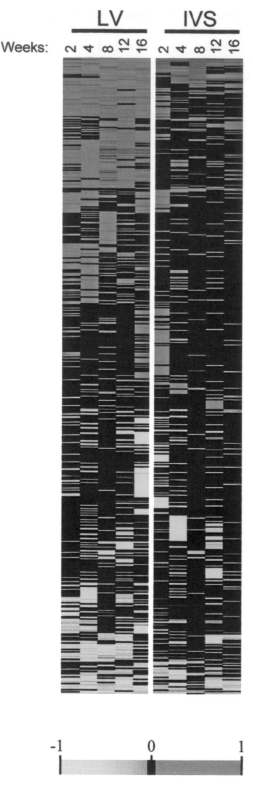

Figure 36.3. Heat map display of clustered gene expression patterns of 731 clones that display differential expression in rat myocardial infarction (MI). Each row represents a different cDNA, and columns pertain to data collected at five time points (weeks) after surgery from left ventricular infarct zone (LV) and non-infarct zone (IVS). Normalized data values, displayed in shades of red and blue, represent elevated and repressed expression, respectively, in MI tissue relative to control tissue (scale is shown at bottom). Insignificant differential expression values between 1.4 and −1.4 (before normalization) are set to 0 and shown in black. Genes with similar expression patterns are clustered together in 58 different clusters, and the clusters were arranged by nearest similarity to other clusters. Reproduced with permission from Stanton LW, Garrard LJ, Damm D et al. Altered patterns of gene expression in response to myocardial infarction. Circ Res 2000; 86: 939–45. See color plate section.

study is that, in contrast to transcriptional changes during embryonic development, changes in mRNA abundance for dysregulated genes in this and other acquired cardiomyopathies, rarely exceed 3-fold differences compared with normal controls.

Transcriptional profiling has also been employed to examine the myocardial responses to another type of insult, namely microbial infection. Taylor et al. evaluated time-dependent responses to coxsackievirus B3 infection in mice, and reported sequential host responses of injury, inflammation and healing, that are quite similar to the responses observed following myocardial infarction.[23] As with infarction, genes from a variety of functional categories, including cell signaling, cell structure, metabolism, transcription regulation and translation regulation, are altered in a variety of temporal patterns during the host response to infection.[23] Myocardial transcriptional

responses to microbial infection also include alterations of genes involved with cell defense and cell division that have been less prominent in reports focusing on ischemic myocardial infarction. Thus, the window provided by transcription profiling reveals the tremendously complex dynamics within the injured heart following myocardial infarction or microbial infection. The facets of this complexity include varied temporal patterns of molecular responses, marked regional differences between infarct, periinfarct and distant myocardial regions, and the involvement of a wide variety of resident and itinerant cell types. Ultimately, the composite of these multifaceted and intricately interwoven dynamics produces the necessary host responses of wound healing and functional adaptation within an organ that must continue its incessant function. Insofar as possible, Fig. 36.4, from Taylor et al. illustrates these complex dynamics.[23]

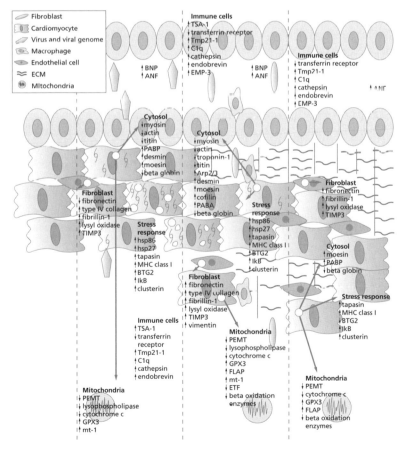

Figure 36.4. *Caricature of host gene expression occurring across the period of myocarditic disease progression. Potentially important gene regulatory events and their cellular and temporal contexts are shown, reflecting insights gained in the current analysis. Reproduced with permission from Taylor LA, Carthy CM, Yang D et al. Host gene regulation during coxsackievirus B3 infection in mice: assessment by microarrays. Circ Res 2000; 87: 328–34.*

Transcription profiling of failing human hearts

Because of the success in sequencing the human genome and interest stemming from diverse fields of inquiry, some of the highest quality and best annotated microarrays are those designed for high-throughput profiling of human tissues. Because gaining insight into human heart failure is the chief objective, it might seem self-evident that applying high-throughput profiling techniques to human myocardial tissues would provide the ideal way to acquire such insights. In reality, a variety of logistical and technical factors tend to undermine the benefits of this approach. First and foremost, virtually all studies obtain failing human ventricular myocardium at the time of cardiac transplantation. This approach precludes the multiple time points of sampling that has proved so informative in studies using animal models. In addition, myocardial profiling from failing hearts is primarily focused on most advanced end of the disease spectrum, where many transcriptional changes are likely to represent responses to, rather than causes of, heart failure. Another major problem confronting human myocardial profiling is the difficulty of obtaining suitable control tissue. Because of rapid tissue and mRNA degradation, samples obtained at necropsy are not suitable. The most 'normal' hearts derived from brain-dead donors are usually used for transplantation rather than research. In addition to factors leading to rejection for allografting, catecholamine fluxes and other interventions surrounding the brain-dead patient may alter their myocardial transcription. Finally, because of differences in disease etiology, disease duration and tempo, host factors and treatments, myocardial tissues from failing human hearts are far more heterogeneous than tissues from animal models. Indeed, Boheler et al. demonstrated that most of the genes dysregulated in failing hearts compared with non-failing controls demonstrate age-related and sex-related variation within groups, independent of the disease-related alterations.[24] Together, these various sources of heterogeneity tend to reduce the statistical power, and potentially confound the interpretation of any tissue-based inquiries including transcriptional profiling efforts.

The above considerations notwithstanding, the use of DNA microarrays does provide an opportunity to visualize a broad-based molecular portrait of heart failure by simultaneously visualizing thousands of genes and multiple pathways that would be too cumbersome to study using a gene-by-gene approach. One example of molecular distinctions defined through profiling relates to comparisons between hypertrophic and dilated cardiomyopathies. Most cases of familial hypertrophic cardiomyopathy (HCM) in humans are caused by mutation in one of sarcomeric proteins, while cytoskeletal mutations can lead to a familial dilated cardiomyopathy phenotype (DCM).[25,26] However, recent evidence indicates that the same mutation can be associated with either DCM or HCM phenotype in same family.[27] In a microarray-based transcriptional profile of heart failure, Hwang et al. demonstrated sets of genes that were abnormal and differentially expressed among the patients with DCM and HCM, despite the common endpoint of severe heart failure.[28] Among upregulated genes, there were more genes related to immune responses and cell/organism defense in DCM than in HCM, while there was a greater representation of genes related to protein synthesis in HCM. Conversely, there were more metabolism genes downregulated in DCM than in HCM, while genes related to cell signaling/communication and cell structure/motility tended to be reduced in HCM more than DCM. Thus, transcriptional profiling supports the concept that distinct etiologies of cardiomyopathy, progressing through different patterns of remodeling, involve distinct molecular dynamics, despite the common clinical outcome of congestive heart failure.

On the other hand, there is considerable transcriptional convergence as failing human hearts progress towards end-stage DCM. For example, comparing DCM and non-failing hearts, Barrans et al. observed dysregulation of more than 100 genes in the failing hearts.[29] In addition to classic markers of hypertrophy such as increases in ANF and β-MHC, these investigators observed failure-associated upregulation of genes encoding numerous sarcomeric and cytoskeletal proteins, regulators of transcription and genes involved with energy metabolism. Conversely, there was downregulation of many signaling genes and calcium signaling and calcium cycling proteins. In applying a hierarchical clustering algorithm, six of the seven DCM hearts and four of the five non-failing hearts clustered together based on their global gene expression patterns, further attesting to the convergence of transcriptional patterns in these hearts. Using similar methodologies in another small group of samples from failing and non-failing human hearts, Tan et al. reported similar findings including far more transcriptional convergence than divergence among the different diseased hearts.[30] Interestingly, many of the genes displaying consistent dysregulation in end-stage failing human hearts, including the genes also expressed during fetal development, are homologs to genes exhibiting dysregulation in the advanced, rather than the early, myopathic stages of hearts from genetically modified mouse models and rodent models of acquired heart disease. This homology supports the conclusion that many of the transcriptional changes observed in severely failing human hearts obtained at the time of transplantation are responses to, rather than causes of, sustained myocardial stress and overall the heart failure phenotype.[31]

Effects of therapeutic interventions on transcriptional profiles in failing hearts

In recent years, it has become increasingly apparent that therapeutic interventions have the ability to induce regression of the pathological phenotype of failing hearts. In both animal models and clinical settings, this phenomenon of so-called 'reverse remodeling' has been observed via both medical interventions and surgical interventions. The cellular and organ level mechanisms that drive the process of myocardial recovery and reverse remodeling are even more enigmatic than the processes that drive the progression of cardiomyopathy in diseased hearts. Accordingly, the use of transcriptional profiling, enabled by high-throughput techniques, is a theoretically attractive way of generating new insights into the molecular biology of myocardial reverse remodeling.

Among medical interventions, the most

dramatic instances of reverse remodeling of cardiac structure and function have been observed with administration of β-adrenergic blockers. In one study examining the manner in which β-adrenergic blockers attenuate the pathological phenotype in $G_{s\alpha}$-overexpressing mice, Gaussin et al. showed that pharmacological treatment attenuated increases in UCP2 and FHL1 compared with untreated controls, further supporting the role of these molecules as participants in the cardiomyopathic process in these mice.[8] In one of the rare studies in which human tissue profiling was performed before end-stage advanced heart failure had developed, Lowes et al. reported that DCM patients who had an improvement in the left ventricular ejection fraction (LVEF) with β-blocker had increases in sarcoplasmic reticulum Ca^{2+} ATPase (SERCA) and α-MHC mRNA that were not observed in individuals with functional improvements.[32] Moreover, the SERCA expression was not increased in a placebo-treated group, even when there was an increase in LVEF. Using real-time polymerase chain reaction analysis of endomyocardial biopsies, Yasumura et al. also observed increases in SERCA and phospholamban abundance and decreases in β-MHC and sodium–calcium exchanger, providing further evidence that β-blocker treatment affects expression of sarcomeric proteins and calcium regulatory proteins.[33] Together these data indicate that transcriptional profiling may indeed support existing hypotheses or provide novel clues to mechanisms of myocardial adaptations observed during pharmacological therapies.

In recent years, some of the most dramatic examples of reverse-remodeling have been observed following placement of left ventricular assist devices (LVADs) in patients with medically refractory heart failure awaiting heart transplantation.[34] Mechanical LVADs have become a reliable means of sustaining medically refractory patients with heart failure awaiting cardiac transplantation. After their implantation, LVADs induce profound and immediate cardiac unloading, and more delayed decreases in systemic neurohormonal activation.[35–37] Because left ventricular tissue is removed at the time of device placement and removal, LVAD support also provides an opportunity to study reverse remodeling at the tissue level. Exploiting this opportunity, studies to date have reported that LVADs result in favorable adaptations to the phenotype of advanced human cardiomyopathy at multiple levels, including reduced cardiomyocyte hypertrophy, improved contractility and contractile reserve, faster rates of relaxation, shorter action potential duration and reduced rates of apoptosis. Among other things, these studies demonstrate that the potential for myocardial reverse remodeling is retained across a wide spectrum of disease severity, including some of the most myopathic hearts.[34]

Exploiting the paired human myocardial specimens available at the time of LVAD implantation and removal, and the analytical power of cDNA microarrays, several investigators have examined broad trends in myocardial gene expression following LVAD support. In one study, Blaxall et al. demonstrated that LVAD support induces significant changes in the gene expression, and demonstrated a distinct separation between the pre- and post-LVAD groups consistent with a unique gene signature associated with reverse remodeling.[38] These investigators also detected disease-related differences in the response to LVAD support, such that the divergence of molecular phenotype between pre-LVAD and post-LVAD was greater for patients with non-ischemic cardiomyopathy than it was for patients with heart failure due to coronary artery disease. In a second study, highlighted in Fig. 36.5, Chen et al. observed that genes related to transcription, cell growth/apoptosis/ DNA repair, structural proteins, metabolism and cell signaling were, in general, upregulated following LVAD support; while genes related to cytokines were downregulated.[39] In general, these observations from transcriptional profiling are consistent with several previous proteomic and functional observations concerning LVAD-associated changes in cytokine expression, signaling, apoptosis and metabolism.[34,40–43]

Interestingly, despite the tendency to focus on LVAD-associated changes in gene expression, recent studies from our laboratory indicate that dysregulated genes in severely failing hearts are far more likely to remain abnormal than they are to recover towards a normal phenotype after LVAD

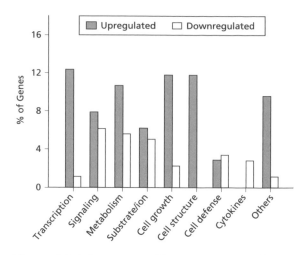

Figure 36.5. *Percentage of known genes in each functional category that were upregulated (filled bars) or downregulated (open bars). Transcription, transcriptional factors; signaling, cell signaling/communication; substrate/ion, substrate/ion transport; cell growth, cell growth/apoptosis/DNA repair; cell structure, cell structure/extrcellular matrix; cell defense, cell/organism defense. Reproduced with permission from Chen Y, Park S, Li Y et al. Alterations of gene expression in failing myocardium following left ventricular assist device support. Physiol Genomics 2003; 14: 251–60.*

support.[44] One possible interpretation of the general discordance between the microarray data indicating persistent pathological abnormalities, and the structural and functional data suggesting significant improvements following LVAD support, is that many key pathological processes are regulated at a post-transcriptional level. It is also conceivable that key transcriptional responses mediating phenotypic adaptations to LVAD support are transient, occur relatively early after cardiac unloading, and are missed by samples obtained at a later time. Another possibility is that many of the transcriptional changes mediating recovery do not necessarily recapitulate the transcriptional profile of the normal heart, a phenomenon we refer to as hysteresis. This latter hypothesis is consistent with the studies of Friddle et al. during induction and regression of hypertrophy during exposure to and removal of isoproterenol and angiotensin II.[16] Overall, it appears that further characterization of failing myocardium with microarray technology, before and after medical and surgical interventions, is likely to generate new insights and hypotheses concerning the biology of reverse remodeling.

Conclusions

Studies profiling transcription patterns in tissues from animal models and humans with myocardial failure have already generated a variety of novel observations that help inform our understanding of myocardial biology. Profiling studies in genetically modified mice and humans with familial cardiomyopathies reveal the integrated nature of myocardial cell biology, by demonstrating how single gene mutations trigger a variety of transcriptional adaptations. These studies further illustrate how early transcriptional adaptations, in advance of severe hypertrophy or dysfunction, are more likely to represent model-specific changes rather than non-specific responses to the development of heart failure and myocardial stress or injury. Conversely, in genetically driven and acquired cardiomyopathies, transcriptional adaptations observed in the presence of more advanced structural and functional defects are more likely to be a consequence, rather than a cause, of heart failure.

Transcriptional profiling of various animal models of acquired hypertrophy and failure support these general observations, and also highlight that the integrated biology of acquired cardiomyopathies is far more complex than in myopathies driven by a single point mutation. Even with the homogeneity of host and insult, potential for serial tissue sampling and correlative phenotypic analysis afforded by well-established animal models, interactions of time-, location- and cell-type dependent variability provides a formidable challenge to simple interpretations of transcriptional data. Nevertheless, carefully executed transcriptional profiling using clinically relevant animal models of acquired myocardial disease still provides a powerful opportunity to identify previously undiscovered molecular interactions contributing to myocardial adaptations to external stress. Considering the bio-

logical heterogeneity, various confounding factors, and difficulties obtaining normal and less-diseased human myocardial specimens for time-course analysis, it seems likely that animal models will provide much better opportunities than human myocardium for elucidating robust pathophysiological insights through transcriptional profiling.

Of course, transcriptional profiling in heart failure will continue to evolve. Advances in microarray technology and analysis routines will make this powerful assay technique more cost-effective, less cumbersome and available to more investigators. Transcriptional profiling will be complemented by increasing capacity for high-throughput assessments of protein abundance and post-translational modifications. Beyond technical advances, further comparison between existing and new experimental models will dramatically enhance the pace of new discoveries. Further enrichment of myocardial profiling will derive from more extensive clarification of distinctions between adaptations (transcriptional and otherwise) within different cell types, between forward remodeling and reverse remodeling, and between different therapeutic interventions. With these extensions, myocardial profiling will occupy an increasingly important role as a data-driven means of developing novel hypotheses to guide future mechanistic inquires.

References

1. Schneider MD, Schwartz RJ. Chips ahoy: gene expression in failing hearts surveyed by high-density microarrays. Circulation 2000; 102: 3026–7.
2. Cook SA, Rosenzweig A. DNA microarrays: implications for cardiovascular medicine. Circ Res 2002; 91: 559–64.
3. Shaughnessy J, Jr. Primer on medical genomics. Part IX: scientific and clinical applications of DNA microarrays—multiple myeloma as a disease model. Mayo Clin Proc 2003; 78: 1098–109.
4. Quackenbush J. Microarray data normalization and transformation. Nat Genet Dec 2002; 32 (Suppl): 496–501.
5. Quackenbush J. Computational analysis of microarray data. Nat Rev Genet Jun 2001; 2: 418–27.
6. Hoffmann R, Seidl T, Dugas M. Profound effect of normalization on detection of differentially expressed

genes in oligonucleotide microarray data analysis. Genome Biol 2002; 3: RESEARCH0033.

7. Tang Z, McGowan BS, Huber SA et al. Gene expression profiling during the transition to failure in TNF-alpha over-expressing mice demonstrates the development of autoimmune myocarditis. J Mol Cell Cardiol 2004; 36: 515–30.

8. Gaussin V, Tomlinson JE, Depre C et al. Common genomic response in different mouse models of beta-adrenergic-induced cardiomyopathy. Circulation 2003; 108: 2926–33.

9. Yun J, Zuscik MJ, Gonzalez-Cabrera P et al. Gene expression profiling of alpha(1b)-adrenergic receptor-induced cardiac hypertrophy by oligonucleotide arrays. Cardiovasc Res 2003; 57: 443–55.

10. Aronow BJ, Toyokawa T, Canning A et al. Divergent transcriptional responses to independent genetic causes of cardiac hypertrophy. Physiol Genomics 2001; 6: 19–28.

11. Blaxall BC, Spang R, Rockman HA, Koch WJ. Differential myocardial gene expression in the development and rescue of murine heart failure. Physiol Genomics 2003; 15: 105–14.

12. Depre C, Shipley GL, Chen W et al. Unloaded heart in vivo replicates fetal gene expression of cardiac hypertrophy. Nat Med 1998; 4: 1269–75.

13. Ueno S, Ohki R, Hashimoto T et al. DNA microarray analysis of in vivo progression mechanism of heart failure. Biochem Biophys Res Commun 2003; 307: 771–7.

14. Ding JH, Xu X, Yang D et al. Dilated cardiomyopathy caused by tissue-specific ablation of SC35 in the heart. EMBO J 2004; 23: 885–96.

15. Okuda T, Sumiya T, Iwai N, Miyata T. Pyridoxine 5′-phosphate oxidase is a candidate gene responsible for hypertension in Dahl-S rats. Biochem Biophys Res Commun 2004; 313: 647–53.

16. Friddle CJ, Koga T, Rubin EM, Bristow J. Expression profiling reveals distinct sets of genes altered during induction and regression of cardiac hypertrophy. Proc Natl Acad Sci USA 2000; 97: 6745–50.

17. Pfeffer JM, Pfeffer MA. Angiotensin converting enzyme inhibition and ventricular remodeling in heart failure. Am J Med 1988; 84: 37–44.

18. Stanton LW, Garrard LJ, Damm D et al. Altered patterns of gene expression in response to myocardial infarction. Circ Res 2000; 86: 939–45.

19. Sehl PD, Tai JT, Hillan KJ et al. Application of cDNA microarrays in determining molecular phenotype in cardiac growth, development, and response to injury. Circulation 2000; 101: 1990–9.

20. Hadjiargyrou M, Lombardo F, Zhao S et al. Transcriptional profiling of bone regeneration. Insight into the molecular complexity of wound repair. J Biol Chem 2002; 277: 30177–82.

21. Iyer VR, Eisen MB, Ross DT et al. The transcriptional program in the response of human fibroblasts to serum. Science 1999; 283: 83–7.

22. Lopaschuk GD, Saddik M. The relative contribution of glucose and fatty acids to ATP production in hearts reperfused following ischemia. Mol Cell Biochem 1992; 116: 111–16.

23. Taylor LA, Carthy CM, Yang D et al. Host gene regulation during coxsackievirus B3 infection in mice: assessment by microarrays. Circ Res 2000; 87: 328–34.

24. Boheler KR, Volkova M, Morrell C et al. Sex- and age-dependent human transcriptome variability: implications for chronic heart failure. Proc Natl Acad Sci USA 2003; 100: 2754–9.

25. Hwang JJ, Dzau VJ, Liew CC. Genomics and the pathophysiology of heart failure. Curr Cardiol Rep. May 2001; 3: 198–207.

26. Bowles NE, Bowles KR, Towbin JA. The 'final common pathway' hypothesis and inherited cardiovascular disease. The role of cytoskeletal proteins in dilated cardiomyopathy. Herz 2000; 25: 168–75.

27. Schonberger J, Seidman CE. Many roads lead to a broken heart: the genetics of dilated cardiomyopathy. Am J Hum Genet 2001; 69: 249–60.

28. Hwang JJ, Allen PD, Tseng GC et al. Microarray gene expression profiles in dilated and hypertrophic cardiomyopathic end-stage heart failure. Physiol Genomics 2002; 10: 31–44.

29. Barrans JD, Allen PD, Stamatiou D, Dzau VJ, Liew CC. Global gene expression profiling of end-stage dilated cardiomyopathy using a human cardiovascular-based cDNA microarray. Am J Pathol 2002; 160: 2035–43.

30. Tan FL, Moravec CS, Li J et al. The gene expression fingerprint of human heart failure. Proc Natl Acad Sci USA 2002; 99: 11387–92.

31. Vikstrom KL, Bohlmeyer T, Factor SM, Leinwand LA. Hypertrophy, pathology, and molecular markers of cardiac pathogenesis. Circ Res 1998; 82: 773–8.

32. Lowes BD, Gilbert EM, Abraham WT et al. Myocardial gene expression in dilated cardiomyopathy treated with beta-blocking agents. N Engl J Med. May 2 2002; 346: 1357–1365.

33. Yasumura Y, Takemura K, Sakamoto A, Kitakaze M, Miyatake K. Changes in myocardial gene expression associated with beta-blocker therapy in patients with chronic heart failure. J Cardiac Fail 2003; 9: 469–74.

34. Margulies KB. Reversal mechanisms of left ventricular remodeling: lessons from left ventricular assist device experiments. J Card Fail 2002; 8(6 Suppl): S500–505.

35. Altemose GT, Gritsus V, Jeevanandam V, Goldman B, Margulies KB. Altered myocardial phenotype after mechanical support in human beings with advanced cardiomyopathy. J Heart Lung Transplant 1997; 16: 765–73.

36. James KB, McCarthy PM, Thomas JD et al. Effect of the implantable left ventricular assist device on neuroendocrine activation in heart failure. Circulation 1995; 92(9 Suppl): II191–5.

37. Milting H, A ELB, Kassner A et al. The time course of natriuretic hormones as plasma markers of myocardial recovery in heart transplant candidates during ventricular assist device support reveals differences among device types. J Heart Lung Transplant 2001; 20: 949–55.

38. Blaxall BC, Tschannen-Moran BM, Milano CA, Koch WJ. Differential gene expression and genomic patient stratification following left ventricular assist device support. J Am Coll Cardiol 2003; 41: 1096–106.

39. Chen Y, Park S, Li Y et al. Alterations of gene expression in failing myocardium following left ventricular assist device support. Physiol Genomics 2003; 14: 251–60.

40. Flesch M, Margulies KB, Mochmann HC et al. Differential regulation of mitogen-activated protein kinases in the failing human heart in response to mechanical unloading. Circulation 2001; 104: 2273–6.

41. de Jonge N, van Wichen DF, van Kuik J et al. Cardiomyocyte death in patients with end-stage heart failure before and after support with a left ventricular assist device: low incidence of apoptosis despite ubiquitous mediators. J Heart Lung Transplant 2003; 22: 1028–36.

42. Torre-Amione G, Stetson SJ, Youker KA et al. Decreased expression of tumor necrosis factor-alpha in failing human myocardium after mechanical circulatory support: a potential mechanism for cardiac recovery. Circulation 1999; 100: 1189–93.

43. Razeghi P, Young ME, Ying J et al. Downregulation of metabolic gene expression in failing human heart before and after mechanical unloading. Cardiology 2002; 97: 203–9.

44. Margulies KB, Matiwala S, Cornejo C, Olsen H, Craven WA, Bednarik D. Mixed messages: transcription patterns in failing and recovering human myocardium. Circ Res 2005; 96: 592–9.

CHAPTER 37

Animal models of hypertrophic cardiomyopathy: relevance to human disease

Christopher Semsarian, J G Seidman

Introduction

Advances in molecular techniques particularly over the last decade have allowed significant improvements in our understanding of basic cellular and biochemical processes in health and disease. The cornerstone of these advances has been the development of whole-animal models of human disease. Given the high sequence homology between mice and humans at a genomic level, and the practicalities of housing and breeding small animals, genetically modified mouse (murine) models of human disease have proven invaluable. Furthermore, development of other animal models including rats and rabbits, have contributed to further our knowledge about disease processes. Animal models have not only confirmed primary pathological processes caused by specific gene defects, but have also provided a system whereby basic molecular, cellular, biochemical and cytological processes can be studied.

Genetically modifying the animal genome and studying the consequences clearly has important implications for understanding the basis of human disease. For example, studying the phenotype of a mouse in which a specific gene has been manipulated not only gives important information about the structure and function of the encoded protein, but it can also confirm a primary etiology as well as elucidate pathogenic processes and signaling events that ultimately lead to the phenotype. The consequences of the specific genetic manipulation can be studied from many perspectives. Furthermore, the role of environmental effects can be studied in detail, e.g. a particular gene may only become physiologically important during stress, exercise, or when diet is altered. Finally understanding the function of proteins and the molecular events involved in pathogenesis forms a platform for the identification of potential molecular targets for therapeutic intervention.

This chapter will discuss the development of animal models of hypertrophic cardiomyopathy (HCM). A particular focus will involve how these models differ both genotypically and phenotypically, how these models have contributed to our understanding of disease pathogenesis, and the utility and relevance of such models in specifically understanding human HCM.

Genetics of hypertrophic cardiomyopathy in humans

HCM is a primary cardiac disorder characterized by hypertrophy, usually of the left ventricle, in the absence of other loading conditions, such as hypertension. The disorder has a wide clinical spectrum, ranging from a benign, asymptomatic course, to symptoms of heart failure and sudden cardiac death.[1] HCM is the commonest structural cause of sudden cardiac death in individuals aged less than

35 years, including competitive athletes.[2,3] HCM was the first cardiovascular disorder for which the genetic basis has been identified, and as such, has acted as a paradigm for the study of a genetic cardiac disorder. HCM is an inherited disorder and is transmitted as an autosomal dominant trait. Since 1989, major advances have been made in understanding the molecular basis for HCM. Indeed, HCM is a genetically heterogeneous disease with at least ten causative genes now identified, the majority of which encode sarcomere proteins (Table 37.1). These include the cardiac β-myosin heavy chain (β-MHC), cardiac troponin T (cTnT), α-tropomyosin (α-TM), myosin-binding protein C (MyBP-C), cardiac troponin I (cTnI), essential and regulatory myosin light chains, and more recently, titin and actin genes (Fig. 37.1; Table 37.1).

While much is known about the clinical aspects of HCM as well as the underlying causative gene mutations, many questions remain unanswered. How does the mutation in the sarcomere lead to the observed phenotype? How do the mutations perturb sarcomere function? What are the signaling pathways which lead from the gene defect to disease? Why is it that affected siblings can have such diverse phenotypes, i.e. asymptomatic versus sudden death, if they carry the same gene mutation? What factors influence the expression of the mutant gene and hence modify the end phenotype? What are the mechanistic triggers that predispose

to sudden death in HCM and are there any treatments that can prevent disease?

Searching for answers to these questions is limited in human studies, due to a variety of factors including diverse genetic backgrounds, environmental stimuli which may vary between individuals such as diet and exercise, small numbers of individuals with the same mutation, and the relative difficulty in getting human samples for study. For these reasons, the development of animal models in HCM have been particularly useful, where there is effectively an unlimited supply of 'patients' with the same mutation, where genetic and environmental backgrounds can be controlled, and where access to tissue samples is essentially unlimited.

Animal models of hypertrophic cardiomyopathy

Many animal models have been developed over the last decade in an attempt to reproduce the human HCM phenotype. Following is an abbreviated description of some of these models as a representation of the more common genes known to cause human HCM, and to illustrate how these models mimic the human disease very closely. Many other animal models of HCM currently exist which will not be covered here.

Table 37.1. *Causal genes in HCM*

HCM gene	Symbol	Chromosome locus	% of all HCM
β-MHC	*MYH7*	14q12	35
Myosin-binding protein C	*MYBPC3*	11p11.2	30–35
Cardiac troponin T	*TNNT2*	1q32	10–15
α-tropomyosin	*TPM1*	15q22.1	< 5
Cardiac troponin I	*TNNI3*	19q13.4	< 5
Myosin light chains			
essential	*MYL3*	3p21	< 1
regulatory	*MYL2*	12q24.3	< 1
Actin	*ACTC*	15q14	< 0.5
Titin	*TTN*	2q24.3	< 0.5
α-MHC	*MYH6*	14q12	< 0.5

HCM, hypertrophic cardiomyopathy; MHC, myosin heavy chain.

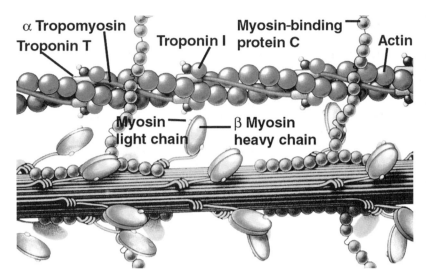

α **Tropomyosin**
Troponin T
Troponin I
Myosin-binding protein C
Actin
Myosin light chain
β **Myosin heavy chain**

Figure 37.1. *Schematic representation of cardiac sarcomere. Diagram shows key components of the thick and thin filaments of the sarcomere. Modified from Arad M, Seidman JG, Seidman CE. Phenotypic diversity in hypertrophic cardiomyopathy. Hum Mol Genet 2002; 11: 2499–506.*

Myosin heavy chain mutations

Mutations in the β-MHC gene comprise approximately 35% of all gene defects found in families with HCM, and these were the first to be identified in HCM.[4,5] The myosin heavy chain is the major component of the thick filament in the sarcomere and provides the key motor function to enable cardiac contraction to proceed. There are two isoforms of the myosin heavy chain in the heart, termed α and β. The β isoform is the predominant isoform in the human heart, while the α isoform predominates in the mouse heart. The α-MHC[403/+] mouse model was the first animal model of HCM developed and has been the most extensively studied.[6] The model illustrates how human HCM is replicated in mice. The mouse model was generated by introducing an Arg403Gln mutation into the α-cardiac MHC gene by gene targeting and homologous recombination (knock-in/knock-out). The Arg403Gln mutation is well characterized in humans with HCM, is associated with high penetrance (>90% express the phenotype by age 20 years) and early sudden death, with approximately 50% of affected heterozygous individuals carrying this mutation dying by the age of 45 years.[5] The heterozygous mouse genetically recapitulates the human situation, i.e. the mutation is present in one allele only, is under endogenous regulatory mechanisms, and is expressed at physiological levels. Analysis of the phenotype in these heterozygous α-

MHC[403/+] mice has revealed features that parallel the human disease (summarized in Table 37.2). α-MHC[403/+] mice develop diastolic dysfunction by age 12 weeks, classical histopathological changes of HCM (myocyte hypertrophy, disarray and fibrosis) by age 15 weeks (Fig. 37.2), and echocardiographically detectable left ventricular hypertrophy by 30 weeks.[7] This mimics the human disease very closely (Table 37.2).

More recently, this model has enabled studies to be performed to elucidate underlying cellular mechanisms involved in triggering myocyte growth and death, and therefore likely to provide novel avenues for intervention and prevention, e.g. mechanisms related to Ca^{2+} handling. In the α-MHC[403/+] model of HCM, myofibrillar protein extract, immunohistochemical and ryanodine receptor phosphorylation studies suggest an important early cellular event in HCM is dysregulation of the release of Ca^{2+} from the sarcoplasmic reticulum, possibly secondary to Ca^{2+} becoming 'trapped' in the mutated sarcomere.[8,9] Such studies have led to the identification of pharmacological agents which can both promote severe hypertrophy and sudden death, and prevent disease. For example, ciclosporin and minoxidil have separately been shown to severely exacerbate cardiac hypertrophy (independent of any blood pressure effects) in α-MHC[403/+] HCM mice, associated with a high incidence of premature death.[8] In contrast,

Table 37.2. *Comparison of Arg403Gln mutation in mice and humans*

HCM feature	Arg403Gln in mice	Arg403Gln in humans
Hypertrophy	Present in all by age 30 weeks	Present in 90% by age 20 years
Histopathology	Myocyte hypertrophy, disarray and fibrosis	Myocyte hypertrophy, disarray and fibrosis
Cardiac function	Systolic function supernormal, diastolic function impaired early	Systolic function supernormal, diastolic function impaired early
Ventricular arrhythmias	Common (~60%)	Common (>50%)
Sudden death	Rare except with vigorous exercise, e.g. swimming	Common (survival at age 45 years is ~50%)
Phenotype heterogeneity	Yes	Yes
Others	Gender effects observed, exercise capacity decreased	Gender effects observed, exercise limitations unclear

HCM = hypertrophic cardiomyopathy.

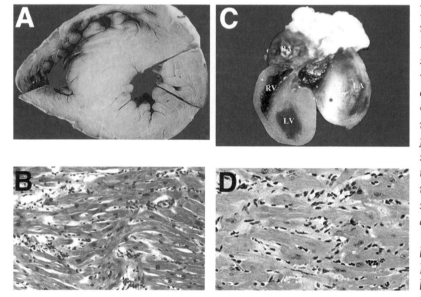

Figure 37.2. *Comparison of mouse and human HCM. Postmortem human heart sample showing gross hypertrophy and reduction in left ventricular chamber size (A) associated with myocyte hypertrophy, myofiber disarray and interstitial fibrosis (B). Postmortem heart sample form mouse expressing the Arg403Gln mutation in the myosin heavy chain gene, showing myocardial hypertrophy and left atrial enlargement (C) associated with typical histopathological features of HCM (D). LA, left atrium; LV, left ventricle; RV, right ventricle. See color plate section.*

administration of the L-type calcium channel inhibitor diltiazem early in life prevents disease in approximately 50% of α-MHC[403/+] mice.[9] Of further mechanistic interest is the observation that when these α-MHC[403/+] mice are bred to homozygosity, they develop severe dilated cardiomyopathy and die either *in utero* or within 7 days of birth, suggesting a gene dosage effect.[10]

This model of HCM has allowed investigators to begin to evaluate the role of modifiers, primarily

due to the unique ability in mice to control environmental influences, as well as to alter the genetic background by breeding the mutant mice in different mouse strains. Indeed, breeding of the α-MHC[403/+] mice in different genetic backgrounds, both inbred and outbred, has led to the identification of phenotypic differences in terms of hypertrophy, exercise capacity and histopathology, indicating the presence of a gene modifier in HCM.[11] Definition of such modifying factors by gene mapping strategies has great significance for defining novel targets for therapeutic interventions in human HCM.

The same mutation indicated above has also been introduced in a transgenic rabbit model of HCM.[12] This gene overexpression model expresses a human β-MHC cDNA with either the Arg403Gln mutation or the wild-type cDNA (as a control). The main advantage of the rabbit model is that the mutation is introduced into the β-isoform of MHC (in contrast to α-isoform in mice) which is the same as the human isoform. The larger size of rabbits makes many of the physiological investigations easier and more reliable. In addition, the heart rate of rabbits is closer to that of the human than with mice. In rabbits carrying the mutant transgene, a significant increase in ventricular septal wall thickness and left ventricular mass was seen, as well as myocellular disarray, and an increase in interstitial collagen.[12] Premature death was also more common in mutant transgenic rabbits. Taken together, these features are consistent with the human form of HCM. While rabbit models generally have been of value, the expense in generating rabbit lines, the costs of housing, longer breeding times, and relative lack of genomic information indicate mouse models will continue to be the primary source of transgenic models.

Transgenic mouse lines have also been made where the α-MHC gene (MYH6) has been completely removed.[13] Homozygous α-MHC null mice died in utero of gross heart defects, whereas heterozygous mice survived with normal external appearance. Histopathological analysis in these mice revealed fibrosis and disrupted sarcomeric structure, associated with severe impairment of contractility and relaxation. This has led to the conclusion that both α-MHC alleles are required for normal cardiac development and function.

Myosin-binding protein C mutations

Mutations in the MyBP-C gene (MYBPC3) comprise approximately 30–35% of all gene defects found in families with HCM.[14,15]. MyBP-C binds to both myosin and titin filaments and plays an important role in the assembly and overall stability and architecture of the sarcomere. A MyBP-C mouse model has been generated by introducing a neomycin gene into exon 30 of MYBPC3 using gene targeting and homologous recombination techniques, resulting in a 30-amino acid truncation of the MyBP-C protein.[7] The truncation causes disruption of the myosin- and titin- binding domains. This mutant protein is identical to a truncation which occurs in humans with HCM. Analysis of the phenotype of these heterozygous MyBP-C mice has identified features consistent with the human disease. In patients with HCM due to MyBPC3 mutations, the clinical course is characterized by late penetrance (>40 years) and a generally benign course. These mice develop hypertrophy very late, i.e. only a third of heterozygote MyBPC3 mice have hypertrophy by age 30 weeks (compared to 100% in α-MHC$^{403/+}$ mice at the same age), but all have hypertrophy by age two years.[7] Histopathological features are also consistent with human HCM, while electrophysiological testing suggests that there is a significantly increased vulnerability to ventricular arrhythmias, and therefore sudden death. Comparison of this MyBP-C model with the α-MHC$^{403/+}$ model will be invaluable in addressing the issues of disease heterogeneity and penetrance, as well as the effects of aging in the progression of cardiomyopathy.

In an analogous way to the α-MHC$^{403/+}$ model, interesting results have arisen by breeding the heterozygote MyBP-C mouse line to homozygosity. Homozygosity for the truncated MyBP-C protein leads to a dilated cardiomyopathy.[16] In contrast to the Arg403Gln homozygote mice, where dilated cardiomyopathy occurs in neonates and all mice die of heart failure by age 7 days, homozygous MyBPC3 mice develop dilated cardiomyopathy by age 3 weeks, but subsequently develop compensatory hypertrophy and indeed have a normal lifespan.[16] The ability to breed these mice to both heterozygosity and homozygosity has resulted in clinically relevant models of human HCM and dilated cardiomyopathy and provides a platform for further studies both to understand pathogenesis, and to potentially identify therapeutic options and targets.

Most recently, humans with HCM have been identified who carry two HCM-causing gene mutations.[15,17] Some of these compound heterozygote individuals have more severe clinical disease.

Cross-breeding of mouse models of HCM to produce mice that carry two mutations to form compound heterozygotes (e.g. Arg403Gln combined with truncated *MyBPC3*) will allow studies to be performed to understand how multiple gene defects may interact and alter clinical outcome.

Troponin T mutations

Mutations in the cardiac troponin T (cTnT) gene (*TNNT2*) cause 5–15% of HCM. cTnT is the tropomyosin-binding component of the troponin complex.[18] The cTnT protein plays an important role in both positioning the troponin complex along the thin filament and in conferring calcium sensitivity to the inhibitory function of the troponin complex on ATPase. In contrast to other genes causing HCM, individuals with mutations in *TNNT2* have less cardiac hypertrophy but remain predisposed to both symptoms as well as sudden cardiac death.[18] Transgenic mouse lines modeling human mutations in the *TNNT2* have been described. The first such model mimicked the product of a splice donor site mutation (in intron 15), leading to a truncated cTnT protein where exon 16 is deleted.[19] This region contains a tropomyosin-binding domain. Mouse lines expressing low levels (4–6%) of the cTnT truncated protein were viable, but exhibited a smaller left ventricular size, severe diastolic dysfunction and mild systolic abnormalities. Histopathological analysis of these hearts showed myocellular disarray, but no fibrosis and no induction of hypertrophy, mimicking the human phenotype. Breeding these mice to double transgene expression resulted in a severe phenotype with affected mice dying within 24 hours of birth, reflecting the effect of gene dosage and the severity of this allele. Interestingly, a rat model expressing a truncated human TnT protein (8–10%), again missing exon 16, has also been described, and showed diastolic dysfunction similar to the mouse model and increased susceptibility to arrhythmias in an isolated working heart preparation.[20]

A second cTnT transgenic mouse model has been studied in which the Arg92Gln missense mutation was overexpressed in the heart.[21] Lines were established expressing 30%, 67% and 92% of their total TnT protein as the missense allele. Like the previous truncated cTnT model, these mice also had a small left ventricle, but in contrast had significant fibrosis and induction of the hypertrophic markers in the absence of myocyte hypertrophy. Isolated cardiac myocytes showed sarcomeric activation, impaired relaxation and shorter sarcomere length, while isolated working heart preparations showed hypercontractility and diastolic dysfunction. These findings are distinct from a transgenic mouse model with the same mutation which has been studied by a different group, with only 1% to 10% of the total TnT protein being from the mutated allele.[22] These mice had normal left ventricular dimensions and systolic function, but diastolic dysfunction, as well as the typical histopathological features. The differences between these two models probably reflect variations in the amount of expression of the mutant allele as a proportion of total cTnT expression.

Troponin I mutations

Mutations in the cTnI gene account for less than 5% of genetically defined HCM patients.[23] cTnI is the inhibitory component of the troponin complex, and prevents effective actin–myosin interaction. The human disease-causing mutation Arg145Gly in the cTnI gene has been overexpressed in a cardiac-specific manner in mice.[24] These mice develop cardiomyocyte hypertrophy and interstitial fibrosis associated with upregulation of RNA markers for hypertrophy. Functionally, these mice develop hypercontractility and impairment of diastolic function, with increased calcium hypersensitivity demonstrated in isolated muscle fibers. While these features are consistent with the human form of HCM, sudden death was observed in a significant proportion of transgenic mice, and survival overall was reduced in mice which did not die suddenly. This is different from the more common phenotype observed in humans with cTnI mutations, and may reflect differences in overall mutant protein expression in this mouse model, or differential responses to calcium changes specifically in the mouse heart, which beats ten times faster than in humans.

Tropomyosin mutations

Mutations in the α-TM gene (*TPM1*) account for less than 5% of genetically defined HCM patients.[25]

α-TM is the predominant isoform in the human heart, and is involved in both stabilizing the thin filament and in blocking the myosin-binding site of actin when calcium is absent. Both mouse and rat models of *TPM1* have been developed. Introduction of the Asp175Asn mutation in the *TPM1* gene caused increased calcium sensitivity in a mouse model.[26] These mice showed myocellular disarray and hypertrophy, with impairment of both systolic and diastolic function. In a second model, heterozygous inactivation of the *TPM1* gene resulted in normal protein levels, despite the presence of only a single functional allele,[27,28] suggesting that the null-allele hypothesis for the *TPM1* gene is less likely.

At least two different rat models with mutations in *TPM1* have recently also been described.[29] One model harbors the same mutation as indicated above, while the second transgenically overexpresses the Glu180Gly mutation. Both these rat models did not develop hypertrophy, but showed myocellular disarray, induction of hypertrophic markers such as α-skeletal actin, diastolic dysfunction and arrhythmias. Comparison of these phenotypes caused by different mutations in two different rodent systems will enable pathogenic mechanisms to be studied in detail.

Naturally occuring animal models

There are several other animals in which spontaneous, naturally occurring mutations lead to a HCM phenotype. There is a hamster model for HCM, in which a naturally-occurring deletion of exon 1 of the δ-sarcoglycan gene (*SGCD*) has been described.[30] This genetic defect can lead to either HCM, or dilated cardiomyopathy, or DCM depending on the subline (BIO14.6 versus TO-2). The hamsters with the HCM subline develop fibrosis, myofiber disarray, and left ventricular hypertrophy. To date, no human mutations in dystrophin-associated glycoproteins have been shown to cause HCM, although other cytoskeletal proteins such as actin have been described as a rare cause of HCM (Table 37.1). The findings in this hamster model may shed light on alternate mechanisms of how defects in cytoskeletal proteins may lead to cardiomyopathies in humans.

Naturally-occurring HCM has also been described in a family of Maine Coon cats.[31] This strain of cats have a disorder that resembles human disease, exhibiting both morphological and histopathological features of human HCM. Interestingly, almost 50% of affected cats die of heart failure or sudden cardiac death. The genetic basis of HCM in these cats has not yet been elucidated. However it is clear that this model may prove very useful in understanding the pathogenesis of the most severe complications of HCM, i.e. heart failure and sudden death.

Approaches to therapy in animal models of HCM

To date, there are no human studies which definitively show that therapy in HCM can result in either regression or prevention of onset of disease. While treatment with β-blockers, calcium antagonists, and other anti-arrhythmic agents is widely used, none of these therapies have been shown to improve survival in randomized controlled trials. To this end, studies in animal models of HCM have been important in elucidating potential therapies that may alter the natural history of this disease, with a particular emphasis on hypertrophy and fibrosis. Three particular recent examples have illustrated how pharmacological agents with different properties may be of therapeutic benefit in HCM.

In the first of these studies, treatment of transgenic mice expressing mutant cTnT with the angiotensin II inhibitor, losartan, resulted in reversal of interstitial fibrosis, associated with a 50% reduction in transforming growth factor-β, a mediator of the angiotensin II pro-fibrotic effect.[32] In a second study, treatment of transgenic rabbits expressing mutant β-MHC gene (*MYH7*) with the HMGcoA reductase inhibitor, simvastatin, resulted in regression of both cardiac hypertrophy (reduction in left ventricular mass by 37%) and interstitial fibrosis (reduction in collagen volume fraction by 44%), associated with improvement in cardiac function.[33] In contrast to these two regression studies, a more recent study provides evidence that early preclinical treatment can prevent the onset of phenotypic changes of HCM. Specifically, treatment of the gene-targeted mice expressing mutant α-MHC (*MYH6*) with the L-type calcium channel inhibitor, diltiazem, results in prevention of hypertrophy and fibrosis in approximately 50% of mice.[9] Taken together, these three studies illustrate how animal models in HCM will be an important tool in elucidating different mechanisms of disease pathogenesis and in determining how three therapeutic agents, with vastly different pharmacological properties, can result in significant clinical benefit. It is likely that such studies in animal models will provide the basis for initiating clinical trials in genotyped humans with HCM.

Future directions

Understanding how sarcomere mutations perturb biophysical events of muscle contraction and cell signaling pathways within the myocyte, will inevitably allow advances in future treatment interventions. An exponential growth in our knowledge of gene defects that cause HCM will undoubtedly continue. Essentially any human gene of interest can now be modified or knocked out in the mouse, and the subsequent effects of the genetic manipulation studied in detail, both in terms of the overall phenotype, and in the dissection of molecular, biochemical and structural events at a cellular level. Continued study and development of animal models of HCM should further enable studies of the integrative physiology of multiple organ systems involved in the development of HCM.

Development of animal models in which the targeted gene can be manipulated during life, i.e. conditional models where the mutant gene can be turned 'on' or 'off',[34] will no doubt further improve the utility of such animal models. These conditional or bigenic models are currently being developed. One such technique is the use of a binary tetracycline-controlled system in which gene expression is regulated by the presence or absence of tetracycline in the drinking water.[35,36] Conditional trangenesis will therefore allow the researcher not only to determine whether a particular gene defect causes disease, but whether these changes can be reversed when the transgene is turned off. Elucidation of signaling events and physiological responses leading from mutant protein to clinical phenotype, and the identification of factors, both genetic and environmental that modify the response to mutations, may provide new insights that can enhance our fundamental understanding of the pathogenesis of HCM, and therefore improve therapeutic intervention.

References

1. Seidman JG, Seidman CE. The genetic basis for cardiomyopathy: from mutation identification to mechanistic paradigms. Cell 2001; 104: 557–67.
2. Maron BJ. Sudden death in young athletes. N Engl J Med 2003; 349: 1064–75.
3. Maron BJ, Shirani J, Poliac LC et al. Sudden death in young competitive athletes. Clinical, demographic, and pathological profiles. JAMA 1996, 276: 199–204.
4. Geisterfer-Lowrance AA, Kass S, Tanigawa G et al. A molecular basis for familial hypertrophic cardiomyopathy: a beta-cardiac myosin heavy chain gene missense mutation. Cell 1990; 62: 999–1006.
5. Watkins H, Rosenzweig A, Hwang DS et al.. Characteristics and prognostic implications of myosin missense mutations in familial hypertrophic cardiomyopathy. N Engl J Med 1992; 326: 1108–14.
6. Geisterfer-Lowrance AAT, Christe ME, Conner DA et al. A mouse model of familial hypertrophic cardiomyopathy. Science 1996; 272: 731–4.
7. McConnell BK, Fatkin D, Semsarian C et al. Comparison of two murine models of familial hypertrophic cardiomyopathy. Circ Res 2001; 88: 383–9.
8. Fatkin D, McConnell BK, Mudd JO et al. An abnormal Ca^{2+} response in mutant sarcomere protein mediated familial hypertrophic cardiomyopathy. J Clin Invest 2000; 106: 1351–9.
9. Semsarian C, Giewat M, Georgakopoulos D et al. The L-type calcium-channel inhibitor diltiazem prevents cardiomyopathy in a mouse model. J Clin Invest 2002; 109: 1013–20.
10. Fatkin D, Christe ME, Aristazabal O et al. Neonatal cardiomyopathy in mice homozygous for the Arg403Gln mutation in the α-cardiac myosin heavy chain gene. J Clin Invest 1999, 103: 147–53.
11. Semsarian C, Fatkin D, Healey MJ et al. A polymorphic modifier gene alters the hypertrophic response in a murine model of familial hypertrophic cardiomyopathy. J Mol Cell Cardiol 2001; 33: 2055–60.
12. Marian AJ, Wu Y, Lim D-S et al. A transgenic rabbit model for human hypertrophic cardiomyopathy. J Clin Invest 1999; 104: 1683–92.

13. Jones WK, Grupp IL, Doetschmann T et al. Ablation of the murine α-myosin heavy chain gene leads to dosage effects and functional deficits in the heart. J Clin Invest 1996; 98: 1906–17.

14. Niimura H, Bachinski LL, Sangwatanaroj S et al. Mutations in the gene for cardiac myosin-binding protein C and late-onset familial hypertrophic cardiomyopathy. N Engl J Med 1998; 338: 1248–57.

15. Richard P, Charron P, Carrier L et al. Hypertrophic cardiomyopathy: distribution of disease genes, spectrum of mutations, and implications for a molecular diagnosis strategy. Circulation 2003; 107: 2227–32.

16. McConnell BK, Jones KA, Fatkin et al. Dilated cardiomyopathy in homozygous myosin-binding protein-C mutant mice. J Clin Invest 1999; 104: 1235–44.

17. Jeschke B, Uhl K, Weist B et al. A high risk phenotype of hypertrophic cardiomyopathy associated with a compound genotype of two mutated beta-myosin heavy chain genes. Human Genet 1998; 102: 299–304.

18. Watkins H, McKenna WJ, Thierfelder L et al. Mutations in the genes for cardiac troponin T and alpha-tropomyosin in hypertrophic cardiomyopathy. N Engl J Med 1995; 332: 1058–64.

19. Tardiff JC, Factor SM, Tompkins BD et al. A truncated cardiac troponin T molecule in transgenic mice suggests multiple cellular mechanisms for familial hypertrophic cardiomyopathy. J Clin Invest 1998; 101: 2800–11.

20. Frey N, Franz WM, Gloeckner K et al. Transgenic rat hearts expressing a human cardiac troponin T deletion reveal diastolic dysfunction and ventricular arrhythmias. Cardiovasc Res 2000; 47: 254–64.

21. Tardiff JC, Hewett TE, Palmer BM et al. Cardiac troponin T mutations result in allele-specific phenotypes in a mouse model for hypertrophic cardiomyopathy. J Clin Invest 1999; 104: 469–81.

22. Oberst L, Zhao G, Park JT et al. Dominant-negative effect of a mutant cardiac troponin T on cardiac structure and function in transgenic mice. J Clin Invest 1998; 102: 1498–1505.

23. Kimura A, Harada H, Park JE et al. Mutations is the cardiac troponin I gene associated with hypertrophic cardiomyopathy. Nat Genet 1997; 16: 379–82.

24. James J, Zhang Y, Osinska H et al. Transgenic modeling of a cardiac troponin I mutation linked to familial hypertrophic cardiomyopathy. Circ Res 2000; 87: 805–11.

25. Thierfelder L, Watkins H, MacRae C et al. Alpha-tropomyosin and cardiac troponin T mutations cause familial hypertrophic cardiomyopathy: a disease of the sarcomere. Cell 1994; 77: 701–12.

26. Muthuchamy M, Pieples K, Rethinasamy P et al. Mouse model of a familial hypertrophic cardiomyopathy mutation in α-tropomyosin manifests cardiac dysfunction. Circ Res 1999; 85: 47–56.

27. Blanchard E, Iizuka K, Christe M et al. Targeted ablation of the murine α-tropomyosin gene. Circ Res 1997; 81: 1005–10.

28. Rethinasamy OP, Muthachamy M, Hewett T et al. Molecular and physiological effects of α-tropomyosin ablation in the mouse. Circ Res 1998; 82: 116–23.

29. Wernicke D, Thiel C, Isac C et al. Characterization of a transgenic rat model of familial hypertrophic cardiomyopathy with missense mutations Asp175Asn or Glu180Gly in α-tropomyosin. Circulation 1999; 100: 1–268.

30. Sakamoto A, Ono K, Abe M et al. Both hypertrophic and dilated cardiomyopathies are caused by mutation of the same gene, delta-sarcoglycan, in hamster: an animal model of disrupted dystrophin-associated glycoprotein complex. Proc Natl Acad Sci USA 1997; 94: 13873–8.

31. Kittleson MD, Meurs KM, Munro MJ et al. Familial hypertrophic cardiomyopathy in Maine Coon cats – an animal model of human disease. Circulation 1999; 99: 3172–80.

32. Lim DS, Lutucuta S, Bachireddy P et al. Angiotensin II blockade reverses myocardial fibrosis in a transgenic mouse model of human hypertrophic cardiomyopathy. Circulation 2001; 103: 789–91.

33. Patel R, Nagueh SF, Tsybouleva N et al. Simvastatin induces regression of cardiac hypertrophy and fibrosis and improves cardiac function in a transgenic rabbit model of human hypertrophic cardiomyopathy. Circulation 2001; 104: 317–24.

34. Redfern CH, Degtyarev MY, Kwa AT et al. Conditional expression of a Gi-coupled receptor causes ventricular conduction delay and a lethal cardiomyopathy. Proc Natl Acad Sci USA 2000; 97: 4826–31.

35. Bujard H. Controlling genes with tetracyclines. J Gene Med 1999; 1: 372–4.

36. Fishman GI, Kaplan ML, Buttrick PM. Tetracycline-regulated cardiac gene expression *in vivo*. J Clin Invest 1994; 93: 1864–8.

CHAPTER 38

Clinical proteomics: technologies to define and diagnose heart disease

Brian A Stanley, Jennifer E Van Eyk

The proteome is the fingerprint of disease

Proteomics is the global analysis of the protein species derived from a genome (the proteome). Proteins are made up of 20 amino acids, each encoded for by three nucleotides. Similar to genomics, proteomics utilizes the unique sequence of amino acids in each protein to identify them. Unlike genomics, where the relative complexity of the analysis is determined by the presence of alternative gene forms, proteomics must deal with virtually limitless differences in protein chemistry and potential modifications. Therefore, the complexity of proteomics is generally underappreciated by the casual observer. A typical gene for instance may have multiple different isoforms, which can be generated by alternative splicing of the transcript, generating two different proteins. Now consider the potential magnification in complexity of these proteins (Fig. 38.1). During translation through the secretory pathway for instance, there may be different degrees of folding in the endoplasmic reticulum. Although folding is not a post-translational modification (PTM), it can affect whether the potential sites for the addition of PTMs are available. Upon proper folding, the protein is now able to travel through the Golgi body, where it can be glycoslyated. Since there can be different degrees of glcyosylation, each one of these now qualifies as a new protein (for the sake of

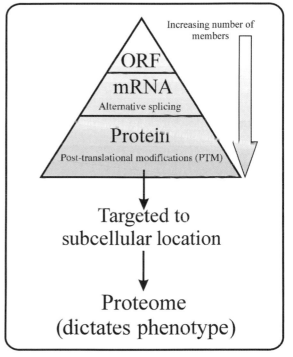

Figure 38.1. *A schematic diagram illustrating the increasing complexity that occurs as a single open reading frame (ORF) is transcribed into many functional proteins.*

proteomics), since each of these species may have a unique function or give a unique representation of a disease. Further export of a protein can reveal the cytoplasmic tail of the protein, which can now be acylated, ubiquinated or phosphorylated, or a combination of all (or some) of the above, in addition to many other PTMs. Considering only three potential differences at each of these steps, we now have at least $2^{(3n)}$ unique proteins where n represents the number of steps. Following translation, each different form of the protein can be modified by as many as 500 known PTMs,[1] producing a staggering array of protein products. PTMs are modifications to proteins by the addition of a modifying group to one or more amino acids (Table 38.1), or proteolytic cleavage which can affect protein activity, interactions, and cellular localization. Translocation of proteins to specific subcellular regions allows the same proteins to exert different biological effects or act as an alternative form of protein control, often resulting in an observable cellular phenotype. In contrast to the genome, which is a static blueprint, the proteome is a dynamic system of production, modification and degradation. It is the dynamic nature of the proteome which determines the phenotype of the cell. The changes in the proteome also characterize the cell's response to change in the environment that induces disease. Thus, the characterization of a cell's proteome allows one to understand the disease processes.

Traditionally, the goal of proteomics has been the unbiased analysis of a proteome through the identification and quantification of every protein species present (both modified and unmodified) in a cell. Since the proteome is in constant flux, any single time-point analysis provides only a snapshot of the proteome at a given time. Consequently, when comparing a diseased tissue proteome to that of control tissue, any difference in protein quantity of modified forms can be functionally benign, advantageous, or detrimental to the cell. To determine which changes are results of disease progression, one needs to observe multiple time points through a disease process. These changes can be found in tissue or other body fluids such as a patient's serum (due to cellular necrosis occurring during the disease progression or in response to disruptions to the cardiovascular system). Regardless of location or cause, any protein change can be considered part of the global 'fingerprint' of the disease state, and may be an invaluable tool in the diagnosis/treatment of the disease.

Clincial proteomics: proteins for the prognosis and prevention of disease

Disease is a change in the normal phenotype of a patient, and reflects alterations in the proteome (Fig. 38.2). In theory, the complete analysis of a patient's proteome can provide a direct quantifica-

Table 38.1. *Common post-translational modifications to proteins*

PTM	Residues modified
Phosphorylation	Ser/Thr/Tyr
Acylation	N-terminal Met
N-linked glycosylation	Asp
O-linked glycosylation	Ser/Thr
Methylation	Lys
Sulfation	Tyr
Oxidation	Met/Cys
Nitrosylation	Cys/Tyr

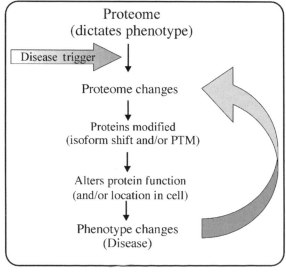

Figure 38.2. *A flow chart illustrating the cellular changes that can occur following a disease trigger and the resulting feedback.*

tion of a disease state. However, even partial proteomic analysis may reveal diagnostic markers (or a diagnostic pattern) which are a reflection of the disease state giving a physician an insight into a disease before any overt clinical changes occur. The field of clinical proteomics can be considered the direct application of proteomic technologies to the medical field.[2] The goals of clinical proteomics differ from traditional proteomics, in that the primary focus is altered towards the development of diagnostic markers and/or therapeutic reagents (Fig. 38.3) for direct clinical application. To date, the area of clinical proteomics has concentrated almost exclusively on biomarker discovery, since without an effective diagnostic to define a disease

state it is not possible to intervene or design targeted therapeutics specifically for the condition. For example, in order to develop a clinical intervention strategy for the early stages of heart failure (HF), clinicians need to first distinguish presymptomatic from healthy individuals. This requires the development of a specific prognostic biomarker(s) to identify patients in the population that are at risk for developing HF within the next 1–5 years.

Strategies for biomarker/therapeutic discovery

Biomarker development can take two routes, depending upon the type of prognostic marker. If a general serological marker associated with a disease state is the goal, then an unbiased analysis of serum or plasma should be undertaken. Alternatively, if one is interested in determining a tissue-specific disease marker, then tissue samples should be examined prior to the examination of body fluids. Tissue can be obtained from either an animal model which simulates the human form of the disease or, if possible, from clinical patients. Although patient samples directly correlate to the disease process, they exhibit a much higher variability in their proteome compared to inbred animals which only approximate the human disease process. Biomarkers are then obtained by identifying proteins altered in the disease state, and comparing proteomic fingerprints from a diseased tissue with those from normal tissue. Tissue-specific biomarkers may be detectable in serum if they are secreted or released following cellular necrosis or apoptosis, but due to the low concentrations of tissue specific markers in sera compared to other proteins, it is unlikely that the unbiased analysis of serum will reveal tissue specific biomarkers.

Once a model has been determined, a defined set of studies of increasing levels of difficulty can then be undertaken (Fig. 38.4). The first study that should be undertaken is to develop a 'database' of the identity of the proteins in the model system. A database, simply stated, is a list of all of the proteins found in the model system. The process of developing a database generally follows the four steps illustrated in Fig. 38.4. These steps are (1) selection

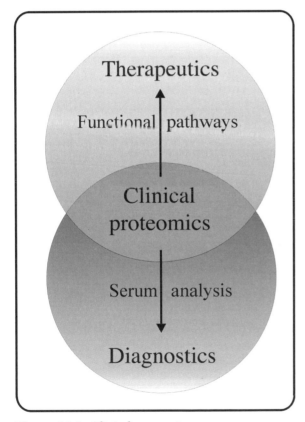

Figure 38.3. *Clinical proteomics can generate information on the functional pathway in a cell, which allows for the generation of clinical therapeutics (top). Conversely, a serum biomarker may also be generated which allows for the development of diagnostic tests (bottom).*

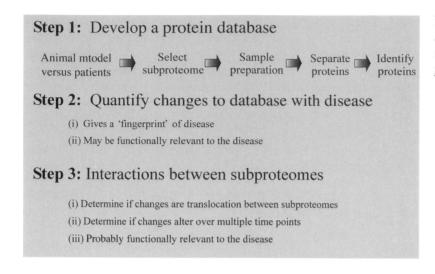

Step 1: Develop a protein database

Animal mtodel versus patients → Select subproteome → Sample preparation → Separate proteins → Identify proteins

Step 2: Quantify changes to database with disease

(i) Gives a 'fingerprint' of disease

(ii) May be functionally relevant to the disease

Step 3: Interactions between subproteomes

(i) Determine if changes are translocation between subproteomes

(ii) Determine if changes alter over multiple time points

(iii) Probably functionally relevant to the disease

Figure 38.4. *A strategy for the development of biomarkers or therapeutics by clinical proteomics.*

of a subproteome if desired; (2) sample preparation; (3) separation of proteins; and (4) identification of the proteins. The simplest database is one composed of the extract from a single tissue, or whole serum/plasma. This type of extraction has had some success in characterizing proteins as shown by the generation of multiple cardiac databases in the 1990s.[1,3] However, due to technological limitations of the time, only the most abundant cellular proteins were observed and minimal characterization of PTMs was carried out. To observe the proteome in greater depth, a more complex cellular extraction is necessary. This is accomplished through subproteome fractionation. A subproteome is a collection of proteins that can be consistently fractionated based on a common property. This can be based on cellular location (i.e. organelle), or pathway specific, in the form of isolating protein-binding partners involved in a biochemical signal pathway. For example, Taylor et al. have developed a database of proteins from the human cardiac mitochondria,[4] whereas Edmondson et al. have characterized the binding partners of protein kinace Cε (PKCε),[5] a protein important in cardiac preconditioning, through co-immunprecipitation. A subproteome may also be based on one of the intrinsic properties of the proteins, such as their ability to be solubilized under different pH conditions. Using such a method, Neverova et al. were able to obtain a subproteome from cardiac myocytes enriched in myofilaments while maintaining normal protein-binding stoichiometries.[6]

Body fluid subproteomes: physiological systems monitoring

In theory, proteomic analysis of any body fluid or tissue sample can be carried out using proteomic technologies for biomarker discovery. For applicability, the fluid should be easily available in sufficient quantities and be affected by a particular disease process. Table 38.2 lists the bodily fluids that have been used for this purpose to date. The most commonly used body fluid has been serum/plasma, due to the ease of obtaining it from patients. However, it is a challenging fluid to analyze because a large dynamic range exists between the highest and lowest concentration of proteins present. For example, in blood, albumin and immunoglobulin G (IgG) have concentrations between 30 and 50 mg/mL, whereas biomarkers like cardiac troponin I (cTnI; when present) are in the range of 0.1–2 ng/mL, and various cytokines are present at pg/mL concentrations. In fact, 22 proteins comprise approximately 98% of the protein content of plasma (or serum), and the remaining 2% comprise the 1000s of low-abundance proteins that are likely to be useful as biomarkers. Furthermore, in-depth analysis is still compounded by both the lack of obvious serum subproteomes, and the limited utility of current protein separation and identification technologies when applied to

Table 38.2. *List of bodily fluids used for biomarker discovery*

Body fluid	Marker specificity	Availability	Reference numbers
Mucosa	Specific	Minimal	7, 8
Bronchoalveolar lavage	Systemic, lung specific	Minimal	9–11
Breath condensation	Systemic	Minimal	12, 13
Breast aspirant	Specific	Minimal	14
Plasma	Systemic	High	15, 16
Serum	Systemic	High	17–19
Spinal fluid	Systemic	Minimal	20–22
Tear drops	Specific	Minimal	23
Renal ultrafiltrate	Systemic, specific	High	24
Tissue	Specific	Variable	1, 25, 26

serum. Differences in sample processing, handling, and long-term storage can introduce several artifacts which make the analysis of body fluids technically challenging. Additionally, natural variability in cohorts, including different drug treatment and confounding diseases, makes clinical phenotyping extremely difficult. Several approaches can be used to minimize these variables – longitudinal studies or cohorts on patients that are carefully defined. Thus, it is not currently clear whether the protein changes observed through proteomics will give rise to specific markers that are unique to a disease, nor is it possible to observe the low-abundance proteins like TnI, and cytokines using classical proteomic methods. Therefore, the probability of detecting an organ-specific biomarker is reduced when analyzing serum or plasma directly. Rather, one will probably be detecting changes in proteins and peptides peripherally related to disease, due to the changes in the cardiovascular (CV)/pulmonary system as a result of the disease. For example, C-reactive protein (CRP) is increased in acute heart disease, chronic heart failure, and a host of other diseases due to inflammation. CRP reflects the 'sickness' of the cardiovascular (CV) system, and hence the declining health of the patient. Monitoring CRP in the CV system adds value and is therefore worthwhile. However, it is more difficult to discern specificity, and it is easier to have protein biomarkers that only change due to specific disease states. This illustrates that discovery is an imperfect process, and the weeding out of 'dirty' markers in the validation

stage will be necessary to choose correct markers or panels that truly represent disease.

Developing databases: protein separation and identification technologies

The technical difficulty in identifying and characterizing proteins is proportional to the complexity of the sample being analyzed. Even by limiting analysis to subproteomes, there are often 100s to 1000s of proteins that need to be separated. Consequently, protein separation methods should ideally be able to separate protein mixtures into single protein species at sufficient quantity for identification and characterization of PTMs. This separation is usually based on one of the three intrinsic properties of proteins: charge, mass, or hydrophobicity. Orthogonal separation based on more than one independent physical parameter is often necessary, depending upon the complexity of the sample. Based on the mathematical theory of Giddings,[27] when samples are orthogonally separated, the number of uniquely resolved proteins is the product of each stage of separation. For instance, 100 proteins could be resolved using one step that separates 100 proteins, or 10 000 proteins could be separated using two orthogonal steps. The human proteome has been estimated to be of the order of 100 000 protein species (includes PTMs, isoforms, etc), with as many as 30 000 expressed at a given time.[28] These are many more products than can be separated using single-step separation technologies. It would be possible to resolve these 30 000 distinct protein species using multiple orthogonal separation steps. For instance, the entire proteome could be separated using two steps resolving 173 proteins, or three orthogonal steps with each step resolving 31 proteins. This means that in theory, the entire human proteome could be resolved by three orthogonal separation stages. In practice, no more than two stages of protein separation are simultaneously undertaken on a routine basis. The two most common protein separation methods are gel electrophoresis and liquid chromatography. A number of different forms of liquid chromatography have been developed over the years, which can successfully separate proteins or peptides based on either mass, charge, or hydrophobicity. A summary of these methods as well as gel electrophoresis is shown in Table 38.3.

Gel electrophoresis
SDS polyacrylamide gel electrophoresis (SDS-PAGE)

SDS polyacrylamide gel electrophoresis (SDS-PAGE) is a cornerstone of proteomics,[29] and is still the most common method of protein separation due to its ease of use and its relatively inexpensive start-up costs. SDS-PAGE separates proteins based on their molecular weight (Fig. 38.5). SDS is an anionic detergent that binds proteins at approximately one molecule per every two amino acids or 1.4 g SDS/g protein.[30,31] SDS–protein complexes are almost always soluble in water, and the incorporation of SDS onto a protein masks the inherent charge of the protein, allowing the protein to migrate based solely on its molecular weight when in a polyacryalamide gel exposed to an electric field. Some proteins, however, may migrate differently than their actual molecular weight, due to altered binding of SDS. For instance, the contractile protein troponin I (TnI) will bind more SDS than normal, which results in the protein migrating at an apparent molecular weight of 32 kDa on a gel rather than its true molecular weight of 23 kDa.

Complete separation of a protein mixture can be accomplished using a single dimension SDS-PAGE (1-DE) if the number of proteins is small. Protein separation can further be optimized by altering the concentration of acrylamide present in the gel, such that either high- or low-molecular weight proteins are resolved. However, if a given protein mixture is too complex, each band resolved by SDS-PAGE will probably contain multiple protein species possessing a range of charges and hydrophobicity. For such complex mixtures two-dimensional gel electrophoresis (2-DE) is a more

Table 38.3. *Summary of the different methods of separation used in proteomics*

Type of separation	Separates by:	Binding motif
Liquid chromatography		
Size exclusion	Size (≈molecular weight)	Size-dependent migration in matrix
Anion exchange	Charge	Negatively charged residues
Cation exchange	Charge	Positively charged residues
Immobilized metal affinity	Charge	Metal-binding sites or negatively charged residues (i.e. PO_4^{2-})
Antibody affinity (biospecificity)	Immunodetection Antibody-binding site on protein	
Reverse phase	Hydrophobicity	Hydrophobic residues and peptide backbone
Gel electrophoresis		
SDS-PAGE	Molecular weight	Size-dependent migration in gel matrix
IEF	Charge (pI)	Will migrate in field until pH = pI

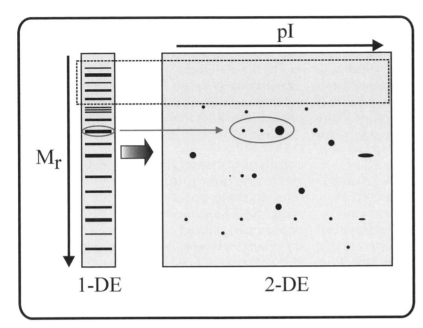

Figure 38.5. *One- and two-dimensional gel electrophoresis (1-DE, 2-DE) demonstrate different abilities to separate proteins. 1-DE allows for the detection of higher-molecular weight (M_r) proteins, not observed by 2-DE (broken lines). 2-DE allows for the separation of multiple protein isoforms as observed by changes in pI.*

appropriate method, as it separates proteins based on two properties, isoelectric point (pI) and mass.

In 2-DE, the first dimension uses isoelectric focusing (IEF), whereas the second dimension utilizes SDS-PAGE (Fig. 38.5). This allows a more extensive separation of different proteins as well as allowing PTMs and multiple isoforms to be resolved. For instance, protein phosphorylation will result in an acidic shift in a protein's pI, such that the phosphorylated and non-phosphorylated forms of proteins will be separated by ~0.2 pH units but have the same molecular weight (Fig. 38.5). A benefit of 2-DE is seen by its superior ability to separate proteins with as many as 10 000 species reported.[32] Unfortunately, high-molecular weight proteins often do not resolve in the first dimension, resulting in under-representation. Highly hydrophobic proteins are also under-represented in 2-DE, due in part to their inability to enter the first dimension, and to difficulties in solubilizing these proteins.

Theory of IEF and sample preparation requirements

Proteins are amphoteric molecules with positive and negative regions distributed throughout the protein, depending on the pH of the solution. The net charge of a protein is the sum of all the charged amino acid side chains as well as the unmodified amino- and carboxy-termini. This charge is a function of the pH of the medium. When the pH of the medium is such that the number of positive and negative charges on the proteins is equal, it is known as the isoelectric point (pI). Therefore, in solution, proteins are positively charged at a pH below their pI, and negatively charged at a pH above their pI. Any PTM that affects the overall charge of the protein can alter its pI by shifting this charge ratio.

Similar to 1-DE, the first dimension of 2-DE functions by the migration of proteins through an acrylamide gel exposed to an electric field. With IEF, a pH gradient is constructed in the gel prior to the addition of protein. When proteins enter the gel they will either be negatively or positively charged, and as such will migrate due to the electric field until they reach the pH equal to their pI, at which point they become uncharged and stop migrating. The degree of protein resolution is determined by the range of the pH gradient and the distance over which the gradient is formed. The resolving power of IEF is illustrated by its ability to separate proteins with a pI difference of 0.0016 units when used on a large format gel, focused on a single pH unit gradient.[33] IEF was initially performed within tube polyacrylamide gels in which the pH gradients were created using hundreds of carrier ampholytes (CA-IEF).[34,35] In modern 2-DE, the pH gradient is no longer dependent upon the focusing of small molecules in solution, but determined by the concentration of specific buffering groups covalently bonded in place to the acrylamide matrix. This is known as an immobilized pH gradient (IPG). The stability of the pH gradient generated within the IPG allowed 2-DE analysis of basic pH ranges up to pH 12.[36] A drawback to 2-DE is the requirement that the charge of the proteins must not be altered from their natural state during IEF. Consequently, buffers need to avoid strongly charged detergents such as SDS, because they will alter this charge and inhibit focusing when present at concentrations greater than 0.25%.[37] The incompatibility of charged detergents such as SDS with IEF has also necessitated the adoption of zwitterionic detergents such as CHAPS to enhance solubility of hydrophobic/membrane proteins; however, these proteins are still under-represented on 2-DE gels.

These drawbacks can discourage researchers from using 2-DE, however, the technology is the most robust method of separating proteins that is currently available and deserves a premier position in a proteomic investigation. Another drawback with 2-DE, which makes its use as the main tool for large-scale analysis unlikely, is the necessity that proteins be removed from the gel prior to identification. It would be preferable if the proteins could remain in the liquid phase throughout the separation process. This can be accomplished by separating proteins using liquid chromatography (LC).

Liquid chromatography separation

During liquid chromatography (LC), a protein mixture is passed over a column containing a matrix that exhibits an affinity for one of the intrinsic characteristics/properties of proteins. In cation exchange chromatography (CEC) the protein mixture is passed over a negatively charged matrix which 'captures' proteins through its affinity for residues containing positively charged side chains, such as lysine or arginine. Proteins can then be variably eluted from the column through the addi-

tion of increasing quantities of a substance (e.g. imidazole) that will compete with the protein for the matrix-binding site. Historically, these methods have had enormous success for the purification of individual proteins or simple mixtures for analysis by protein biochemists. It is only recently that these methods have been more widely applied to proteomics.

Multidimensional separation of proteins using liquid chromatography

Two different approaches are used to resolve a proteome through liquid chromatography. The first is to separate the proteins intact (top-down separation), and the second method is to cleave the proteins through enzymatic digestion and then separate the fragments (bottom-up separation). The benefit of separating proteins instead of peptides is that small differences in the proteins from PTM or small isoform shifts can result in a modified affinity for the separating matrix. This will allow each form of the protein to be identified. 2-DLC of proteins has found some early success using size exclusion chromatography (SEC) coupled to a reverse phase high-pressure liquid chromatography (HPLC) column. Using this combination, Opiteck et al. have observed low-abundance proteins from E. coli crude extracts,[38] whereas Zhang et al. have separated different forms of β-crystallin, which would not normally resolve as distinct spots by 2-DE, from human eye tissue.[39] Significant success with protein 2-DLC has been accomplished using IEF with a pH gradient on an ion exchange column in the first dimension, followed by reversed phase column (RP-HPLC) in the second dimension,[40,41] to separate HCT-116 colon adenocarcinoma cell proteins of drug-treated and untreated cells. Using 0.2 pH unit increments followed by RP-HPLC for separation, they could observe over 1000 distinct protein peaks and quantify changes in abundance by examining ultraviolet (UV) absorption. Unfortunately, 0.2 pH units may be too broad a pH range to observe PTMs such as phosphorylation, which typically result in a ~0.2 pH unit shift. Consequently, the mechanism of separation during the first dimension needs to be refined to allow finer pH separation.

Multidimensional separation of peptides using liquid chromatography

The other method of LC separation is to enzymatically digest the proteins of interest (generally with trypsin) and separate these fragments by 2-DLC. This method has the benefit that individual peptides are more likely to maintain their solubility throughout the separation as compared to whole proteins. Additionally, identification can be accomplished by directly analyzing the effluent from the column directly using a mass spectrometer (MS). Alternatively, the fractions can be pooled and analyzed later via a process termed offline analysis. Due to buffer restrictions for mass spectrometry the final column prior to the MS is required to be RP-HPLC for online analysis. The first apparent limitation to peptide separation is the large number of fragments present. For instance, with 25 tryptic fragments per protein on average, one would have to separate as many as 500 000 peptides from a crude lysate to observe every species – an incredibly difficult feat. However, in order to identify a protein, only a small handful of these peptides are necessary. This means that the increased number of species is actually a benefit, as the odds are higher that a novel peptide will elute at a distinct position, which will increase the number of proteins identified. Unfortunately, the analysis of lower-abundance proteins is still limited using this method.

Work by Moore and Jorgenson has demonstrated the utility of this method using SEC coupled to RP-HPLC and capillary zone electrophoresis with success.[42] However, the most common method for peptide 2-DLC utilizes ion exchange chromatography followed by RP-HPLC. This method has been termed multidimensional protein identification technology (MudPIT) by Washburn et al.[43–45]. Through this method they were able to identify 1484 proteins from a yeast protein extract. PTMs are difficult to observe by this method. To address this, MacCoss et al. have demonstrated a method of determining PTMs using MudPIT,[46] by increasing sequence overlap through the use of multiple proteases; this method has yet to be widely adopted.

Identification of proteins

Although the ability has existed since the 1970s to separate thousands of proteins by 2-DE, the only methods available to identify proteins at the time was through N-terminal sequencing or immunoblotting. These methods require large quantities of protein and a specific antibody for the latter. The difficulty in using these methods is highlighted by the attempts by Baker et al.[3] to characterize the human myocardial proteome. Although Baker could separate the extract into over 1500 protein spots, only about 50 of these could be identified. These limitations were overcome by technical developments in mass spectrometry (MS) instrumentation in the 1990s. The main components of a MS are (a) a method of protein ionization; (b) a mass filter; and (c) a detector. Different forms of each of these components have been developed and these can be used in different combinations in different MS machines. The development of electrospray ionization (ESI) and matrix-assisted laser desorption ionization (MALDI) has allowed for the routine, highly accurate, determination of peptide masses in proteomic investigations.[47,48] ESI functions by ionizing small quantities of sample from solution, whereas MALDI ionizes peptides in a crystalline matrix (via laser pulses) on a solid support. Because of the differences in ionization modes, ESI is generally used following liquid chromatography separation, while MALDI is used to analyze proteins following 2-DE. A mass spectrometer functions by ionizing a species and determining its mass with high precision, through the spectrometer's mass analyzer (a term which can refer to any of four current technologies). These mass analyzers can either be a 'time-of-flight' (TOF), quadrupole (Q), ion trap, or Fourier transform mass spectrometer (FT-MS). Regardless, of the mechanism of ionization or mass analyzer, the most common species being analyzed in a mass spectrometer are peptide fragments generated by the enzymatic digestion of the protein sample. This is primarily due to the limitations of detectors, resulting in an increased resolution observed for lower masses, as well as the ability to determine the identity of the protein either indirectly through peptide mass fingerprinting (PMF) or with much higher confidence through amino acid sequencing using tandem mass spectrometers (MS/MS).

Identification of proteins using peptide mass fingerprinting and MALDI-TOF MS

The most common initial method of identifying proteins is known as peptide mass fingerprinting (PMF) (Fig. 38.6A). In this procedure, a protein of interest is either excised from a portion of a gel or obtained from a HPLC fraction. This protein is then enzymatically digested using a protease that specifically cleaves the protein at known positions (e.g. trypsin cleaves following R or K). This generates a reproducible and unique set of peptide fragments whose mass can then be determined using MALDI–TOF MS. Individually, this information is of very little use. However, these masses can be compared to theoretical digests of proteins obtained from genomic sequencing projects, to determine the protein identity. Unfortunately, instead of a single protein, search algorithms generate a list of possible proteins corresponding to the submitted peptides, with different statistical rankings. This can lead to false identification if a different protein happens to have a similar digest pattern with slight differences in mass. Additionally, not every peptide from the protein is observed in the spectra, due to variations in ionization efficiency. Consequently, observing 40% of the tryptic fragments from a protein is often considered sufficient to confirm the presence of that protein. In order to increase the confidence on the identity of the peptide, a more precise method of identifying proteins using tandem mass spectrometry (MS/MS) can be utilized (Fig. 38.6B). This method allows one to directly determine the peptide sequence, rather than inferring the identity through the fingerprinting method.

Determining protein identity using MS/MS

An alternative method to identify a protein of interest is to directly determine the amino acid sequence of the peptide through MS/MS. MS/MS analysis involves two mass analyzers, often separated by a collision cell which contains an inert gas. The two mass analyzers are generally any combination of Q or TOF mass analyzers. Table 38.4 lists some of the tandem mass spectrometers commercially available, and their combinations of mass analyzers. The first mass analyzer allows a profile of the peptides present and the selection of a single fragment based on its mass. This peptide alone is allowed to pass into the collision cell which contains an inert gas. When the peptide of n amino acids collides with the inert gas, the peptide will fragment, generating a list of fragments of varying size corresponding to n, $n - 1$, $n - 2$, etc amino acids. These smaller fragments from the original peptide will then pass into the second mass analyzer, allowing one to derive the amino acid sequence of the peptide. By comparing the masses

Table 38.4. *Some of the commercially available mass spectrometers*

Manufacturer	Name of MS	Ionization	Mass analyzer
Agilent Technologies	Agilent 1100 Series	ESI	Ion Trap
Amersham Pharmacia	Ettan MALDI-TOF MS	MALDI	TOF
	Ettan LC-MS	ESI	TOF
Applied Biosystems	Voyager DE-PRO	MALDI	TOF
Bruker Daltonics	BioTOF II	ESI	TOF
	BioTOF-Q	ESI	Quadrupole/TOF
	AutoFlex II	MALDI	TOF/TOF
Ciphergen Biosystems	ProteinChip® System	SELDI	TOF
IonSpec Corp	Ultima FTMS	MALDI-ESI	FTMS
Kratos Analytical	Axima-CFR	MALDI	TOF
MicroMass-Walters	M@LDI LR	MALDI	TOF
	Q-Tof Ultima	ESI/MALDI	Quadrupole/hexapole/TOF

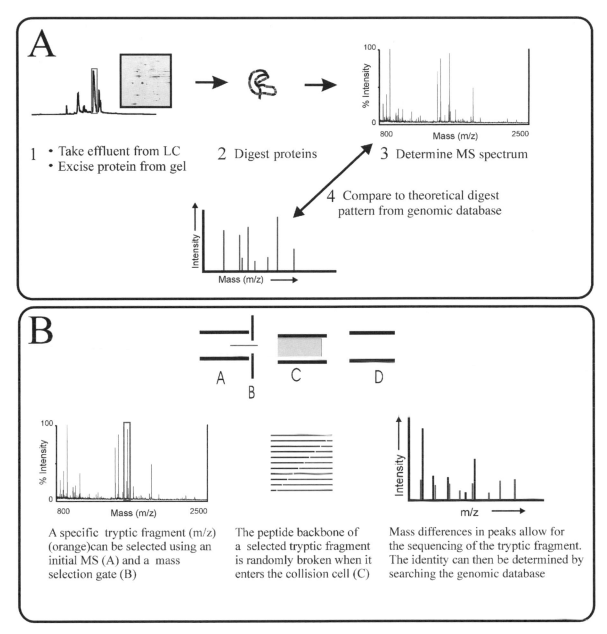

Figure 38.6. (A) *An overview of protein identification through peptide mass fingerprinting (PMF) using a MALDI-TOF MS.* (B) *An overview of amino acid sequencing of a single peptide by MS/MS.*

of closely spaced fragments differing by one amino acid (i.e. *n* versus *n* − 1), the mass difference will give the mass of the additional amino acid. The complete sequence of the peptide can then be determined using multiple fragments. However, it is impossible to distinguish between leucine and

isoleucine, because their masses are identical, or between lysine and glutamine because their masses differ by only 0.03%. MS/MS sequencing can also reveal the presence of PTM, as modified amino acids have a defined increase in mass (e.g. +80 Da for phosphorylation). Many of these peaks may not be available for peptide sequencing, however, as the presence of a PTM can often lead to an inability to ionize, and consequently they will not be detected. The increased power of MS/MS comes at the expense of a much higher equipment cost, which often results in MS/MS becoming a secondary form of identification for proteins of higher interest, following PFM.

Does the database change with disease? Quantification of protein species from gels and LC

After establishing the database, the next step is to determine which proteins are altered during a disease state. This necessitates a method of quantifying the amount of protein present. The methods of quantification currently used depend on the type of separation. In gel electrophoresis, proteins are visualized using a general protein stain such as Coomassie brilliant blue (CBB), silver, or fluorescence. CBB is a visible dye which functions by binding to arginine and hydrophobic residues on proteins, and is by far the most commonly used stain owing to its linear increase in absorbance over two orders of magnitude. The limitation of CBB is that it has a poor sensitivity. Conversely, silver stain resolves some of these issues in that it possesses a 100× greater sensitivity than Coomassie, but unfortunately its drawback is that it has a small linear dynamic range (<1 order of magnitude).[49,50] Silver-stained gels also suffer from the drawback that protein spots cannot be identified through MS, due to the use of glutaraldehyde in the staining process which crosslinks proteins to the gel. This can be overcome by removing glutaraldehyde during staining, but this can significantly reduce the sensitivity of the stain.[51] Fluorescent stains tend to be much better than either of the other stains, in that they exhibit a linear response over three orders of magnitude,[52] and possess a sensitivity that is slightly less than that of silver. The major drawback to the use of fluorescent dyes is that they are considerably expensive, and require imaging scanners that are also relatively expensive. Fluorescent dyes can be used to stain proteins in-gel similar to CBB (Sypro Ruby, Molecular Probes; Deep Purple, GE Healthcare), or covalently attached to the proteins prior to sep-

aration (cyanine dyes, GE Healthcare). It must be emphasized that staining is an indirect observation of a protein (with the exception of cyanine dyes), and variations in binding ability of stains can result in different proteins being observed by different stains. For example, the contractile protein troponin I is known to be visible by Coomassie, yet by the more sensitive silver stain the protein is only weakly visible.[53]

The quantification of proteins by liquid chromatography separation is relatively limited compared to gel electrophoresis, in that proteins are directly observed through their UV absorbance. By measuring the absorbance of the effluent, one can determine the concentration of protein present in individual fractions. Specifically, absorbance at 210 nm is due to the absorption of the peptide backbone, whereas the aromatic side chains will absorb light at 280 nm. By integrating over the area of the peak, the total quantity of protein can then be determined. Because one cannot distinguish different proteins by UV absorbance, a given absorbance is a sum of all the proteins eluting at a given time. Thus, eluting proteins off a column should be done in such a fashion that individual proteins are eluted per fraction. This can be approximated by allowing the absorbance to return to baseline between each peak.

Alternative MS-based quantification technologies

In addition to the methods listed for gel and liquid chromatographic quantification, alternate methods have been developed to allow protein quantification using MS. MS is generally not considered a quantitative assay due to the fact that different peptides of equal abundance will exhibit largely different signals due to variations in their ability to ionize. However, it is relatively easy to obtain a relative quantity (ratio) between two states by labeling individual peptides with either a heavy or light isotope, so that they can be distinguished by a small mass shift. Because the difference in the peptide due to the heavy isotope will have a negligible effect on its ability to ionize, the ratio of signal should be equal to the relative abundance of the two species. Multiple methods have been developed utilizing this principle, but for the scope of this chapter only two methods will be discussed (other methods are reviewed in references 54 and 55). By far the most commonly used method is isotope-coded affinity tags (ICAT), developed by Aebersold's group.[56] In ICAT, either a heavy or light biotin-labeled probe (depending upon whether the sample is control or experimental) is covalently attached to a protein via free cysteine residues. The protein mixtures can then be combined and enzymatically digested to obtain fragments suitable for MS, and fragments enriched for using a biotin affinity (avidin) column. Early work using ICAT suffered due to heavy probe design and non-specific binding to columns, which inhibited similar peptides co-eluting. However, these issues have been resolved with the commercially available ICAT reagents (Applied Biosystems, Framingham, MA), which utilize ^{13}C instead of ^{2}H in the heavy probe, and contain an acid-cleavable linker which allows the removal of the biotin affinity tag prior to MS.

Instead of attaching a probe to the proteins of interest, the incorporation of the heavy or light isotopes can be accomplished during proteolysis. During proteolysis, an oxygen atom from the solvent is incorporated into the C terminus of the protein. Thus, by incubating the protein in either heavy or light oxygen water during the enzymatic digestion, selective incorporation of two ^{18}O will occur. Thus, the ratio of abundance between two states can be obtained by the ratio of the two peaks separated by 4 Da. This small mass shift does make the analysis technologically challenging, and may require separate experiments in which each of the two states are labeled with ^{18}O to confirm the results.[57]

The methods discussed so far rely on quantifying identified proteins to generate a cellular fingerprint. However, profiles using unknown proteins (primarily serum proteins) can also be used for diagnosis. An approach which is gaining popularity for this purpose is the ProteinChip® surface-enhanced laser desorption-ionization (SELDI) MS technology.[58] SELDI utilizes a chromatographic surface such as an ion exchange or a reversed phase matrix to selectively bind low-mass proteins or peptides. This surface can then be analyzed directly by MS, to generate a profile of the proteins bound to the surface. A diseased tissue profile can be determined using a small cohort of patients with a particularly well-defined disease state. This profile pattern should then be compared against another cohort to determine specificity and sensitivity of the test. The applicability of profiling is unclear as it is only in its infancy. Adding to the uncertainty is whether one can resolve differences in diseased versus normal states, by observing only the highly abundant proteins. Additionally, resolving clinically relevant changes is complicated due to the fact that serum has a very dynamic proteome, which is influenced by both physiological and pathophysiological conditions. Currently, no cardiac biomarkers have been developed through this method. However, Qu et al. have determined six different biomarkers for prostate cancer using the ProteinChip®/SELDI MS.[59] Although four of these biomarkers were previously known, two novel biomarkers were obtained that demonstrated increased abundance in the cancer cells.

Conclusions: choosing the right tool depends upon the goal

The proteomic tools overviewed in the previous section demonstrates that there are many different techniques that are available for clinical biomarker discovery. In practice it is always wise to choose the simplest tool/approach that match the goals of the investigator. For instance, if one had a relatively small subproteome and was not interested in analyzing PTMs then 1-DE or 1-DLC would be perfectly suited to their goals. However, if one is interested in examining thousands of hydrophobic proteins, then 2-DLC or ICAT would be more suited. Conversely, if one is interested in determining whether changes to protein PTM have occurred, then 2-DE is the tool of choice. A summary of the methods available for a biomarker discovery is shown in Table 38.5.

The key for developing effective therapy is to understand the underlying molecular mechanism driving the disease process. This entails detailed proteomic analysis using the complete set of tools available. As well, it will be essential to determine the functional effects of an altered proteome. This

Table 38.5. *A summary of the different technologies used in proteomics and their limitations*

Technologies	Quantify?	PTM?	Protein difficulties?	Labor intensive?
Electrophoresis				
1-DE	No	No	No	No
2-DE	Yes[a]	Yes	Yes[b]	Yes: gel and MS
Liquid chromatography				
1-DLC	No	No	No	No
2-DLC – peptide	No	No[c]	No	Yes: MS
2-DLC – protein	Yes[d]	Yes	No	Yes: MS
MS quantification				
ICAT	Yes[e]	No	Yes[f]	Yes: MS
Isotope labeled culture	Yes[e]	No	No	Yes: MS
SELDI	Yes[e]	Yes	Yes[g]	No

[a]Restricted by linear range of dye used to visualize proteins.
[b]Limited in the ability to separate high molecular weight, hydrophobic proteins.
[c]Possible by method of MacCoss et al., 2002.[46]
[d]Method can quantify if single protein species elutes at a time.
[e]Only ratios of proteins/peptides can be obtained.
[f]Will only detect proteins which contain a free cysteine residue.
[g]Limited to the analysis of low molecular weight proteins.

is not necessary if one is solely interested in developing one or more markers that predict with high sensitivity/specificity for a disease state. As proteomics evolves as a scientific discipline, with new technologies and approaches, a greater number of proteins will be observed with increasing ease. However, it will be the understanding of whether a protein change is predictive or an extraneous event which will be the next challenge facing clinical proteomics.

Acknowledgements

The authors would like to thank David Graham, Lesley Kane and Todd McDonald for editing of this manuscript. Financial assistance was provided by the Daniel P. Amos Family Foundation, the Donald W. Reynolds Foundation and the NHLBI proteomics grant (Contract#HV-28180).

References

1. Jager D, Junghlut PR, Muller-Werdan U. Separation and identification of human heart proteins. J Chromatogr B Analyt Technol Biomed Life Sci 2002; 771: 131–53.

2. Granger CB, Van Eyk JE, Mockrin SC et al. National Heart, Lung, And Blood Institute Clinical Proteomics Working Group report. Circulation 2004; 109: 1697–703.

3. Baker CS, Corbett JM, May AJ et al. A human myocardial two-dimensional electrophoresis database: protein characterisation by microsequencing and immunoblotting. Electrophoresis 1992; 13: 723–6.

4. Taylor SW, Fahy E, Zhang B et al.. Characterization of the human heart mitochondrial proteome. Nat Biotechnol 2003; 21: 281–6.

5. Edmondson RD, Vondriska TM, Biederman KJ et al. Protein kinase C epsilon signaling complexes include metabolism- and transcription/translation-related proteins: complimentary separation techniques with LC/MS/MS. Mol Cell Proteomics 2002; 1: 421–33.

6. Neverova I, Van Eyk JE. Application of reversed phase high performance liquid chromatography for subproteomic analysis of cardiac muscle. Proteomics 2002; 2: 22–31.

7. Li X, Zhang W, Wang B et al. Preliminary establishment and optimization of two-dimensional gel electrophoresis for proteomics of gastric mucosa. Jiefangjun Yixue Zazhi 2003; 28: 438–40.

8. Baek HY, Lim JW, Kim H et al. Oxidative-stress-related proteome changes in Helicobacter pylori-infected human gastric mucosa. Biochem J 2004; 379: 291–9.

9. Neumann M, von Bredow C, Ratjen F et al. Bronchoalveolar lavage protein patterns in children with malignancies, immunosuppression, fever and pulmonary infiltrates. Proteomics 2002; 2: 683–9.

10. He C. Proteomic analysis of human bronchoalveolar lavage fluid: expression profiling of surfactant-associated protein A isomers derived from human pulmonary alveolar proteinosis using immunoaffinity detection. Proteomics 2003; 3: 87–94.

11. Wattiez R, Hermans C, Cruyt C et al. Human bronchoalveolar lavage fluid protein two-dimensional database: study of interstitial lung diseases. Electrophoresis 2000; 21: 2703–12.

12. Garey KW, Neuhauser MM, Robbins RA et al. Markers of inflammation in exhaled breath condensate of young healthy smokers. Chest 2004; 125: 22–6.

13. Griese M, Noss J, Von Bredow C. Protein pattern of exhaled breath condensate and saliva. Proteomics 2002; 2: 690–6.

14. Petricoin EE, Paweletz CP, Liotta LA. Clinical applications of proteomics: proteomic pattern diagnostics. J Mammary Gland Biol Neoplasia 2002; 7: 433–40.

15. Pieper R, Su Q, Gatlin CL et al. Multi-component immunoaffinity subtraction chromatography: an innovative step towards a comprehensive survey of the human plasma proteome. Proteomics 2003; 3: 422–32.

16. Anderson NL, Polanski M, Pieper R et al. The human plasma proteome: a nonredundant list developed by combination of four separate sources. Mol Cell Proteomics 2004; 3: 311–26.

17. Adkins JN, Varnum SM, Auberry KJ et al. Toward a human blood serum proteome: analysis by multidimensional separation coupled with mass spectrometry. Mol Cell Proteomics 2002; 1: 947–55.

18. Pieper R, Gatlin CL, Makusky AJ et al. The human serum proteome: display of nearly 3700 chromatographically separated protein spots on two-dimensional electrophoresis gels and identification of 325 distinct proteins. Proteomics 2003; 3: 1345–64.

19. Tirumalai RS, Chan KC, Prieto DA et al. Characterization of the low molecular weight human serum proteome. Mol Cell Proteomics 2003; 2: 1096–103.

20. Puchades M, Hansson SF, Nilsson CL et al. Proteomic studies of potential cerebrospinal fluid protein markers for Alzheimer's disease. Brain Res Mol Brain Res 2003; 118: 140–6.

21. Hammack BN, Owens GP, Burgoon MP et al. Improved resolution of human cerebrospinal fluid proteins on two-dimensional gels. Mult Scler 2003; 9: 472–5.

22. Wenner BR, Lovell MA, Lynn BC. Proteomic analysis of human ventricular cerebrospinal fluid from neurologically normal, elderly subjects using two-dimensional LC-MS/MS. J Proteome Res 2004; 3: 97–103.

23. Zhou L, Huang LQ, Beuerman RW et al. Proteomic analysis of human tears: Defensin expression after ocular surface surgery. J Proteome Res 2004; 3: 410–16.

24. Ward RA, Brinkley KA. A proteomic analysis of proteins removed by ultrafiltration during extracorporeal renal replacement therapy. Contrib Nephrol 2004; 141: 280–91.

25. Corbett JM, Wheeler CH, Baker CS et al. The human myocardial two-dimensional gel protein database: update 1994. Electrophoresis 1994; 15: 1459–65.

26. De Souza AI, McGregor E, Dunn MJ et al. Preparation of human heart for laser microdissection and proteomics. Proteomics 2004; 4: 578–86.

27. Giddings JC. Concepts and comparisons in multidimensional separation. J High Resolut Chromatogr Chromatogr Commun 1987: 319–23.

28. Harrison PM, Kumar A, Lang N et al. A question of size: the eukaryotic proteome and the problems in defining it. Nucleic Acids Res 2002; 30: 1083–90.

29. Laemmli UK. Cleavage of structural proteins during the assembly of the head of bacteriophage T4. Nature. 1970; 227: 680–5.

30. Nelson CA. The binding of detergents to proteins. I. The maximum amount of dodecyl sulfate bound to proteins and the resistance to binding of several proteins. J Biol Chem 1971; 246: 3895–901.

31. Reynolds JA, Tanford C. Binding of dodecyl sulfate to proteins at high binding ratios. Possible implications for the state of proteins in biological membranes. Proc Natl Acad Sci U S A 1970; 66: 1002–7.

32. Klose J, Kobalz U. Two-dimensional electrophoresis of proteins: an updated protocol and implications for a functional analysis of the genome. Electrophoresis 1995; 16: 1034–59.

33. Gunther S, Postel W, Wiering H et al. Acid phosphatase typing for breeding nematode-resistant tomatoes by isoelectric focusing with an ultranarrow immobilized pH gradient. Electrophoresis 1988; 9: 618–20.

34. O'Farrell PH. High resolution two-dimensional electrophoresis of proteins. J Biol Chem 1975; 250: 4007–21.

35. Klose J. Protein mapping by combined isoelectric focusing and electrophoresis of mouse tissues. A novel approach to testing for induced point mutations in mammals. Humangenetik. 1975; 26: 231–43.

36. Gorg A. IPG-Dalt of very alkaline proteins. Methods Mol Biol 1999; 112: 197–209.

37. Harder A, Wildgruber R, Nawrocki A et al. Comparison of yeast cell protein solubilization procedures for two-dimensional electrophoresis. Electrophoresis 1999; 20: 826–9.

38. Opiteck GJ, Ramirez SM, Jorgenson JW et al. Comprehensive two-dimensional high-performance liquid chromatography for the isolation of overexpressed proteins and proteome mapping. Anal Biochem 1998; 258: 349–61.

39. Zhang Z, Smith DL, Smith JB. Multiple separations facilitate identification of protein variants by mass spectrometry. Proteomics 2001; 1: 1001–9.

40. Wall DB, Kachman MT, Gong S et al. Isoelectric focusing nonporous RP HPLC: a two-dimensional liquid-phase separation method for mapping of cellular proteins with identification using MALDI-TOF mass spectrometry. Anal Chem 2000; 72: 1099–111.

41. Yan F, Subramanian B, Nakeff A et al. A comparison of drug-treated and untreated HCT-116 human colon adenocarcinoma cells using a 2-D liquid separation mapping method based upon chromatofocusing PI fractionation. Anal Chem 2003; 75: 2299–308.

42. Moore AW, Jr., Jorgenson JW. Comprehensive three-dimensional separation of peptides using size exclusion chromatography/reversed phase liquid chromatography/optically gated capillary zone electrophoresis. Anal Chem 1995; 67: 3456–63.

43. Washburn MP, Ulaszek RR, Yates JR 3rd. Reproducibility of quantitative proteomic analyses of complex biological mixtures by multidimensional protein identification technology. Anal Chem 2003; 75: 5054–61.

44. Washburn MP, Ulaszek R, Deciu C et al. Analysis of quantitative proteomic data generated via multidimensional protein identification technology. Anal Chem 2002; 74: 1650–7.

45. Washburn MP, Wolters D, Yates JR, 3rd. Large-scale analysis of the yeast proteome by multidimensional protein identification technology. Nat Biotechnol 2001; 19: 242–7.

46. MacCoss MJ, McDonald WH, Saraf A et al. Shotgun identification of protein modifications from protein complexes and lens tissue. Proc Natl Acad Sci USA 2002; 99: 7900–5.

47. Fenn JB, Mann M, Meng CK et al. Electrospray ionization for mass spectrometry of large biomolecules. Science 1989; 246: 64–71.

48. Karas M, Hillenkamp F. Laser desorption ionization of proteins with molecular masses exceeding 10,000 daltons. Anal Chem 1988; 60: 2299–301.

49. Giometti CS, Gemmell MA, Tollaksen SL et al. Quantitation of human leukocyte proteins after silver staining: a study with two-dimensional electrophoresis. Electrophoresis 1991; 12: 536–43.

50. Switzer RC 3rd, Merril CR, Shifrin S. A highly sensitive silver stain for detecting proteins and peptides in polyacrylamide gels. Anal Biochem 1979; 98: 231–7.

51. Shevchenko A, Wilm M, Vorm O et al. Mass spectrometric sequencing of proteins silver-stained polyacrylamide gels. Anal Chem 1996; 68: 850–8.

52. Patton WF. A thousand points of light: the application of fluorescence detection technologies to two-dimensional gel electrophoresis and proteomics. Electrophoresis 2000; 21: 1123–44.

53. Labugger R, McDonough JL, Neverova I et al. Solubilization, two-dimensional separation and detection of the cardiac myofilament protein troponin T. Proteomics 2002; 2: 673–8.

54. Lill J. Proteomic tools for quantitation by mass spectrometry. Mass Spectrom Rev 2003; 22: 182–94.

55. Aebersold R, Mann M. Mass spectrometry-based proteomics. Nature 2003; 422: 198–207.

56. Gygi SP, Rist B, Gerber SA et al. Quantitative analysis of complex protein mixtures using isotope-coded affinity tags. Nat Biotechnol 1999; 17: 994–9.

57. Wang S, Regnier FE. Proteomics based on selecting and quantifying cysteine containing peptides by covalent chromatography. J Chromatogr A 2001; 924: 345–57.

58. Reddy G, Dalmasso EA. SELDI ProteinChip(R) Array Technology: Protein-Based Predictive Medicine and Drug Discovery Applications. J Biomed Biotechnol 2003; 2003: 237–41.

59. Qu Y, Adam B-L, Yasui Y et al. Boosted decision tree analysis of surface-enhanced laser desorption/ionization mass spectral serum profiles discriminates prostate cancer from noncancer patients. Clin Chem 2002; 48: 1835–43.

CHAPTER 39

Animal models of dilated cardiomyopathy and relevance to human disease

Arundathi Jayatilleke, Howard A Rockman

Introduction

Dilated cardiomyopathy (DCM) is one of the most common diseases of the myocardium, and a leading cause of heart failure in the United States. It comprises a group of diseases of the heart muscle that is characterized by both cardiac chamber dilation and systolic dysfunction. Myocardial infarction and coronary artery disease are common causes of DCM in the United States, but DCM may arise from a wide variety of etiologies, including hormonal disorders such as diabetes mellitus and thyroid disease, ingestion of certain substances or heavy metals, bacterial or viral infections, or connective tissue diseases. The etiology of DCM is often multifactorial, with both genetic and environmental components, although a number of genetic mutations have been identified in hereditary forms of DCM. The development of animal models of DCM has been important, not only in dissecting the pathways and mechanisms leading to heart failure, but also in identifying new molecular targets for therapeutic strategies, and providing physiological models of human heart failure in which to test these interventions.

Genetic models of DCM

A wide variety of genetic models of DCM have been developed in recent years, most notably in the mouse. The use of genetic models and inbred animals allows us to avoid the variability in genetic background that may affect the DCM phenotype and its penetrance, as well as to better control and select for the phenotype being examined. By targeting overexpression or knockout of specific genes and pathways, these models have been critical in furthering our understanding of the molecular mechanisms of DCM.

Sarcomeric proteins

One of the first described genetically engineered murine models of DCM is the muscle LIM protein (MLP) knockout mouse. MLP regulates myogenic differentiation in the development of both skeletal and cardiac muscle,[1] while the LIM domain targets MLP to the actin cytoskeleton.[2] While mice heterozygous for disruption of the MLP gene were phenotypically the same as wild-type littermates, homozygous MLP knockout mice developed DCM with hypertrophy after birth, with disruption of myofibrillar organization evident on histological examination of cardiomyocytes.[3] About half of MLP[-/-] homozygotes died within two weeks of birth, but those animals that survived to adulthood displayed a cardiac phenotype similar to that of humans with DCM, with left ventricular (LV) enlargement, thinning of chamber walls, and decreased LV function.[3] The histological features of adult MLP knockout hearts, including interstitial cell infiltration and myocardial fibrosis, also resemble those of human patients with DCM.[3] While these phenotypic similarities alone do not prove that MLP is involved in human DCM, a mutation in a conserved region of human *MLP* (K69R) was recently identified in a patient with DCM.[4] The W4R mutation associated with DCM was found to result in a defect of titin interaction,[5] and several different *MLP* mutations were identified in patients with familial HCM.[6]

The MLP knockout model has also been useful in studying novel therapeutic strategies in heart failure through crossbreeding experiments with other genetically modified mice. For example, expression of β-adrenergic receptor kinase-1 (βARK1) is known to be elevated in human heart failure, and MLP knockout mice overexpressing the C-terminal 194 amino acids of βARK1 (βARKct), which acts as a βARK1 inhibitor, had normal LV chamber size,[7] and increased LV contractility,[8] compared to MLP knockout mice

(Fig. 39.1). Another crossbreeding experiment showed that deficiency in phospholamban, resulting in Ca^{2+} reuptake in cardiomyocytes, prevented expression of the DCM phenotype in double knockout mice.[9] Lastly, short-term administration of recombinant human growth hormone to MLP knockout mice improved LV systolic and diastolic function, without causing a significant increase in cardiac and body weights,[10] although the benefits of long-term growth hormone therapy in humans with DCM have not yet been proven. These studies not only provide evidence that defects in sarcomeric proteins may act to cause heart failure, but also shed new light on possible treatment strategies.

Mutations in the gene encoding cardiac myosin-binding protein-C (MyBP-C) are among the most common genetic alterations associated with familial hypertrophic cardiomyopathy (FHC), accounting for about 15% of cases.[11] MyBP-C

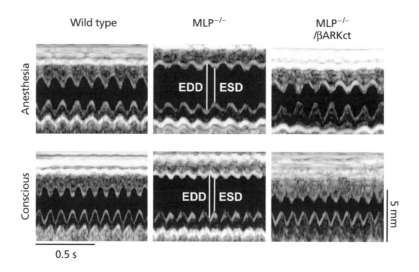

Figure 39.1. *Overexpression of βARKct in MLP knockout (MLP^(−/−)) mice attenuates the DCM phenotype. βARKct inhibits βARK-1, which mediates βAR phosphorylation, demonstrating the importance of βAR downregulation and desensitization in the pathogenesis of heart failure. Representative echocardiographic tracings are shown for the same mouse under anesthesia and in the conscious state. LV dimensions are indicated by the white lines: EDD, end-diastolic dimension; ESD, end-systolic dimension. MLP^(−/−) mice (center) display marked chamber dilation and depressed cardiac function compared to WT mice (left) under both conscious and anesthetized conditions, while MLP^(−/−)/ βARKct mice (right) display normal cardiac performance and chamber diameter. From Esposito G, Santana LF, Dilly K et al. Cellular and functional defects in a mouse model of heart failure. Am J Physiol Heart Circ Physiol 2000; 279: H3101–12, with permission.*

localizes to the A bands of sarcomeres,[12] and is thought to regulate positive inotropy in a phosphorylation-dependent manner.[13] Transgenic mice homozygous for the truncated MyBP-C peptide had increased LV systolic and diastolic diameters, and decreased LV fractional shortening compared to wild-type mice. Mutant mice recapitulated the fetal gene expression characteristic of murine DCM, including increased expression of α-skeletal actin, brain natriuretic peptide (BNP), and β-myosin heavy chain (β-MHC).[14] Despite truncation of the MyBP-C peptide and decreased MyBP-C mRNA expression levels, homozygous mutant mice had nearly normal sarcomeres with only somewhat indistinct M bands.[14]

The role of MyBP-C deficiency in the pathogenesis of DCM is unclear. In humans, heterozygosity for an allele of the cardiac MyBP-C gene which encodes a truncated MyBP-C peptide leads to FHC,[15] in contrast to the DCM phenotype exhibited by transgenic mice. In addition, while the truncated MyBP-C peptide was expressed in the cardiac tissue of homozygous mice, one human study of an FHC-associated MyBP-C mutation found no expression of truncated MyBP-C peptides in cardiac tissue.[16] However, while this model of murine DCM does not directly correlate to a similar phenotype in humans, it may be useful in further studies to analyze the role of MyBP-C in the development of DCM.

Tropomodulin is a thin filament protein that caps the pointed ends of actin filaments. Overexpression of tropomodulin (between 7- and 11-fold above normal levels) in the myocardium, using the cardiac-specific α-MHC promoter, leads to a phenotype of DCM with loss of myofibril organization and myocyte apoptosis. Echocardiographic analysis revealed LV dilation, wall thinning rather than hypertrophy, and impaired systolic performance in tropomodulin-overexpressing transgenic (TOT) mice.[17] Expression of β-MHC and atrial natriuretic factor (ANF) mRNA was increased, similar to other murine models of DCM.[18] A crossbreeding experiment using mice overexpressing insulin-like growth factor-1 (IGF-1), which is known to inhibit myocyte apoptosis under oxidative stress, showed that hearts from mice overexpressing both proteins (TIGFO mice)

had normal mass, normal wall thickness, and decreased cardiomyocyte apoptosis compared to TOT mice, which was associated with some improvement in cardiac function.[19] In addition, pharmacological inhibition of calcineurin, a Ca^{2+}-dependent protein phosphatase prevented ventricular dilation and loss of myofibrillar organization, though fractional shortening remained impaired.[20]

Mutations in tropomodulin are not yet known to be associated with human cases of DCM. However, analysis of the mechanisms of DCM in TOT mice at both the cellular and molecular levels should prove useful in understanding the similarities between mouse models and human cases of DCM, and in the testing of novel therapeutic strategies for DCM.

Extrasarcomeric cytoskeletal proteins

The cardiomyopathic Syrian hamster (CM hamster) strain was a well-known and often-used model of DCM, long before the gene responsible for the cardiomyopathic phenotype was identified. The responsible gene is now known to encode δ-sarcoglycan, a component of the sarcoglycan–dystroglycan complex that associates with sarcospan,[21] and anchors dystrophin, a cytoskeletal protein in the sarcolemma of striated muscle cells.[22] In CM hamsters, the δ-sarcoglycan gene mutation is a deletion of 27.4 kb of genomic material, including the first exon of δ-sarcoglycan. The CM hamster develops DCM with myocardial necrosis that begins within 10–12 weeks after birth, and causes death within one year.[23] Recent studies in δ-sarcoglycan-deficient mice showed they also developed severe DCM and myocardial necrosis progressing to fibrosis with age,[24] and hearts of transgenic mice lacking γ-sarcoglycan, another component of the sarcoglycan complex, developed fibrosis at 20 weeks of age.[25]

Only a few putative disease-causing mutations in the gene encoding δ-sarcoglycan have been identified in screens of DCM patients.[26–28] However, because the CM hamster model has been in use for many years, it has proven a useful model for the study of the molecular pathophysiology of DCM, including alterations in calcium handling, as well as testing new therapeutic modalities for DCM.

Dystrophin, as mentioned above, is associated with the sarcoglycan complex. In humans, mutations that cause premature termination and dystrophin deficiency cause Duchenne muscular dystrophy (DMD), which is characterized by skeletal muscle wasting and progressive cardiomyopathy. The X chromosome-linked muscular dystrophy (mdx) mouse is a model of DMD that also lacks dystrophin due to premature termination of translation; however, mdx mice do not display severe cardiomyopathy and have less skeletal muscle damage than humans with DMD, possibly due to greater regenerative capabilites.[29] On the other hand, the combined mdx:MyoD[−/−] mutant, which lacks both dystrophin and the skeletal muscle-specific transcription factor MyoD, displays skeletal muscle wasting as well as cardiomyopathy, with LV dilation, increased heart-to-body weight ratio, and myocardial fibrosis.[30] Because MyoD is found in skeletal muscle and not cardiac muscle, the mechanism by which it causes the observed cardiomyopathic changes is uncertain.

The mdx:MyoD[−/−] mouse is useful because it provides a model for human DMD cardiomyopathy previously unavailable in the single loss-of-function mdx mouse. While the exact mechanism by which MyoD deficiency leads to cardiomyopathy is unknown, the presence and severity of skeletal muscle damage may be a contributory factor in mice, and may be involved in the pathogenesis of DMD-associated cardiomyopathy in humans.

An important process that has recently been shown to involve disruption of the sarcoglycan-dystrophin complex is coxsackievirus B (CVB) infection. CVB, an enterovirus known to cause viral myocarditis, is thought to be an etiologic agent in idiopathic DCM. The coxsackieviral protease 2A cleaves and prevents proper localization of dystrophin, leading to acquired cardiomyopathy (Fig. 39.2).[31] Transgenic mice with cardiac-specific expression of the CVB3 genome, a strain of CVB known to cause persistent infection, developed DCM with significant fibrosis and impaired systolic function.[32] Further studies in cultured cardiomyocytes and in the heart showed that the sarcolemmal localization of dystroglycan and sarcoglycans was disrupted by infection with CVB3.[33] These studies are especially interesting

because they use a genetic model to highlight the importance of the sarcoglycan–dystroglycan–dystrophin complex in acquired DCM, and because it may be applicable to the development of hereditary DCM caused by mutations in the dystrophin complex.

Structural proteins outside of the dystrophin–sarcoglycan complex also seem to be involved in the pathogenesis of DCM. Intermediate filaments involving desmin, a muscle-specific protein, are thought to link myofibrils to each other and other cell structures, and are important in maintaining myocyte structural integrity. Desmin knockout mice develop abnormalities in cardiac, skeletal, and smooth muscle in knockout mice; cardiac muscle degeneration was most severe, possibly related to higher activity and repair rates.[34] In addition, desmin knockout mice hearts displayed extensive areas of interstitial fibrosis, large calcified plaques, hypertrophy, and ventricular dilation with depressed systolic function.[34] While the phenotype of DCM was not explored further in these mice, the model remains of interest because of the possible application to human disease. The absence of desmin and intact intermediate filaments disrupts cardiac function, and may help explain the pathophysiology of cardiomyopathy in human desminopathies. Absence of desmin in humans has not been observed, but missense mutations in desmin are associated with human familial DCM.[35]

Integrins are cellular adhesion receptors which link the actin cytoskeleton with the extracellular matrix,[36] and are required for normal cardiac development.[37] Both integrin ablation in ventricular myocytes,[38] and transgenic expression of a dominant negative chimera of integrin β_{1A} and interleukin-2- (IL-2) receptor domains[39] leads to LV dilation, decreased LV contractility, and myocardial fibrosis in response to pressure overload. Moreover, knockout of the striated muscle-specific integrin-binding protein melusin in transgenic mice leads to LV dilation, systolic dysfunction, and decreased survival in response to pressure overload, but without a fibrotic response.[40]

The adherens junction in cardiac myocytes is also involved in cell–cell adhesion as well as force transmission via the intercalated disc, to which

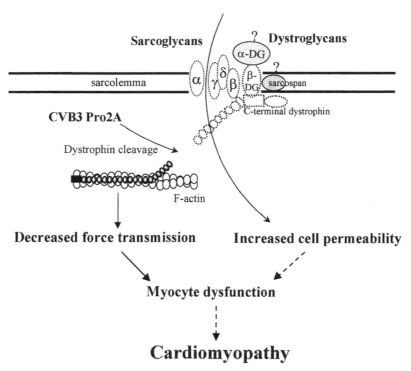

Figure 39.2. *Model of the pathogenic role of the dystrophin–glycoprotein complex in acquired coxsackievirus B cardiomyopathy. This model demonstrates the interactions between environment and cellular pathways of DCM. The dystrophin–glycoprotein complex in cardiac muscle comprises dystrophin, β-dystroglycan (DG), and α-, β-, γ- and δ-sarcoglycans, connects the actin cytoskeleton to the extracellular matrix, and is thought to play an important role in force transmission. In this schematic, cleavage of dystrophin by the viral protease 2A leads to loss of sarcoglycans and β-dystroglycan from the sarcolemmal membrane, thus causing functional impairment of the dystrophin–glycoprotein complex and acquired DCM. Adapted from Lee GH, Badorff C, Knowlton KU. Dissociation of sarcoglycans and the dystrophin carboxyl terminus from the sarcolemma in enteroviral cardiomyopathy. Circ Res 2000; 87: 489–95, with permission.*

myofibrils are attached. Cadherins, cell adhesion receptors, are an important part of the adherens junction, and mediate the interaction between adjacent myocytes.[41] Targeted ectopic expression of E-cadherin (the epithelial isoform) and over-expression of N-cadherin (the muscle isoform) led to DCM in transgenic mice, possibly due to stoichiometric or structural changes interfering with adherens junction organization.[42] Unlike several of the other cytoskeletal-based DCM phenotypes, cadherin misexpression does not have a known human equivalent. However, it lends support to the hypothesis that mutations in cytoskeletal and myofibrillar proteins cause DCM via impaired force transmission.

Calcium regulation and signaling

Calsequestrin is a Ca^{2+}-binding protein located in the junctional sarcoplasmic reticulum. Voltage-dependent influx of Ca^{2+} through L-type Ca^{2+} channels in the plasma membrane activates cardiac ryanodine receptors (RyR2), which open and allow calsequestrin to release Ca^{2+} into the cytosol, thus coupling cardiac myocyte excitation and contraction. Calsequestrin-overexpressing (CSQ) mice have an initial cardiac phenotype of LV hypertrophy and decreased systolic function, that progresses to LV dilation, severely impaired cardiac function and premature death.[43] CSQ mice also showed altered Ca^{2+} signaling, with reduced Ca^{2+}-induced Ca^{2+} release.[44,45]

Hearts from CSQ mice demonstrate classic abnormalities in βAR signaling, including decreased βAR density and reduced agonist responsiveness (downregulation and desensitization).[43] βARKct overexpression was used to test whether blocking βAR desensitization would prevent the development of heart failure in CSQ mice. CSQ mice crossed with transgenic mice overexpressing βARKct resulted in significantly improved survival in double transgenic mice, from 9 weeks in CSQ mice, to 15 weeks in CSQ/βARKct mice.[46] In addition, treatment of CSQ/βARKct mice with metoprolol, a β-blocker, further improved survival to 25 weeks.[46] Moreover, the CSQ model has been used to identify genetic modifier loci that alter the heart failure phenotype,[47,48] and may lead to the identification of novel candidate genes that affect prognosis in human heart failure. Thus, the CSQ model may not only provide insight into the altered signaling pathways and genetics involved in the pathogenesis of DCM in humans, but also lead to identification of potential therapeutic strategies such as inhibition of βARK1 in heart failure.

Phospholamban plays an important role in intracellular Ca^{2+} handling, by inhibiting the sarcoplasmic reticulum Ca^{2+} ATPase pump (SERCA2a), and is itself inactivated via phosphorylation by protein kinase A (PKA) and Ca^{2+}/calmodulin kinase II (CaMKII). The activity of SERCA2a, which promotes cardiac relaxation, is known to be decreased in human heart failure,[49] while ablation of phospholamban, as mentioned above, rescues heart failure in MLP knockout mice, suggesting that increased or unopposed phospholamban activity contributes to heart failure. Indeed, transgenic mice overexpressing wild-type phospholamban had decreased contractility (although without ventricular dilation).[50] In addition, transgenic mice with 2-fold overexpression of a superinhibitory valine-49 to glycine phospholamban mutation showed systolic dysfunction and interstitial fibrosis of the myocardium. In females, this led to cardiac hypertrophy, while males developed progressive dilation of the heart, fetal gene expression, and congestive heart failure leading to death within seven months of age.[51] The mechanism by which this particular phospholamban

mutant increases inhibition of sarcoplasmic reticulum (SR) Ca^{2+} uptake is unknown, but may be related to increased association between phospholamban and SERCA2a.

A recent study in transgenic mice heterozygous for an arginine to cystine mutation in the conserved ninth residue of phospholamban (PLN[R9C]), showed that PLN[R9C] mice displayed rapidly progressing DCM as well as physiological symptoms of heart failure.[52] Cell culture studies revealed that the mutant PLN[R9C] inhibited phosphorylation of wild-type (WT) phospholamban by sequestering PKA. Thus, WT phospholamban remained active, resulting in constitutive inhibition of SERCA2a and prolonged decay of Ca^{2+} transients in cardiac myocytes. Interestingly, this murine model of DCM was engineered using reverse genetic techniques: screening of individuals with inherited DCM for mutations in phospholamban revealed a cytosine to thymidine substitution resulting in the arginine to cystine mutation at position 9; this mutation caused DCM which was inherited in an autosomal dominant manner in the proband's family (Fig. 39.3).[52] The engineered PLN[R9C] mice developed a phenotype very similar to that in humans, with decreased LV contractility, myocyte enlargement without myofibril disarray, and extensive interstitial fibrosis.[52] The prevalence of this mutation in humans is not yet known, but this mouse model is particularly noteworthy because few animal models correlate so well to their human genetic counterparts.

Inhibition of phospholamban using a recombinant adeno-associated virus (rAAV) to transfer a pseudophosphorylated phospholamban mutant (PLNS16E) into the hearts of cardiomyopathic hamsters has been shown to improve cardiac function.[53] A similar gene transfer study performed in rats after myocardial infarction showed improved post-infarction cardiac contractility and suppressed LV remodeling in rats treated with rAAV-PLNS16E compared to control animals.[54] Thus, phospholamban inhibition may prove to be another therapeutic strategy in the treatment of heart failure.

Protein kinase C (PKC) is involved in the regulation of the phosphorylation state of phospholamban. PKC phosphorylates protein phosphatase

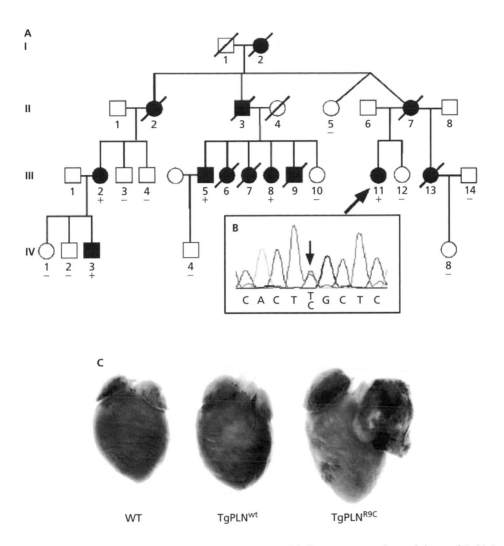

Figure 39.3. *PLNR9C and dilated cardiomyopathy. This model illustrates an inherited form of DCM which was identified in humans and reverse-engineered in transgenic mice. (A) Pedigree of a large family displaying clinical status (circle, female; square, male; solid, affected (DCM); clear, unaffected; slash, death) and genetic status (+, PLN^{R9C/+}; −, PLN^{wt}) shows cosegregation of PLN^{R9C} and DCM. Individual III-11 was the proband. (B) DNA sequencing revealed a cytosine to thymidine substitution at position 25. (C) Hearts of WT, PLN^{wt} transgenic (TgPLN^{wt}), and PLN^{R9C} transgenic (TgPLN^{R9C}) mice at 4 months, showing enlargement of the PLN^{R9C} heart. From Schmitt JP, Kamisago M, Asahi M et al. Dilated cardiomyopathy and heart failure caused by a mutation in phospholamban. Science 2003; 299: 1410–3, with permission.*

inhibitor-1 (I-1), which inhibits protein phosphatase-1 (PP-1), which in turn dephosphorylates phospholamban,[55] preserving phospholamban in its activated state. Overexpression of WT PKC-α causes a progressive reduction of cardiac contractility in transgenic mice, while transgenic mice which lack PKC-α display hypercontractility without decrement over a period of 10 months.[56] PKC-α overexpression indirectly leads to hypophosphory-lation of phospholamban in the hearts of transgenic mice,[56] promoting inhibition of SERCA2a and alter-ation of Ca^{2+} signaling. These studies show that PKC-α may provide another novel therapeutic tar-get to improve cardiac contracility in heart failure.

Another class of molecules involved in trans-ducing Ca^{2+} signals is that of the Ca^{2+}/calmodulin-dependent protein kinases (CaMKs). CaMKII, a multimeric enzyme, phosphorylates a variety of sar-coplasmic reticulum proteins, including RyR2, phospholamban, L-type Ca^{2+} channels, and cAMP response element-binding protein (CREB). The subunits of CaMKII have tissue-specific distribu-tions, with the δ isoform predominating in the heart.[57] CaMKIIδ expression is also increased in the failing heart,[58] and two transgenic mouse studies examined the possible role of CaMKII in heart failure. Cardiac-specific overexpression of CaMKIIδB, a splice variant of CaMKIIδ that local-izes to the nucleus, caused cardiomegaly associated with mild ventricular dilation: 27% increase in LV-to-body mass ratio by 12 weeks of age, and 14% increase in the LV end-diastolic diameter.[59] However, the CaMKIIδB isoform is unlikely to regulate Ca^{2+} handling, being targeted to the nucleus.

A similar study in transgenic mice overexpress-ing CaMKIIδC, a cytoplasmic splice variant of CaMKIIδ, showed that a 3-fold increase in myocardial levels of CaMKIIδC caused markedly enlarged hearts and increased LV-to-body mass ratio by 8 weeks of age compared to wild-type levels.[60] CaMKIIδC overexpression was associated with increased phosphorylation of both phospho-lamban and RyR2, at levels 2.3- and 2.5-fold above wild-type, respectively.[60] Because the increased phospholamban phosphorylation would be expected to release SERCA inhibition and enhance contractility, the full effects of increased CaMKIIδC are still unclear; however, *SERCA* mRNA levels were decreased in transgenic mice, and may instead lead to decreased contractility. The increased CaMKIIδ levels in the failing heart may mediate in part the complex alterations in phosphorylation and activity of Ca^{2+}-signaling pro-teins, and thus play a role in the contractile dys-function associated with DCM.

Adrenergic signaling

Increased adrenergic activation in the failing heart is important to maintain cardiac function acutely; however, chronic adrenergic stimulation con-tributes to myocardial dysfunction and ventricular remodeling. Several transgenic mouse models have been created in order to explore the roles of differ-ent aspects of βAR signaling pathways in heart failure.

In the failing human heart, β_1AR density is reduced, while the remaining β_1ARs are desensi-tized.[61] Overexpression of β_1ARs, the most abun-dant βAR subtype in the heart, at levels 20- to 40-fold over wild-type produced a distinct pheno-type of myocyte hypertrophy, myofibrillar disarray, increased myocyte apoptosis, and LV dilation in young animals, and progressive deterioration in cardiac function with age.[62]

On the other hand, the level of β_2AR is not significantly changed in the failing heart,[61] result-ing in an altered β_1AR:β_2AR ratio. Transgenic mice overexpressing β_2ARs exhibited enhanced contractility and cardiac function without decreased mortality;[63] however, very high expres-sion levels (350-fold above wild-type) led to a rap-idly progressing DCM.[64] In contrast to the β_1AR-overexpressing mice, high-expressing β_2AR transgenic mice showed maximal β_2AR activation in the absence of agonist, and did not respond to treatment with propanolol, a β-blocker,[64] unless the β-blocker had inverse agonist activity.[65] Thus, it seems that intrinsic receptor signaling activity, perhaps due to an increased number of sponta-neously isomerized β_2ARs, was deleterious to car-diac function. Interestingly, disruption of G_i signaling by pertussis toxin in β_2AR-overexpressing mice led to rescue of cAMP signaling and contrac-tility in response to agonist administration,[66] suggesting that coupling of β_2AR to G_i impairs contractility in the failing heart.

Phosphoinositide 3-kinase-γ (PI3Kγ) is a pro-tein and lipid kinase which acts downstream of the βAR to promote βAR internalization. Upon agonist stimulation, it is recruited by βARK2 to activated βARs, where it generates phosphatidyli-nositol-3,4,5-tri-phosphate (PtdIns-3,4,5-P3) lead-ing to βAR endocytosis.[67,68] Transgenic mice overexpressing a catalytically inactive mutant of

PI3Kγ (PI3Kγ$_{inact}$) that displaces endogenous PI3K from βARK1 did not display βAR downregulation or desensitization in response to chronic agonist stimulation, while wild-type mice displayed markedly attenuated βAR responsiveness.[69] Moreover, these PI3Kγ$_{inact}$ mice show less cardiac dysfunction and increased survival after pressure overload-induced heart failure compared to WT mice.[69] This model demonstrates that disruption of the βARK1/PI3K interaction prevents βAR desensitization and downregulation, and may be critical in maintaining cardiac function in the setting of pressure overload.

In response to βAR signaling, protein kinase A (PKA) increases cardiac contractility by phosphorylating target molecules such as phospholamban (see above), releasing SERCA2a inhibition and increasing SR Ca^{2+} uptake. PKA also phosphorylates and activates the L-type voltage-gated Ca^{2+} channel, increasing Ca^{2+} influx, and RyR2, increasing Ca^{2+}-dependent Ca^{2+} release. Transgenic mice overexpressing the catalytic subunit of PKA in the heart were initially phenotypically normal, but by 13 weeks had developed dilated hearts with mild fibrosis and symptoms of heart failure.[70] Contractility was impaired in transgenic mice, and arrhythmias increased. Both survival and chamber dilation correlated to the degree of PKA overexpression.

PKA may be involved in the altered contractility associated with chronic βAR signaling. PKA overexpression caused hyperphosphorylation of RyR2 and dissociation of FKBP12.6, which normally binds to RyR2 and prevents Ca^{2+} release. These changes have been shown to lead to uncoupling of RyR2 from each other, and decreased gain in excitation–contraction coupling.[71] The resultant alteration of Ca^{2+} release from the SR may play a role in the development of heart failure in PKA-overexpressing transgenic mice. Likewise, as PKA phosphorylation of RyR2 may be increased in human heart failure,[71] PKA may, in part, mediate the effects of abnormal βAR signaling in the failing myocardium.

G proteins

G$_{sα}$ is a G protein that transduces the stimulatory signal from the agonist-occupied βAR, activating adenylyl cyclase and resulting, in the normal heart, in increased contractility and heart rate.[72] However, chronic adrenergic stimulation results in reduced βAR responsiveness,[61] and may contribute to the progression of decompensated heart failure in humans. Although G$_{sα}$ is thought to be in excess of adenylyl cyclase, transgenic mice overexpressing G$_{sα}$ under the control of the cardiac-specific α-MHC promoter, initially exhibited more rapid activation of adenylyl cyclase, as well as increased heart rate and contractility in response to catecholamine stimulation.[73] However, the older mice demonstrated features of DCM: decreased ejection fraction, significantly increased incidence of ventricular arrhythmias, and histological evidence of cardiomyopathy, including myocardial necrosis and fibrosis.[74] Interestingly, the number of apoptotic myocytes was also increased in mature transgenic mice, suggesting that apoptosis contributes to the progression of heart failure in a model of heightened βAR signaling.[74]

Cardiac-specific overexpression of G$_{αq}$, which is involved in the downstream signaling pathways of α$_1$-adrenergic and angiotensin II receptors, also results in a cardiomyopathic phenotype in transgenic mice.[75] Hypertrophy with systolic dysfunction is the predominant feature in mice with 4-fold G$_{αq}$ overexpression.[75] Activation of PKCε, a signaling protein known to act downstream of G$_{αq}$, and decreased βAR-dependent cAMP production were also noted, once again pointing to the importance of βAR signaling in the pathogenesis of heart failure.[75] Further experiments showed that G$_{αq}$-overexpressing transgenic mice subjected to hemodynamic pressure overload using transverse aortic constriction (TAC) developed left heart failure,[76] while female mice were susceptible to peripartum heart failure with cardiomyocyte apoptosis.[77] Treatment of mice with IDN-1965, a caspase inhibitor, rescued G$_{αq}$ peripartum myopathy,[78] once again highlighting the importance of apoptosis and suggesting that the transition between LV hypertrophy and DCM is mediated in part by apoptotic pathways in the heart.

Mice with cardiac-directed overexpression of adenylyl cyclase type VI were crossbred with G$_{αq}$-overexpressing mice; double transgenic mice had histologically normal hearts, improved βAR responsiveness, and increased agonist-dependent cAMP production. LV contractility and fractional shortening were also returned towards normal levels, effects which did not diminish over a period of several weeks.[79] Thus, while chronic adrenergic stimulation is known to have deleterious results in the heart, adenylyl cyclase overexpression in $G_{\alpha q}$-overexpressing mice led to increased adrenergic responsiveness without chronic adrenergic stimulation or desensitization, a finding that may lead to new heart failure treatment strategies specifically involving the activity of adenylyl cyclase.

The G_i signaling pathway, in contrast, is an inhibitory pathway that is coupled to A1 adenosine receptors and M2 muscarinic receptors, and decreases cAMP levels by inhibiting adenylyl cyclase, resulting in decreased contractility and heart rate. Conditional and cardiac-specific expression of a G_i-coupled receptor (Ro1, a modified receptor activated only by synthetic ligand, based on the κ opioid receptor) provided a model of DCM without hypertrophy.[80] Transgenic mice had enlarged ventricles and decreased fractional shortening under basal Ro1 signaling, while stimulation of the Ro1 receptor led to marked bradycardia.[80] However, suppression of Ro1 expression in transgenic mice led to partial reversal of some cardiac changes, such as improved fractional shortening, without reversal of structural changes such as ventricular dilation.[80] While increased G_i signaling is not known to cause cardiomyopathy in humans, LV levels of G_i protein and mRNA are increased in patients with idiopathic DCM, and probably does play a role in βAR desensitization.[81] Thus, whereas the previously-held belief that increased G_i is compensatory to the increased sympathetic nervous system activity in heart failure, the Ro1 receptor model suggests a possible causative role for G_i and related receptors in DCM.

Mitogen-activated protein kinase (MAPK) cascades

MAPK pathways transduce a variety of signals and lead to changes in cell growth and differentiation patterns. In the heart, MAPK cascades are of particular interest with respect to cardiac hypertrophy and ventricular remodeling,[82] and are activated in response to pressure overload.[83] Transgenic mice that overexpress a constitutively active upstream kinase of p38 (MKK3bE) die prematurely with signs of heart failure and myocardial fibrosis.[82] LV function was compromised in these transgenic mice, with decreased cardiac output and increased LV volume compared to WT mice.[82] Interestingly, these mice did not develop hypertrophy, suggesting that ventricular hypertrophy is mediated by another arm of the MAPK pathway.

Rac1 is one of a number of GTP-binding proteins that couple G protein-coupled receptors (GPCRs) to MAPK pathways by activating upstream protein kinases. Transgenic mice expressing constitutively active Rac1 at normal levels in the heart (RacET mice) had one of two phenotypes: severe DCM leading to death within the first two weeks of life, or cardiac hypertrophy with hypercontractility and normalization of heart-to-body weight ratio.[84] The DCM phenotype was not characterized by myofibrillar disarray; decrease in cardiac function may instead be due to Rac1 activation of PAK, a serine/threonine kinase associated with loss of focal adhesion complexes.[85] As with the cadherin-overexpressing mouse described above, the resultant decrease in cellular adhesion may cause impaired force transmission that leads to DCM.

RhoA is another GTP-binding protein thought to regulate actin cytoskeleton organization, and transduce signals from a subset of GPCRs to stress fiber formation and assembly of focal adhesion complexes.[86] Overexpression of wild-type RhoA in the heart led to DCM, sinus and atrioventricular nodal conduction defects, ventricular dilation and interstitial fibrosis.[87] Interestingly, marked bradycardia was noted in all transgenic mice before the development of ventricular dysfunction, suggesting that RhoA leads to heart failure through impaired conduction. The mechanism behind these conduction abnormalities is still unknown; however, this model should be useful in elucidating the role of RhoA in heart failure and its association with sinus and atrioventricular nodal dysfunction.

Protein phosphatase-2A (PP2A) is a serine/threonine phosphatase involved in cellular signal transduction, including cross-talk between MAPK and extracellular signal-regulated kinase (ERK) pathways, and modulation of apoptotic pathways.[88] Muscle-specific expression of a dominant negative mutant of the regulatory A subunit of PP2A led to a phenotype of DCM without hypertrophy in transgenic mice.[89] The mutant subunit prevents proper assembly of the PP2A holoenzyme, causing instead an increase in core enzymes, such that PP2A is altered to favor core enzyme activity instead of holoenzyme activity.[89] Altered PP2A signaling has been implicated in several human diseases, including colon and lung carcinomas, but its role in DCM has not yet been fully explored.

Cytokines

Circulating levels of tumor necrosis factor-α (TNF-α), an inflammatory cytokine, are increased in patients with heart failure, with some correlation between the TNF-α level and the severity of disease.[90] Circulating TNF-α derives from cardiac myocytes, and leads to decreased contractility and LV remodeling.[91] Transgenic mice overexpressing TNF-α in the heart developed DCM with histopathological changes including grossly enlarged hearts, interstitial fibrosis, and apoptotic cell death.[92] Expression of ANF mRNA, consistent with activation of the heart failure-associated fetal gene program, was also elevated in transgenic mice.[92] These findings argue that elevated levels of TNF-α play a causative role in DCM, rather than simply being a heart failure-associated phenomenon.

Further experiments on the same TNF-α-overexpressing mice showed that inhibition of matrix metalloproteinases, known to be active in collagen deposition and extracellular matrix remodeling, reduced collagen expression and prevented ventricular hypertrophy in young mice.[93] Anti-TNF-α therapy, in this case via adenovirus-mediated gene transfer of a soluble recombinant TNF receptor, also improved certain aspects of the phenotype, most notably collagen content of the heart.[94] Nonetheless, treatment of heart failure patients with anti-TNF-α agents has yielded inconclusive results, and has not yet proven to be of benefit in patients.[95]

Interleukin-6 (IL-6) is another proinflammatory cytokine whose levels are elevated in heart failure.[96] The IL-6 family of cytokines shares a common receptor known as gp130, overexpression of which can lead to cardiac hypertrophy.[97] Mice with a cardiac-specific knockout of gp130 did not show overt cardiac dysfunction or hypertrophy up to six months of age. However, use of transverse aortic constriction (TAC) to generate pressure overload in knockout mice led to decreased contractility, chamber dilation and increased mortality, while control mice were observed to develop only cardiac hypertrophy.[98] In control mice, TAC induced activation of STAT3, a gp130-dependent pathway, while knockout mice were unable to activate the STAT3 pathway.[98] Taken together, these results suggest that the gp130 pathway may be critical in preventing cell death leading to DCM in response to biomechanical stress.

Growth factors and their receptors

Platelet-derived growth factors (PDGFs) are mitogenic proteins thought to be involved in replacement fibrosis in a variety of organs. Overexpression of PDGF-C targeted to the heart using the α-MHC caused sex-dependent phenotypes in transgenic mice.[99] Both sexes displayed interstitial fibrosis of the myocardium, and altered vasculature with decreased capillary density. However, by three months of age, male transgenic mice had developed fibrosis and hypertrophy, while females had developed DCM.[99] The severity of the disease was also greater in female transgenic mice, and few were alive at six months of age, despite similar expression levels of PDGF-C and activation of its receptor in both sexes.[99] The mechanism behind these sex-related differences is as yet unknown, as is their significance to human DCM. However, the observed fibrotic reaction and vascular changes implicate PDGF-C-dependent mitogenic pathways in the pathological changes observed in DCM and heart failure.

Targeted knockout of the gene encoding vascular endothelial growth factor A (VEGF-A) in cardiomyocytes also results in contractile dysfunction and thin-walled, dilated hearts in transgenic mice.[100] While the cardiac macrovasculature of these mice was unchanged, capillary density was reduced, suggesting a role for hypovascularity and ischemia in the development of this phenotype. However, signs of severe prolonged ischemia such as cardiomyocyte necrosis and fibrosis were not observed. The targeted VEGF-A knockout may instead serve as a genetic model for cardiac dysfunction resulting from prolonged mild ischemia, and may lead to refinement of therapeutic strategies involving stimulation of angiogenesis in ischemic heart disease.

Angiotensin II (Ang II), a component of the renin–angiotensin–aldosterone system, is thought to adversely affect ventricular remodeling in the setting of heart failure. Type 1 Ang II (AT1) receptors mediate vasoconstriction in response to Ang II stimulation, while the role of type 2 receptors (AT2), which appear to be chronically upregulated in human heart failure,[101] is less well understood. Transgenic mice overexpressing AT2 receptors in the ventricles via the ventricle-specific myosin light chain (MLC-2v) promoter showed LV systolic and diastolic dysfunction corresponding to the level of AT2 transgene expression, while LV dilation, wall thinning, and fibrosis were only observed in high-expressing mice.[102] Phosphorylation of PKC-α, PKC-β and phospho-ERK12 was also increased in high-expressing mice, indicating activation of growth pathways. Thus, upregulation of AT2 receptors in chronic DCM may lead to activation of pathways involved in ventricular remodeling, and the AT2 receptor may be a potential target for treatment of chronic heart failure.

Endothelin-1 (ET-1) is a small vasoconstrictive peptide and mitogenic factor whose levels are increased in patients with CHF,[103] although its role in the pathogenesis of heart failure is not well defined. Non-targeted overexpression of ET-1 does not result in a cardiac phenotype, perhaps due to embryonic lethality. However, conditional overexpression of ET-1 in the adult mouse heart caused inflammatory infiltration of the myocardium, LV dilation and fibrosis, leading to overt symptoms of heart failure and premature death within weeks of transgene induction.[104] Interestingly, ET-1 overexpression also caused nuclear translocation of nuclear factor-κB (NF-κB) and inflammatory cytokine expression, suggesting that ET-1 may initiate a proinflammatory cascade which leads to myocardial damage and dysfunction, and, ultimately, heart failure. Non-selective ET-1 receptor blockade in transgenic mice resulted in a small improvement in survival,[104] and further research

may show blocking ET-1 pathways to be of therapeutic benefit in DCM.

Transcription factors

The cAMP response element binding protein (CREB) is a basic leucine zipper transcription factor which, in its phosphorylated state, responds to cAMP levels by binding to DNA and activating transcription. CREB^{A133}, a dominant negative form of CREB, can bind DNA but not activate transcription, and is thought to act by displacing WT CREB from associated proteins and DNA. Expression of CREB^{A133} in cardiomyocytes led to progressive DCM in transgenic mice: interstitial fibrosis, biventricular dilation, and decreased systolic function compared to non-transgenic controls were noted.[105] Transgenic mice also demonstrated increased mortality, with over 40% dead by 20 weeks of age. Apoptotic and inflammatory cells were not increased in the hearts of transgenic mice.

Mutations in the CREB gene have not been identified as candidate genes in inherited DCM in humans. However, the possibility remains that other genes in the same transcriptional pathway may be involved in idiopathic DCM. In addition, CREB is downregulated by chronic βAR stimulation in the rat,[106] suggesting that the hyperadrenergic state seen in human heart failure may act in part through downregulation of the CREB transcriptional pathway, to lead to cardiac dysfunction, and that therapies that increase CREB activity may be beneficial in treating heart failure.

Calcineurin, also known as protein phosphatase 2B, is, like PP2A, a serine/threoneine phosphatase involved in intracellular signaling pathways. It is activated by the binding of Ca^{2+} and calmodulin (CaM) to the regulatory and catalytic subunits, respectively, and in turn dephosphorylates and activates the nuclear factor of activated T cells-3 (NF-AT3), a transcription factor which induces T cell activation.[107] Transgenic mice expressing an activated form of calcineurin (lacking regulatory requirements for Ca^{2+} and CaM) in the heart developed marked hypertrophy within the first three weeks of life.[108] The ventricles became dilated with age, and transgenic mice had an increased incidence of sudden death due to heart failure compared to non-transgenic control animals. These findings suggest that activation of calcineurin, perhaps in response to elevated intracellular Ca^{2+} concentrations, results in cardiac hypertrophy and later DCM through activation of NF-AT3 and other signaling pathways.

These findings suggested the possible therapeutic use of calcineurin inhibitors in treating heart failure. Cyclosporin A (CsA), a common immunosuppressant drug and calcineurin inhibitor, was used to treat transgenic animals. While animals treated with vehicle developed hypertrophic and dilated hearts, the hearts of animals treated with CsA were not significantly larger than those of non-transgenic animals (Fig. 39.4).[108] Unfortunately, repeated studies in transgenic mice, as well as clinical studies in humans, failed to provide conclusive evidence for the use of CsA or FK506, another calcineurin inhibitor, in the treatment of cardiac hypertrophy or DCM, possibly due to confounding of the data by technical variability and drug toxicity.[109] Thus, while calcineurin inhibitors have not yet been proven beneficial in the treatment of heart failure, calcineurin and its associated signaling pathways may provide further targets for pharmacotherapy.

Apoptosis

The role of apoptosis in DCM is the subject of much research, because of findings of increased levels of myocyte apoptosis in failing human hearts,[110] and several of the animal models of DCM described in this chapter also exhibit increased myocyte apoptosis. The details of apoptotic involvement in cardiovascular disease are described elsewhere in this text (Chapter 3), but mention must be made of recent genetic mouse models of DCM in which apoptotic pathways are specifically activated. One such model used a conditionally active form of caspase-8, a cysteine protease which cleaves cellular proteins and causes cell death, to examine the role of myocyte apoptosis in the development of DCM.[111] Acute activation of caspase-8 in transgenic mice (via administration of an exogenous ligand) led to rapid myocyte apoptosis and death within 18 hours.[111]

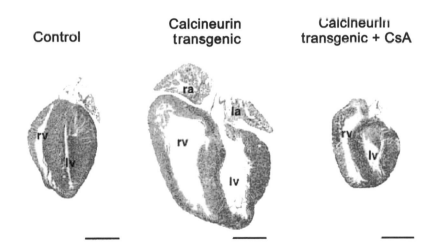

Figure 39.4. *Inhibition of calcineurin signaling prevents the development of DCM in activated calcineurin-expressing transgenic mice. This transgenic model implicates calcineurin signaling as a novel pathway for the development of cardiac hypertrophy and DCM. Hematoxylin and eosin-stained histological section of hearts from non-transgenic mice (left) and calcineurin transgenic mice treated with vehicle (center) or CsA (right). Mice expressing activated calcineurin show DCM at 25 days (center), while CsA treatment attentuates cardiac chamber dilation. la, left atrium; lv, left ventricle; ra, right atrium; rv, right ventricle. From Molkentin JD, Lu JR, Antos CL et al. A calcineurin-dependent transcriptional pathway for cardiac hypertrophy. Cell 1998; 93: 215–28, with permission. See color plate section.*

However, even without the exogenous ligand, low levels of cardiomyocyte apoptosis were noted in high-expressing transgenic mice, possibly due to spontaneous auto-activation of caspase-8. These mice displayed biventricular dilation, interstitial fibrosis and systolic dysfunction, along with a cardiomyocyte apoptotic frequency roughly 15 times that of WT mice, and lower than that observed in failing human hearts.[111] Inhibition of caspase-8 activity in transgenic mice by the polycaspase inhibitor IDN led not only to inhibition of apoptosis but also normalization of LV size, improvement in systolic function and decreased fibrosis, supporting the theory that caspase-8 and associated apoptotic pathways have a causal role in the development of DCM.

Another transgenic mouse model was created by overexpressing WT mammalian sterile 20-like kinase 1 (Mst1) in the heart. Mst1 is a ubiquitous serine/threonine kinase which is cleaved and activated by caspases in response to apoptotic stimuli, and is thought to sensitize cells to apoptosis.[112] Overexpression of the WT Mst1 in transgenic mouse hearts caused impaired systolic function and ventricular dilation.[113] Histological analysis of the myocardium revealed that fibrosis, apoptotic cells, and caspase activity were increased, while myocyte density was decreased. Conversely, transgenic mice expressing a dominant negative form of Mst1 did not activate apoptotic pathways in response to ischemia–reperfusion, while non-transgenic control mice did, presumably through endogenous Mst1, implicating Mst1 in the apoptotic response to reperfusion injury.

In another study exploring the function of inactive mutants of 14-3-3η, a phosphoserine-binding protein that acts in a variety of intracellular signaling pathways, transgenic mice were found to have normal phenotypes until subjected to TAC, when they rapidly developed LV dilation, marked cardiomyocyte apoptosis, and heart failure.[114] 14-3-3, then, may act through modulation of cell signaling to block apoptosis such that inactivation of 14-3-3 renders animals susceptible to stress-induced heart failure.

These models underscore the importance of myocyte apoptosis in the development of DCM, not only through mutations in proteins which directly participate in apoptic pathways, but also through the interplay of diverse cell signaling pathways. Targeted inhibition of apoptotic factors may be useful in the treatment of DCM in humans.

Mitochondrial dysfunction

Muscle-specific knockout of the mitochondrial transcription factor A (Tfam), which regulates transcription of mitochondrial DNA, leads to DCM with heart block and death at two to four weeks of age.[115] Unsurprisingly, electron microscopy of cardiomyocytes revealed abnormal mitochondria. While such a defect does not exist in humans, the model may still be relevant to other human mitochondrial cardiomyopathies, including Kearns–Sayre syndrome, in which portions of the mitochondrial DNA genome are deleted.[116]

Other genetic models of DCM

Several rat strains have been used as models for cardiac hypertrophy and dilated cardiomyopathy, including the Dahl salt-sensitive (DS) rat, which develops hypertension and LV hypertrophy after being fed a high-salt diet.[117] The spontaneously hypertensive (SHR) rat develops hypertension which progresses to heart failure between 18 and 24 months of age, while the related SHHF rat develops heart failure before 18 months of age.[118] Mapping is being carried out to identify the genes responsible for the hypertensive and cardiomyopathic phenotypes in the DS and SHHF rats. Because of their similarities to human heart failure phenotypes, these rats have proven useful in pharmacological testing, and in analysis of Ca^{2+} cycling in failing myocytes.[119]

Ventricle-specific overexpression of a constitutively active retinoid receptor in mice leads to DCM with wall thinning, and systolic dysfunction varying with transgene copy number.[120]

Other known genetic models of DCM do not target specific pathways or genes implicated in the development of DCM. One such model is the overexpression of polyomavirus large T-antigen, thought to be an immortalizing gene that can remove normal controls on cell growth. Transgenic mice developed cardiomyopathy whose severity varied with gene expression levels.[121] Another model examines expression of the Epstein–Barr virus nuclear antigen-leader protein (EBNA-LP), which is required for B cell transformation by Epstein–Barr virus (EBV). Transgenic mice expressing EBNA-LP had no initial cardiac phenotype, but developed symptoms of heart failure, including LV dilation, after several months.[122]

Non-genetic models of DCM

Non-genetic models of DCM have also proven useful in the study of heart failure, particularly in the mechanical simulation of conditions which lead to human heart failure. For example, conditions of pressure overload, myocardial ischemia, and volume overload can be surgically created in animal models to mimic the progression of DCM in humans. Non-surgical models have also been used to study specific etiologies of DCM, including viral and drug-induced myocardiopathy.

Chronic pressure overload has proven to be an extremely important method of mechanically inducing progression to LV failure and evaluating rescue models. Subjecting mice to TAC[123] causes progressive deterioration of cardiac function and LV dilation.[124] In contrast, targeted inhibition of G_q signaling in transgenic mice which overexpress the carboxy-terminal peptide of $G_{\alpha q}$ (TgG_{ql}) caused a reduction in TAC-induced hypertrophy,[125] and cardiac dysfunction,[124] highlighting the role of $G_{\alpha q}$ in the hypertrophic response to pressure overload. Further, mice that do not express dopamine β-hydroxylase ($Dbh^{-/-}$) also showed less hypertrophy and no deterioration in cardiac function in response to chronic pressure overload.[124] These studies contradict the classic dogma that adaptive hypertrophy acts to preserve cardiac function, but rather suggest that limiting detrimental hypertrophic signaling pathways reduces development of cardiac dysfunction.[124]

Myocardial infarction (MI) resulting from coronary artery ligation is an extensively used model of heart failure. Incomplete ligation of the coronary arteries in Sprague-Dawley rats may simulate the development of heart failure in response to chronic ischemia,[126] while complete ligation leads to infarcts of varying size and a corresponding range in the degree of ventricular dilation and dysfunction.[127] Larger infarcts led to early dilation of the left ventricle and chronic decrease in contractility.[127] Similar experiments have been performed in dogs, and also in the mouse in order to facilitate the use of transgenic mice in elucidating the mechanism of post-MI changes in the heart. As in the rat, ventricular remodeling and LV dilation were observed six weeks after MI.[128] The MI model is now routinely being used to simulate ischemic cardiomyopathy and the neurohumoral changes seen in heart failure, as well as to study post-infarct ventricular remodeling due to myocardial necrosis.

Surgical creation of volume overload has also been used to induce heart failure. Surgical creation of an aortocaval fistula, or disruption of the mitral valve in dogs,[129,130] aortic regurgitation in rabbits,[131] and creation of an aortocaval fistula in the mouse[132] are associated with varying degrees of cardiac dysfunction.

Tachycardia pacing has been used in dogs and numerous other species to produce heart failure. Otherwise healthy dogs paced for three weeks at a heart rate of 250 beats per minute develop biventricular dilation and decreased ejection fraction without hypertrophy.[133].

Viral myocarditis can also lead to acquired DCM. Mice infected with encephalomyocarditis virus (EMCV) developed myocardial necrosis and inflammatory infiltrates, as well as biventricular dilation.[134]

Experimental models of chronic heart failure due to drug-induced cardiomyopathy have also been developed. Adriamycin, a cancer chemotherapeutic drug, causes dose-dependent cardiomyopathy in humans, and models of adriamycin-induced cardiomyopathy have been reported in rabbits,[135] dogs[136] and rats.[137]

Table 39.1. *Genetic Models of DCM*

Model	Major phenotypic characteristics	Relevance
Sarcomeric proteins		
MLP knockout[3]	LV dilation Systolic dysfunction Myofibrillar disarray Premature death	Rescued by βARKct overexpression[8] Rescued by phospholamban knockout[9] Rescued by growth hormone therapy[10] Mutations associated with human DCM[4,5] and FHC[6]
MyBP-C knockout[14]	LV dilation Systolic dysfunction Fetal gene expression	Mutations associated with human FHC[11]
Tropomodulin overexpression[17]	LV dilation Systolic dysfunction Wall thinning Fetal gene expression[18]	Rescued by calcineurin inhibition[20] Rescued by IGF-1 overexpression[19]
Extrasarcomeric cytoskeletal proteins		
δ-sarcoglycan knockout (hamster,[23] mouse[24])	LV dilation Systolic dysfunction Myocardial necrosis Premature death	Mutations associated with human DCM[26-28]
Dystrophin/myoD combined knockout[30]	LV dilation Systolic dysfunction Myocardial fibrosis Skeletal muscle wasting	Model of Duchenne muscular dystrophy
Overexpression of CVB3 genome (disruption of sarcoglycan complex)[32]	LV dilation Systolic dysfunction Myocardial fibrosis	Mechanism of coxsackie virus B cardiomyopathy involves cytoskeletal integrity
Desmin knockout[34]	LV dilation Mutations associated with human DCM[35] Systolic dysfunction Cardiac hypertrophy Myocardial fibrosis Calcified plaques Myocardial degeneration	

Continued

Table 39.1. *Genetic Models of DCM – continued*

Integrin ablation[38]	LV dilation, systolic dysfunction, and myocardial fibrosis in response to TAC	Supports role for integrins in the pathogenesis of DCM
Dominant negative integrin expression[39]	LV dilation, systolic dysfunction, and myocardial fibrosis in response to TAC	Supports role for integrins in the pathogenesis of DCM
Melusin deficiency[40]	LV dilation, systolic dysfunction, and decreased survival in response to TAC	Supports role for integrins in the pathogenesis of DCM
Ectopic E-cadherin expression[42]	LV dilation Systolic dysfunction	Supports role for cadherins and impaired force transmission in the pathogenesis of DCM
N-cadherin overexpression[42]	LV dilation Systolic dysfunction	Supports role for cadherins and impaired force transmission in the pathogenesis of DCM
Ca²⁺ signaling and regulation		
Calsequestrin overexpression[44]	Initial LV hypertrophy LV dilation Severe cardiac dysfunction Premature death	Rescued by βARKct overexpression[46]
Overexpression of phospholamban V49G[51]	Systolic dysfunction Myocardial fibrosis Cardiac hypertrophy in females LV dilation, fetal gene expression and premature death in males	Supports involvement of altered Ca²⁺ signaling in the pathogenesis of DCM
Expression of PLN^R9C [52]	Myocardial fibrosis Myocyte enlargement	Same mutation was found in inherited DCM in humans[52]
Overexpression of PKC-α[56]	Reduced contractility	Contributes to the understanding of the roles of altered Ca²⁺ signaling and phospholamban in DCM
Overexpression of CaMKIIδB[59]	Mild ventricular dilation	CaMKIIδ expression elevated in heart failure
Overexpression of CaMKIIδC[60]	Marked cardiomegaly	CaMKIIδ expression elevated in heart failure
Adrenergic signaling		
Overexpression of β₁AR[62]	LV dilation Systolic dysfunction Cardiac hypertrophy Myofibrillar disarray Increased apoptosis	Chronic adrenergic stimulation contributes to myocardial dysfunction in human heart failure

Continued

Table 39.1. *Genetic Models of DCM – continued*

Model	Major phenotypic characteristics	Relevance
Overexpression of β_2AR (350-fold)[64]	LV dilation Systolic dysfunction Rapidly progressing DCM	Rescued by pertussis toxin treatment[66]
Overexpression of PKA catalytic subunit[70]	LV dilation Systolic dysfunction Mild fibrosis Increased arrhythmias	May be involved in altered Ca^{2+} signaling and phospholamban phosphorylation in DCM
G-proteins		
Overexpression of $G_{s\alpha}$[73]	Systolic dysfunction Myocardial necrosis Myocardial fibrosis Increased arrhythmias Increased apoptosis	Supports involvement of apoptotic pathways in heart failure induced by heightened βAR signaling
Overexpression of $G_{\alpha q}$[75]	Cardiac hypertrophy Heart failure in males subjected to TAC[76] Peripartum heart failure with myocyte apoptosis in females[77]	Rescued by adenylyl cyclase type VI overexpression[79] Peripartum cardiomyopathy rescued by caspase inhibition[78]
Expression of Ro1[80]	Biventricular dilation Systolic dysfunction Premature death	Systolic function rescued by Ro1 suppression[80]
MAPK cascades		
Expression of MKK3bE-activated mutant[82]	LV dilation Systolic dysfunction Myocardial fibrosis	Activation of MAPK pathways can cause cardiac remodeling
Expression of constitutively active Rac1[84]	Severe heart failure Premature death	May be involved in coupling GPCR activation to MAPK pathways in DCM
Overexpression of RhoA[87]	LV dilation Myocardial fibrosis Bradycardia Nodal conduction defects	Model of heart failure associated with nodal dysfunction
Expression of PP2A dominant negative mutant[89]	LV dilation Systolic dysfunction	May be associated with altered MAPK and ERK signaling in heart failure

Continued

Table 39.1. *Genetic Models of DCM – continued*

Model	Phenotype	Notes
Cytokines		
Overexpression of TNF-α[92]	LV dilation Systolic dysfunction Myocardial fibrosis Increased apoptosis	Rescued by anti-TNF-α therapy[94] Rescued by matrix metalloprotease inhibition[93]
gp130 knockout[98]	Heart failure and increased apoptosis after TAC	IL-6 levels are elevated in human heart failure,[96] gp130 receptor overexpression can lead to cardiac hypertrophy[97]
Growth factors and receptors		
Overexpression of PDGF-C[99]	Myocardial fibrosis Systolic dysfunction and increased mortality in females Fibrosis and hypertrophy in males	PDGF-dependent mitogenic pathways may be involved in replacement fibrosis in DCM
VEGF-A knockout[100]	LV dilation Systolic dysfunction Wall thinning Microvascular changes	Model for cardiac dysfunction in the setting of prolonged ischemia
Overexpression of AT2 receptors[102]	Systolic and diastolic dysfunction LV dilation Wall thinning Myocardial fibrosis in high-expressing mice	AT2 receptors upregulated in human heart failure[101]
Overexpression of ET-1[104]	Inflammatory infiltrate Myocardial fibrosis Heart failure Premature death	ET-1 levels are increased in human heart failure[103]
Transcription factors		
Expression of CREB^{A133} dominant negative mutant[105]	Biventricular dilation Decreased systolic function Myocardial fibrosis Premature death	CREB is downregulated by chronic βAR stimulation in the rat[106]
Expression of activated calcineurin[108]	Biventricular dilation Cardiac hypertrophy Increase in sudden death	Rescued by cyclosporin A and FK-506 therapy[108]

Continued

Table 39.1. *Genetic Models of DCM – continued*

Model	Major phenotypic characteristics	Relevance
Apoptosis		
Conditional caspase-8 activation[111]	Biventricular dilation Myocardial fibrosis Increased apoptosis Death within 18 hours of caspase activation	Rescued by polycaspase inhibitor treatment[111]
Overexpression of Mst1[113]	LV dilationSystolic dysfunction Myocardial fibrosis Increased apoptosis	Increased apoptosis is noted in human heart failure[113]
14-3-3η knockout[114]	LV dilation Systolic dysfunction Increased apoptosis after TAC	May sensitize animals to stress-induced heart failure
Mitochondrial dysfunction		
Tfam knockout[115]	Heart failure Heart block Premature death	Model for human mitochondrial cardiomyopathies
DS rat[117]	Hypertension and hypertrophy on high-salt diet	Similar to human phenotype of hypertension and cardiomyopathy
SHR and SHHF rat[118]	Hypertension and heart failure by 18 (SHHF) or 24 (SHR) months	Similar to human diabetes mellitus type II phenotype
Expression of constitutively active retinoid receptor[120]	Biventricular dilation Systolic dysfunction Wall thinning	Retinoids are important in cardiac morphogenesis
Expression of polyomavirus large T-antigen[121]	LV dilation Systolic dysfunction	Disruption of cell growth pathways
Expression of Epstein–Barr virus nuclear antigen-leader protein[122]	LV dilation Systolic dysfunction	Disruption of cell growth pathways

Conclusion

Animal models have greatly aided us in understanding the molecular and physiological mechanisms of DCM. Each model examined (see Table 39.1) not only highlights a unique factor in DCM but, perhaps more importantly, also stresses the features common to several models, including dysregulation of adrenergic signaling, alteration of Ca^{2+} regulation, and sarcomeric dysfunction. Both genetic and non-genetic models of DCM have provided different advantages in the study of heart failure, and the two strategies have been combined with great success in the study of transgenic models which display attenuation of cardiac dysfunction in response to mechanical stressors such as TAC. These models have been critical in the study of the pathogenesis of DCM as well as the identification of new therapeutic targets. Hopefully, as more transgenic models and therapeutic strategies are developed, we will be better able to translate these findings to humans, and develop treatment strategies for patients.

References

1. Arber S, Halder G, Caroni P. Muscle LIM protein, a novel essential regulator of myogenesis, promotes myogenic differentiation. Cell 1994; 79: 221–31.

2. Arber S, Caroni P. Specificity of single LIM motifs in targeting and LIM/LIM interactions in situ. Genes Dev 1996; 10: 289–300.

3. Arber S, Hunter JJ, Ross J Jr. et al. MLP-deficient mice exhibit a disruption of cardiac cytoarchitectural organization, dilated cardiomyopathy, and heart failure. Cell 1997; 88: 393–403.

4. Mohapatra B, Jimenez S, Lin JH et al. Mutations in the muscle LIM protein and alpha-actinin-2 genes in dilated cardiomyopathy and endocardial fibroelastosis. Mol Genet Metab 2003; 80: 207–15.

5. Knoll R, Hoshijima M, Hoffman HM et al. The cardiac mechanical stretch sensor machinery involves a Z disc complex that is defective in a subset of human dilated cardiomyopathy. Cell 2002; 111: 943–55.

6. Geier C, Perrot A, Ozcelik C et al. Mutations in the human muscle LIM protein gene in families with hypertrophic cardiomyopathy. Circulation 2003; 107: 1390–5.

7. Rockman HA, Chien KR, Choi DJ et al. Expression of a beta-adrenergic receptor kinase 1 inhibitor prevents the development of myocardial failure in gene-targeted mice. Proc Natl Acad Sci USA 1998; 95: 7000–5.

8. Esposito G, Santana LF, Dilly K et al. Cellular and functional defects in a mouse model of heart failure. Am J Physiol Heart Circ Physiol 2000; 279: H3101–12.

9. Minamisawa S, Hoshijima M, Chu G et al. Chronic phospholamban-sarcoplasmic reticulum calcium ATPase interaction is the critical calcium cycling defect in dilated cardiomyopathy. Cell 1999; 99: 313–22.

10. Hongo M, Ryoke T, Schoenfeld J et al. Effects of growth hormone on cardiac dysfunction and gene expression in genetic murine dilated cardiomyopathy. Basic Res Cardiol 2000; 95: 431–41.

11. Niimura H, Bachinski LL, Sangwatanaroj S et al. Mutations in the gene for cardiac myosin-binding protein C and late-onset familial hypertrophic cardiomyopathy. N Engl J Med 1998; 338: 1248–57.

12. Gilbert R, Kelly MG, Mikawa T et al. The carboxyl terminus of myosin binding protein C (MyBP-C, C-protein) specifies incorporation into the A-band of striated muscle. J Cell Sci 1996; 109: 101–11.

13. Hartzell HC. Effects of phosphorylated and unphosphorylated C protein on cardiac actomyosin ATPase. J Mol Biol 1985; 186: 185–95.

14. McConnell BK, Jones KA, Fatkin D et al. Dilated cardiomyopathy in homozygous myosin-binding protein-C mutant mice. J Clin Invest 1999; 104: 1235–44.

15. Bonne G, Carrier L, Richard P et al. Familial hypertrophic cardiomyopathy: from mutations to functional defects. Circ Res 1998; 83: 580–93.

16. Rottbauer W, Gautel M, Zehelein J et al. Novel splice donor site mutation in the cardiac myosin-binding protein-C gene in familial hypertrophic cardiomyopathy. Characterization of cardiac transcript and protein. J Clin Invest 1997; 100: 475–82.

17. Sussman MA, Welch S, Cambon N et al. Myofibril degeneration caused by tropomodulin overexpression leads to dilated cardiomyopathy in juvenile mice. J Clin Invest 1998; 101: 51–61.

18. Vikstrom KL, Bohlmeyer T, Factor SM et al. Hypertrophy, pathology, and molecular markers of cardiac pathogenesis. Circ Res 1998; 82: 773–8.

19. Welch S, Plank D, Witt S et al. Cardiac-specific IGF-1 expression attenuates dilated cardiomyopathy in tropomodulin-overexpressing transgenic mice. Circ Res 2002; 90: 641–8.

20. Sussman MA, Lim HW, Gude N et al. Prevention of cardiac hypertrophy in mice by calcineurin inhibition. Science 1998; 281: 1690–3.

21. Crosbie RH, Lebakken CS, Holt KH et al. Membrane targeting and stabilization of sarcospan is mediated by the sarcoglycan subcomplex. J Cell Biol 1999; 145: 153–65.

22. Jung D, Yang B, Meyer J et al. Identification and characterization of the dystrophin anchoring site on beta-dystroglycan. J Biol Chem 1995; 270: 27305–10.

23. Sakamoto A, Ono K, Abe M et al. Both hypertrophic and dilated cardiomyopathies are caused by mutation of the same gene, delta-sarcoglycan, in hamster: an animal model of disrupted dystrophin-associated glycoprotein complex. Proc Natl Acad Sci USA 1997; 94: 13873–8.

24. Coral-Vazquez R, Cohn RD, Moore SA et al. Disruption of the sarcoglycan-sarcospan complex in vascular smooth muscle: a novel mechanism for cardiomyopathy and muscular dystrophy. Cell 1999; 98: 465–74.

25. Hack AA, Ly CT, Jiang F et al. Gamma-sarcoglycan deficiency leads to muscle membrane defects and apoptosis independent of dystrophin. J Cell Biol 1998; 142: 1279–87.

26. Tsubata S, Bowles KR, Vatta M et al. Mutations in the human delta-sarcoglycan gene in familial and sporadic dilated cardiomyopathy. J Clin Invest 2000; 106: 655–62.

27. Sylvius N, Duboscq-Bidot L, Bouchier C et al. Mutational analysis of the beta- and delta-sarcoglycan genes in a large number of patients with familial and sporadic dilated cardiomyopathy. Am J Med Genet 2003; 120A: 8–12.

28. Karkkainen S, Miettinen R, Tuomainen P et al. A novel mutation, Arg71Thr, in the delta-sarcoglycan gene is associated with dilated cardiomyopathy. J Mol Med 2003; 81: 795–800.

29. Coulton GR, Morgan JE, Partridge TA et al. The mdx mouse skeletal muscle myopathy: I. A histological, morphometric and biochemical investigation. Neuropathol Appl Neurobiol 1988; 14: 53–70.

30. Megeney LA, Kablar B, Perry RL et al. Severe cardiomyopathy in mice lacking dystrophin and MyoD. Proc Natl Acad Sci USA 1999; 96: 220–5.

31. Badorff C, Lee GH, Lamphear BJ et al. Enteroviral protease 2A cleaves dystrophin: evidence of cytoskeletal disruption in an acquired cardiomyopathy. Nat Med 1999; 5: 320–6.

32. Wessely R, Klingel K, Santana LF et al. Transgenic expression of replication-restricted enteroviral genomes in heart muscle induces defective excitation-contraction coupling and dilated cardiomyopathy. J Clin Invest 1998; 102: 1444–53.

33. Lee GH, Badorff C, Knowlton KU. Dissociation of sarcoglycans and the dystrophin carboxyl terminus from the sarcolemma in enteroviral cardiomyopathy. Circ Res 2000; 87: 489–95.

34. Milner DJ, Weitzer G, Tran D et al. Disruption of muscle architecture and myocardial degeneration in mice lacking desmin. J Cell Biol. 1996; 134: 1255–70.

35. Li D, Tapscoft T, Gonzalez O et al. Desmin mutation responsible for idiopathic dilated cardiomyopathy. Circulation 1999; 100: 461–4.

36. Calderwood DA, Shattil SJ, Ginsberg MH. Integrins and actin filaments: reciprocal regulation of cell adhesion and signaling. J Biol Chem 2000; 275: 22607–10.

37. Fassler R, Rohwedel J, Maltsev V et al. Differentiation and integrity of cardiac muscle cells are impaired in the absence of beta 1 integrin. J Cell Sci. 1996; 109: 2989–99.

38. Shai SY, Harpf AE, Babbitt CJ et al. Cardiac myocyte-specific excision of the beta1 integrin gene results in myocardial fibrosis and cardiac failure. Circ Res 2002; 90: 458–64.

39. Keller RS, Shai SY, Babbitt CJ et al. Disruption of integrin function in the murine myocardium leads to perinatal lethality, fibrosis, and abnormal cardiac performance. Am J Pathol 2001; 158: 1079–90.

40. Brancaccio M, Fratta L, Notte A et al. Melusin, a muscle-specific integrin beta1-interacting protein, is required to prevent cardiac failure in response to chronic pressure overload. Nat Med 2003; 9: 68–75.

41. Soler AP, Knudsen KA. N-cadherin involvement in cardiac myocyte interaction and myofibrillogenesis. Dev Biol 1994; 162: 9–17.

42. Ferreira-Cornwell MC, Luo Y, Narula N et al. Remodeling the intercalated disc leads to cardiomyopathy in mice misexpressing cadherins in the heart. J Cell Sci 2002; 115: 1623–34.

43. Cho MC, Rapacciuolo A, Koch WJ et al. Defective beta-adrenergic receptor signaling precedes the development of dilated cardiomyopathy in transgenic mice with calsequestrin overexpression. J Biol Chem 1999; 274: 22251–6.

44. Jones LR, Suzuki YJ, Wang W et al. Regulation of Ca^{2+} signaling in transgenic mouse cardiac myocytes overexpressing calsequestrin. J Clin Invest 1998; 101: 1385–93.

45. Wang W, Cleemann L, Jones LR et al. Modulation of focal and global Ca^{2+} release in calsequestrin-

overexpressing mouse cardiomyocytes. J Physiol 2000; 524: 399–414.

46. Harding VB, Jones LR, Lefkowitz RJ et al. Cardiac beta ARK1 inhibition prolongs survival and augments beta blocker therapy in a mouse model of severe heart failure. Proc Natl Acad Sci USA 2001; 98: 5809–14.

47. Le Corvoisier P, Park HY, Carlson KM et al. Multiple quantitative trait loci modify the heart failure phenotype in murine cardiomyopathy. Hum Mol Genet 2003; 12: 3097–107.

48. Suzuki M, Carlson KM, Marchuk DA et al. Genetic modifier loci affecting survival and cardiac function in murine dilated cardiomyopathy. Circulation 2002; 105: 1824–9.

49. Kubo H, Margulies KB, Piacentino V, 3rd et al. Patients with end-stage congestive heart failure treated with beta-adrenergic receptor antagonists have improved ventricular myocyte calcium regulatory protein abundance. Circulation 2001; 104: 1012–8.

50. Kadambi VJ, Ponniah S, Harrer JM et al. Cardiac-specific overexpression of phospholamban alters calcium kinetics and resultant cardiomyocyte mechanics in transgenic mice. J Clin Invest 1996; 97: 533–9.

51. Haghighi K, Schmidt AG, Hoit BD et al. Superinhibition of sarcoplasmic reticulum function by phospholamban induces cardiac contractile failure. J Biol Chem 2001; 276: 24145–

52. Schmitt JP, Kamisago M, Asahi M et al. Dilated cardiomyopathy and heart failure caused by a mutation in phospholamban. Science 2003; 299: 1410–3.

53. Hoshijima M, Ikeda Y, Iwanaga Y et al. Chronic suppression of heart-failure progression by a pseudophosphorylated mutant of phospholamban via in vivo cardiac rAAV gene delivery. Nat Med 2002; 8: 864–71.

54. Iwanaga Y, Hoshijima M, Gu Y et al. Chronic phospholamban inhibition prevents progressive cardiac dysfunction and pathological remodeling after infarction in rats. J Clin Invest 2004; 113: 727–36.

55. MacDougall LK, Jones LR, Cohen P. Identification of the major protein phosphatases in mammalian cardiac muscle which dephosphorylate phospholamban. Eur J Biochem 1991; 196: 725–34.

56. Braz JC, Gregory K, Pathak A et al. PKC-alpha regulates cardiac contractility and propensity toward heart failure. Nat Med 2004; 10: 248–54.

57. Baltas LG, Karczewski P, Krause EG. The cardiac sarcoplasmic reticulum phospholamban kinase is a distinct delta-CaM kinase isozyme. FEBS Lett 1995; 373: 71–5.

58. Kirchhefer U, Schmitz W, Scholz H et al. Activity of cAMP-dependent protein kinase and Ca^{2+}/calmodulin-dependent protein kinase in failing and nonfailing human hearts. Cardiovasc Res 1999; 42: 254–61.

59. Zhang T, Johnson EN, Gu Y et al. The cardiac-specific nuclear delta(B) isoform of Ca^{2+}/calmodulin-dependent protein kinase II induces hypertrophy and dilated cardiomyopathy associated with increased protein phosphatase 2A activity. J Biol Chem 2002; 277: 1261–7.

60. Zhang T, Maier LS, Dalton ND et al. The deltaC isoform of CaMKII is activated in cardiac hypertrophy and induces dilated cardiomyopathy and heart failure. Circ Res 2003; 92: 912–9.

61. Bristow MR, Ginsburg R, Minobe W et al. Decreased catecholamine sensitivity and beta-adrenergic-receptor density in failing human hearts. N Engl J Med 1982; 307: 205–11.

62. Engelhardt S, Hein L, Wiesmann F et al. Progressive hypertrophy and heart failure in beta1-adrenergic receptor transgenic mice. Proc Natl Acad Sci USA 1999; 96: 7059–64.

63. Milano CA, Allen LF, Rockman HA et al. Enhanced myocardial function in transgenic mice overexpressing the beta 2-adrenergic receptor. Science 1994; 264: 502–6.

64. Liggett SB, Tepe NM, Lorenz JN et al. Early and delayed consequences of beta(2)-adrenergic receptor overexpression in mouse hearts: critical role for expression level. Circulation 2000; 101: 1707–14.

65. Bond RA, Leff P, Johnson TD et al. Physiological effects of inverse agonists in transgenic mice with myocardial overexpression of the beta 2-adrenoceptor. Nature 1995; 374: 272–6.

66. Xiao RP, Avdonin P, Zhou YY et al. Coupling of beta2-adrenoceptor to Gi proteins and its physiological relevance in murine cardiac myocytes. Circ Res 1999; 84: 43–52.

67. Naga Prasad SV, Barak LS, Rapacciuolo A et al. Agonist-dependent recruitment of phosphoinositide 3-kinase to the membrane by beta-adrenergic receptor kinase 1. A role in receptor sequestration. J Biol Chem 2001; 276: 18953–9.

68. Naga Prasad SV, Laporte SA, Chamberlain D et al. Phosphoinositide 3-kinase regulates beta2-adrenergic receptor endocytosis by AP-2 recruitment to the receptor/beta-arrestin complex. J Cell Biol 2002; 158: 563–75.

69. Nienaber JJ, Tachibana H, Naga Prasad SV et al. Inhibition of receptor-localized PI3K preserves car-

diac beta-adrenergic receptor function and ameliorates pressure overload heart failure. J Clin Invest 2003; 112: 1067–79.

70. Antos CL, Frey N, Marx SO et al. Dilated cardiomyopathy and sudden death resulting from constitutive activation of protein kinase a. Circ Res 2001; 89: 997–1004.

71. Marx SO, Reiken S, Hisamatsu Y et al. PKA phosphorylation dissociates FKBP12.6 from the calcium release channel (ryanodine receptor): defective regulation in failing hearts. Cell 2000; 101: 365–76.

72. Rockman HA, Koch WJ, Lefkowitz RJ. Seven-transmembrane-spanning receptors and heart function. Nature 2002; 415: 206–12.

73. Gaudin C, Ishikawa Y, Wight DC et al. Overexpression of Gs alpha protein in the hearts of transgenic mice. J Clin Invest 1995; 95: 1676–83.

74. Iwase M, Bishop SP, Uechi M et al. Adverse effects of chronic endogenous sympathetic drive induced by cardiac GS alpha overexpression. Circ Res 1996; 78: 517–24.

75. D'Angelo DD, Sakata Y, Lorenz JN et al. Transgenic Galphaq overexpression induces cardiac contractile failure in mice. Proc Natl Acad Sci USA 1997; 94: 8121–6.

76. Sakata Y, Hoit BD, Liggett SB et al. Decompensation of pressure-overload hypertrophy in G alpha q-overexpressing mice. Circulation 1998; 97: 1488–95.

77. Adams JW, Sakata Y, Davis MG et al. Enhanced Galphaq signaling: a common pathway mediates cardiac hypertrophy and apoptotic heart failure. Proc Natl Acad Sci USA 1998; 95: 10140–5.

78. Hayakawa Y, Chandra M, Miao W et al. Inhibition of cardiac myocyte apoptosis improves cardiac function and abolishes mortality in the peripartum cardiomyopathy of Galpha(q) transgenic mice. Circulation 2003; 108: 3036–41.

79. Roth DM, Gao MH, Lai NC et al. Cardiac-directed adenylyl cyclase expression improves heart function in murine cardiomyopathy. Circulation 1999; 99: 3099–102.

80. Baker AJ, Redfern CH, Harwood MD et al. Abnormal contraction caused by expression of G(i)-coupled receptor in transgenic model of dilated cardiomyopathy. Am J Physiol Heart Circ Physiol 2001; 280: H1653–9.

81. Flesch M, Schwinger RH, Schnabel P et al. Sarcoplasmic reticulum Ca^{2+}ATPase and phospholamban mRNA and protein levels in end-stage heart failure due to ischemic or dilated cardiomyopathy. J Mol Med 1996; 74: 321–32.

82. Liao P, Georgakopoulos D, Kovacs A et al. The in

vivo role of p38 MAP kinases in cardiac remodeling and restrictive cardiomyopathy. Proc Natl Acad Sci USA 2001; 98: 12283–8.

83. Esposito G, Prasad SV, Rapacciuolo A et al. Cardiac overexpression of a G(q) inhibitor blocks induction of extracellular signal-regulated kinase and c-Jun NH(2)-terminal kinase activity in in vivo pressure overload. Circulation 2001; 103: 1453–8.

84. Sussman MA, Welch S, Walker A et al. Altered focal adhesion regulation correlates with cardiomyopathy in mice expressing constitutively active rac1. J Clin Invest 2000; 105: 875–86.

85. Manser E, Huang HY, Loo TH et al. Expression of constitutively active alpha-PAK reveals effects of the kinase on actin and focal complexes. Mol Cell Biol 1997; 17: 1129–43.

86. Mackay DJ, Hall A. Rho GTPases. J Biol Chem 1998; 273: 20685–8.

87. Sah VP, Minamisawa S, Tam SP et al. Cardiac-specific overexpression of RhoA results in sinus and atrioventricular nodal dysfunction and contractile failure. J Clin Invest 1999; 103: 1627–34.

88. Liu Q, Hofmann PA. Protein phosphatase 2A-mediated cross-talk between p38 MAPK and ERK in apoptosis of cardiac myocytes. Am J Physiol Heart Circ Physiol 2004.

89. Brewis N, Ohst K, Fields K et al. Dilated cardiomyopathy in transgenic mice expressing a mutant A subunit of protein phosphatase 2A. Am J Physiol Heart Circ Physiol 2000; 279:H1307–18.

90. Levine B, Kalman J, Mayer L et al. Elevated circulating levels of tumor necrosis factor in severe chronic heart failure. N Engl J Med 1990; 323: 236–41.

91. Feldman AM, Combes A, Wagner D et al. The role of tumor necrosis factor in the pathophysiology of heart failure. J Am Coll Cardiol 2000; 35: 537–44.

92. Kubota T, McTiernan CF, Frye CS et al. Dilated cardiomyopathy in transgenic mice with cardiac-specific overexpression of tumor necrosis factor-alpha. Circ Res 1997; 81: 627–35.

93. Li YY, Kadokami T, Wang P et al. MMP inhibition modulates TNF-alpha transgenic mouse phenotype early in the development of heart failure. Am J Physiol Heart Circ Physiol 2002; 282:H983–9.

94. Kubota T, Bounoutas GS, Miyagishima M et al. Soluble tumor necrosis factor receptor abrogates myocardial inflammation but not hypertrophy in cytokine-induced cardiomyopathy. Circulation 2000; 101: 2518–25.

95. Chung ES, Packer M, Lo KH et al. Randomized, double-blind, placebo-controlled, pilot trial of inflix-

imab, a chimeric monoclonal antibody to tumor necrosis factor-alpha, in patients with moderate-to-severe heart failure: results of the anti-TNF Therapy Against Congestive Heart Failure (ATTACH) trial. Circulation 2003; 107: 3133–40.

96. Testa M, Yeh M, Lee P et al. Circulating levels of cytokines and their endogenous modulators in patients with mild to severe congestive heart failure due to coronary artery disease or hypertension. J Am Coll Cardiol 1996; 28: 964–71.

97. Hirota H, Yoshida K, Kishimoto T et al. Continuous activation of gp130, a signal-transducing receptor component for interleukin 6-related cytokines, causes myocardial hypertrophy in mice. Proc Natl Acad Sci USA 1995; 92: 4862–6.

98. Hirota H, Chen J, Betz UA et al. Loss of a gp130 cardiac muscle cell survival pathway is a critical event in the onset of heart failure during biomechanical stress. Cell 1999; 97: 189–98.

99. Ponten A, Li X, Thoren P et al. Transgenic overexpression of platelet-derived growth factor-C in the mouse heart induces cardiac fibrosis, hypertrophy, and dilated cardiomyopathy. Am J Pathol 2003; 163: 673–82.

100. Giordano FJ, Gerber HP, Williams SP et al. A cardiac myocyte vascular endothelial growth factor paracrine pathway is required to maintain cardiac function. Proc Natl Acad Sci USA 2001; 98: 5780–5.

101. Tsutsumi Y, Matsubara H, Ohkubo N et al. Angiotensin II type 2 receptor is upregulated in human heart with interstitial fibrosis, and cardiac fibroblasts are the major cell type for its expression. Circ Res 1998; 83: 1035–46.

102. Yan X, Price RL, Nakayama M et al. Ventricular-specific expression of angiotensin II type 2 receptors causes dilated cardiomyopathy and heart failure in transgenic mice. Am J Physiol Heart Circ Physiol 2003; 285: H2179–87.

103. Stewart DJ, Cernacek P, Costello KB et al. Elevated endothelin-1 in heart failure and loss of normal response to postural change. Circulation 1992; 85: 510–7.

104. Yang LL, Gros R, Kabir MG et al. Conditional cardiac overexpression of endothelin-1 induces inflammation and dilated cardiomyopathy in mice. Circulation 2004; 109: 255–61.

105. Fentzke RC, Korcarz CE, Lang RM et al. Dilated cardiomyopathy in transgenic mice expressing a dominant-negative CREB transcription factor in the heart. J Clin Invest 1998; 101: 2415–26.

106. Muller FU, Boknik P, Horst A et al. cAMP response element binding protein is expressed and phosphory-lated in the human heart. Circulation 1995; 92: 2041–3.

107. Flanagan WM, Corthesy B, Bram RJ et al. Nuclear association of a T-cell transcription factor blocked by FK-506 and cyclosporin A. Nature 1991; 352: 803–7.

108. Molkentin JD, Lu JR, Antos CL et al. A calcineurin-dependent transcriptional pathway for cardiac hypertrophy. Cell 1998; 93: 215–28.

109. Olson EN, Molkentin JD. Prevention of cardiac hypertrophy by calcineurin inhibition: hope or hype? Circ Res 1999; 84: 623–32.

110. Olivetti G, Abbi R, Quaini F et al. Apoptosis in the failing human heart. N Engl J Med 1997; 336: 1131–41.

111. Wencker D, Chandra M, Nguyen K et al. A mechanistic role for cardiac myocyte apoptosis in heart failure. J Clin Invest 2003; 111: 1497–504.

112. Lee KK, Ohyama T, Yajima N et al. MST, a physiological caspase substrate, highly sensitizes apoptosis both upstream and downstream of caspase activation. J Biol Chem 2001; 276: 19276–85.

113. Yamamoto S, Yang G, Zablocki D et al. Activation of Mst1 causes dilated cardiomyopathy by stimulating apoptosis without compensatory ventricular myocyte hypertrophy. J Clin Invest 2003; 111: 1463–74.

114. Xing H, Zhang S, Weinheimer C et al. 14-3-3 proteins block apoptosis and differentially regulate MAPK cascades. EMBO J 2000; 19: 349–58.

115. Wang J, Wilhelmsson H, Graff C et al. Dilated cardiomyopathy and atrioventricular conduction blocks induced by heart-specific inactivation of mitochondrial DNA gene expression. Nat Genet 1999; 21: 133–7.

116. Hubner G, Gokel JM, Pongratz D et al. Fatal mitochondrial cardiomyopathy in Kearns-Sayre syndrome. Virchows Arch A Pathol Anat Histopathol 1986; 408: 611–21.

117. Inoko M, Kihara Y, Morii I et al. Transition from compensatory hypertrophy to dilated, failing left ventricles in Dahl salt-sensitive rats. Am J Physiol 1994; 267: H2471–82.

118. McCune SA, Park S, Radin MJ et al. SHHF/Mcc-facp rat model: a genetic model of congestive heart failure. In: Singal PK, ed. Mechanisms of heart failure. Boston: Kluwer Academic Publishers; 1995: xxiv, 453.

119. Gomez AM, Valdivia HH, Cheng H et al. Defective excitation-contraction coupling in experimental cardiac hypertrophy and heart failure. Science 1997; 276: 800–6.

120. Colbert MC, Hall DG, Kimball TR et al. Cardiac

compartment-specific overexpression of a modified retinoic acid receptor produces dilated cardiomyopathy and congestive heart failure in transgenic mice. J Clin Invest 1997; 100: 1958–68.

121. Chalifour LE, Gomes ML, Wang NS et al. Polyomavirus large T-antigen expression in heart of transgenic mice causes cardiomyopathy. Oncogene 1990; 5: 1719–26.

122. Huen DS, Fox A, Kumar P et al. Dilated heart failure in transgenic mice expressing the Epstein-Barr virus nuclear antigen-leader protein. J Gen Virol. 1993; 74: 1381–91.

123. Rockman HA, Ross RS, Harris AN et al. Segregation of atrial-specific and inducible expression of an atrial natriuretic factor transgene in an in vivo murine model of cardiac hypertrophy. Proc Natl Acad Sci USA 1991; 88: 8277–81.

124. Esposito G, Rapacciuolo A, Naga Prasad SV et al. Genetic alterations that inhibit in vivo pressure-overload hypertrophy prevent cardiac dysfunction despite increased wall stress. Circulation 2002; 105: 85–92.

125. Akhter SA, Luttrell LM, Rockman HA et al. Targeting the receptor-Gq interface to inhibit in vivo pressure overload myocardial hypertrophy. Science 1998; 280: 574–7.

126. Kajstura J, Zhang X, Reiss K et al. Myocyte cellular hyperplasia and myocyte cellular hypertrophy contribute to chronic ventricular remodeling in coronary artery narrowing-induced cardiomyopathy in rats. Circ Res 1994; 74: 383–400.

127. Pfeffer JM, Pfeffer MA, Fletcher PJ et al. Progressive ventricular remodeling in rat with myocardial infarction. Am J Physiol Heart Circ Physiol 1991; 260: H1406–14.

128. Patten RD, Aronovitz MJ, Deras-Mejia L et al. Ventricular remodeling in a mouse model of myocar-

dial infarction. Am J Physiol Heart Circ Physiol 1998; 274: H1812–20.

129. McCullagh WH, Covell JW, Ross J, Jr. Left ventricular dilatation and diastolic compliance changes during chronic volume overloading. Circulation 1972; 45: 943–51.

130. Kleaveland JP, Kussmaul WG, Vinciguerra T et al. Volume overload hypertrophy in a closed-chest model of mitral regurgitation. Am J Physiol Heart Circ Physiol 1988; 254: H1034–41.

131. Magid NM, Opio G, Wallerson DC et al. Heart failure due to chronic experimental aortic regurgitation. Am J Physiol Heart Circ Physiol 1994; 267: H556–62.

132. Tanaka N, Dalton N, Mao L et al. Transthoracic echocardiography in models of cardiac disease in the mouse. Circulation 1996; 94: 1109–17.

133. Armstrong PW, Stopps TP, Ford SE et al. Rapid ventricular pacing in the dog: pathophysiologic studies of heart failure. Circulation 1986; 74: 1075–84.

134. Matsumori A, Kawai C. An animal model of congestive (dilated) cardiomyopathy: dilatation and hypertrophy of the heart in the chronic stage in DBA/2 mice with myocarditis caused by encephalomyocarditis virus. Circulation 1982; 66: 355–60.

135. Arnolda L, McGrath B, Cocks M et al. Adriamycin cardiomyopathy in the rabbit: an animal model of low output cardiac failure with activation of vasoconstrictor mechanisms. Cardiovasc Res 1985; 19: 378–82.

136. Hanai K, Takaba K, Manabe S et al. Evaluation of cardiac function by echocardiography in dogs treated with doxorubicin. J Toxicol Sci 1996; 21: 1–10.

137. Mettler FP, Young DM, Ward JM. Adriamycin-induced cardiotoxicity (cardiomyopathy and congestive heart failure) in rats. Cancer Res 1977; 37: 2705–13.

CHAPTER 40

Clinical implications for molecular genetics of cardiomyopathy

Christine E Seidman, J G Seidman, Barry J Maron

Introduction

Cardiomyopathies are primary disorders of the myocardium that are classified as dilated, hypertrophic, or restrictive, according to the changes produced in cardiac morphology and physiology. Cardiomyopathy is an important cause of heart failure, and the most common diagnosis in patients referred for cardiac transplantation. Both adverse clinical outcomes and diminished survival from cardiomyopathy have provided considerable anatomical and histopathological information about these disorders over the past several decades. In concert with hemodynamic profiling, investigations have defined profound structural remodeling and functional changes that reflect both adaptive and maladaptive responses to triggers of cardiomyopathy. While such studies have defined natural history and provided important insights that shape management, these data have yielded few insights into the causes of primary cardiomyopathies. However, with the recognition of familial incidence and the delineation of Mendelian patterns of transmission of primary cardiomyopathies, the strategies used to study these intriguing disorders shifted from pathophysiology to human molecular genetics. Definition of autosomal dominant, recessive and X-linked inheritance of cardiomyopathies has fostered genome-wide mapping studies and, more recently, the discovery of cardiomyopathy genes. This productive line of research has caused

a fundamental change in conventional paradigms about human cardiomyopathies: these 'idiopathic' disorders are now recognized as resulting from heritable gene mutations.

For the researcher, fundamental discovery of disease-causing mutations has provided unparalleled opportunities to uncover the mechanisms by which genetic defects trigger cardiac remodeling.[1] Molecular engineering of human mutations into animal models provides the reagents necessary to examine signals evoked by gene mutations that produce myocardial responses that evolve over many years in human patients, but within only months in mouse models (see Chapters 37 and 39). Concurrent with this progress to genetically manipulation mice, came miniaturization of cardiac investigational devices, so that comprehensive analyses of morphological, hemodynamic, and electrophysiological functions in mice could provide longitudinal assessment of the consequence of human mutations. When coupled with molecular profiling, these genetically engineered animal models of human cardiomyopathy are important resources for elucidating mechanisms of disease, and hold great promise for enabling evaluations of novel therapeutic targets to attenuate or prevent disease development.

The benefits for elucidating the molecular basis for human cardiomyopathies are not limited to researchers. Indeed, for clinicians, knowledge about the genetic causes of human cardiomyopathy is of immediate and practical value. Precise information about genetic causes has considerable ramifications for diagnosis, profoundly changes screening strategies for at-risk families' members, and increasingly provides insights into prognosis that may alter patient management. While translation of the discovery of molecular causes of human cardiomyopathies into robust clinical tools remains an evolving effort, current strategies are appropriate for integration of this important molecular information into standards of clinical practice.

Like many advances in cardiology, molecular genetics methodologies, when applied to cardiomyopathy, are both technically demanding and costly. The expense of gene-based diagnosis reflects both biological and technological issues. Molecular discoveries of the causes of human cardiomyopathy, even when classified into the major subtypes of hypertrophic or dilated pathologies, indicate that these are each heterogeneous groups of genetic disorders. For example, although hypertrophic cardiomyopathy (HCM) is usually regarded as a discrete clinical entity, ten different HCM disease genes (see Chapter 30) are now known: β and α myosin heavy chain, α tropomyosin, cardiac actin, troponin I and T, essential and regulatory myosin light chains, myosin-binding protein-C and titin. While mutations in the β myosin heavy chain, myosin-binding protein-C, cardiac troponin I, or cardiac troponin T account for over 50% of HCM cases, even in these four HCM genes hundreds of distinct human mutations have been reported. While the diversity of HCM disease genes can be understood by appreciating the related function of the encoded constituent thin and thick sarcomere filaments proteins, the myriad of genetic causes for other cardiomyopathies remain an enigma. Dilated cardiomyopathy (DCM) can arise by human mutations in genes (see Chapters 31–34) that encode proteins involved in many distinct myocyte processes, including contractile force production (β myosin heavy chain cardiac myosin, troponin T, α tropomyosin), force transmission (titin, actin, desmin and sarcoglycan proteins), calcium cycling (phospholamban, cardiac ryanodine receptor) as well as proteins of unknown function (lamin A/C and a transcriptional co-activator). In addition to the over 15 known disease-causing DCM genes,

there are multiple linkage studies that have defined a chromosome location but not yet the gene responsible for DCM. Collectively these data indicate that the genetics of DCM will be exceedingly complex.[2]

This biological heterogeneity has considerable impact on the translation of basic discoveries into clinical ascertainment of genetic cause of human cardiomyopathies. First, without information that specifically implicates one or even a subset of disease genes, identification of the causal mutation requires the study of many genes. This is not a trivial undertaking, since most cardiomyopathy genes are both large and organized into exons that span thousands of base pairs (bp); for example the β myosin heavy chain gene is encoded by 6008 bp and is organized into 40 exons that span over 16 000 bp.[3] Second, most cardiomyopathies are dominant disorders (Fig. 40.1 A and C),[4] and affected individuals are heterozygous, with one normal gene and one mutant gene. As such DNA analyses must detect a mutation in the presence of wild-type sequence (Fig. 40.1 B and D). Third a single nucleotide substitution in a disease gene is often sufficient to cause cardiomyopathy. Collectively these issues indicate that genetic technologies to define human cardiomyopathy mutations must allow high throughput analyses of multiple large genes and be sufficiently robust to detect a single nucleotide change in one of two gene copies.

While several indirect methodologies have been developed to increase the efficiency of mutation screens in multiple genes, direct DNA sequence analyses remains the gold standard for accurate mutation identification. Early DNA sequence technologies used high-specific activity radioactive reagents that only allowed determination of tens to a few hundred base pairs of sequence in a single experiment. Subsequent to the development of di-deoxy chain termination reactions, and production of fluorescent-labeled nucleotides, the safety, efficiency and capacity for rapid DNA sequence determination has soared.[5] Recent modifications that employ capillary electrophoresis allow the determination of 800–1000 bp of DNA sequence in a single reaction at the cost of only about $5.

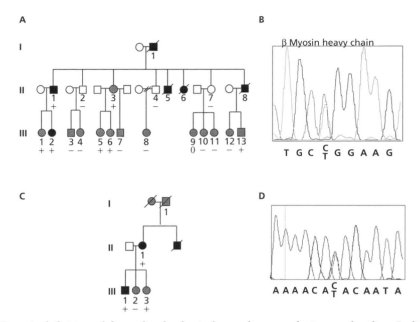

Figure 40.1. *Genetic definition of the molecular basis for cardiomyopathy in two families. Pedigrees indicated sex (circles denote women, boxes denote men) and clinical status (solid fill, affected; open, unaffected, hatched uncertain status). Genotypes are provided: + mutation present; – mutation absent. (A) Sudden death in multiple individuals (I-1, II-5, -6, -8) prompted clinical evaluation of surviving family members. Clinical diagnosis was uncertain in II-3 who had prior thoracic surgery and poor imaging studies, and in younger family members (generation IV). Genetic analyses of affected individual II-1 defined a disease-causing mutation that allowed definitive genotype assignment of all family members. Note that genotype clarified diagnosis of II-3 and identified individuals at risk death from HCM. (B) An arginine to glutamine substitution at residue 719 in the β myosin heavy chain gene causes HCM. DNA traces show both the normal and mutant myosin sequences. (C) A small family with cardiac hypertrophy (II-1 and III-1) and electrographic abnormalities (II-1, first degree atrioventricular block and bradycardia; III-1 ventricular pre-excitation pattern). Pedigree structure is consistent with dominant inheritance. (D) DNA analyses of the PRKAG2 gene revealed a tyrosine to histidine missense mutation in residue 477, indicating glycogen storage as the cause of cardiac hypertrophy. Individual II-1 subsequently developed conduction system disease necessitating pacemaker implantation. See color plate section.*

Still newer advances are on the horizon. For example, adaptation of nanotechnologies allow *in situ* amplification and contact replication of multiple individual DNA molecules, coupled with fluorescent *in situ* sequencing, so as to provide high-fidelity nucleic acid sequence analyses of multiple genes,[6] at costs that are markedly reduced from traditional approaches. This innovative methodology and others are ideally suited to the genetic complexity of human cardiomyopathies, and should enhance the diagnostic and investigative tools available to cardiovascular clinicians and researchers.

The intersection of research aimed at defining the molecular causes of cardiomyopathies and advances in DNA sequencing platforms, makes information about the precise gene mutation that causes cardiomyopathy in a single patient an emerging and powerful clinical tool. It is therefore timely to review the implication of molecular diagnostics on clinical management of genetic cardiomyopathies, using HCM as an example. Today, DNA sequence analyses of sarcomere protein genes to define the disease-causing mutation in HCM can be performed from a peripheral blood sample of 7 ml from an affected individual, in less than one month.[7] Because strategies used for HCM are likely to be broadly applicable to other cardiomyopathies, clinicians should expect that molecular analyses of other important genetic cardiovascular disorders will also be available in the near future.

Genetic information: implication for diagnosis

DNA analyses that define a mutation in a known disease gene provide unequivocal diagnosis, even when clinical findings are ambiguous, or when interpretation of diagnostic information is confounded by co-existing conditions. In HCM, the benefits for gene-based diagnosis are considerable. With the prevalent use of non-invasive imaging, left ventricular hypertrophy (LVH) is commonly identified by echocardiography (Fig. 40.2 A and B). In the absence of another cardiac disease capable of producing the magnitude of LVH that is evident

(e.g. systemic hypertension or aortic valvular stenosis), abnormally increased LV wall thickness (usually ≥ 15 mm) in association with non-dilated ventricular chambers represents the contemporary clinical diagnosis for HCM.[8] Borderline LV wall thicknesses of 13 mm or 14 mm in an adult patient are regarded as possible evidence of HCM, but often elicit a differential diagnosis with the physiologically based athlete's heart or systemic hypertension, which may be difficult to resolve on clinical grounds alone.[9,10] The frequency with which unexplained LVH is identified,[11,12] approximately 0.2% (1 in 500) of young adults in the general population as assessed in a recent epidemiological studies, indicates the potential magnitude for an HCM diagnosis to be 500 000 in the United States.[12] An alternative interpretation is that a definitive clinical diagnosis of HCM is difficult. Demonstration that an individual has a sarcomere gene mutation provides irrefutable evidence for the diagnosis of HCM (Fig. 40.1B).

Clinical diagnosis that combines imaging

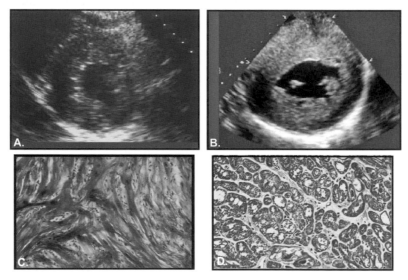

Figure 40.2. *Genetic causes of hypertrophy: HCM and PRKAG2 cardiomyopathy. Comparison of cross-sectional echocardiogram images in HCM (A) and PRKAG2 cardiomyopathy (B). Note the comparable and significant LVH in each. Images were obtained late in diastole. (C) Mason trichrome stain of LV tissue reveals classifc HCM histopathology with myocyte (red) enlargement and disarray, and also significant interstitial fibrosis (blue). (D) Histopathology in PRKAG2 cardiomyopathy also shows myocyte hypertrophy but without disarray and with prominent vacuoles that appear empty, due to the loss of glycogen during tissue fixation. Note the minimal interstitial fibrosis compared to HCM specimen (C). See color plate section for (C) and (D).*

studies with personal family history certainly improves diagnostic accuracy in HCM, but even in the setting of heritable cardiac hypertrophy, genetic analyses have an important role. Recently, other genes responsible for inherited LVH with the clinical presentation of HCM (Fig. 40.2B) have also been identified. Unlike previously described sarcomere protein mutations, these defects alter proteins involved in cardiac metabolism. The first of these is in the gene encoding the gamma-2-regulatory subunit of the AMP-activated protein kinase (*PRKAG2*).[13–15] Clinical manifestations include familial LVH in association with electrophysiological manifestations. Early in the course of disease, ventricular pre-excitation is present, but over time the conduction system disease progresses and patients often develop complete atrioventricular block, necessitating pacemaker implantation. Remarkably, the histopathology of patients harboring *PRKAG2* mutations can be distinguished from those with sarcomere protein gene mutations (Fig. 40.2 C and D) by the absence of typical HCM manifestations of myocyte disarray, and by the presence of distinct glycogen accumulations within myocytes.[15] While these data imply that examination of myocardial tissue could differentiate between these two heritable forms of cardiac hypertrophy, practical issues (particularly patient risk) limit the appropriateness for cardiac biopsy to improve clinical diagnosis. Genetic analyses that define either *PRKAG2* or a sarcomere mutation (Fig. 40.1B and D) provide a non-invasive resolution of this dilemma in differential diagnosis.

Attention has recently focused on two other genetic causes of cardiac hypertrophy. Two genes α-galactosidase (*GLA*) and the lysosome-associated membrane protein 2 (*LAMP2*) that are encoded on the X chromosome are recognized to cause systemic human disorders, Fabry disease,[16,17] and Danon disease,[18] respectively. Remarkably, some mutations in *GLA*[19] and *LAMP2*[20] can also produce predominantly cardiac disease, with clinical and morphological manifestations that mimic HCM. Male gender and the absence of family history are typical in both Fabry and Danon disease, but these are insufficient parameters to be clinically useful in the differential diagnosis. Some women who carry mutations in these genes also develop cardiac

hypertrophy, and indeed may have far fewer systemic manifestations than men. The ages of onset for cardiac hypertrophy in *GLA* and *LAMP2* mutations overlaps with that observed in HCM caused by sarcomere gene mutations. Mutations in *GLA* most often result in late-onset hypertrophy, as do myosin binding-protein C mutations.[21,22] *LAMP2* mutations cause cardiac hypertrophy with onset during childhood,[23] a scenario that might suggest HCM due to a β myosin heavy chain gene mutation; however findings of massive hypertrophy (left ventricular (LV) wall thickness over 30 mm) with electrophysiological abnormalities of ventricular pre-excitation are more likely to indicate a *LAMP2* gene defect.[20] The histopathology of both *GLA* and *LAMP2* gene defects shows glycogen storage in myocytes, thus suggesting a common mechanism may cause hypertrophy in the Fabry disease, Danon disease and PRKAG2 cardiomyopathy. Molecular diagnosis establishes whether unexplained hypertrophy is HCM and reflects a sarcomere gene mutation, or a glycogen storage cardiomyopathy and caused by *PRKAG2*, *GLA*, or *LAMP2* mutation.

Establishing a precise diagnosis of HCM or of a metabolic cardiomyopathy is not only of academic interest, but is important for patient management. Effective enzyme replacement in Fabry disease is currently available, and improves cardiac function; presumably enzyme therapy would also benefit patients who have predominantly cardiac manifestations of *GLA* mutations. While specific treatments are not available for most other genetic causes of cardiac hypertrophy, recognition of the different clinical courses associated with HCM versus glycogen storage cardiomyopathy underscores the need for accurate molecular diagnosis. Despite an increased risk for sudden death in HCM,[8] the natural history of sarcomere gene mutations often enables long-term survival,[24] with progression to severe heart failure occurring in only a small subset of individuals. Prognosis in *PRKAG2* mutations is also good, with appropriate longitudinal assessment and management of progressive conduction system disease.[13–15] In contrast, *LAMP2* mutations that cause predominantly cardiac Danon disease[20] predispose to progressive heart failure early in adult life, with current management therapies limited to transplantation. For physicians to provide appropriate counseling to patients and families, knowledge of the genetic cause of cardiac hypertrophy is essential.

Genetic information: implication for risk stratification

The second clinical advance provided by gene-based definition of causes in HCM relates to the important issue of stratification of risk for future disease complications. Sudden and unexpected death is the most common mode of demise, and the most devastating and unpredictable complication of HCM.[8,24-27] Within the broad disease spectrum of HCM, overall annual mortality rate is about 1%,[24] however high-risk subsets at a much greater risk (perhaps 5% per year or more),[25-27] constitute an important minority of all HCM patients. The family pedigree in Fig. 40.1A illustrates this high-risk subset. Historically, a complex challenge has been the identification of such patients,[25,28,29] in advance of life-threatening events. For example, sudden death can be the initial clinical manifestation of HCM, often in the absence of prior symptoms.[8] Sudden death occurs most commonly in children and young adults aged ≤ 30 years,[30] but uncommonly in infants and young children aged less than 12 years.[31] Risk for sudden death extends across a wide range of years, through mid-life and beyond;[30] indeed achieving a particular age does not itself confer immunity to sudden catastrophe. Sudden death occurs most commonly during mild exertion or sedentary activities, but not infrequently is related to vigorous physical exertion.[32] Indeed, HCM is the most common cause of cardiovascular sudden death in young people, including trained competitive athletes – most commonly in basketball and football and frequently in African-Americans.[33,34]

The majority of HCM patients (i.e. 55%) do not demonstrate any of the acknowledged risk factors in this disease, and it is exceedingly uncommon for such patients to die suddenly;[24,35] the subset at increased risk appears to comprise about 10–20% of the patient population.[37] Highest risk for sudden death in HCM has been associated with any of the following non-invasive clinical markers:[8,28,28] (1) prior cardiac arrest or spontaneous

sustained ventricular tachycardia (VT); (2) family history of premature HCM-related death, particularly if sudden, in close relatives, or multiple (Fig. 40.1A); (3) syncope, particularly when exertional or recurrent, in young patients or, when documented, arrhythmia-based or clearly unrelated to neurocardiogenic mechanisms; (4) multiple and repetitive, or prolonged, bursts of non-sustained VT on serial ambulatory (Holter) electrocariogram (ECG) recording; (5) hypotensive blood pressure response to exercise, particularly in patients <50 years old; and (6) extreme (massive) LV hypertrophy with maximum wall thickness ≥ 30 mm, of particular relevance in adolescents and young adults.

Presentation of HCM in young children, while uncommon, nevertheless usually creates a clinical dilemma often due to fortuitous diagnosis so early in life, because of the uncertainty regarding future risk over such long periods of time.[8,24,31] The role of invasive strategies such as electrophysiological testing with programmed ventricular stimulation (to provoke arrhythmias) in detecting the substrate for VF in individual HCM patients is unresolved, and has been largely abandoned.[8,28] Limitations include the infrequency with which monomorphic VF is provoked and non-specificity of rapid polymorphic VT and ventricular fibrillation (VF), as well as the uncertainty as to how frequently re-examination is necessary over the long risk period characteristic of HCM patients.[8]

No single risk factor has yet been defined that can identify all HCM patients at unacceptable risk for a lethal arrhythmia or heart failure death. Yet the recognition of HCM patients who are at particularly high risk for sudden death has become particularly relevant in the last several years, given the application of the implantable cardioverter-defibrillator to selected patients with this disease, for the primary (and secondary) prevention of sudden death.[8,36] Based on genotype–phenotype correlations, knowledge of the specific mutation that causes HCM represents another stratifying marker for sudden death and heart failure risk. Some gene defects convey either favorable or adverse prognosis (i.e. high- and low-risk mutations designated in the literature as 'malignant' or 'benign').[25] For example, a formulated hypothesis suggests that some β-myosin heavy chain mutations (e.g. Arg403Gln and Arg719Gln; Fig. 40.1B) and some troponin-T mutations may be associated with higher frequency of premature death and/or heart failure, while other mutations including many defects in myosin-binding protein C convey a more favorable prognosis, and are frequently associated with normal life expectancy.[21,37] While caution is warranted before drawing strong conclusions regarding prognosis based solely on the limited genetic epidemiological data (that may be skewed by virtue of patient selection bias toward high-risk families), inclusion of mutation data in the overall clinical assessment of risk has merit. Acquisition of considerably more genotype–phenotype data in HCM will probably expand this clinical application.

Genetic information: implications for family screening

The value of gene-based diagnosis extends well beyond the single affected patient. While diagnosis of a primary cardiomyopathy (e.g. hypertrophic, dilated or restrictive) already warrants clinical screening of family members, appropriate evaluations also require longitudinal investigations by cardiologists, internists and pediatricians over many years. A traditional clinical strategy for screening relatives in HCM families calls for such evaluations on a 12- to 18-month basis, usually beginning at about age 12 years. In the event that such investigations do not show evidence of the HCM phenotype with LVH (one or more LV segments with abnormally increased wall thickness) by the time full growth and maturation is achieved at the age of about 18 to 21 years, it was previous customary practice to conclude that a HCM-mutant gene was probably absent. However more recent recognition of delayed, late-onset LVH in adult relatives has unavoidably led to a new proposed paradigm for family screening. Therefore, clinical evaluations for HCM in relatives may need to be extended past adolescence and into midlife.[38,39] Cost, inconvenience and inadequate commitment by both family members and the physicians who care for them, often results in incomplete and inadequate screening.

Abnormal 12-lead ECG patterns have been demonstrated before LVH is detectable on echocardiogram, thereby potentially providing early clinical evidence for mutant HCM genes. However, while a distinctly abnormal ECG can in some cases be regarded as a surrogate clinical marker for the HCM phenotype,[38,40] relatively minor ECG alterations often represent non-specific or normal variants completely unrelated to HCM. In addition, recent imaging studies with reduced load-dependent tissue Doppler echocardiography provide evidence that diastolic dysfunction may also precede the appearance of the HCM phenotype on echocardiogram.[41,42] Nevertheless, in the absence of LV hypertrophy, neither the scalar ECG nor tissue Doppler echocardiography convey sufficient sensitivity or specificity as reliable diagnostic alternatives to laboratory DNA testing.

By definitive ascertainment of genotype, regardless of age at ascertainment, DNA-based laboratory diagnosis can resolve clinical ambiguities in diagnosis. Sarcomere gene mutations, like most cardiomyopathy disease genes, are inherited as dominant traits (Fig. 40.1A). As such, both male and female offspring of an affected individual are at risk for disease development. In contrast, the finding that a family member does not carry a gene mutation eliminates the potential risk of HCM in all offspring. Genetic screening of families with HCM and other genetic cardiomyopathies is therefore likely to recognize affected individuals who would otherwise be unaware of their disease status. Those with clinical manifestations require appropriate risk assessment and management. Those without clinical manifestations of HCM or another cardiomyopathy require longitudinal clinical evaluation.

Gene-based diagnosis has particular relevance for screening relatives in high-risk families in which HCM-related sudden deaths may have already occurred (Fig. 40.1A). It should be emphasized, however, that while sudden death events occurring in HCM in advance of obvious phenotypic expression have been reported, at this time they appear to be quite rare.[8] While conclusive data linking genotype in the absence of phenotype to any particular prognosis are lacking, longitudinal follow-up of this newly recognized emerging subset of genotype-positive, phenotype-negative (Fig. 40.1A individuals III-1, -5, -6, -13) population is essential, and limitations on competitive athletic participation is probably prudent.

Concluding perspectives

Definition of the disease-causing mutation in a clinically affected individual requires technically demanding DNA analyses that are labor intensive and expensive. To make this investment worthwhile, the probability for mutation identification should be high. In HCM and several other forms of inherited cardiac hypertrophy, definition of the spectrum of disease genes is currently sufficient for genetic analyses to be productive. As such, DNA-based diagnosis of the genetic causes of LVH should no longer be viewed as a research endeavor, but rather as a robust clinical tool. In contrast, genetic analyses of DCM remain almost exclusively within research laboratories that are involved in the discovery of specific disease genes. While substantially more effort is required to define the full spectrum of cardiomyopathy disease genes, the clinical knowledge gained by this investment is considerable. With genetic data, diagnosis is certain, and shapes the framework for appropriate management strategies and reflective counseling. Also with genetic information, assignment of disease status in all family members is accurate, efficient and inexpensive. Continued progress in understanding the mechanisms by which gene mutations cause cardiomyopathy will also undoubtedly occur, knowledge that will probably provide new therapies that are predicated on genetic cause in cardiomyopathy. As such, physicians, patients and family members can expect that the definition of a disease-causing mutation will substantially benefit all aspects of clinical care in cardiomyopathy, today by improving diagnosis and management, and in the future by directing therapy.

Acknowledgement

This chapter is adapted, in part, from Maron BJ, Seidman JG, Seidman CE. Proposal for contemporary screening strategies in families with hypertrophic cardiomyopathy. J Am Coll Cardiol 2004; 44: 2125–32; with permission from the American College of Cardiology (and Elsevier Inc., New York, NY).

References

1. Seidman JG, Seidman CE. The genetic basis for cardiomyopathy: from mutation identification to mechanistic paradigms. Cell 2001; 104; 557–67.

2. Schonberger J, Seidman CE. Many roads lead to a broken heart: the genetics of dilated cardiomyopathy. Am J Hum Genet 2001; 69: 249–60.

3. Jaenicke T, Diederich KW, Haas W et al. The complete sequence of the human beta-myosin heavy chain gene and a comparative analysis of its product. Genomics 1990; 8: 194–206.

4. Online Mendelian Inheritance in Man: www.ncbi.nih.gov (accessed 17 May 2005).

5. Galas DJ, McCormack SJ. An historical perspective on genomic technologies. Curr Issues Mol Biol 2003; 5: 123–7.

6. Shendure J, Mitra RD, Varma C, Church GM. Advanced sequencing technologies:methods and goals. Nat Rev Genet 2004; 5: 335–44.

7. Harvard-Partners Center for Genetics and Genomics, Laboratory of Molecular Medicine: www.hpcgg.org/LMM/tests.html (accessed 17 May 2005).

8. Maron BJ, Danielson GK, Kappenberger et al. American College of Cardiology/European Society of Cardiology Clinical Expert Consensus Document on Hypertrophic Cardiomyopathy. A report of the American College of Cardiology Task Force on Clinical Expert Consensus Documents and the European Society of Cardiology Committee for Practice Guidelines and Policy Conference. J Am Coll Cardiol 2003; 42: 1687–713.

9. Maron BJ, Pelliccia A, Spirito P. Cardiac disease in young trained athletes: insights into methods for distinguishing athlete's heart from structural heart disease with particular emphasis on hypertrophic cardiomyopathy. Circulation 1995; 91: 1596–601.

10. Lewis JF, Maron BJ. Diversity of patterns of hypertrophy in patients with systemic hypertension and marked left ventricular wall thickening. Am J Cardiol 1990; 65: 874–81.

11. Hada Y, Sakamoto T, Amano K et al. Prevalence of hypertrophic cardiomyopathy in a population of adult Japanese workers as detected by echocardiographic screening. Am J Cardiol 1987; 59: 183–4.

12. Maron BJ, Peterson EE, Maron MS et al. Prevalence of hypertrophic cardiomyopathy in an outpatient population referred for echocardiographic study. Am J Cardiol 1994; 73: 577–80.

13. Gollob MH, Green MS, Tang AS-L et al. Identification of a gene responsible for familial Wolff-Parkinson-White syndrome. N Engl J Med 2001; 344: 1823–31.

14. Blair E, Redwood C, Ashrafian H et al. Mutations in the gama(-2) subunit of AMP-activated protein kinase cause familial hypertrophic cardiomyopathy: evidence for the central role of energy compromise in disease pathogenesis. Hum Mol Genet 2001; 10: 1215–20.

15. Arad M, Benson DW, Perez-Atayde AR et al. Constitutively active AMP kinase mutations cause glycogen storage disease mimicking hypertrophic cardiomyopathy. J Clin Invest 2002; 109: 357–62.

16. Bernstein HS, Bishop DF, Astrin KH et al. Fabry disease: six gene rearrangements and an exonic point mutation in the alpha-galactosidase gene. J Clin Invest 1989; 83: 1390–9.

17. Kornreich R, Bishop DF, Desnick RJ. Alpha-galactosidase A gene rearrangements causing Fabry disease: identification of short direct repeats at breakpoints in an Alu-rich gene. J Biol Chem 1990; 265: 9319–26.

18. Nishino I, Fu J, Tanji K et al. Primary LAMP 2 deficiency causes X-linked vacuolar cardiomyopathy and myopathy (Danon disease). Nature 2000; 406: 906–10.

19. Nagao Y, Nakashima H, Fukuhara Y et al. Hypertrophic cardiomyopathy in late-onset variant of Fabry disease with high residual activity of alpha-galactosidase A. Clin Genet 1991; 39: 233–7.

20. Arad M, Maron BJ, Gorham JM et al. Glycogen storage diseases presenting as hypertrophic cardiomyopathy. N Engl J Med 2004; 352: 362–72.

21. Niimura H, Bachinski LL, Sangwatanaroj S et al. Mutations in the gene for human cardiac myosin-binding protein C and late-onset familial hypertrophic cardiomyopathy. N Engl J Med 1998; 338: 1248–57.

22. Niimura H, Patton KK, McKenna WJ et al. Sarcomere protein gene mutations in hypertrophic cardiomyopathy of the elderly. Circulation 2002; 105: 446–51.

23. Ackerman MJ, Van Driest SL, Ommen SR et al. Prevalence and age-dependence of malignant mutations in the beta-myosin heavy chain and troponin T genes in hypertrophic cardiomyopathy: a comprehensive outpatient perspective. J Am Coll Cardiol 2002; 39: 2042–8.

24. Maron BJ, Olivotto I, Spirito P et al. Epidemiology of hypertrophic cardiomyopathy-related death. Revisited in a large non-referral-based patient population. Circulation 2000; 102: 858–64.

25. Watkins H. Sudden death in hypertrophic cardiomyopathy. N Engl J Med 2000; 372: 422–23.

26. Wigle ED, Sasson Z, Henderson MA et al. Hypertrophic cardiomyopathy. The importance of the site and the extent of hypertrophy. A review. Prog Cardiovasc Dis 1985; 28: 1–83.

27. Schwartz K, Carrier L, Guicheney P et al. Molecular basis of familial cardiomyopathies. Circulation 1995; 91: 532–40.

28. Elliott PM, Poloniecki J, Dickie S et al. Sudden death in hypertrophic cardiomyopathy: Identification of high-risk patients. J Am Coll Cardiol 2000; 36: 2212–18.

29. Elliott PM, Gimeno JR, Mahon NG et al. Relation between severity of left-ventricular hypertrophy and prognosis in patients with hypertrophic cardiomyopathy. Lancet 2001; 357: 420–4.

30. Maron BJ, Olivotto I, Spirito P et al. Epidemiology of hypertrophic cardiomyopathy-related death. Revisited in a large non-referral-based patient population. Circulation 2000; 102: 858–64.

31. Skinner JR, Manzoor A, Hayes AM et al. A regional study of presentation and outcome of hypertrophic cardiomyopathy in infants. Heart 1997; 77: 229–33.

32. Maron BJ, Mitchell JH. 26th Bethesda Conference. Recommendations for determining eligibility for competition in athletes with cardiovascular abnormalities. J Am Coll Cardiol 1994; 24: 845–99.

33. Maron BJ, Shirani J, Poliac LC et al. Sudden death in young competitive athletes: Clinical, demographic and pathological profiles. JAMA 1996; 76: 199–204.

34. Maron BJ, Carney KP, Lever HM et al. Relationship of race to sudden cardiac death in competitive athletes with hypertrophic cardiomyopathy. J Am Coll Cardiol 2003; 41: 974–80.

35. Maron BJ, Casey SA, Poliac LC et al. Clinical course of hypertrophic cardiomyopathy in a regional United States cohort. JAMA 1999; 281; 650–5.

36. Maron BJ, Estes NAM III, Maron MS et al. Primary prevention of sudden death as a novel treatment strategy in hypertrophic cardiomyopathy. Circulation 2003; 107: 2872–5.

37. Watkins H, McKenna W, Thierfelder L et al. Mutations in the genes for cardiac troponin T and α-tropomyosin in hypertrophic cardiomyopathy. N Eng J Med 1995; 332: 1058–64.

38. Rosenzweig A, Watkins H, Hwang D-S et al. Preclinical diagnosis of familial hypertrophic car-

diomyopathy by genetic analysis of blood lymphocytes. N Eng J Med 1991; 325: 1753–60.

39. Maron BJ, Seidman JG, Seidman CE. Proposal for contemporary screening strategies in families with hypertrophic cardiomyopathy. J Am Coll Cardiol 2004; 44: 2125–32.

40. Montgomery JV, Gohman TE, Harris KM et al. Electrocardiogram in hypertrophic cardiomyopathy revisited: Does ECG pattern predict phenotypic expression and left ventricular hypertrophy or sudden death? J Am Coll Cardiol 2002; 39(Suppl A): 161A.

41. Ho CY, Sweitzer NK, McDonough B et al. Assessment of diastolic function with Doppler tissue imaging to predict genotype in preclinical hypertrophic cardiomyopathy. Circulation 2002; 105: 2992–7.

42. Nagueh SF, Bachinski LL, Meyer D et al. Tissue Doppler imaging consistently detects myocardial abnormalities in patients with hypertrophic cardiomyopathy and provides a novel means for an early diagnosis before and independently of hypertrophy. Circulation 2001; 104: 128–30.

CHAPTER 41

Future directions for heart failure research

Richard A Walsh

Prediction is very hard . . . especially regarding the future (Neils Bohr).

In his landmark book on the structure of scientific revolutions, Thomas S Kuhn challenged the empiricist's view of science as an orderly objective progression toward the truth.[1] Scientific investigation is, in his view, a series of peaceful periods which are intermittently eclipsed by dramatic revolutions in which one scientific paradigm is replaced by another. It is instructive to view progress in our understanding of the causes and consequences of heart failure over the past half century in this light. A hemodynamic view of heart failure was replaced by a neurohormonal model. Genetic, molecular and cellular approaches now dominate mechanistic studies of the causes and potential treatment of heart failure. We are now on the verge of a new systems biology approach to this and other chronic diseases.

Despite these shifting paradigms, congestive heart failure continues to have a major adverse impact on public health in developed countries. There is an estimated prevalence of over 5 million individuals with a clinical diagnosis, and an incidence of over 500 000–800 000 new cases of congestive heart failure per year. It is a condition that affects 1–2% of our overall population, and 6–10% of those individuals over the age of 65. It is the only cardiovascular disorder with an increasing incidence and prevalence, and, despite contemporary

pharmacotherapy, mortality ranges from 5–40% per year. This equals or exceeds that of many common malignancies. It is facile to think that control of known environmental risk factors such as hypertension, diabetes, lipid disorders and tobacco use will minimize the importance of this condition. The aging of our population will permit genetic and environmental factors to operate over more protracted time. Societal, physician and patient focus on prevention will remain imperfect, despite the acknowledgement and importance of the quality movement in medicine. Individual attitudes regarding healthy behavior and risk taking will continue to challenge primary and secondary prevention in healthy and diseased populations respectively. Finally, as developing countries improve their economies, diabetes, lipid disorders and other modifiable risk factors will similarly affect a larger segment of their populations. Given these probabilities, focus on heart failure prevention, novel therapies, and the resulting economic impact of improved diagnostic and therapeutic technology will continue to challenge healthcare.

Increasing focus on single nucleotide polymorphisms that will enhance susceptibility to the development of pathological hypertrophy and failure, and, coupled with pharmacogenetics and genomics, will provide important new therapeutic insights. In contrast to other tissues such as the bone marrow, liver and vascular endothelium, the role of regenerative medicine in the prevention and management of heart failure will remain challenging. The application of systems biology and computational analysis will be necessary to understand and integrate an increasingly complex information base into coherent strategies for targeted therapeutics. Heart failure, like any chronic disease, results from complex interactions among polygenic, environmental and other factors. The search for new therapies will, therefore, prove challenging. We have attempted to set the scene in this book to encourage basic, translational and clinical investigators to take fresh approaches and to excite scientists from other disciplines to focus on this important public health problem.

Reference

Kuhn TS. The Structure of Scientific Revolutions (2e, enlarged). Chicago, IL, USA; London, UK: University of Chicago Press; 1970.

Index